Petroleum Engineering Handbook

Petroleum Engineering Handbook

Larry W. Lake, Editor-in-Chief

I **General Engineering** *John R. Fanchi, Editor*
II **Drilling Engineering** *Robert F. Mitchell, Editor*
III **Facilities and Construction Engineering** *Kenneth E. Arnold, Editor*
IV **Production Operations Engineering** *Joe Dunn Clegg, Editor*
V **Reservoir Engineering and Petrophysics** *Edward D. Holstein, Editor*
VI **Emerging and Peripheral Technologies** *H.R. Warner Jr., Editor*
VII **Indexes and Standards**

Petroleum Engineering Handbook

Larry W. Lake, Editor-in-Chief
U. of Texas at Austin

Volume V - Part A

Reservoir Engineering and Petrophysics

Edward D. Holstein, Editor
Consultant

Society of Petroleum Engineers

Printed in the United States of America.

ISBN 978-1-959025-55-9 (print)
ISBN 978-1-55563-335-6 (digital)

Society of Petroleum Engineers
222 Palisades Creek Drive
Richardson, TX 75080-2040 USA

http://store.spe.org/
service@spe.org
1.972.952.9393

Foreword

This 2006 version of SPE's *Petroleum Engineering Handbook* is the result of several years of effort by technical editors, copy editors, and authors. It is designed as a handbook rather than a basic text. As such, it will be of most benefit to those with some experience in the industry who require additional information and guidance in areas outside their areas of expertise. Authors for each of the more than 100 chapters were chosen carefully for their experience and expertise. The resulting product of their efforts represents the best current thinking on the various technical subjects covered in the *Handbook.*

The rate of growth in hydrocarbon extraction technology is continuing at the high level experienced in the last decades of the 20th century. As a result, any static compilation, such as this *Handbook*, will contain certain information that is out of date at the time of publication. However, many of the concepts and approaches presented will continue to be applicable in your studies, and, by documenting the technology in this way, it provides new professionals an insight into the many factors to be considered in assessing various aspects of a vibrant and dynamic industry.

The *Handbook* is a continuation of SPE's primary mission of technology transfer. Its direct descendents are the "Frick" *Handbook*, published in 1952, and the "Bradley" *Handbook*, published in 1987. This version is different from the previous in the following ways:

- It has multiple volumes in six different technical areas with more than 100 chapters.
- There is expanded coverage in several areas such as health, safety, and environment.
- It contains entirely new coverage on Drilling Engineering and Emerging and Peripheral Technologies.
- Electronic versions are available in addition to the standard bound volumes.

This *Handbook* has been a monumental undertaking that is the result of many people's efforts. I am pleased to single out the contributions of the six volume editors:

General Engineering—John R. Fanchi, Colorado School of Mines
Drilling Engineering—Robert F. Mitchell, Landmark Graphics Corp.
Facilities and Construction Engineering—Kenneth E. Arnold, AMEC Paragon
Production Operations Engineering—Joe D. Clegg, Shell Oil Co., retired
Reservoir Engineering and Petrophysics—Ed Holstein, Exxon Production Co., retired
Emerging and Peripheral Technologies—Hal R. Warner, Arco Oil and Gas, retired

It is to these individuals, along with the authors, the copy editors, and the SPE staff, that accolades for this effort belong. It has been my pleasure to work with and learn from them.

—Larry W. Lake

Preface

This volume consists of 27 chapters that deal with the many aspects of reservoir engineering. The chapters were assembled to provide information on acquiring and interpreting data that describe reservoir rock and fluid properties; acquiring, understanding, and predicting fluid flow in the reservoir; interpreting measurements of well performance; calculating the factors that impact both primary and improved recovery mechanisms from oil and gas reservoirs; estimating reserves and calculating project economics; simulating reservoir performance; and structuring and measuring the effectiveness of a reservoir management system.

These chapters have been written as a handbook and, as such, assume that the reader has a familiarity with fundamentals and some experience in the production of hydrocarbons and will use this publication as a refresher or to expand knowledge in certain areas of technology. Extensive references in each chapter indicate the amount of material that has been considered and distilled.

This updated version of SPE's *Petroleum Engineering Handbook* contains information on many of the subjects covered by the 1962 and 1987 versions. All chapters in this version are new and greatly expanded; chapters existing in the older versions have been revised extensively or rewritten completely, and new chapters have been added on geophysics; geology; petrophysics; production logs; chemical tracers; foam, polymer, and resin injection; miscible processes; valuation; and reservoir management.

The high quality of this volume is a credit to the authors of each chapter; they were chosen for their extensive knowledge and experience. Special recognition and thanks go to E.C. Thomas, who is responsible for the quality and thoroughness of the chapters contained in the petrophysical section. The resulting product contains many good tips and guidelines that represent lessons learned in the practice of applying various technologies to a wide range of actual field conditions. Numerous field examples have been included to illustrate these applications.

—Edward D. Holstein

Contents

Chapter 1
Reservoir Geology

F. Jerry Lucia, SPE, U. of Texas at Austin

1.1 Introduction

The efficient extraction of oil and gas requires that the reservoir be visualized in 3D space. Engineers need a conceptual model of reservoirs, an integral part of the decision-making process, whether it be selecting perforations or forecasting future production. However, most engineering measurements made on reservoirs have little or no spatial information. For example, a core measurement has no dimensional information, wireline logs and continuous core measurements are 1D, and production data and pressure information are volumetric but with unconstrained spatial information. Geologic information, on the other hand, contains valuable spatial information that can be used to visualize the reservoir in 3D space. Therefore, engineers should understand the geologic data that can improve their conceptual model of the reservoir and, thus, their engineering decisions.

The first and most important geologic information is the external geometry of the reservoir, which is defined by seals or flow barriers that inhibit the migration of hydrocarbons, forming a hydrocarbon trap. The buoyancy force produced by the difference in density between water and hydrocarbons drives migration. Migration will cease, and a hydrocarbon reservoir will form, only where hydrocarbons encounter a trap. Traps are composed of top, lateral, and bottom seals; the geometry of traps can have structural, sedimentary, or diagenetic origins.

The second most important geologic information is the internal reservoir architecture. A reservoir is composed of rock types of varying reservoir quality that are systematically stacked, according to stratigraphic and diagenetic principles. The lateral distribution of depositional textures is related to depositional environments, and the vertical stacking of textures is described by stratigraphy, which is the geological study of the form, arrangement, geographic distribution, chronologic succession, classification, and correlation of rock strata. Diagenesis, changes that happen to the sediment after deposition, can also control the lateral continuity and vertical stacking of reservoir rock types. This fact is most important in carbonate reservoirs, in which the conversion of limestone to dolostone and the dissolution of carbonate have a large effect on internal reservoir architecture.

The most basic concern for most engineers is the spatial distribution of petrophysical properties, such as porosity, permeability, water saturation, and relative permeability. To visualize the reservoir in petrophysical terms, the engineer must be able to equate measurements (log, core, or production) with geologic models because the measurements themselves do not contain spatial information. Linking engineering measurements with geologic descriptions is best done at the rock-fabric level because rock fabric controls pore-size distribution, which, in turn, controls porosity, permeability, and capillary properties. Rock fabrics can be tied directly to stratigraphic models and, thus, to 3D space.

1.2 External Geometry—Reservoir Traps

1.2.1 Introduction. Hydrocarbons are formed by anaerobic decomposition of organic matter that accumulates from the deposition of plankton in deep ocean basins. Oil and gas are generated as the sediments are buried and the temperature rises. Oil is the first hydrocarbon to be generated, followed by wet gas, and lastly by dry gas. Once generated, oil and gas flow vertically and laterally through overlying sediments because of the density difference between hydrocarbons and formation water and they migrate through permeable formations until they encounter a reservoir trap in which oil and gas accumulate. Oil will fill the traps first because it is first to be generated. Higher temperatures resulting from continued burial cause gas to be generated. Migrating gas will displace oil from the traps because gas has a lower density. The displaced oil will migrate further updip and fill any trap encountered.[1]

Traps filled with hydrocarbons are often referred to as pools. However, engineers normally use the term reservoir instead of pool for an oil and gas accumulation, and *reservoir* will be used throughout this chapter. A field is composed of one or more reservoirs in a single area. A trap is defined by the geometry of its seals, which are formations with very low permeability and very small pores that will impede or stop the flow of hydrocarbons. To trap migrating hydrocarbons, seals must contain flow in 3D: the seals must form a closure. In the simplest terms, a trap is similar to a box with its bottom removed. The box is the seal composed of top and lateral seals. A trap may also contain a bottom seal. Imagine a smaller box inserted into the base of the original box. The smaller box is also a seal and confines the reservoir to a layer within the larger box.

Seals may be in the form of impermeable lithologies or faults. The simplest traps are convex structures in which the sealing layer dips in all directions from a central structural high, forming domes or doubly dipping anticlines. More complex structural traps are formed when convex structures are truncated by faults or when faulting occurs around a piercement structure. Many traps are combinations of structural uplift, faulting, and stratigraphy, such as an updip pinchout of a sand body into an impermeable shale. A purely stratigraphic trap may form when deposition creates a topographic high that is encased by impermeable lithology, such as shale or salt.

The volume of oil and gas that can accumulate is defined partly by the height of the trap because any additional hydrocarbons will spill out the bottom. The base of the trap is therefore called the spill point (see **Fig. 1.1**). The trap may not be full because the height of the oil column will be controlled by the capacity of the seal to impede flow and the volume of oil that migrates to the trap. In addition, oil/water contacts need not be horizontal because subsurface fluids are rarely static, and the flow can cause the oil/water contact to tilt in the direction of flow.

1.2.2 Traps. Geologists commonly describe traps on the basis of their origin. Structural traps are closures formed by structural movements within the Earth, and stratigraphic traps are closures formed by sedimentation and diagenesis, without the need for structural movements. Structural/stratigraphic traps are closures formed by patterns of reservoir rock that impinge upon a structure. This organization will be used hereafter, even though new groupings based on sealing surfaces were suggested by Milton and Bertram.[2]

Structural traps are formed most commonly by structural uplift and differential compaction. Typical structural traps are structural domes and doubly plunging anticlines (see Fig. 1.1).

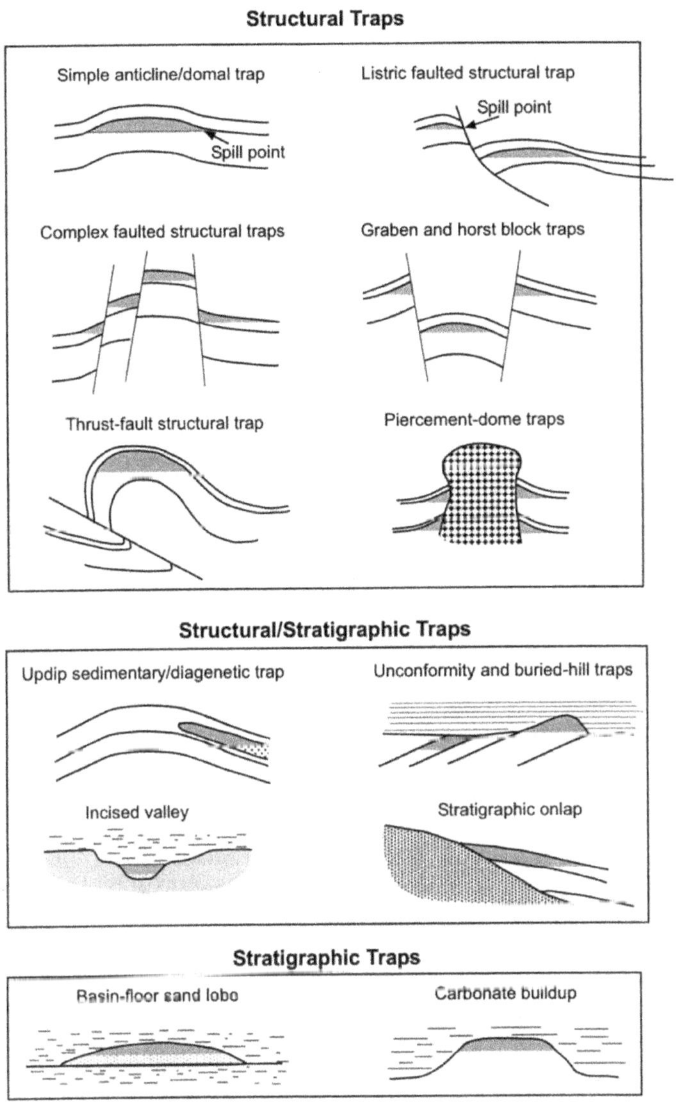

Fig. 1.1—Diagram description of hydrocarbon traps.

These traps have a structural high and quaquaversal dips (the seal dips away from a structural high in all directions). The bulk of the world's oil is found in these four-way-closure traps,[3] which were the first type to be exploited by surface mapping. Many major oil fields in the world were discovered by using surface mapping to locate domal structures.

A more complex method of forming a structural trap is by faulting and structural uplift (see Fig. 1.1). Faulted structures can vary from a simple faulted anticline to complex faulting around piercement structures and domal uplifts. Faulted structures are very common and form some of the most complex reservoirs known. Types of faults include normal, listric, reverse, and thrust, which are related to the stress fields generated during structural movement. Piercement traps (diapirs) are formed typically by salt moving up through a stack of sediment driven by the density difference between salt and quartz or carbonate. Closure is achieved by the uplift of sediments juxtaposed to the piercement dome, by the top seal being an overlying impervious bed and the lateral seals being formed by structural dip, by sealing faults, or by the

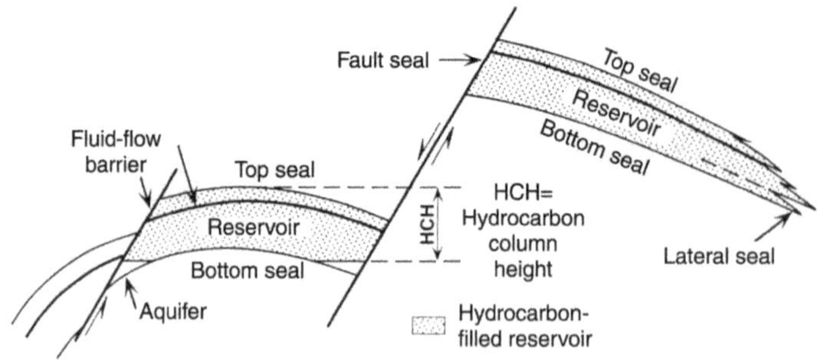

Fig. 1.2—Some typical seals; HCH is the hydrocarbon-column height that the weakest seal will hold. Figure courtesy of Sneider Exploration Inc.

piercement salt. Faulted reservoirs commonly have a bottom seal formed by the lower contact of sand with shale. The bottom seal, along with the oil/water contact within the sand body, forms the base of the reservoir.

Structural/stratigraphic traps are formed by a combination of structure, deposition, and diagenesis. The most common form, the updip pinchout of reservoir lithology into a sealing lithology (see Fig. 1.1), is found in the flanks of structures. The top and updip seal is normally an impervious rock type, and the lateral seals are formed by either structural dip or the lateral pinchout of reservoir rock into seal material. The base of the reservoir is defined by a bottom seal composed of impervious rock and by an oil/water contact. During relative sea-level fall, streams may erode deep valleys, thus forming lateral seals for fluvial sediments. Onlap of sand onto a paleotopographic high during relative sea-level rise can produce an updip seal for a sand body. Unconformity traps are formed by the truncation of dipping strata by overlying bedded sealing lithology. The reservoir rock may be found in the form of buried hills formed by erosion during the time of the unconformity. The oil/water contact forms the base of these reservoirs. A stratigraphic trap may be partly related to diagenetic processes; for example, the updip seal for the supergiant Coalinga field, California, is tar- and asphalt-filled sandstone and conglomerate. Many traps in the Permian reservoirs of west Texas, are formed by lateral changes related to stratigraphy from porous to dense dolomite in an updip direction.

Stratigraphic traps are formed by depositional processes that produce paleotopographic highs encased in impermeable material, such as evaporite or shale (Fig. 1.1). Closure occurs when there is contact between seal material and underlying sediment. The most common type is a carbonate buildup, usually erroneously called a "reef." Piles of sand deposited on the seafloor by density currents often form broad topographic highs that, in turn, form stratigraphic traps. Structure may also play a part in the geometry of stratigraphic traps, although the defining characteristic is that structure is not required to form the trap.

1.2.3 Seals. A seal is a low-permeable to impermeable rock or immobile fluid, such as tar, with a capillary entry pressure large enough to dam up or trap hydrocarbons.[4] Typical seals include top, bottom, lateral, and fault, as shown in **Fig. 1.2**. Faults may be sealing or nonsealing, depending on whether the sand offsets another sand (nonsealing) or shale (sealing).[5] Any lithology can be a seal or flow barrier. The requirement is that the minimum capillary displacement pressure of the seal or flow-barrier material be greater than the buoyancy pressure of the hydrocarbons in the accumulation. The continuous, small, pore-throat sizes create a barrier to moving hydrocarbons, causing them to dam up or become trapped. Therefore, the size of the continuous pore throats and the density of the hydrocarbons and water are critical elements in evaluating a seal or flow barrier.

Porosity and permeability are not the best criteria for evaluating seal and flow-barrier behavior. The seal capacity of a rock can be best evaluated with a mercury porosimeter that can inject mercury into material using pressures as high as 60,000 psi. The key equation used in capillary pressure/saturation evaluation of reservoir rocks, seals, and flow barriers is:

$$h_c = \left(\frac{p_c}{\rho_w - \rho_{hc}}\right) 0.433 \text{ ft}, \quad (1.1)$$

in which h_c = the maximum hydrocarbon column held, p_c = the capillary entry pressure, ρ_w= the density of water, and ρ_{hc} = the density of the hydrocarbon. The capillary pressure used is usually not the capillary entry pressure but the capillary pressure at a mercury saturation of between 5 and 10% because of closure effects.

Effective hydrocarbon seals for exploration plays and reservoirs must be laterally continuous. Some typical seal lithologies, illustrated in **Fig. 1.3**, have entry pressures ranging from 14 to 20,000 psi. With data from more than 3,000 seals, we can group the data into Classes A through E to categorize the typical lithologies listed in Fig. 1.3 from most to least ductile. Fig. 1.3 also illustrates the hydrocarbon column that can be held, assuming that the fluid is 35°API oil and saline water. Evaporite and kerogen-rich shale can hold the greatest oil column—from 1,000 to more than 5,000 ft. Clay-mineral-rich shale, silty shales, and dense mudstones can hold between 500 and 1,000 ft of oil column. Sandy shales are ranked next, with a 100- to 500-ft capacity, whereas very shaly siltstone and sandstone, anhydrite-filled dolostones, and cemented sandstone each have between 50 and 100 ft of capacity. In addition, immobile fluids, such as tar, bitumen, and asphalt, can be effective seals and barriers. For example, the updip seal for the supergiant Coalinga field, California, is tar- and asphalt-filled sandstone and conglomerate.

1.3 Reservoir Base

Whereas structure and stratigraphy most often define the reservoir trap or top of the reservoir, factors controlling the base of a petroleum reservoir include seal capacity, spill point, capillary forces, and hydrodynamics.[6,7] The reservoir base is defined as the zero capillary pressure level, also referred to as the free-water level. Reservoir height is determined by the height from seal to spill point, if seal capacity is large enough. If the height is less than that from seal to spill point, seal capacity or hydrocarbon charge will determine the position of the reservoir base.

Subsurface groundwater is seldom static. Differences in water density, structural tilting, tectonic forces, and other factors combine to create a difference in hydrodynamic potentials that result in the movement of fluids in the subsurface. Fluid movement is controlled by the fluid potential as defined by Hubbert[6] and illustrated by the following formula[8]:

$$H = Z + \frac{P}{\rho_w g}, \quad (1.2)$$

in which Z = elevation relative to a datum (sea level), P = measured static pressure, and ρ = density of the fluid (water).

A potentiometric map is a map that connects points of equal fluid potential within an aquifer. If the potentiometric surface is not horizontal, the aquifer will flow in the direction of lowest potential. Calculating fluid potential requires accurate subsurface pressure measurements in the aquifer. The flow of water under a reservoir will cause the zero-capillary-pressure level to tilt, referred to as a tilted water table. The degree of tilt can be estimated from the following equation.

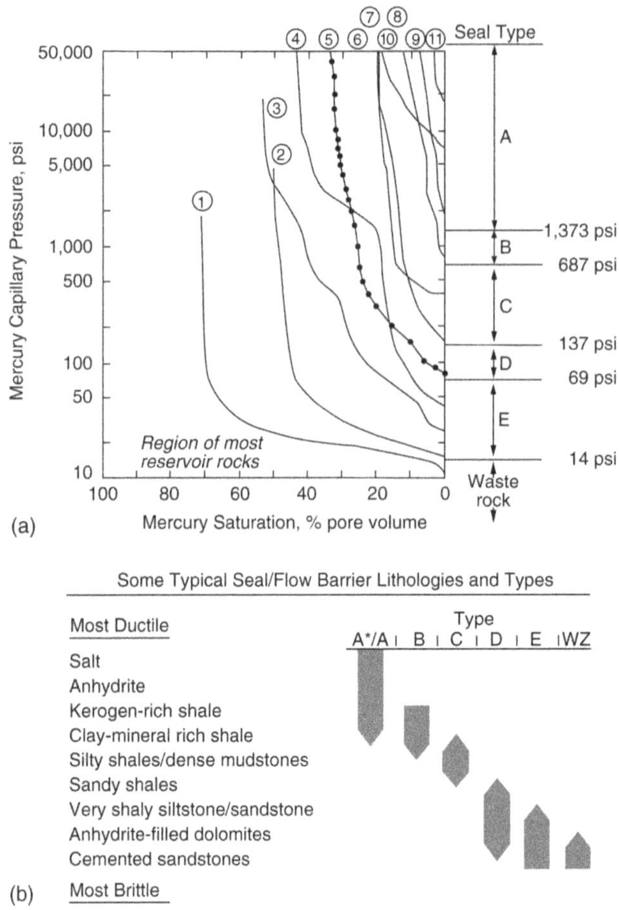

(c)

Seal - Flow Barrier Type	35° API Oil Column Held: Meters	Feet
A*	≥ 1500	≥ 5,000
A	≥ 300 to < 1,500	≥1,000 to < 5,000
B	≥ 150 to < 300	≥ 500 to < 1,000
C	≥ 30 to < 150	≥ 100 to < 500
D	≥ 15 to < 30	≥ 50 to < 100
E	< 15	< 50
F	Waste Zones Rocks	

Fig. 1.3—(a) Air/mercury capillary pressure curves of seal lithologies. The ordinate is a log scale. (b) Typical seal lithologies. (c) Sneider *et al.*[3] classification of seals and flow barriers. The hydrocarbon column that can be held assumes that the fluids in the reservoirs are 35°API oil and saline water. Types are keyed to Figs. 3a and 3b. Figure courtesy of Sneider Exploration Inc.

$$\tan\,\theta = \frac{\Delta Z}{x} = \left(\frac{\rho_W}{\rho_w - \rho_o}\right)\frac{dh}{dx}, \quad (1.3)$$

in which $\frac{\Delta Z}{x}$= change in reservoir height for distance x (tilt in water table), ρ_w= density of water in aquifer, ρ_o = density of hydrocarbon, and $\frac{dh}{dx}$ = change in potentiometric surface for distance x.

During primary development, the economic base of the reservoir is normally defined as the producing oil/water contact, or the level at which oil and water are first coproduced. This level is generally assumed to be at approximately 50% water saturation, according to relative permeability considerations. During tertiary development, the economic base of the reservoir may be defined as the level of zero oil saturation. However, because pore size can vary with stratigraphy, 50 or 100% water saturation may not occur at the same height throughout the reservoir. The economic base of the reservoir will not be horizontal.

Defining the base of a reservoir is often made difficult by the presence of residual-oil and tar zones below the producing oil/water contact. Residual-oil and tar zones can be as thick as 300 ft and are thought to form by a variety of processes, including biodegradation of hydrocarbons, flushing of part of the oil column as a result of hydrodynamic forces, and remigrating of hydrocarbons because of leaky seals and structural tilting. The presence of this material may indicate that the reservoir is in an imbibition rather than a drainage mode. Estimates of the original oil in place will depend on which capillary pressure model is assumed. An incorrect model can lead to large errors in estimates of the original oil in place.

1.4 Internal Geometry—Reservoir Architecture

1.4.1 Introduction. Information that defines the external reservoir geometry, including trap configuration, seal capacity, and the base of the reservoir, is of primary importance during exploration and initial development of a reservoir. As development continues, reservoir architecture becomes key to predicting the distribution of reservoir quality so that primary- and secondary-development programs can be planned. Reservoir architecture is important because it provides a basis for distributing petrophysical properties in 3D space. In most cases, this operation is done by relating lithofacies to petrophysical properties because lithofacies can be directly linked to depositional processes for prediction.

We commonly correlate lithofacies from one well to the next by assuming a degree of horizontality and continuity of similar facies. This approach leads to images with highly continuous lithofacies and porosity zones. Many depositional facies, however, are known to be highly discontinuous laterally and vertically, and correlating similar lithofacies from one well to the next can lead to unrealistic displays of reservoir architecture. Modern correlation methods rely more on the chronostratigraphic approach, one that uses time stratigraphy rather than lithostratigraphy to determine continuity between wells. This approach is referred to as sequence stratigraphy and provides a basis for correlating time surfaces between which lithofacies are distributed systematically in a predictable pattern.

1.4.2 Sequence Stratigraphy. Sequence stratigraphy is a chronostratigraphic method of correlation. It groups lithofacies into time-stratigraphic units between chronostratigraphic surfaces, which are sometimes defined by unconformities and facies shifts. A key premise is that the surfaces are formed in response to eustatic sea-level changes of various scales and periodicity (eustatic refers to worldwide sea-level changes affecting all oceans). It is thought that eustatic sea-level changes can be linked to climatic changes and to eccentricities in the Earth's orbit. The Russian astronomer Milanovitch defined cyclic variation in the shape of the Earth's orbit and in the tilt and wobble of the axis. These Earth cycles are precession (19,000 to 23,000 years), obliquity (41,000 years), and eccentricity (1,000,000 to 4,000,000 years) and are thought to cause changes in the Earth's climate, resulting in more or less water trapped as ice at the poles. The trapping or release of water from the ice caps is thought to result in sea-level rise and fall, referred to as eustasy.

Sequence stratigraphy is important for reservoir modeling because a chronostratigraphic surface is present in every well in the reservoir. This fact provides geologists with a powerful tool for correlating packages of lithofacies between wells. A more realistic image of reservoir archi-

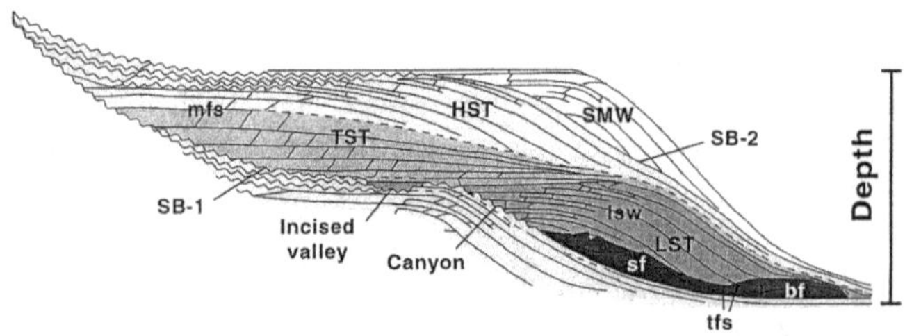

Fig. 1.4—Exxon idealized depositional-sequence cross-sectional model for siliciclastic sediments. Sequence boundary (SB), LST composed of basin-floor fan (bf), slope fan (sf), and lowstand prograding wedge (LSW), TST, HST, and shelf-margin wedge (SMW). Maximum flooding surface (mfs) separates the TST from the HST. Taken from Kerans and Tinker.[10]

tecture can, therefore, be constructed by distributing lithofacies and petrophysical properties within a detailed sequence-stratigraphic framework.

The terminology of sequence stratigraphy, like most geologic terminology, is complex and constantly evolves as concepts and ideas change.[9,10] It is the intent here to present a basic overview of the terminology to provide the reader with sufficient understanding to communicate with reservoir geologists. The classic Exxon model (see **Fig. 1.4**) shows the terminology used in siliciclastic stratigraphy. The terms used in carbonate stratigraphy, although similar, have important differences because carbonates are organic in origin and clastics are terrigenous in origin. The terminology used in carbonate stratigraphy is illustrated in **Fig. 1.5**.

The smallest time-stratigraphic unit is the high-frequency cycle (HFC), or parasequence, a unit composed of genetically related lithofacies deposited during one basic sea-level rise and fall. Assuming a constant rate of subsidence, each cycle begins with a flooding event as sea level rises. The flooding event is also referred to as transgression or retrogradation, the backward and landward movement or retreat of a shoreline or coastline. The sea transgresses the land, the shoreline retreats, and the space for sediment to accumulate increases. The space created by the transgression is referred to as accommodation space. Sea-level rise is followed by a stillstand, during which sediment completely or partly fills the accommodation space. The buildup of sediment by deposition is referred to as aggradation. The stillstand is followed by a relative sea-level fall during which accommodation space is reduced, forcing sediment to be transported into the basin and resulting in progradation of the sediment body. Progradation refers to the building forward and outward toward the sea of a shoreline or body of sediment. During sea-level fall, the most-landward sediment may become subaerially exposed, forming an unconformity. Farther basinward the water is deeper, and the shallowing event is identified by a facies shift in the vertical stacking of lithofacies. The next sea-level rise produces another flooding event, and the depositional cycle is repeated. Flooding events approximate chronostratigraphic surfaces and define the HFC as a time-stratigraphic unit.

Repeated eustatic sea-level cycles result in the vertical stacking of HFC. Cycles are stacked vertically into retrogradational cycles, aggradational cycles, and progradational cycles. Retrogradational cycles are formed when the eustatic sea-level rise for each cycle is much more than the fall. The shoreline will move farther landward with each successive cycle, a pattern described as back stepping or transgression. The sediments are said to be deposited in the transgressive system's track (TST). Aggradational cycles are formed when eustatic rise and fall are equal, and the resulting facies will stack vertically. These cycles are defined as part of the highstand system's tract (HST). Progradational cycles form when the eustatic fall for each cycle is greater that the rise. The shoreline for each successive cycle will move seaward, a pattern described as progradation or regression, and the sediments are said to be deposited in the HST. Sediments deposited when relative sea level is lowest are said to be deposited in the

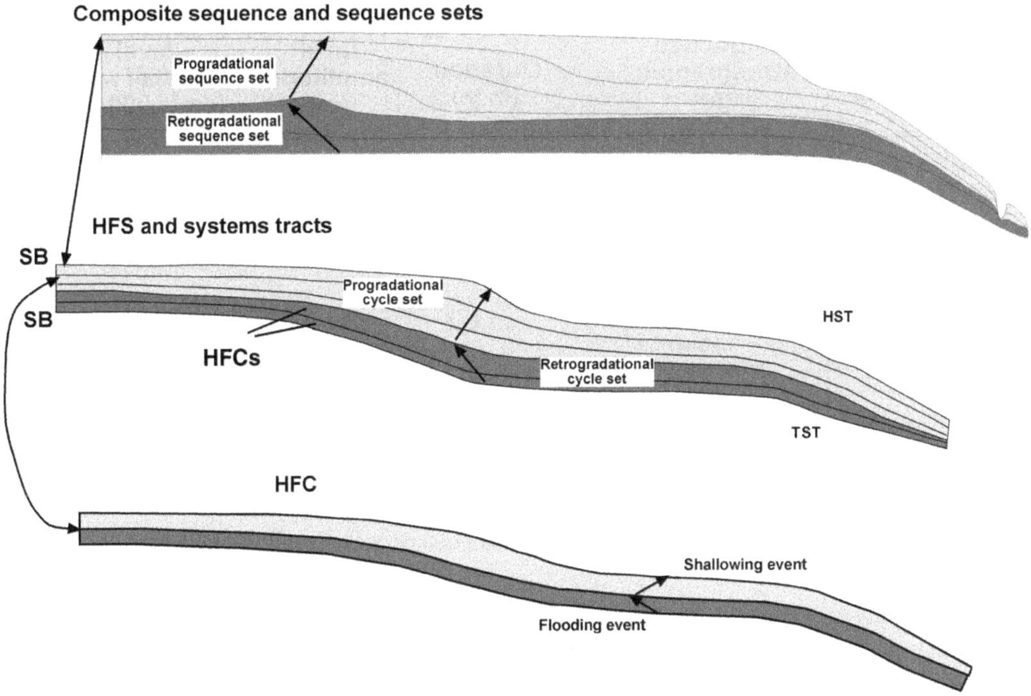

Fig. 1.5—Hierarchy of carbonate sequence-stratigraphic terminology in a cross-sectional view. An HFC is composed of a flooding event (transgression) and a shallowing event (prograding). HFSs are composed of a retrogradational cycle set that makes up the TST and a prograding cycle set that makes up the HST. Composite sequences are composed of a basal retrogradational set of HFSs and an upper set of progradational HFSs. The dark shading indicates transgressive units, and the light shading indicates progradational units. See Kerans and Tinker.[10]

lowstand system's track (LST). The sequence from TST to HST to LST defines a larger-scale sea-level signal referred to as a high-frequency sequence (HFS). The turnaround from transgression to aggradation and progradation is termed the maximum flooding surface (MFS). HFSs can be packaged into longer-term signals called composite sequences on the basis of the observation that they tend to stack vertically into transgressive, progradational, and lowstand sequences.

The terminology and duration of the cycle hierarchies estimated by Goldhammer[11] are shown in **Fig. 1.6**. HFC, HFS, and composite sequences are commonly referred to as fifth-, fourth-, and third-order cycles, respectively, with characteristic durations ranging from 0.01 to 10 million years (m.y.). First- and second-order cycles, or supersequences, have much longer durations, from 10 to more than 100 m.y., and are related more to structural movements than to eustasy. These major sequences are useful not only for regional but also for worldwide correlations. The durations of all these cycles and sequences are approximate and are based on radiogenic dates extrapolated to the numbers of cycles and sequences of various scales.

1.5 Carbonate Reservoirs

1.5.1 Introduction. A basic overview of carbonate-reservoir model construction was presented by Lucia,[12] and much of what is presented herein is taken from that book. Carbonate sediments are commonly formed in shallow, warm oceans either by direct precipitation out of seawater or by biological extraction of calcium carbonate from seawater to form skeletal material. The result is sediment composed of particles with a wide range of sizes and shapes mixed together to

Tectono-eustatic/ eustatic cycle order	Sequence stratigraphic unit	Duration (m.y.)	Relative sea-level amplitude (m)	Relative sea-level rise/fall rate (cm/1,000 yr)
First		>100		<1
Second	Supersequence	10 to 100	50 to 100	1 to 3
Third	Depositional sequence composite sequence	1 to 10	50 to 100	1 to 10
Fourth	High-frequency sequence, parasequence and cycle set	0.1 to 1	1 to 150	40 to 500
Fifth	Parasequence, high-frequency cycle	0.01 to 0.1	1 to 150	60 to 700

(after Goldhammer and others, 1991)

Fig. 1.6—Terminology of cycle hierarchies and order of cyclicity from Goldhammer.[11]

form a multitude of depositional textures. The sediment may be bound together by encrusting organisms or, more commonly, deposited as loose sediment subject to transport by ocean currents.

Depositional textures are described using a classification developed by Dunham.[13] The Dunham classification divides carbonates into organically bound and loose sediments (see **Fig. 1.7**). The loose sediment cannot be described in simple terms of grain size and sorting because shapes of carbonate grains can vary from spheroid ooids to flat-concave and high-spiral shells having internal pore space. The grain content of a grain-supported sediment composed of shells can be as little as 30% of the bulk volume because the shells occupy less space than spheroids. Loose sediment is, therefore, described on the basis of the concept of mud vs. grain support. Mud refers to mud-size carbonate particles, not to mud composed of clay minerals. Grain-supported textures are grainstone, which lacks carbonate mud, and packstone, which contains mud. Mud-supported textures are referred to as wackestone, which contains more than 10% grains, and mudstone, which contains less than 10% grains. To complete the description, generic names are modified according to grain type such as "fusulinid wackestone" or "ooid grainstone."

Dunham's boundstone class was further divided by Embry and Klovan[14] because carbonate reefs are commonly composed of large reef-building organisms, such as corals, sponges, and rudists, which form sediments composed of very large particles. They introduced the terms bafflestone, bindstone, and framestone to describe autochthonous (in-place) boundstone reef material. Floatstone and rudstone are used to describe allochthonous, (transported) reef sediment with particles larger than 2 mm in diameter. Rudstone is grain-supported, whereas floatstone is mud-supported sediment.

Enos and Sawatsky[15] measured the porosity and permeability of modern carbonate sediments. The average porosity and permeability of grainstone are approximately 45% and 10 darcies, respectively, whereas the average porosity and permeability of a wackestone are approximately 65% and 200 md, respectively. The higher porosity in mud-supported sediments is caused by the needle shape of small aragonite crystals that make up the carbonate mud, and the decrease in permeability is caused by the small pore size found between mud-sized particles. An important observation based on this data is that all carbonate sediments have sufficient porosity and permeability to qualify as reservoir rocks.

Depositional texture					Depositional texture not recognizable
Original components not bound together during deposition				Original components were bound together during deposition as shown by intergrown skeletal matter, lamination, or sediment-floored cavities that are roofed over by organic or questionable organic matter and are too large to be interstices.	(Subdivided according to classifications designed to bear on physical texture or diagenesis)
Contains mud			Lacks mud and is grain supported		
Mud supported		Grain supported			
Less than 10% grains	More than 10% grains				
Mudstone	Wackestone	Packstone	Grainstone	Boundstone	Crystalline carbonate

Boundstone	Bafflestone	Bindstone	Framestone
Sediment Grains > 2 mm	Mud supported	Grain supported	
	Floatstone	Rudstone	

Fig. 1.7—Dunham[13] and Embry and Klovan[14] classification of carbonate sediments.

With modifications, the Dunham approach can be used to characterize the petrophysical properties of carbonate rocks. The classification must be modified, however, because diagenesis significantly alters the depositional texture, and a rock-fabric classification that incorporates diagenetic overprints and that can be linked to petrophysical properties is required. The classification proposed by Lucia[16] is designed for this purpose (see **Fig. 1.8**). All pore space is divided into interparticle (intergrain and intercrystal) and vuggy (pore space within grains/crystals and much larger than grains/crystals). Interparticle pore space is classified with the Dunham classification approach. Instead of grain support vs. mud support, however, *grain-* and *mud-dominated* are used as a basic division. Grain-dominated fabrics include grainstone and grain-dominated packstone. Mud-dominated fabrics include mud-dominated packstone, wackestone, and mudstone. The packstone class is divided into grain- and mud-dominated packstone because the petrophysical properties of grain-dominated packstone are according to grain size, whereas mud size controls the properties of mud-dominated packstone. Diagenetic reductions in porosity by cementation and compaction are reflected in the amount of interparticle porosity.

Dolostones are classified similarly if the precursor limestone fabric can be determined. The principal petrophysical difference between limestones and dolostones occurs in mud-dominated fabrics. Limestone-mud-dominated fabrics have mud-sized particles (< 20 μm) and very small pores. Dolomitized mud-dominated fabrics have crystal sizes ranging from 10 μm or less to more than 200 μm, with corresponding pore sizes. Dolomitization must, therefore, improve reservoir quality by increasing particle and pore size.

The classification of vuggy pore space is an important aspect of rock-fabric classification that is not found in the classification of depositional textures. Vuggy pore space is divided into two groups on the basis of how the pore space is connected. Separate vugs are connected to each other through interparticle pore space, and touching vugs are connected directly to one another. Selective dissolution of grains, such as ooids or skeletal material, and intrafossil poros-

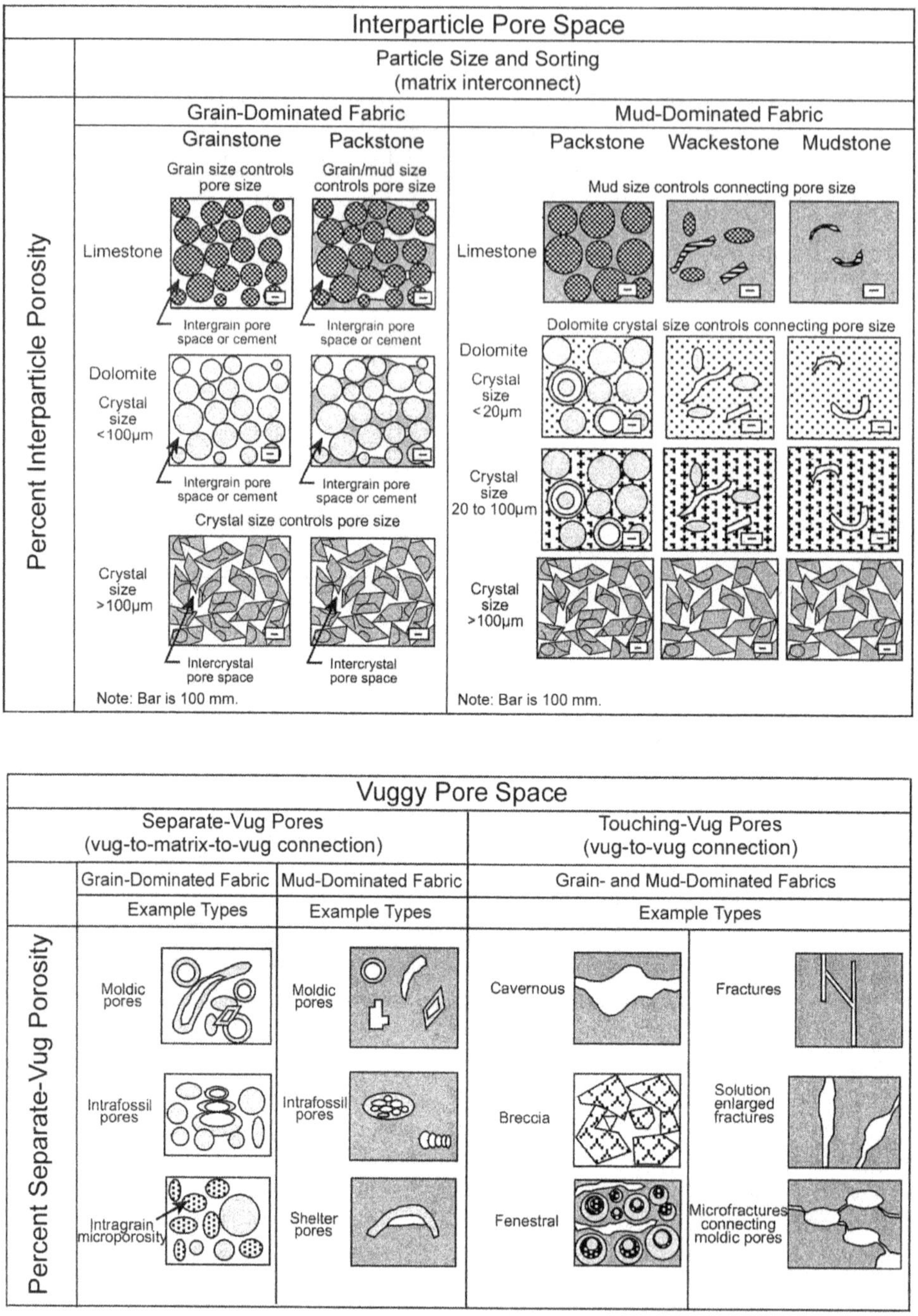

Fig. 1.8—Lucia[16] classification of carbonate pore space.

ity are types of separate vugs. Because separate-vug porosity is poorly connected, it contributes less to permeability than would be expected if the porosity were located between the particles. Touching vugs are commonly formed by mass dissolution and fracturing. These processes can form reservoir-scale vuggy pore systems that dominate the performance of carbonate reservoirs.

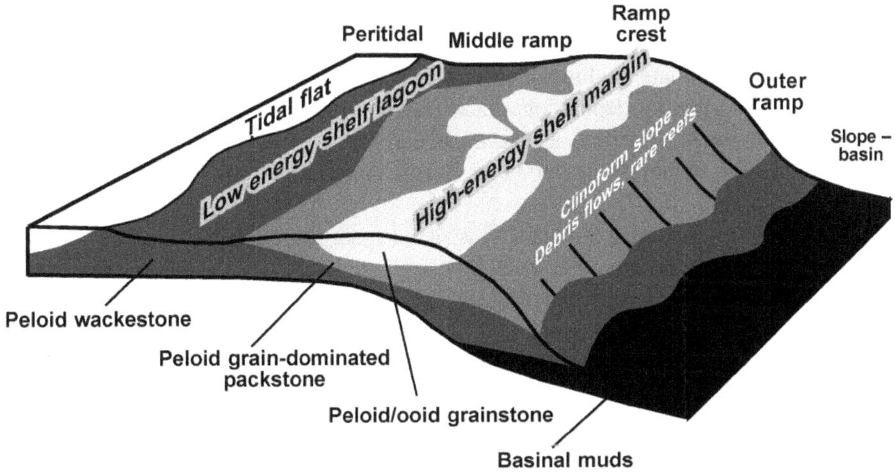

Fig. 1.9—Generalized block diagram Illustrating carbonate-ramp facies patterns and topography.

1.5.2 Depositional Environments. Carbonate sediments accumulate in depositional environments that range from tidal flats to deepwater basins. Most carbonate sediments originate on a shallow-water platform, shelf, or ramp and are transported landward and basinward. "Platform" is a general term for the shallow-water environment, whereas "shelf" and "ramp" refer to topography—shelves with flat platform tops and steep foreslopes and ramps having gently dipping platform tops and slightly steeper foreslopes.

The lateral distribution of depositional environments reflects energy levels, topography, and organic activity. These changes can be related to the geometry of the carbonate platform. Ocean currents are produced by tides and waves and are concentrated at major topographic features, such as ramp and shelf margins, islands, and shorelines. Grainstones and boundstones are concentrated in the areas of highest energy, commonly at ramp and shelf margins. Sediment is transported from the shelf edge onto the shelf slope and into the basin environment. This transport occurs primarily during highstand and results in progradation of the shelf margin. Calcareous plankton is deposited in the basinal environment as well. Sediment is also transported landward onto the shoreline, creating tidal-flat deposits that prograde, primarily during regression. Transgressive sediments are generally wackestones and mudstones at all locations because rising sea level typically creates a low-energy depositional environment.

The combination of organic activity, ocean currents, topography, and eustasy produces a typical facies progression from land to basin during highstand: peritidal, middle ramp, ramp crest, ramp shelf or slope, and basin, as shown in **Fig. 1.9**. The peritidal facies, composed of tidal-flat-capped cycles, normally defines the most landward position of an HFC. The cycles are formed by filling accommodation space and depositing sediment above sea level by transporting carbonate sediment onto the mud flat with tidal and storm currents. Tidal-flat sediments are key indicator facies because they define sea level. The tidal-flat environment is divided into the intertidal zone overlain by the supratidal zone. Sediment in the intertidal zone is characterized by burrowed, pelleted, muddy sediment. Algal laminates are concentrated at the boundary between the intertidal and supratidal zones. The supratidal zone is easily identified by its characteristic irregular lamination, pisolites, mud cracks, intraclasts, and fenestral fabrics. The supratidal zone is sometimes called a "sabkha" environment, referring to the extensive evaporitic flats on the western shore of the Persian Gulf.

In arid climates, evaporite deposits may form by precipitation of gypsum ($CaSO_4{\cdot}2H_2O$) or anhydrite ($CaSO_4$) from evaporation of seawater trapped on or in the supratidal zone. Halite

(NaCl) is normally found in isolated basins similar to the Dead Sea. Sulfate minerals are found as deposits in hypersaline lakes and as beds and crystals within the peritidal sediments. Sulfates found within carbonate sediments are properly classified as diagenetic minerals and cannot be used to describe the depositional environment, but sulfate deposited out of a standing body of water, is properly classified as sediment and is characteristic of the depositional environment as well as the climate. For sulfate to precipitate from seawater, three conditions must be met:

1. The body of seawater must be highly restricted from the ocean.
2. The hypersaline water must be able to escape either by returning to the ocean or by seeping into the underlying sediment (*seepage reflux*), otherwise large volumes of Halite will precipitate forming a bed of salt.
3. The climate must be sufficiently arid to allow the seawater to evaporate to at least one-third its original volume.

The middle-ramp facies is characterized by quiet-water deposits typically composed of skeletal wackestones and mudstones. Burrowing organisms churn the muddy sediment and produce fecal pellets that, together with skeletal material, comprise the grain fraction of the sediment. During highstand, accommodation space may be reduced and water depth lessened to the point at which wave and storm energy increase, lime mud is winnowed out, and a packstone texture is produced. The increase in grain content, possibly capped by packstone, is used to define sea-level changes in this environment.

The ramp-crest facies is characterized by high-energy deposits, typically grainstones and packstones. The classic upward-shoaling succession of wackestone to packstone and grainstone typifies this environment. Typical high-energy deposits are as follows:

- Shelf-margin, tidal-bar, and marine-sand belts.
- Back-reef sands associated with landward transport of sediment for fringing reefs.
- Local middle-shelf deposits associated with gaps between islands or tidal inlets forming lobate tidal deltas.

Packstones are typically churned by burrowing organisms and show no evidence of current transport. Grainstones are commonly crossbedded, often in multiple directions, indicating deposition out of tidal currents. Reefs are also found in the ramp-crest facies. The term reef has been much misused in the petroleum industry. At one time, all carbonate reservoirs were referred to as reefs, and the term is commonly used today to describe any carbonate buildup. However, the term should be restricted to carbonate bodies composed of bindstone, bafflestone, and associated float- and rudstones.

The outer-ramp, or slope, facies is formed by transport of shelf-margin and inner-shelf sediment onto the shelf slope. Sediments are typically wackestones and mudstones, along with occasional packstones and grainstones, in channels associated with density flows into the basin. On steep slopes, sediments may be dominated by sedimentary breccias and debris flows produced by the collapse of a steep shelf margin. The basin facies is typically composed of thin-bedded, quiet-water lime muds that contain planktonic organisms. Wackestones are often punctuated by debris and grain flows. Classic turbidite textures and cycles are also found in basinal carbonate deposits.

1.5.3 Diagenetic Environments. Because all carbonate-reservoir rocks have undergone significant diagenesis, understanding their diagenetic history can be as important as understanding their depositional history. Modern carbonate sediments have sufficient porosity and permeability to qualify as reservoir rocks. Many ancient carbonates, however, lack the porosity and permeability needed to produce hydrocarbons economically. Loss of reservoir quality occurs when sediment changes after deposition. The processes that cause these changes are referred to as diagenetic processes, and the resulting fabric is often referred to as the diagenetic overprint.

Carbonate diagenetic processes include calcium-carbonate cementation; mechanical and chemical compaction; selective dissolution; dolomitization; evaporite mineralization; and massive dissolution, cavern collapse, and fracturing. Whereas sedimentation is a one-time event, diagenesis is a continuing process, and diagenetic processes interact with one another in time and space. Thus, a sequence of diagenetic events may be extremely complicated and the pattern of diagenetic products difficult to predict if they are not related to depositional patterns.

The process of diagenetic overprinting of depositional textures must be understood to predict the distribution of petrophysical properties in a carbonate reservoir. To this end, diagenetic processes are grouped according to their conformance to depositional patterns. Calcium-carbonate cementation, compaction, and selective dissolution form the first group. These processes have the highest conformance to depositional patterns. Reflux dolomitization and evaporite mineralization form the second group. Although these processes depend greatly on geochemical and hydrological considerations, they are often predictable because they can be related to tidal-flat and evaporite depositional environments. Massive dissolution, collapse brecciation and fracturing, and late dolomitization form the third group. These processes have the lowest conformance to depositional patterns, and their products are quite unpredictable.

Calcium-carbonate cementation, compaction, and selective dissolution can often be linked to depositional textures. Because calcium-carbonate cementation begins soon after deposition, it is often connected to the depositional environment. It continues as the sediment is buried, so the distribution of late cements is often unpredictable. Cementation fills pore space from the pore walls inward, reducing both pore size and porosity in proportion to the amount of cement. Compaction and associated cementation are a function of depositional texture and the time-overburden history. Compaction is both a physical and a chemical process resulting from increased overburden pressure caused by burial. Textural effects include porosity loss; pore-size reduction; grain penetration, breaking, and deformation; and microstylolites. Compaction does not require the addition of material from an outside source and is often related to depositional textures. Experiments and observations have shown that mud-supported sediments compact more readily than those that are grain-supported.

Selective dissolution occurs when one fabric element is selectively dissolved in preference to others. Carbonate sediments are composed of three varieties of calcium carbonate—low-magnesium calcite, high-magnesium calcite (magnesium substituted for some calcium in the crystal lattice), and aragonite. Aragonite, in particular, is an unstable form and is rarely found in carbonate rocks. Grains composed of aragonite tend to be dissolved, and the carbonate is deposited as calcite cement. This distribution of aragonite grains can be predicted on the basis of depositional models.

Dolostone (a rock composed of dolomite) is an important reservoir rock. The composition of dolomite is $CaMg(CO_3)_2$, and it is formed by replacement of calcite and by occlusion of pore space. In the following dolomitization equation, x = the amount of carbonate added to the rock in excess of the amount in the sediment.

$$(2-x)CaCO_3 + Mg^{+2} + xCO_3^{-2} = CaMg(CO_3)_2 + (1-x)Ca^{+2} \; . \qquad (1.4)$$

A main source of magnesium is thought to be modified seawater circulating through the sediment in response to various hydrodynamic forces, including density, elevation, and temperature differences. Many pore volumes of dolomitizing fluid are needed to convert a limestone to a dolostone. Therefore, the hydrologic system must be understood for the distribution of dolostone to be predicted. The hypersaline reflux model can be used to predict dolomite patterns because it can be linked to an evaporitic environment. In an arid climate, seawater is trapped in tidal-flat sediment and hypersaline lakes and is concentrated through evaporation, producing a dolomitizing fluid. A hydrodynamic potential is created because the evaporated fluid is denser than seawater or groundwater and the tidal flats are at a slightly higher elevation than sea level. As a result, the hypersaline fluid will reflux down through the underlying sediment, converting it to dolomite. The geometries of dolostone bodies formed by this mechanism can be predicted if the distribution of evaporitic tidal-flat facies is known.

The hypersaline reflux model also accounts for the addition of $CaSO_4$, commonly an evaporite mineral in carbonate reservoirs. $CaSO_4$ is most commonly formed near the Earth's surface in its hydrous form, gypsum ($CaSO_4{\cdot}2H_2O$). However, at higher temperatures, the stable form is anhydrite $CaSO_4$, which is the form most commonly found in carbonate reservoirs. In some locations, tectonics has uplifted carbonate strata into a cooler temperature, and anhydrite has hydrated, forming gypsum.

Four types of anhydrite are commonly found in dolostone reservoirs. Pore-filling anhydrite is typically composed of large crystals filling interparticle and vuggy pore space. Poikilotopic anhydrite is found as large crystals with inclusions of dolomite scattered throughout the dolostone. They are both replacive and pore filling. Nodules of anhydrite are composed of microcrystalline anhydrite, often showing evidence of displacing sediment. They make up a small percentage of the bulk volume and have little effect on reservoir quality. Bedded anhydrite is found as beds composed of both coalesced nodules and laminations. Anhydrite beds are flow barriers and seals in reservoirs.

Massive dissolution, collapse brecciation and fracturing, and late dolomitization are the most unpredictable diagenetic processes. Massive dissolution refers to nonfabric selective dissolution, including cavern formation at any scale, collapse brecciation and fracturing, solution enlargement of fractures, and dissolution of bedded evaporites. This process is thought to be most commonly related to the flow of near-surface groundwater, referred to as meteoric diagenesis but often included under the general heading of karst. The products of this diagenetic environment are controlled by precursor diagenetic events, tectonic fracturing, and groundwater flow and show little relationship to depositional environments. Reservoirs of this type are, therefore, difficult to model.

1.6 Siliciclastic Reservoirs

1.6.1 Introduction. Siliciclastic rocks are composed of terrigenous material formed by the weathering of pre-existing rocks, whereas carbonate rocks are composed principally of sediment formed from seawater by organic activity. Clastic sediments are composed of grains and clay minerals, and siliciclastic sediments are first classified according to grain type. The three basic grain types are quartz, feldspar, and rock fragments, and the end members are quartz sandstone, arkosic sandstone, and lithic sandstone, as shown in **Fig. 1.10a**. Second, siliciclastics are described in terms of grain size (Fig. 1.10b). Grain-size classes include gravels (boulder size to 2 mm in diameter), sands (2 to 0.0625 mm), and mud, which includes silts (0.0625 to 0.004 mm) and clay (< 0.004 mm). Mixtures are described with a modifying term for a less-abundant size, such as clayey sandstone, sandy siltstone, or muddy sandstone (Fig. 1.10c). Mudstone, composed of clay and silt, is not to be confused with carbonate mudstone. In this classification, mud and clay are terms used to indicate size, not mineralogy.

The porosity and permeability of unconsolidated siliciclastic sediments were measured by Beard and Weyl.[17] Porosity varies from 45% for well-sorted sands to 25% for very poorly sorted sands and does not vary with changes in grain size for well-sorted media. Permeability ranges from 400 darcies in well-sorted, coarse-grained sands to 0.1 darcies (100 md) in very poorly sorted, fine-grained sands. Permeability varies with grain size and sorting because it is controlled by pore-size distribution. Most modern sands are reservoir-quality rock. Modern claystones and mudstones, which are composed primarily of clay minerals, have little permeability and are not reservoir quality.

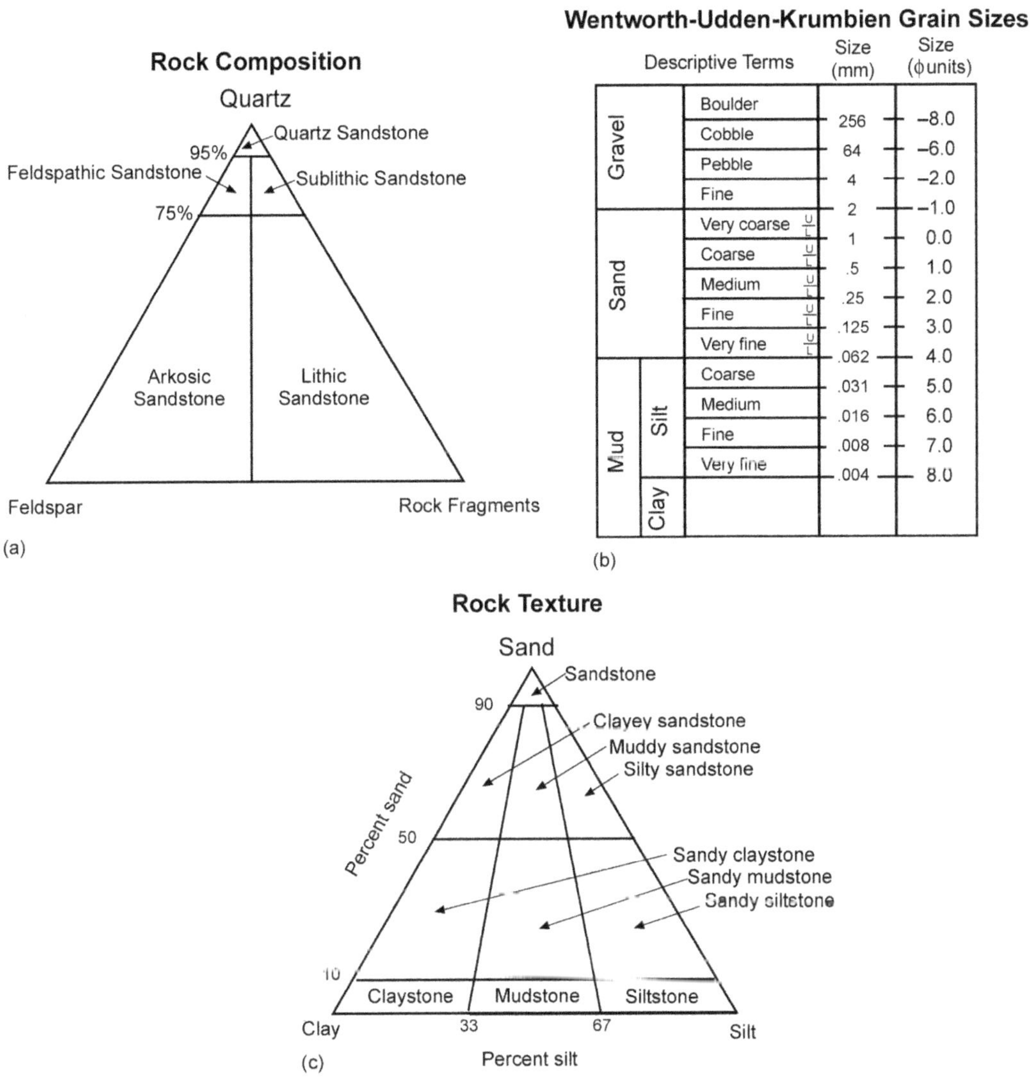

Fig. 1.10—Classification of siliciclastic rocks by (a) composition, (b) grain size, and (c) texture. The ϕ scale is defined as $\phi = -\log_2 (size)$.

The type, amount, and habit of clay minerals in siliciclastic rocks are important characteristics (see **Fig. 1.11**). Clay minerals are sheet-structure silicates that have a profound impact on the petrophysical and production properties of sandstones. They can be deposited as muddy sediment or formed during burial by diagenetic processes (sometimes referred to as authigenic clay). Common clay minerals are kaolinite [$Al_2Si_2O_5(OH)_4$], illite [$KAl_3Si_3O_{10}(OH)_2$], chlorite [$(Al, Mg, Fe)Si_4O_{10}(OH)_2$], and smectite or montmorillonite [$(Al, Mg)Si_4O_{10}(OH)_2$].

The mineralogy of the clay minerals has a great effect on pore size and petrophysical properties.[18] For example, kaolinite-cemented sandstones are more permeable than are illite-cemented sandstones because kaolinite tends to form boolets that reduce pore size and porosity, whereas illite tends to form thin threads that reduce pore size with little effect on porosity. Clay minerals are also known to hinder inflow into the wellbore. Smectite, for example, tends to swell, reducing permeability when in contact with fresh water. Kaolinite is known to get

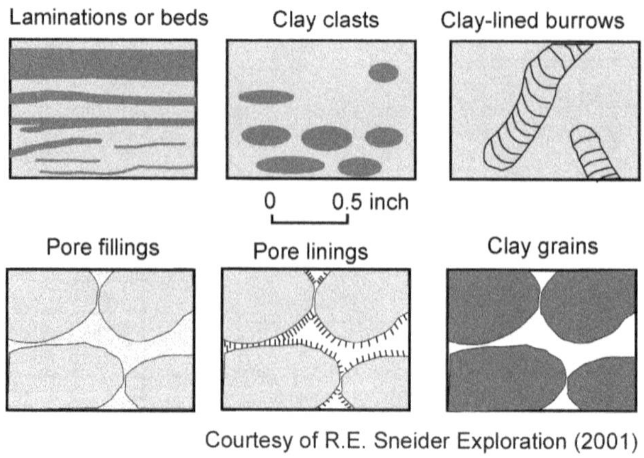

Fig. 1.11—Diagrammatic illustration of the basic, different types of clay habit in siliciclastic rocks. Figure courtesy of Sneider Exploration Inc.

dislodged by high-velocity flow and plug pore throats near the wellbore, reducing permeability. The iron in chlorite is commonly released during acid treatments, plugging perforations.

1.6.2 Depositional Environments. The following discussion is taken primarily from Galloway and Hobday.[19] Grain type, size, and sorting, as well as other characteristics of siliciclastic reservoirs are most commonly controlled by the depositional environment. Many siliciclastic reservoirs are geologically young, and the sediment has undergone only moderate compaction and cementation. Therefore, diagenesis is not a major factor, and petrophysical properties can be predicted on the basis of sedimentology.

Siliciclastic sediments are transported and deposited by wind and flowing water. On land, clastics are deposited by wind and stream flow. In the marine environment, they are transported by tidal, wave, ocean, and density currents. Land-based environments are grouped into alluvial-fan, fluvial, and eolian systems. Ocean-center environments include delta systems and barrier bars, which are transitional between land and marine environments, and shelf, slope, and basinal systems, which are marine (see **Fig. 1.12**).[20]

Alluvial fans are conical, lobate, or arcuate accumulations of predominately coarse-grained clastics extending from a mountain front or escarpment across an adjacent lowland. Some fans terminate directly in lakes or ocean basins as fan deltas, which generally show some degree of distal modification by currents or waves. Most sediment is deposited by stream and debris flow. Stream flow is commonly confined to one or two channels but may spread across the fan as sheet-flow. Debris flows result when clay and water provide a low-viscosity medium of high yield strength capable of transporting larger particles under gravity. Wave and tidal currents modify the distal terminations of fans that build into lakes or the ocean, improving sorting and reservoir quality.

Eolian deposits are typically fine- to medium-grained, well-sorted, quartzose sand with pronounced crossbedding. The sand is transported and deposited by wind currents, which are the most effective agents for sorting clastic particles. Hot, arid regions are the most favored locales for eolian accumulation. Eolian environments can be divided into dune and interdune facies. Dunes are large bed forms that come in an array of forms. Barchans, barchanoid ridges, and transverse dunes form in response to essentially unidirectional winds. Longitudinal dunes arise from varying wind directions. Draas comprise large stellate rosettes with a high central peak and radiating arms and form in response to intense, multidirectional wind systems. The interdune environment is generally a broad, featureless plain covered by lag gravels resulting from

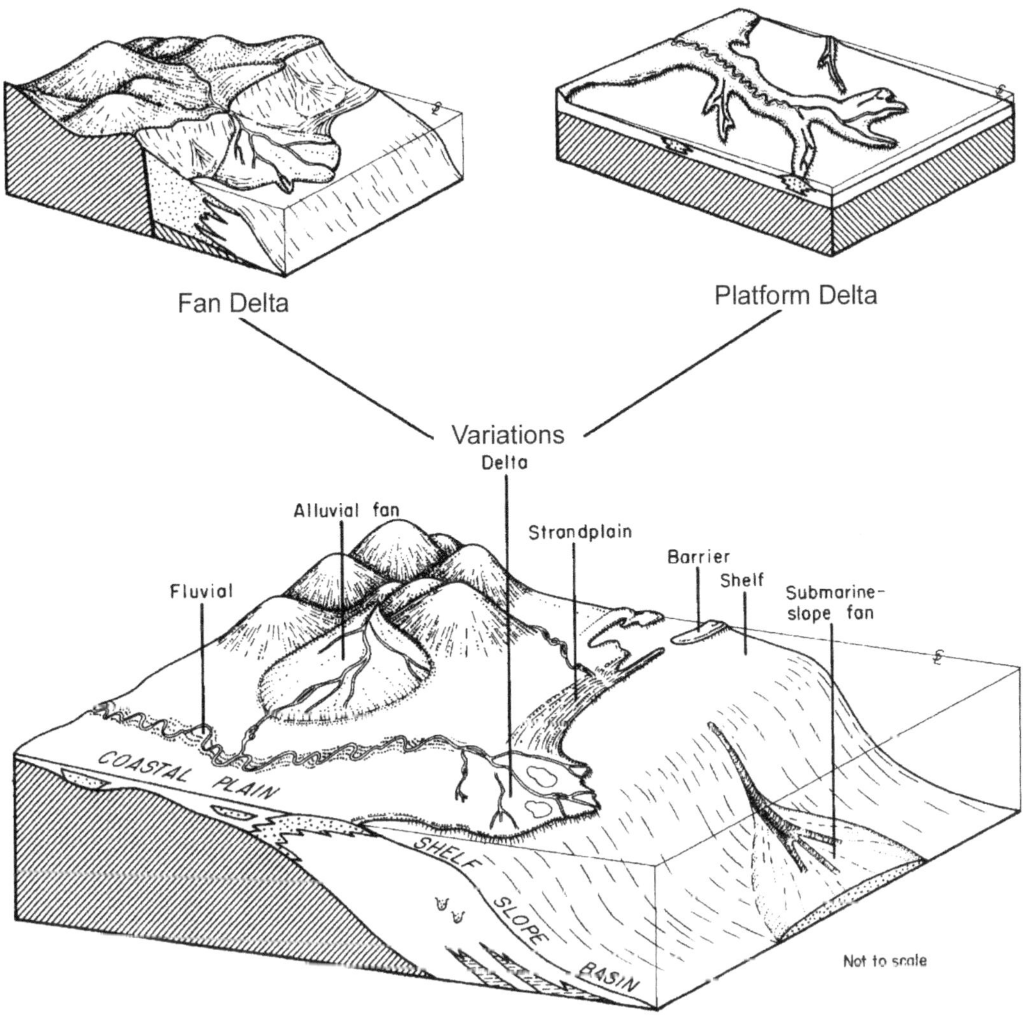

Fig. 1.12—Range of sandstone depositional systems that typically host hydrocarbon resources.[20]

deflation (erosion). Deposition in the interdune area results from rainfall in desert highlands, promoting ephemeral streams that deposit sediment in streambeds and small alluvial fans. Flooding may produce interdune braided-stream deposits. Ponding of water between dunes can create lakes that can precipitate evaporite minerals if the groundwater is sufficiently saline.

Fluvial systems are a collection of stream channels and their floodplains. The channels are sinuous (meandering), with the degree of sinuosity increasing seaward. Braided streams are the result of sand-rich channels. Channel deposits are composed of sand bars and lag deposits. The point bar, a major feature of a high-sinuosity channel, forms by lateral accretion of sediment in the lower-energy, leeward side of a meander. Deposition normally occurs during the ebbing phase of a flood. The highest energy, found in the channel proper, erodes the channel bank, causing the channel to shift constantly; lag deposits are characteristic of the channel. Abandoned channels are commonly clay filled.

Floodplain deposits are deposited as levees, crevasse splays, and flood-basin sediments. Levee deposits are fine sand, silt, and clay deposited along the margins of the channels, when decelerating water rich in suspended sediment spills over the banks during flood stage. Crevasse splays are formed when local breaches in the levees funnel floodwater into near-channel parts of the flood plain. These sediments tend to be highly heterogeneous, composed of sand of variable size, plant debris, and mud clasts. Flood-basin deposits are broad, clay-rich sediments that have been reworked by burrowing animals, plant growth, and pedogenic (soil-forming) processes.

Delta systems form when a river transporting sediment enters a standing body of water, commonly an ocean or a lake, and consist of both fluvial and marine sediments. The depositional architecture of a delta system is characteristically progradational and may fill a small basin. The combination of fluvial and marine processes creates a unique facies assemblage and reservoir architecture. Deltaic sediments are deposited as channel fills, channel-mouth sands, crevasse splays, and delta-margin sand sheets. Together, these facies compose a delta lobe, which is a fundamental building block of a delta system. Delta systems are divided into fluvial-, wave-, and tidal-dominated deltas according to major energy type. Each system has a unique depositional architecture.

Shore-zone systems, excluding deltas, compose a narrow transitional environment that extends from wave base ($\approx$50 ft of water) to the seaward edge of the alluvial coastal plain. They include shoreface, beach, barrier, lagoon, and tidal-flat facies. These systems are supplied principally by onshore transport of river-derived and shelf sediments. Sands are concentrated in barrier-island complexes and tidal sand bodies, with finer sediment landward. Accretion of beach ridges seaward can form a sheetlike sand body referred to as a strandplain sand. The "shoreface facies" refers to that part of the shore zone that is below the zone of wave swash. It is commonly divided into lower-, middle-, and upper-shoreface deposits partly on the basis of water depth and associated energy levels, the highest energy level being the surf zone (upper shoreface). Beach facies includes wave swash and dune zone, all deposited above mean tide. The barrier is formed by aggradation or by progradation of shoreface sands seaward. The lagoon facies, located behind the barrier, is generally composed of clay and fine sand. The barrier may be breached during storms, allowing tidal currents to transport coarser sediment from the ocean into the lagoon, forming tidal deltas.

Shelf systems are broad, deepwater platforms covered by terrigenous sediment. Sediment distribution is controlled by ocean currents, including tidal, wave, storm surge, and density. Facies are defined by bed form and include sand ribbon, wave, ridge, storm, and mud.

Slope and basin systems are found in the relatively deep water beyond the shelf break. Deposition is characterized by the dominance of sediment transport by gravity and density flow, although pelagic settling also occurs. The upper slope is typically a zone of sand remobilization and bypass, with characteristic erosion and channel cutting; the lower slope and basin floor are sites of deposition. Regionally, grain size is the coarsest in the upper slope and decreases in the basin-floor direction. Slope and basin systems are typically distinguished from other systems by their fining-upward-graded bedding, which results from grain settling from a suspended sediment load. Submarine fans are typical slope and basin-floor deposits. Fed from point sources, such as river mouths or submarine canyons, they receive the bulk of their sediments from turbidity currents, a density current produced by sediment-rich water. The upper-fan environment is characterized by feeder channels or canyons that serve as sediment conduits, and sediments are typically coarse gravels. The midfan is characterized by a series of bifurcating, distributary, or braided channels that accumulate massive and pebbly sands showing lenticular bedding, and the lower fan is a smooth, gently sloping surface that received slowly deposited, suspended sediment punctuated by pulses of fine-grained to silt-sized sand. The resulting graded beds are thin, laterally persistent, and monotonously repetitive, commonly through a considerable thickness.

1.6.3 Diagenetic Environments. Sandstones are less susceptible to diagenetic change than carbonates. Common diagenetic processes in sandstones are quartz overgrowth cement, carbonate (calcite and dolomite) cement, compaction, grain dissolution and associated formation of clay

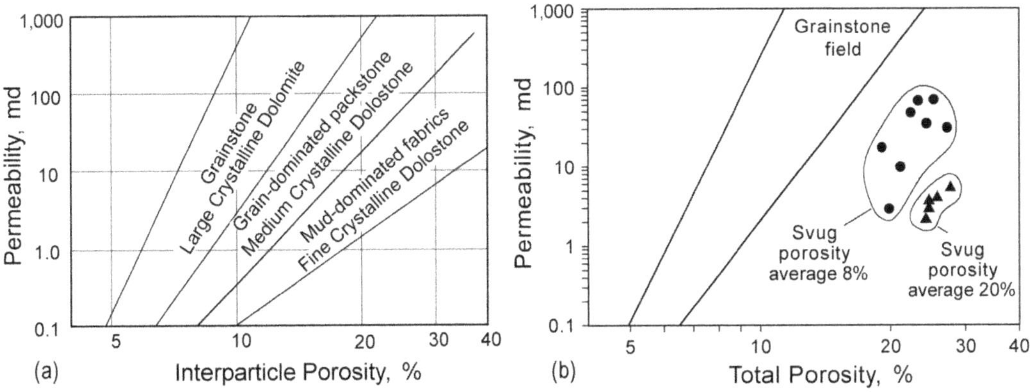

Fig. 1.13—Relationship between carbonate rock-fabric facies, porosity, and permeability.[12] (a) Crossplot of interparticle porosity and permeability showing rock-fabric fields. (b) Crossplot of total porosity and permeability showing effect of separate-vug (Svug) porosity on permeability of grainstone.

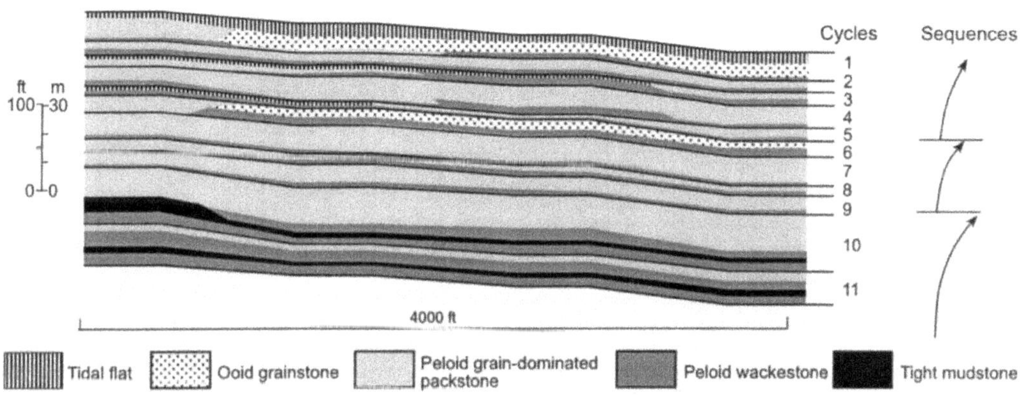

Fig. 1.14—Rock-fabric, sequence-stratigraphic carbonate reservoir suitable for petrophysical quantification.

minerals, and alteration of sedimentary clay minerals. Many of these products can be related to the burial history. Pore space is reduced by mechanical and chemical compaction, resulting in more closely spaced grains and smaller pores, and by quartz overgrowths, which are commonly sourced from chemical dissolution of quartz grains during burial. Carbonate cements are formed by dissolution and precipitation of indigenous carbonate shell material and by importation of carbonate from a more distant source. Iron-rich, pore-filling dolomite is not uncommon.

Feldspar minerals found in rock fragments are commonly unstable in the burial environment and are susceptible to dissolution, forming grain molds similar to those in carbonate rocks. Clay minerals (commonly chlorite) are deposited in the intergrain spaces associated with this dissolution process. Chlorite linings of pore space are thought to inhibit burial cementation and compaction and preserve porosity at depth. Clay minerals are altered during burial diagenesis, and authigenic (diagenetic) clay minerals are formed. Clay-mineral diagenesis causes large increases in surface area and microporosity that, in turn, have large effects on reservoir performance and log analysis.

1.6.4 Reservoir Models. Reservoir models are constructed by distributing petrophysical properties in 3D space with geologic models as a template. Geologic models are constructed by

(a) Porosity, %

Sorting	Grain Size: Coarse	Medium	Fine	Very Fine
Extremely well sorted	43	41	42	43
Very well sorted	41	40	40	41
Well sorted	38	38	39	40
Moderately sorted	33	34	34	34
Poorly sorted	28	31	30	30
Very poorly sorted	27	25	28	30

(b) Permeability, darcies

Sorting	Grain Size: Coarse	Medium	Fine	Very Fine
Extremely well sorted	350	89	22	5
Very well sorted	350	86	21	5
Well sorted	226	57	14	3
Moderately sorted	82	21	5	2
Poorly sorted	34	9	3	0.7
Very poorly sorted	10	2	0.6	0.1

Fig. 1.15—Relationship between siliciclastic texture, porosity, and permeability in unconsolidated sediments.[17] (a) Porosity varies with sorting and not with grain size. (b) Permeability varies with grain size and sorting.

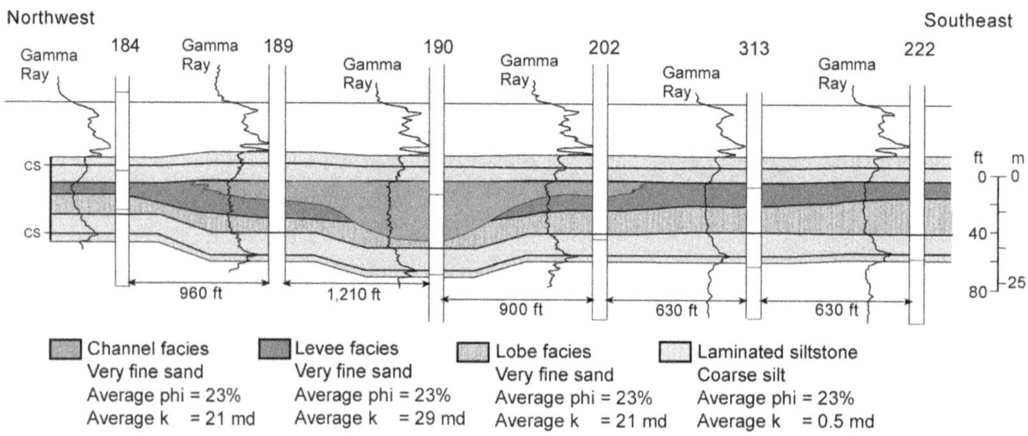

Fig. 1.16—Depositional facies model in a basin-floor fan showing grain size, average porosity, and average-permeability data.[21]

distributing facies within a sequence stratigraphic framework using the systematic distribution of facies within a depositional model as a guide. There are many types of facies, and facies selection is normally based on the question asked. Water depth and changes in sea level are key questions when building a sequence stratigraphic model, and fossil and other grain types together with depositional textures are keys to estimating water depth. Thus, numerous "depositional" facies are commonly described from core material. Once a sequence model is built, however, the problem is to convert the geologic model into a reservoir model by populating the geologic model with petrophysical data. This problem is best resolved by linking petrophysical measurements to rock fabric or texture, and the number of "rock-fabric" facies needed to quantify the geologic model is generally much lower than depositional facies.

Examples of carbonate rock-fabric facies include grain-dominated packstone, medium crystalline dolowackestone, and moldic grainstone. In carbonate reservoirs, rock fabrics include diagenetic overprints as well as depositional textures because diagenesis plays an important role in forming most carbonate pore space. An example of how rock-fabric facies relate to porosity and permeability is illustrated in **Fig. 1.13.**[12] Particle size, sorting, interparticle porosity, and vuggy porosity are used to characterize pore-size distribution. Petrophysical quantification of a carbonate stratigraphic model is accomplished by mapping rock-fabric facies and interparticle porosity and calculating permeability and initial water saturation from rock-fabric-specific relationships to porosity. An example of a rock-fabric reservoir model is illustrated in **Fig. 1.14.**

Petrophysical properties of siliciclastic rocks are often directly related to facies characterized by grain size and sorting because pore space is generally located between grains and the variability of porosity within a facies is small. Porosity generally decreases with decreasing sorting and remains constant with changes in average grain size (see **Fig. 1.15**). However, porosity will vary with cementation and compaction. Permeability decreases with decreasing sorting and grain size (Fig. 1.15), and as porosity decreases even though grain size and sorting remain constant. The finest grain size is found in shales (mudstones) that typically have little permeability. Once the depositional facies are distributed in 3D space, the model can be quantified with petrophysical properties by using grain size, sorting, and porosity characteristics of the depositional facies. As illustrated in **Fig. 1.16**[21] (a geologic model from a deepwater sediment fan), depositional facies commonly have similar petrophysical properties that reduce the number of textural facies needed to quantify the geologic model. The distribution of shales (mudstone) is important for reservoir modeling because they typically have little permeability and act as flow barriers in the reservoir.

Fracture porosity can have a large influence on performance of a carbonate or siliciclastic reservoir. Characterizing and modeling fracture porosity is difficult and is beyond the scope of this discussion. The book on fractured reservoirs by Aguilera[22] is suggested as a good general overview of the problem.

Nomenclature

g = acceleration of gravity, ft/sec^2
h_c = maximum hydrocarbon column, ft
H = fluid potential, ft
P = measured static pressure, psia
p_c = capillary entry pressure, psia
x = horizontal distance, ft
Z = elevation relative to a datum, sea level, ft
θ = angle between horizontal and dip of water table, degrees
ρ_w = density of water, g/cm^3
ρ_{hc} = density of the hydrocarbon, g/cm^3
ρ_o = density of oil, g/cm^3
ϕ = porosity, fraction
k = permeability, md

References

1. Hunt, J.M.: *Petroleum Geochemistry and Geology,* W.H. Freeman and Co., San Francisco (1979) 617.
2. Milton, N.J. and Bertram, G.T.: "Trap Styles—A New Classification Based on Sealing Surfaces," *AAPG Bull.* (1992) **76,** No. 7, 983.
3. Demaison, G. and Huizinga, B.J.: "Genetic Classification of Petroleum Systems," *AAPG Bull.* (1991) **75,** No. 10, 1626.

4. Sneider, R.M. *et al.*: "Comparison of Seal Capacity Determinations: Conventional Cores Versus Cuttings," *Seals, Traps, and the Petroleum System*, R.C. Surdam (ed.), *AAPG Memoir 67* (1997) 12.
5. Smith, D.A.: "Theoretical Considerations of Sealing and Non-Sealing Faults," *AAPG Bull.* (1966) **50,** No. 2, 363.
6. Hubbert, M.K.L.: "Entrapment of Petroleum under Hydrodynamic Conditions," *AAPG Bull.* (1953) **37,** No. 8, 1954.
7. Schowater, T.T.: "Mechanics of Secondary Hydrocarbon Migration and Entrapment," *AAPG Bull.* (1979) **63,** No. 5, 723.
8. North, F.K.: *Petroleum Geology*, Unwin Hyman Inc., Boston, Massachusetts (1985) 631.
9. Von Wagoner, J.C.: "Overview of Sequence Stratigraphy of Foreland Basin Deposits: Terminology, Summary of Papers, and Glossary of Sequence Stratigraphy," *Sequence Stratigraphy of Foreland Basin Deposits; Outcrop and Subsurface Examples from the Cretaceous of North America*: J.C. Van Wagoner and G.T. Bertram (eds.), *AAPG Memoir 64* (1995) 4.
10. Kerans, C. and Tinker, S.W.: "Sequence Stratigraphy and Characterization of Carbonate Reservoirs," *SEPM Short Course Notes No. 40* (1997) 130.
11. Goldhammer, R.K.: "Hierarchy of Stratigraphic Forcing: Example from Middle Pennsylvanian Shelf Carbonates of the Paradox Basin," *Sedimentary Modeling: Computer Simulations and Methods for Improved Parameter Definition,* E.K. Franseen, *et al.* (eds.) *Kansas Geological Survey Bull. 233* (1991) 361.
12. Lucia, F.J.: *Carbonate Reservoir Characterization,* Springer, New York (1999) 226.
13. Dunham, R.J.: "Classification of Carbonate Rocks According to Depositional Texture," *Classification of Carbonate Rocks—A Symposium,* W.E. Ham (ed.), *AAPG Memoir No. 1* (1962) 108.
14. Embry, A.J. and Klovan, R.E.: "A Late Devonian Reef Tract of Northeastern Banks Island, N.W.T.," *Bulletin of Canadian Petroleum Geology* (1971) **19,** No. 4, 730.
15. Enos, P. and Sawatsky, L.H.: "Pore networks in Holocene carbonate sediments," *J. of Sedimentary Petrology* (1981) **51,** No. 3, 961.
16. Lucia, F.J.: "Rock-Fabric/Petrophysical Classification of Carbonate Pore Space for Reservoir Characterization," *AAPG Bull.* (1995) **79,** No. 9, 1275.
17. Beard, D.C. and Weyl, P.K.: "Influence of Texture on Porosity and Permeability of Unconsolidated Sand," *AAPG Bull.* (1973) **57,** No. 2, 349.
18. Neasham, J.W.: "The Morphology of Dispersed Clay in Sandstone Reservoirs and Its Effect on Sandstone Shaliness, Pore Space and Fluid Flow Properties," paper SPE 6858 presented at the 1977 SPE Annual Technical Conference and Exhibition, Denver, 9–12 October.
19. Galloway, W.E. and Hobday, D.K.: *Terrigenous Clastic Depositional Systems; Applications to Petroleum, Coal, and Uranium Exploration,* Springer-Verlag, New York (1983) 423.
20. Galloway, W.E. *et al.*: *Atlas of Major Texas Oil Reservoirs,* U. of Texas at Austin, Bureau of Economic Geology (1983) 139.
21. Dutton, S.P., Barton, M.D., Asquith, G.B., Malik, M.A., Cole, A.G., Gogas, J., Guzman, J.I., and Clift, S.J.: *Geologic and Engineering Characterization of Turbidite Reservoirs, Ford Geraldine Unit, Bell Canyon Formation, West Texas,* The U. of Texas at Austin, Bureau of Economic Geology Report of Investigations No. 255 (1999) 88.
22. Aguilera, R.: *Naturally Fractured Reservoirs*, PennWell Publishing Co., Tulsa (1995).

SI Metric Conversion Factors

°API	141.5/(131.5 + °API)		= g/cm^3
ft ×	3.048*	E − 01	= m
g/cm^3 ×	1.0*	E + 03	= kg/m^3
mm ×	2.54*	E + 01	= in.
psi ×	6.894 757	E + 00	= kPa

*Conversion factor is exact.

Chapter 2
Fundamentals of Geophysics

Bob A. Hardage, U. of Texas at Austin

2.1 Introduction

Geophysics is a broad subject that encompasses potential field theory (gravity and electromagnetic fields) and seismic technology. Potential field data are valuable in many studies, but seismic data are used in more reservoir characterization and reservoir management applications. This chapter focuses on seismic fundamentals and does not consider gravity, magnetic, or electrical concepts.

Seismic data have been used for many years to guide exploration. More recently, seismic data have been used to support reservoir characterization in field development planning and subsequent reservoir management. As the technology in equipment and interpretation techniques has advanced, so has the ability to define the size, shape, fluid content, and variation of some petrophysical properties of reservoirs. This chapter provides insight into the fundamentals of seismic data acquisition, interpretation techniques, and the types of information that can be derived. See the chapter on reservoir geophysics in the Emerging and Peripheral Technologies section of this *Handbook* for information on emerging technologies that apply geophysical data.

Most seismic data are acquired with surface-positioned sources and receivers. For the first 4 or 5 decades that seismic-reflection data were acquired, sources and receivers were deployed along the same straight line to create 2D seismic profiles. Two-dimensional seismic data do not yield a correct image of subsurface stratigraphy when a 2D seismic line crosses a complex subsurface structure because the acquisition geometry cannot distinguish reflections that originate from outside the profile plane from reflections that occur within the 2D vertical image plane.

This imaging deficiency of 2D seismic profiling has been remedied by the implementation of 3D seismic data acquisition, which allows data processing to migrate reflections to their correct image coordinates in 3D space. Industry largely abandoned 2D seismic profiling in the 1990s and now relies almost entirely on 3D seismic data acquisition. This chapter covers the basics of 3D seismic technology, but does not consider 2D seismic technology involving surface-positioned equipment.

In some reservoir applications, seismic data are acquired with downhole sources and receivers. If the receiver is stationed at various depth levels in a well and the source remains on the surface, the measurement is called vertical seismic profiling (VSP). This technique produces a high-resolution, 2D image that begins at the receiver well and extends a short distance (a few tens of meters or a few hundred meters, depending on the source offset distance) toward the source station. This image, a 2D profile restricted to the vertical plane passing through the source and receiver coordinates, is useful in tying seismic responses to subsurface geologic and engineering control.

If the source is deployed at various depth levels in one well and the receiver is placed at several depth stations in a second well, the measurement is called crosswell seismic profiling (CSP). Images made from CSP data have the best spatial resolution of any seismic measurement used in reservoir characterization because a wide range of frequencies is recorded. CSP data are useful for creating high-resolution images of interwell spaces and for monitoring fluid movements between wells. However, a CSP image is also a 2D profile with the image limited to the vertical plane that passes through the source and receiver coordinates. This chapter includes brief descriptions of the fundamentals of subsurface VSP and CSP technologies to complement the descriptions of surface-positioned seismic technology.

2.2 Impulsive Sources

A variety of seismic sources exist that can apply vertical impulse forces to the surface of the ground. These devices are viable energy sources for onshore seismic work. Included in this source category are gravity-driven weight droppers and other devices that use explosive gases or compressed air to drive a heavy pad vertically downward. Refs. 1 through 3 describe these types of sources.

Chemical-explosive energy sources are popular for onshore seismic surveys but are prohibited at some sites because of environmental conditions, cultural restrictions, or federal and state regulations. Chemical explosives are no longer used as marine energy sources for environmental and ecological reasons.

Field tests should always be made before an extensive seismic program is implemented. First, it should be determined whether the selected impulsive source creates adequate energy input to provide data with an appropriate signal-to-noise ratio and a satisfactory signal bandwidth at appropriate offset distances. Second, it is important to determine whether an impulsive source causes unwanted reverberations in shallow strata.

2.3 Vibrators

Vibroseis™ energy sources are some of the more popular seismic source options for onshore hydrocarbon exploration. The generic term vibrator refers to these types of seismic sources. Vibrators have several features that make them attractive for seismic data acquisition. They are quite mobile and allow efficient and expeditious illumination of subsurface targets from many different shotpoint locations. Also, the frequency content of a vibrator signal often can be adjusted to better meet resolution requirements needed for a particular target. In addition, the magnitude of the energy input into the Earth can be tailored for optimal signal-to-noise conditions by varying the size and number of vibrators or by altering the output drive of individual vibrators. For these reasons, vibrators are one of the most versatile onshore seismic energy sources.

Vibrators work on the principle of introducing a user-specified band of frequencies, known as the sweep, into the Earth and then crosscorrelating that sweep function with the recorded data to define reflection events. The parameters of a vibrator sweep are start frequency, stop frequency, sweep rate, and sweep length. A vibrator can do an upsweep that starts with a frequency as low as 8 to 10 Hz and stops at a high value of 80, 100, or 120 Hz. Alternatively, vibrators can do a downsweep that starts with a high frequency and finishes with a low frequency. Most Vibroseis data are generated with upsweeps.

Sweep rate can be linear or nonlinear. A linear rate causes the vibrator to dwell for the same length of time at each frequency component. Nonlinear sweeps are used to emphasize higher frequencies because the vibrator dwells longer at higher frequencies than it does at lower frequencies.

Sweep length defines the amount of time required for the vibrator to transverse the frequency range between the start and stop frequencies. As sweep length is increased, more energy is put into the Earth because the vibrator dwells longer at each frequency component. Sweep length is usually in the range of 8 to 14 seconds.

If a vibrator sweep is 12 seconds long, then each reflection event also spans 12 seconds in the raw, uncorrelated data. It is not possible to interpret uncorrelated Vibroseis data because all reflection events overlay each other and individual reflections cannot be recognized. The data are reduced to an interpretable form by a crosscorrelation of the known input sweep with the raw data recorded at the receiver stations. Each time the correlation process finds a replication of the input sweep, it produces a compact symmetrical correlation wavelet centered on the long reflection event. In this correlated form, Vibroseis data exhibit a high signal-to-noise ratio, and reflection events are robust wavelets spanning only a few tens of milliseconds.

As a general observation, if an area is plagued by random noise, vibrators are an excellent energy source because the correlation process used to reduce the vibrator sweep to an interpretable form discriminates against noise frequencies that are outside the source sweep range. If several sweeps are summed, unorganized noise within the sweep range is attenuated. However, if coherent noise with frequencies within the vibrator sweep frequency range is present, then the correlation process may accentuate these noise modes. Refs. 4 through 13 are a portion of the available literature that describes the operating characteristics of vibrators and their ability to optimize signal-to-noise ratios.

Probably the most important improvement in vibrator operations has been the development of ground-force phase-locking technology.[13] Application of this technology results in the same ground-force function (i.e., the same basic seismic wavelet) being generated during hundreds of successive sweeps by vibrators positioned over a wide range of ground-surface and soil conditions and by all vibrators in a multivibrator array. All aspects of seismic data processing benefit when a source generates consistent output wavelets throughout a seismic survey, hence the appeal of vibrators as the source of choice for most onshore surveys.

2.4 Shear Wave Sources

All seismic energy sources generate compressional (P) and shear (S) body waves. To study the physics and exploration applications of S-waves, it is often necessary to increase the amount of S-wave energy in the downgoing wavefield and to produce a shear wavefield that has a known vector polarization. These objectives can be accomplished with sources that apply horizontally directed impulses to the Earth or by vibrators that oscillate their baseplates horizontally rather than vertically. In either case, a heavy metal pad is used to impart horizontal movement to the Earth by means of cleats on the bottom side of the pad that project into the Earth. A specific design for a horizontal shear-wave vibrator can be found in a patent issued to Fair.[14]

Horizontal vibrators have also been improved with the introduction of ground-force phase-locking technology that results in more consistent shear wavelets from sweep to sweep as horizontal vibrators move across a prospect. Surface damage has been minimized by reducing the size of the cleats underneath the baseplate so that they make only shallow ground depressions.

2.5 Marine Air Guns

Air guns are now the primary energy sources used in offshore seismic profiling. Chemical explosives are no longer used for safety reasons and because of their adverse effects on marine biology. Modern seismic vessels tow multiple arrays of air guns, and each array sometimes has 10 or more air guns. The size and position of each air gun in the array are engineered so that the output wavelet has minimal bubble oscillations and optimal peak-to-peak amplitude. **Fig. 2.1** shows map and section views of the deployment of air guns from a seismic vessel. The air guns and hydrophone cable are positioned at proper lateral offsets from each other by steering

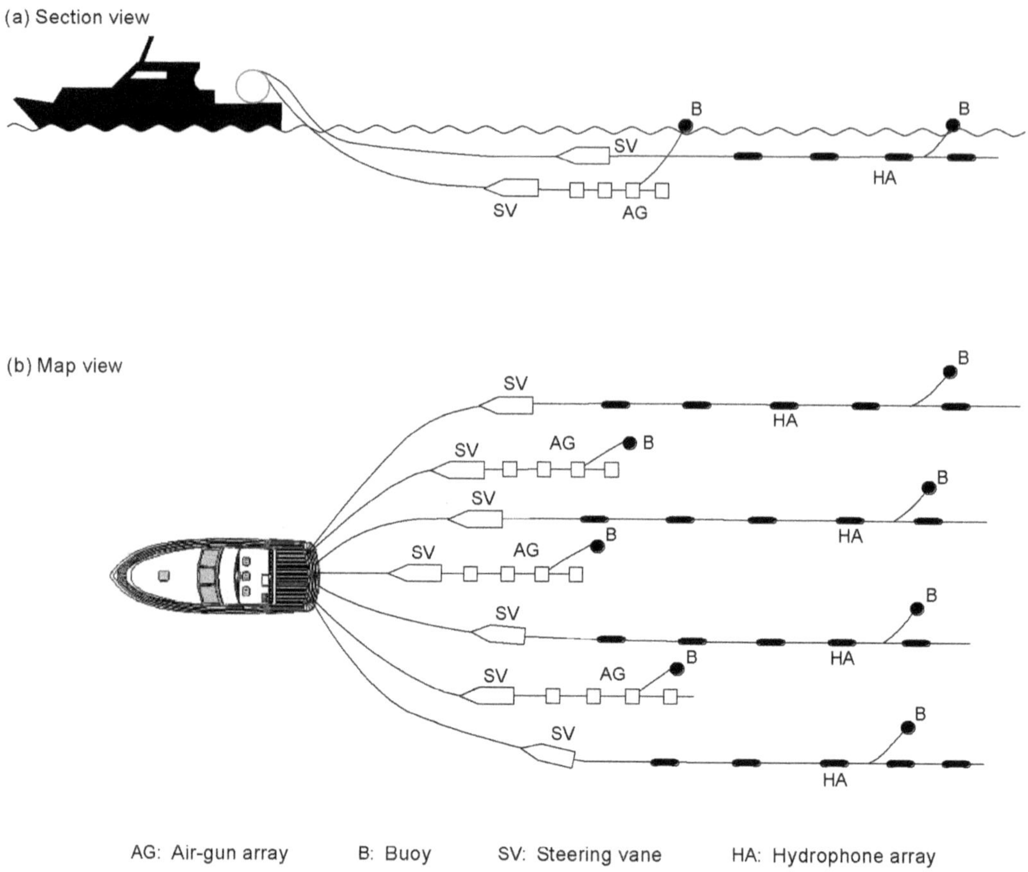

Fig. 2.1—Equipment deployed by marine seismic vessel with towed-cable technology.

vanes. Combinations of depth-control vanes and surface buoys keep the air guns and hydrophones at a constant depth as they are towed across a survey area.

The air-gun arrays are powered by large onboard compressors that allow the guns to fire and repressure (to approximately 2,000 psi) at intervals of 8 to 10 seconds as the vessel steams along a prescribed course at constant speed. These high-repetition firing rates create shotpoints at regular spacings of 20 to 100 m, depending on boat speed, along the length of the source line. In wide-line profiling, a vessel tows several air-gun arrays spaced 50 to 100 m apart laterally to create parallel source lines in one traverse of the vessel across a prospect area (Fig. 2.1).

Fig. 2.2 shows the raypaths involved in air-gun illumination of a geologic target. The seismic energy produced by an air-gun shot propagates up and down from the source array. The downgoing raypath creates the primary arrival. The upgoing raypath reflects from the water surface (where the reflection coefficient is –1 for an upgoing pressure wave) and travels downward as a time-delayed ghost event. Reduced-amplitude versions of the primary and ghost events follow at later times as the air bubble created by the shot oscillates and decays. These three components—primary arrival, ghost arrival, and bubble oscillations—combine to form the air-gun source wavelet.

The effectiveness of air-gun array parameters is tested in deepwater environments in which a hydrophone can be positioned at a deep, far-field station to record the source output wavelet, as shown in Fig. 2.2. The term far field refers to that part of wavefield propagation space that

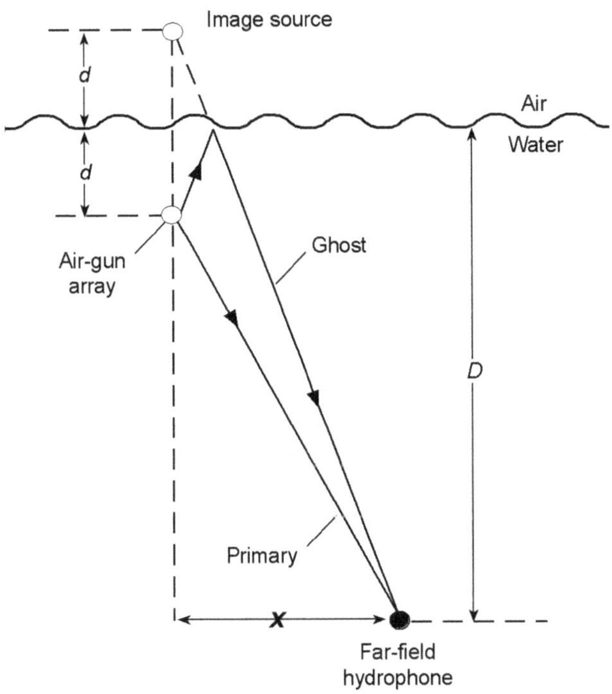

Fig. 2.2—Seismic raypaths involved in air-gun illumination of geologic targets.[15]

is several seismic wavelengths away from the source. This far-field requirement means that the hydrophone depth, *D,* shown in Fig. 2.2 is several hundreds of meters.

Air-gun arrays are designed to create source wavelets that are as compact in time as possible and that have minimal bubble oscillations. Compact wavelets are desired because such wavelets have wide signal-frequency spectra; minimal bubble oscillations are desired so that the signal spectrum will be as smooth as possible.

Fig. 2.3 shows an ideal source wavelet. The vertical dash line marks the arrival time of the bulk of the energy that travels the primary raypath (Fig. 2.2). The energy that travels the ghost raypath (Fig. 2.2) arrives at a time delay, $2d/v,$ where v is the velocity of the pressure pulse in seawater and d is the depth of the air-gun array. The ghost event has a polarity opposite that of the primary arrival because the reflection coefficient for an upgoing pressure wave at the air/water interface is –1. The primary and ghost arrivals define the peak-to-peak amplitude of the source wavelet. Small-amplitude events occur at later times in the source wavelet because the air bubble produced by the air-gun discharge oscillates as it decays (Fig. 2.3). The number, sizes, and relative separations of the guns in the array control the amplitude of these residual bubble oscillations.

The two wavelet properties of greatest interest are its peak-to-peak strength (PTP) and its primary-to-bubble ratio (PBR). The objectives of air-gun array design are to maximize the PTP property of an air-gun wavelet, which is the difference, A–B, in **Fig. 2.4**, and to minimize the PBR parameter, which is the ratio (A–B)/(C–D).

Many factors, such as the number of guns in the array, the volume of the guns, and the depth of the guns, affect the amplitude, phase, and frequency character of an air-gun array wavelet. **Table 2.1** summarizes Dragoset's[15] analysis of air-gun array parameters and their effects on the fundamental PTP and PBR properties of air-gun wavelets. A key point in this

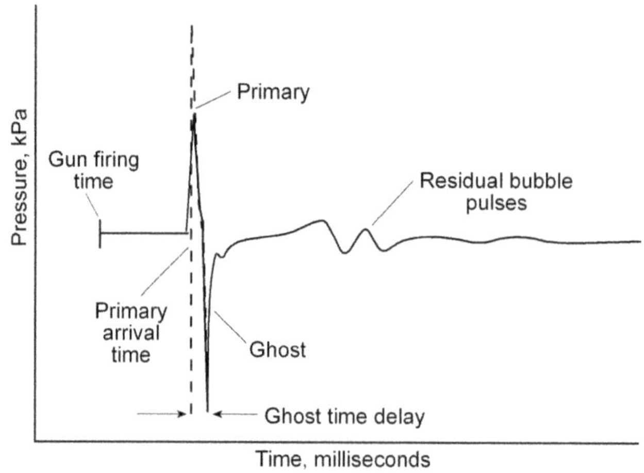

Fig. 2.3—Key features of an air-gun wavelet.[15]

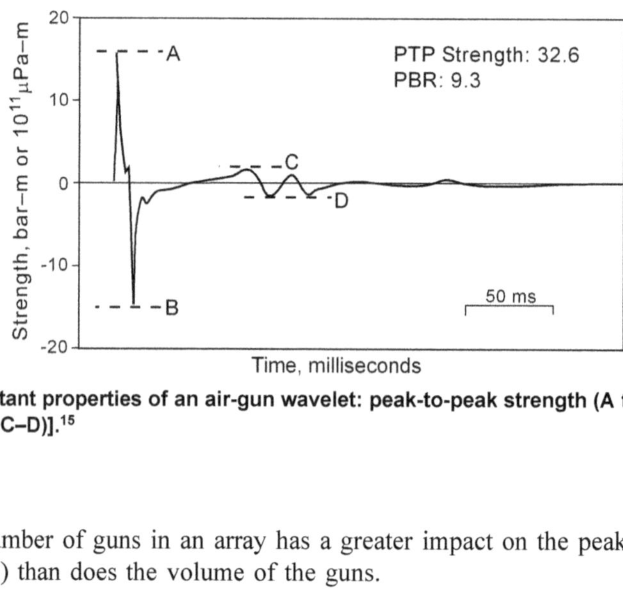

Fig. 2.4—Two important properties of an air-gun wavelet: peak-to-peak strength (A to B) and primary-to-bubble ratio [(A–B)/(C–D)].[15]

table is that the number of guns in an array has a greater impact on the peak-to-peak amplitude (or wavelet energy) than does the volume of the guns.

2.6 Seismic Sensors

Two classes of seismic sensors are used to acquire seismic data: scalar sensors and vector sensors. A scalar sensor measures the magnitude of Earth motion created by a seismic disturbance but does not indicate the direction of that motion. A hydrophone is an example of a popular scalar sensor used throughout the seismic industry. Hydrophones measure pressure variations (scalar quantities) associated with a seismic disturbance. A hydrophone cannot distinguish a pressure variation caused by a downgoing wavefield from a pressure change created by an upgoing wavefield. Hydrophones provide no directional (vector) information about a propagating seismic event.

A vector sensor indicates the direction that a seismic event causes the Earth to move. The classic example of a vector sensor is the moving-coil geophone that has been used for decades to record onshore seismic data. The principal of a moving-coil geophone is that a lightweight coil, with several hundred turns of thin copper wire, is suspended by springs that are, in turn,

TABLE 2.1—AIR-GUN SPECIFICATIONS AND ARRAY PARAMETERS (after Dragoset[15])

	PTP Strength	Primary-to-Bubble Ratio
Number of guns	Proportional to number of guns	No direct relation
Gun volume	Proportional to cube root of gun volume	Increases as volume increases
Gun depth	Increases at first, then decreases as depth increases	Decreases as gun depth increases
Initial firing pressure	Increases as firing pressure increases	Increases as firing pressure increases
Port area	Proportional to square root of port area	Increases as port area increases
Port-closure pressure	No effect if closure pressure < 0.5 of firing pressure	Decreases as port-closure pressure increases
Gun synchronization	Decreases as spread in firing time increases	Decreases as spread in firing times increases
Horizontal separation of gun	Decreases as guns move closer together	Increases dramatically when gun bubbles coalesce

attached to the case of the geophone. The springs are designed to allow the geophone case and the lightweight coil to move independently of each other over a frequency band of interest. Permanent magnets are attached to the geophone case to create a strong internal magnetic field. When the case is moved by a seismic disturbance, an electrical voltage is created as the coils cut the magnetic lines of force. The magnitude of the voltage output is proportional to the number of magnetic lines of force cut per unit time; thus, geophone response indicates the velocity of the geophone case, which, in turn, is proportional to Earth particle velocity at the geophone station.

The polarity of the geophone output voltage depends on the direction that the electrical conductors are moving as they cut across the magnetic lines of force. If an upward movement creates a positive voltage, a downward movement produces a negative voltage. Thus, a geophone is a vector sensor that defines not only the magnitude of Earth motion, but also the direction of that motion.

Because geophones are directional sensors and can distinguish between vertical and horizontal Earth motions, they are used to record multicomponent seismic data. Three-component (3C) geophones are used to record compressional and shear seismic data onshore. Shear waves do not propagate in fluids. In marine environments, geophones have to be placed in direct contact with the Earth sediment on the seafloor, with data-recording cables connected to surface-positioned ships or telemetry buoys. Four-component (4C) sensors used for this service are encased in large, robust, watertight enclosures that include a hydrophone and a 3C geophone. **Fig. 2.5** illustrates a segment of an ocean-bottom cable (OBC) used for deploying 4C marine seismic sensors. In this cable design, a 4C sensor station is positioned at intervals of 50 m along the 150-m cable segment. A large number of these segments, each containing three receiver stations, are connected end to end to make a continuous OBC receiver line several kilometers long. The exact length of the receiver line is determined by the depth of the target that is to be imaged.

Fig. 2.6 shows an OBC line deployed on the seafloor and connected to a stationary data-recording vessel. A second boat towing an air-gun array traverses predesigned source lines either parallel to, or orthogonal to, the OBC cable. The 4C sensors remain motionless on the seafloor as data are recorded, just as onshore geophones do when onshore seismic data are acquired. Seafloor sensors are not towed as are the conventional marine hydrophone cables shown in Fig. 2.1.

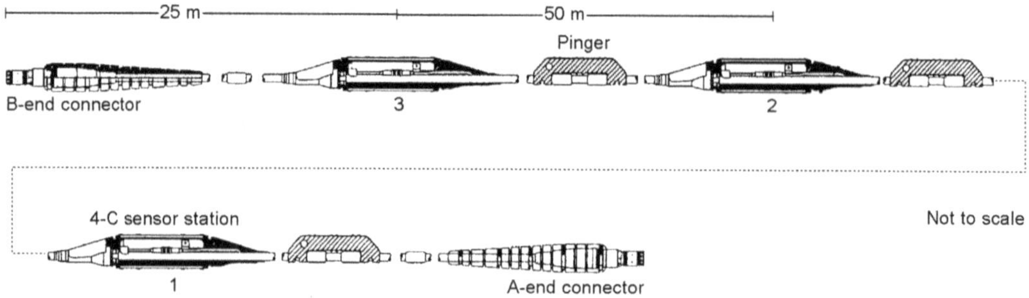

Fig. 2.5—Segment of ocean-bottom cable with three 4C sensor stations spaced 50 m apart.

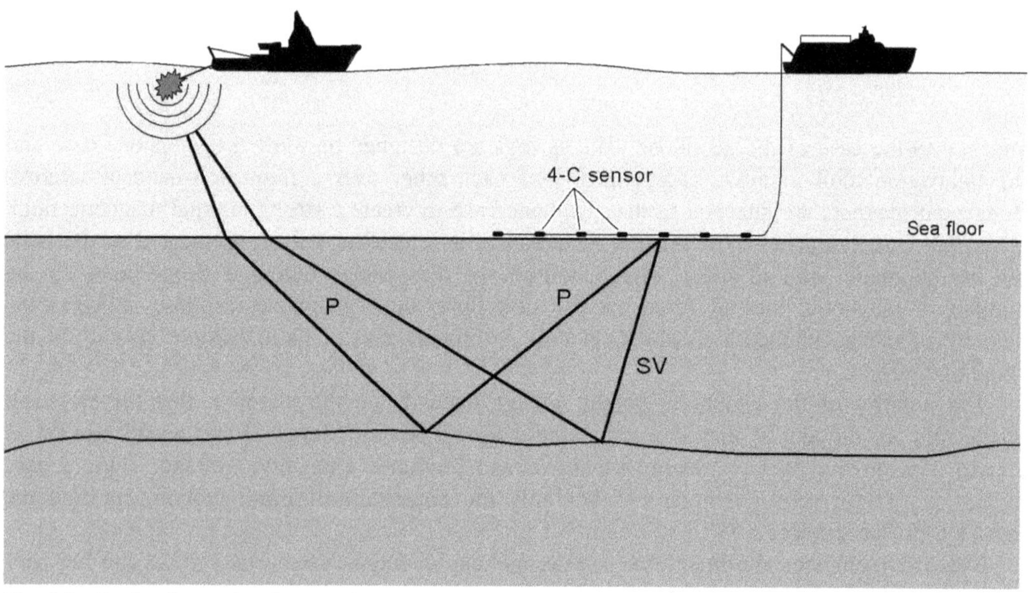

Fig. 2.6—Basic elements of ocean-bottom cable data acquisition. P denotes a seismic compressional wave; SV is a converted shear mode.

2.7 Seismic Wave Propagation

The full elastic seismic wavefield that propagates through an isotropic Earth consists of a P-wave component and two shear (SV and SH) wave components. Marine air guns and vertical onshore sources produce reflected wavefields that are dominated by P and SV modes. Much of the SV energy in these wavefields is created by P-to-SV-mode conversions when the downgoing P wavefield arrives at stratal interfaces at nonnormal angles of incidence (Fig. 2.6). Horizontal-dipole sources can create strong SH modes in onshore programs. No effective seismic horizontal-dipole sources exist for marine applications.

A principal difference among P, SV, and SH wavefields is the manner in which they cause rock particles to oscillate. **Fig. 2.7** illustrates the relationships between propagation direction and particle-displacement direction for these three wave modes. A compressional wave causes rock particles to oscillate in the direction that the wavefront is propagating. In other words, a P-wave particle displacement vector is perpendicular to its associated P-wave wavefront. In contrast, SV and SH waves cause rock particles to oscillate perpendicular to the direction that the wavefront is moving, with the SH and SV displacement vectors orthogonal to each other.

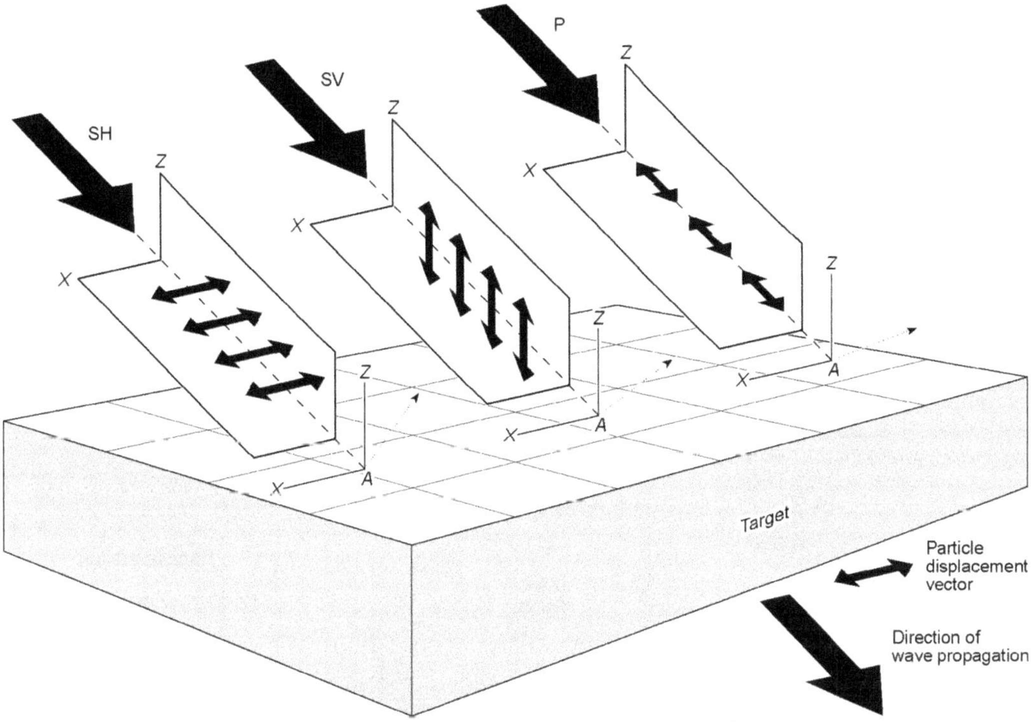

Fig. 2.7—Distinction between the three components of an elastic wavefield.

A shear-wave particle-displacement vector is thus tangent to its associated wavefront. In a flat-layered isotropic Earth, the SH displacement vector is parallel to stratal bedding, and SV displacement is in the plane that is perpendicular to bedding.

To create optimal images of subsurface targets, a seismic wavefield must be segregated into its P, SV, and SH component parts so that a P-wave image can be made that has minimal contamination from interfering SV and SH modes. Likewise, an SV image must have no interfering P and SH modes, and an SH image must be devoid of P and SV contamination.

A P wave travels at velocity V_p in consolidated rocks, which is approximately two times faster than velocity V_s of either the SH or SV wave. In carbonates, the velocity ratio (V_p / V_s) tends to be approximately 1.7 or 1.8. In siliciclastics, V_p / V_s varies from approximately 1.6 in hard sandstones to approximately 3 in some shales. This velocity difference aids in separating interfering P and S wave modes during data processing. An equally powerful technique for separating a seismic wavefield into its component parts is to use data-processing techniques that concentrate on the distinctions in the particle displacements associated with the P, SH, and SV modes (Fig. 2.7).

The P, SH, and SV particle displacements shown in Fig. 2.7 form an orthogonal coordinate system. The fundamental requirement of multicomponent seismic imaging is that reflection wavefields must be recorded with orthogonal 3C sensors that allow these P, SH, and SV particle motions to be recognized. To date, most exploration seismic data have been recorded with single-component sensors that emphasize P-wave modes and do not capture SH or SV wave modes.

2.8 Body Waves and Surface Waves

Seismic wavefields propagate through the Earth in two ways: body waves and surface waves. Body waves propagate in the interior (body) of the Earth and illuminate deep geologic targets.

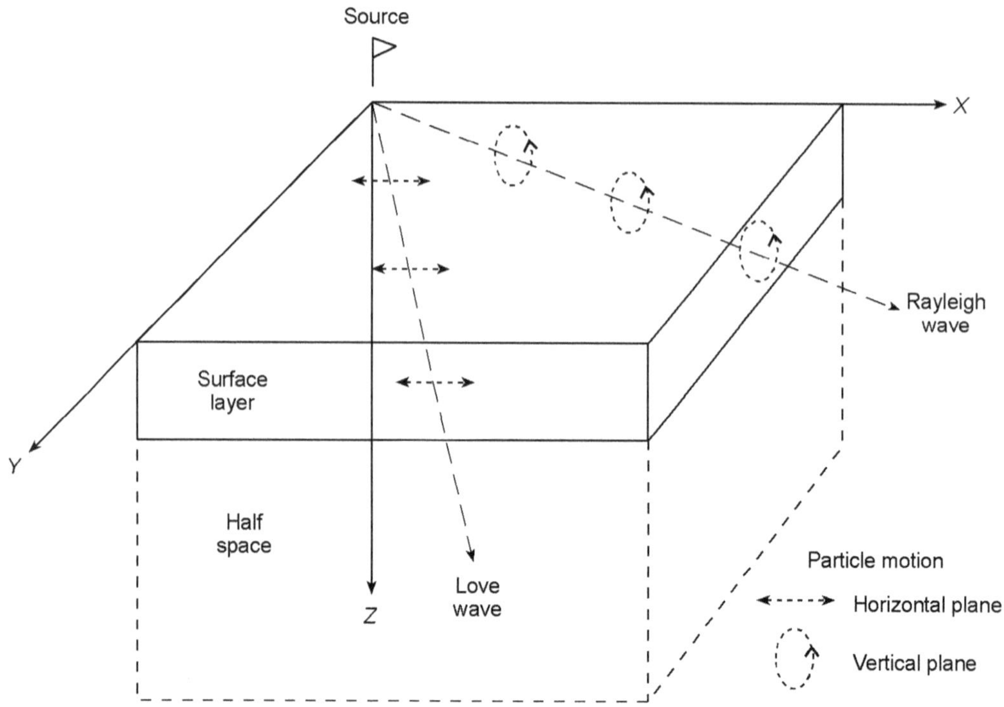

Fig. 2.8—Particle motions produced by the two principal seismic surface-wave noise modes: the Love wave and the Rayleigh wave.

These waves generate the reflected P, SH, and SV signals that are needed to evaluate prospects and to characterize reservoirs. Reflected (or scattered) body waves are the fundamental signals sought in seismic data-acquisition programs.

Surface waves travel along the Earth/air interface and do not illuminate geologic targets in the interior of the Earth. Surface waves are noise modes that overlay the desired body-wave reflections. Surface waves can be a serious problem in onshore seismic surveys. Surface waves do not affect towed-cable marine data because they require some shear-wave component to propagate, and shear waves cannot propagate along the air/water interface. An exception in the marine case is sometimes encountered when data are recorded with ocean-bottom sensors (OBS) because interface waves can propagate along the water/sediment boundary and become a type of surface-wave noise that degrades OBS marine seismic data.

There are two principal surface waves: Love waves and Rayleigh waves (**Fig. 2.8**). Love waves are an SH-mode surface wave and do not affect conventional P-wave seismic data. Love waves are a serious noise mode only when the objective is to record reflected SH wavefields. The more common surface wave is the Rayleigh wave, which combines P and SV motions and is referred to as ground roll on P-wave seismic field records. Love waves create particle displacements in the horizontal plane; Rayleigh wave displacements are in the vertical plane (Fig. 2.8).

Much of the field effort in onshore seismic programs concentrates on designing and deploying receiver arrays that can attenuate horizontally traveling surface waves (ground-roll noise) and, at the same time, amplify upward-traveling reflection signals. The most effective field technique is to deploy 10, 12, 16, or more geophones at a uniform spacing at each receiver station so that the distance from the first geophone to the last geophone is the same as the dominant wavelength of the ground-roll event. All geophone responses are then summed to create a single output response at that receiver station. The idea is to create a sensor array length such that half of the geophones are moving up and half are moving down as the horizontally traveling ground roll passes the receiver station. The summed output of the geophone array is essentially zero because of the passage of the ground-roll event. In contrast, upward-traveling reflections arriving at this same receiver array are not attenuated because such events cause all geophones to move up and down in unison. The summed output of the array for an upward-traveling reflection wavefield is thus a strong voltage signal.

2.9 Seismic Impedance

The concept of acoustic (or seismic) impedance is critical to understanding seismic reflectivity. Seismic impedance controls the seismic reflection process in the sense that seismic energy is reflected only at rock interfaces in which there are changes in impedance across the interface. Seismic impedance is defined as

$$I = \rho V, \quad \text{(2.1)}$$

where I = impedance, ρ = the bulk density of the rock, and V = the velocity of seismic wave propagation through the rock. V is set to V_p if the wave mode of interest is a P wave; it is set to V_s if S-wave reflectivity is being considered. Any alteration in rock properties that causes ρ and/or V to change can be the genesis of a seismic reflection event; therefore, areal and vertical variations in seismic reflectivity can be used to infer spatial distributions of rock types and porosity trends.

2.10 Reflection Coefficients

Seismic reflectivity is best explained with a simple two-layer Earth model in which Layer 1 is above Layer 2 (**Fig. 2.9**). The seismic reflection coefficient, R, for a downgoing particle-velocity wave mode that arrives perpendicular to the interface between the two layers is

$$R = \frac{I_1 - I_2}{I_1 + I_2} = \frac{\rho_1 V_1 - \rho_2 V_2}{\rho_1 V_1 + \rho_2 V_2}. \quad \text{(2.2)}$$

A negative algebraic sign has to be inserted on the right side of Eq. 2.2 if the downgoing wavefield is a pressure wavefield (hydrophone measurement) rather than a particle-velocity wavefield (geophone measurement). The velocity parameters, V_1 and V_2, are P-wave velocities if P-wave reflectivity is being calculated; they are S-wave velocities if S-wave reflectivity is to be determined. At any interface, R can be positive, negative, or zero, depending on the impedance contrast $(\rho_1 V_1 - \rho_2 V_2)$ across the interface.

A seismic reflection/transmission process is indicated in Fig. 2.9 by the raypaths labeled A_o, A_r, and A_t. For nonnormal incidence angles, the expression for reflection coefficient involves trigonometric functions that ensure that horizontal slowness (the inverse of horizontal velocity) is conserved and is a more complex expression than that given in Eq. 2.2. Ref. 16 gives a detailed mathematical treatment. The seismic reflection, A_r, is given by

$$A_r = RA_o, \quad \text{(2.3)}$$

and the transmitted seismic event, A_t, is given by

$$A_t = (1 + R)A_o. \quad \text{(2.4)}$$

The magnitude and algebraic sign of A_r depends on R and, in turn, the basic control on R is the variation of impedance ρV across the interface (Eq. 2.2). **Table 2.2** lists the common geo-

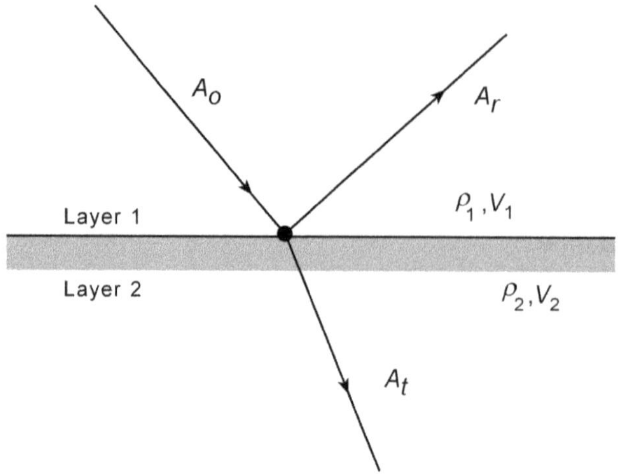

Fig. 2.9—Two-layer Earth model describing seismic reflectivity parameters.

TABLE 2.2—GEOLOGICAL INFLUENCES ON ACOUSTIC IMPEDANCE

Geologic Condition	Impedance Effect
Lithological interfaces	A common cause of impedance contrasts. Generally ρ and V change whenever there is a change in lithology.
Porosity variations	ρ and V are porosity dependent. Gradual changes in porosity generate modest impedance contrasts. Abrupt changes in porosity can create large reflection coefficients.
Changes in pore fluid	Pore-fluid density affects bulk density ρ and velocity V. A change of pore fluid from water to oil creates a small impedance contrast that can be detected seismically only in ideal signal-to-noise conditions. A change from liquid (either water or oil) to gas can produce large impedance contrasts and robust seismic reflections.
Overpressure	ρ and V decrease in overpressure zones. If the onset of overpressure is gradual, the impedance contrasts may be too small to create detectable seismic reflections. Abrupt onsets of overpressure can produce strong reflections.
Cementation	Cementation affects the mechanical strength of rocks. V generally increases as mechanical strength increases (assuming a constant rock type). In some instances, variations in cementation can create impedance contrasts sufficient to result in seismic reflections.

logic conditions that often create impedance contrasts that result in nonzero reflection coefficients at rock interfaces.

Two types of petrophysical properties control the value of acoustic impedance in individual rock layers: elastic properties of the rock matrix and properties of the fluid in the pore spaces of the rock. P-waves travel through elastic materials and fluids; thus, any change in either the rock matrix (such as a change in mineralogy or porosity) or in the type of fluid occupying the pore spaces will create a discontinuity in the P-wave seismic impedance of the rock system.

Fig. 2.10 illustrates the relationships between petrophysical conditions that occur at an impendance boundary and the existence of P and S reflections at that boundary. A P-wave reflection will occur at boundaries at which there is a change in either the rock matrix or the pore fluid, or both. In constrast, S-waves are not affected by changes in pore fluid or are only weakly affected. Consequently, a change in the properties of the rock matrix can create a re-

ROCK AND FLUID PROPERTIES THAT CREATE P AND S REFLECTIONS

Boundary conditions	P-reflection	S-reflection
Rock matrix 1, Pore fluid 1 Rock matrix 1, Pore fluid 2	Yes	No (may be weak)
Rock matrix 1, Pore fluid 1 Rock matrix 2, Pore fluid 1	Yes	Yes
Rock matrix 1, Pore fluid 1 Rock matrix 2, Pore fluid 2	Yes	Yes

Fig. 2.10—Relationships between petrophysical conditions that occur at an impedance boundary and the existence of P and S reflections at that boundary.[17]

flecting boundary for S-waves, but a change in pore fluid will create only a small (usually negligible) S-wave reflection boundary (Fig. 2.10). If a small, nonzero S-wave reflection coefficient occurs at a fluid boundary, that reflection coefficient usually exists because the bulk density of the rock system varies across the fluid boundary.

These wave physics provide valuable geologic insights when both P and S reflection data are acquired across a prospect area. When P and S reflections occur at the same depth coordinate, the reflecting boundary at that depth is associated with a change in the rock matrix (that is, with a lithological change). There may or may not be a change in pore fluid at that boundary. When a P reflection occurs at a boundary but there is no S reflection, that boundary quite likely marks a change in pore fluid and not a change in rock matrix. (That is, the lithology probably does not change at that depth, but the type of pore fluid does.)

2.11 Seismic Attributes

The fundamental properties of processed seismic data that are used in interpretation are temporal and spatial variations of reflection amplitude, reflection phase, and wavelet frequency. Structural and stratigraphic interpretations of 3D seismic data are inferences of geologic conditions made by analyzing areal patterns of these three seismic attributes—amplitude, phase, and frequency—across selected seismic time surfaces. Any procedure that extracts and displays these seismic parameters in a convenient, understandable format is an invaluable interpretation tool.

Taner and Sheriff [18] and Taner *et al.*[19] began using the Hilbert transform to calculate seismic amplitude, phase, and frequency instantaneously, meaning that a value of amplitude, phase, and frequency is calculated for each time sample of a seismic trace. Since that introduction, numerous Hilbert transform algorithms have been implemented to calculate these useful seismic attributes (e.g., Hardage[20]).

2.12 Complex Seismic Trace

Fig. 2.11 illustrates the concept of a complex seismic trace in which $x(t)$ represents the real seismic trace and $y(t)$ is the Hilbert transform of $x(t)$. In this discussion, we ignore what a Hilbert transform is and how the function $y(t)$ is calculated. Most modern seismic data-process-

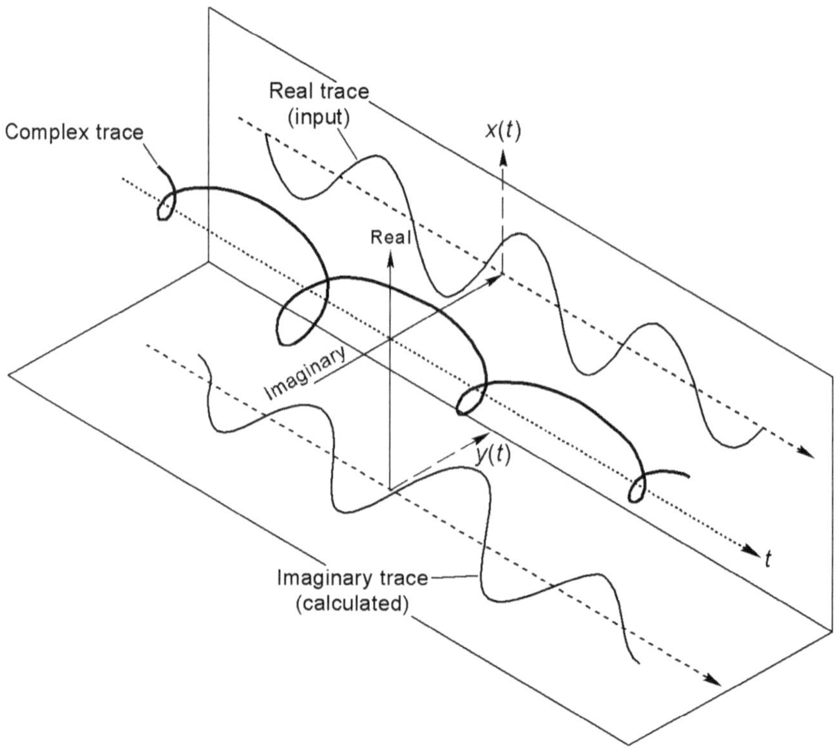

Fig. 2.11—A complex seismic trace.[21]

ing software packages provide Hilbert transform algorithms and allow processors to create the function $y(t)$ shown in Fig. 2.11 easily. These two data vectors are displayed in a 3D (x, y, t) space in which t is seismic traveltime, x is the real data plane, and y is the imaginary plane. In this complex trace format, the actual seismic trace, $x(t)$, is confined to the real x plane, and $y(t)$, the Hilbert transform of $x(t)$, is confined to the imaginary y plane. When $x(t)$ and $y(t)$ are added vectorally, the result is a complex seismic trace, $z(t)$, in the shape of a helical spiral extending along, and centered about, the time axis t. The projection of this complex function $z(t)$ onto the real plane is the real seismic trace, $x(t)$, and the projection of $z(t)$ onto the imaginary plane is $y(t)$, the calculated Hilbert transform of $x(t)$.

Fig. 2.12 illustrates the reason for converting the real seismic trace, $x(t)$, into what first appears to be a more mysterious complex seismic trace, $z(t)$, in which the attributes known as instantaneous seismic amplitude, instantaneous phase, and instantaneous frequency are introduced. At any point on the time axis of this complex seismic trace, a vector $a(t)$ can be calculated that extends away from the t axis in a perpendicular plane to intersect the helically shaped complex seismic trace, $z(t)$. The length of this vector is the amplitude of the complex trace at that particular instant, hence the term instantaneous amplitude. This amplitude value is calculated with the equation for $a(t)$ shown in Fig. 2.12.

The orientation angle, $\phi(t)$, of the amplitude vector, $a(t)$, at time t, which is generally measured relative to the positive axis of the real x-plane, is defined as the phase of $z(t)$ at that moment in time. Numerically, the phase angle is calculated from the equation for $\phi(t)$ defined in Fig. 2.12. As seismic time progresses, vector $a(t)$ moves along the time axis and rotates continually about the time axis to maintain contact with the spiraling complex trace, $z(t)$. Each full rotation of the vector around the time axis increases the phase value by 360°.

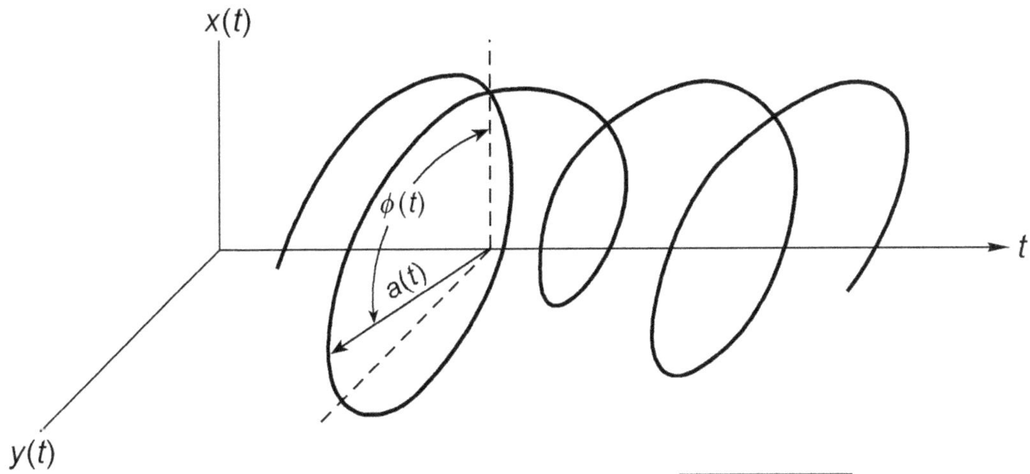

Instantaneous amplitude $a(t) = \sqrt{x^2(t) + y^2(t)}$

Instantaneous phase $\phi(t) = \tan^{-1}\left(\frac{y(t)}{x(t)}\right)$

Instantaneous frequency $\omega(t) = \frac{d}{dt}\left[\phi(t)\right]$

Fig. 2.12—Key seismic attributes (instantaneous amplitude, instantaneous phase, and instantaneous frequency) that can be calculated once a complex seismic trace is created.[21]

In any oscillating system, and specifically for a seismic trace, frequency can be defined as the time rate of change of the phase angle. This fundamental definition describes the frequency of the complex seismic trace so that the instantaneous frequency, $\omega(t)$, at any seismic time sample is given by the time derivative of the phase function specified by the equation in Fig. 2.12.

2.13 Instantaneous Phase and Instantaneous Frequency Calculations

Fig. 2.13 illustrates calculations of the instantaneous phase associated with a typical seismic trace. The figure's bottom panel shows the actual seismic trace, and the center panel shows the real and imaginary components of the associated complex trace. Applying the second equation of Fig. 2.12 to the real and imaginary components of the complex seismic trace (center panel) produces the instantaneous phase function at the top of Fig. 2.13. The phase behavior at times t_1, t_2, t_3 is critical to understanding the geologic significance of anomalous frequencies. Although phase is a positive function that monotonically increases in magnitude with seismic time, it is customarily plotted as a repetitive, wraparound function with plot limits of 0 to 360° (or −180° to +180°). Each wraparound of 360° corresponds to a full rotation of vector $a(t)$ around the seismic time axis while the vector stays in contact with the spiraling complex seismic trace $z(t)$ (see Fig. 2.12).

The top panel of **Fig. 2.14** shows the instantaneous frequencies calculated for the seismic trace presented in Fig. 2.13. Although the calculated frequency values at times t_1 and t_2 shown by the solid-line curve are physically impossible because they are negative, these anomalous frequency behaviors are some of the most useful seismic attributes that an interpreter can use. They should be preserved because, when displayed in an eye-catching, contrasting color, they

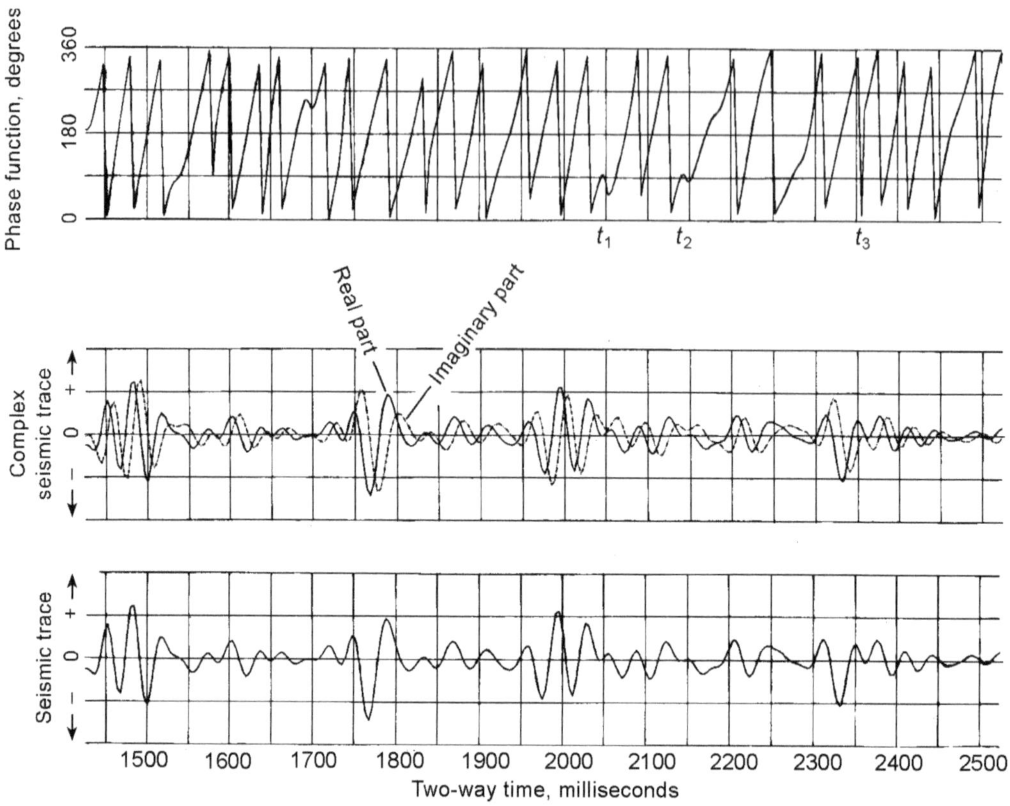

Fig. 2.13—Instantaneous phase seismic attribute function.[21]

serve to shift an interpreter's attention quickly to subtle structural and stratigraphic discontinuities in a seismic image.

Comparing the time coordinates of these anomalous frequency values with the time coordinates of the instantaneous phase function in Fig. 2.13 shows that rather than exhibiting its typical, monotonically increasing behavior in these time intervals, the phase momentarily decreases in magnitude. This causes the time rate of change of phase (or the slope of the phase function), which is the instantaneous frequency, to be negative at time samples t_1 and t_2. Numerical algorithms should not camouflage these unrealistic frequency values, which some algorithms do by arbitrarily reversing the algebraic sign of any negative frequency. The dashed-curve segments in the enlargement show how some software algorithms change the algebraic sign of negative frequency values to positive values. Although this seems like a logical correction, it should be avoided because it reduces the interpretive value of a 3D volume of instantaneous frequencies.

At times t_1, t_2, and t_3, some type of wavelet interference (that is, a wavelet distortion) occurs in the seismic trace (bottom panel). As a result, the reflection waveform at times t_1, t_2, and t_3 is slightly distorted (bottom panel) because of the destructive interference of two or more overlapping wavelets, demonstrating that anomalous frequencies tend to coincide with, and emphasize, distorted wavelets such as are produced at structural and stratigraphic discontinuities. High positive frequency values also occur at 1580, 1650, 1740, and 1840 milliseconds, but these values are not anomalous in the sense that they do not exceed the Nyquist limit (125 Hz). However, they too coincide with distorted wavelets (bottom panel).[21]

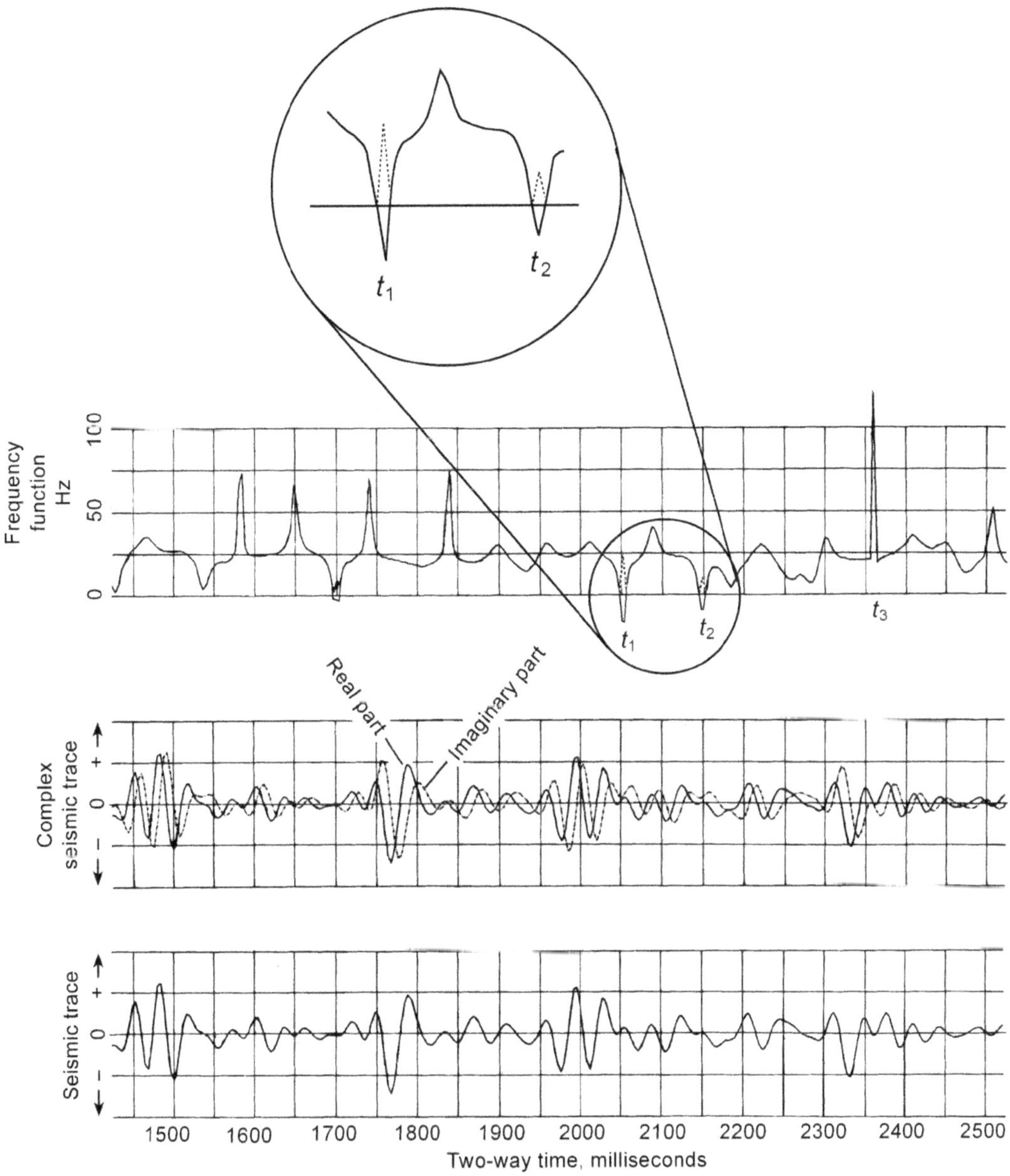

Fig. 2.14—Instantaneous frequency seismic attribute function calculated for the same seismic trace discussed in Fig. 2.13.

2.14 Applications of Seismic Attributes

All instantaneous seismic attributes (amplitude, phase, frequency) can be used in interpretation. In practice, most interpreters use instantaneous amplitude, or some variation of an amplitude attribute, as their primary diagnostic tool. Amplitude is related to reflectivity, which in turn is related to subsurface impedance contrasts. Thus, amplitude attributes provide information about all the rock, fluid, and formation-pressure conditions listed in Table 2.2.

Instantaneous phase is useful for tracking reflection continuity and stratal surfaces across low-amplitude areas where it is difficult to see details of reflection waveform character. In general, instantaneous phase is the least used of the seismic attributes.

Instantaneous frequency sometimes aids in recognizing changes in bed thickness and bed spacing. Anomalous values of instantaneous frequency (negative values or unbelievably high positive values) are particularly useful for recognizing edges of reservoir compartments, subtle faults, and stratigraphic pinchouts. Hardage[22] demonstrated these applications of instantaneous frequency.

2.15 Seismic Interpretation

A stratal surface is a depositional bedding plane: a depositional surface that defines a fixed geologic time. A siliciclastic rock deposited in a high-accommodation environment contains numerous vertically stacked stratal surfaces. A fundamental thesis of seismic stratigraphy is that a seismic reflection event follows an impedance contrast associated with a stratal surface; that is, a seismic reflection is a surface that represents a fixed point in geologic time.[23,24] The term chronostratigraphic defines this type of seismic reflection event. Because lithology varies across the area spanned by a large depositional surface, the implication of this interpretation principle is that an areally pervasive seismic reflection event does not necessarily mark an impedance contrast boundary between two fixed rock types as that reflection traverses an area of interest. The application of this fundamental concept about the genetic origin of seismic reflections to seismic interpretation is referred to as stratal-surface seismic interpretation.

Tipper[25] illustrated and discussed situations in which a seismic reflection can be either chronostratigraphic or diachronous (meaning that the event moves across depositional time surfaces), depending on the vertical spacings between beds, the lateral discontinuity between diachronous beds, and bed thickness. The conclusion that a seismic reflection is chronostratigraphic or diachronous needs to be made with caution because the answer depends on the local stratigraphy, the seismic bandwidth, and the horizontal and vertical resolution of the seismic data.

If two seismic reflection events, A and B, are separated by an appreciable seismic time interval (a few hundred milliseconds) yet are conformable to each other (that is, they parallel each other), then the uniform seismic time thickness between these two events represents a constant and fixed period of geologic time throughout the seismic image space spanned by reflectors A and B. An implication of seismic stratigraphy that can be invoked in such an instance is that any seismic surface intermediate to A and B, which is also conformable to A and B, is also a stratal surface.

A key first step in seismic interpretation is to use well logs and cores to identify the three types of stratal interfaces that exist in geologic intervals of interest: flooding surfaces, maximum flooding surfaces, and erosion surfaces. Flooding surfaces are widespread interfaces that contain evidence of an upward, water-deepening facies dislocation, such as contact between rooted, unfossiliferous floodplain mudstones and overlying fossiliferous marine shale. A ravinement surface is a specific type of flooding surface that suggests that transgressive passage of a surf zone has eroded underlying shallower-water facies.

Maximum flooding surfaces are interfaces that contain evidence of a widespread, upward, water-deepening facies dislocation that is associated with the inferred, deepest water facies encountered in a succession of strata. A maximum flooding surface is commonly represented by a thin condensed section, typically a black, organic-rich shale with a low-diversity fossil assemblage representing deepwater, sediment-starved conditions.

Maximum flooding surfaces bound and define upward-coarsening facies successions that are called genetic sequences by Galloway.[26] These genetic sequences are similar to cycles or cyclothems in other terminology.[27,28]

An erosion surface is an interface in which there is evidence of a facies offset that indicates that an abrupt decrease in water depth occurred. If an erosion surface is widespread, truncation of older strata can be documented on well log cross sections. Some of these surfaces may be disconformities representing downcutting during periods of subaerial exposure caused by allocyclic (extrabasinal) mechanisms, such as eustatic sea-level changes. These major chronostratigraphic surfaces are often manifested as mappable seismic reflections. All 3D seismic data volumes should be calibrated with mappable, key surfaces recognized from cores and well logs, with priority given to flooding surfaces, maximum flooding surfaces, and erosion surfaces. See the chapter on reservoir geology in this volume for more discussion on sequence stratigraphy and depositional environments.

2.16 Structural Interpretation

The original use of seismic reflection data (circa 1930 through 1960) was to create maps depicting the geometry of a subsurface structure. Because many of the world's largest oil and gas fields are positioned on structural highs, structural mapping has been, in a historical sense, the most important application of exploration seismic data. When the seismic industry converted from analog to digital data recording in the mid-1960s, digital technology increased the dynamic range of reflected seismic signals and allowed seismic data to be used for applications other than structural mapping, such as stratigraphic imaging, pore-fluid estimation, and lithofacies mapping. These expanded seismic applications have led to the discovery of huge oil and gas reserves confined in subtle stratigraphic traps, and seismic exploration is now no longer limited to just "mapping the structural highs." However, even with the advances in seismic technology, structural mapping is still the first and most fundamental step in interpretation. When 3D seismic data are interpreted with modern computer workstations and interpretation software, structural mapping can be done quickly and accurately.

Different seismic interpreters use different approaches and philosophies in their structural interpretations. The technique described here is particularly robust and well documented.[29] The first step of the procedure is to convert the 3D seismic data volume that has to be interpreted to a 3D coherency volume. Coherency is a numerical measure of the lateral uniformity of seismic reflection character in a selected data window. As the waveform character of side-by-side seismic traces becomes more similar, the coherency value for the traces approaches a value of +1.0; as the traces become more dissimilar, the coherency of the traces approaches zero. All modern seismic interpretation software can perform the numerical transform that converts 3D seismic wiggle-trace data into a 3D coherency volume.

Fig. 2.15 shows an example of a horizontal time slice through a 3D coherency volume from the Gulf of Mexico. This figure is also discussed in the chapter on reservoir geophysics in the Emerging Technologies volume of this *Handbook*. The narrow bands of low coherency values that extend across this time slice are created by faults that disrupt the lateral continuity of reflection events. Fault mapping is a major component of structural mapping, and this type of coherency display can be used to create fast, accurate fault maps. Coherency technology has evolved into the optimal methodology for detecting and mapping structural faults in 3D seismic image space.

The second step of the structural interpretation procedure is to transfer the fault pattern defined by coherency data to the associated 3D seismic wiggle-trace data volume. **Fig. 2.16** illustrates the projection of the faults in Fig. 2.15 onto a vertical profile through 3D seismic image space. The coherency time slice in Fig. 2.15 defines the X, Y coordinates of each intersected fault at one constant, image-time coordinate across the image space. Additional coherency time slices are made at image-time intervals of 100 or 200 milliseconds to define the X, Y coordinates of each fault as a function of imaging depth. This procedure causes the orientations and vertical extents of faults transferred to a 3D seismic wiggle-trace volume to be quite accurate.

The first-order fault labeled in Fig. 2.16 extends through the entire stratigraphic column and create large vertical displacements of strata. The second-order faults have less vertical extent and cause less vertical displacement than the first-order faults. Other structural and stratigraphic features that are common in Gulf of Mexico geology are labeled. These features are identified to indicate the imaging capabilities of seismic data. Rollover indicates fault-related flexing of bedding, which results in structural trapping of hydrocarbons. The bright spot is an

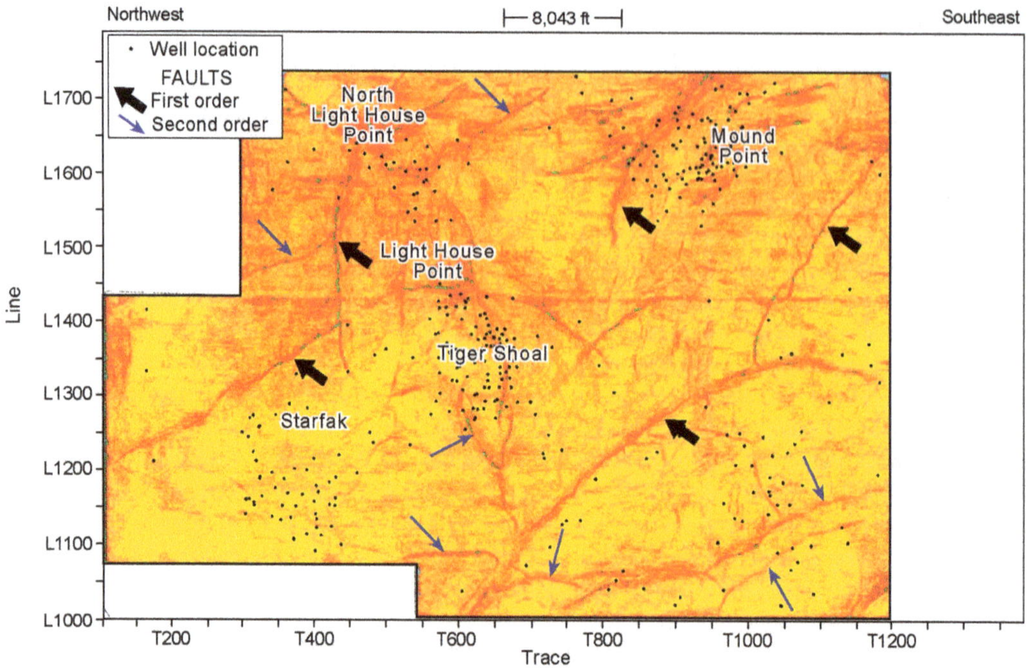

Fig. 2.15—Horizontal slice through a 3D coherency volume imaging a producing area in the Gulf of Mexico.[29]

example of reflection amplitude reacting as a direct hydrocarbon indicator (see changes in pore fluid in Table 2.2). The velocity sag feature is a false structural effect caused by anomalously low seismic propagation velocity that delays reflection arrival times, leaving the misleading appearance of a structural sag.

The third step of this approach to structural mapping is to interpret a series of chronostratigraphic surfaces across the seismic image space. These surfaces can be any of the chronostratigraphic surfaces (flooding surfaces, maximum flooding surfaces, and erosion surfaces) described in Sec. 2.15, depending on the amount and quality of subsurface well control available to the interpreter. If there is no well control, interpreters must use their best judgment as to how to correlate equivalent strata across a seismic image space and then adjust their interpretation, if necessary, as wells are drilled.

When a selected stratal surface is extended across the complete seismic image space, the geometrical configuration of that chronostratigraphic surface can be displayed as a structure map. The structure map in **Fig. 2.17** is one of the chronostratigraphic surfaces interpreted across this Gulf of Mexico prospect with the fault geometry information defined by coherency slices (Fig. 2.15) and vertical slices (Fig. 2.16). The producing fields shown in the map are positioned on local structural highs associated with one or more first-order faults.

In the lower left of the map in Fig. 2.17, an arbitrary profile XX′ is shown crossing the fault swarm. **Fig. 2.18** displays a vertical section along this profile to demonstrate the degree to which faults compartmentalize producing strata. This expanded view of the seismic reflection character also reveals critical stratigraphic features, such as lowstand wedges, that are embedded in the faulted structure. (A lowstand wedge is a sedimentary wedge deposited during a period of low sea level.) This type of seismic interpretation allows stratigraphers to construct detailed models of the internal architecture of targeted reservoir systems.

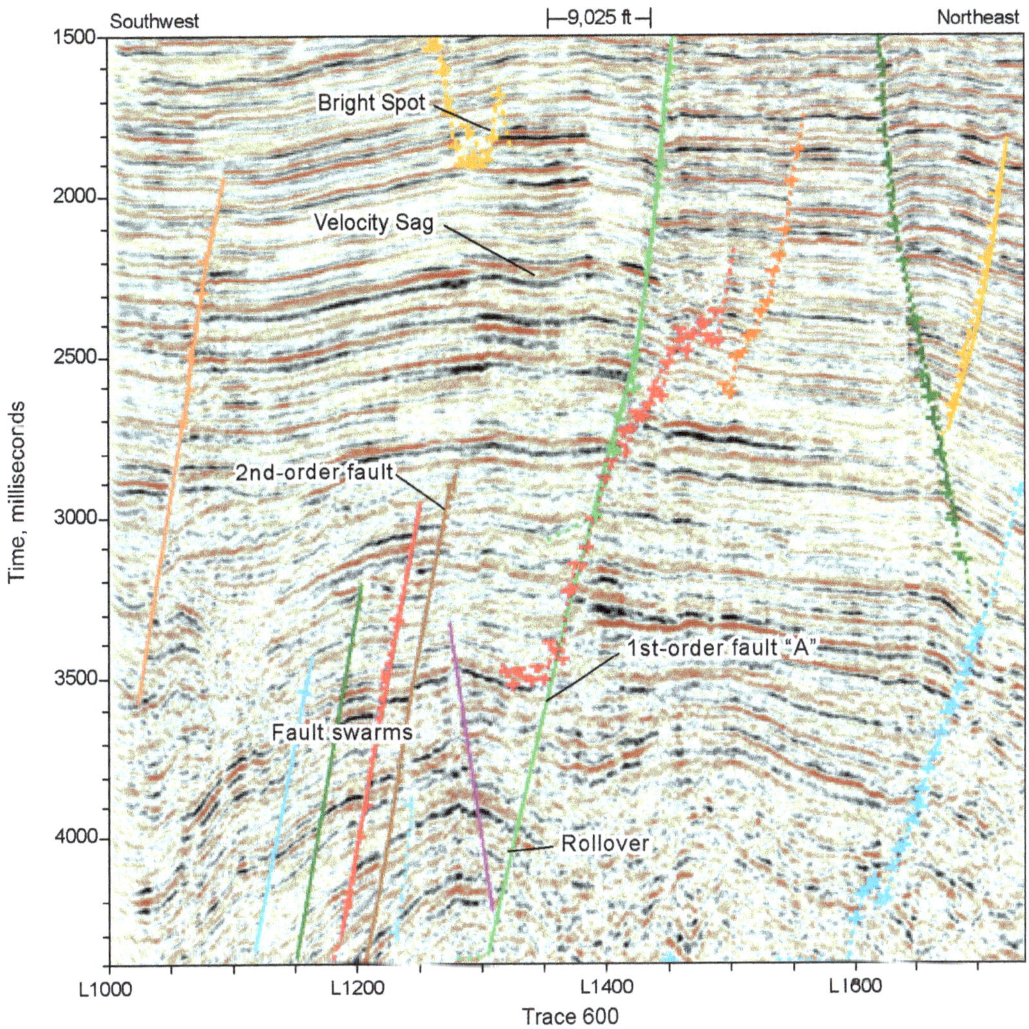

Fig. 2.16—Vertical seismic slice along crossline T600 of Fig. 2.15.[29]

Fig. 2.19 shows a second structural map constructed from a shallower chronostratigraphic surface to illustrate that less fault compartmentalization is in shallow reservoirs than in the deeper reservoirs associated with the structure shown in Fig. 2.17. The first-order faults still displace strata at this shallow level, but most second-order faults have terminated at deeper depths and no longer cause reservoir compartmentalization.

The structure maps shown in Figs. 2.17 and 2.19 are time-structure maps. These maps can be converted to depth maps once seismic propagation velocities are determined through the stratigraphic column.

2.17 Imaging Reservoir Targets

Fig. 2.20 shows a data window from a vertical slice of a 3D seismic data volume that is centered on a targeted channel system. These data include a good-quality reflection peak labeled "reference surface." The reference surface is a reference seismic stratal surface used to construct additional stratal surfaces that pass through the targeted thin-bed interval.[30] The fluvial system is embedded in the reflection peak that occurs at 0.73 seconds at inline coordinate 120.

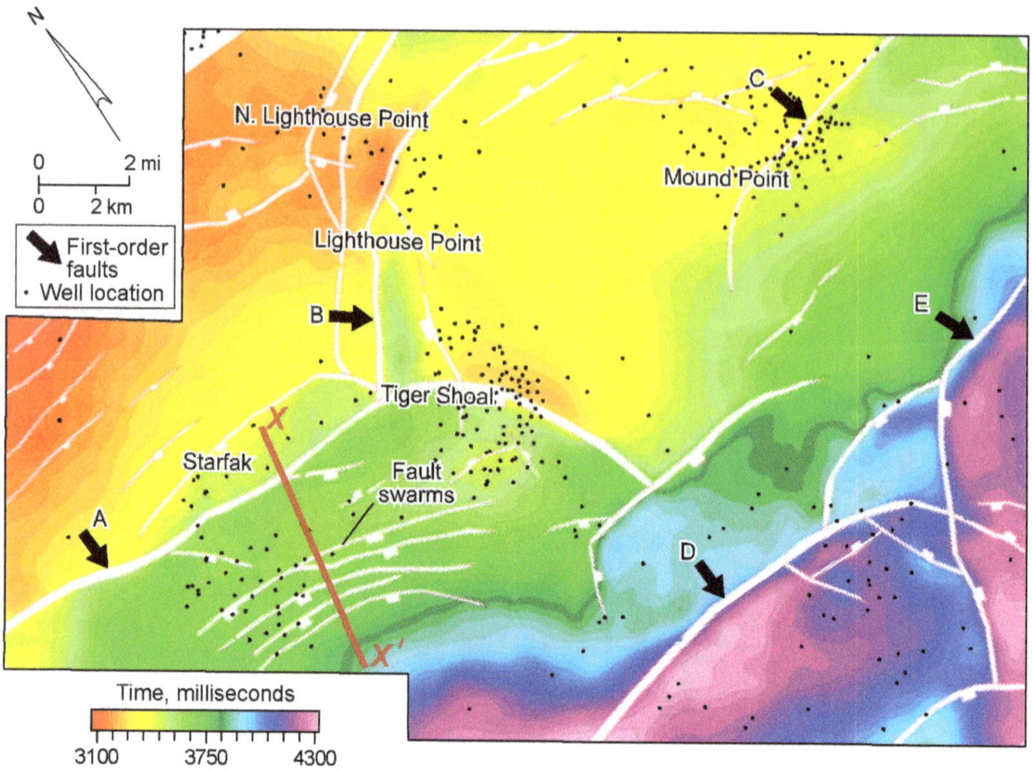

Fig. 2.17—Time-structure map of a deep reservoir system exhibiting considerable fault-induced compartmentalization.[29]

This particular reflection peak satisfies the fundamental criteria required of a reference stratal surface used to study thin-bed sequences: the event extends over the total 3D image space and has a high signal-to-noise character; the event is reasonably close to the targeted thin-bed sequences that need to be studied (i.e., the strata related to the anomalous reflection waveforms labeled "Channel 1" approximately 90 milliseconds above the reference surface); and the event is conformable to (i.e., parallel to) this targeted thin-bed sequence. The third criterion is the most important requirement for any seismic stratal surface that is to be used as a reference surface. Because this reference surface follows the apex of an areally continuous reflection peak, the basic premise of seismic stratigraphy is that this reference surface follows an impedance contrast that coincides with a stratal surface.

Fig. 2.21 displays this crossline section view with four conformable surfaces (A, B, C, and D) that pass through the targeted thin-bed interval added to the profile. These four surfaces are, respectively, 92, 90, 88, and 86 milliseconds above—and conformable to—the reference surface. Visual inspection of the reflection events above and below surfaces A, B, C, and D shows that all these reflection peaks and/or troughs are reasonably conformable to the reference surface event. Surfaces A, B, C, and D can thus be assumed to be stratal surfaces, or constant-depositional-time surfaces, because they are conformable to a known stratal surface (the reference surface) and are embedded in a 200-millisecond seismic window in which all reflection events are approximately conformable to the selected reference surface. The highlighted data window encircles subtle changes in reflection waveform that identify the seismic channel facies.[30]

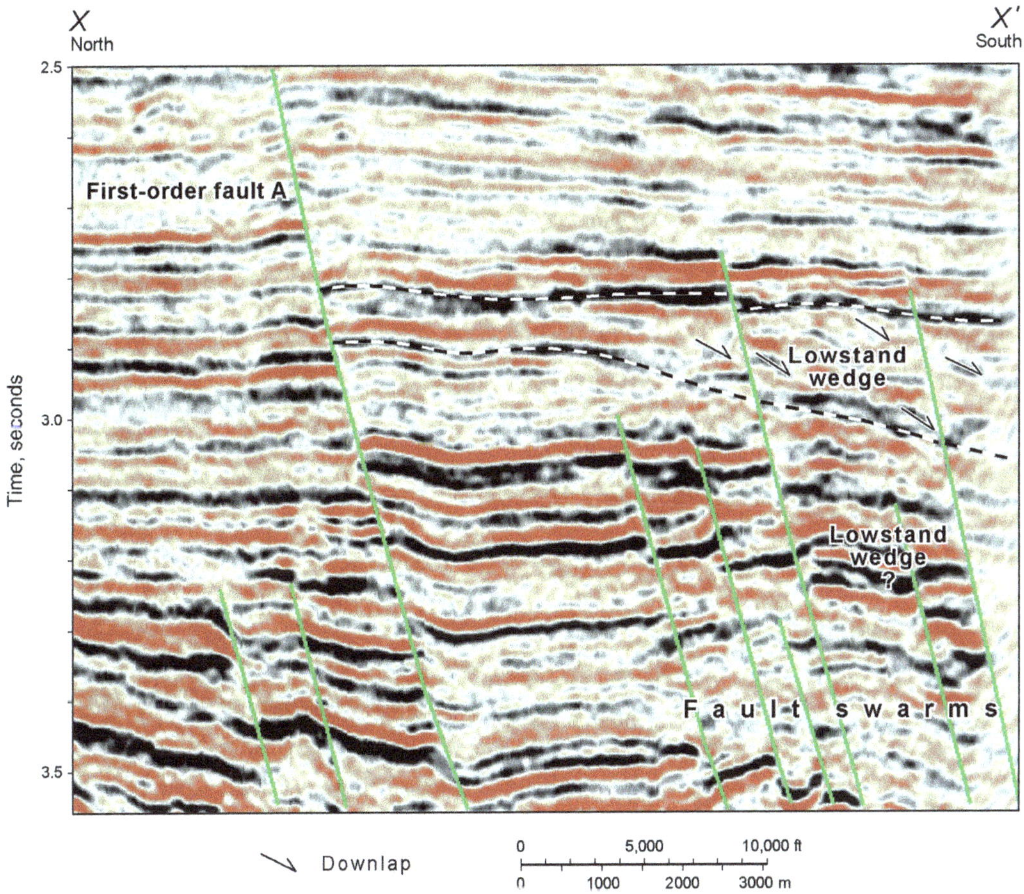

Fig. 2.18—Vertical seismic slice along profile XX′ (Fig. 2.17) showing faulted stratigraphic features.[29]

The circled features in Fig. 2.21 identify locations where stratal surfaces A, B, C, and D intersect obvious variations in reflection waveform. These waveshape changes are the critical seismic reflection character that distinguishes channel facies from nonchannel facies, as can be verified by comparing the inline coordinates spanned by the circled features (coordinates 60 to 70) with these same inline coordinates where this crossline (number 174) intersects the channel features labeled "Channel 1" in **Fig. 2.22**.

Fig. 2.22 shows reflection-amplitude behavior on stratal surface B, which is 90 milliseconds above, and conformable to, the selected seismic reference surface. This surface shows portions of the Channel 1 system in the lower right quadrant of the image. A second channel system (Channel 2) is located in the upper right quadrant.[30] The channel-system image shown in Fig. 2.22 is a surface-based image; that is, the seismic attribute that is displayed (which is reflection amplitude in this instance) is limited to a data window that vertically spans only one data sample. When a 1-point-thick data window is a good approximation of a stratal surface that passes through the interior of a targeted thin-bed sequence, then the seismic attributes defined on that surface can be important depictions of facies distributions within the sequence, as the image in this figure demonstrates.

An alternate, and usually more rigorous, way of determining facies distributions within a thin-bed sequence is to calculate seismic attributes in a data window that spans several data points vertically, yet is still confined (approximately) to only the thin-bed interval that needs to be

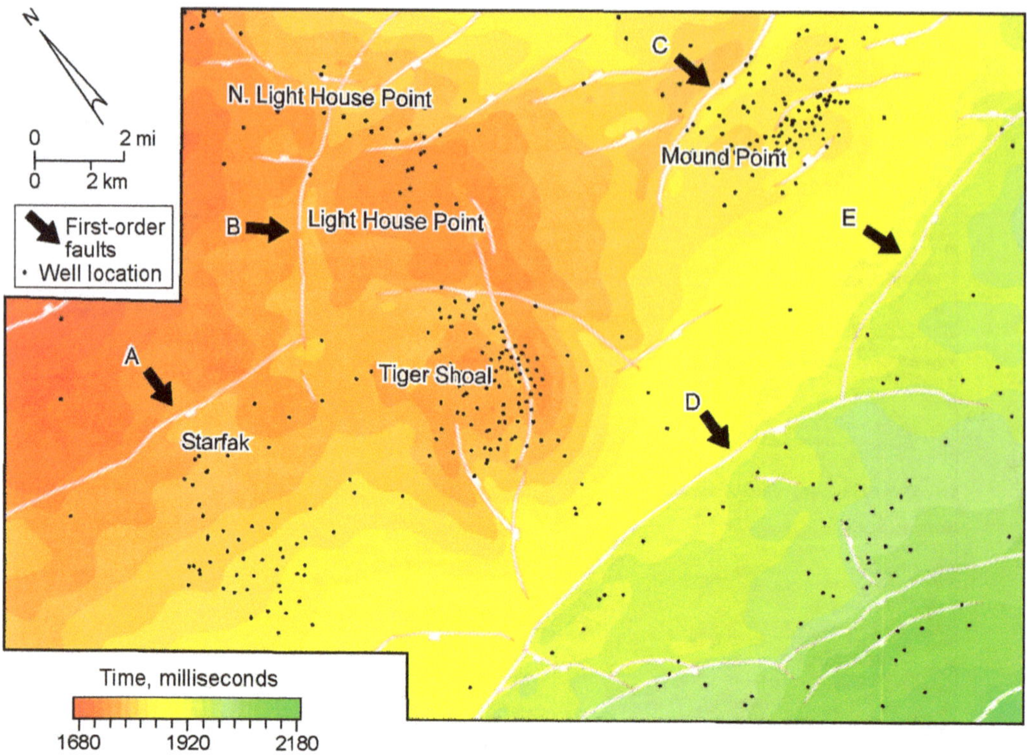

Fig. 2.19—Shallow time-structure map showing reduced influence of second-order faults compared with the deeper structure of Fig. 2.17.[29]

studied. The bottom stratal surface of this data window must reasonably coincide with the onset depositional time of the sequence, and the top stratal surface must be a good approximation of the shutoff depositional time of the sequence. Such a data window is called a stratal-bounded seismic analysis window.

Stratal surfaces A and D shown in the section view in Fig. 2.21 are examples of surfaces that define a stratal-bounded data analysis window that spans a targeted thin-bed sequence, specifically a thin-bed fluvial channel system that was the interpretation objective of this 3D seismic program. In this instance, the analysis window is 4 data points (8 milliseconds) thick. As stated in the discussion of Fig. 2.21, surfaces A, B, C, and D are good-quality stratal surfaces because each horizon images a significant part of the thin-bed fluvial system that was deposited over a "short" geological time period. Because each of these four seismic horizons is a good approximation of a constant-depositional-time surface, the four surfaces collectively are a good representation of the facies distribution within the total thin-bed sequence that they span.

One way to evaluate facies-sensitive seismic information spanned by surfaces A and D is to calculate some type of an averaged seismic attribute in each stacking bin (the concept of a stacking bin is described in Sec. 2.18.1) of the 4-point-thick data analysis window bounded by horizons A and D. For example, the average peak amplitude between A and D could be used to show an alternate image of the total channel system.

In thin-bed interpretations such as the fluvial channel system considered here, it is important to try to define two seismic reference surfaces that bracket the thin-bed system to be interpreted: one reference surface below the interpretation target and the second reference surface above the target. By creating conformable reference stratal surfaces above and below a thin-

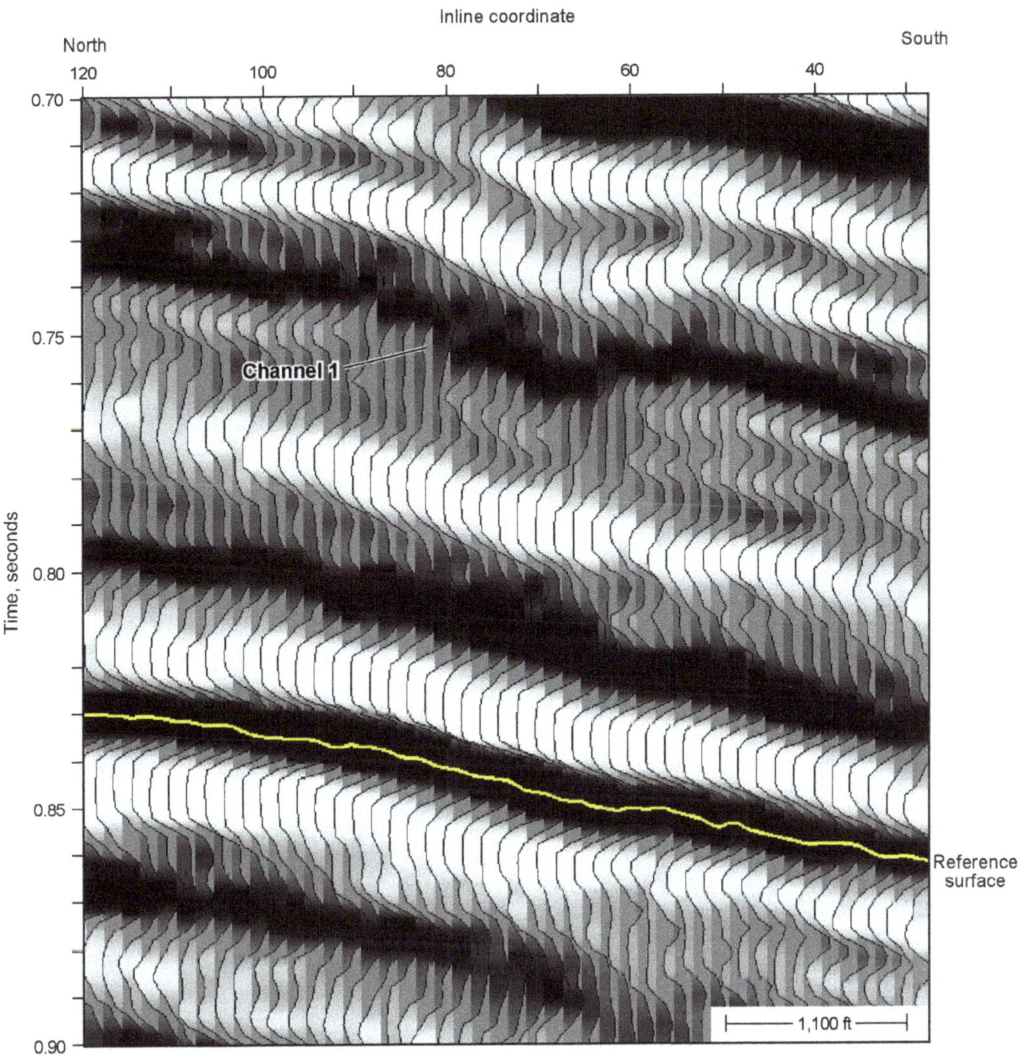

Fig. 2.20—Data window from a vertical slice of a 3D seismic data volume that is centered on a targeted channel system (Channel 1).

bed system, an interpreter can extend a series of conformable seismic stratal surfaces from two directions to sweep across a thin-bed target. A set of seismic stratal surfaces extended across an interval from above the interval is often a better approximation of constant-depositional-time surfaces within a targeted thin-bed sequence than is a set of stratal surfaces extended across the interval from below the interval (or vice versa). The more accurate set of surfaces will produce more reliable images of facies patterns within the thin-bed unit.

To illustrate the advantage of this opposite-direction convergence of seismic stratal surfaces onto a thin-bed target, a second reference surface was interpreted above (and, in this case, closer to) the targeted fluvial system studied in Figs. 2.20 through 2.22. Specifically, this second stratal reference surface followed the apex of the reflection troughs immediately above the thin-bed channels. **Fig. 2.23** shows the location of reference surface 2 on a second vertical slice (crossline 200) through the 3D seismic data volume. Reference surface 1 is the horizon labeled "reference surface" in Fig. 2.20. Reference surface 2 is an alternate seismic stratal surface posi-

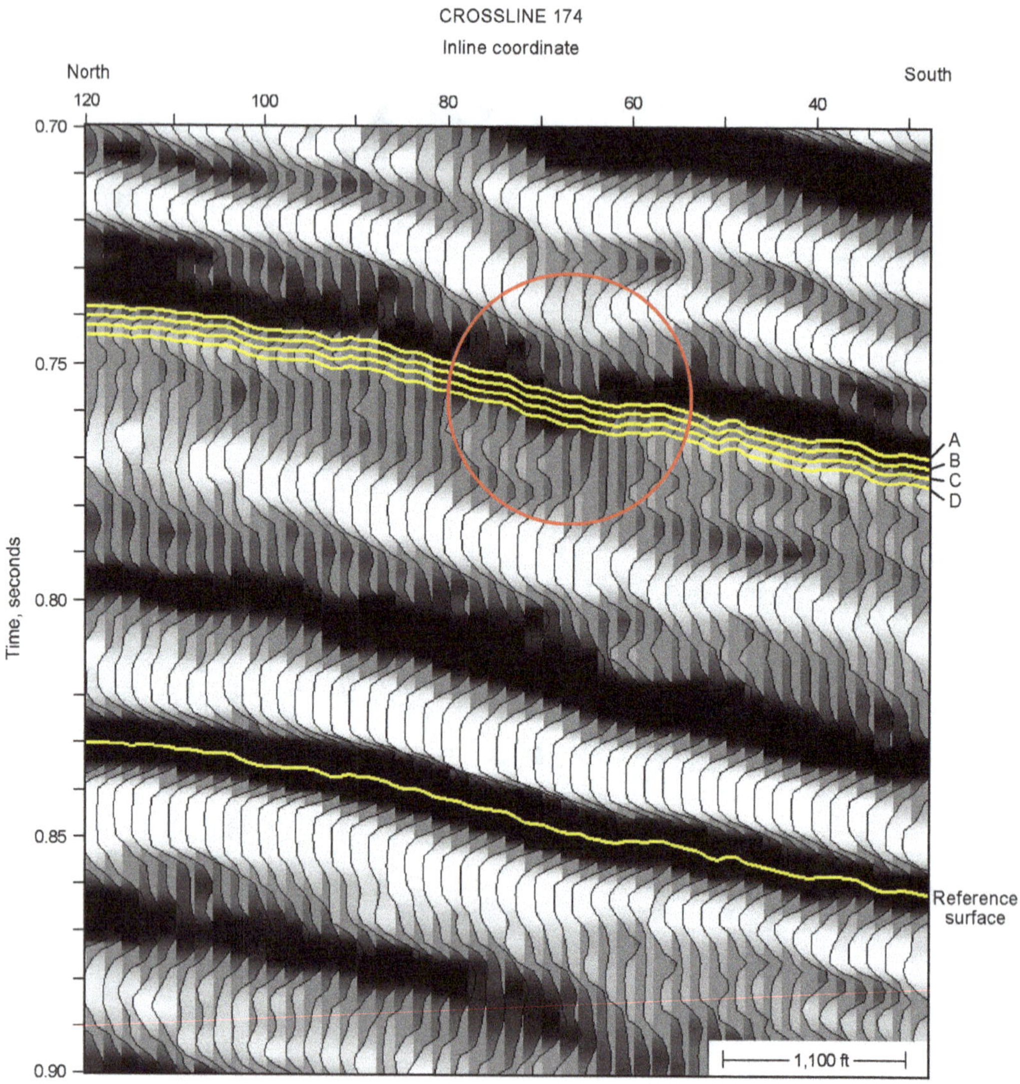

Fig. 2.21—Data window from the selected vertical slice showing stratal surfaces that traverse the channel system.[30]

tioned above the Channel 1 thin-bed target.[30] The targeted fluvial system referred to as Channel 1 is approximately 24 to 30 milliseconds below this second reference surface.

Fig. 2.24 displays the reflection-amplitude response across the channel systems observed on a stratal surface 26 milliseconds below and conformable to reference surface 2, as defined in Fig. 2.23. An improved channel image occurs, when compared with the image in Fig. 2.22, because in this case stratal surfaces that are conformable to the overlying seismic stratal surface happen to be better approximations of constant-depositional-time surfaces for this channel system than are stratal surfaces that are conformable to the deeper reference surface. This result illustrates that the combination of upward and downward extrapolations of conformable stratal surfaces across a thin-bed target is a good interpretation procedure, especially in those instances in which valid stratal reference surfaces can be interpreted both above and below the targeted thin bed.

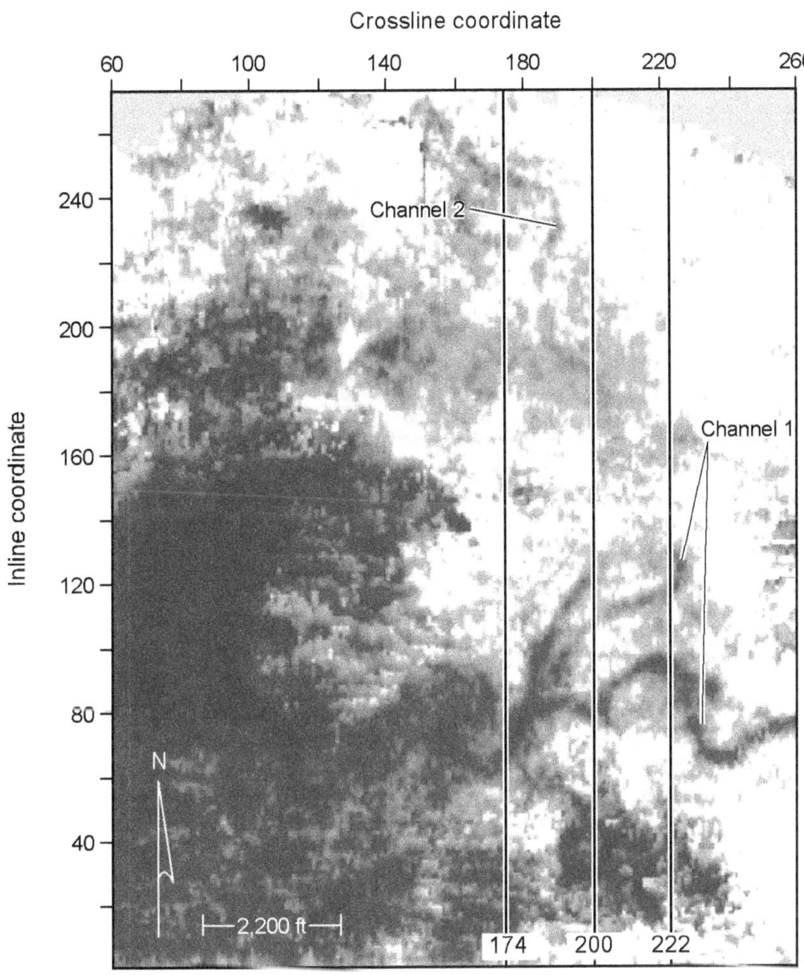

Fig. 2.22—Reflection-amplitude behavior on stratal surface B, which is 90 milliseconds above, and conformable to, the selected seismic reference surface.[30]

In summary, a good technique for interpreting thin-bed targets in 3D seismic data volumes is to interpret a reference surface that is conformable to the areal geometry of the thin-bed sequence and then to create seismic stratal surfaces conformable to this reference surface that pass through the thin-bed target. If the seismic stratal surfaces constructed according to this logic are satisfactory approximations of constant-depositional-time surfaces that existed during the deposition of the thin-bed sequence, the seismic attributes across these stratal surfaces are usually valuable indicators of facies distributions within the sequence.

A second technique is to expand the application of this stratal-surface concept by calculating seismic attributes inside a thin, stratal-bounded analysis window that is centered vertically on the thin-bed target. Facies-sensitive attributes extracted from carefully constructed stratal-bounded windows are often better indicators of facies distributions within a thin-bed target than are attributes that are restricted to a 1-point-thick stratal surface that passes through the target. This fact implies that the geologic time interval during which a thin-bed sequence is deposited can sometimes be portrayed satisfactorily by a stratal-bounded data window, whereas a fixed geologic time during the thin-bed deposition is not well approximated by a 1-point-thick seismic stratal surface. Interpreters have to try both approaches to determine an optimal procedure.

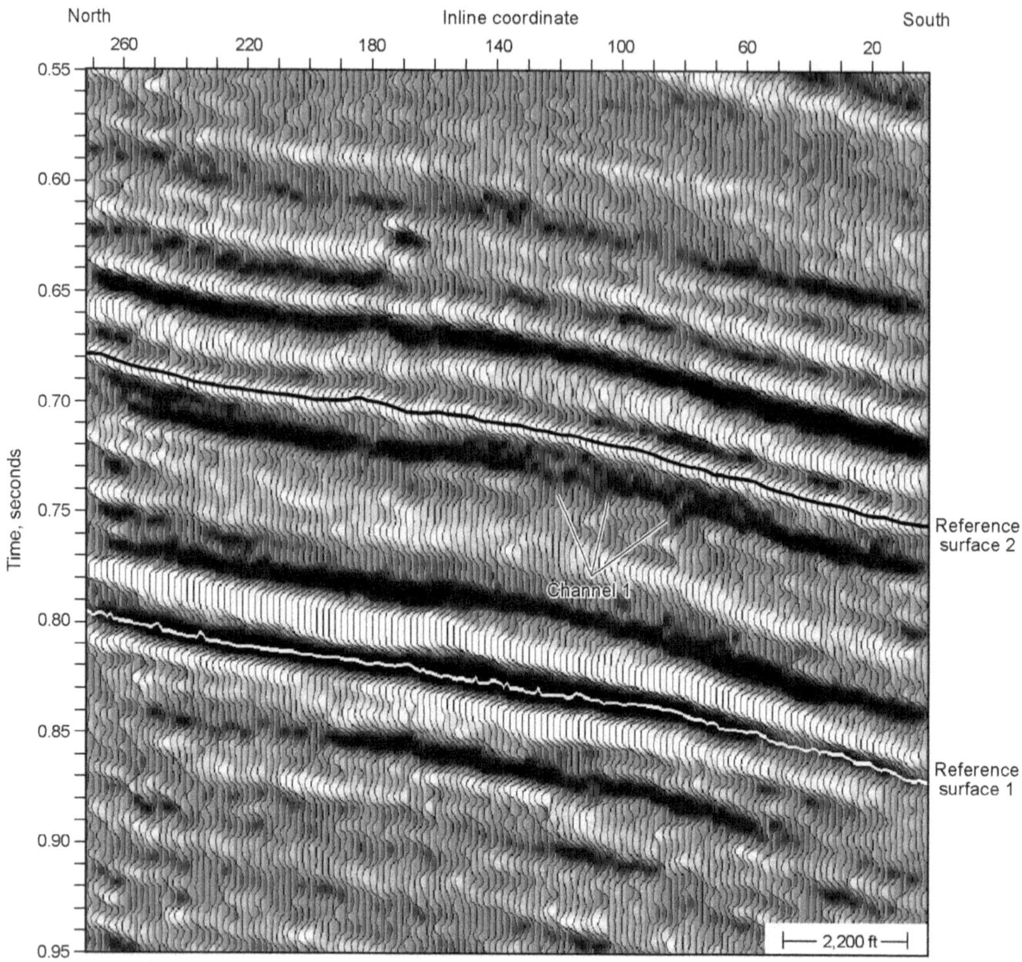

Fig. 2.23—Location of reference surface 2 on a second vertical slice (crossline 200) through the 3D seismic data volume.[30]

A third technique is to extend a series of conformable seismic stratal surfaces and stratal-bounded windows onto the thin-bed target from opposite directions, that is, from both below and above the thin-bed target. The logic in this dual-direction approach is that one of the seismic reference surfaces may be more conformable to the thin-bed sequence than the other reference surface and that this improved conformability will lead to improved attribute imaging of facies distributions within the thin bed.

2.18 Three-Dimensional Seismic Survey Design

2.18.1 Stacking Bins. The horizontal resolution a 3D seismic image provides is a function of the trace spacing within the 3D data volume. As the separation between adjacent traces decreases, horizontal resolution increases. At the conclusion of 3D data processing, the area spanned by a 3D seismic image is divided into a grid of small, abutted subareas called stacking bins. Each trace in a 3D seismic data volume is positioned so that it passes vertically through the midpoint of a stacking bin.

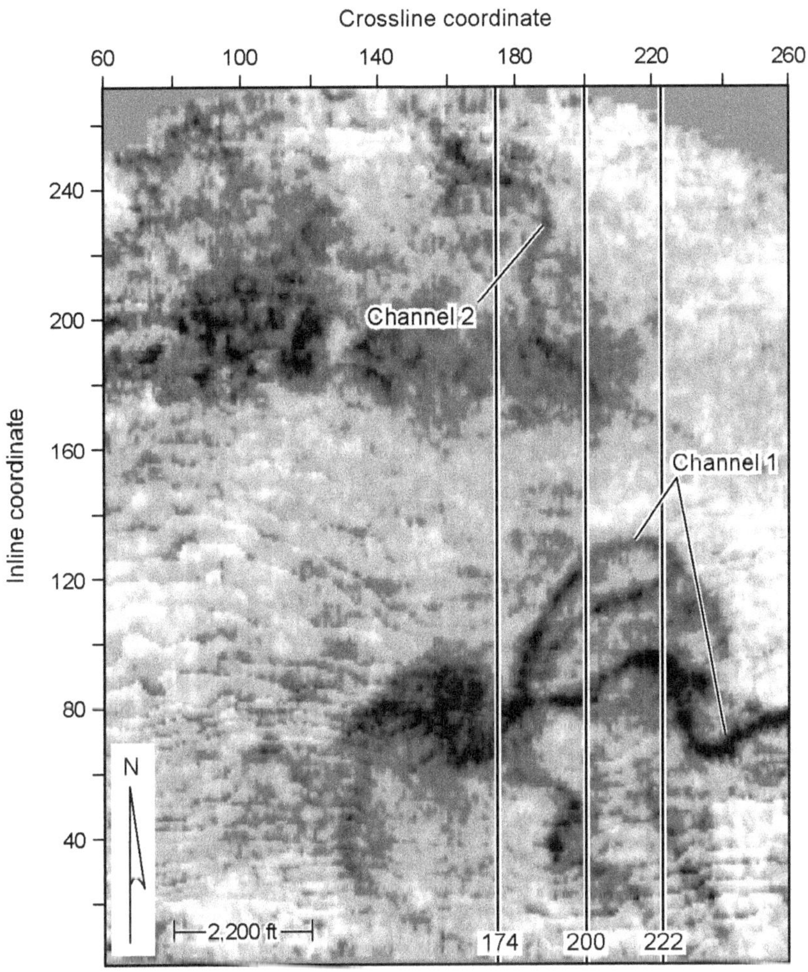

Fig. 2.24—Three-dimensional seismic image of the targeted thin-bed fluvial channel system.[30]

In **Fig. 2.25**, each stacking bin has lateral dimensions of Δx and Δy. The horizontal separations between adjacent processed traces in the 3D data volume are also Δx and Δy. The term inline is defined as the direction in which receiver cables are deployed, which is north/south in this example. Inline coordinates increase from west to east as shown. Crossline refers to the direction that is perpendicular to the orientation of receiver cables; thus, the crossline coordinates increase from south to north.[31] These stacking bins can be square or rectangular, depending on an interpreter's preferences. The dimension of the trace spacing in a given direction across a 3D image is the same as the horizontal dimension of the stacking bin in that direction. As a result, horizontal resolution is controlled by the areal size of the stacking bin.[32]

The imaging objective dictates how small a stacking bin should be. Smaller stacking bins are required if the resolution of small stratigraphic features is the primary imaging requirement. As a general rule, there should be a minimum of three stacking bins, and preferably at least four bins, across the narrowest stratigraphic feature that needs to be resolved in the 3D data volume. This imaging principle causes the targeted stratigraphic anomaly to be expressed on three or four adjacent seismic traces.

As **Fig. 2.26** illustrates, the critical parameter to be defined in 3D seismic design is the smallest (narrowest) horizontal dimension of a stratigraphic feature that must be seen in the 3D

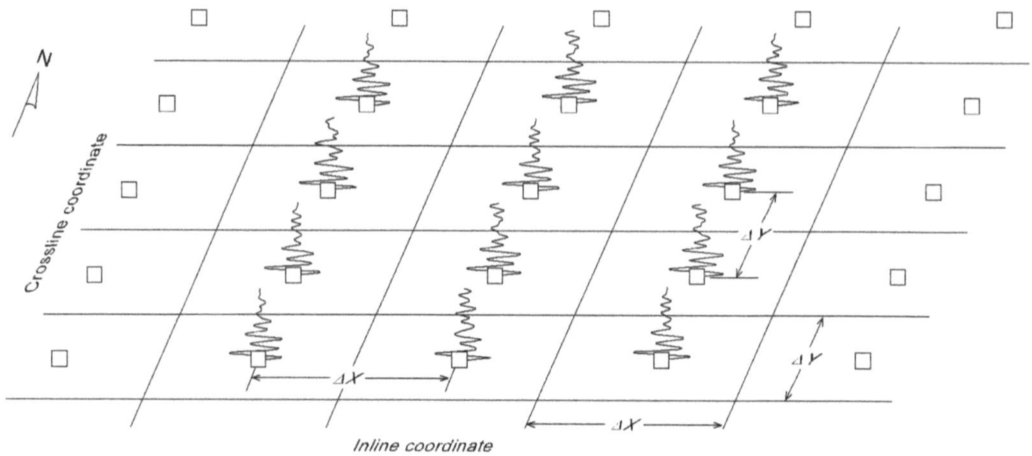

Fig. 2.25—Oblique view of 3D stacking bins.

data volume. For purposes of illustration, it is assumed that the narrowest feature to be interpreted is a meander channel. At least three, and ideally four, stacking bins (that is, seismic traces) must lie within the narrowest dimension, W, of this channel if the channel is to be reliably seen in the seismic image during workstation interpretation. Once W is defined, the dimensions of the stacking bins are also defined. The bin dimensions should be no wider than $W/3$. Ideally, they should be approximately $W/4$.[31] A variation in seismic reflection character on three to four adjacent traces is usually noticed by most interpreters, whereas anomalous behavior on fewer traces tends to be ignored or may not even be seen when a 3D data volume is viewed.

For example, if the interpretation objective is to image meandering channels that are as narrow as 200 ft, then the stacking bins should have lateral dimensions of approximately 50 ft (Fig. 2.26). This would cause a 200-ft channel to affect four adjacent traces. One of the first 3D design parameters to define, therefore, is the physical size of the stacking bin to be created. The bin size, in turn, can be determined by developing a stratigraphic model of the target that is to be imaged and then using that model to define the narrowest feature that needs to be seen. Once this minimum target dimension is defined, stacking bins with lengths and widths that are approximately one-fourth the minimum target width must be created if the target is to be recognized in a 3D data volume. Conversely, once a stacking-bin size is established, the narrowest stratigraphic feature that most interpreters can recognize will be a facies condition that spans at least three or four adjacent stacking bins.

2.18.2 Station Spacings. The distance between adjacent source points along a seismic line is the source-station spacing; the distance between adjacent receiver arrays along that same line is the receiver-station spacing. Previous publications on the topic of seismic acquisition[33,34] show that the trace spacing (i.e., the stacking-bin dimension) along a 2D seismic profile is one-half the receiver-station spacing (assuming the usual condition that the source-station spacing along the line is equal to or greater than the receiver-station spacing). Applying this principle to 3D seismic design leads to the following: the dimension of a 3D stacking bin in the direction in which receiver lines are deployed in a 3D grid is one-half the receiver-station spacing along these receiver lines, and the dimension of the stacking bin in the direction in which source lines are oriented is one-half the source-station spacing along the source lines.

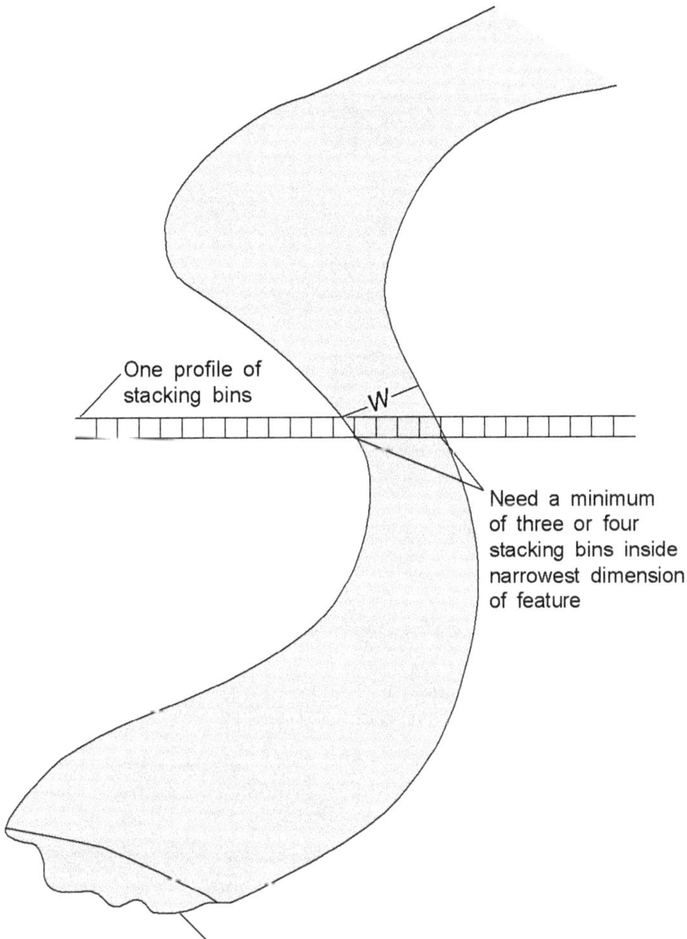

Fig. 2.26—Example of the narrowest feature that must be seen in a 3D image.[31]

As stated previously, once a decision has been made about the narrowest target that must be imaged, the required size of a stacking bin is automatically set at one-third or one-fourth that target dimension (Fig. 2.26). As a result, the source-station and receiver-station spacings are also defined because source-station spacing is twice the horizontal dimension of the chosen stacking bin in the source-line direction, and receiver-station spacing is twice the dimension of the stacking bin in the receiver-line direction. Stated another way, the source-station and receiver-station spacings should be one-half the narrowest horizontal dimension that needs to be interpreted from the 3D data.

When the geology involves steep dips or large changes in rock velocity across a fixed horizontal plane, rigorous calculations of station spacing (or bin size) should be made with commercial 3D seismic design software rather than by following the simple relationships described here.

2.18.3 Stacking Fold. The stacking fold associated with a particular 3D stacking bin is the number of field traces that are summed during data processing to create the single image trace positioned at the center of that bin (Fig. 2.25). In other words, the stacking fold is the number

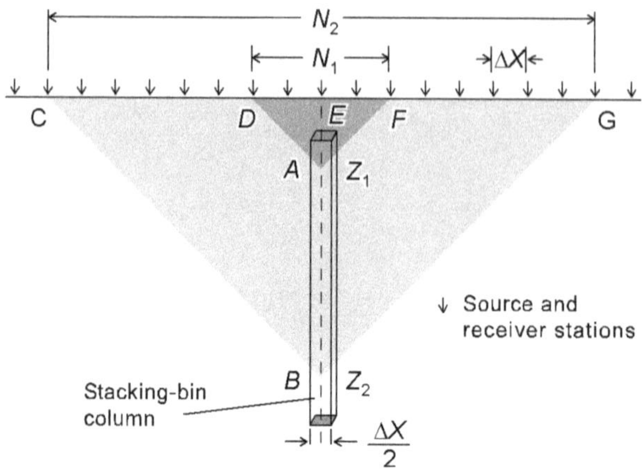

Fig. 2.27—Vertical variation in stacking fold.[31]

of distinct reflection points that are positioned inside a bin because of the particular source-receiver grid that is used.

At any given stacking-bin coordinate, the stacking fold inside that bin varies with depth. **Fig. 2.27** illustrates vertical variation in stacking fold. The source-station and receiver-station spacings along this 2D profile both have the same value for Δx, which results in a stacking bin width of $\Delta x/2$. The vertical column shows the coordinate position of one particular stacking bin. For a deep target at depth Z_2, the stacking fold in this bin is a high number because there is a large number, N_2, of source-receiver pairs that each produce a raypath that reflects from subsurface point B. Only one of these raypaths, CBG, is shown. For a shallow target depth, Z_1, the stacking fold is low because there is only a small number, N_1, of source-receiver pairs that can produce individual raypaths that reflect from point A. One of these shallow raypaths, DAF, is shown. When a 3D seismic data volume is described as a 20-fold or 30-fold volume, the designers are usually referring to the maximum stacking fold created by the 3D geometry, which is the stacking fold at the deepest target.[31]

In Fig. 2.27, when the stacking bin is centered around deep reflection point B, the stacking fold is at its maximum because the largest number of source and receiver pairs can be used to produce individual reflection field traces that pass through the bin. The number of source-receiver pairs that can contribute to the image at B is typically confined to those source and receiver stations that are offset horizontally from B by a distance that is no larger than depth Z_2 to reflection point B. Thus, the distances CE and EG are each equal to Z_2.

With this offset criterion to determine the number of source-receiver pairs that can contribute to the seismic image at any subsurface point, we see that the stacking fold at depth Z_2 would be N_2, as Fig. 2.27 shows, because N_2 unique source-receiver pairs can be found that produce N_2 distinct field traces that reflect from point B. When the stacking bin is kept at the same x and y coordinates but moved to shallower depth, Z_1, the stacking fold decreases to the smaller number, N_1. Only N_1 source-receiver pairs generate field traces that reflect from A and still satisfy the geometrical constraint that these pairs are offset by distance DE (or EF) that does not result in critical wavefield refractions at interfaces above A. When critical refraction occurs, the transmitted raypath, bent at an angle of 90°, follows a horizontal interface rather than continuing to propagate downward and illuminating deeper targets.

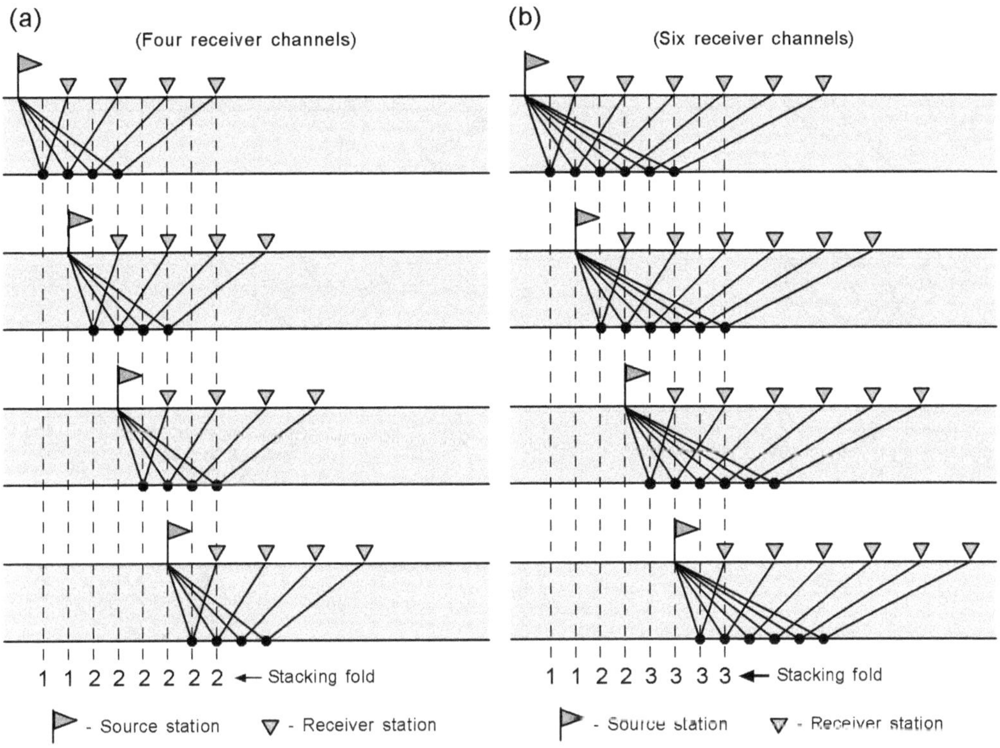

Fig. 2.28—Effect of number of active receivers on inline stacking fold.[31]

In 2D acquisition geometry, the inline stacking fold, F_{IL}, is a function of two geometrical properties: the number of active receiver channels and the ratio between the source-station interval and the receiver-station interval. The raypath diagrams in **Figs. 2.28 and 2.29** illustrate the manner in which each of these geometrical parameters affects inline stacking fold. Fig. 2.28 establishes the principle that inline stacking fold is one-half of the active receiver stations when the source-station interval equals the receiver-station interval.

$$F_{IL} = \frac{n_c}{2} \quad \text{(2.5)}$$

The raypaths in Fig. 2.28a show the distribution of reflection points (the solid circles on the subsurface interface) when there are four active receiver channels and the source-station interval is the same as the receiver-station interval. The vertical dashed lines pass through successive reflection points. The stacking-fold numbers at the bottom of the diagram define the number of distinct source-receiver pairs that create a reflection image at each subsurface point, that is, the number of reflection points that each vertical dashed line intersects.

The maximum stacking-fold for this four-receiver situation is 2. The raypaths in Fig. 2.28b show the distribution of reflection points and the stacking fold that results when there are six-receiver channels. The maximum stacking fold for this six-receiver geometry is 3.[31]

Fig. 2.29 expands the inline stacking-fold analysis to show that for geometries in which the source-station interval does not equal the receiver-station interval, then

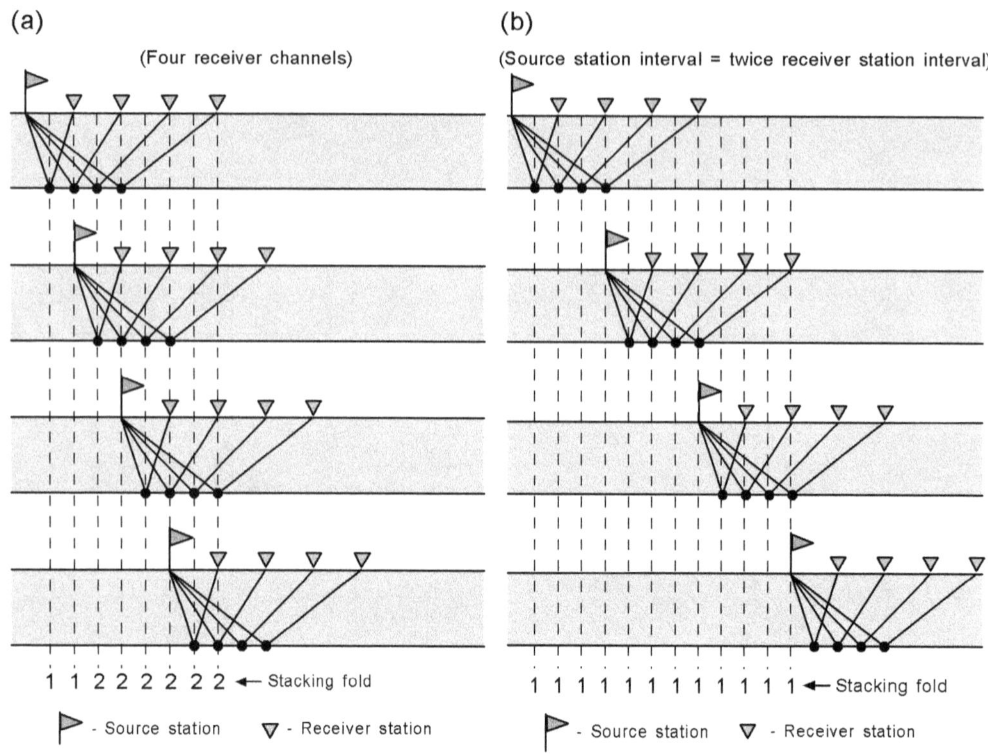

Fig. 2.29—Effect of source-station spacing on inline stacking fold.[31]

$$F_{IL} \sim \frac{i_r}{i_s}, \qquad (2.6)$$

where ~ means "is proportional to."

The raypath diagram in Fig. 2.29a shows the distribution of subsurface reflection points (the solid circles on the subsurface interface) when there are four active receiver channels and the source-station spacing equals the receiver-station spacing. The inline stacking fold is the number of independent reflection points that occur at the same subsurface coordinates, which is the same as the number of reflection points intersected by each vertical dashed line. The stacking fold is shown by the sequence of numbers at the base of the diagram and, in this geometry, the maximum fold is 2.

The raypath picture in Fig. 2.29b shows the distribution of reflection points when the number of active receiver channels is the same as in Fig. 2.29a; that is, there are four receiver groups, but the source-station spacing is now twice the receiver-station spacing. (Note that the incremental movement of the source-station flag in Fig. 2.29b is two times greater than the flag movement in Fig. 2.29a.) The resulting stacking fold is shown by the number written below each vertical dashed line, which is the number of reflection points intersected by each of those lines. The maximum stacking fold in this geometry is only 1. These two diagrams establish the principle that inline stacking fold is proportional to the ratio of the receiver-station interval to the source-station interval.

Combining Eqs. 2.5 and 2.6 leads to the design equation for inline stacking fold:

$$F_{IL} = \frac{1}{2}(n_c) \times \left[\frac{i_r}{i_s}\right]. \quad (2.7)$$

In 2D seismic profiling, the source-station interval is usually the same as the receiver-station interval, making the ratio term in the brackets in Eq. 2.7 equal to 1. However, in 3D profiling, the source-station spacing along a receiver line is the same as the source-line spacing, which is several times larger than the receiver-station spacing. Crossline fold, F_{XL}, is given by

$$F_{XL} = \frac{1}{2}n_l. \quad (2.8)$$

In a 3D context, the stacking fold is the product of the inline stacking fold (the fold in the direction in which the receiver cables are deployed) and the crossline stacking fold (the fold perpendicular to the direction in which the receiver cables are positioned). This principle leads to the important design equation:

$$F = F_{IL} \times F_{XL}. \quad (2.9)$$

To build a high-quality 3D image, it is critical not only to create the proper stacking fold across the image space but also to ensure that the traces involved in that fold have a wide range of offset distances and azimuths. Eq. 2.9 provides no information about the distribution of either the source-to-receiver offset distances or azimuths that are involved in the stacking fold. When it is critical to know the magnitudes and azimuth orientations of these offsets, commercial 3D seismic design software must be used. Offset analysis is a technical topic that goes beyond the scope of this discussion. Ref. 35 describes the parameters involved in onshore 3D seismic survey design.

2.19 Vertical Seismic Profiling

In vertical seismic profiling (VSP), a seismic sensor is lowered to a sequence of selected depths in a well by wireline. **Fig. 2.30** shows the source-receiver geometry involved in VSP. A wall-locked seismic sensor is manipulated downhole by wireline so that the receiver occupies a succession of closely spaced vertical stations. This receiver records the total seismic wavefield, both downgoing and upgoing events, produced by a surface-positioned energy source. Only 6 receiver stations are indicated here for simplicity, but a typical VSP consists of 75 to 100 receiver stations. The vertical spacing between successive stations is a few tens of feet. A common receiver spacing is 50 ft (15 m). The horizontal distance, X, between the surface source and the downhole receiver is the offset and can assume different magnitudes, depending on the specific VSP imaging application. Fig. 2.30 depicts a VSP measurement made in a vertical wellbore, but the VSP technique can also be implemented in deviated wells.[37]

Because the receiver stations are aligned vertically, the data-recording procedure is called VSP to distinguish the technique from conventional horizontal seismic profiling, in which seismic receivers are deployed across the surface of the Earth. In horizontal seismic profiling along the Earth surface, only upgoing seismic wavefields are recorded. The crucial information of the downgoing wavefields is not available to assist seismic data processors and interpreters. Seismic data recorded with a vertical receiver array have many valuable applications,[36,37] but the only uses stressed here are the abilities of such data to calibrate stratigraphic depth to specific waveform features of surface-recorded seismic reflection data and to provide an independent, high-resolution image of the subsurface in close proximity to the VSP receiver well.

Velocity check-shot data are recorded with the same source-receiver geometry used for VSP data recording (Fig. 2.30). However, the vertical distance between successive receiver sta-

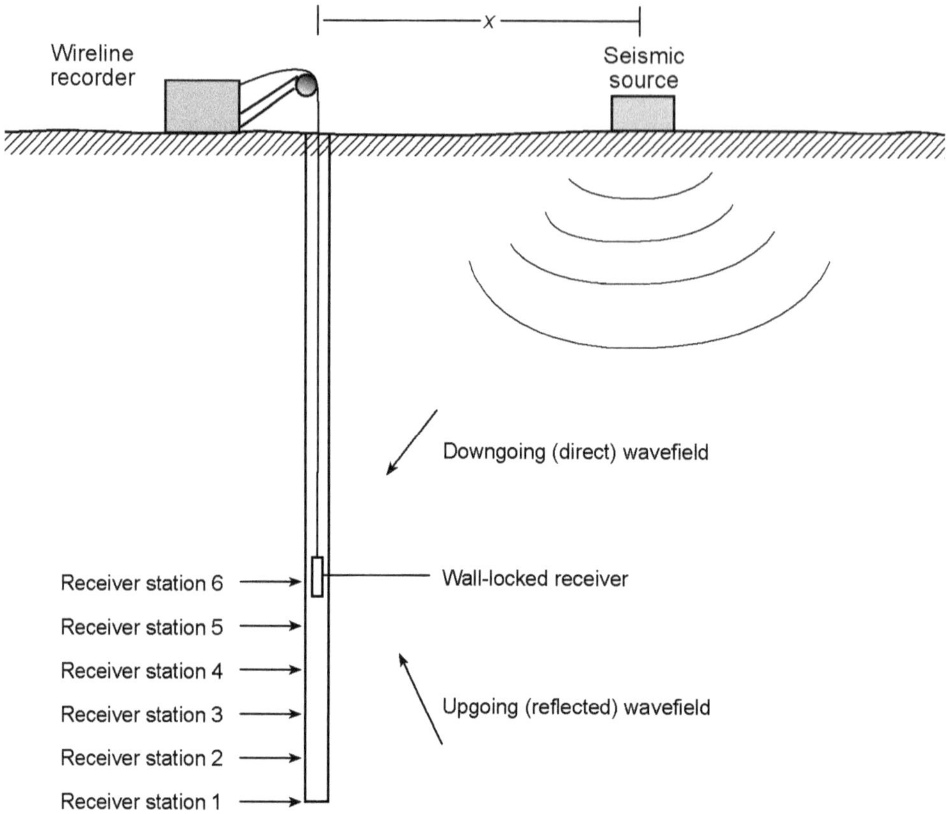

Fig. 2.30—Source-receiver geometry used in vertical seismic profiling.[38]

tions is on the order of 500 ft (150 m) or more, compared with a smaller station spacing of approximately 50 ft (15 m) used to record VSP data. This order-of-magnitude difference in the spatial sampling of subsurface seismic wavefields is the principal difference between VSP and velocity check-shot data. The primary use of velocity check-shot data is to create a rigorous relationship between stratigraphic depth coordinates and seismic image-time coordinates. These depth-to-time relationships are critical for transforming log data and engineering data from the depth domain to the seismic image-time domain. This coordinate transformation allows critical geologic and engineering information to be associated with proper data windows in the seismic image.

Because equivalent source-receiver recording geometries are used, velocity check-shot data can provide a rigorous relationship between stratigraphic depth and seismic travel-time, just as VSP data do. One shortcoming of check-shot data, however, is that they do not provide an independent seismic image that can be correlated with surface-recorded seismic reflection images. Such a correlation can verify the precise amount of time shift that should be imposed to bring subsurface stratigraphy into exact phase agreement with a surface-recorded 3D seismic image. To a seismic interpreter, two images are in phase agreement when the peaks and troughs of the two sets of wiggle-trace data occur at the same time coordinates over a window of interest and the waveshapes of key events in the two images are similar over that window. In contrast to check-shot data, VSP data provide an independent seismic image, and this VSP image is the unique feature of the VSP technique that allows subsurface stratigraphy to be inserted into 3D seismic image volumes at precise seismic travel-time coordinates.[38]

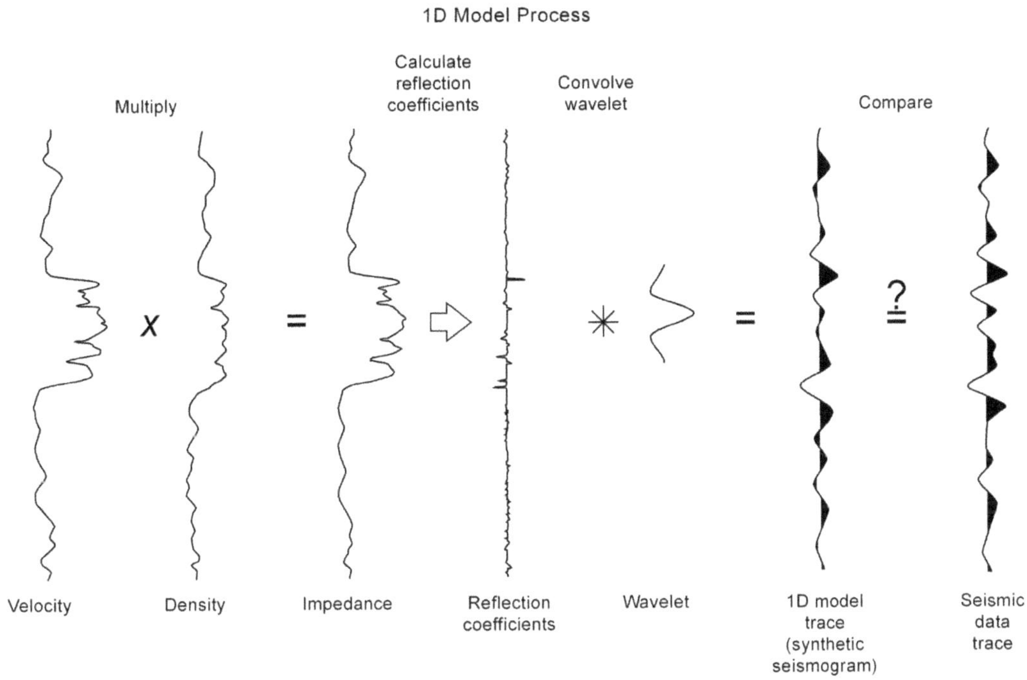

Fig. 2.31—Computational steps involved in calculating a synthetic seismogram.

2.19.1 Synthetic Seismograms. Some seismic interpreters argue that a synthetic seismogram made from sonic and density log data can provide an independent image that can be used to determine the proper time shift between surface-recorded seismic data and check-shot-positioned stratigraphy encountered in the check-shot well. **Fig. 2.31** illustrates the steps taken to create a synthetic seismogram and to use that synthetic model in interpretation.

First, sonic log data and density log data recorded in a chosen calibration well are multiplied to create a log of the layer impedances penetrated by the well (left three curves of Fig. 2.31). Eq. 2.1 describes this calculation. Either before or after this multiplication, these log data have to be converted from functions of depth to functions of vertical seismic travel time. Such a transformation is done with a simple equation:

$$D = v \times t \text{ . } \qquad (2.10)$$

The velocity function in this equation is provided by the sonic log used in the calculation. Sonic log data usually have to be adjusted by small percentage amounts so that the integrated sonic log time agrees with seismic check-shot time. With Eq. 2.2, the time-based layer impedance wave is converted to a time series of reflection coefficients, and an estimated seismic wavelet is convolved with this reflectivity series. The result is the synthetic seismogram trace shown in Fig. 2.31. The interpretation step is done by comparing the synthetic seismogram with real seismic traces near the calibration well (last step of Fig. 2.31). During this comparison, the synthetic trace is shifted up and down in time to determine what time shift, if any, is required to create an optimal alignment of reflection peaks and troughs between the synthetic and real traces. Most geophysicists would describe the wiggle-trace alignment shown in Fig. 2.31 as a good phase tie.

There are instances in which synthetic seismograms are a poor match to seismic data. When there are only a few well penetrations, this can be a problem best addressed with VSP data. As the number of wells increase and greater areal coverage is provided, poor synthetic data can be eliminated and reliable synthetic seismograms can be used to leverage a limited number of VSP surveys. There are several reasons that synthetic seismograms sometimes fail to provide the reliability needed for calibrating thin-bed stratigraphy with seismic reflection character.[39] The more common failures are usually related to one or more of the following factors:

- The log-determined velocity and density values used in a synthetic seismogram calculation represent petrophysical properties of rocks that have been mechanically damaged by drilling and altered by the invasion of drilling fluids. In addition, irregular changes in borehole diameter sometimes induce false log responses. As a result, well log determinations of rock velocity and density, which are the fundamental data used to produce the reflection coefficients needed for a synthetic seismogram calculation, may not represent the velocity and density values in undrilled rocks near the logged well, which are the fundamental rock properties that determine the reflection waveshape character of seismic data recorded at the wellsite.
- A synthetic seismogram represents an estimate of the seismic image that would result if the imaging raypath traveled vertically downward from a source and then reflected vertically upward along that same travel path to a receiver located exactly at the source position. In contrast, each trace of an actual seismic profile is a composite of many field traces that represent wavefield propagation along a series of oblique raypaths between sources and receivers that are laterally displaced from each other, with each of these raypaths reflecting from the same subsurface point. These two images (synthetic seismogram and actual seismic trace) thus involve raypaths that travel through different portions of the Earth.
- Even when log-determined velocity and density values (and any synthetically calculated seismic reflectivity derived from these log data) represent the correct acoustic impedances of a stratigraphic succession, that stratigraphy may be localized around the logged well and not be areally large enough to be a reflection boundary for a surface-generated seismic wavefield. This situation may be more common in heterogeneous rock systems than many interpreters may appreciate.
- The effects of the near surface are not included in a synthetic seismogram calculation because logs are not recorded over shallow depths. At some wellsites, the near surface can induce significant effects into the waveform character of surface-recorded seismic data. In contrast, near-surface effects, such as peg-leg multiples, frequency absorption, and phase shifting, are included in VSP data because VSP wavefields propagate through the total stratigraphic section, including the near surface, just as surface-recorded seismic wavefields do.

2.20 Calibrating Seismic Image Time to Depth

VSP recording geometry causes the stratigraphy at the VSP well, where sequence boundaries are known as a function of depth from well log and core control, to be locked to the VSP image. This stratigraphy, in turn, is also known as a function of VSP reflection time. This fixed relationship between stratigraphy and the VSP image results because VSP receivers are distributed vertically through the seismic image space. This data-recording geometry allows both stratigraphic depth and seismic traveltime to be known at each receiver station. The dual-coordinate domain (depth and time) involved in a VSP measurement means that any geologic property known as a function of depth at the VSP well can be accurately positioned on, and rigidly welded to, the time-coordinate axis of the VSP image.

Fig. 2.32 illustrates the VSP depth-to-time calibration. VSP data are unique in that they are the only seismic data that are recorded simultaneously in the two domains critical to geologic interpretation: stratigraphic depth and seismic reflection time (Fig. 2.32a). As a result, specific stratigraphic units, known as a function of depth from well log data, can be positioned precisely in their correct VSP-image time windows (Fig. 2.32b). Each numbered stratigraphic unit shown in Fig. 2.32b is a thin-bed reservoir penetrated by the VSP well. When the VSP image is shifted up or down to correlate better with a surface-recorded seismic reflection image that

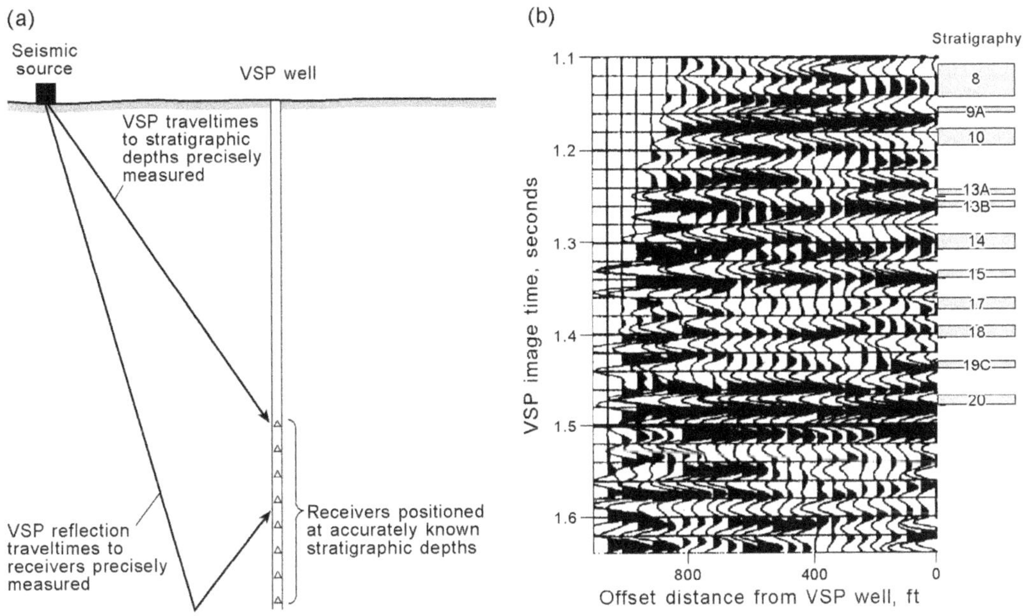

Fig. 2.32—Concept of VSP depth-to-time calibration.[38]

crosses the VSP well, the VSP-defined time window that spans each stratigraphic unit should be considered to be welded to the VSP data. This causes the stratigraphy to move up and down in concert with the VSP image during the VSP-to-surface seismic correlation process. The seismic time scale involved in the depth-to-time calibration illustrated here is VSP image time, which may be different from the image time for surface-recorded reflection data. **Fig. 2.33** illustrates the transformation of stratigraphy from VSP image time to 3D seismic image time.

The reverse situation is also true; that is, the VSP image could be positioned on, and welded to, the depth-coordinate axis of the stratigraphic column at the VSP wellsite. This option of transforming a VSP image to the stratigraphic depth domain is not often done because the common objective of seismic interpretation is to insert stratigraphy into 3D seismic data volumes that are defined as functions of seismic traveltime, not as functions of stratigraphic depth.

The concept of a welded bond between a VSP image and the stratigraphy at the VSP wellsite means that whenever a VSP image is moved up to better correlate with a 3D seismic image, the stratigraphy moves up by that same amount of time in the 3D image. If the VSP image has to be moved down to create a better waveform character match with the 3D data, then the stratigraphy shifts down by the same amount in the 3D data volume. The fact that VSP data provide an independent image that can be moved up and down to find an optimal match between VSP and 3D reflection character is the fundamental property of the VSP-to-seismic calibration technique, which establishes the correct time shift between 3D seismic reflection time and VSP reflection time.

When the time shift between these two images is determined, the correct time shift between the 3D seismic image and the stratigraphy at the VSP-calibration well is also defined because that stratigraphy is welded to the VSP traveltime scale and moves up and down in concert with the VSP image time coordinate. Fig. 2.33 shows a specific example of a VSP-based stratigraphic calibration of a 3D data volume. The rigid welding of stratigraphic depth to VSP traveltime as described in Fig. 2.32 is repeated here as Fig. 2.33a. In this example, the VSP image must be advanced (moved up) by 18 milliseconds to optimally align with the 3D seismic image at

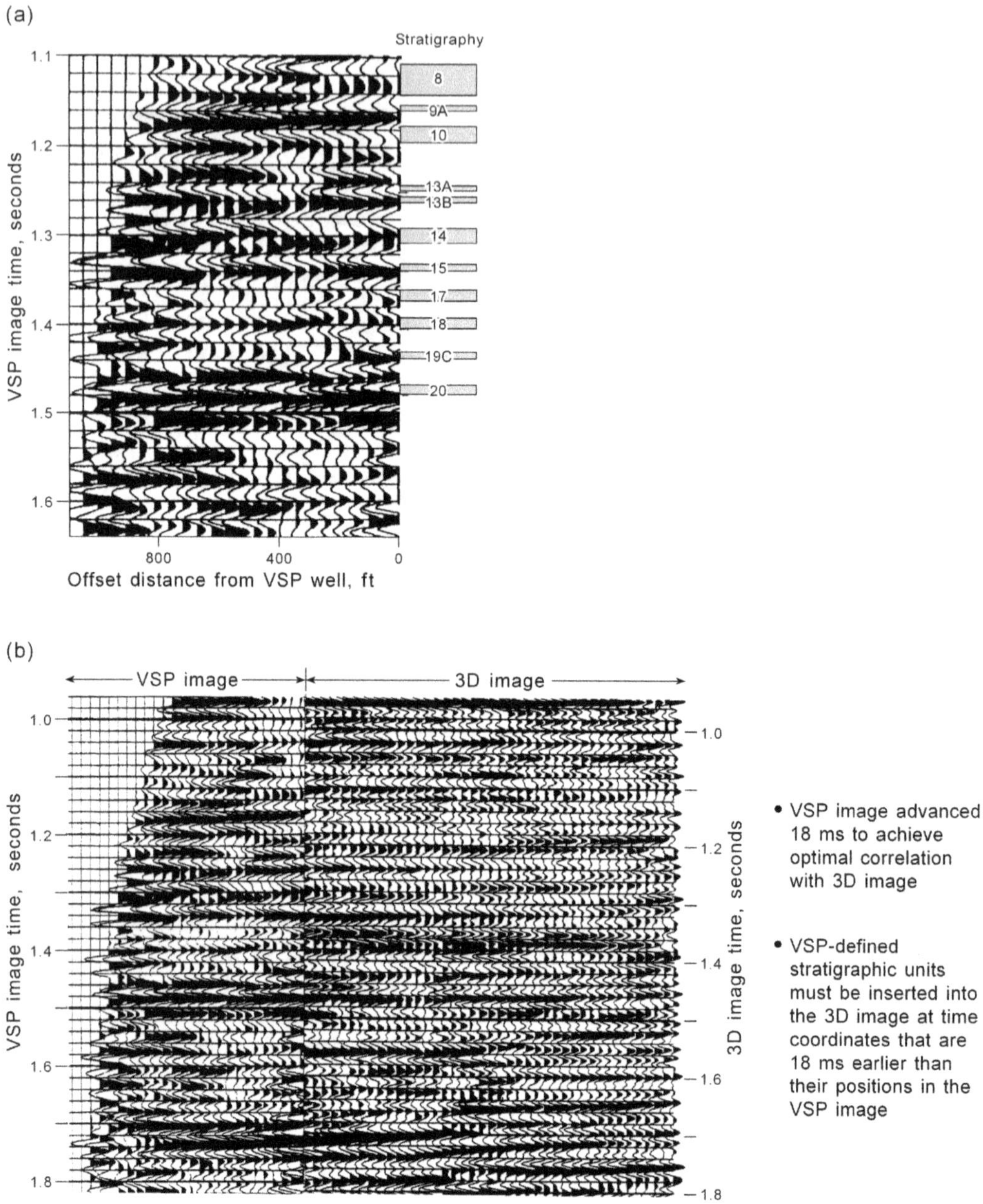

Fig. 2.33—VSP-based calibration of thin-bed stratigraphy in 3D seismic images.[38]

the VSP well (Fig. 2.33b). Because the stratigraphy penetrated by the VSP well is welded to the VSP image, the positions of the targeted thin-bed time windows in the 3D image also move up by 18 milliseconds to align with their positions in the VSP image. The VSP technique provides not only a time-vs.-depth calibration function but also an independent reflection image that can be time shifted to correlate with a surface-recorded image in the manner shown here. This is the unique feature that makes a VSP calibration of stratigraphy to 3D seismic image time more reliable than a check-shot-based stratigraphic calibration.[38]

This VSP image was produced from a large-offset VSP survey in which the offset distance, X (Fig. 2.30), was a little more than 2,000 ft (600 m). In Fig. 2.33, the VSP-based interpretation procedure leads to the conclusion that although the tops of thin-bed reservoirs 19C and 15 are positioned at VSP travel times of 1.432 and 1.333 seconds, they have to be inserted into the 3D data volume at 3D seismic travel times of 1.414 and 1.315 seconds.

Three-dimensional VSP data can be acquired when many source stations encircle a receiver well. Technically, there is no barrier to 3D VSP imaging. The major industry objection to 3D VSP technology is the relatively high cost of data acquisition and processing compared with the cost of conventional 3D surface-based seismic imaging. In special cases that have justified the cost, 3D VSP imaging has been done to create high-resolution images around a receiver well. To date, only a few such surveys have been done worldwide.

2.21 Crosswell Seismic Profiling

Fig. 2.34 shows distinctions among the source-receiver geometries involved in vertical seismic profiling (VSP), reverse vertical seismic profiling (RVSP), and crosswell seismic profiling (CSP). Fig. 2.34a shows the field geometry used in conventional VSP. Source S is positioned on the surface of the Earth, and seismic receiver R is lowered into the well where the data are to be recorded. The direct arrival path is SR, and the reflected travel path is SPR. The position of reflection point P can be varied by moving either source S or receiver R. If the source is directly above the receiver, the measurement is called a zero-offset VSP. If the source is not directly above the receiver, the measurement is called an offset VSP.

In RVSP, the positions of the source and receiver are exchanged. As Fig. 2.34b shows, receiver R is on the surface for an RVSP, and source S is located in the well. The offset in this diagram has the same meaning as it does for a conventional VSP. (Offset is the lateral distance between a vertical line passing through the source position and a vertical line passing through the receiver position.) In a vertical well, offset can be measured relative to the wellhead, if desired. In nonvertical wells, offset must be measured strictly between the coordinates of the source and the receiver. Three-dimensional RVSP data can be acquired at rather low cost because it is not difficult to distribute a large number of receiver stations on the Earth's surface in an areal pattern around a source well.

In a CSP measurement, both the source and the receiver are below the surface and in separate wells, as Fig. 2.34c shows. The direct travel path is again SR, and the reflected path is SPR. One of the attractions of CSP data is that no part of either path SR or path SPR passes through the near-surface weathered layer, as occurs when VSP and RVSP data are recorded. As a result, crosswell data do not suffer a significant loss of the higher-frequency components of the source wavefield. These components are usually attenuated as they pass through the surface weathered layer to complete any of the VSP-type travel paths. Because spatial resolution improves as the frequency content of the signal is increased, crosswell data reveal greater reservoir detail than do either type of VSP measurement.

In crosswell data acquisition, two types of source-receiver offsets can be considered, depending on whether the direct or the reflected wavefield is being analyzed. These two offsets are transmission offset and reflection offset, respectively. Transmission offset in a crosswell geometry (Fig. 2.34c) is measured orthogonal to the direction in which reflection offset (Figs. 2.34a and 2.34b) is measured and can be defined as the vertical distance between a horizontal line passing through the source position and a horizontal line passing through the receiver position.

There are three techniques by which the interwell space of a reservoir system can be investigated using CSP data: attenuation tomography, for which the basic measurement is the amplitude of the direct seismic arrival wavelet; velocity tomography, for which the principal measurement is the traveltime required for the direct seismic arrival to propagate across the interwell space; and elastic wavefield imaging.

In velocity and attenuation tomography, the only information in the crosswell wavefield that is used are the travel times and amplitudes of the seismic direct arrival. In elastic wave-

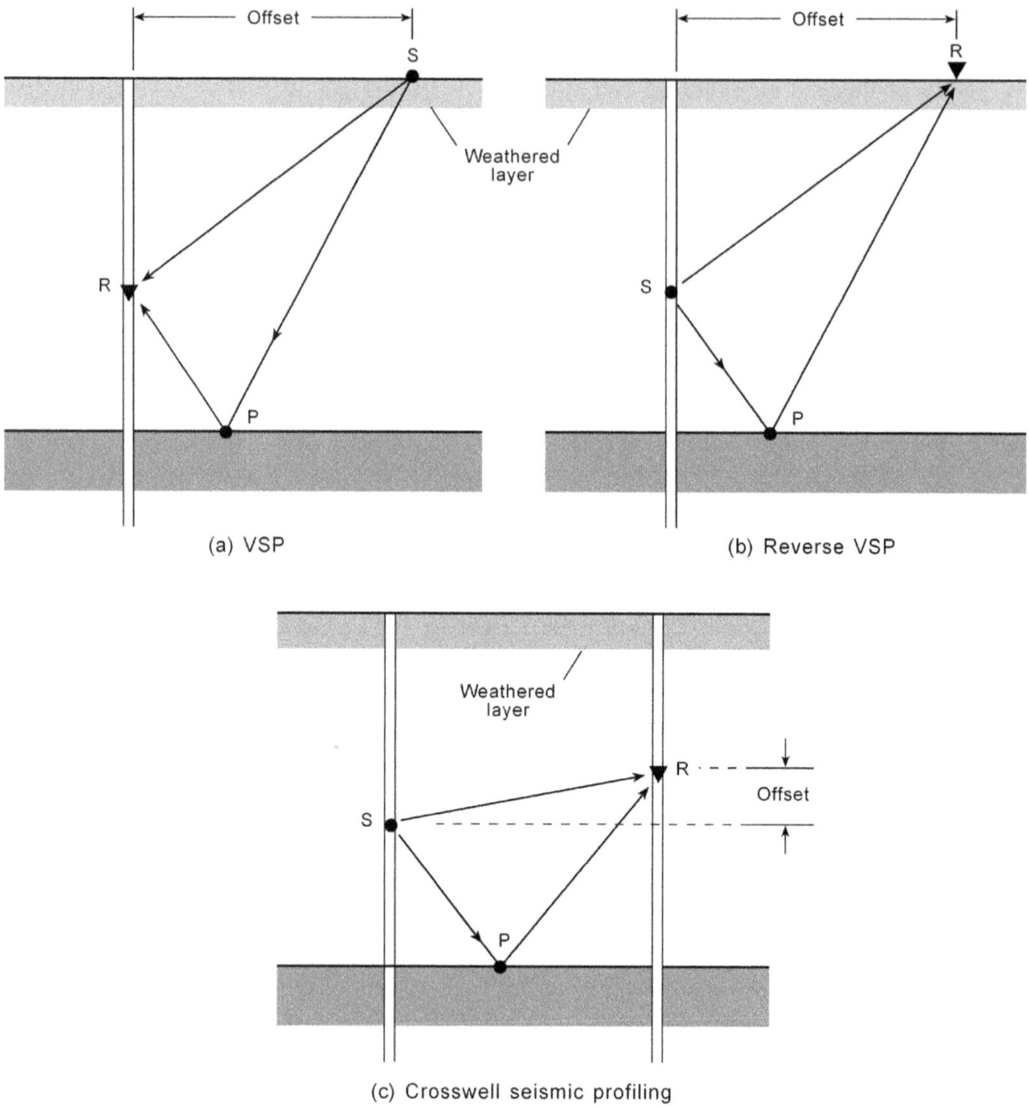

Fig. 2.34—Source-receiver geometries involved in VSP, RVSP, and CSP data acquisition.[40]

field imaging, the complete seismic wavefield is used. The major imaging contributions come from the scattered wavefield that occurs after the direct arrival.

Tomographic data are used to infer spatial distributions of rock and fluid properties in interwell spaces. Velocity tomograms are more widely used than are attenuation tomograms. Table 2.2 lists several reservoir properties (lithological variations, porosity, pore fluid) that affect seismic wave velocity. In concept, crosswell velocity tomograms can define the spatial patterns of these properties in the 2D vertical plane passing through the source and receiver wells.

Elastic wavefield imaging of CSP data provides more information about interwell conditions than do velocity tomograms because the images are presented in wiggle-trace format similar to surface-recorded seismic data. Interpreters can use standard seismic interpretation software to analyze these images, calculate amplitude and frequency attributes, and map stratal surfaces.

Because CSP technology provides data with signal frequencies as high as 1000 to 2000 Hz, some CSP data have dominant wavelengths as short as 3 m [10 ft]. Thus, CSP technology provides a better spatial resolution of reservoir properties than does surface-based seismic technology. By acquiring CSP data in a time-lapse sequence (usually 12 to 15 months between surveys), engineers can often track fluid movements in interwell spaces to determine if secondary recovery processes are performing as planned.

Fig. 2.35 gives a visualization of the portions of a crosswell wavefield that are involved in these approaches to CSP imaging. In this measurement, a source was kept at the depth labeled "Source" in a well that was 1,800 ft [550 m] away from the receiver well in which the data were recorded. A wall-clamped 3C geophone was then positioned in the receiver well at depth stations 25 ft apart, starting at a depth of 6,100 ft and extending up to a depth of 500 ft. Fig. 2.35 displays the response of the vertical geophone in the top wavefield, and the bottom wavefield shows the summed response of both horizontal geophones. It is probably not wise to sum the responses of the two horizontal geophones into a single wavefield because then the SV and SH shear modes (see Fig. 2.7) cannot be distinguished. As a result of this summation, all shear events in Fig. 2.35 are labeled as S, not as SH or SV.

The compressional (P) wavefield arrives first and is followed by the shear (S) wavefield. The arrival times of these wavefields are labeled on the shallow geophone trace. The S wavefront has more curvature than the P wavefront because S velocity is less than P velocity. CSP data record both downgoing reflection events (when the reflecting interface is above the receiver depth) and upgoing reflection events (when the reflecting interface is below the receiver depth). The opposite traveling reflection events create a crisscross pattern in the data, an effect that is pronounced in the S wavefield. The depth at which each S reflection occurs can be determined by extending each of these crisscrossing events back to its point of origin on the S first-arrival wavefront. Many P reflection events exist in the data at times later than the P first-arrival wavefront, but they are difficult to see in these unprocessed data. The labeled linear events sloping up and down behind the P first-arrival wavefront are SV events created by P-to-SV mode conversions at stratal interfaces. These events are better seen on the display of the horizontal-geophone data. The depth at which a reflection occurs can be determined by extrapolating a linear event to intersect the P-wave first arrival. The interpreted reflector depths can then be compared with the depths of rock and fluid interfaces defined by logs recorded in the receiver well and with the formation depths calculated from surface-recorded seismic data.

For a velocity tomography analysis of the interwell space illuminated by the wavefields in Fig. 2.35, the P and S first-arrival times can be picked at each depth station. These travel times then can be used to synthesize the interwell velocity structure by some type of iterative travel-path reconstruction technique. The particular downhole source used in this instance was a vibrator that produced a symmetrical wavelet. In this example, the data are not deconvolved to reduce the wavelet side lobes; thus, the arrival times would be the center point of the long, ringing, symmetrical direct arrivals.

To produce an estimate of the spatial distribution of seismic attenuation properties of the interwell space, amplitudes of the P and S direct arrivals have to be analyzed with other factors such as the consistency of the receiver couplings, the shot-to-shot energy levels, and the geometric shapes of the source radiation and receiver antenna patterns.

To produce P and S seismic images of the interwell space, the reflection portions of the wavefields that are noted need to be processed with interwell velocities determined by the velocity tomography analysis to position each reflection wavelet at its subsurface point of origin. The vertical axis of images created from CSP data is true stratigraphic depth, not image time, because the source and receiver stations are distributed over known depth coordinates. CSP images can be correlated to surface-based seismic images only if the surface data are transformed from the image-time domain to the depth domain.

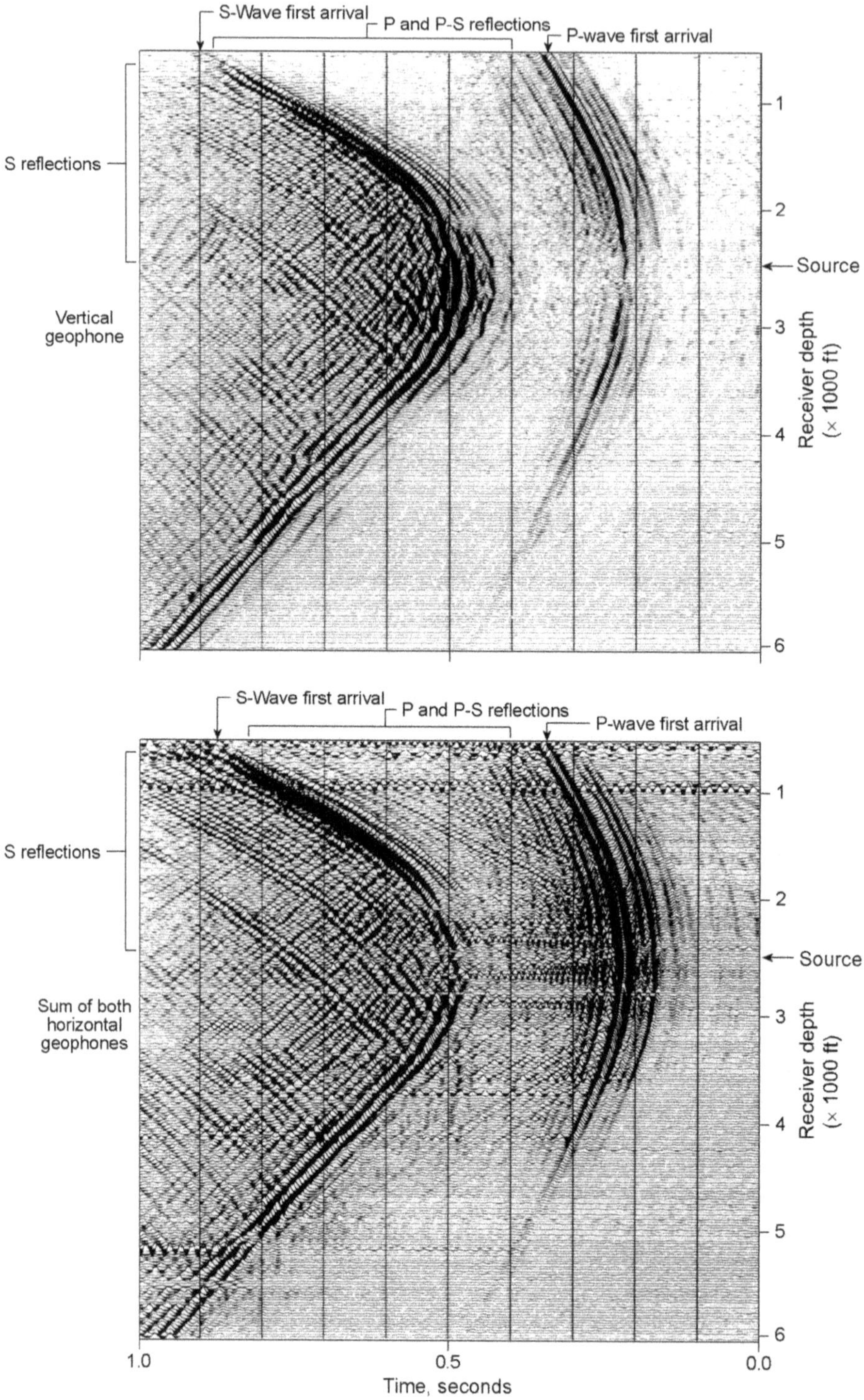

Fig. 2.35—Crosswell seismic wavefield that allows velocity tomograms, amplitude attenuation tomograms, and elastic wavefield (P and S) images of the interwell space between a source well and a receiver well to be constructed.[40]

Nomenclature

$a(t)$ = seismic trace amplitude value
A_o = incident seismic amplitude
A_r = seismic reflection amplitude
A_t = transmitted seismic amplitude
d = depth of air-gun array, L, ft or m
D = depth, L, ft or m
F = 3D stacking fold
F_{IL} = inline stacking fold
F_{XL} = crossline stacking fold
i_r = receiver-station interval, L, ft or m
i_s = source-station interval, L, ft or m
I = seismic impedance, (g·m)/(cm^3· sec)
n_c = number of receiving channels
n_l = number of receiver lines in the recording patch
N_1 = number of source-receiver pairs that image target at depth Z_1
N_2 = number of source-receiver pairs that image target at depth Z_2
R = seismic reflection coefficient or seismic receiver
t = seismic traveltime, t, second
v = velocity, L/t
V = velocity of seismic wave propagation, L/t
V_p = propagation velocity of a compressional (P) wave, L/t
V_s = propagation velocity of a shear (S) wave, L/t
W = minimum target width, L, ft or m
x = real data plane
$x(t)$ = real seismic trace
X = offset distance, L, ft or m
y = imaginary plane
$y(t)$ = Hilbert transform of $x(t)$
$z(t)$ = complex seismic trace
Z_1 = shallow target depth, L, ft or m
Z_2 = deep target depth, L, ft or m
Δx = inline dimension of stacking bin, L, ft or m
Δy = crossline dimension of stacking bin, L, ft or m
ρ = bulk density of the rock, m/L^3, g/cm^3
$\phi(t)$ = phase angle
$\omega(t)$ = instantaneous frequency, cycles/sec

References

1. Dobrin, M.B.: *Introduction to Geophysical Prospecting,* McGraw-Hill Book Co., New York City (1976) 630.
2. Telford, W.M. *et al.: Applied Geophysics,* Cambridge U. Press, Cambridge, U.K. (1976) 860.
3. Cholet, J. and Pauc, A.: "Device for Generating Seismic Waves by Striking a Mass Against a Target Member," U.S. Patent No. 4,205,731 (1980).
4. Crawford, J.M., Doty, W.E.N., and Lee, M.R.: "Continuous Signal Seismograph," *Geophysics* (1960) **25,** 95.
5. Geyer, R.L.: "Vibroseis Parameter Optimization," *Oil & Gas J.* (1971) **68,** No. 15, 116, and (1971) **68,** No. 17, 114.

6. Geyer, R.L.: "The Vibroseis System of Seismic Mapping," *Canadian J. of Exploration Geophysics* (1970) **6,** No.1, 39.
7. Seriff, A.J. and Kim, W.H.: "The Effect of Harmonic Distortion in the Use of Vibratory Surface Sources," *Geophysics* (1970) **35,** 234.
8. Goupillaud, P.L.: "Signal Design in the Vibroseis Technique," *Geophysics* (1976) **41,** 1291.
9. Bernhardt, T. and Peacock, J.H.: "Encoding Techniques for the Vibroseis System," *Geophysical Prospecting* (1978) **26,** 184.
10. Waters, K.H.: *Reflection Seismology—A Tool for Energy Resource Exploration*, John Wiley & Sons, New York City (1978).
11. Cunningham, A.B.: "Some Alternate Vibrator Signals," *Geophysics* (1979) **44,** 1901.
12. Edelmann, H.A.K., and Werner, H.: "The Encoded Sweep Technique for Vibroseis," *Geophysics* (1982) **47,** 809.
13. Sallas, J.J.: "Seismic Vibrator Control and the Downgoing P-Wave," *Geophysics* (1984) **49,** 732.
14. Fair, D.W.: "Shear Wave Transducer," U.S. Patent No. 3,159,232 (1964).
15. Dragoset, W.H.: "Air-Gun Array Specs—A Tutorial," *The Leading Edge* (1990) **9,** No. 1, 24.
16. Aki, K. and Richards, P.G.: *Quantitative Seismology—Theory and Methods*, W.H. Freeman and Co., New York City (1980).
17. Hardage, B.A.: "Combining P-Wave and S-Wave Seismic Data To Improve Prospect Evaluation," Bureau of Economic Geology Report of Investigations, No. 237, U. of Texas, Austin, Texas (1996).
18. Taner, M.T. and Sheriff, R.E.: "Application of Amplitude, Frequency, and Other Attributes to Stratigraphic and Hydrocarbon Determination," *American Assn. of Petroleum Geologists Memoir* (1977) **26,** 301.
19. Taner, M.T., Koehler, F., and Sheriff, R.E.: "Complex Seismic Trace Analysis," *Geophysics* (1979) **44,** 1041.
20. Hardage, B.A.: *Seismic Stratigraph,* Geophysical Press, London (1987).
21. Hardage, B.A.: "Instantaneous Frequency—A Seismic Attribute Useful in Structural and Stratigraphic Interpretation," technical summary, GRI-96/0019, Bureau of Economic Geology (1996).
22. Hardage, B.A. *et al.*: "3-D Instantaneous Frequency Used as a Coherency/Continuity Parameter To Interpret Reservoir Compartment Boundaries Across an Area of Complex Turbidite Deposition," *Geophysics* (1998) **63,** 1520.
23. Mitchum, R.M. Jr., Vail, P.R., and Thompson, S. III: "Seismic Stratigraphy and Global Changes in Sea Level, Part 2, The Depositional Sequence as a Basic Unit for Stratigraphic Analysis," *Seismic Stratigraphy—Applications to Hydrocarbon Exploration*, C.E. Payton (ed.), American Assn. of Petroleum Geologists, Memoir 26 (1977) 53.
24. Vail, P.R. and Mitchum, R.M. Jr.: "Seismic Stratigraphy and Global Changes of Sea Level, Part 1, Overview," *Seismic Stratigraphy—Applications to Hydrocarbon Exploration,* C.E. Payton (ed.), American Assn. of Petroleum Geologists, Memoir 26 (1977) 51–52.
25. Tipper, J.C.: "Do Seismic Reflections Necessarily Have Chronostratigraphic Significance?" *Geology* (1993) **130,** 47.
26. Galloway, E.W.: "Genetic Stratigraphic Sequences in Basin Analysis I: Architecture and Genesis of Flooding-Surface Bounded Depositional Units," *American Assn. of Petroleum Geologists Bull.* (1989) **73,** No. 2, 125.
27. Wanless, H.R. and Weller, J.M.: "Correlation and Extent of Pennsylvanian Cyclothems," *Geol. Soc. Am. Bull.* (1932) **43,** 1003.
28. Moore, R.C. *et al.*: "*Correlation of Pennsylvanian Formations in North America*," *Geol. Soc. Am. Bull.* (1944) **55,** No.6, 657.
29. De Angelo, M.V. and Wood, L.J.: "3-D Seismic Detection of Undrilled Prospective Areas in a Mature Province, South Marsh Island, Gulf of Mexico," *The Leading Edge* (2001) **20,** 1282.
30. Hardage, B.A. and Remington, R.L.: "3-D Seismic Stratal Surface Concepts Applied to the Interpretation of a Fluvial Channel System Deposited in a High-Accommodation Environment," *Geophysics* (1999) **64,** 609.
31. Hardage, B.A.: "Principles of Onshore 3-D Seismic Design," *Geological Circular 97–5*, Bureau of Economic Geology, U. of Texas, Austin, Texas (1997).

32. Ebrom, D., Li, X., and McDonald, J.: "Bin Spacing in Land 3-D Seismic Surveys and Horizontal Resolution in Time Slices," *The Leading Edge* (1995) **14,** No. 1, 37.
33. Stone, D.G.: *Designing Seismic Surveys in Two and Three Dimensions*, Soc. of Exploration Geophysicists, Geophysical References Series (1994).
34. Cordsen, A., Galbraith, M., and Peirce, J.: *Planning Land 3-D Seismic Surveys*, Geophysical Developments, Soc. of Exploration Geophysicists, Tulsa (2000) **9.**
35. Galbraith, M.: *Land 3-D Survey Design by Computer*, Exploration Geophysics, Australian Soc. of Exploration Geophysicists (1994) **25,** 71–77.
36. Balch, A.H. and Lee, M.W.: *Vertical Seismic Profiling—Technique, Applications, and Case Histories,* Intl. Human Resources Development Corp., Boston, Massachusetts (1984).
37. Hardage, B.A.: *Vertical Seismic Profiling—Principles*, third edition, Amsterdam, Elsevier (2000).
38. Hardage, B.A.: "A Practical Use of Vertical Seismic Profiles—Stratigraphic Calibration of 3-D Seismic Data," Geological Circular 97–4, Bureau of Economic Geology, U. of Texas, Austin, Texas (1997).
39. Geotz, J.F., Dupal, L., and Bowler, J.: "An Investigation Into Discrepancies Between Sonic Log and Seismic Check-Shot Velocities," *Australian Petroleum Exploration Assoc. J.* (1979) **19,** part 1, 131–141.
40. Hardage, B.A.: *Crosswell Seismology and Reverse VSP,* Geophysical Press, London (1992).

Glossary

2D (two-dimensional) is an adjective used to describe seismic data acquisition when a single vertical plane can pass through all source and receiver stations. The resulting image is restricted to the vertical plane passing through the source and receiver stations.

3D (three-dimensional) is an adjective used when seismic receivers are distributed across an expansive area surrounding a seismic source station. The point of origin of a reflection event recorded by this areal receiver patch can then be positioned correctly in *x, y, z* space.

3C (three-component) is an adjective used to describe seismic data acquired with three orthogonal sensors at each receiver station. These sensors measure Earth movement as a vector quantity in *x, y, z* coordinate space.

4C (four-component) is an adjective used to describe seismic data acquired on the seafloor. These seafloor receivers consist of a 3C geophone that measures vector movement of the seafloor in *x, y, z* space plus a hydrophone that measures scalar pressure variations at the seafloor.

Air gun is a marine seismic energy source that releases a high-pressure (~2,000 psi) pulse of compressed air to produce a robust, high-resolution seismic wavelet.

Body wave is a seismic wave that propagates in the interior (body) of the Earth. See surface wave.

Bright spot is a dramatic increase in seismic reflection amplitude that makes a subsurface object the dominant feature on any amplitude-based data display. A "true" bright spot is caused by gas replacing liquid in the pore spaces of a reservoir rock. A "false" bright spot is caused by a change in the rock matrix, such as a lateral transition from sand to coal.

Check-shot survey is a seismic measurement involving a seismic source on the Earth surface and a seismic sensor suspended by wireline in a well. The objective is to measure the travel time required for a seismic wavelet to travel from the Earth's surface to the downhole receiver. Check-shot data allow interpreters to convert seismic image times to stratigraphic depth (or vice versa).

Chronostratigraphic is an adjective meaning "time layer" (chrono=time, strata=layer). Interpreters must define the context in which they are using the term "time" (i.e., geologic time or seismic image time).

Coherency is a numerical measure of the similarity of reflection wave shapes in a user-specified data window. Coherency values are scaled to the numerical range +1 to –1. A value near +1 means the comparison wave shapes are identical; a value near –1 means the wave shapes are identical but have opposite polarities. A value near zero means the comparison wave shapes have little similarity.

Complex seismic trace is the result of applying a Hilbert transform to a seismic trace. A complex seismic trace consists of a real part (the input seismic trace) and an imaginary part produced by the Hilbert transform. The reason for transforming seismic data from the "real" domain to the "complex" domain is that reflection amplitude, phase, and frequency can then be calculated at each time sample point of the seismic wiggle trace.

Crossline is the direction that is perpendicular to seismic receiver lines. See inline.

Crosswell seismic profiling (CSP) data are acquired with a downhole seismic source in one well and downhole seismic sensors in a second well. CSP data provide high-resolution 2D images of geologic conditions across interwell spaces.

Depth-structure map is a seismic-derived map showing the geometry of subsurface structure in terms of depth coordinates. See time-structure map.

Diachronous is a term used to describe a surface or a seismic reflection that cuts across geologic time (dia=across, chrono=time). See chronostratigraphic and stratal surface.

Erosion surface (ES) is a subsurface interface marking an ancient erosional event that has removed portions of one or more stratigraphic units.

Far field is that portion of a seismic propagation medium that is a distance of at least 3 or 4 wavelengths away from a source station.

Flooding surface is a depositional surface marking the transgression of a flooding event across an area. Deeper-water fauna occur above a flooding surface; shallower-water fauna exist below that surface.

Genetic sequence is a package of stratigraphic units that are genetically related. Genetic sequences are bounded by erosional surfaces, flooding surfaces, or maximum flooding surfaces.

Ground force phase locking is a technology that ensures each vibrator in an array of vibrators generates the same source wavelet regardless of variations in soil conditions beneath the vibrator base plates.

Ground roll is a robust, high-amplitude wave produced by onshore seismic sources that travels along the Earth/air interface. A ground-roll wave does not propagate in the interior (body) of the Earth. See Rayleigh wave.

Impedance is the product of bulk density and seismic propagation velocity in the medium in which the wavefield propagates.

Inline is the direction in which receiver lines are deployed. See crossline.

Instantaneous amplitude is the amplitude of a complex seismic trace at a specified time coordinate along that trace. Instantaneous amplitude is not the same as trace amplitude. See complex seismic trace.

Instantaneous frequency is the frequency of a complex seismic trace at a specified time coordinate along that trace. Instantaneous frequency is the time derivative of instantaneous phase. See complex seismic trace.

Instantaneous phase is the phase of a complex seismic trace at a specified time coordinate along that trace. Instantaneous phase is the inverse tangent of the ratio: real part of the complex trace divided by imaginary part of the complex trace. See complex seismic trace.

Love wave is a surface wave that propagates along the Earth/air interface and creates a particle displacement that is tangent to the Earth surface and also perpendicular to the direction of wave propagation. A Love wave does not propagate in the interior of the Earth. See body wave, ground roll, Rayleigh wave, and surface wave.

Low-stand wedge is an asymmetrical accumulation of sediment and strata occurring during a period of lower sea level.

Maximum flooding surface (MFS) is a depositional surface that marks conditions associated with the deepest water depth occurring in a geologic time period of interest.

Ocean-bottom cable (OBC) is a cable-based seismic receiver system that is positioned on the seafloor so shear-wave data can be acquired in addition to compressional wave data. OBC technology usually involves 4C sensors. See 4C.

P wave is a compressional wave. It is sometimes called a primary (P) wave because it is the portion of a seismic wavefield that arrives first at an observation point. A P wave causes rock particles to oscillate in a direction that is perpendicular to its wavefront.

Peak-to-peak (PTP) is a parameter used to describe air-gun performance. An air-gun wavelet consists of a high-amplitude peak followed by a high-amplitude trough. The parameter PTP defines the magnitude of the source wavelet amplitude measured from the apex of the leading peak of the wavelet to the apex of the following trough. A high PTP value indicates a high energy output.

Primary-to-bubble ratio (PBR) is a parameter used to define the performance of a marine airgun array. "Primary" refers to the amplitude of the wavelet created by the output pulse of high-pressure air. "Bubble" refers to the amplitude of the wavelet created by the subsequent collapse of the air bubble in the water column. A high PBR value (~30) is desired.

Rayleigh wave is the correct name for a ground-roll wave. A Rayleigh wave is a surface wave that propagates along the Earth/air interface, not in the body of the Earth. A Rayleigh wave creates an elliptical motion of Earth particles along its propagation path. The horizontal particle displacement associated with this ellipse is oriented in the direction of wave propagation. The horizontal displacements associated with Love waves and Rayleigh waves are orthogonal to each other.

Reflection coefficient is a parameter that defines the amplitude of the wave that reflects from an interface. The magnitude of a reflection coefficient at an interface is linearly proportional to the difference in seismic impedance across that interface. The algebraic sign of a reflection coefficient defines the polarity of the reflection event.

Rollover is a downward bending of strata that often forms a structural trap for oil and gas. The bending movement is initiated by tectonic forces.

S wave is a shear wave. It is sometimes called a secondary (S) wave because it arrives later than the primary wave (see P wave). The term S wave needs to be used carefully because there are several types of shear waves. S waves include SH and SV modes and, in a complex anisotropic Earth, each of these modes (SH and SV) divides into a fast and slow component. Converted SV waves (called C waves) that result when P waves arrive at interfaces at nonnormal angles of incidence are another type of S wave. The term "S wave" spans a large family of distinct wave types.

SH is a shear wave mode that has its particle-displacement vector oriented perpendicular to the vertical plane passing through the source and receiver stations. In a flat-layered isotropic Earth, an SH mode has its particle-displacement vector oriented parallel to horizontal interfaces. See SV.

Stacking bin is the smallest definable area within a 3D seismic image space. The number of stacking bins in a 3D seismic volume is the same as the number of seismic traces in that volume. At the conclusion of 3D data processing, one image trace passes vertically through the center of each stacking bin. The lateral dimension of a stacking bin is the same as the spacing between adjacent traces in the 3D volume.

Stacking fold is a number that specifies how many seismic field traces are summed to create a final image trace in a 3D seismic volume. A stacking fold of 20 means that 20 field traces were summed to create one stacked trace.

Stratal surface is a depositional surface associated with a fixed geologic time. Geologic time is constant along a stratal surface, but the rock types above and below the surface can vary.

Surface wave is a seismic wave that trends along an interface, particularly along the Earth/air interface. Surface waves do not enter the body of the Earth and image deep targets. See ground roll, Love wave, and Rayleigh wave.

SV is a shear-wave mode that has its particle-displacement vector oriented in the vertical plane passing through the source and receiver stations. An SV displacement vector is orthogonal to an SH displacement vector. See SH.

Synthetic seismogram is a mathematical construction of a seismic wavefield. Synthetic seismograms can be calculated in 1D, 2D, or 3D data space. Either downgoing wavefields, upgoing wavefields, or both can be included in the calculation.

Thin bed is a sedimentary layer with a thickness less than one-fourth of the length of the dominant wavelength in the illuminating seismic wavefield. Typical dominant wavelengths can be 200 to 300 ft (~65 to 100 m); therefore, many thin beds have thicknesses of 50 to 75 ft (~15 to 22 m). The top and base of a thin bed cannot be resolved in a seismic image.

Time slice is a horizontal slice through a 3D seismic data volume. Seismic image time is constant across a time slice, but geologic time is not, unless the stratigraphy is perfectly flat.

Time-structure map is a map identifying the seismic image times at which subsurface structure is located. Time-structure maps can be converted into depth-structure maps if seismic propagation velocity can be defined throughout 3D seismic image space. See depth-structure maps.

Tomograms are popular products produced by crosswell seismic profiling. A CSP velocity tomogram shows the spatial distribution of seismic propagation velocities in the interwell space between a source well and a receiver well.

Vibrator is a popular onshore seismic energy source. Vibrators are large vehicles weighing 60,000 lbs or more. They transmit seismic energy into the Earth through a heavy baseplate that is pressed to the ground and then vibrated over a prescribed frequency range.

Vertical seismic profile (VSP) data are acquired with an energy source on the Earth's surface and a vertical array of closely spaced receiver stations in a well. VSP receiver stations are positioned vertically at increments of approximately 50 ft (15 m). This small spatial sampling allows all wave modes (downgoing, upgoing, shear, compressional) to be separated from the raw data.

SI Metric Conversion Factors

ft	×	3.048*	E − 01 = m
mile	×	1.609 344*	E + 00 = km
psi	×	6.894 757	E + 00 = kPa

*Conversion factor is exact.

Chapter 3A
Petrophysics

E.C. Thomas, Bayou Petrophysics

3A.1 Introduction

3A.1.1 Definitions. The term "petrophysics" was coined by G.E. Archie and J.H.M.A. Thomeer in a quiet bistro in The Hague.[1] By their definition, petrophysics is the study of the physical and chemical properties of rocks and their contained fluids. It emphasizes those properties relating to the pore system and its fluid distribution and flow characteristics. These properties and their relationships are used to identify and evaluate hydrocarbon reservoirs, hydrocarbon sources, seals, and aquifers.

The petrophysicist or petrophysical engineer practices the science of petrophysics as a member of the reservoir management team (RMT). (See the chapter on reservoir management in this section of the *Handbook.*) The petrophysicist provides answer products needed and used by team members, as well as physical and chemical insights needed by other teammates. The reservoir and fluid characteristics to be determined are thickness (bed boundaries), lithology (rock type), porosity, fluid saturations and pressures, fluid identification and characterization, permeability (absolute), and fractional flow (oil, gas, water).

It is easy to define these characteristics and to appreciate their part in the assessment of reserves. The difficult part comes in determining their actual value at a level of certainty needed to make economic decisions leading to development and production. The seven characteristics listed are interdependent (i.e., to properly determine porosity from a wireline log, one must know the lithology, fluid saturations, and fluid types). The science of petrophysics is then used to unscramble the hidden world of rock and fluid properties in reservoirs from just below the Earth's surface to ones more than four miles deep. The petrophysicist then takes on many characteristics of the fictional sleuth Sherlock Holmes to extrapolate, from the most meager of clues, the true picture of the subsurface reservoir using dogged determination to wrest all possible information from the available data, all the while enjoying the thrill of the hunt.

How does the petrophysicist solve this difficult problem? Archie's general method is to subdivide the problem into smaller segments and iterate using all data until all data agree. One starting point is to determine rock types (petrofacies) wherein we identify pore type, pore size distribution, pore throat type, and pore throat distribution. When coupled with fluid type, one can establish a capillary pressure model that will lead to understanding in-situ fluid saturations and fluid flow. The discussion of the process is lengthy and is covered here and throughout the

TABLE 3A.1—TOOL CHOICES FOR A PARTICULAR RESERVOIR PROPERTY*

Reservoir Property	Primary Tool	Secondary Tool	Backup
Thickness (bed boundaries)	Gamma ray log (borehole image log when available)	Electric log with SP log	Density/neutron log
Lithology/rock type	Core and core analysis	Mud log	Gamma ray spectroscopy log
Porosity	Density log (given rock type and fluid identification)	Neutron log (given rock type and fluid ID)	Core and core analysis
Saturations (oil, gas, water)	Electric log (given porosity, rock type, and fluid ID) calibrated with a capillary pressure model	Carbon-oxygen log (given favorable borehole/formation environment)	Core and core analysis (given appropriate drilling environment and reservoir position)
Fluid ID/characterization	Wireline formation tester	Mud log	Core and core analysis
Permeability (absolute)	Core and core analysis	Wireline formation tester	Drillstem test
Fractional flow (oil, gas, water)	Core and core analysis	Wireline formation tester	Drillstem test

*The choices in this table are subjective and can change depending on the investment risk a particular client company can tolerate and the geographical circumstances involved.

Handbook. However, the tools available to the petrophysicist are mud logging (solids, liquids, gasses, volumes, rates, concentrations, and temperature); measurements while drilling (MWD) and logging while drilling (LWD); wireline logging (open- and cased-hole); core sampling [(wireline (percussion and drilled) and whole] and core analysis; and fluid sampling (wireline and/or drillstem tests). This list is arranged in order of the usual acquisition sequence. The economics of a given evaluation may restrict the application of any of these tools.

These topics are discussed in further detail in this and other sections of the *Handbook*. These chapters explain how to make the necessary measurements and how to ensure that the resulting data are accurate enough. We will now demonstrate how to choose the appropriate tools to determine the required reservoir and fluid characteristics.

3A.1.2 Choosing Tools. The choices of tools for the evaluation program are usually made by the RMT with recommendations from the petrophysicist. Cost and accuracy are usually highly correlated; thus, the team's choices quickly become one of finding the maximum allowable risk for the minimum cost (time and money) of the evaluation. **Table 3A.1** summarizes the typical evaluation program choices and can be used as a reference duringsubsequent discussions on determination of various reservoir properties.

3A.1.3 Determining Layer Thickness. Bed boundaries are usually the easiest of all reservoir properties to measure; however, there are some fatal traps that await the unwary. The geologist's knowledge of the rock types in the play, in general (and in the well of interest, in particular) can be used. By convention, reservoir thickness is expressed in feet and is rounded to the nearest foot, even though most modern wireline logs are recorded digitally every 6 in. Many of the world's reservoirs are logged in metric units, and thickness is expressed in meters. In this regime, thickness is rounded to the nearest 0.1 m.

Sands and Shales. The term sand is used generically and can also refer to sandstone or other siliciclastic formations. The term shale is used generically and can also refer to mudrock or claystone. When the reservoir beds are mostly sand [typically low gamma ray (GR)] and shale (typically high GR), then the GR log can usually be used to select bed boundaries. The inflection point of the GR count rate expressed in American Petroleum Institute (API) units is selected as the bed boundary. (See examples in the chapter on nuclear logging in this section of the *Handbook.*) The choice of which bed thickness is to be determined is usually made by the geologist largely on the basis of pattern recognition skills developed during the play definition. Not all sand beds have low GR levels. If the sand bed contains sizable amounts of potassium feldspar, mica, or volcanic debris, the sands may be as radioactive as the shales and difficult to tell apart. In this case, the spontaneous potentials (SP) log is often used if the well is drilled in water-based mud. Again, the inflection point of the log is used to denote the bed boundary. (See examples in the chapter on resistivity and SP logging in this section of the *Handbook.*) However, in low-porosity and high-resistivity environments, the SP is suppressed and cannot be relied on as a bed boundary indicator.

Another tool that can be used to mark bed boundaries is the electric (resistivity or conductivity) log. This log is most useful when the shales are water-bearing and the sands are shale-free and hydrocarbon bearing. However, the bed boundary is no longer at the inflection point on the log, but is offset by one-half of the electrode spacing of the tool used to measure resistivity. Each service company has charts to aid the petrophysicist in determining the correct offset, while many wellsite computer products take this offset into account automatically. The finest resolution of all is afforded by electric or acoustic borehole image logs, but these logs are not routinely run because of their expensive acquisition and processing costs. The chapter on specialized logging topics in this section of the *Handbook* shows examples of borehole image logs. If the well is drilled in oil-based mud, then the SP log is not available and the density/neutron log can be used to define sand from shale. If the density/neutron log is a modern digitally sampled tool, then the inflection point can be used as the bed boundary. If, however, the density/neutron log is an older analog tool, the bed boundary can be offset because of drag settings. Vendor publications are available to aid the user in determining the offset. Acoustic borehole image logs are an option in small boreholes and lightweight oil-based muds. Nuclear magnetic resonance (NMR) logs are also useful in distinguishing zones that have movable fluids from those containing bound fluids (examples are shown in the chapter on NMR logging in this section of the *Handbook*).

Carbonates. When the reservoir beds are generally composed of carbonates (i.e., limestones and/or dolomites), bed definition becomes more complex. In this case, the presence and absence of porosity defines the reservoir from the seal; thus, the GR log may not be useful in demarking bed boundaries. Likewise, carbonates are generally in a higher resistivity regime, and the SP log is of little use. Thus, in this environment, one relies on a tool sensitive to porosity to delineate the bed boundaries, such as the density, neutron, or acoustic log. Borehole image logs are also useful when the borehole is relatively smooth. Large vugs and washouts invalidate readings from these tools. (The chapter on specialized logging topics in this section of the *Handbook* includes an example of a borehole image log in a carbonate formation.)

What To Count. Determining gross bed thickness is straightforward; however, determining what part of a given bed contains producible hydrocarbons is tricky. At present, there is no single standard sanctioned by SPE or API as to the definition of, or the method for determining, net pay. One of the more common complications is a result of the small thickness of layers containing hydrocarbons. Some common depositional environments can result in sands and shales being laid down into zones thinner than the resolving power of almost all wireline logs. One cannot count directly what one cannot see. In this case, we settle for thicker averages that mathematically equate to correct fluid volumes (i.e., calculate that from depth X to depth X+30 ft, the ratio of sand to shale is 0.5; thus, it contains 15 ft of reservoir-quality rock in layers too thin to individually resolve). Finally, one can cut a continuous core from the top through the bottom of the reservoir in question and, given the good fortune to recover 100% of the core at the surface, one can use sophisticated core-analysis techniques to determine the thickness of the layers with a precision and accuracy better than any other part of the reservoir remaining in the Earth. Because of the high cost of cutting and analyzing whole core, this method is seldom used, unless no other method can be proven to work.

3A.1.4 Determining Lithology and Rock Type. The identification of a bed's lithology is fundamental to all reservoir characterization because the physical and chemical properties of the rock that holds hydrocarbons and/or water affect the response of every tool used to measure formation properties.

Definitions. The term lithology is used as a gross identification for a rock layer in the subsurface and uses familiar names such as sandstone (or sand), limestone, dolostone (or dolomite), claystone (or clay), chert, coal, shale (or mudrock), diatomite, halite, anhydrite, gypsum, and tuff. (The preceding list is not exhaustive. For detailed lists, see Ref. 2.) The term rock type is a more detailed description than lithology because it reflects the natural groupings of pore systems that produce recognizable properties used to predict flow properties, volumes, and fluid saturations. Lithology focuses on grains, while rock type focuses on pores. The list of rock types contains more than 250 classifications.* (See the chapter on petroleum geology in this section of the *Handbook.*) Another term used in the literature is the Greek equivalent "petrofacies." (Another term that is finally dropping from use is the incorrectly named "electrofacies.")

Direct Determination. Obtaining a physical sample of the reservoir is the surest way to unambiguously determine lithology and rock type, but obtaining this physical sample is not always easy. Mud logs are the first choice in wildcat wells, but exact assignment of a rock fragment to a particular depth is not without error. The size of the individual rock sample examined at the surface is rather small because it is limited by the size of drill cuttings and rock strength. Weak rocks, ones without cement, are often reduced to original detrital grain size by the drilling process, making it difficult to determine rock type, but still possible to determine lithology. Once the well is drilled and logged and rock layers are marked for further study, rock samples can be obtained through the use of wireline core takers or sidewall core drills.

Recovery of sidewall samples is not always a sure thing, and we often fail to get rock samples from zones of interest. In a wildcat well with a zone of high interest identified with logs, rock samples, and fluid samples, we can plug back the well several hundred feet, set a whipstock, perform a parallel sidetrack (called a bypass hole), and then take a whole core across the zone of interest. Because the two holes are parallel, we know exactly at which depth to swap out the drill bit with a core barrel. The resulting whole core can be sampled, and sophisticated core analysis can be used to identify the rock type of each zone and determine petrophysical parameters, which are used to refine the formation evaluation from the log data.

Indirect Determination. If no direct rock sample is available in a given zone, log responses must be used alone to simultaneously determine lithology, porosity, and fluid saturations. Rock typing is much more involved and requires the use of commercially available catalogs of analog data[3,4] or locally collected data to pinpoint petrophysical properties that can be used to refine porosity and fluid-saturation calculations. If, in a given wellbore, one is lucky enough to encounter parts of the desired reservoir below the free-water level, then fluid saturation is known and gross lithology can be determined from routine wireline logs such as density, neutron, and acoustic tools. All wireline service companies provide charts and answer products that convert wireline logging data into lithology; however, few beds are simple homogeneous layers. (See the chapter on Reservoir Geology in this section of the *Handbook.*) If some of the layers are beneath the resolving thickness of the tool, some average response of the layers, depending on bed and borehole geometry, can be observed. Single layers of salt (NaCl), anhydrite, gypsum, coal, or tuft are easily identified when several feet thick or thicker; however, salt-

* Personal communication with R.M. Sneider, R.M. Sneider Exploration Inc., Houston (1990).

plugged sandstone can mimic gas-filled clean sandstone, or a very shaly rock type can hide the presence of gas.

Mica or potassium feldspar, when abundant in sandstones, can confuse lithology determination on the basis of the GR response. The more sophisticated GR spectroscopy tools can help to identify the chemical species in a given rock and often can lead to deducing its lithology. (See the chapter on nuclear logging in this section of the *Handbook.*) However, no single wireline log or set of logs can determine pore size distribution and pore throat size distribution. These data are used to establish rock type, resulting fluid flow characteristics, and other petrophysical parameters needed to determine fluid saturation. If every reservoir in our wildcat well is hydrocarbon filled, the problem of quantification becomes more difficult, particularly when formation water resistivity, R_w, is unknown. In difficult areas, a follow-up well and a whole core, drilled with oil-based mud that is taken high in the hydrocarbon column, must be drilled to guarantee no in-situ water displacement. Subsequent sophisticated core analysis can determine the value of in-situ water salinity. (See the chapter on petrophysical applications in this section of the *Handbook*). If coring is not an option, a well will have to make a significant amount of water to test. In pressure-depletion reservoirs that never produce much water, the accurate determination of water salinity may never happen if the reservoir is drilled with oil-based mud, negating the use of the SP log.

3A.1.5 Determination of Porosity. The determination of porosity is paramount because it determines the ultimate volume of a rock type that can contain hydrocarbons. The value and distribution of porosity, along with permeability and saturation, are the parameters that dictate reservoir development and production plans. Determination of porosity from a wireline log is only part of the problem, because the values determined in one well must be upscaled into the space between wells. To extrapolate correctly, the team must identify depositional environments and rock types and then have access to analog data sets. Only then can the correct statistical distributions be extrapolated across the reservoir. (See the chapter on geostatistics in the Emerging and Peripheral Technology section of the *Handbook.*)

Direct Determination. If one has access to an undamaged whole core from the reservoir in question, direct measurement of the porosity is possible, if care is taken. Some profess that all cores are damaged by the coring process, thus no accurate assessment of porosity is possible by coring. This author does not agree with that premise. X-ray computerized tomography scans, thin sections, and scanning electron microscope examination can verify that grain contacts are unmoved, authigenic pore linings are undamaged, and heavy-weight drilling fluids and/or particles are absent. Thus, given a satisfactory core, porosity can be determined accurately by several different methods specified in API *RP 40.*[5]

There are a number of pitfalls to avoid. One pitfall is that cleaning the core of crude and brine must be both thorough and gentle. One must remove all the heavy ends of the crude (typically asphaltenes) but not damage the authigenic minerals. NaCl crystals left from the in-situ brine when the water is removed must also be removed, but the authigenic clay minerals must not be removed or disturbed. Tar must not be removed if it occurs naturally in situ, and NaCl pore linings must not be removed if it occurs naturally in situ. Gypsum, when present, must not be dehydrated during the determination of porosity, and hydroxyl water of clay minerals must not be removed and counted as part of the pore space. Pore space must not be created during the cleaning process by flowing large volumes of fluid through rocks with soluble grains, such as gypsum, anhydrite, limestone, or salt.

Another pitfall to avoid is that some rock samples are mechanically weak and uncompact when brought to the surface and freed from the overburden load. These rock samples must be returned to in-situ conditions of effective stress to return the rock sample to its in-situ value of porosity. Correct procedures to make these measurements have been published by several authors.[6,7] Typically, stressed porosity measurements are more time consuming and more costly. If the core is subdivided into rock types, empirical correlations between stressed and unstressed measurements on a given rock type can be used to extrapolate to a larger data set.

An additional pitfall is that some sandstones contain fibrous pore-bridging clay minerals. These clay types are fragile and are damaged by most routine methods used to clean the core before porosity determination. When this type of clay is determined to be present, all subsequent core cleaning must be done using critical point drying methods.[8] In addition to the methods defined in API *RP 40,* NMR spectroscopy methods can be used to determine porosity on small core samples, When care is taken to ensure that the core sample is 100% saturated with water and or liquid hydrocarbon, the NMR method yields an accurate value of porosity. The NMR method for porosity determination can also be used in the borehole using wireline NMR tools with excellent results. The chapter on NMR logging in this section of the *Handbook* shows examples.

Indirect Determination. Direct determination of porosity by core analysis is the "gold standard" and is used where available to calibrate all indirect measurements. Indirect methods allow leveraging of limited core data to provide more information on areal and vertical variations in porosity when coring is too expensive and/or only partial cores are recovered.

Sands. Patchett and Coalson[9] determined that the density log is the most accurate method to determine porosity when one has knowledge of grain density and fluid density. While this method is standard in production wells, these parameters are often unknown for wildcats. Grain density can change rapidly along the borehole as lithology changes. Fluid types and saturations change more slowly, except at fluid contacts. Thus, we have four unknowns: porosity, grain density, hydrocarbon saturation, and water saturation, and one measurement: bulk density. A statistical method would be used to combine the environmentally corrected log readings from the density, neutron, acoustic, and GR to solve for the four unknowns. Often, a shallow resistivity log is included in the mix, and the acoustic log is dropped. Every logging vendor provides an answer product that uses this type of method, and most larger oil companies have published their own method.[10] If we are certain that we are below the free water level and if we assume that we know the rock type is sand with a grain density of 2.65 g/cm^3, the measured density log reading of bulk density can be converted into porosity, as the chapter on nuclear logging in this section of the *Handbook* shows.

Another simplified method is to use the density-neutron crossplot provided by each of the logging vendors. Patchett and Coalson[9] found no benefit to using the density-neutron crossplot over a density log if one used known and variable grain density. The novice often uses a simple method with little regard for changes in rock type, fluid type, or borehole conditions, and the result is considerable error in the determination of porosity. The density log can be quite accurate when logged in ideal to semi-ideal borehole conditions. However, in rugose boreholes, extremely thick mudcakes, or unusual weighting materials in the mudcake (e.g., hematite), the bulk density readings seen on the log will no longer reflect those of the borehole wall, and these readings must be discarded and the value of porosity determined by other methods. The borehole caliper and density correction curves are used to validate the quality of the bulk density readings.

Heterogeneity. The foregoing discussion on sands provides an answer for porosity that is correct, but it reflects the average porosity over the depth of resolution of the tools (i.e., approximately 3 ft). Sophisticated digital processing can increase the resolution to approximately 1 ft. Thus, when the reservoir is heterogeneous on a scale smaller than 1 ft, one must use other methods to deconvolve the resulting averages into values that reflect the true porosity of the individual rock types. One of the most common heterogeneous reservoirs is the laminated sand-shale sequence, in which the shale layers are often less than 1 in. thick. One published method used to determine the porosity of the sand layers free of the unresolved shale layers is the Thomas-

Stieber method.[11] In a sand-shale reservoir in which the shale laminations have a porosity lower than the sand layers, one will consistently understate the value of reservoir porosity if the unresolved shale laminations are not properly accounted for. Furthermore, if one is using porosity-to-permeability transforms, the value of permeability will be underpredicted.

Carbonates. Determination of porosity in carbonates is generally straightforward unless the rock type is one with large vugs (i.e., fist-sized or larger) or fractures. Density-neutron and neutron-acoustic crossplot have been historically useful and accurate when calibrated to core measurements. When the rock types become complex and numerous, then statistical, multiple-log methods that match the number of unknowns to independent log measurements are required. Every logging vendor provides an answer product to produce a reasonable value of porosity when all logs are environmentally corrected and validated. Large vugs can be spotted with borehole image logs (see the chapter on specialized logging topics in this section of the *Handbook*) and with large diameter cores (see the chapter on relative permeability and capillary pressure in the General Engineering section of the *Handbook*). With appropriate sampling, the borehole readings can be corrected for the effects of large vugs. Logging tools that investigate larger volumes are given higher weights in the analysis.

Fractured Reservoirs. The dual-porosity system that exists in a fractured matrix reservoir provides a challenge in the opposite direction in that often the overall value of porosity is quite low (2 to 3%). Because most wireline porosity logs have random statistical error of 1 to 2%, the error is as big as the value being measured. Under these conditions, reservoir simulation and history match is the most reliable method to determine storage capacity and reserves, and porosity becomes moot. Borehole image logs are used to locate the fractures and provide probable production intervals.

3A.1.6 Oil, Gas, and Water Saturation. The determination of in-situ saturations relies on interpretations of logging devices that read far from the borehole and away from any fluid alterations caused by invasion during drilling.

The workhorse tools are the deep induction and deep laterolog. All other tools such as density, neutron, acoustic, NMR, shallow laterologs, and GR provide readings from the flushed zone that has altered saturations. The most important transform that converts resistivity readings into water saturation is the well-known Archie relationship, which is discussed in the chapter on resistivity and SP logging in this section of the *Handbook*. The Archie relationship has many unknowns [i.e., porosity (which may have three or four unknowns itself), R_w (resistivity of the in-situ water), m (an empirical fitting parameter between porosity and resistivity, often called Archie's cementation exponent), and n (an empirical fitting parameter between water saturation and resistivity often called Archie's saturation exponent)], and one measured parameter, the formation resistivity, often called the true resistivity. Thus, we have six or seven unknowns and one measurement, which is why a standard is needed (but a standard is rarely available).

In most cases, the saturations in a core have been flushed by the mud filtrate and are not representative of the in-situ reservoir value. Only when we drill with oil-based mud and core high in the oil column where the relative permeability to water is quite low (practically zero) do we recover a core with a value of water saturation at in-situ conditions, after we correct for blowdown from dissolved gas and stress effects on the pore volume of the core sample. The result is an accurate value of in-situ water saturation for zones with less than S_w=50%.[12,13] Even under these ideal conditions, we have to make empirical corrections to our gold standard. The other case that can be used to calibrate Archie's relationship is when heavy crude or tar does not move when we core. These cores usually have little or no blowdown because there is little or no dissolved gas. The only correction that is needed is for a stress correction to the pore volume; the result is an accurate value of in-situ oil saturation.

Calibration. Because the Archie method relies on so many adjustable and often unknown parameters, a calibration step is required to ensure that saturation values are accurate. The method of choice is calibration to a capillary model, which uses multiple core samples for each rock type to provide statistical precision. For a given rock type, several capillary pressure curves are averaged to provide a capillary pressure vs. saturation relation. With some information about the type of hydrocarbon in the reservoir, this relation can be converted into saturation vs. height above the free water level relation. With this model, we can predict, for a given rock type, the hydrocarbon saturation at any elevation in the reservoir and compare it with that computed from the Archie method. Discrepancies between the methods must be resolved by further study, but typically they result in adjustment to one of the many parameters in the Archie method.

Invaded Zone Saturations. The near-wellbore environment is usually altered by the drilling process in several ways, one of which is mud filtrate invasion as a result of overbalance and/or imbibition. The size of the invaded zone depends on many parameters. Some are overbalance magnitude, mud-fluid-loss parameters, mudcake permeability, formation porosity, formation permeability, and in-situ fluid viscosity. The exact shape of the invaded zone is unknown but is assumed to be cylindrical. (This cylindrical assumption is not as robust when the borehole encounters dipping beds or is drilled as a deviated hole.)

The radial extent of this invaded zone can be determined with multiple-spaced resistivity tools if the invasion process has altered resistivity, unless the depth of invasion is beyond the zone of investigation of the resistivity tool. If one is comfortable that the shallow-reading resistivity device responds solely from the invaded zone, then one can use the Archie relationship to compute water saturation, as discussed previously. One must take care to determine if the n parameter has changed because the invaded zone is on the imbibition cycle, rather than the drainage cycle usually observed deep within the reservoir. NMR is another method that can be used to determine the invaded zone saturation. Wireline NMR tools do not see deeply into the formation and usually read invaded zone values. The chapter on NMR logging in this section of the *Handbook* shows examples of these responses. Generally, filtrate invasion results from the use of either water-based or oil-based mud. Filtrate invasion from water-based mud into the water leg of a reservoir makes no change in S_w. It remains at 100%.

Water-based mud filtrate invasion into the hydrocarbon leg of the reservoir can dramatically decrease the hydrocarbon saturation caused by imbibition, viscous stripping, and gas dissolution. The measured value of fluid saturations in the invaded zone should not be used to predict the residual hydrocarbon saturation by water displacement within the body of the reservoir because the dynamics of displacement are too different.

In the case of oil-based mud filtrate invasion, water saturation will remain immobile if S_w is less than 50% and mild surfactants are used in the mud. There will be no change in oil saturation in the invaded zone when compared with that deeper in the reservoir, unless extreme overbalance is being used. Shallow resistivity logs and NMR logs can be used to determine oil saturation in the invaded zone. These values can then be extrapolated to similar rock types throughout the reservoir.

3A.1.7 Fluid Identification and Characterization. Detection of a change in fluid type in the rocks while drilling is usually straightforward with the use of gas and chromatographic measurements. The chapter on mud logging in this section of the *Handbook* discusses these methods. Gas shows and oil shows while drilling are time-honored indicators of zones that need further investigation through logs, testers, and cores. In the rare case of gas-bearing, high-permeability rock drilled with high overbalance, gas will be flushed from the rock ahead of the bit, will not be circulated to the surface in the mud, and will not produce a gas show.

Water-Based Drilling Mud. Because hydrocarbons are not always part of a water-based-mud formulation, sophisticated analytical chemical techniques can be used on the oil and gas samples circulated to the surface and captured to determine the properties of hydrocarbons in a given zone penetrated by the drill bit. Occasionally, small amounts of hydrocarbons are added to water-based mud to reduce friction on the drillpipe. This hydrocarbon response must be accounted for in any analysis. Sidewall samples will contain some remaining hydrocarbons when brought to the surface. If promptly wrapped, frozen, and analyzed, the crude remaining in the core can be used to characterize density, gas/oil ratio, viscosity, and other fluid properties. Companion sidewall samples can be analyzed at the wellsite for odor, color, fluorescence, and cut-fluorescence, which can be used to distinguish oil/water and gas/oil contacts. If one is uncomfortable predicting reservoir fluid properties from such a small sample, the zone in question can be sampled with a wireline formation tester or ultimately subjected to a drillstem test. Wireline formation testers are often unsuccessful in obtaining a hydrocarbon sample because of the extent of the invaded zone. The low oil saturation in the invaded zone has a low relative permeability resulting in little oil production and little, if any, movement while being sampled by a wireline tester tool until all of the invaded, water-based fluid has been removed.

The other fluid in the rock of interest is water. The value of the water resistivity in the virgin reservoir is needed to interpret the deep-reading resistivity tools. However, the water-based mud filtrate dilutes the in-situ water, making its characterization problematic. Some authors have proposed a method to tag the mud filtrate with radioactive tritium while coring the reservoir. The resulting core is sampled, its water extracted and analyzed for tritium. The measured concentration of tritium is used with the measured porosity to compensate the fluid properties for the effect of dilution from mud filtrate. Other methods use an uncommon anion as the tagged material to avoid radiation-based methods. The formation tester tool can also be used to obtain a water sample in the water leg of the reservoir. Although the zone is invaded with mud filtrate, we can use the pump-out feature available on all vendors' tools to pump fluid from the invaded zone into the wellbore until the entire invasion has been removed. At this point, a water sample can be taken. The tester tool has an onboard water resistivity cell that is used to monitor the change in resistivity as the invasion fluid is pumped. When this value no longer changes, we switch over and take a water sample for analysis.

Oil-Based Drilling Mud. The overwhelming presence of oil in this type of mud dilutes the hydrocarbon in the invaded zone to such a point that analysis of the hydrocarbon in sidewall samples can no longer be used to predict hydrocarbon fluid properties. On the other hand, the water phase is now intact and can be sampled and analyzed with usual methods.

To determine hydrocarbon properties, wireline formation testers must be used. Two methods are useful. One method requires that many judiciously placed pressure tests be taken in the wellbore. This will permit the determination of the fluid-pressure gradient and density, which can be used to identify gas, oil, or water. Another method requires the use of the pump-out feature available on all wireline vendor tester tools to pump invaded zone fluid into the wellbore until the invading fluid has been reduced to a sufficiently low value, and then switches the pump to capture fluids in a removable, transportable fluid chamber. The feature of these tester tools that makes this possible is onboard spectrometers that have the ability to discern oil-based mud filtrate from in-situ hydrocarbons that guide the changeover from pumping into the wellbore to the sample chamber. Pressure/volume/temperature laboratories for the required fluid properties can then analyze these fluids. The operation of wireline formation testers is neither cheap nor without risk of tool sticking; thus, some operators choose not to run these wireline tools and opt instead for a drillstem test. All wireline tester tools can be used to take a water sample in the water leg, and this practice is encouraged. Not only are errors in water resistivity reduced, but possible incompatibilities between in-situ water and completion fluids are discovered before completion attempts.

3A.1.8 Determination of Absolute Permeability. The "gold" standard for permeability is to make measurements on core samples and to determine permeability with the methods outlined in API *RP 40*.[5] All other techniques are calibrated back to core measurements. However, because core measurements sample such a minute part of the reservoir, we must rely on techniques that can be applied in a widespread fashion across the reservoir. These methods rely on measurements on sidewall samples, correlation to wireline logging responses, interpretation of NMR logs, wireline formation tester pressure responses, and drillstem tests.

Sidewall Samples. This technique is valid for slightly to unconsolidated sandstone rock types. Carbonate rock types are generally too heterogeneous for small samples to provide any meaningful reservoir-wide value for permeability. Sidewall samples of sandstone rock types are inherently contaminated with drilling mud particles and are of little use for direct measurement of permeability. However, we can inspect the rock sample with a binocular microscope to estimate median grain size, sorting, and degree of consolidation, and to characterize pore fills. With these data, we can develop correlations to permeability on the basis of whole core measurements. An alternative is to disaggregate the sample and determine a grain size analysis with laser light scattering, which can then be correlated to permeability on the basis of whole core analysis.

Wireline Logging Correlations. Permeabilities measured in cores can be correlated to wireline measurements taken in the cored borehole. At various times and places, almost every wireline log has been used to correlate to permeability. The porosity-permeability crossplot is, perhaps, the most used; however, it is subject to considerable error. In select basins, the GR log response can be used to correlate to permeability while, in other basins, the neutron log or acoustic log seems to provide the correlation with least statistical scatter.

NMR Logs. Interpretation of NMR logging responses provides a volumetric distribution of pore sizes. If the pores are assumed to be spherical in shape, a value for permeability can be computed. These size-dependent data have been coupled with NMR pore volumes and NMR fluid saturations to produce an NMR permeability log. The chapter on NMR logging in this section of the *Handbook* shows examples of these techniques.

Wireline Formation Testers. All wireline tester vendors provide answer products that take the drawdown and buildup pressure vs. time responses and compute mobility. Mobility can be converted into permeability if a value of fluid viscosity is assumed. This permeability must be used with some caution. First, the pressure measurements are made on the borehole wall that has suffered possible drilling damage and pore throat plugging from mud solids. Second, one must take note if the measurement is in an invaded zone with two phases and, hence, the permeability determined is an effective permeability, not an absolute permeability. Depending on rock type and fluid saturations, the effective permeability may be an order of magnitude too small. The chapter on fluid sampling in the General Engineering section of this *Handbook* presents examples of wireline formation tester responses and derived permeability and the use of these pressure measurements to determine fluid gradients.

Drillstem Tests. These measurements are covered in the chapter on data acquisition and interpretation in the Drilling section of this *Handbook*.

3A.1.9 Fractional Flow. To predict the fractional flow from a given zone in a wellbore, the absolute and relative permeability of the rock types open to flow in the wellbore must be characterized. Determining the complete relative permeability function vs. fluid saturations measured on cores maintained at reservoir conditions of temperature, pressure, and the use of live reservoir crude is a daunting task. Most reservoirs never have these measurements performed on them. Usually, we take small snippets of these curves and attempt to characterize their entire response with various models. Proof of success is often taken to be a successful history match; others prefer to look for agreement between future predictions and future performance.

Regardless of our ability to measure these parameters in the laboratory, the scaleup to a heterogeneous, faulted reservoir is a challenge of immense proportions. These subjects are covered in the chapter on reservoir simulation in this section and the chapter on relative permeability and capillary pressure in the General Engineering section of this *Handbook.*

3A.1.10 How To Put It Into Use. This chapter is designed to show the reader how to use petrophysics in a general sense—how the many facets of petrophysics are tied together and how they relate to other elements of petroleum engineering. However, there is nothing like examples to show new practitioners how petrophysics really works and how they may put it to use themselves. The chapter on petrophysical applications provides these examples.

Nomenclature

m = Archie's cementation exponent
n = Archie's saturation exponent
R_w = resistivity of the in-situ water, ohm m^2/m
S_w = water saturation, fraction of pore volume
X = a specific depth

References

1. Thomas, E.C.: "50th Anniversary of the Archie Equation: Archie Left More Than Just an Equation," *The Log Analyst* (May–June 1992) 199.
2. Deeson, A.F.L.: *The Collector's Encyclopedia of Rocks & Minerals,* Clarkson N. Potter Inc., New York City (1973).
3. *A Catalog of Petrophysical and Geological Properties of Typical Reservoir Rocks*, Shell Oil Co., Houston (1995).
4. *The World Wide Rock Catalog,* Reservoirs Inc., Houston (1990).
5. *RP 40, Recommended Practices for Core Analysis,* second edition, API, Washington, DC (1998).
6. Swanson, B.F. and Thomas, E.C.: "The Measurement of Petrophysical Properties of Unconsolidated Sand Cores," *The Log Analyst* (September–October 1980) 22.
7. Wei, K.K., Morrow, N.R., and Brower, K.R.. "The Effect of Fluid, Confining Pressure, and Temperature on Absolute Permeabilities of Low-Permeability Sandstones," *SPEFE* (August 1986) 413.
8. Wawak, B.E. and Campbell, W.L.: "Characterization of Clay Fabric Using Critical Point Drying to Preserve Clay Texture and Morphology," *Scanning Electron Microscopy* (1986) **4,** 1323.
9. Patchett, J.G. and Coalson, E.B.: "The Determination of Porosity in Sandstone: Part Two, Effects of Complex Mineralogy and Hydrocarbons," paper T presented at the 1982 Annual Soc. of Professional Well Log Analysts Symposium, Corpus Christi, Texas, 6–9 July.
10. Peeters, M. and Visser, R.: "A Comparison of Petrophysical Evaluation Packages: LOGIC, FLAME, ELAN, OPTIMA and ULTRA," *The Log Analyst* (1991) **32,** No. 4, 350.
11. Thomas, E.C. and Stieber, S.J.: "The Distribution of Shale in Sandstones and Its Effect on Porosity," paper presented at the 1975 Annual Soc. of Professional Well Log Analysts Symposium, New Orleans, 4–7 June.
12. Richardson, J.G. *et al.*: "Investigation of As-Received Oil-Base-Core Water Saturation From the Ivishak Reservoir, Prudhoe Bay Field," paper SPE 28592 presented at the 1994 SPE Annual Technical Conference and Exhibition, New Orleans, 25–28 September.
13. Holstein, E.D. and Warner, H.R. Jr.: "Overview of Water Saturation Determination for the Ivishak (Sadlerochit) Reservoir, Prudhoe Bay Field," paper SPE 28573 presented at the 1994 SPE Annual Technical Conference and Exhibition, New Orleans, 25–28 September.

SI Metric Conversion Factors

bbl ×	1.589873	E − 01 = m^3
ft ×	3.048*	E − 01 = m

*Conversion factor is exact.

Chapter 3B
Resistivity and SP Logging

T.D. Barber, A. Brie, and **B.I. Anderson,** Schlumberger

3B.1 Fundamentals

Resistivity logging is an important branch of well logging. Essentially, it is the recording, in uncased (or, recently, even cased) sections of a borehole, of the resistivities (or their reciprocals, the conductivities) of the subsurface formations, generally along with the spontaneous potentials (SPs) generated in the borehole. This recording is of immediate value for geological correlation of the strata and detection and quantitative evaluation of possibly productive horizons. The information derived from the logs may be supplemented by cores (whole core or sidewall samples of the formations taken from the wall of the hole).

As will be explained later, several types of resistivity measuring systems are used that have been designed to obtain the greatest possible information under diverse conditions (e.g., induction devices, laterolog, microresistivity devices, and borehole-imaging devices). Many service companies offer resistivity-logging services, and most offer a Web-based catalog that describes each service.

3B.1.1 Earth Resistivity. Formation resistivity is a key parameter in determining hydrocarbon saturation. An electric current can pass through a formation because it contains water with enough dissolved ions to be conductive. With a few rare exceptions, such as metallic sulfides and graphite, dry rock matrix is a good electrical insulator. However, perfectly dry rocks seldom occur below ground level, so nearly all subsurface formations have finite, measurable resistivities because of the water in their pores, adsorbed onto their grain surfaces, or absorbed into a clay structure.

The resistivity of a formation depends on the resistivity of the formation water, the amount of water present, and the structure and geometry of the pores. The resistivity (specific resistance) of a substance is the electrical resistance measured between opposite faces of a unit cube of the substance at a specified temperature or, generally,

$$R = r\frac{A}{L}, \quad \text{(3B.1)}$$

where R = resistivity in ohm·m, r = resistance in ohm, A = area in m^2, and L = length in m.

Conductivity, σ, is the reciprocal of resistivity, expressed in Siemens/m. To avoid the excessive use of small decimal numbers in well logging, conductivity is expressed in milliSiemens/m (mS/m), where 1000 mS/m = 1 Siemen/m, so

$$\sigma = \frac{1000}{R} \quad \text{(3B.2)}$$

Formation resistivities are usually in the range of 0.2 to 1000 ohm·m. Resistivities higher than 1000 ohm·m are uncommon in most permeable formations but are observed in impervious, low-porosity formations such as evaporites. A few low-porosity hydrocarbon-bearing formations with almost no formation water can have resistivities as high as 20 000 ohm·m.

Formation resistivities are measured either by passing a known current through the formation and measuring the electrical potential (electrode or galvanic devices) or by inducing a current distribution in the formation and measuring its magnitude (induction devices).

Because resistivities cannot be read accurately over the entire measurement range when displayed on a linear scale, all resistivity logs are now presented on logarithmic grids, usually in 4 decades across two log tracks. This allows the display of readings from 0.2 to 2000 ohm·m, with a single curve covering the useful range of nearly all logs. A backup curve is used in the exceptional cases of readings outside that range. As resistivity data are used more and more digitally, and the log plots are mostly for reference, other formats are in wide use.

Most wireline resistivity-logging tools also have the ability to measure and record small differences in electrical potential that occur spontaneously in conductive muds as a continuous SP curve. The SP curve records the electrical potential differences generated by the interaction of formation water, conductive drilling fluid, and ion-selective shales. It has a wide range of formation-evaluation applications, including differentiating potentially porous and permeable reservoir rocks from impermeable clays and shales, defining bed boundaries, correlating nearby wells, indicating the shaliness of shaly formations, and estimating formation-water resistivity, R_w.

3B.2 The Logging Environment

Measuring the resistivity of formations of interest is complicated by the invasion of drilling fluids into permeable rocks. The invaded fluid can displace some or all of the connate water or hydrocarbon present. The resistivity of the borehole is often much less than the formations of interest, sometimes by orders of magnitude. In addition, the formation often consists of rock layers with widely varying resistivity. **Fig. 3B.1** shows these factors.

To measure the uninvaded portion of the rock from the borehole, a resistivity device must include a large volume of formation—there must be adequate sensitivity to the region of interest. Over the years, the main developments in resistivity-logging tools have been targeted at eliminating response from unwanted parts of the formation and recovering the resistivity of the uninvaded portion of a single layer. The response from unwanted parts of the formation is lumped under the term "environmental effects."

The main objectives of resistivity logs are the determination of R_t and R_{xo} and, for the newer imaging devices, the mapping of resistivity profiles into and around the borehole. This is accomplished by incorporating devices with at least three, and preferably more, depths of investigation. The deep-reading focused devices include the deep laterolog device (LLd) and the deep induction (ILd) device. Medium-depth devices include the shallow laterolog (LLs) and the medium induction (ILm). The shallow devices include the microresistivity device or the spherically focused log (SFL). The latest laterolog and induction tools include arrays focused at many depths of investigation—from five to eight depths—measured simultaneously. These al-

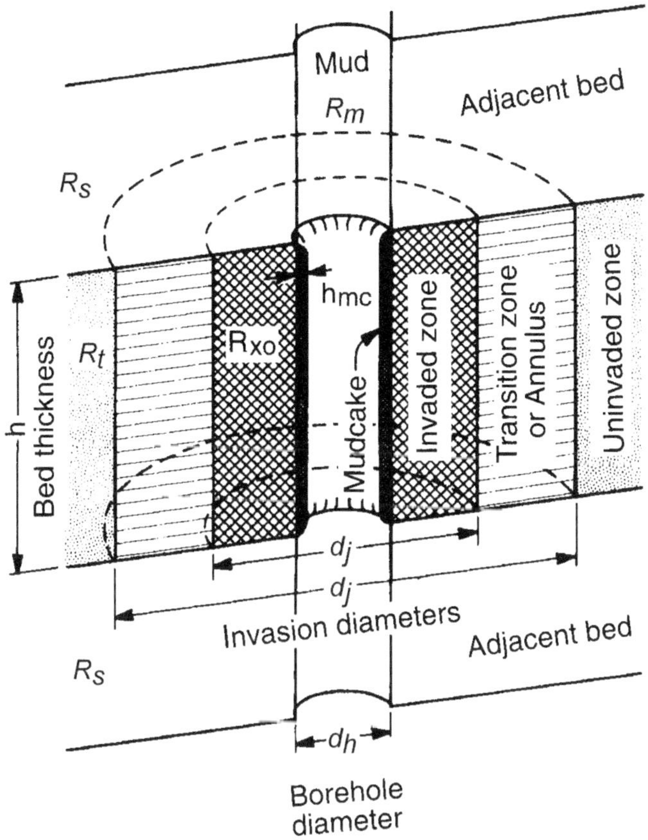

Fig. 3B.1—The borehole/formation environment for well logging.

low a better description of the invaded zone and allow for interpretation of complex invasion profiles.

3B.3 Electrode Resistivity Devices

3B.3.1 Normal and Lateral Devices. During the first quarter century of well logging, the only electrical surveys (ES) available were the resistivity logs made with so-called lateral and normal devices plus the SP. Thousands of them were run each year all over the world. Since then, new logging methods have been developed to measure values much closer to R_{xo} and R_t. Nevertheless, the conventional ES logs (consisting of SP; 16-in. normal; 64-in. normal; and 18-ft, 8-in. lateral) are stored in log archives all over the world. Because new information can often be obtained by reinterpreting old ES logs, this chapter includes discussion of the principles and responses of the ES measurements.

The first resistivity devices were the normals and laterals.[1–3] These were, in concept, extensions to laboratory four-terminal resistivity-measuring cells. Current is injected in the formation from a single electrode and returned to a point remote from the well. The current near the injection electrode spread out radially from the electrode. Two voltage–measuring electrodes (M and N) on the sonde approximated the measurement of a constant-voltage spherical shell around the injection electrode. The measurements of voltage and current are converted to a resistivity measurement.

For normal devices (**Fig. 3B.2**), the distance AM is small: 1 to 6 ft as compared with MN, MB, and BN. In practice, N or B may be placed in the hole at a large distance above A and M

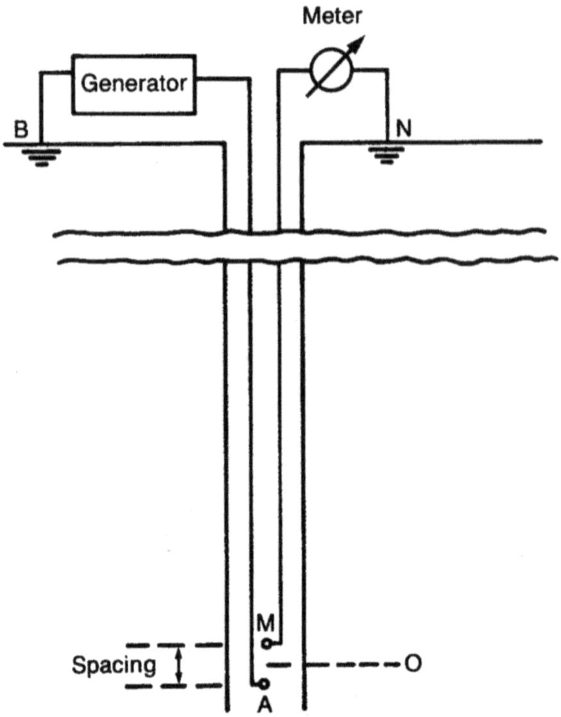

Fig. 3B.2—Schematic representation of the normal device.

[the voltage measured is practically the potential of M (because of current from A), referred to an infinitely distant point]. The distance AM of a normal device is its spacing. The point of measurement is midway between A and M. The most common normal spacings were 16 and 64 in.

The lateral device was designed to provide a deeper resistivity measurement than the normal tools, while at the same time improving the detection of thin beds. For lateral devices (**Fig. 3B.3**), measuring electrodes M and N are close to each other and located several feet below current electrode A. Current-return electrode B is at a great distance above A or at the surface. The voltage measured is approximately equal to the potential gradient at the point of measurement 0, midway between M and N. The distance AO is the spacing of the lateral device.

In an alternate version of the lateral, the positions of the current and voltage electrodes are interchanged. A and B are moved to M and N, and N and M are moved to B and A. This tool is called the "inverse," and it records the same resistivity values as the lateral by reciprocity. The inverse arrangement made it more practical to record measurements by the two normals and the lateral simultaneously.

Interpretation of laterals and normals is very complicated because the response is a complicated function of the formation being measured.[4] **Fig. 3B.4** shows a computed response of the 16- and 64-in. normals for a series of beds with and without invasion. The separation is not a clear function of invasion, but is also a function of bed thickness. **Fig. 3B.5** shows the 18-ft, 8-in. lateral tool in the same series of beds. The relation of the curve to the bed is not clear at all. Many charts (called departure curves) were published to aid in interpretation of the ES logs. Modern interpretation methods include 2D inversion (after the curves are digitized) and iterative forward modeling for when they are not digitized.[5]

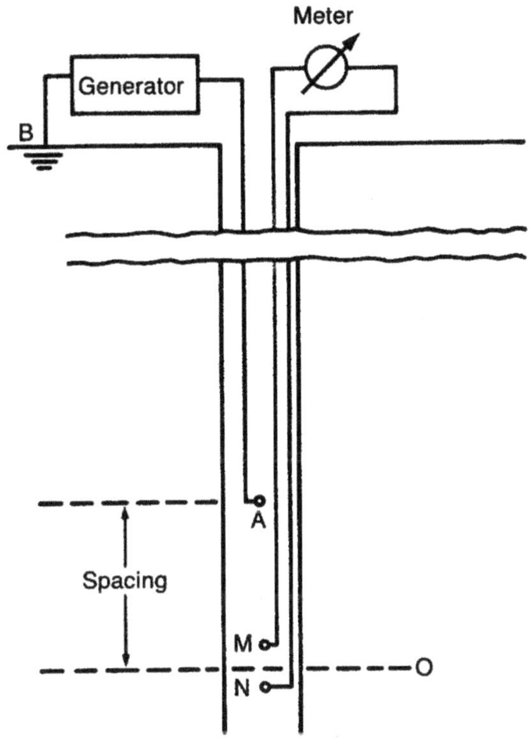

Fig. 3B.3—Schematic representation of the lateral device.

3B.3.2 Laterologs. Laterolog devices are designed to minimize the influence of near-tool effects (i.e., the borehole and the invaded zone) for a deep laterolog measurement, as well as to reduce the response from adjacent beds.[6]

To estimate R_t under a variety of different logging conditions and in different formations, a simple three-parameter, step-profile invasion model is often used. This model consists of a flushed zone of resistivity R_{xo} and a sharp boundary at diameter d_i, with the uninvaded zone of resistivity R_t. Three independent, borehole-corrected resistivity measurements with appropriately chosen depths of investigation contain enough information from the formation to reliably solve for R_t using this model. Measurements with the following features should be chosen: small, correctable borehole effects; similar vertical resolutions; and well-distributed radial depths of investigation—one reading as deep as practical, one very shallow reading, and one intermediate reading.

In conductive muds, the Dual Laterolog (DLL) Resistivity–R_{xo} combination tool provides simultaneous measurements suitable for evaluating R_t, R_{xo}, and d_i.[7] It should be said that the value of R_t in a given bed is an interpreted parameter, and is almost never measured. As long as the formation is invaded, assumptions about the invasion profile must be made to estimate R_t.

Dual Laterolog Resistivity Measurements. **Fig. 3B.6** shows the electrode array used for deep and shallow laterolog measurements (LLd and LLs, respectively). Both logs share the same electrodes and have the same current-beam thickness, but different focusing currents give them different depths of investigation. The measure current (I_0) is emitted from the central A0 electrode, returning to an "infinitely distant" electrode, usually at the surface. The focusing current (I_a) flows from the A1 and A2 (and A1′ and A2′) electrodes to a distant electrode for the

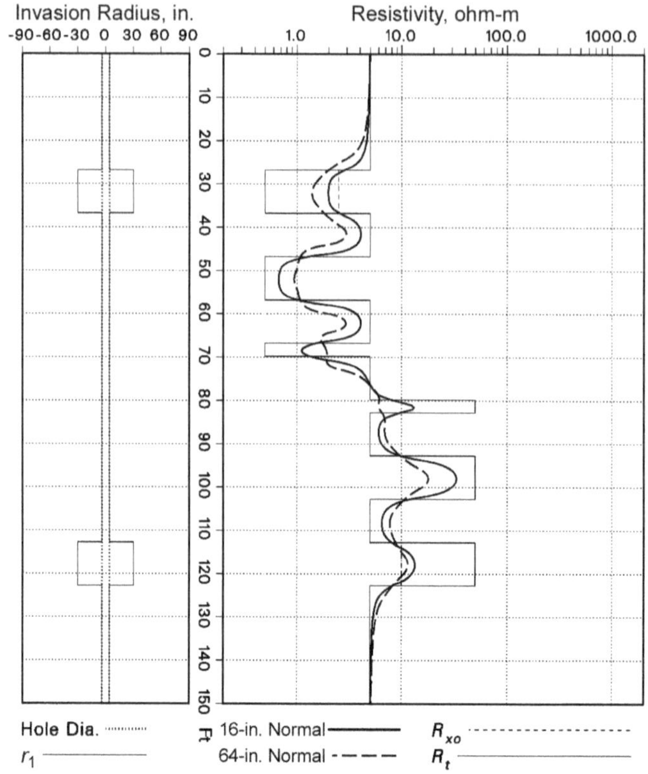

Fig. 3B.4—Computed response of 16- and 64-in. normals in beds of varying thickness.

LLd measurement and from A1 to A2 (and A1′ to A2′) for the LLs measurement. The focusing current is adjusted so that electrodes M1 and M2 (and M1′ and M2′) remain at the same potential.

A constant-power measuring system ensures measurement accuracy over a wide range of resistivities (from 0.2 to 40,000 ohm·m). Both the measure current and measure voltage (V_0) are varied and measured, but their product power I_0V_0 is kept constant.

Long guard electrodes are required to achieve the desired depth of investigation and measurement range. For electrode tools focused using guard electrodes, the depth of investigation increases only as the square root of the length of the guard electrodes. This requirement results in the 28-ft [8.5-m] total length of the LLd electrode array. A beam thickness of only 28 in. [0.7 m], however, ensures good vertical resolution.

The LLs measurement shares most of the electrodes with the deeper measurement. This is achieved by operating LLs and LLd at different frequencies. The LLs and LLd measurements have the same vertical resolution, but the LLs device uses a less constrained focusing condition in which the focusing current returns to electrodes on the array instead of to a remote electrode. The LLs measurement therefore has a shallower depth of investigation and responds more strongly to the region around the borehole that is usually affected by invasion.

Laterolog Anomalies. The Groningen effect was named after the large Dutch gas field where the anomaly was first identified. The effect[8] is an anomalously high resistivity reading that occurs for approximately 100 ft [30 m] below a thick, highly resistive bed such as the thick evaporitic Zechstein caprock at the Groningen field. The effect is maximum around 1 ohm·m. Because the DLL measure current is AC (albeit very low frequency), skin effect re-

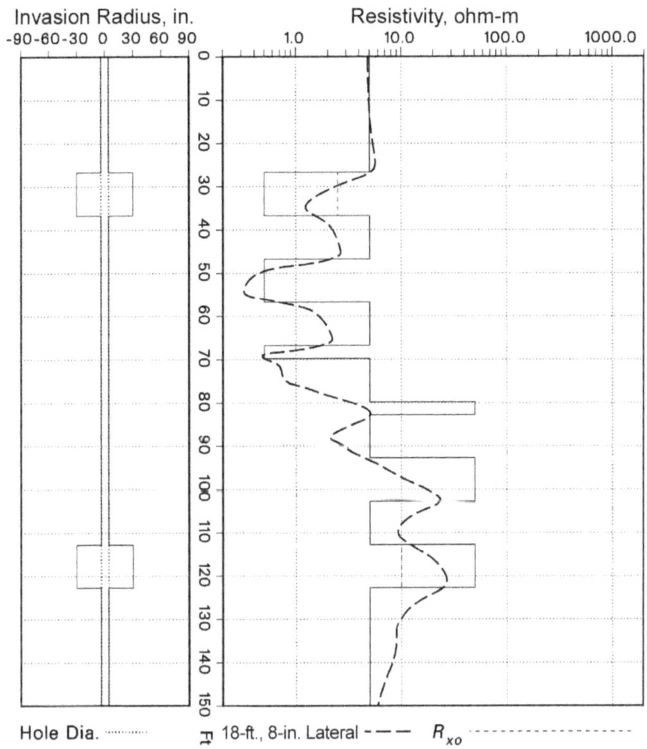

Fig. 3B.5—The 18-ft, 8-in. lateral in the same series of beds as in Fig. 3B.4.

duces the volume around the well where the measure and focusing current can flow. Little of the current is able to return to the remote electrodes through the highly resistive formation, with the majority flowing in the conductive mud in the borehole. This creates a negative potential at the far reference electrode used as the potential reference for the laterolog measurement. If casing has been set in or below the resistive zone, it accentuates the "short circuit" effect of the borehole, and the Groningen effect is more pronounced. Drillpipe conveyance produces the same effect, with the drillpipe becoming the "short circuit." This problem severely limits the use of drillpipe conveyance of the DLL in high-angle or horizontal wells in many reservoirs.

A mild Groningen effect may be difficult to identify from the LLd curve alone. The Schlumberger DLL has a modified-geometry measurement that can also be recorded. This provides an LLg curve that separates from the LLd curve when Groningen effect is present. If the Groningen effect is positively identified, an estimate of its magnitude can be made by analyzing the signal phases in the tool, and an approximate correction can be applied to the log. The LLs measurement uses different current paths and does not suffer from the Groningen effect. The array laterolog (see the following) is not affected by Groningen effect.

Azimuthal Dual Laterologs. In the early 1990s, a new dual laterolog that had an additional azimuthally segmented current electrode was introduced.[9] The Schlumberger ARI[*] Azimuthal Resistivity Imager has a set of segmented azimuthal electrodes incorporated in a conventional dual laterolog array. The tool records azimuthal resistivity variations around the borehole and

[*] Throughout this chapter, tool names are service marks of the referenced companies.

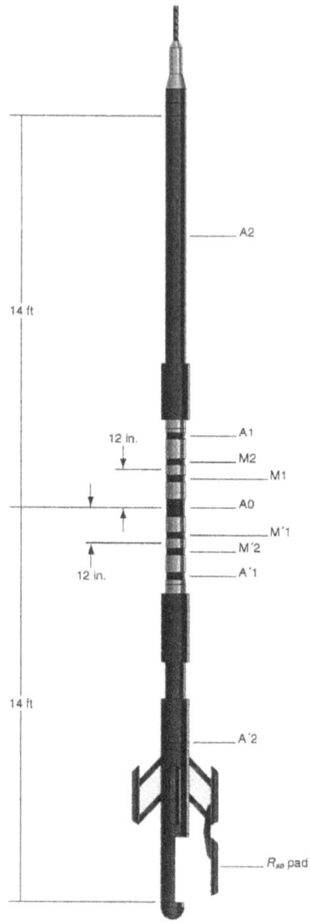

Fig. 3B.6—The dual laterolog array layout.

produces an image of the variations. The azimuthal electrodes are placed at the center of the A2 electrode of the DLL tool and do not interfere with the standard LLd and LLs measurements.

The deep azimuthal measurement operates at the same frequency as the deep laterolog measurement, and the currents flow from 12 azimuthal current electrodes to the surface. They are focused by the current from the A2 electrode's upper and lower portions and by currents from the other current electrodes. In addition, the current from each azimuthal electrode is focused passively by the currents from its neighbors. The resulting operation of the azimuthal array has no effect on the LLd and LLs measurements.

Twelve azimuthal resistivities are computed, and from their sum, a high-resolution resistivity measurement, LLhr, is derived. This is equivalent to replacing the azimuthal electrodes with a single cylindrical electrode of the same height.

The high-resolution LLhr curve reads almost as deeply into the formation as a deep laterolog LLd curve, particularly when R_{yo} is less than R_t. An LLhr log can therefore replace an LLd log for interpretation, especially where its vertical resolution is an advantage. Individually selected azimuthal resistivities can be used in the same manner where the logged interval is azimuthally anisotropic or includes highly dipping thin beds.

The azimuthal resistivity measurements are sensitive to tool eccentering in the borehole and to irregular borehole shape. Auxiliary measurements are made that are very shallow, with current paths close to the tool. Most of the current returns to the A2 electrode near the azimuthal array. Because the borehole is generally more conductive than the formation, the current tends to stay in the mud, and the measurement responds primarily to the volume of mud in front of each azimuthal electrode. The measurement is therefore sensitive to borehole size and shape and to eccentering of the tool in the borehole. However, for best use of the azimuthal measurements, the tool should be well centralized in the borehole.

A High-Resolution Azimuthal Laterolog Sonde (HALS) resembles an ARI or a dual laterolog array.[10] Although it is shorter than either of these tools, the HALS is not just a scaled-down version. Its dimensions are optimized to achieve similar performance as an ARI sonde with a tool that is only approximately one-half its overall length.

Like the ARI tool, the azimuthal array of the HALS makes deep and shallow resistivity measurements around the borehole with a 1- or 2-ft [0.3- or 0.6-m] vertical resolution. (In this chapter, vertical resolution is defined as the 90% width of the vertical response function.) Formation resistivity images can be derived from either the deep or the shallow measurements. Mud resistivity and tool standoff are also measured. In addition to providing a visual image of formation lamination and anisotropy, the azimuthal images can be used to estimate the gross formation dip and to correct deep resistivity measurements in dipping beds.

Because the HALS is shorter than ARI and DLL sondes, the borehole effect of the shallow measurement is larger. Combined with the slightly reduced depth of investigation of the deep array, this reduces the precision of the invasion correction in cases of invasion where $d_i > 50$ in. [1.3 m]. The inherent vertical resolution is sharper—24 in. [0.6 m] for the HALS deep and shallow resistivity curves (HLLD and HLLS, respectively) compared with 40 in. [1 m] for the LLd and LLs curves of the DLL log. The HALS provides high-resolution deep and shallow curves (HRLd and HRLs, respectively) with the same 12-in. vertical resolution as the LLhr curve of an ARI log.

Real-time corrections can be made for Groningen effect, electrical path changes imposed by tough logging conditions (TLC) logging in which the logging tool is transported on drillpipe, and borehole effects. A two-parameter inversion model can also be used in real time to solve for R_t and d_i, with R_{xo} provided by the microcylindrically focused log (MCFL) measurements of the Platform Express tool.

3B.3.3 Array Electrode Tools. The most recent development in electrode tools is the array laterolog or array lateral tools. These combine multiple depths of investigation with 2D inversion of the data to give much improved response in invaded thin beds with conductive mud.

The Schlumberger high-resolution laterolog array (HRLA) tool consists of five laterolog arrays with different depths of investigation.[11] All current is returned to electrodes above and below the array, so no bridle is used. Because it has no bridle, it does not suffer from Groningen effect.

Borehole and shoulder effects are minimized by the use of laterolog-style focusing. Focusing involves injecting current from guard or bucking electrodes to ensure that the current from the central measure electrode flows into the formation rather than along the borehole. By having all currents return to the tool body rather than surface, Groningen effect is eliminated and shoulder-bed effect reduced. More importantly, the surface current return and insulating bridle are no longer needed. All signals are measured at the same time and logging position. This avoids horns or oscillations caused by irregular tool motion and ensures that the measurements are always exactly depth-aligned.

The HRLA tool uses segmented bucking electrodes and multifrequency operation (ranging from 75 to 270 Hz) to acquire six simultaneous measurements. The six modes are focused by a combination of hardware and software focusing. The hardware injects the currents in a way that is as close to focused as possible. The shallowest mode, RLA0, is mostly sensitive to the borehole and is used to estimate the mud resistivity. The apparent resistivities RLA1 through RLA5 are all sensitive to the formation, becoming progressively deeper in investigation.

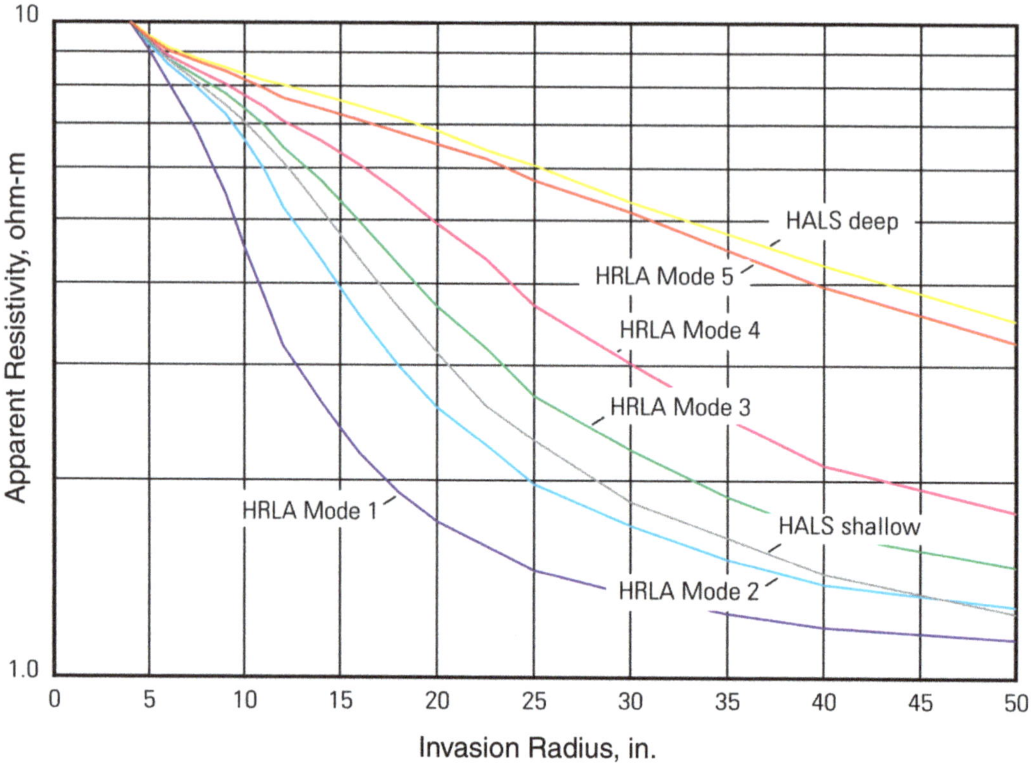

Fig. 3B.7—Computed radial response of the HRLA logs compared to the HALS, HLLD, and HLLS.

Fig. 3B.7 shows the radial response of the optimized HRLA tool compared to the HLLd and HLLs measurements from the HALS tool. **Fig. 3B.8** shows the HRLA logs compared to the DLL, LLd, and LLs logs.

The shoulder-bed and invasion responses of laterolog tools are combined in the tool response in ways that are difficult to separate. For this reason, some sort of 2D inversion is necessary to determine the formation parameters accurately in the case of invaded thin beds. An automatic 2D inversion is available that yields the best results from the data in invaded thin beds. **Fig. 3B.9** shows the field logs, 1D inversion results, and 2D inversion results in the same well.

The Baker Atlas high-definition lateral log (HDLL) tool[12] acquires 8 potential and 16 first differences, and computes 14 second differences. These are used as inputs to a 2D inversion to solve for formation resistivities. The inversion process begins with the raw data input and an initial estimate of a parametric model describing the formation resistivity distribution. Simulated logs or synthetic tool responses for each of the array sensors are derived for the selected initial formation parameters as a solution to Maxwell's equations.

The inversion process is initiated by using the shallow measurements (associated with short-spacing sensors) to identify and evaluate the shallow-formation resistivity structure. Deep measurements (associated with the longest sensor spacings) are used in evaluating the uncontaminated formation resistivity structure. The intermediate measurements are used to derive the radial-invasion profile. Using inversion processing, measurements of different vertical resolution and depths of investigation are combined in a single interpretation process to provide an accurate resistivity distribution image.

Final inversion results consist of the estimated resistivity structure satisfying all HDLL data and the statistical quality indicators represented in terms of the importance of the formation

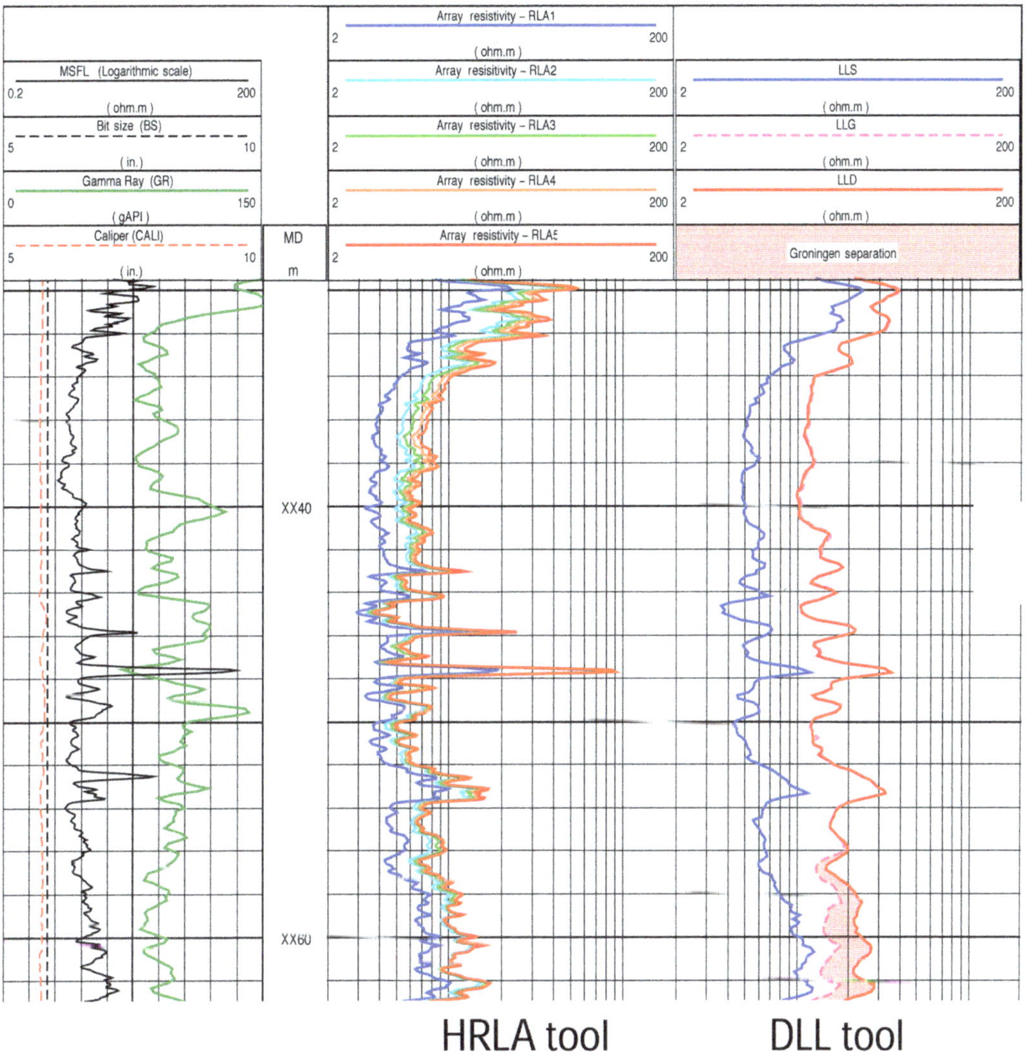

Fig. 3B.8—A log showing the high resolution of the HRLA vs. DLL.

parameter along with its corresponding error bounds. **Fig. 3B.10** shows a comparison of HDLL R_t and R_{xo} inversion results with conventional Dual Laterolog (RS is DLL shallow and RD is DLL deep) and Micro Laterolog (MLL) measurements.

3B.3.4 SFLs. The SFL device measures the conductivity of the formation near the borehole. It uses small electrodes that can be combined with the dual-induction tool to provide shallow-investigation data for invasion evaluation.

The SFL device uses two independent current systems. A focusing current system establishes constant-potential spherical "shells" around the current electrode, even in the presence of a conductive borehole, and the I_0 survey current flows through the volume of investigation.

The SFL electrode array consists of current-emitting, current-return, and measure electrodes. Two equipotential spheres are established around the I_0 survey current electrode; the first sphere is approximately 9 in. [0.2 m] from the electrode, and the other is approximately 50 in. [1.3 m] from it. The volume of formation between these two surfaces is constant, and

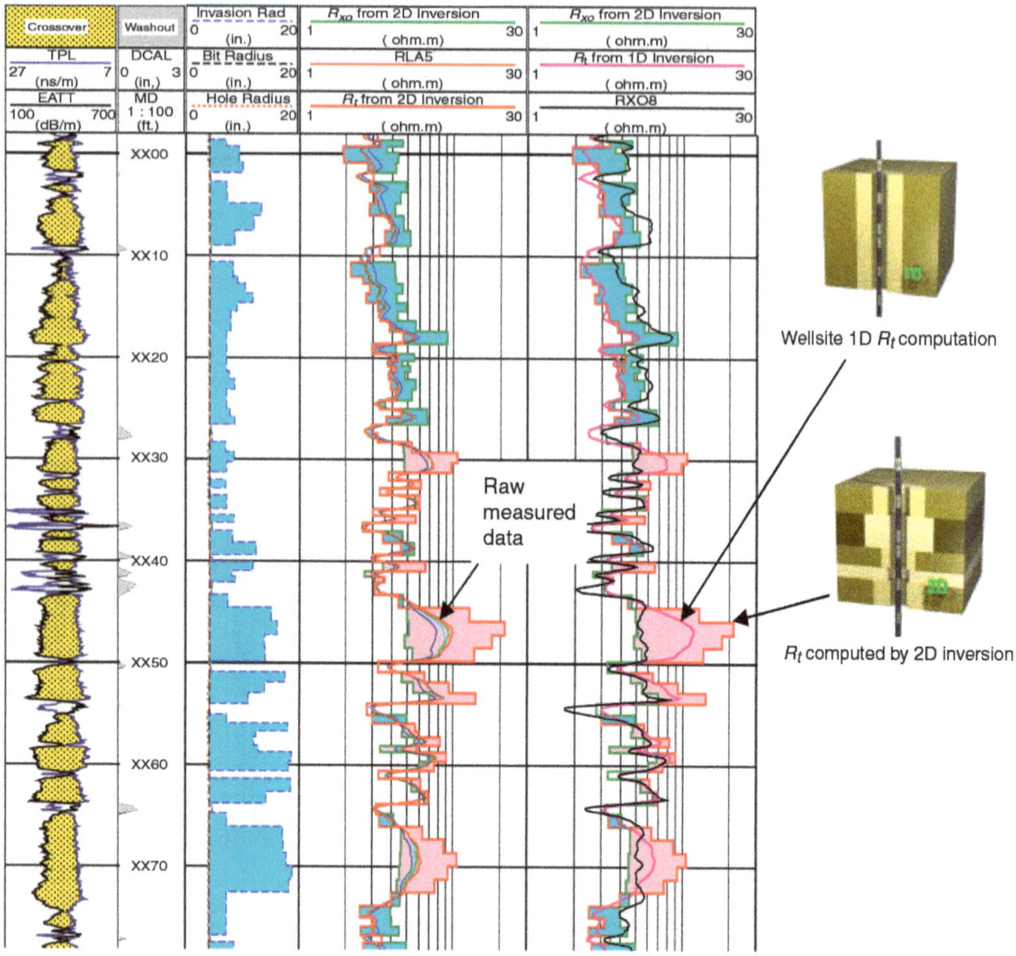

Fig. 3B.9—HRLA field logs, 1D real-time inversion results, and 2D inversion results.

because a potential difference of 2.5 mV is maintained between the spheres, the conductivity of this volume of formation is proportional to the I_0 survey current intensity.

3B.3.5 Cased-Hole Resistivity Tools. Despite the apparent paradox of measuring formation resistivity through the highly-conductive steel casing, tools are now available that can measure the formation resistivity to considerable accuracy. The idea originated in the 1930s[13] and was revisited by Kaufman[14] and Vail[15] in the late 1980s. Commercial tools were introduced in 2000 by Schlumberger[16] and Baker Atlas.[17] Both of these tools operate on the Kaufman-Vail principles.

The Schlumberger cased-hole formation resistivity (CHFR) tool has three sets of four arms that contain electrodes that are forced into contact with the inside of the casing. A current generator on the surface is connected to an electrode at the top of the tool. The current is injected into the casing and returns to an electrode in the earth some distance from the casing. Although most of the current returns through the casing, some small fraction of it will leak off from the outside of the casing and will return through the earth. This leakoff current forms the basis for the CHFR measurement.

The leakoff current is determined by measuring the voltage drop along a section of the casing. The double-differenced voltage Δ contains both the leakoff term and the voltage drop

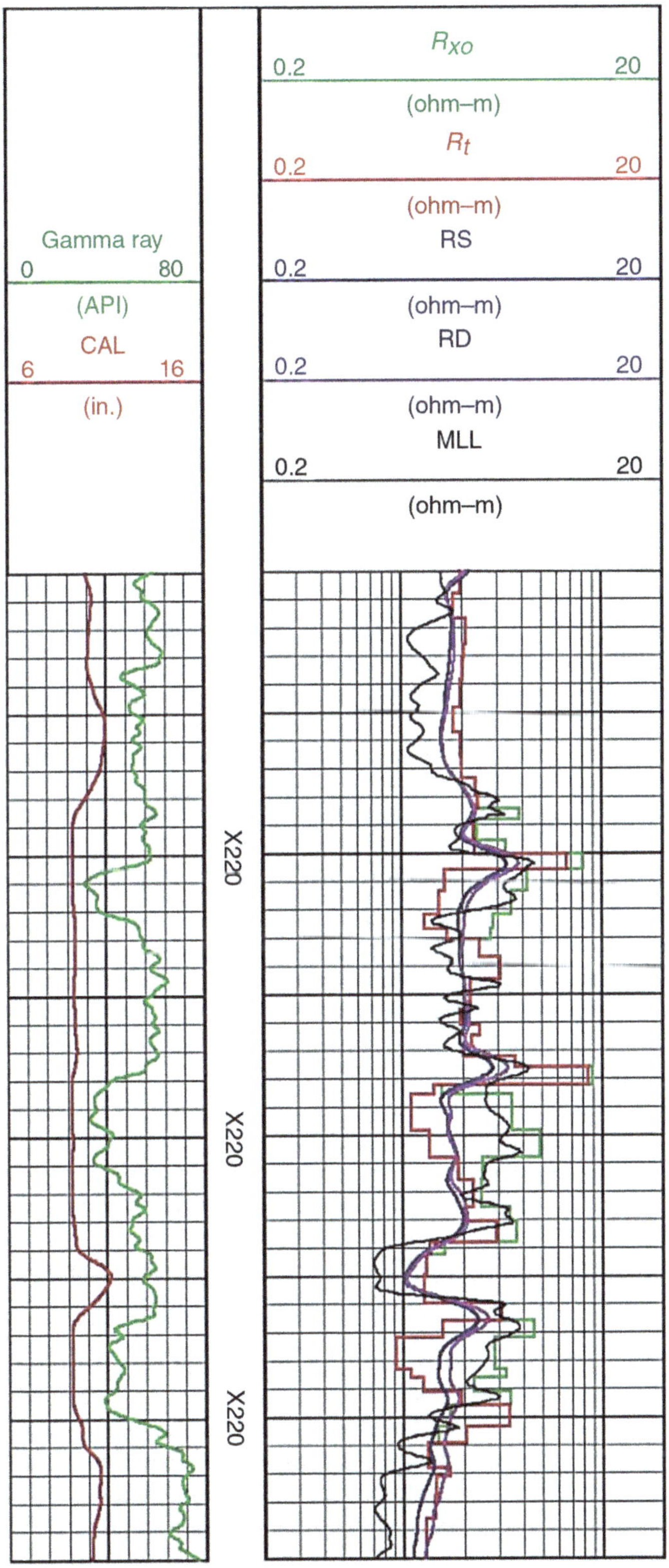

Fig. 3B.10—HDLL 2D inversion results compared with DLL logs.

produced by the current flowing through the casing and the resistance, R, of the casing. The current switch is changed to position 2. Now a current from inside the tool is sent from the upper electrode to a lower electrode. The voltage difference is now measuring the resistance of the casing, $R1$. All of the measurements are combined in the equation

$$I_f = \left(\frac{1}{R_1}\right)\left[\Delta - E\left(\frac{I_2}{I_0}\right)\right] \qquad (3B.3)$$

to produce a formation resistivity measurement. The measurements are taken while the tool is stationary and take approximately a minute per station.

Fig. 3B.11 shows a log of the CHFR in a newly cased well compared with open-hole HALS and AIT logs. The comparison is very good, and in the zone from 865 to 900 ft, with $R_{xo} > R_t$, the CHFR agrees well with the AIT 90-in. log, showing the great depth of investigation of the CHFR log.

Cased-hole resistivity is becoming accepted for applications including contingency logging, reservoir monitoring, and evaluation of old producing wells. One application combines CHFR with pulsed-neutron logs to do cased-hole formation evaluation.

The Baker Atlas Through-Casing Resistivity (TCR) tool operates on a similar measurement principle as the CHFR. It is smaller in diameter (2⅛ vs. 3⅜ in. for the CHFR).

GVR Resistivity-at-the-Bit Tool. The Schlumberger Geovision Resistivity (GVR) tool[18] is an electrode resistivity tool that measures five resistivity values—bit, ring, and three button resistivities—as well as gamma-ray and shock measurements.

A 1500-Hz alternating current passes through the toroidal-coil lower transmitter that is 1 ft from the bottom of the tool, inducing a voltage in the collar below. Current flows through the collar and bit and into the formation in front of the bit, returning to the collar farther up the drillstring. The resistivity at the bit is derived from the axial current, which is measured by a ring monitor toroid, and the induced voltage, which is a function of the transmitter current.

When the GVR tool is positioned directly above the bit, the resistivity measurement has a resolution of approximately 2 ft [61 cm], which is usually adequate for "geostopping"—stopping drilling precisely at casing or coring depths.

Focused Multidepth Resistivity. There are four focused-resistivity measurements incorporated in the RAB tool. These include the ring electrode measurement, with a depth of investigation of approximately 9 in. [0.23 m], and three button electrode measurements, with depths of approximately 1, 3, and 5 in. [2.5, 8, and 13 cm, respectively] into the formation.

The button measurements are radial, acquiring azimuthal resistivity profiles as the tool rotates in the borehole. A rotational speed of at least 30 rpm is required for full profile recording, with each button recording 56 resistivity measurements per rotation. The data are usually stored downhole for later retrieval, although a compressed image and selected button data may be transmitted to the surface in real time together with the ring and bit resistivities and gamma-ray measurements. **Fig. 3B.12** shows a recorded GVR image compared with an image from the wireline FMI borehole resistivity image tool (see the following for a description of this tool).

All four focused resistivities use the same measurement principle: Current from the upper transmitter flows down the collar and out into the formation, leaving the collar perpendicular to its surface and returning to the collar above the transmitter. Low-impedance circuits measure the current at each button electrode, and the axial current flowing down the collar is measured at the ring electrode by the ring monitor toroid and at the lower transmitter by the lower monitor toroid. These resistivity measurements are repeated using current from the lower transmitter.

In a homogeneous formation, the equipotential surfaces near the button and ring electrodes on the RAB tool are cylindrical. However, in layered formations, there is a tendency for cur-

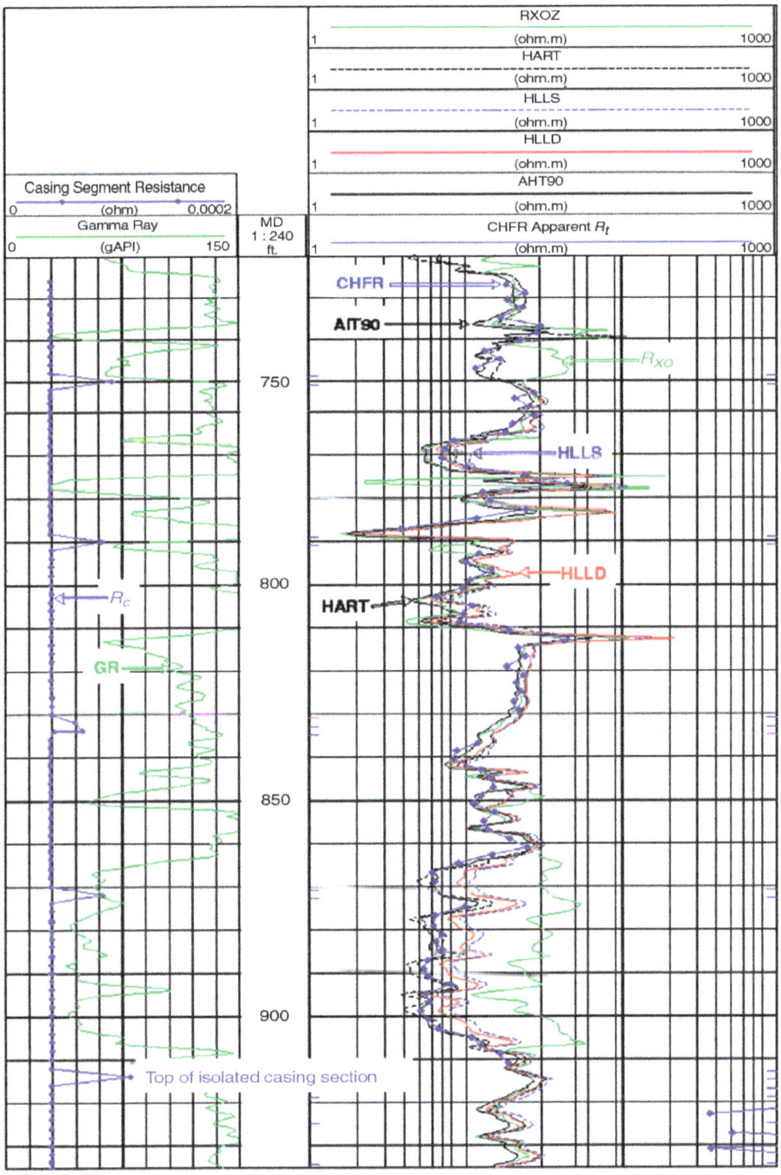

Fig. 3B.11—CHFR log vs. HALS laterolog and array induction openhole logs.

rent to flow preferentially in the more conductive beds and avoid the more resistive beds. This effect is known as "squeeze" for conductive beds and "antisqueeze" for resistive beds, and it leads to horn-like distortion of resistivity readings at bed boundaries.

A cylindrical focusing technique (CFT) is used to measure and compensate for this distortion by restoring the cylindrical geometry of the equipotential surfaces in front of the measurement electrodes. This is achieved by regulating the currents generated by the upper and lower transmitters for zero axial current flow at the ring monitor electrode, which avoids current flow along the borehole and focuses the ring current into the formation. This focusing technique produces a response very similar to that of a wireline laterolog.

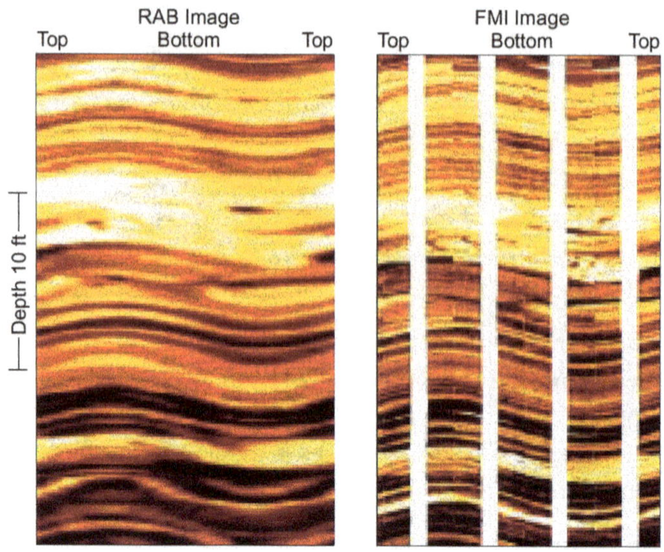

Fig. 3B.12—GVR button measurements image compared with wireline FMI tool images.

3B.4 Environmental Effects on Laterolog Tools. Laterolog and SFL log readings are influenced by the borehole mud, adjacent shoulder beds, and the invaded zone as well as the uninvaded formation. If automatic corrections are not available, log-interpretation charts provided by the service company are used to manually correct the log readings for these influences. The borehole corrections must always be made first, followed by bed-thickness corrections and finally invasion corrections of the determination of R_t, R_{xo}, and d_i.

Invasion Corrections. The "geometric factor" relates the effect of a portion of formation on the logging tool reading to its position relative to the tool in an infinite homogeneous medium. It has a particular application to induction logging tools, but pseudogeometrical factors are a useful comparative tool for other resistivity devices.

Fig. 3B.13 is a plot of integrated pseudogeometrical factors for several focused resistivity logs. It graphically compares the relative contributions of the invaded zone to the tool responses and their relative depths of investigation. The good spread in radial characteristics of the LLd and LLs measurements enables accurate resistivity analysis over a wide range of invasion conditions.

To evaluate the three unknowns of the simple step-profile invasion model (R_{xo}, R_t, and d_i), a combination of at least three carefully chosen resistivity measurements is required. LLd and LLs curves, with a very shallow resistivity measurement that reads R_{xo} directly, may be sufficient. See the section on invasion interpretation for more details on the determination of R_t, R_{xo}, and d_i.

3B.4 Induction Logging

Induction logging[19] was originally developed to measure formation resistivities in boreholes containing oil-based muds and in air-drilled boreholes because electrode devices could not work in these nonconductive boreholes. However, because the tools were easy to run and required much less in the way of chart corrections than laterals or normals, induction tools were used in a wide range of borehole salinity soon after their introduction.

3B.4.1 Principles. Commercial induction tools consist of multiple coil arrays designed to optimize vertical resolution and depth of investigation. However, to illustrate induction-tool funda-

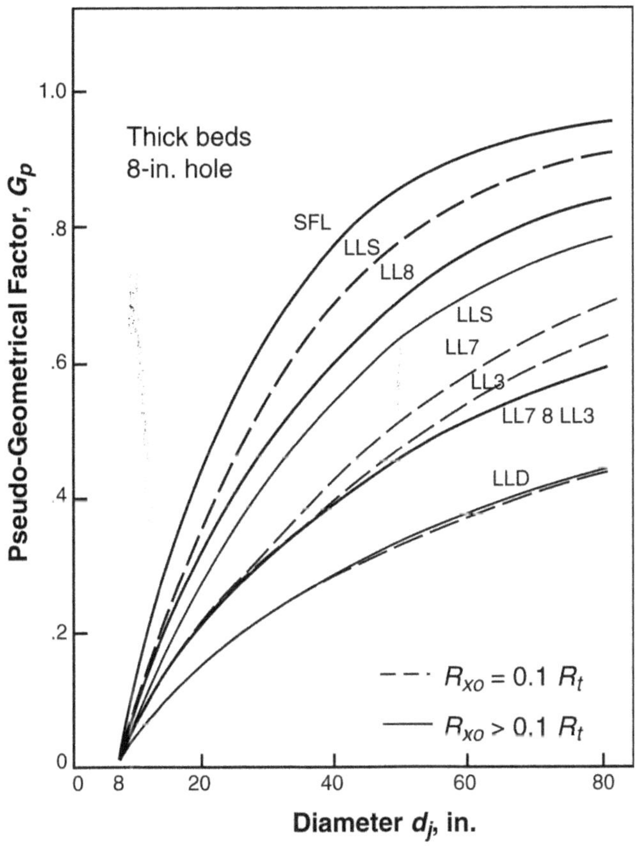

Fig. 3B.13—Pseudogeometrical factors, or radial response functions, for several laterolog tools.

mentals, it is instructive to first examine the basic building block of multiple-coil arrays, the two-coil sonde.

Fig. 3B.14 shows that a two-coil sonde consists of a transmitter and receiver mounted coaxially on a mandrel. Typical coil separations range from 1 to 10 ft apart. In practice, each coil can consist of from several to 100 or more turns, with the exact number of turns determined by engineering considerations. The operating frequency of commercial induction tools is in the tens to hundreds of kilohertz range, with 20 kHz being the most commonly used frequency before 1990.

The induction transmitter coil is driven by an alternating current that creates a primary magnetic field around the transmitter coil. The primary magnetic field causes eddy currents to flow in a continuous circular distribution (often mistakenly called "ground loops") centered around the borehole axis. The color contours in Fig. 3B.14 show the current distribution. These eddy currents are proportional to the formation conductivity, and they in turn generate a secondary magnetic field, which induces an alternating voltage in the receiver coil. This receiver voltage is first-order proportional to the conductivity of the formation.[20]

Because the transmitter current is alternating, there is a phase shift between the transmitter current and the current density in the formation. This phase shift is not the same in all parts of the formation—it increases with distance into the formation (Fig. 3B.14). Similarly, the phase in the receiver is even further shifted. At very low conductivities, the total phase shift is approximately 180° and increases with increasing formation conductivity. Induction tools have always

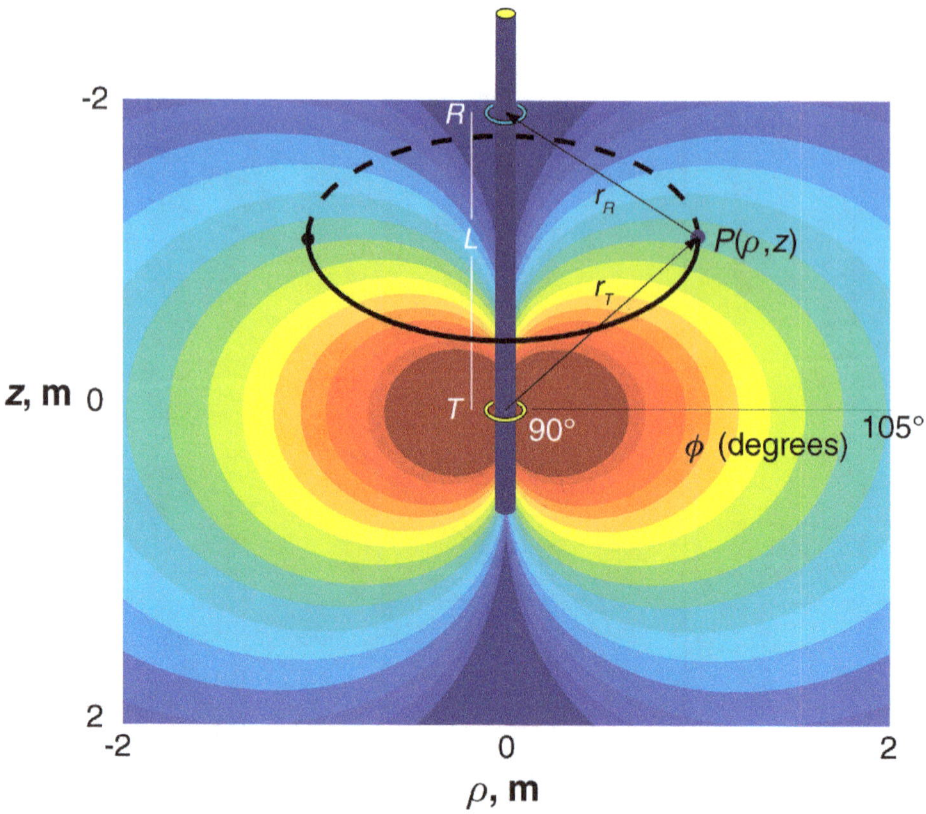

Fig. 3B.14—Schematic representation of a two-coil induction array showing the distribution of the currents induced in the formation by the transmitter coil.

measured the part of the voltage that is exactly 180° phase-shifted from the transmitter current (called the R-signal). As the conductivity increased, and the phase shifted, the voltage was a bit less than expected from a linear relationship. This difference is called skin effect. Modern induction tools make an additional measurement at a phase shift of 270° from the transmitter current (called the X-signal). These two measurements, being in quadrature, allow precise phase and amplitude measurement of the receiver voltage.

3B.4.2 Induction Response. To produce adequate sensitivity to the uninvaded zone, induction tools perforce include signals from a large volume of formation. The challenge is to determine exactly where the measurement is coming from in the formation. Doll defined the geometrical factor as a 2D function $g(\rho,z)$, which defines the part of the total signal that comes from an infinitesimally thin loop around the borehole. This definition is valid only at very low conductivities. Moran defined a modification of the geometrical factor that is valid in low contrast formations at any conductivity. This response is known as the Born response.[21]

The response to formation layers is given by the vertical response function $g_V(z)$, which is defined as the integral of the 2D response function $g(\rho,z)$ over radius ρ. The response to radial variations in a thick bed is given by the radial response function $g_R(\rho)$, which is defined as the integral of $g(\rho,z)$ over z. The response of the array to invasion in a thick bed is characterized by the integrated radial response G_R, which is the cumulative integral of $g_R(\rho)$ over radius.

3B.4.3 Multicoil Arrays. Because the direct transmitter-receiver mutual coupling of a two-coil array can produce a voltage several thousand times that from a formation, two-coil arrays are

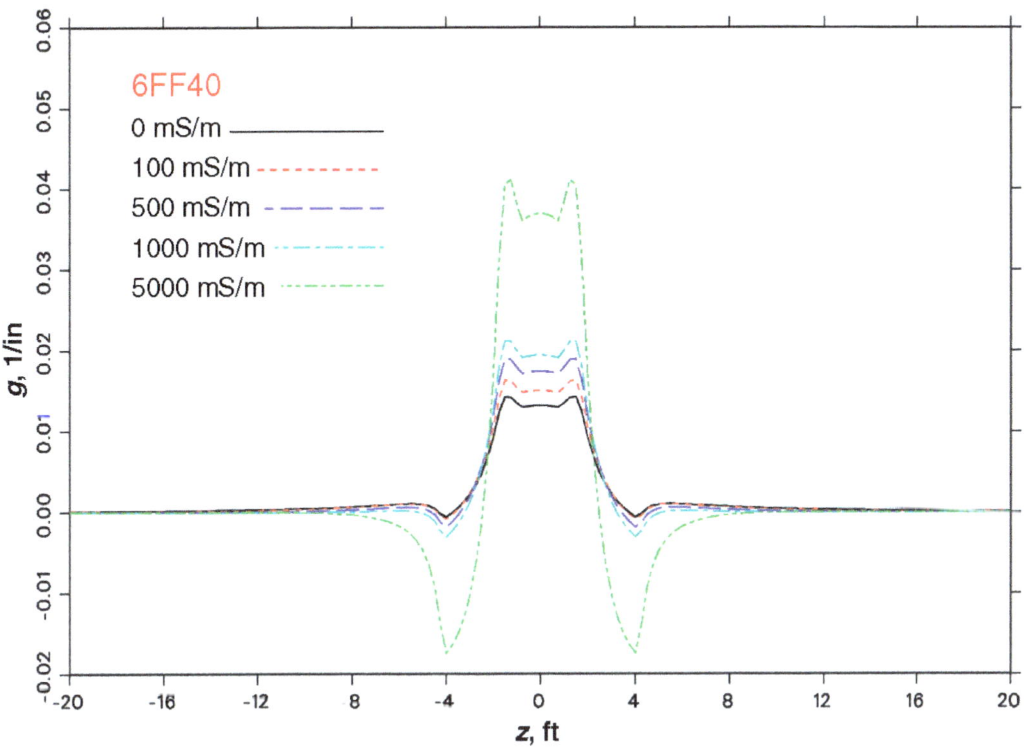

Fig. 3B.15—Vertical response function *G*(*R*) for the 6FF40 array at several formation conductivities.

not practical. The simplest practical array is a three-coil array with a transmitter and two receivers. The second receiver is placed between the transmitter and main receiver, and is wound oppositely so that the voltages in the two receivers exactly cancel when the array is in free space. The response is the sum of the coil pair responses.

One of the most successful induction arrays was the 6FF40 array introduced in 1960.[22] It had three transmitters and three receivers, with a symmetric Born response *g*. **Figs. 3B.15 and 3B.16** show its vertical and radial responses. The array was designed to achieve deep investigation, reasonable vertical resolution, and a low borehole effect. However, the large peaks in the 2D response along the tool resulted in sensitivity to borehole washouts, called cave effect.

3B.4.4 Dual-Induction Tools. One of the challenges of measuring formation resistivity is to sort out the resistivity of the invaded zone from that of the virgin zone. The earliest concept to successfully solve the problem (at least in thick beds with uncomplicated invasion profiles) was the dual-induction tool. This tool combined a 6FF40 array as a deep-induction measurement (ID or ILD) with a set of receivers that worked with the 6FF40 transmitters to produce a shallower measurement. This was referred to as the medium-depth induction (IM or ILm).[23,24]

Because there are three parameters in the simplest step-profile invasion model, at least three measurements are required to solve for these parameters. The shallow measurement was a shallow laterolog (LL8 or SFL) colocated with the induction arrays. The radial response function involves very complicated mathematics, and the solution offered to users of the dual induction logs was a graphical solver called the tornado chart.

The ILD-ILM-SFL logs separate when there is invasion, and this separation is what allows interpretation for invasion parameters. **Fig 3B.17** shows the modeled response of the dual induction-SFL tool (DIT) in a typical Gulf of Mexico pay zone with a transition over a water zone.

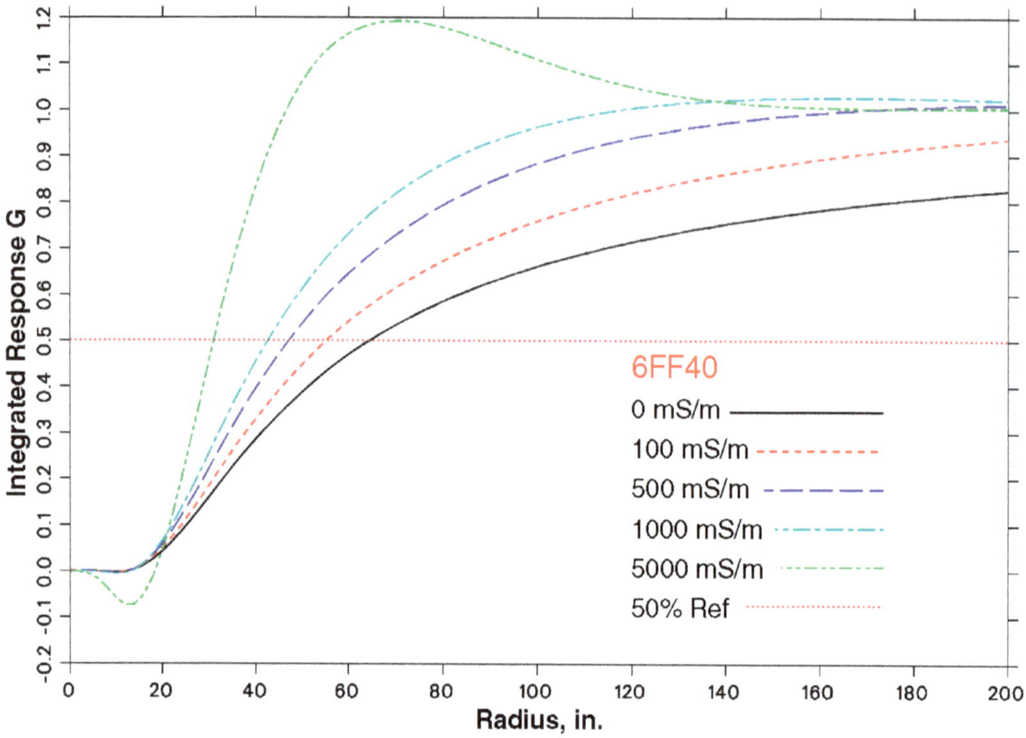

Fig. 3B.16—Integrated radial response function *G*(*R*) for the 6FF40 at several conductivities.

The ILm and ILd measurements do not respond linearly to the formation conductivity. This nonlinearity is closely related to the changes in the response shape and depth of investigation with increasing conductivity. This nonlinear response of an induction array is called skin effect because it is related to the "skin depth" effect of AC current flowing in conductors.

Some sort of function must be applied to the tool voltages to correct for this nonlinearity. The processing applied to the Schlumberger DIT[25] consisted of a skin-effect function ("boost") applied to the measured R-signals from the induction arrays. This was based on computations of the response in an infinite homogeneous medium. The ILd was further processed using a three-station deconvolution filter to slightly sharpen the bed-boundary transition and to correct for shoulder effect over a limited resistivity range (1 to 10 ohm·m). At other formation-resistivity ranges, the response either produced horns or large shoulder effects. **Fig. 3B.18** shows the DIT logs in a set of formation layers with the same shoulder-bed contrasts, but centered on 1, 10, and 100 ohm·m.

Borehole correction was also hand-applied to the induction and SFL logs. The borehole correction chart was derived from measurements made with a DIT in plastic pipes full of salt water. The 6FF40-based dual induction-shallow electrode tool was offered by most service companies.

3B.4.5 Phasor Induction. The DIT tool became the standard resistivity tool and remained virtually unchanged for more than 20 years. However, as its application moved from the original Gulf of Mexico formation contrasts to higher-resistivity formations, the shoulder-effect problem became much worse. Although shoulder-correction charts were provided for high resis-

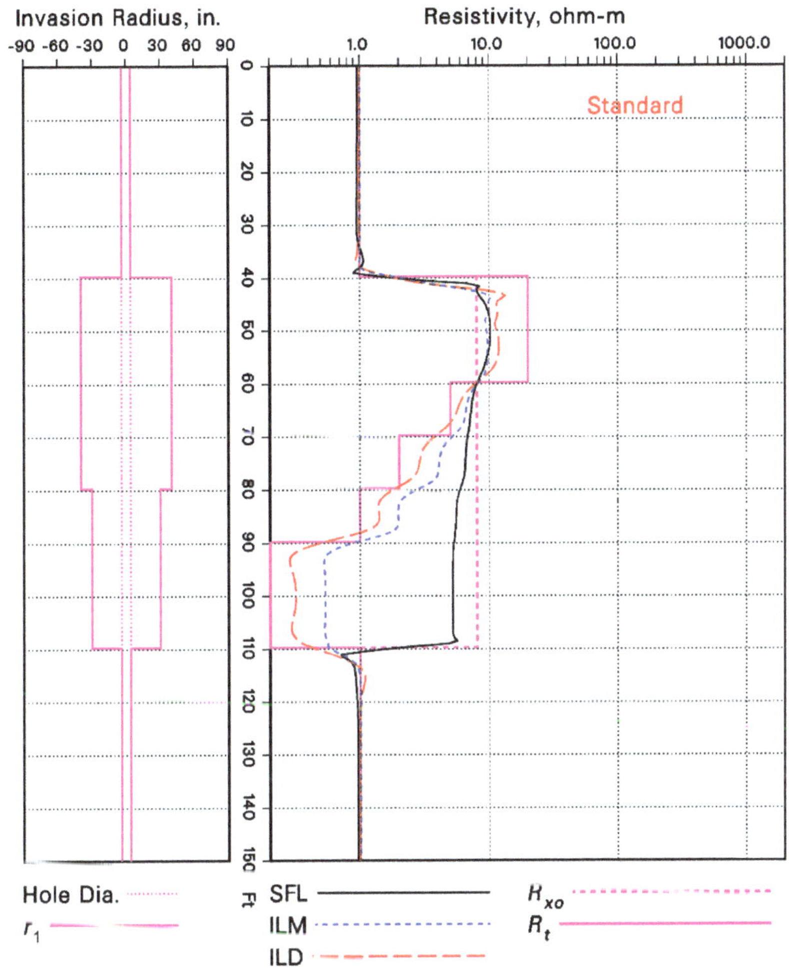

Fig. 3B.17—Modeled DIL logs in a typical Gulf of Mexico formation in a pay zone, a transition zone, and a water zone.

tivity, they mainly indicated that the problem was bad rather than serving as a usable correction mechanism.

The fundamental problem in induction log interpretation is to isolate the response of a thin bed and the virgin zone from the shoulders and the invaded zone after the measurement process has thoroughly mixed them. The Phasor induction tool was introduced in the mid-1980s and was the first tool to automate the environmental corrections. It uses a linear deconvolution function to correct for shoulder effect and uses the X-signal measurement to correct for skin effect. This algorithm was the basis for Phasor Processing.[26] It can be shown that a filter fitted at low conductivity works well at low conductivity but produces large errors at high conductivity. The error is, however, a slowly varying function closely related to the X-signal. An algorithm applied to the X-signal to match it to the skin-effect error allows a single FIR filter to correct for shoulder effect over a wide range of conductivities.

Fig. 3B.19 shows the results of Phasor processing in the formation models of Fig. 3B.18. The induction logs are fully shoulder-effect-corrected at all conductivity levels. Phasor logs in the Gulf of Mexico simulation of Fig. 3B.17 are not very different from the DIT logs. This is

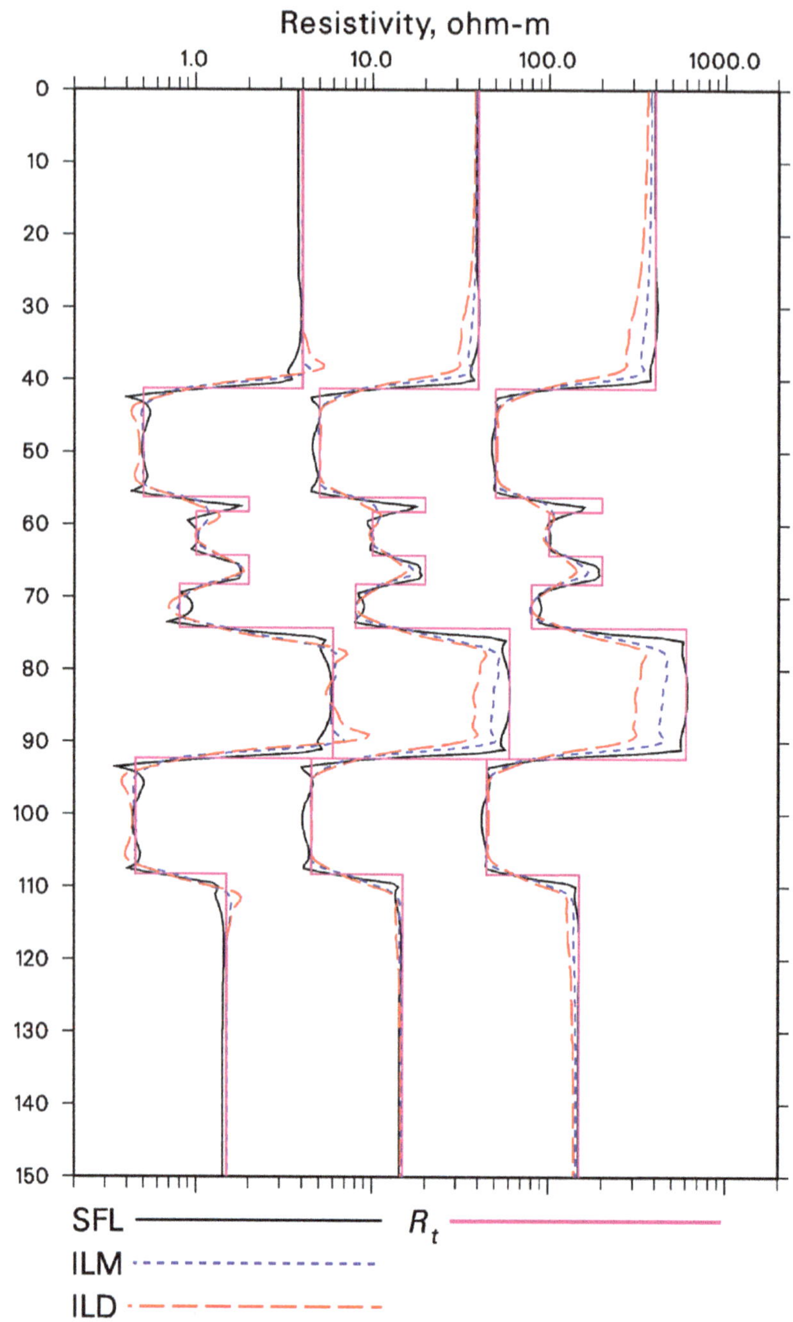

Fig. 3B.18—Modeled DIT logs in a set of formation layers with the same shoulder-bed contrasts, but centered on 1, 10, and 100 ohm-m.

in part because this formation is where the DIT logs were designed to work well. Although tornado charts were published for the Phasor induction logs, the invasion parameters are computed in real time at the wellsite. Borehole corrections are based on computer models of an eccentered tool in a wide range of borehole salinities and formation conductivities.[27] Borehole corrections are applied in real time at the wellsite. The Phasor induction tool was the first in-

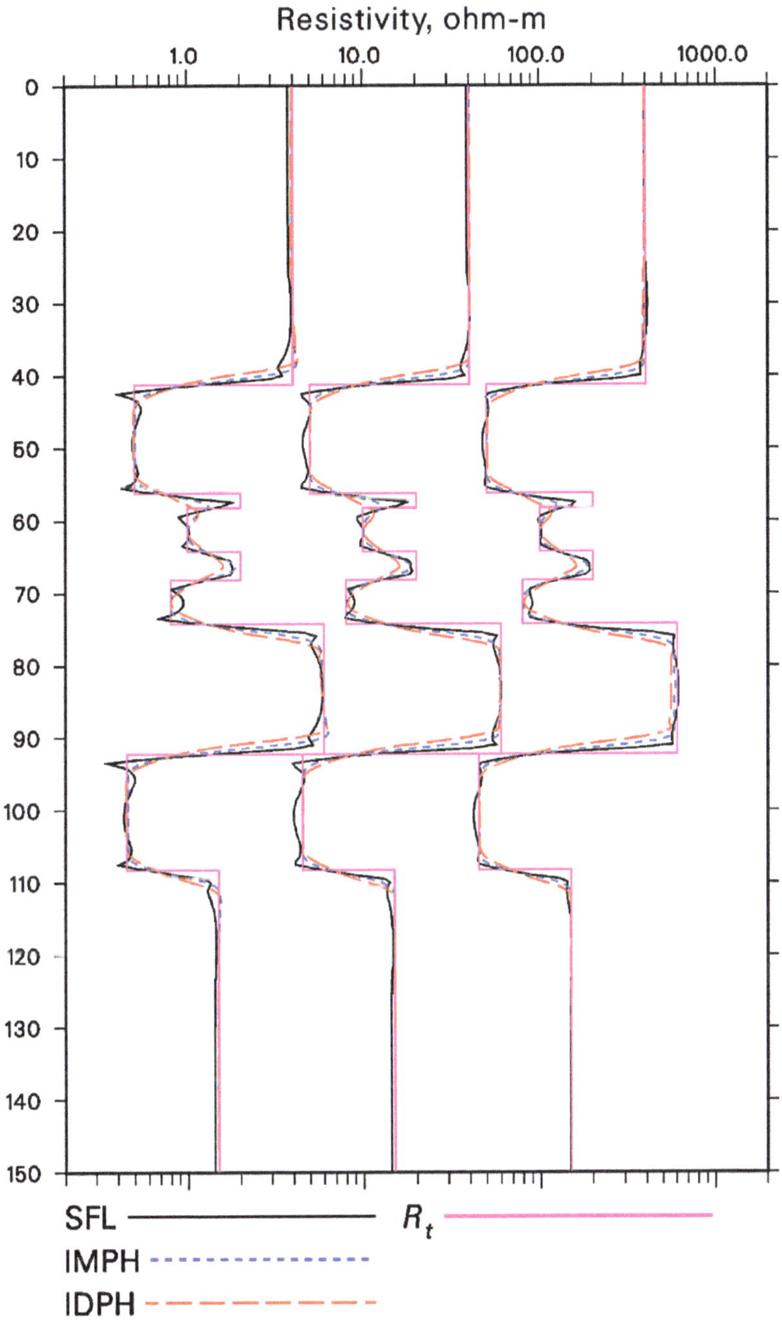

Fig. 3B.19—Phasor logs in the cases of Fig. 3B.18.

duction tool that could provide full environmental correction and invasion parameter determination at the wellsite. In 1987, changes to the deconvolution filters allowed induction logs with a 2-ft vertical resolution (compared with 5 ft for ILm and 8 ft for ILd).

Dual-induction tools that measured both R and X signals and applied automatic shoulder-effect corrections were introduced by Atlas (the Dual-Phase Induction Tool, or DPIL) and Gearhart (the High-Resolution Induction Tool, or HRI tool). The HRI tool also achieved a ver-

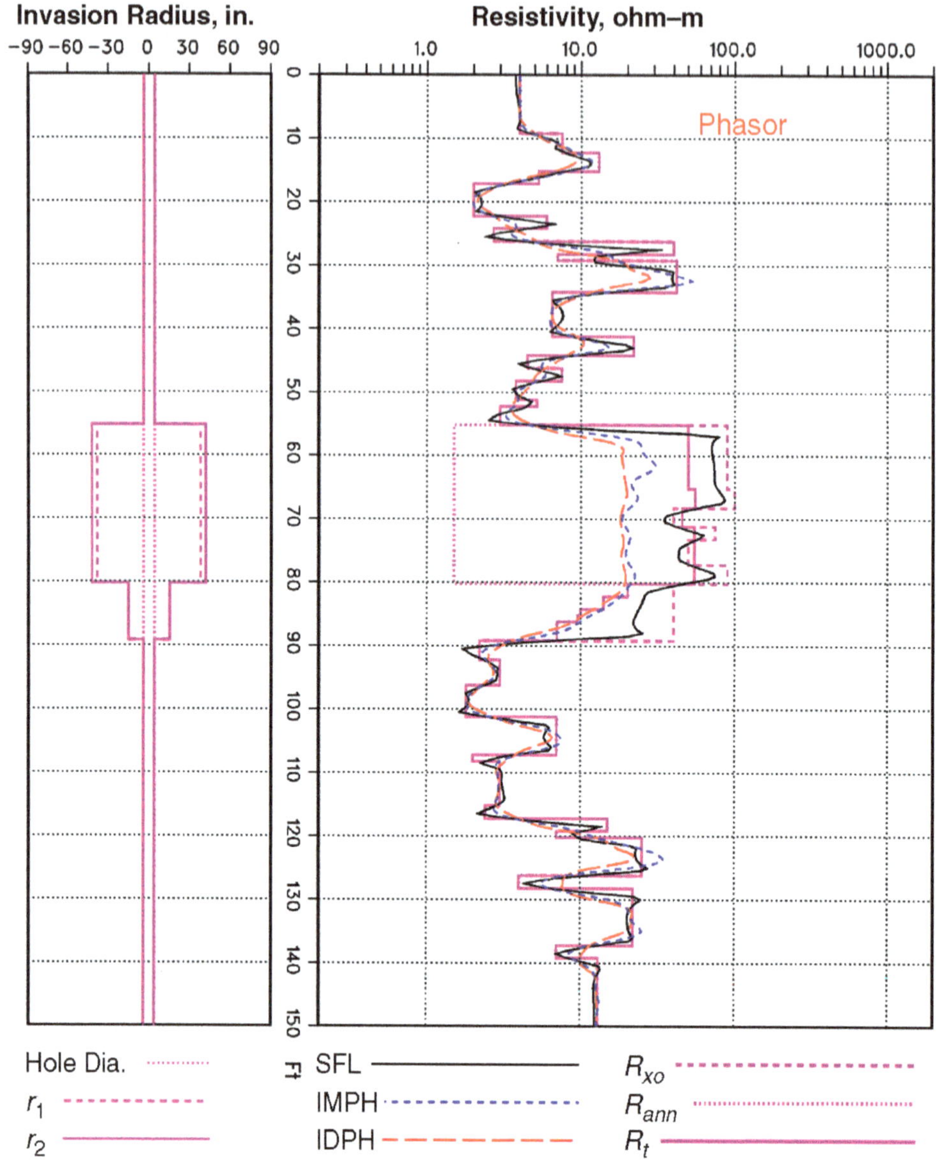

Fig. 3B.20—Phasor logs in a simulation of a gas-bearing formation where an annulus has formed.

tical resolution of 2 ft. It was also the first dual-induction tool to be based on a different array from the 6FF40. Its deep array had a median depth of investigation of approximately 90 in. After the breakup of Gearhart, Halliburton acquired the HRI tool and commercialized it.

However, all of these tools are based on two induction arrays—a shallow array and a deep array. Performance in complex invasion profiles is limited by the small number of measurements. **Fig. 3B.20** shows the Phasor logs in a simulation taken from a field log in a gas reservoir. Here an annulus has developed, and the deep log reads much less than R_t. In this well, the three-parameter invasion model will not return the correct value of R_t.

In the case of oil-based mud (OBM), the SFL is not usable. Separation between the medium and deep logs is only a qualitative indication of invasion and is not quantitatively interpretable.

3B.4.6 Array-Induction Tools. With the Phasor Induction tool, the dual-induction concept had reached its limits. In particular, improvements were needed in better estimates of R_t in the presence of deep-invasion or complex transition zones. As the grosser environmental distortions were corrected by Phasor Processing or similar processing, annulus profiles and other transitions were encountered more often.

These response problems, coupled with an increasing use of OBM, led to the concept of using several induction arrays with different depths of investigation. With the problems of applying linear deconvolution filters solved, then Doll's approach of using a simple array was applicable. The Schlumberger AIT was designed with eight simple three-coil arrays ranging in length from 6 in. to 6 ft.[28]

Array-Induction Principles. The first step in log formation in the AIT family of tools is to correct all raw array signals for borehole effects. This process is based on a forward model of the arrays in a circular borehole, and it includes an exact description of the tool in the model.[29]

The signal measured by an induction sonde eccentered in a borehole can be described mathematically as a function of four parameters. These are the borehole radius r, the mud conductivity σ_m, the formation conductivity σ_f, and the tool position x with respect to the borehole wall (commonly referred to as the "standoff").

The correction algorithm is designed to solve for some of these parameters by minimizing the difference between the modeled and actual logs from the four shortest arrays. The information content of these measurements is not sufficient to solve for all the borehole parameters at the same time. In practice, two of the four parameters can be reliably determined by this method. The other two parameters have to be either measured or fixed. The equivalent homogeneous formation conductivity σ_f must always be solved for because no measurement is closely enough related to it. This leaves one of the other parameters to be determined, and the remaining two parameters must be entered as measurements. This leads to the three borehole correction methods to compute mud resistivity, hole diameter, and standoff. All of the AITs except the original AIT-B have integral mud resistivity sensors,[30] and "compute standoff" is the default borehole-correction method in water-based mud (WBM).

A method was developed[31] to combine these array measurements to focus the resulting log to the desired depth of investigation, while at the same time doing so with a high vertical resolution and minimizing cave effect. The log-formation process is described by the equation

$$\sigma_{\log}(z) = \sum_{n=1}^{N} \sum_{Z=Z_{\min}}^{Z_{\max}} w_n(z')\sigma_a^{(n)}(z - z') \; . \qquad (3B.4)$$

In this equation, $\sigma_{\log}$ is the recorded AIT log, $\sigma_a^{(n)}$ is the measured log from the nth channel, and N is the total number of measure channels. This process produces a log that is different from that produced by any of the individual arrays. It is still characterized by a response function. This response function is a weighted sum of the response functions of each of the individual channels n. Skin effect is handled in a manner similar to the Phasor tool.

The result of this equation is a combination of the logs from the eight array that "distills" the radial information from the eight arrays into five independent logs with depths of investigation of 10, 20, 30, 60, and 90 in. Each of these five logs is available at a resolution of 1, 2, and 4 ft. The radial profile is identical at all resolutions, and the vertical resolution is identical for all radial depths. The set of weights w in Eq. 3B.1 determines which log is produced.

If the mud is very salty, or if the borehole is very large, the signal in the AIT arrays from the borehole will be very large. With salty mud, even normal variations in the borehole surface from the drilling operation can cause "wiggles" on the short array data. Several years of practice have shown that these can affect the final logs, especially the 1-ft logs. This experience has shown that in an 8-in. borehole with 1½-in. standoffs, the 1-ft logs are normally usable at R_t / R_m contrasts up to 100; the 2-ft logs are usable up to a contrast of 450, and the 4-ft logs are usable up to contrasts of 1000. Algorithms based on the real-time use of a chart and "road-noise" analysis of the 6-in. array allow real-time selection of the appropriate resolution based on actual logging conditions.

AIT logs separate in invaded zones and give a good visual indication of invasion, even with OBM. **Fig. 3B.21** shows the AIT logs in the same formation as Fig. 3B.17 (left). In the annulus case, Fig. 3B.21 (right), the AIT logs are "out of order," clearly indicating the nonstep nature of the invasion profile.

With the additional curves, the invasion profile parameters can be solved for using inverse methods. AIT invasion processing has three models that can be selected: step, ramp, and annulus. All are available at the computing center, while the ramp profile is used for real-time logs.

Interpretation of logs in deviated wells or where the apparent dip is high is considerably complicated. First, one has to recognize that the logs are at high apparent dip. **Fig. 3B.22** shows AIT logs in a formation with an apparent dip of 85°. Although this is high, the characteristics that appear here—horns and strange log order—appear in logs with dips as low as 40°. In many fields, faults and slumping of young sediments can produce high apparent dips that are not detectable on seismic profiles. Merlin processing has been developed to produce logs fully corrected for dip effect.[32] Recently, real-time high-angle processing was made available.[33] This processing produces logs that are independent of dip angle. However, the resulting logs are also shallow. **Fig. 3B.23** shows dip-invariant processing (Grimaldi) on the right.

Users of induction logs should be very careful making quantitative analyses in wells that are deviated, or if the formation is dipping. If the shoulder-bed contrast is 20 or less, then the minimum angle where dip correction is needed is approximately 30°. At shoulder-bed contrast of over 100, the logs will need correction at dips as low as 10°.

3B.4.7 Field-Log Examples. A few field-log examples will illustrate the richness of information available in array-induction logs. The first example, **Fig. 3B.24**, is a comparison of AIT and Phasor induction logs in a gas zone from Canada. The AIT shows a nonmonotonic curve order, indicating an annulus profile. If the data from this zone is inverted into an annulus profile using material-balance constraints to determine the thickness of the annulus, then the complete annulus parameters can be recovered (**Fig. 3B.25**).

The next example is from south Texas, and is in a well drilled with oil-based mud. **Fig. 3B.26** shows the AIT logs compared to the Phasor logs in this well. Because of the OBM, only the induction logs from the DIT are available. In the lower zone, the AIT logs show a conductive invasion profile, suggesting that the OBM has broken down as it has invaded into the formation.[34] The invasion profile is much clearer with the AIT logs.

The final example is from a deviated well in Canada in a 4¾-in. borehole drilled with OBM (**Fig. 3B.27**). The well was air-drilled and turned horizontal. This part was at a deviation of approximately 60°—going "round the bend." The 2-ft field logs in the left track have a scrambled curve order—shallow logs are higher than deep in the resistive beds and then reversed in the conductive beds. This is a signature of high-dip response. The dip-invariant logs in the center track do not exhibit the curve scrambling. The Merlin logs also show no scrambling and high resolution as well.

3B.4.8 Other Array Induction Tools. Baker Atlas introduced its High-Definition Induction Log (HDIL) array induction tool in 1996.[35] It is a seven-array tool that operates at eight frequencies. This information can be processed in a variety of ways, depending on the environment. The multiple frequencies are used for skin-effect correction. This algorithm is developed by computing the R signal measured by a given array at each frequency at a wide range of

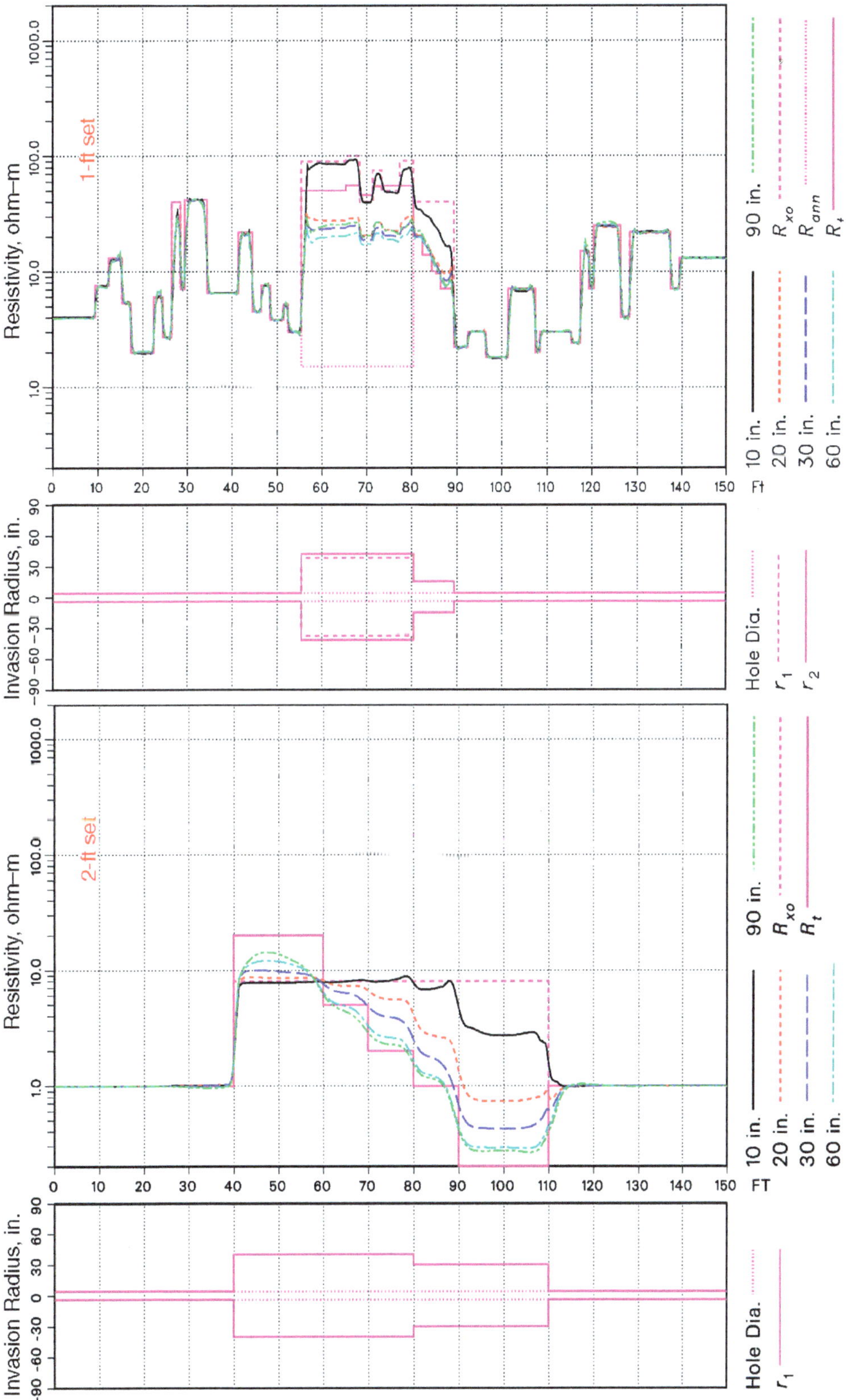

Fig. 3B.21—Phasor logs in the cases of Fig. 3B.18.

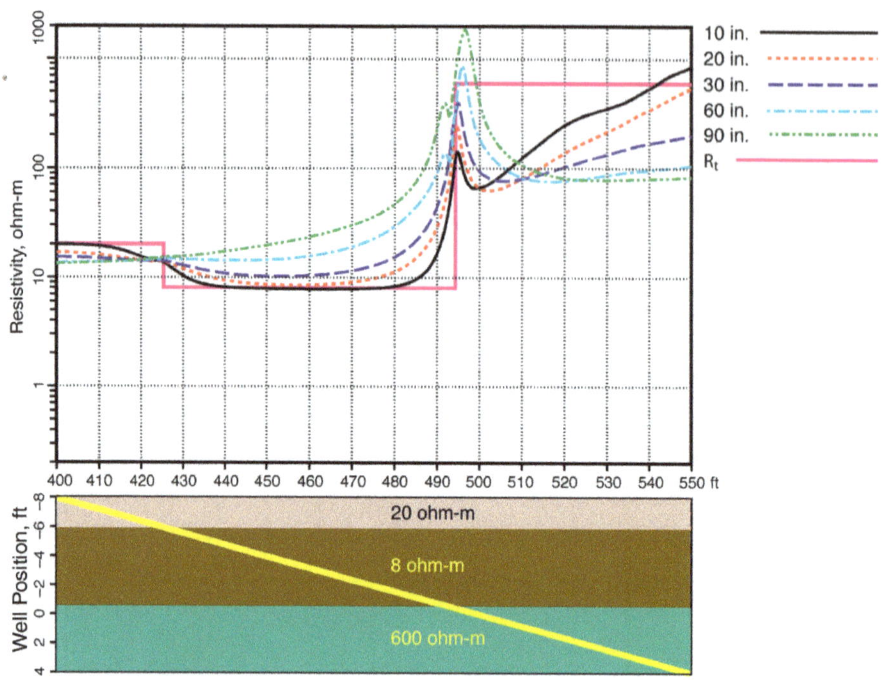

Fig. 3B.22—Modeled AIT logs in a formation with a relative dip of 85°.

formation conductivities. The data at each frequency are fitted to the true formation conductivity. The resulting function is used for the skin-effect correction.

The skin-effect-corrected conductivities are then deconvolved with filters to form six logs at depths of 10, 20, 30, 60, 90, and 120 in., and at three matched resolutions of 1, 2, and 4 ft. An additional presentation is the "true resolution" log set. This has the same six depths, but the resolution of each depends on the depth of investigation. This presents the resolution information content that actually comes from the formation region near the midpoint of the integrated radial response function. Dip correction is provided at the computing center. **Fig. 3B.28** shows an example. **Fig. 3B.29** shows a 2D inversion available at the computing center.

Halliburton introduced its High-Resolution Array Induction (HRAI) tool in 2000.[36] It is a six-array tool based on the array layout of the HRI. Standard HRAI tool logs present resistivities at vertical resolutions of 4, 2, and 1 ft, each with six depths of investigation (10, 20, 30, 60, 90, and 120 in.). The log resistivities are inverted to yield true resistivities of the formation in the virgin zone, R_t, and in the invaded zone, R_{xo}, near the borehole. Invasion diameters (D_i) corresponding with R_t and R_{xo} are also presented. HRAI tool answer products are available in real time while logging.

A variant on the array induction principle was introduced by Weatherlord (previously Reeves Wireline).[37] The array induction donde (AIS) combines four simple induction arrays with a shallow focused-electrode array. The induction data have been presented in two ways. Originally, the four arrays were combined in software to match the response of the ILd and ILm arrays. Later, the Vectar processing was introduced to produce a higher-resolution log.[38] Data from each array is skin-effect-corrected and then resolution-matched to the shortest array. Up to six curves are presented from the four arrays.

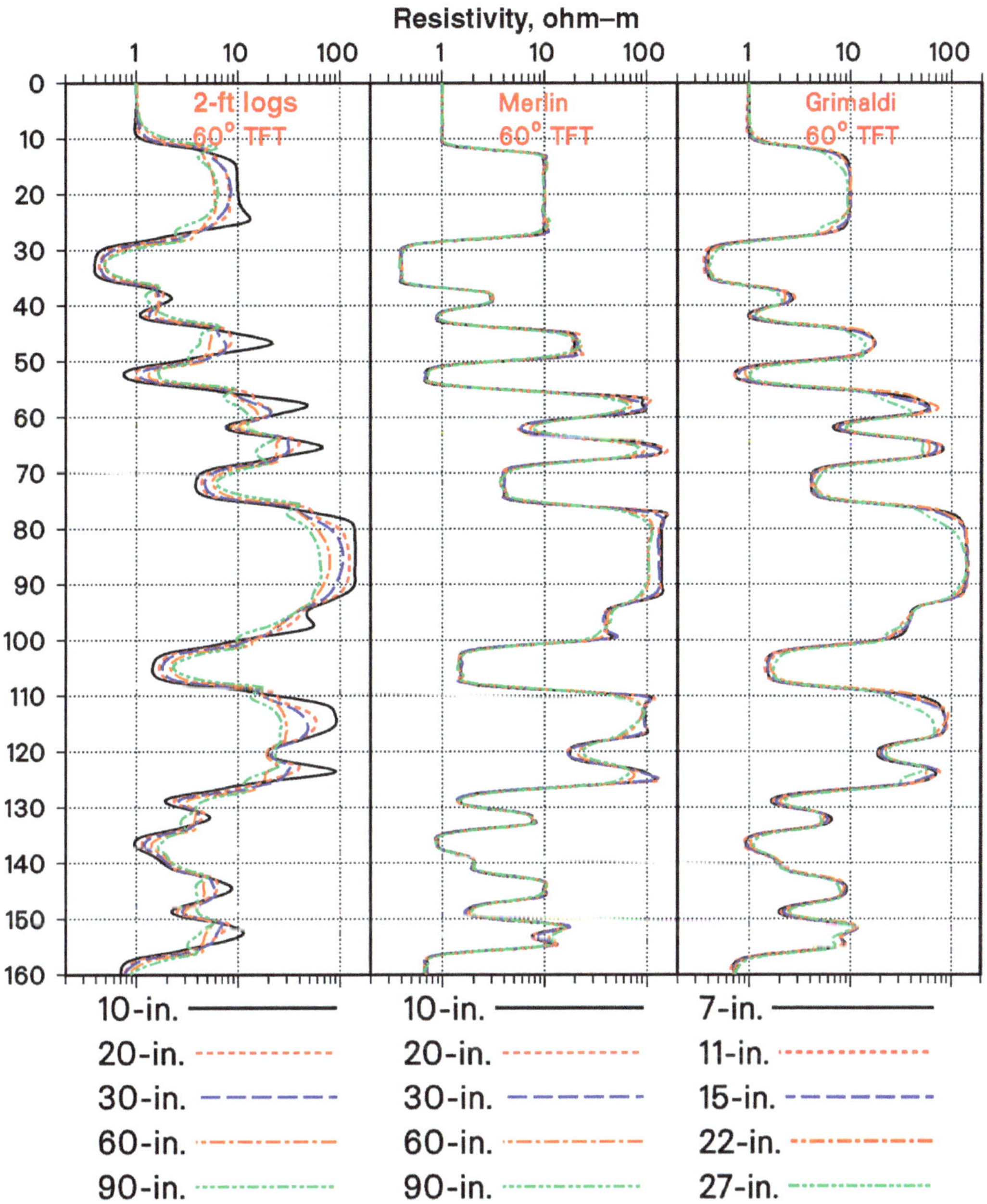

Fig. 3B.23—AIT dip-invariant logs.

3B.5 LWD Induction Tools

3B.5.1 Induction Tools on the Drillpipe. Commercial resistivity measurements made while drilling first became available in the late 1970s. Because the drilling environment is much more adverse than the wireline logging environment, a simple short normal tool mounted behind the drill bit was used as the first LWD resistivity tool.[39] However, short normal tools were only able to provide information for basic interpretation, such as correlation of geological markers and estimation of gross water saturation,[40] because of their shallow depth of investiga-

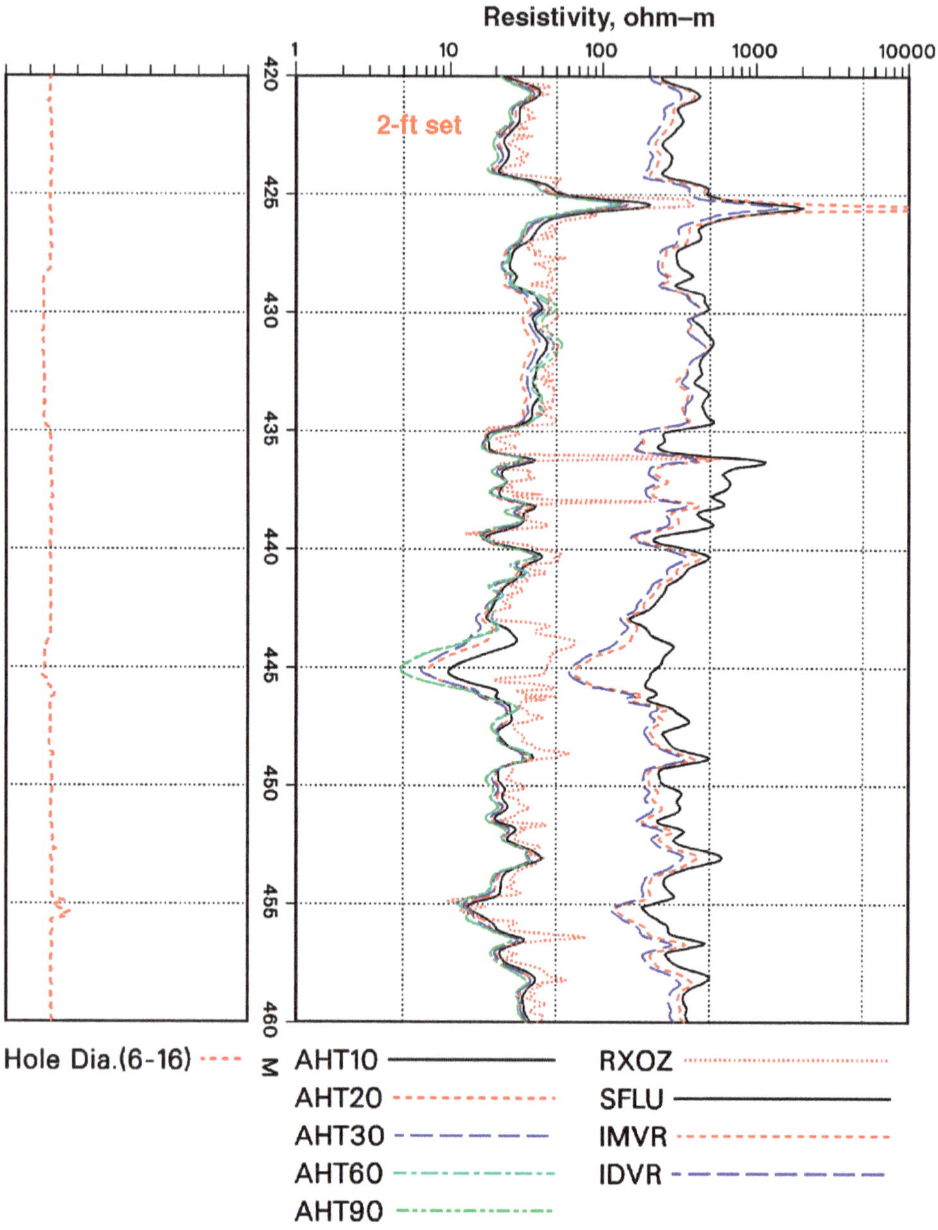

Fig. 3B.24—Comparison of AIT and Phasor logs in a gas zone in Canada.

tion and relatively poor vertical resolution. Being DC electrode devices, normal tools are also limited to conductive mud environments.

To expand the LWD resistivity market to OBM environments, induction-type propagation measurements were introduced in the early 1980s. The first commercial device was the electromagnetic wave resistivity (EWR)[41,42] tool from NL Information Services (later merged with Sperry-Sun). Shortly after this, Schlumberger introduced the compensated dual resistivity (CDR)[43,44] tool. All LWD propagation tools are run with a gamma ray tool for lithology estima-

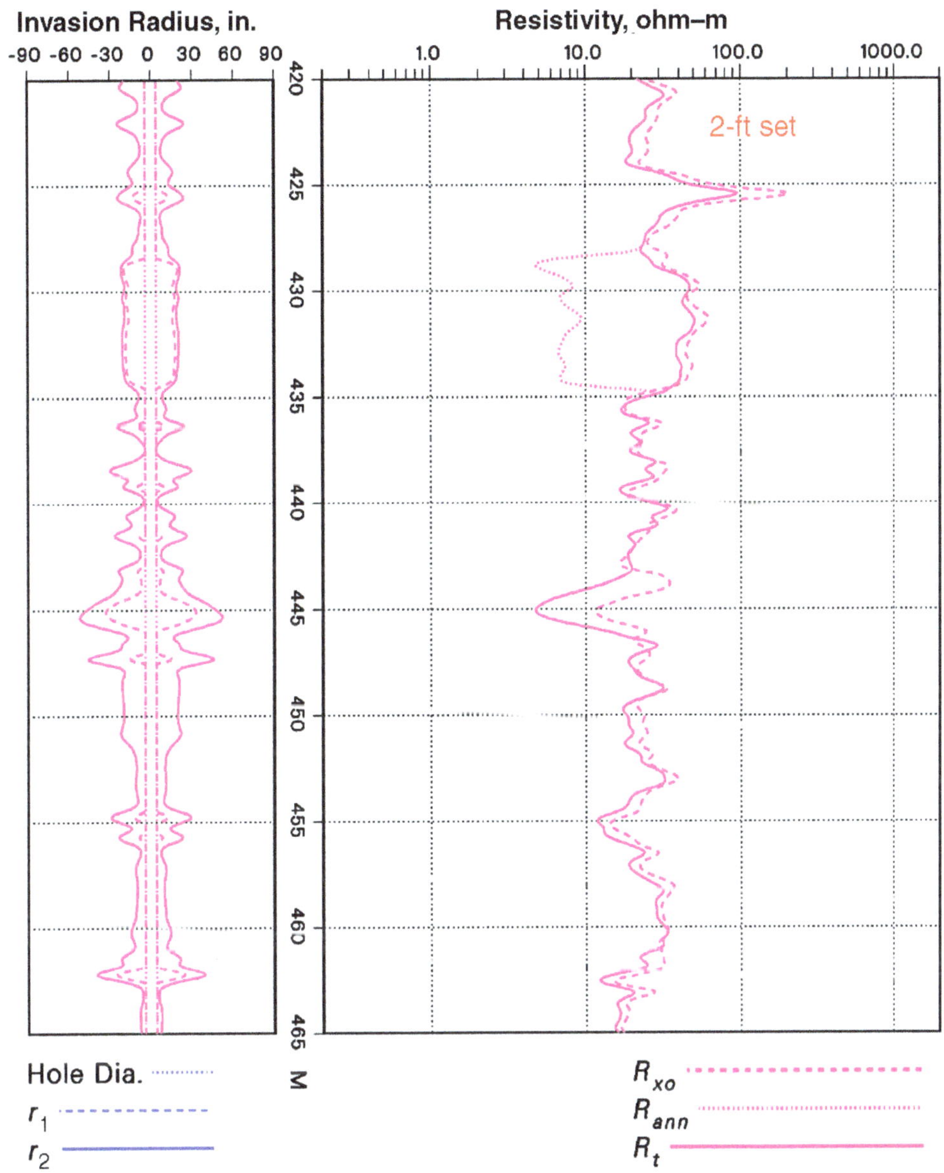

Fig. 3B.25—Radial inversion of the AIT logs in Fig. 3B.24 for invasion parameters using the annulus model.

tion and correlation. Log data are transmitted uphole in real time using mud-pulse telemetry. Downhole memory and batteries allow raw and processed data to be stored for later retrieval.

3B.5.2 Principles of the Propagation Measurement. Because the CDR is a simple tool that measures both attenuation and phase shift, it is used to demonstrate the basic concepts of propagation measurements. Conventional induction measurements are made with mutually balanced arrays of transmitter and receiver coils operating in the kilohertz frequency range and that phase-lock measurement electronics. Because it was difficult to engineer this type of measurement on a steel drill collar using the technology of the early 1980s, a higher-frequency propagation measurement was considered to be more practical for LWD. A frequency of 2 MHz was chosen

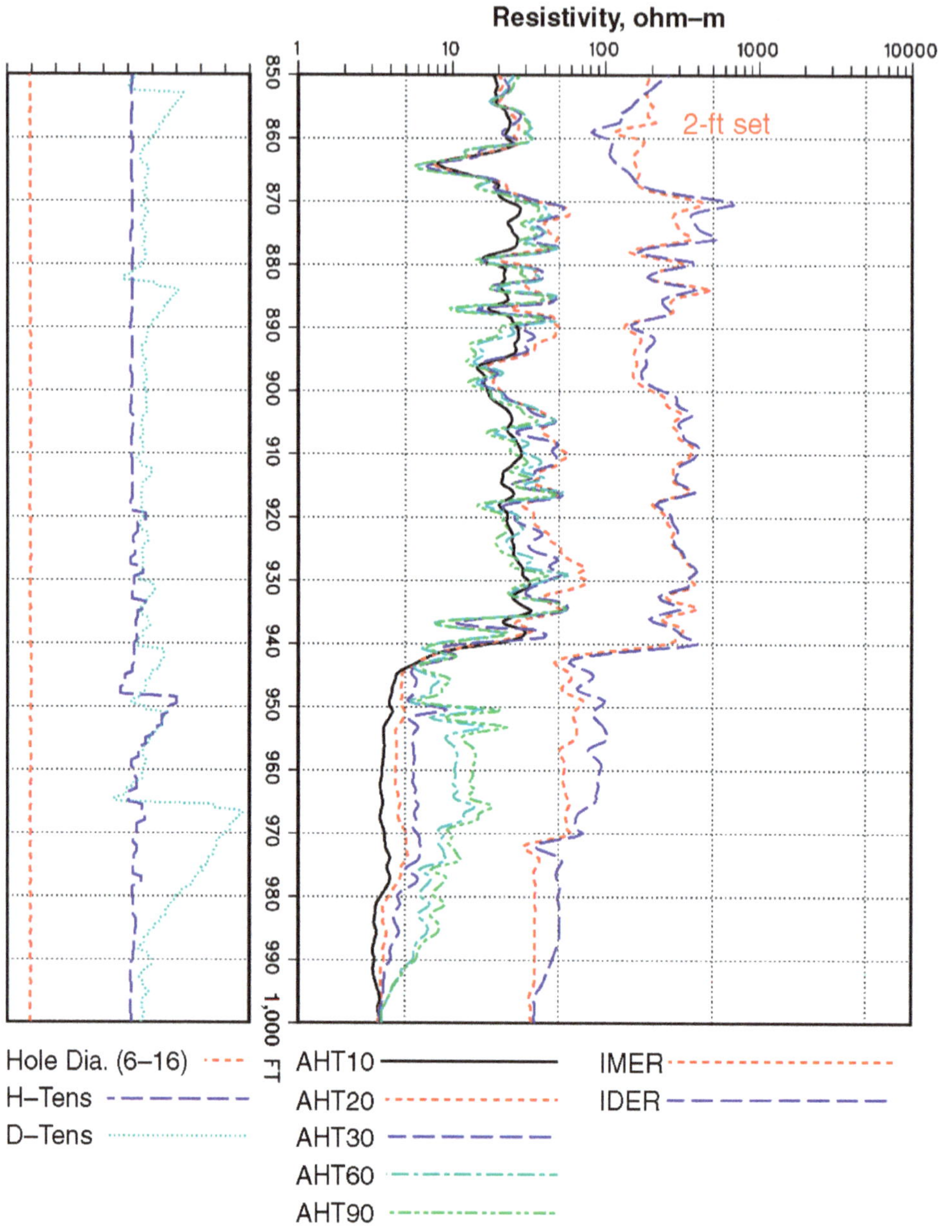

Fig. 3B.26—AIT and Phasor logs in a south Texas well drilled with OBM. Note the conductive invasion profile despite the OBM.

because it was the lowest frequency at which accurate propagation measurements could be made on a drill collar at that time.

The CDR tool broadcasts a 2-MHz electromagnetic wave. A propagation measurement is made by taking the difference between the phases (phase shift) and amplitudes (attenuation) of the voltages recorded at the two receivers. Attenuation increases as a function of increasing conductivity, while the wavelength decreases as conductivity increases. Thus, the two measurements are proportional to formation conductivity and can be used to generate resistivity logs.

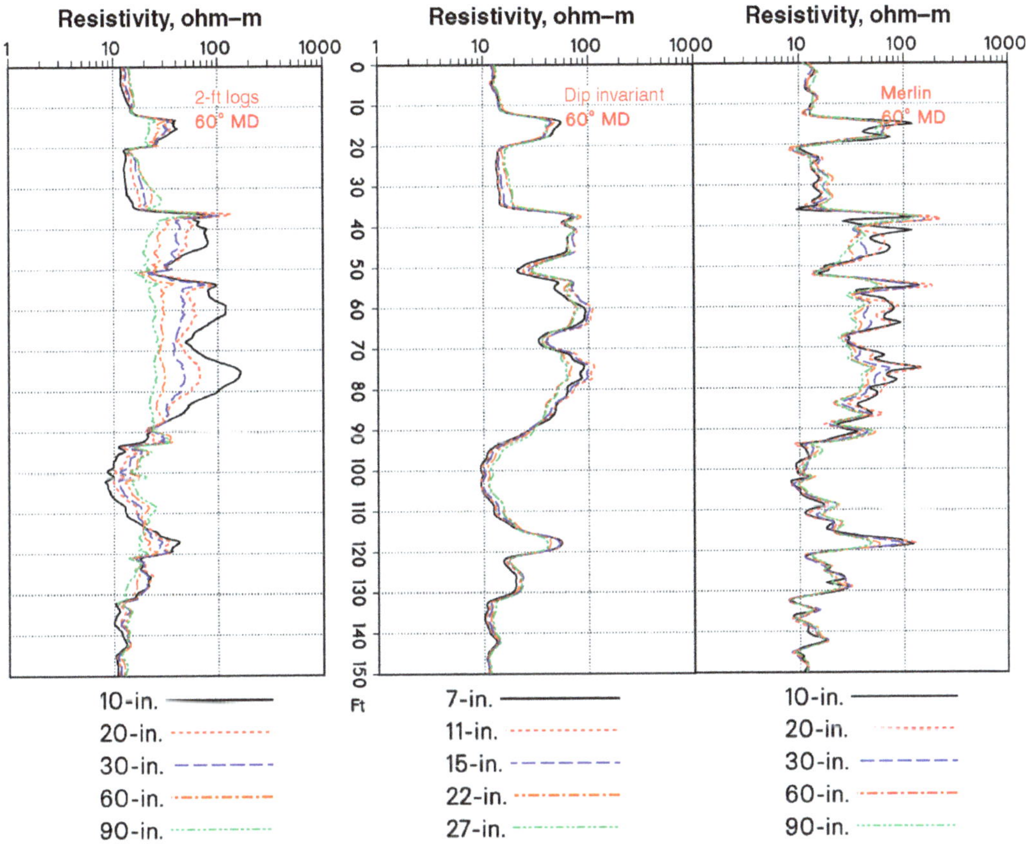

Fig. 3B.27—Slim-access AIT logs from a well in Canada. Relative dip is approximately 60°. Field processing, dip-invariant processing, and Merlin logs are shown.

Note that the CDR tool has two transmitters. The phase-shift and attenuation measurements generated by the upper transmitter between the two receivers, and by the lower transmitter between the two receivers, are averaged to symmetrize the tool response. This averaging is known as borehole compensation because it also reduces the effect of borehole rugosity. The transmitter-to-receiver spacings for the CDR tool are 25 and 31 in.

The phase-shift and attenuation measurements are transformed to two independent resistivities, which for the CDR tool are known as RPS (phase shift, shallow) and RAD (attenuation, deep). Because at 2 MHz dielectric effect can be significant at high resistivity levels, a dielectric correction is performed before the raw data are converted to apparent resistivity. Most service companies have developed their own proprietary algorithms to perform dielectric correction. Joint inversion for both resistivity and dielectric constant is also possible with today's multiarray propagation tools. The dielectric-corrected phase shift and attenuation are converted to resistivity using a table look-up algorithm based on computed tool response in homogeneous isotropic media of known resistivity, R_t.

The phase-shift and attenuation measurements are both relatively insensitive to borehole size and mud resistivity. Borehole correction is only necessary in conductive holes with large washouts when the R_t/R_m contrast is greater than 100 to 1.

Invasion is usually quite shallow at the time of drilling when most LWD logs are run. However, LWD logs may also be recorded each time that the drillstring is pulled to replace the

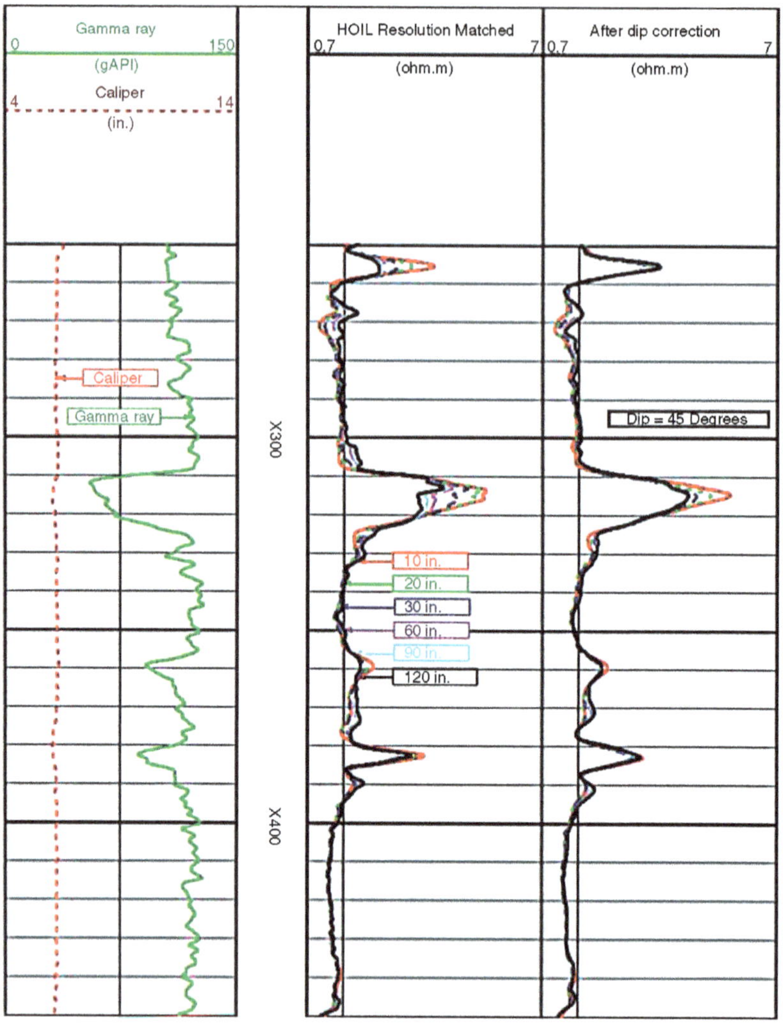

Fig. 3B.28—HDIL logs in a formation with 45° relative dip before and after dip correction.

drill bit. At these later times, invasion can be much deeper. The two resistivity values, RPS and RAD, provide two independent depths of investigation as an indication of invasion.

The reason that two depths of investigation can be obtained from a single measurement is made clearer by examining the behavior of the electromagnetic field. The surfaces of constant phase are spheres because the wave travels with the same speed in all directions. The surfaces of constant amplitude are toroids because the wave is stronger in the radial direction than in the vertical direction, which is a normal characteristic of vertical magnetic dipole antennas. If we compare the phase and amplitude contour lines passing through the receivers, the amplitude extends to a significantly deeper region of the formation than the phase.

Depth of investigation can also be studied by modeling tool response in invaded formations. **Fig. 3B.30** shows CDR radial response for a case with $R_{xo} > R_t$, and **Fig. 3B.31** shows the radial response for $R_{xo} < R_t$. In both figures, RPS and RAD are plotted as a function of increasing invasion radius. In Fig. 3B.30, RPS reads consistently closer to R_{xo}, indicating that RPS is the shallower of the two measurements. In Fig. 3B.31, RPS is again consistently shallower than RAD. In this case, the RPS curve extends below the value of R_{xo} between radii of

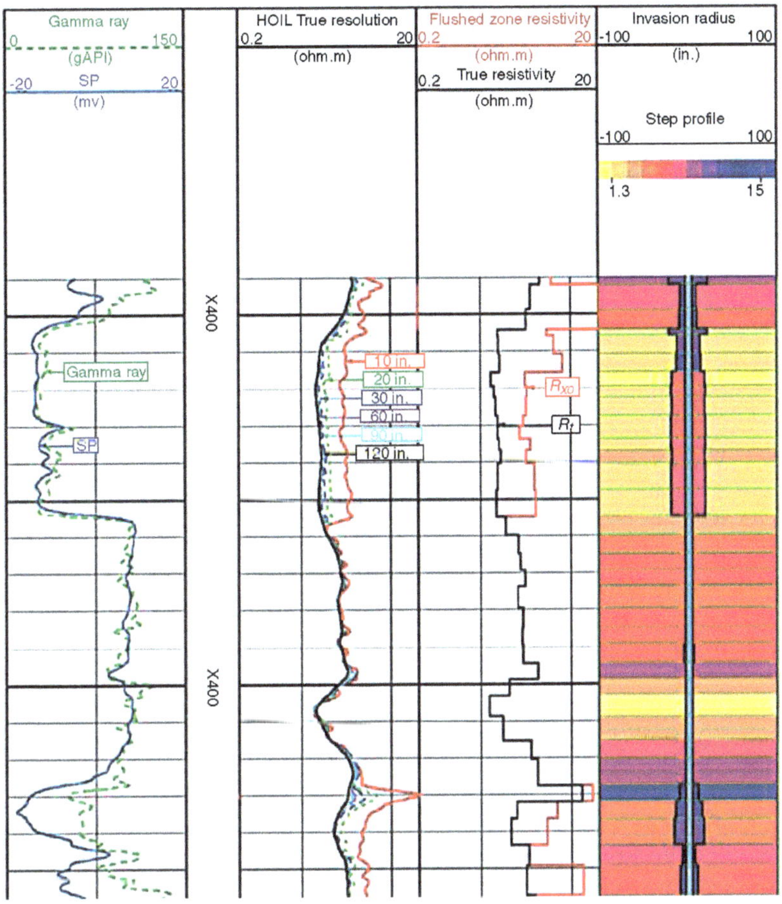

Fig. 3B.29—HDIL logs with 2D inversion for invasion parameters.

30 and 50 in. because of wave reflection at the invasion front. In general, the depth of investigation of RPS is 10 to 20 in. shallower than that of RAD.

The depth of investigation and vertical resolution of 2-MHz tools is influenced to a significant degree by skin effect. 2D tool response can be characterized using Born response functions,[45] which are geometrical factors that take skin effect into account. They show the amount of tool signal coming from a specific volume of the formation. The depths of investigation of both the phase-shift and attenuation measurements are shallower at lower formation-resistivity levels (higher conductivities) because the electromagnetic signal is more attenuated because of skin effect. In general, 2-MHz tools have a shallower depth of investigation and higher vertical resolution in low-resistivity formations, and a deeper depth of investigation and lower vertical resolution in high-resistivity formations.

Vertical resolution is characterized in more detail in **Fig. 3B.32**, which compares CDR and Phasor induction logs. In this low-resistivity formation, the vertical resolution of CDR log is slightly sharper than that of the Phasor log. (The SFL log indicates that shallow invasion has taken place at wireline time.) However, the vertical resolution of 2-MHz logs deteriorates as the formation resistivity level increases. Because conventional induction logs always undergo vertical processing, while 2-MHz logs are seldom processed because their depth sampling is irregular, care must be taken when comparing wireline induction and 2-Mhz logs in resistive

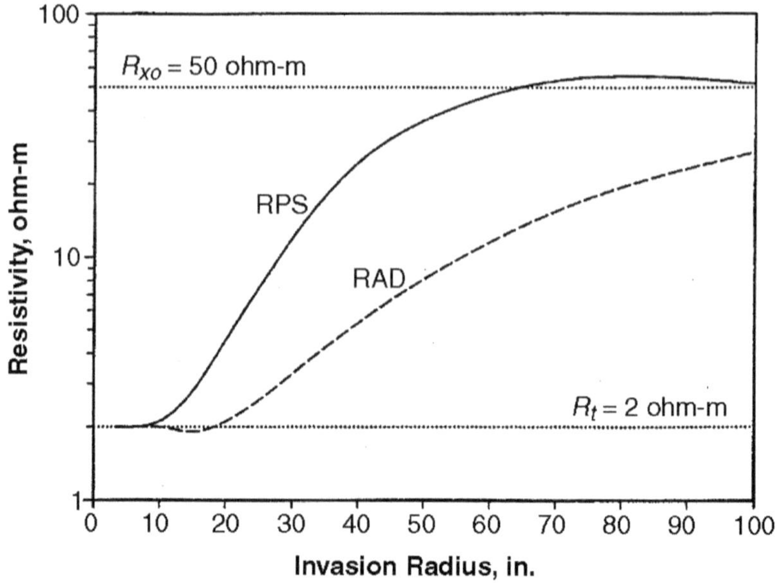

Fig. 3B.30—CDR radial response for $R_{xo} < R_t$.

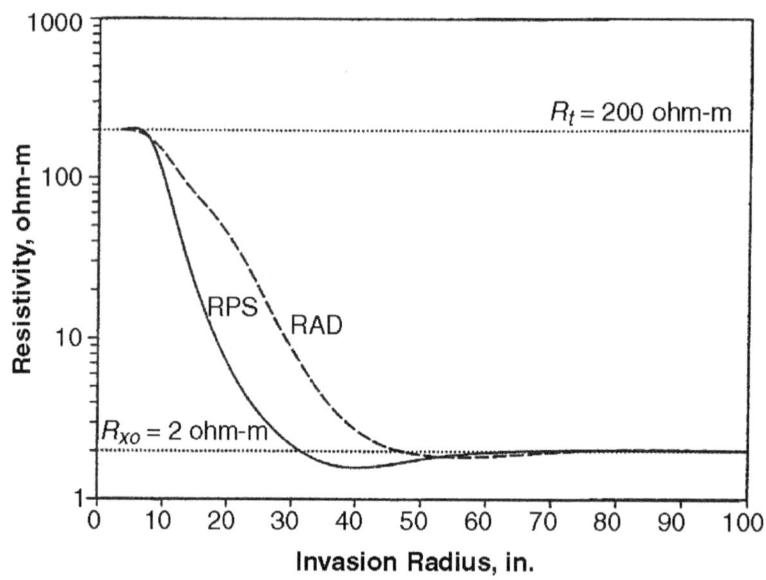

Fig. 3B.31—CDR radial response for $R_{xo} > R_t$.

formations. **Fig. 3B.33** shows the variation in CDR vertical resolution as a function of resistivity level.

3B.5.3 Multiarray Propagation Tools. During the 1990s, all major service companies developed multiarray versions of 2-MHz tools. Schlumberger introduced the array resistivity compensated tool (ARC5)[46] as a replacement for the CDR tool and to accommodate the increasing

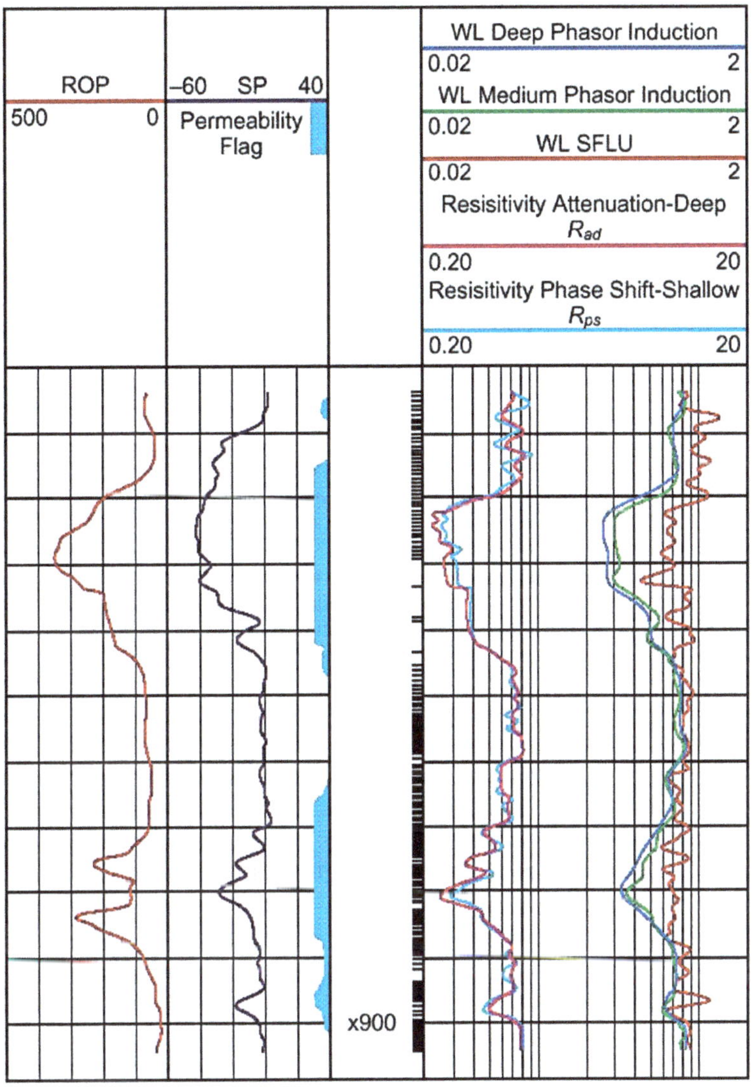

Fig. 3B.32—Comparison of CDR and Phasor induction wireline logs in a Gulf of Mexico well at low resistivity. The "permeability flag" is a function of separation between the phase and attenuation resistivities.

number of slimholes being drilled. The ARC5 tool makes five independent phase-shift and attenuation measurements. The number of measurements was deliberately chosen to be the same as that of wireline array-induction tools to allow the sharing of interpretation methods for analyzing complex invasion profiles and estimating R_t.

The ARC5 antenna configuration has five transmitters and two receivers. The phase shift and attenuation of the signal broadcast by each transmitter is measured between the two receivers for a total of five raw phase shifts and five raw attenuations. Because the transmitters are not arranged symmetrically above and below the receivers, conventional borehole compensation cannot be performed. Instead, the ARC5 relies on linear combinations of three transmitters to provide "mixed borehole compensation."[46] This process results in five calibrated phase-shift and attenuation-resistivity logs which are characterized by the antenna spacings: 10, 16, 22, 28, and 34 in. The 28-in. spacing yields a log identical to a CDR tool.

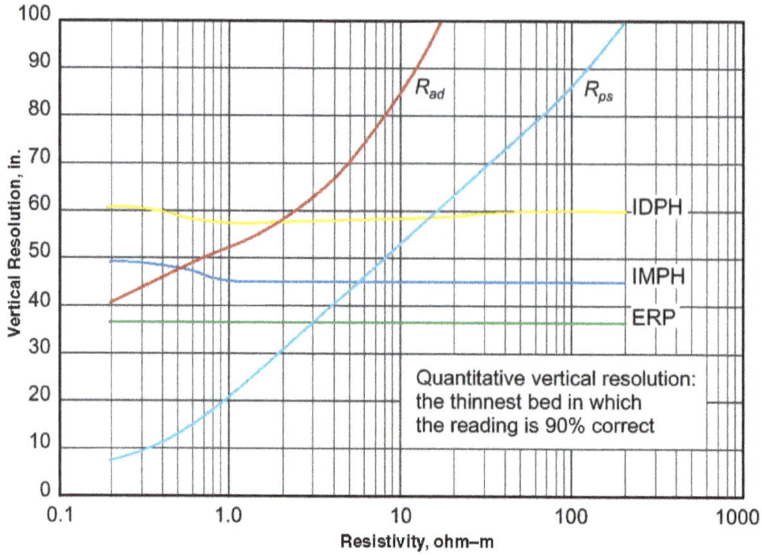

Fig. 3B.33—Vertical resolution of CDR measurements as a function of formation resistivity compared with Phasor and enhanced Phasor induction. The highly nonlinear response of propagation measurements makes resolution-matching and shoulder-effect correction difficult.

Vertical resolution is related to the 6-in. receiver spacing that is common to all the measurements, but the depth of investigation increases as the transmitter spacing increases. The result is five phase-shift resistivity logs with different depths of investigation, and five deeper attenuation-resistivity logs, also with different depths of investigation. The phase-shift logs are matched in vertical resolution in conductive beds, but not in resistive beds. Similarly, the attenuation logs are better matched in conductive beds than resistivity beds. This difference in apparent vertical resolution is shown in Fig. 3B.32.

The ARC5 makes a set of measurements at 2 MHz. Larger versions of the tool (6¾ and 8¼ in.) make measurements at both 2 MHz and 400 kHz. The 400-kHz measurement provides higher signal level in conductive formations (less than 1 ohm·m) and is less sensitive to borehole conditions, particularly in formations where $R_t < 1$ ohm·m and with OBM. It also has a deeper depth of investigation than the 2-MHz measurement in conductive formations. However, 400-kHz measurements have less sensitivity to R_t in resistive formations than 2-MHz measurements.

Fig. 3B.34 compares ARC5 phase shift and wireline AIT logs. The vertical resolution of both sets of logs is similar, with the deeper AIT 90-in. curve giving a slightly higher value for R_t in the resistive beds. **Fig. 3B.35** shows a comparison with the ARC and the ARI dual laterolog in very nonconductive mud. In the bottom track, the ARC logs are inverted for invasion parameters.

In addition to the tools described here, the following LWD resistivity services are also available. References describing early versions of current tools are listed because old logs are sometimes used to interpret existing reservoirs. In 1991, Sperry-Sun introduced a version of the EWR tool with three phase-shift measurements and three attenuation measurements.[47] In 1993, it introduced the EWR-Phase 4[48] tool with four phase-shift measurements and four attenuation measurements, giving a total of eight apparent-resistivity curves. The three shortest measurements are made at 2 MHz, while the longest measurement is made at 1 MHz. Both versions of the EWR are run without borehole compensation.

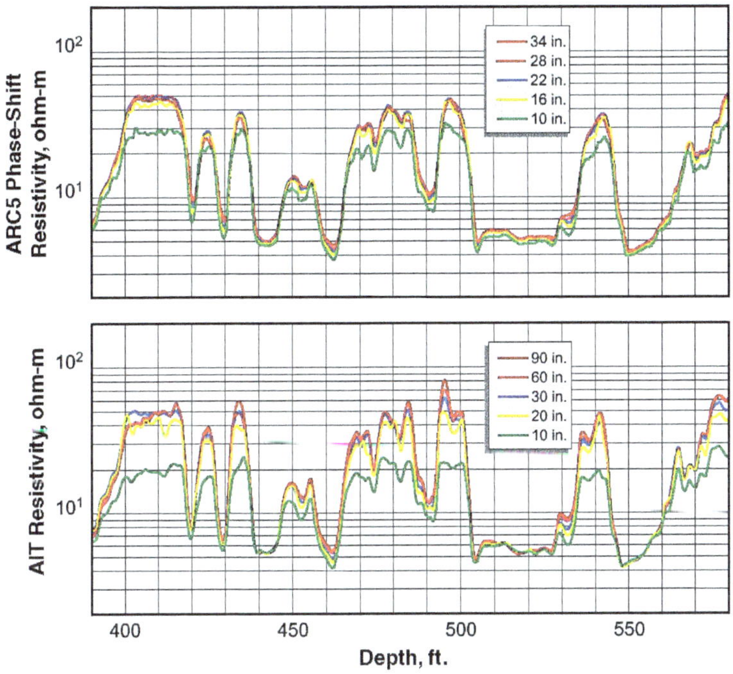

Fig. 3B.34—Comparison of ARC5 phase shift and wireline AIT logs at 2-ft vertical resolution.

In 1989, Teleco introduced the 2-MHz Dual Propagation Resistivity (DPR)[49] tool. This tool measured the phase shift and attenuation at receivers located 27 and 35 in. from a single transmitter (borehole compensation was not used). Teleco was taken over by Baker Hughes, and in 1993 the DPR tool was replaced by the multiple propagation-resistivity (MPR)[50] tool. The MPR configuration consists of upper and lower long- and short-spaced transmitters surrounding a central receiver pair. Antenna spacings range from 23 to 35 in. The two receivers measure the phase shift and attenuation of 2-MHz and 400-kHz signals broadcast by each transmitter. Borehole compensation is performed by averaging the measurements from the symmetrically opposed long and short transmitter pairs. This yields a total of eight logs (long-spaced and short-spaced phase shift and attenuation at 2 MHz and 400 kHz). Resistivity is calculated and displayed either as "apparent" or "borehole-corrected" (hole size and mud-resistivity corrected). Further processing of the logs is available in the MPRteq processing. This processing corrects for a variety of environmental effects. **Fig. 3B.36** compares the processed logs with unprocessed logs.

In 1993, Halliburton introduced the 2-MHz compensated wave-resistivity (CWR)[51] tool. After Halliburton's acquisition of Sperry-Sun, the existing Halliburton business was spun off as PathFinder. The CWR tool makes a set of shallow and deep phase-shift and attenuation measurements with borehole compensation. The transmitter to receiver spacing is approximately 40 in. for the deep mode and 20 in. for the shallow mode.

Geosteering With LWD Propagation Tools. Because 2-MHz resistivity measurements are made behind the bit and can be sent uphole in real time, they are often used to steer the drilling of horizontal wells. Before drilling a horizontal well, potential hydrocarbon-bearing zones are first identified using vertical exploration wells. Then the horizontal well is drilled toward a target bed, with marker beds used to maintain the wellbore trajectory. 2-MHz resistivity logs recorded behind the bit are compared to logs from the exploration wells to identify the

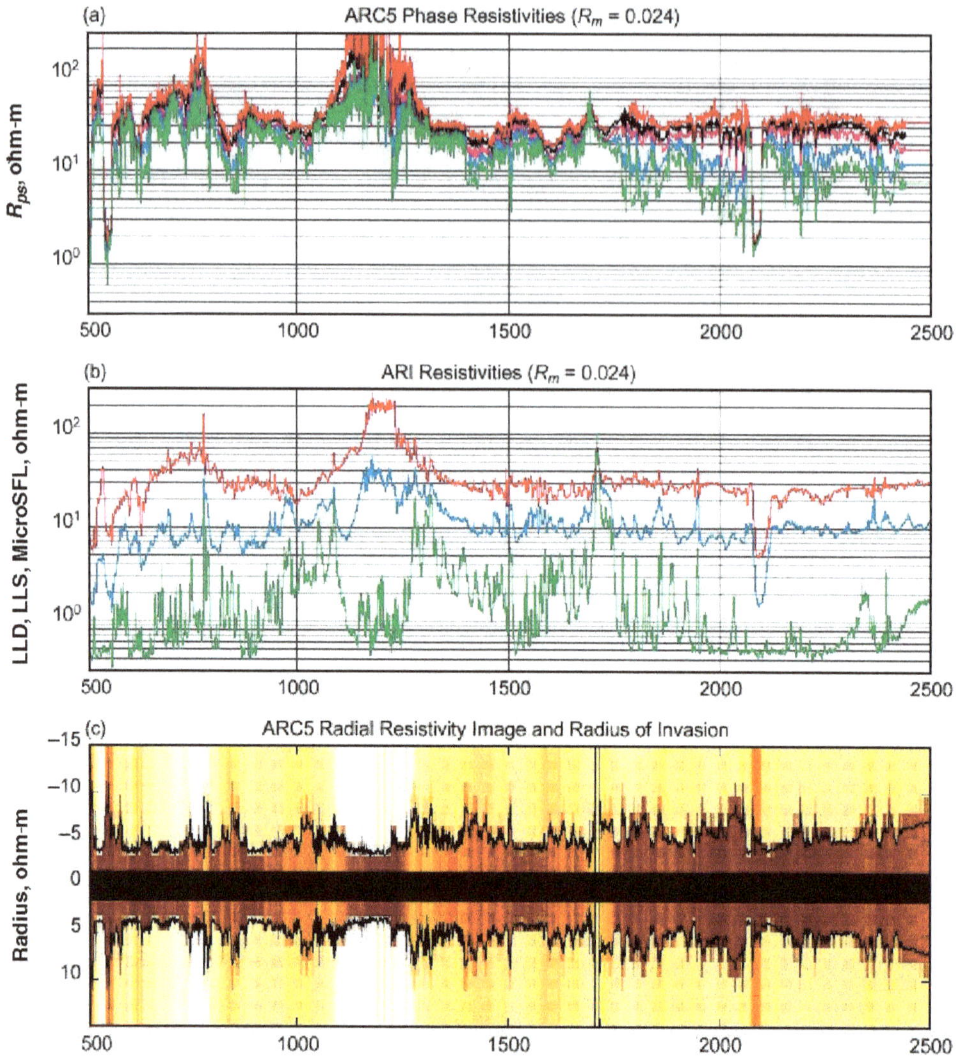

Fig. 3B.35—ARC logs compared with wireline ARI dual laterolog logs. In the bottom track, the ARC logs are inverted for invasion parameters.

marker beds. Computer modeling of predicted resistivity-tool response at different well deviation angles is used to modify the well path. This process is called geosteering.

When comparing resistivity logs in a horizontal well to logs from a vertical exploration well in the same zone, the resistivity value often differs in shales and in laminated formations. This difference is caused by anisotropy (the variation of resistivity with direction). In addition to particle-size anisotropy, formations consisting of a series of thin beds of different lithology (such as sequences of sand and shales) also behave anisotropically if a logging tool is significantly longer than the bed thickness. In vertical wells, resistivity tools (conventional induction, 2-MHz and laterologs) read the effective horizontal resistivity, R_h, which can be calculated from the volume average of the layer conductivities (inverse resistivities),

$$\sigma_h = V_{\text{sand}}\sigma_{\text{sand}} + V_{\text{shale}}\sigma_{\text{shale}}, \quad \text{(3B.5)}$$

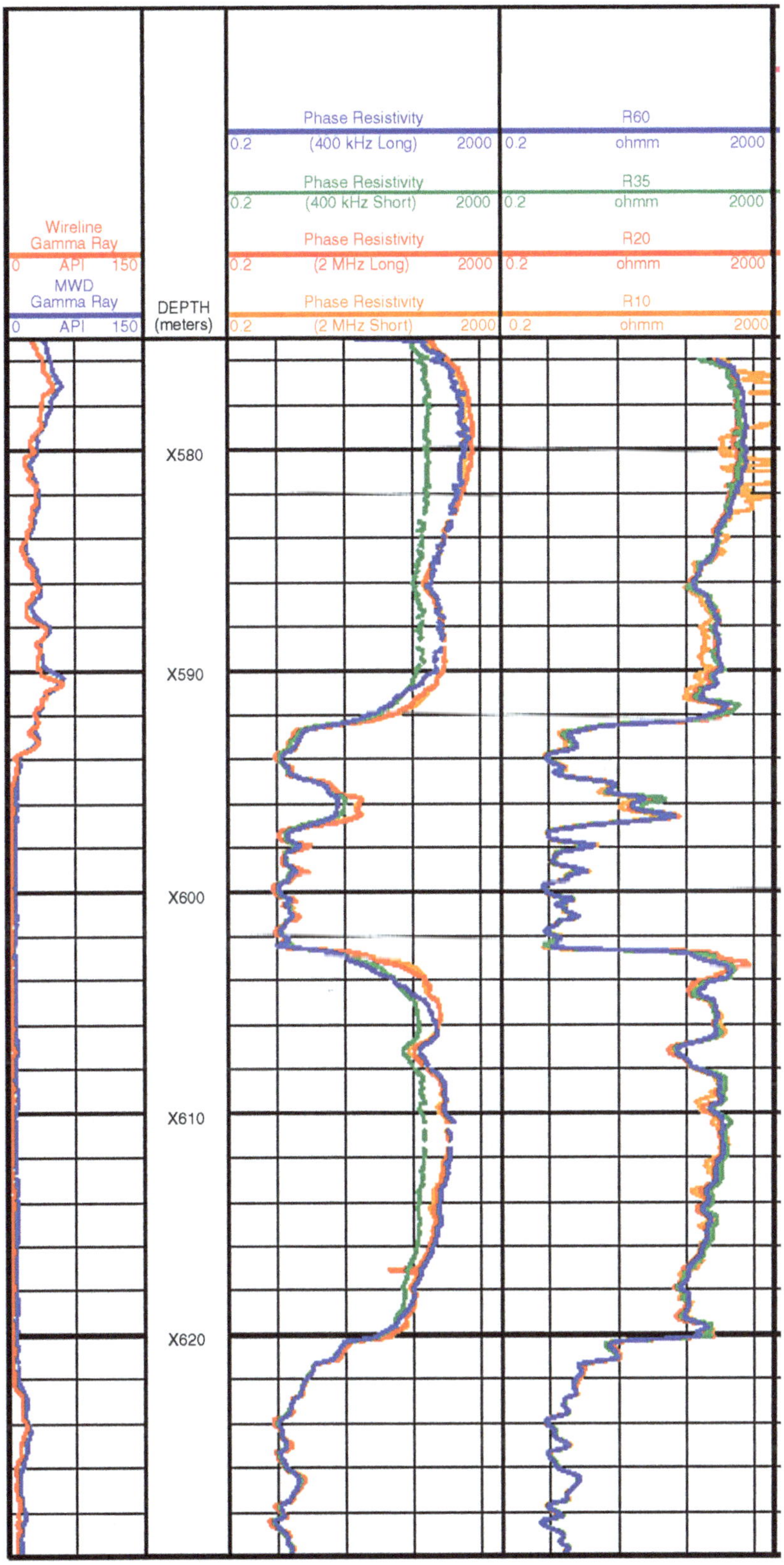

Fig. 3B.36—Comparison of unprocessed MPR logs and MPRteq processed logs (courtesy of Baker Inteq).

where conductivities are expressed in mS/m. V_{sand} and V_{shale} are the bulk volume fractions (percentages) distributed throughout the layered region (layers are all assumed to be approximately uniform in thickness). The effective vertical resistivity, R_v, can be calculated in a similar manner from the volume average of the layer resistivities,

$$R_v = V_{sand}R_{sand} + V_{shale}R_{shale} \text{ . } \quad (3B.6)$$

In deviated wells, the apparent resistivity R_a in anisotropic media can be calculated using the approximation

$$R_a = \frac{R}{\sqrt{\sin^2\alpha + \lambda^2\cos^2\alpha}}, \quad (3B.7)$$

where α is the angle between the tool axis and vertical,

$$\lambda = \sqrt{\frac{R_v}{R_h}}, \quad (3B.8)$$

and

$$R = \sqrt{R_v R_h} \text{ . } \quad (3B.9)$$

For α = 90° (horizontal wells), R_a = R. For α = 0° (vertical wells), R_a = R_h. Thus, the vertical resistivity cannot be detected at all by conventional resistivity logging tools in vertical wells.

Fig. 3B.37 shows a modeled tool response illustrating differences caused by anisotropy between CDR logs in a vertical well (0° dip) and in a highly deviated well (80° dip). At 0° dip, the CDR log reads R_h. At 80°, the two CDR curves increase in the direction of R_v in the anisotropic bed, with the phase-shift resistivity reading higher than the attenuation resistivity.

Induction and 2-MHz tools both generate azimuthally polarized electric fields that induce current loops that are tilted with respect to the bedding anisotropy. These tilted current loops sense a weighted average of R_v and R_h that depends on dip angle. Extensive modeling and analysis of 2-MHz tool response has demonstrated that radiation effects control the phase-shift measurement more strongly than the attenuation measurement. Thus, separation between 2-MHz phase-shift and attenuation logs provides a good indication of anisotropy (in the absence of invasion and shoulder-bed effect). There is sufficient sensitivity to invert for R_h and R_v only at considerable distances from bed boundaries. Also, note the polarization horn that occurs near the bed boundary at 80°. Polarization horns are a common occurrence at high dip angles, and are an indication in geosteering that the well has crossed into a target bed.

3B.6 Induction vs. Laterolog Measurements

Laterolog and induction logging tools each have unique characteristics that favor their use in specific situations and applications.

The induction log is generally recommended for holes drilled with only moderately conductive drilling muds or nonconductive muds (e.g., OBM) and for empty or air-drilled holes. The laterolog is generally recommended for holes drilled with very conductive drilling muds (i.e., salt muds).

Induction tools are conductivity-sensitive devices, which are most accurate in low- to medium-resistivity formations. Laterolog tools are resistivity devices, which are most accurate in

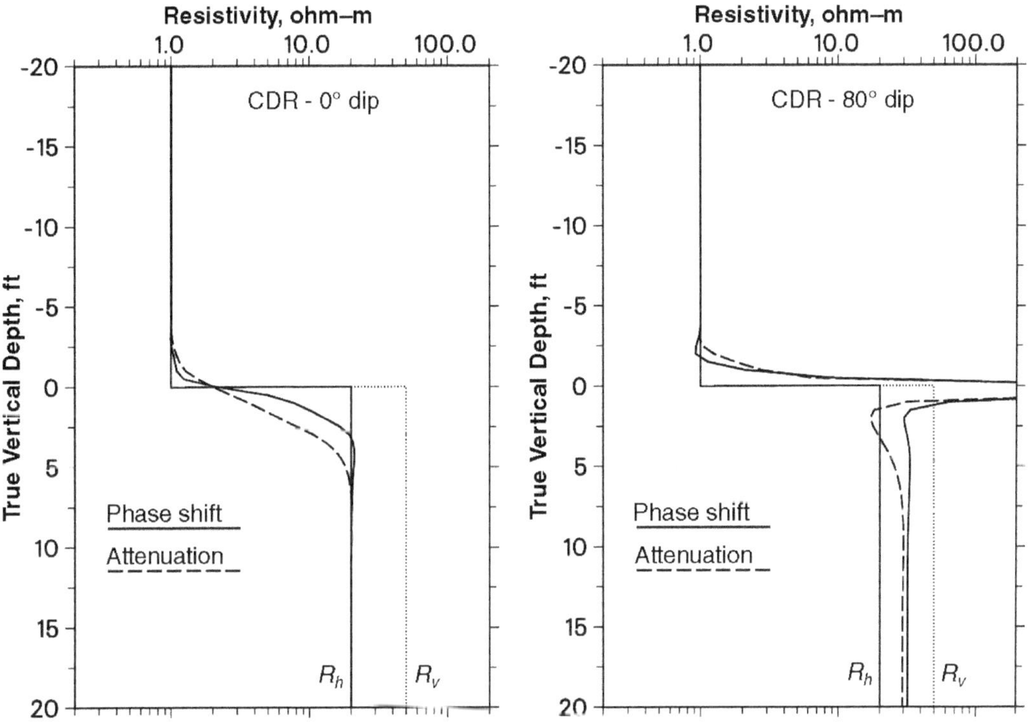

Fig. 3B.37—CDR response at different dips as the tool logs an isotropic bed above an anisotropic bed.

medium- to high-resistivity formations. In practice, both modern laterolog and induction-logging tools are suitable for most logging conditions, and it is no longer practical to make a specific recommendation for one type in preference to the other, except in extreme conditions.

Laterolog devices see the more resistive zones, and induction tools see the more conductive zones. Therefore, when R_{xo} is greater than R_t, an induction tool is preferred for R_t determination because laterolog tools will be affected mostly by R_{xo}. Conversely, a laterolog tool is preferred when R_{xo} is less than R_t. Conductivity in the borehole has a strong influence on an induction measurement, but little influence on a laterolog measurement.

Starting with the Phasor Induction tool, borehole-corrected logs for induction tools have been available at the wellsite. A caliper and estimate of mud resistivity is essential for induction-borehole correction, either by hand using a chart or automatically.

The AITs have only automatic borehole correction—no charts exist. An analytic forward model was used to compute thousands of cases for the AIT covering the range of each of these parameters. At the wellsite, a caliper and accurate measurement of R_m are used as inputs, and the other two parameters are solved for in a least-squares inversion through the computed table. This method is essential to produce an accurate 10-in. log over a wide range of borehole sizes and mud resistivities.

The following are guidelines for running induction logs, especially array-induction tools:

1. A caliper is required *in the same toolstring* as the induction tool.
2. R_m must be measured adequately, preferably downhole, using an accurate sensor. There can be large errors in values of mud resistivity based on surface measurements.
3. Adequate standoff is essential. Never run slick.

If these guidelines are followed, modern AITs can give accurate estimates of R_t even when R_{xo}/R_t is as low as 0.2.

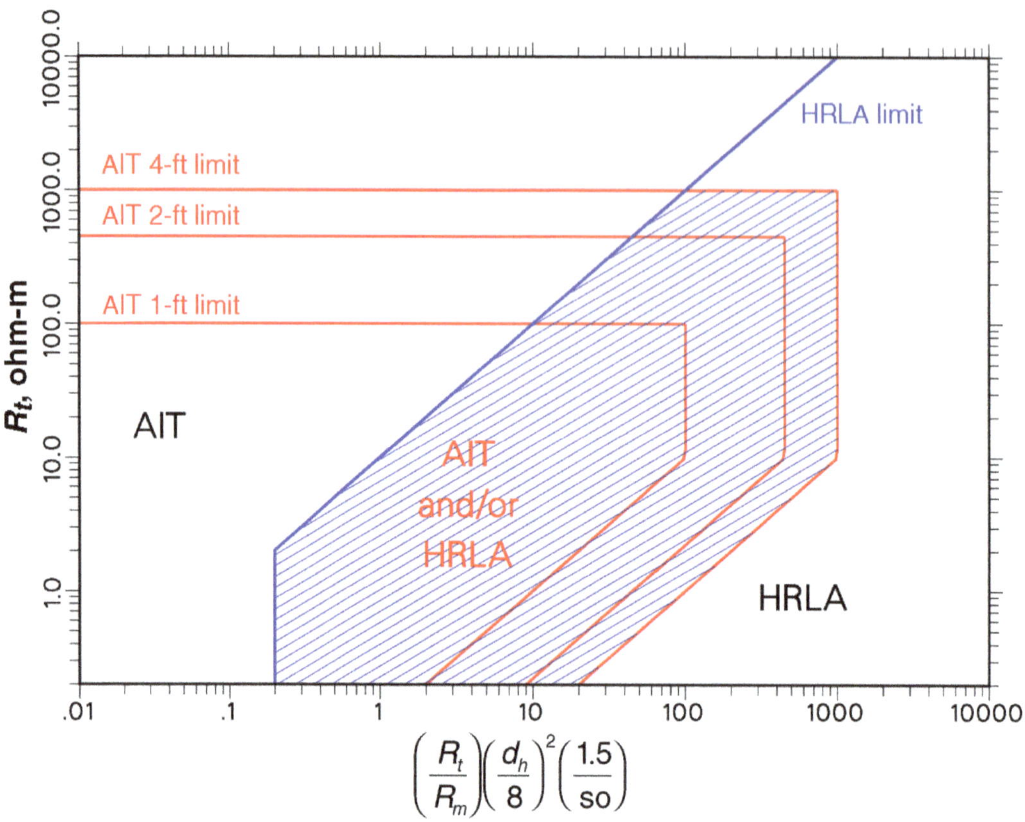

Fig. 3B.38—Chart showing the usable range of the AIT and HRLA logs.

The HRLA array laterolog tool with its inversion of the five array logs has extended the usability of laterolog tools further into the $R_{xo} > R_t$ region. The AITs, again with inversion of the logs, have extended the induction range in the $R_{xo} < R_t$ region. **Fig. 3B.38** shows the range of usability of the AIT and HRLA tools. In the broad overlap region, both tools can be used. In this region, the HRLA array laterolog tool can be combined with the AIT to determine anisotropic resistivity—R_v and R_h in vertical wells.[52]

When looking at both induction and laterolog logs from the same well or the same field, do not expect the logs to overlay. The R_t values for both tools should be close, but the logs themselves, uncorrected for environmental effects, can be quite different.[53]

When working with older logs, one must keep in mind that both laterolog and induction measurements are influenced by the borehole and by surrounding beds. Surprisingly, thick beds may have some effect on their measurements, depending on shoulder-bed contrast. The measurements of both devices should always be corrected for borehole and surrounding bed effects. Although these corrections are in many cases small, it is good practice to make them routinely. This will ensure that they are not overlooked in the larger number of cases where they are significant.

With either laterolog or induction deep-resistivity measurements, it is essential to record at least three resistivity-log curves with different depths of investigation. With fewer than three competent measurements, it is not possible to make an estimate of the invasion parameters, and R_t and R_{xo} become guesses. Array-induction and array-laterolog tools make a sufficient number of measurements to use the more rigorous inversion solutions, deriving even more reliable values of R_t and R_{xo}.

3B.7 Microresistivity Logs

Microresistivity devices measure the resistivity of the flushed zone and delineate permeable beds by detecting the presence of mudcake.

When invasion is moderate to deep, knowledge of R_{xo} is required to derive R_t from the deep-resistivity measurement. To evaluate a formation with logs, the R_{xo}/R_t ratio is required for some saturation-estimation methods. In clean formations, a value of the formation resistivity factor F can be computed from R_{xo} and R_{mf} if S_{xo} is known or can be estimated.

Tools designed to measure R_{xo} have a very shallow depth of investigation, because the flushed zone may extend only a few inches beyond the borehole wall. To avoid the effect of the borehole, a sidewall-pad tool is used. The pad, carrying an array of closely spaced electrodes, is pressed against the formation to minimize the short-circuiting effect of the mud. Currents from the electrodes on the pad must pass through any mudcake to reach the flushed zone.

Microresistivity readings are affected by mudcake; the effect depends on the mudcake resistivity and thickness (h_{mc}). Mudcakes are usually anisotropic, with the resistivity parallel to the borehole wall lower than the resistivity across the mudcake. This increases the mudcake effect on microresistivity readings to make the effective, or electrical, mudcake thickness greater than the physical thickness indicated by the caliper.

Microresistivity measurements have evolved from the first microlog, through the obsolete microlaterolog[54] and proximity-log devices, to the current MicroSFL and Platform Express MCFL microresistivity measurements.

3B.7.1 Microlog. The microlog[55] is still used qualitatively for its ability to detect permeable intervals with a fine vertical resolution, but not for the evaluation of R_{xo}. The measurement comprises two short-spaced devices with different depths of investigation, providing resistivity measurements of small volumes of mudcake and formation adjacent to the borehole. The presence of a mudcake, identified by a "positive" separation of the two curves, indicates an invaded and, therefore, permeable formation. As a qualitative log, the microlog is usually presented on a linear grid.

Principles. The flexible oil-filled microlog pad is pressed against the borehole wall by arms and springs. The face of the pad has three small in-line electrodes spaced 1 in. [2.5 cm] apart. The electrodes record a 1×1-in. "microinverse" log and a 2-in. "micronormal" log simultaneously.

In an invaded permeable zone, R_{mc} is usually significantly lower than R_{xo}. The 2-in. micronormal device has a greater depth of investigation than the microinverse. It is, therefore, less influenced by the mudcake and reads a higher resistivity when mudcake is present. In impermeable formations, the two curves read approximately the same resistivity or have a small negative separation, and the resistivities are usually much greater than in permeable formations.

Positive separation occurs in a permeable zone. Although the microlog curves identify permeable formations, quantitative inferences of permeability are not possible.

Under favorable circumstances, R_{xo} values can be derived from microlog measurements using charts provided by the service companies. R_{mc} values for this purpose can be measured directly or estimated from other charts, and h_{mc} is obtained from comparing the caliper curve to bit size. The limitations of the method are as follows:

1. The ratio R_{xo}/R_{mc} must be less than approximately 15 (porosity more than 15%).
2. The value of h_{mc} must be no greater than 0.5 in. [1.3 cm].
3. Depth of invasion must be greater than 4 in. [10 cm]; otherwise, the microlog readings are affected by R_t.

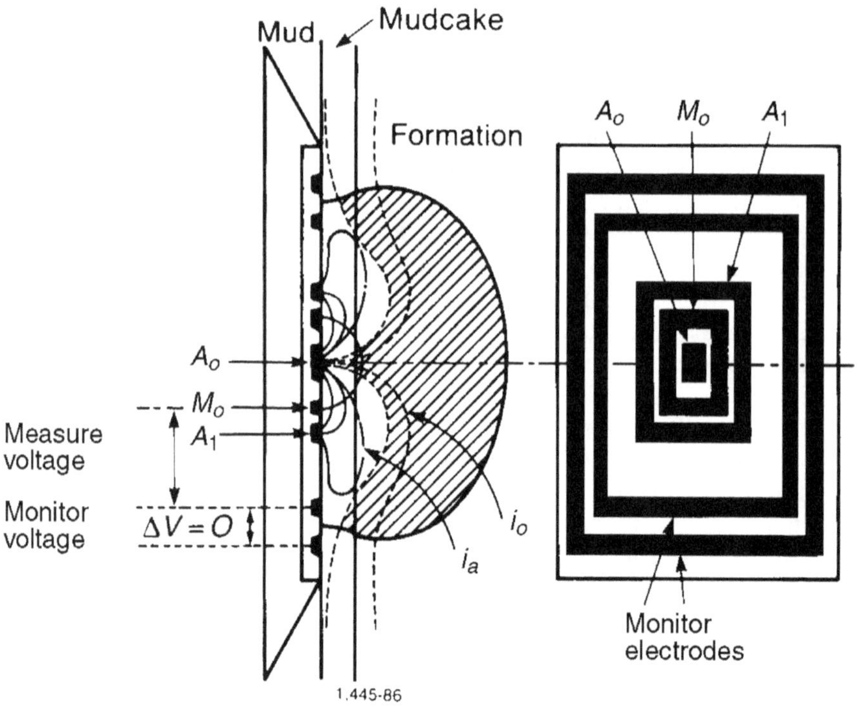

Fig. 3B.39—Pad layout of the MicroSFL tool.

3B.7.2 MicroSFL Log. The MicroSFL (MSFL) device[56] is a pad-mounted spherically focused logging sensor with two distinct advantages over the microlaterolog and proximity tools it replaced. Unlike these earlier R_{xo} devices, the MSFL tool is combinable with other logging tools, which eliminates the necessity of a separate logging run to obtain R_{xo} information. The MSFL log also performs better in shallow invaded zones in the presence of mudcake.

Fig. 3B.39 shows the electrode arrangement (right) and current patterns (left) of the MSFL tool. The surveying current flows outward from a central electrode, A0. Bucking currents, passing between electrodes A0 and A1, flow in the mudcake and the formation. The measuring current, I_0, is confined to a path directly into the formation, where it quickly spreads and returns to a remote electrode. By forcing the measure current to flow initially directly into the formation, the effect of mudcake resistivity on the tool response is minimized, yet the tool still has a very shallow depth of investigation.

Synthetic microlog curves (microinverse and micronormal) can be computed from MSFL parameters, because the measure current sees mostly flushed zone and the bucking current sees primarily mudcake.

Environmental Corrections. Although the influence of mudcake on the readings is relatively small, MSFL measurements must be corrected for thickness. Mudcake thickness is normally deduced from a comparison of the actual borehole size, as measured with the caliper, to the known bit size.

3B.7.3 MCFL Log. The Schlumberger MCFL microresistivity measurements[57] made with Platform Express tool strings are different from previous measurements in several respects:

1. Electrodes are mounted on a rigid, mostly metal pad that is not deformed by the borehole wall, allowing a more consistent standoff measurement.

2. Survey currents are independently focused in planes parallel and perpendicular to the tool axis, reducing the sensitivity to borehole geometry.

3. Three measurements are made with different depths of investigation, which allows a more reliable resolution of mudcake and formation properties with independent response equations.

4. The microresistivity sensors are interlaced with the density sensors, so both measurements sample the same volume of formation at the same time.

A vertical resolution of the raw measurements better than 1 in. [2.5 cm] is achieved, and R_{xo}, R_{mc}, and h_{mc} are solved simultaneously. Two curves can be displayed in a microlog-like presentation. When the two curves are superimposed, they both read R_{xo}. Any separation indicates pad standoff from the formation, which is usually caused by mudcake and indicates a permeable formation.

Because R_{xo}, R_{mc}, and h_{mc} are obtained directly from the Platform Express microresistivity measurements by inversion processing, no mudcake thickness corrections are required. The values of R_{xo} can be used directly with medium and deep-resistivity measurements (or array-resistivity measurements) to derive R_t.

3B.8 Formation-Resistivity Determination

Resistivity is the one of the most difficult formation parameters to measure accurately because of the complex changes that occur during and after drilling a well and that may still be occurring during logging. The various components of the downhole environment may have strongly contrasting resistivities, some of which cannot be measured directly, and their physical dimensions may not be readily available. Fig. 3B.1 shows an idealized relationship of the main environmental components. The resistivities and dimensions of all these "layers" (mud, mudcake, flushed zone, and zone of transition) influence all deep-reading resistivity measurements. There is no direct measurement of R_t. It must be inferred from the multiple-depth resistivity measurements.

In a permeable formation, mud resistivity is commonly 1 to 3 orders of magnitude lower than the formation resistivity, or, in the case of OBM, it can be much higher. The downhole mud resistivity can be estimated approximately by measuring the resistivity of a surface sample taken just before mud circulation was stopped before logging and adjusting it for the difference in temperature using an appropriate chart or equation. The shape and size of the borehole and the position of the logging tool in the hole have an influence on resistivity measurements that will not be apparent from a single hole-size measurement.

Attempts to measure the resistivity of an artificially formed sample of mudcake is unlikely to represent in-situ mudcake resistivity accurately. Mudcake thickness cannot be directly measured with current logging tools; it can only be estimated with some uncertainty.

The resistivity of the flushed zone can be measured with reasonable precision if the depth of invasion is greater than the depth of investigation of the R_{xo} logging tool, but the depth of invasion and the resistivity profile and geometry of the zone of transition are difficult to estimate.

3B.9 Traditional R_t Methods

Evaluation of the uninvaded formation resistivity R_t is sometimes referred to as invasion correction. It is usually performed assuming a simple three-parameter, step-profile invasion model consisting of a flushed zone of uniform resistivity R_{xo} with a sharp boundary at the diameter of invasion d_i to resistivity R_t. This is clearly not realistic, but it allows a complex problem to be solved relatively simply, usually with acceptable accuracy, by using a minimum of three resistivity measurements with different depths of investigation.[5]

A shallow microresistivity measurement, such as the MicroSFL log, is corrected for the influence of the mudcake by using the best available estimates of mudcake thickness and resistivity. It is assumed that d_i is greater than the depth of investigation of the MicroSFL, so that the MicroSFL log reads only the flushed zone and the mudcake-corrected MicroSFL reading is R_{xo}.

Next, the deep- and medium-resistivity measurements are corrected for environmental effects using the charts for the tool used. These effects are always corrected in the following order: borehole effect, bed thickness, and shoulder effect. Then the invasion parameters (R_{xo}, R_t, d_i) can be found for induction tools using tornado charts. For laterolog tools, butterfly charts are used. Some service companies offer charts for R_t and d_i when R_{xo} is known from a microresistivity device.

3B.9.1 Inversion for Invasion Parameters. The latest array-induction and array-electrode tools all use some form of inversion (rather than charts) to estimate the invasion parameters—invasion radius or diameter, d_i, R_{xo}, and R_t. There are two types of inversion used: 1D and 2D. The formation geometry of each is shown in Fig. 3B.9.

Inversion means first building a parametric model of the formation, then estimating from the log values a "first guess" of the parameter values. Then, a modeling code is used to compute the log response to the model. This is compared with the actual logs. The difference between modeled and measured is used to pick a new set of model parameters, and the log response computed. Again, the modeled logs are compared to the actual logs. This continues until the difference reaches some preset minimum, and then the model parameters are output as the invasion parameters.

Induction response is mostly a function only of coil spacings, with a weak response to the large-scale average formation conductivity. If we can assume that each formation layer is more or less uniformly invaded, then an inversion through a 1D radial forward model[58] will give a close estimate of invasion parameters R_{xo} and R_t. This is most often true when $R_{xo} > R_t$.

Laterolog tools are much more affected at the same time by both bed thickness and invasion parameters in that bed. For this reason, when the formation of interest consists of thin (< 30 ft [10 m]) beds, 2D inversion is necessary for accurate estimation of R_{xo}, R_t, and d_i. 1D inversion tends to be used in real-time processing, and 2D inversion is used at the computing center. Some commercial 2D-inversion applications allow very sophisticated choices of parametric models, including transition zones and annulus models.

Inversion methods also return "goodness-of-fit" criteria. At the same time, the modeled logs can be compared with the actual logs as a quality control. Inversion methods also allow the parametric model to be selected to better fit the situation at hand. A four-parameter inversion model, with a zone of transition defined from d_i to the diameter of the limit of invasion at d_j, gives reliable answers in a much wider range of conditions than the traditional three-parameter model. It also generates "quality-of-fit" parameters that indicate when the log readings are not consistent with the model.

Annulus formation is a common phenomenon, caused by the sweeping of conductive ions from the formation by the invading borehole fluid. Fluid-flow models predict annulus formation in a wide range of formation/borehole fluid conditions. Annulus formation has been observed with both WBM and OBM. Even with five or six depths of investigation, the information content is not sufficient to solve for the five annulus parameters (R_{xo}, R_{ann}, R_t, r_1, and r_2) independently. However, by invoking the constraint of material balance[59] and supplying an estimate of R_w, the annulus problem can be reduced to three parameters: R_{xo}, r_1, and R_t. r_2 is tied to r_1 through the material-balance constraint, and R_{ann} can be estimated from R_{mf}, R_{xo}, and R_w.

Invasion also changes with time, sometimes rapidly. For this reason, the combination of data taken at different times (e.g., wireline and LWD, or LWD time-lapse), must be done with care. One method that has shown promise is to use a consistent parametric model (step or annulus) and assume R_{xo} and R_t are constant, allowing only for the invasion radius to change.

Before performing an inversion for invasion parameters, be sure the cause of curve separation is actually invasion. Causes of curve separation include tool-response effects such as shoulder effect, not matching the resolution of the curves, and improper borehole correction. Other formation effects that can cause curve separation include dipping beds and drilling-induced fractures.

The introduction of array-resistivity tools has clearly delineated invasion profiles that were not as expected, even after years of logging in a region. In some regions, it was assumed that the formation was uninvaded because the ILd, ILm, and SFL logs all were very close in pay zones. Logs made by an array-induction tool showed an $R_{xo} < R_t$ profile. The discrepancy exists because the SFL, being a laterolog, can have a depth of investigation as deep as ILd under $R_{xo} < R_t$ conditions. The ILm was long considered an inferior measurement to ILd because it would lie lower in resistivity than either SFL or ILd. Modeling shows that either an annulus or $R_{xo} < R_t$ will produce this curve order.

A powerful method for handling the fundamentally underdetermined problem of invasion correction of resistivity logs is iterative forward modeling.[5] Complex formation geometries and invasion profiles can be worked out by building a model of the best estimate of formation parameters from logs, field knowledge, and petrophysical constraints; modeling the resistivity logs; and varying the parameters until a fit is obtained.

Finally, keep in mind that R_t cannot be measured directly, but must be inferred from multiple-depth resistivity measurements.

3B.9.2 Resistivity Imaging. Many modern resistivity- and microresistivity-logging tools have arrays of sensors that make multiple measurements, enabling the creation of 2D images of formation resistivity. The images represent resistivity variations with the azimuth around the borehole or with distance away from the borehole.

Inspired by the images produced by early acoustic borehole televiewer tools, borehole resistivity imaging was developed to see actual formation variations, rather than the surface effects that the acoustic images depicted. The first practical resistivity images were produced by an array of closely spaced, shallow-reading button electrodes applied to the borehole wall.

The borehole coverage of this high-resolution microresistivity image was increased on later tools, and deeper reading tools with imaging capabilities have been subsequently developed.

Microresistivity Images. The first microelectrical imaging tool had an array of button electrodes mounted on an enlarged section of the tool mandrel. Electrodes were then placed on two of the four pads of the Dual Dipmeter tool. The Fullbore Formation MicroImager (FMI) tool[62] has button electrodes on all four pads, with extension flaps on each pad to increase the borehole coverage.

FMI image data have sufficient resolution and character to allow using selected electrode signals for conventional dipmeter computations, and the tool's borehole coverage provides a detailed visual appreciation of geologic features, **Fig. 3B.40**. Formation dips and fracture orientations can also be derived directly from the images.

Seeing the shape of formation-resistivity variations often provides understanding of the lack of coherent dip found by a dipmeter computation program.

Halliburton's Electromagnetic MicroImager (EMI) tool is a six-arm resistivity borehole imager tool. Its principle of operation is similar to the FMI tool. Baker Atlas' Simultaneous Acoustic and Resistivity (STAR) Imager tool integrates resistivity and acoustic borehole imaging sensors into one instrument. The resistivity imager is a six-arm device with powered centralization to keep the acoustic transducer centered. The acoustic sensor also works in OBM where the resistivity imager performance is poor.

ARI Images. The formation resistivity around the borehole is displayed in ARI images as a 2D azimuthal image, with the same dimensions of well depth and azimuthal angle around the

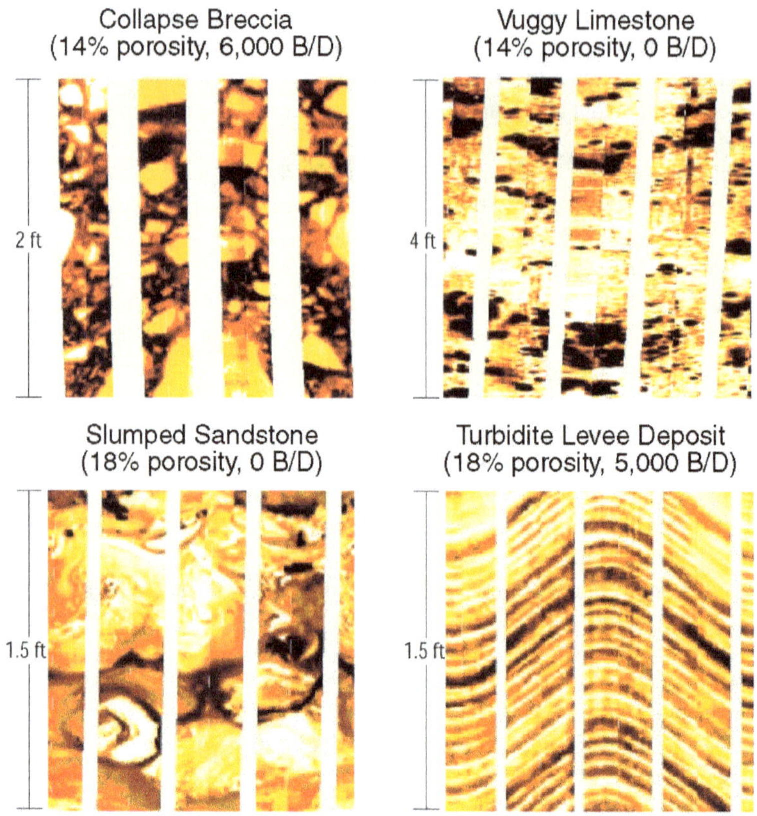

Fig. 3B.40—FMI images in four different geological settings.

well as FMI images. This image has much lower spatial resolution than acoustic or microelectrical images from the UBI and FMI tools, but it complements them well because of its sensitivity to features beyond the borehole wall and its lower sensitivity to shallow features.

AIT Images. The AIT provides images of variations in formation resistivity or conductivity with distance away from the borehole. This capability brings a new dimension to formation image data, because the image contains invasion information useful for understanding how deeply the formation can be invaded.

Radial response functions are used to invert the set of matched vertical resolution logs in the four-parameter invasion model, producing a detailed description of the radial resistivity. Introducing other petrophysical parameters, such as F, R_w, and R_t, and a suitable saturation equation (see Saturation Determination section) allows imaging-computed virgin and invaded-zone saturations.

3B.9.3 LWD Resistivity Images. The GVR tool incorporates three 1-in.-diameter azimuthal button electrodes that produce borehole resistivity images during rotary drilling by recording 56 resistivity measurements per rotation with each electrode. The data are processed and recorded downhole for later retrieval.

Because the GVR button electrodes are larger than FMI electrodes and are not in contact with the formation, GVR images are less sharp than FMI images, as seen in Fig. 3B.12. However, often the timeliness of the images more than makes up for the resolution. A compression algorithm allows the images to be sent up in real time for geosteering.

Formation Dip From LWD Images. Dip computation by conventional dipmeter data processing is most effective when the apparent dips (i.e., dips relative to tool inclination) are less than approximately 70°, which is suitable for most vertical and normally deviated wells. LWD has major applications (e.g., geosteering) in highly deviated and horizontal wells, where apparent dips are commonly greater than 70°. A dip-computation process that returns dip from the GVR in real time was developed for high-angle wells.[61] The dip-azimuth and magnitude computations are performed by a robust algorithm in the downhole tool, allowing real-time transmission of the dip data. Although the well deviation is accurately measured by instruments in the drill collar, the relative dip with respect to the bedding is important for geosteering wells along the bedding planes. This method allows improved well placement.

The confidence in computed GVR dips is increased by using data from all three electrodes. Because the electrodes are at fixed distances from each other, irregular tool movement in the hole is unimportant.

3B.10 SP Log

The SP curve is a continuous recording vs. depth of the electrical potential difference between a movable electrode in the borehole and a surface electrode.[1] Adjacent to shales, SP readings usually define a straight line known as the shale baseline. Next to permeable formations, the curve departs from the shale baseline; in thick permeable beds, these excursions reach a constant departure from the shale baseline, defining the "sand line." The deflection may be either to the left (negative) or to the right (positive), depending on the relative salinities of the formation water and the mud filtrate. If the formation-water salinity is greater than the mud-filtrate salinity (the more common case), the deflection is to the left.

The relevant features of the SP curve are its shape and the size of its departure from the shale baseline. Because the absolute reading and position of the shale baseline on the log are irrelevant, the SP sensitivity scale and shale-baseline position are selected by the logging engineer for convenience. The SP log is typically scaled at 100 mV per log track. If the resistivities of the mud filtrate and formation water are similar, the SP deflections are small and the curve is rather featureless. An SP curve cannot be recorded in holes filled with nonconductive muds, such as OBMs.

3B.10.1 Origin of the SP.

Deflections of the SP curve are the result of electrochemical and electrokinetic potentials in the formations that cause electric currents to flow in the mud in the borehole.

Electrochemical Component. *Membrane Potential.* The structure of clay minerals in shales and the concentration of negative electric charges on the clay particle surfaces give shales a selective permeability to electrically charged ions. Most shales act as "cationic membranes" that are permeable to positively charged ions (cations) and impermeable to negative ions (anions).[62]

The upper part of **Fig. 3B.41** shows saline formation water in a sandstone formation and mud in the borehole separated by a shale. Sodium chloride, which is usually present in both the formation water and the drilling mud, separates into charged ions (Na^+ and Cl^-) in solution in water. The Na^+ and Cl^- ions tend to migrate from a more-concentrated to a less-concentrated solution, but because the intervening shale is a cationic membrane, impervious to Cl^- ions, only the Na^+ ions can migrate. If, as usual, the formation water is a more concentrated NaCl solution than the mud, there is a net flow of positive ions through the shale from the sandstone to the borehole. This corresponds to a positive electric current in the same direction (indicated by the curved arrow) driven by an electric potential, or electromotive force (EMF), across the shale. Because the shale acts as an ion-selective membrane, the electric potential is known as the membrane potential.

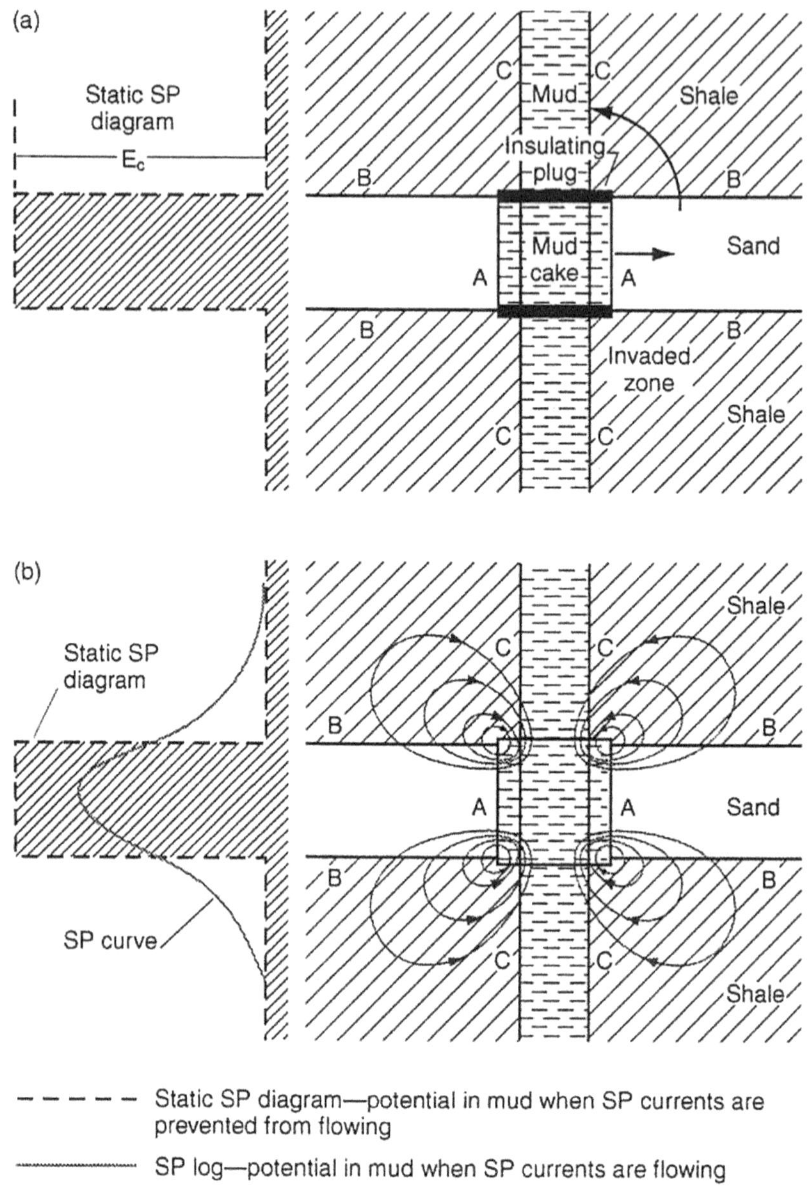

Fig. 3B.41—Schematic representation of potential and current distribution in and around a permeable bed.

Liquid-Junction Potential. At the edge of the invaded zone, where the mud filtrate and formation water are in direct contact, Na^+ and Cl^- ions can move freely from one solution to the other. But Cl^- ions are smaller and have greater mobility than Na^+ ions, so the net diffusion of ions from the more-concentrated formation water to the less-concentrated mud filtrate includes a greater number of Cl^- ions than Na^+ ions. This is equivalent to a positive current flow in the opposite direction (indicated by the straight arrow at A in Fig. 3B.41.)

The current flowing across the junction between solutions of different salinity is driven by an EMF called the liquid-junction potential. The magnitude of the liquid-junction potential is only approximately one-fifth of the membrane potential.

If the solutions contain substantial amounts of salts other than NaCl, the value of K at 77°F may not be 71. If the permeable formation contains some shale or dispersed clay, the total electrochemical potential, and therefore the SP deflections, is reduced.

Electrokinetic Component. An electrokinetic potential (E_k, also called the streaming potential or electrofiltration potential) is produced when an electrolyte flows through a permeable medium.[63,64] The size of the electrokinetic potential is determined mainly by the differential pressure producing the flow and the resistivity of the electrolyte.

In the borehole, the electrokinetic potential E_{kmc} is produced by the flow of mud filtrate through the mudcake deposited on the borehole wall opposite permeable formations. Little or no electrokinetic potential is generated across the permeable formation itself because the differential pressure is usually low. The electrokinetic potential E_{ksh} may, however, be produced across a shale if it has any permeability.

Typically, E_{kmc} and E_{ksh} are similar in magnitude, and the net electrokinetic contribution to the SP deflection is negligible. If the formation water is fairly saline (resistivity less than 0.1 ohm·m) and the differential pressure is in the normal range of only a few hundred psi, the contribution of the electrokinetic potential can usually be ignored.

Electrokinetic effects may be significant in highly depleted formations or when heavy drilling muds are used because of unusually large differential pressures. Significant electrokinetic effects may also occur in very-low-permeability formations, where an appreciable part of the pressure differential occurs in the formation itself, especially if little or no mudcake is formed. If the formation water is brackish, the mud is resistive, and the low-permeability formation is clean and has some porosity, the electrokinetic effect could be as large as –200 mV.

3B.10.2 SP and Permeability. The movement of ions, essential to develop an SP, is possible only in formations with some permeability, however small—a small fraction of a millidarcy is sufficient. There is no direct relationship between the magnitude of the SP deflection and the value of either the formation's permeability or its porosity.

3B.10.3 Static SP. The lower part of Fig. 3B.41 shows SP currents in the borehole and formations. The current directions indicated correspond to the more usual case of formation-water salinity greater than mud-filtrate salinity, producing a potential by the permeable bed lower than the potential by the shale. This corresponds to a deflection to the left on the SP log by the permeable bed.

If the mud-filtrate salinity is greater than the formation-water salinity, the currents flow in the opposite direction, producing positive SP deflections. If the salinities of the mud filtrate and formation water are similar, no SP is generated.

The SP currents flow through four different media: borehole fluid, the invaded zone, the uninvaded part of the permeable formation, and surrounding shales. The SP log measures only the potential drop from the SP currents in the borehole fluid, which may not represent the total SP because there are also potential drops in the formation. If the currents could be interrupted by hypothetical insulating plugs (see the upper part of Fig. 3B.41), the potential observed in the mud would be the total spontaneous potential. This idealized SP deflection is called the static SP (or SSP). The SP deflection practically reaches the SSP in a thick, clean formation.

The borehole presents a much smaller cross-sectional area to current flow than the formations around it, so the resistance of the borehole part of the SP current loop is much higher than the formation part. Nearly all the SP potential drop, therefore, occurs in the borehole if formation resistivities are low-to-moderate and formation beds are thick, so, in practice, the recorded SP deflection approaches the static SP value in thick, permeable beds.

Determination of SSP. To determine the SSP, a sand line is drawn through the maximum (usually negative) excursions of the SP curve adjacent to the thickest permeable beds. A shale

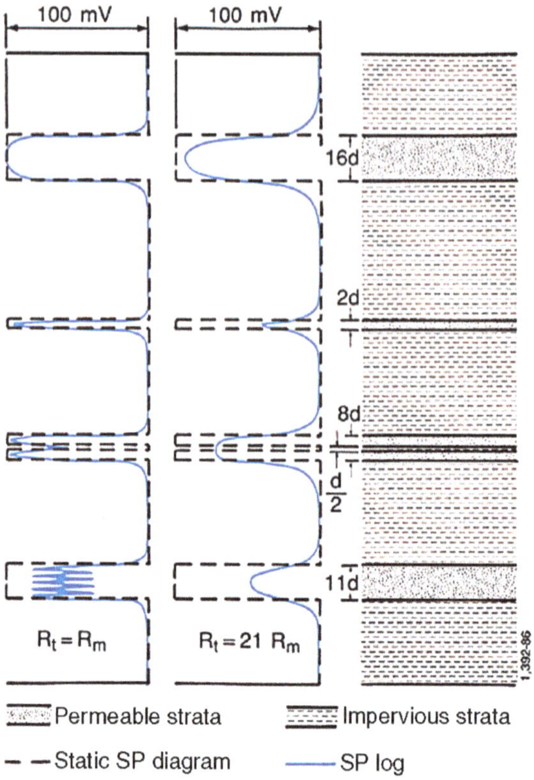

Fig. 3B.42—SP curve in beds of different thickness for $R_t = R_m$ (left) and $R_t = 21\ R_m$ (center).

baseline is drawn through the SP through the intervening shale beds. The separation of the sand line from the shale baseline, measured in mV, is the SSP. Any SP anomalies are discounted.

If there are no thick, clean, permeable invaded beds in the zone under study, the SP reading can be corrected for the effects of bed thickness and invasion to estimate the SSP by using charts available from service companies.

Shape of the SP Curve. The slope of the SP curve is proportional to the intensity of the SP currents in the borehole at that depth. Because the current intensity is highest at the boundaries of the permeable formation, the slope of the SP curve is at a maximum, and an inflection point occurs at these bed boundaries.

The shape of the SP curve and the amplitude of its deflection in permeable beds depend on the following factors: thickness and true formation resistivity of the permeable bed, resistivity of the flushed zone (R_{xo}) and diameter d_i, resistivity of the adjacent shale bed (R_s), and resistivity of the mud and the diameter of the borehole (d_h).

Fig 3B.42 shows examples of SP curves computed for $R_t = R_s = R_m$ (on the left) and $R_t = R_s = 21R_m$ (in the center). In the first case ($R_t = R_s = R_m$), the SP curve gives a much sharper definition of the boundaries of the permeable beds, and the SP deflections approach the SSP value more closely than in the case where the formation-to-mud resistivity ratio is 21.

SP Anomalies. The SP curve may be difficult to interpret and use for R_w determination because it does not always behave ideally. The following are a few cases of apparently anomalous SP responses.

Highly Resistive Formations. Highly resistive formations between some shales and permeable beds can significantly alter the distribution of the SP currents and change the expected

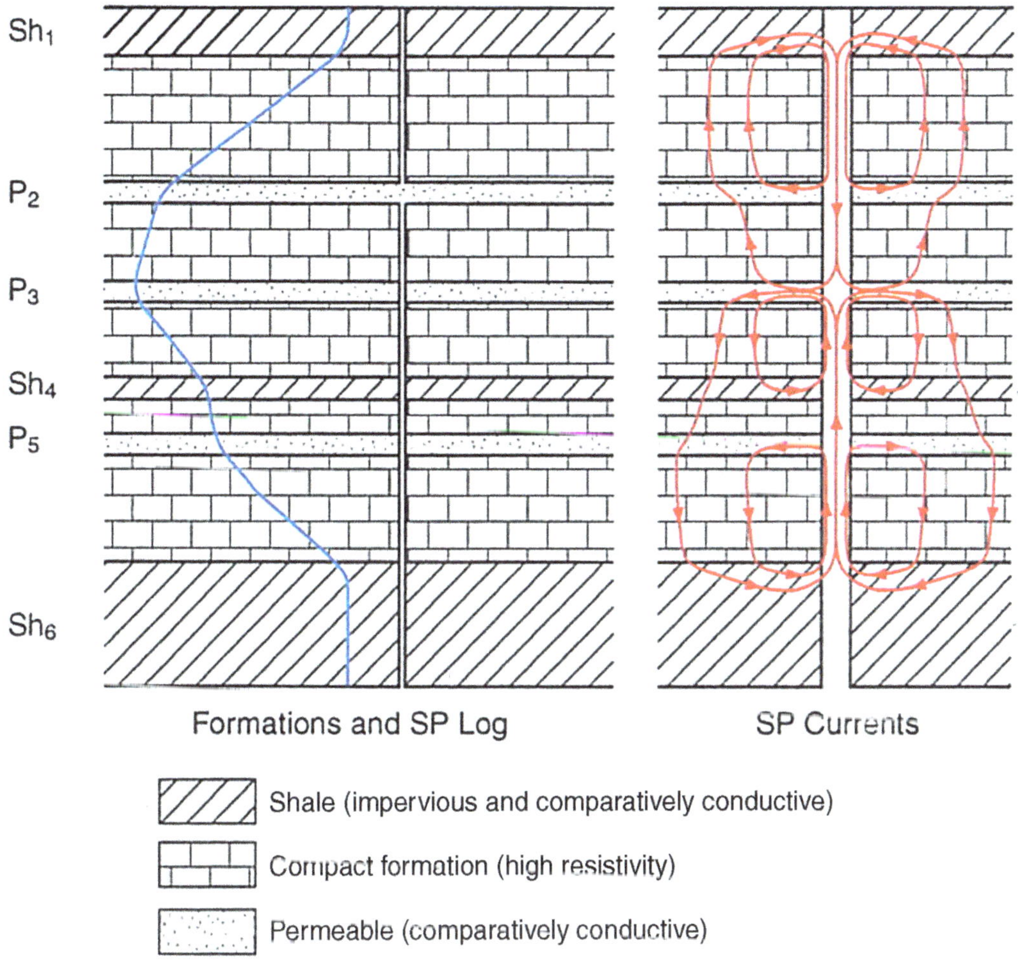

Fig. 3B.43—Schematic representations of SP phenomena in highly resistive formations.

shape of the SP curve. The currents shown flowing from shale bed Sh_1 toward permeable bed P_2 in **Fig. 3B.43** are largely confined to the borehole by the high resistivity of the formation separating Sh_1 and P_2. The current in the borehole over this interval is constant, so for a constant borehole diameter, the SP curve is a straight line inclined to the shale baseline.

The SP curve consists of straight portions adjacent to the high-resistivity zones with a change of slope at each more conductive permeable interval (the SP curve is concave toward the shale line) and opposite every shale bed (the SP curve is convex toward the shale line). Defining permeable bed boundaries from the SP log is difficult in the vicinity of highly resistive formations.

Shale-Baseline Shifts. A shift of the shale baseline can occur when formation waters of different salinities are separated by a shale bed that is not a perfect cationic membrane.[65] **Fig. 3B.44** shows an SP log recorded in a series of sandstones (B, D, F, and H) separated by thin shales or shaly sandstones (C, E, and G). The SSP of Sandstone B is –42 mV. Shale C is not a perfect cationic membrane, and the SP curve does not return to the shale baseline defined by Shale A. A new shale baseline defined by Shale E gives SP deflections of +44 mV in Sandstone D and –23 mV in Sandstone F.

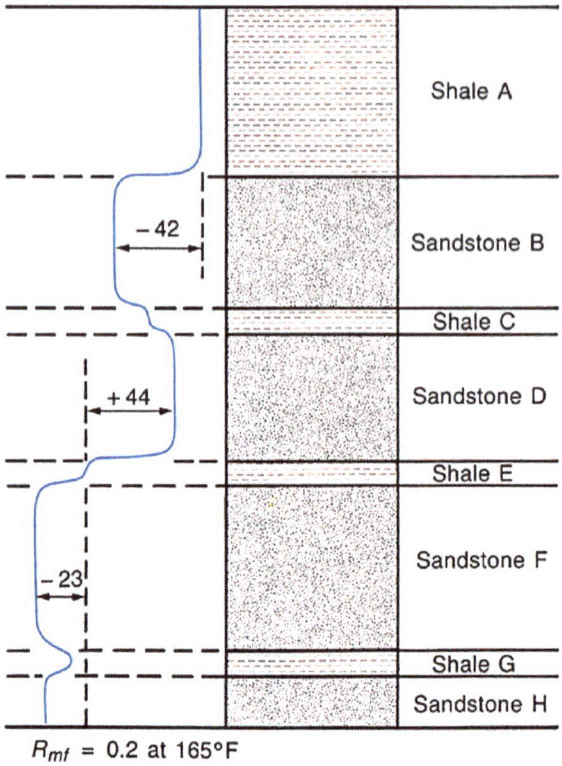

Fig. 3B.44—SP baseline shift.

Baseline shifts also occur when formation waters of different salinities are separated by an impermeable layer that is not a shale. In this case, the SP curve shows little or no variation at the level of the change in salinity, but the deflections at the upper and lower shale boundaries are different and may even have different polarities.

Invasion-Related Anomalies. If the mud filtrate and the formation water have significantly different salinities, and therefore different densities, gravity-induced fluid migration can cause SP anomalies in highly permeable formations, as shown in Fig. 3B.44. Invasion is very shallow near the lower boundary of each permeable interval and deeper near the upper boundary.

The SP curve is rounded at the upper boundary because of the deep invasion, and it may have a sawtooth profile at thin, impervious shale streaks in which the SP deflection exceeds the SSP above the shale streak. A reading greater than the SSP is caused by the accumulation of filtrate below the shale streak. Encircling the hole is a horizontal disk of shale sandwiched between salt water and fresher mud filtrate that acts like a battery cell. The EMF of this cell is superimposed on the normal SSP, producing the sawtooth profile.

Noisy SP Logs. SP measuring circuits are sensitive and therefore prone to recording spurious electrical noise superimposed on the SP curve. Occasionally, the source of noise cannot be eliminated during logging, and a noisy log is recorded. However, this does not always render the log unusable.

A regular sine-wave signal may be superimposed on the SP curve when some part of the logging winch is magnetized. An intermittent contact between the casing and cable armor may also cause spurious spikes on the SP curve. In these situations, the SP curve can usually be read so that the sine-wave amplitude or noise spikes are not added to or subtracted from the authentic SP deflection.

Direct currents flowing through formations near the SP electrode can cause erroneous SP readings, particularly where formation resistivities are high. These currents may be caused by "bimetallism," when the logging tool has exposed metal housings. The currents are small and have a significant effect on the SP only in highly resistive formations. If an SP curve looks questionable in highly resistive formations, it should be relied on only in lower-resistivity intervals.

The offshore logging environment is notorious for its ample supply of sources of electrical noise, such as wave motion, cathodic protection systems, rig welding, onboard generators, and leaky power sources. On land, proximity to power lines and pumping wells may have a similar effect on the SP curve, but the effects can usually be minimized by carefully choosing the ground-electrode location.

3B.11 Uses and Interpretation of Well Logs

3B.11.1 Determination of Saturation. Water saturation is the fraction of the pore volume of the reservoir rock that is filled with water. It is generally assumed, unless otherwise known, that the pore volume not filled with water is filled with hydrocarbons. Determining water and hydrocarbon saturation is one of the basic objectives of well logging.

Clean Formations. All water saturation determinations from resistivity logs in clean (non-shaly) formations with homogeneous intergranular porosity are based on Archie's water saturation equation, or variations thereof.[66,67] The equation is

$$S_w^n = \frac{F R_w}{R_t} \quad \text{(3B.10)}$$

R_w is the formation water resistivity, R_t is the true formation resistivity, and F is the formation resistivity factor. F is usually obtained from the measured porosity of the formation through the relationship

$$F = a / \phi^m \quad \text{(3B.11)}$$

For S_{xo}, the water saturation in the flushed zone, a similar expression exists:

$$S_{xo}^n = \frac{F R_{mf}}{R_{xo}}, \quad \text{(3B.12)}$$

where R_{mf} is the mud filtrate resistivity and R_{xo} is the flushed zone resistivity.

For simplicity, the saturation exponent n is usually taken as 2. Laboratory experiments have shown that this is a reasonable value for average cases. For more exacting work, electrical measurements on cores will produce better numbers for n, a, and m. When core measured values are unavailable, the values of a and m in Eq. 3B.7 can be estimated as follows: in carbonates, $F = 1/\phi^2$ is usually used; in sands, $F = 0.62/\phi^{2.15}$ (Humble formula), or $F = 0.81/\phi^2$ (a simpler form practically equivalent to the Humble formula). These equations are easily programmed into spreadsheets and are available in most log interpretation software.

The accuracy of the Archie equation, Eq. 3B.10 and its derivatives, depends in large measure, of course, on the accuracy of the fundamental input parameters: R_w, F, and R_t. The deep resistivity measurement (induction or laterolog) must be corrected, therefore, for borehole, bed thickness, and invasion (see Sections 3B.8 and 3B.9). It is almost never safe to make the assumption "deep = R_t." The most appropriate porosity log (sonic, neutron, density, magnetic resonance, or other) or combination of porosity and lithology measurements must be used to obtain porosity, and the proper porosity-to-formation factor relationship must be used. Finally, the R_w value should be verified in as many ways as possible: calculation from the SP curve, water catalog, calculation from nearby water-bearing formation, and/or water sample measurement.

Alternate methods for determining water saturation include analysis of cores cut with low-invasion OBMs and single-well chemical-tracer tests (described in the chapter on single-well chemical-tracer testing in this section of the *Handbook*). These independent methods can be used to calibrate log analyses.

Resistivity-vs.-Porosity Crossplots. Combining Eqs. 3B.10 and 3B.11, the Archie saturation equation may be written

$$S_w^n = \frac{aR_w}{\phi^m R_t} . \qquad (3B.13)$$

If n and m are equal to 2, and $a = 1$, then

$$\phi S_w = \sqrt{R_w / R_t} . \qquad (3B.14)$$

Eq. 3B.14 shows that for R_w constant, ϕS_w is proportional to $1/\sqrt{R_t}$; ϕS_w is the quantity of water per unit volume of formation. To emphasize the proportionality between ϕ and $1/\sqrt{R_t}$, Eq. 3B.14 may be rewritten:

$$\phi = \frac{\sqrt{R_w}}{S_w}\frac{1}{\sqrt{R_t}} . \qquad (3B.15)$$

For a 100% water-saturated formation, $S_w = 1$ and $R_t = R_0$. If R_0 for water-saturated formations is plotted on an inverse square-root scale vs. ϕ, all points should fall on a straight line given by $\phi = \sqrt{R_w / R_0}$.

Furthermore, the points corresponding to any other constant value of S_w will also fall on a straight line, because in Eq. 3B.14 the coefficient is constant for constant values of R_w and S_w.

Fig. 3B.45 shows several points plotted over an interval in which formation-water resistivity is constant (as indicated by constant SP deflections opposite the thick, clean permeable beds). Assuming that at least some of the points are from 100% water-bearing formations, the line for $S_w = 1$ is drawn from the pivot point ($\phi = 0$, $R_t = \infty$) through the most northwesterly plotted points. The slope of this line defines the value of R_w as shown on Fig. 3B.45, for ϕ = 10%, R_0 = 6.5 ohm·m. For this formation, the most appropriate F–ϕ relation is $F = 1/\phi^2$. Thus, for ϕ = 10%, F = 100. Because $R_w = R_0/F$, R_w = 0.065 ohm·m, as shown.

For other S_w values, R_t and R_0 are related by the equation $R_t = R_0 / S_w^2$. For S_w = 50%, and $1/S_w^2 = 4$, and $R_t = 4\ R_0$. This relation establishes the line for S_w = 50%. Other S_w lines may be defined in a similar manner.

If the matrix composition remains constant over the formations under investigation, the basic measurement from the sonic, density, or neutron logs can be plotted directly vs. R_t with similar results.[68] This is possible because of the linear relationship between porosity and bulk density, sonic transit time, or neutron-hydrogen index response. An example of a sonic-induction crossplot is shown in **Fig. 3B.46**. The transit time has been plotted against the induction resistivity for several levels. The northwesterly points define the 100% water saturation line.

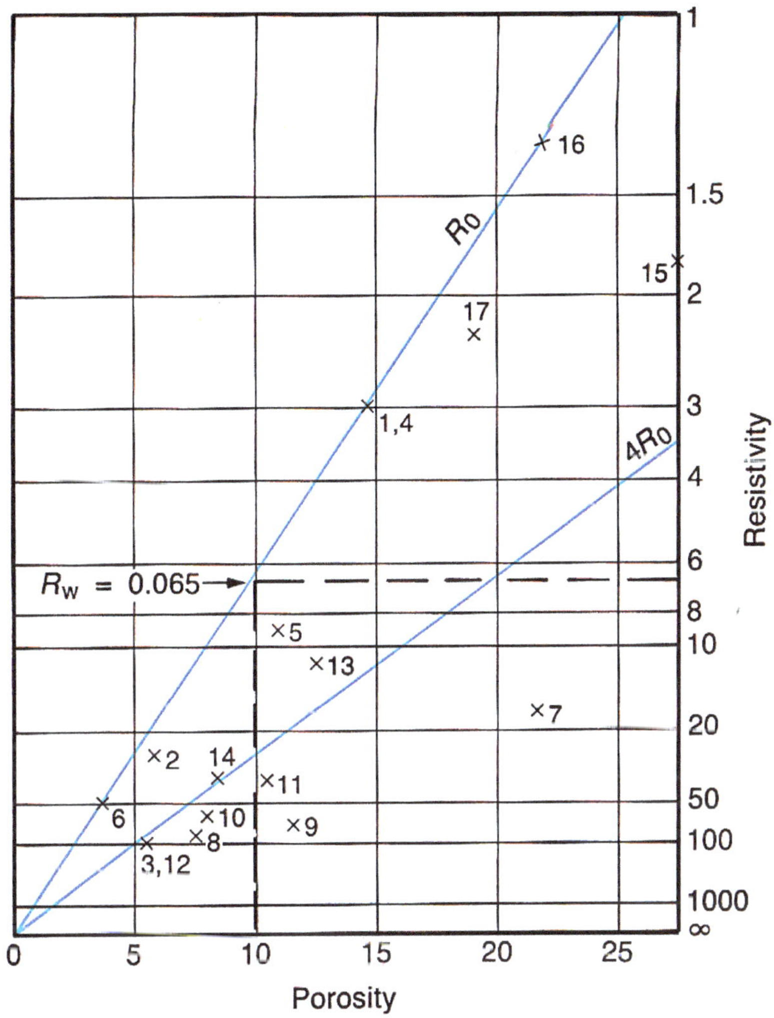

Fig. 3B.45—Resistivity-porosity crossplot for determining R_w and S_w.

The transit-time value at the point where this line intersects the horizontal line of infinite resistivity is the matrix-transit time, t_{ma} In Fig. 3B.46, t_{ma} is found to be approximately 47.5 µs/ft (156 µs/m). This corresponds to a matrix velocity of 21,000 ft/sec (6,400 m/s).

By knowing t_{ma}, a porosity scale, a scale of formation factor (e.g., from $F = 1/\phi^2$) can be easily derived. A vertical line drawn through $F = 100$ (or $\phi = 10$) intersects the water line at $R_0 = 5$ ohm·m; accordingly, R_w (= R_0/F) is 0.05 ohm·m.

The lines for other S_w values are straight lines, determined as previously described, radiating out from the $R_t = \infty$, $t_{\text{ma}} = 47.5$ pivot point.

Density and neutron logs can be crossplotted against resistivity in a manner identical to the sonic logs. For density logs, the intersection of the 100% water line with the infinite-resistivity line yields the matrix-density value, ρ_{ma}. For neutron logs, the intersection defines the matrix-hydrogen index, or apparent matrix porosity. Knowledge of matrix density or hydrogen index permits the ρ_B or ϕ_N scale to be rescaled in ϕ and F units. With the F scale defined, R_w can

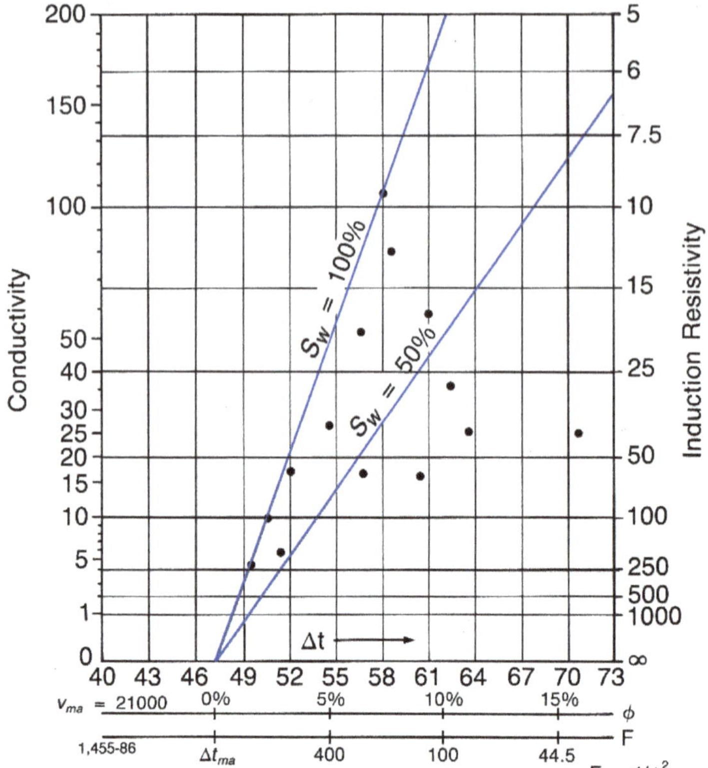

Fig. 3B.46—Acoustic-induction crossplot.

be calculated as for the sonic-resistivity crossplot, and lines of constant water saturation can be constructed in a similar manner.

These resistivity-vs.-porosity crossplots require that formation water resistivity be constant over the interval plotted, that lithology be constant, that invasion not be deep, and that the measured-porosity log parameter (i.e., t, ρ_B, or ϕ_N) can be linearly related to porosity. This last condition implies that the time-average transform for the conversion of t into porosity is appropriate.

The neutron-resistivity crossplot is not as satisfactory in gas-bearing formations as are the sonic- or density-resistivity crossplots. The apparent porosity measured by the neutron log in gas zones is often much too low. This results in overstated S_w values in gas zones. Indeed, in a gas zone, the neutron resistivity may indicate a porous gas-bearing zone to be near zero porosity and 100% water bearing. In contrast, the sonic- or density-resistivity tends to be slightly optimistic in gas zones (i.e., porosities may be slightly high and water saturations slightly low).

Microresistivity-vs.-Porosity Crossplots. This method is particularly useful for older logs or cases in which the analyst has only a paper copy of the log. A resistivity-porosity plot can also be made using the values from a shallow-investigation resistivity log such as the microlaterolog, MSFL, or MCFL log. If the microresistivity log reads approximately R_{xo}, then a line through points of mud-filtrate-saturated formations (S_{xo} = 1) should have a slope related to R_{mf}. R_{mf} is an important parameter, and this check of its value by means of a sonic-microresistivity or density-microresistivity crossplot is often useful.

These plots are also valuable for improved determinations of matrix parameters (either t_{ma} or ρ_{ma}), particularly in cases where the sonic-resistivity or density-resistivity plot does not give

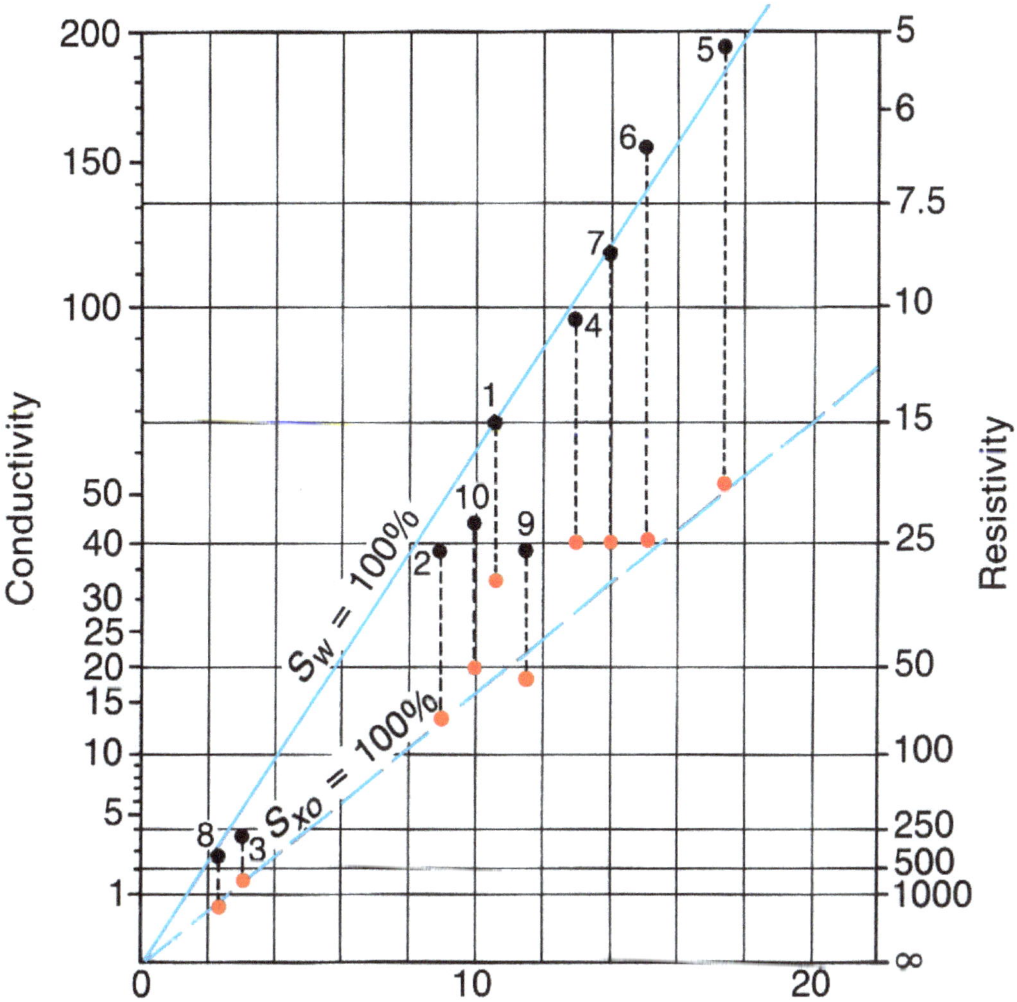

Fig. 3B.47—Resistivity-porosity crossplot showing points from deep-induction and microlaterolog. S_w = 1 and S_{xo} = 1 lines are shown.

a clear answer because of hydrocarbon saturation. The $F\ R_{mf}$ line should be easier to determine because S_{xo} is usually fairly high even in hydrocarbon-bearing formations.

Fig. 3B.47 shows a resistivity-porosity plot in which both the deep induction reading and the microlaterolog at the same levels are plotted in a series of water-bearing formations. The porosity values were derived in this case from a neutron-density crossplot. The plots from the two logs define two trends corresponding respectively to S_w = 1 (using deep induction) and S_{xo} = 1 (using microlaterolog data). The points not in these trends can be divided into two groups:

1. Points whose microlaterolog readings fall on the S_{xo} = 1 line but whose deep induction log readings fall below the S_w = 1 line (Points 2, 9, and 10) are probably the result of either deep invasion or adjacent-bed effect in which deep resistivity is greater than R_t.

2. Points whose induction log readings fall on the S_w = 1 line but whose microlaterolog points fall above the S_{xo} = 1 line are possibly a result of shallow invasion in which RMLL is lower than R_{xo}.

Resistivity-porosity plots are thus often more informative if the short-spaced resistivity or medium-induction values are also plotted. Not only does this permit an appreciation of invasion effects, but it may also indicate moved oil.

3B.11.2 R_{wa} Comparison. If water saturation is assumed to be 100%, the Archie water saturation equation (Eq. 3B.10) reduces to

$$R_{wa} = \frac{R_t}{F} \approx \frac{R_{ID}}{F} \text{ . ... (3B.16)}$$

The term R_{wa} is used in Eq. 3B.16 rather than R_w to indicate that this is an apparent formation water resistivity. It is only equal to R_w in 100% water-bearing formations. In hydrocarbon-bearing formations, R_{wa} computed from Eq. 3B.12 will be greater than R_w. Indeed, by combining Eqs. 3B.16 and 3B.10, the relationship between S_w, R_{wa}, and, R_w can be shown to be

$$S_w = \sqrt{R_w / R_{wa}} \text{ . ... (3B.17)}$$

The R_{wa} technique can, therefore, be useful for identifying potential hydrocarbon-bearing zones and for obtaining R_w values.

In practice, R_{wa} is obtained by simply dividing the deep induction resistivity (or deep laterolog resistivity) by the formation factor obtained from a porosity log or a combination of porosity logs. Today, a continuous R_{wa} computation is made over a long interval of the borehole in real time. If one has only paper logs, many individual manual computations are made so as to approximate a continuous computation.

Resistivity-Ratio Methods. In resistivity-ratio methods, it is assumed that a formation is divided into two distinct regions—a flushed zone and a noninvaded zone. Both zones have the same F, but each contains water of a distinct resistivity (R_{mf} in the invaded zone and R_w in the noninvaded zone). The resistivities of the two zones must be measurable or derivable from logs, and methods for determining the resistivity of the water in each zone must be available.

Because of the necessary assumptions, the resistivity-ratio methods have limitations, but when no porosity or formation factor data are available, they are sometimes the only choice. The principal limitation arises from the inability of any resistivity device to measure either R_x or R_t totally independent of the other. Simply put, invasion must be deep enough to allow a shallow investigating resistivity device to measure R_{xo} but not so deep that a deep-resistivity device cannot measure R_t.

Another difficulty appears when hydrocarbons are present. In this case, some knowledge or assumption of the value of the flushed or invaded zone saturation is necessary.

Flushed-Zone Method. If n = 2 is assumed and Eq. 3B.10 is divided by Eq. 3B.12,

$$\left(\frac{S_w}{S_{xo}}\right)^2 = \frac{R_{xo} / R_t}{R_{mf} / R_w} \text{ . ... (3B.18)}$$

This equation gives the ratio of S_w to S_{xo}, and no knowledge of formation factor or porosity is needed. R_{xo} may be found from a microresistivity log, R_t from an induction or laterolog, and R_{mf}/R_w from measured values or from the SP curve.

The ratio is valuable in itself as an index of oil movability. If S_w / S_{xo} = 1, then no hydrocarbons have been moved by invasion, whether or not the formation contains hydrocarbons. If S_w /

S_{xo} is approximately 0.7 or less, movable hydrocarbons are indicated. The value of S_w / S_{xo}, along with ϕ and S_{wo}, is useful in evaluating reservoirs.

To determine S_w from Eq. 3B.18, S_{xo} must be known. For moderate invasion and average residual oil saturation, an empirical relation between S_w and S_{xo} has been found useful: $S_{xo} = S_w^{1/5}$. Inserting this into Eq. 3B.18 gives:

$$S_w = \left(\frac{R_{xo} / R_t}{R_{mf} / R_w} \right)^{0.625} \quad \text{.......... (3B.19)}$$

Service companies provide charts for graphical solution of this equation, or it can be easily programmed into a spreadsheet.

Invaded-Zone Method. The invaded-zone method is useful for water saturation determination when only an ES, IES, or other early-resistivity log is available and no porosity-log or formation-factor data exist. (This section also uses some early nomenclature.) For application of the method, R_i / R_m must be at least 10.

Archie's equation for the invaded zone is

$$S_i^2 = \frac{F R_z}{R_t}, \quad \text{.......... (3B.20)}$$

where R_z is the resistivity of the water in the invaded zone. Because of incomplete flushing, R_z is a mixture of mud filtrate, R_{mf}, and formation water, R_w.

Studies of many logs suggest that S_i and S_w are related by

$$S_i = \sqrt{S_w} \quad \text{.......... (3B.21)}$$

Dividing the noninvaded-zone water saturation equation (Eq. 3B.10) by Eq. 3B.20 and using the relationship presented in Eq. 3B.21 yields an expression for S_w:

$$S_w = \frac{R_i / R_t}{R_z / R_w} \quad \text{.......... (3B.22)}$$

To use Eq. 3B.22, R_t is taken from a deep resistivity device such as a deep induction or deep laterolog (corrected for borehole effect and bed thickness). R_i is taken from a shallow resistivity device such as a Laterolog 8, 16-in. normal, or SFL (corrected for borehole effect and bed thickness).

R_z is given by the relationship

$$\frac{1}{R_z} = \frac{z}{R_w} + \frac{(1-z)}{R_{mf}}, \quad \text{.......... (3B.23)}$$

where z is the fraction of the invaded zone pore water, which is formation water, and $1 - z$ is the fraction that is mud filtrate. Experience has indicated that z varies from 0.075 in cases of normal invasion to 0.035 in cases of deep invasion or vuggy formations.

Fig. 3B.48 solves Eq. 3B.22. It is entered with R_{mf} / R_w on the appropriate z scale and R_i / R_t (oblique lines) to determine S_w. When R_i / R_t is close to unity, some caution is required. The formation may be extremely invaded or there may be little invasion, or it may be dense and

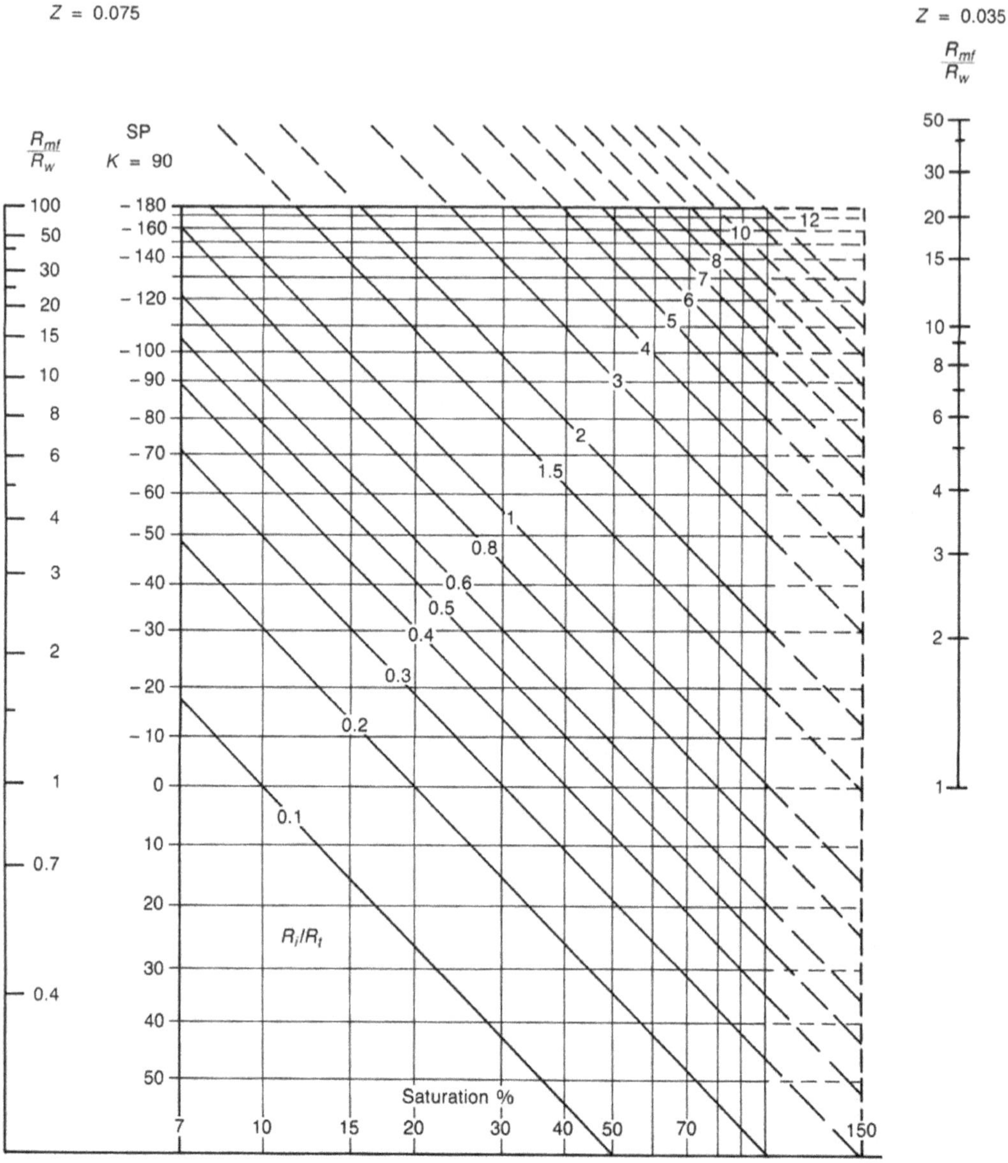

Fig. 3B.48—Empirical resistivity-ratio method.

impermeable. On the other hand, many good hydrocarbon-bearing reservoirs will have $R_i/R_t \approx 1$.

3B.11.3 R_{xo}/R_t Quicklook. The R_{xo}/R_t quicklook method can be used to identify hydrocarbon-bearing formations and to indicate hydrocarbon movability (producibility). When S_w/S_{xo} is 1 in a permeable zone, the zone will produce water or be nonproductive regardless of water saturation. A value S_w/S_{xo} significantly less than 1 indicates that the zone is permeable and contains some hydrocarbons, and that the hydrocarbons have been flushed (moved) by invasion. Thus, the zone contains producible hydrocarbon.

Eq. 3B.18 can be written as

$$\frac{S_w}{S_{xo}} = \left[\frac{R_{xo}/R_t}{(R_{mf}/R)_{SP}}\right]^{1/2}, \quad \text{(3B.24)}$$

which shows that an indication of S_w/S_{xo} can be obtained by comparing R_{xo}/R_t with R_{mf}/R_w, where the subscript SP emphasizes that R_{mf}/R_w is derivable from the SP. Equivalently, the comparison can be between log R_{xo}/R_t and the SP curve for an indication of log S_w/S_{xo}.

The value of log R_{xo}/R_t is computed from solving the three or more resistivity logs for invasion parameters. It is used as an overlay comparison curve with the SP. Separations between the log R_{xo}/R_t curve, properly scaled to match the SP, and the SP curve provide a quick-look location of producible hydrocarbons.

Originally, log R_{xo}/R_t was computed from R_{LL8}/R_{ID} or R_{SFL}/R_{ID}. Use was made of the observation that over a wide range of invasion diameters (from approximately 20 to 100 in.), R_{xo}/R_t depends primarily on the value of R_{LL8}/R_{ID} or R_{SFL}/R_{ID}. The relationship used for the LL8 device was

$$R_{xo}/R_t = 1.85(R_{LL8}/R_{ID}) - 0.85. \quad \text{(3B.25)}$$

For the SFL device, it was

$$R_{xo}/R_t = 1.45(R_{SFL}/R_{ID}) - 0.45. \quad \text{(3B.26)}$$

Much more sophisticated algorithms are now used to obtain R_{xo}/R_t. These values are output in real time as separate logs.

To interpret the R_{xo}/R_t quick-look curve, the impermeable zones must be eliminated by reference to the SP, GR, or microlog curves or by resistivity ratios. Then, if the SP and R_{xo}/R_t (actually –K log R_{xo}/R_t) curves coincide in a permeable zone, the zone will most probably produce water. If, however, the R_{xo}/R_t curve reads appreciably lower (i.e., to the right) than the SP, the zone should produce hydrocarbons. An R_{xo}/R_t value less than the SP amplitude indicates movable hydrocarbons are present.

The R_{xo}/R_t quick-look technique is applicable to fresh mud conditions ($R_{xo} > R_t$) in formations where invasion falls within the limits demanded by the R_{xo}/R_t computation. For the simpler computation technique using Eq. 3B.25 and –25, that is for d_i 30 to 70 in.; for the more sophisticated techniques, that is, between 20 and 120 in. Even in the more restrictive case, however, any errors are optimistic. In other words, water zones may appear to be hydrocarbon-productive. This constitutes a safeguard against overlooking pay zones, and it is considered a desirable feature in any quick-look approach.

The R_{xo}/R_t technique efficiently handles variations in formation water resistivity, R_w, and in shaliness. Any change in R_w is reflected similarly into both the computed R_{xo}/R_t and the SP amplitude. Thus, comparing the two curves still permits formation-fluid identification. Shaliness also affects the two curves in a similar manner. All other things remaining constant, shaliness reduces the R_{xo}/R_t value and the SP amplitude. Finally, the R_{xo}/R_t quick-look technique does not require porosity data, nor use of any F–ϕ relationships.

Fig. 3B.49 is an example of a shaly gas sand at 3,760 through 3,788 ft and several water-productive sands with varying amounts of shaliness. The productive-gas sand is identified by the separation between the R_{xo}/R_t and SP curves. Water-productive zones are shown by lack of separation. In shaly water zones, the variation in the SP curve is essentially the same as the variation in the R_{xo}/R_t ratio—a result of the same shale. Therefore, the comparison is not significantly disturbed by shaliness. Neither is it disturbed by variations in R_w.

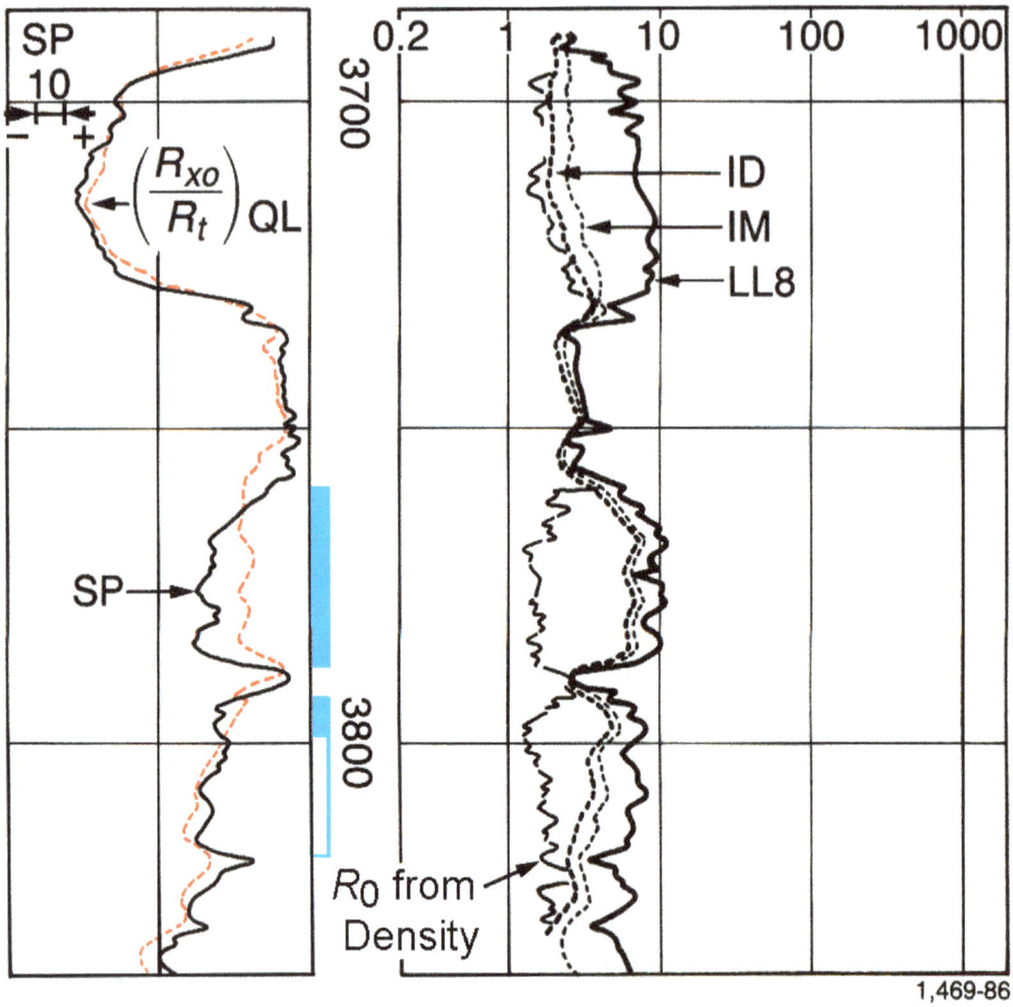

Fig. 3B.49—Examples of R_{xo}/R_t quick-look curve used for comparison with SP to identify zones with movable hydrocarbons.

Estimates of water saturation and saturation ratio in clean formations can be made by comparing the R_{xo}/R_t and SP curves. Eq. 3B.24 permits S_w/S_{xo} to be estimated, and then Eq. 3B. 19 allows S_w to be estimated.

3B.12 Shaly Formations

Shaly-sand analysis has been the subject of much continuing work over the past 20 years, and a detailed coverage of that work is far beyond the scope of this *Handbook*. We will present the problem to be solved and a couple of methods that can be applied to older logs where detailed information is not available.

Shales are one of the more important common constituents of rocks in log analysis. Aside from their effects on porosity and permeability, this importance stems from their electrical properties, which have a great influence on the determination of fluid saturations.

Shales are loose, plastic, fine-grained mixtures of clay-sized particles or colloidal-sized particles and often contain a high proportion of clay minerals. Most clay minerals are structured in sheets of alumina-octahedron and silica-tetrahedron lattices. There is usually an excess of negative electrical charges within the clay sheets. The substitution of Al^{+++} by ions of lower valence is the most common cause of this excess; the structure of the crystal remains the same. This local electrical imbalance must be compensated to maintain the electrical neutrality of the clay particle. The compensating agents are positive ions—cations or counterions—which cling to the surface of the clay sheets in a hypothetical dry state. The positive surface charge is usually measured in terms of milli-ions equivalents per 100 grams of dry clay minerals and is called the cation exchange capacity (CEC). When the clay particles are immersed in water, the Coulomb forces holding the positive surface ions are reduced by the dielectric properties of water. The counterions leave the clay surface and move relatively freely in a layer of water close to the surface (the electrical balance must be maintained so that the counterions remain close to the clay water interface) and contribute to the conductivity of the rock.

The Archie water saturation equation, which relates rock resistivity to water saturation, assumes that the formation water is the only electrically conductive material in the formation. The presence of another conductive material (i.e., shale) requires either that the Archie equation be modified to accommodate the existence of another conductive material, or that a new model be developed to relate rock resistivity to water saturation in shaly formations. The presence of clay also complicates the definition or concept of rock porosity. The layer of closely bound surface water on the clay particle can represent a very significant amount of porosity. However, this porosity is not available as a potential reservoir for hydrocarbons. Thus, a shale or shaly formation may exhibit a high total porosity, yet a low effective porosity as a potential hydrocarbon reservoir.

The way shaliness affects a log reading depends on the amount of shale and its physical properties. It may also depend on the way the shale is distributed in the formation. Shaly material can be distributed in the formation in three ways:

• Shale can exist in the form of laminae between which are layers of sand. The laminar shale does not affect the porosity or permeability of the sand streaks themselves. However, when the amount of laminar shale is increased and the amount of porous medium is correspondingly decreased, overall average effective porosity is reduced in proportion.

• Shale can exist as grains or nodules in the formation matrix. This matrix shale is termed structural shale; it is usually considered to have properties similar to those of laminar shale and nearby massive shales.

• The shaly material can be dispersed throughout the sand, partially filling the intergranular interstices. The dispersed shale may be in accumulations adhering to or coating the sand grains, or it may partially fill the smaller pore channels. Dispersed shale in the pores markedly reduces the permeability of the formation.

All these forms of shale can, of course, occur simultaneously in the same formation.

Over the years, a large number of models relating resistivity and fluid saturations have been proposed. Many have been developed assuming the shale exists in one of the three specific geometric forms. All these models are composed of a clean sand term, described by the Archie water saturation equation, plus a shale term. The shale term may be fairly simple or quite complex; the shale term may be relatively independent of, or it may interact with, the clean sand term. All the models reduce to the Archie water saturation equation when the fraction of shale is zero; for relatively small amounts of shaliness, most models and methods yield quite similar results.

Only a very few of these models will be reviewed here to provide some flavor and understanding for the evolution of shaly-sand interpretation logic.

3B.12.1 Laminated Sand/Shale—Simplified Model. In this laminar shale model, R_t, the resistivity in the direction of the bedding planes, is related to R_{sh} (the resistivity of the shale laminae) and to R_{sd} (the resistivity of the clean sand laminae) by a parallel resistivity relationship,

$$\frac{1}{R_t} = \frac{1 - V_{\text{lam}}}{R_{\text{sd}}} + \frac{V_{\text{lam}}}{R_{\text{sh}}}, \quad \text{(3B.27)}$$

where V_{lam} is the bulk-volume fraction of the shale, distributed in laminae, each of more-or-less uniform thickness.

For clean-sand laminae, $R_{\text{sd}} = F_{\text{sd}} R_w / S_w^2$, where F_{sd} is the formation resistivity factor of the clean sand. Because $F_{\text{sd}} = a / \phi_{\text{sd}}^2$ (where ϕ_{sd} is the sand-streak porosity) and $f = (1 - V_{\text{lam}})\phi_{\text{sd}}$ (where ϕ is the bulk-formation porosity), then

$$\frac{1}{R_t} = \frac{\phi^2 S_w^2}{(1 - V_{\text{lam}}) a R_w} + \frac{V_{\text{lam}}}{R_{\text{sh}}} . \quad \text{(3B.28)}$$

To evaluate S_w by the laminated model, R_t, R_w, ϕ, V_{lam}, and R_{sh} must be determined.

For the determination of R_t, the problem is the same as for clean formations. If R_w is not known, its determination usually involves looking at a nearby clean sand and solving for R_w using the SP measurement. If the formation is water-bearing, the resistivity and porosity measurements can be used.

For the determination of ϕ and V_{lam}, a combination of porosity logs can be used. For example, as illustrated in **Fig. 3B.50**, a crossplot of ϕ_N and ϕ_B from a density log is effective. The triangle of the figure is defined by the matrix point, water point, and shale point. In this example, the matrix point is at $\phi_N = 0$ (the neutron log was scaled in apparent sandstone porosity) and $\phi_{\text{ma}} = 2.65$ g/cm^3 (quartz matrix). The shale point is at $\phi_N = 50$ p.u. and $\phi_{\text{sh}} = 2.45$ g/cm^3. These values were taken in a nearby thick shale bed; it is assumed that shale laminae in the shaly sand under investigation are similar to the nearby massive shale beds. The water point is, of course, located at $\phi_N = 100$ p.u. and $\phi_B = 1$ g/cm^3. The matrix-water line and shale-water line are linearly divided into porosity; the matrix-shale line and water-shale line are linearly divided into shale percentages.

Point A, plotted as an example, corresponds to log readings of $\phi_B = 2.2$ g/cm^3 and $\phi_N =$ 33 p.u. Interpretation by the lines on the plot yields 23% and V_{sh} (or V_{lam}) = 16 %.

Direct use of this crossplot assumes 100% water saturation in the zone investigated by the tools. Because oil has a density and hydrogen content normally not greatly different from water, this crossplot technique can be used with acceptable accuracy in oil-bearing formations. The presence of gas or light hydrocarbon, however, decreases ϕ_N and decreases ϕ_B. This would cause the point to shift in a northwesterly direction. When gas or light hydrocarbons are present, an additional shaliness indicator, such as GR or SP data, is needed to evaluate the amount of the shift.

Using the laminated model, an equation for R_{xo} analogous to Eq. 3B.28 could be written. S_{xo} would replace S_w, and R_{mf} would replace R_w. The other terms (ϕ, V_{lam}, and R_{sh}) remain the same in the two equations. Assuming $S_{xo} = S_w^{1/5}$ (as in the flushed-zone ratio method) and the ratio of the PSP (SP deflection in the shaly sand) to the SSP (SP deflection in a nearby clean sand of similar formation water) is a measure of shaliness, V_{lam}, water saturation could be calculated from R_{xo} / R_t and PSP in the shaly sand and SSP (or R_{mf} / R_w) in a nearby clean sand.

3B.12.2 Dispersed Clay. In this model, the formation conducts electrical current through a network composed of the pore water and dispersed clay. As suggested by de Witte,[67] it seems acceptable to consider that the water and the dispersed shale conduct an electrical current like a mixture of electrolytes. Development of this assumption yields

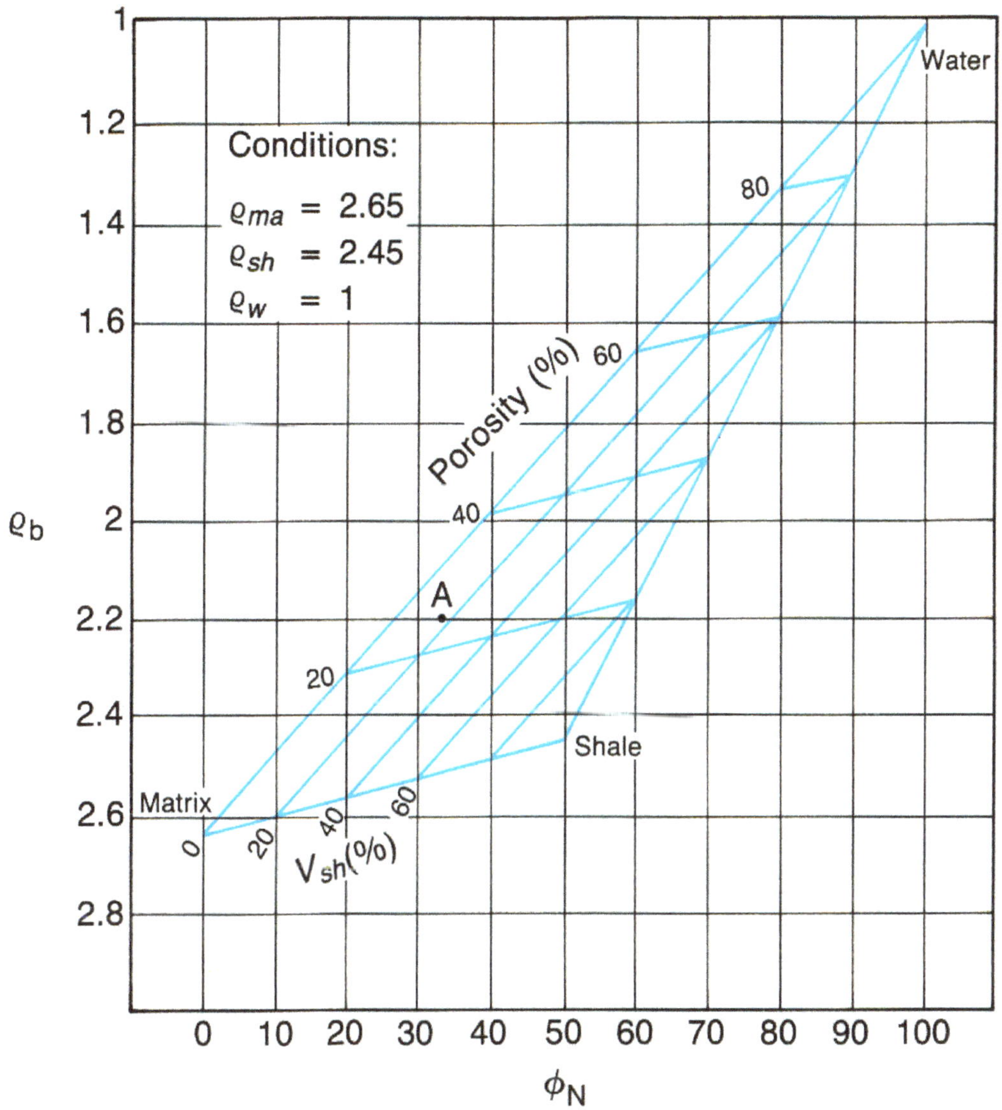

Fig. 3B.50—Neutron-density crossplot showing matrix, water, and shake points, scaled for determination of V_{shale} and porosity.

$$\frac{1}{R_t} = \frac{\phi_{im} S_{im}}{a}\left(\frac{q}{R_{shd}} + \frac{S_{im}}{R_w}\right), \qquad (3B.29)$$

where ϕ_{im} = intermatrix porosity, which includes all the space occupied by fluids and dispersed shale; S_{im} = the fraction of the intermatrix porosity occupied by the formation-water, dispersed-shale mixture; q = the fraction of the intermatrix porosity occupied by the dispersed shale; and R_{shd} = the resistivity of the dispersed shale. Also, it can be shown that $S_w = (S_{im} - q)/(1 - q)$, where S_w is the water saturation in the fraction of true effective formation porosity.

Combining these relations and solving for S_w yields

$$S_w = \frac{\sqrt{\frac{aR_w}{\phi_{im}^2 R_t} + \left[\frac{q(R_{shd} - R_w)}{2R_{shd}}\right]^2} - \frac{q(R_{shd} + R_w)}{2R_{shd}}}{1 - q} \quad \text{(3B.30)}$$

Usually, ϕ_{im} can be obtained directly from a sonic log because dispersed clay in the rock pores is seen as water by the sonic measurement. The value of q can be obtained from a comparison of a sonic and density log. Indeed, if $\rho_{shd} \cong \rho_{ma}$, then $qsv\ \phi\ (\phi_{SV} - \phi_D)/\phi_{SV}$, where ϕ_{SV} and ϕ_D are the sonic and density derived porosities, respectively. In this case, ϕ_D approximates ϕ, the effective porosity available for fluid saturation.

The value of R_{sh} is more difficult to evaluate. It is usually taken as equal to R_{sh} in nearby shale beds. Fortunately, its value is not too critical if it is at least several times greater than R_w. In fact, when R_w is small compared to R_{sh} and the sand is not too shaly, Eq. 3B.30 can be simplified to a form independent of R_{sh}:

$$S_w = \frac{\sqrt{\frac{aR_w}{\phi_{im}^2} + \frac{q^2}{4}} - \frac{q}{2}}{1 - q} \quad \text{(3B.31)}$$

3B.12.3 Total Shale Relationship. Based upon the previously described ideas, laboratory investigations, and field experience, it has been found that a simple relationship of the following form works well for many shaly formations independent of the distribution of the shale and over the range of S_w values encountered in practice:

$$\frac{1}{R_t} = \frac{\phi^2 S_w^2}{(1 - V_{sh})aR_w} + \frac{V_{sh}S_w}{R_{sh}} \quad \text{(3B.32)}$$

In using this equation, R_{sh} is taken equal to the resistivity of the adjacent shale beds, and V_{sh} is the shale fraction as determined from a total shale indicator.

Before the Waxman-Smits formulation, equations of the form of Eq. 3B.29 and 3B.32 gained wide acceptance in the evaluation of shaly sands. These equations have a general form of

$$\frac{1}{R_t} = \alpha S_w^2 + \gamma S_w, \quad \text{(3B.33)}$$

where α denotes a predominant sand term that is dependent on the amount of sand, its porosity, and the resistivity of the saturating water. The sand term always reduces to Archie's water saturation equation when the shale fraction is zero. γ denotes a predominant shale term that depends on the amount and resistivity of the shale.

3B.13 Dual Water Models

In 1968, Waxman and Smits proposed, based on extensive laboratory work and theoretical study, a saturation-resistivity relationship for shaly formations that related the resistivity contribution of the shale (to the overall resistivity of the formation) to the CEC of the shale.[69] The Waxman-Smits relationship is

TABLE 3B.1—DUAL-WATER MODEL

Solids			Fluids		
Matrix	Silt	Dry clay	Bound water	Free water	Hydrocarbons
Matrix		Shale		Effective porosity	
				Total porosity	

$$\frac{1}{R_t} = \frac{S_w^2}{F^* R_w} + \frac{B Q_v S_w}{F^*}, \qquad (3B.34)$$

where F^* is the formation factor of the interconnected porosity, S_w also relates to the interconnected pores, B is the equivalent conductance of the sodium clay-exchange cations as a function of the formation water conductivity, and Q_v is the CEC of the rock per unit pore volume.

Unfortunately, a continuous in-situ measurement of rock CEC was not available when this study was presented. As a result, the dual water model was developed as a practical solution.[70] The dual water method is based on three premises:

- The conductivity of clay is because of its CEC.
- The CEC of pure clays is proportional to the specific surface area of the clay.
- In saline solutions, the anions are excluded from a layer of water around the surface of the grain. The thickness of this layer expands as the salinity of the solution (below a certain limit) decreases, and the thickness is a function of salinity and temperature.

Therefore, because CEC is proportional to specific area (area per unit weight) and to the volume of water in the counter-ion exclusion layer per unit weight of clay. Consequently, the conductivity of clay is proportional to the volume of the counter-ion exclusion layer, this layer being "bound" to the surface of the clay grains. For clays, this very thin sheet of bound water is important because of the large surface areas of clays relative to sand grains (several magnitudes greater). Therefore, in the dual water model, a clay is modeled as consisting of two components: bound water and clay minerals.

The clay minerals are modeled as being electrically inert; the clay electrical conductivity is modeled as being derived from the conductivity of the bound water, C_{wb}. C_{wb} is assumed to be independent of clay type (from the second postulate described previously). The amount of bound water varies according to clay type, being higher for the finer clays (with higher surface areas), such as montmorillonite, and lower for coarser clays, such as kaolinite. Salinity also has an effect; in low-salinity waters (roughly < 20,000 ppm NaCl), the diffuse layer expands.

The bound water is immovable under normal conditions; therefore, the volume it occupies cannot be displaced by hydrocarbons. Because the clay minerals (dry colloids) are considered electrically inert, they may be treated just as other minerals. Schematically, shaly formations are modeled with the dual water model, as illustrated in **Table 3B.1**.

For most rocks (except for conductive minerals such as pyrite, which cannot be treated in this way) only the porous part needs to be considered when discussing electrical properties, and it is treated according to the Archie water-saturation equation. The equation becomes

$$\sigma_t = \frac{\phi_t^m S_{wt}^m}{a} \sigma_{we}, \qquad (3B.35)$$

where *a, m,* and *n* have the usual Archie connotations. σ_t is the conductivity of the noninvaded, virgin formation ($1/R_t$), and σ_{we} is the equivalent conductivity of the waters in the pore space.

Note that ϕ_t and S_{wt} refer to total pore volume; this includes the pore volumes saturated with the bound water and the formation connate water (sometimes called the "free" water). The equivalent water conductivity, σ_{we}, is

$$\sigma_{we} = \frac{V_w \sigma_w + V_{wb} \sigma_{wb}}{V_w + V_{wb}}, \quad \text{(3B.36)}$$

where V_w and V_{wb} are the bulk volumes of formation water and bound water, respectively, and σ_w and σ_{wb} are their conductivities.

In terms of saturation, Eq. 3B.36 becomes

$$\sigma_{we} = \frac{\phi_t \sigma_w \left(S_{wt} - S_{wb}\right) + \phi_t \sigma_{wb} S_{wb}}{\phi_t \left(S_{wt} - S_{wb}\right) + \phi_t S_{wb}}, \quad \text{(3B.37)}$$

or

$$\sigma_{we} = \sigma_w \frac{S_{wt} - S_{wb}}{S_{wt}} + \sigma_{wb} \frac{S_{wb}}{S_{wt}}, \quad \text{(3B.38)}$$

or

$$\sigma_{we} = \sigma_w + \left(\sigma_{wb} - \sigma_w\right) \frac{S_{wb}}{S_{wt}}, \quad \text{(3B.39)}$$

where S_{wb} is the bound water saturation (i.e., the fraction of the total pore volume occupied by the bound water).

Eq. 3B.39 describes the equivalent-water conductivity as a function of the formation water conductivity plus the bound-water conductivity. The saturation equation (Eq. 3B.35) becomes

$$\sigma_t = \frac{\phi_t^m S_{wt}^n}{a} \left[\sigma_w + \frac{S_{wb}}{S_{wt}} \left(\sigma_{wb} - \sigma_w\right) \right]. \quad \text{(3B.40)}$$

The porosity and water saturation of the sand (clean formation) phase (that is, the nonclay phase) of the formation is obtained by subtracting the bulk-volume fraction of bound water (ϕ_t S_{wb}). Therefore, the effective porosity is

$$\phi = \sigma_t \left(1 - S_{wb}\right), \quad \text{(3B.41)}$$

and the water saturation is

$$S_w = \frac{S_{wt} - S_{wb}}{1 - S_{wb}}. \quad \text{(3B.42)}$$

To evaluate a shaly formation using the dual water model, four parameters must be determined. They are σ_w (or R_w), σ_{wb} (or R_{wb}), ϕ_t, and S_{wb}. A neutron-density crossplot provides a good value of ϕ_t. S_{wb} is obtainable from a variety of shale-sensitive measurements (SP, GR,

ϕ_N, R_t, $\phi_N - \rho_B$, $t - \rho_B$, etc.). R_{wb} and R_w are usually determined by the log analyst and entered as input parameters.

3B.14 Summary

This chapter presents the fundamentals of the various logging tools used to measure formation resistivities, conductivities, and naturally occurring currents that normally exists in wellbores containing conductive fluids. The borehole and formation environments are described, along with their effect on log response and what can be done to determine true formation resistivity. Simplified manual methods for interpreting log responses are presented, along with a description of the expanded amount of information that can now be generated with sophisticated computer programs. It has been the objective to provide sufficient insight to use existing log data and to indicate what tools are available to capture data in new wells. Other sections in this petrophysics chapter contain information on other types of logs that are needed in conjunction with resistivity logs to obtain an understanding of reservoir rock and fluid properties.

Service companies, oil companies, and many third parties have developed software to calculate $S_{w,}$ in most situations. Planning logging jobs will always be a balance of the types of tools available, the data needed, and the costs of acquisition. In the long run, it is best to include all tools that provide sufficient data for the formation at hand.

Nomenclature

a_{mf} = mud filtrate chemical activity
a_w = formation water chemical activity
A = area, m^2
d_i = diameter of invasion (in., m)
E_k = electrokinetic potential
E_{kmc} = electrokinetic potential of the mudcake
E_{ksh} = electrokinetic potential of shale
F = formation factor relating resistivity to porosity
g = induction-response function
G = induction integrated radial-response function
h_{mc} = mudcake thickness
I = electrical current, Amperes
L = length, m
r = resistance, ohm
R = resistivity (ohm·m)
R_{ann} = resistivity of the annulus
R_h = resistivity in the horizontal direction (ohm·m)
R_m = resistivity of the mud column (ohm·m)
R_{mc} = resistivity of the mudcake
R_{mf} = resistivity of the mud filtrate
R_{xo} = resistivity of the invaded zone
R_t = resistivity of the uninvaded formation
R_v = resistivity in the vertical direction (ohm·m)
R_w = resistivity of the formation connate water (ohm·m)
R_{wa} = apparent water resistivity from deep resistivity and porosity
S_{xo} = water saturation of the invaded zone
S_w = water saturation in the uninvaded zone
t = acoustic travel time (μs/ft)
t_{ma} = acoustic travel time of the rock matrix(μs/ft)

V = electrical voltage, volts
V_{sd} = fraction of the total formation volume that is sand
V_{sh} = fraction of the total formation volume that is shale
ρ = density
ρ_{ma} = density of the rock matrix
σ = conductivity, mS/m
σ_m = conductivity of the mud column, mS/m
ϕ = porosity

References

1. Schlumberger, C., Schlumberger, M., and Leonardon, E.G.: "A New Contribution to Subsurface Studies by Means of Electrical Measurements in Drill Holes," *Trans.*, AIME (1933) **103.**
2. Schlumberger, C., Schlumberger, M., and Leonardon, E.G.: "Electrical Coring; A Method of Determining Bottom-Hole Data by Electrical Measurements," *Trans.*, AIME (1932).
3. Schlumberger, C., Schlumberger, M., and Leonardon, E.G.: "Some Observations Concerning Electrical Measurements in Anisotropic Media and Their Interpretation," *Trans.*, AIME (1934).
4. Lynch, E.J.: "Formation Evaluation," *Harper's Geoscience Series*, Harper and Row, New York City (1962).
5. Anderson, B.I.: *Modeling and Inversion Methods for the Interpretation of Resistivity Logging Tool Response*, Delft U. Press, Delft, The Netherlands (2001).
6. Doll, H.G.: "The Laterolog: A New Resistivity Logging Method With Electrodes Using an Automatic Focusing System," *Trans.*, AIME (1951) **192,** 305.
7. *The Dual Laterolog*, Schlumberger, Houston (1970).
8. Woodhouse, R.: "The Laterolog Groningen Phantom Can Cost You Money," paper R presented at the 1978 SPWLA Annual Logging Symposium.
9. Davies, D.H. *et al.*: "Azimuthal Resistivity Imaging: A New Generation Laterolog," paper SPE 24676 presented at the 1992 SPE Annual Technical Conference and Exhibition, Washington, DC, 4–7 October.
10. Smits, J.W. *et al.*: "High Resolution From a new Laterolog With Azimuthal Imaging," paper SPE 30584 presented at the 1995 SPE Annual Technical Conference and Exhibition, Dallas, 22–25 October.
11. Smits, J.W. *et al.*: "Improved Resistivity Interpretation Utilizing a New Array Laterolog Tool and Associated Inversion Processing," paper SPE 49328 prepared for presentation at the 1998 SPE Annual Technical Conference and Exhibition, New Orleans, 27–30 September.
12. Iskovick, G.B. *et al.*: "High Definition Lateral Log," paper PP presented at the 1998 SPWLA Annual Logging Symposium, Keystone, Colorado, 6–9 June.
13. Alpin, L.: "The Method of Electric Logging in the Borehole With Casing," U.S. Patent 56,026 (1939).
14. Kaufman, A.A.: "Conductivity Determination in a Formation Having a Cased Well," U.S. Patent 4,796,186 (1989).
15. Vail, W.B.: "Method and Apparatus for Measurement of Resistivity of Geological Formations From Within Cased Boreholes," U.S. Patent 4,820,989 (1989).
16. Béguin, P. *et al.*: "Recent Progress on Formation Resistivity Through Casing," paper CC presented at the 2000 SPWLA Annual Logging Symposium, Dallas, 4–7 June.
17. Maurer, H.M. and Hunziker, J.: "Early Results of Through-Casing Field Tests," paper DD presented at the 2000 SPWLA Annual Logging Symposium, Dallas, 4–7 June.
18. Rosthal, R.A. *et al.*: "Formation Evaluation and Geological Interpretation From the Resistivity-at-the-Bit Tool," paper SPE 30550 presented at the 1995 SPE Annual Technical Conference and Exhibition, Dallas, 22–25 October.
19. Doll, H.G.: "Introduction to Induction Logging and Application to Wells Drilled With Oil Base Mud," *JPT* (June 1949) 148; *Trans.*, AIME, **186.**

20. Moran, J.H. and Kunz, K.S.: "Basic Theory of Induction Logging and Application to Study of Two-Coil Sondes," *Geophysics* (December 1962) **44,** No. 7, 829.
21. Anderson, B. and Barber, T.: *Induction Logging,* Schlumberger, Houston (1995).
22. Tanguy, D.R.: "Induction Well Logging," U.S. No. Patent 3,067,383 (1962).
23. Tanguy, D.R.: "Methods and Apparatus for Investigating Earth Formations Featuring Simultaneous Focused Coil and Electrode System Measurements," U.S. Patent No. 3,329,889 (1967).
24. Attali, G.: "Methods and Apparatus for Investigating Earth Formations Including Measuring the Resistivity of Radially Different Formation Zones," U.S. Patent No. 3,453,530 (1969).
25. Anderson, B.I. and Barber, T.D.: "Deconvolution and Boosting Parameters for Obsolete Schlumberger Induction Tools," *The Log Analyst* (1999) **40,** No. 1, 133.
26. Barber, T.: "Introduction to the Phasor Dual Induction Tool," *JPT* (September 1985) 1699.
27. Kienitz, C. *et al.:* "Accurate Logging in Large Boreholes," paper III presented at the 1986 SPWLA Annual Logging Symposium, Houston, 9–13 June.
28. Hunka, J.F. *et al.*: "A New Resistivity Measurement System for Deep Formation Imaging and High-Resolution Formation Evaluation," paper SPE 20559 presented at the 1990 SPE Annual Technical Conference and Exhibition, New Orleans, 23–26 September.
29. Grove, G.P. and Minerbo, G.N.: "An Adaptive Borehole Correction Scheme For Array Induction Tools," paper P presented at the 1991 SPWLA Annual Logging Symposium, Midland, Texas, 16–19 June.
30. Barber, T.D. *et al.*: "A Multiarray Induction Tool Optimized for Efficient Wellsite Operation," paper SPE 30583 presented at the 1995 SPE Annual Technical Conference and Exhibition, Dallas, 22–25 October.
31. Barber, T.D. and Rosthal, R.A.: "Using a Multiarray Induction Tool To Achieve Logs With Minimum Environmental Effects," paper SPE 22725 presented at the 1991 SPE Annual Technical Conference and Exhibition, Dallas, 6–9 October.
32. Barber, T.D.: "Interpretation of AIT-Family Logs in Invaded Formations at High Relative Dip Angles," paper A presented at the 1998 SPWLA Annual Logging Symposium, Keystone, Colorado, 26–29 May.
33. Barber, T.D. and Minerbo, G.N.: "An Analytic Method of Producing Multiarray Induction Logs That Are Free of Dip Effect," paper SPE 77718 presented at the 2002 SPE Annual Technical Conference and Exhibition, San Antonio, Texas, 29 September–2 October.
34. La Vigne, J., Barber, T., and Bratton, T.: "Strange Invasion Profiles: What Multiarray Induction Logs Can Tell Us About How Oil-based Mud Affects the Invasion Process and Wellbore Stability," paper B presented at the 1997 SPWLA Annual Logging Symposium, Houston, 15–18 June.
35. Beard D.R., Zhou, Q., and Bigelow, E.L.: "Practical Application of a New Multichannel and Fully Digital Spectrum Induction System," paper SPE 36504 presented at the 1996 SPE Annual Technical Conference and Exhibition, Denver, 6–9 October.
36. Beste, T. *et al.:* "A New High Resolution Array Induction Tool," paper C presented at the 2000 SPWLA Annual Logging Symposium, Dallas, 4–7 June.
37. Martin, D.W., Spencer, M.C., and Patel, H.K.: "The Digital Induction—A New Approach To Improving the Response of the Induction Measurement," paper M presented at the 1984 SPWLA Annual Logging Symposium, New Orleans, 10–13 June.
38. Samworth, J.R. *et al.:* "The Array Induction Tool Advances Slim-hole Logging Technology," paper Y presented at the 1994 SPWLA European Formation Evaluation Symposium, Aberdeen, 11–13 October.
39. Tanguy, D.R. and Zoeller, W.A.: "Applications of Measurements While Drilling," paper SPE 10324 presented at the 1981 SPE Annual Technical Conference and Exhibition, San Antonio, Texas, 5–7 October.
40. Jan, Y.M. and Campbell, R.L.: "Borehole Correction of MWD Gamma Ray and Short Normal Resistivity Logs," paper PP presented at the 1984 SPWLA Annual Logging Symposium, New Orleans, 10–13 June.
41. Rodney, P.F. *et al.:* "The Electromagnetic Wave Resistivity MWD Tool," *SPEDE* (October 1986) 337.

42. Coope, D., Shen, L.C., and Huang, F.S.: "The Theory of 2 MHz Resistivity Tool and Its Application to Measurement While Drilling," *The Log Analyst* (May/June 1984) 35.
43. Clark, B. *et al.*: "Electromagnetic Propagation Logging While Drilling: Theory and Experiment," *SPEFE* (September 1990) 263.
44. Clark, B. *et al.*: "A Dual Depth Resistivity Measurement for Formation Evaluation While Drilling," paper A presented at the 1988 SPWLA Annual Logging Symposium, San Antonio, 5–8 June.
45. Habashy, T. and Anderson, B.: "Reconciling Differences in Depth of Investigation between 2-MHz Phase Shift and Attenuation Resistivity Measurements," paper E presented at the 1991 SPWLA Annual Logging Symposium, Midland, Texas, 16–19 June.
46. Bonner, S.D. *et al.*: "New 2-MHz Multiarray Borehole-Compensated Resistivity Tool Developed for MWD in Slimholes," paper SPE 30547 presented at the1995 SPE Annual Technical Conference and Exhibition, Dallas, 22–25 October.
47. Rodney, P.F. *et al.*: "An MWD Multiple Depth of Investigation Electromagnetic Wave Resistivity Sensor," paper D presented at the 1991 SPWLA Annual Logging Symposium, Midland, Texas, 16–19 June.
48. Oberkircher, J., Steinberger, G., and Robbins, B.: "Applications for a Multiple Depth of Investigation MWD Resistivity Measurement Device," paper OO presented at the 1993 SPWLA Annual Logging Symposium, Calgary, 13–16 June.
49. Fredericks, P.D., Hearn, F.D., and Wisler, M.M.: "Formation Evaluation While Drilling With a Dual Propagation Resistivity Tool," paper SPE 19622 presented at the 1989 SPE Annual Technical Conference and Exhibition, San Antonio, Texas, 8–11 October.
50. Meyer, W.H. *et al.*: "A New Slimhole Multiple Propagation Resistivity Tool," paper NN presented at the 1994 SPWLA Annual Logging Symposium, Tulsa, 19–22 June.
51. Gianzero, S. *et al.*: "New Developments in 2 MHz Electromagnetic Wave Resistivity Measurements," paper MM presented at the 1994 SPWLA Annual Logging Symposium, Tulsa, 19–22 June.
52. Griffiths, R., Barber, T., and Faivre, O.: "Optimal Evaluation of Formation Resistivities Using Array Induction and Array Laterolog Tools," paper BBB presented at the 2000 SPWLA Annual Logging Symposium, Dallas, 4–7 June.
53. Crary, S. *et al.*: "Effect of Resistive Invasion on Resistivity Logs," paper SPE 71708 presented at the 2001 SPE Annual Technical Conference and Exhibition, New Orleans, 30 September–3 October.
54. Doll, H.G.: "The Microlaterolog," *JPT* (January 1953) 17; *Trans.*, AIME, **198.**
55. Doll, H.G.: "The MicroLog: A New Electrical Logging Method for Detailed Determination of Permeable Beds," *Trans.*, AIME (1950) **189,** 155.
56. Suau, J. *et al.*: "The Dual Laterolog-Rxo Tool," paper SPE 4018 presented at the 1972 SPE Annual Meeting, Dallas, 8–11 October.
57. Eisenmann, P. *et al.*: "Improved R_{xo} Measurements Through Semi-Active Focusing," paper SPE 28437 presented at the 1994 SPE Annual Technical Conference and Exhibition, New Orleans, 25–28 September.
58. Howard, A.Q.: "A New Invasion Model for Resistivity Log Interpretation," *The Log Analyst* (March–April 1992) **33,** No. 2, 96.
59. Singer, J. and Barber, T.: "The Effect of Transition Zones on the Response of Induction Logs," paper L presented at the 1988 SPWLA Annual Logging Symposium, San Antonio, Texas, 5–8 June.
60. Ekstrom, M. *et al.*: "Formation Imaging With Microelectrical Scanning Arrays," *The Log Analyst* (March 1987) 294.
61. Rosthal, R. *et al.*: "Real-Time Formation Dip From a Logging-While-Drilling Tool," paper SPE 38647 presented at the 1997 SPE Annual Technical Conference and Exhibition, San Antonio, Texas, 5–8 October.
62. Gondouin, M., Hill, H.J., and Waxman, M.H.: "A Tri-Chemical Component of the SP Curve," *JPT* (March 1962) 301.
63. Gondouin, M. and Scala, C.: "Streaming Potential and the SP Log," *JPT* (August 1958) 170; *Trans.*, AIME, **213.**

64. Wyllie, M.R.J.: "A Quantitative Analysis of the Electro-Chemical Component of the SP Curve," *JPT* (1949) 17; *Trans.*, AIME, **186.**
65. Pied, B. and Poupon, A.: "SP Base Line Shifts in Algeria," presented at the 1966 SPWLA Annual Logging Symposium, Tulsa, 9–11 May.
66. Archie, G.E.: "Electrical Resistivity Log as an Aid in Determining Some Reservoir Characteristics," *Trans.*, AIME (1942) **146.**
67. de Witte, L.: "Relations Between Resistivities and Fluid Contents of Porous Rocks," *Oil & Gas J.* (24 August 1950).
68. Lindley, R.H.: "Use of Differential Sonic-Resistivity Plots to Find Movable Oil in Permian Formations," *JPT* (August 1961) 749.
69. Waxman, W.H. and Smits, L.J.M.: "Electrical Conductivities in Oil-Bearing Sands," *SPEJ* (June 1968) 107.
70. Clavier, C., Coates, G., and Dumanoir, J.: "Theoretical and Experimental Bases for the Dual-Water Model for Interpretation of Shaly Sands," *JPT* (April 1984) 153.

SI Metric Conversion Factors

cycles/sec	×	1.0*	E + 00	= Hz
ft	×	3.048*	E − 01	= m
ft^2	×	9.290 304*	E − 02	= m^2
in.	×	2.54*	E + 00	= cm
$in.^3$	×	1.638 706	E + 01	= cm^3

*Conversion factor is exact.

Chapter 3C
Acoustic Logging

Doug Patterson, Baker Hughes and **Stephen Prensky,** SPE, Consultant

3C.1 Introduction

Petroleum applications of acoustic-wave-propagation theory and physics include both surface- and borehole-geophysical methods. These data-acquisition methods cover a broad range of scales from millimeters to hundreds of meters (**Fig. 3C.1**). Acoustic logging is a subset of borehole-geophysical acoustic techniques. This chapter provides an overview of borehole acoustic-logging theory, modern tool design, data processing methods, and data applications. **Table 3C.1** lists other common surface- and borehole-geophysical methods. The chapter on the Fundamentals of Geophysics, in this volume, and the chapter on Reservoir Geophysics, in the Emerging and Peripheral Technologies volume of this *Handbook,* also discuss these methods.

A virtual explosion in the volume of acoustic research conducted over the past 20 years has resulted in significant advances in the fundamental understanding of downhole acoustic measurements. These advances, in turn, have greatly influenced practical logging technology by allowing logging-tool designs to be optimized for specific applications.[1]

Borehole acoustic-logging measurements are used in a wide variety of geophysical, geological, and engineering applications and play an important role in evaluating reservoirs, reducing exploration and production risks, selecting well locations, designing completions, and increasing hydrocarbon recovery (**Table 3C.2**).

Modern logging tools include conventional borehole-compensated (BHC) monopole devices as well as the newer array devices—both monopole and multipole (monopole/dipole)—and logging-while-drilling (LWD) acoustic services. These logging tools provide acoustic measurements in all borehole mud types (but not in air- or foam-filled boreholes) in vertical, deviated, and horizontal wells, in both open and cased hole. They are combinable with other logging devices and are available in a variety of sizes to accommodate a range of borehole and casing diameters. Specialized tool designs are used for cement and casing evaluation and borehole imaging.

Historically, the primary and the most routine uses of acoustic logs in reservoir engineering have been porosity determination, identification of gas-bearing intervals, and cement evaluation. Continuing developments in tool hardware and in interpretation techniques have expanded the utility of these logs in formation evaluation and completion (fracture) design and evaluation. This chapter discusses the potential applications of these logs to allow the reader to evaluate the appropriate applications for a particular well or field. However, site-specific assess-

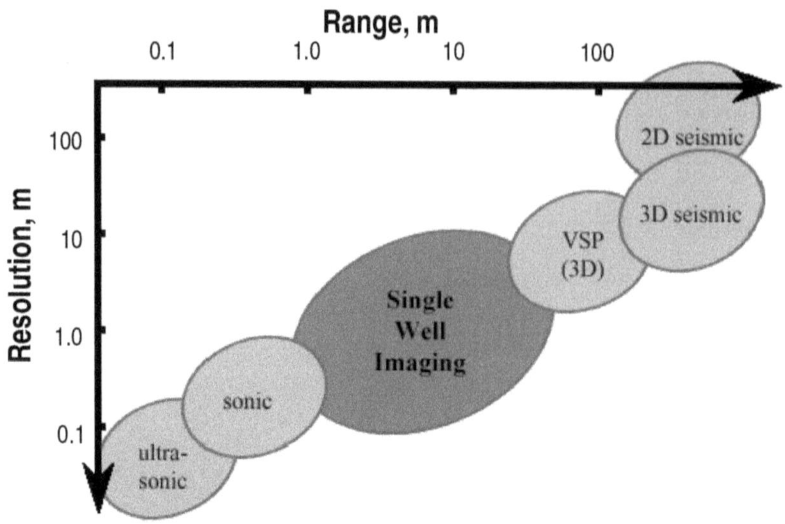

Fig. 3C.1—Diagram showing the maximum and minimum ranges vs. the resolution for various acoustic methods.[211] Acoustic logging includes the sonic and ultrasonic ranges (courtesy of SPE).

ments are required to determine whether acoustic logs, with the proper planning, can provide the desired results in a cost-effective manner.

3C.2 Acoustic Theory and Wave Propagation

The principles of borehole acoustic logging (and surface seismic methods) are based on the theory of wave propagation in an elastic medium, as detailed in Refs. 2 through 5. The oscillating motion generated by a sound source (transducer) in an elastic medium (rock formation) is called an elastic wave or acoustic wave (also called head or body waves). Wave theory predicts how an acoustic signal propagates through the borehole and formation. Snell's law explains how the acoustic signal behaves at the velocity boundary separating the borehole and the formation, that is, how it is transmitted into the formation and back to the receivers. Elasticity is the property of matter that causes it to resist deformation in volume or shape. It is the elastic nature of rock formations that permits wave propagation. Acoustic waves have four measurable properties: velocity, amplitude, amplitude attenuation, and frequency. Acoustic logging tools are designed to measure one or more of these properties, with velocity (slowness) being the most common.

The waveform recorded at the logging tool's receivers is a composite signal containing different energy modes, each with a different frequency, velocity, and amplitude. For borehole logging, the modes of primary interest are (in order of arrival) compressional, shear, and Stoneley (tube) waves (**Fig. 3C.2**). The waveform is recorded as acoustic amplitude as a function of time.

These waves are transmitted through the medium some distance from the origin of displacement. The particles of the medium do not travel with the wave, but only vibrate around their mean central position. Acoustic waves are classified according to the direction of particle displacement with respect to the direction of wave propagation as either longitudinal (i.e., particle displacement is parallel to the direction of propagation) or transverse (i.e., direction of particle displacement is perpendicular to the direction of propagation). In acoustic logging, the longitudinal wave is known as the compressional wave and the transverse wave is known as the shear wave. The presence of the borehole excites two additional acoustic energy modes, called guided waves: normal (pseudo-Rayleigh) and tube (Stoneley) waves.

TABLE 3C.1—GEOPHYSICAL ACOUSTIC RESERVOIR EVALUATION METHODS

Acquisition Mode	Geophysical Method
Surface	2D and 3D
Reflection seismic	Time-lapse reservoir monitoring (4D)
	Amplitude variation with offset (AVO)
Borehole	Acoustic logging
Wireline	Vertical seismic profiling (VSP)
	Amplitude variation with offset (AVO)
	Continuity logging
	Crosswell surveys
	Fracture surveys (microseismicity)
Wireline/LWD	Seismic while drilling (Reverse VSP)
Permanent sensors	Reservoir monitoring

TABLE 3C.2—APPLICATIONS OF BOREHOLE ACOUSTIC DATA TO SPECIFIC DISCIPLINES

Discipline Application	Geophysics	Geology	Formation Evaluation	Production Engineering	Reservoir Engineering	Well Completion	Drilling	Enhanced Recovery
Synthetic seismogram	X	X	X					
VSP analysis	X	X	X					
AVO calibration	X	X	X		X			X
Porosity evaluation	X	X	X		X	X		X
Lithology estimate	X	X	X		X	X		X
Saturation evaluation		X	X	X	X			X
Gas detection	X	X	X	X	X	X	X	X
Hydrocarbon typing	X	X	X		X		X	X
Fracture analysis	X	X	X		X	X	X	X
Permeability index		X	X		X	X		X
Abnormal pressure		X	X	X	X	X	X	
Wellbore stability		X	X	X		X	X	
Perforation stability			X	X	X	X		X
Anisotropy	X	X	X	X	X	X	X	X
Cement evaluation			X	X	X	X		X

Acoustic-wave velocity is controlled by a number of factors: lithology (mineralogy), cementation, clay content, texture, porosity, pore-fluid composition and saturation, overburden-and pore-fluid pressure (stress), and temperature. The rock's mechanical properties, elastic dynamics, and density are a constant for a particular homogeneous and isotropic material. Acoustic-wave velocity can be related to rock elastic properties through three constants of proportionality, elastic moduli (e.g., Young's, shear, and bulk), and Poisson's ratio. This serves as the basis for mechanical-property evaluation by acoustic logs (see the additional discussion under Geomechanical Applications—Rock Mechanical Properties). In reality, most petroleum reservoirs contain varying pore sizes, pore fill (e.g., clays), fractures, etc. and consequently, are neither truly isotropic nor homogeneous. Furthermore, in fluid-saturated rocks, these acoustic properties also depend on the type and volume of fluids present.

3C.2.1 Compressional Waves. Compressional (P, primary, or pressure) waves are longitudinal waves that are transmitted through an elastic formation by compression or pressure. Particle motion is parallel to the direction of wave propagation (**Fig. 3C.3**). They can travel through solids, liquids, and gases and are the fastest wave type—they represent the acoustic first arrival. Of all acoustic wave types, they are least affected by faults, unconsolidated formations, and borehole fluids, and are, therefore, the most reliable. The wave is transmitted by both the

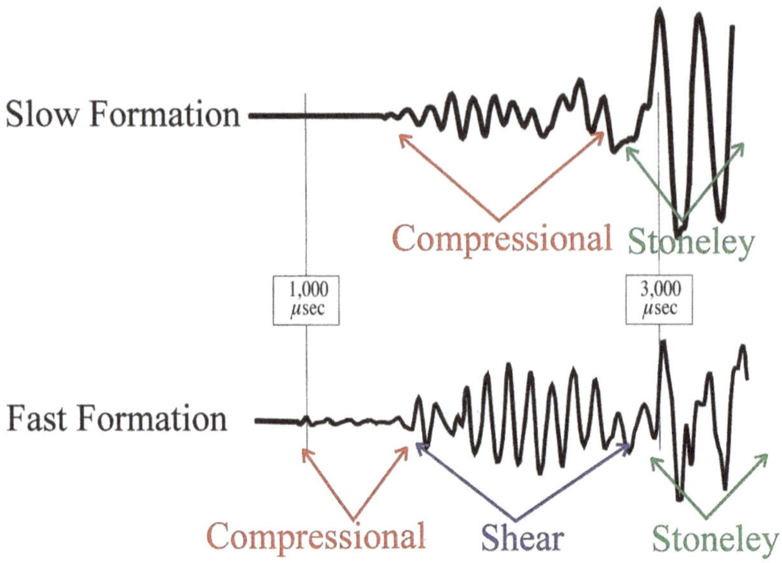

Fig. 3C.2—Generalized acoustic waveform showing the wave signature of different energy modes (courtesy of Baker Atlas).

rock matrix (i.e., the framework) and the fluid present in the pore throats. A compression, together with an adjacent rarefaction preceding or following it, constitutes a complete cycle. The distance between complete cycles is called the wave length and the number of cycles propagating through a point in the medium per unit time is the frequency. The velocity of elastic-wave propagation in an isotropic homogeneous medium can be derived from a combination of the theory of elasticity with Newton's law of motion. Compressional-wave velocity (or travel time) is a function of the density and elasticity of the medium and is a constant for a given material.

3C.2.2 Shear and Borehole Flexural Waves. Shear (S, secondary) waves are transverse waves that are transmitted by lateral displacement of particles in a rigid elastic formation. Particle motion is perpendicular to the direction of motion (Fig. 3C.3). Normally, shear waves are the second arrival in an acoustic wave train. In most reservoir rocks, shear waves generally have higher amplitudes than compressional waves but lower velocities, by as much 40 to 50%. There are two types of borehole shear waves—direct and indirect, also known as refracted or induced. Indirect shear waves are induced in a formation through a process known as mode conversion in which some of the compressional energy is transferred from the borehole fluid into the rock formation. Monopole transmitters generate these indirect shear waves while dipole transmitters generate direct shear waves by inducing a flexural (asymmetric mode) in the borehole. Shear-wave propagation requires a medium that has shear strength (rigidity). Consequently, shear waves can only travel in solids, not in liquids or gas. In liquids and gas, the shear head-wave generated within the formation is converted into a compressional wave and propagated back across the borehole fluid to the acoustic receivers as a later-arriving compressional wave.

Unconsolidated or poorly consolidated sandstones ("soft" or "slow" rocks) are less rigid and more compressible than well-consolidated ("hard" or "fast") rocks. When the formation shear-wave velocity is less than the acoustic velocity of the borehole fluid ($V_s < V_f$), a rock formation is called "slow." There is no refracted shear-wave from monopole devices in slow formations and low-frequency dipole transmission and reception is required to adequately de-

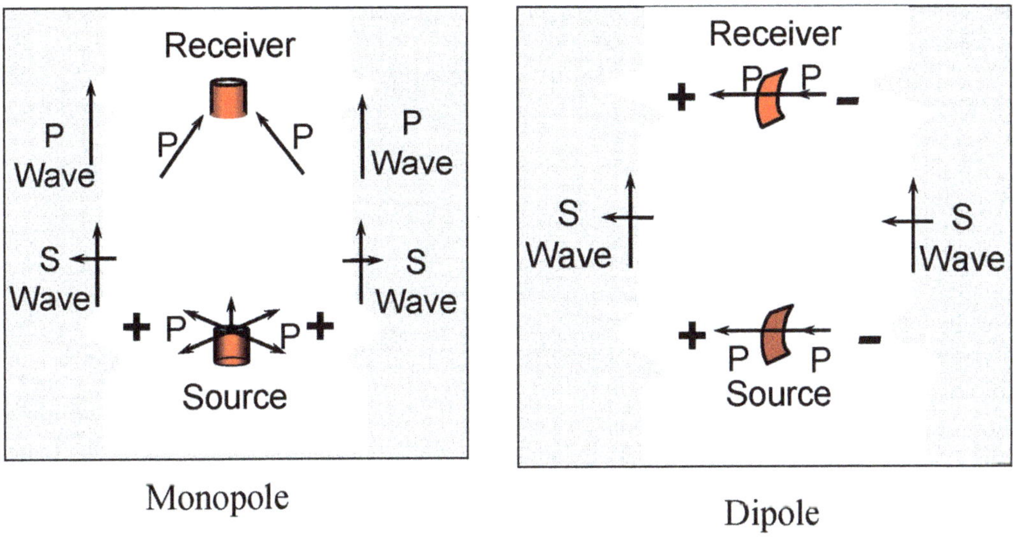

Fig. 3C.3—Direction of particle motion in compressional and shear waves and wave motion generated by monopole and dipole transmitters (courtesy of Baker Atlas).

tect low-frequency flexural arrivals for the shear-wave slowness determination. However, if a monopole-array tool is used in these conditions, a shear-wave slowness can be estimated from Stoneley-wave velocity dispersion.[6] In very slow formations, where $V_c < V_f$, special processing may be required to extract the formation compressional signal.[7]

Flexural-wave velocity varies with frequency—a phenomenon called dispersion. In contrast, the compressional and shear headwaves generated by monopole sources are generally not dispersive. At very low frequencies, the flexural wave travels at the formation shear velocity. This dispersion effect diminishes as the wavelength of the flexural-wave increases and is generally minimal when the wavelength is at least three times the borehole diameter. Fast formations exhibit a center frequency slightly greater than 3 kHz, while slow formations exhibit a center frequency of ≈ 1 kHz, or less. The received frequency spectrum of a dipole array is a function of transmitter frequency, rock properties, and borehole size. Modern dipole transmitters are broadband transmitters, i.e., they operate over a range of frequencies, to account for dispersion and to accommodate different formation types.

3C.2.3 Stoneley Waves. Stoneley (tube) waves are high-amplitude guided waves that are generated by a radial (symmetric) flexing of the borehole as the acoustic energy passes from the borehole fluid into the rock formation. They propagate at low frequencies along the fluid-rock interface at the borehole wall; hence, they are sensitive to the rock properties adjacent to the borehole wall. They are the slowest acoustic mode. They can be measured in both open and cased boreholes, but in cased holes Stoneley-wave features are primarily controlled by the casing rigidity. Similarly to shear waves, Stoneley waves are also dispersive; i.e., wave velocity varies with frequency—the amount of dispersion is related to formation rock properties. However, Stoneley waves are notable for several special properties: there is no cut-off frequency; dispersion is very mild; for all frequencies, Stoneley-wave velocity is less than fluid velocity; and group velocity nearly equals phase velocity over the frequency range.

All acoustic waves undergo attenuation, a reduction in signal amplitude away from the source. For logging devices this means radially away from the borehole wall. Signal attenuation results from the geometric spread of energy through reflection, refraction, and scattering, and through absorption by the medium through which the acoustic energy travels. Attenuation, usually expressed in dB/ft, is characteristic of different materials and increases with frequency of the acoustic wave. Generally, attenuation is large in slow formations and very small to negligible in fast formations. Because of these features, Stoneley waves are used to identify acoustic leakage away from the borehole that may be caused by formation permeability or the presence of fractures.

3C.3 Acoustic-Logging Tools

Acoustic-logging devices are comprised of transmitters (sources), receiver arrays, and accompanying electronics. They have been designed to measure one or more acoustic-wave properties. Acoustic sources (transmitters) generally consist of piezoelectric transducers that generate the acoustic signal by converting electrical signals into a sonic vibration that travels through the borehole and adjacent rock formations. Monopole (axisymmetric) transducers generate omnidirectional acoustic waves around the tool circumference, while dipole (nonaxisymmetric) transducers generate azimuthally oriented acoustic waves (Fig. 3C.3).

Modern receivers are piezoelectric crystals that transform the received (measured) acoustic signal back into electrical signals. Different logging tools use different piezoelectric materials and operate at different frequencies to measure different energy modes (wave types). The pressure variation produced by an acoustic wave displaces the piezoelectric material causing it to ring or oscillate. When the receiver oscillates, it develops a small voltage. This voltage is amplified and the raw data are processed downhole and sent to the surface-acquisition system in the form of a waveform signal (wireline) or acoustic slowness (wireline and LWD). Digital data are stored, either at the surface (wireline) or downhole (LWD), for wellsite log presentations and post-acquisition processing and playback.

The energy (amplitude) of acoustic waves is lost (attenuated or dispersed) primarily by travel through the borehole fluid and rock matrix. Additional attenuation may result from a number of other factors that include internal particle friction within the propagation medium, changes in acoustic impedance [the product of density (ρ) and acoustic velocity (V)] at interfaces (boundaries) between different mediums, borehole rugosity, and signal cancellation resulting from tool eccentering. In general, the largest signal possible occurs when the instrument is in the center of the hole; a dipole tool in an extremely large borehole may be exception to this rule. For high frequencies, the monopole signal is reduced by as much 50% of the centralized value by displacing the instrument only inches from the center of the hole. Consequently, whenever possible, centralizers should be used with acoustic-logging tools.

3C.3.1 Critical Spacing. The acoustic wave travels through the formation much faster than it does through the borehole fluid and reaches the receiver by the longer formation route first. This is true as long as the critical spacing is less than actual spacing. Critical spacing is the transmitter-to-receiver spacing at which the fluid-signal and formation-signal arrive at the receiver at the same time. The critical-spacing value depends on the diameter of the logging sonde, the diameter of the borehole, the time of travel through the fluid, and the time of travel through the rock.

In soft formations, and under certain conditions, the critical spacing can exceed actual spacing and the fluid arrival, which exists for a noncentralized tool, can interfere with the acoustic signal. When this occurs, data-processing methods can exclude this fluid arrival, thereby reducing or possibly eliminating interference from the fluid arrival. Longer transmitter-receiver spacings are used to minimize this occurrence.

The transmitter-to-receiver spacing in modern monopole tools is set to enable separation of the compressional- and shear-energy packets to allow for accurate measurement of both in fast formations. When logging with an array-acoustic device, the receivers nearest the transmitter see stronger acoustic waves than the more distant receivers.

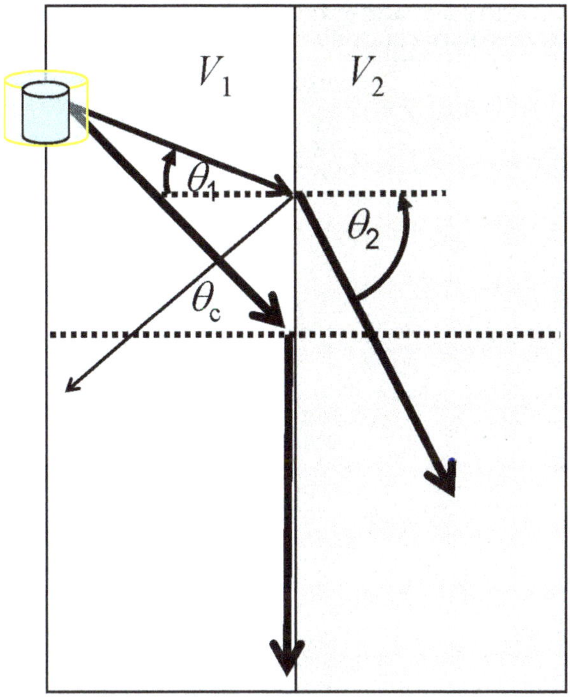

Fig. 3C.4—Illustration of Snell's law showing how acoustic waves are refracted (wave A), reflected (wave C), or converted (wave B) as they encounter a boundary separating media having different velocities (indices of refraction) (courtesy of Baker Atlas).

3C.3.2 Monopole Excitation. The transmitter emits acoustic energy uniformly around the tool. In fast formations, this energy excites three waves that travel down the borehole wall: compressional, shear, and Stoneley. The compressional wave travels away from the transmitter with a velocity, V_f, in the mud. When these waves reach the borehole face, they are reflected, refracted, and converted according to Snell's law (**Fig. 3C.4**).

For angles of incidence less than the compressional-wave critical angle (θ_c, Fig. 3C.4), part of the energy is transmitted into the formation in the form of compressional wave (Wave A, Fig. 3C.4), another part is converted as a (refracted) shear wave (Wave B, Fig. 3C.4), and the remainder is reflected back into the mud as a compressional wave (Wave C, Fig. 3C.4). The transmitted waves travel at velocity V_p and V_s in the formation, close and parallel to the borehole wall, while continuously radiating energy back into the mud as converted compressional-waves, at the same compressional-wave critical angle at which it entered. It is this radiated energy that is detected by the receivers.

If the formation shear-wave velocity is slower than borehole-fluid compressional velocity ($V_s < V_f$), shear waves cannot be refracted along the borehole wall, and no shear wave is measured. Beyond the shear-wave critical angle, all the incident energy is reflected back into the mud to form the guided waves. The Stoneley wave travels at approximately the velocity of compressional waves in the borehole fluid. Compressional and Stoneley arrivals are always present. In the absence of a refracted shear-wave arrival (i.e., in formations in which $V_p < V_f$), the Stoneley wave can be used to estimate formation shear-wave velocity when a formation bulk-density measurement is available, using certain assumptions. However, because of the uncertainty associated with these assumptions, dipole shear-wave measurements are recommended in slow formations. Stoneley-wave amplitude decreases (attenuates) significantly at high frequen-

TABLE 3C.3—VELOCITY AND ACOUSTIC SLOWNESS (TRANSIT TIME) VALUES FOR COMMON RESERVOIR FLUIDS AND LITHOLOGIES (after Carmichael[9])

Fluid	V_f (ft/sec)	Δt_f (μsec/ft)	
Water with 20% NaCl	5,300	189	
Water with 15% NaCl	5,000	200	
Water with 10% NaCl	4,800	208	
Water (pure)	4,600	218	
Oil	4,200	238	
Methane	1,600	626	
Air	1,100	910	
Lithology (matrix)	V_{ma} (ft/sec)	Compressional Δt_{ma} (μsec/ft)	Shear Δt_{ma} (μsec/ft)
Sandstone			
Unconsolidated	17,000 or less	58.8 or more	93
Semiconsolidated	18,000	55.6	92.9
Consolidated	19,000	52.6	92.9
Limestone	21,000	47.6	
Dolomite	23,000	43.5	72
Shale	6,000 to 16,000	167 to 62.5	
Salt (halite)	15,000	67	116
Coal			
Lignite	7,150	140	
Bituminous	8,300	120	
Anthracite	9,500	105	
Granite	20,000	50	
Minerals			
Calcite	22,000	46	89
Quartz	18,000	51	74
Evaporites			
Anhydrite	20,000	54	98
Gypsum	19,000	53	
Trona		65	
Iron Minerals			
Limonite		57	103
Pyrite		38	59
Siderite		44	85
Hematite		46	72
Micas			
Biotite		51	224
Muscovite		47	79

cies and modern tools use low-frequency transmitters (< 1 to 12 kHz) to ensure acquisition of the Stoneley arrival in slow formations.[8]

3C.3.3 Dipole Excitation. The dipole transmitter exerts a differential pressure on one side of the transmitter element that creates a flexural wave in the borehole, much like the wave produced when a vertical rope is shaken from side to side. The flexural wave is dispersive, but at low frequencies this wave travels down the borehole at the formation shear velocity. Receivers, sensitive only to differential pressures, are used to detect this flexural wave. Because the receivers are not sensitive to axially symmetric pressure fields, both the compressional head wave and the Stoneley waves are suppressed. This is desirable because it simplifies data processing. The desired output is the velocity of the formation shear wave. If the wavelength of the flexural wave is at least three times the diameter of the borehole, the flexural wave travels at very nearly the formation bulk-shear velocity. However, because this is a dispersive mode, if

TABLE 3C.4—DEVELOPMENT OF ACOUSTIC-LOGGING DEVICES

Application	Date	Device
Velocity		
Wireline		
	1950s	Single receiver Dual receiver
	1960s	Borehole-compensated log Long-spaced sonic
	1970s	Monopole array Circumferential sonic
	1980s	Digital array
	1990s	Dipole array Multipole array Crossed-dipole array
LWD		
	Early 1990s	Monopole array
	Late 1990s–early 2000s	Multipole array
Openhole Borehole Imaging		
	1960s	Borehole televiewer (BHTV)
	1980s–1990s	Improved designs
Cement Evaluation		
	1960s	Amplitude (CBL)
	1980s	Attenuation rate (e.g., CBT, RBT)
	1980s	Radial (pad, ultrasonic) (e.g., SBT, PET, CET)
	1990s	Reflection (pulse/echo) (e.g., USI)
Casing Evaluation		
	1980s	Ultrasonic (e.g., CET, PET)
	1990s	Reflection (pulse/echo) (e.g., USI)

the wavelength is shorter (because of higher frequency), this flexural mode will travel slower than the shear velocity and dispersion corrections are needed.

3C.3.4 Logging Documentation. All types of well logs, both open- and cased-hole, should be accompanied by complete documentation to ensure good-quality logs and sound interpretation. It is important to remember that any acoustic analysis represents an interpretation of measured acoustic waves. The wellsite engineer or geologist must ensure that all data pertaining to a particular log run, including borehole information (bit and casing sizes and depths), tool configuration, borehole fluid, formation parameters, and tool centralization are recorded in the well-log header for future reference. Cased-hole logs should also contain information on cement composition, casing weight and thickness, if the log was run with pressure, and the amount of that pressure.

3C.4 Evolution of Acoustic-Logging Tools

The most commonly used acoustic-wave property acquired in borehole logging is the compressional-wave velocity. Modern velocity-logging tools measure the time, Δt, required for a compressional or shear wave to travel through a fixed distance of formation; it is recorded as a function of depth. This parameter, Δt, referred to as the interval transit time, transit time, travel time, or slowness, is the reciprocal of the velocity of the compressional waves, $\Delta t = 1/V_p$. For

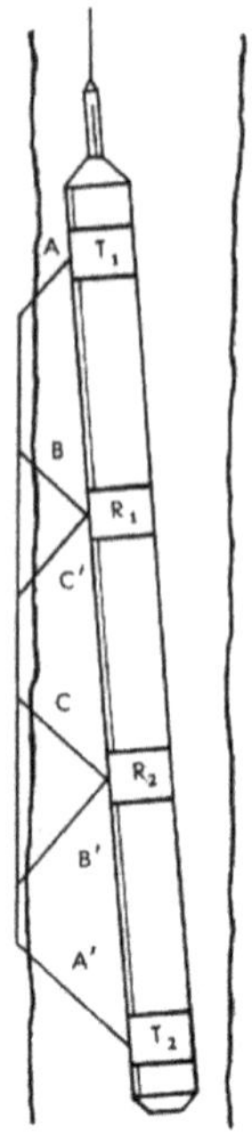

Fig. 3C.5—Two-transmitter, two-receiver scheme used for borehole compensation (courtesy of Baker Atlas).

the formations typically encountered in acoustic logging, travel times range from 40 to 250 μsec/ft, corresponding to velocities ranging from 25,000 to 4,000 ft/sec (**Table 3C.3**).

The resolution of any acoustic method is a function of the signal wavelength; the lower limit is one-quarter of the propagating wavelength. Seismic-reflection exploration methods typically operate in the frequency range of 10 to 50 Hz. In typical petroleum reservoir rocks, the resolution of these methods is approximately 30 to 160 ft (10 to 50 m), depending on depth (signal strength). In contrast, conventional BHC logging tools operate within the frequency range of 10 to 40 kHz, while newer array devices operate at even lower frequencies, 1 to 12 kHz. Logging-tool resolution is also a function of both the array aperture and the methodology employed for the array processing. Consequently, acoustic-logging tools typically have resolutions on the order of 1.0 to 4.0 ft (0.3 to 1.2 m) (see the section on Resolution Enhancement in this chapter).

3C.4.1 Velocity/Porosity Logging. Acoustic well logging developed out of the need for downhole velocity (time-depth) measurements to improve the accuracy (calibrate) of surface seismic measurements. Surface seismic maps out sub-surface structures referenced to time and borehole acoustic tools provide a bridge to understand how time is related to depth. Downhole geophones were introduced in the 1930s to provide acoustic travel times to the surface, and continuous-velocity-logging tools were introduced in the 1950s (**Table 3C.4**). Soon after the introduction of the continuous-velocity log, it was recognized that these data also provided an excellent means for stratigraphic correlation, lithologic identification, and for evaluation of formation porosity.

The first acoustic-logging tools used a single monopole transmitter and a single receiver. Tool designs rapidly evolved to improve the accuracy of the velocity measurement by minimizing or eliminating influences related to borehole effects, i.e., fluid, geometry, and tool position

TABLE 3C.5—FACTORS AFFECTING BHC SLOWNESS

Factor	Cause	Effect on Slowness
Borehole diameter	Signal attenuation dependent on TR spacing	Reads mud Δt if TR > critical spacing
Noise	Intermittent; far receiver more susceptible	Underestimated or erratic
Cycle skip	Near receiver not triggered because of signal attenuation. Possible causes: unconsolidated formations, fractures, gas-cut mud, excessively large, rugose, or washed-out borehole.	Overestimated
Δt stretch	Signal attenuation; can produce serious errors in integrated travel time.	Overestimated
Velocity inversion	Increase in slowness near borehole because of invasion or formation alteration	Underestimated
Gas effect	Gas present in flushed (measurement) zone	Overestimated
High apparent dip	Acoustic signal may travel along dip planes rather than along borehole. May result in false thickness in fast formations	Underestimated
Signal splitting	Signal cancellation because of tool eccentricity (reduces signal-to-noise ratio)	Same as noise

(tilt) within the borehole (see the next section of this chapter); and to measure additional acoustic-wave properties.

These simple devices were soon replaced by two receiver designs that had the advantage of eliminating the need to correct for travel time in the drilling mud (for more on the historical development of acoustic logging, consult Refs. 10 through 13). Modern borehole-compensated designs, introduced in the 1960s, use two transmitters and two receivers to compensate for variations in borehole diameter and tool position within the borehole (**Fig. 3C.5**). Two separate values of two receiver (R_1 and R_2) and two transmitters (T_1 and T_2) interval-transit times are provided, and an average of the two effectively compensates for any problems (Eq. 3C.1).

$$\Delta t = \frac{(t_1 - t_2) + (t_4 - t_3)}{2X_{12}}, \quad \text{(3C.1)}$$

where t_1 = travel time between T_2 and R_1; t_2 = travel time between T_2 and R_2; t_3 = travel time between T_1 and R_1; t_4 = travel time between T_1 and R_2; and X_{12} = distance between R_1 and R_2. Fig. 3C.5 illustrates travel paths that show that the averages of AA′, BB′, and CC′ are essentially equal.

The BHC tools' velocity measurements may be affected by a variety of factors including borehole diameter, signal noise, cycle skipping, Δt stretch, velocity inversion, gas effect, and dip angle, with respect to the borehole (**Table 3C.5**). Modern digital-circuitry and array-tool designs reduce or eliminate many of these problems, but they may still be present on older logs.

Conventional BHC monopole-acoustic logs, with their short transmitter-to-receiver spacing, have shallow depths of investigation and they largely measure mud filtrate that fills the pore space in the invaded (flushed) zone around the borehole. Long-spaced and array devices can acquire measurements beyond the filtrate and altered zone.

3C.4.2 Long-Spaced Acoustic Tools. Typical BHC devices have a transmitter-to-receiver spacing (TR) of 3 to 5 ft. These work well in many circumstances; however, in cases which borehole enlargement prevents acquisition of reliable data, due to the increase of the critical spacing, or in which the drilling process damages or alters the shales surrounding the borehole, long-spacing tools (TR = 8 to 15 ft) may be necessary, or advised, to obtain accurate measurements. In contrast to conventional BHC devices, in which the transmitters and receivers are

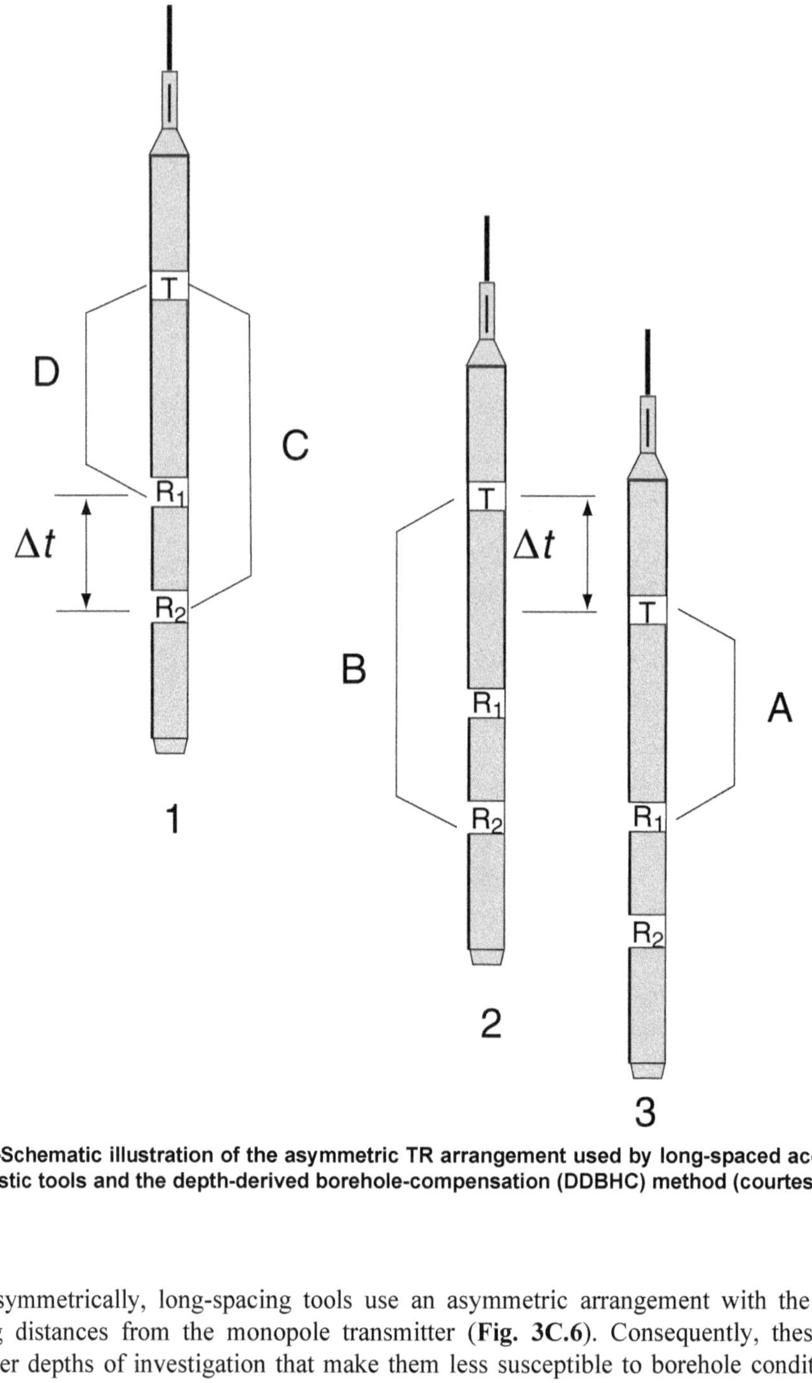

Fig. 3C.6—Schematic illustration of the asymmetric TR arrangement used by long-spaced acoustic and array acoustic tools and the depth-derived borehole-compensation (DDBHC) method (courtesy of Baker Atlas).

arranged symmetrically, long-spacing tools use an asymmetric arrangement with the receivers at varying distances from the monopole transmitter (**Fig. 3C.6**). Consequently, these devices have deeper depths of investigation that make them less susceptible to borehole conditions such as enlargement and shale alteration. These tools operate in both open and cased holes.

Because long-spaced and array tools use an asymmetric configuration of transmitters and receivers, borehole compensation is achieved through a process called depth-derived borehole compensation (DDBHC)[14] (Fig. 3C.6). The processing is accomplished downhole by either the tool's electronics or in the surface recording system and uses a depth-based delay to create synthetic transmitter arrays from multiple tool positions. The compensated travel-time measurement, Δt, is determined through the following procedure. At position three (Fig. 3C.6), the transmitter (T) is at the depth where the far receiver (R_2) will be when the tool is moved to position one. The interval-transit time (A) between the transmitter (T) and near receiver (R_1), which includes mud and formation signals, is recorded and delayed (memorized). At position two (Fig. 3C.6), the transmitter (T) is at the same depth the near receiver (R_1) will be when the tool moves to position one. The interval transit time or waveform (B) between T and R_2 is recorded and delayed. When the tool reaches position one, the two interval transit times (A and B) are equal to the interval transit time that would result if a second transmitter were below the receivers. The correct compensated value of Δt is obtained by combining the two delayed values of transit time (recorded at positions two and three) with transit times C and D, recorded at position one. Compensated transit time (Δt) is then correctly represented by

$$\Delta t = \frac{(B - A) + (C - D)}{2x}, \quad \text{(3C.2)}$$

where x = the distance between R_1 and R_2.

Analog recording of full acoustic waveforms, called "amplitude" logging, was developed in the 1960s. However, it was not until digital technology, instrumentation, and signal-processing methods were introduced in the late 1970s that the recording of full-waveform data became routine. While these techniques enabled the extraction of shear-wave data from conventional BHC acoustic data, determination of shear arrivals using conventionally spaced BHC tools (TR = 3 to 5 ft) suffers from interference between late-arriving compressional waves and shear-wave arrivals. The use of long-spaced tools reduced this problem by allowing greater temporal separation between the different wave packets and provided accurate estimates of shear–wave slowness.[15–18]

3C.4.3 Monopole Array Devices. Modern array tools are a natural outgrowth of the long-spaced tool design. Additional receivers (4 to 13) were added to provide the statistical redundancy needed to enhance extraction of wave arrival times; some designs also include multiple transmitters. The monopole transmitters in these tools use lower frequencies (e.g., 1 to 12 kHz vs. 20 to 40 kHz) and have broader frequency ranges than earlier tools to permit acquisition of high-quality acoustic waveforms. These devices are typically comprised of several sections or subs that house the tool components: electronics, receiver array, acoustic isolator, and transmitters. The acoustic isolator, placed between the transmitter and receiver sections, prevents or minimizes and delays direct sound transmission between transmitters and receivers. The electronics section provides timing and control for the transmitter and receiver sections, digitizes the received acoustic waves, analyzes the acquired waveforms, and transmits the data to the surface-data-acquisition system, all in real time.

Array tools can record full waveforms: compressional, shear, and Stoneley arrivals. These tools operate in either a single or a variety of combination-acquisition modes that include full waveform, compression Δt, and cement-bond logging. The number of modes that can be activated during a single log run is a function of the logging speed. Acquisition of full waveforms permits the use of waveform-correlation techniques for waveform amplitude, coherent slowness from the coherent-wave moveout, arrival-time processing, and most importantly, allows for detailed post-acquisition processing that improves the interpreted results. Because these techniques are insensitive to cycle skipping, they are particularly effective in gas-saturated, rugose, and washed-out boreholes.

3C.4.4 Through-Casing Acoustic Measurements. The extended transmitter-to-receiver offset (6 to 19 ft) provided by array instruments allows them to quantify formation-compressional and shear-wave energy through casing. Successful cased-hole operation requires a good cement bond to provide the necessary acoustic coupling to the formation and to minimize or eliminate the casing arrival.[19–21] New processing techniques may allow valid acoustic evaluation even in cases of poor bonding.[22]

3C.4.5 Dipole and Multipole Array Devices. In hard (fast) formations, monopole array tools can acquire a refracted shear wave but not in soft and unconsolidated (slow) formations. Before the introduction of dipole transmitters, processing techniques were developed to derive shear-wave transit time from Stoneley-wave data.[1,23,24]

Tools using dipole transmitters were conceived as early as the 1960s,[25] but were not actually developed until the 1980s.[26] In contrast to monopole logging tools, dipole acoustic devices can excite a low-frequency flexural wave in the borehole at shear velocity. Low-frequency (< 1 kHz) dipole sources allow for shear-velocity determination that is much closer to seismic shear waves and permits acquisition of direct-shear velocities in slow and fast formations. However, increased noise (i.e., a lower signal-to-noise ratio) is one limitation of low-frequency operation. Noise has been reduced through improved acquisition electronics, the use of semi-rigid tool designs, and by choosing the operational mode of the dipole source. A semi-rigid tool body not only reduces the influence of the tool body on the measurement but also permits operation in deviated wells.

At high frequencies, or when the borehole diameter is large, flexural-mode propagation is slower and a dispersion correction is needed to obtain the shear velocity from the measured flexural velocity. This dispersion correction is a function of mud compressional velocity, formation compressional and shear velocities, the ratio of formation and mud densities, and the product of borehole diameter and processing frequency. Few, if any corrections are required if the flexural wavelength (velocity/frequency) is at least three times the borehole diameter, which is why low frequencies (< 1 kHz) are used. Where a correction is necessary, it is typically only a few percent, but can be higher under certain conditions.

The latest commercial tool designs are multipole array devices that operate in open and cased hole. These tools typically integrate multiple transmitters (monopole and dipole) and one or more arrays of monopole and dipole receivers. Multiple monopole transmitters, or a single "programmable" transmitter,[19,27,28] provide the preferred frequency-range for optimal acquisition of conventional BHC and long-spaced compressional, shear, low-frequency Stoneley modes along with cement-bond logs. The dipole transmitter, more recently a variable-frequency (wide bandwidth) transmitter, provides the crossed-dipole mode.[28–32] The receivers are positioned axially along the length of the tool, for short- and long-spaced measurements, and may be interlaced or independent. At each axial position a group of receivers (e.g., 4 or 8) is positioned azimuthally around the circumference of the sonde. The receivers are oriented with the transmitter to allow for alignment of directional (multimode) source excitation and data acquisition. This allows radial imaging of acoustic parameters and measured properties.[33,34] The dipole receivers may be aligned inline with the dipole transmitters or orthogonal (crossed) to them for crossed-dipole analysis.

Multipole tools provide enhanced Stoneley-wave and crossed-dipole shear-wave data for analysis of formation permeability and anisotropy (e.g., stress and fractures) (see later discussion).

3C.4.6 Logging While Drilling. Acoustic slowness measurements are a relatively recent addition to the suite of LWD measurements.[35,36] Similar to wireline devices, LWD tools consist of transmitter, isolator, and receiver-array sections contained in either a single or separate drill-collars and use monopole-type (axisymmetric) transmitters. However, unlike wireline devices, which are small relative to the borehole size, the rigid collar, structural design, and large diameter of acoustic LWD drill collars may actually interfere with the physics of acoustic-energy propagation, making it more difficult to decouple the transmitter-to-receiver signal traveling along the tool body and complicating generation of borehole guided modes, such as Stoneley and flexural.

TABLE 3C.6—ADVANTAGES AND DISADVANTAGES OF LWD ACOUSTIC ACQUISITION	
Advantages	Disadvantages
Real-time Δt	Relies on downhole processing; waveforms stored in memory
Preinvasion	Increased noise from drilling
Compressional and shear measurement in fast formations	Time based, depths can be irregular
Time-lapse studies	Tool centralization may be poor
	Pure dipole-shear measurement not possible

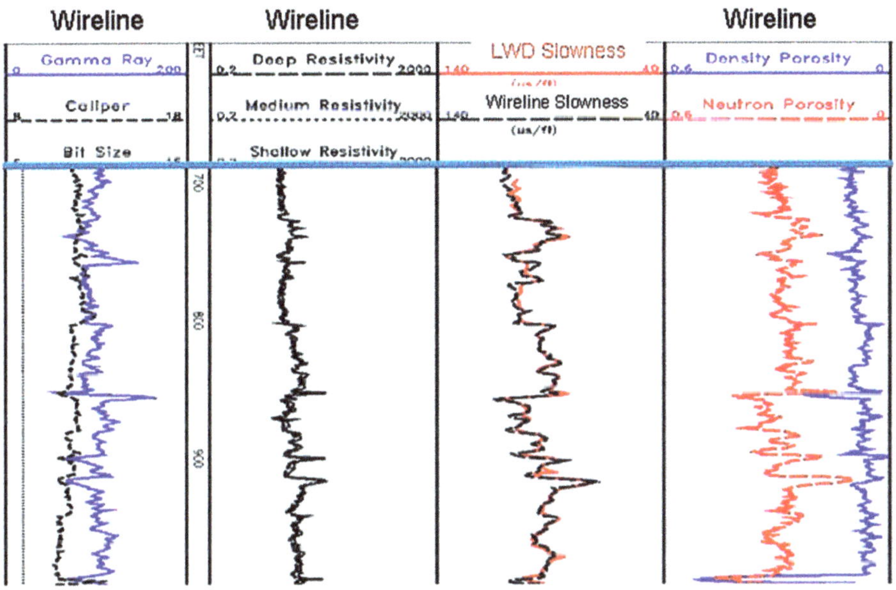

Fig. 3C.7—Comparison of LWD (red, Track 3) and wireline (blue, Track 3) acoustic slowness in a fast formation (courtesy of Baker Atlas).

Downhole processing provides acoustic slowness data in real time and also allows for storage of raw waveform data in downhole memory that can be downloaded during bit trips. The receiver array provides data redundancy and is coupled with a much narrower recording bandwidth to enhance the signal-to-noise ratio in the presence of the drilling-related noise. This results in more accurate measurements of acoustic slowness. As with all types of LWD tools, the primary advantage of LWD acoustic measurements is the acquisition of data before significant fluid invasion or alteration of the formation can occur, and providing the data in sufficient time to influence drilling decisions for improved safety, well placement, and well productivity.[37,38] **Table 3C.6** summarizes the advantages and disadvantages of LWD acoustic measurements. LWD acoustic data recorded in casing during bit trips offers the potential for cased-hole evaluation, e.g. formation velocity and cement-evaluation, during bit trips.[39]

Multipole tools (monopole and pseudo-dipole or quadrupole) have recently been introduced and continue to be developed.[40–43] These devices use the monopole transmitter for compressional-slowness and shear-slowness in fast formations. Similar to wireline tools, these tools may include an azimuthal receiver array.[42,43] With these tools, the compressional-slowness measurement is obtained in fast and slow formations by summing the azimuthal measurements at each receiver level. **Fig. 3C.7** illustrates the excellent agreement between LWD data and wireline

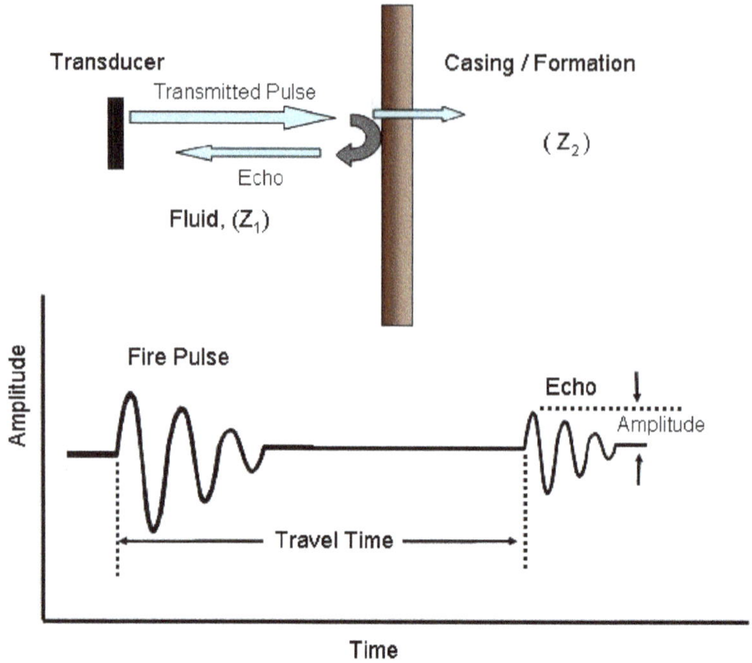

Fig. 3C.8—A simplified representation of the path of the televiewer pulse-echo signal (upper) and the measurements made (lower) (courtesy of Baker Atlas).

data. Measurement of compressional and shear slowness in slow formations is especially impacted as the diameter of the LWD drill collar approaches the borehole diameter. Because the drilling environment precludes propagation of a true dipole flexural mode, different methodologies are being used to obtain a pseudo-shear measurement using other borehole-guided modes, either through tool design, data processing, or both.[41,44–51] Current solutions include multiple-frequency and higher-frequency dipole-type transmitters, and quadrupole transmitters. These measurements require corrections for frequency dispersion and eccentering.[52–55]

Acoustic LWD can provide a limited real-time seismic-while-drilling capability in combination with a surface source.[56–58] There is also great interest among drillers and geophysicists to develop a viable drilling-based system in which the drillbit acts as the acoustic source and the sensors (ruggized geophones) are located in the borehole, as part of an LWD bottomhole assembly. Both types of systems provide checkshot surveys for time-depth correlation with pre-drill surveys and look-ahead seismic capability.[59]

3C.4.7 Ultrasonic Reflection (Pulse/Echo) Acoustic Devices. Reflection (pulse/echo) acoustic devices were introduced in 1967 with the borehole televiewer (BHTV).[60] In contrast to conventional acoustic-logging devices, which record the transmission of acoustic waves through the formation, pulse/echo devices record the travel time, amplitude, and azimuth of ultrasonic acoustic pulses (echoes) that are reflected off the formation wall in openhole or off the casing or cement in cased hole (**Fig. 3C.8**). The difference between the acoustic impedance of the borehole fluid and formation determines the magnitude (amplitude) of the transmitted ultrasound pulse that is reflected off the formation and back to the transducer. A portion of the transmitted signal is not reflected and continues to travel into the formation (openhole) or casing (cased hole) (Fig. 3C.8).

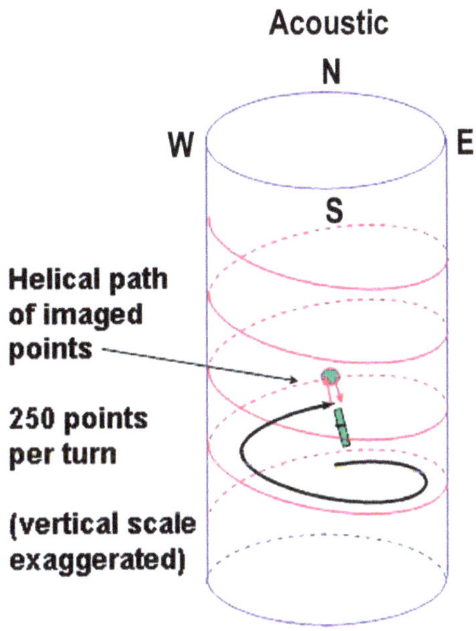

Fig. 3C.9—Figure illustrating the helical-scan data-acquisition path used by pulse/echo tools (courtesy of Baker Atlas).

Televiewer-type devices use a rotating transducer that acts as both transmitter and receiver, to acquire as many as 250 samples per revolution. Because the tool is moving uphole continuously, data are acquired as a helical scan along the borehole (**Fig. 3C.9**). Magnetometers provide azimuthal information for each scan. Acoustic devices operate in both conductive (water-based) and nonconductive (oil-based) muds. Televiewer-type tools cannot operate in air- or gas-filled boreholes.

The ultrasonic pressure pulses transmitted from the front face of the transducer form a beam pattern that defines the resolution and detection capabilities of the tool. While image resolution is directly proportional to signal frequency, the operating frequency (200 to 650 kHz) is a compromise between the desire for high image resolution and the need to operate in high-density muds in which signal attenuation is directly related to the frequency. In openhole, in which high-resolution, focused transducers are used, resolution may be as fine as 0.2 in. Image resolution in cased hole, in which unfocused transducers may be used, is generally lower.

The peak amplitude of the reflected signal is used to generate a 360° image of the borehole wall and the travel-time measurements are used as a caliper to provide a measurement of borehole geometry or casing corrosion. The primary factors that contribute to the measured pulse/echo amplitude are borehole-fluid ultrasonic attenuation, borehole-fluid/formation reflection coefficient, physical features of the formation, and transducer-beam angle of incidence on the formation. Loss in signal amplitude (image quality) results from conditions that either scatter, absorb, or spread the acoustic energy, such as tool eccentering, irregularities in the borehole shape and surface, high-density and some oil-based drilling muds, and contrasts in acoustic impedance between the borehole fluid and borehole wall or casing (**Table 3C.7**). A separate transducer, commonly in the "mud-sub," continually measures the borehole-fluid velocity for the caliper measurement.

In openhole, amplitude images of the borehole wall are used for fracture analysis, stress analysis (borehole breakouts), and formation evaluation. Cased-hole applications include cement evaluation (distribution and quality), casing evaluation (wear and damage), and perfora-

TABLE 3C.7—FACTORS AFFECTING PULSE/ECHO AMPLITUDE

Factor	Effect
Fluid attenuation	Primary factor. High-fluid attenuation or large-diameter boreholes can result in a complete loss of useful information. OBMs (synthetic or otherwise) generally produced more signal loss than water-based mud of the same weight. For any particular fluid, attenuation usually increases with increasing fluid density, temperature, oil/water ratio, and any increase in particle size or gas content. Fluid attenuation generally decreases with greater concentrations of CaCl. High ultrasound attenuation in large boreholes will produce speckles (similar to snow on a weak television signal), fuzziness, and streaks.
Eccentering	Tool decentralization and/or borehole ellipticity can cause vertical dark streaks or bands.
Reflection coefficient	The difference between the acoustic impedance of the fluid and formation
Physical features	e.g., rugosity, fractures, or pitting
Beam angle of incidence (non-normal)	e.g., tool tilting and angular deformations on surface

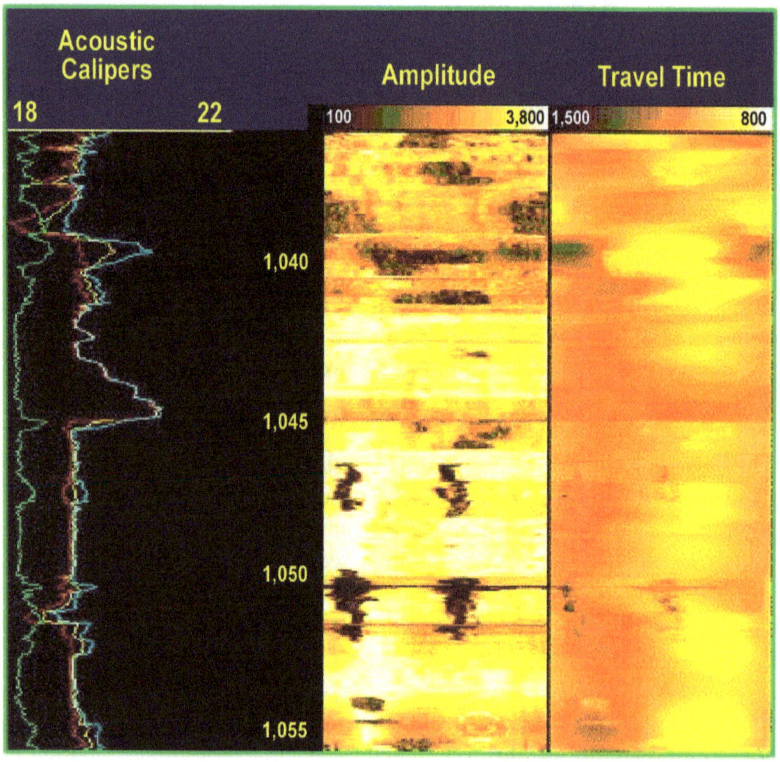

Fig. 3C.10—Typical televiewer "open-cylinder" image presentation; Track 1 contains the gamma ray curve and acoustic caliper; Track 2 is the reflected acoustic amplitude; and Track 3 is acoustic travel time (courtesy of Baker Atlas).

tion control.[61] Smooth casing or borehole walls produce high-amplitude reflections (light), whereas openings (e.g., fractures or vugs) or imperfections (e.g., rugosity or pitting) in the borehole wall or casing result in low-amplitude reflections (dark). These images are displayed in the typical open-cylinder image presentation (**Fig. 3C.10**).

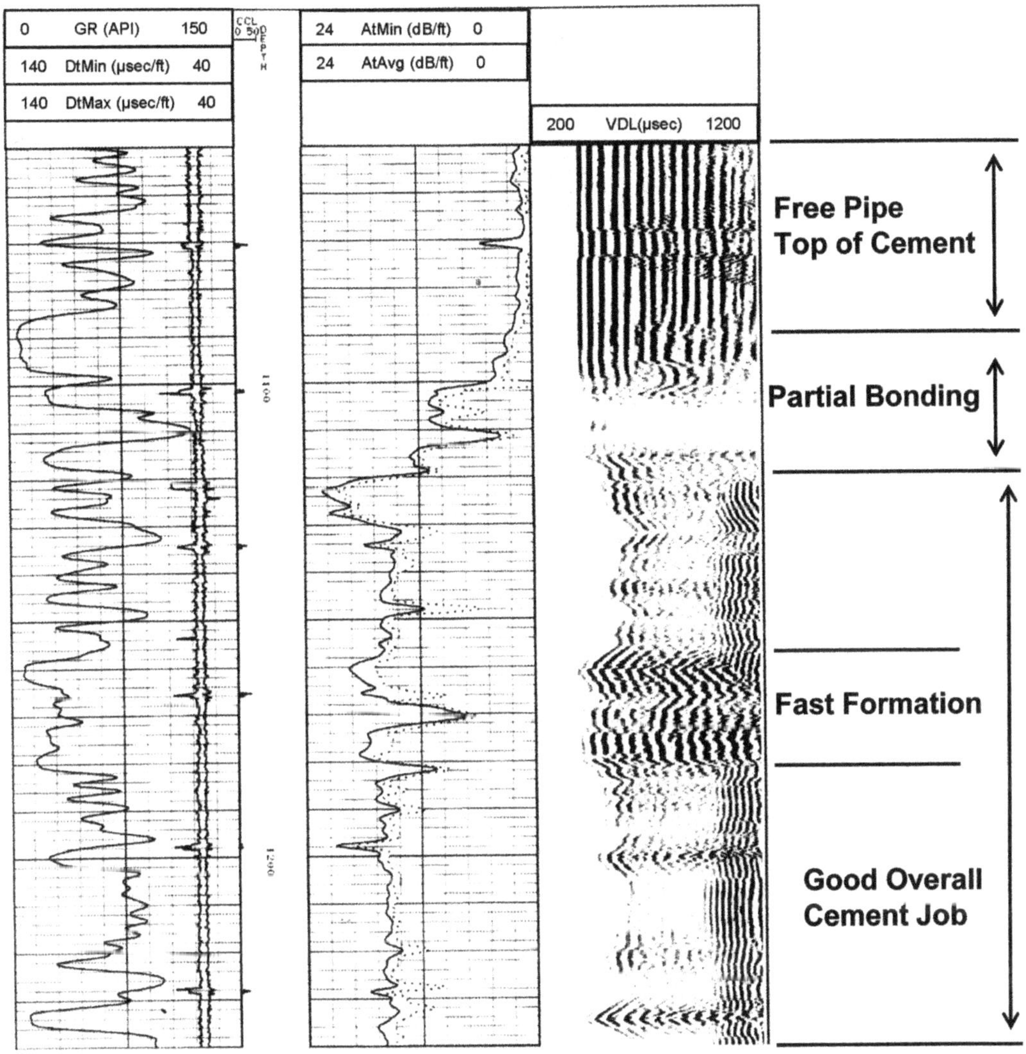

Fig. 3C.11—Typical cement-bond log presentation (courtesy of Baker Atlas).

3C.4.8 Cement-Bond Logging. Proper cement placement between the well casing and the formation is essential to support the casing (shear bond), to prevent fluid from leaking to the surface, and for isolating producing zones from water-bearing zones (hydraulic bond). Acoustic logs provide the primary means for evaluating the mechanical integrity and quality of the cement bond.[62–66] Acoustic logs do not measure cement quality directly, rather, this value is inferred from the degree of acoustic coupling of the cement to the casing and to the formation. Properly run and interpreted, cement-bond logs (CBL) provide highly reliable estimates of well integrity and zone isolation. Just as filtrate invasion and formation alteration may produce changes in formation acoustic properties, and thus variation in acoustic logs over time,[67–69] so too, cement-bond logs may vary over time as the cement cures and its properties change.

Modern acoustic cement-evaluation (bond) devices are comprised of monopole (axisymmetric) transmitters (one or more) and receivers (two or more). They operate on the principle that acoustic amplitude is rapidly attenuated in good cement bond but not in partial bond or free pipe. These cased-hole wireline tools measure compressional-wave travel time (transit time),

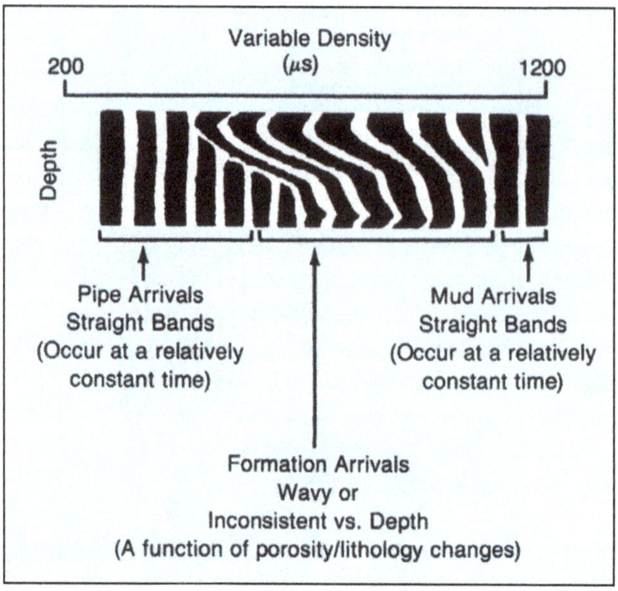

Fig. 3C.12—Identification of important features on a variable-density log (courtesy of Baker Atlas).

amplitude (first pipe arrival), and attenuation per unit distance. Conventional CBL tools provide omnidirectional measurements, while the newer radial cement-evaluation tools provide azimuthally sensitive measurements for channel evaluation.

When the acoustic wave generated by the transmitter reaches the casing, part is refracted down the casing (amplitude and travel-time measurement), part travels through the mud (fluid arrival), and other parts are refracted into the annulus and the formation and received back (formation arrival). Amplitude, measured directly or as an attenuation ratio, is the primary bond measurement and is used to provide quantitative estimations of cement compressive strength, bond index, and qualitative interpretation of the cement-to-formation interface. Tool response depends on the acoustic impedance of the cement, which, in turn is function of density and velocity. On the basis of empirical data, the log can be calibrated directly in terms of cement compressive strength. However, in foamed cements or when exotic additives are used, these calibrations can be inaccurate. In these situations, users are advised to consult with the logging service company regarding the appropriate calibrations.

A typical cement-log presentation includes a correlation curve (gamma ray), travel time (μsec), amplitude (mV), and attenuation (dB/ft) curves, and a full-waveform display (μsec). Presentation of the full acoustic waveform assists in resolving bond ambiguities arising from use of an amplitude measurement alone and provides qualitative information about the cement-to-formation bond. Waveform displays may be in variable density (VDL) or intensity (also called microseismograms) formats, oscilloscope waves (also known as x-y or "signature"), or both (**Fig. 3C.11**).

Variable density is a continuous-depth time display of full-waveform amplitude presented as shades of black and white. Positive waveform amplitudes are shown as dark bands and negative amplitudes as gray or white bands; contrast is proportional to amplitude. On a variable-density log, free pipe and fluid arrivals (if present) are easily identified as straight dark and light lines (indicating homogenous acoustic properties) at either side of the display (**Fig. 3C.12**). The zigzag, wavy, or chevron pattern between these two arrivals is the formation signal (indicating varying acoustic transit time). In cases of poor bonding, casing-collar signals may also be identified as "w" patterns (anomalies) (**Fig. 3C.13**).[70]

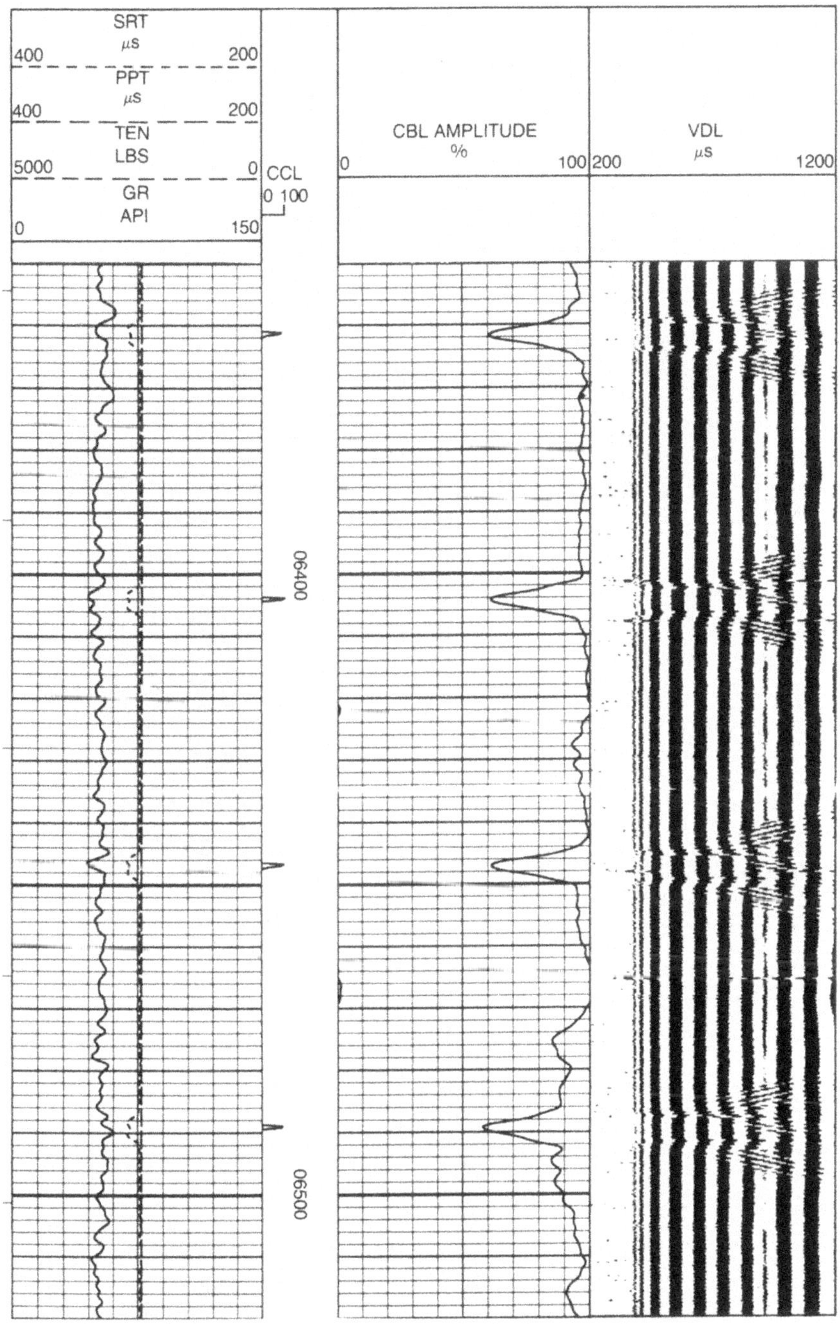

Fig. 3C.13—Casing-collar arrivals indicated on peak amplitude curve (Track 2) and variable density (Track 3). The height of the collar disturbance is a function of measurement TR spacing (amplitude, 3-ft interval; variable density, 5-ft interval) (courtesy of Baker Atlas).

Conventional Cement-Bond Devices. Early CBL designs (1960s) used a single transmitter and single receiver for an amplitude measurement. In an evolution similar to that of openhole acoustic logs (Table 3C.4), new designs were subsequently introduced that measured signal am-

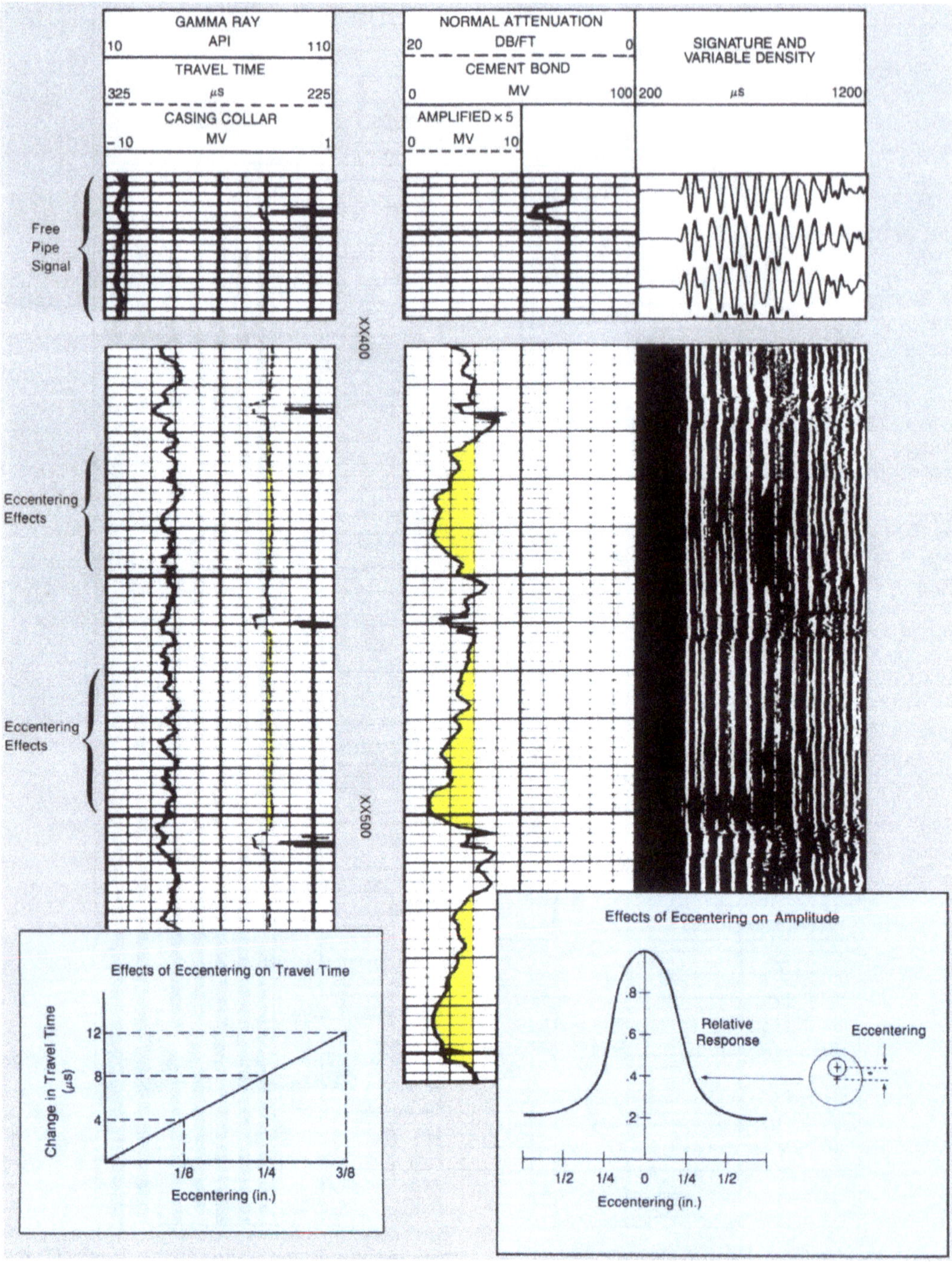

Fig. 3C.14—Effects of tool eccentering (courtesy of Baker Atlas).

plitude at a near receiver and a full waveform from a far receiver. Eventually, borehole-compensated devices using dual transmitters and dual receivers were introduced in the 1980s, and today most commercial devices use multiple transmitters and receivers in a variety of arrangements to provide compensated measurements. These devices measure the attenuation between two transmitters and receivers as a way of eliminating, or at least minimizing, the

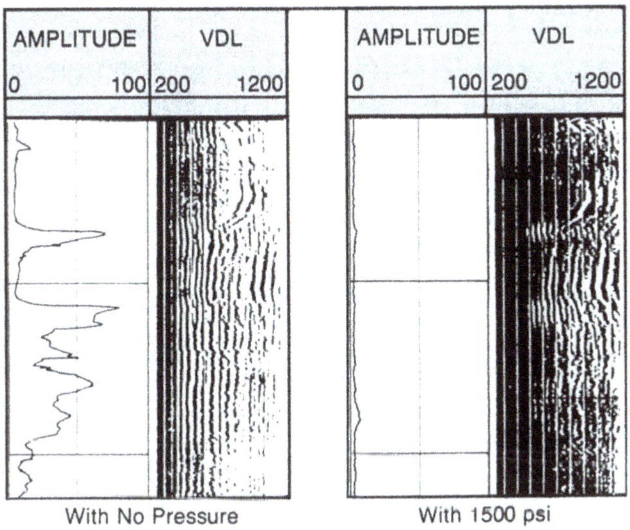

Fig. 3C.15—Field example showing microannulus effect on amplitude and VDL log displays (courtesy of Baker Atlas).

effects of tool eccentering, fluid attenuation, receiver sensitivity, temperature drift, and calibration. In addition to specialized cement-bond devices, modern openhole array tools are designed to also provide conventional cement-evaluation measurements in cased hole. The cement-bond instrument sleeve is typically slotted to suppress and delay the tool signal that might otherwise be confused with the important casing signals.

TR spacing typically ranges from 3 to 5 ft. The shorter spacing (e.g., 3 ft) provides optimum signal level and resolution at high attenuation rates and is normally used for amplitude and travel-time (TT) measurements. A longer spacing, commonly ≥ 5 ft, is used for the full-waveform recording because longer TR spacing provides greater separation of the casing and formation-signal arrival times. This separation allows for easier analysis of the formation-signal strength and is used to monitor cement-to-formation bonding.

These tools typically operate at higher frequencies than conventional openhole tools—between 20 and 30 kHz. As with openhole tools, cement-bond tools require centralization to ensure accurate measurements. Centering in the cased hole is more critical because the higher-operating frequencies (i.e., shorter wavelengths) and the tool measurement are based on signal amplitude. Tool eccentering reduces signal amplitude and travel time (**Fig. 3C.14**).

Cement-bond logging tools use gated systems to measure the specific parts of the acoustic waveform needed for the primary bond-amplitude measurement. Gates are time periods during which measurements are made—they can be either fixed or sliding (floating). Fixed-gate systems are commonly used for amplitude measurements and floating gates for travel-time measurements. Fixed gates are set (generally at the wellsite) to open, remain open, and close at designated times; opening time for the gate is a function of the casing size and the borehole-fluid velocity. If the gate opening is too large, there may be interference between early and late-arriving signals. Floating gates remain open, but recording is only triggered by an amplitude value greater than a designated threshold value.

A casing cement job can result in one or more of the following situations: free pipe, good bond, bond to casing only, and partial bond. In the first scenario, free pipe, there is no cement bond between the casing and cement. Consequently, there is no acoustic coupling with the formation and most of the transmitted acoustic energy is confined to the casing and the borehole

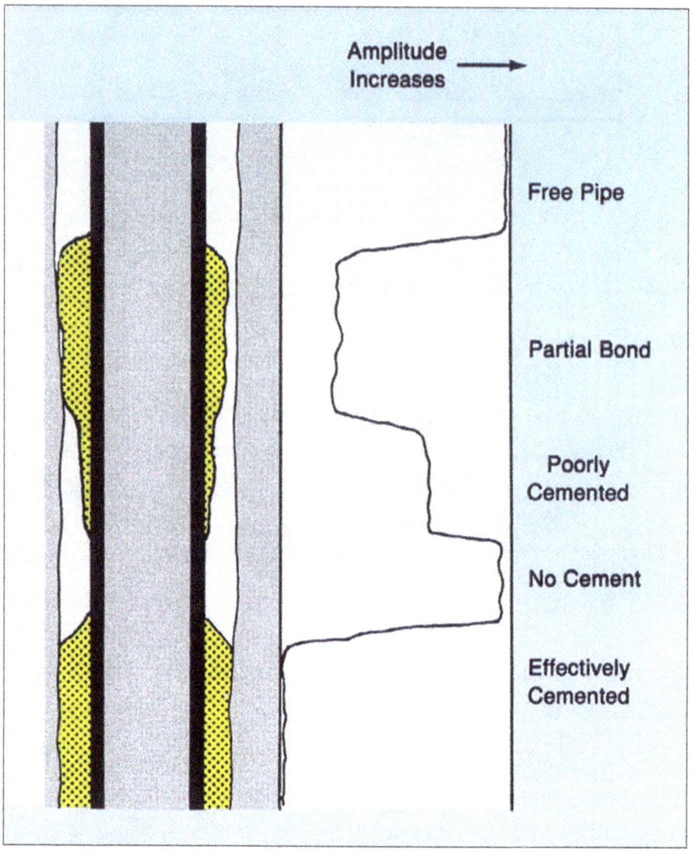

Fig. 3C.16—Summary of qualitative interpretation based on the amplitude curve (courtesy of Baker Atlas).

TABLE 3C.8—FACTORS AFFECTING BOND INTERPRETATION FROM AMPLITUDE RESPONSE[62,71]

Factor	Description
Detection method	Fixed or floating gates
Gate width	Wide or narrow (multiple casing strings)
Measurement threshold	Cycle skipping
Eccentering	Tool or casing
Presence of gas	In borehole fluid or the cement
Casing variability	Size and weight (thickness)
Cement variability	Thickness [< 0.75 in. (2 cm)], voids (foam)
Formation type	In fast formations (e.g., low-porosity carbonates and evaporites) the formation signal may arrive at the same time or before the casing signal
Borehole-fluid variability	Density and viscosity; gas cut

fluid. As a result, a free-pipe acoustic signal is long-lived, high-amplitude, and of uniform frequency.

In the second scenario, good bond, cement is properly bonded to casing and to the formation. This provides good acoustic coupling and most of the acoustic energy is transmitted to the formation, resulting in little (weak) to no casing signals and little amplitude until the arrival of the strong formation signal.

TABLE 3C.9—LIMITATIONS OF CONVENTIONAL CEMENT-BOND LOGS

Operating Factor	Limitation
Centralization	Required
Detecting small channels	Difficult to distinguish between the case of high-strength cement with a channel vs. an even distribution of low-strength (foam) cement (similar amplitude response)
Shear coupling	Required (microannulus may have a free-pipe signal)
Tool calibration	May not apply in fast formations in which the formation arrival precedes the casing arrival
Pressurization of casing	Recommended to eliminate microannulus effect

The third scenario, bond to casing only, is a common condition in which cement is bonded to the casing but not to the formation. This can occur because the mudcake dries and shrinks away from cement, or because the cement did not bond with mudcake in poorly consolidated formations. In this situation, energy traveling through the casing is attenuated drastically because of the highly attenuating cement sheath. At the same time, the annulus outside the cement sheath provides poor acoustic coupling. The result is that little energy is transferred to the annular fluid and virtually none is transferred to the formation. This condition is indicated by the lack of later-arriving formation energy. A similar response can be caused by the presence of formation gas in shallow, high-porosity zones.

In the last scenario, partial bond, a space exists within an otherwise well-bonded casing. This may occur with the presence of a microannulus or channels within the cement. The resulting waveform is comprised of a casing signal and a formation signal; the casing signal arrives first, followed by the formation signal.

When channeling occurs, it is generally localized and nonuniform; that is, it occurs over relatively short intervals and can frequently be identified by variations in the amplitude response. Channeling is significant because it prevents a hydraulic seal. In contrast, a microannulus (a small gap between the casing and cement sheath) may extend over long sections of casing but may not prevent a hydraulic seal. Microannulus may result from thermal expansion or contraction of the pipe during cementing or to the presence of contaminants, such as grease or mill varnish, on the casing's exterior surface. A common practice is to run cement-bond logs with the casing under pressure to expand the casing against the cement, thereby decreasing any microannulus that might exist. If the initial log run was not under pressure and the log indicates poor bond, the presence of a microannulus can be evaluated by running a second bond log under pressure to see if there is a difference. Pressuring the casing improves the acoustic coupling to the formation and the casing signal will decrease and the formation signal will become more obvious (**Fig. 3C.15**). However, if only channeling exists, pressuring the casing will not significantly change the log. When conducting a cement evaluation, information on the type of cement used is essential. For example, foam cements, which intentionally create void spaces in the cured cement, can be misinterpreted as partial bond if normal cement is assumed. **Fig. 3C.16** summarizes this discussion, and **Table 3C.8** lists additional factors that may affect interpretation of bond quality from the amplitude response.

One caveat regarding the use of the amplitude curve for bond evaluation: pipe amplitude represents the quality of the bond of pipe to cement but provides no indication of the quality of the bond between the cement and the formation. Whenever possible, amplitude data should be used in conjunction with the other measurements presented on the log (e.g., travel time or full waveform) for a more-reliable bond evaluation. For example, the presence of shear-wave amplitudes on the full-waveform display is an indication of good acoustic coupling to the formation. **Table 3C.9** lists the limitations of conventional cement-bond logs.

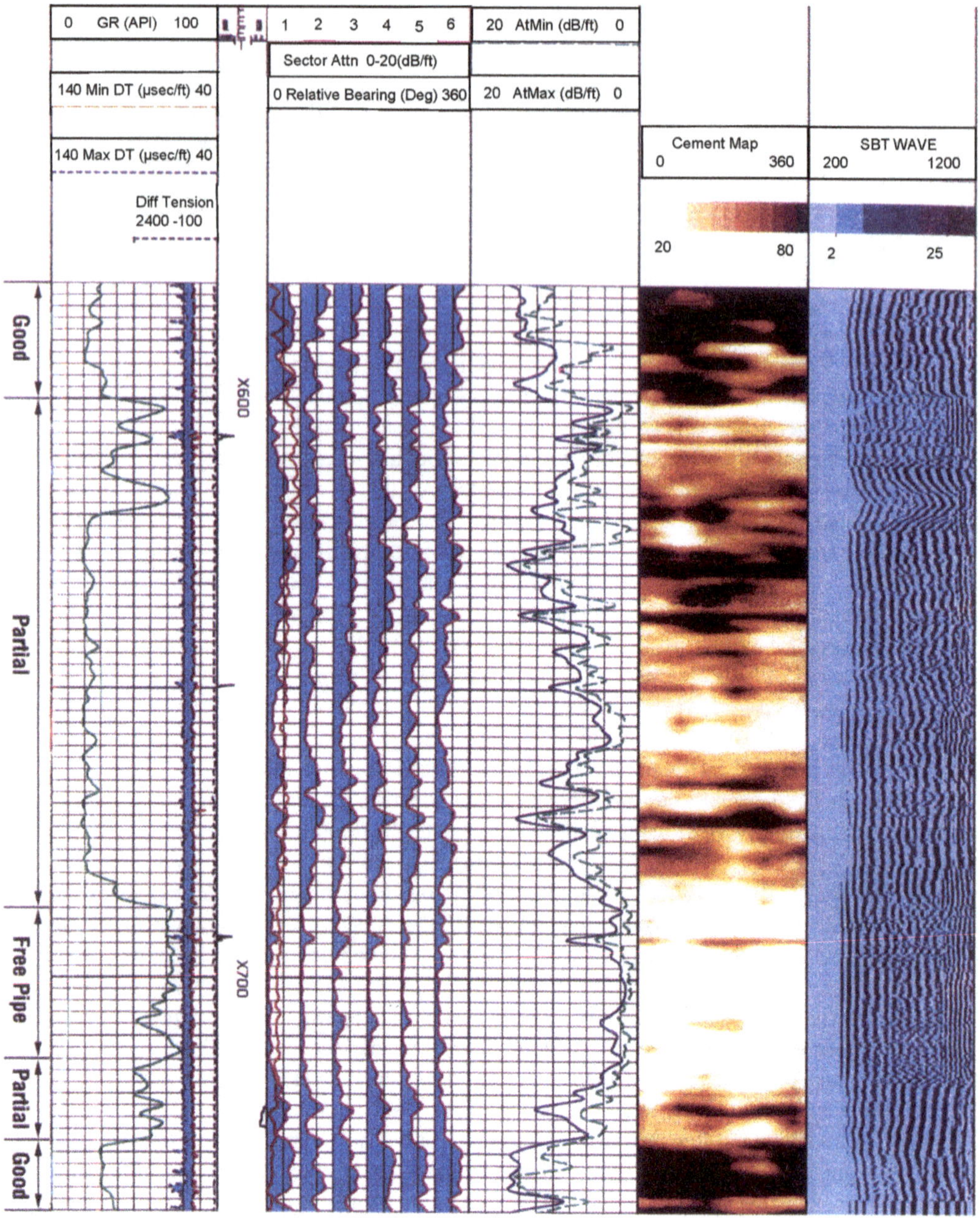

Fig. 3C.17—Log presentation for the Baker Atlas SBT tool containing individual log curves, cement map, and VDL display (courtesy of Baker Atlas).

Formations with transit times less than casing (57 μsec/ft) can cause problems when interpreting the amplitude curves because the formation signal can arrive in the amplitude gate. The VDL should be examined to ensure that the formation arrival is impacting the amplitude curve.

The bond index (BI) is a qualitative measure of cement bond based on signal amplitude. This dimensionless quantity is the ratio of measured attenuation to maximum attenuation:

$$\text{Attenuation} = 20/TR \ \log(\text{measured amplitude}/\text{free-pipe amplitude}) \text{ . ... (3C.3)}$$

TABLE 3C.10—CAPABILITIES AND GUIDELINES FOR RUNNING CEMENT-EVALUATION LOGS[63]

	Conventional CBL	Radial Ultrasonic	Radial Pad
Amplitude/attenuation	Yes	Yes (different)	Yes
Travel time (transit time)	Yes	Yes (different)	Yes
Full waveform	Yes	No	Yes
Good bond identification	Yes (if shear arrival)	Yes	Yes
Partial bond identification	Yes	Yes	Yes
Channel identification	No	Yes	Yes
Microannulus effect	Strongly affected (repeat log under pressure)	Moderately affected	Yes
Repeat (calibration) runs	Yes	Yes	Yes
Centralization	Essential (Use attenuation-ratio log where centralization is difficult)	Less critical	Not a factor
Free-pipe calibration (above top of cement)	Desirable	Not necessary	Not a factor
Heavy mud	Moderately affected	Strongly affected	Not a factor
Gas in microannulus	Strongly affected	Strongly affected	Strongly affected
Casing geometry	No	Yes	No
Casing thickness	No	Yes	No
Casing corrosion	No	Yes	No

A BI value of 1.0 represents a perfect cement bond. A value of less than 1.0 indicates an incomplete bond. This technique requires attenuation measurements in zones with 100% bond and in free pipe.

Radial-Cement Evaluation. Radial-cement-evaluation devices were developed to overcome some limitations of conventional cement-bond tools and to permit more accurate evaluation of cement distribution behind casing by providing the precise location of partial bond and channeling. These tools use one or more azimuthally sensitive transducers to evaluate cement quality around the circumference of the casing. Data from these tools are presented as individual log curves or as azimuthal images ("maps") of cement quality generated by interpolating between the individual azimuthal measurements (**Fig. 3C.17**). In addition, each tool design also provides a conventional 5-ft VDL waveform measurement to provide information about the cement-to-formation bond.

There are four radial-evaluation-tool designs in current use: televiewer-types that use a single rotating ultrasonic transducer,[72,73] circular ultrasonic pulse/echo transducers arranged in a fixed helical pattern around the sonde,[74,75] a multipad device that provides six compensated attenuation measurements,[76] and an array of eight TR pairs, arranged azimuthally around the sonde, that provide compensated CBL amplitude.[77,78]

The ultrasonic tools compute the acoustic impedance of the material beyond the casing. To do this, repeated acoustic pulses are directed at the casing to make it resonate in its thickness mode and the energy level (attenuation) of the decaying reflected wave is measured. Good cement bond to casing produces a rapid damping (higher impedance) of this resonance; poor cement bond results in longer resonance decay (lower impedance). Measurements from these devices are influenced by the same factors as openhole televiewer devices.

The pad device makes multiple short-spaced, compensated, azimuthal-attenuation measurements. Because the pads are in direct contact with the casing—in contrast to ultrasonic measurements—measurements are unaffected by gas in the borehole, fast formations, heavy-mud conditions, or minor tool eccentricity. The attenuation in each segment is measured in two directions using a pair of acoustic receivers and two transmitters. The two measurements are

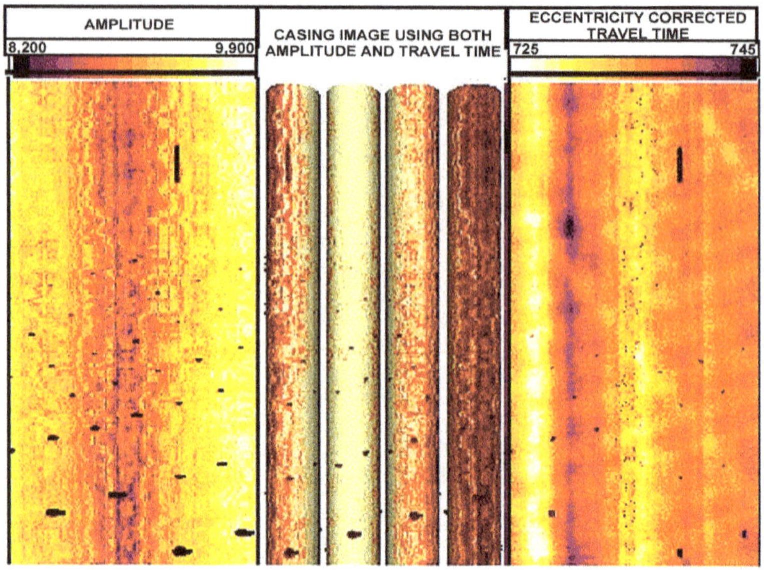

Fig. 3C.18—Casing-evaluation log display. Holes in the casing are visible in the series of ultrasonic images that are based on amplitude (left) and corrected travel time (right). The center 3D images show the pipe in 90° quadrants. The image shading is generated from the amplitude data[86] (courtesy of SPE).

combined to form a result that compensates for surface roughness and the effects of minor residual cement on the inside of the casing. Transmitting elements and the firing sequence are controlled to direct (steer) and enhance the acoustic-energy output of both the pad transmitters and the VDL transmitter. This has the effect of improving the signal strength of both the casing and cement-to-formation arrivals, respectively. This technique improves VDL interpretation, particularly in soft formations in which the standard VDL may wash out.

The use of new high-performance low-density, foam, and complex cements is increasing. However, the presence of gas in cement slurries, as an inert component or as contamination, may seriously affect ultrasonic-tool interpretation. New interpretation methods integrate ultrasonic and attenuation measurements from conventional tools to provide improved cement evaluation in these conditions.[79–82] The latest ultrasonic tool has a conventional pulse-echo transducer plus a flexural transmitter and two flexural receivers that provide greater depth of investigation. Interpretation techniques combining these different measurements provide improved evaluation in lightweight cements, especially in the annulus, beyond the casing-cement bond.[83] **Table 3C.10** summarizes the capabilities and guidelines for running the different types of cement-bond evaluation tools.

Casing Evaluation. Preventing casing failure caused by deformation, physical wear, or corrosion is critical to maintaining well production. Ultrasonic radial-cement-evaluation devices and modified openhole-imaging devices are also used to evaluate casing for indications of potential collapse, thinning, and internal or external metal loss.[73,84–86] Echo amplitude and travel time provide images of the condition of the inside casing surface (e.g., buildup, defects, and roughness such as pitting and gouges) (**Fig. 3C.18**), and travel-time and resonant-frequency analysis provide casing thickness (**Fig. 3C.19**).

The acoustic caliper generated from the pulse/echo travel time provides the casing inside diameter (an average of all transducers or a single circumferential scan). An estimate of casing ovality is obtained using only the maximum and minimum measurements. Then, if the nominal value of the outside casing diameter is assumed, changes in thickness can be calculated and internal defects identified. Frequency analysis determines the casing resonant frequency from

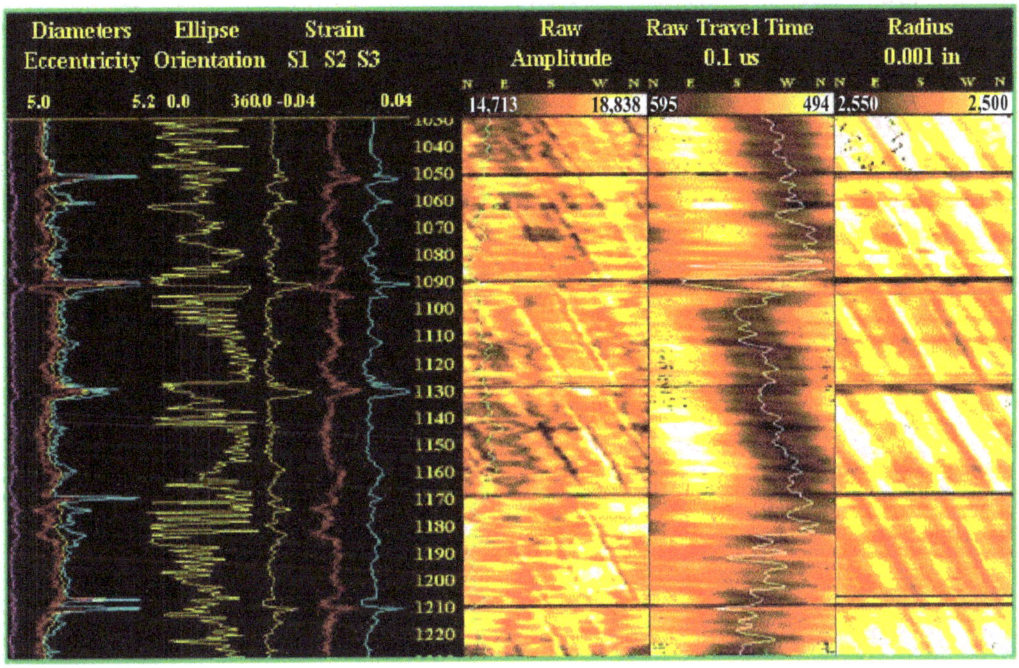

Fig. 3C.19—Ultrasonic casing-evaluation display. In this example, casing radius and shape are presented as log curves and image maps and deformed casing is easily identified (courtesy of Baker Atlas).

the acoustic waveform; casing thickness is inversely related to the resonant frequency. By combining travel time and resonant-frequency measurements and using data from all available transducers (or a single scan), presentations showing casing cross sections are used to highlight casing damage such as thinning, corrosion metal loss, and collapse (Fig. 3C.19).

3C.5 Conventional Applications

Interpretation of acoustic-log data begins with the slowness values obtained from processing the recorded waveforms. Slowness, or porosity derived from slowness, may be corrected for additional factors (discussed below) before use in applications. Today, log analysis and interpretation is routinely performed by computers during data acquisition (in real time) or in post-acquisition processing in offices and computing centers. Modern interpretation software is built on algorithms derived from the empirical relationships originally used to create a variety of graphical solutions contained in crossplots and nomograms.[87] Because porosity tools vary in their sensitivity to lithology, porosity, and fluid properties, the combination of different measurements allows more-accurate determination of porosity, petrophysical, and geological characteristics. The logging service companies issue chartbooks with tool-specific charts to facilitate rapid manual determination of porosity, lithology, shaly-sand analysis, saturation, mechanical properties, and cement-bond evaluation. These charts, which are accompanied by instructions for their use, form the basis for quick-look interpretations and for quality control of computer interpretations.[88,89]

The earliest applications of acoustic-logging measurements made use of compressional slowness, used alone or in combination with resistivity logs and other porosity logs. This group of applications, which includes velocity, porosity, gas identification, lithology, fluid saturation, and rock strength, still comprises the basic set of applications for acoustic-log data. In the 1960s, laboratory studies indicated that shear-wave data could also prove useful in formation evaluation.[90] The development and introduction of reliable shear-wave acquisition and full-wave-

TABLE 3C.11—ACOUSTIC-LOG APPLICATIONS BY WAVE TYPE

Wave Type (Energy Mode)	Acoustic Property	Application
Transmission		
Compressional		
	Velocity	Porosity evaluation
	Velocity	Lithology identification
	Velocity	Gas identification
	Velocity	Fluid saturation
	Velocity	Rock-strength determination
	Velocity	Seismic time/depth correlation
	Velocity	Synthetic seismic
	Velocity	Geopressure prediction
	Attenuation	Cement evaluation
Shear		
	Velocity	Porosity evaluation
	Velocity	Lithology evaluation
	Velocity	Hydrocarbon typing (combined with compressional)
	Velocity, amplitude	Azimuthal anisotropy analysis
	Velocity	Rock-strength determination
	Amplitude, attenuation	Fracture identification and evaluation
	Velocity	Seismic correlation and calibration, AVO, VSP
	Velocity	Geopressure prediction
	Velocity	Completion design (hydrofracture evaluation)
	Velocity	Sand control
	Attenuation	Permeability
Stoneley (tube)	Velocity, attenuation	Permeability analysis
	Amplitude	Fracture evaluation
Reflection		
Ultrasonic		
	Velocity/amplitude	Borehole imaging
	Velocity/amplitude	Cement evaluation
	Velocity/amplitude	Casing evaluation

form logging in the late 1970s and 1980s resulted in improved accuracy in these conventional applications.[91] Further advances throughout the 1980s and 1990s, in both basic and applied research as well as in tool technology, led to new and advanced applications, e.g., permeability estimation, anisotropy determination, and direct hydrocarbon indicators, that use shear- and Stoneley-wave amplitude and attenuation (**Table 3C.11**).

3C.5.1 Petrophysical Applications. ***Porosity Evaluation.*** The compressional velocity of sound in fluid is less than the velocity in rock. If there is pore space in the rock, and it is fluid-filled, the acoustic energy will take longer to get from the transmitter to the receiver (i.e., low velocity indicates high porosity). The recorded velocity or travel time represents the sum of the velocity of the solid part or framework of the rock (i.e., the rock matrix), the rock lining the pores, and the fluid filling the pore space. In turn, travel time in the rock matrix, Δt_{ma}, is influenced by variations in lithology (i.e., the chemical composition) and confining pore pressure (i.e., compaction).[92] These factors are related through an empirical relationship known as the Wyllie time-average equation.[93] When the velocity (transit time, Δt, or travel time, t) of the rock matrix and borehole fluids are known, porosity can be computed the following ways (Eq. 3C.4 to Eq. 3C.7).

- In terms of velocity, v:

TABLE 3C.12—GUIDELINES FOR SELECTING MATRIX TRANSIT TIME

Δt_{sh} (μsec/ft)	V_{ma} (ft/sec)	Δt_{ma} (μsec/ft)
70 to 80	20,000	50
80 to 90	19,000	52.5
90 to 100	18,000	55.5
> 100	Use compaction correction (C_p) adjustments	

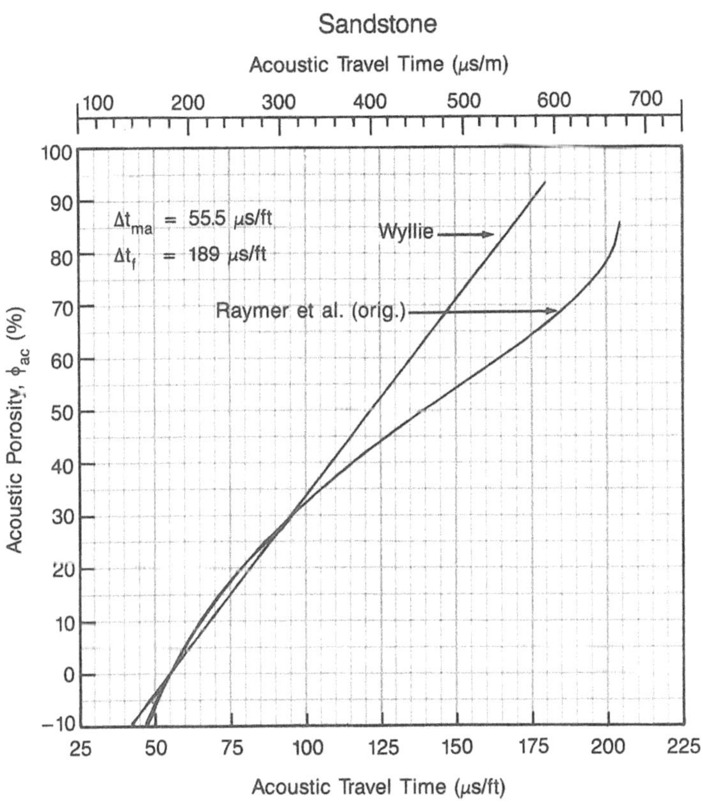

Fig. 3C.20—Graphical solutions of the Wyllie and Raymer velocity-porosity relationships (sandstone matrix) (courtesy of Baker Atlas).

$$\frac{1}{v} = \frac{\phi}{v_f} + \frac{(1-\phi)}{v_{ma}}, \quad \text{(3C.4)}$$

where ϕ = fractional porosity of the rock, v = velocity of the formation (ft/sec), v_f = velocity of interstital fluids (ft/sec) and, v_{ma} = velocity of the rock matrix (ft/sec).

- In terms of transit time (Δt):

$$\Delta t = \phi \Delta t_f + (1-\phi)\Delta t_{ma}, \quad \text{(3C.5)}$$

or

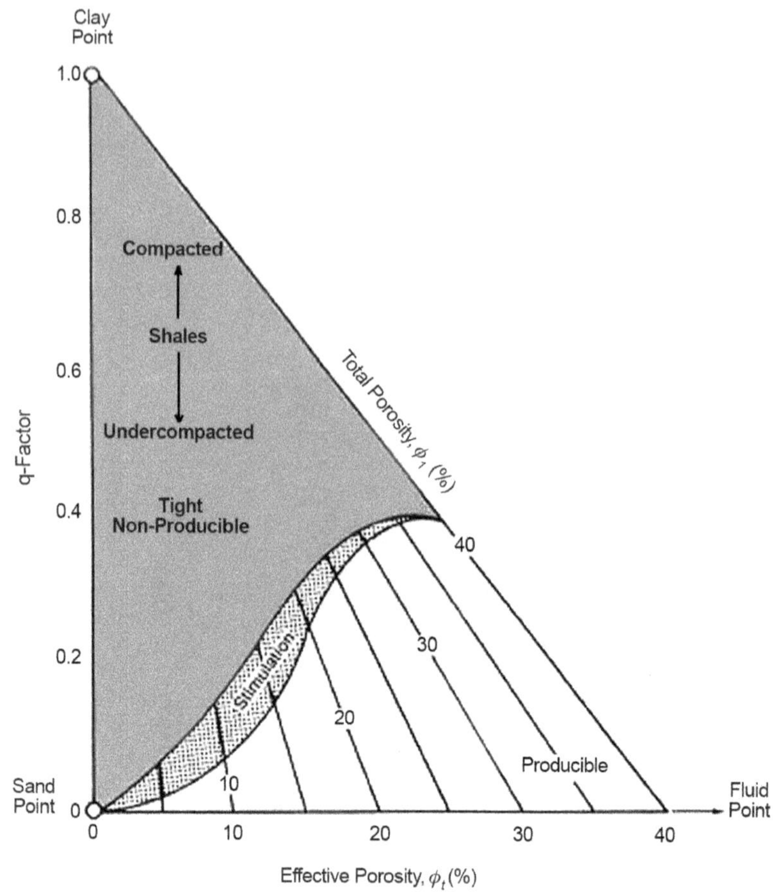

Fig. 3C.21—Chart used to predict productivity in shaly-sand reservoirs (courtesy of Baker Atlas).

$$\phi = \frac{\Delta t - \Delta t_{ma}}{\Delta t_f - \Delta t_{ma}}, \quad \text{(3C.6)}$$

where Δt = acoustic transit time (µsec/ft), Δt_f = acoustic transit time of interstitial fluids (µsec/ft), and Δt_{ma} = acoustic transit time of the rock matrix (µsec/ft). (See Table 3C.3 for typical values of Δt_{ma} and Δt_f.)

- In terms of travel time:

$$t_{log} = t_{f\phi} + t_{ma}(1 - \phi)$$

$$t_{log} = t_{f\phi} + t_{ma} - t_{ma\phi}$$

$$t_{log} - t_{ma} = t_{f\phi} - t_{ma\phi}$$

$$t_{log} - t_{ma} = (t_f - t_{ma})\phi$$

$$\therefore$$

$$\phi = \frac{t_{log} - t_{ma}}{t_f - t_{ma}}. \quad \text{(3C.7)}$$

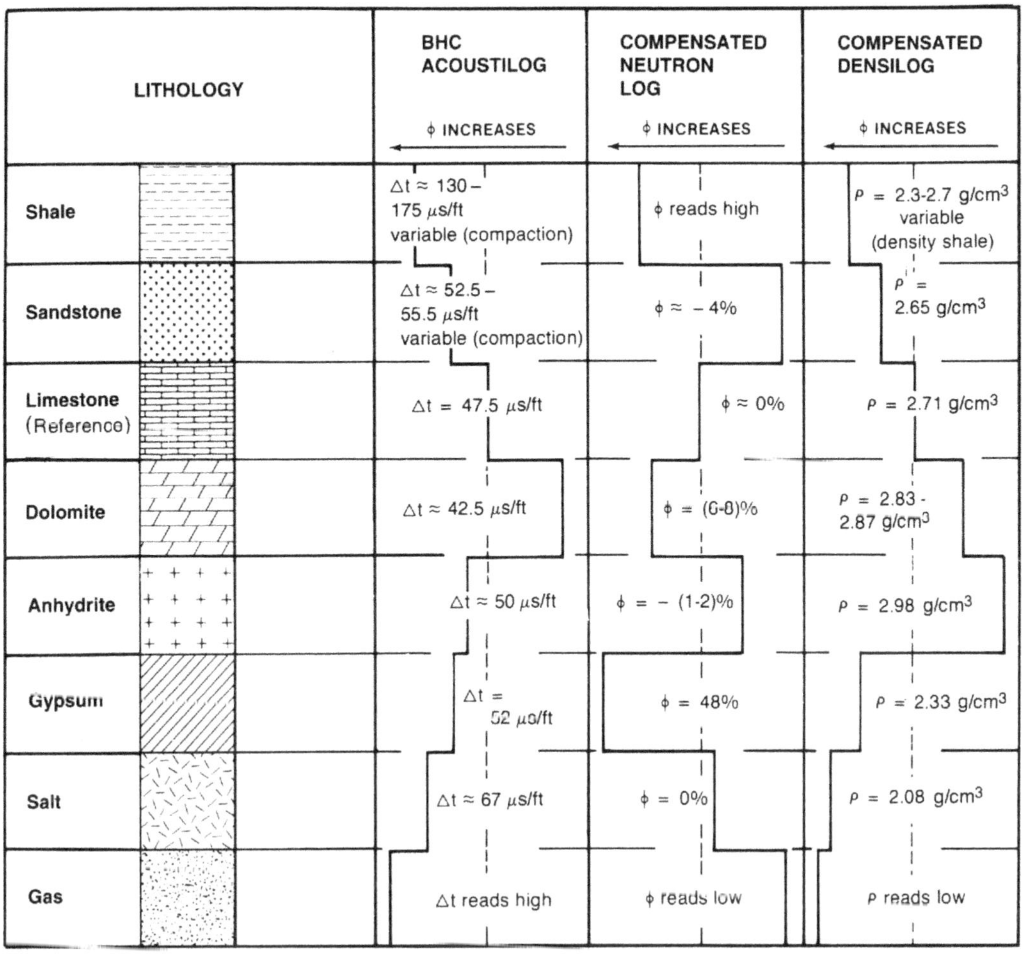

Fig. 3C.22—Generalized comparison of acoustic-, neutron-, and density-log responses in common reservoir lithologies and to gas (courtesy of Baker Atlas).

The velocity of most borehole and reservoir fluids (except gas) does not vary greatly and a fluid velocity (Δt_f) of 189 μsec/ft (5,300 ft/sec) is generally assumed for fresh drilling fluids; a slightly lower value, 185 μsec/ft, is used for salt muds. Fluid type becomes more of a concern when oil-based mud (OBM) is used if the formation of interest is not invaded or if invasion is very shallow. The lithology must be known or estimated in order to select the appropriate matrix velocity.

The Wyllie equation represents consolidated and compacted formations. In poorly consolidated or unconsolidated rocks, a correction factor is necessary (Eq. 3C.8). Also, the presence of shale or clay within the sand matrix will increase Δt by an amount proportional to the bulk-volume fraction of the clay. An empirical equation is used for calculating porosity in sandstones in which adjacent shale values (Δt_{sh}) exceed 100 μsec/ft (Eq. 3C.9):

$$\phi = \frac{\Delta t - \Delta t_{ma}}{\Delta t_f - \Delta t_{ma}} \times \frac{1}{C_p}, \qquad (3C.8)$$

where the compaction correction factor C_p is

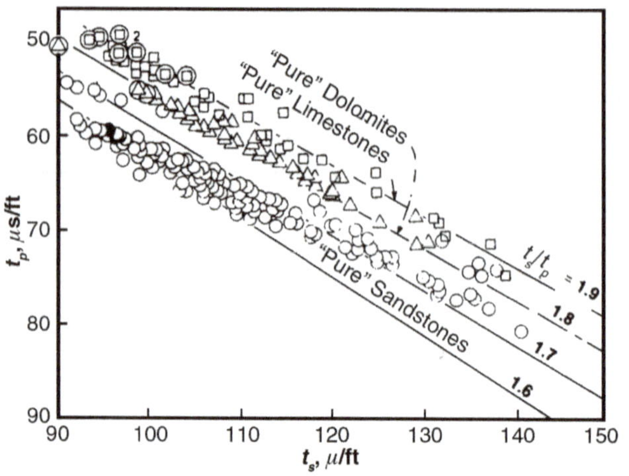

Fig. 3C.23—Δt_p vs. Δt_s crossplot demonstrates sensitivity to variations in lithology[90] (courtesy of SPE).

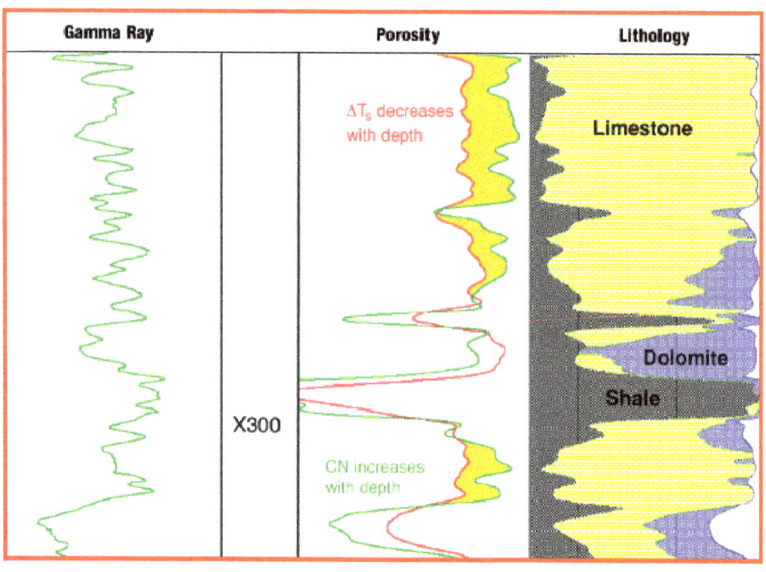

Fig. 3C.24—Shear slowness as a lithology indicator. The cased-hole apparent compensated-neutron porosity (Track 2) appears to increase with depth, while the dipole tool shear-velocity data decrease with depth. This indicates a lithology change rather than a porosity change, which is confirmed in the lithology track (courtesy of Baker Atlas).

$$C_p = \frac{\Delta t_{sh}(C)}{100}, \quad \text{(3C.9)}$$

where Δt_{sh} = specific acoustic transit time in adjacent shales (μsec/ft), and 100 = acoustic transit time in compacted shales (μsec/ft). The shale compaction coefficient (C) generally ranges from 1.0 to 1.3, depending on the regional geology.

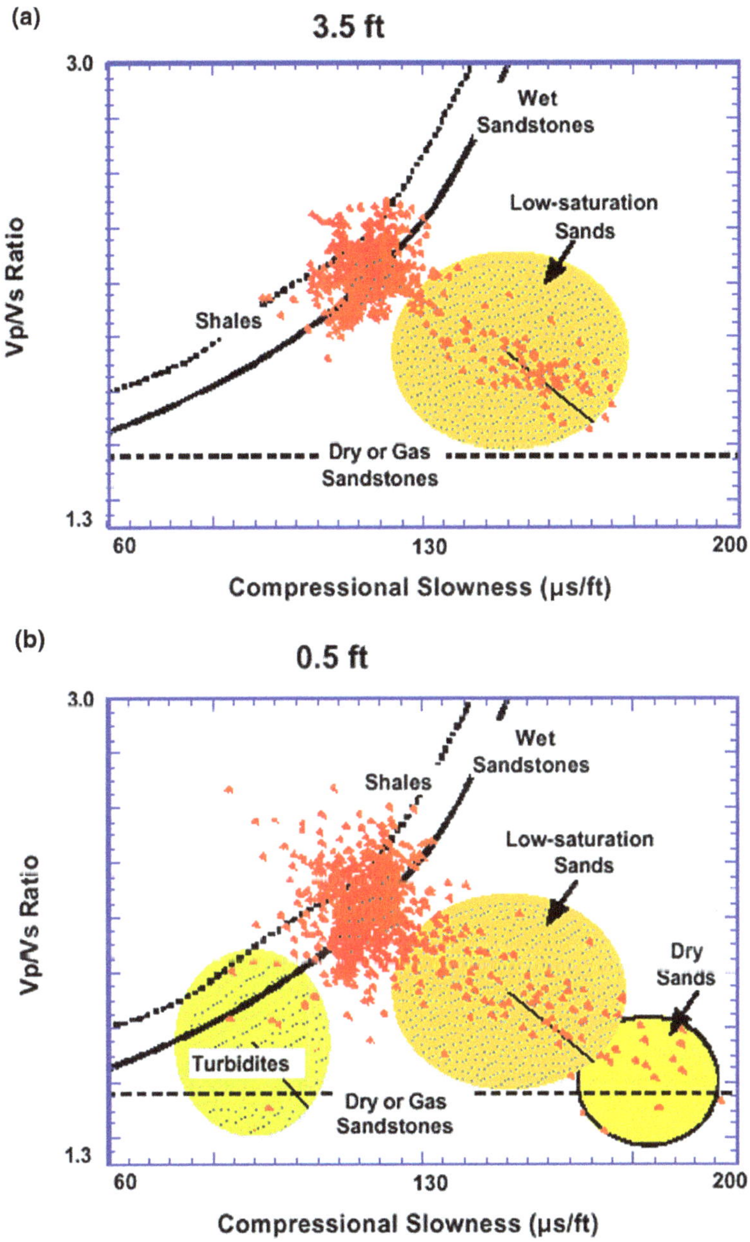

Fig. 3C.25—V_p / V_s vs. Δt_p crossplot. In addition to the low-saturation sands identified in (a), the high-resolution result (b) also resolves turbidite and dry sand intervals[224] (courtesy of SPWLA).

The highest velocities observed in sandstones approach 20,000 ft/sec (50 μsec/ft), but most sandstones have a lower matrix velocity. Velocities in adjacent shales are used to adjust the matrix velocity for sands with velocities lower than 18,000 ft/sec. **Table 3C.12** provides guidelines for selecting the appropriate value of Δt_{ma}. If the lithology of carbonate rocks can be reasonably estimated and if the porosity distribution is fairly uniform, the Wyllie time-average formula can provide reliable determination of porosity for this group.

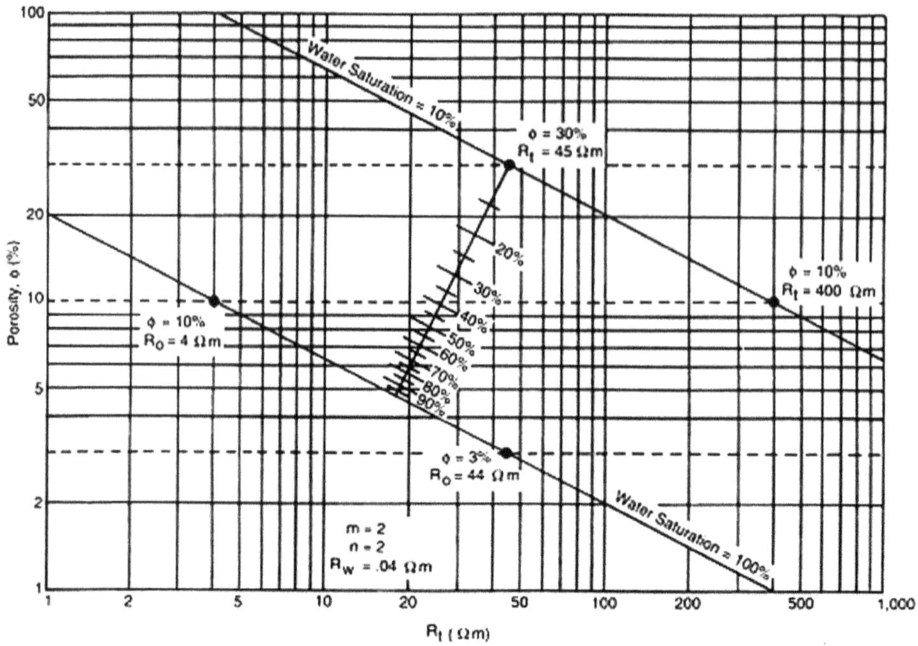

Fig. 3C.26—Example of a Pickett plot, including a saturation scaler (courtesy of Baker Atlas).

A second empirical velocity/porosity relationship, the Raymer-Hunt-Gardner equation,[94] was introduced to correct for observed anomalies and shortcomings of the Wyllie time-average formula (Eq. 3C.10). It provides improved porosity correlation over the entire porosity range and is applicable to both consolidated and some unconsolidated formations, thus eliminating the need for a compaction correction. However, in high-porosity, unconsolidated, and uncemented (slow) rocks, neither of the empirical velocity/porosity transforms may be adequate[95]

$$\phi_{\log} = -\alpha - \left[\alpha^2 + \left(\Delta t_{ma} / \Delta t_{\log}\right) - 1\right]^{0.5}, \quad \text{(3C.10)}$$

where $\alpha = (\Delta t_{ma} / 2\Delta t_f) - 1$.

Graphical solutions for both algorithms for a sandstone matrix are compared in **Fig. 3C.20.** One caveat regarding the use of empirically derived porosity transforms: they do not account for all the factors influencing acoustic velocity. Consequently, these relationships may not be valid for all reservoirs.

In fast formations, the shear velocity can also be used for porosity evaluation in a manner similar to that described above for compressional velocity.[91] Further, the combination of compressional and shear slowness can provide an enhanced porosity determination.[96] Using shear velocity for porosity determination offers several distinct advantages because shear velocity is: generally more sensitive to porosity than compressional velocity, insensitive to gas effects, less affected by borehole washout, can be used to replace nuclear porosity in some situations,[97] and porosity evaluation can be conducted in cased hole using dipole tools (or in some cases, monopole-array tools using special processing).[98,99]

These velocity/porosity methods are for clean (shale free), water-filled formations. The calculated apparent porosity must still be corrected for the volume of pore-filling material (shale). If the formation contains shale or dispersed-clay particles, or is hydrocarbon bearing and invaded to only a very shallow depth, corrections to the basic log data are necessary before reasonable porosity values can be calculated.

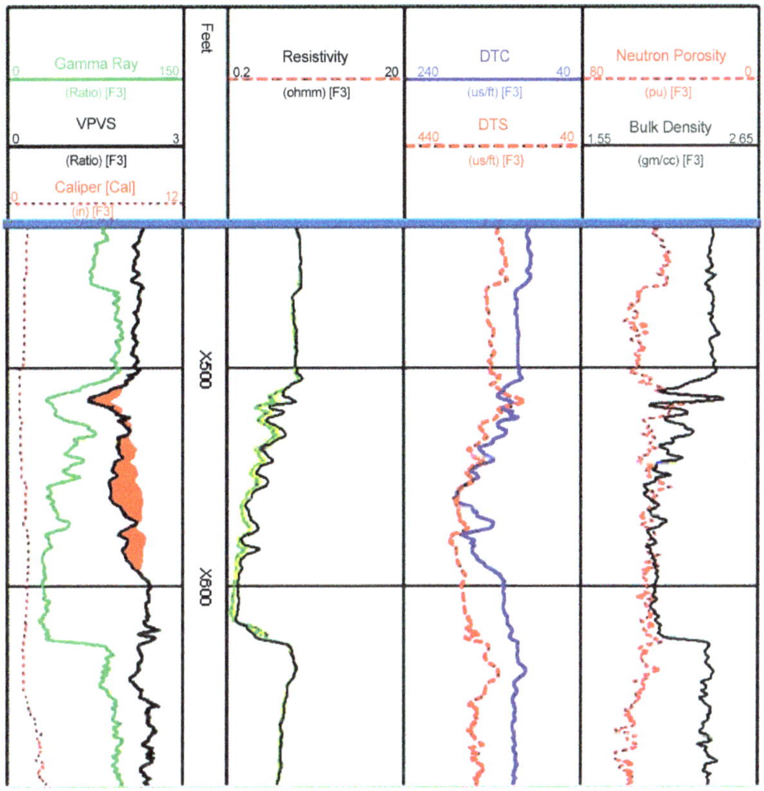

Fig. 3C.27—Log presentation illustrating the use of Δt_c and Δt_s curves as a quick-look gas indicator (courtesy of Baker Atlas).

Because shale transit times range from 62 to 167 μsec/ft, failure to correct for the presence of shale may result in overly optimistic porosity calculations. The acoustic measurement is also influenced by the way the shale is distributed within the sandstone reservoirs. The fraction of shale, or shale volume, can be estimated using a combination of log measurements that are influenced by shale, such as neutron porosity, density, gamma ray, or spontaneous potential (SP). Chartbook nomograms developed for porosity determination include graphical solutions for both undercompaction and shale volume.

In some producing regions, producibility indexes based on the volume of shale in producing sandstone reservoirs have been developed. The fraction of total porosity occupied by dispersed clay (q factor) is empirically related to effective and total porosity and production characteristics. Local experience is used to create permeability cutoffs using the q factor (**Fig. 3C.21**).[87]

Acoustic travel time in gas and oil is higher than in water. The presence of unflushed hydrocarbons in an interval can result in high values of apparent formation porosity. Commonly used correction factors are 0.9 in oil zones and 0.7 in gas zones.[92] More recently, a gas-zone porosity-correction technique using shear slowness has been developed.[100]

An additional empirical velocity-porosity predictive model has recently been proposed and is still in the experimental phase.[101,102]

Carbonate and complex lithology reservoirs are generally comprised of varying proportions of limestone, dolomite, chert, quartzite, and occasionally, evaporites. The primary influences on porosity in these rocks are lithology and pore type. Generally, any shale present is in dispersed form and in small amounts that do not significantly impact porosity calculations. Acoustic poros-

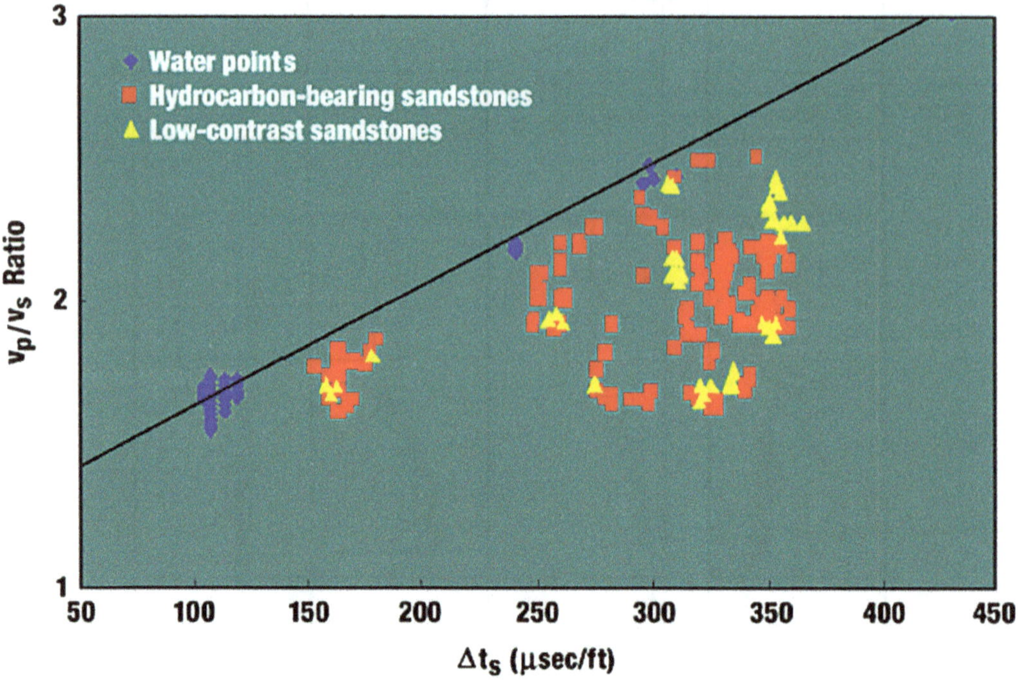

Fig. 3C.28—Crossplot of shear slowness vs. V_p / V_s serves as a quick-look hydrocarbon indicator. Compressional-slowness travel times increase where the rock pore space is occupied by more compressible fluid[81] (modified from original, courtesy of Baker Atlas).

ity is a measure of the primary or intergranular (matrix) porosity. In dual-porosity reservoirs, the secondary porosity (e.g., isolated pores, vugs, and fractures) may significantly influence the rock-pore distribution, but may be overlooked by acoustic-log measurements. This topic is the subject of ongoing research.[103] In contrast, nuclear-porosity devices, such as density and neutrons, measure total porosity. The difference between the nuclear-porosity and acoustic-porosity measurements is an approximation of the secondary porosity.

Lithology Identification. Acoustic velocity is primarily a function of the rock matrix and can be used to identify different lithologies and for stratigraphic correlations. A variety of crossplot techniques, using acoustic measurements alone, or in combination with other porosity logs (neutron and density), have been devised to assist in lithologic identification (**Fig. 3C.22**). In particular, the M-N and mineral-identification-plot (MID) techniques use all three porosity logs in different combinations.[104,105] Before lithology determination, the individual log measurements must be corrected for influences of gas effect, secondary porosity, bad hole conditions, and shaliness. In general, the MID plot is more sensitivity to lithology, gas, and secondary porosity and provides superior results to M-N plots. Crossplots, using a variety of log measurements or combinations of measurements can be used to resolve specific lithologic problems related to local or regional geology.[87,106]

The ratio of compressional to shear velocity, V_p / V_s, is an effective lithology indicator because each lithology exhibits a defined trend that is independent of porosity and depth (**Fig. 3C.23**).[90,107,108] However, because the V_p / V_s ratio is affected by formation anisotropy, the ratio values may not be absolute indicators of a particular lithology.[109]

The addition of shear slowness to lithology identification provides a more robust result that can be particularly useful in cased-hole evaluations where density logs are not available. In **Fig. 3C.24**, the combined use of shear slowness and cased-hole neutron porosity results in en-

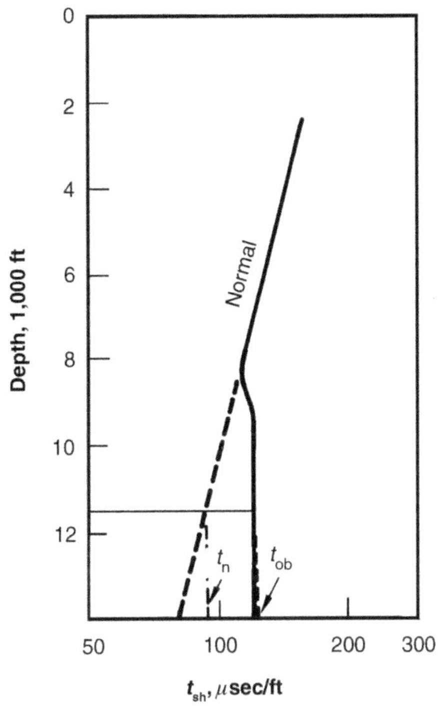

Fig. 3C.29—Semi-log plot illustrating acoustic detection of geopressure by use of shale slowness[148] (courtesy of SPE).

hanced-porosity determination in a complex lithology. Crossplots of V_p/V_s ratio vs. compressional travel time, Δt_c, facilitate identification of lithology trends with respect to porosity and lithology (**Fig. 3C.25**). Recent studies of complex carbonate reservoirs indicate that the V_p/V_s ratio is a function of porosity and in some cases, can differentiate higher-permeability facies.[110,111]

Saturation Determination. The potential of acoustic velocity for determination of fluid saturation was recognized soon after development of the first logging devices and quickly became a staple input for log-interpretation methods and software. Acoustic-derived porosity serves as one variable, together with resistivity, in a variety of graphical and empirical methods used for solving the Archie saturation equation. Hingle and Pickett plots are the two most commonly used resistivity-porosity plots. Depending on the data available, the Hingle plot[112,113] can solve for R_w, Δt_{ma}, and water saturation (S_w). Pickett plots[114,115] (**Fig. 3C.26**) solve for Archie parameters a and m, formation factor, resistivity index, and S_w. A new technique using an empirical equation based on acoustic- and resistivity-log parameters is reported to successfully estimate water saturation in clean and shaly sandstones.[116]

Hydrocarbon Identification. *Gas.* Acoustic coupling between solid and gas or fluid and gas is poor, resulting in a high loss of energy. A sudden loss of energy (amplitude) in the measured acoustic signal, primarily in the compressional wave and only secondarily in the shear wave: (e.g., the cycle skipping or high slowness values) may indicate gas-filled pore space (gas effect). In the cased of gas-filled porosity, the acoustic-neutron crossplot can be useful for this purpose because neutron porosity is lower than acoustic porosity in gas zones.

Compressional velocities are affected (slowed) by the compressive fluids in the pore space, while shear velocity is affected only by the rock matrix. Consequently, the presence of gas is especially noticeable in compressional-wave slowness. The combination of compressional and

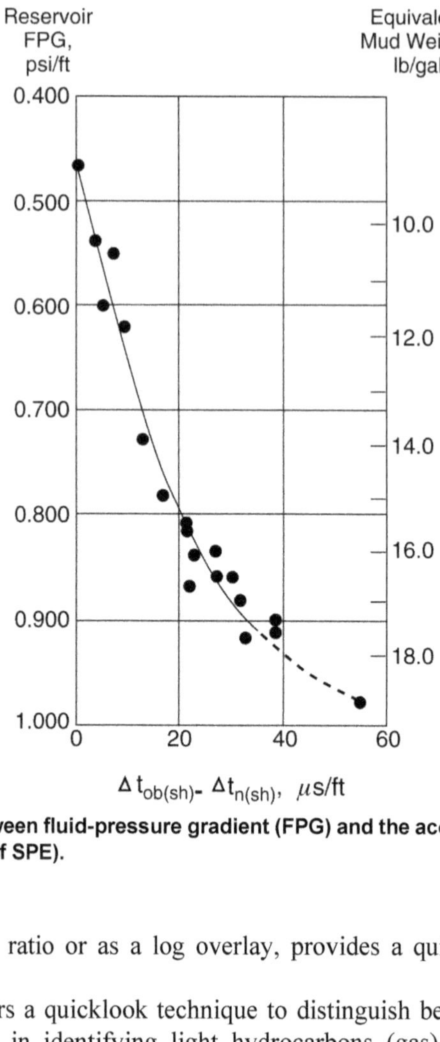

Fig. 3C.30—Relationship between fluid-pressure gradient (FPG) and the acoustic slowness difference for U.S. Gulf Coast[148] (courtesy of SPE).

shear slowness, either as a ratio or as a log overlay, provides a quick-look gas indicator (**Fig. 3C.27**).

The ratio of V_p/V_s offers a quicklook technique to distinguish between reservoir fluids[117–120] and is especially effective in identifying light hydrocarbons (gas).[121,122] When Δt_s is plotted against the V_p/V_s ratio, water-bearing sands and shales show a linear relationship and points falling below matrix lines result from the slowing effect of Δt_p in light hydrocarbons (**Fig. 3C.28**). These relationships are used for correcting porosity in gas-bearing intervals.[100] Further, laboratory investigations suggest that the V_p/V_s ratio can be used for saturation determination in some situations[123] as well as time-lapse changes accompanying steam flood of heavy-oil reservoirs.[124]

Oil. Recent work suggests that the V_p/V_s ratio may also serve as an indicator of bypassed oil in cased wells.[125,126] Research on the acoustic properties of heavy oils indicates that under the proper conditions of temperature and viscosity, these oils may behave as solids and generate shear waves that may be detectable at logging-tool frequencies.[127]

3C.5.2 Geophysical Applications. The higher operating frequency of acoustic-logging tools and the smaller TR distances allows for higher-quality velocity data and finer vertical resolution than surface reflection techniques. Acoustic-velocity logs were originally developed for calibrating surface seismic velocities and reflectors. Acoustic-log interval travel time or transit time, Δt, can be summed, i.e., integrated, over the entire logged interval to provide the equiva-

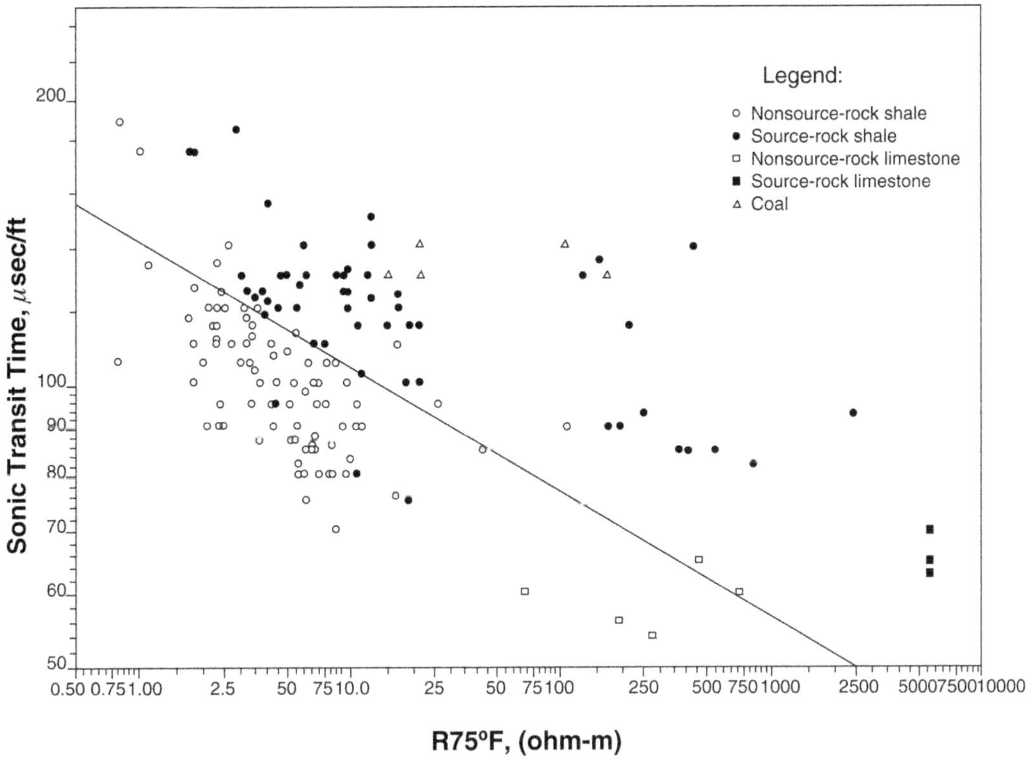

Fig. 3C.31—Identification of source-rock potential using a sonic transit-time-resistivity crossplot—points plotted above the line are source rocks, while those below are not.[181] R75°F is well-log resistivity corrected to 75°F (courtesy of AAPG).

lent of seismic one-way time which is compared to borehole seismic surveys and reflection seismic two-way time.

Acoustic-log data are commonly calibrated using checkshot (velocity) or vertical seismic profile (VSP) surveys prior to use in geophysical applications. Data from these surveys, which use downhole receivers and surface acoustic sources, are used to adjust the log data for drift and borehole conditions and result in improved time-depth correlation. Acoustic-log data are combined with density-log data, to generate an impedance log that in turn is used to produce a synthetic seismogram. Synthetic seismograms are artificial seismic records that tie seismic time to log depth and are also used to match well-log quantities to seismic attributes for distinguishing primary seismic events (geologic structure and stratigraphy). It is possible, however, that a synthetic seismogram may not provide a very good match to the seismic field data. Disagreements commonly result from the differences scale and acquisition physics used in seismic and well-log measurement; for example, operational frequency (wavelength), borehole condition, and angle of measurement (particularly in the presence of anisotropy).[128–132] Acoustic-log data provide a fundamental and essential element of modern seismic reservoir characterization.[133] The chapter on Fundamentals of Geophysics in this volume of the *Handbook* contains more information on the determination and use of these types of analyses.

3C.5.3 Drilling and Reservoir Engineering Applications. ***Pore Pressure and Overpressure Detection and Evaluation.*** Abnormal pressure is defined as any departure from normal hydrostatic pressure at a given depth.[134] Abnormal subsurface pressures, either overpressure (geopressure) or underpressure, are encountered in hydrocarbon basins throughout the world in

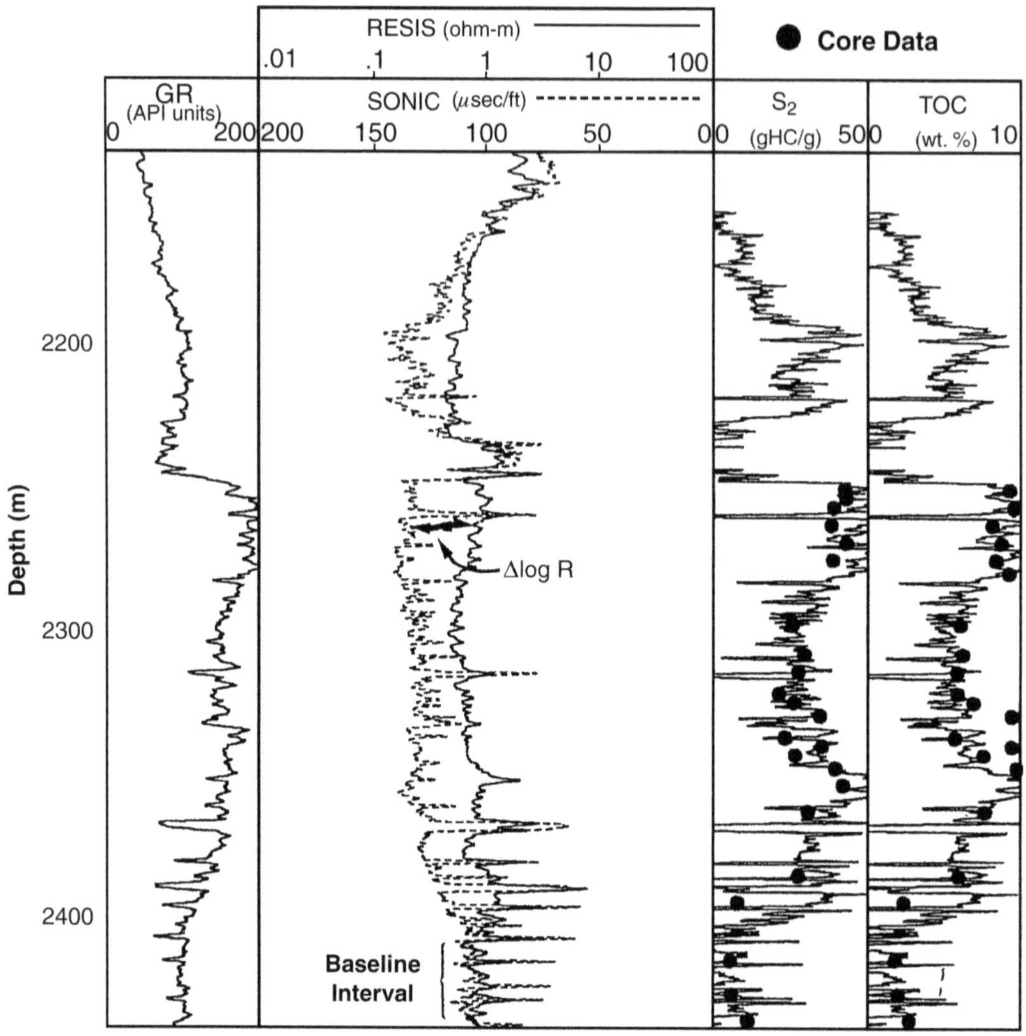

Fig. 3C.32—Source-rock identification and assessment using an acoustic/resistivity overlay technique. Separation of the acoustic and resistivity curves (Track 2), labeled Δ log R, indicates organic-rich intervals as shown by the core analyses in Tracks 3 and 4. The amount of separation is directly related to the amount of total organic carbon and is a function of thermal maturity[182] (courtesy of AAPG).

all lithologies, all geologic ages, and at all depths.[135] Early and reliable detection of geopressure is vital to avoid or mitigate potential drilling and safety hazards, e.g., shallow water flow, blowouts, and shale instability. During drilling, advanced warning of approaching geopressuring enables the mud weight to be adjusted to avoid well and reservoir damage and to determine casing points. This is a particular concern in deepwater wells in which the pressure difference; i.e., the operating window, between the hydrostatic gradient and the fracture gradient can be very narrow.

Geopressuring in hydrocarbon reservoirs may result from a variety of geologic and tectonic processes.[135–137] Borehole-acoustic detection methods using compressional and shear slowness can identify abnormally pressured zones before they are drilled and can quantify pressure gradients. These methods, discussed below, are used in conventional borehole logging (wireline and LWD), new seismic-while-drilling techniques, and more recently, surface seismic data.[37,56,136,138–147]

TABLE 3C.13—DEFINITIONS OF THE ELASTIC MODULI AND POISSON'S RATIO

Elastic Constants	Basic Equations	Interrelation of Equations	Equations in Well Logging Terms
Young's modulus	$E = \dfrac{9Kpv_s^2}{3K + pv_s^2}$	$E = \dfrac{3K\mu}{3K+\mu} = 2\mu(1-\sigma) = 3K(1-2\sigma)$	$E = \left(\dfrac{p}{\Delta t_s^2}\right)\left(\dfrac{3\Delta t_s^2 - 4\Delta t_c^2}{\Delta t_s^2 - \Delta t_c^2}\right) \times 1.34 \times 10^{30}$
Bulk modulus	$K = pv_p^2 - \frac{4}{3}v_s^2$	$K = \dfrac{E\mu}{3(3\mu) - E} = \mu\dfrac{2(1+\sigma)}{3(1-2\sigma)} = \dfrac{E}{3(1-2\sigma)}$	$K = p\left(\dfrac{3\Delta t_s^2 - 4\Delta t_c^2}{3\Delta t_s^2 \Delta t_c^2}\right) \times 1.34 \times 10^{30}$
Shear modulus	$\mu = pv_s^2$	$\mu = \dfrac{3KE}{9K \times E} = 3K\dfrac{1-2\sigma}{2+2\sigma} = \dfrac{E}{2+2\sigma}$	$\mu = \dfrac{p}{\Delta t_s^2} \times 1.34 \times 10^{10}$
Poisson's ratio	$\sigma = \frac{1}{2}\dfrac{\left(\frac{v_p^2}{v_s^2}\right) - 2}{\left(\frac{v_p^2}{v_s^2}\right) - 1}$	$\sigma = \dfrac{3K - 2\mu}{2(3K+\mu)} = \left(\dfrac{E}{2\mu} - 1\right) = \dfrac{3K - E}{6K}$	$\sigma = \frac{1}{2}\left(\dfrac{\Delta t_s^2 - 2\Delta t_c^2}{\Delta t_s^2 - \Delta t_c^2}\right)$

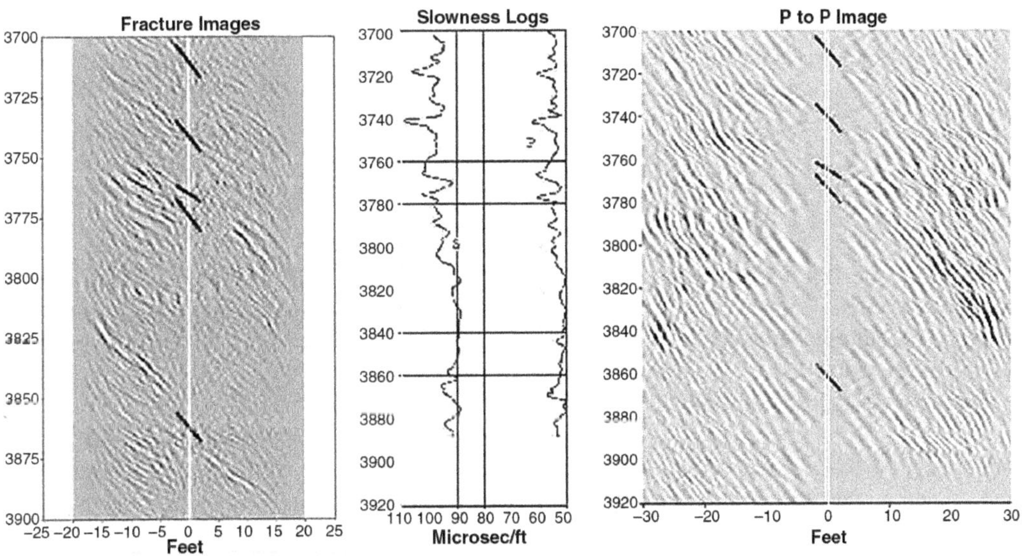

Fig. 3C.33—Near-well acoustic image of a fractured reservoir using direct (compressional) (right panel) and converted (shear) (left panel) acoustic modes. The fractures indicated in the converted mode data are more apparent than in the direct data and also correlate better with fractures seen on the televiewer image[211] (not shown) (courtesy of SPE).

Undercompaction is the primary mechanism for creating overpressure, particularly in deltaic basins in which high rates of deposition commonly prevent the escape of pore water trapped in shales. Undercompacted shales have higher acoustic transit times (i.e., higher apparent porosity) than normally pressured shales at the same depth.[148–151] With the onset of overpressuring, a semi-logarithmic plot of acoustic slowness with depth will diverge from a normal (hydrostatic) straight-line trend of decreasing slowness (increasing velocity) with depth (**Fig. 3C.29**).

The "normal," or hydrostatic, trend for the well, which may vary with different geologic provinces, is defined by plotting slowness values for shale beds (> 10-ft thickness) in the well. A constant overburden gradient of 1.0 psi/ft is generally assumed. The difference in acoustic slowness between the normal and abnormal trends can be converted to an equivalent fluid-pres-

TABLE 3C.14—TECHNIQUES USED IN ACOUSTIC SLOWNESS PROCESSING

Method	Output	Comments
Data stacking		Improves signal-to-noise ratio
Single-mode analysis	First break; wave slowness, filtered wave signatures	Slowness of individual energy mode (e.g., compressional or shear)
Multimode analysis	First break; wave slowness, filtered wave signatures	Two or more energy modes in an acoustic array; recommended for openhole shear and cased-hole compressional slowness
First-motion analysis	DDBHC acoustic slowness	Travel time of first coherent amplitude change on each array signature. Quick-look technique; affected by tool noise, signal amplitude, and cycle skipping; openhole only
Cross correlation	DDBHC acoustic slowness	Statistical correlation method based on first motion over a fixed range of slowness; rapid with 6-in. resolution; not affected by cycle skipping or signal amplitude; openhole only
Semblance	Slowness, quality curves	Statistical correlation; high confidence; openhole and cased hole; resolution limited to 3.5 ft
*N*th-root stacking	Slowness, quality curves	Statistical correlation; high confidence; openhole and cased hole; resolution limited to 3.5 ft
Waveform matching	Slowness, quality curves	Subarray processing, resolution < 3.5 ft (to 1 ft)

sure gradient and formation pressure for a given depth (**Fig. 3C.30**). This method may not be applicable or effective in all areas or where undercompaction is not the mechanism for overpressuring.[152,153] More-general approaches to determination of pore pressure and fracture gradient use an effective-stress, rock-mechanics approach.[152,154,155]

Recent investigations into the effects of pressure on shale porosity suggest that the relationship is more complex than previously thought. While additional study is necessary, the results to-date suggest that it may be necessary to reconsider or revise the well-log methods currently used in pore pressure and exhumation analysis (see section on Geological Applications below).[147,156–158]

Fracture Identification. Locating fractures, recognizing fracture morphology, and identifying fluid-flow properties in the fracture system are important criteria in characterizing reservoirs that produce predominantly from fracture systems. However, fracture identification and evaluation using conventional resistivity and compressional-wave acoustic logs is difficult, in part because fracture recognition is very dependent on the dip angle of fractures with respect to the borehole.

Fractures are physical discontinuities that generate acoustic reflection, refraction, and mode conversion—all of which contribute to a loss of transmitted acoustic energy. In particular, compressional- and shear-wave amplitude and attenuation and Stoneley-wave attenuation are significantly affected by the presence of fractures. Compressional waves are primarily affected by oblique fractures—those with dip angles between 15° and 85°—while shear waves are primarily affected by horizontal or near-horizontal fractures.[159] On conventional-velocity logs, fracture-induced attenuation may be evidenced as cycle skipping, variations in the V_p/V_s ratio and on VDL presentations, or as chevron (crisscross) patterns caused by mode-conversion interference.[160,161] Borehole-televiewer-type imaging devices provide a higher degree of success in identifying fractures and determining whether or not they are open (producible) or closed.

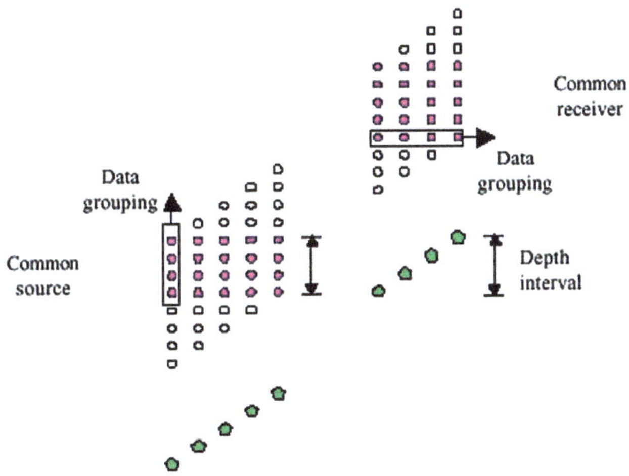

Fig. 3C.34—Diagram showing the gather of array acoustic data into either common-source subarrays or common-receiver subarrays. These subarrays cover the same depth interval. Data grouping is in the vertical direction for the source gather, and in the horizontal direction for the receiver gather[224] (courtesy of SPWLA).

The development of reliable full-waveform shear- and borehole-imaging devices enabled enhanced fracture identification and evaluation.[61,162] Aguilera[163] summarizes the use of conventional acoustic-log methods for fracture identification.

Recently developed anisotropy-analysis methods use crossed-dipole shear, Stoneley-wave, and acoustic-imaging data—individually or in combination—to provide reliable identification and evaluation of in-situ and induced fractures (see the Crossed-Dipole Anisotropy Analysis section of this chapter).

3C.5.4 Geological Applications. ***Estimates of Erosion and Uplift (Exhumation).*** The amount of erosion that has occurred in a region that has been uplifted can be estimated from the degree of shale compaction measured by acoustic travel time.[164–166] This technique assumes that shale compaction is irreversible and that the shale retains the degree of compaction it gained at its maximum burial depth. Uplift and erosion will result in lower porosities than expected for the current burial depth (i.e., a shale will appear to be overcompacted).[147,167–177]

Determination of Organic Richness and Source-Rock Potential. Acoustic slowness, used alone or in conjunction with formation resistivity, can provide qualitative indications and quantitative determination of source-rock potential (when calibrated to laboratory data). The identification of potential petroleum-source rocks and characterizing the thermal maturity of these rocks is important for assessing petroleum potential (risking) and for basin modeling.

Studies of coals and organic-rich shales have demonstrated that acoustic velocity is reduced by the presence of organic material, that changes in velocity are proportional to the volume of organic material present, and that increases in thermal maturity (largely a function of burial depth and temperature) are accompanied by increases in acoustic velocity (decreases in transit time, Δt). Total organic carbon and the level of organic thermal maturation, expressed in terms of vitrinite reflectance, are two key parameters used for determining the potential of a formation to source hydrocarbons, and each can be mathematically related to Δt.[178–180] Because acoustic velocity is influenced by a number of factors in addition to organic carbon content, a combination of log measurements can provide improved results when other factors do not mask responses. In particular, acoustic-resistivity crossplot techniques (**Fig. 3C.31**) and log overlays (**Fig. 3C.32**) have proved successful.[181–183]

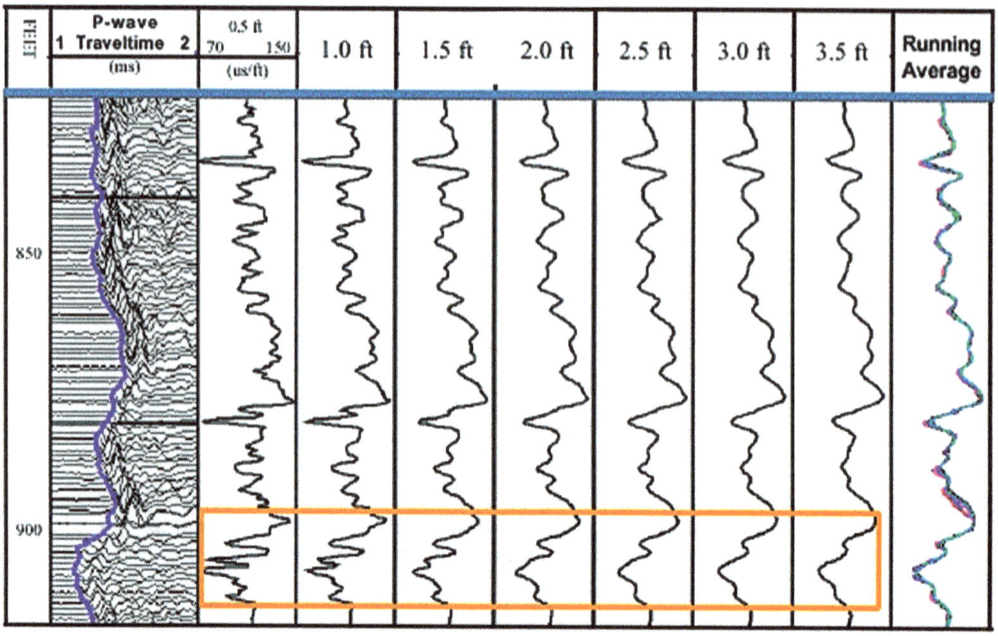

Fig. 3C.35—Compressional-wave-slowness curves obtained for various configurations (apertures) of possible overlapping subarrays based on an array tool having four transmitters and eight receivers. Note the increasing resolution of the curves with decreasing subarray aperture. Track 9 is a consistency check obtained by averaging the curves to 3.5-ft aperture and overlaying the results[224] (courtesy of SPWLA).

3C.5.5 Geomechanical Applications. ***Rock Mechanical Properties.*** The determination of a reservoir's mechanical properties is critical to reducing drilling risk and maximizing well and reservoir productivity. Estimates of rock mechanical properties are central to drilling programs, well placement, and well-completion design.[184] Mechanical properties include the elastic properties (Young's modulus, shear modulus, bulk modulus, and Poisson's ratio) and the inelastic properties (fracture gradient and formation strength).

Elasticity is the property of matter that causes it to resist deformation in volume or shape. Hooke's law describes the behavior of elastic materials and states that for small deformations, the resulting strain is proportional to the applied stress. Stress is the force applied per unit area, and strain is the fractional distortion that results because of the acting force. The modulus of elasticity is the ratio of stress to strain. Depending on the mode of the acting geological force and type of geological media the force is acting upon, three types of deformation can result as well as three elastic moduli that correspond to each type of deformation. Young's modulus, E, is the ratio of uniaxial compressive (tensile) stress to the resultant strain. Bulk modulus, K, is the change in volume under hydrostatic pressure (i.e., the ratio of stress to strain). (K is the reciprocal of compressibility.) Shear modulus, μ, is the ratio of shearing (torsional) stress to shearing strain. An additional parameter, Poisson's ratio, σ, is a measure of the geometric change of shape under uniaxial stress. These four elastic parameters are interrelated such that any one can be expressed in terms of two others and can also be expressed in terms of acoustic-wave velocity and density (**Table 3C.13**).

The data needed to compute mechanical rock properties are compressional and shear velocities (slowness) and density. Shear and compressional velocities are a function of the bulk modulus, shear modulus, and density of the formation being measured. The V_p/V_s ratio, combined with formation density, ρ, is used to calculate Poisson's ratio, Young's modulus, the bulk modulus, and the shear modulus. Whenever possible, log-derived, dynamic rock properties

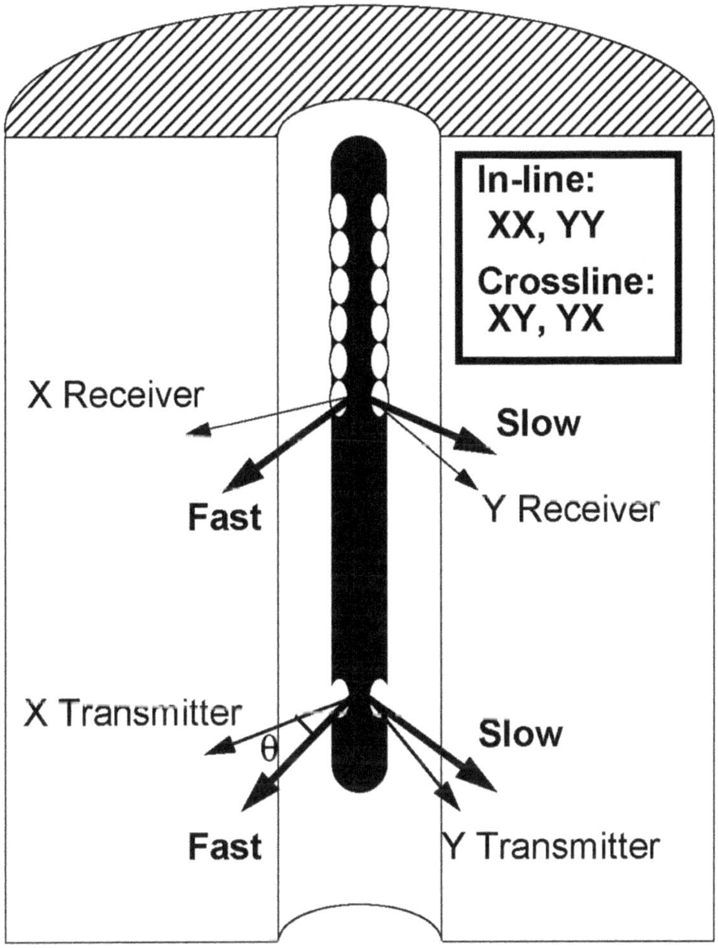

Fig. 3C.36—Diagram illustrating the principle and configuration of crossed-dipole logging: two orthogonal dipole transmitter and receiver array systems. The tool acquires four array data sets, two in-line, XX and YY; two crossed-line, XY and YX, and uses the physics of shear-wave splitting in azimuthally anisotropic rocks to determine the fast-shear polarization azimuth (courtesy of Baker Atlas).

should be calibrated to core-derived static (laboratory) properties, because the static measurements more accurately represent the in-situ reservoir mechanical properties.[185–189] Rock mechanical properties can be determined using conventional empirical charts[190] or computer programs. The elastic moduli and Poisson's ratio are used in a variety of applications.[191] These applications include predictions of formation strength,[192–196] well stimulation (fracture pressure and fracture height),[197–200] borehole and perforation stability,[201] sand production and drawdown limits in unconsolidated formations,[202–204] coal evaluation,[205] and determining the roof-rock-strength index for underground mining operations.[206,207] Rock-mechanics applications of modern multipole tools are discussed in the Anisotropy Analysis section following.

3C.5.6 Near-Well Imaging. Acoustic data acquired using modern array tools can provide high-resolution (0.5 m), microscale "seismic" 2D and 3D images of structural features in the near-borehole region (10 to 15 m). Conventional seismic-processing techniques, including filtering and migration, are used to extract compressional and shear reflections from the acoustic data. The reflections are then used to image geological features near the borehole. This technique

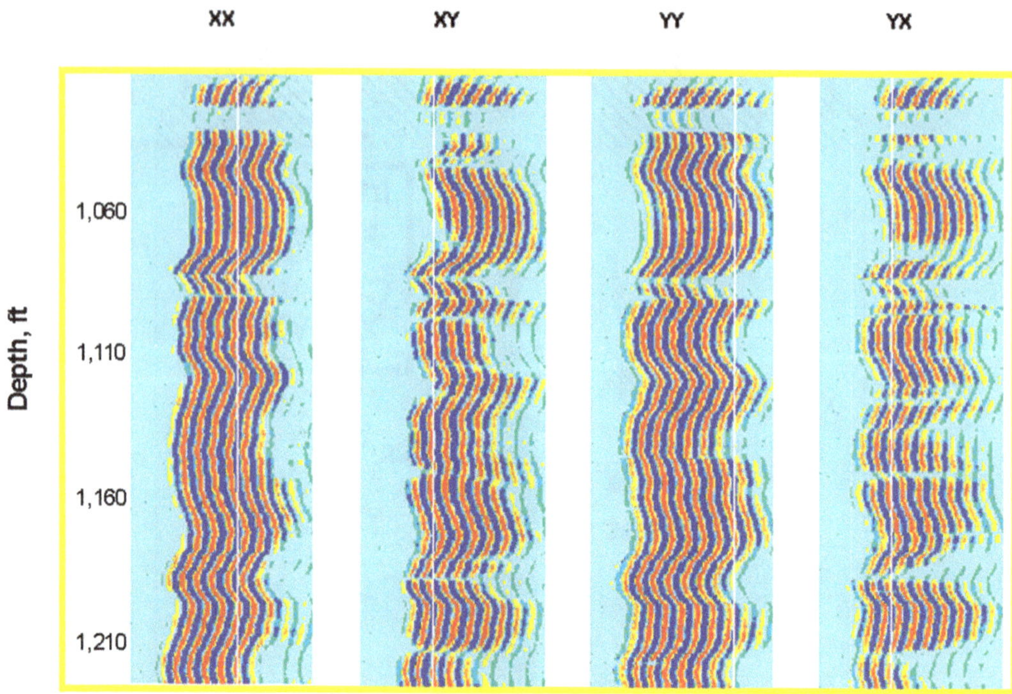

Fig. 3C.37—Example of four-component, crossed-dipole waveform log data showing shear-wave splitting caused by formation anisotropy (courtesy of Baker Atlas).

allows the imaging of bed boundaries, thin beds (stringers), fractures, and faults in openhole and cased wells (**Fig. 3C.33**).[32,208–215]

3C.6 Advanced Data Analysis and Applications

Processing acoustic data downhole as well as at the surface is necessary to transform the raw acoustic signals recorded by modern logging instruments into data suitable for interpretation and analysis. Data processing takes place during acquisition, in the logging tool itself and in the surface acquisition unit, and also in post-acquisition processing at computing centers. There are a variety of sources of noise in the downhole environment that contaminate the recorded acoustic signal: tool ("road") noise, measurement error, reflection and scattering from rough borehole or bed boundaries, mode conversion, and interferences that occur in the downhole environment. The goal of acoustic-data processing is to minimize the data noise while maximizing the petrophysical information.[216] Data preprocessing reduces the influences of these sources, thus allowing extraction of the true formation signal.

Following the rapid theoretical advances in acoustic-wave propagation made during the 1980s and 1990s, significant advances in data processing provided improved quality in slowness measurements and enabled a number of new applications using Stoneley and dipole-shear wave in open and cased holes. The combined interpretation of Stoneley and dipole-shear acoustic measurements with NMR and borehole imaging enhances formation evaluation.

3C.6.1 Slowness Analysis. One of the primary goals in borehole acoustic logging is to obtain formation slowness from array measurements. Accurate slowness analysis is vital to most petrophysical, geophysical, and seismic applications. A variety of techniques are used for computing slowness from array tools (**Table 3C.14**).

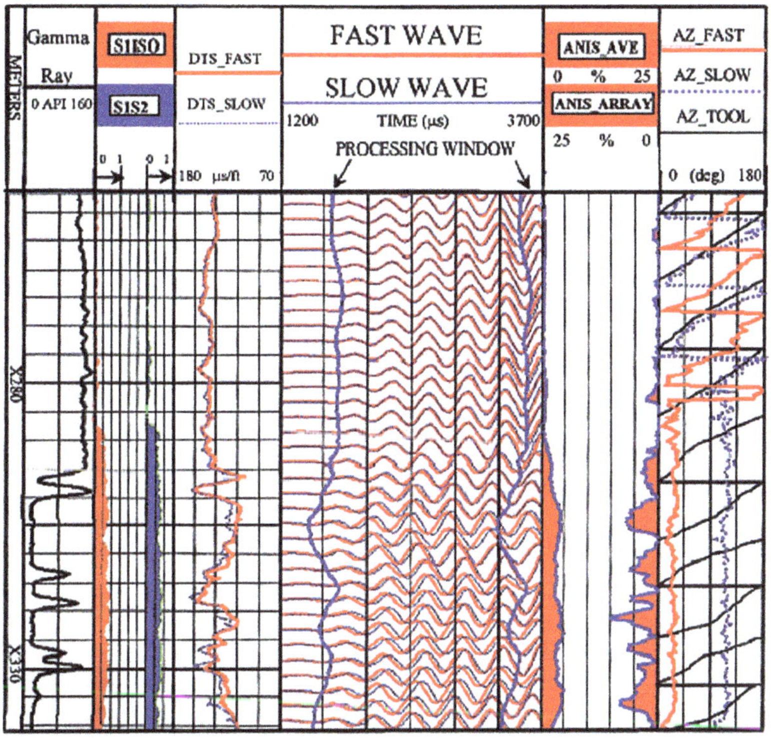

Fig. 3C.38—Post-processing results from crossed-dipole analysis indicating the magnitude and azimuth of anisotropy (courtesy of Baker Atlas).

Semblance and Nth-Root Stacking. The two most commonly used techniques to determine slowness from borehole acoustic array are semblance[217–219] and *N*th-root stacking[220,221]; both are cross-correlation, coherency techniques that compare signatures in an acoustic array and find the similarities that correspond to coherent wave types. Semblance has a direct physical interpretation, whereas *N*th-root stacking is a purely mathematical solution. Although the semblance method is faster, the *N*th-root stacking method is more tolerable to noise and ignores amplitude variations across the array and, in general, provides better results. Collecting the array data, either as receiver or transmitter (source) arrays (**Fig. 3C.34**), enhances the slowness output from both of these techniques.

The primary application of the receiver- and transmitter-array-derived slowness curves is to provide compensation (DDBHC) to the monopole compressional and shear slowness for borehole irregularities such as washouts or cave-ins. Borehole compensation is achieved by averaging the slowness estimates from the receiver and transmitter arrays. The dipole-shear wave does not require borehole compensation because the flexural-shear wave is not as sensitive to the borehole geometry although it is often used to improve results in rough boreholes.

The objective of the semblance technique is to find the slowness that maximizes the coherence (time domain) among the wave power spectra over the receiver array. The *N*th-root stacking technique is very similar to the semblance technique, except that the waveform amplitudes are modified in a different way to produce a so-called "pseudo" power spectrum instead of the true power spectrum.

Although the algorithms used in semblance and *N*th-root stacking differ, the processing procedure is similar. A time window of fixed length is set up on each signature in the array. The windows are offset by a specified time interval on the successive signatures and a correlation

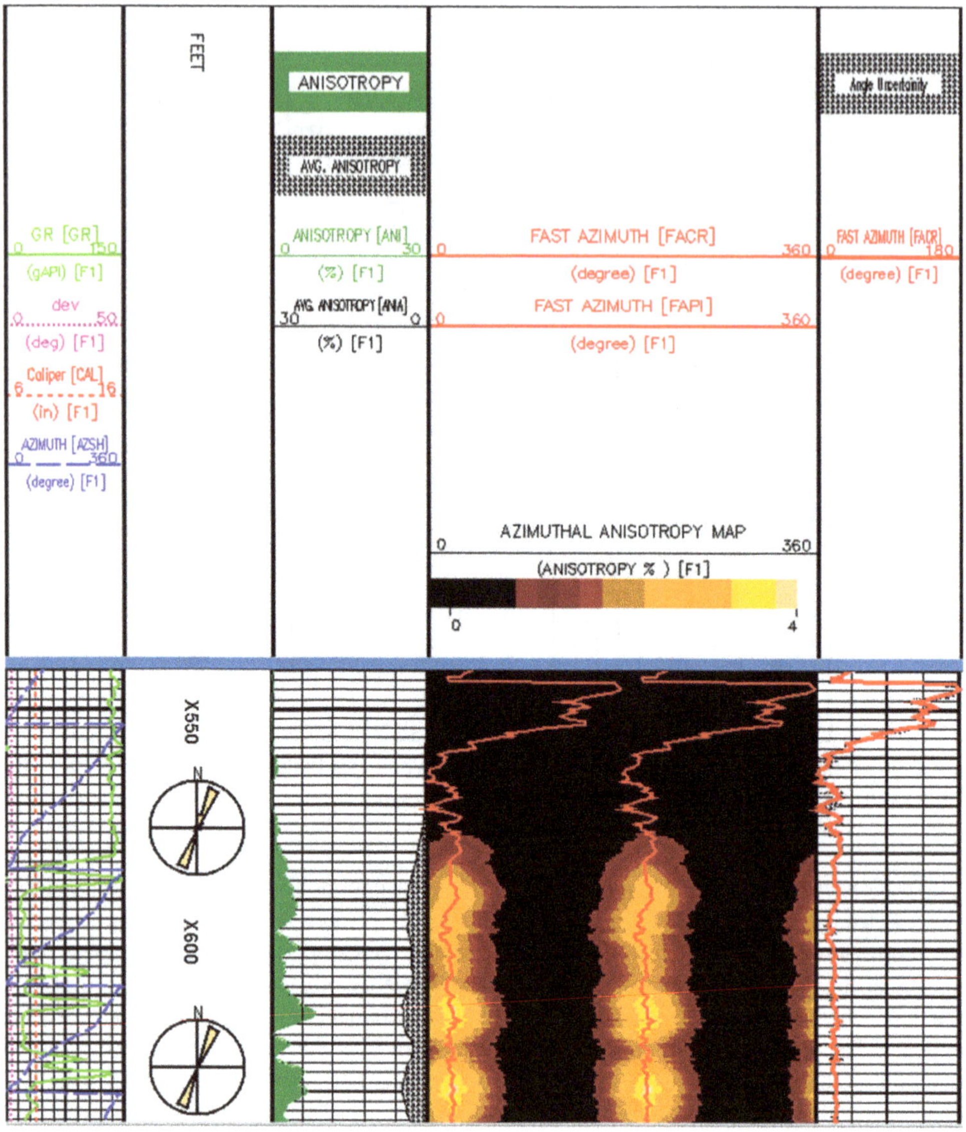

Fig. 3C.39—The results of anisotropy analysis displayed as an anisotropy map (courtesy of Baker Atlas).

is computed over these windows. The windows on all but the first signature are then stepped a certain distance out in time and another correlation value is computed. The process continues for the range of anticipated slowness in the well. This process is repeated as the window is progressively stepped on the first signature.

The group of correlation values obtained is known as a correlogram, in which the waveform coherence or correlation values either from semblance or *N*th root method are displayed as a function of slowness and arrival time creating a 3-D surface. The correlation values in a range of arrival times are further combined to form a "combined correlogram," which is a projection of the 3-D surface on the slowness axis. The correlogram peak at each measured-depth level is used to obtain the Δt slowness for the wave type. These individual slowness values are then displayed as a continuous log curve. Filtering for the compressional wave or for the dispersive shear wave is normally required before using either method. As a quality check, the

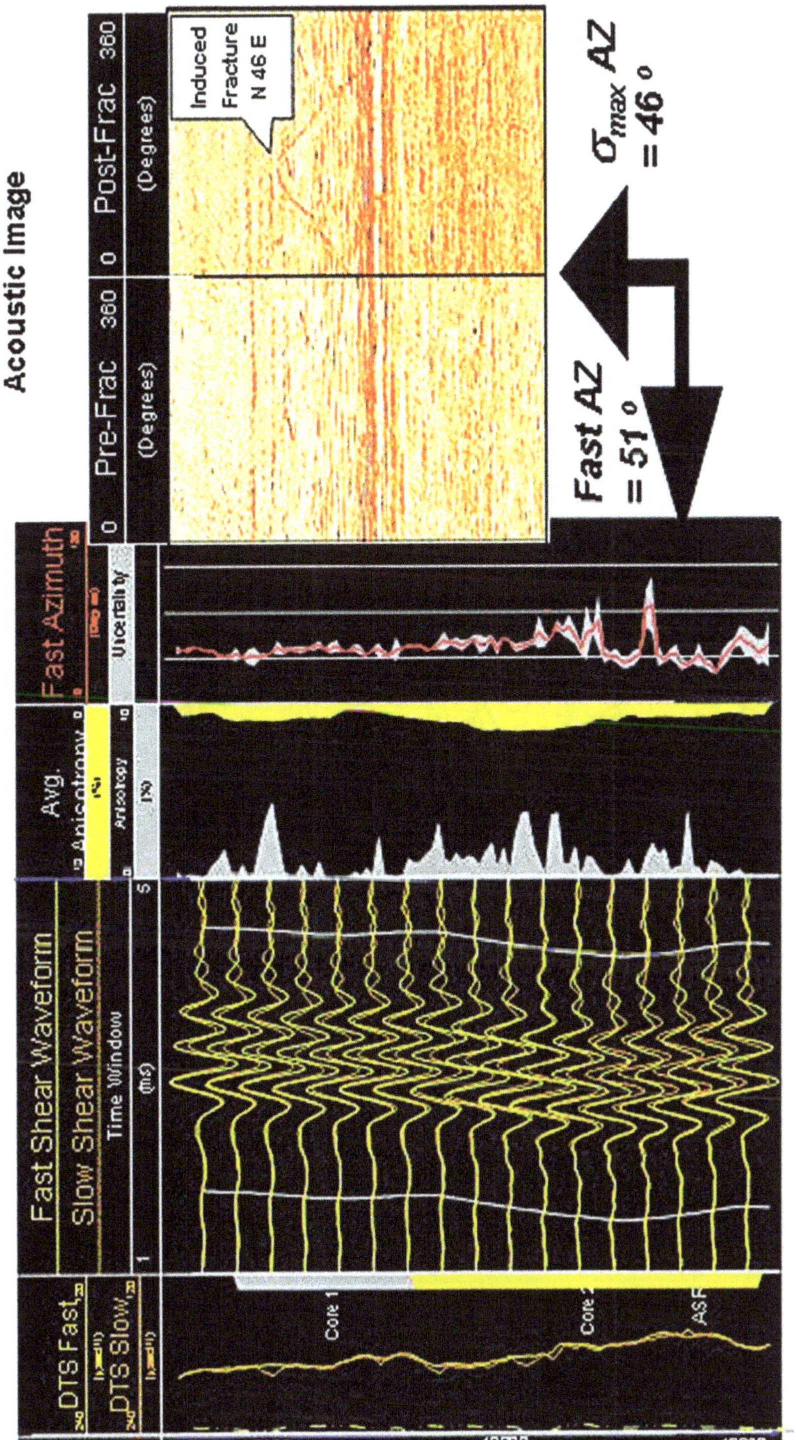

Fig. 3C.40—Crossed-dipole logging determines a fast shear azimuth of 51° before fracture treatment. The maximum stress direction of 46°, determined after the fracture job, is in good agreement with the stress direction from the crossed-dipole log results (courtesy of Baker Atlas).

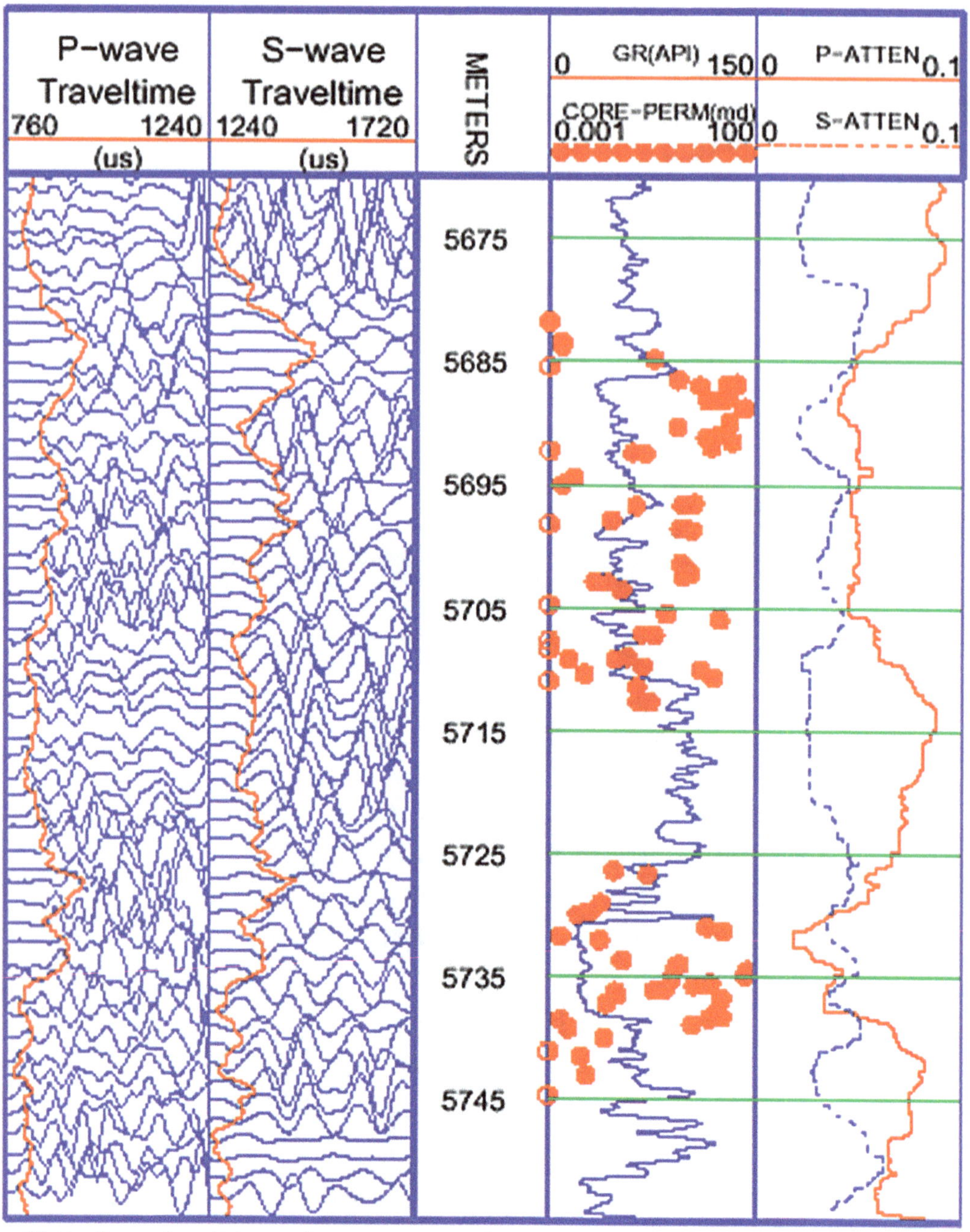

Fig. 3C.41—Comparison between shear-wave attenuation logs and core permeability in a permeable oil zone. The two attenuation curves agree in high-permeability zones[278] (courtesy of SEG).

slowness results of semblance or *N*th-root stack processing can also be plotted in the combined correlogram together with a "computed travel time" which is plotted against the waveform. The computed travel time represents the sum of the transit-time measurements from the transmitter to the receiver along with the mud travel time. Dipole shear-slowness processing must

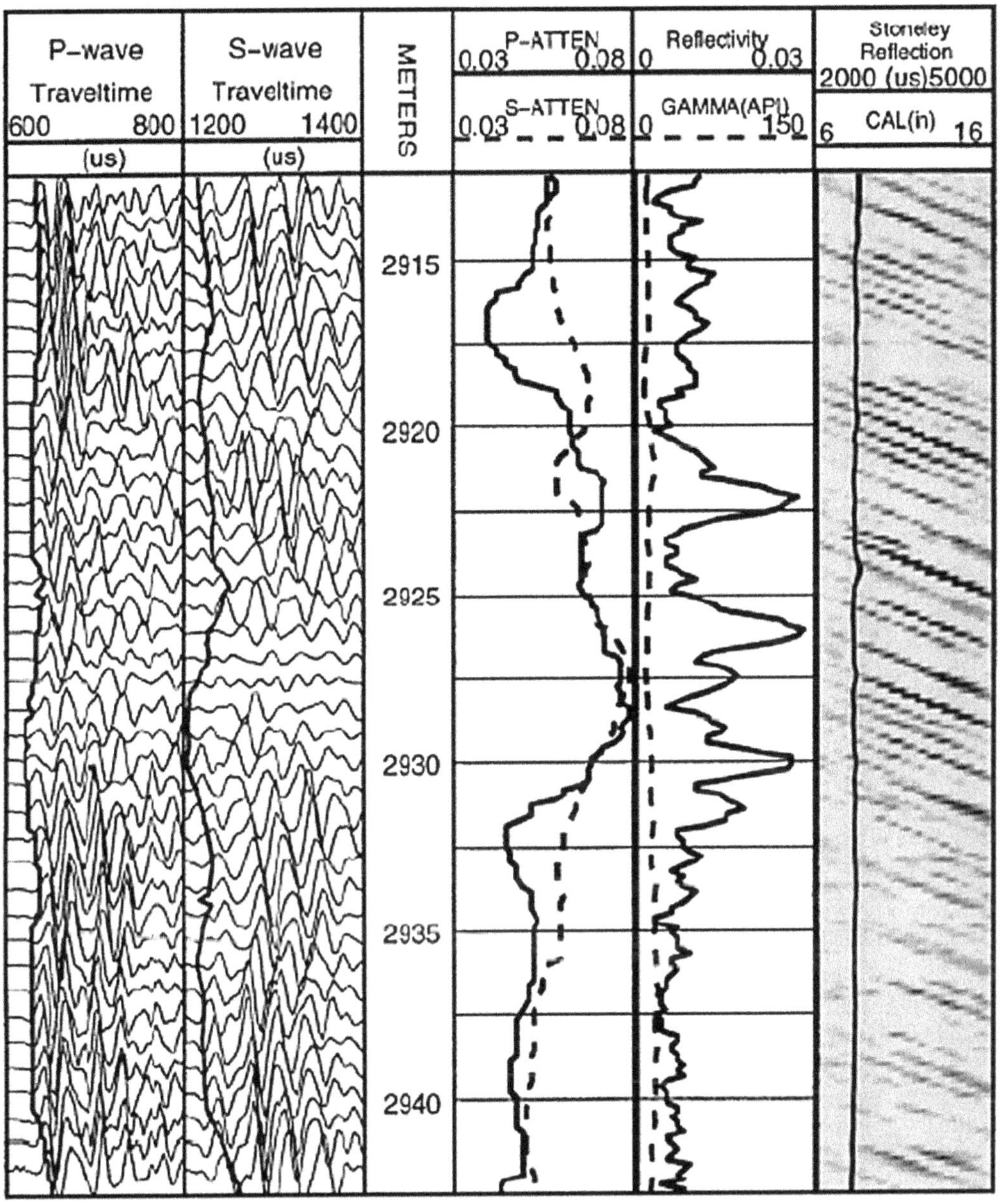

Fig. 3C.42—Correlation between compressional- and shear-wave attenuation and fracture location. The interval of 2920 to 2930 m in which P- and S-wave attenuation curves overlay corresponds to the interval of high acoustic reflectivity (Track 4) and fractures identified by the down-going Stoneley-wave reflectivity (Track 5)[278] (courtesy of SEG).

also take into account dispersion of the flexural wave.[222] Depending on the frequency of the flexural mode, a correction may be required to obtain shear-wave slowness.

Resolution Enhancement. The aperture (vertical resolution) of array tools is typically 3.5 ft. This means that a bed must be at least that thickness to measure true acoustic slowness, although such tools can detect (resolve) beds down to 2-ft thickness. To meet the need for the higher resolution necessary for thin-bed evaluation, waveform-matching (multishot) processing techniques use the redundant information contained in overlapping receiver subarrays to improve vertical resolution (Fig. 3C.34).[223] Recently introduced processing techniques reduce the aperture to 0.5 ft and thus achieve a true vertical resolution of 1.0 ft (**Fig. 3C.35**).[224] However, reducing the aperture makes the measurements more susceptible to noise.

3C.6.2 Anisotropy Analysis. Formation anisotropy—the directional variation of physical properties—can be the result of depositional processes (intrinsic) or tectonic processes (stress-induced). Formation anisotropy is evidenced through variations in permeability, rock strength, fractures, and borehole failure. In acoustic/seismic terms, intrinsic anisotropy is structural in nature and is commonly seen as transverse isotropy (TI or vertical transverse isotropy, VTI), in which properties differ in the vertical or horizontal planes, such as in shales or thinly bedded intervals.[225] Stress-induced anisotropy is known as azimuthal anisotropy (or horizontal transverse isotropy, HTI), in which acoustic parameters in a vertical borehole vary with azimuthal orientation, such as the case of fractures parallel to the borehole.

Analyses of in-situ anisotropy (primarily stress induced) are made using direct or derived shear-wave velocity and provide the magnitude and azimuth of anisotropy (i.e., direction of the maximum and minimum horizontal stresses as well as an indication of their difference). Anisotropy analysis has been widely used in solid-earth seismology, geothermal studies, and more recently, in exploration geophysics (see summaries in Crampin and Chastin[226] and Helbig and Thomsen[227]). In the petroleum industry, these results are used in well design and well placement for optimum reservoir drainage,[228] to detect and characterize faults and fractures in openhole and cased hole,[229] to predict borehole instability and sand production,[230] and for optimizing the design and evaluation of well completions (perforations and hydraulic fracturing) (see following section on Crossed-Dipole Anisotropy Analysis).

Compressional-, shear- and Stoneley-wave properties are each affected, to some degree, by the presence and type of formation anisotropy. While the shear-wave response to azimuthal anisotropy (see the following section on Crossed-Dipole Anisotropy Analysis) and Stoneley waves to VTI anisotropy is well known, the effect on compressional-wave energy is less well characterized.[231–234] In some situations, information from additional measurements, such as borehole images, dip logs, or both, may be necessary for relating the measured anisotropy to geological features.

Borehole acoustic azimuthal and VTI anisotropy analysis is an advance made possible by the recent introduction of new inversion methods.[235–242] These methods use the crossed-dipole shear to derive azimuthal anisotropy and the Stoneley wave to derive TI anisotropy in slow formations, or a combination of these modes in deviated wells. A reasonable shear velocity can be derived using inversion techniques with low-frequency Stoneley-wave dispersion which is sensitive to the horizontal shear (in contrast to the dipole's sensitivity to the vertical shear).[6,23,243]

Several new applications have been made possible by anisotropy analysis: identification of formation alteration using dipole-shear dispersion,[33,237,244,245] stress estimation,[239] and distinguishing between intrinsic and stress-induced anisotropy using dispersion crossover.[246–249]Anisotropy analysis can also be conducted using shear-wave parameters derived from Stoneley-wave dispersion.[242,250]

Crossed-Dipole Anisotropy Analysis. In anisotropic media, shear waves (both monopole refracted and dipole flexural) split into orthogonally polarized components having different velocities. This is known as shear-wave splitting, shear-wave birefringence, or shear-wave velocity anisotropy.[251–254] The difference in fast and slow shear-wave slowness provides a measure of the magnitude of anisotropy. Shear-wave splitting is useful for evaluating fractures, faults, bedding planes that intersect the borehole at an angle, and unbalanced tectonic stresses perpendicular to the borehole.

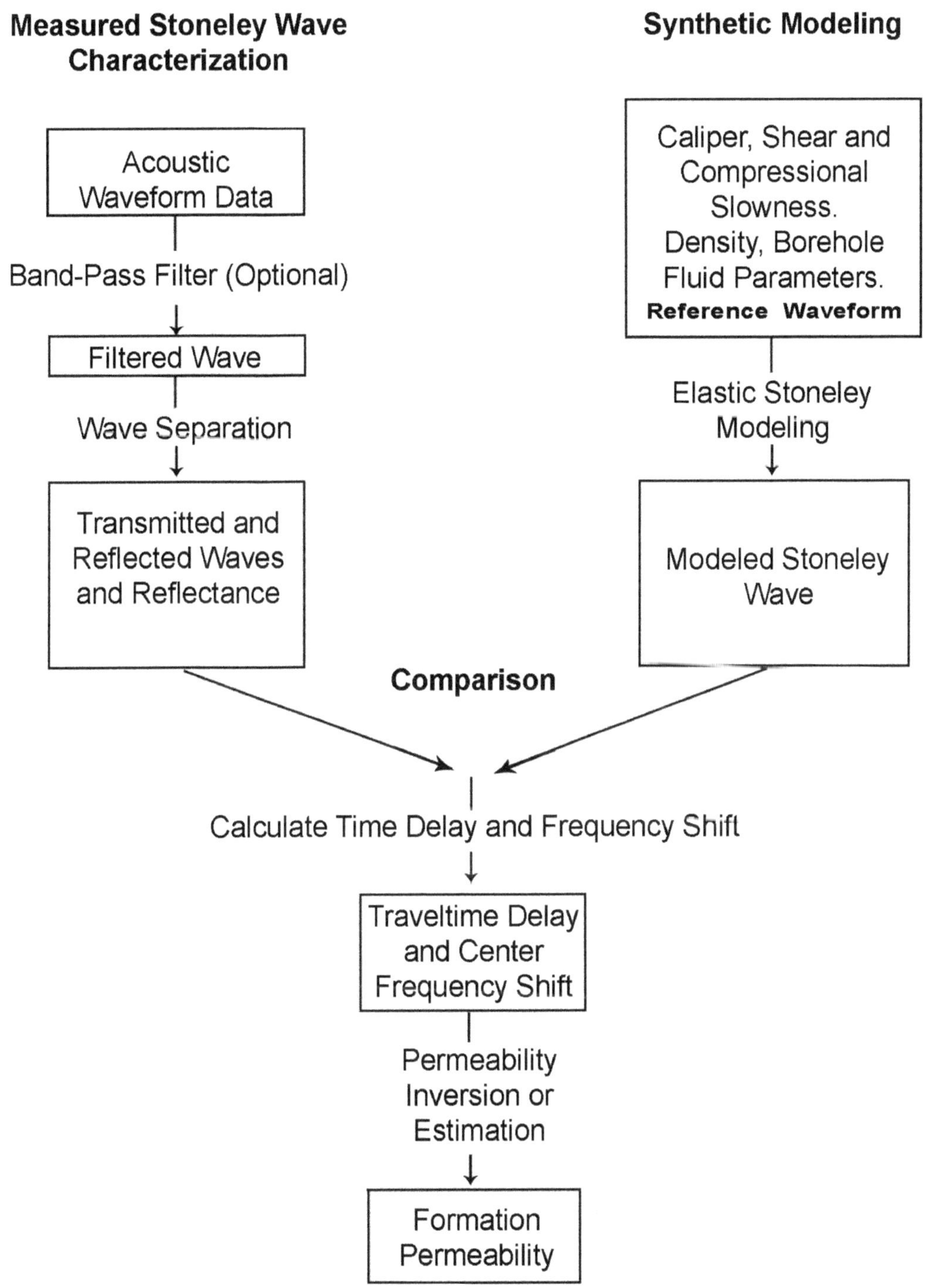

Fig. 3C.43—Stoneley-derived-permeability processing flow (courtesy of Baker Atlas).

Because the flexural waves induced by dipole-acoustic sources have a directional component, the use of mutually perpendicular (crossed) pairs of dipole transmitters and receivers can detect and measure dipole shear-wave splitting.[255–258] In isotropic formations, both receiver com-

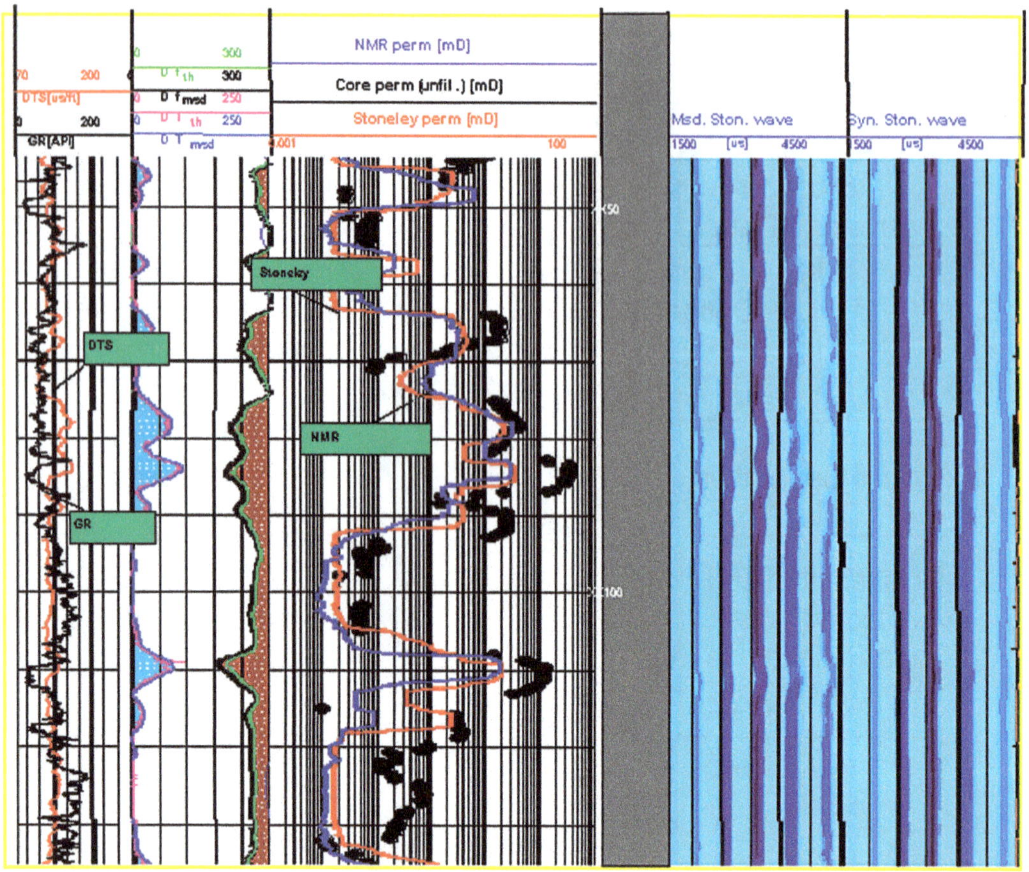

Fig. 3C.44—Comparison of Stoneley-, NMR- and core-permeability profiles. The agreement of these fundamentally different measurements confirms the validity of each (courtesy of Baker Atlas).

ponents (X and Y) will measure the same shear-wave arrival time. However, in an azimuthally anisotropic formation, the shear wave measured at the X-component of the receiver pair will be different from the one measured at the Y-component; one is called the fast shear component, the other, the slow shear component (**Fig. 3C.36**). To measure the fast and slow shear-wave slowness, four components are measured: two inline components, X-X and Y-Y, and two crossline components, XY and YX (**Fig. 3C.37**).

The dipole flexural shear mode is affected by a variety of factors including formation anisotropy. Significant borehole ellipticity (the result of borehole failure or breakouts)[259] and high relative inclination between the borehole and formation[260,261] may result in erroneous interpretation of dipole-derived anisotropy and must be accounted for during data processing.[262] Additionally, the presence of shale anisotropy in high-angle and horizontal wells, can significantly influence compressional velocity, which must be corrected for this effect.[263]

Initially, mathematical rotation methods originally developed for use in processing surface seismic data[264] were used to determine the fast-wave direction together with the slowness values for the fast and slow shear wave. More recently, inversion methods[265] are being applied to the four data sets to simultaneously determine the azimuth and magnitude of the anisotropy (**Fig. 3C.38**).

An anisotropy map (**Fig. 3C.39**) combines the derived average anisotropy and its azimuth to generate an azimuthal image. This display facilitates interpretation by allowing the analyst to

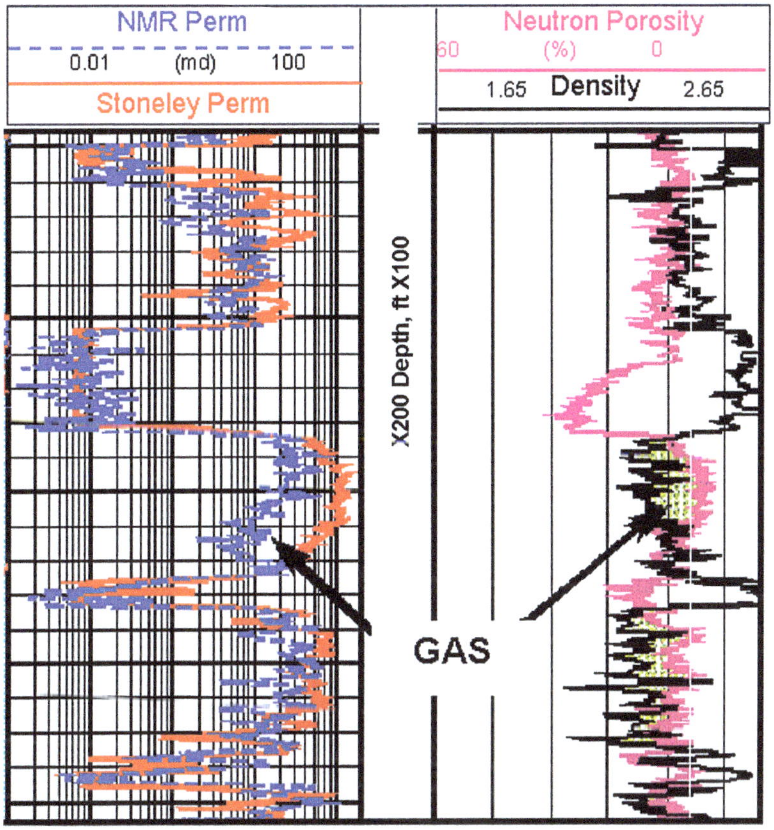

Fig. 3C.45—Stoneley-NMR-permeability gas indicator (courtesy of Baker Atlas).

quickly assess depth intervals of interest by looking at the brightness, direction, and continuation of the features on the map. The map also facilitates comparison with borehole-image logs. Rose diagrams (Track 3) provide an accurate indication of the fast shear azimuth over each labeled depth interval. The integration of monopole and dipole measurements yields improved estimates of anisotropy, magnitude of anisotropy, and permeability.[243,266]

Horizontal Stress and Hydraulic Fracturing. Fractures, both natural and hydraulically induced, develop in relation to regional or localized stress patterns and play a major role in optimizing production and reservoir drainage. An estimate of the magnitude and azimuth of the horizontal stresses surrounding a borehole is needed for accurately placing wells to take advantage of existing fracture patterns and for artificially inducing fracture patterns during well completion through hydraulic stimulation.[228,267]

The stresses operating on a rock formation are described by a triaxial coordinate system that consists of two principal horizontal stresses, σ_x and σ_y, and a vertical stress component, σ_z, which is the overburden. In a borehole, these downhole stresses are expressed as radial components at the borehole wall: the vertical component, σ_z, the radial component, σ_r, and the tangential component, σ_θ , and the (azimuthal) shear component, $\sigma_{r\theta}$.

Unbalanced-formation-stress components produce distortion around the borehole (stress-induced anisotropy) that results in shear-wave splitting. The azimuth of the fast-shear wave parallels the direction of maximum horizontal stress and the azimuth of the slow-shear wave parallels the direction of the minimum horizontal stress. Hydraulic-fracture azimuth is parallel to the direction of maximum horizontal stress.[268]

Hydraulic well stimulation consists of perforating an interval, then packing and pressuring these perforations to create fractures behind casing to allow increased production. Crossed-dipole anisotropy logging can estimate the vertical extent of the formation-stress fracture along the borehole and its azimuth in the formation (**Fig. 3C.40**).[34,239–241,258,269–274]

3C.6.3 Attenuation Analysis. Acoustic-wave attenuation correlates with a variety of petrophysical parameters, including formation lithology and pore-fluid permeability, the degree of fluid saturation, and fractures type.[275–278] A recent study suggests that the combination of compressional- and shear-wave attenuation logs may provide a potential formation-evaluation tool (**Figs. 3C.41 and 3C.42**).[278]

3C.6.4 Stoneley-Wave Analysis. ***Permeability.*** Stoneley-wave velocity and attenuation are sensitive to formation and fracture permeability, particularly at low frequencies.[2,279–281] Stoneley-wave velocity decreases, and its attenuation increases, as permeability increases. Initial efforts (begun in the 1970s) to derive permeability information from Stoneley data were unsuccessful because neither the necessary low-frequency tools nor the appropriate processing methods had been developed. The parallel development of modern multipole array tools and sophisticated semblance- and inversion-processing methods enable computation of continuous profiles of formation permeability from monopole Stoneley-wave data.[282–284] Typically, these methods first model the nonpermeability effects using the elastic-wave theory and then relate differences between the modeled and the measured data to formation permeability. One approach to Stoneley-wave processing is comprised of three parts[284,285]: slowness analysis, reflectance mapping, and permeability estimation (**Fig. 3C.43**).

Stoneley-wave data are typically presented as an interval transit time, Δt_t and as a ratio of the amplitudes for the two receivers used. Both traces have been shown to correlate well with permeability changes and compare well with core data, when it is available. Stoneley-wave amplitude can be computed and used in conjunction with the slowness and the signatures to analyze dispersion and attenuation characteristics. Use of the attenuation (center-frequency shift) and dispersion (travel-time delay) provide good permeability indication and better quality control for permeability estimation.

Wave-separation processing minimizes the effects of nonpermeability-related influences (e.g., road noise and borehole scattering) and yields reflectance logs for the direct and reflected Stoneley-wave data. The center-frequency log for the reflected wave data characterizes the Stoneley-wave attenuation and can be used to indicate fractures, vugs, and bed boundaries. The center-frequency log for the direct (transmitted) data is used to estimate formation permeability. Knowledge of the formation-fluid properties (viscosity and compressibility from core or NMR) enables quantitative estimates. Without this information log-derived permeability estimates are only qualitative.

These models require sophisticated computer processing. A simplified, field-oriented technique based on Stoneley amplitude[286] has so far provided good results in ideal conditions and when calibrated to core or NMR data.

Stoneley-derived formation permeability compares well with estimates obtained by nonacoustic methods, including core analysis and NMR logs (**Fig. 3C.44**).[287,288] Additional improvement in permeability estimation is possible when anisotropy information is incorporated in the process.[289] Formation evaluation is further enhanced when these diverse measurements are integrated through joint interpretation.[290] As indicated by the plots in Fig. 3C.44, calibration of log responses to core data would improve log-predicted values in noncored intervals.

The Stoneley wave measures total permeability and NMR measures vuggy permeability. A comparison of these two measurements in carbonate formations makes it possible to evaluate the permeability contributions arising from fractures and vugs.[290] A combination of these two

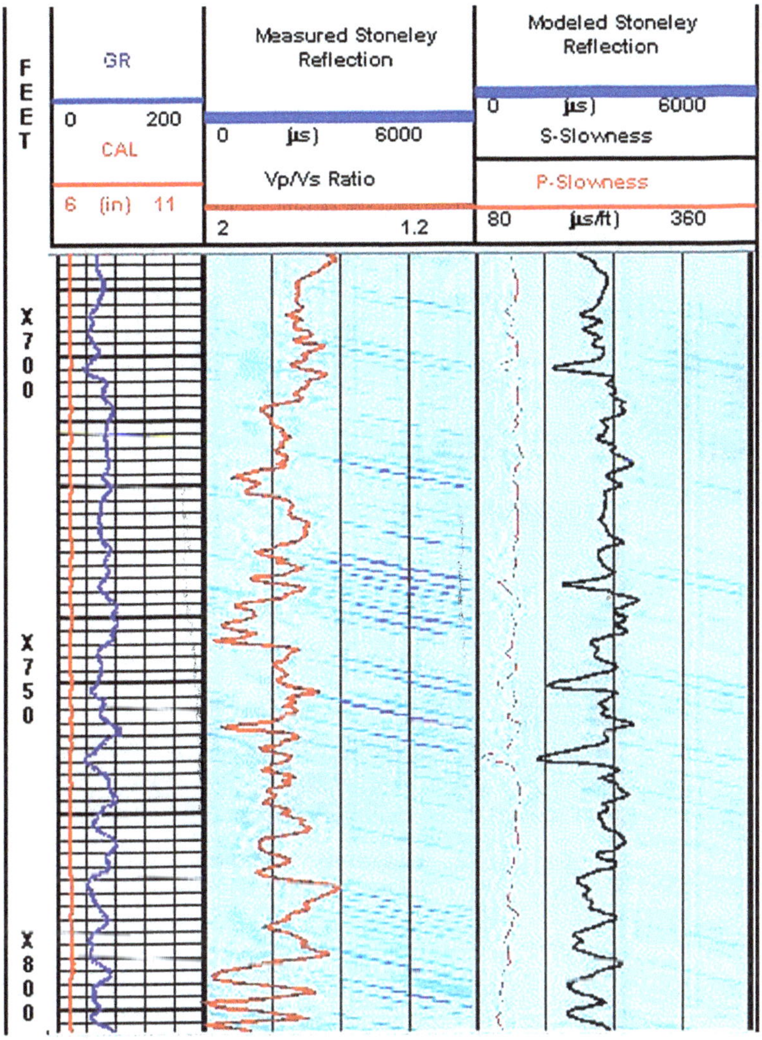

Fig. 3C.46—Gas detection in a high-porosity, laminated formation using thin-bed slowness and Stoneley reflection. The increase in P-wave slowness (Track 2) across the interval of thin gas sands, X740 to X800 m, is accompanied by low density values (Track 2) and numerous Stoneley-wave reflections (Track 3) (courtesy of Baker Atlas).

measurements calibrated using wireline formation-tester data enables improved permeability estimation in these reservoirs.[291]

Gas Detection. Stoneley-wave properties, used in combination with other data, facilitate identification of gas-bearing intervals. A gas-saturated formation interval has drastically different fluid mobility and compressibility compared to those of the surrounding formations. In the presence of gas, Stoneley-derived permeability is overestimated because of increased pore-fluid mobility (decreased viscosity) and compressibility, and NMR permeability may be underestimated because of a decreased hydrogen index. If first calibrated in a nongas-bearing interval, separation of the two permeability curves indicates the presence of gas (**Fig. 3C.45**).[290] The plot also shows the neutron-density crossplot is effective in detecting the gas zone and may be more sensitive in some intervals.

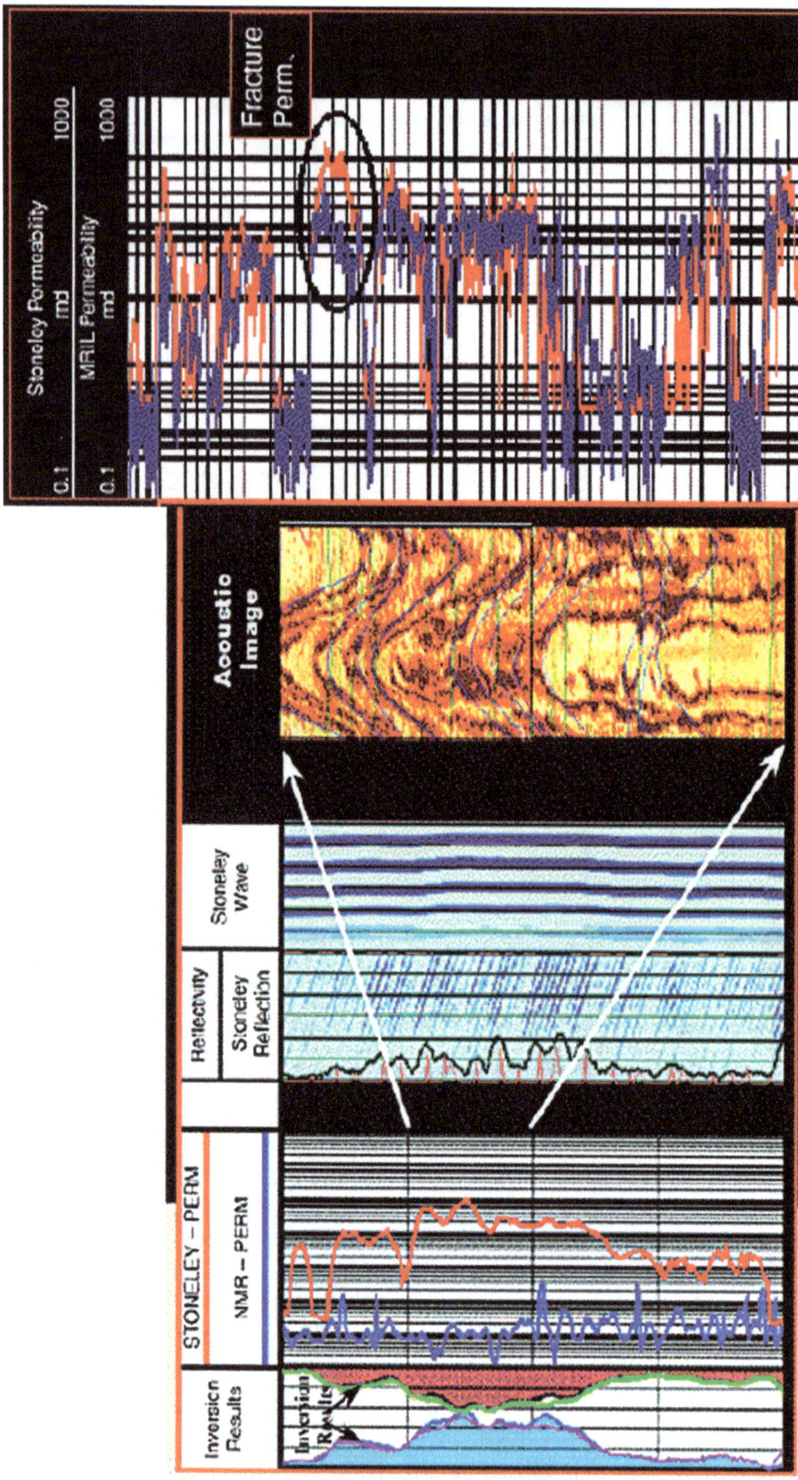

Fig. 3C.47—Stoneley-wave and NMR permeability profiles in the presence of borehole fractures. Stoneley-wave results are dominated by the fracture-system contribution, while the NMR results are dominated by the matrix-based permeability. The fractured sandstone interval, indicated by separation of the two permeability curves (Tracks 2 and 6), is confirmed by Stoneley-wave reflection data (Track 3) and the acoustic image log (Track 5) (courtesy of Baker Atlas).

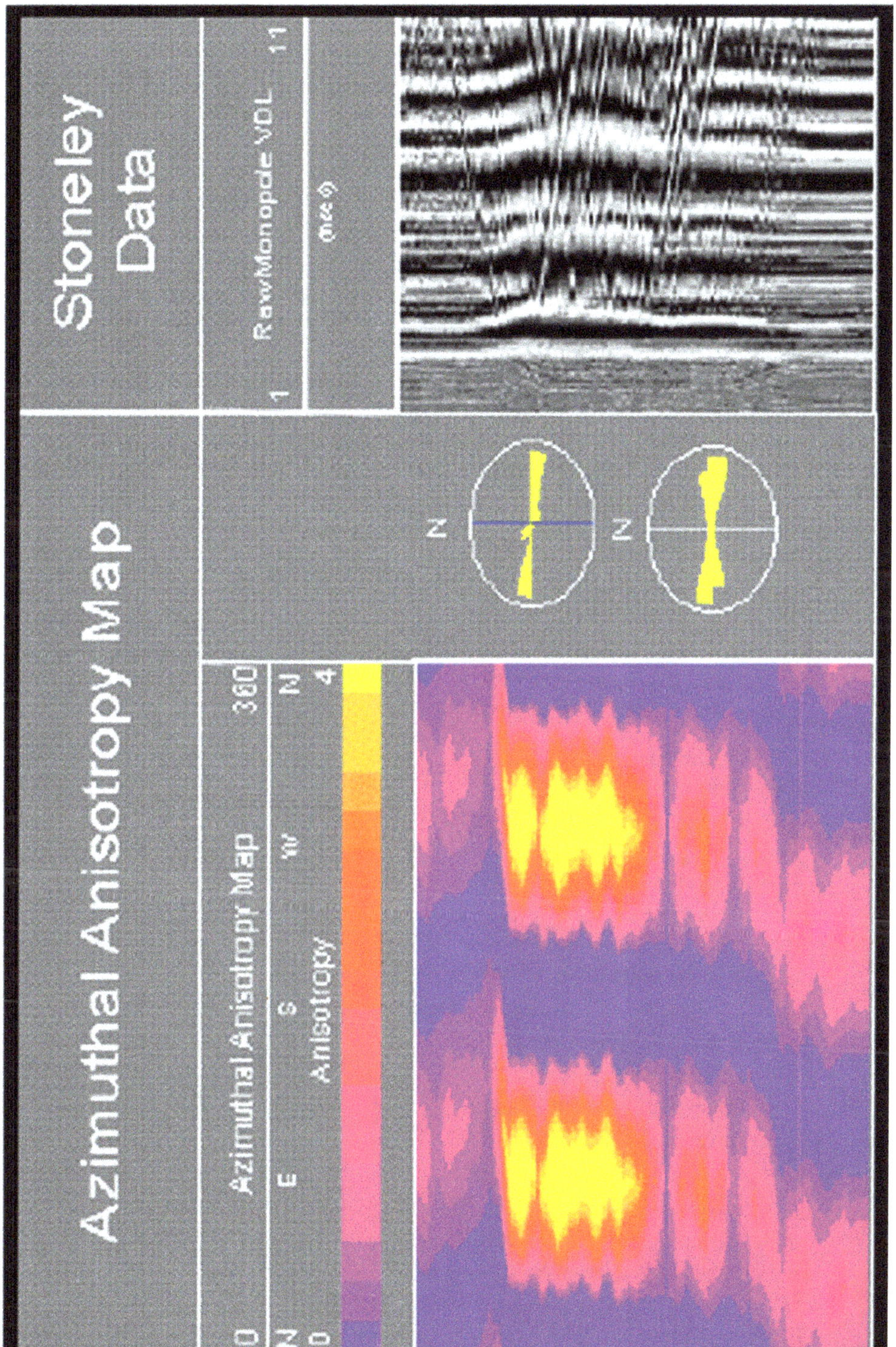

Fig. 3C.48—Correlation between azimuthal anisotropy and Stoneley reflections. A fracture zone in a U.S. mid-continent well causes both shear-wave splitting and Stoneley-wave reflections—in the anisotropy map (left), brighter colors represent higher anisotropy. The azimuth of the fast shear arrival is east/west (center). The fractures intersecting the borehole cause significant up- and down-going Stoneley-wave reflections seen in Stoneley-waveform amplitudes (right) (courtesy of Baker Atlas).

The significant contrast in fluid mobility and compressibility in a gas zone also generates a measurable Stoneley-wave reflection. The presence of these reflections, together with an increase in compressional slowness or a decrease in the V_p/V_s ratio, identifies a gas-bearing zone (**Fig. 3C.46**).[292] This type of joint interpretation is particularly effective in thin-bed evaluation.

Fracture Evaluation. The pressure-driven, fluid-borne nature of Stoneley waves makes them sensitive to fluid movement, and thus, to open fractures. The effects of open fractures on Stoneley waves are amplitude reduction, an increase in Stoneley slowness, the occurrence of mode conversion, and the occurrence of Stoneley reflections.[162,293,294]

The Stoneley wave responds to fracture permeability while NMR does not, thus, the Stoneley-wave permeability is greater in the presence of fractures than NMR permeability. The combined use of Stoneley- and NMR-derived permeability provides a fracture indicator in both sandstones (**Fig. 3C.47**) and carbonates (**Fig. 3C.48**).[290]

3C.7 Summary

This chapter has summarized the characteristics and capabilities of acoustic-logging tools. Although determination of geophysical (seismic) and petrophysical (porosity) properties have traditionally been the most widely used applications for acoustic-log data, new and advanced applications in geomechanics (anisotropy, fracture, and stress evaluation) and petrophysics (permeability) have been made possible by the multiarray tools and inversion processing methods now available. Interpretation methods that combine acoustic data with core and other log data, such as resistivity, nuclear, NMR, and borehole images, facilitate determination of fluid saturation and typing when logging conditions provide reliable data. For all applications of acoustic logging careful planning is critical to ensure successful acquisition of the desired data.

Nomenclature

a = Archie parameter
C = shale compaction coefficient
C_p = compaction correction factor
E = Young's modulus
K = bulk modulus
m = Archie parameter
R = resistivity, ohm-m
R_1 = receiver one in a two-receiver tool configuration
R_2 = receiver two in a two-receiver tool configuration
S_w = water saturation, fraction
t = travel time
t_1 = travel time between T_2 and R_1
t_2 = travel time between T_2 and R_2
t_3 = travel time between T_1 and R_1
t_4 = travel time between T_1 and R_2
T = transmitter
TR = transmitter-to-receiver spacing (ft)
T_1 = transmitter one in a two-transmitter tool (BHC) configuration
T_2 = transmitter two in a two-transmitter tool (BHC) configuration
v = velocity of the formation
V = velocity, ft/sec
x = distance between R_1 and R_2
X_{12} = distance between R_1 and R_2
α = porosity correction factor
Δt = transit time, sec

Δt_c = compressional travel time
ϕ = fractional porosity of rock
θ = angle, degrees
ρ = formation density, lbm/ft^3
μ = shear modulus
σ = Poisson's ratio
σ_r = borehole radial stress
σ_x = principal horizontal stress x-direction
σ_y = principal horizontal stress y-direction
σ_z = principal vertical stress z-direction
σ_θ = borehole tangential stress

Subscripts

c = compressional wave
c' = critical
f = fluid
ma = rock matrix
p = compressional wave
s = shear wave
sh = shale
w = water

References

1. Paillet, F.L., Cheng, C.H., and Pennington, W.D.: "Acoustic-Waveform Logging—Advances in Theory and Application," *The Log Analyst* (1992) **33,** No. 3, 239.
2. Paillet, F.L. and Cheng, C.H.: *Acoustic Waves in Boreholes,* CRC Press, Boca Raton, Florida (1991) 1–264.
3. Mavko, G., Mukerji, T., and Dvorkin, J.: *The Rock Physics Handbook—Tools for Seismic Analysis in Porous Media*, Cambridge U. Press, Cambridge, England (1998) 1–329.
4. Hearst, J.R., Nelson, P.H., and Paillet, F.L.: "Acoustic Logging," *Well Logging for Physical Properties*, second edition, Wiley and Sons Inc., New York City (2000) Chap. 8, 257–303.
5. Tang, X.-M. and Cheng, A.: "*Quantitative Borehole Acoustic Methods*," *Handbook of Geophysical Exploration,* Seismic Exploration **24**, Pergamon Press (Elsevier) (2004).
6. Cheng, C.H. and Toksoz, M.N.: "Determination of Shear Wave Velocities in 'Slow' Formation," paper V, *Trans.,* 1983 Annual Logging Symposium, Soc. of Professional Well Log Analysts (SPWLA) 1–18.
7. Valero, H.P. *et al.*: "Processing of Monopole Compressional in Slow Formation," paper BG 2.5, in *Expanded Abstracts,* 2004 Annual Meeting Technical Program, Soc. of Exploration Geophysicists (SEG) 318–321.
8. Wu, P.T., Darling, H.L., and Scheibner, D.: "Low-Frequency P-Wave Logging for Improved Compressional Velocity in Slow Formation Gas Zones," paper BG1.3, *Expanded Abstracts,* 1995 Annual Meeting Technical Program, SEG, 9–12.
9. *Handbook of Physical Properties of Rocks*, R.S. Carmichael (ed.) CRC Press Inc., Boca Raton, Florida (1982) **2,** 1–228.
10. Leonardon, E.G.: "Logging, Sampling, and Testing," *History of Petroleum Engineering*, American Petroleum Inst., Div. of Production, Dallas (1961) Chap. 8, 493–578.
11. Snyder, D.D. and Fleming, D.B.: "Well-Logging—A Twenty-Five-Year Perspective," *Geophysics* (1985) **50,** No. 12, 2,504–2,529.
12. Jorden, J.R. and Campbell, F.L.: *Well Logging II—Electric and Acoustic Logging*, Monograph Series, SPE, Richardson, Texas (1986) **10,** 95–151.

13. Timur, A.: "Acoustic Logging," *Petroleum Engineering Handbook*, H.B. Bradley (ed.), SPE, Richardson, Texas (1987) Chap. 7, 51-1–51-52.
14. Liu, O.Y.: "The Sources of Errors in Slowness Measurements and an Evaluation of Full Waveform Compensation Techniques," paper SPE 16772 presented at the 1987 SPE Annual Technical Conference and Exhibition, Dallas, 27–30 September.
15. Aron, J., Murray, J., and Seeman, B.: "Formation Compressional and Shear Interval Transit Time Logging by Means of Long Spacings and Digital Techniques," paper SPE 7446 presented at the 1978 SPE Annual Technical Conference and Exhibition, Houston, 1–3 October.
16. Koerperich, E.A.: "Shear Wave Velocities Determined from Long- and Short-Spaced Borehole Acoustic Device," *SPEJ* (October 1980) 317.
17. Morris, C.F., Little, T.M., and Letton, W. III: "A New Sonic Array Tool for Full Waveform Logging," paper SPE 13285 presented at the 1984 SPE Annual Technical Conference and Exhibition, Houston, 16–19 September.
18. Williams, D.M. *et al.*: "The Long Spaced Acoustic Logging Tool," paper T, *Trans.*, 1984 Annual Logging Symposium, SPWLA, 1–15.
19. Chen, S.T. and Eriksen, E.A.: "Compressional- and Shear-Wave Logging in Open and Cased Holes Using a Multipole Tool," *Geophysics* (1991) **56,** No. 4, 550–557.
20. Georgi, D.T. *et al.*: "Application of Shear and Compressional Transit-Time Data to Cased-Hole Carbonate Reservoir Evaluation," *The Log Analyst* (1991) **32,** No. 3, 129.
21. Valero, H.P., Skelton, O., and Cao, H.: "An Overview of Sonic Slowness Evaluation Behind Casing," paper I, *Trans.*, 2000 Well Logging Symposium Japan, SPWLA, Japan Chapter, 1–8.
22. Tang, X., and Patterson, D.: "Analyzing and Processing Acoustic Logging Data for Poorly Bonded Cased Boreholes," paper OO, *Trans.*, 2005 Annual Logging Symposium, SPWLA, 1–11.
23. Stevens, J.L. and Day, S.M.: "Shear Velocity Logging in Slow Formations Using the Stoneley Wave," *Geophysics* (1986) **51,** No. 1, 137–147.
24. Liu, O.Y.: "Stoneley Wave-Derived Delta-T Shear Log," paper ZZ, *Trans.*, 1984 Annual Logging Symposium, SPWLA, 1–14.
25. White, J.E.: "The HULA Log—A Proposed Acoustic Tool," paper I, *Trans.*, 1967 Annual Logging Symposium, SPWLA, 1–30.
26. Zemanek, J. *et al.*: "Continuous Acoustic Shear Wave Logging," paper U, *Trans.*, 1984 Annual Logging Symposium, SPWLA, 1–14.
27. Kessler, C. *et al.*: "New Development in Monopole Acoustic Logging," paper Z, *Trans.*, 2004 Annual Logging Symposium, SPWLA, 1–12.
28. Harrison, A.R. *et al.*: "Acquisition and Analysis of Sonic Waveforms from a Borehole Monopole and Dipole Source for the Determination of Compressional and Shear Speeds and Their Relation to Rock Mechanical Properties and Surface Seismic Data," paper SPE 20557 presented at the 1990 SPE Annual Technical Conference and Exhibition, New Orleans, 23–26 September.
29. Tello, L.N. *et al.*: "A Dipole Array Sonic Tool for Vertical and Deviated Wells," paper SPE 56790 presented at the 1999 SPE Annual Technical Conference and Exhibition, Houston, 3–6 October.
30. Oden, C.P., Stowell, J.R., and LoCoco, J.J.: "Variable Frequency Monopole-Dipole Sonic Logging for Shear Velocity—Applications and Test Results," *Proc.*, 2000 Intl. Symposium on Borehole Geophysics for Minerals, Geotechnical, and Groundwater Applications, Minerals, and Geotechnical Logging Soc., SPWLA, 71–76.
31. Kessler, C. *et al.*: "A New Generation Crossed Dipole Logging Tool: Design and Case Histories," paper SPE 71740 presented at the 2001 SPE Annual Technical Conference and Exhibition, New Orleans, 30 September–3 October.
32. Pistre, V. *et al.*: "A Modular Wireline Sonic Tool for Measurements of 3D (Azimuthal, Radial, and Axial) Formation Acoustic Properties," paper P, *Trans.*, 2005 Annual Logging Symposium, SPWLA, 1–13.
33. Plona, T. *et al.*: "Mechanical Damage Detection and Anisotropy Evaluation Using Dipole Sonic," paper F, *Trans.*, 2002 Annual Logging Symposium, SPWLA, 1–14.
34. Sinha, B. *et al.*: "Optimal Well Completions Using Radial Profiling of Formation Shear Slownesses," paper SPE 95837 presented at the 2005 SPE Annual Technical Conference and Exhibition, Dallas, 9–12 October.
35. Aron, J. *et al.*: "Sonic Compressional Measurements While Drilling," paper SS, *Trans.*, 1994 Annual Logging Symposium, SPWLA, 1–17.

36. Minear, J. *et al.*: "Compressional Slowness Measurements While Drilling," paper VV, *Trans.*, 1995 Annual Logging Symposium, SPWLA, 1–12.
37. Hsu, K. *et al.*: "Interpretation and Analysis of Sonic While Drilling Data in Overpressured Formations," paper FF, *Trans.*, 1997 Annual Logging Symposium, SPWLA, 1–14.
38. Hashem, M. *et al.*: "Seismic Tie Using Sonic-While-Drilling Measurements," paper I, *Trans.*, 1999 Annual Logging Symposium, SPWLA, 1–13.
39. Market, J., Schmitt, D., and Deady, R.: "LWD Sonic Logging in Cased Holes," paper U, *Trans.*, 2004 Annual Logging Symposium, SPWLA, 1–14.
40. Varsamis, G. *et al.*: "A New MWD Full Wave Dual Mode Sonic Tool Design and Case Histories," paper F, *Trans.*, 1999 Annual Logging Symposium, SPWLA, 1–12.
41. Varsamis, G.L. *et al.*: "LWD Shear Velocity Logging in Slow Formations—Design Decisions and Case Histories," paper O, *Trans.*, 2000 Annual Logging Symposium, SPWLA, 1–13.
42. Joyce, B. *et al.*: "Introduction to a New Omni-Directional Acoustic System for Improved Real Time LWD Sonic Logging—Tool Design and Field Test Results," paper SS, *Trans.*, 2001 Annual Logging Symposium, SPWLA, 1–14.
43. Leggett, J.V. III *et al.*: "Field Test Results Demonstrating Improved Real-Time Data Quality in an Advanced LWD Acoustic System," paper SPE 71732 presented at the 2001 SPE Annual Technical Conference and Exhibition, New Orleans, 30 September–3 October.
44. Tang, X. *et al.*: "Shear-Velocity Measurement in the Logging-While-Drilling Environment—Modeling and Field Evaluations," *Petrophysics* (2003) **44,** No. 2, 79–90.
45. Market, J. *et al.*: "Processing and Quality Control of LWD Dipole Sonic Measurements," paper PP, *Trans.*, 2002 Annual Logging Symposium, SPWLA, 1–14.
46. Dubinsky, V. *et al.*: "Engineering Aspects of LWD Quadrupole Measurements and Field Test Results," paper SPE 84248 presented at the 2003 SPE Annual Technical Conference and Exhibition, Denver, 5–8 October.
47. Goldberg, D. *et al.*: "Analysis of LWD Sonic Data in Low-Velocity Formations," paper BH 1.5, *Expanded Abstracts,* 2003 Annual Meeting Technical Program, SEG, 301–304.
48. Boonen, P. and Yogeswaren, E.: "A Dual-Frequency LWD Sonic Tool Expands Existing Unipolar Transmitter Technology to Supply Shear Wave Data in Soft Formations," paper X, *Trans.*, 2004 Annual Logging Symposium, SPWLA, 1–11.
49. Endo, T., Yoneshima, S., and Valero, H.-P.: "Analysis of LWD Sonic Data in Very Slow Formation—Compressional Processing and Indirect V_p/V_s Estimation," paper G, *Proc.*, 2004 Formation Evaluation Symposium of Japan, SPWLA Japan Chapter, 1–8.
50. Tang, X., Zheng, Y., and Dubinsky, V.: "Logging While Drilling Acoustic Measurement in Unconsolidated Slow Formations," paper R, *Trans.*, 2005 Annual Logging Symposium, 1–13.
51. Zhu, Z. *et al.*: "Experimental Studies of Multipole Acoustic Logging with Scaled Borehole," paper BG 2.6, *Expanded Abstracts,* 2005 Annual Meeting Technical Program, SEG, 376–379.
52. Tang, X. *et al.*: "Logging-While-Drilling Shear and Compressional Measurements in Varying Environments," paper II, *Trans.*, 2003 Annual Logging Symposium, SPWLA, 1–13.
53. Wang, T., and Tang, X.: "Investigation of LWD Quadrupole Shear Measurement in Real Environments," paper KK, *Trans.*, 2003 Annual Logging Symposium, SPWLA, 1–12.
54. Haugland, S.: "Frequency Dispersion Effects on LWD Shear Sonic Measurements in Acoustically Slow Environments," paper SPE 90505 presented at the 2004 SPE Annual Technical Conference and Exhibition, Houston, 26–29 September.
55. Huang, X., Zheng, Y., and Toksoz, N.M.: "Effects of Tool Eccentricity on Acoustic Logging While Drilling (LWD) Measurements," paper BG 1.4, *Expanded Abstracts,* 2004 Annual Meeting Technical Program, SEG, 290–293.
56. Underhill, W. *et al.*: "Demonstrations of Real-Time Borehole Seismic from an LWD Tool," paper SPE 71365 presented at the 2001 SPE Annual Technical Conference and Exhibition, New Orleans, 30 September–3 October.
57. Haldorsen, J.B.U. *et al.*: "Optimizing the Well Construction Process: Full-Waveform Data from While-Drilling Seismic Measurements in the South Caspian Sea," paper SPE 79844 presented at the 2003 SPE/IADC Drilling Conference, Amsterdam, 19–21 February.
58. Esmersoy, C. *et al.*: "Seismic MWD: Drilling in Time, on Time, It's About Time," *The Leading Edge* (2005) **24,** No. 1, 56–62.

59. Althoff, G. *et al.:* "New Concepts for Seismic Surveys While Drilling," paper SPE 90751 presented at the 2004 SPE Annual Technical Conference and Exhibition, Houston, 26–29 September.
60. Zemanek, J. *et al.:* "The Borehole Televiewer: A New Logging Concept for Fracture Location and Other Types of Borehole Inspection," *JPT* (June 1969) 762.
61. Prensky, S.: "Advances in Borehole Imaging Technology and Applications," *Borehole Imaging —Applications and Case Histories,* M.A. Lovell, G. Williamson, and P.K. Harvey (eds.), Geological Soc., London, Special Publication No. 159 (1999) 1–44.
62. Bigelow, E.L.: *Cement Evaluation*, Atlas Wireline Services, Houston (1990) 1–142.
63. Hill, A.D.: "Cement-Quality Logging," *Production Logging—Theoretical and Interpretative Elements*, Monograph Series, SPE, Richardson, Texas (1990) **14,** 123–140.
64. Jutten, J. and Morriss, S.L.: "Cement Job Evaluation," *Well Cementing,* E.B. Nelson (ed.), Elsevier, Amsterdam, Developments in Petroleum Science No. 28 (1990) Chap. 16, 16-1–16-44.
65. Rouillac, D.: *Cement Evaluation Logging Handbook*, Editions Technip, Paris (1994) 1–176.
66. Smolen, J.J.: *Cased Hole and Production Log Evaluation*, PennWell Publishing Co., Tulsa (1996) Chaps. 10 and 11, 161–213.
67. Baker, L.J.: "The Effect of the Invaded Zone on Full Wavetrain Acoustic Logging," *Geophysics* (1984) **49,** No. 6, 796–809.
68. Alberty, M.: "The Influence of the Borehole Environment upon Compressional Sonic Logs," *The Log Analyst* (1996) **37,** No. 4, 30–44.
69. Chi, S. *et al.:* "Assessment of Mud-Filtrate Invasion Effects on Borehole Acoustic Logs and Radial Profiling of Formation Elastic Parameters," paper SPE 90159 presented at the 2004 SPE Annual Technical Conference and Exhibition, Houston, 26–29 September.
70. Walker, T.: "Origin of the 'W' Pattern on Cased Hole Micro-Seismogram Logs," *The Log Analyst* (1968) **9,** No. 2, 30–32.
71. Pilkington, P.E.: "Cement Evaluation—Past, Present, and Future," *JPT* (February 1992) 132.
72. Hayman, A.J., Hutin, R., and Wright, P.V.: "High-Resolution Cementation and Corrosion Imaging by Ultrasound," paper KK, *Trans.,* 1991 Annual Logging Symposium, SPWLA, 1–25.
73. Graham, W.L. *et al.:* "Cement Evaluation and Casing Inspection with Advanced Ultrasonic Scanning Methods," paper SPE 38651 presented at the 1997 SPE Annual Technical Conference and Exhibition, San Antonio, Texas, 5–8 October.
74. Froelich, B. *et al.:* "Cement Evaluation Tool: A New Approach to Cement Evaluation," *JPT* (August 1982) 1835.
75. Sheives, T.C. *et al.:* "A Comparison of New Ultrasonic Cement and Casing Evaluation Logs with Standard Cement Bond Logs," paper SPE 15436 presented at the 1986 SPE Annual Technical Conference and Exhibition, New Orleans, 5–8 October.
76. Bigelow, E.L., Domangue, E.J., and Lester, R.A.: "A New Innovative Technology for Cement Evaluation," paper SPE 20585 presented at the 1990 SPE Annual Technical Conference and Exhibition, New Orleans, 23–26 September.
77. Schmidt, M.G.: "The Micro CBL—A Second Generation Radial Cement Evaluation Instrument," paper Z, *Trans.,* 1989 Annual Logging Symposium, SPWLA, 1–16.
78. Tello, L.N. *et al.:* "New Efficient Method for Radial Cement Bond Evaluation," *Canadian Well Logging Soc. J.* (1995) **20,** 60–69.
79. Butsch, R.J.: "Overcoming Interpretation Problems of Gas-Contaminated Cement Using Ultrasonic Cement Logs," paper SPE 30509 presented at the 1995 SPE Annual Technical Conference and Exhibition, Dallas, 22–25 October.
80. Frisch, G.J., Graham, W.L., and Griffith, J.: "Assessment of Foamed-Cement Slurries Using Conventional Cement Evaluation Logs and Improved Interpretation Methods," paper SPE 55649 presented at the 1999 SPE Rocky Mountain Regional Meeting, Gillette, Wyoming, 15–18 May.
81. Frisch, G., Graham, L., and Griffith, J.: "A Novel and Economical Processing Technique Using Conventional Bond Logs and Ultrasonic Tools for Enhanced Cement Evaluation," paper EE, *Trans.,* 2000 Annual Logging Symposium, SPWLA, 1–14.
82. Frisch, G. *et al.:* "Advances in Cement Evaluation Tools and Processing Methods Allow Improved Interpretation of Complex Cements," paper SPE 97186 presented at the 2005 SPE Annual Technical Conference and Exhibition, Dallas, 9–12 October.

83. Van Kuijk, R. *et al.*: "A Novel Ultrasonic Cased-Hole Imager for Enhanced Cement Evaluation," paper IPTC 10546 presented at the 2005 International Petroleum Technology Conference, Doha, Qatar, 21–23 November.
84. Dumont, A., Patin, J.-B., and Le Floch, G.: "A Single Tool for Corrosion and Cement Evaluation," paper SPE 13140 presented at the 1984 SPE Annual Technical Conference and Exhibition, Houston, 16–19 September.
85. Roberts, D.L. and Richards, J.W.: "Casing Corrosion Evaluation Using Ultrasonic Techniques: A Unique Approach for West Texas Wells," *Proc.,* 1987 Annual Short Course, Southwestern Petroleum Short Course Association, Lubbock, Texas, 20–33.
86. Frisch, G. and Mandal, B.: "Advanced Ultrasonic Scanning Tool and Evaluation Methods Improve and Standardize Casing Inspection," paper SPE 71399 presented at the 2001 SPE Annual Technical Conference and Exhibition, New Orleans, 30 September–3 October.
87. Fertl, W.H.: "Openhole Crossplot Concepts: A Powerful Technique in Well Log Analysis," *JPT* (March 1981) 535.
88. Bigelow, E.L.: *Introduction to Wireline Log Analysis*, Baker Atlas, Houston (1995) 1–312.
89. Boyer, S. and Mari, J.-L.: *Seismic Surveying and Well Logging Oil and Gas Exploration Techniques*, Editions Technip, Paris (1997) 1–192.
90. Pickett, G.R.: "Acoustic Character Logs and Their Applications in Formation Evaluation," *JPT* (June 1963) 659.
91. Medlin, W.L. and Alhilali, K.A.: "Shear-Wave Porosity Logging in Sands," *SPEFE* (March 1992) 106.
92. Tixier, M.P., Alger, R.P., and Doh, C.A..: "Sonic Logging," *JPT* (1959) 106; *Trans.,* AIME, **216.**
93. Wyllie, M.R.J., Gregory, A.R., and Gardner, L.W.: "Elastic Wave Velocities in Heterogeneous and Porous Media," *Geophysics* (1956) **21,** No. 1, 41–70.
94. Raymer, L.L., Hunt, E.R., and Gardner, J.S.: "An Improved Sonic Transit Time-to-Porosity Transform," paper P, *Trans.,* 1980 Annual Logging Symposium, SPWLA, 1–12.
95. Dvorkin, J. and Nur, A.: "Time-Average Equation Revisited," *Geophysics* (1998) **63,** No. 2, 460–464.
96. Krief, M. *et al.*: "A New Petrophysical Interpretation Using the Velocities of P and S Waves (Full Waveform Sonic)," *The Log Analyst* (1990) **31,** No. 6, 355–369.
97. Spears, R., and Nicosia, W.: "Porosity Estimation from Shear Wave Interval Transit Time in the Norphlet Aeolian Jurassic Sandstone of Southern and Offshore Alabama," paper N, *Trans.,* 2003 Annual Logging Symposium, SPWLA, 1–12.
98. Valero, H.-P., Skelton, O., and Almeida, C.M.: "Processing of Monopole Sonic Waveforms Through Cased Hole," paper BH 1.1, *Expanded Abstracts,* 2003 Annual Meeting Technical Program, SEG, 285–288.
99. Tello, L.N. *et al.*: "Through-Tubing Hostile-Environment Acoustic Logging Tool," paper Y, *Trans.,* 2004 Annual Logging Symposium, SPWLA, 1–11.
100. Smith, J., Fisburn, T., and Bigelow, E.: "Acoustic Porosity Corrections for Gas and Light Hydrocarbon-Bearing Sandstones," paper AAA, *Trans.,* 1996 Annual Logging Symposium, SPWLA, 1–9.
101. Knackstedt, M.A., Arns, C.H., and Pinczewski, W.V.: "Velocity-Porosity Relationships, Part 1: Accurate Velocity Model for Clean Consolidated Sandstones," *Geophysics* (2003) **68,** No. 6, 1822–1834.
102. Knackstedt, M.A., Arns, C.H., and Pincsewski, W.V.: "Velocity-Porosity Relationship: Predictive Velocity Model for Cemented Sands Composed of Multiplemineral Phases," *Geophysical Prospecting* (2005) **53,** No. 3, 349–372.
103. Kazatchenko, E., Markov, M., and Mousatov, A.: "Determination of Primary and Secondary Porosity in Carbonate Formations Using Acoustic Data," paper SPE 84209 presented at the 2003 SPE Annual Technical Conference and Exhibition, Denver, 5–8 October.
104. Burke, J.A., Campbell, R.L. Jr., and Schmidt, A.W.: "The Litho-Porosity Cross Plot—A Method of Determining Rock Characteristics for Computation of Log Data," paper SPE 2771 presented at the 1969 SPE Illinois Basin Regional Meeting, Evansville, Indiana, 30–31 October.
105. Clavier, C. and Rust, D.H.: "MID Plot: A New Lithology Technique," *The Log Analyst* (1976) **17,** No. 6, 16–24.

106. Robert, H.V. and Campbell, R.L. Jr.: "Applications of CORIBAND to the Evaluation of Sandstones Containing Mica," *The Log Analyst* (1976) **17,** No.1, 33–40.
107. Nations, J.F.: "Lithology and Porosity from Acoustic Shear and Compressional Wave Transit Time Relationships," *The Log Analyst* (1974) **15,** No. 6, 3–8.
108. Eastwood, R.L. and Castagna, J.P.: "Interpretation of V_p/V_s Ratios from Sonic Logs," *Shear-Wave Exploration*, S.H. Danbom and S.N. Domenico (eds.), SEG, Geophysical Development Series No. 1 (1987) 139–153.
109. Johnston, J.E. and Christensen, N.I.: "Compressional to Shear Velocity Ratios in Sedimentary Rocks," *Intl. J. of Rock Mechanics and Mining Sciences and Geomechanical Abstracts* (1993) **30,** No. 7, 751–754.
110. Assefa, S., McCann, C., and Sothcott, J.: "Velocities of Compressional and Shear Waves in Limestone," *Geophysical Prospecting* (2003) **51**, No. 1, 1–13.
111. Tsuneyama, F. *et al.:* "V_p/V_s Ratio as a Rock Frame Indicator for a Carbonate Reservoir," *First Break* (2003) **21**, No. 7, 53–58.
112. Hingle, A.T. Jr.: "The Use of Logs in Exploration Problems," *Technical Program,* 1959 Annual Meeting, SEG, 38.
113. Berry, J.E.: "Acoustic Velocity in Porous Media," *JPT* (1959) 262; *Trans.,* AIME, **216.**
114. Pickett, G.R.: "A Review of Current Techniques for Determination of Water Saturation from Logs," *JPT* (November 1966) 1425.
115. Pickett, G.R.: "Pattern Recognition as a Means of Formation Evaluation," *The Log Analyst* (1973) **14,** No. 4, 3–11.
116. Kamel, M.H., and Mabrouk, W.M.: "An Equation for Estimating Water Saturation in Clean Formations Utilizing Resistivity and Sonic Logs: Theory and Application," *J. of Petroleum Science and Engineering* (2002) **36**, No. 3–4, 59–168.
117. Souder, W.W.: "Using Sonic Logs to Predict Fluid Type," *Petrophysics* (2002) **43,** No. 5, 412–419.
118. Chardac, O., Brie, A., and Chouker, A.C.: "Correlations of Shear vs. Compressional in Shaly Sands and Application to Quicklook Hydrocarbon Detection," paper SPE 84205 presented at the 2003 SPE Annual Technical Conference and Exhibition, Denver, 5–8 October.
119. Hamada, G.M.: "V_p/V_s Crossplots Identify Reservoir Fluids," *Oil & Gas J.* (28 June 2004) **102,** No. 24, 41–44.
120. Khazanehdari, J., and McCann, C.: "Acoustic and Petrophysical Relationships in Low-Shale Sandstone," *Geophysical Prospecting* (2005) **53,** No. 4, 447–462.
121. Williams, D.M.: "The Acoustic Log Hydrocarbon Indicator," paper W, *Trans.,* 1990 Annual Logging Symposium, SPWLA, 1–22.
122. Brie, A. *et al.:* "Shear Sonic Interpretation in Gas-Bearing Sands," paper SPE 30595 presented at the 1995 SPE Annual Technical Conference and Exhibition, Dallas, 22–25 October.
123. Rogen, B. *et al.:* "Ultrasonic Velocities of North Sea Chalk Samples: Influence of Porosity," *Geophysical Prospecting* (2005) **53,** No. 4, 481–496.
124. Watson, I.A., Lines, L.R., and Brittle, K.F.: "Heavy-Oil Reservoir Characterization Using Elastic Wave Properties," *The Leading Edge* (2002) **21,** No. 8, 736–739.
125. Gutierrez, M., Dvorkin, J., and Nur, A.: "In-Situ Hydrocarbon Identification and Reservoir Monitoring Using Sonic Logs, La Cira-Infantas Oil Field (Colombia)," paper RPB 2.6, *Expanded Abstracts,* 2000 Annual Meeting Technical Program, SEG, 1727–1730.
126. Gutierrez, M.A., Dvorkin, J., and Nur, A., "Theoretical Rock Physics for Bypassed Oil Detection Behind the Casing—La Cira-Infantas Oil Field," *The Leading Edge* (2001) **20,** No. 2, 192–198.
127. Batzle, M. *et al.:* "Heavy Oils: Seismic Properties," paper RP P2.2, *Expanded Abstracts,* 2004 Annual Meeting Technical Program, SEG, 1762–1765.
128. Thomas, D.H.: "Seismic Applications of Sonic Logs," *The Log Analyst* (1978) **19,** No. 1, 23–32.
129. Liner, C.L.: "Interpreting Seismic Data," *Exploring for Oil and Gas Traps, Treatise of Petroleum Geology, Handbook of Petroleum Geology* E.A. Beaumont, and N.H. Foster (eds.), AAPG (1999) Chap. 12, 12-13–12-17.
130. Bork, J. and Wood, L.C.: "Seismic Interpretation of Sonic Logs," paper INT 1.4, *Expanded Abstracts,* 2001 Annual Meeting Technical Program, SEG, 510–513.
131. Box, R., and Lowrey, P.: "Reconciling Sonic Logs with Check-Shot Surveys; Stretching Synthetic Seismograms," *The Leading Edge* (2003) **22,** No. 6, 510–517.

132. White, R.E.: "Tying Well-Log Synthetic Seismograms to Seismic Data: The Key Factors," paper W-3.7, *Expanded Abstracts,* 2003 Annual Meeting Technical Program, SEG, 2449–2452.
133. Walls, J., Dvorkin, J., and Carr, M.: "Well Logs and Rock Physics in Seismic Reservoir Characterization," paper OTC 16921 presented at the 2004 Offshore Technology Conference, Houston, 3–6 May.
134. Bruce, B. and Bowers, G.: "Pore Pressure Terminology," *The Leading Edge* (2002) **21,** No. 2, 170–173.
135. Fertl, W.H., Chapman, R.E., and Holz, R.F. (eds.): *Studies in Abnormal Pressure*, Developments in Petroleum Science No. 38, Elsevier, Amsterdam (1994) 1–454.
136. Bowers, G.L.: "Detecting High Pressure," *The Leading Edge* (2002) **21,** No. 2, 174–177.
137. Chilingar, G.V., Serebryakov, V.A. and Robertson, J.O. Jr. (eds.): *Origin and Prediction of Abnormal Formation Pressures*, Developments in Petroleum Science No. 50, Elsevier, Amsterdam (2002) 1–390.
138. Evans, B.J.: "Developments in Pore Pressure Prediction Using Seismic-While-Drilling Technology," *Australian Petroleum Production and Exploration Assn. J.* (1999) **39,** Part 1, 461–474.
139. Walls, J. *et al.:* "Use of Compressional and Shear Wave Velocity for Overpressure Detection," paper OTC 11912 presented at the 2000 Offshore Technology Conference, Houston, 1–4 May.
140. Badri, M.A. *et al.:* "Pore Pressure Prediction Data Using Seismic Velocities and Log Data in the Offshore Nile Delta, Egypt," paper SPE 68195 presented at the 2001 SPE Middle East Oil Show, Bahrain, 17–20 March.
141. Carcione, J.M. and Tinivella, U.: "The Seismic Response to Overpressure—A Modelling Study Based on Laboratory, Well and Seismic Data," *Geophysical Prospecting* (2001) **49,** No. 5, 523–539.
142. Huffman, A.R.: "The Future of Pore-Pressure Prediction Using Geophysical Methods," paper OTC 13041 presented at the 2001 Offshore Technology Conference, Houston, 30 April–3 May.
143. Sayers, C.M. and Woodward, M.J.: "Seismic Pore-Pressure Prediction Using Reflection Tomography and 4-C Seismic Data," *The Leading Edge* (2002) **21,** No. 2, 188–192.
144. Citta, F. *et al.:* "Deepwater Hazard Avoidance in a Large Top-Hole Section Using LWD Acoustic Data," *The Leading Edge* (2004) **23**, No. 6, 566–573.
145. Althoff, G. *et al.:* "New Concepts for Seismic Surveys While Drilling," paper SPE 90751 presented at the 2004 SPE Annual Technical Conference and Exhibition, Houston, 26–29 September.
146. Mancilla-Castillo, J., Mendez-Hernandez, E., and Santana-Fernandez, J.: "Geopressure Evaluation from Seismic Data and Its Application for Exploratory Wells in Mexico," paper OTC 15250 presented at the 2003 Offshore Technology Conference, Houston, 5–8 May.
147. Storvoll, V., Bjorlykke, K., and Mondol, N.M.: "Velocity-Depth Trends in Mesozoic and Cenozoic Sediments from the Norwegian Shelf," *AAPG Bulletin* (2005) **89**, No. 3, 359–381.
148. Hottman, C.E. and Johnson, R.K.: "Estimation of Formation Pressures from Log-Derived Shale Properties," *JPT* (June 1965) 717; *Trans.,* AIME, **234.**
149. Ham, H.H.: "A Method of Estimating Formation Pressures from Gulf Coast Well Logs," *Trans., Gulf Coast Assn. of Geological Soc.* (1966) **16,** 185–197.
150. Japsen, P.: "Overpressured Cenozoic Shale Mapped from Velocity Anomalies Relative to a Baseline for Marine Shale, North Sea," *Petroleum Geoscience* (1999) **5,** 321–336.
151. Draou, A. and Osisanya, S.O.: "New Methods for Estimating Formation Pressures and Fracture Gradients from Well Logs," paper SPE 63263 presented at the 2000 SPE Annual Technical Conference and Exhibition, Dallas, 1–4 October.
152. Bowers, G.L.: "Pore Pressure Estimation from Velocity Data: Accounting for Overpressure Mechanisms Besides Undercompaction," *SPEDC* (June 1995) 89.
153. Swarbrick, R.E.: "Pore-Pressure Prediction—Pitfalls in Using Porosity," paper OTC 13045 presented at the 2001 Offshore Technology Conference, Houston, 30 April–3 May.
154. Eaton, B.A.: "The Equation for Geopressure Prediction from Well Logs," paper SPE 5544 presented at the 1975 SPE Annual Technical Conference and Exhibition, Dallas, 28 September–1 October.
155. Eaton, B.A. and Eaton, T.L.: "Fracture Gradient Prediction for the New Generation," *World Oil* (1997) **218,** No. 10, 93–94 and 96–100.

156. Lahann, R.W., McCarty, D.K., and Hsieh, J.C.C.: "Influence of Clay Diagenesis on Shale Velocities and Fluid-Pressure," paper OTC 13046 presented at the 2001 Offshore Technology Conference, Houston, 30 April–3 May.
157. Domnesteanu, P., McCann, C., and Sothcott, J.: "Velocity Anisotropy and Attenuation of Shale in Under- and Overpressured Conditions," *Geophysical Prospecting* (2002) **50**, No. 5, 487–503.
158. Dewhurst, D.N., Siggins, A.F., and Raven, M.D.: "Influence of Pore Pressure, Composition, and Microstructure on the Acoustic Properties of Shales," paper SPE 78197 presented at the 2002 SPE/ISRM Rock Mechanics Conference, Irving, Texas, 20–23 October.
159. Morris, R.L., Grine, D.R., and Arkfeld, T.E.: "Using Compressional and Shear Acoustic Amplitudes for the Location of Fractures," *JPT* (June 1964) 623.
160. Minne, J.-C. and Gartner, J.: "Fracture Detection in the Middle East," paper SPE 7773 presented at the 1979 SPE Middle East Technical Conference and Exhibition, Bahrain, 25–28 February.
161. Cheung, P.S.V.: "Fracture Detection Using the Sonic Tool," paper 42, *Trans.*, 1984 Intl. Formation Evaluation Symposium, SPWLA, Paris Chapter (SAID) 1–8.
162. Paillet, F.L.: "Qualitative and Quantitative Interpretations of Fracture Permeability by Means of Acoustic Full-Waveform Logs," *The Log Analyst* (1991) **32,** No. 3, 256–270.
163. Aguilera, R.: "Formation Evaluation by Well Log Analysis," *Naturally Fractured Reservoirs*, second edition, PennWell Publishing Co., Tulsa (1995) Chap. 3, 181–315.
164. Magara, K.: "Thickness of Removed Sedimentary Rocks, Paleopore Pressure, and Paleotemperature, Southwestern Part of Western Canada Basin," *AAPG Bulletin* (1976) **60,** No. 4, 554–565.
165. Magara, K.: "Thickness of Erosion," *Geological Models of Petroleum Entrapment,* Elsevier Applied Science Publishers, London (1986) Chap. 7, 129–151.
166. Bulat, J., and Stoker, J.S.: "Uplift Determination from Interval Velocity Studies, UK Southern North Sea," *Petroleum Geology of North West Europe,* Brooks, J. and Glennie, K. (eds.), Graham and Trottman, London (1987) **1,** 293–305.
167. Hillis, R.R.: "Quantifying Erosion in Sedimentary Basins from Sonic Velocities in Shales and Sandstones," *Exploration Geophysics* (1993) **24,** 561–566.
168. Hillis, R.R., Thomson, K., and Underhill, J.R.: "Quantification of Tertiary Erosion in the Inner Moray Firth Using Velocity Data from the Chalk and the Kimmeridge Clay," *Marine and Petroleum Geology* (1994) **11,** No. 3, 283–293.
169. Hansen, S.: "A Compaction Trend for Cretaceous and Tertiary Shales on the Norwegian Shelf Based on Sonic-Transit Times," *Petroleum Geoscience* (1996) **2,** No. 1, 59–68.
170. Heasler, H.P. and Kharitonova, N.A.: "Analysis of Sonic Well Logs Applied to Erosion Estimates in the Bighorn Basin, Wyoming," *AAPG Bulletin* (1996) **80,** No. 5, 630–646.
171. Evans, D.J.: "Estimates of the Eroded Overburden and the Permian-Quaternary Subsidence History of the Area West of Orkney," *Scottish J. of Geology* (1997) **33,** Part 2, 169–181.
172. Japsen, P.: "Regional Velocity-Depth Anomalies, North Sea Chalk—A Record of Overpressure and Neogene Uplift and Erosion," *AAPG Bulletin* (1998) **82,** No. 11, 2,031–2,074.
173. Densley, M.R., Hillis, R.R., and Redfearn, J.E.P.: "Quantification of Uplift in the Carnarvon Basin Based on Interval Velocities," *Australian J. of Earth Sciences* (2000) **47,** No. 1, 111–122.
174. Fuh, S.-C.: "Magnitude of Cenozoic Erosion from Mean Sonic Transit Time, Offshore, Taiwan," *Marine and Petroleum Geology* (2000) **17,** No. 9, 1,011–1,028.
175. Al-Chalabi, M.: "The Use of Instantaneous Velocity in Uplift Investigation," *Geophysical Prospecting* (2001) **49**, 645–655.
176. Ware, P.D. and Turner, J.P.: "Sonic Velocity Analysis of the Tertiary Denudation of the Irish Sea Basin," *Exhumation of The North Atlantic Margin; Timing, Mechanisms and Implications for Petroleum Exploration*, Dore, A.G., Cartwright, J.A., Stoker, M.S., Turner, J.P., and White, N.J. (eds.), Special Publication 196, Geological Society, London (2002) 355–370.
177. Mavromatidis, A., and Hillis, R.: "Quantification of Exhumation in the Eromanga Basin and Its Implications for Hydrocarbon Exploration," *Petroleum Geoscience* (2005) **11,** No. 1, 79–92.
178. Stocks, A.E. and Lawrence, S.R.: "Identification of Source Rocks from Wireline Logs," *Geological Applications of Wireline Logs*, A. Hurst, M.A. Lovell, and A.C. Morton (eds.), Geological Soc. of London Special Publication No. 48 (1990) 241–252.

179. Herron, S.L.: "In Situ Evaluation of Potential Source Rocks by Wireline Logs," *Source and Migration Processes and Evaluation Techniques*, Treatise of Petroleum Geology, R.K. Merrill (ed.), *Handbook of Petroleum Geology,* AAPG (1991) Chap. 13, 127–134.
180. Lang, W.H.: "The Determination of Thermal Maturity in Potential Source Rocks Using Interval Transit Time/Interval Velocity," *The Log Analyst* (1994) **35,** No. 6, 47–59.
181. Meyer, B.L., and Nederlof, M.H.: "Identification of Source Rocks on Wireline Logs by Density/Resistivity and Sonic Transit Time/Resistivity Crossplots," *AAPG Bulletin* (1984) **68,** No. 2, 121–129.
182. Passey, Q.R. *et al.:* "A Practical Model for Organic Richness from Porosity and Resistivity Logs," *AAPG Bulletin* (1990) **74,** No. 12, 1,777–1,794.
183. Huang, Z. *et al.:* "Cyclicity in the Egret Member (Kimmeridgian) Oil Source Rock, Jeanne d'Arc Basin, Offshore Eastern Canada," *Marine and Petroleum Geology* (1996) **13,** No. 1, 91–105.
184. Fjaer, E. *et al.: Petroleum Related Rock Mechanics*, Developments in Petroleum Science No. 33, Elsevier, Amsterdam (1992) 1–338.
185. Montmayeur, H. and Graves, R.M.: "Prediction of Static Elastic/Mechanical Properties of Consolidated and Unconsolidated Sands from Acoustic Measurements: Basic Measurements," paper SPE 14159 presented at the 1985 SPE Annual Technical Conference and Exhibition, Las Vegas, Nevada, 22–26 September.
186. Montmayeur, H. and Graves, R.M.: "Prediction of Static Elastic/Mechanical Properties of Consolidated and Unconsolidated Sands from Acoustic Measurements: Correlations," paper SPE 15644 presented at the 1986 SPE Annual Technical Conference and Exhibition, New Orleans, 5–8 October.
187. Holt, R.M., Ingsoy, P., and Mikkelsen, M.: "Rock Mechanical Analysis of North Sea Reservoir Formations," *SPEFE* (March 1989) 33.
188. Gatens, J.M. III *et al.*: "In-Situ Stress Tests and Acoustic Logs Determine Mechanical Properties and Stress Profiles in the Devonian Shales," *SPEFE* (September 1990) 248.
189. Yale, D.P. and Jamieson, W.H. Jr.: "Static and Dynamic Rock Mechanical Properties in the Hugoton and Panoma Fields, Kansas," paper SPE 27939 presented at the 1994 SPE Mid-Continent Gas Symposium, Amarillo, Texas, 22–24 May.
190. Kowalski, J.J.: "Formation Strength Parameters from Well Logs," paper N, *Trans.,* 1978 Annual Logging Symposium, SPWLA, 1–19.
191. Sethi, D.K.: "Well Log Applications in Rock Mechanics," paper SPE 9833 presented at the 1981 SPE/DOE Low Permeability Gas Reservoirs Symposium, Denver, 27–29 May.
192. Tixier, M.P., Loveless, G.W., and Anderson, R.A.: "Estimation of Formation Strength from the Mechanical Properties Log," *JPT* (March 1975) 283.
193. Stein, N.: "Mechanical Properties of Friable Sands from Conventional Log Data," *JPT* (July 1976) 757.
194. Onyia, E.C.: "Relationships Between Formation Strength, Drilling Strength, and Electric Log Properties," paper SPE 18166 presented at the 1988 SPE Annual Technical Conference and Exhibition, Houston, 2–5 October.
195. Stein, N.: "Sonic Log Data Help Determine Formation Strength," *Oil & Gas J.* (28 December 1992) 96.
196. Raaen, A.M. *et al.:* "FORMEL—A Step Forward in Strength Logging," paper SPE 36533 presented at the 1996 SPE Annual Technical Conference and Exhibition, Denver, 6–9 October.
197. Anderson, T. and Walker, T.: "Log-Derived Rock Properties for Use in Well Stimulation Design," paper SPE 4095 presented at the 1972 SPE Annual Meeting, San Antonio, Texas, 8–11 October.
198. Anderson, R.A., Ingram, D.S., and Zanier, A.M.: "Determining Fracture Pressure Gradients from Wells Logs," *JPT* (November 1973) 1259.
199. Newberry, B.M., Nelson, R.F., and Ahmed, U.: "Prediction of Vertical Hydraulic Fracture Migration Using Compressional and Shear Wave Slowness," paper SPE 13895 presented at the 1985 SPE/DOE Low Permeability Gas Reservoirs Symposium, Denver, 19–22 May.
200. Stein, N.: "How to Calculate Fracture Pressures from Well Logs," *Petroleum Engineer Intl.* (1988) **60,** No. 8, 36–38.
201. Bruce, S.: "A Mechanical Stability Log," paper IADC/SPE 19942 presented at the 1990 IADC/SPE Drilling Conference, Houston, 27 February–2 March.

202. Stein, N. and Hilchie, D.W.: "Estimating the Maximum Production Rate Possible from Friable Sandstones Without Using Sand Control," *JPT* (September 1972) 1157.
203. Edwards, D.P., Sharma, Y., and Charron, A.: "Zones of Sand Production Identified by Log-Derived Mechanical Properties—A Case Study," paper S, *Trans.,* 1983 European Formation Evaluation Symposium, SPWLA, London Chapter, 1–23.
204. Ong, S. *et al.:* "An Accurate Characterization of Sand Strength in Weak and Unconsolidated Formations Aids Offshore Production Test Designs—A Bohai Bay Case Study," paper SPE 64738 presented at the 2000 SPE International Oil and Gas Conference and Exhibition in China, Beijing, 7–10 November.
205. Fertl, W.H. and DeVries, M.R.: "Coal Evaluation Using Geophysical Well Logs," paper F, *Trans.,* 1997 Formation Evaluation Symposium, Canadian Well Logging Society, 1–17.
206. Bond, L.O., Alger, R.P., and Schmidt, A.W.: "Well Log Applications in Coal Mining and Rock Mechanics," *Trans.,* AIME (1971) **250,** 355–362.
207. Kowalski, J. and Fertl, W.H.: "Application of Geophysical Well Logging to Coal Mining Operations," *Energy Sources* (1977) **3,** No. 2, 133–147.
208. Hornby, B.E.: "Imaging of Near-Borehole Structure Using Full-Waveform Sonic Data," paper SPE 19872 presented at the 1989 SPE Annual Technical Conference and Exhibition, San Antonio, Texas, 8–11 October.
209. Hornby, B.E.: "Use of Full-Wave Sonic Data to Image Near-Borehole Structural Features," *Petroleum Geoscience* (1995) **1,** No. 2, 109–114.
210. Esmeroy, C. *et al.:* "Sonic Imaging—A Tool for High-Resolution Reservoir Description," paper BH 2.7, *Expanded Abstracts,* 1997 Annual Meeting Technical Program, SEG, **1,** 278–281.
211. Coates, R. *et al.:* "Single-Well Sonic Imaging: High-Definition Reservoir Cross-Sections from Horizontal Wells," paper SPE 65457 presented at the 2000 SPE/PS-CIM International Conference and Exhibition on Horizontal Well Technology, Calgary, 6–8 November.
212. Chabot, L. *et al.:* "Single-Well Imaging Using the Full Waveform of an Acoustic Sonic," paper BH 3.7, *Expanded Abstracts,* 2001 Annual Meeting Technical Program, SEG, 420–423.
213. Yamamoto, H. *et al.:* "Borehole Acoustic Reflection Survey Experiments in Horizontal Wells for Accurate Well Positioning," paper SPE 65538 presented at the 2000 SPE/CIM International Conference on Horizontal Well Technology, Calgary, 6–8 November.
214. Tang, X.M.: "Imaging Near-Borehole Structure Using Directional Acoustic-Wave Measurement," *Geophysics* (2004) **69**, No. 6, 1378–1386.
215. Zheng, Y., and Tang, X.: "Imaging Near-Borehole Structure Using Acoustic Logging Data with Prestack F-K Migration," paper BG 2, *Expanded Abstracts,* 2005 Annual Meeting Technical Program, SEG, 360–363.
216. Mari, J.-L., Glangeaud, F., and Coppens, F.: *Signal Processing for Geologists and Geophysicists*, Editions Technip, Paris (1999) 1–458.
217. Neidell, N.S. and Taner, M.T.: "Semblance and Other Coherency Measures for Multichannel Data," *Geophysics* (1971) **36,** No. 3, 482–497.
218. Kimball, C.B. and Marzetta, T.M.: "Semblance Processing of Borehole Acoustic Array Data," *Geophysics* (1984) **49,** No. 3, 274–281.
219. Kimball, C.V.: "Shear Slowness Measurements by Dispersive Processing of the Borehole Flexural Mode," *Geophysics* (1998) **63,** No. 2, 337–344.
220. McFadden, P.L., Drummond, B.J., and Kravis, S.: "The *N*th-Root Stack—Theory, Applications, and Examples," *Geophysics* (1986) **51,** No. 10, 1,879–1,892.
221. Kravis, S.: "The *N*th Root Slant Stack—A New Method of Coherency Enhancement," *First Break* (1990) **8,** No. 9, 339–344.
222. Brie, A. *et al.:* "Shear Slowness Determination from Dipole Measurements," paper F, *Trans.,* 1997 Annual Logging Symposium, SPWLA, 1–14.
223. Hsu, K. and Chang, S.K.: "Multiple-Shot Processing of Array Sonic Waveforms for High Resolution Sonic Logs," *Geophysics* (1987) **52,** No. 10, 1,376–1,390.
224. Zhang, T., Tang, X.M., and Patterson, D.: "Evaluation of Laminated Thin Beds in Formations Using High-Resolution Acoustic Slowness Logs," paper XX, *Trans.,* 2000 Annual Logging Symposium, SPWLA, 1–14.
225. Wang, Z.: "Seismic Anisotropy in Sedimentary Rocks, Part 1—A Single-Plug Laboratory Method; Part 2—Laboratory Data," *Geophysics* (2002) **67,** No. 5, 1415–1440.

226. Crampin, S. and Chastin, S.: "A Review of Shear Wave Splitting in the Crack-Critical Crust," *Geophysical J. Intl.* (2003) **155**, 221–240.
227. Helbig, K. and Thomsen, L.: "75-plus Years of Anisotropy in Exploration and Reservoir Seismics: A Historical Review of Concepts and Methods," *Geophysics* (2005) **70,** No. 6, 9ND–23ND.
228. Franco, J.L.A.: "Using Shear-Wave Anisotropy to Optimize Reservoir Drainage and Improve Production in Low-Permeability Formations in the North of Mexico," paper SPE 96808 presented at the 2005 SPE Annual Technical Conference and Exhibition, Dallas, 9–12 October.
229. Klimentos, T.: "Shear-Wave Anisotropy Applications for Perforation Strategy and Production Optimization in Oil Bearing Porous Sands," paper LL, *Trans.,* 2003 Annual Logging Symposium, SPWLA, 1–12.
230. Klimentos, T., Farghaly, A., and Qleibo, M.: "Finding Faults with Shear-Wave Anisotropy," paper F, *Trans.,* 2003 Annual Logging Symposium, SPWLA, 1–12.
231. Furre, A.-K. and Brevik, I.: "Characterization of Angle Dependency in Sonic Logs," paper BH 2.1, *Expanded Abstracts,* 1998 Annual Meeting Technical Program, SEG, 292–295.
232. Hornby, B., Howie, J., and Ince, D.: "Anisotropy Correction for Deviated Well Sonic Logs—Application to Seismic Well Tie," paper BH/RP 4.7, *Expanded Abstracts,* 1999 Annual Meeting Technical Program, SEG, 112–115.
233. Wang, Z.: "Seismic Anisotropy in Sedimentary Rocks," paper RP 2.4, *Expanded Abstracts,* 2001 Annual Meeting Technical Program, SEG, 1740–1743.
234. Sun, Y. *et al.:* "Effects of Stress-Induced Anisotropy on Monopole and Dipole Logging," paper RBG P1.2, *Expanded Abstracts,* 2003 Annual Meeting Technical Program, SEG, 1282–1285.
235. Koster, K. *et al.:* "Applied Production Geophysics Using Shear-Wave Anisotropy—Production Applications for the Dipole Shear Imager and the Multicomponent VSP," paper DP1.1, *Expanded Abstracts,* 1994 Annual Meeting Technical Program, SEG, 233–235.
236. Esmersoy, C. *et al.:* "Fracture and Stress Evaluation Using Dipole-Shear Anisotropy Logs," paper J, *Trans.,* 1995 Annual Logging Symposium, SPWLA, 1–12.
237. Tang, X.: "Processing Dipole Waveform Logs for Formation Alteration Identification," paper BG 3.4, *Expanded Abstracts,* 1996 Annual Meeting Technical Program, SEG, **1,** 162–165.
238. Huang, X. *et al.:* "Formation Stress Estimation Using Standard Acoustic Logging," paper BH/RP 2.6, *Expanded Abstracts,* 1999 Annual Meeting Technical Program, SEG, 53–56.
239. Tang, X., Cheng, N., and Cheng, A.: "Identifying and Estimating Formation Stress from Borehole Monopole and Cross-Dipole Acoustic Measurements," paper QQ, *Trans.,* 1999 Annual Logging Symposium, SPWLA, 1–14.
240. Badri, M., Sousa, S., and Klimentos, T.: "Shear Anisotropy Applications in Production Optimization, Western Desert, Egypt," paper RPB 1.5, *Expanded Abstracts,* 2000 Annual Meeting Technical Program, SEG, 1695–1698.
241. Tang, X., Patterson, D., and Markovic, M.: "Fracture Measurements in Open/Cased Holes Using Cross-Dipole Logging—Theory and Field Results," paper RPB 1.6, *Expanded Abstracts,* 2000 Annual Meeting Technical Program, SEG, 1699–1702.
242. Tang, X.M.: "Determining Shear-Wave Transverse Isotropy from Borehole Stoneley Waves," *Geophysics* (2003) **68**, No. 1, 118–126.
243. Cheng, N. and Cheng, C.H.: "Estimations of Formation Velocity, Permeability, and Shear-Wave Velocity Using Acoustic Logs," *Geophysics* (1996) **61,** No. 2, 437–443.
244. Murray, D., Plona, T., and Valero, H.P.: "Case Study of Borehole Sonic Dispersion Curve Analysis," paper BB, *Trans.,* 2004 Annual Logging Symposium, SPWLA, 1–14.
245. Sinha, B. *et al.:* "Near-Wellbore Alteration and Formation Stress Parameters Using Borehole Sonic Data," paper SPE 95841 presented at the 2005 SPE Annual Technical Conference and Exhibition, Dallas, 9–12 October.
246. Plona, T.J. *et al.:* "Measurement of Stress Direction and Mechanical Damage Around Stressed Boreholes Using Dipole and Microsonic Techniques," paper SPE 47234 presented at the 1998 SPE/ISRM Rock Mechanics in Petroleum Engineering Conference, Trondheim, Norway, 8–10 July.
247. Plona, T. *et al.:* "Stress-Induced Dipole Anisotropy—Theory, Experiment and Field Data," paper RR, *Trans.,* presented at the 1999 Annual Logging Symposium, SPWLA, 1–14.
248. Plona, T.J. *et al.:* "Using Acoustic Anisotropy," paper H, *Trans.,* 2000 Annual Logging Symposium, SPWLA, 1–12.

249. Sinha, B.K., Kane, M.R., and Frignet, B.: "Dipole Dispersion Crossover and Sonic Logs in Limestone Reservoir," *Geophysics* (2000) **65,** No. 2, 390–407.
250. Xu, S., Wu, X., Huang, X., and Yin, H.: "Evaluation of Anisotropic Rock Properties in Sedimentary Rocks from Well Logs," paper OTC 17251 presented at the 2005 Offshore Technology Conference, Houston, 2–5 May.
251. Crampin, S.: "Evaluation of Anisotropy by Shear-Wave Splitting," *Geophysics* (1985) **50,** No 1, 142–152.
252. Rai, C.S. and Hanson, K.E.: "Shear-Wave Velocity Anisotropy in Sedimentary Rocks—A Laboratory Study," *Geophysics* (1988) **53,** 800–806.
253. Yale, D.P. and Sprunt, E.S.: "Prediction of Fracture Direction Using Shear Acoustic Anisotropy," *The Log Analyst* (1989) **30,** No. 2, 65–70.
254. Winkler, K.W.: "Azimuthal Velocity Variations Caused by Borehole Stress Concentrations," *J. of Geophysical Research* (1996) **101,** No. B4, 8,615–8,621.
255. Esmersoy, C. *et al.:* "Dipole Shear Anisotropy Logging," paper SL 3.7, *Expanded Abstracts,* 1994 Annual Meeting Technical Program, SEG, 1,139–1,142.
256. Hatchell, P.J. *et al.:* "Quantitative Comparison Between a Dipole Log and VSP in Anisotropic Rocks from Cymric Oil Field, California," paper BG1.4, *Expanded Abstracts,* 1995 Annual Meeting Technical Program, SEG, 13–16.
257. Beckham, W.E.: "Seismic Anisotropy and Natural Fractures from VSP and Borehole Sonic Tools —A Field Study," *Geophysics* (1996) **61,** No. 2, 456–466.
258. Walls, J.D., Dvorkin, J., and Mavko, G.: "In-Situ Stress Magnitude and Azimuth from Well Log Measurements, Final Report," Report GRI-95/0356, Gas Research Inst., Chicago (1996) 1–147.
259. Grandi, S.K. *et al.:* "In Situ Stress Modeling at a Borehole—A Case Study," paper BH 1.4, *Expanded Abstracts,* 2003 Annual Meeting Technical Program, SEG, 297–300.
260. De, G.S. and Schmitt, D.P.: "Issues with Shear-Wave Azimuthal Anisotropy in Highly Deviated Wells," paper OTC 17647 presented at the 2005 Offshore Technology Conference, Houston, 2–5 May.
261. Tang, X.M. and Patterson, D.: "Characterizing Seismic Anisotropy Using Cross-Dipole Measurement in Deviated Wells," paper BG 2.5, *Expanded Abstracts,* 2005 Annual Meeting Technical Program, SEG, 372–375.
262. Patterson, D.J. and Tang, X.: "Pit Falls in Dipole Logging—Anisotropy: Cause of Discrepancy in Borehole Acoustic Measurements," paper OTC 17644 presented at the 2005 Offshore Technology Conference, Houston, 2–5 May.
263. Hornby, B.E., Howie, J.M., and Ince, D.W.: "Anisotropy Correction for Deviated-Well Sonic Logs—Application to Seismic Well Tie," *Geophysics* (2003) **68**, No. 2, 464–471.
264. Alford, R.M.: "Shear Data in the Presence of Azimuthal Anisotropy—Dilley, Texas," paper S9.6, *Expanded Abstracts,* 1986 Annual Technical Meeting Program, SEG.
265. Tang, X.M. and Chunduru, R.K.: "Simultaneous Inversion of Formation Shear-Wave Anisotropy Parameters from Cross-Dipole Acoustic Array Waveform Data," *Geophysics* (1999) **64,** No. 5, 1,502–1,511.
266. Gelinsky, S. *et al.:* "Anisotropic Permeability in Fractured Reservoirs from Integrated Acoustic Measurements," paper RC 3.5, *Expanded Abstracts,* 1998 Annual Meeting Technical Program, SEG, 956–959.
267. Smith, M.B.: "Effect of Fracture Azimuth on Production with Application to the Wattenburg Gas Field," paper SPE 8298 presented at the 1979 SPE Annual Technical Conference and Exhibition, Las Vegas, Nevada, 23–26 September.
268. Hubbert, M.K. and Willis, D.G.: "Mechanics of Hydraulic Fracturing," *Trans.,* AIME (1957) **210,** 153–166.
269. Cipolla, C.L., Liu, D., and Kyte, D.G.: "Practical Application of In-Situ Stress Profiles," paper SPE 28607 presented at the 1994 SPE Annual Technical Conference and Exhibition, New Orleans, 25–28 September.
270. Cipolla, C.L.: "Hydraulic Fracture Technology in the Ozona Canyon and Penn Sands," SPE 35196 presented at the 1996 SPE Permian Basin Oil and Gas Recovery Conference, Midland, Texas, 27–29 March.
271. De, G.S. *et al.:* "Predicting Natural or Induced Fracture Azimuths from Shear-Wave Anisotropy," *SPEREE* (August 1998) 311.

272. Garg, A., Desai, A.M., and Towler, B.F.: "Horizontal Stresses in Anisotropic Formation and Prediction of Hydraulic Fracture Direction," paper SPE 38342 presented at the 1997 SPE Western Regional Meeting, Long Beach, California, 25–27 June.
273. Aguila, F.J.: "Using Cross-Dipole Sonic-Anisotropy Data to Improve Reservoir Understanding in the Southern/Marine Areas of Mexico," paper SPE 84204 presented at the 2004 SPE International Symposium and Exhibition on Formation Damage Control, Lafayette, Louisiana, 18–20 February.
274. Barajas, J.: "Case History—Cased Hole Dipole Sonic Applications in Mexico," paper SPE 90703 presented at the 2004 SPE Annual Technical Conference and Exhibition, Houston, 26–29 September.
275. Klimentos, T. and McCann, C.: "Relationships Between Compressional Wave Attenuation, Porosity, Clay Content, and Permeability in Sandstones," *Geophysics* (1990) **55,** 998–1,014.
276. Klimentos, T.: "Attenuation of P- and S-Waves as a Method of Distinguishing Gas from Oil and Water," *Geophysics* (1995) **60,** No. 2, 447–458.
277. Akbar, N., Dvorkin, J., and Nur, A.: "Relating P-Wave Attenuation to Permeability," *Geophysics* (1993) **58,** No. 1, 20–29.
278. Sun, X. *et al.*: "P- and S-Wave Attenuation Logs from Monopole Sonic Data," *Geophysics* (2000) **65,** No. 3, 755–765.
279. Burns, D.R.: "Predicting Relative and Absolute Variations of In-Situ Permeability from Full-Waveform Acoustic Logs," *The Log Analyst* (1991) **32,** No. 3, 246–255.
280. Tang, X.M., Cheng, C.H., and Toksoz, M.N.: "Dynamic Permeability and Borehole Stoneley Waves—A Simplified Biot-Rosenbaum Model," *J. of the Acoustical Soc. of America* (1991) **90,** No. 3, 1,632–1646.
281. Tang, X., Cheng, C.H., and Toksoz, M.N.: "Stoneley-Wave Propagation in a Fluid Filled Borehole with a Vertical Fracture," *Geophysics* (1991) **56,** 447–460.
282. Saxena, V.: "Permeability Quantification from Borehole Stoneley Waves," paper T, *Trans.,* 1994 Annual Logging Symposium, SPWLA, 1–22.
283. Kimball, C.V. and Endo, T.: "Quantitative Stoneley Mobility Inversion," paper BH 1.1, *Expanded Abstracts,* 1998 Annual Meeting Technical Program, SEG, 252–255.
284. Tang, X.M. *et al.*: "Permeability from Borehole Acoustic Logs—An Overview with Recent Advances," paper BH 2.6, *Expanded Abstracts,* 1997 Annual Meeting Technical Program, SEG, 274–277.
285. Sinha, A. *et al.*: "A New Method for Deriving Permeability from Borehole Stoneley Waves and Its Application in the North Mongas Field of Eastern Venezuela," paper P, *Trans.,* 1998 Annual Logging Symposium, SPWLA, 1–12.
286. Canaday, W., Spooner, P., and Vasquez, R.: "Permeability Estimation from Stoneley Amplitude, Corrected for Borehole Geometry and Rugosity," paper SPE 96598 presented at the 2005 SPE Annual Technical Conference and Exhibition, Dallas, 9–12 October.
287. Geerits, T. *et al.*: "Comparison Between Stoneley, NMR, and Core-Derived Permeabilities," paper TT, *Trans.,* 1999 Annual Logging Symposium, SPWLA, 1–14.
288. Qobi, L. *et al.*: "Permeability Determination from Stoneley Waves in the Ara Group," paper SPE 63140 presented at the 2000 SPE Annual Technical Conference and Exhibition, Dallas, 1–4 October.
289. Tang, X.M., and Patterson, D.: "Estimating Formation Permeability and Anisotropy from Borehole Stoneley Waves," paper W, *Trans.*, 2004 Annual Logging Symposium, SPWLA, 1–14.
290. Tang, X.M., Altunbay, M., and Shorey, D.: "Joint Interpretation of Formation Permeability from Wireline Acoustic NMR, and Image Log Data," paper KK, *Trans.,* 1998 Annual Logging Symposium, SPWLA, 1–13.
291. Aladeddin, E., Tchambaz, M., and Al-Adani, N.: "Combining NMR and Stoneley Analysis for a Better Estimation of Permeability in Carbonate Reservoirs," paper E, *Proceedings,* 2004 Formation Evaluation Symposium of Japan, Society of Petrophysicists and Well Log Analysts, Japan Chapter, 7 p.
292. Tang, X.M. and Patterson, D.: "Detecting Thin Gas Beds in Formations Using Stoneley Wave Reflection and High-Resolution Slowness Measurements," paper OO, *Trans.,* 2001 Annual Logging Symposium, SPWLA, 1–14.

293. Hornby, B.E. *et al.:* "Fracture Evaluation Using Reflected Stoneley-Wave Arrivals," *Geophysics* (1989) **54,** No. 10, 1,274–1,288.
294. Medlin, W.L.: "Fracture Diagnostics with Tube-Wave Reflection Logs," *JPT* (March 1994) 239.

SI Metric Conversion Factors

dB/ft	×	3.048*	E − 01 = dB/m
ft	×	3.048*	E − 01 = m
ft/sec	×	3.048*	E − 01 = m/sec
in.	×	2.54*	E + 00 = cm
psi/ft	×	2.262 059	E + 01 = kPa/m
μsec/ft	×	3.280 840	E + 00 = μsec/m

*Conversion factor is exact.

Chapter 3D
Nuclear Logging

Gary D. Myers, ConocoPhillips Inc.

3D.1 Introduction

Nuclear logging has been used in some form since the late 1920s to provide information on lithology and rock characteristics. Continued technological advances have provided improved methods for analyzing the measurements of natural and induced nuclear readings. Even with better tool designs, the long-standing problem remains that logging tools do not directly measure the formation properties that engineers, geologists, and petrophysicists need to describe a reservoir. The goal of log analysis is to map out the downhole values of reservoir characteristics chiefly as porosity, fluid saturations, and permeability. Unfortunately, nuclear-logging tools only measure gamma ray or neutron count rates at cleverly positioned detectors. Inference, empiricism, experience, and alibis bridge these count rates to the rocks and fluids in the reservoir. Nuclear-log interpretation rests on smarter processing of these tool readings. Understanding what the tools really measure is the key to better log analysis.

Consider some of the limitations of the current technology. Grouping nuclear logs according to their underlying nuclear physics highlights the blurry relationship between what they measure and what we expect from them. **Table 3D.1** summarizes such a classification scheme. Two types of problems skew tool measurements away from their targets. First, because a nuclear tool averages over a shallow bulk volume, the borehole often represents a major part of the tool's response. Second, even if all borehole effects can be removed, the fact remains that nuclear tools do not respond directly to reservoir properties. Sometimes, the reservoir parameter of interest does not even dominate the underlying physics of the tool. Historically, such problems have been addressed with calibrations at a few points accessible in the laboratory; these are then generalized into correction charts. Two books[1,2] serve as excellent general introductions to the convoluted physics of logging tools.

Nuclear logs work because gamma rays and neutrons are penetrating radiation. Unlike visible light, they can traverse tool housings and boreholes and still sample a significant volume of the formation. They can even penetrate casing, giving them a near monopoly in cased-hole formation evaluation. They also respond to different properties than resistivity logs, which merely measure the conductivity of a formation.

To exploit a reservoir, the engineer must characterize it. That basically means building an understanding of two things: the amount and distribution of hydrocarbons and the recoverabili-

TABLE 3D.1—SELECTION OF COMMON NUCLEAR-LOGGING TOOLS: HOW THEY WORK AND WHAT LOG ANALYSTS REALLY WANT FROM THEM TO DESCRIBE HYDROCARBON RESERVOIRS

Tool	Physics Exploited	Reads	Desired Information
Gamma ray	Natural gamma rays	Gamma ray counts, K-U-Th counts	Clay type and bedding
Gamma density	Gamma ray scattering	Bulk Compton cross section	Formation density, porosity
Neutron porosity	Neutron slowing	Migration length	Porosity, lithology
Pulsed neutron	Neutron lifetime	Bulk capture cross section	S_w, porosity
Carbon/oxygen, geochemical	Neutron-induced gamma ray	Elemental count rates	S_0, lithology

ty of those hydrocarbons. Amount and distribution starts with a description of hydrocarbon volume in place. To the first order, this means bulk volume hydrocarbon:

$$V_{bh} = \phi\left(1 - S_w\right)h_V A\ . \quad \text{(3D.1)}$$

Estimating porosity requires detailed knowledge of rock fabric, one of the primary uses of nuclear logs. Rock-fabric information runs the gamut from primary lithology (e.g., sandstone vs. limestone) to diagenesis to clay volume and distribution. Nuclear logs can provide estimates of bulk formation properties such as density and hydrogen content. With some geologic insight and simplifications, these bulk properties can be related to reservoir fabric through simple bulk mixing laws. One of the virtues of nuclear tools is that they are bulk-averaging devices. They average formation properties over a volume on the order of 1 ft^3. While nuclear logs are much less sensitive to the difference between water and liquid hydrocarbons than resistivity logs, they are very sensitive to the difference between liquid- and gas-phase fluids.

The second class of reservoir properties, the actual production of hydrocarbons, is less directly accessible to nuclear measurements. Producibility depends on two broad factors: heterogeneity and permeability. At some level, log measurements can give insight into heterogeneity. Laterally, multiple wells can be correlated and overlaid on seismic cross sections to track the continuity of layers of similar properties. Logs certainly can speak to vertical continuity, at least within the limitations of their vertical resolutions. Nuclear logs do provide the best vertical resolution of any of the standard suite of log measurements (as fine as 6 in. for gamma ray-based measurements). Bedding thinner than that can be assessed only with some sort of special borehole-imaging log. In electrical logs, thin bedding may manifest itself as anisotropy, but nuclear logs' bulk-averaging nature removes most sensitivity to the detailed internal structure of the volume they investigate. This also means that nuclear logs cannot speak directly to permeability because that does depend on the microscopic details of grain shape and size, the arrangement of the grains, clay minerals, and their distributions, nor can nuclear logs discriminate secondary porosity in vugs or fractures from primary intergranular porosity. In the end, nuclear-log interpretation is a matter of model choice as much as tool reading.

3D.2 The Physics of Nuclear Logs

3D.2.1 Nuclear Measurements and Statistics. For logging purposes, all nuclear radiation behaves as particles, and all nuclear-log measurements are particle-counting experiments. There is randomness to the arrival of the particles, so accurate count-rate measurements need to be very long-term averages. As a result, fluctuations in radioactive logs may be statistical rather than the result of a real change in formation properties. This accounts for the apparent chatter in most nuclear logs. Because of this, nuclear logs almost always set the maximum logging speed of a simple quad-combo logging suite [i.e., a tool string with no specialty logs like borehole images, nuclear magnetic resonance (NMR), or waveform sonic].

Counting experiments obey Poisson statistics. A Poisson distribution, as **Fig. 3D.1** shows, characterizes such experiments. If the expected value of the number of counts received in a given period of time is μ, then the probability of obtaining a particular number of counts, x, on a given repetition of the experiment is given by

$$P_x = \mu^x \frac{e^{-\mu}}{x!} \quad \text{(3D.2)}$$

This is the familiar bell-shaped curve, with its width or standard deviation, σ, given by

$$\sigma = \sqrt{\mu} \quad \text{(3D.3)}$$

(Generally, a Poisson distribution applies to discrete events like nuclear counting experiments, while a Gaussian or normal distribution applies to continuous properties like the length of a rod; shapes and moments are nearly identical.) This means that unlike sonic- or resistivity-log measurements, nuclear-log measurements do not have a fixed precision. The precision of a nuclear measurement depends on the number of counts received. If the expected value of the nuclear counts is N, then approximately 32% of all attempts to measure N will fall outside the range $N \pm \sqrt{N}$. Realizing that the number of counts is simply the count rate times the integration time, to improve the precision of a nuclear-log measurement by a factor of two, one must count four times longer. Log measurements are depth-based (i.e., measurements are made approximately twice per foot). To count four times longer, one needs to log four times slower (e.g., 5 ft/min instead of 20).

3D.3 Nuclear Radiation Transport

Nuclear logs are based on the interaction of nuclear radiation with matter—materials like sand, clay, water, and hydrocarbons that together make up a reservoir. For logging, the interactions are primarily particle-scattering interactions. Even though gamma rays are usually discussed as electromagnetic radiation, for nuclear logging they are treated as photons—classical particles. Even in well logging, quantum mechanics rears its head.

Well logs exploit two types of nuclear radiation: gamma rays and neutrons. Depending on the type and energy of the particle, different scattering processes predominate. **Fig. 3D.2** shows a beam of particles impinging on a slab of formation from the left. Particle beams are characterized in terms of their flux, which has units of particles per unit area per unit time. The incoming flux is labeled ϕ_i. For scattering, the slab is characterized by the number density of potential scattering particles within it (in other words, the number of atoms per cm^3). Suppose the slab in the figure has an average of N_p atoms/cm^3. If the slab is h cm thick, a beam of unit area will encounter and have a chance to interact with $N_p \times h$ atoms as it passes through the slab. The actual probability of a given radiation particle interacting with a given atom in the slab of formation depends on a number of factors, including the nature and energy of the radiation and the characteristics of the target atom. Physicists lump these probabilities as cross sections, typically labeled $\sigma(E)$, where E refers to the energy dependence of the cross section. A cross section has the units of area because it corresponds to the apparent size of the scattering target as seen by the incoming particle. Because these are atomic-scale interactions, the apparent target sizes are on the order of 10^{-29} cm^2. Because humans relate best to numbers

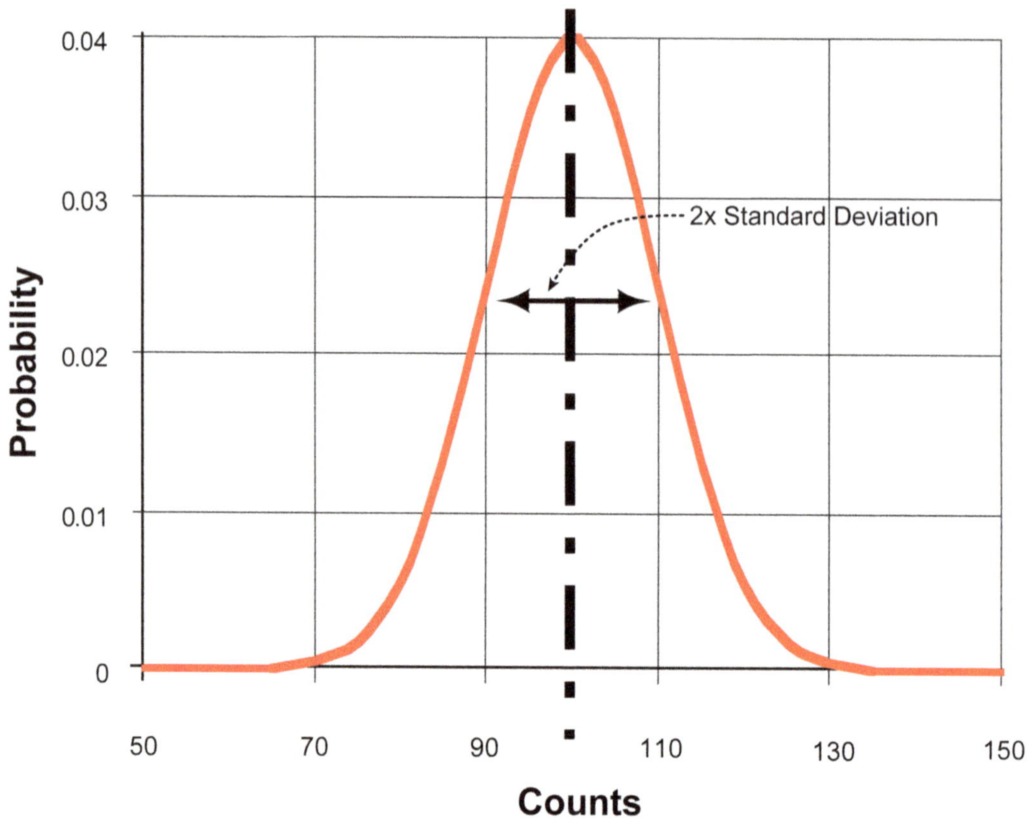

Fig. 3D.1—The Poisson distributions capture how the statistical nature of nuclear counting measurements influences the precision of such measurements.

they can count on their fingers, a special unit of cross section, the barn, equal to 10^{-24} cm^2, was created.

The number of particles that will be scattered out of the original beam of radiation as it passes through the slab of formation can be written as

$$\delta\phi = \phi_i N_p \sigma(E) h \; . \quad \text{(3D.4)}$$

$\phi_i N_p \sigma$ is interpreted as a reaction rate per unit volume of formation resulting from the incident beam of radiation. Number density can be calculated from bulk density, ρ_b, average atomic weight, A, and Avogadro's number, N_A, according to the formula

$$N_p = \frac{N_A}{A}\rho_b \; . \quad \text{(3D.5)}$$

In practice, cross sections are measured experimentally and tabulated as a function of energy for various reaction types and target nuclei. The discussion above is simplified to a slab of formation made up of a single type of atom. For real formations with a variety of atoms, the actual amount of scattering is just a volume-weighted sum of the various atoms in the formation. Many times, as **Tables 3D.2 through 3D.4** show, only one type of atom will account for the vast majority of the scattering. This is in fact the basis of nuclear logging.

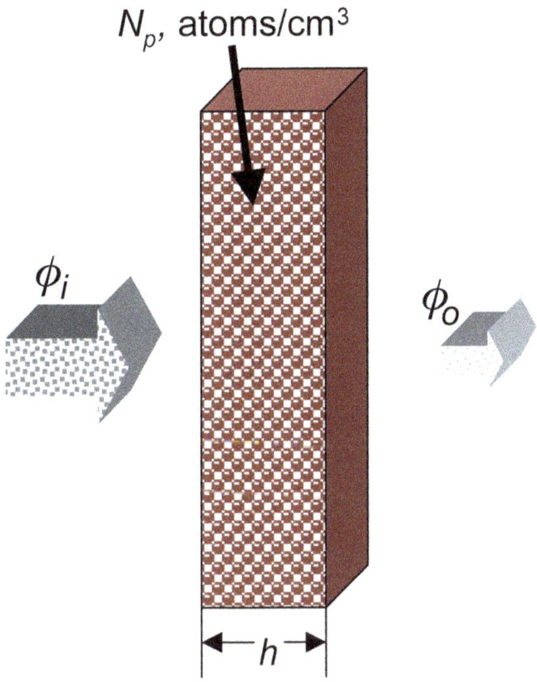

Fig. 3D.2—A schematic of a beam of radiation passing through a uniform slab of material.

TABLE 3D.2—NEUTRON SLOWING-DOWN POWER OF COMMON RESERVOIR ELEMENTS, EXPRESSED AS THE NUMBER OF COLLISIONS REQUIRED TO THERMALIZE A NEUTRON (i.e., TO SLOW IT DOWN FROM 4 MeV TO LESS THAN 0.1 MeV)

Element	Number of Collisions to Thermalize
Hydrogen	18
Carbon	114
Oxygen	150
Silicon	257
Chlorine	329
Calcium	368

Integrating the differential flux equation, Eq. 3D.4, produces the unscattered flux that emerges through a thickness of formation, *h*, as Fig. 3D.2 shows.

$$\phi_o = \phi_i e^{-N_p \sigma h} \quad \text{(3D.6)}$$

This leads naturally to the concept of mean free path. The mean free path, λ, is the thickness of formation that will reduce a beam of radiation to $1/e$ (approximately 37%) of its original value.

TABLE 3D.3—CAPTURE CROSS SECTION FOR SLOW (THERMAL-ENERGY) NEUTRONS BY SOME COMMON FORMATION ELEMENTS

Element	Capture Cross Section (barns)
Gadolinium	40,000
Boron	771
Lithium	71
Chlorine	33.4
Manganese	13.3
Titanium	6.4
Copper	3.8
Potassium	2.2
Iron	2.56
Calcium	0.44
Hydrogen	0.332
Aluminum	0.232
Silicon	0.16
Magnesium	0.064
Carbon	0.0034
Oxygen	0.0002

TABLE 3D.4—HYDROGEN CONTENT OF SOME COMMON EARTH ELEMENTS

Material	Hydrogen Density (atm/cm^3 × 10^{23})	Hydrogen Index
Water (60°F, 14.7 psi)	0.669	1.000
Water (200°F, 2,000 psi)	0.667	1.000
Brine (200 kppm NaCl)	0.614	0.920
Methane (60°F, 14.7 psi)	0.001	0.002
Methane (200°F, 2,000 psi)	0.329	0.490
n-octane (60°F, 14.7 psi)	0.667	1.000
n-octane (200°F, 2,000 psi)	0.639	0.960
Gypsum	0.325	0.490
Anthracite	0.268	0.400
Kaolinite	0.250	0.370
Chlorite	0.213	0.320
Illite	0.059	0.090

$$\lambda = \frac{1}{N_p \sigma} \quad \text{(3D.7)}$$

It depends on the amount of material in the formation and its cross section. The mean free path of radiation in a formation determines its depth of investigation and its vertical resolution.

Density tools and neutron-porosity tools simply measure the drop-off in radiation with distance from the source of radiation. While this can be done with a single detector at an appropriate distance from a source of known strength, all modern tools use at least two detectors at different distances from the source. These designs are referred to as compensated. With a near and a far detector, it is possible for the tool designer to compensate for the borehole effects and variations in source strength.

Nuclear-logging tools exploit only two types of radiation: gamma rays and neutrons. Both follow the basic scattering principles defined above but have unique reaction types and cross sections.

3D.4 Single-Log Interpretation

Nuclear log interpretation is simply the practice of solving tool-response mixing-law equations with the judicious application of some assumptions and constraints.

3D.4.1 The Generalized Interpretation Process. All interpretation is an approximate model. As more factors are taken into account, the interpretation usually improves, but the model becomes more complicated.

For the neutron-porosity log, the simplest interpretation model is to naively accept the raw log reading. This is acceptable in a clean, water-filled reservoir, with a single known matrix lithology:

$$\phi_o = \phi_{\text{CNL}} \text{ .} \qquad (3\text{D}.8)$$

If the reservoir is shaly, or if the fluid density is not the same as water, a hydrogen-index linear-mixing law will generally do.

$$\begin{aligned}\phi_{\text{CNL}} &= \phi_{\text{CNLma}} \times \left(1 - V_{\text{sh}} - \phi_e\right) + \phi_{\text{CNLsh}} \times V_{\text{sh}} \\ &+ \phi_e \times \left[I_{H\_hc} \times \left(1 - S_w\right) + I_{H\_w} \times S_w\right] \text{ .}\end{aligned} \qquad (3\text{D}.9)$$

This equation can be solved for ϕ_e given the neutron tool response to each of the various formation components. If S_w is not known and the hydrogen index of the hydrocarbon and connate water phases differ appreciably, it may be necessary to solve for porosity and fluid content simultaneously or at least iteratively.

Invasion adds the next level of complexity. Depending on the degree of invasion, the analyst may need to generate an average fluid hydrogen index, weighted by the invasion diameter and the corresponding radial geometric function for the logging tool. If the porosity is gas-filled, the simple bulk mixing law fails because the neutron response becomes nonlinear. In that case, the next level of interpretation sophistication accounts for this nonlinearity through an "excavation effect," another term added to Eq. 3D.9.

As the lithology becomes more complex, the analyst can move to a macroparameter approach characterized by a neutron migration length. Finally, if the standard environmental correction charts do not cover borehole effects, a full-blown Monte Carlo model may be needed. Fortunately, this is a very rare circumstance.

3D.4.2 Single-Log Interpretation—More Details. Even though log analysts' primary interest is only in virgin formation properties, there are at least three regions that contribute to a nuclear log measurement: the borehole (including borehole size, composition, temperature, and even the tool body), the invaded zone, and the real target, the virgin formation. The tool will respond to a weighted average of all three. For nuclear tools, the weights given to the borehole and invaded zone may be large because they are closer to the sources and detectors.

Log analysts typically divide the problem into three corresponding parts and attempt to handle them sequentially. First, one compensates for borehole effects by applying semiempirical environmental corrections, either from chart books or equivalent computerized correction formulas. Second, one finesses the invaded-zone problem if possible. For liquids, the density and hydrogen index of the mud filtrate may be similar enough to those of the formation fluids that the effect of invasion can be ignored, and average fluid properties can be assigned with little damage to the accuracy of the calculated porosity. If all of the potential fluid properties are not similar enough, analysts frequently assume that a log responds primarily to either the invaded zone or the deeper, uninvaded zone. The gamma-gamma scattering density log's depth of invasion may be less than 6 in. Because of this very shallow penetration, it is usually safe to assume that the density tool responds only to the invaded zone. The neutron log is often problematic. With shallow invasion, it may primarily see the virgin formation. With deep invasion, even the neutron log may be responding only to the invaded zone. If the formation fluid is gas, the fluid density and hydrogen index differ wildly from those of a typical drilling mud. This is especially the case at shallow depths, where gas density is low. In this case, partial saturation may also feed into the solution of the log's response. While iterative solutions for light hydrocarbons frequently work well enough, the simultaneous solution of all the log responses (including resistivity logs and saturation equations) gives the best result in these cases. While there are numerous commercial simultaneous-solver computer programs, an interesting treatment, including the effects of invasion, was given by Patchett and Wiley.[3]

To put depths of investigation for nuclear tools in perspective, it is useful to introduce the concept of integrated radial geometric factors, or *J*-factors.[4] This is a method of standardizing the data from tools with different depths of investigation. The function is a measure of what fraction of a tool's response comes from inside a certain radius, *x,* defined by

$$J(x) = \frac{(U_x - U_T)}{(U_i - U_T)} \quad \text{(3D.10)}$$

Here, U_x refers to the tool response integrated out to some distance x into the formation, U_T is the tool's full response out to infinity in the absence of invasion, and U_i is the tool response to a fully invaded formation. Depth of investigation is commonly defined as the radial distance into the borehole wall at which the tool response reaches 90% of the final value.

Fig. 3D.3 compares the radial geometric response functions for the three basic nuclear logs. The radial geometric function is a quick, approximate contrivance for determining whether tool response is predominantly coming from the invaded zone or the virgin formation. The curve labeled "gr-reservoir" corresponds to a bulk density of 2.35 g/cm^3. For comparison, a deep induction log does not reach its 50% response point until approximately 150 radial in. The base case shown is for a 20% porosity limestone. Obviously, the depth of investigation varies with formation composition, which is, after all, the principle on which density and neutron tools are based. The depth of investigation of a density log ranges from 4.4 in. at 5% porosity to 5.0 in. at 40%. For compensated neutron logs, depth of investigation ranges from 9.5 in. at 40% porosity limestone to 16 in. at 2.5 p.u. (porosity unit or % porosity). Note that increasing porosity increases the depth of investigation of the density log but decreases the depth of investigation of the neutron log. This makes sense in light of the different dominant scattering processes for neutrons (as opposed to gamma rays). A passive gamma ray response function for 100% water is shown for comparison. Even though it is stretched somewhat, as expected, the difference is not nearly as large as between nuclear measurements, and a deep induction log with a 90% response point may be deeper than 20 ft.

No matter what the approach, the trick is estimating the invasion diameter so that it can be compared to the tool's depth of investigation. Invasion is a complicated function of mud

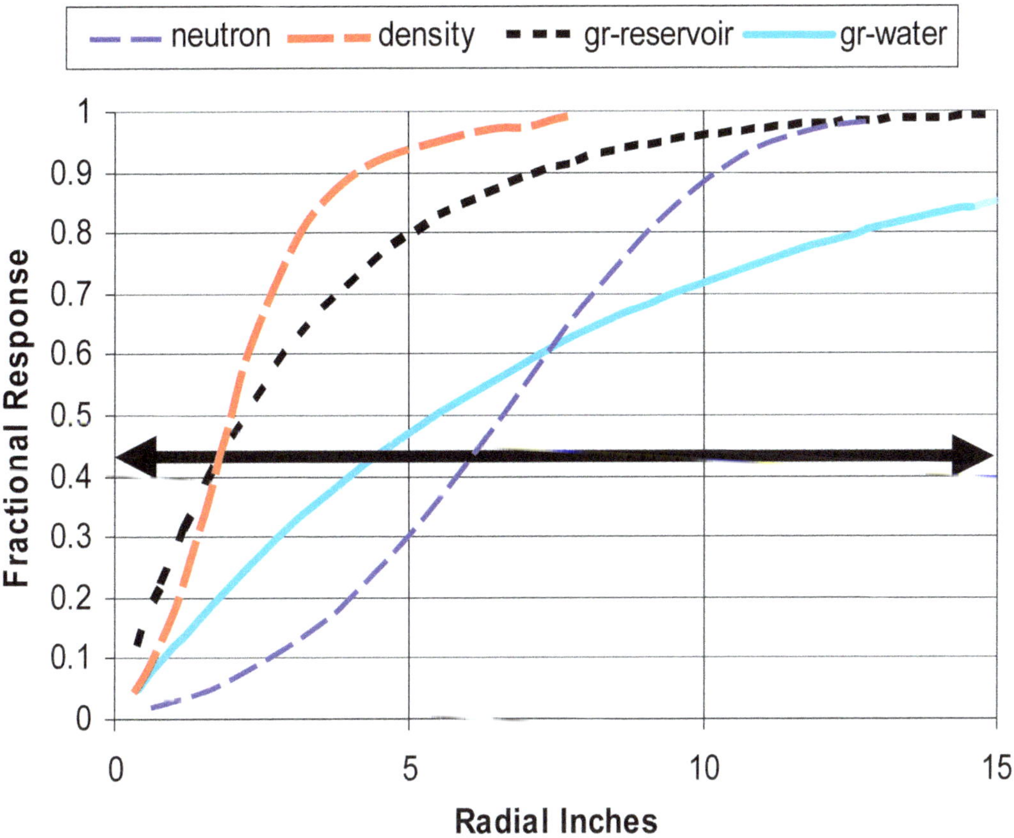

Fig. 3D.3—Nuclear radial geometric functions: comparison of radial geometric functions for various nuclear logs in a 20-p.u. limestone.

weight, mud composition, formation pressure, porosity, and permeability, so a quick and dirty estimate of its extent is rarely possible. If the standard three resistivity logs with three different depths of investigation have been run, a rough estimate of the diameter of invasion can be made from so-called tornado-chart calculations. In addition to refined estimates of the true formation resistivity and the invaded-zone resistivity, a diameter of invasion also will be derived. This is based on the assumption of piston displacement of formation fluids by drilling mud, as suggested schematically in **Fig. 3D.4**. This step-function invasion model is far from physically correct, but at least it is a step in the right direction. As a first hurdle, this diameter of invasion can be compared to the appropriate integrated radial geometric function to confirm (or contradict) the validity of assumptions that a log's response is predominantly from the invaded zone or the unadulterated formation. While modern array resistivity tools can produce much more detailed descriptions of the invaded zone, the response to an equivalent step-invasion profile is much more tractable. If a log's response includes significant elements of both zones (i.e., the diameter of invasion lies somewhere between the 10 and 90% points on the J-function), then its response can be weighted to the two regions. For step invasion, this is simply

$$R = J(x)I + [1 - J(x)]T, \qquad (3D.11)$$

where R = the tool reading, I = the tool's bulk response in the case of complete invasion, and T = the likely bulk response to no invasion.

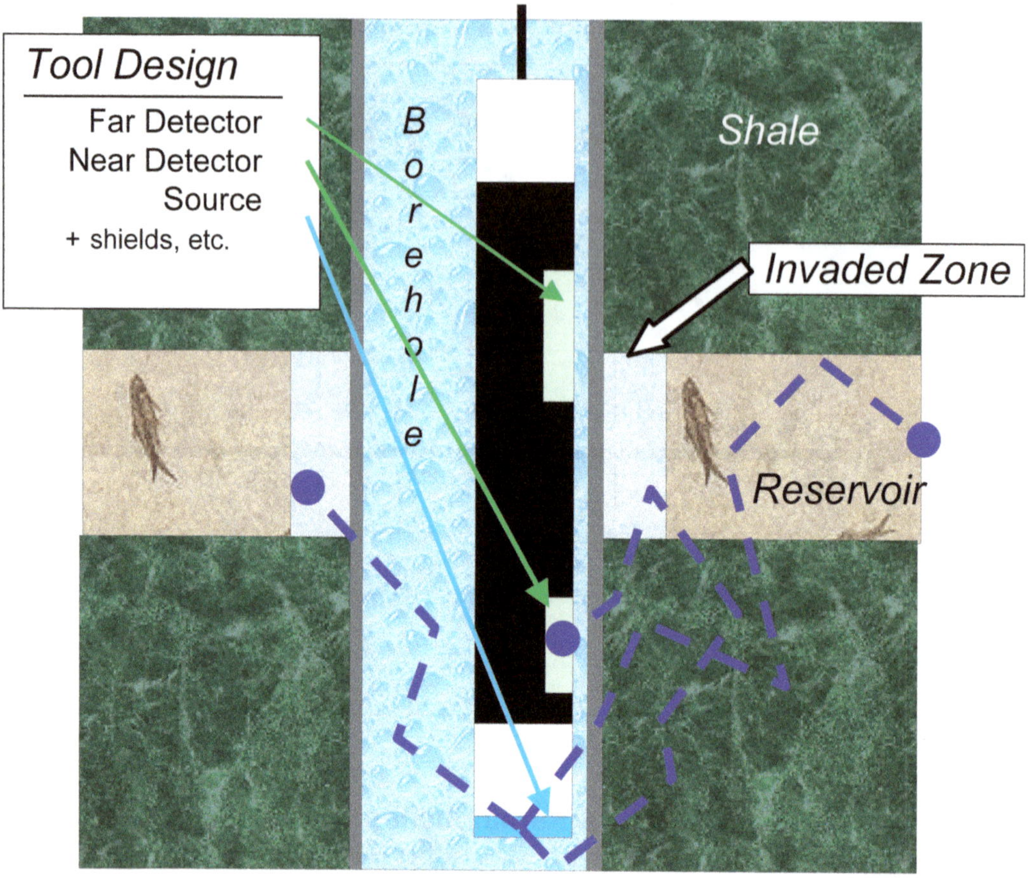

Fig. 3D.4—Monte Carlo modeling predicts tool response from a detailed description of geometry and composition.

The third and final step, and the one on which the log analyst spends the most time, is the understanding of the tool's bulk response to a formation, the T in Eq. 3D.11. Consider a small cube of formation. While the cube is heterogeneous on the microscopic scale, it can be treated as uniform on the scale of nuclear scattering. At the scale of 10 cm, the exact arrangement of crystalline rock and fluids does not matter, only the bulk average number of scattering centers of a given type. Tool response is determined by these bulk averages. Fortunately, these can be related to bulk average formation properties. **Fig. 3D.5** summarizes the process. Mathematically, this bulk averaging corresponds to a linear mixing law:

$$R = \sum_{i=1}^{n} V_i R_i, \quad \text{(3D.12)}$$

where R is the bulk tool response, V_i is the fractional volume of the ith material, R_i is the response to that material only, and n is the total number of materials present. For example, if water occupies 10% of the pore space and the porosity is 20%, the fractional volume of water (also known as bulk volume water) is $0.1 \times 0.2 = 0.02$. For a density, log R_i is the density of the ith component. Pure water has a density of 1.0 g/cm^3, so R of water is 1.0 g/cm^3. The

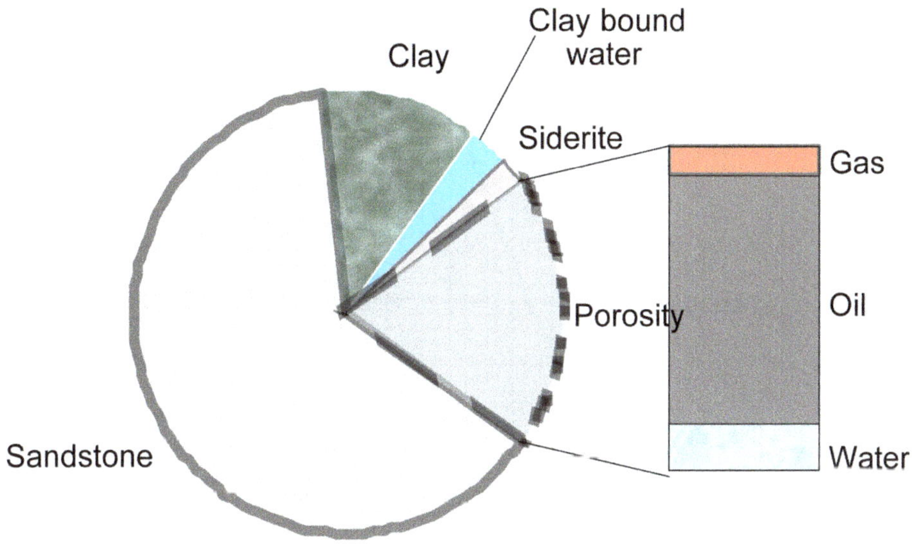

Fig. 3D.5—Bulk formation volumes and their properties are the key to the interpretation of nuclear logs.

contribution of that water to the tool response is $V_i \times R_i = 0.02 \times 1.0 = 0.02$ g/cm^3. Similarly, for a neutron-porosity log, R_i is the hydrogen index of material *i*.

More generally, the responses may not be linear, but there is still an equation or mixing law:

$$R = R(V_i,\ R_i)\ . \qquad (3D.13)$$

An example of this would be the nonlinear response of a neutron-porosity log to gas, referred to as the excavation effect or any of the numerous *ad hoc,* nonlinear gamma ray models.

Single-log interpretation amounts to the assumption that considering just two components (or at least only two at a time) can capture the log's response to a formation. Examples include the determination of shale volume or total porosity from a single log curve. As an example, consider calculating shale volume from a gamma ray curve. According to the linear mixing-law equation, the gamma ray tool's response can be written as

$$\gamma = \gamma_{sh} V_{sh} + \gamma_{ns} V_{ns}\ . \qquad (3D.14)$$

This seems to be one equation in two unknowns, but there is another, implicit equation, namely that the formation is composed entirely of shale and nonshale (or reservoir or clean sand); that is,

$$V_{sh} + V_{ns} = 1. \qquad (3D.15)$$

This really amounts to an assumption. Most importantly, it assumes that there is such a thing as "reservoir rock" characterized by a single bulk gamma ray response. Even in the simplest case, reservoir rock consists of matrix- and water-filled porosity.

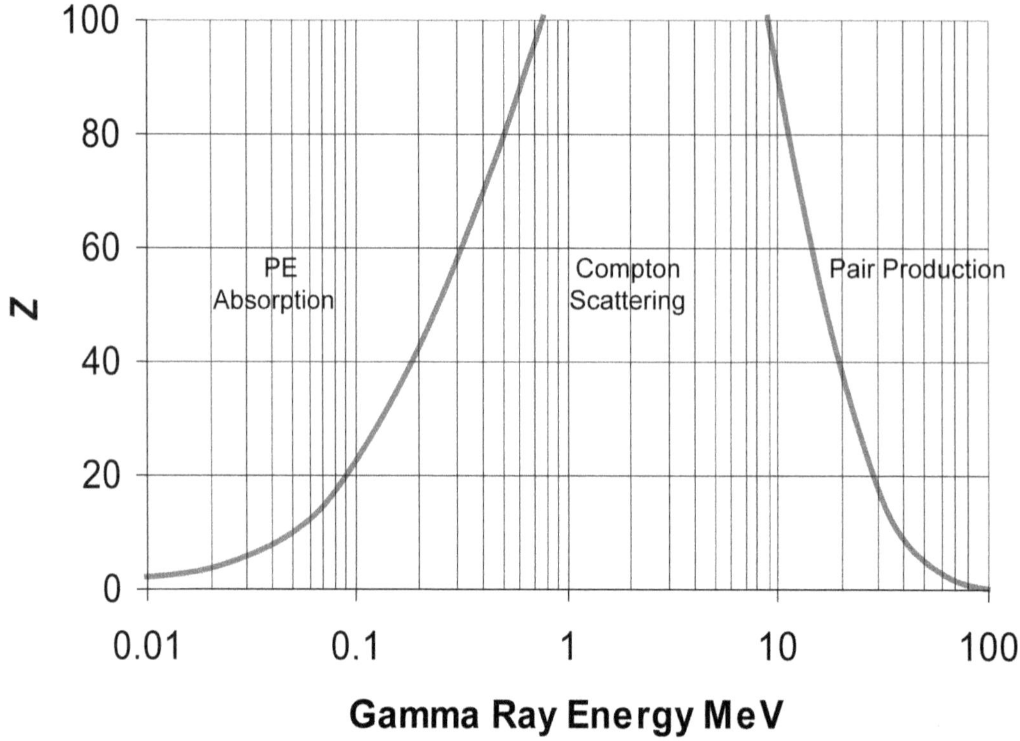

Fig. 3D.6—Illustration of dominant gamma-ray interactions with a formation as a function of gamma-ray energy.

3D.5 Logs Exploiting Gamma Rays and Gamma Ray Transport

3D.5.1 Gamma Ray Interactions With Formations. Gamma rays interact with formations in three different ways: photoelectric absorption, Compton scattering, and, to a limited extent, pair production. One of these will dominate depending on the energy of the gamma ray, as **Fig. 3D.6** shows.

Compton Scattering. The most important interaction for logging measurements involving gamma rays is Compton scattering, which dominates in the middle energy range. The density log itself is designed to exploit Compton scattering. Compton scattering also controls the transport of natural gamma rays through a formation to the standard gamma ray tool. Compton scattering is scattering off an atomic electron. In the process, the gamma ray loses some of its energy to the electron. Compton scattering is the dominant form of gamma ray interaction with a formation, from several hundred keV (kilo electron volts, a unit of energy) all the way to 10 MeV (mega electron volts). The cross section for Compton scattering changes very little with energy. The loss of gamma rays is proportional to

$$\sigma_{co}\frac{N_A}{A}\rho_b Z, \quad \text{(3D.16)}$$

where Z is the average atomic number of the formation. The attenuation law for gamma ray intensity falloff is then

TABLE 3D.5—VALUES OF BULK DENSITY, ELECTRON DENSITY, AND APPARENT DENSITY (AS READ BY A LOGGING TOOL) FOR SOME COMMON MINERALS AND FLUIDS

Material	Formula	Bulk Density (ρ_b), g/cm^3	Electron Density (ρ_e), g/cm^3	Apparent Density(ρ_a), g/cm^3
Quartz	SiO_2	2.654	2.650	2.648
Calcite	$CaCO_3$	2.710	2.708	2.710
Dolomite	$CaCO_3$*Mg	2.870	2.863	2.876
Fresh water	H_2O	1.000	1.110	1.000
Salt water	200 kppm	1.146	1.237	1.135
Oil	$n(CH_2)$	0.850	0.970	0.850
Gas	$C_{1.1}H_{4.2}$	ρ_g	1.238 ρ_g	1.325 ρ_g–0.188

$$\phi_o = \phi_i e^{\rho_b \frac{Z}{A} N_A \sigma_{co} h} \quad \text{(3D.17)}$$

For Compton scattering to be a true measure of bulk density, ρ_b, Z/A must be a constant. For almost all formation elements, $Z/A = ½$, and a measurement of gamma ray attenuation in the 1- to 10-MeV range can indeed be calibrated to bulk density. The notable exception is hydrogen, for which $Z/A=1$. **Table 3D.5** lists some density values for comparison.

Photoelectric (PE) Absorption. Not surprisingly, the PE log is based on the photoelectric absorption of gamma rays, the scattering process that dominates at low energy. In this process, the incoming gamma ray is absorbed by an atomic electron, giving up all its energy to the electron in the process. If the gamma ray is energetic enough, the added energy causes the electron to break free from its atom. As another electron falls into the vacancy, a characteristic X-ray, generally less than 100 keV, is emitted. These X-rays are too low in energy to contribute to logging measurements.

The PE cross section falls off very strongly as the energy of the incoming gamma ray increases. The cross section is proportional to

$$\frac{Z^{4.6}}{E_{GR}^{3.15}} \quad \text{(3D.18)}$$

It is a significant factor in gamma ray scattering only for energies less than 100 keV. This means that it is easy to separate the effects of PE absorption from those of Compton scattering by simply windowing the energies of the gamma rays detected. The same tool can make both measurements simultaneously. By examining the falloff of low-energy gamma ray flux, a logging tool can be calibrated to measure the PE factor (PEF). The PEF, in turn, is primarily sensitive to the average atomic number, Z, of the formation. Because hydrocarbons and water have very low Z values, they contribute very little to the average PE of a formation. Conversely, because the major rock matrices have very different Zs, the PE factor is a nearly porosity-independent lithology indicator.

Pair Production. The final process by which gamma rays interact with a formation needs only a passing comment because its impact on logging measurements is minimal. This process, pair production, occurs only at very high gamma ray energies. It is another absorption process, with a threshold of 1.022 MeV. The incoming gamma ray interacts with the electric field of the nucleus and is absorbed if it has enough energy. This generates an electron-positron pair. The positron (actually just an antimatter electron) is quickly annihilated, yielding two 511-keV

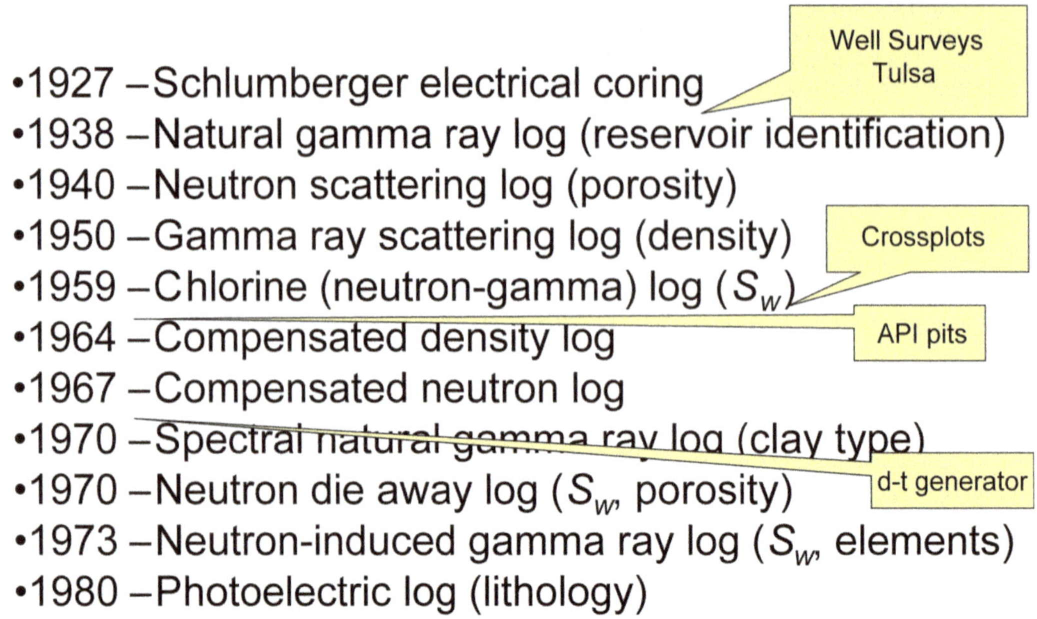

Fig. 3D.7—A timeline of nuclear logging highlights the introduction and evolution of commercial nuclear-logging measurements.

gamma rays. This has little impact on passive gamma ray or gamma-scattering density measurements but does play a role in the appearance and analysis of gamma ray spectra from neutron-induced gamma ray logs.

3D.6 Passive Gamma Ray Tools

Conceptually, the simplest tools are the passive gamma ray devices. There is no source to deal with and generally only one detector. They range from simple gross gamma ray counters used for shale and bed-boundary delineation to spectral devices used in clay typing and geochemical logging. Despite their apparent simplicity, borehole and environmental effects, such as naturally radioactive potassium in drilling mud, can easily confound them.

The gamma ray tool was the first nuclear log to come into service, around 1930 (see **Fig. 3D.7**). Gamma ray logs are used primarily to distinguish clean, potentially productive intervals from probable unproductive shale intervals. The measurement is used to locate shale beds and quantify shale volume. Clay minerals are formed from the decomposition of igneous rock. Because clay minerals have large cation exchange capacities, they permanently retain a portion of the radioactive minerals present in trace amounts in their parent igneous micas and feldspars. Thus, shales are usually more radioactive than sedimentary rocks. The movement of water through formations can complicate this simple model. Radioactive salts (particularly uranium salts) dissolved in the water can precipitate out in a porous formation, making otherwise clean sands appear radioactive.

There are only three radioactive elements that occur naturally: potassium, uranium, and thorium. Thorium and uranium both decay to daughters that are also radioactive, and those daughter elements in turn decay to other radioactive daughters, and so on for several generations. Most of these decays result in gamma rays. The energies of the gamma rays are characteristic of the element decaying. This leads to a characteristic pattern or spectrum of gamma ray energies for thorium and uranium, as shown in **Fig. 3D.8**. Potassium, for example, decays directly to a stable daughter, argon, emitting a single gamma ray with energy 1.46 MeV.

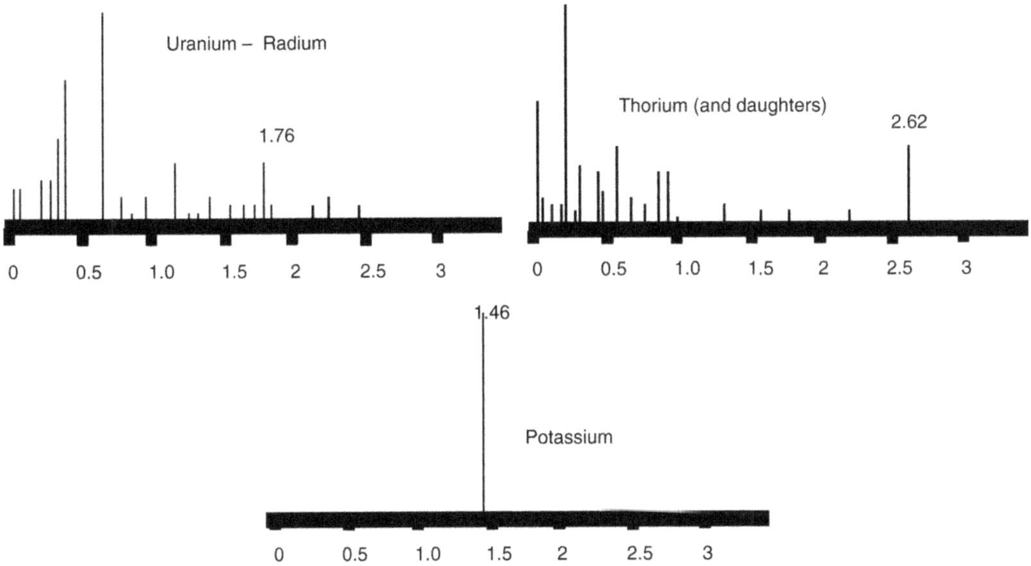

Fig. 3D.8—The spectra of gamma rays emitted by the three naturally occurring radioactive measurements (MeV) and, in the case of uranium and thorium, their daughters.

Before getting into how to use the log readings, let us consider the workings of the tool. Unlike all other nuclear tools (and, in fact, all other logging measurements), it is completely passive. It emits no radiation. Instead, it simply detects incoming gamma rays from the formation and (unfortunately) the borehole. Gamma rays are electromagnetic radiation, generally in the energy range 0.1 to 100 MeV. As light, this would correspond to very short wavelengths indeed. The difference between gamma rays and X-rays is largely semantic because they overlap in energy.

Originally, the detector was a Geiger-Müeller tube, just as in the Geiger counter. More recently, the detectors have been switched to solid-state scintillation crystals such as NaI. When a gamma ray strikes such a crystal, it may be absorbed. If it is, the crystal produces a flash of light. This light is "seen" by a photomultiplier staring into the end of the crystal. The photomultiplier shapes the light into an electrical pulse that is counted by the tool. Hence, like all nuclear tools, the raw measured quantity in a gamma ray log is counts. As discussed above, this means that the precision of gamma ray log measurements is determined by Poisson statistics. The precision is the square root of the total number of counts recorded at a given depth. Counts recorded are basically proportional to the volume of the detector crystal times its density (which determine the probability that a gamma ray will be captured within the crystal) times the length of time counted. As with all nuclear-logging measurements, the only part of this that the logger controls is the counting time. Because log measurements are depth driven, the length of time the logger counts is inversely proportional to the logging speed.

Historically, gamma ray sondes have recorded the total flux of gamma radiation integrated over all energies emanating from a formation as a single count rate, the gamma ray curve. Logging tools are not uniform in their energy sensitivity. No detector responds to all the gamma rays that impinge on it. Many pass through with no effect. The sizes of a detector, the solid angle it subtends, and its thickness, as well as its composition (particularly its density), all affect its efficiency for detecting gamma rays. The tool housing around the detector, the casing, and even the density of the borehole fluid can all filter the gamma rays coming from the formation. All these factors not only lower the overall tool efficiency, they also lead to variations in efficiency for gamma rays of different energies. In short, the count rate recorded in a particular radioactive shale bed is not a unique property of the shale. It is a complex function of tool design and borehole conditions as well as the actual formation's radioactivity. Even though gamma ray readings are generally used only in a relative sense, with reservoir (clean) and shale values determined in situ, there are advantages to a common scale. In the U.S. and most places outside the former Soviet Union, gamma ray logs are scaled in American Petroleum Inst. (API) units. This harkens back to a desire to compare logs from tools of different designs. Tools with different detector sizes and compositions will not have the same efficiency and thus will not give the same count rate even in the same hole over the same interval. To provide a common scale, API built a calibration facility at the U. of Houston. It consists of a concrete-filled pit, 4 ft in diameter, with three 8-ft beds penetrated by a 5½-in. hole cased with 17-lbm casing. The top and bottom beds are composed of extremely-low-radioactivity concrete. The middle bed was made approximately twice as radioactive as a typical midcontinent U.S. shale, resulting in the zone containing 13 ppm uranium, 24 ppm thorium, and 4% potassium. The gamma ray API unit is defined as 1/200 of the difference between the count rate recorded by a logging tool in the middle of the radioactive bed and that recorded in the middle of the nonradioactive bed.

While it has served fairly well for more than 40 years, this is a poor way to define a fundamental unit. Different combinations of isotopes, tool designs, and hole conditions may give the same count rate, so the calibration does not transfer very far from the calibration-pit conditions. In contrast, Russian gamma ray logs are typically scaled in microroentgens (μR)/hr, which does correspond to a specific amount of radiation. Converting this to API units is a bit vaguely defined, but it is often suggested that the conversion factor is 1 μR/hr = 10 API units for Geiger tube detectors, but 15 μR/hr = 10 API for scintillation detectors. This falls in with the previous discussion of the many factors that can affect gamma ray readings. As will be seen later, the problem is further aggravated in logging-while-drilling (LWD) measurements. The API unit provides a degree of standardization, but despite the best efforts of tool designers, one cannot expect tools of different designs to read exactly the same under all conditions. Fortunately, none of this is very important because gamma ray measurements are generally used only in a relative way.

Because we use gamma ray logs as relative measures, precise calibration is not very important except as a visual log display feature. Environmental effects are much more important. Consider a radioactive volume of rock traversed by a borehole. Referring back to the bit of nuclear physics above, gamma rays are absorbed as they pass through the formation. For typical formations, this limits the depth of investigation to approximately 18 in. Considering only the geometry, the count rate opposite a given rock type will be much lower in a larger borehole in which the detector is effectively farther from the source of gamma rays. In an open hole, borehole size almost always has the greatest effect on the count-rate calibration. This problem can go well beyond changes in bit size. Especially if shales or sands are selectively washed out, borehole size can imprint itself of the expected gamma ray contrast between shales and sands. If the borehole is large enough, the density of the fluid filling the borehole can also impact the calibration by absorbing some of the gamma rays before they get to the tool.

Barite in the mud is another complication, filtering the incoming gamma rays. Thus, the gamma ray borehole size and fluid corrections are often very important and should be made if at all possible. Obviously, casing absorbs a large fraction of the gamma rays traversing it on their way to the borehole, so if the tool is run in a cased hole, casing corrections are very important. Tool design has a large impact on environmental corrections. The housing and location of the detectors all filter the incoming gamma rays. It is important to use the right environmental corrections for the tool being run. This is especially true for LWD tools that may consist of multiple detectors embedded in large, heavy drill collars that filter the incoming gamma rays in unique ways.

Now that we know how the tools work, we are ready to discuss how gamma ray logs are used in log analysis. While the gamma ray log traditionally has been used primarily for well-to-well correlation, it also plays a role in quantitative log analysis. As mentioned at the outset, gamma ray logs are used primarily to define and quantify productive intervals. As discussed above, there are only three naturally occurring radioactive elements—potassium, uranium, and thorium (or K, U, and Th by their elemental symbols)—and all of these tend to be associated with shales, not clean matrix minerals (e.g., quartz sand, SiO_2, limestone $CaCO_3$).

The most common interpretation method is the simple bulk linear mixing law presented previously.

$$\gamma = \sum_i V_i \gamma_i \; . \qquad (3D.19)$$

Even though we know that the distribution of clays in shales and reservoir rocks is quite complex, to first order, log analysts frequently simplify the linear bulk mixing law to the determination of shale volume:

$$\gamma = \gamma_{sh} V_{sh} + \gamma_{ma} V_{ma} + \gamma_f V_f \; . \qquad (3D.20)$$

Standard log analysis separates the log-analysis problem into a series of sequential, independent steps. Because shale-volume determination is usually the first step in the sequential process of formation evaluation from logs, porosity and fluid volumes are not yet known. As a result, the equation is further simplified to

$$\gamma = \gamma_{sh} V_{sh} + \gamma_{cn} V_{cn} \; . \qquad (3D.21)$$

Adding closure,

$$V_{sh} + V_{ma} = 1 \qquad (3D.22)$$

leads to the familiar formula for calculating shale volume from a borehole-corrected gamma ray log:

$$V_{sh} = \frac{\gamma_{log} - \gamma_{cn}}{\gamma_{sh} - \gamma_{cn}} , \qquad (3D.23)$$

where the "clean" terms represent the lumped response to the matrix grains and the fluids in the porosity. Further complications arise because the shale values are taken from overlying shale beds. The clays distributed in the reservoir rock are almost certainly not simply dispersed versions of the shales, unless they occur as thin laminations. At the very least, there will be differences between shale, made up of clay minerals, clay bound water, and silt-size particles, and the clay minerals alone distributed in the matrix. Worse, because of differences in the processes at work when the shales were laid down vs. the shaly sands, the clay minerals in the sands may not be the same as those in the matrix. To compensate for this, numerous nonlinear relationships have been proposed. These have geologically significant-sounding names like Larinov older rocks but are simply empirical and have no physical basis.[5] They are used to improve the correlation between gamma ray-derived shale volumes and other estimates of the shale volume, especially from core. The equations all start with the linear gamma ray index discussed above and reduce the intermediate values from there. **Fig. 3D.9** lists a few of the

$$\text{Linear:}\quad I_{GR} = \frac{GR_{log} - GR_{clean}}{GR_{sh} - GR_{clean}}$$

$$\text{Larinov\_older:}\quad V_{sh} = 0.33\times(2^{2\times I_{GR}} - 1)$$

$$\text{Larinov\_tertiary:}\quad V_{sh} = 0.083\times(2^{3.7\times I_{GR}} - 1)$$

$$\text{Clavier:}\quad V_{sh} = 1.7 - [3.38 - (I_{GR} + 0.7)^2]^{0.5}\; V_{sh} = 1.7 - \left[3.38 - (I_{GR} + 0.7)^2\right]^{0.5}$$

$$\text{Stieber:}\quad V_{sh} = \frac{I_{GR}}{3 - 2\times I_{GR}},\; V_{sh} = \frac{I_{GR}}{2 - I_{GR}},\text{etc.}$$

Fig. 3D.9—A summary of various nonlinear shale-volume models used to reduce the amount of shale below the linear, bulk mixing-law prediction.

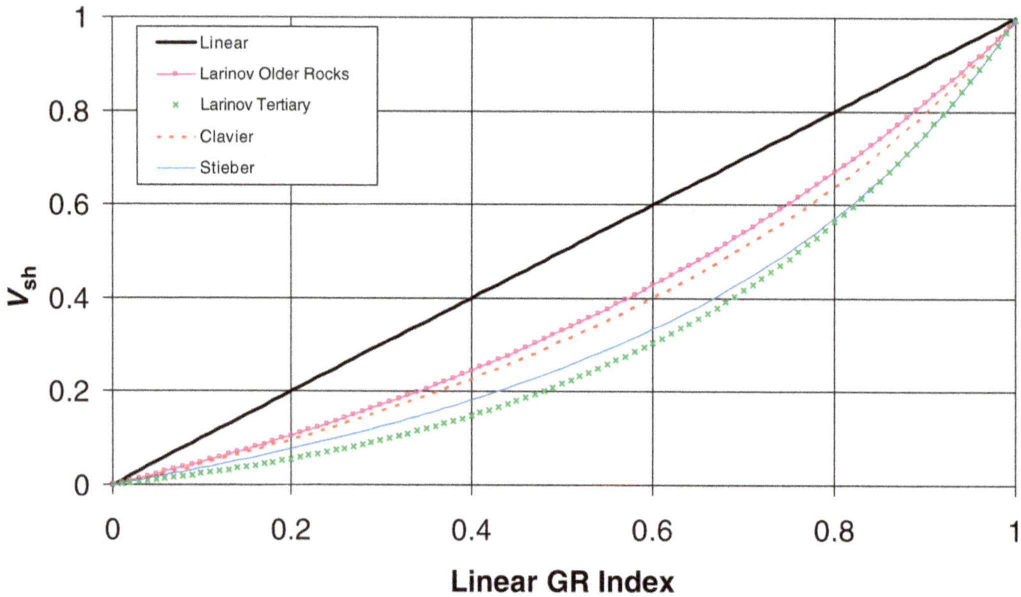

Fig. 3D.10—Illustration of the amount of shale reduction predicted by various nonlinear shale-volume models listed in Fig. 3D.9.

more common equations. **Fig. 3D.10** illustrates the degree of shale reduction that the various models afford. If one of these models must be used, select the one that best fits other available estimates of clay volume.

One disadvantage of the various empirical, nonlinear models is that they generally require core data for calibration or at least justification. This is a generic problem with more complex models; they require more parameters to characterize them. To set or calibrate those parameters in turn requires more independent log or core measurements.

It is also assumed that the clean reservoir material (the sum of the pore fluids and matrix minerals) has a fixed amount of radiation associated with it. As long as the gamma ray reading associated with the clean reservoir material is small compared to the shales, this assumption is safe. As the sands become hotter (more radioactive), lumping the fluids and matrix together becomes problematic, particularly if the porosity is large.

The latest variant on the gamma ray log is the spectral gamma ray log. This starts out exactly the same as the standard gross gamma ray count log discussed above. Gamma rays from the formation are counted in a detector system. However, there is an added level of sophistication. The energy of the gamma ray captured by the detector is proportional to the brightness of the light pulse it produces, and this brightness, in turn, determines the size of the electrical pulse produced by the photomultiplier. By sorting the pulses from the photomultiplier into bins by their size, a spectrum equivalent to the energy spectrum of the incoming gamma rays is produced. As noted above, the energy of the gamma rays is determined by which element emitted them.

Spectral gamma ray measurements offer several advantages. They can help with clay typing. Variations of the relative amounts of potassium, thorium, and uranium are associated with specific shale minerals. As is so often the case in log analysis, crossplots are used to highlight these differences. Different clay minerals may (sometimes) array themselves in the pie slices of a thorium/potassium crossplot. Clays of different types also may plot in different regions on a crossplot of potassium or the thorium/potassium ratio against P_e, the PE factor. For the mathematically inclined, the same relationships may be captured in an equation of the form

$$\gamma_{\text{API}} = A[\text{Th}]$$

$$\gamma_{\text{API}} = A\left[\text{Th}_{(\text{ppm})}\right] + B\left[\text{U}_{(\text{ppm})}\right] + C\left[\text{K}_{(\text{wt \%})}\right] . \quad (3\text{D}.24)$$

For a typical shale, the coefficients are in the ratio of A:B:C=1:2:4. Uranium is more often associated with fluid movement in porous rocks than shale minerals. At the very least, the effects of uranium can be removed. In other cases, potassium may be associated with feldspar rather than shale. Differences in the ratios between the overlying reference shale and the shaly sands may highlight the problems with carrying clay properties from the overlying shale into the shaly-sand interval.

3D.6.1 Precision. Consider briefly some details of how a standard, gross-count rate gamma ray tool works. Most modern tools (in nuclear logging, "modern" means within the past 25 years) use a solid-state scintillator crystal (most often sodium iodide, NaI) to detect gamma rays. When a gamma ray strikes the crystal, there is some probability that it will be captured. That probability is mostly proportional to the size and density of the crystal. If it is captured, it gives off a flash of light. A photomultiplier mounted on one end of the crystal converts that light to an electrical pulse, which is then fed to an electronic pulse counter. To measure a count rate with a given precision in the laboratory, one counts until enough counts are registered to give the desired level of precision (see the discussion of counting statistics above). Then, one divides that number of counts by the time it took to get that many to obtain a count rate. Unfortunately, in a logging tool, all measurements are depth-based. To measure a count rate, the tool counts for the length of time it takes the tool to move ½ ft (or whatever the depth increment is), then divides by the length of time it took the tool to move that distance. This means that the precision of a nuclear-logging measurement in a given lithology is proportional to one over the square root of the logging speed. Remember that the number of counts received crossing a clean ½ ft will be much less than the number when crossing a shaly ½ ft.

3D.6.2 Environmental Effects. The simple consideration of the discussion of radiation transport above helps clarify which environmental effects most seriously distort the gamma ray log. Imagine what happens as borehole size increases. There is less of the radiating radioactive material near the detector, and the measured count rate goes down, even though the actual level of radioactivity in the formation remains the same. Further imagine the rather typical case in which the shales are eroded and broken out while the sands remain in gauge. This would suppress the apparent gamma ray count rate in the eroded shales much more than in the sands, suppressing the gamma ray contrast between eroded shales and sands. This is typically one of the largest environmental effects on the gamma ray count rate. Again from the discussion of radiation transport, heavier materials in the path that the gamma rays must follow from the formation through the detector will absorb more gamma rays than lighter materials (as will be seen in a later section, this is the basis for the bulk density log, but that is another story and a different log). Worse yet, barite is a big absorber of gamma rays. The lesson to carry away is that borehole size and fluid corrections are almost always important when running the gamma ray log.

3D.6.3 Spectral Gamma Ray Logs. As we saw earlier, the energy of a gamma ray depends on its source. Each of the standard naturally occurring radioactive elements (K, U, and Th) gives off a gamma ray of a unique energy when it decays. Potassium gives off only a gamma ray. The other elements give off a gamma ray, then decay to other elements called daughters, which, because they are still radioactive, give off other gamma rays, and so on. This gives rise to the pattern of gamma ray energies in Fig. 3D.8. These are called spectra of the elements and are as unique as fingerprints. It is not surprising that the brightness of a light pulse produced in a scintillator crystal is proportional to the energy of the gamma ray. The amount of current in the electrical pulse from the photomultiplier is in turn proportional to the brightness of the pulse of light. It is a simple matter to sort the pulses coming out of the photomultiplier into bins according to their pulse size before counting them. This is called pulse-height spectrum analysis and gives rise to a histogram of count rates such as those in Fig. 3D.8 instead of a single count rate. Common scintillators lack the resolution to break the gamma rays into fine enough bins to reproduce the spectra in Fig. 3D.8, so some sophisticated mathematical deconvolution is needed to infer proportions of uranium, potassium, and thorium from broadly windowed pulse-height spectra.

There are several new things that can be done once we have K, U, and Th count rates rather than just total gamma ray. The most important is that we can produce a count rate only because of potassium and thorium. This is very useful because these elements most often tag only clays, while uranium salts can be associated with moved water. These uranium salts can be precipitated out in porous reservoir rock, especially at the wellbore, where pressure changes may occur. This uranium can produce what appear to be hot sands on a gross gamma ray log. Using the uranium-free gamma ray curve from a spectral tool (CGR, in Schlumberger's mnemonics) can circumvent this problem and improve sand/shale discrimination in such environments. Occasionally, the ratio of thorium to potassium can be exploited in clay typing. The downside of spectral gamma ray curves is reduced count rate and the accompanying reduced precision. By dividing the spectra into three components, the count rate for any one component may be less than one-third that of the total gamma ray measurement. Further errors occur in the math of deconvolution. If high-precision spectral gamma ray measurements are needed, reduced logging speed is required. The service companies have charts and computer programs that can help in the selection of logging speeds to achieve specific precisions.

3D.6.4 LWD. LWD gamma ray tools typically embed the detector inside of a drill collar. Two to three inches of steel are interposed between the detectors and the formation. That steel acts as an energy-cutoff filter, passing high-energy gamma rays better than lower-energy ones. As a result, these tools are more sensitive to the high-energy potassium gamma rays than the lower-energy uranium and thorium. The API gamma ray unit defined in the U. of Houston facility fails to recognize that different gamma ray energy distributions (arising from different relative concentrations of potassium, uranium, and thorium in the formation, as well as different borehole conditions and detector response functions) can cause the same counting rate at the detector in the borehole. In addition, the borehole diameter of the calibration pit is too small to accommodate most measurement-while-drilling (MWD) tools. To allow direct comparison with familiar wireline gamma ray logs, MWD contractors have attempted to transfer the API unit to the new (and larger-diameter) spectral gamma ray calibration pits, also at the U. of Houston. Because of the differences in spectral response between wireline and MWD tools, there is no unambiguous way to transfer the API unit to MWD tools. This problem is not unique to MWD tools, but because of their suppressed low-energy sensitivity, it is particularly severe for them.

To offset these effects, most MWD gamma ray tools are spectral gamma ray tools that divide the spectrum into 256 channels downhole. The precise use of these windows is still evolving, but they clearly can be used for K-U-Th determination. In one case, the shielding provided by the drill collar is turned to an advantage to produce a directional gamma ray log. As the tool rotates in the hole, it looks in different directions. At a dipping bed boundary with gamma ray contrast, such as when entering or leaving a shale bed, the gamma ray reading oscillates as the tool first sees the bed on the top of the hole and then the bottom. This fact can be used to estimate the angle at which the drillstring is striking the bed and to keep drilling within a bed.

3D.6.5 Other Applications. Gamma ray logs have a number of other niche applications. For example, injected fluids can be tagged with radioactive tracers and their progress through a field monitored with gamma ray logs in wells adjacent to the injection site.

Spectral natural gamma ray systems designed for K-U-Th logging have been applied to evaluate stimulations and completions.[6] One or more radioactive isotopes tag the various materials sent downhole. From a spectral log that separates the different isotopes, engineers establish the vertical zones of each of the different phases of the treatment. By examining peak-to-Compton-background ratios from the spectra, it is also possible to discriminate material inside the borehole from that outside the borehole. The same data yield a feeling for how far into the formation (remembering that gamma rays penetrate reservoir rocks only approximately 6 in.) the materials extend. By applying directional gamma ray detection schemes, it is also possible to infer fracture direction.

3D.6.6 Gamma-Gamma Scattering Density Tools. A density-logging tool sends gamma rays into a formation and detects those that are scattered back. Typical logging sondes use a Cesium-137 source, which emits gamma rays of 0.66MeV. At this high energy level, Compton scattering dominates. The average electron density in the volume of formation probed by the tool controls the scattered gamma ray count rates at the detectors. As we saw above, average electron density, in turn, correlates strongly (but not perfectly) with bulk density. Because the gamma rays cannot penetrate far into the formation, the volume of investigation is small. Mudcake and tool standoff have particularly strong effects on this measurement. For less energetic gamma rays, PE absorption controls the observed count rates. Here, the average atomic number (which correlates with rock type) sets the amount of PE absorption that a formation exhibits. Again, all the bulk average effects accrue, with special problems posed by barium-weighted mud.

The depth of investigation of a density tool decreases with increasing density and never exceeds 6 in., as the pseudogeometric factor in Fig. 3D.3 shows. The log almost always measures the invaded zone, at least in porous, permeable formations. As is typical of the nuclear scattering family of measurements, the density tool uses two detectors at progressively longer distances from the source. The distance between the near and far detectors sets the vertical resolution, approximately 10 in. typically. Correction is by the spine-and-ribs technique. The spine is the normal calibrated relationship between the density measured by the near-spaced and far-spaced detectors in the absence of any gap between the tool and the borehole wall. Gaps cause departures from this spine and lead to density corrections that are applied to the density from the long-space detector. This correction is presented as a curve with the density log. Above an inch or so of standoff, the compensation fails. There is no universal correction threshold, but corrections greater than 0.05 to 0.1 g/cm^3 are suspicious. Unfortunately, a low correction does not guarantee a good measurement. Connecting a particular rib back to the correct location on the spine (and, hence, to the correct bulk density) requires that the density pad be parallel to the borehole wall. Likewise, very small water-filled gaps can give rise to large corrections that are perfectly correct.

3D.6.7 LWD. This is the one case in which the difference in design between wireline and LWD tools is significant. Wireline tools use a skid-mounted pad that is pressed directly against the borehole wall. The pad follows hole rugosity, at least at the 1- to 2-ft vertical frequency level, minimizing the tool standoff. Because this is a very shallow measurement, minimizing the thickness of the layer of mud and borehole fluid between the tool and the formation is very important. The location of the sensor in the rotating drill collar precludes pad mounting of the LWD gamma ray density sensor. Thus, direct contact with the formation is eliminated, and a large and variable mud layer is introduced into the volume of investigation. Two different approaches are used to compensate for this mud effect. One design places the source and detector in stabilizer blades. These blades displace the drilling mud, providing a nearly direct path to the formation for the probing gamma rays. Because a stabilizer can steer the bit, tool subs using them must be run several joints behind the bit to maintain directional control. The blades are also subject to wear that can affect the tool calibration. A focused source and careful detector-window design further minimize borehole effects. Measurements generally use tool rotation to correct for hole-size variations. Spectral detection of scattered gamma rays is exploited in combination with low-density windows to produce a PE measurement. In an alternate approach to solving the mud problem, multiple detectors are placed radially around the collar, and the results are averaged to remove mud effects when the mud density is known. Of course, the tool may undergo complex motion, requiring a more sophisticated algorithm. This procedure can be misled by tool precession in nearly vertical holes.

3D.6.8 Density-Porosity Interpretation. In the standard sequential interpretation process, the analyst determines porosity directly from the density log. It is conceptually the easiest of the porosity logs to interpret because, if ever a tool obeyed a linear bulk mixing law, it is the density logs.

In a simple clean reservoir, the interpretation model is:

$$\rho_b = \phi \rho_{\text{fl}} + (1 - \phi)\rho_{\text{ma}} \text{ . } \qquad \text{(3D.25)}$$

Solved for porosity, this yields

$$\phi_T = \frac{\rho_{\text{ma}} - \rho_b}{\rho_{\text{ma}} - \rho_{\text{fl}}} \text{ . } \qquad \text{(3D.26)}$$

This porosity is subscripted with a T for total porosity because it draws no distinction between pore fluid and fluid possibly bound in shales. As mentioned in the general interpretation discussion, the density log rarely sees past the invaded zone, so $\rho_{\text{fl}} = \rho_{mf}$, the mud-filtrate density. Even if shale is present, the interpretation merely requires an additional term:

$$\rho_b = \phi_e \rho_{\text{fl}} + (1 - \phi_e - V_{\text{sh}})\rho_{\text{ma}} + V_{\text{sh}} \rho_{\text{sh}}, \qquad \text{(3D.27)}$$

and the corresponding shale correction to turn total porosity into effective porosity is straightforward. Again, more sophisticated formation descriptions require more parameters, which in turn require more measured data for calibration.

In known lithology, the grain density, ρ_{ma}, can be chosen from a table like Table 3D.5. Sensitivity can be analyzed directly by taking the partial derivative of the response equation with respect to grain density. For a 30-p.u., water-filled sandstone, an error of 0.05 g/cm^3 in the supposed grain density will only alter the calculated porosity by 2 p.u. A similar analysis of the sensitivity to fluid density shows that a variation of more than 0.1 g/cm^3 in the fluid density corresponds to a similar 2-p.u. error in the calculated porosity.

This is fortunate because fluid density within the density tool's volume of investigation may be very difficult to estimate. As already discussed, because of its shallow depth of investigation, the density log is commonly a flushed-zone device, and the fluid density that it sees may be taken to be that of the mud filtrate. The effect of a small amount of immovable hydrocarbon or connate water can, more often than not, be ignored. If gas is present, its low density will produce an apparent increase in porosity. Here, we could have a case in which, to calculate density porosity, we must first know the average flushed-zone fluid saturation.

As discussed in the section about radiation transport, the biggest hitch in gamma-gamma scattering density evaluation is the difference between bulk density and electron density. The device measures electron density. As Table 3D.5 shows, this matters only for fluids. Knowing that the tool will be used in fluid-filled rocks, the service companies transform the electron density to a water-filled, porous limestone, calibrated bulk density. The apparent density as read by the tool is thus altered to read:

$$\rho_{b\_log} = 1.0704\rho_e - 0.188. \qquad (3D.28)$$

In the principal lithologies of interest, this transformed density departs less than 0.004 g/cm^3 from the true bulk density. The case of a high-porosity reservoir filled with gas (and with minimal invasion) may require additional correction beginning with the removal of this "calibration." In those cases, the analyst should use the apparent density rather than the true bulk density in any mixing-law equation.

3D.6.9 PE. In addition to gamma ray scattering, modern density tools also analyze the low-energy region of the scattered gamma ray spectrum separately. These low-energy gamma rays are subject to photoelectric absorption, which is controlled by the atomic number, *Z. Z,* in turn, strongly correlates with lithology (see **Fig. 3D.11**). The length of the lines represents a variation from 0 (top) to 40 p.u. Note how effectively the lithologies are discriminated independent of porosity.

The PE absorption cross section, in barns (10^{-24} cm^2), is strongly dependent on the energy of the gamma rays, *E,* as well as the average atomic number, *Z.*

$$\sigma = 12.1E^{-3.15}Z^{-4.6} \; . \qquad (3D.29)$$

This means that low-energy gamma ray flux is attenuated according to

$$\Phi = \Phi_o e^{-n\tau x} \; . \qquad (3D.30)$$

To suppress this energy dependence, the PE log is scaled as a PE index or factor:

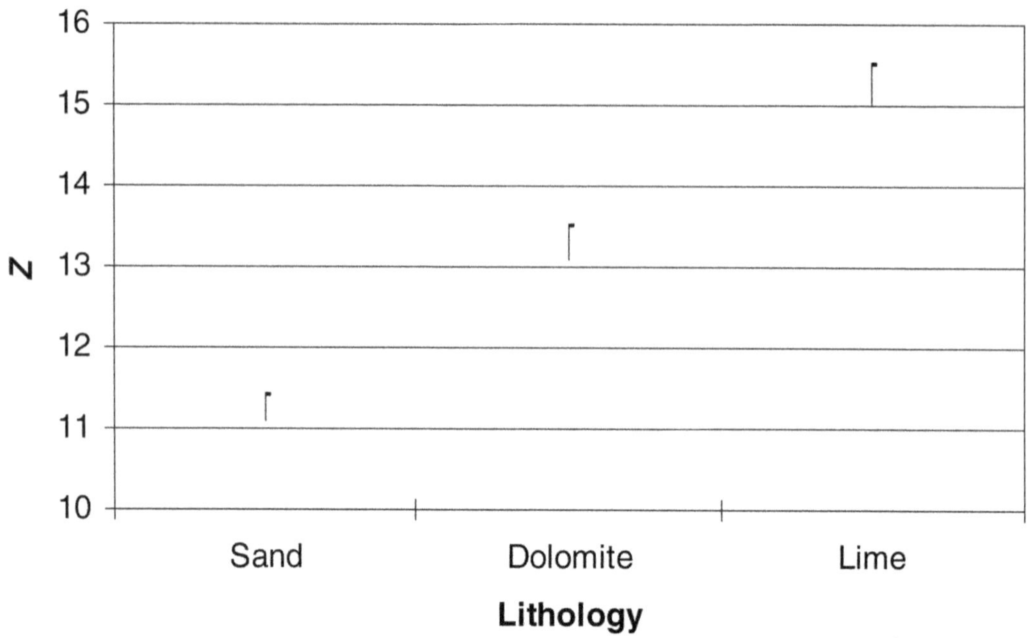

Fig. 3D.11—*Z* and, hence, the PEF measurement, discriminates lithologies, largely independent of porosity. The line lengths represent ranges of porosity from 0 to 40% in the respective lithologies. (There are three vertical lines corresponding to the porosity, ranging from 0 p.u. at the top to 40 p.u. at the bottom for each of the three common reservoir matrix lithologies. This illustrates that *Z* and, hence, PE is nearly porosity independent while strongly discriminating lithology. The *x*-axis takes on only three discrete values for each of the three lithologies.)

$$P_e \equiv F_{pe} \equiv \left[\frac{Z}{10}\right]^{3.6} . \quad \text{(3D.31)}$$

So, in terms of P_e, the attenuation (which is what the tool actually measures) of low-energy gamma rays is simply

$$\Phi \propto \Phi_o e^{-n_e P_e x}, \quad \text{(3D.32)}$$

where n_e is the electron-number density.

3D.6.10 *U*. Unfortunately, P_e does not obey a linear, volumetric mixing law on which log analysis thrives. To get around this, a new parameter, U, was developed.

$$U = F_{pe} \rho_e = \frac{F_{pe}(\rho_b + 0.1883)}{1.0704}, \quad \text{(3D.33)}$$

where ρ_b is the formation density in g/cm^3. In terms of multiple components,

$$U_T = \sum_i P_{e,i} \rho_{e,i} V_i = \sum_i U_i V_i, \quad \text{(3D.34)}$$

TABLE 3D.6—PE FACTORS AND *U* VALUES FOR SOME COMMON FORMATION CONSTITUENTS

Material	PEF (barns/electron)	*U* (barns/cm^3)
Quartz	1.80	4.80
Calcite	5.10	13.80
Dolomite	3.10	9.00
Kaolinite	1.80	4.40
Illite	3.50	8.70
Chlorite	6.30	17.00
Fresh water	0.36	0.40
Brine (120 kppm NaCl)	0.81	0.96
Oil	0.12	0.12

which is a linear bulk mixing equation. **Table 3D.6** shows typical values of PEF and *U* for some common formation constituents.

3D.7 Neutron Logs

3D.7.1 Neutron-Scattering Porosity Tools. By far, the most difficult nuclear logs to interpret are those that exploit neutron scattering to estimate porosity. The log targets the average hydrogen density of the volume investigation. If all the hydrogen in the formation is in the form of porosity-filling liquid, in particular water or oil, the hydrogen index will track the porosity. A modern, compensated tool actually estimates the size of a cloud of neutrons around a source by measuring the ratio of count rates at two different distances from the source. The straight-line distance that the average neutron travels away from the source before collisions with formation atoms slow it down to thermal energy sets the size of the neutron cloud. Once a neutron slows to thermal energy and is thus in equilibrium with the rock matrix, it diffuses only very slowly away from the source. This forms the static cloud of neutrons whose size the tool measures. This characteristic distance is called the slowing-down length. Neutrons slow down through elastic, billiard-ball-type collisions. Conservation of momentum requires that they lose the most speed in collisions with nuclei of nearly their own mass (e.g., hydrogen). Obviously, collisions with other nuclei also slow neutrons, but less effectively than those with hydrogen. Table 3D.2 summarizes just how dominant hydrogen is at slowing down neutrons compared to the other common formation elements. Ideally, a porosity tool would count only epithermal neutrons because the slower thermal neutron population depends as much on the absorption cross section of the formation as it does on the slowing-down length. To get count rates high enough for statistical accuracy, logging tools typically count all neutrons, epithermal and thermal. Thus, the tool reading must be corrected for the effects of neutron capture in the formation. **Fig. 3D.12** outlines the conversion from tool reading to porosity.

The neutron-porosity log first appeared in 1940. It consisted of an isotopic source, most often plutonium-beryllium, and a single detector. Many variations were produced exploiting both thermal and epithermal neutrons. In most of the early tools, neutrons were not detected directly. Instead, the tools counted gamma rays emitted when hydrogen and chlorine capture thermal neutrons. Because hydrogen has by far the greatest effect on neutron transport, the borehole effects on such a tool are large. Until recently, this type of tool was still in common use in the former Soviet Union. The now-standard compensated neutron-porosity logging (CNL) tool, in common use since the 1970s, is still a very simple tool. Like a density tool, it

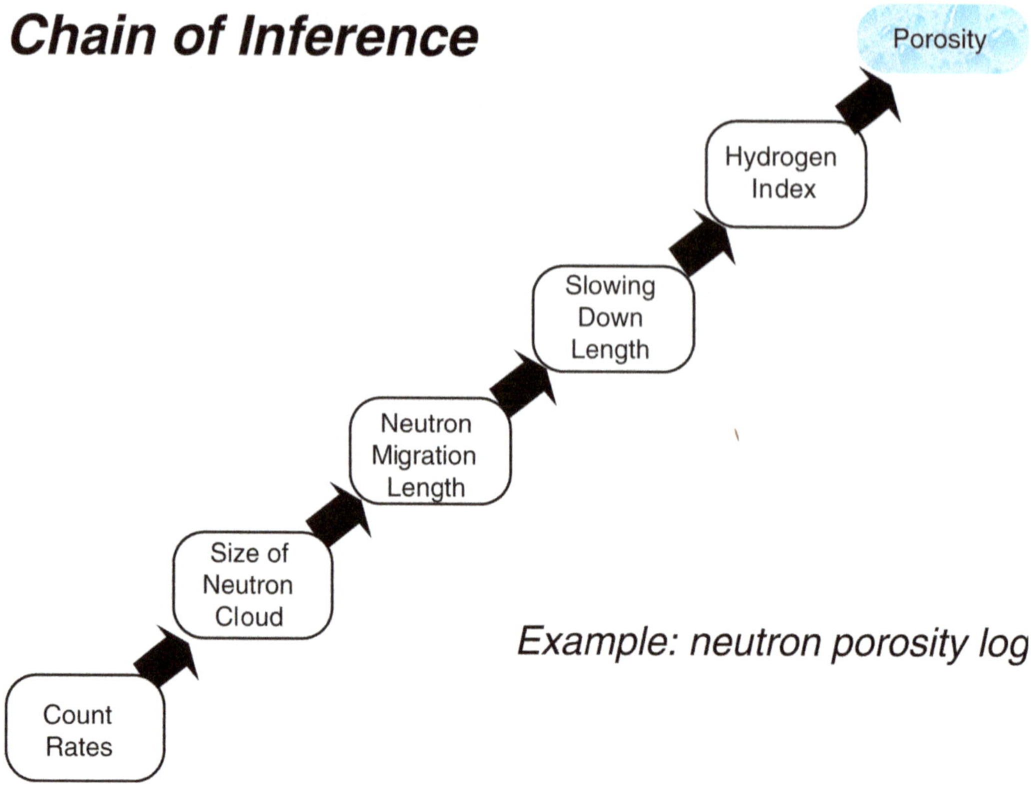

Fig. 3D.12—The nuclear-log-interpretation process can be seen as a chain of inferences, as illustrated by the neutron-porosity log.

consists of an isotopic source (now most often americium-beryllium, although at least one tool uses an accelerator source) and two neutron detectors. The tool measures the size of the neutron cloud by characterizing the falloff of neutrons between the two detectors. Because neutrons penetrate considerably further than gamma rays, the design is much simpler than that of a density tool. It requires little collimation and does not need to be pressed against the borehole wall. The size of the fluid-filled borehole is obviously an important environmental effect that must be taken into account. As a result, even "raw" CNL porosities are reported with a borehole-size correction already applied.

3D.7.2 LWD. As expected, some of the most interesting variations in design occur for the most troublesome measurement, the LWD tools. Although all MWD devices share the basic wireline configuration of a neutron source and two differently spaced detectors, the drillstring environment forces changes. Again, a pad-mounted tool is not possible, increasing potential borehole effects. He-3 detectors with long central wires, the standards in wireline tools, are sensitive to the effects of vibration that can cause false counts. One service company uses a new neutron detector, an Li-6 scintillator. Because this detector can respond to gamma rays as well as neutrons, spectral processing is required to strip out the gamma ray counts that show up as low-height pulses. The detector absorbs essentially all incident thermal neutrons, resulting in a high counting efficiency, but the metal hatch over the detector acts as a filter, giving a substantial epithermal character to the response, which lies somewhere between the thermal and epithermal responses of a wireline tool. In another novel approach, multiple Geiger Marsden tubes arrayed around the circumference of the collar detect capture gamma rays. In principle, most detected gamma rays come not from neutron capture in the formation, but from capture in the iron of the collar. They thus reflect the neutron population near the detector as if they were neutron detectors. In practice, this tool can exhibit lithology and salinity effects. In formations containing siderite, some correlation between porosity and grain density has been observed from this design. This indicates that some of the gamma rays recorded by the detector do indeed come from the formation. A third design takes a wireline-like approach. The primary measurement uses banks of He-3 detectors at two different spacings from an americium-beryllium source. The source is centered in the drill collar rather than near the formation. Although this makes the source separately retrievable, it also filters and lowers the neutron flux at the formation. Source-to-detector spacings are similar to wireline tools. Porosity is calculated much as it is for wireline logs. A near-to-far count-rate ratio is taken. Shop calibration factors, borehole diameter, mud weight, salinity, temperature, pressure, and matrix corrections are applied to the ratio before finally calculating porosity. This procedure is superior to correcting porosities for these effects in that it reflects perturbations on what the tool measures: neutron slowing-down length, not porosity. Because of absorption of thermal neutrons by the drill collar, the measurement has an epithermal flavor. Each bank of detectors consists of three detectors distributed radially around the circumference of the drill collar. Examining the count rates from the three detectors allows for correction for tool position in the borehole.

Like the gamma ray API unit, some historical baggage accompanies the presentation and scaling of neutron-porosity logs. The curve is presented as porosity, frequently without reference to the matrix, even though the matrix does matter. Although the curve is most often scaled as apparent limestone porosity, as we have seen earlier, it is actually a measurement of the distance required for a neutron to slow down, referred to by physicists as the slowing-down length, L_s. Neutrons produced by an americium-beryllium ($Am^{241}Be^{9}$) isotopic source have an energy of approximately 4.3 MeV, corresponding to a speed of more than 2000 cm/μs (44 million mph). Above approximately 0.1 eV (a mere 6,000 mph), neutrons slow down primarily through elastic collisions with the nuclei of atoms in the formation. Elastic collisions are like billiard-ball collisions: the nearer the nucleus struck is to the mass of the neutron, the more energy the neutron loses in the collision. This means that hydrogen, the nucleus of which has only a single proton (and which has altogether the same mass as the incoming neutron), is by far the most effective atom at slowing down a neutron (see Table 3D.2). Neutron slowing down by elastic collisions (thermalization) may be visualized as a random walk (see **Fig. 3D.13**). The average straight-line distance that a neutron manages to get from the source before it comes to thermal equilibrium with the reservoir is the slowing-down length, L_s. Note how much longer the slowing-down length for limestone is than for water, as shown in the figure.

Logging tools measure the size of the neutron cloud by looking at the falloff in neutron flux between two detectors. The falloff is inferred from a ratio of near-to-far neutron counting rates. As shown in the top half of **Fig. 3D.14**, this ratio can be uniquely related to slowing-down length. The data points in the plot represent a variety of lithologies and porosites, but all fall along the same trend line. Although it depends on details of lithology and fluid composition, conversion of slowing-down length to porosity is straightforward, as will be seen in the discussion of macroparameters later.

The standard neutron-porosity measurement counts all neutrons, not just the unthermalized or epithermal ones. They are designed this way because the epithermal count rate represents only a small fraction of the neutrons at a distance from the source. To get the count rate (and the precision) up, all available neutrons are counted. This is a textbook example of trading accuracy for precision. While elastic scattering and, thus, the hydrogen content of the formation control epithermal neutron flux, thermal neutrons can interact with the formation in other ways. In particular, they can be captured by other elements in the formation. Neutron capture is not dominated by hydrogen, as **Table 3D.3** shows. This means that the standard thermal

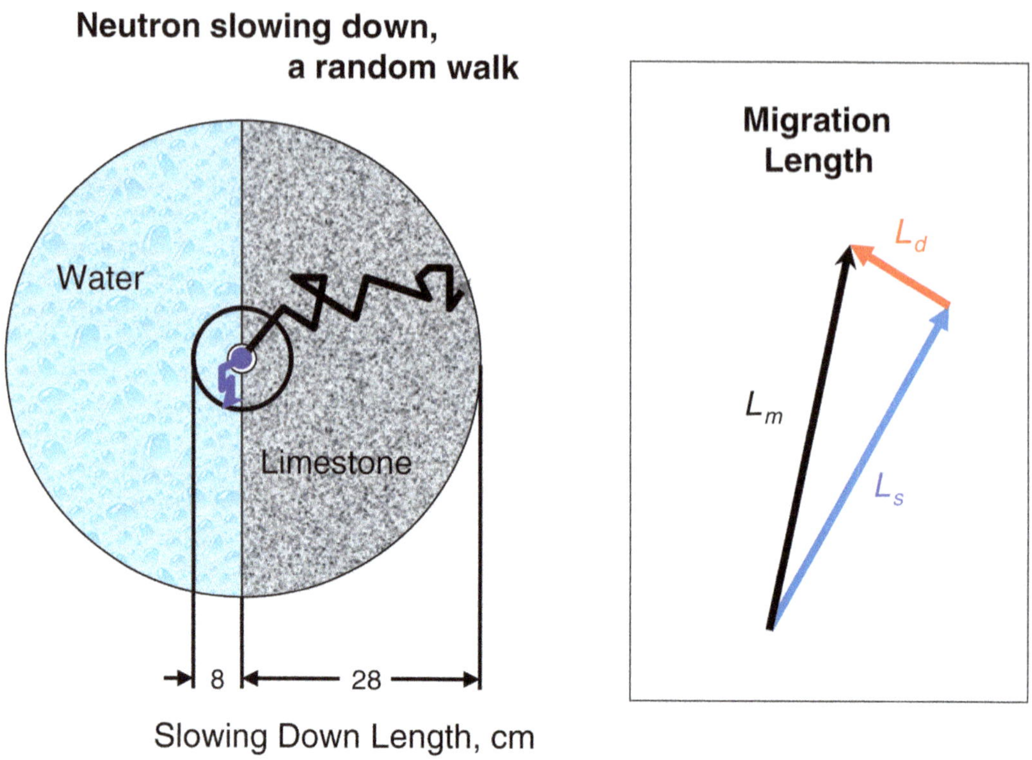

Fig. 3D.13—Neutron scattering determines the size of a cloud of neutrons surrounding a source embedded in a formation.

neutron porosity is contaminated by the subsequent diffusion distance, L_d, that the thermalized neutrons travel before they are finally captured. This is illustrated by the vector sum in Fig. 3D.13. This total distance the neutron travels, slowing down plus diffusion, is called the migration length, L_m. Because of the enduring confusion around the matrix assumed for a particular display of neutron porosity and the tenuous relationship between what the tool measures and porosity, values would be better reported in migration length than apparent porosity. It is as if density logs were always displayed as density porosity without always reporting the assumed matrix and fluid densities.

These problems with the interpretation of the standard compensated thermal neutron log have started a search for a new neutron-porosity tool with a more epithermal character. Accelerator neutron sources have begun to appear in commercial porosity tools. These tools take a more sophisticated approach than the simple CNL-like count-rate ratios used by current pulsed-neutron lifetime tools to obtain porosity. For example, Schlumberger has fielded a neutron generator-based porosity sonde that measures several different "neutron porosities" with both thermal and epithermal characters. These different measurements of porosity can be compared and combined to improve the final porosity estimate. While the measurement may be a better hydrogen-index porosity measurement, it is not the same as a standard compensated thermal neutron-porosity log.

3D.7.3 Log Interpretation. Because we are stuck with values reported in apparent neutron porosity, that is how we typically interpret them. Most interpretation schemes assume that the neutron porosity is scaled in apparent limestone units; that means a limestone matrix and water-filled porosity. If the neutron matrix is not known for certain, but the actual formation matrix

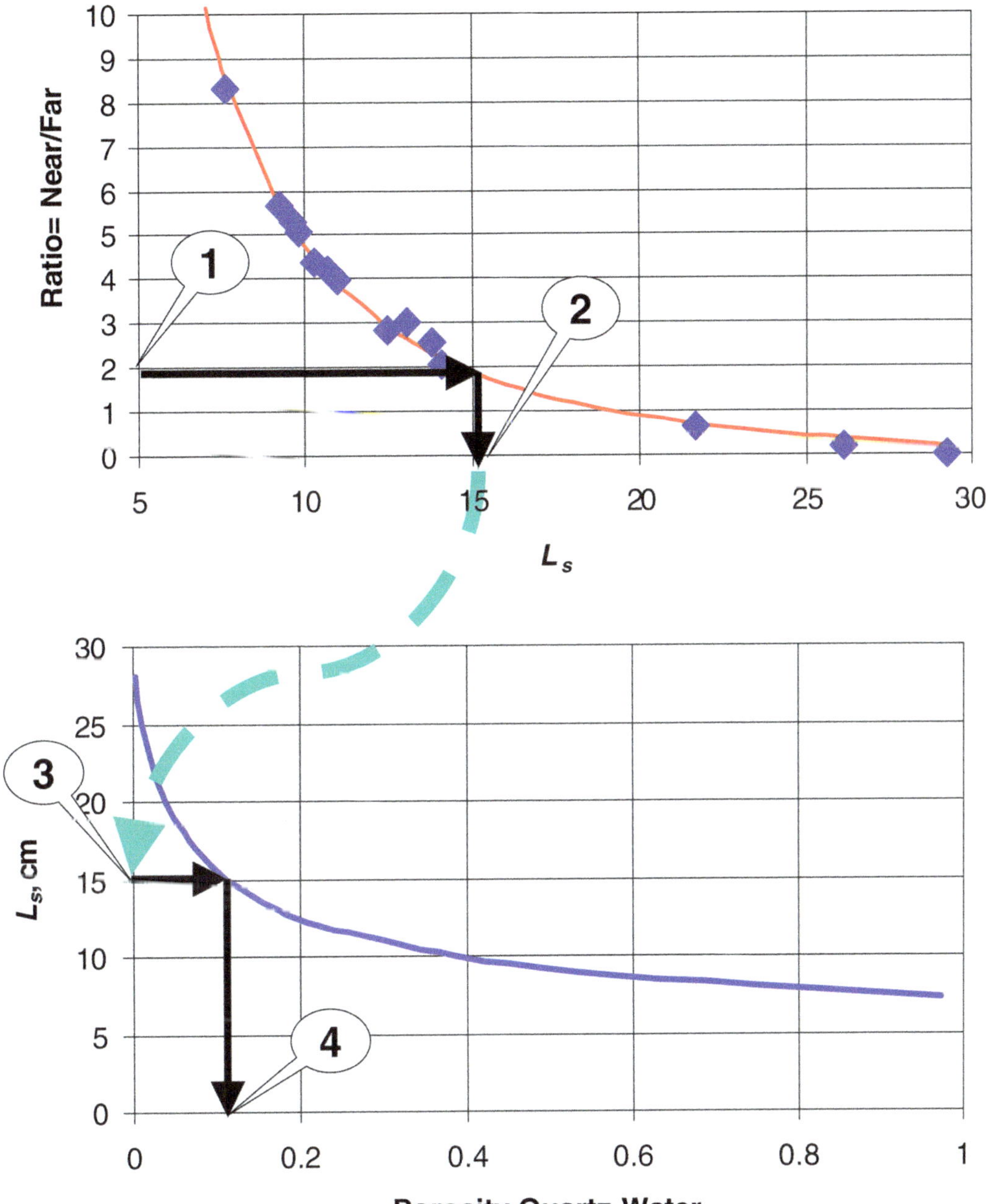

Fig. 3D.14—In an alternate neutron-porosity log-interpretation process, the intermediate response parameter, slowing-down length, relates better to basic tool response. The data points in the top plot represent a variety of lithologies and porosities, but all fall on a single trend.

is, the matrix on which the neutron-porosity log was recorded can be verified by making a density-neutron crossplot. **Fig. 3D.15** shows a schematic example. If the points fall along the overlay line for the actual formation matrix, the neutron log is most likely in limestone (calcite) units. If the points fall along the calcite overlay line, the log matrix is the same as the formation matrix. In particular, if the points fall along the limestone line and the reservoir is known to be

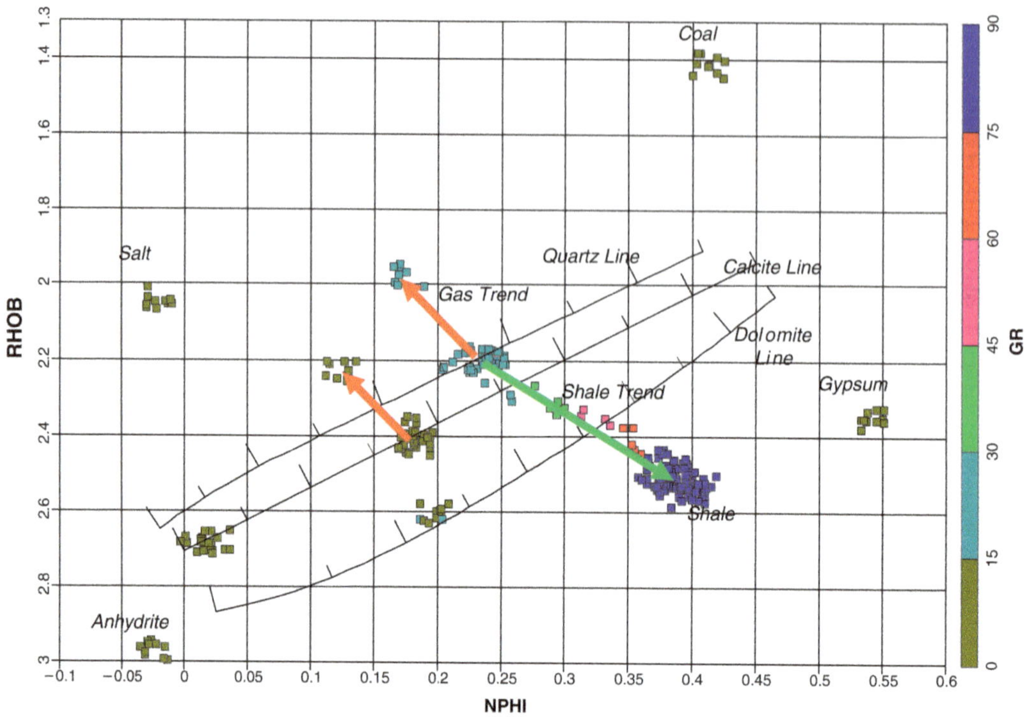

Fig. 3D.15—Neutron-density crossplot showing where the common lithologies in Fig. 3D.16 fall.

sandstone, the neutron log is in sandstone units and should be transformed to limestone units before proceeding with interpretation. As the schematic shows, gas and shale can obscure these trends.

To first order, once the log has been environmentally corrected, its reading can be characterized by a linear mixing law

$$\phi_{\text{CNL}} = \sum_i V_i \phi_{\text{app}i} \approx \sum_i V_i I_{H\_i} \ . \quad \text{(3D.35)}$$

In the case of a purely epithermal neutron log, the approximately equal sign can be replaced by an equal sign. Because most logs encountered are thermal neutron logs, the rest of the discussion will center on thermal neutron interpretation.

The linear mixing law implies that matrix and shale effects can be handled by a simple apparent porosity of a 0-p.u. mineral. Some examples of apparent neutron porosity (on a limestone matrix) for a number of materials are given in **Table 3D.7**. As can be seen from the matrix overlays on the density-neutron crossplot (Fig. 3D.15), this mixing law is not quite linear, and these apparent porosities vary with true porosity. While they are correct at 0 p.u., by the time true porosity reaches 10 p.u., quartz's apparent porosity has climbed to –4 p.u. and dolomite's to +6 p.u. These values also can vary somewhat with tool design. In a known, single clean lithology, it is best to use the contractor's more elaborate transforms. However, in shaly sands or dolomitized limestone, canned transforms will not exist, and the analyst falls back on the linear mixing law. In the shaly sand case, the mixing law is of the form

$$\phi_{\text{CNL}} = \left(1 - V_{\text{sh}} - \phi_e\right)\phi_{\text{ma,app}} + V_{\text{sh}}\phi_{\text{sh,app}} + \phi_e I_{H\_pf} , \quad \text{(3D.36)}$$

TABLE 3D.7—APPARENT THERMAL NEUTRON POROSITIES OF SOME COMMON RESERVOIR MATERIALS

Material	Apparent ϕ_{CNL}
Quartz	–0.020
Calcite	0.000
Dolomite	0.020
Siderite	0.120
Glauconite	0.380
Coal	>0.40
Kaolinite	~0.370
Illite	~0.300
Anhydrite	–0.020
Water	1.000
Brine (200 kppm)	0.920
Gas (STP)	0.002
Gas (reservoir)	0.540
Oil	~0.900

Note: "Gas (STP)" refers to gas at standard temperature and pressure (20°F, 14.7 psia), and "Gas (reservoir)" refers to gas at the nominal reservoir conditions of 200°F and 7,000 psia.

where ϕ_{app} and I_H are used interchangeably because both are calibrated to the apparent neutron porosity of pure water.

Fluid Effects. The hydrogen index of the pore fluid (see Table 3D.4) and its equivalent apparent neutron porosity (Table 3D.7) can have a much bigger effect. The difference between pure water, most brines, and typical oils is small, but as the table shows, gas can have much different neutron response properties. While the presence of gas increases the apparent porosity seen by a density log, it decreases the apparent porosity seen by the neutron log. This is the source of "gas crossover" on neutron density-log displays (see Fig. 3D.16). Moreover, the shallow-reading density log frequently is an invaded-zone measurement, completely masking the gas effect on it. Because the neutron porosity is deeper reading, it is often the only log that can be used for gas detection. Even when not completely reading the invaded zone, the neutron-porosity log probably reads a mixture of invaded and virgin formation. This leads to a very complex response equation, even in a clean reservoir:

$$\phi_{\text{CNL}} = (1-\phi_e)\phi_{\text{ma,app}} + f(r_i)[\phi_e S_{xo} I_{H\_mf} + (1-S_{xo})\phi_e I_{H\_hc}]$$
$$+[1-f(r_i)][\phi_e S_w I_{H\_w} + (1-S_w)\phi_e I_{H\_hc}], \quad \text{(3D.37)}$$

where $f(r_i)$ is the radial geometric function discussed above, r_i is the step-invasion profile approximation for the radius of invasion, *mf* refers to mud filtrate, *hc* refers to hydrocarbon, and *w* refers to formation water. Of course, shaly or multimineral interpretations add additional terms.

Furthermore, the response equation becomes decidedly nonlinear when gas is introduced. To compensate for this, an additional term was introduced to the response equation. This artifact of the gas is labeled the excavation effect.

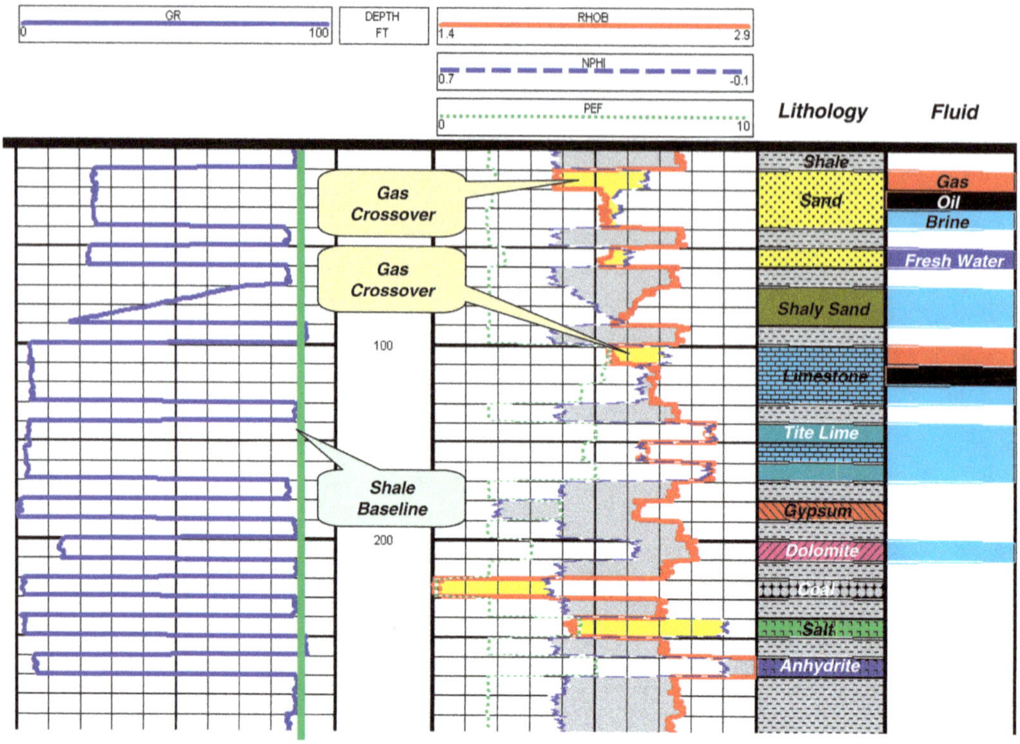

Fig. 3D.16—Schematic nuclear-log responses for some common lithologies.

Advanced Processing. The effect of formation absorption has long been recognized. Chartbook corrections for environmental effects on the thermal neutron log are extensive but confusing. This situation reflects more on the futility of attempts to handle every downhole situation with a handful of correction charts rather than any real error by service companies or log analysts. Clearly, a new approach is needed.

New log-processing methods that extend laboratory benchmark data with a more detailed mathematical description of the tool's response have been developed. Such methods replace one-size-fits-all correction charts. Effectively, log analysts using such techniques generate custom correction charts that exactly match their downhole situations. In lithologies that depart significantly from the standard limestone/sandstone/dolomite triplet, most particularly those with high capture cross-sectional minerals or fluids, the results can be dramatically different from chartbook values. With such model-based processing, corrections need not be made serially, nor are they limited to a few cases. Using laboratory benchmarked forward modeling, analysts can generate a broader range of corrections for complex lithologies and fluids. These procedures permit access to temperature and pressure regimes unattainable in laboratory formation models. This is especially important for neutron-porosity tools, whose response to porosity is both complex and tenuous (Fig. 3D.12).

Unfortunately, this approach requires iterative, forward modeling. Logging-tool response is rarely unique; many different lithology/fluid/borehole combinations can produce the same log reading. The availability of other well information and the judgment of the analyst becomes important. Such other reservoir knowledge limits the inputs to the forward model and reduces the number of trial-and-error cycles required to interpret logs with this method.

For neutron tools, there is a need to rethink what the tools measure and how we parameterize the measurement. The tools measure the size of a neutron cloud, expressed as a function of a neutron macroparameter, migration length. In the past, several analysts have formulated the use of forward tool-response modeling to improve neutron-log interpretation in greater detail and include field examples. They range from mixing-law treatments[7] to more elaborate use of neutron transport properties.[8]

There are two broad approaches to forward modeling of nuclear-tool response: macroparameters such as neutron migration length and Monte Carlo modeling. Macroparameters characterize tool response to bulk average formation composition (Fig. 3D.5), while Monte Carlo models treat geometry as well as composition (Fig. 3D.4).

Macroparameters. For porosity tools, macroparameters include such things as slowing-down length (when only epithermal neutrons are considered) and migration length (when thermal neutrons are considered as well). They are averaged over composition, geometry, and energy. This approach uses simplified but physically realistic theory to calculate bulk tool response. It is important to recognize that porosity tools measure neutron migration length, not porosity. A macroparameter model calculates migration length from average porosity, matrix, and fluid types. Macroparameter methods are fast, particularly compared to Monte Carlo methods, and are tractable as part of the routine interpretation process.[9] Schlumberger published the SNUPAR program,[10] which generates macroparameters for a variety of neutron and gamma ray transport tools. Once the macroparameters are understood, it is necessary to map them into the count-rate-ratio response of a particular tool design to complete the analysis. This mapping can be accomplished by regression analysis of laboratory data taken with the particular tool.[9]

Monte Carlo Modeling. When the detailed effects of geometry cannot be ignored, service companies (and, occasionally, even log analysts) resort to Monte Carlo modeling. It can account for borehole effects, standoff, invasion, thin beds, and tool design. All effects are calculated simultaneously as they occur physically and account for interactions and interdependencies that are ignored in the serial chart-book approach. The problem is not artificially divided into independent, noninteracting regimes, and no effects need be ignored.

In its most straightforward form, analog Monte Carlo modeling simulates millions of particle trajectories, tracing the progress of every particle emitted by the source. It begins with their emission at the source and follows their movement in straight-line segments. Probability distributions for interactions are accessed with random numbers generated by the computer (hence the name Monte Carlo, like a roll of the dice). These simulate particle collisions (i.e., mean distance between collisions, what the particle collides with, and its direction and speed after the collision). It is a brute-force, but fairly intuitive, approach. The models are limited primarily by the quality of the input data, particularly the nuclear cross sections of the materials involved. For real-world tools, the considerable amount of information about tool design that must be included in the model may also limit the accuracy of the Monte Carlo method.

The problem with analog Monte Carlo is that very few of the particles traced end up at the detector. Therefore, enormous numbers of particle histories must be followed to score enough counts at the detector to be statistically significant; even on fast computers, days of computations may be required. The answer is a mixed bag of tricks for ignoring some particles, steering others toward the detectors, and counting still others more than once. This approach can be mathematically valid but is quite tricky, especially with general-purpose codes like Los Alamos' Monte Carlo Nuclear Parameters (MCNP).

The biggest problem is that the models run very slowly, far too slowly to be used as the foot-by-foot forward model in an iterative interpretation process. Geometric effects are still separated from the interpretation process into the environmental-correction process. Monte Carlo modeling is generally confined to creating special-purpose correction charts. Indeed, most current service-company correction charts are generated by Monte Carlo modeling benchmarked to a few lab measurements.

3D.7.4 Pulsed-Neutron-Lifetime (PNL) Devices. PNL logs measure the die-away time of a short-lived neutron pulse. They probe the formation with neutrons but detect gamma rays. Chlorine has a particularly large capture cross section for thermal neutrons. If the chlorine in the formation brine dominates the total neutron capture losses, a neutron-lifetime log will track chlorine concentration and, thus, the bulk volume of water in the formation. For constant porosity, the log will track water saturation, S_w. The neutrons are little affected by steel casing, so this is the standard cased-hole saturation tool. Like other nuclear tools, modern PNL tools incorporate two detectors for borehole compensation. These detectors also permit the calculation of a ratio porosity. This ratio porosity is similar, but not identical, to that of a compensated neutron-porosity tool. They differ because the energy of the neutrons from the pulsed accelerator source is higher than the energy from the isotopic source used in compensated neutron logging. Also, the neutron-lifetime tools detect capture gamma rays rather than direct neutrons.

The basis of operation is similar to the other nuclear radiation transport tools in that the tool infers a cross section. In this case, the tool measures the time required for a pulse of neutrons to be absorbed by a formation. The mechanism by which the neutrons disappear is primarily thermal neutron capture. The time evolution of a pulse of neutrons follows the usual exponential decay law:

$$N_t = N_o e^{-\frac{\Sigma_{\mathrm{abs}}}{4550}t}, \quad \text{(3D.38)}$$

where Σ_{abs} is the total neutron capture cross section of the formation expressed in capture units (c.u. = 1000 × cm²/cm³, which has units of cross-sectional area per unit volume). The total capture cross section for a formation follows the standard linear volumetric mixing law discussed above:

$$\Sigma_{\mathrm{abs}} = \sum_{i=1}^{n} V_i \Sigma_i, \quad \text{(3D.39)}$$

where V_i is the volume of a particular constituent (mineral or fluid) of a formation and Σ_i is the capture cross section of that constituent. Because the corrected tool reads the total capture cross of the formation, this equation forms the basis of interpretation. For example, in the case of a clean sand with a porosity that is filled with oil and water, the tool reading will be

$$\Sigma_{\mathrm{abs}} = (1-\phi)\Sigma_{\mathrm{quartz}} + \phi S_w \Sigma_w + \phi(1-S_w)\Sigma_o . \quad \text{(3D.40)}$$

If porosity is known either from openhole logging or the ratio porosity measured by the pulsed neutron tool itself, the various cross sections can be looked up in a table, and it is a simple matter to solve for S_w. **Table 3D.8** lists the capture cross sections for several materials that commonly make up reservoirs. Several things can be observed on the basis of this table and the response equation above. First and foremost, for this measurement to be very sensitive to replacing oil by water in the pore space, the water salinity needs to be higher than 50,000 ppm NaCl. Otherwise, the oil and water capture cross sections are so similar that the measured formation cross section will not change perceptibly when one is substituted for the other. These sensitivities can be evaluated easily by setting up a simple spreadsheet and varying the values in the equation above. Eq. 3D.40 also suggests the value of running this log in a baseline monitor mode or time-lapse mode. Differences between successive logging runs will depend only on differences in fluid volumes because the terms involving the unchanging rock matrix will

TABLE 3D.8—THEORETICAL MACROSCOPIC CAPTURE CROSS SECTIONS OF A FEW MATERIALS OF RESERVOIR INTEREST

Mineral	Capture Cross Section (c.u.)
Quartz	4.3
Calcite	7.3
Dolomite	4.8
Glauconite	25
Chlorite	25
Shale	35–55
Fresh water	22
Gas	0–12
Oil	18–22
Brine (50 kppm NaCl)	35–40
Brine (200 kppm NaCl)	95–100

subtract out. In this way, no explicit knowledge of the capture cross sections of the minerals and clays is required to interpret saturation changes.

$$S_{w1} - S_{w2} = \frac{\Sigma_1 - \Sigma_2}{\phi(\Sigma_w - \Sigma_o)} \quad \text{(3D.41)}$$

While it is rarely done, this method is particularly valuable if a baseline run is made early in the production history of the well before S_w has had a chance to change significantly.

Other Applications. Pulsed-neutron logging tools have been applied in nonconventional ways to solve several production-logging problems. Not all of these applications make use of neutron die-away time. Instead, they monitor gamma rays from neutron activation of specific elements that can be thought of as tracers. In other applications, the pulsed-neutron ratio porosity can be used for excavation-effect-style gas interpretations.

3D.7.5 Log-Inject-Log Measurements for Residual Oil Saturation. In a variation on the baseline monitor mode of operation suggested above, residual oil saturation can be determined by a log-inject-log procedure. In this procedure, a pulsed neutron log is run over a zone of interest to get a baseline reading. Then, a brine of contrasting salinity is injected into the formation while logging pass after pass with the pulsed neutron tool. Over time, all of the movable, original formation fluids are displaced by the new brine until S_{orw} (residual oil saturation to waterflood) is achieved. Differences between the preinjection-pass and the final-pass formation capture cross sections give direct access to S_{orw}. As a bonus, changes in the capture cross-section profile over time during injection highlight permeability variations in the formation.

3D.7.6 Mechanical Integrity. Oxygen-activation flow logging may be used as a test of well integrity and zonal isolation. This is a stopwatch measurement. The neutron generator activates the oxygen in a slug of water. The time it takes the slug to move from its birthplace at the generator until it is opposite one of several remote detectors is measured. The flow velocity is just the distance from source to detector divided by the transit time. Because of the short half-life of oxygen, a particular source-to-detector spacing will be optimal only for a narrow range of flow rates. This procedure works equally well for flow inside and behind pipe. In principle, similar measurements can discern distance to the flow.

Boron has a very high neutron-absorption cross section that greatly reduces the neutron lifetime measured by a pulsed-neutron tool. This makes it a useful tracer when used in conjunction with pulsed-neutron logging. It has been exploited in mechanical-integrity testing by injecting borated water into a well. Any place to which the boron-tagged water finds its way will stand out on the pulsed-neutron log.

Gravel-Pack Logs. In another example, silicon activation is used to evaluate gravel-pack quality.[11] Gravel packs are placed in oil and gas wells to prevent the accumulation of formation material that otherwise would clog wellbores and production facilities. In the conventional logging method for gravel-pack evaluation, a nonfocused density tool detects the density contrast between packing material and completion fluid. When a pulsed-neutron log is used, it detects activation gamma rays from silicon and aluminum in the packing material that have a half-life of approximately 2.24 minutes. Of the other common downhole elements, oxygen has a much shorter half-life (7.2 seconds), and chlorine, sodium, and iron have half-lives of 30 minutes or longer. Thus, a judicious choice of logging speed can maximize sensitivity to silicon and aluminum. Because the threshold for silicon activation is high (4 to 5 MeV), the measurement is very shallow, maximizing sensitivity to the gravel-pack region.

3D.7.7 Induced Gamma Ray Spectroscopy Tools. A final class of neutron logs, neutron-induced gamma ray logging, records the energy spectra of the induced gamma rays. Because elements excited by neutrons emit characteristic gamma rays, such spectra can be analyzed for elemental concentrations. Most commonly, carbon and oxygen concentrations are used to determine oil saturation, although more detailed geochemical information lies buried in the spectra.

3D.7.8 Carbon/Oxygen (C/O) Logs. Recompletion of existing wells and the search for bypassed oil in established fields require knowledge of the current oil saturation behind pipe. In fields with connate water salinity > 20,000 ppm chlorides, PNL logs provide a convenient measurement of water saturation through tubing and casing. If the salinity is low, the neutron lifetime is not determined by the chlorine concentration in the formation. If the salinity is variable, the chlorine concentration does not track the water saturation. In both cases, a PNL log fails to give useful fluid saturations. C/O logging was developed for these situations. The tools exploit inelastic scattering of high-energy neutrons off carbon and oxygen to induce gamma rays. Spectral analysis of the resulting gamma rays yields the amounts of oxygen and carbon in the volume of investigation. Unfortunately, the carbon sensitivity of the measurement is low. The depth of investigation is extremely shallow (only a few inches into the formation). Such a small volume necessarily includes a large percentage of borehole compared to the amount of formation. It is also true that although carbon is present in oil but not in water, and oxygen is present in water but not in oil, the rock matrix (particularly carbonates) may contain significant amounts of both. Together, these result in substantial borehole and formation effects that must be accounted for in the C/O log-interpretation process. To obtain a lithology compensation, most C/O tools also record neutron capture spectra in which elements such as calcium, silicon, and iron reveal themselves. Neutron capture only occurs shortly after the neutrons have slowed down to thermal energies. Buffer timing separates inelastic C/O spectra (during the neutron burst) from capture spectra (slightly after the burst). Experience with C/O logging has been uneven at best.

3D.7.9 Geochemical Logs. Geochemical logging is still struggling to find applications. Schlumberger's latest incarnation is called the environmental capture sonde (ECS).[12] Applications lie primarily in rock and clay typing for reservoir description. The goal is to add additional elemental concentrations to the formation model. Natural gamma ray spectroscopy measurements provide potassium, uranium, and thorium data. The capture spectroscopy tool detects silicon, calcium, iron, sulfur, gadolinium, titanium, chlorine, and hydrogen.

In analyzing the data, elemental concentrations are derived and processed to obtain a complete mineralogical description. There are several sources of ambiguity. Most importantly, uncertainties in the measurements of the elemental concentrations are not explicitly accounted for. Because elemental concentrations derive from least-squares deconvolution of unresolved gamma ray spectra, they are not determined independently. Furthermore, sensitivities to different elements vary by orders of magnitude and often are very small, requiring large integration times. Finally, element-to-mineral mapping is not sufficiently unique. For best results, a limited suite of minerals must be selected before analyzing the data. The correct choice of a mineral suite depends on knowledge of local mineralogy from other experiences, such as core in an offset well. This arises in part from incomplete and inaccurate elemental analysis and partly from the nearly infinite variety of minerals and the small number of elements. As an example, quartz and opal contain the same elements but are quite different in their impact on a reservoir.

Mineralogy can in turn be related to such properties as permeability, porosity, and cation/ion-exchange capacity. Unfortunately, the minerals-to-petrophysical-properties inversion is not unique either, partly because the tool has no information about the physical configuration of the minerals (for example, grain size or fractures). Even when geochemical logging can give accurate elemental abundances, conversion of those numbers to mineralogy and petrophysical parameters such as permeability still requires a locally calibrated interpretation model. With limited goals and careful local calibration, geochemical logs do provide useful information. The logs remain tied to a local database and *ad hoc* knowledge to relate mineralogy to petrophysical properties.

3D.8 Multiple-Log Interpretation

3D.8.1 Visualization—The Multitrack Log Display. Multiple-log interpretation began with, and still revolves around, simple, quick-look visual displays. The familiar third track on a standard log display is the best example. At first glance, it may seem strange and the product of hidebound tradition. The familiar and seemingly arcane display and scales provide a great deal of information at a glance. The three principal porosity logs (density, neutron, and sonic) are scaled and placed so that they overlay and track one another in a particular clean matrix. Violated assumptions as to matrix and fluids stand out when the logs do not overlay. A particular matrix must, of course, be chosen. Most often, all logs are displayed on a limestone matrix. The standard scales are 1.95 to 2.95 g/cm^3 for the density log and 0.45 to −0.15 for the neutron log. Running the density/porosity response equation presented above for water-filled limestone confirms that 1.95 g/cm^3 does indeed correspond to 0.45 porosity, and 2.95 corresponds to the curious number −0.15; 140 and 40 μs/ft also correspond, but that is the subject of another chapter.

Fig. 3D.16 shows how the various nuclear-log curves would overlay on these scales for a variety of common lithologies. Note the several porosity units (1 p.u. = 1% = 0.01 fractional) of mismatch between the density and neutron logs when the matrix is sandstone instead of limestone, as is implicit in this choice of scales. This positive crossover corresponds to the apparent density porosity being larger than the apparent neutron porosity. The word "apparent" is belabored on purpose. As discussed above, logs do not read porosity until they have been passed through an interpretation model, and that interpretation model requires assumptions about rock and fluid type. In these examples, the logs are plotted as if they were water-filled limestone. If they are not, assumptions have been violated, and logs will not overlay. The 5 to 7 p.u. of matrix crossover in sandstones plotted as if they were limestones is often mistaken for a gas effect. As the simulated logs show, gas effect is in the same direction but should result in even more crossover depending on the gas properties. In gas crossover, the violated assumptions are fluid properties. The pore space is filled with gas, not water. Gas has a lower density than water, so apparent density porosity will calculate higher than the true porosity. Neutron porosity calculates lower because gas also has a lower hydrogen index than water. Thus, apparent density and neutron porosity are "off" in opposite directions in a gas zone, resulting in large crossover in the conventional log display. As in the discussion of gas response with the individual logs, invasion often confuses the matter. The shallow penetrating density log may be reading entirely from the invaded zone, where mud filtrate (not gas) fills the porosity. With luck, the neutron log may be seeing at least some distance beyond the invasion front, so at least part of its response includes gas-filled porosity. In any case, gas crossover may be largely a neutron-log artifact, and the amount often will be less than expected. Depth (or, more precisely, pressure) also suppresses the gas crossover because gas fluid properties depend strongly on pressure.

In a known sandstone reservoir, the third track is sometimes displayed on a sandstone matrix. In this case, the density log is scaled from 1.65 to 2.65 g/cm^3, and the neutron log is scaled from 60 to 0% porosity, in sandstone units. As above, 1.65 g/cm^3 does indeed correspond to 60% porosity, assuming a quartz matrix density (2.65 g/cm^3) and freshwater-filled porosity (fluid density equal to 1.0 g/cm^3). In similar fashion to the limestone display, the curves will overlay each other exactly in clean sandstone. Generally, the neutron porosity will read to the left of the density in shales and to the right in gas-affected intervals.

3D.8.2 Visualization—Crossplots. The crossplot is another method for visualizing petrophysical data. A clever crossplot can reveal even more about a formation than a standard log-depth display. In a crossplot, the analyst plots one log value on the x-axis against a different log value, at the same depth, on the y-axis. This is repeated for all depths of interest, creating a scatterplot such as that shown in Fig. 3D.15. With luck, the location of points on such a plot can discriminate underlying mineralogy and reveal trends such as shaliness or porosity. Each pure mineral will plot as a single point. The power to discriminate depends on the independence and uniqueness of log responses to the lithologies of interest. Crossplots frequently include calculated overlay points and lines. The points locate various lithologic endpoints of interest, while the lines track the simultaneous solution of the response equations for the two logs over a range variable such as porosity, or percentage of one mineral vs. another. These response equations are simply the linear mixing-law response equations discussed in the sections above on the individual logs. With only two variables—the two logs—only two unknowns can be extracted. For example, one could determine matrix type (and its associated endpoint-log readings) and the amount of water-filled porosity.

In crossplots, nuclear logs have a clear advantage over sonic or resistivity laws. As we have seen, nuclear logs generally obey simple, linear, bulk mixing laws that have a firm basis in physics. The mixing laws for sonic and resistivity measurements are not only nonlinear but also largely empirical, with only weak connections to theory. Nonlinear terms in a mixing law show up on crossplots as curved lines (the simultaneous solution for a given set of conditions corresponds to a line). In this section, the discussion will be confined to crossplots involving only nuclear logs, although many other useful combinations are possible.

A third variable is sometimes displayed as a z-axis in the form of a color scale. In the example, the color of each point represents its gamma ray log reading according to the key along the right side. This highlights the location of shales and facilitates the selection of shale properties. This highlights the location of shales and facilitates the selection of shale properties needed in further log analysis. For example, the shale density and apparent neutron porosity of the shale can be read off the plot as the values corresponding to the cluster of shale points (in this case, approximately 2.5 g/cm^3 and 40 p.u.).

Perhaps the most useful crossplot in log analysis is an old standard, the neutron-density crossplot. An example based on the synthetic-type logs in Fig. 3D.16 is shown in Fig. 3D.15.

By convention (and convention is very important to quick-look, visual techniques), the neutron log, expressed in limestone porosity units, is plotted on the x-axis against the density log in g/cm^3 on the y-axis, with the scales reversed (i.e., from highest to lowest density). Ideally, because both are porosity logs, points of a given porosity in a pure lithology will fall along a diagonal line. Such a line represents the simultaneous solution of the density and neutron mixing laws as a function of varying porosity. Three such lines are generally plotted as overlays on this crossplot. They correspond to a calcite, dolomite, or quartz matrix with water-filled porosity. If the neutron log were a true hydrogen index log, the lines would extend from a y-intercept corresponding to the grain density of the particular lithology (the zero-porosity limit) to a common upper-right point corresponding to 100% water (i.e., 1.0 g/cm^3 density and 100% neutron porosity). While this is largely true, neutron logs are not perfect hydrogen index measures.

As discussed above, the most commonly run compensated neutron log actually measures neutron migration length, which is a mixture of a large hydrogen index-controlled term and a smaller term controlled by neutron capture that is matrix- and fluid-type dependent. The mix of the two terms in a given tool is design dependent. For example, epithermal neutron porosity is a nearly perfect hydrogen index log. The more commonly used thermal neutron porosity includes some capture effect. This superimposes a linear, matrix-dependent term on the neutron response and a small amount of nonlinearity when hydrogen index is low, such as in gas. Because tool design affects the relative contribution of these terms, each service company generates its own, slightly different overlays for the neutron-density crossplot. This also explains apparent differences between wireline and LWD neutron-porosity measurements.

Returning to the example in Fig. 3D.15, the location of points on the neutron-density crossplot can be mapped to specific lithologies, a number of which are shown on the figure. Other lithology points can be plotted from their neutron- and density-log readings taken from Tables 3D.7 and 3D.5, respectively. Edmundson and Raymer[13] present a more complete tabulation of pure mineral-log readings, as do most service-company chart books. Lines connecting two points on a crossplot represent the mixing of the two lithologies. Remember that water can be used as a lithology endpoint on a crossplot. This creates a porosity trend line from the pure, 0% porosity point for a given matrix to the 100% water point. Lines and points on the crossplot represent specific, simultaneous solutions of the neutron and density mixing for specific supposed lithologies. Cross-cutting lines may represent lithology trends—changes from one lithology to another or simultaneous changes in lithology and porosity. Violated assumptions can be especially revealing. A given formation thought to be a limestone may actually lie along the dolomite line, indicating that it is a dolomite or a sand plot to the lower right of the sand line and, thus, may not be as clean as hoped. The most commonly violated assumption is that the pore space is filled with a liquid (specifically water, although liquid hydrocarbons do not fall very far from the water-filled porosity line). If it were filled with gas instead, the points on the crossplot would move to the upper left, away from the water-filled porosity line. This is the same effect demonstrated by neutron-density crossover on a standard log display. More subtly, a neutron-density crossplot can flag diagenesis. For instance, dolomitization of a limestone might reveal itself as a trail of points scattering from the tight end of the limestone line to the moderate-porosity region of the dolomite one. This can be a very beneficial process, increasing the porosity of the formation. If this process were missed and the formation treated as a pure limestone, much lower porosity would be calculated, and the reservoir might be bypassed. Examination of the neutron-density crossplot should often be one of the first steps in reconnaissance log analysis. A crossplot can help the analyst identify rock types and porosity ranges and guide the selection of facies and zones.

By exploiting the principal of closure (the fact that the volume percentages of all the constituents of a formation must add up to exactly 1), three components can be extracted from a 2D crossplot. Consider a three-component system composed of sand, shale, and water-filled

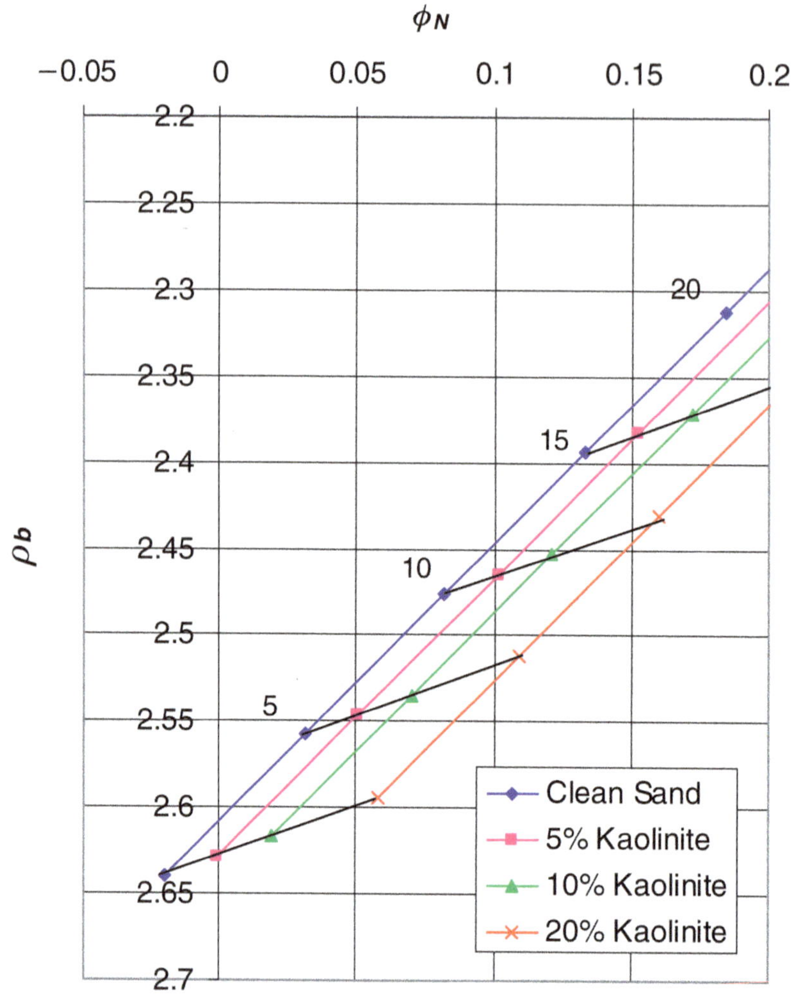

Fig. 3D.17—The neutron-density crossplot can be used to quantify clay volume and porosity in sand/shale mixtures using simple linear mixing laws to plot lines for given bulk properties (e.g., shale volumes).

porosity. Qualitatively, the shaly sand progression beginning at a single clean-sand porosity is sketched in Fig. 3D.15 as a trend line. Even if not done quantitatively, this process can indicate the direction that points would move in the presence of a change in composition. As this suggests, the neutron-density crossplot can be a useful alternative to simple gamma ray interpretation for the determination of shale volume. **Fig. 3D.17** is a neutron-density crossplot overlaid with a grid of lines. The grid is calculated from the density and neutron response equations, varying relative amounts of sand, water-filled porosity, and clay.

An example of a different, less commonly used nuclear-log crossplot is shown in **Fig. 3D.18**. As in the neutron-density example, the sample data from the logs in Fig. 3D.16 are plotted as small squares. This display crossplots synthetic variables, not raw logs. On the *x*-axis is the *U* matrix apparent. As discussed above, this transformation converts the nonlinear P_e log to U_{maa}, a characteristic that obeys linear volumetric mixing. On the *y*-axis is apparent grain density from the neutron and density logs. Somewhat simplified, this is the grain density needed to produce the neutron-log porosity from the density-log reading, assuming water-filled porosity. The blue, ternary grid shows the generic endpoints for sandstone, calcite, and

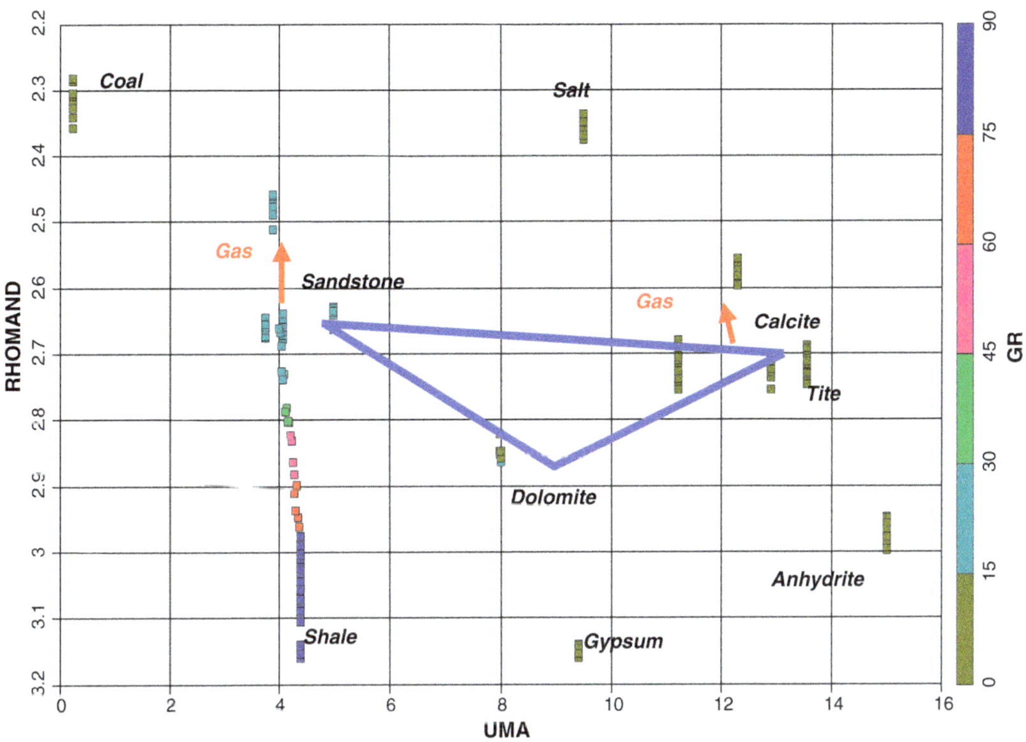

Fig. 3D.18—Example MID plot for the logs in Fig. 3D.16.

dolomite. The various labels (e.g., coal and anhydrite) mark the locations at which those minerals should ideally fall on the plot. This technique, sometimes called the matrix-identification (MID) plot, is especially useful in unwinding multicomponent lithologies, as the widely separated overlay points suggest. It gets much of its power from the fact that P_e is largely porosity independent. This accounts for the near-vertical trends in much of the overlaid data from Fig. 3D.16. As in all crossplots, uncorrected environmental effects may show up as misplaced points, the hallmark of a violated assumption. For instance, because the P_e is a very shallow measurement, barite (with its high iron content) in the mud can cause a wholesale shift of the data cloud to the right.

These are but two examples of the visualizations possible with petrophysical crossplots. Other derived parameters useful in crossplotting incorporate sonic logs. These include the *n*lith and *m*lith crossplot (where *m*lith and *n*lith are derived from combining density, neutron, sonic, and PE logs) and the crossplot of apparent matrix density (from the neutron-density crossplot) vs. apparent matrix travel time (from the neutron-sonic crossplot). These procedures can reduce the simultaneous solution of more than two log responses to an *x-y* plot visualization. Much of this can, of course, be done mathematically by solving multiple-log response equations simultaneously. Crossplot visualizations, however, may set limits on the possible formation constituents and define the input parameters to the formation model before attempting a mathematical solution.

3D.9 Conclusions

Despite the absence of any fundamentally new nuclear-logging measurements on the horizon, the future holds significant promise. LWD continues to mature. By getting the measurements made before significant invasion occurs, LWD mitigates one of the most serious problems of nuclear logs: their limited depths of investigation. New array resistivity measurements better map the invaded zone. This better understanding of the invaded zone should benefit nuclear-log interpretation by letting the analyst better deconvolve the invaded-zone and virgin-zone portions of the tool response. More generally, the evolution and application of simultaneous solver techniques is accelerating. It may soon be common to solve the invasion problem explicitly, simultaneously getting all the invaded and virgin formation fluids correct while interpreting porosity and ensuring consistent solution of all of the log-response equations. This is, after all, where environmental corrections, multiple-log overlays, and crossplots have been heading all along. Nuclear logging is about tools and response models, and for the foreseeable future, progress may come more from more sophisticated handling of the response models than from better measurements. Several once-promising nuclear-log measurements failed to gain traction over the last decade. The passive spectral gamma ray measurement remains a niche tool. Geochemical logging (neutron-induced gamma ray spectroscopy) failed to mature.

Nomenclature

A = area, L^2
A = atomic weight
e = natural logarithm base
E = energy dependence of the cross section
E_{GR} = gamma ray energy
F_{pe} = photoelectric factor
h = thickness traversed
h_V = vertical thickness, L^2
I_H = hydrogen index
I_{H_hc} = hydrogen index hydrocarbon
I_{H_pf} = hydrogen index pore fluid
I_{H_w} = hydrogen index water
J = radial geometric function
L_d = neutron diffusion length
L_m = neutron migration length
L_s = neutron slowing-down length
n_e = electron-number density
N = a counting number
N_A = Avogadro's number = 6.02×10^{23} molecules/gram molecular weight
N_p = particle number density
P_e = the photoelectric factor
P_x = Poisson probability distribution
R = tool reading
R_i = tool reading for pure material
S_o = oil saturation
S_{orw} = residual oil saturation to waterflood
S_w = water saturation
S_{xo} = flush zone water saturation
U = density-weighted F_{pe}
U_i = tool response to flushed zone
U_{maa} = U matrix apparent
U_T = tool response integrated to infinity
U_x = tool response integrated out to a radial distance$x$

V_{bh} = bulk volume of hydrocarbon
V_{cn} = volume of clean formation
V_f = volume of fluid
V_i = volume of a particular constituent (mineral or fluid) of a formation
V_{ma} = volume fraction of matrix mineral
V_{ns} = nonshale volume
V_{sh} = volume of shale
x = particular number of counts
Z = average atomic number
γ = gamma ray tool reading in API units
γ_{cn} = gamma ray flux from 100% clean formation component
γ_f = gamma ray flux from 100% fluid
γ_{ma} = gamma ray flux from 100% matrix
γ_{ns} = gamma ray tool reading in nonshale
γ_{sh} = gamma ray tool reading in 100% shale
$\delta\phi$ = a change in flux
λ = mean free path
μ = mean or expected value of a quantity
ρ_b = bulk density
ρ_e = electron number density
ρ_{fl} = bulk density of fluid
ρ_g = gas density
ρ_{ma} = bulk density of matrix mineral
ρ_{mf} = bulk density of mud filtrate
ρ_{sh} = shale density
σ = standard deviation of a Poisson distribution
$\sigma(E)$ = energy-(E) dependent scattering cross section
σ_{co} = Compton scattering cross section
Σ_{abs} = total neutron capture cross section
Σ_i = capture cross section of ith formation component
Σ_w = capture cross section of water
τ = decay time
ϕ = porosity
$\phi_{\text{app}i}$ = apparent porosity measured by a CNL in lithology i
ϕ_{CNL} = porosity measured by a compensated neutron-logging tool
$\phi_{\text{CNL}x}$ = apparent porosity measured by a CNL in lithology x(e.g., shale)
ϕ_e = effective porosity
ϕ_i = initial particle flux
ϕ_{ma} = apparent matrix porosity
ϕ_N = neutron porosity
ϕ_o = unscattered particle flux
ϕ_{sh} = shale porosity
ϕ_T = total porosity

Subscripts

GR = gamma ray

hc = hydrocarbon
i = item count or index
t = time

References

1. Hearst, J.R., Nelson, P.H., and Paillet, F.L.: *Well Logging for Physical Properties,* second edition, John Wiley & Sons, New York City (2000).
2. Ellis, D.V.: *Well Logging for Earth Scientists,* Elsevier Science Publishing, New York City (1987).
3. Patchett, J.G. and Wiley, R.: "Inverse Modeling Using Full Nuclear Response Functions Including Invasion Effects Plus Resistivity," paper H presented at the 1994 SPWLA Annual Logging Symposium, Tulsa, 19–22 June.
4. Sherman, H. and Locke, S.: "Effect of Porosity on Depth of Investigation of Neutron and Density Sondes," paper SPE 5510 presented at the 1975 SPE Annual Technical Conference and Exhibition, Dallas, 28 September–1 October.
5. Katahara, K.: "Gamma Ray Log Response in Shaly Sands," *The Log Analyst* (1995) **36,** No. 4, 50.
6. Gadeken, L.L. *et al.:* "The Interpretation of Radioactive Tracer Logs Using Gamma ray Spectroscopy Measurements," *The Log Analyst* (1991) **32,** No. 1, 24.
7. Dahlberg, K.E.: "A Practical Model for Analysis of Compensated Neutron Logs in Complex Formations," paper PP presented at the 1989 SPWLA Annual Logging Symposium, Denver, June 1989.
8. Wiley, R. and Pachett, J.G.: "CNL (Compensated Neutron Log) Neutron Porosity Modeling, A Step Forward," *The Log Analyst* (1990) **31,** No. 3, 133.
9. Tittle, C.W.: "Prediction of Compensated Neutron Response Using Neutron Macroparameters," *Nuclear Geophysics* (1988) **2,** No. 2, 95.
10. McKeon, D.C. and Scott, H.D.: "SNUPAR (Schlumberger nuclear parameters)—A nuclear parameter code for nuclear geophysics applications," *Nuclear Geophysics* (1988) **2,** No. 4, 215.
11. Olesen, J.R., Hudson, T.E., and Carpenter, W.W.: "Gravel Pack Quality Control by Neutron Activation Logging," paper SPE 19739 presented at the 1989 SPE Annual Technical Conference and Exhibition, San Antonio, Texas, 8–11 October.
12. Herron, S.L. and Herron, M.M.: "Quantitative Lithology—An Application for Open and Cased Hole Spectroscopy," paper E presented at the 1996 SPWLA Annual Logging Symposium.
13. Edmundson, H.N. and Raymer, L.L.: "Radioactive Logging Parameters for Common Minerals," *The Log Analyst* (1979) **19,** No. 1, 38.

General References

This is a small bibliography of papers and books that compose a basic library in nuclear well logging.

Bibliographies

Mills, W.R., Stromswold, D.C., and Allen, L.S.: "Advances in Nuclear Oil Well Logging," *Nuclear Geophysics* (1991) **5,** No. 3, 209.

Myers, G.D.: "A Review of Nuclear Logging," *The Log Analyst* (1992) **33,** No. 3, 228.

Density Logs

Bertozzi, W., Ellis, D.V., and Wahl, J.S.: "The Physical Foundations of Formation Lithology Logging with Gamma Rays," *Geophysics* (October 1981) **46,** No. 10, 1439.

Tittman, J. and Wall, J.S.: "Formation Density Logging (Gamma-Gamma) Principles and Practice," *Society of Professional Well Log Analysts; Gamma Ray, Neutron and Density Logging Reprint Volume,* SPWLA (1978) paper D.

Gamma Ray Logs

Wahl, J.S.: "Gamma ray Logging," *Geophysics* (1983) **48,** No. 11, 1536.

General

Ellis, D.V.: "Neutron and Gamma Ray Scattering Measurements for Subsurface Geochemistry," *Science* (5 October 1990) **250,** 82.

Patchett, J.G. and Wiley, R.: "The Effects of Invasion on Density/Thermal Neutron Porosity Interpretation," paper G presented at the 1994 SPWLA Annual Logging Symposium, Tulsa, 19–22 June.

Tingey, J.C., Nelson, R.J., and Newsham, K.E.: "Comprehensive Analysis of Russian Petrophysical Measurements," paper S presented at the 1995 SPWLA Annual Logging Symposium, Paris, June 1995.

Tittle, C.W.: "A History of Nuclear Well Logging in the Oil Industry," *Nuclear Geophysics* (1989) **32,** No. 2, 75.

Neutron-Induced Gamma Ray Logs

Oliver, D.W., Frost, E., and Fertl, W.H.: "Continuous Carbon/Oxygen (C/O) Logging—Instrumentation, Interpretive Concepts and Field Applications," paper TT presented at the 1981 SPWLA Annual Logging Symposium, Mexico City, June 1981.

Woodhouse, R. and Kerr, S.A.: "The Evaluation of Oil Saturation Through Casing Using Carbon/Oxygen Logs," *The Log Analyst* (1992) **33,** No. 1, 1.

Youmans, A.H. *et al.:* "Neutron Lifetime, A New Nuclear Log," *Society of Professional Well Log Analysts; Pulsed Neutron Logging Reprint Volume, revised edition,* SPWLA (1979) 3.

Neutron-Porosity Logs

Ellis, D.V.: "Some Insights on Neutron Measurements," *IEEE Trans. on Nuclear Science* (1990) **37,** No. 2, 959.

Tittle, C.W.: "Porosity—An Improved Porosity Code for Processing CNL Data on Desktop Computers," *The Log Analyst* (1994) **35,** No. 1, 27.

Textbooks

Doveton, J.H.: *Geologic Log Analysis Using Computer Methods*, AAPG, Tulsa (1994).

Tittman, J.: *Geophysical Well Logging*, Academic Press, New York City (1986).

SI Metric Conversion Factors

barns ×	1.0	E − 24 = cm²
ft ×	3.048*	E − 01 = m
in. ×	2.54*	E + 00 = cm
in.^{2} ×	6.451 6*	E + 00 = cm^2
in.^{3} ×	1.638 706	E + 01 = cm^3
psi ×	6.894 757	E + 00 = kPa

*Conversion factor is exact.

Chapter 3E
Nuclear Magnetic Resonance Applications in Petrophysics and Formation Evaluation

Stephen Prensky, SPE, Consultant and **Jack Howard,** SPE, Halliburton

3E.1 Introduction

Nuclear magnetic resonance (NMR) has been, and continues to be, widely used in chemistry, physics, and biomedicine and, more recently, in clinical diagnosis for imaging the internal structure of the human body. The same physical principles involved in clinical imaging also apply to imaging any fluid-saturated porous media, including reservoir rocks. The petroleum industry quickly adapted this technology to petrophysical laboratory research and subsequently developed downhole logging tools for in-situ reservoir evaluation (see the next section of this chapter).

NMR logging, a subcategory of electromagnetic logging, measures the induced magnet moment of hydrogen nuclei (protons) contained within the fluid-filled pore space of porous media (reservoir rocks). Unlike conventional logging measurements (e.g., acoustic, density, neutron, and resistivity), which respond to both the rock matrix and fluid properties and are strongly dependent on mineralogy, NMR-logging measurements respond to the presence of hydrogen protons. Because these protons primarily occur in pore fluids, NMR effectively responds to the volume, composition, viscosity, and distribution of these fluids (i.e., oil, gas, and water). NMR logs provide information about the quantities of fluids present, the properties of these fluids, and the sizes of the pores containing these fluids. From this information, it is possible to infer or estimate the volume (porosity) and distribution (permeability) of the rock pore space, rock composition, type and quantity of fluid hydrocarbons, and hydrocarbon producibility (**Fig. 3E.1**).

NMR logging provides measurements of a variety of critical rock and fluid properties in varying reservoir conditions (e.g., salinity, lithology, and texture), some of which are unavailable using conventional logging methods (Fig. 3E.1) and without requiring radioactive sources (**Table 3E.1**). Whether run independently as a standalone service or integrated with conventional log and core data for advanced formation and fluid analyses, NMR logging has significantly contributed to the accuracy of hydrocarbon-reservoir evaluation. During the past decade, a new generation of wireline-logging devices has been introduced into commercial service. In the past few years, logging-while-drilling (LWD) devices and downhole NMR spectrometers have also been introduced.

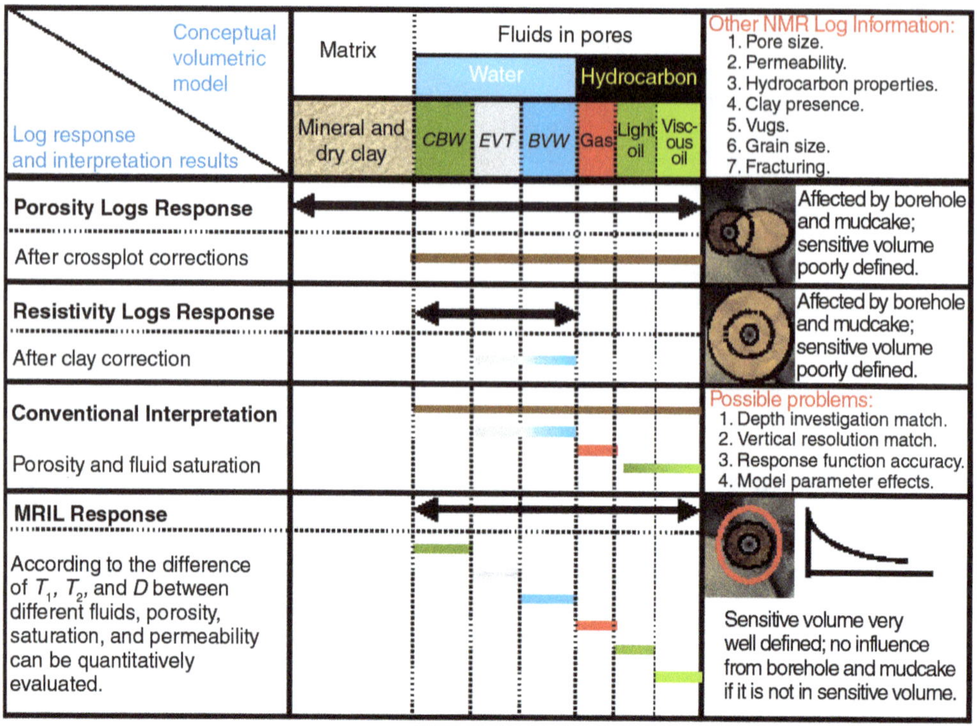

Fig. 3E.1—NMR logging-tool response compared to conventional logging tools. NMR porosity is independent of matrix minerals, and the total response is very sensitive to fluid properties. Differences in relaxation times and/or fluid diffusivity allow NMR data to be used to differentiate clay-bound water, capillary-bound water, movable water, gas, light oil, and viscous oils. NMR-log data also provide information concerning pore size, permeability, hydrocarbon properties, vugs, fractures, and grain size.

3E.2 Historical Development

Within a few years after the first successful observations of NMR in 1946, and the demonstration of free-precession NMR in the earth's magnetic field in 1948, the petroleum industry recognized the potential of NMR measurements for evaluating reservoir rocks, pore fluids, and fluid displacement (flow).

In the early 1950s, several companies—particularly California Research (Chevron), Magnolia (Mobil), Texaco, Schlumberger, and Shell—began extensive investigations to understand the NMR properties of fluids in porous media for the purpose of characterizing reservoir rocks (porosity, permeability, and fluid content).[1–3] In addition to laboratory research, these investigations included proposals for logging devices and the development of well-logging methods to permit formation evaluation in situ.[1,4] Although a number of patents for logging tools were issued in the 1950s, it was not until Chevron completed an experimental Earth's field nuclear-magnetic-log (NML) logging device in 1958 that a functioning device was actually developed.[1,5] Limited commercial service of these devices was introduced in 1962 by Atlas, using the Chevron centralized design, and followed in 1965 by Schlumberger, using a pad-type tool of its own design. An improved version of the Schlumberger tool was introduced in 1978. Although the potential applications for this measurement were significant, particularly in the shallow, heavy-oil fields of the San Joaquin Valley,[6] in general, they did not live up to expectations and were not commercially successful.[7] Tool reliability and operational limitations proved to be major obstacles: the tool was not combinable, it required high (surface) power; the signal

TABLE 3E.1—APPLICATIONS OF NMR-LOGGING DATA FOR RESERVOIR CHARACTERIZATION

Application
Petrophysical Analyses
Porosity and porosity distribution
Determination of effective porosity, ϕ_e
Determination of total porosity, ϕ_t
Evaluation for clay-bound water (CBW) and bulk-volume irreducible (BVI) water in shaly sands
Evaluation of permeability (k)
Hydrocarbon Identification, Characterization, and Typing
for hydrocarbon vs. water
for gas vs. oil
for heavy vs. light oil
Estimating viscosity
Reserves Estimates
Evaluation of low-resistivity/low-contrast pay
Evaluation of low porosity/low permeability, "tight" formations
Determinations of residual-oil saturations
Evaluation of producible (movable) fluids by means of the free-fluid index (FFI)
Characterization of Lithologies in Formations
Determinations of grain size and sorting
Mineralogy
Other Applications
Enhancing stimulation design by selection of "best sites" for well paths and perforations

level varied geographically and was generally very low as a result of the low-operational frequency (2 kHz); and the borehole had to be doped with powdered magnetite to suppress the proton signal from the mud.[1,2,8] The final version of the Schlumberger NML tool—a centralized tool introduced in 1984—proved reliable and commercially successful and was in service until the advent of modern pulse-echo tools in 1994.

In 1978, the Los Alamos Natl. Laboratory developed a logging tool that employed permanent magnets and used a pulsed radio frequency (RF) (pulse-echo) NMR method. Although this particular design had serious limitations—such as a low signal-to-noise ratio (S/N) that prohibited continuous, nonstationary logging—the concept set the stage for the development of modern commercial NMR tools. This advance was soon followed by improvements in magnet and coil design that enabled continuous logging. During the 1980s, while developing a commercial logging tool, industry also carried out laboratory experiments to further understand NMR behavior in fluid-filled porous media and to develop petrophysical interpretations from these data. Ultimately, two wireline tools using different magnet and coil configurations emerged from these efforts: Numar's mandrel device (MRIL) and Schlumberger's skid (sometimes called "pad") design [Combinable Magnetic Resonance (CMR) tool]. Commercial logging began with these tools in 1991 and 1995, respectively. These wireline-tool designs continue to evolve (see section on Tool Design in this chapter). Recent improvements allow simultaneous acquisition of more measurements, operation in a wider range of borehole conditions, and faster logging speeds. Detailed accounts of the historical development of NMR and NMR logging are available in several published references.[4,9–11]

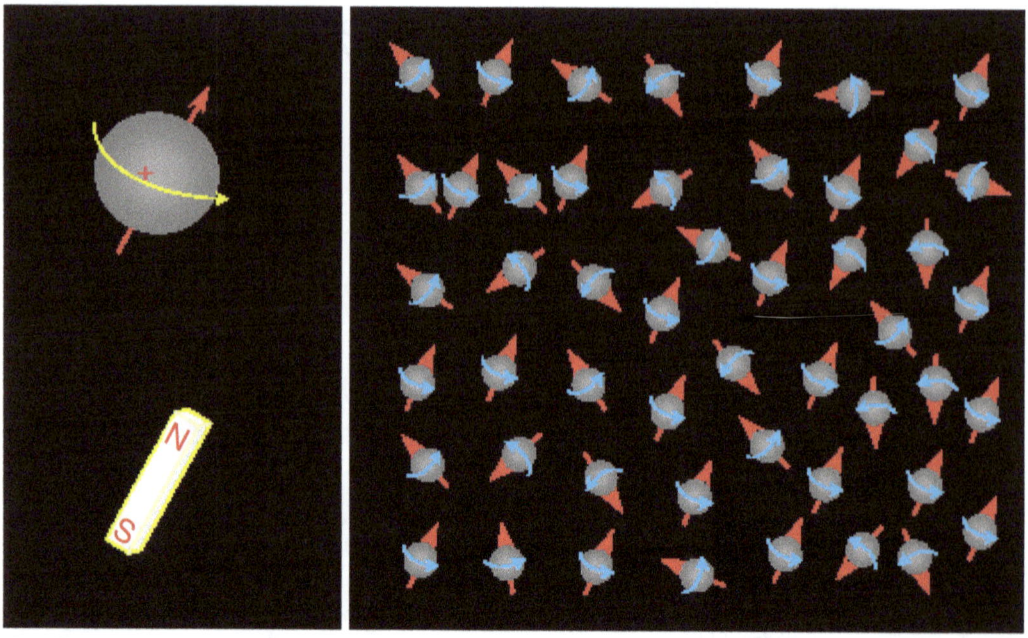

Fig. 3E.2—Because of their inherent nuclear magnetism, hydrogen nuclei (left) behave as though they are tiny bar magnets aligned with the spin axes of the nuclei. In the absence of an external magnetic field, the nuclear-magnetic axes (right) are randomly aligned.

In addition to improvements in wireline tools, new acquisition schemes and processing methods have improved the resolution, quality, and utility of the acquired data and enabled enhanced interpretation methods and data analysis. Concurrent with wireline improvements, LWD NMR logging services were being developed and have been introduced in the past few years. In a related development, a downhole NMR spectrometer is now available for use with a formation-testing tool for in-situ fluid analysis.

3E.3 NMR Physics

Atomic nuclei spin, and this angular moment produces a magnetic moment (i.e., a weak magnetic field). The NMR technique measures the magnetic signal emitted by spinning protons (hydrogen nuclei are the protons of interest in NMR logging) as they return to their original state following stimulation by an applied magnetic field and pulsed RF energy. These signals, which are observed (measured) as parallel or perpendicular to the direction of the applied magnetic field, are expressed as time constants that are related to the decay of magnetization of the total system.

NMR devices—both laboratory spectrometers and logging tools—use strong magnets to create a static magnetic field, B_0, that aligns (polarizes) the protons in the pore fluid from their resting (random) state to the direction of the imposed magnetic field (**Fig. 3E.2**).

Polarization is not instantaneous—it grows with a time constant, which is called the longitudinal relaxation time, denoted as T_1. Once full polarization (magnetic equilibrium) has been achieved, the applied static magnetic field, B_0, is turned off.

The protons begin to lose energy as the imposed magnetization, M_0, decays and the protons fall out of alignment, back to their original orientation and low-energy state. The protons' angular momentum causes them to behave like tiny gyroscopes, and the loss of energy occurs during a wobbling or axial rotation (called precession) in the direction of the applied magnetic field. M_0, also known as the bulk magnetization, provides the signals measured by NMR de-

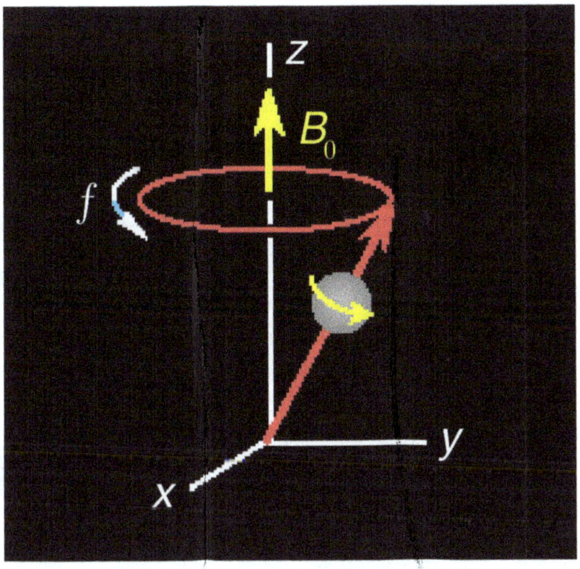

Fig. 3E.3—In an external magnetic field, the precessional frequency (*f*) of a nucleus depends on the gyromagnetic ratio (*γ*) of the nucleus and the strength of the external field (B_0).

vices. The frequency at which the energy is emitted or is initially absorbed, *f*, called the Larmor or resonance frequency, is proportional to the strength of the external magnetic field, B_0, (**Fig. 3E.3**). The Larmor frequency is used to tune a NMR probe, permitting it to image very thin slices of a sample at different distances from the tool.

An antenna detects and records the decaying magnetic field generated by the precessing nuclei. At any given time, *t*, the strength of this magnetic field, M_z, is proportional to the number of protons, the magnitude of B_0, and the inverse of the absolute temperature (Eq. 3E.1):

$$M_z(t) = M_0\left(1 - e^{\frac{-t}{T_1}}\right), \quad \text{(3E.1)}$$

where $M_z(t)$ = the magnitude of magnetization at *t*, M_0 = the final and maximum magnetization at a given magnetic field, and *t* = the time that the protons are exposed to the B_0 field.

The signal recorded parallel to the direction of the applied magnetic field (*z* plane) is called T_1, or longitudinal (spin-lattice) relaxation. T_1 describes how quickly the protons align within the static magnetic field. The T_1 curve is an exponential curve that characterizes the rate of change of the proton magnetization (**Fig. 3E.4**).

T_1 is the time at which the magnetization reaches 63% of its final value, and three times T_1 is the time at which 95% polarization is achieved. Full polarization of typical reservoir-pore fluids may take several seconds. Large values of T_1 (measured in milliseconds) correspond to weak coupling between the fluid and its surrounding environment and a slow approach to magnetic equilibrium, whereas, small T_1 values represent strong coupling and a rapid approach to equilibrium.[1] Different fluids, such as water, oil, and gas, have very different T_1 values. T_1 is directly related to pore size and viscosity.

Pulse NMR devices use precisely timed bursts (pulse sequences) of RF energy that generate an oscillating magnetic field (B_1) that tilts or "tips" the aligned protons perpendicular (*x-y* plane) to the direction of the applied magnetic field. The application of B_1 results in a change

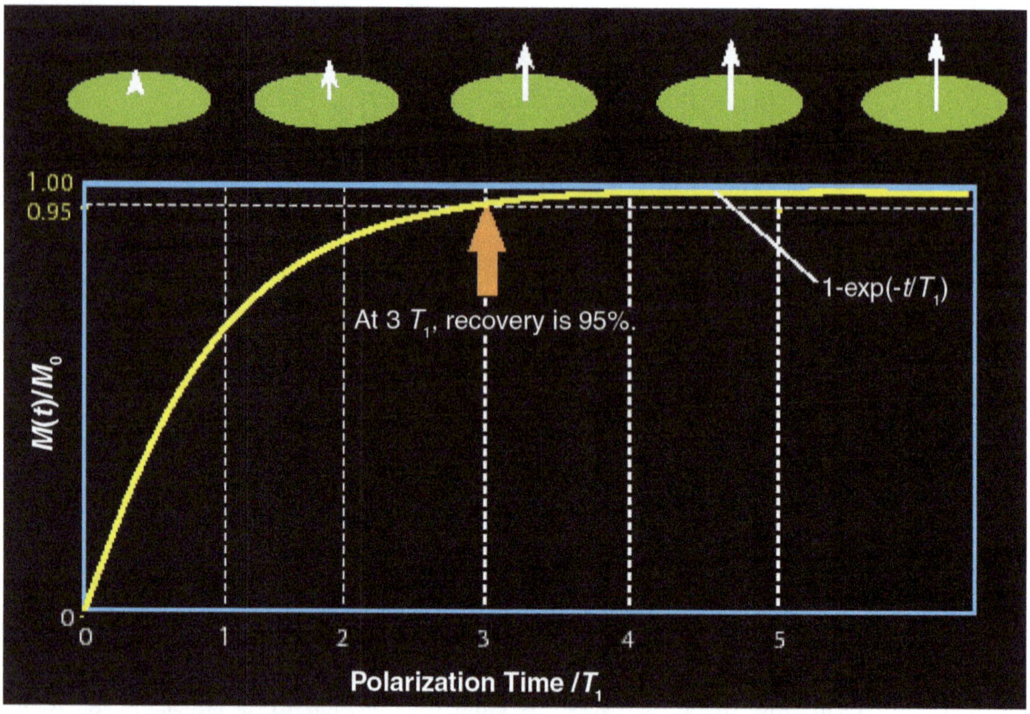

Fig. 3E.4—T_1-relaxation (polarization) curves indicate the degree of proton alignment, or magnetization, as a function of the time that a proton population is exposed to an external magnetic field.

in energy state that causes the protons to precess in phase to one another. These changes are known as NMR.

When the B_1 field is turned off, the precessions of the protons are no longer in phase with one another, and the net magnetization decreases. In this situation, a receiver coil (antenna) that measures magnetization in the transverse direction will detect an exponential decaying signal called free-induction decay (FID); see **Fig. 3E.5**. NMR-logging tools use the same antenna to transmit the RF pulse (kilowatt scale) and receive the decay signal (nanovolt scale).

The FID signal measured in the x-y plane is called T_2—the transverse or spin-spin relaxation. In contrast to T_1, T_2 of hydrocarbons is much shorter (see **Table 3E.2**).

The primary objectives in NMR logging are measuring T_1 signal amplitude (as a function of polarization), T_2 signal amplitude and decay, and their distributions. The total signal amplitude is proportional to the total hydrogen content and is calibrated to give formation porosity independent of lithology effects. Both relaxation times can be interpreted for pore-size information and pore-fluid properties, especially viscosity.

In the laboratory, T_1 is generally measured by either of two pulse sequences: inversion recovery or saturation recovery. Inversion recovery consists of a 180° spin inversion followed by a variable recovery time and then a 90° read pulse. The magnetization vector is entirely in the longitudinal range and, thus, has a higher dynamic range than the other method. Saturation recovery uses a 90° pulse, followed by a 90° read pulse. Saturation recovery is generally considered the more robust and efficient method. Although the actual T_1 sampling sequence is very short—involving several short echoes trains, each of which requires only a few milliseconds—the total amount of time required to obtain the number of samples sufficient to define the T_1 spectrum is significantly greater.

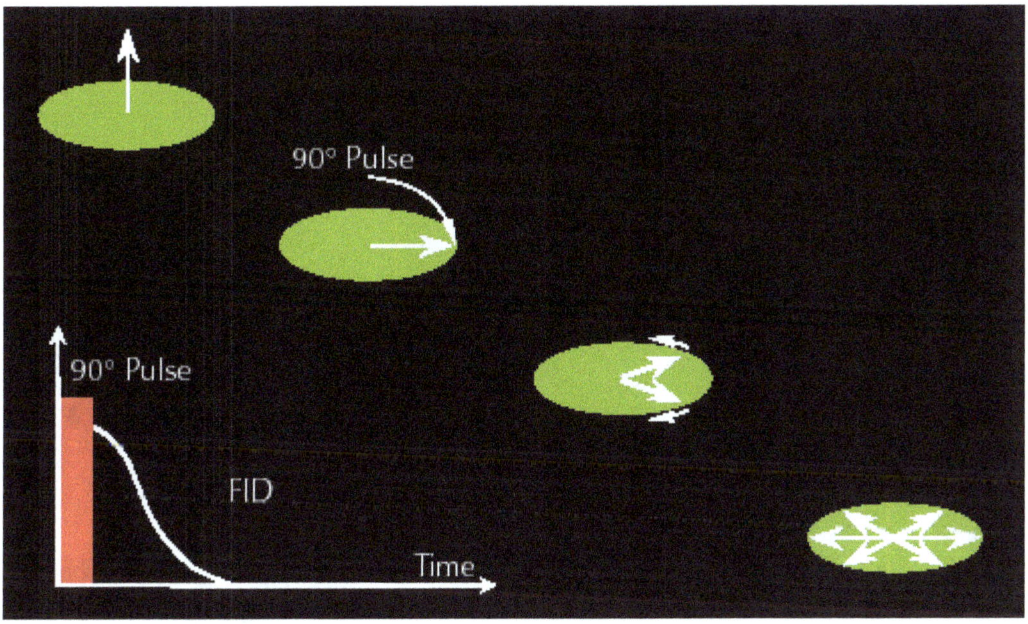

Fig. 3E.5—After application of a 90° pulse, the proton population dephases and an FID signal can be detected.

TABLE 3E.2—TYPICAL NMR-RELATED FLUID PROPERTIES FOR GULF OF MEXICO SANDSTONES[31]

Pore Fluid	T_1 (ms)	T_2 (ms)	$\xi(T_1/T_2)$	HI	η (cp)	$D_0 \times 10^{-5}$ (cm^2/s)
Brine	1–500	1–500	2	1	0.2–0.6	4–14
Oil	3,000–4,000	300–1,000	4	1	1–1,000	3–9
Gas	4,000–5,000	30–60	80	0.2–0.4	0.011–0.014	80–100

Depending on the activation used, the computation of a T_1 spectrum requires at least 25% more, and sometimes double, the time needed for the computation of a T_2 spectrum. In NMR logging, T_1 measurement initially required either a stationary mode or very slow logging speeds. With the latest multifrequency tools, a technique used for speeding up T_1 measurements is to make simultaneous measurements of the individual steps observed during a T_1 recovery experiment in adjacent volumes; at least two such volumes are required. This technique enables T_1 acquisition in less time, thereby permitting faster logging speeds.

T_2 measurement uses the spin-echo technique,[12] in which the protons are first tipped into the transverse (x-y) plane by a 90° RF pulse and then inverted (flipped) by a subsequent 180° RF pulse at a fixed-time interval to rephase the dephasing protons. Rephasing the protons creates a detectable signal called a spin echo (**Fig. 3E.6**).

In practice, a sequence of pulses is used to generate a series of spin echoes (echo train) in which echo amplitude decreases exponentially with the time constant, T_2. A variety of multiple-echo pulse sequences have been developed for different purposes.[11] In well logging and petrophysical studies, the most widely used is the Carr-Meiboom-Purcell-Gill (CMPG) sequence.[13,14] A polarization period is followed by a 90° tip pulse, which in turn is followed by

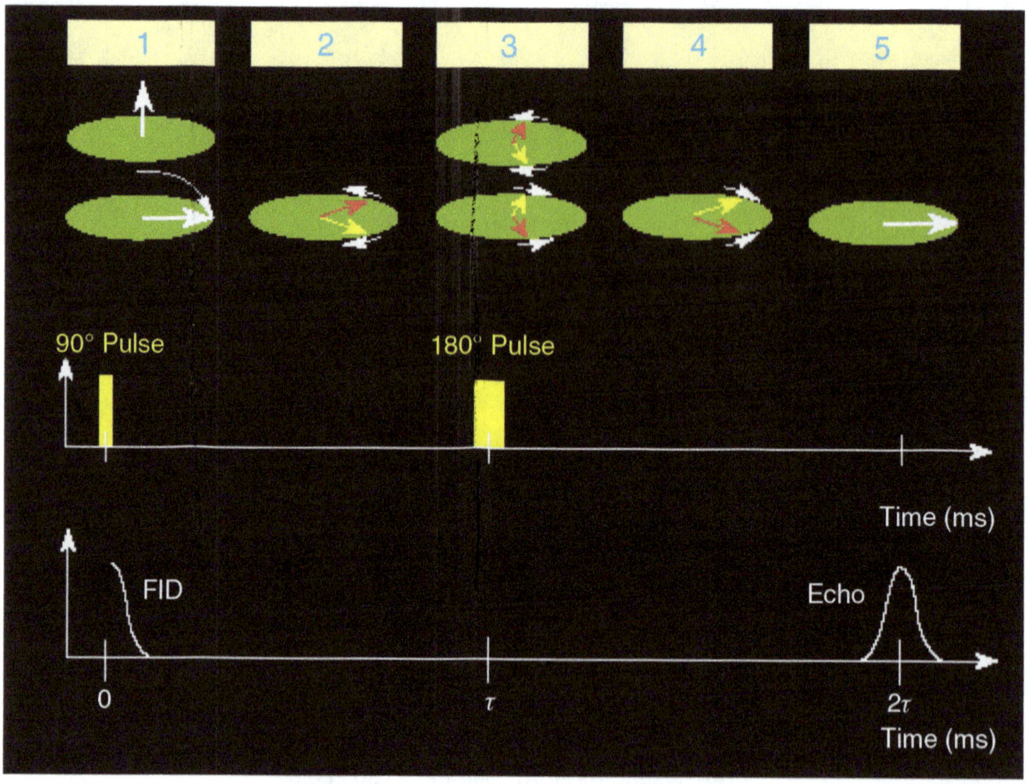

Fig. 3E.6—NMR spin echo: (1) to generate a spin echo, a 90° B_1 pulse is first applied; (2) after cessation of the 90° pulse, dephasing starts; (3) at τ, a 180° B_1 pulse is applied to reverse the phase angles and, thus, initiate rephasing; (4) rephasing proceeds; and (5) rephasing is complete, and a measurable signal (a spin echo) is generated at 2τ.

a series of alternating RF pulses and measurements of echo amplitudes detected by the logging-tool antenna. Successive 180° pulses are applied at a fixed-time interval (echo spacing, T_E), and the echoes are recorded between the pulses (**Fig. 3E.7**). By recording an echo train, T_2 can be calculated from the decay in the height (amplitude) of successive echoes[11] using Eq. 3E.2:

$$M_x(t) = M_{0x} e^{\frac{-t}{T_2}}, \quad \text{(3E.2)}$$

where $M_x(t)$ = the amplitude of the transverse magnetization (i.e., the amplitude of the spin-echo train) at time t, and M_{0x} = the magnitude of the transverse magnetization when t = 0 (i.e., the time at which the 90° pulse stops).

A single T_2-pulse sequence may involve several hundred or thousand echoes. Only the amplitude (peak) of each spin echo is measured and stored. A series of echo trains is recorded and the signals stacked to improve S/N, especially at shorter relaxation times.

When recording multiple CMPG sequences, the time period between spin-echo recovery and the next 90° CMPG excitation—during which the protons are repolarized by the static magnetic field—is called the wait time, T_W (**Fig. 3E.8**). Each CMPG sequence may use a different wait time, echo spacing, and number of echoes. An additional advantage of the CMPG sequence is that a small echo spacing, T_E, in the CMPG sequence can minimize the diffusion

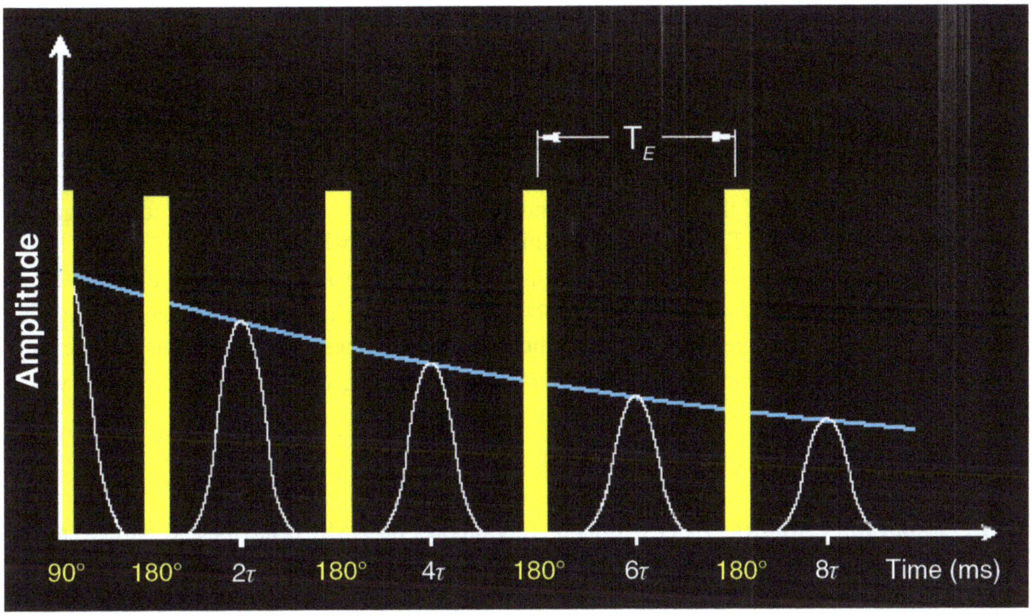

Fig. 3E.7—The CPMG pulse sequence is used in NMR logging to generate a spin-echo train; a 90° B_1 pulse is followed by a sequence of 180° B_1 pulses. Spin echoes of decreasing amplitude follow the 180° B_1 pulses.

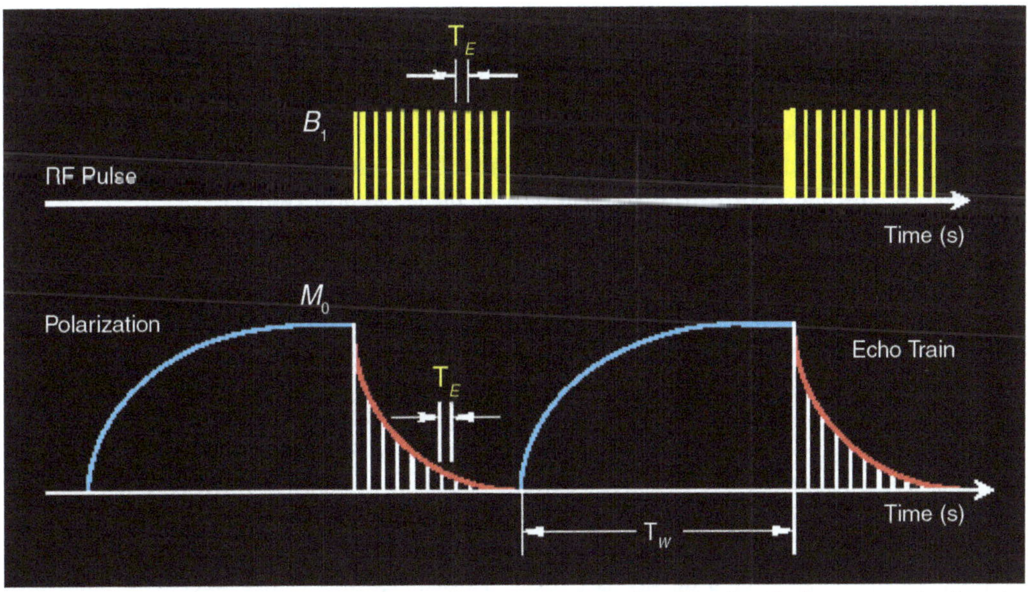

Fig. 3E.8—CPMG pulse sequences (top) and alternating polarization (T_1-relaxation) curves and spin-echo train acquisition (bottom).

effect on T_2. CMPG measurement sets are always collected in phase-alternate pairs (PAP) to preserve the signal and to eliminate low-frequency electronic offsets. In general, pulse NMR offers better methods to measure relaxation times and quantify liquid displacement in rock.[15]

3E.3.1 In-Gradient Diffusion. FID is caused by inhomogeneities in the magnetic field that are primarily caused by the existence of magnetic-field gradients. Gradients in the magnetic field occur, in part, because of the distance from the magnet to the sensitive (measurement) volume. For a given geometry, the gradient is inversely related to magnetic-field strength. Compared to laboratory and clinical NMR devices, NMR-logging tools produce a relatively weak and inhomogeneous static magnetic field. In the case of reservoir rocks, differences between the magnetic properties of the rock matrix and pore fluids may also contribute to a magnetic-field gradient. T_2, but not T_1, is affected by this phenomenon, which is called diffusion. In the presence of high magnetic-field gradients, diffusion effects make T_2 interpretation difficult. However, because the gradients produced by NMR-logging tools are relatively constant, they can be accounted for in T_2 interpretation. In fact, the existence of these field gradients has actually proved beneficial in NMR logging. Magnetic resonance imaging (MRI) is the process by which NMR measurements are obtained in a gradient magnetic field.

3E.4 NMR Petrophysics

3E.4.1 Laboratory Studies. Extensive laboratory studies on NMR behavior and on the properties of fluid-saturated porous media have been conducted since the inception of NMR and throughout the development of NMR-logging tools. The results from these investigations have provided the petrophysical foundation for understanding the logging measurements and for developing interpretation models and applications.

Low-field, bench-top pulse-NMR spectrometers were developed concurrently with logging tools so that wellbore measurements could be duplicated on core samples in the laboratory.[16,17] These instruments operate and record data in the same manner as NMR-logging tools.[18] Because NMR analysis is nondestructive, NMR and conventional capillary-pressure measurements can be performed on the same samples, in both the saturated and partially saturated states. Low-field spectrometers provide the ability to make repeatable measurements of rock- and fluid-NMR properties. This ability, in turn, permitted correlation and calibration of laboratory and field measurements and also permitted direct transfer of interpretation models developed in the laboratory to logging data. Where core is unavailable for NMR-log calibration, new technology and methods now allow NMR petrophysical measurements on drill cuttings.[19]

Laboratory NMR studies are routinely conducted for the following purposes:

- Verifying formation porosity.
- Evaluating textural effects, such as microporosity, on NMR-log response.
- Determining formation-specific models that enhance the accuracy of determining bulk-volume-irreducible (BVI) water, free-fluid index (FFI), and, ultimately, permeability.
- Developing models to identify and quantify hydrocarbons, including residual oil.
- Developing models to predict changes in pore size (facies).

Much of this work is summarized in Kenyon,[20] Murphy,[21] Woessner,[3] and Dunn *et al.*[11] The most recent laboratory studies have suggested that some established NMR core-log relationships should be further investigated to better account for data scatter.[22] A related area of study not dealt with here is NMR imaging of fluid flow in core.[23]

As in NMR logging, data quality is critical. To achieve the desired objectives, laboratory NMR studies should include a preplanning phase similar to that used in logging. (See the Job Planning section of this chapter).

3E.4.2 Petrophysical Properties. The basic petrophysical parameters of porosity, permeability, and producibility can be determined from either T_1 or T_2 echo-decay data. Until low-field spectrometers were developed, T_1 was the preferred acquisition method in the laboratory, where time is not a concern.[18,20] Because T_1 measurement requires more time than T_2, T_2 became the

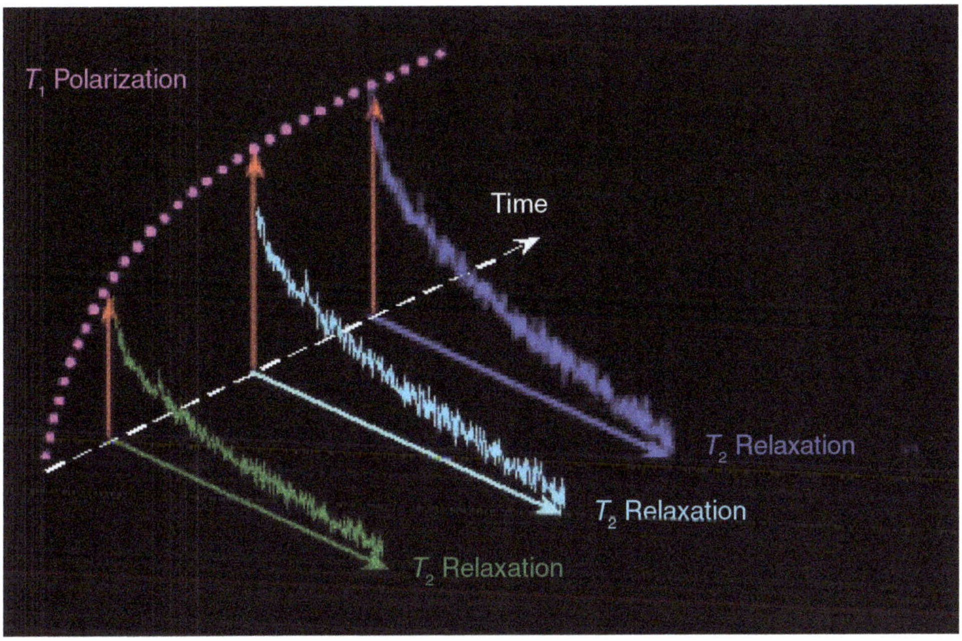

Fig. 3E.9—T_1 can be determined from a series of multiple wait-time T_2 measurements. The peak amplitudes from the T_2 measurements trace the outline of the T_1 curve (dotted).

primary acquisition mode in pulse-NMR logging because it allowed logging at speeds that were commercially viable. Fortunately, there is a correlation between T_1 and T_2[24]; and T_1 can be estimated from T_2 data by extrapolating the T_2 decay-obtained-polarization pulses of different lengths (i.e., using different values of T_W; see **Fig. 3E.9**).

Modern logging tools are capable of operating in either T_1- or T_2-acquisition modes. The logging mode is dictated by operational factors and job objectives and may, in fact, switch back and forth, as needed. (See further discussion in the LWD Tool section of this chapter.)

3E.4.3 NMR Properties of Fluids. T_1 relaxation occurs when the precessing proton system transfers energy to its surroundings. T_2 relaxation occurs through a similar transfer in energy and also through dephasing. Consequently, transverse relaxation, T_2, is always faster than longitudinal relaxation, T_1. The emphasis of the proton NMR techniques used in formation evaluation is on the NMR fluid response from the pore fluids where $T_2 \leq T_1$. The NMR response in solids (i.e., the non-shale/clay component of the rock matrix) is very short compared to the pore-fluid signal, and is generally not measured by laboratory or logging devices.

NMR relaxation of fluids depends on whether the fluid is measured in bulk form as a wetting-pore fluid within a rock matrix or in a gradient magnetic field. Bulk relaxation is the intrinsic relaxation property of a fluid that is controlled by viscosity, chemical composition, temperature, and pressure: $T_{2\text{bulk}} \cong T_{1\text{bulk}}$. Fluids contained within pores have different relaxation characteristics, namely those of surface relaxation.

Surface relaxation occurs at the fluid-solid interface between a wetting-pore fluid and the rock-pore walls (**Fig. 3E.10**) and is different from the relaxation in either the solid or the fluid, individually. Surface relaxation dramatically decreases both T_2 and T_1 and is the dominant component contributing to T_2. When a nonwetting fluid (e.g., oil) is also present in rock pores, the nonwetting fluid may continue to relax at its bulk-relaxation rate.

Surface relaxation is expressed by the following equations (Eqs. 3E.3 and 3E.4):

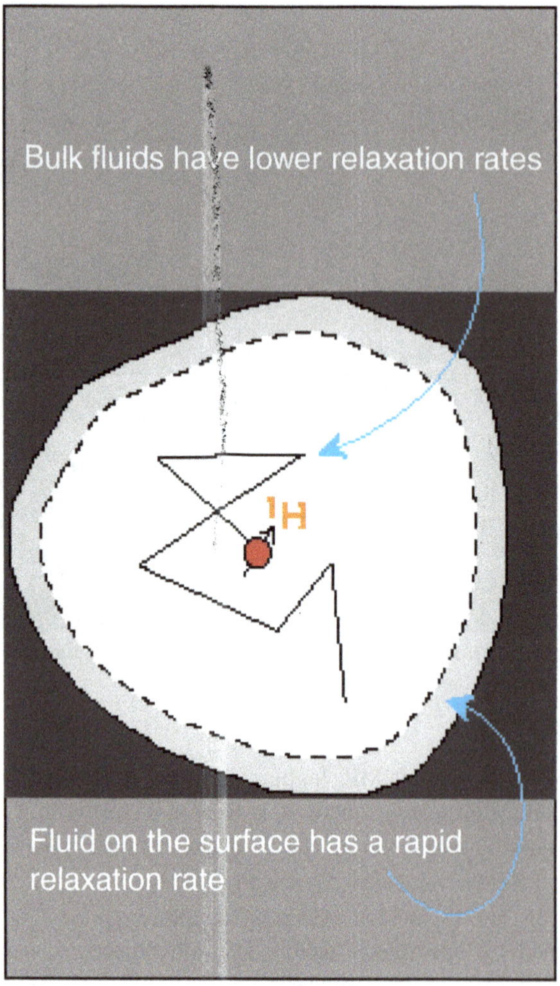

Fig. 3E.10—Schematic illustrating surface relaxation.

$$\frac{1}{T_{2\text{surface}}} = \rho_2 \times \left(\frac{S}{V}\right)_{\text{pore}} \quad \text{(3E.3)}$$

and

$$\frac{1}{T_{1\text{surface}}} = \rho_1 \times \left(\frac{S}{V}\right)_{\text{pore}}, \quad \text{(3E.4)}$$

where ρ_2 = T_2 surface relaxivity (i.e., T_2 relaxing strength of the grain surfaces); ρ_1 = T_1 surface relaxivity (i.e., T_1 relaxing strength of the grain surfaces); and $(S/V)_{\text{pore}}$ = ratio of pore surface to fluid volume.

Surface relaxivity varies with mineralogy; and, for simple pore shapes (S is the surface area of a pore and V is the volume of the same pore), S/V is a measure of pore size. In a brine-wet rock, T_2 in smaller pores will be less than T_2 in large pores; consequently, identical pore water in different rocks can have a wide range of relaxation times because of variations in surface

TABLE 3E.3—PRIMARY T_2 RELAXATION MECHANISM IN WATER-WET ROCKS

Fluid in Pores of Water-Wet Rock	Dominant T_2 Relaxation Mechanism	Application of Capability
Brine	$T_{2surface}$	Evaluate clay- and capillary-bound water
Heavy oil	T_{2bulk}	Identify heavy oil as T_2 is notably less than light oil and water
Intermediate viscosity and light oil	T_{2bulk} and $T_{2diffusion}$ (viscosity dependent)	Fluid typing
Gas	$T_{2diffusion}$	Fluid typing

relaxivity. Laboratory studies have demonstrated that in water-wet rocks, the surface and volume ratio (S/V) is also a measure of permeability.

Fluids controlled by surface relaxation exhibit T_2 values that are not dependent on temperature and pressure. For this reason, laboratory NMR measurements made at room conditions are commonly used to calibrate formulas used to estimate petrophysical parameters such as permeability and bound water.[25,26]

Diffusion-induced relaxation occurs when a significant gradient exists in the static magnetic field. Molecular diffusion in this gradient causes additional dephasing that contributes to increased T_2 relaxation. In addition to the magnetic-field gradient, diffusion is also controlled by inter-echo spacing and fluid diffusivity, viscosity, molecular composition, temperature, and pressure.

Bulk-fluid processes and surface relaxation affect both T_1 and T_2, while diffusion only affects T_2 relaxation. All three processes are independent and act in parallel according to the following equations (Eqs. 3E.5 and 3E.6):

$$\frac{1}{T_2} = \frac{1}{T_{2\text{bulk}}} + \frac{1}{T_{2\text{surface}}} + \frac{1}{T_{2\text{diffusion}}} \quad \text{(3E.5)}$$

and

$$\frac{1}{T_1} = \frac{1}{T_{1\text{bulk}}} + \frac{1}{T_{1\text{surface}}} \ . \quad \text{(3E.6)}$$

The relative importance of the three diffusion-relaxation mechanisms depends on the fluid type in the pores (e.g., water, oil, or gas), the sizes of the pores, the strength of the surface relaxation, and the wettability of the rock surface (see **Table 3E.3**).

3E.4.4 T_2 Decay. Eq. 3E.2 states that the T_2 decay associated with a single pore size in water-saturated rocks is proportional to the pore size.[20] In fact, because reservoir rocks typically comprise a distribution of pore sizes and frequently contain more than one fluid type, a CMPG T_2 spin-echo train actually consists of a distribution of T_2 decays, rather than a single T_2 decay. In these cases, the exponential decay is described by Kenyon *et al.*[27] as follows:

$$M(t) = \sum M_i(0) e^{-\frac{t}{T_{2i}}}, \quad \text{(3E.7)}$$

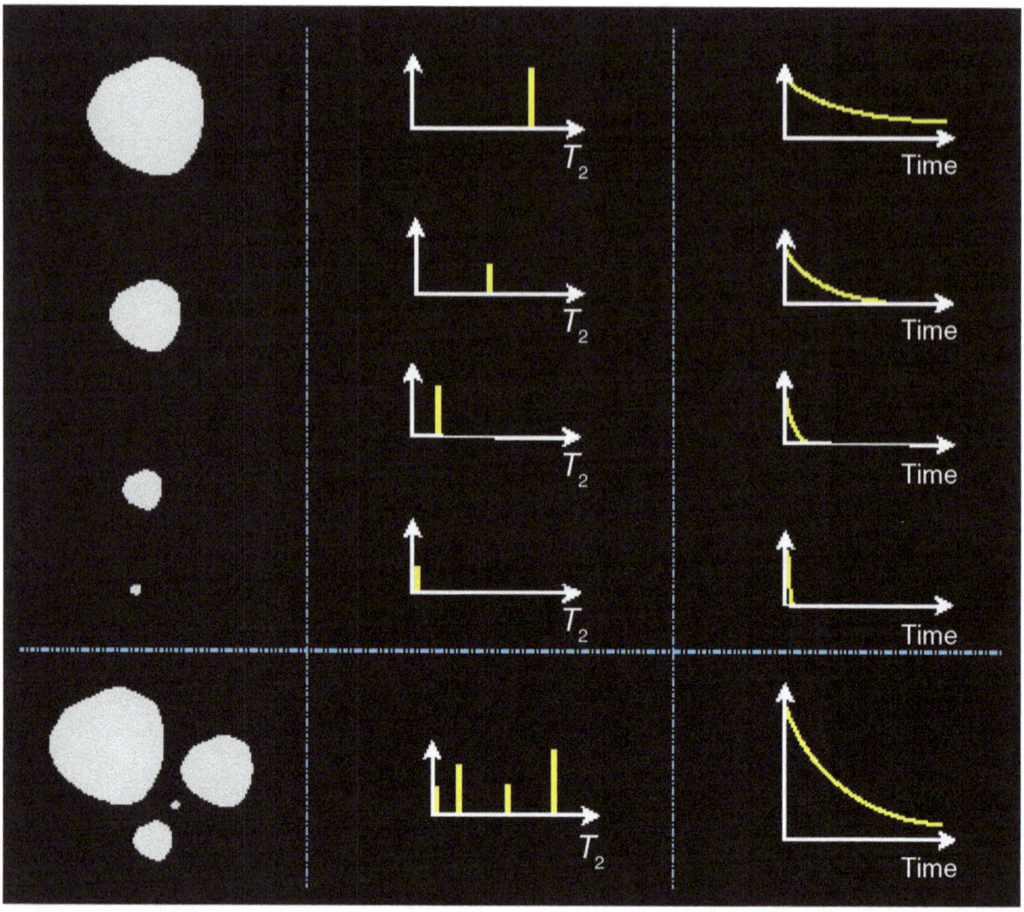

Fig. 3E.11—A 100% water-saturated pore (upper left) has a single T_2 value (upper center) that depends on pore size and, thus, its spin-echo train exhibits a single exponential decay (upper right) that also depends on pore size. Multiple pores at 100% water saturation (bottom left) have multiple T_2 values (bottom center) that depend on the pore sizes and, thus, their composite spin-echo train exhibits multiexponential decay (bottom right) that also depends on the pore sizes.

where $M(t)$ = measured magnetization at t; $M_i(0)$ = initial magnetization from the ith component of relaxation; and T_{2i} = decay constant of the ith component of transverse relaxation. The summation is over the entire sample (i.e., all pores and all different types of fluid).

Fig. 3E.11 illustrates the multiexponential decay character of a porous medium containing pores of different sizes and a single wetting phase. Surface relaxation dominates when a short inter-echo spacing is used and the formation is only brine saturated. Under this condition, T_2 is directly proportional to pore size. When all pores are assumed to have similar geometric shape, the largest pores (see Fig. 3E.11, left column) have the lowest S/V and, thus, the longest T_2. Medium-size pores have smaller S/V, yielding shorter T_2 values. The smallest pores have the highest S/V and the shortest T_2 values.

Eq. 3E.7 can also be expressed as follows:[28]

$$M(t) = \sum M_{0i} e^{-\rho_2 \left(\frac{S}{V}\right)_i t}, \qquad (3E.8)$$

where $(S/V)_i$ is the surface-to-volume ratio of the ith pore. When $t = 0$, the following is true:

$$M(0) = \sum M_{0i} \;. \qquad \text{(3E.9)}$$

If the measured magnetization for 100% bulk water, $M_{100\%}(0)$, is known, then $M(0)$ and M_{0i} can be calibrated to porosity by the following equation:

$$\phi = \frac{M(0)}{M_{100\,\%}(0)} = \frac{\sum M_{0i}}{M_{100\,\%}(0)} = \sum \frac{M_{0i}}{M_{100\,\%}(0)} = \sum \phi_i, \qquad \text{(3E.10)}$$

where ϕ = calibrated porosity of the formation, ϕ_i = calibrated porosity associated with all pores of the ith pore size (also known as the incremental porosity). Therefore, the T_2 distribution (in the form of the amplitudes, M_{0i}, associated with the time constants, T_{2i}, is calibrated to the porosity distribution (i.e., the individual pores ϕ_i with the associated time constants T_{2i}).

If the rock is water-wet and the pores are partially saturated (i.e., the pores contain oil and/or gas in addition to water), then the total signal comprises contributions from each component in the following equation:

$$M(t) = \sum M_{0i} e^{-\rho_2 \left(\frac{S}{V}\right)_i t} + M_{\text{oil}} e^{-\frac{t}{T_{2\text{oil}}}} + M_{\text{gas}} e^{-\frac{t}{T_{2\text{gas}}}}, \qquad \text{(3E.11)}$$

where M_{oil} = magnetization produced by oil protons in the pores, M_{gas} = magnetization produced by gas protons in the pores, $T_{2\text{oil}}$ = T_2 of oil measured with a CMPG sequence, and $T_{2\text{gas}}$ = T_2 of gas measured with a CMPG sequence.

In a brine-saturated rock, the T_2 decay spectrum represents a pore-size distribution. However, when nonwetting fluids (e.g., oil or gas) are present, the T_2 spectrum includes a bulk response from the nonwetting fluid, in addition to the pore-size response. Pores containing the nonwetting fluid either appear in the spectrum at a decay time that is faster than is normally associated with the pores, or do not appear at all if the surface layer is too thin. This behavior affects the appearance of the T_2 spectrum and associated T_2 distribution but not the total signal amplitude (i.e., porosity).

3E.4.5 Data Fit—Inversion. The raw data recorded by an NMR device are a series of spin-echo amplitudes (echo train) as a function of time, usually at fixed time increments (bins). These NMR measurements are statistical, and stacking is required to improve the precision of the log outputs. The precision depends on formation-fluid properties, activation (acquisition) type, and the number of pulses stacked.[29] The data are mathematically inverted (mapped) by use of a best-fit curve to produce a distribution of T_2 values as a function of relaxation time (**Figs. 3E.12 and 3E.13**). Initially, biexponential-fitting algorithms were used[30]; however, T_2 decay in fluid-saturated rocks is multiexponential, because of this, multiexponential models were developed and are commonly used for inverting the data.[11] The inverse solution (T_2 distribution) is a function of both the measured echo data and the chosen smoothness (regularization) for the inversion. However, because regularization is controlled in part by the S/N, the fit that is actually used is not unique; that is, there can be a number of differently shaped T_2 distributions that fit the original echo-decay curve. In general, the area under the T_2-distribution curve (porosity) and the general location in time of the high-porosity bins are robust; however, caution is advised when interpreting the fine details of the distribution.[31] Because the spin echoes are measured over a very short time, an NMR tool travels no more than a few inches

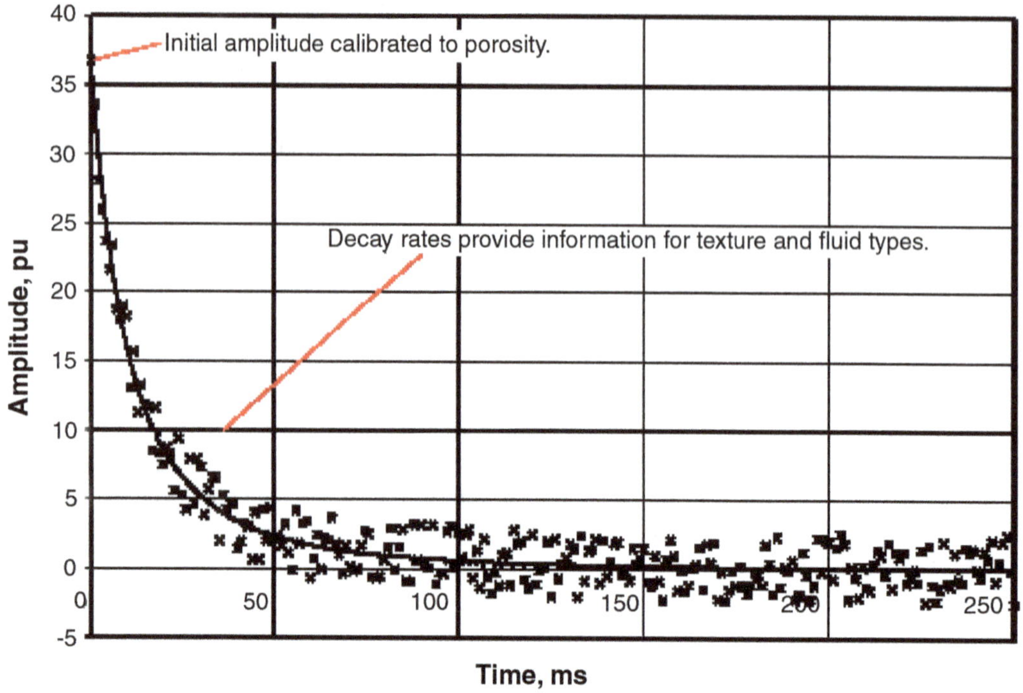

Fig. 3E.12—Spin-echo-decay train. The amplitudes of the spin echoes are recorded over time. A best fit line is applied to the data before inversion. In this example, the raw spin-echo measurements recorded at T_E = 1 ms are shown together with a data-fitting curve (solid).

along the wellbore while recording the spin-echo train; thus, the recorded spin-echo data can be displayed on a log as a function of depth.

3E.4.6 T_2 Distribution. The mathematical statement (Eq. 3E.10 and Eq. 3E.11) that T_2 distribution observed in water-saturated rock represents the pore-size distribution and porosity of the rock has been confirmed using mercury-injection capillary pressure (MICP) methods (see **Figs. 3E.14 and 3E.15**).[32,33]

NMR responds to pore-body size, and MICP responds to pore-throat size.[32,34–36] In clastic rocks in which there is a good correlation between pore-body and pore-throat size, there is often good qualitative agreement between NMR and MICP data, as Figs. 3E.14 and 3E.15 illustrate. The NMR porosity, reflected in the T_2 distribution, is a spectrum comprising rock matrix and fluid components (see **Fig. 3E.16**).

Although matrix minerals and dry clay may contain hydrogen atoms in the form of hydroxyl groups, the T_1 relaxation times of these nuclei are too long to be polarized by a moving NMR-logging tool, and their T_2 relaxation times are too short to be recorded.[37] The fluid component is subdivided into bound and free subcomponents. The hydrogen nuclei of clay-bound water are adsorbed on the surfaces of clay grains. These hydrogen protons can be polarized by NMR-logging tools and recorded when a sufficiently short T_E is used.[38] Similarly, hydrogen protons in capillary-bound water and movable fluids (e.g., free water, mud filtrates, oil, and gas), are polarized and recorded by NMR-logging tools with appropriate values for T_E and T_W.

The porosity and pore-size information from NMR measurements can be used to estimate producible porosity (i.e., the movable fluids) and permeability and for hydrocarbon identification (see the Applications section of this chapter).

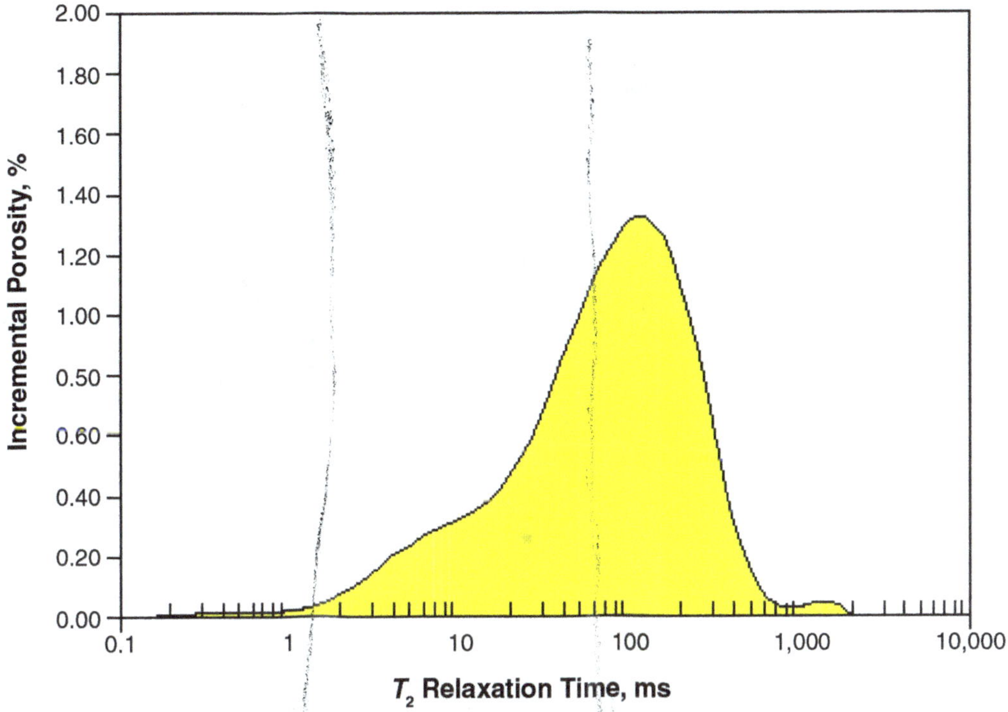

Fig. 3E.13—T_2 distribution. Mathematical inversion is used to convert the spin-echo-decay data to a T_2 distribution. This distribution is the most likely distribution of T_2 values that produce the echo train. With proper calibration, the area under the T_2-distribution curve is equal to the porosity. When the rock is water-saturated, this distribution will correlate with pore-size distribution. The presence of hydrocarbons will affect the T_2 distribution depending on the hydrocarbon type, viscosity, and saturation.

3E.5 Logging Tools

Compared to laboratory or clinical NMR, the dimensions of a typical borehole and the nature of continuous logging impose severe constraints on the physics, equipment, and operation of NMR-logging tools. Unlike laboratory devices, logging measures an external sample by use of a weaker magnetic field while in motion relative to the sample.

3E.5.1 NML Tool. The NML tool—the first generation of NMR logging (1960–1994)—was an Earth's field device that measured the free-induction decay in the Earth's magnetic field. Proton polarization (alignment) was achieved using a magnetic field produced by a coil energized with a strong direct current. Each experiment required several seconds to allow complete polarization. The power was turned off, and the same coil was used to receive the free-induction signal.

There were a number of problems inherent in this technology. NML devices measured the proton signal in the borehole fluid as well as in the surrounding formation. Magnetite powder was circulated into the mud system to cancel the borehole hydrogen signal thereby preventing it from overwhelming the formation signal. This was a time-consuming process. The low intensity of the Earth's magnetic field (0.5 gauss) resulted in a low S/N. The NML had a large dead time (i.e., time between the cessation of the static magnetic field and the first measurement), approximately 20 ms; pore fluids with relaxation times less than the dead time were not measurable. The initial signal amplitude had to be extrapolated backwards from the subsequent free-induction decay data. The NML could not distinguish between oil and water or measure total

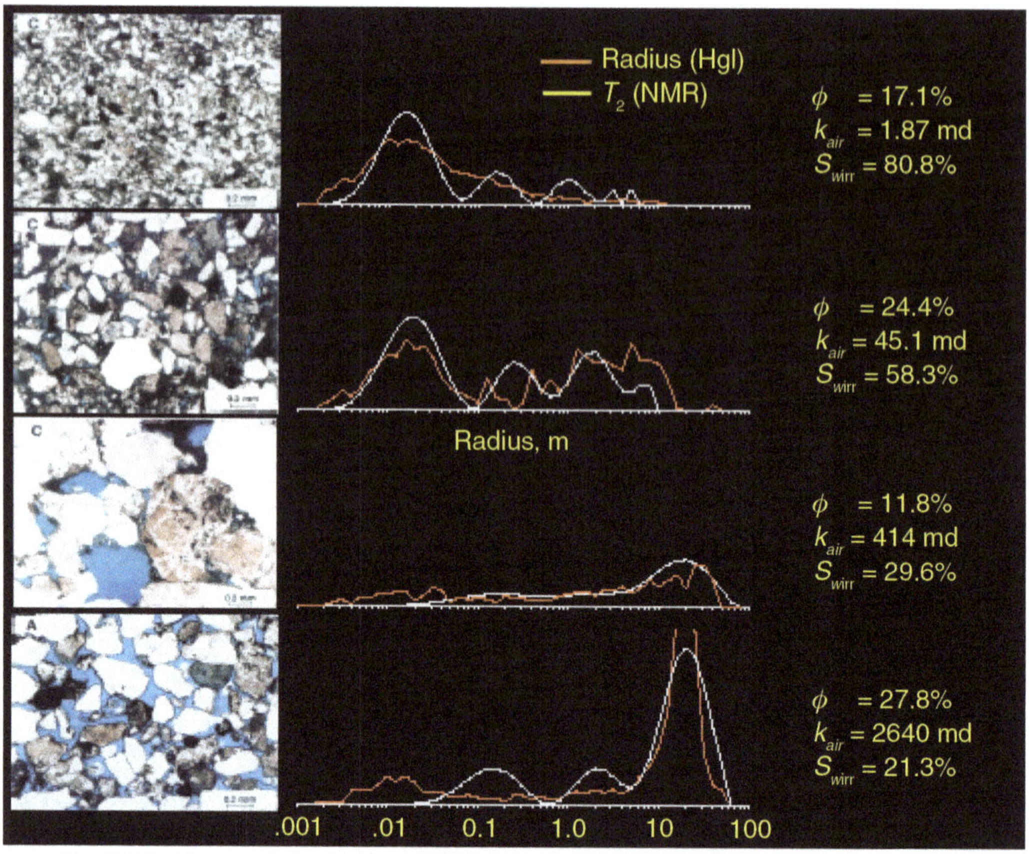

Fig. 3E.14—The correlation between MICP and T_2 distribution in water-saturated sandstones.

porosity; however, the dead time was close to the T_2 cutoff established for irreducible water saturation. The NML could detect movable fluids and provided a measurement called the Free Fluid Index (FFI).[11] An expanded discussion of the NML technology can be found in Brown and Neuman[39] and Brown.[1]

3E.5.2 Pulse NMR. Unlike conventional magnetic resonance imaging (MRI) devices in which the sample is placed inside the stationary coil, a borehole-logging device investigates a sample (rock volume) that is outside the device itself while moving along the borehole. This setup has been called the "inside-out" NMR problem because the geometry of the magnets and coils is inverse to that used in laboratory-NMR spectrometers.[40–44] To obtain useful measurements, a logging tool must generate a large, radially symmetric static magnetic field and also a high-frequency oscillating magnetic field. Each must be capable of penetrating one or more inches into the formation surrounding the borehole. The diameter of a typical well limits the size of the permanent magnets that can be used and, therefore, the strength of the magnetic field that can be generated by a logging device. In comparison with laboratory-NMR and clinical-NMR devices, which may operate at 10 MHz and the static magnetic fields of which are commonly in the range 1 to 2 Tesla (high field), modern logging tools and laboratory spectrometers designed for petroleum investigations are considered low-field devices. They operate at or below 2 MHz and generate relatively weak (< 200 gauss) and inhomogeneous magnetic fields (i.e., gradients up to 20 gauss/cm). By comparison, the NML operating frequency was only 2 kHz,

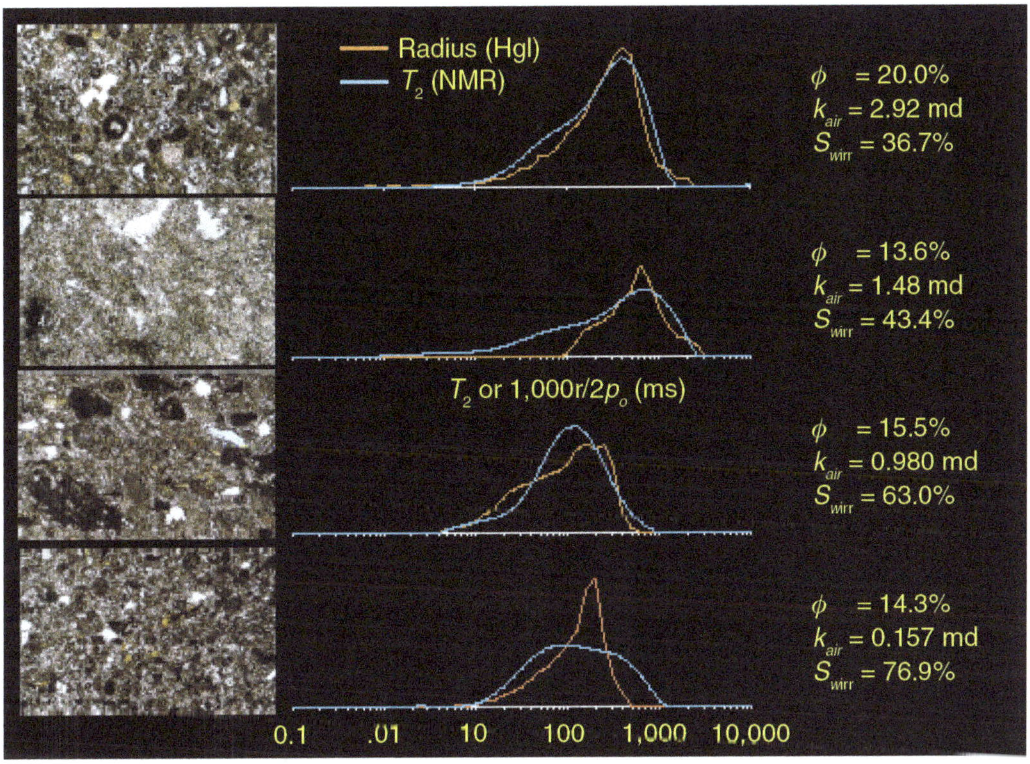

Fig. 3E.15—The correlation between MICP and T_2 distribution in water-saturated carbonates.

and the Earth's magnetic field is only 0.5 gauss. These factors limit borehole investigation to protons (hydrogen) and the use of relaxation data only. Chemical shifts (widely used in the biological field) are not observable.[7] Furthermore, to compensate for the lower S/N that results from low-field intensity, logging tools must acquire more echoes and/or stack data to improve S/N. As mentioned previously, with the introduction of pulse-echo tools, T_2 became the primary acquisition mode because it permitted faster logging—a major factor in high-cost wells.

This new design provides a number of operational advantages over the NML:

• Using permanent magnets rather than electromagnets reduces the surface-power requirement.

• Focusing the sensitive region of the magnetic field at some distance into the formation eliminates the requirement for suppressing the mud signal.

• Using an RF pulse from a coil tuned to the Larmor frequency ensures that only nuclei in the sensitive region are in resonance.

• Controlling the pulse duration means shorter dead times that, in turn, allow a better estimate of initial decay amplitude (porosity) measurement for short T_2 components (bound-fluid evaluation).

• Enabling more sophisticated pulse sequences allows for measurement of additional rock and fluid properties.[44]

One initially unwanted product of this design is the creation of gradients in the static magnetic field that causes molecular diffusion. The strength of the magnetic field gradient, G, is controlled by tool design and configuration (i.e., tool size and tool frequency); by environmental conditions such as formation temperature; and by internal gradients induced by the applied field, B_0. Subsequent characterization of these gradients has enabled the in-gradient diffusion to be used for hydrocarbon typing.

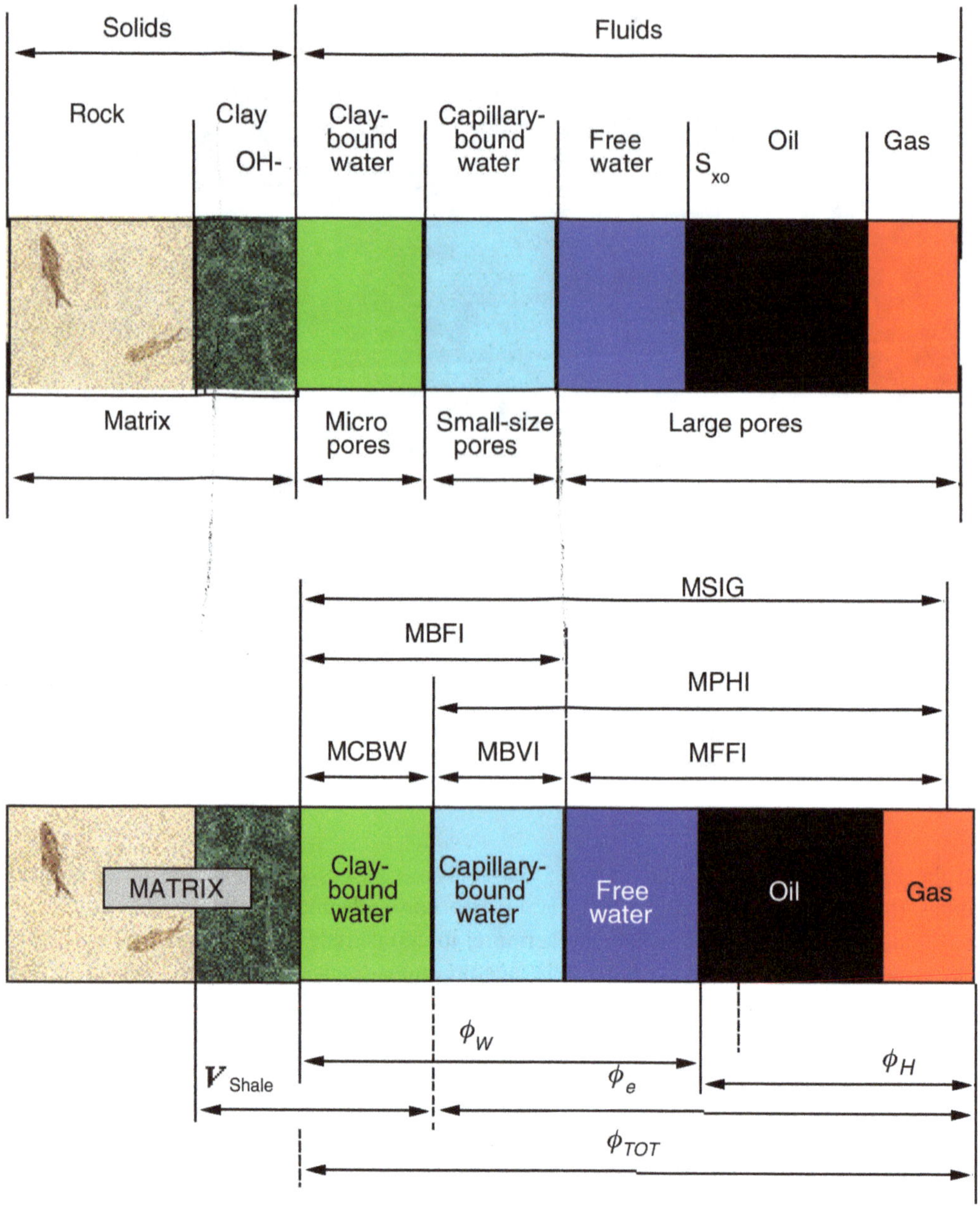

Fig. 3E.16—NMR-porosity model. These figures illustrate the distribution of pore size and pore fluids used in NMR logging. The suggested NMR mnemonics are Halliburton's. Because NMR tools respond to the invaded zone, a mud-filtrate component is added between movable water and oil. NMR-logging tools are sensitive to fluids but not to matrix materials and dry clay. NMR porosity is affected by HI, T_W, and T_E.

3E.5.3 Wireline-Tool Designs. From the beginning, a debate in NMR logging has been whether to use a centralized or eccentered tool design. There are two different wireline designs in current commercial service: (1) the Numar (now a part of Halliburton) magnetic resonance imaging tool (MRIL), a centralized mandrel device[30,45–48]; and (2) the Schlumberger CMR tool[49–55] and the Baker Atlas MREX[56] eccentered devices, both of which require contact with the borehole wall. The general designs of both the MRIL (mandrel-type) and CMR (pad-type) have evolved over the past decade with the addition of new capabilities and faster logging

TABLE 3E.4—CAPABILITIES AND ADVANTAGES OF T_1 AND/OR T_2 MEASUREMENTS

Measurement Capability			Relative Influence on Measurement		
Quantity	T_1	T_2	Parameter	T_1	T_2
Porosity	Yes	Yes	Field sensitivity	Low	High
Permeability	Yes	Yes	Signal decay	No	Yes
Producibility	Yes	Yes	Experiment time	Long	Short
Viscosity	No	Yes	S/N (at logging frequency)	High	Low
Diffusion effect	No	Yes*	Motion sensitivity		
Signal amplitude	Yes	Yes	Vertical	High	Low
			Lateral	Low	High**
			Bed-boundary effects	High	Low

* constrained by echo spacing
** constrained by tool design

speeds, made possible by improved electronics and improved data acquisition and processing. Computalog is field testing a prototype mandrel device: the Nuclear Magnetic Resonance Tool (NMRT) developed by NPF Karotazh.[57]

A centralized NMR-logging tool like the MRIL must be longer than a pad device, simply to contain magnets of sufficient strength to project the required magnetic field, through the borehole and into the formation. This factor results in a greater sensitivity to borehole salinity than a pad device, which can exclude the mud from its measurement. The MRIL design generates relatively thin (1 to 2 mm) sensitive volumes, but the reduced S/N that accompanies these volumes is compensated by the vertical size of the sensitive area. It also generates a relatively high magnetic gradient. In contrast, a contact device, such as the CMR, can use smaller magnets and electronics, which provide higher vertical resolution but a shallower depth of investigation (DOI) and greater sensitivity to borehole rugosity. In addition to the standard permanent magnets, some designs now include "prepolarization" magnets, which are added to ensure full polarization at typical logging speeds.

The latest wireline NMR-logging tools operate simultaneously in several RF frequencies to measure (image) multiple sample volumes. In the presence of a gradient magnetic field, pulses with different frequencies will cause protons in different (and parallel) regions of space (i.e., measurement or sensitive volumes) to resonate.[31] Cycling through several frequencies excites protons in different cylindrical volumes, allowing measurements to be made more quickly. If the frequencies of multifrequency measurements are very close together, then the sensitive volumes are very close together; and, for practical purposes, the rocks sampled can be considered to be the same. This principle is the same as that used for slice selection in medical MRI imaging. These tools acquire multiple echo trains using different values of T_W, T_E, and variable magnetic gradients (G) in a single logging pass. The time between measurements made at multiple frequencies can be as little as the time of an echo train, and the time between measurements made at a single frequency is essentially the time needed to repolarize (T_W). The thickness of these sensitive volumes may be as small as 1 mm. Furthermore, recent advances in tool design now permit cost-effective T_1 acquisition.[58] **Table 3E.4** summarizes the capabilities, advantages, and disadvantages of T_1 and T_2 acquisition. These differences influenced the designs of the LWD tools discussed in the next section. Multifrequency operation provides measurements at multiple DOIs (typically 1 to 4 in.). This allows invasion effects to be accounted for in the data interpretation, thus enabling determination of near-wellbore fluid saturation and oil properties at high resolution.[54,58–60]

3E.5.4 LWD Tools. The latest entries into NMR logging are LWD tools. The development of LWD-NMR is ongoing and significant changes in hardware design, as well as significant changes and improvements in data acquisition and processing, can be expected in the next few years. The general benefits of LWD have been discussed elsewhere—in particular, NMR-LWD offers a nonradioactive alternative for porosity measurement, an NMR alternative to wireline in high-risk and high-cost wells, and enables high-resolution fluid analysis in thin beds and laminated reservoirs.[61]

By definition, logging tools operating in the drilling environment are built into drill collars and are, therefore, mandrel devices. In contrast to wireline tools, they must be capable of making measurements when the drillstring is stationary, sliding or rotating, centered or eccentered.

LWD-tool measurements are either omnidirectional or azimuthal, depending on tool design. The incorporation of magnetometers allows binning of the data into azimuthal sectors. Omnidirectional measurements can be generated from azimuthal data, but not the reverse. Although current LWD-NMR services provide only omnidirectional data, patents have been issued for tools with azimuthal capability.[62]

A major concern introduced with the advent of LWD-NMR is the effect of drillstring lateral motion on the basic NMR measurement.[63] NMR measurements are not instantaneous, they involve both polarization and decay, time-dependent components. Lateral movement of a wireline tool or LWD drillstring shifts the polarization volume and the sensitive-measurement area, relative to one another, and these shifts may result in incomplete polarization or incomplete measurement of the decay. Low-velocity motion affects only the decay, but high-speed motion can also affect the initial decay amplitude.

The currently available LWD tools offer different solutions to this concern. While both tool designs can operate in either T_1 or T_2 acquisition mode and both incorporate accelerometers and magnetometers for detecting lateral motion for quality control of the NMR measurements, they differ in their choice of primary measurement mode. Operational factors, such as the slower logging speed (i.e., rate of penetration in drilling), compared with typical wireline logging, also affect the choice of measurement mode.

The NMR acquisition sequences are programmable and interchangeable with those used in wireline tools. Switching between these acquisition modes is accomplished by a variety of methods, including elapsed time, counting measurements, and differentiating between drilling and nondrilling conditions. T_1-mode records echo amplitudes as a function of time. Data output consists of porosity, free-fluid, and bound-water volumes, and can provide a quick-look permeability estimate. T_2-mode is a multifrequency mode that records multiple wait-time CMPG spin-echo amplitudes and is capable of using all wireline-pulse sequences. T_2-mode output includes porosity, free fluid, clay- and capillary-bound water, and differences in the multiple wait-time data are used for hydrocarbon indication.

The Halliburton tool (MRI-LWD)[64–68] uses T_1 as its preferred acquisition mode. Halliburton considers T_1 more robust for determining porosity and free-fluid volume. The anticipated maximum rates of penetration for LWD—1 to 3 ft/min—allow T_1 acquisition. T_1 is motion tolerant compared with T_2. A sequence of interleaved measurements made at different recovery times is used to construct the T_1 relaxation decay (buildup). As long as the sensitive volume (shell) is contained within the much larger volume reached by the saturation pulse, the measurement is valid. During post-processing, drilling and nondrilling periods are identified, and the invalid T_1 data recorded during drilling are discarded.

Schlumberger's tool (ProVision LWD-NMR)[69–71] uses T_2 for its primary measurement. While both companies agree that T_1 is motion tolerant, Schlumberger considers T_1 to be less robust for estimating porosity, bound-fluid, and free-fluid volumes because of the poor S/N resulting from the longer time require for equivalent-quality T_1 measurements. A reduced S/N impacts data quality (e.g., statistical repeatability and vertical resolution), logging speed, and, ultimately, the results. T_2 measurement also enables rapid calibration and correlation with the large body of wireline-NMR data. The multiple wait-time acquisition includes a fully polarizing (3 to 12 seconds), a partially polarizing (normally 0.6 to 1 seconds), and a very fast wait time (typically 0.08 seconds). Porosity and bound-fluid volume are calculated for both the long- and the medium-wait-time measurements, and significant differences between the long-wait-time and medium-wait-time porosity provide real-time hydrocarbon indication. When necessary, logs are corrected for incomplete polarization of the hydrogen nuclei, and T_1 distributions are estimated from the measured T_2 relaxation.

Baker Atlas is presently field testing a new LWD device (MagTrak) that operates in T_2 acquisition mode and has a vertical resolution of less than 3 in.[72] This tool design achieves a motion-tolerant T_2 measurement through a combination of a very-low gradient magnetic field, circuitry that permits a T_E as low as 0.6 ms, and the use of special stabilizers.

3E.5.5 Downhole NMR Spectrometer. Contamination of hydrocarbon reservoirs by oil-based mud (OBM) and synthetic oil-base mud (SOBM) is a significant problem for accurate reservoir and fluid analysis in wells in which OBMs are used, especially offshore. Direct knowledge or an estimate of the OBM's NMR characteristics is required to distinguish it from connate oil in fluid samples obtained by formation testers. Acquisition of T_1, T_2, and D_0 (the self-diffusion coefficient) permits evaluation of connate oil, oil viscosity, and gas/oil ratio.[31,73] Furthermore, in costly drilling environments, real-time acquisition of NMR properties permits immediate, rather than delayed, fluid evaluation. Information on NMR fluid properties is also valuable for the interpretation of NMR wireline and LWD logs.

The recently introduced downhole NMR spectrometer, incorporated into a formation-testing tool, can obtain NMR measurements of OBM contamination directly, on live samples at in-situ conditions.[74–77] As the testing tool pumps fluid from the reservoir into the borehole or sample chamber, the spectrometer—using a measurement time of 30 seconds—measures the hydrogen index (HI), T_1, T_2, and diffusion. The T_1 measurement is made while flowing; T_2 and diffusion are static measurements. The T_1 distribution is important in differentiating between highly refined OBM filtrates and native oil. The T_1 characteristics of these common filtrates are measured and cataloged so that the data from oil-based filtrates can be distinguished from native hydrocarbons.

3E.6 Log Presentation

NMR-log data are presented in a variety of formats designed to emphasize specific aspects of the data and thus enable rapid visual interpretation of movable and immovable fluids, porosity, and permeability. Data interpretation is further enhanced when additional log and core information are also included in the log presentation.

The T_2 distribution is typically displayed in waveform presentation, image (variable-density log, VDL) format, and bin-distribution plot. Each T_2 format represents the distribution of the porosity over T_2 values, and hence, over the pore sizes (**Figs. 3E.17 and 3E.18**).

3E.7 NMR Applications

NMR-log data can be analyzed independently or in combination with conventional-log and core data. As an independent logging service, NMR can provide porosity, permeability index, and complete information on fluid type and saturation of the flushed zone. Some data-interpretation methods operate in the echo-decay time domain, while others operate in the T_2-relaxation domain.

3E.7.1 Porosity Determination with NMR. The initial amplitude of the spin-echo train is proportional to the number of hydrogen nuclei associated with the fluids in the pores within the sensitive volume. This amplitude is calibrated in porosity units (see Eq. 3E.10). The accuracy

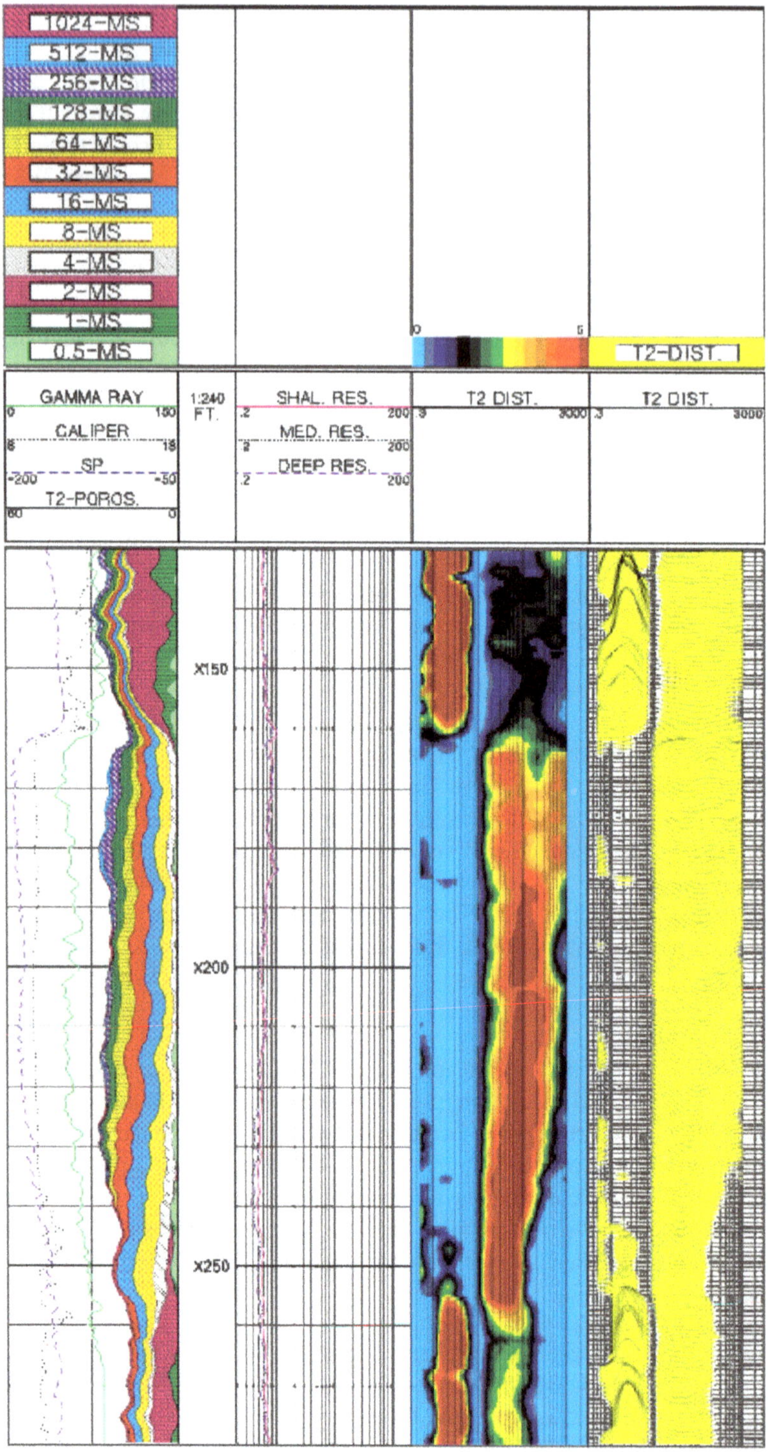

Fig. 3E.17—Log presentation illustrating some of the formats used to present T_2 distributions: Track 1, a plot of the cumulative amplitudes from the binned T_2-distribution; Track 2, shallow, medium, and deep resistivity log curves; Track 3, a color VDL image of the binned T_2 distribution; Track 4, a waveform presentation of the same information. The displayed T_2 distribution typically corresponds to binned amplitudes for exponential decays that may range from 0.5 to 1,024 ms. Logging data are much noisier than laboratory data, and the T_2 distribution shown on well logs is comparatively coarse.

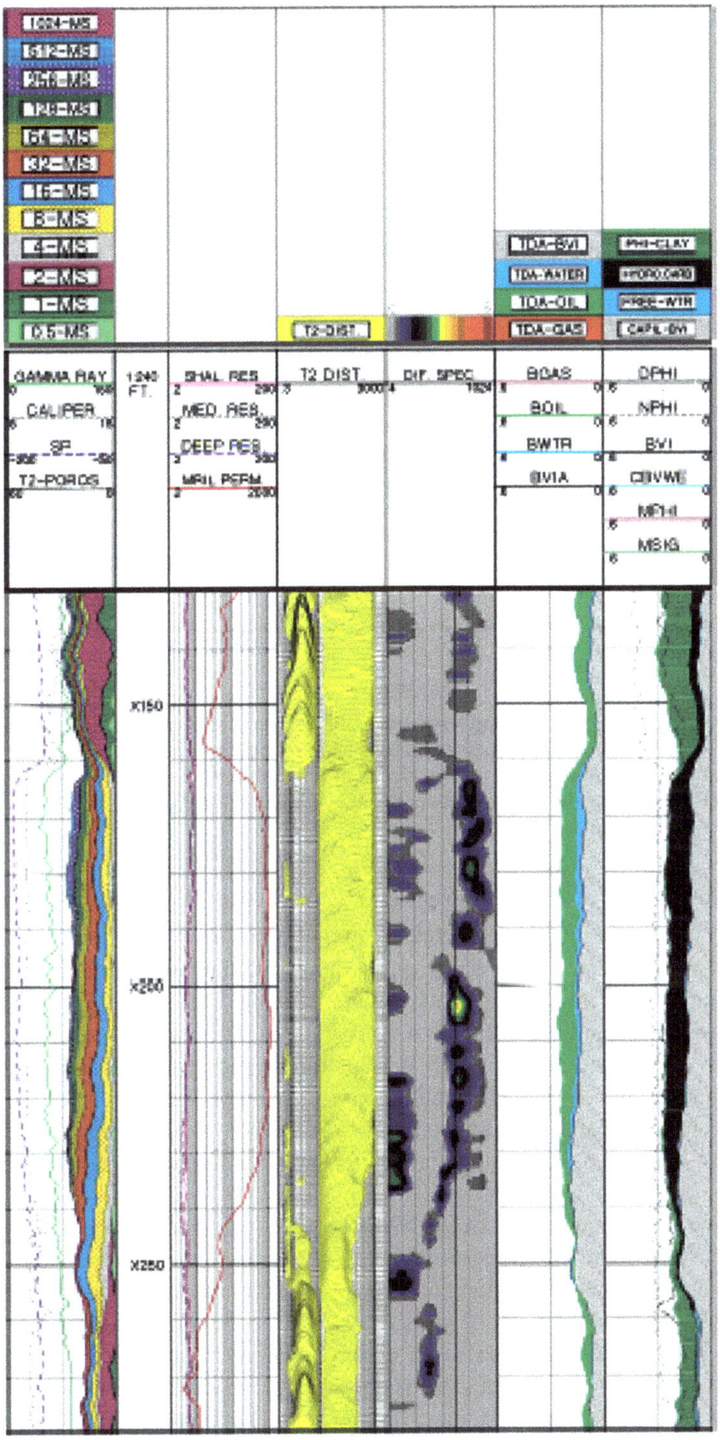

Fig. 3E.18—Integrated NMR-log evaluation. This log presentation includes GR, caliper, and SP (Track 1), resistivity data (Track 2), NMR T_2 distribution (Track 3), DSM-analysis results (Track 4), TDA-analysis results (Track 5), and NMR-resistivity integrated analysis results (MRIAN, Track 6).

of this amplitude measurement depends on three factors: first, a sufficiently long polarization time, T_W, is needed to achieve complete polarization of the hydrogen nuclei in the fluids; sec-

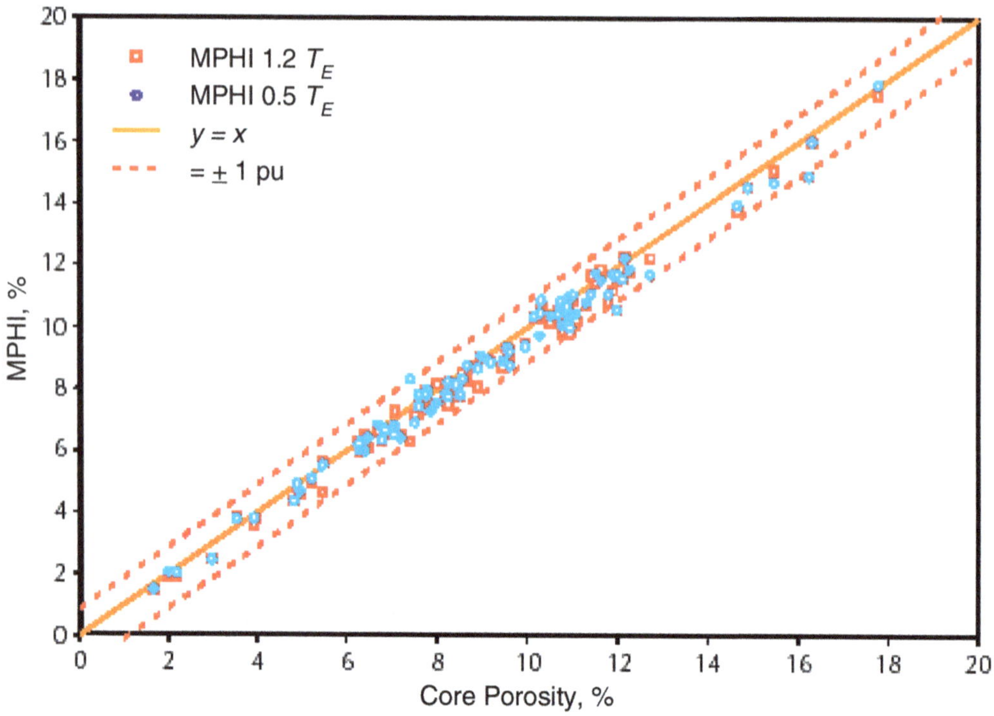

Fig. 3E.19—Comparison of core- and log-NMR porosity. In this clean sandstone example, there is good agreement between porosity derived from laboratory-NMR measurements and porosity derived from conventional core analysis. NMR-porosity values typically fall within ±1 p.u. of the measured core-porosity values. Here, NMR-laboratory data were measured at T_E = 0.5 ms and T_E = 1.2 ms.

ond, a sufficiently short inter-echo spacing, T_E, is needed to record the decays for fluids associated with clay pores and other pores of similar size—if either T_W is too short or T_E too long, NMR porosity will be underestimated; third, the number of hydrogen nuclei in the fluid should equal the number in an equivalent volume of water (i.e., HI = 1); if the HI of any of the pore fluids is significantly less than 1, a correction to porosity is necessary.

Porosity was one of the earliest NMR measurements and is still an important one. Assuming that the logging tool is properly calibrated and functioning normally (see discussion of Quality Control in this chapter), interpretation of the porosity measurement depends on several other factors, including vintage of the logging tool, whether hydrocarbons are present, and fluid type.

There are two major contrasts between NMR porosity and conventional density and neutron-nuclear-porosity logs. First, NMR porosity does not depend on the mineralogy of the matrix, except in cases in which the formation contains significant amounts of ferromagnetic or paramagnetic materials, and in most cases, it is considered a lithology-independent measurement. Second, NMR porosity is not sensitive to either borehole or mudcake and does not require corrections because the measurement zone (i.e., sensitive volume) is focused within the formation, beyond the borehole wall and these influences. The accuracy and precision of NMR-derived porosity has been confirmed through comparisons with core porosity obtained using conventional laboratory measurement methods (**Fig. 3E.19**).

Influence of Tool Version on the Porosity Measurement. The original NML tool had such a large dead time (e.g., minimum echo spacing, T_E) that it could not measure clay- and capillary-bound fluids; thus, it was only capable of measuring free-fluid porosity, which is now called

NMR effective porosity (MPHI). The minimum T_E used in the early versions of pulse-NMR tools was still relatively large, and those tools still could not see all the clay-bound fluids. With improvements in logging technology, most NMR logs recorded after 1997 are considered "total-porosity" logs. Current logging tools use a minimum T_E of 0.3 ms in continuous mode and a T_E as small as 0.1 ms in stationary mode.

In clean, water-filled formations, MPHI should approximately equal neutron-density cross-plot porosity. In shaly sands, MPHI should approximately equal density porosity, calculated with the correct grain density; however, the MPHI may not equal effective porosity because of the effects of HI and long T_1 components (Eq. 3E.12):

$$\text{MPHI} = \phi_e \cdot \text{HI} \cdot \left[1 - e^{-\left(\frac{T_W}{T_1}\right)}\right], \quad \text{(3E.12)}$$

where MPHI is measured by the NMR tool; ϕ_e is effective porosity of the formation; HI is related to the amount of fluid in the effective porosity system; T_W is polarization time used during logging; and T_1 is longitudinal relaxation time of the fluid in the effective porosity system. Older NMR logs may indicate very low porosity in shales.

MPHI is almost always less than NMR total porosity (MSIG):

$$\text{MSIG} = \text{MPHI} + \text{MCBW}, \quad \text{(3E.13)}$$

where MSIG is measured by NMR total-porosity logging, and clay-bound-water porosity (MCBW) is measured by the NMR tool with partial-polarization acquisition. In very clean formations, however, NMR MCBW is virtually zero, and then MPHI equals MSIG (see Fig. 3E.16).

NMR total porosity may approximate density or neutron-density porosity. The reported clay-bound volume should not be considered absolute because the boundary between clay- and capillary-bound porosity components can vary with clay type and distribution. Whenever possible, the relationship of clay type and volume should be confirmed through core analysis and comparison with other log data.

Porosity-Logging Modes. Several modes of porosity logging are currently in use: standard T_2 mode, short T_W (e.g., bound-fluid logging), intermediate T_W (e.g., polarization correction and the "clean, wet formation" method), and long T_W, or multiple-wait-time acquisition (e.g., total-porosity logging).

Total-porosity-logging mode provides data to determine porosity, permeability, and productivity (mobile fluids). Total-porosity acquisition is a multiple-wait-time acquisition that acquires two echo trains to obtain the total porosity. One echo-train acquisition uses a long T_E (e.g., 0.9 or 1.2 ms) and a long T_W (3 to 8 seconds) to achieve complete polarization. This echo train provides the effective porosity. The second echo-train acquisition uses a short T_E (0.2 to 0.6 ms) and a short T_W (20 ms) that is only long enough to achieve complete polarization of the fluids in the small pores (i.e., clay-bound porosity).[46] Although the long T_W acquisition may achieve full polarization in gas zones, the porosity may still require HI correction. The short T_W echo trains are stacked to improve S/N. Total porosity is obtained by combining the two separate T_2 distributions, usually at 4 ms. The standard T_2 acquisition is used in situations in which there is little to no diffusion or T_1 contrast. The benefits of standard T_2 acquisition are increased logging speed without degradation of data quality; or superior data quality at normal logging speeds.

In bound-fluid logging mode, T_W is kept relatively short, ranging from 0.3 to 1.0 seconds. The benefit of using a short T_W is faster logging speeds while determining the clay- and capillary-bound fluid volumes. The drawback is underpolarization of the free-fluid volume.

When this logging mode is used, the free-fluid volume is computed from other log data (e.g., density, neutron-density crossplot, or acoustic). In polarization mode (CMR tool), a correction to the free-fluid volume is made using the observed relationship of T_1 to T_2, on the basis of the T_2 acquired while logging. The magnitude of the polarization correction must be monitored because, in addition to porosity, the T_1/T_2 ratio is also influenced by changes in the types of formation and pore fluid. This method generally works well in clastics, but in conditions of high T_1 and T_2, such as in carbonates or light oil, the correction may actually introduce additional error.[78]

Another mode of operation is to find a clean, wet formation and perform a sweep of different wait times (T_W) ranging from 2 to 5 seconds to determine the one needed for full polarization. Because oil and gas may have very long T_1 values or require an HI correction, the use of intermediate T_W methods does not guarantee that the NMR porosity will be fully reported.

Optimization of the acquisition is critical to accurate NMR measurements and interpretation (see the Job Planning section of this chapter). The effect of incorrect acquisition parameters on the T_2 distribution is illustrated in **Fig. 3E.20**. In this figure, the sequence of graphs shows the effects of T_E, T_W, pulsing time (T_P), and S/N on the measured NMR characteristics of a core sample:

- A: Initial acquisition results show an apparent unimodal distribution.
- B: Increasing the T_P shows that this sample is, in fact, bimodal. There is a significant change in the T_2 distribution, but little to no change in the cumulative porosity.
- C: Decreasing the echo spacing (T_E) reveals fast components in this sample that were previously masked by a T_E that is too long. There is a significant change in the cumulative porosity with a characteristic shift of the fastest components in the distribution to shorter T_2 values.
- D: Increasing the wait time (T_W) has significantly increased cumulative porosity. After correction for a T_W that is too short, there is a characteristic shift in the slow T_2 components to longer T_2 values.
- E: Increasing the S/N results in sharper resolution of the two major components. Although the definition of each component in the T_2 distribution improves, little change is noted in the cumulative porosity.
- F: The final result, in which all acquisition parameters have been optimized (red), is compared with the initial result (blue).

BVI Determination. Determining the BVI of water in a formation is one of the earliest and most widely used applications of NMR logging. BVI refers to the immovable or bound water in a formation. BVI is a function of both the capillary-pressure curve for the rock and the height above free water (**Fig. 3E.21**).

In practice, BVI serves as an indicator of, and is used for, estimating producibility and permeability. There are two methods currently used for BVI determination. One is the cutoff-BVI (CBVI) model, which is based on a fixed-T_2 value ($T_{2\text{cutoff}}$) that divides BVI and FFI. The second is the spectral BVI (SBVI) model, which assumes the coexistence of free and bound fluids in any pore described by the T_2 distribution at water saturation (S_w) = 100%. This coexistence is expressed through a weighting factor that defines the fraction of bound water associated with each pore size.[79] The SBVI method is used primarily for quantifying movable water and, secondarily, for estimating permeability.

CBVI (Cutoff) Model. The T_2 signal from the rock matrix is actually so rapid that even modern logging tools cannot detect it. Consequently, the recorded T_2 distribution represents only the porosity occupied by movable (FFI) water and immovable BVI and clay-bound-water

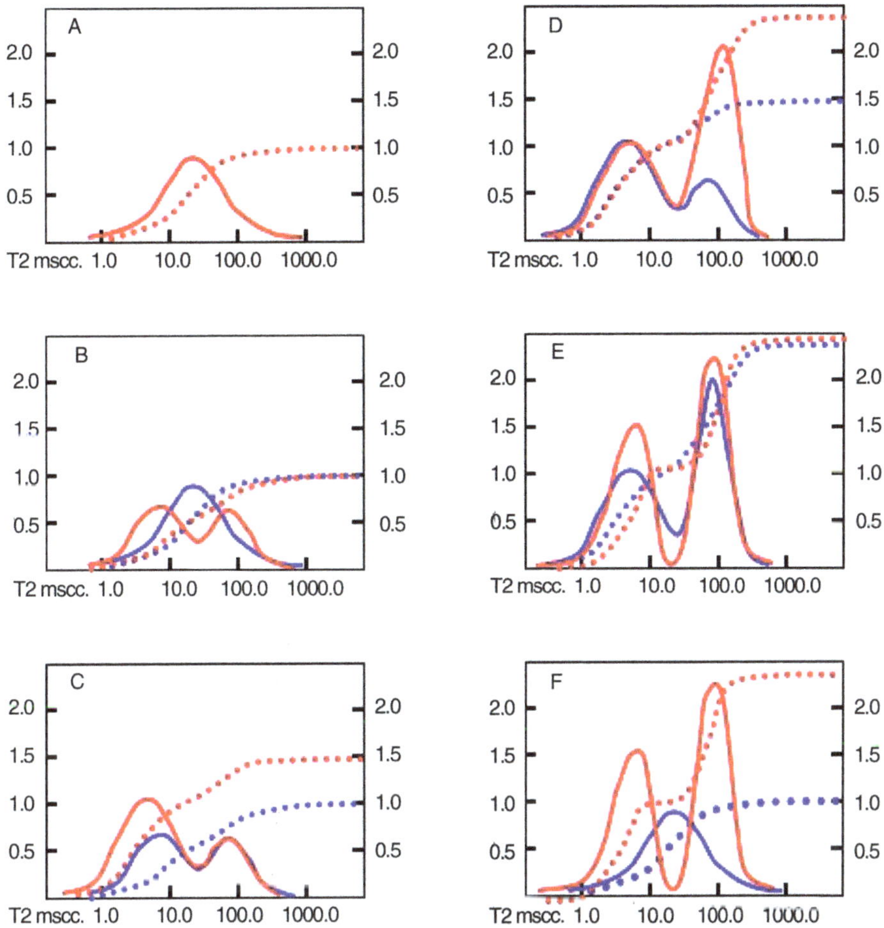

Fig. 3E.20—Dependence of T_2 distribution on NMR-acquisition parameters. In all figures, blue represents previous results, red represents new results, solid lines represent incremental T_2 (read on the left axis), and dashed lines represent cumulative T_2 vs. T_2 relaxation time (read on the right axis).

(CBW) components. Assuming that the mobility of reservoir fluids is primarily controlled by pore size (i.e., the producible fluids reside in large pores and the immobile, or bound, fluids reside in small pores), a fixed-T_2 value can relate directly to a pore size at or below which fluids will not move. This value ($T_{2\text{cutoff}}$) is used to divide the T_2 distribution into movable (i.e., producible or free fluids and FFI) and immovable (i.e., bound-fluid, BVI, and CBVI) components (**Fig. 3E.22**).[80] $T_{2\text{cutoff}}$ is a variable that differs from one formation to another and is influenced by a variety of factors including capillary pressure, lithology, grain size, compaction, and pore characteristics.

In practical use, $T_{2\text{cutoff}}$ is either determined in the laboratory or a default value is assumed. If time and expense permit, accurate $T_{2\text{cutoff}}$ values can be defined by comparing the T_2 distributions obtained on fully and partially water-saturated core samples taken from the logged interval[81,82] (**Fig. 3E.23**).

In the absence of laboratory data, lithology-dependent default values are used for $T_{2\text{cutoff}}$. Default $T_{2\text{cutoff}}$ values are based on local field experience or common practice; for example, typical values used in the Gulf of Mexico are 33 ms for clastics (sandstones) and 92 or 100 ms

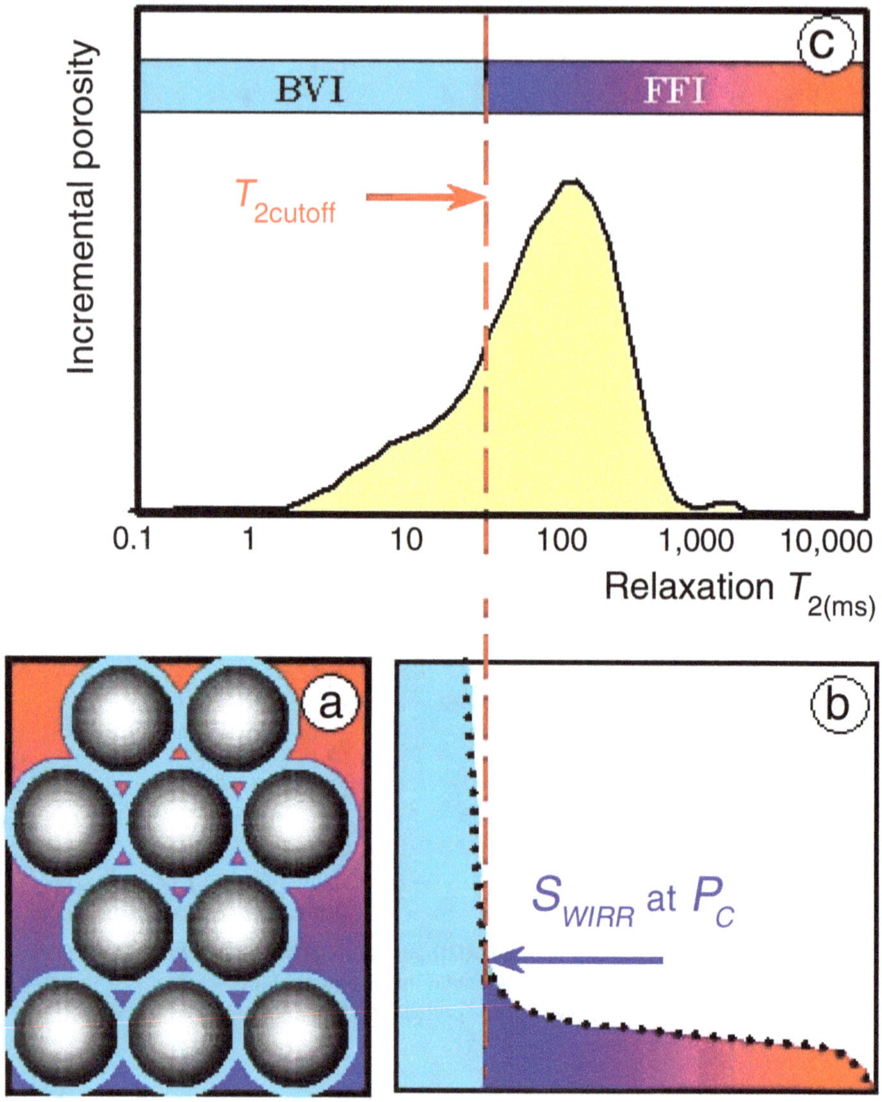

Fig. 3E.21—Correlating NMR $T_{2cutoff}$ to capillary pressure: (a) grains are shown in gray, capillary-bound water (BVI) in light blue, and free fluids in a gradient tone blue/red; (b) capillary pressure curve (black dots) defines capillary pressure (P_c) in which water saturation becomes irreducible (S_{wirr}); (c) a single-porosity cutoff value on the T_2 distribution, corresponding to this pressure.[79]

for carbonates.[83–85] When using a default value, remember that, in fact, the $T_{2cutoff}$ can actually vary within a single lithology because the capillary pressure at which irreducible water saturation (S_{wirr}) is achieved varies between the actual rocks. In addition, $T_{2cutoff}$ is also influenced by pore-wall chemistry, the presence of minor paramagnetic or ferromagnetic components, texture, and pore-throat to pore-body ratios. The use of an incorrect value for $T_{2cutoff}$ may result in underestimation of recoverable reserves or in bypassing a pay zone.

SBVI (Spectral) Model. The SBVI model was developed to address the limitations of the CBVI model in some rocks, particularly in very-well-sorted rocks in which there is a very narrow range of grain and pore sizes. In these rocks, the NMR echo-decay can typically be fit by

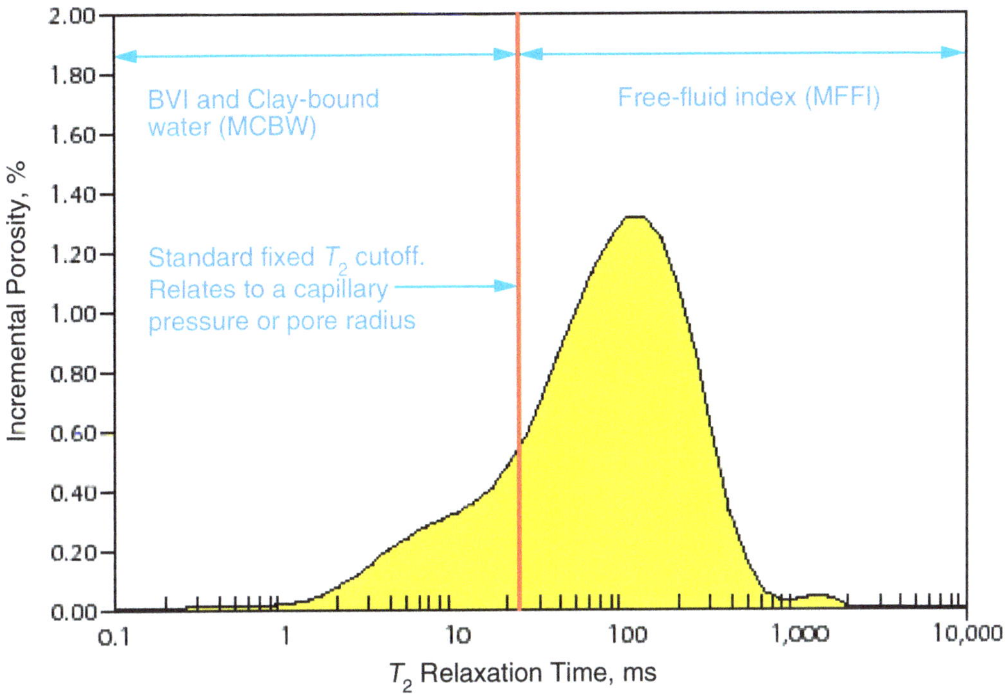

Fig. 3E.22—Illustration of the $T_{2cutoff}$ used in the CBVI model to divide NMR porosity into movable and immovable components.

a single-exponential decay that produces a sharp peak in the T_2 distribution. When a fixed cutoff is used for determining BVI, it may result in significant error because of the presence of immovable water contained in microporosity associated with pore irregularities.[84] This microporosity component is apparent at S_{wirr}, but not at S_w = 100% (**Fig. 3E.24**).

A weighting function defines the fraction of bound water in each pore size that is present in the T_2 distribution at S_w = 100% (**Fig. 3E.25**). Several methods have been proposed for obtaining the weighting function.[31,85,86]

In general, the SBVI method is superior to the CBVI method for determining BVI in rocks in which S_w = 100%. The presence of hydrocarbons (i.e., S_w < 100%), however, introduces complications to the SBVI method. To overcome the limitations of both methods for determining BVI, the recommended practice is to compute two bound-water values—one from the cutoff (CBVI) and one from the spectral (SBVI) method—and take the larger of the two.

3E.7.2 Permeability Estimation. The ability to estimate formation permeability is one of the earliest benefits of NMR logging and remains the most important application. Laboratory studies demonstrate that pore-water relaxation time is inversely related to the S/V ratio of the pore space (**Fig. 3E.26**). The NMR estimate of permeability is based on theoretical and core-based models that show that permeability increases with increasing porosity and pore size (S/V).[80,87–90]

The measurement of formation permeability, in general, is greatly influenced by the method used, the limitations of each, and the scale at which the measurements are made.[91] As stated previously, MICP curves obtained on core samples correlate to pore-throat size, while NMR measures pore-body size.

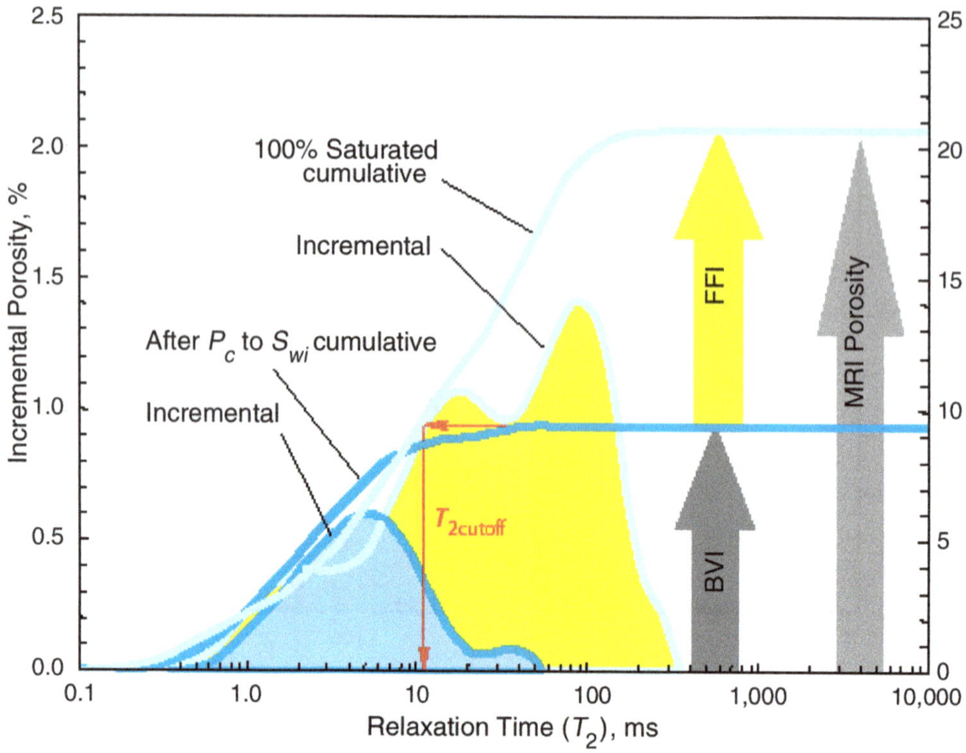

Fig. 3E.23—Determination of $T_{2cutoff}$ from core measurements. NMR measurements on fully saturated (S_w = 100%) core samples and on samples at irreducible saturation (S_{wirr}) can be used to establish a $T_{2cutoff}$ for use in a CBVI model. The T_2 distributions are displayed as incremental porosity and cumulative porosity. The cumulative curves are used to determine $T_{2cutoff}$.

NMR logging does not provide direct and continuous measurement of permeability; rather, a formation-permeability estimate, or index, is calculated from the spectral-porosity measurements using permeability models that are based on a combination of empirical and theoretical relationships. Several permeability models have been developed, and two are in common use: the free-fluid (Timur-Coates or Coates) model and the mean-T_2 [the Schlumberger-Doll-Research (SDR)] model.[27,92–94] The free-fluid model can be applied to water-saturated and hydrocarbon-saturated reservoirs, and the mean-T_2 model can be applied to water-saturated reservoirs.[95] These permeability models assume that a good correlation exists between porosity, pore-body and pore-throat size, and pore connectivity. This assumption is generally valid in clastic (e.g., sand/shale) sequences, but in carbonates or other lithologies, model-derived permeabilities may not be reliable.

Typically, a permeability model is calibrated over a particular zone of interest and verified, wherever possible, by core or well/formation test data. Once this is done, the NMR log can provide a robust continuous-permeability estimate within the zone of interest.

Both models treat permeability as an exponential function of porosity, ϕ^4, and include a parameter to account for the fact that NMR measures pore-body size, not pore-throat size[20] (**Fig. 3E.27**). In the Coates model, the pore-size parameter enters implicitly through $T_{2cutoff}$, which determines the ratio of FFI to BVI. In the SDR model, the size parameter enters through the geometrical mean of the relaxation spectra, T_{2gm}. In water-saturated rocks, both models provide similar and good results; however, when hydrocarbons are present, the SDR model fails because T_{2gm} is no longer controlled exclusively by pore size.[96]

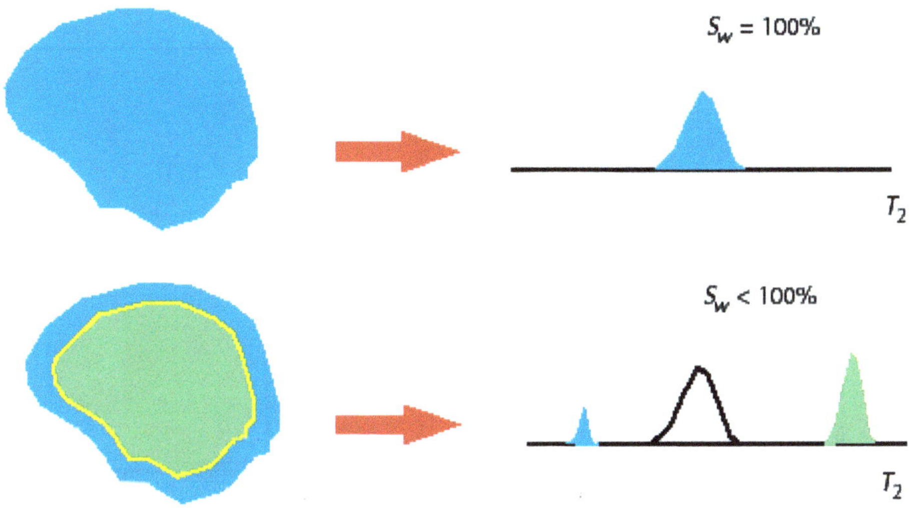

Fig. 3E.24—BVI associated with microporosity in a coarse-grained rock having a narrow range in pore sizes. The T_2 distribution at S_w=100% exhibits a single sharp peak (upper). The T_2 distribution at S_w<100% exhibits two peaks (lower). One appears below the $T_{2cutoff}$ value, the result of irreducible water on the pore surfaces. The other appears above the $T_{2cutoff}$ and represents the oil in the pore fluid. The T_2 value of the second peak is close to the T_2 of bulk oil. This same effect is seen in a North Sea chalk.

Free-Fluid (Timur-Coates or Coates) Model. In the simplest form of the free-fluid model, permeability, k_{Coates}, is expressed as follows (Eq. 3E.14):

$$k_{\text{Coates}} = \left[\left(\frac{\phi}{C}\right)^2 \frac{\text{MFFI}}{\text{MBVI}}\right]^2, \quad \text{(3E.14)}$$

where ϕ is MSIG, MBVI is obtained through the CBVI or SBVI method, MFFI is the difference between MSIG and MBVI (assuming that there is no clay-bound water, see Fig. 3E.16), and C is a formation-dependent variable. The free-fluid model is very flexible and has been calibrated using core data for successful use in different formations.

To calibrate the model to core, Eq. 3E.14 is solved in the form of a straight line, $y = mx + b$:

$$\sqrt{\frac{MFFI}{MBFV}} = m\left(\frac{\sqrt[4]{k}}{MSIG}\right) + b \,. \quad \text{(3E.15)}$$

Assuming $b = 0$ in the equation (3E.15), core permeability is substituted for k. The slope of the line, m (i.e., C value in Eq. 3E.15), is determined using a least-squares regression (**Fig. 3E.28**).

Despite the flexibility of this model there are formation conditions that limit the effectiveness of the model and may require a correction (**Table 3E.5**). The presence of hydrocarbons (i.e., oil, oil filtrate, or gas) in the BVI component may result in an overestimate of BVI by either the CBVI or SBVI methods, leading to an underestimate of permeability. An HI correction can be applied when gas is present. The very short T_2 values associated with heavy oil may be counted in the BVI component and result in an underestimate of permeability.

Mean-T_2 [Schlumberger-Doll-Research (SDR)] Model. Using the SDR model, permeability is expressed as

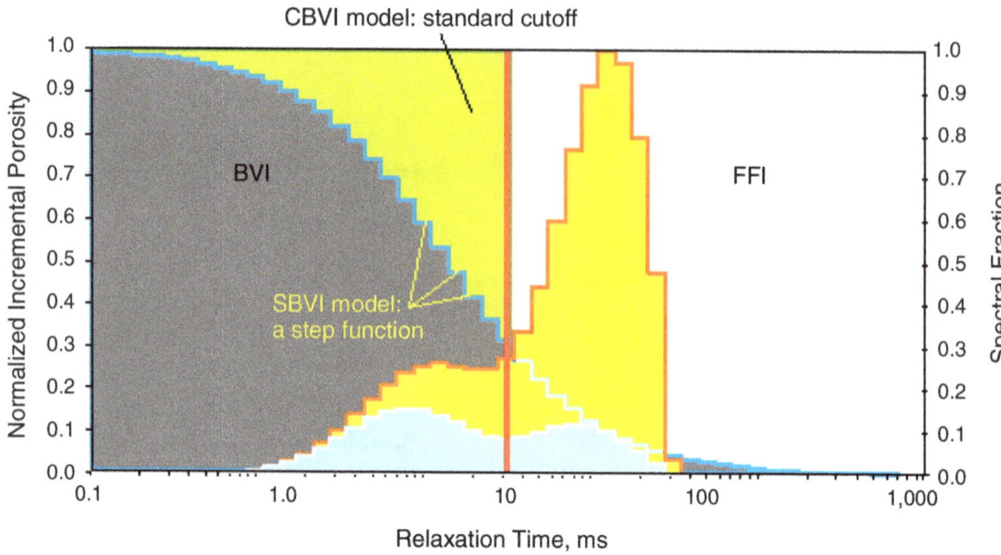

Fig. 3E.25—SBVI model. For a given T_2, the SBVI-weighting factor provides the fractional volume of irreducible fluids in the pores whose size is associated with that T_2. The model shown is a step function and is commonly used to describe the weighting factors.

$$k_{SDR} = C \times T_{2gm}^2 \times \phi^4, \qquad (3E.16)$$

where ϕ is NMR effective porosity (MPHI), T_{2gm} is the geometric mean of the T_2 distribution, and C is a formation-dependent variable. The SDR model works very well in water-saturated zones. In the presence of oil or oil filtrates, the mean T_2 is skewed toward the $T_{2\text{bulk}}$, because of the effects of partial polarization, leading to an overestimate of permeability. In unflushed gas zones, mean-T_2 values are too low relative to the flushed-gas zone; and permeability is underestimated. Because hydrocarbon effects on T_{2gm} are not correctable, the SDR model fails in hydrocarbon-bearing formations. The Coates and SDR models represent matrix permeability and, therefore, are not applicable to estimation of permeability in fractured formations.

Other Comments on NMR-Based Permeability Estimation. **Table 3E.5** compares the Coates and SDR models under different reservoir conditions, and it may be advisable to use both methods in an effort to constrain values for permeability.

There are a number of benefits in having available NMR-derived permeability and BVI. This information enables more-accurate quantification of reservoir heterogeneity and improves estimation of reserves and ultimate recovery. Other applications include optimizing the locations of perforations, well spacing, tailoring completions to maximize recovery rates and efficiencies, and improving primary and secondary recovery design schemes.

3E.7.3 Hydrocarbon (Fluid) Typing. The NMR T_2-porosity relationship in which T_2 is a function of pore size (i.e., S/V ratio, see Eq. 3E.8) holds for water-saturated rocks. The presence of hydrocarbons in water-wet rocks alters the T_2 distribution, thus affecting the porosity interpretation. Despite the variability in the NMR properties of fluids, the locations of signals from different types of fluids in the T_2 distribution can often be predicted or, if measured data are available, identified (**Fig. 3E.29**).

Hydrocarbon typing and prediction of fluid properties by NMR logs is predicated on reliable laboratory correlations between NMR measurements (i.e., relaxation times and diffusion)

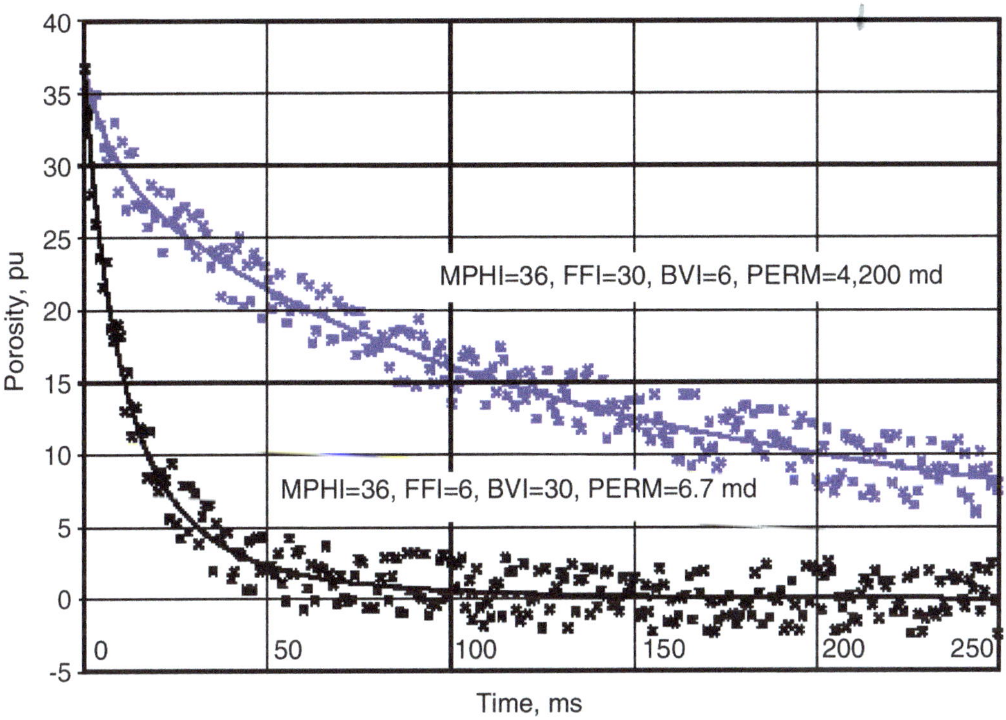

Fig. 3E.26—Variation in T_2 decay with permeability. This plot illustrates the difference between echo trains obtained from formations with similar porosity but different pore sizes. In terms of T_2 distribution, this difference is expressed in different FFI/BVI ratios. The permeabilities were computed using the Timur-Coates model.

and fluid properties [e.g., specific gravity, viscosity, and gas/oil ratio (GOR)]. Early studies were limited to investigations at ambient conditions;[97] however, using the standard correlations derived from these studies may result in seriously underestimating viscosity. More-recent studies have expanded these correlations to oils and mud filtrates at reservoir conditions.[97–99]

NMR logging uses specialized CMPG pulse-acquisition sequences to exploit these differences in pore-fluid NMR properties to achieve specific objectives. The CMPG parameters—wait time (T_W), echo spacing (T_E), the number of echoes (N_E), and the number of sequence repetitions—are selected to take advantage of the wide variation in NMR fluid properties (**Fig. 3E.30**) for estimating total porosity and for hydrocarbon typing. Table 3E.2 illustrates the range of NMR-related properties of fluids for Gulf of Mexico sandstones, for example. **Table 3E.6** lists variations in CMPG pulse sequences for different objectives using time-domain analysis (TDA) and Enhanced Diffusion Method (EDM). The multifrequency tools now in service permit the simultaneous acquisition of multiple measurements on the same rock volume using different acquisition sequences during the same logging run. The general approach is to log in a mode that allows gathering the full spectrum of data. Specialized applications, including some direct fluid-identification methods, involve customized-acquisition sequences that require slower logging and acquisition of more echoes.

Determination of the appropriate interpretation method is largely based on the estimated viscosity of the anticipated hydrocarbons (see **Fig. 3E.31**). These methods will be explained in the following paragraphs.

Selection of the appropriate acquisition sequence and the choice of acquisition parameters depend on the logging objectives and are part of prejob planning. This process considers sever-

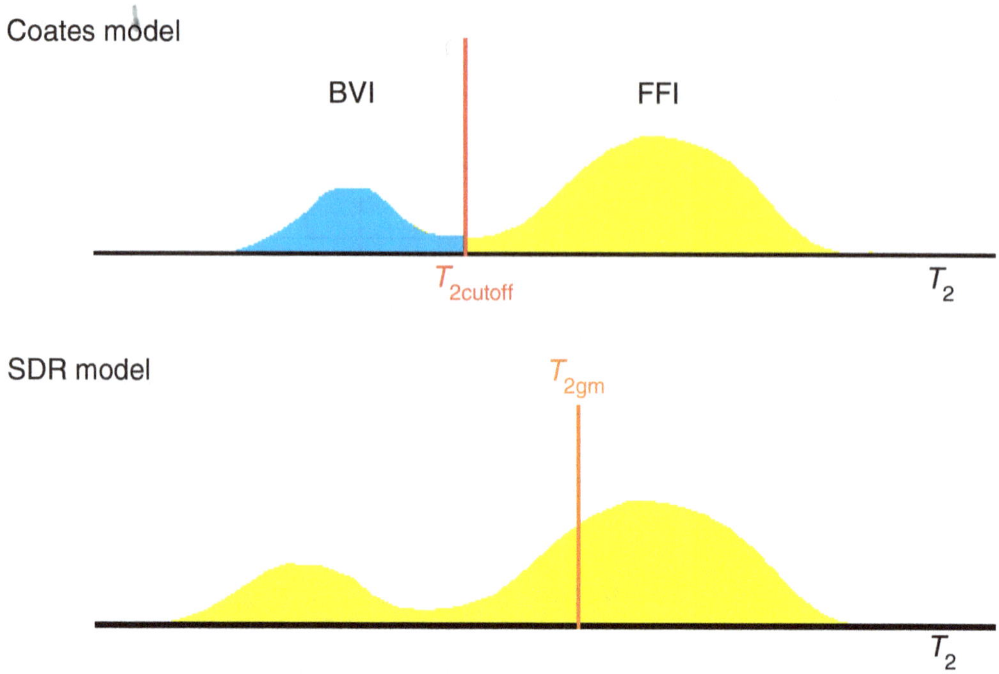

Fig. 3E.27—NMR-permeability models. The free-fluid permeability, Coates, model (top) uses the MFFI/MBVI ratio to describe changes in S/V ratio. The SDR permeability model (bottom) uses an average T_2 value to describe changes in S/V.

al factors regarding the anticipated rock and pore-fluid properties. For example, for characterizing large pores (clastics), clean formations, carbonates, and light oil—all of which are associated with long-T_2 values[103]—a large number of echoes should be acquired, and a long T_W should be used. However, more echoes and longer T_W may require reduced logging speeds. In contrast, characterization of small pores (i.e., low permeability), shaly formations, low porosity, and BVI determination involves the short T_2 component of the NMR-porosity spectrum and often it can be accomplished using fewer echoes and shorter T_W, which may allow normal logging speeds.

Fluid-typing methods fall into two broad categories depending on the NMR properties that are being exploited: (1) T_1-weighting mechanisms take advantage of differences in fluid T_1 values, and (2) diffusion-weighted mechanisms make use of the diffusion-constant differences between oil and water. There are two general types, or sets, of CMPG acquisitions that are associated with each mechanism: dual T_W and dual T_E. These two sets cover the range of major fluid-typing objectives; some interpretation techniques can use one or the other, or both. Each serves specific purposes and is optimized to provide data for specialized analysis programs. In general terms, each consists of at least two echo-train acquisitions in which one or more parameters are varied. The total, or the difference between the echo-train signals, provides an estimate or indicator of porosity, light hydrocarbons, or oil (**Fig. 3E.32**).

Advanced hydrocarbon-typing objectives can involve customized-acquisition sequences[104] that are run at reduced logging speeds or even in stationary mode.

T_1-Weighted Mechanism (Dual-T_W Acquisition). The dual-T_W acquisition method is used primarily to identify and quantify light hydrocarbons (gas and light oil) by separating them from the water signal through T_1 weighting. The dual-T_W acquisition can also provide the standard-T_2 acquisition dataset for determination of spectral NMR porosity, permeability, and

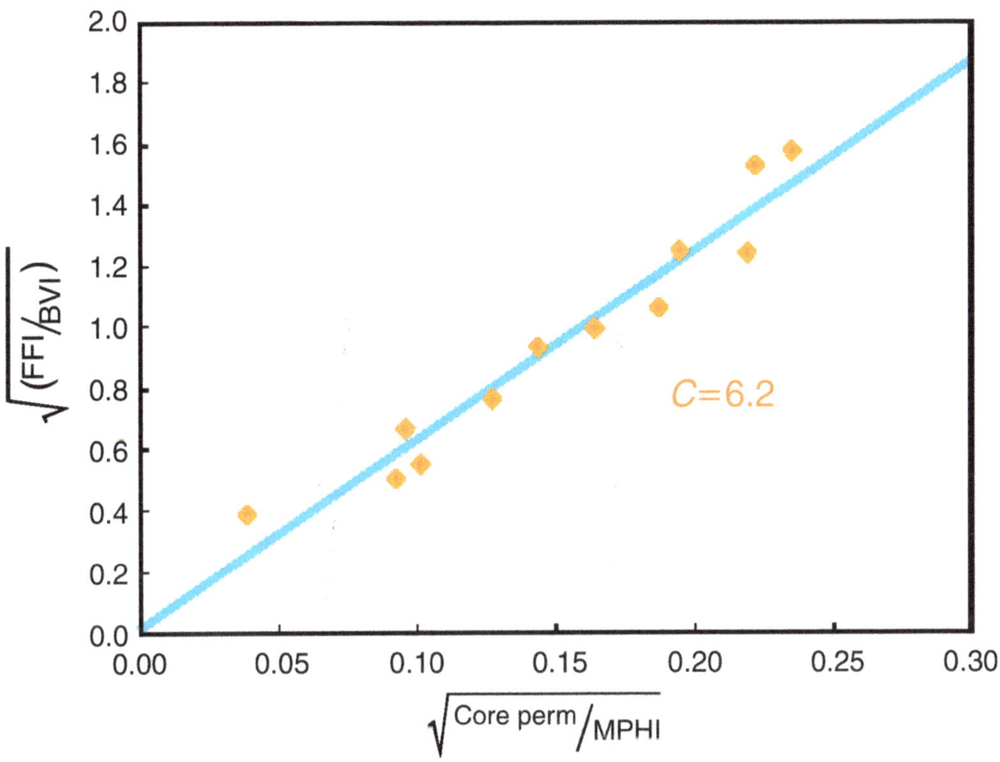

Fig. 3E.28—Crossplot of core and NMR data used for determining the Coates permeability constant, *C*. Assuming *b* = 0 in Eq. 3E.15, core permeability is substituted for *k*, and the slope of the line, *m*, is the *C* value in Eq. 3E.14.

TABLE 3E.5—COMPARISON OF COATES AND SDR MODELS FOR ESTIMATING PERMEABILITY UNDER DIFFERENT FORMATION CONDITIONS

Formation Condition	Free-Fluid (Coates) Model	Mean T_2 (SDR) Model
Water saturated	Accurate	Accurate
Oil and oil filtrates	Affected by filtrates	Underestimates
Unflushed gas	Underestimates	Underestimates
Heavy oil	Underestimates	Underestimates
Fractures	Underestimates	Underestimates

Note: The free-fluid model can be corrected for HI in hydrocarbon-bearing intervals, but the SDR cannot be corrected. Both models represent matrix permeability and underestimate fracture permeability.

productivity (mobile fluids). The T_E is kept the same in both echo-train acquisitions (0.9 or 1.2 ms), and a short T_W (1 second) and long T_W (8 seconds) are used. The water signal is contained in both acquisitions, but because of the light hydrocarbons (which have long T_1 values), the signal is suppressed in the acquisition using the short T_W. Thus, the presence of a signal in the difference of the measurements (differential spectrum) is considered a strong indicator of gas or light oils[105] (**Fig. 3E.33**).

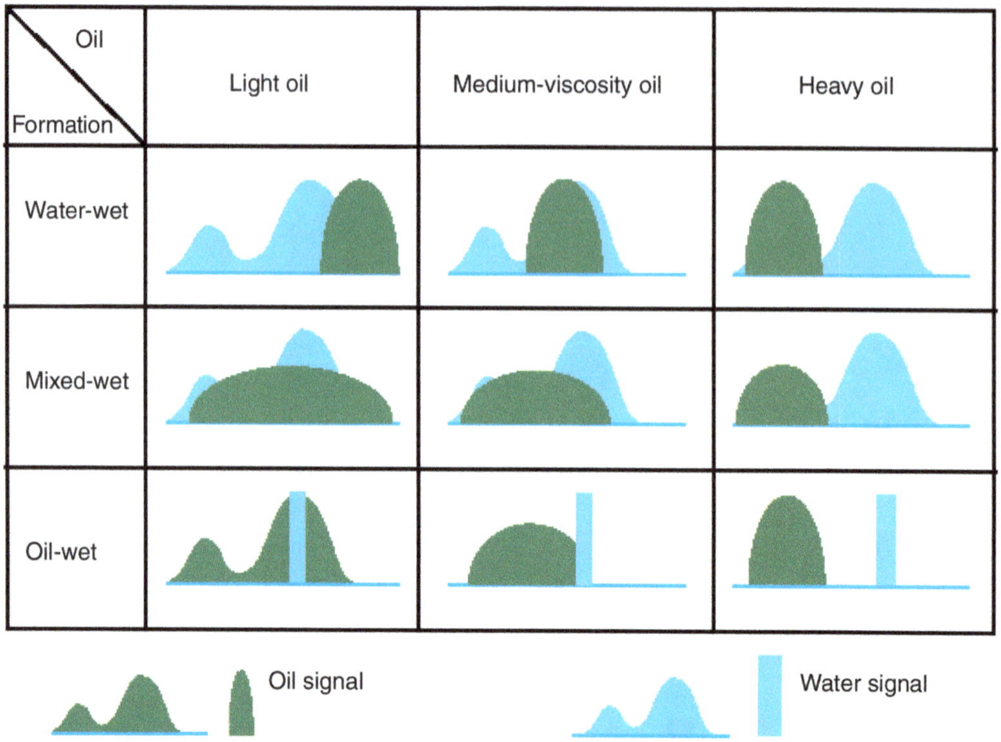

Fig. 3E.29—NMR-oil typing. The position and spread of the oil component in the T_2 distribution depends on oil viscosity and formation wettability. Oil typing is easiest in water-wet formations because of the moderate breadth and distinct positions of the different oil components in the T_2 distribution. Oil typing is most difficult in mixed-wet formations because the oil and water components are broad and overlap one another.

TABLE 3E.6—VARIATIONS IN CMPG PULSE SEQUENCE,[100] FOR SPECIFIC OBJECTIVES USING TDA[101] AND EDA METHODS[102]

CMPG Sequence Variation	Effect	Objective	Method
Short T_E	Emphasize short T_2 species	Distinguish oil from gas	EDA
Long T_E	Increase diffusion-in-gradient effects	Distinguish gas from oil	
Long T_W	Obtain full-porosity response of long T_1 components	Distinguish gas from oil	TDA
Short T_W	Discriminate against long-polarizing components by incomplete polarization	Distinguish gas from oil	

Fluid volumes can be quantified by integrating the difference into T_2 spectra and correcting for the polarization difference between long and short wait times. Fluid typing and quantification are performed through the differential-spectrum (DSM) method and the TDA method. Two conditions must be met to ensure successful DSM interpretation: T_1 contrast between the hydrocarbon and brine phases (i.e., a water-wet formation containing light hydrocarbons), and T_2 contrast between the gas and oil phases. DSM assumes that the hydrocarbon phases relax uniex-

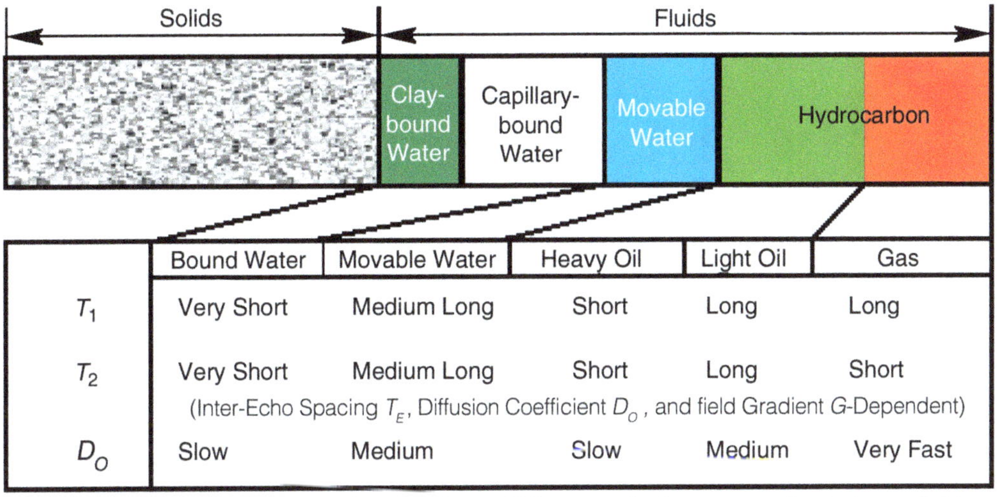

Fig. 3E.30—Typical NMR properties for pore fluids. Differences in these properties between fluids are exploited by methods that use specific acquisition sequences for hydrocarbon typing by use of T_1 or diffusion-weighting mechanisms.

ponentially. DSM is commonly used for: (1) hydrocarbon typing in shaly sands, where the the neutron-density crossover may be suppressed because of the high amount of clay minerals in the rock[106,107] to confirm the presence of light oil in fine-grained rock, and (2) for gas detection in the presence of OBM invasion.[108] Ideally, DSM can be used to compensate for the hydrocarbon effects on NMR measurements and thereby enable correction of NMR total porosity and NMR effective porosity; however, because of S/N requirements, TDA is the preferred technique for correcting NMR logs.

In contrast to the DSM approach, the TDA technique works directly with the echo-decay data (i.e., in the time domain rather than the T_2 domain; see **Fig. 3E.34**). The measured long- and short-wait-time decay trains are subtracted into an echo difference that indicates the presence of a long-T_2 component (usually a hydrocarbon). By working in the time domain, when the T_2 inversion is performed, only the hydrocarbon T_2 component will be present.

TDA—an outgrowth of the DSM technique—is based on the fact that different fluids have different rates of polarization (i.e., different T_1 relaxation times).[109] TDA provides the following results:

- Fluid types in the flushed zone.
- Corrected NMR porosity in gas reservoirs.
- Corrected NMR porosity in light oils.
- Complete fluid-saturation analysis in the flushed zone, using only NMR data.

The TDA technique is a more robust method than DSM, in part because it provides better corrections for underpolarized hydrogen and HI effects[101] (**Figs. 3E.34 and 3E.35**).

Several factors inherent in the dual-T_W acquisition require reduced logging speeds: to achieve full polarization (long T_W) in one acquisition channel, to acquire the small signal amplitudes associated with the short T_W values in another other channel, and to meet the requirement for acceptable S/N levels. A new triple-wait-time method addresses these issues and also allows T_1 acquisition.[110]

Diffusion-Weighted Mechanism (Dual-T_E Acquisition). NMR-logging tools generate relatively large field gradients, which are a function of the operating frequency and tool type. If the pore fluids have high diffusivity constants (D_0), these large gradients may cause diffusion

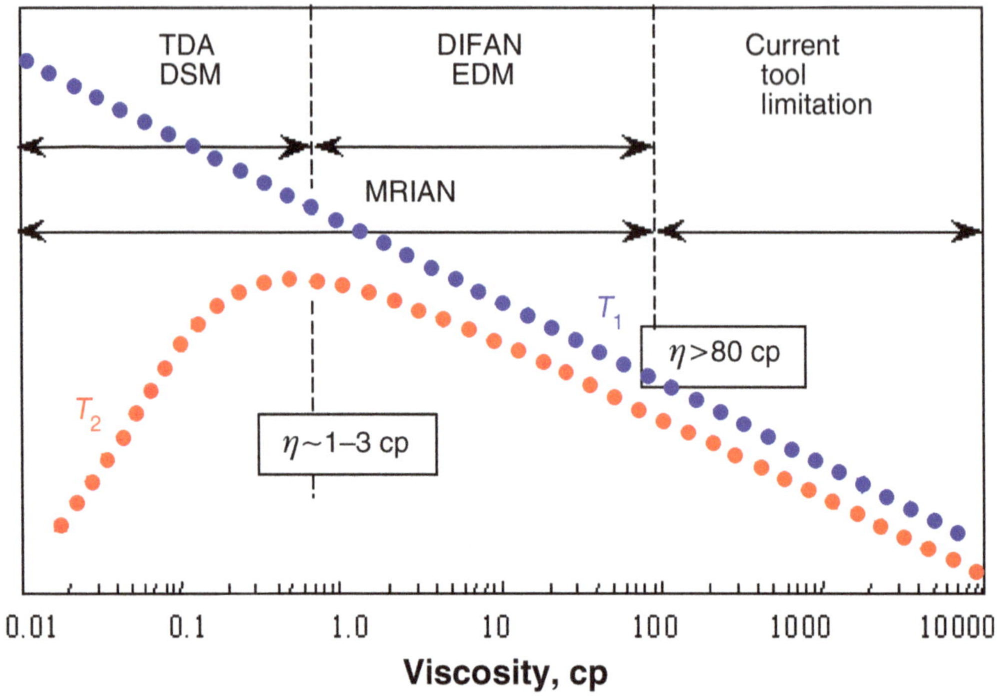

Fig. 3E.31—Graph indicating the current viscosity limits for NMR-interpretation techniques. The resistivity-density technique is limited to gas reservoirs, and the SSM method has been effectively replaced by the newer DSM method. The *y*-axis is the relaxation time.

to become the dominant T_2-relaxation mechanism, even at a reasonably short interecho time. The T_2 contrast between the liquid and gas phases can then be exploited for hydrocarbon typing.

The identification of viscous oil in water-wet rock using T_1-weighting mechanisms (dual-T_W acquisition) is very difficult because there is little T_1 contrast between water and viscous oil. However, a significant difference in the diffusion characteristics of water and viscous oil is used to create a T_2 contrast that separates the two NMR signals, thereby permitting identification.[102] The inter-echo spacing, T_E, is chosen long enough that the water and oil signals are fully separated in the T_2-domain (i.e., water is at lower T_2 than oil; see **Fig. 3E.36**).

Diffusion-weighting techniques, such as shifted-spectrum method (SSM), enhanced-diffusion method (EDM), and diffusion analysis (DIFAN), use the dual-T_E acquisition to create this T_2 contrast. T_E is chosen such that diffusion, rather than surface relaxation, is the dominant relaxation mechanism, and thus, the upper limit of T_2 for pore water is T_2 diffusion, designated T_{DW}. The dual-T_E acquisition set uses long T_W and T_E values, typically 0.9 or 1.2 ms and 3.6 or 4.8 ms, respectively. In this acquisition, the fluid with the larger diffusion constant (water) has a spectrum shifted more to earlier times than the fluid with the smaller diffusion constant (viscous oil). The presence in the spectra of a minimally shifted portion identifies high-viscosity oil in the formation (see the lower portion of **Fig. 3E.37**).[102,111] These techniques are primarily used for identifying the presence of viscous oil. They are also used in carbonates in which a reduced surface relaxation (compared to clastics) may also result in a reduced T_1 contrast.[112,113]

SSM. This method was the first developed using the diffusion-weighted mechanism and was originally developed for use in gas reservoirs as a qualitative technique (see the upper portion of Fig. 3E.37). However, because of signal-processing difficulties and slower logging speeds, this method was effectively replaced by the EDM.[105]

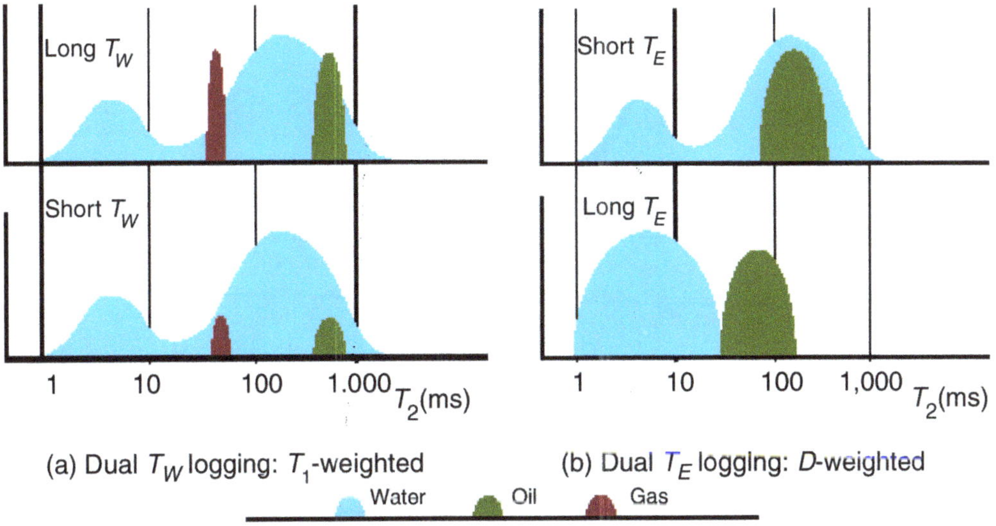

Fig. 3E.32—CPMG acquisition types used for hydrocarbon typing: (a) a T_1-weighted method is used to differentiate hydrocarbons from water, and (b) a diffusivity-weighted method is used to differentiate viscous oil from water or to differentiate gas from liquids.

EDM. This method is the later variation of SSM. Depending on oil-NMR properties and the job objective, EDM data processing can be done in either the T_2 domain or time domain.[102] If the EDM objective is to discriminate pay from nonpay, then a single CMPG measurement with long T_W (for full polarization) and long T_E (for diffusion enhancement) is sufficient; thus, standard-T_2 logging with a long T_E can be used. If the EDM objective is to quantify fluids in a pay zone, then dual-T_E logging will be required. The short-T_E measurements will provide correct MPHI and BVI. If the T_2 contrast over the zone of interest is not expected to be large enough to separate the T_2 values of water and oil, then dual-T_W logging with a single, long T_E may be required to obtain data for TDA processing.

One caveat for users of the EDM: the flushed-zone saturation (S_{xo}) obtained from EDM may not reflect the true volume of residual hydrocarbons in the flushed zone if the faster portion of the hydrocarbon's T_2 spectrum gets fully polarized by both wait-time measurements. A correction to S_{xo} can be applied if the full-hydrocarbon T_1 spectrum can be determined by other means, such as laboratory measurements.

DIFAN. DIFAN is an empirical model used for quantitative-diffusion analysis. This method uses a dual-T_E acquisition and is useful for typing and quantifying oils with viscosities ranging between 0.5 to 35 cp at reservoir conditions (i.e., temperature $\geq$ 200°F and pressure $\geq$ 2,000 psi). It is not recommended for light oils and condensates because the D_0 contrast between hydrocarbons and water is too small.

Neither EDM nor DIFAN is recommended for use in high-viscosity oil (heavy oil) because the difference between the T_2 values of dead oil and irreducible water is too small. The DIFAN model generates two T_2 distributions using the two echo trains generated from dual-T_E logging.

Echo Ratio Method. An apparent diffusion coefficient, calculated from the ratio of long and short T_E echo-train data (time domain), serves as a qualitative gas indicator.[114] This method is accurate and provides a direct measurement of gas volume, but requires high HI values for good results.

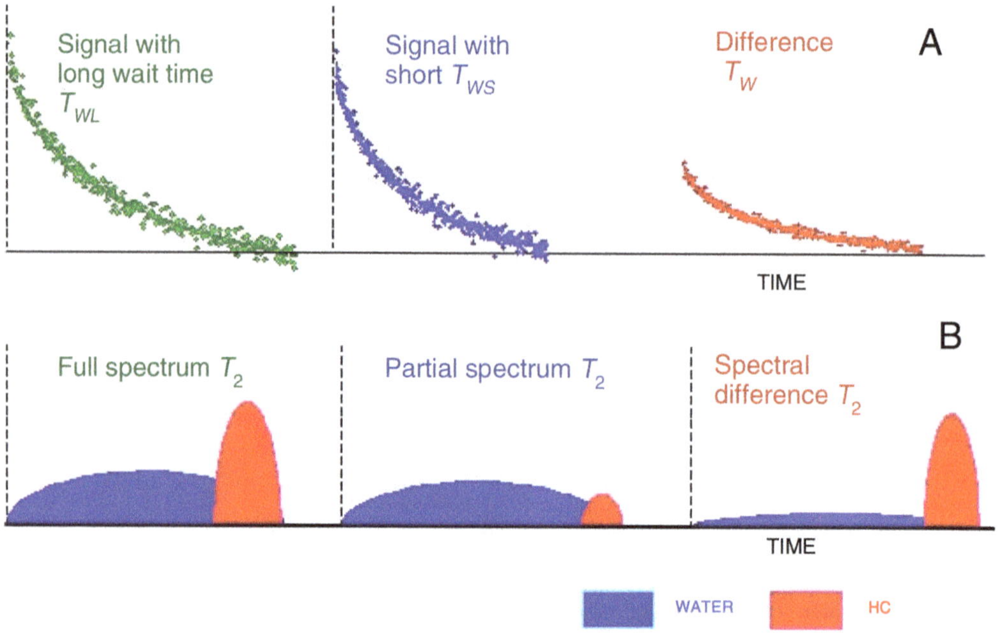

Fig. 3E.33—Methods using T_1-weighting for hydrocarbon typing: (a) in TDA, T_2 echo-train subtraction takes place in the time domain, and (b) in the differential-spectrum method (DSM), T_2 echo-train subtraction takes place in the T_2 domain.

Multifluid (Forward Modeling) Methods. The dual-T_W and dual-T_E techniques described above only provide T_2 distributions and are considered one-dimensional methods.[115] Oil-saturation determination using these methods may be difficult or even impossible because differential diffusion-weighted methods assume that NMR water and hydrocarbon signals can be cleanly separated by subtracting the acquisitions made at different T_E values. In practice, however, partial —or even total—overlap is common.

To avoid these problems, new multifluid-forward models were developed to take advantage of the full suite of data obtained using the latest multifrequency logging tools. These models perform a simultaneous inversion of the NMR fluid properties obtained in multiple parameter-domain dimensions (e.g., T_1, T_2, and D_0).[59,116–120] These methods make use of the enhanced T_2 relaxation in a magnetic-field gradient (i.e., incorporate measurement of the diffusion coefficient, D_0) and the large contrast between diffusion coefficients of oil and water. In effect, they are simultaneously combining both T_1- and diffusion-weighted processing techniques, and thus, provide more accurate and robust results. In addition to porosity and permeability information, the output data from these inversion models are plotted as two-dimensional plots (e.g., D-T_2 plot) in which the oil and water signals are clearly separated.[121] Recent two-dimensional NMR techniques include diffusion editing, diffusion mapping, and relaxation-diffusion 2D.[26,118,122–125]

Multidimensional (2D and 3D) NMR analysis greatly increases the accuracy of fluid typing and saturation determination.[115] These methods are particularly useful for identifying highly diffusive fluids (e.g., gas, condensate, and light oil)[59,125–128] and for identifying wettability alteration caused by OBM-filtrate invasion or chemical surfactants used in enhanced recovery operations.[118,127,129] This information is essential for reservoir simulation, development, and production.[130]

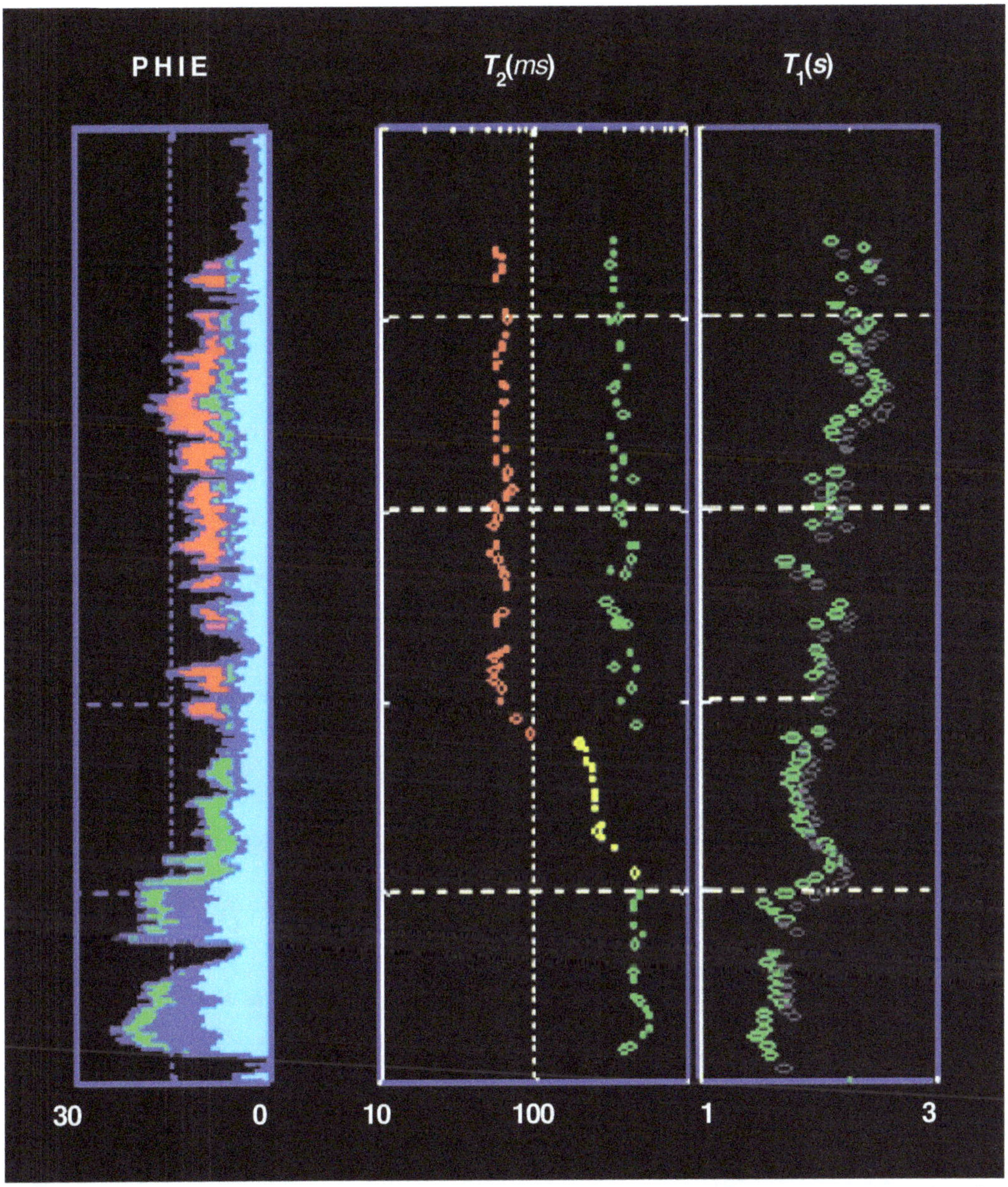

Fig. 3E.34—Example of TDA. Track 1 shows the pore volumes for gas (red), OBM filtrate and/or native oil (green), movable water (dark blue), and capillary-bound water (light blue) obtained from TDA quantitative analysis using an MRIL data set. Tracks 2 and 3 show the T_2 and T_1 values, respectively, of gas and light oil calculated through TDA. The lower part of Track 2 shows the oil/water and gas/oil contacts, while the upper part of the track indicates the presence of gas and oil.

3E.7.4 Residual Oil (S_{xo}) Calculation. Estimation of residual-oil saturation is one of the oldest applications of NMR logging. Unlike resistivity-log analysis, NMR analysis does not rely on formation-water salinity to obtain water saturation. This feature provides NMR logging with a distinct advantage over conventional resistivity analysis in mixed or unknown salinity conditions —an advantage that can be extremely useful in waterflood or steamflood projects for evaluating residual-oil saturation after the flood or to look for bypassed oil. Initially, NMR evaluation

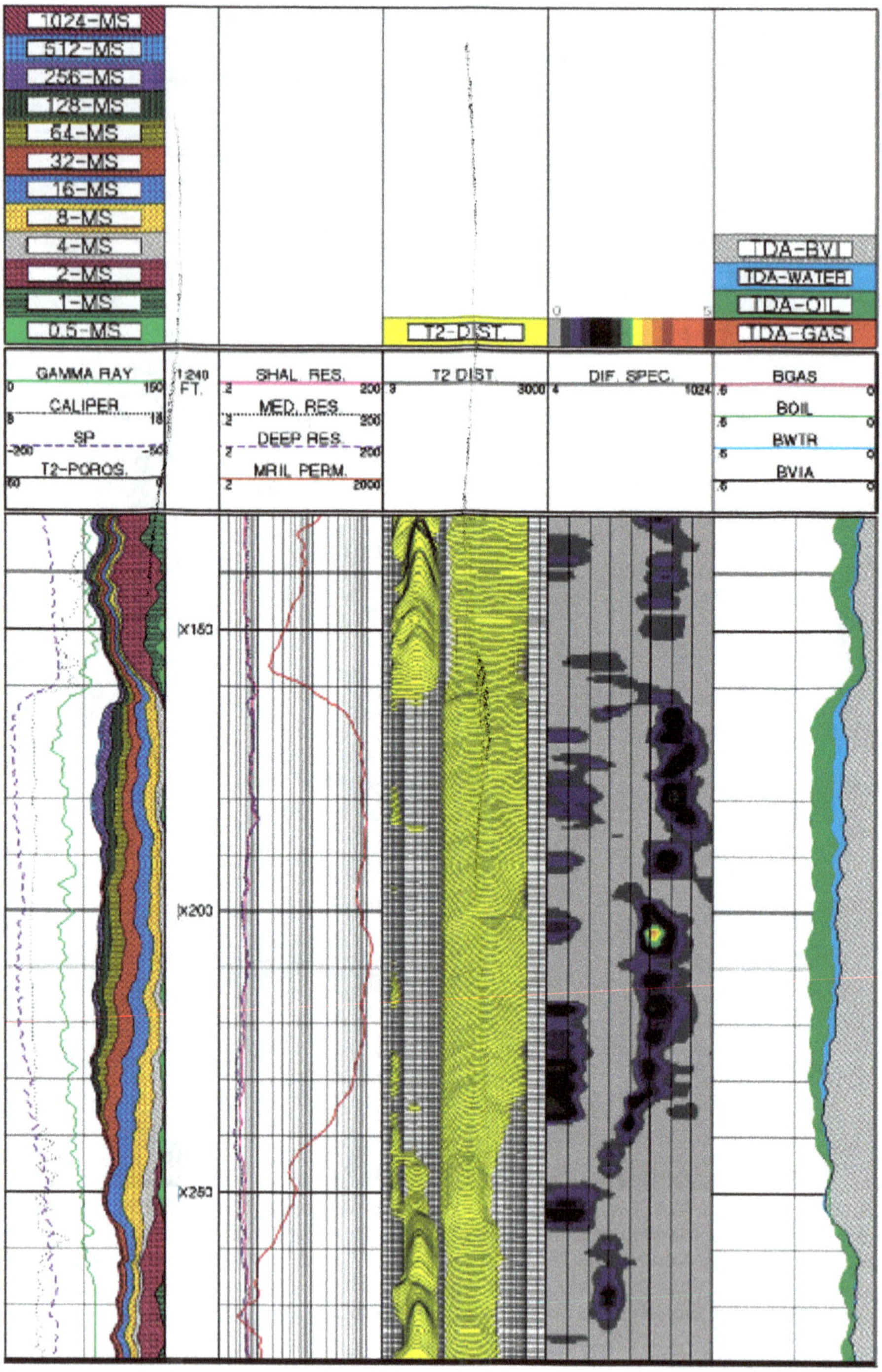

Fig. 3E.35—Log example showing the results of DSM and TDA analysis. Light hydrocarbons can be identified through the subtraction of echo trains obtained at two polarization times. Track 4 displays the differential spectrum obtained from the subtraction of the two separate T_2 distributions derived from echo trains acquired with short- (T_{WL}) times, T_{WS} = 1 second and T_{WL} = 8 seconds. The water signals in each completely cancel, while hydrocarbon signals only partially cancel and remain when the two T_2 distributions are subtracted from one another. Track 5 displays the TDA results. TDA is performed in time domain (as opposed to T_2 domain), and can quantify up to three phases (gas, light oil, and water; gas and water; or light oil and water).

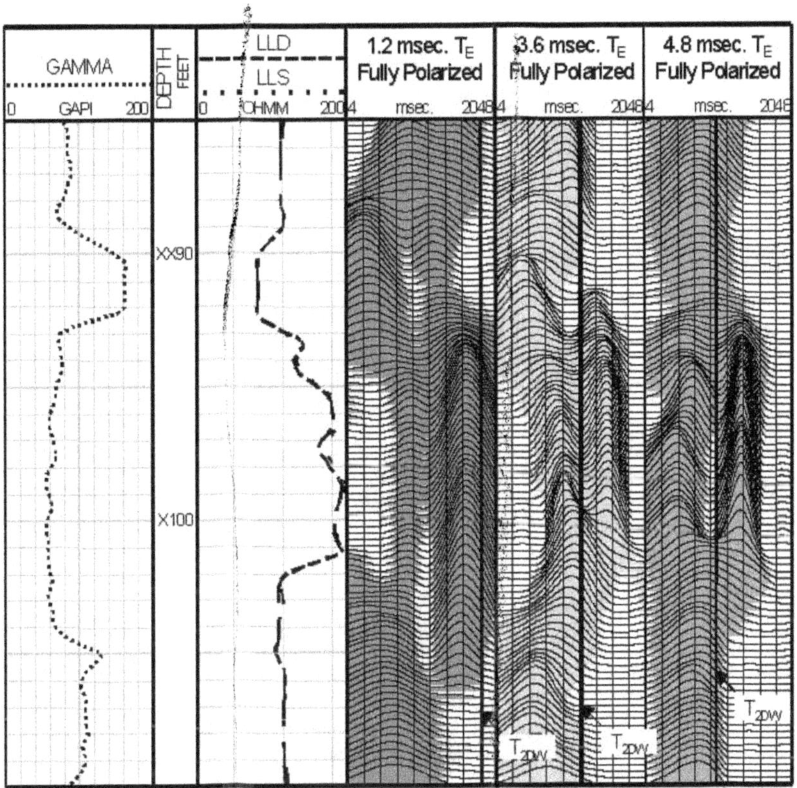

Fig. 3E.36—EDM example. The existence of a signal on the T_2 distribution longer than T_{2DW} indicates the presence of oil in the formation. In this log display, the T_2 distributions from T_E = 1.2, 3.6, and 4.8 ms are shown in Tracks 3, 4, and 5, respectively. In the EDM results, in Track 4 and 5, a significant signal to the right of the T_{2DW} line indicates obvious oil zones. Also note the increased definition and separation in the fluid signal that occurs in Track 5 because of the increased T_E. This observation is used to recognize pay in EDM-log displays.

of residual-oil saturation required use of a dopant (e.g., Mn-EDTA and $MnCl_2$) to flush the invaded zone with paramagnetic ions to shorten the bulk relaxation time of the brine. This process enabled separation of the oil and water signals, leading to direct measurement of S_{xo}. With the combination of modern multifrequency NMR tools and new methods of analysis, such as EDM and TDA, the use of borehole dopant is no longer necessary.[25,111] Standalone NMR interpretation is possible in OBM, but in water-base-mud (WBM), it is possible only when S_{wirr} is known.

3E.7.5 Viscosity Evaluation. In water-wet rocks, the NMR-relaxation spectrum is controlled by viscosity[118,131]; relaxation is directly related to viscosity, η, by Eq. 3E.17:

$$T_1 = 1.2 \times T_K / 298\eta \ . \qquad (3E.17)$$

For oil and water, the diffusion constant, D_0, can be approximated by Eq. 3E.18:

$$D_0 = A \times T_K / 298\eta, \qquad (3E.18)$$

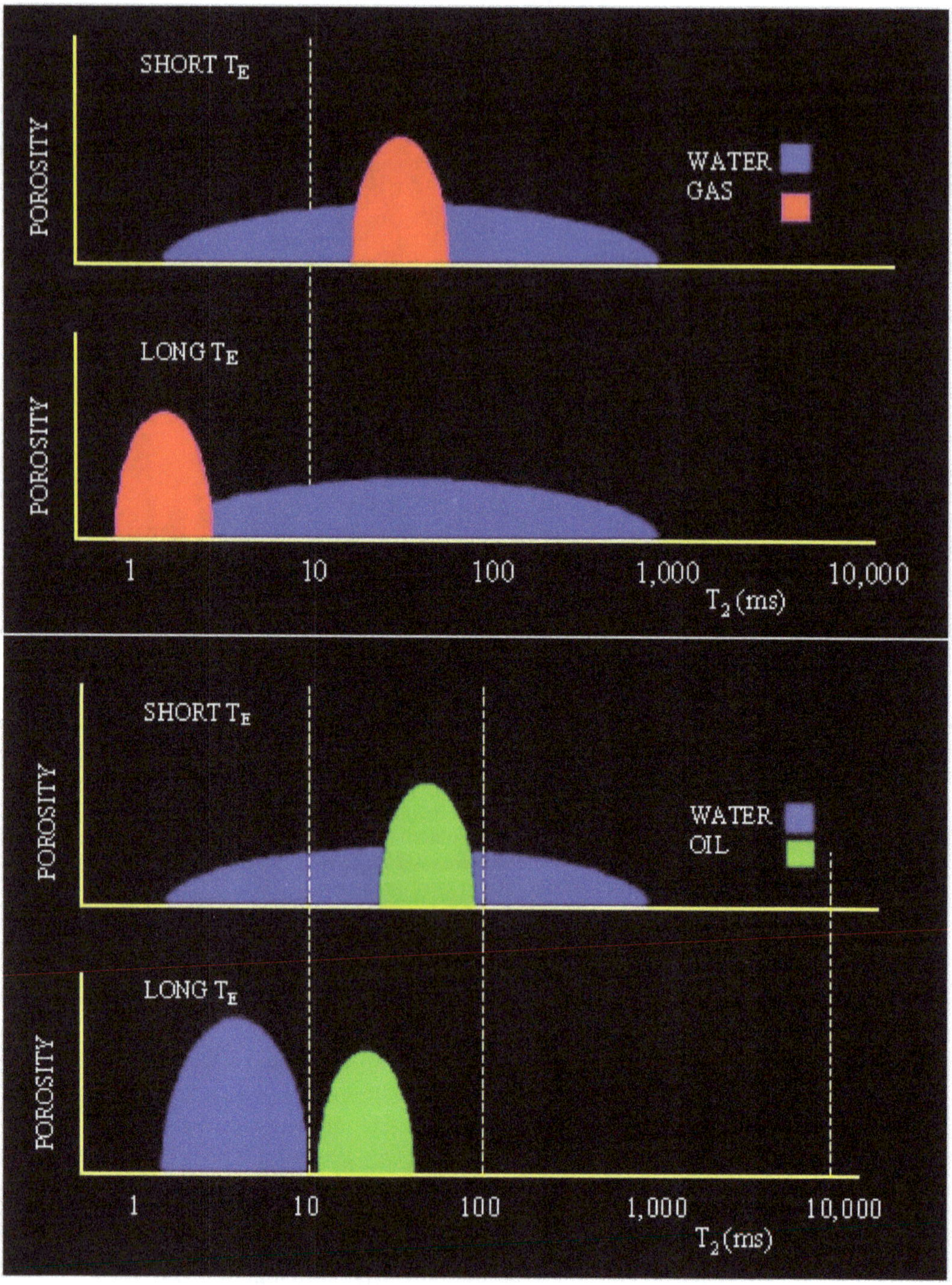

Fig. 3E.37—SSM application in gas (upper) and oil (lower) reservoirs.

where A = 2.5 for water, A = 1.3 for oil, and T_K is temperature in K.

NMR properties of gas can be obtained from published charts that relate viscosity to (1) the center of the relaxation curve, (2) to an American Petroleum Inst. (API) standard value,[105] or (3) from published formulas.[101] These published sources assume that methane is the dominant component of the gas. In the absence of laboratory data at in-situ conditions, reservoir

NMR properties can be estimated by use of Eq. 3E.17. One note of caution: Eqs. 3E.17 and 3E.18 are based on "dead oil" measurements. As mentioned earlier, the relaxation-time dependence on viscosity/temperature of live crude oil may differ significantly from correlations based on hydrocarbon liquids at ambient conditions.[97–99]

3E.7.6 Anisotropy and Geomechanics. NMR T_1 and T_2 relaxometry and derived permeability can be used to determine the Biot elastic constant, which is used for estimating pressures that are critical to sand control, hydraulic fracturing, wellbore stability, and determination of formation stress.[132,133]

3E.7.7 Low-Permeability (Tight) Sandstones. Field experience indicates that invasion or imbibition in tight sands is very shallow. Depending on borehole size, the diameter of invasion, and fluid properties, a mandrel tool may have sufficient DOI to measure beyond the flushed zone. A combined interpretation using both T_1 and T_2 can provide positive identification of fluids in these reservoirs.[134] In addition, in these tight formations, in which formation testers typically may not obtain a fluid sample within a reasonable time period, NMR fluid characterization can separate hydrocarbon from oil filtrate and other pore fluids.[135]

3E.7.8 Heavy Oil, Tar Sands, and Tar Mats. The early acquisition of reliable viscosity information is essential to efficient development of heavy-oil reservoirs. NMR logs offer a viable alternative to downhole fluid sampling for determining viscosity information in heavy-oil reservoirs.[136–138] The presence of tar mats in a reservoir, commonly near the bottom of the oil column, may form vertical permeability barriers and, thereby, isolate the oil leg from the water-drive aquifer. NMR logs, in conjunction with conventional logs, can provide accurate identification of tar-mat levels and viscosity estimation, from empirical relationships.[139] (See the Job Planning section of this chapter.)

3E.7.9 Carbonates and Complex Lithologies. NMR-log evaluation is relatively routine in what might be termed "conventional reservoirs," namely those of homogenous lithology and uniform pore sizes, typically sandstone and chalk reservoirs.[62,140] In contrast, log evaluation of complex and heterogeneous reservoirs, with complex pore geometries, is not straightforward. These reservoirs include, in particular, the highly important Middle East carbonates, as well as other reservoirs comprised of mixed lithologies and mineralogies, or both, in which wettability may also vary. In these reservoirs, there is likely no simple relationship between petrophysical properties and porosity. Instead of a dependence on the volume of pore space, they are dependent on the typically heterogeneous pore distribution, pore types, pore connectivity, and grain sizes. This fundamental difference between siliciclastic and carbonate rocks (primarily the result of diagenetic processes) limits the applicability of routine NMR methods, especially permeability evaluation.[141] Improving NMR evaluation of carbonates has proved challenging and is the subject of a number of recent studies and proposed techniques.[120,141–144] (See the Job Planning section of this chapter.)

3E.7.10 Pseudocapillary-Pressure Curves. As discussed in the Petrophysics section, establishing a correlation between NMR T_2 distribution and MICP is fundamental to NMR interpretation and the computation of S_w. Once such correlations are established for a particular reservoir or field, pseudocapillary-pressure curves can be generated directly from the NMR-log relaxation-time distributions.[145–148]

3E.7.11 Producibility. Accurate determination of BVI enables evaluation of reservoir-fluid (e.g., gas, oil, and water) contacts, production characteristics (producibilty), and the determination of net and recoverable reserves.[149,150] Furthermore, NMR-derived permeability can be used to generate idealized flow profiles across a completion interval. These profiles provide a diagnostic tool for identifying nonflowing portions of the zone and the need for remedial work.[151,152]

3E.8 Combined NMR Applications

NMR tools have shallow depths of investigation and provide results only for the invaded zone. NMR-log data can be integrated with core data and conventional log data (i.e., neutron, density, acoustic, and resistivity) in post-acquisition processing to provide improved determinations of reservoir rock properties, hydrocarbon storage capacity, and reservoir productivity in a variety of environments including gas-bearing and low-resistivity reservoirs. Interpretation models that include NMR data can provide more reliable results than those using only conventional logs.

3E.8.1 NMR-Resistivity Combination. The combination of NMR and deep-resistivity data provides a complete analysis of the fluids in the uninvaded zone. Resistivity measurements alone cannot distinguish between capillary-bound and movable water, but they do represent deep investigations of fluid saturation. Furthermore, resistivity-based methods are often inadequate or unreliable in reservoirs in which salinity and lithology vary. The addition of NMR-derived BVI and CBW from the flushed or invaded zone can significantly enhance the estimation of resistivity-based fluid saturation, both in clastics and carbonates. The addition of NMR data allows the identification and evaluation of water-free production in low-resistivity formations (**Fig. 3E.38**).[107,153,154]

The combination of conventional deep-resistivity data with NMR-derived CBW, BVI, FFI, and MPHI can greatly enhance petrophysical estimations of effective pore volume, water cut, and permeability. It is the preferred technique for identifying low-resistivity pay zones. Fig. 3E.38 presents the results of Halliburton's MRI analysis (MRIAN) service in a turbidite sequence. The sand below XX200 depth has an average resistivity of approximately 0.5 ohm-m (Track 2) and average neutron-density porosity of approximately 38% (Track 4). A quick look, or preliminary analysis, using only the conventional data presented in Tracks 1, 2, and 4, would label this a wet zone. MRIL-NMR data are also presented in Track 1 (T_2-distribution bin data), Track 2 [NMR-derived permeability (Coates)], and Track 3 (VDL presentation of T_2 distribution). The MRIAN results, BVI, and FFI are presented in Track 4. BVI gradually increases with depth, suggesting that the sand is fining downward (i.e., as sand grains become finer, the volume of capillary-bound water that they hold increases). Comparison of the BVI with the resistivity profile (Track 2) shows that the resistivity decreases where the bound water increases. The MRIAN analysis clearly shows that the zone does not contain any movable water and will produce only oil. The interval below XX200 was tested and produced oil with no water.[32]

The combination of NMR fluid identification with resistivity-derived saturation provides a better understanding of hydrocarbon movement in gas monitoring, as seen in the following field example (**Fig. 3E.39**).

NMR and LWD logs were run in a deviated Middle East light-oil carbonate reservoir to monitor gas injection. NMR logs were well suited for gas monitoring by use of the TDA technique and were not affected by formation-water salinity.[31] NMR gas-corrected porosity was in close agreement with core porosity; conventional log-porosity was low because of salinity effects. Combined interpretation of water-saturation analyses from tools with different depths of investigation identified the movable fluid present in the reservoir as WBM filtrate. Track 1 contains T_2 bin data and conventional SP and gamma ray curves; Track 2 shows NMR-permeability derived from the standard Coates model and LWD resistivity; Track 3 contains a T_2 distribution in a variable-density format; Track 4 contains T_2 distributions of data acquired with long T_W and short T_W; Track 5 is the differential spectrum; Track 6 contains TDA results; and Track 7 contains MRIAN results.

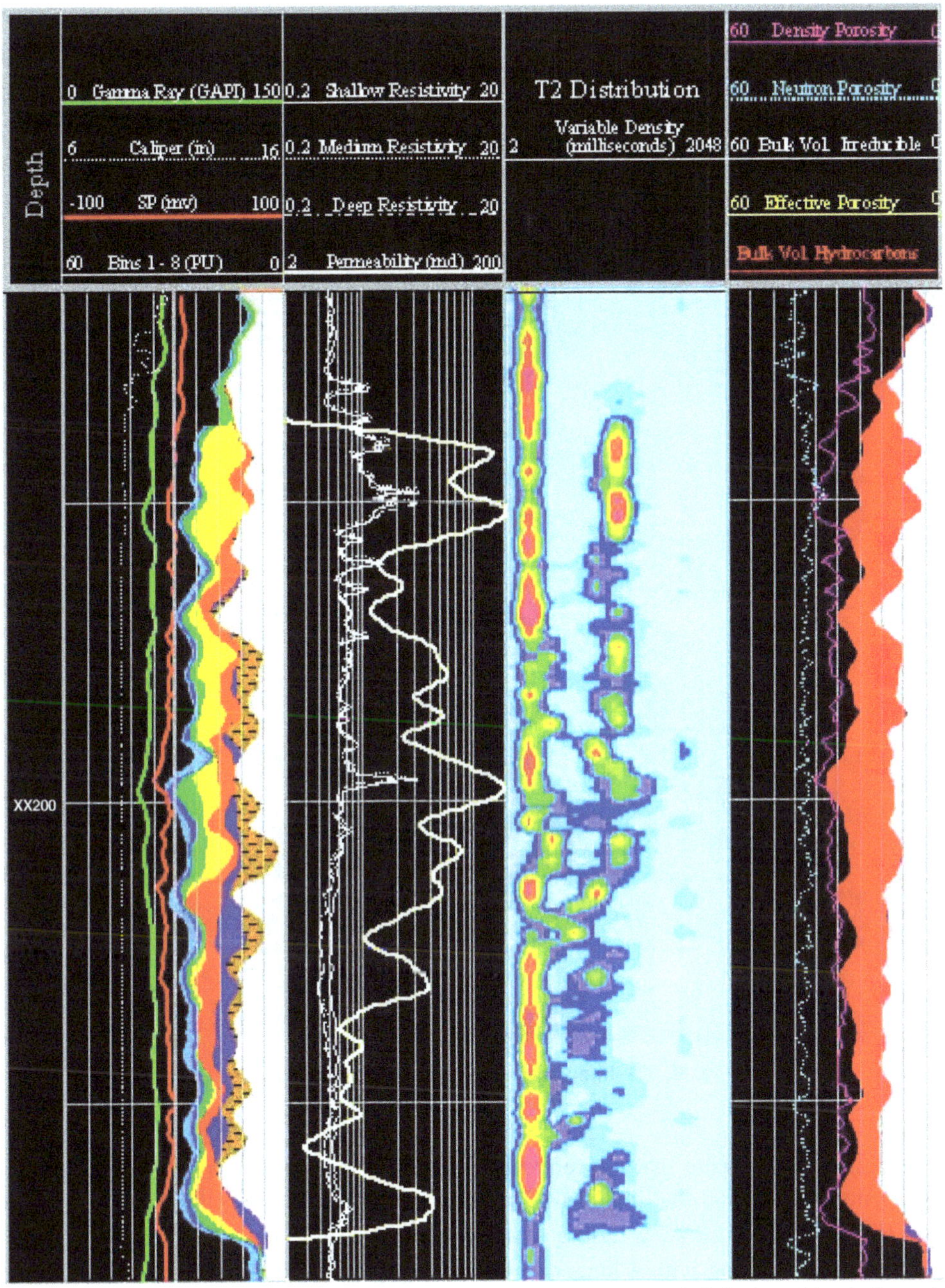

Fig. 3E.38—Integrated-NMR-resistivity shaly sand analysis. Whereas resistivity logs (Track 2) read very low values and indicate that the zone below XX200 is water wet, NMR measurements show that BVI (Track 4, gray) increases with depth. However, T_2 measurements (Tracks 3 and 4) suggested the presence of hydrocarbon, which testing confirmed in what turned out to be a water-free zone.

3E.8.2 NMR Acoustic/Density Combination. Because NMR tools are affected by the lower HI in gas-bearing reservoirs, a moving tool may not fully polarize the gas, which has a long T_1. Conventional acoustic and density logs are not influenced by these factors and, when used jointly with NMR logs, they can provide robust porosity evaluation.[155,156] The basic assump-

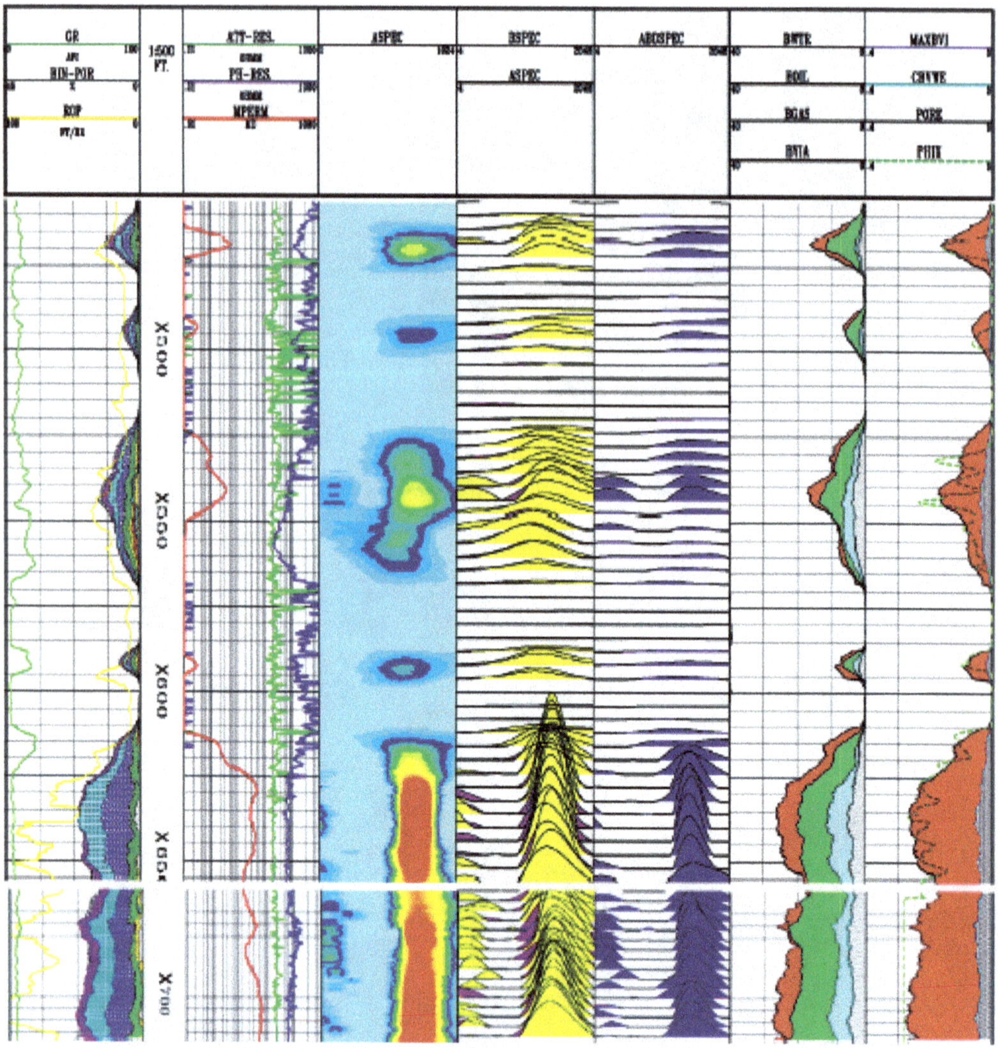

Fig. 3E.39—Integrated-NMR and resistivity analysis.

tions of these techniques are that the reference porosity is measured correctly (e.g., appropriate matrix density is used) and that the NMR T_W is long enough to recover the water but not all of the gas. If the T_W is too short, a false gas signal may be seen. One way to check whether the T_W is appropriate is to log with a longer wait time (e.g., T_W = 2 seconds rather than 1 second) and look for a significant difference in porosity (i.e., the porosity may be slightly higher because of increased polarization of the gas).

3E.9 NMR-Log Quality Control

Quality control (QC) procedures are especially important during the creation of integrated computed products[31,78,157] and to ensure optimal NMR data acquisition. NMR tools have calibration standards and real-time QC indicators. Standard techniques include examination and verification of service company calibrations and QC curves, and calibrations and examination of the repeatability of the computed porosities. Processed results should agree with other data from logs, core, and/or formation test results. NMR QC includes a series of prejob and post-job checks and calibrations.

3E.9.1 Prejob Calibration and Quality Checks During Logging. Several prejob QC steps are necessary to ensure reliable results. These steps are described in this section.

Frequency Check. The amplitude of the NMR measurement is proportional to the square of the magnetic field strength in the measurement region. To attain maximum signal strength, the tool must operate at the correct frequency. Because temperature affects magnetic field strength, a frequency sweep is recorded before the calibration and logging run and is repeated downhole to account for changes in temperature. The CMR tool includes sensors that monitor temperature and changes in the magnetic field strength and adjusts frequency automatically to changes in sonde temperature. The accumulation of metal drilling debris on the magnets can adversely affect the NMR measurements, and the tool may require occasional retuning. The MRIL-gradient field creates a self-correcting effect, and the tool rarely requires retuning after the initial downhole tuning. Drifting off frequency may present itself as a loss of porosity (CMR) or loss of precision (MRIL). Field calibration should be included on the log.

Statistical Calibration. A statistical calibration to 100% porosity (water) should be made at the shop and/or wellsite before and after every logging job for all combinations of T_E frequency and for expected quality factor (Q level); calibration should be done at the shop at least monthly. Primary calibration involves either a flask of fluid (for the CMR tool) or a water tank (for the MRIL tool).

HI. NMR-tool calibration uses water where HI = 1. However, when the HI of the formation pore fluids is expected to be less than one, the NMR-porosity readings will be proportionally lower, and correction is necessary. An NMR tool's processing software will, therefore, compute a correction based on the salinity of the mud filtrate and on the maximum formation pressure and temperature. In general, a mud-filtrate relaxation time of < 200 ms may suppress the long T_2 components in a T_2 distribution. A specific HI value may be used when residual hydrocarbons or OBM filtrates are present.

System Gain. CMR and MRIL tools monitor and calibrate in real time for downhole changes in system gain. A gain measurement is made as a part of each pulse sequence. Gain indicates the amount of loading applied to the NMR tool's transmitter circuit by borehole fluid and formation resistivity. Gain is affected by changes in temperature, mud conductivity, and borehole size. Because gain is frequency dependent, the operating frequency of a tool should be set to achieve maximum gain. Abrupt changes or spikes in gain should not be present. For a particular MRIL tool, the gain value determines the appropriate acquisition power (Q) level.

System Noise (Ringing). Noise contamination of the spin-echo signal, known as ringing, may be a remnant of the RF pulsing process. Noise can interfere with the echo trains and, when present, is usually evident in the NMR porosity readings, either as poor repeatability or lack of agreement with other tools. Both tools (CMR tool and MRIL tool) monitor ringing.

χ. χ is a measure of the quality of fit between the calculated decay curve and the recorded echo amplitudes. Problems with the echo data are likely to be reflected in the χ curve. In general, the value of χ should be less than two but may average slightly higher in certain situations. Spikes in χ may correlate with spikes in porosity[78] and usually indicate tool problems, even if χ remains lower than two. χ serves as is a primary MRIL log-quality indicator and is monitored while logging.

Correction for RF-Tipping Pulse. The strength of the CMPG RF pulse (B_1) that produces proton tipping and rephasing is measured as part of every pulse sequence and requires correction for changes in borehole temperature. If the pulse angles are either < 90° or > 90°, the magnetization will be undertipped or overtipped, respectively. The measured amplitude will, thus, be too small, and porosity will be underestimated. The B_1 curve should be relatively constant but should show some variation with changes in borehole and formation conductivity. B_1 will decrease across conductive washouts and conductive formations. Changes in the B_1 values should track changes in total conductivity and vary together in the same direction as gain. The need for excessive correction (i.e., > 5% of the optimum shop-peak value) may indicate a tool problem and can result in undertipping or overtipping of the protons, a reduced S/N ratio, and a loss of precision in determining porosity. Sudden changes in the B_1 correction curve may also indicate a tool problem.

γ. Regularization methods are used to select a smooth T_2 distribution that is consistent with the spin-echo sequence. These methods require a parameter γ that is automatically computed from the raw echo data. Values of γ are dependent on the S/N ratio and the shape of the underlying T_2 distribution. In high-S/N environments (i.e., medium-to-high-porosity formations), typically $\gamma < 5$; in low-S/N environments (i.e., tight sands and shales), $\gamma > 10$.

Repeatability. Whenever possible, a repeat pass should be recorded with parameters identical to those used in the main pass. These parameters include T_W, N_E, and T_E, as well as computation parameters such as the echoes selected for processing, and for $T_{2cutoff}$. The generally accepted goal for porosity is a standard deviation of 1 porosity unit (p.u.). Repeatability of BVI is usually > 1 p.u., and repeatability of FFI is usually >> 1 p.u.

If repeatability is a concern, decreasing logging speed or increasing the degree of echo data stacking during logging in post-job processing can improve S/N ratio and repeatability. Typically, fluid-typing applications using dual-T_W or dual-T_E methods are more sensitive to data repeatability than a standard NMR log acquired for porosity, bound fluid, and permeability.

A given data set should agree with similar data acquired by other logs, formation tests, and/or core analysis.

Environmental Corrections. NMR logs are similar to neutron logs in that they respond to the hydrogen volume present in a sample. Because the hydrogen volume changes with temperature and pressure, NMR logs require environmental corrections for temperature and pressure similar to those applied to neutron logs. Also, the magnitude of NMR resonance (and S/N ratio) varies inversely with temperature. NMR-logging tools include a temperature sensor to acquire the data needed for this correction. At high temperatures, data stacking can be increased or logging speeds reduced to compensate for a lower S/N ratio.

Porosity Check. NMR porosity can serve as an important diagnostic of data quality. High porosity readings may result from washouts (e.g., borehole fluid in the sensed volume), tool problems, improper environmental corrections or calibration issues, loss of pad contact (for the CMR tool), tool eccentering, or borehole ellipticity (for the MRIL tool). Low porosity readings may result from insufficient wait time, light-hydrocarbon effects, the presence of heavy oil, improper frequency tuning, and/or calibration error.

3E.9.2 Post-Logging Quality Check. NMR-log responses should be checked against conventional logs when they are available (see the Applications section of this chapter).

3E.10 NMR-Log Job Planning

In some complex reservoirs, low-resistivity/low-contrast pay, low-porosity/low-permeability, and medium-to-heavy oil, NMR-log data—independently or in combination with other log data—provide the best and/or only means of accurate formation and fluid evaluation. Because NMR-log data acquisition is complex, job preplanning is essential to ensure optimal selection of acquisition parameters that will result in reliable and accurate data and in the maximum information possible in any given reservoir and logging environment. A clear understanding of the logging job objectives is necessary for optimizing the NMR acquisition parameters to best achieve these objectives.[31,122,157,158] This process must take place before the actual logging.

Typical preplanning consists of three steps:

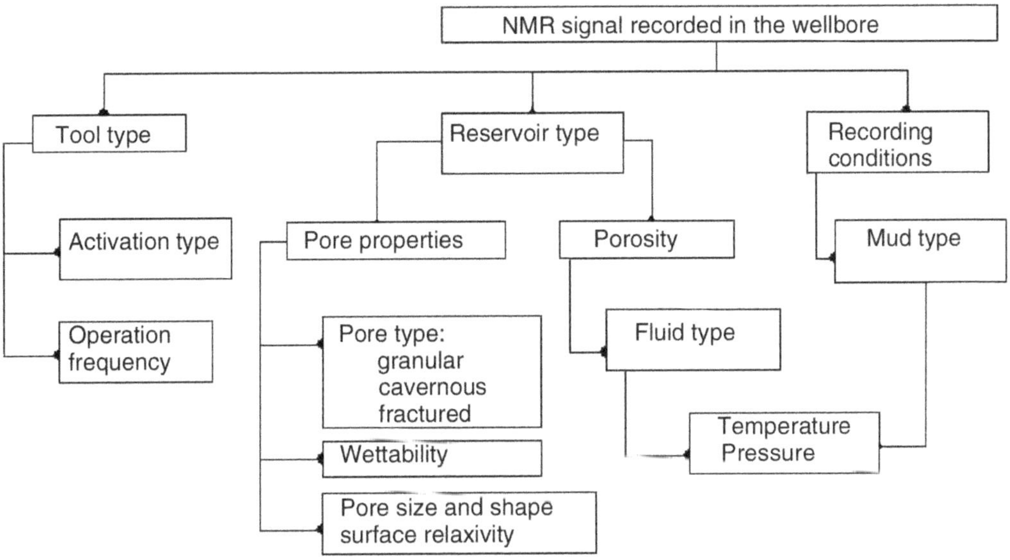

Fig. 3E.40—Reservoir and operational factors influencing the choice of CMPG acquisition.

1. Define the need for NMR measurements.

2. Collect all available borehole (e.g., diameter, mud, salinity, and temperature) and reservoir (e.g., formation and fluid properties) information needed to assess the expected NMR responses in the zone of interest, and understand what can and cannot be resolved with NMR.

3. Select the appropriate tool (on the basis of operational considerations, borehole size, and condition) and acquisition type (i.e., determining the appropriate acquisition parameters, data resolution, and logging speed) that will provide maximum answers for a given job.[31]

Although the actual in-situ reservoir characteristics may be unknown, estimates of the anticipated fluid properties, based on available information such as reports for nearby wells or fields, are used to define and optimize an acquisition sequence that will provide the data needed to meet the job objectives.

In addition to job objectives, determination of the appropriate NMR-acquisition parameters is also influenced by operational considerations and the anticipated in-situ reservoir properties (**Fig. 3E.40**). The most critical factors follow.

3E.10.1 Lithology. Although reservoir lithology generally plays a minor role in NMR-data acquisition, it does play a significant role in data analysis and interpretation. Aspects of reservoir lithology that influence reservoir T_2 values include the following.

Carbonates. Different $T_{2\text{cutoff}}$ values are required because surface relaxivity in carbonates is weaker than in sandstones, resulting in slower relaxation rates (longer T_2). Longer T_1 in carbonates than in sandstones may require longer T_W during acquisition.[103] (See the Application section of this chapter.)

Isolated Pores. The presence of relatively isolated pores (e.g., vugs) will not affect NMR porosity, but it will cause the standard permeability equations (Coates and SDR) to overestimate permeability. (See the Permeability section of this chapter.)

Ferromagnetic and Paramagnetic Minerals. The presence of these minerals may enhance surface relaxation significantly, shifting the T_2 spectrum to very short relaxation times. Depending on the amount of paramagnetic material, relaxation may become too fast to be detected, and the NMR measurement will underestimate porosity. In these cases, standard cutoff values do not apply.[159]

Heavy Oil and Tar Sands. Intervals containing these types of hydrocarbons have very fast relaxation components and may not be detected using conventional acquisition methods. Special methods have been developed for detection and accurate evaluation of these reservoirs.[136,137,160–167] (See the Applications section of this chapter.)

3E.10.2 Wettability. Wettability can have a significant impact on NMR-log response.[168] The use of NMR for determination of wettability has been extensively studied both in laboratory[129,169] and in the field. In general, petrophysical-NMR studies and NMR-logging applications assume reservoir rocks are water wet; however, because mixed-wettability reservoirs do exist (e.g., some carbonates, black shales, and heavy-oil reservoirs),[112] this assumption may lead to incorrect reserve estimates and to unexpected dynamic behavior during waterflood. When a pore is water wet, oil relaxes at its bulk rate. In mixed-wettability reservoirs, the oil and water each relax through a combination of bulk relaxation and surface interaction and depend on the ratio of water-wet surface area to water volume and oil-wet surface area to oil volume. The oil relaxation spectra will be shifted from bulk relaxation into the irreducible-water part, resulting in complex spectra that are difficult to interpret.[170,171] Nevertheless, these shifts provide qualitative wettability indicators that allow NMR logs to provide an early indication of reservoir wetting behavior.[113,118,172–174]

The invasion of OBM or SOBM can alter formation wettability and is a significant concern in NMR logging,[113,131,175] which typically measures fluid and formation properties in the flushed zone. Invasion by these muds can alter strongly water-wet sandstones and carbonates to intermediate-wet or oil-wet rocks. In water-wet reservoirs that have undergone OBM or SOBM invasion, the $T_{2\text{cutoff}}$ model may significantly underestimate $S_{w\text{irr}}$ because wettability alteration changes the water and oil relaxation-time distributions. The magnitude of underestimation depends on the type of OBM surfactants, their concentration in the flushing fluid, and the flushing volume. Controlling the volume of OBM invasion and the concentration of OBM surfactants should minimize the effects of OBM invasion on estimation of $S_{w\text{irr}}$.[176,177]

3E.10.3 Borehole Rugosity. NMR-logging tools have relatively shallow DOIs. Pad-type tools (e.g., the CMR tool) are run eccentered and require good contact with the borehole wall for accurate measurements. Measurements made by pad-type tools can be significantly affected by severe borehole rugosity and washouts, resulting in overestimation of porosity.

Mandrel devices are run centered, and DOI can range from 1 to 4 in., depending on borehole size. The sensitive area is normally beyond minor borehole rugosity. When borehole conditions result in an elliptical borehole (e.g., breakouts or erosion) or otherwise inhibit or prevent centralization (e.g., in highly deviated or horizontal boreholes), contact tools may be a better choice if pad alignment and contact with the borehole wall can be maintained.

3E.10.4 Mud Type. The quality of NMR data acquired in OBM is generally superior to that acquired in WBM. The conductivity of OBM is lower; lower conductivity reduces loading effects on the transmitter/receiver system, resulting in higher S/N ratio. This issue is of greater concern for mandrel tools (e.g., MRIL tool) because, in conductive muds, the power of the RF pulse is reduced as the pulse is transmitted across the borehole. The use of fluid-excluding sleeves can minimize this problem.

Signal dissipation in conductive mud is not a serious concern for pad-type tools (CMR) that maintain contact with the borehole wall. Because NMR tools read the flushed zone, OBM filtrate invasion produces an additional hydrocarbon signal that may significantly complicate log interpretation.[178] Careful prejob planning can reduce interference of the OBM-filtrate signal and the response from the native fluids. The relatively long T_1 relaxation times and diffusivity of OBM make it difficult to differentiate its signal using the shifted-spectrum or differential-spectrum approaches.

3E.10.5 Metal Debris. Metal drilling debris in the borehole fluid may affect NMR-measurement quality by distorting and altering the logging tool's magnetic field. The pad-type tool is more susceptible to field distortion. Metal debris should be removed from the mud, either through the use of the prepolarizing magnets (included in the latest tool designs) or by using magnets at the shale shaker. A new wireline tool uses an autotuning feature to correct the operating frequency for changes in the static magnetic field caused by metallic debris in the borehole.[55]

3E.10.6 Logging Speed and Running Average. The logging speed of an NMR tool is influenced by a number of factors, primarily by tool type (e.g., centered or eccentered, number of operating frequencies, or length of antenna), logging objectives (e.g., acquisition type—T_W, T_E, or N_E—sequence repetitions and vertical resolution), and borehole properties (e.g., diameter and mud resistivity). S/N ratio is primarily controlled by borehole size and mud resistivity. As S/N ratio decreases, the running average (RA) needed to maintain a specified error in porosity increases. The general practice is to require a porosity standard deviation of $\leq$ 1 p.u. The value of RA and the antenna aperture (i.e., length), combined with the logging speed, determines the vertical resolution. Even so, there is always a complex tradeoff in logging speed, accuracy (e.g., S/N ratio or N_E), and job objectives. High accuracy and precision require reduced logging speeds. A method for increasing the overall logging speed is to reduce the vertical resolution in zones of secondary or no interest and also to reduce the vertical resolution in homogeneous intervals.

3E.10.7 Vertical Resolution. The vertical resolution of NMR-logging tools is primarily a function of antenna length (i.e., tool design) and logging speed. The maximum vertical resolution, usually obtained when the tool is at rest (e.g., in stationary mode), is the length of the antenna. During continuous logging, vertical resolution decreases at a rate proportional to logging speed. Contact-logging tools, in general, use smaller sensors and antennae and, thus, have better vertical resolution than centered tools. The contact NMR tool (i.e., the CMR tool) has a resolution advantage when bed thickness is in the range between 0.5 and 5 ft. Outside of this range, both designs deliver similar results. Prejob planning includes selecting a logging speed to obtain the optimum resolution.

Post-job data reprocessing to enhance bed resolution may result in a loss of repeatability. Vertical resolution can also be improved by optimizing the NMR signal through the removal of signal noise during data processing.[179,180]

3E.11 Summary

This chapter has outlined the fundamental properties that are measured by NMR tools and has reviewed the use of these measurements to discern various characteristics of the reservoir rock including its fluid contents. Whether used as a standalone service or in combination with other logs and core data, NMR logs can provide an improved understanding of reservoir petrophysics and producibility. However, NMR logs are the most complex logging service introduced to date and require extensive prejob planning to ensure optimal acquisition of the appropriate data needed to achieve the desired objectives. Review of the references cited in this chapter indicates the rapid pace of advancement in NMR logging research, development, and applications. Users of NMR logging, including engineers, log analysts, petrophysicists, and geologists, should anticipate new developments in this discipline, as reported by SPE, the Soc. of Petrophysicists and Well Log Analysts (SPWLA), and other publications noted in the references.

3E.12 NMR-Tool Mnemonics

Table 3E.7 presents a cross reference of logging-tool output-data mnemonics for the different service companies currently offering NMR-logging services.

TABLE 3E.7—CROSS REFERENCE OF OUTPUT-CURVE MNEMONICS FROM NMR-LOGGING SERVICE PROVIDERS

Measurement	Baker Atlas	Computalog	Halliburton	Schlumberger
Generic porosity	MPHI	MPHI	MPHI	CMPR
Total porosity	MPHS	MSIG	MSIG	TCMR
Effective porosity	MPHE	MPHI	MPHI	CMPR
Free-fluid volume	MBVM	FFI	MFFI	CMFF
Capillary-bound fluid volume	MBVI	MBVI	MBVI	BFV
Clay-bound water	MCBW	MCBW	MCBW	—
Permeability	MPRM	MPERM	MPERM	KTIM/KSDR
Binary/spectral porosities	P01-P0x	BIN1-BIN8	BIN1-BINx or P_1-Px (PR-total)	CBP1-CBP8
Spectral/cutoff bound fluid volume			SBVI/CBVI	
Total bound fluid volume			MBFV	

Nomenclature

A = pore-fluid-specific value used to approximate the diffusion constant
a = constant in the mean-T_2 and viscosity relationship
B_0 = static magnetic field, gauss
B_1 = amplitude of the oscillating magnetic field perpendicular to B_0, gauss
C = coefficient in the Coates permeability model
D_0 = molecular diffusion coefficient, gauss/cm
f = Larmor (precessional) frequency, Hz
G = field-strength gradient, gauss/cm
k = permeability, darcy
k_{Coates} = permeability derived using the Timur-Coates model, darcy
k_{SDR} = permeability derived using the mean-T_2 model, darcy
M = magnetization, gauss/cm^3
M_0 = macroscopic magnetization, gauss/cm^3
M_{0i} = magnitude of the initial magnetization from the ith component, gauss/cm^3
M_{0x} = magnitude of the transverse magnetization at $t = 0$, gauss/cm^3
$M_{100\%}$ = magnitude of the magnetization for 100% bulk water, gauss/cm^3
$M_i(0)$ = magnitude of the initial magnetization from the ith component of relaxation gauss/cm^3
$M(0)$ = magnitude of the initial magnetization, gauss/cm^3
$M(t)$ = measured magnetization at time t, gauss/cm^3
$M_x(t)$ = transverse magnetization at time t, gauss/cm^3
M_z = strength of magnetic field, gauss/cm^3
$M_z(t)$ = longitudinal magnetization at time t, gauss/cm^3
N_E = number of echoes
P_c = capillary pressure, dynes/cm^2
Q = quality factor of a resonant circuit
R_t = true formation resistivity, ohm-m
$(S/V)_i$ = ratio of pore surface (S) to fluid volume (V), of the ith pore, 1/cm
$(S/V)_{\text{pore}}$ = ratio of pore surface (S) to fluid volume (V), 1/cm

S_w = water saturation, %
S_{wirr} = irreducible water saturation, %
S_{wb} = clay-bound water saturation, %
S_{xo} = flushed-zone saturation, %
t = time, seconds
T = temperature, °C
T_1 = longitudinal relaxation time, seconds
$T_{1\text{bulk}}$ = pore-fluid bulk-T_1 relaxation time, seconds
$T_{1\text{surface}}$ = pore-surface T_1 relaxation time, seconds
T_2 = transverse relaxation time, seconds
$T_{2\text{bulk}}$ = pore-fluid bulk-T_2 relaxation time, seconds
$T_{2\text{cutoff}}$ = T_2 cutoff value, seconds
$T_{2\text{diffusion}}$ = pore-fluid T_2 relaxation time in a magnetic field gradient, seconds
T_{2DW} = upper limit of the measured T_2 for water, seconds
T_{2gm} = T_2 geometric mean value, seconds
T_{2i} = pore-fluid surface T_2 relaxation time of the ith component, seconds
$T_{2\text{surface}}$ = pore-fluid surface T_2 relaxation time, seconds
T_{DW} = dual wait time—the value of T_E such that diffusion, rather than surface relaxation, is the dominant relaxation mechanism, seconds
T_E = CMPG interecho spacing, seconds
T_K = absolute temperature, K
T_p = pulsing time, seconds
T_W = polarization (wait) time, seconds
T_{WL} = polarization time, long, seconds
T_{WS} = polarization time, short, seconds
V_{shale} = shale volume, %
x, y, z = cartesian space coordinates
γ = gyromagnetic ratio—the ratio of the magnetic dipole moment to the mechanical angular momentum, Hz/gauss
Δ_{TE} = incremental change in echo spacing, seconds
Δ_{TW} = incremental change in wait time, seconds
η = fluid viscosity, cp
ξ = apparent T_1/T_2 ratio of fluid
ρ_1 = T_1 surface relaxivity, cm/sec
ρ_2 = T_2 surface relaxivity, cm/sec
τ = time over which an oscillating field is applied
ϕ = porosity, %
ϕ_e = effective porosity, %
ϕ_{eff} = effective porosity, %
ϕ_H = hydrocarbon porosity, %
ϕ_i = calibrated porosity associated with all pores of the ith pore size
ϕ_t = total porosity, %
ϕ_w = water porosity, %
χ = goodness of fit

Acknowledgments

The authors of this chapter wish to acknowledge the help provided by the many former and current staff of Halliburton Energy Services who have been involved with the development of NMR logging. All figures except Fig. 3E.21 were provided courtesy of Halliburton Energy Services.

References

1. Brown, R.J.S.: "Earth's Field Nuclear Magnetic Resonance," *Concepts in Magnetic Resonance* (2001) **13,** No. 6, 344.
2. Chandler, R.: "Proton Free Precession (Earth's Field) Logging at Schlumberger (1956–1988)," *Concepts in Magnetic Resonance* (2001) **13**, No. 6, 366.
3. Woessner, D.E.: "The Early Days of NMR in the Southwest," *Concepts in Magnetic Resonance* (2001) **13,** No. 2, 77–102.
4. Brown, R.J.S. *et al.*: "History of NMR Well Logging," *Concepts in Magnetic Resonance* (2001) **13,** No. 6, 340.
5. Brown, R.J.S. and Gamson, B.W.: "Nuclear Magnetism Logging," *JPT* (August 1960) 199; *Trans.*, AIME, **219.**
6. Nikias, P.A. Jr. and Eyraud, L.E.: "Field Examples Of Nuclear Magnetism Logging in the Poso Creek, Kern Front and Kern River Fields of the San Joaquin Valley," *JPT* (January 1963) 23.
7. Kleinberg, R.L. and Jackson, J.A.: "An Introduction to the History of NMR Well Logging," *Concepts in Magnetic Resonance* (2001) **13,** No. 6, 340.
8. Kleinberg, R.L.: "NMR Well Logging at Schlumberger," *Concepts in Magnetic Resonance*, (2001) **13,** No. 6, 396.
9. Becker, E.D., Fisk, C.L., and Khetrapal, C.L.: "The Development of NMR," *Encyclopedia of Nuclear Magnetic Resonance,* D.M. Grant and R.K. Harris (eds.), John Wiley and Sons, New York City (1996) **1,** 1.
10. Kleinberg, R.L.: "Nuclear Magnetic Resonance," *Methods in the Physics of Porous Media, Experimental Methods in the Physical Sciences Series 35,* P. Wong (ed.), Academic Press, San Diego (1999) Chap. 9, 337–385.
11. Dunn, K.-J., Bergman, D.J., and LaTorraca, G.A.: *Nuclear Magnetic Resonance—Petrophysical and Logging Applications, Seismic Exploration,* Series 32, Pergamon Press (2002).
12. Hahn, E.L.: "Spin Echoes," *Physical Review* (1950) **80,** 580.
13. Carr, H.Y. and Purcell, E.M.: "Effects of Diffusion on Free Precession in Nuclear Magnetic Resonance Experiments," *Physical Review* (1954) **94,** 630.
14. Meiboom, S. and Gill, D.: "Modified Spin-Echo Methods for Measuring Nuclear Relaxation Times," *Review of Scientific Instruments* (1958) **29,** No. 8, 699.
15. Woessner, D.E.: "In a Maze in NMR," *Encyclopedia of Nuclear Magnetic Resonance*, D.M. Grant and R.K. Harris (eds.), John Wiley and Sons, New York City (1996) **1,** 700.
16. Griffith *et al.:* "Low Frequency NMR Spectrometer," *Measurement Science and Technology* (1993) **4,** 968.
17. Taicher, Z. *et al.:* "A Comprehensive Approach to Studies of Porous Media (Rocks) Using a Laboratory Spectrometer and Logging Tool With Similar Operating Characteristics," *Magnetic Resonance Imaging* (1994) **12,** No. 2, 285.
18. Straley, C. *et al.:* "Core Analysis by Low Field NMR," *The Log Analyst* (1997) **38,** No. 2, 84.
19. Mirotchnik, K., Kryuchkov, S., and Strack, K.: "A Novel Method to Determine NMR Petrophysical Parameters from Drill Cuttings," paper MM presented at the 2004 Soc. of Petrophysicists and Log Analysts Annual Logging Symposium, Noordwijk, The Netherlands, 6–9 June.
20. Kenyon, W.E.: "Petrophysical Principles of Applications of NMR Logging," *The Log Analyst* (1997) **38,** No. 2, 21.
21. Murphy, D.P.: "NMR Logging and Core Analysis—Simplified," *World Oil* (1995) **216,** No. 4, 65.

22. Lonnes, S., Guzman-Garcia, A., and Holland, R.: "NMR Petrophysical Predictions on Cores," paper DDD presented at the 2003 Soc. of Professional Well Log Analysts Annual Logging Symposium, Galveston, Texas, 22–25 June.
23. Chang, C.T.P., Watson, A.T., and Edwards, C.M.: "NMR Imaging of Fluids and Flow in Porous Media," *Methods in the Physics of Porous Media, Experimental Methods in the Physical Sciences,* P. Wong (ed.), Academic Press, San Diego, California (1999) 387–423.
24. Kleinberg *et al.*: "T_1/T_2 Ratio and Frequency Dependence of NMR Relaxation in Porous Sedimentary Rocks," *J. of Colloid Interface Science* (1993) **158,** No. 1, 195.
25. Flaum, C., Kleinberg, R., and Bedford, J.: "Bound Water Volume, Permeability and Residual Oil Saturation From Incomplete Magnetic Resonance Logging Data," paper UU presented at the 1998 Soc. of Professional Well Log Analysts Annual Logging Symposium, Keystone, Colorado, 26–29 May.
26. Freedman, R. *et al.*: "A New NMR Method of Fluid Characterization in Reservoir Rocks: Experimental Confirmation and Simulation Results," *SPEJ* (December 2001) 452.
27. Kenyon, W.E. *et al.*: "A Three-Part Study of NMR Longitudinal Relaxation Properties of Water-Saturated Sandstones," *SPEFE* (September 1988) 622.
28. Kenyon, W.E. *et al.*: "Pore-Size Distribution and NMR in Microporous Cherty Sandstones," paper LL presented at the 1989 Soc. of Professional Well Log Analysts Annual Logging Symposium, Denver, 11–14 June.
29. Freedman, R. and Morriss, C.E.: "Processing of Data From an NMR Logging Tool," paper SPE 30560 presented at the 1995 SPE Annual Technical Conference and Exhibition, Dallas, 22–25 October.
30. Miller, M.N. *et al.*: "Spin Echo Magnetic Resonance Logging: Porosity and Free Fluid Index Determination," paper SPE 20561 presented at the 1990 SPE Annual Technical Conference and Exhibition, New Orleans, 23–26 September.
31. Coates, G.R., Xiao, L.Z., and Prammer, M.G.: *NMR Logging: Principles and Applications,* Halliburton Energy Services, Houston (1999) 234.
32. Marschall, D. *et al.*: "Method for Correlating NMR Relaxometry and Mercury Injection Data," paper SCA-9511 presented at the 1995 Soc. of Core Analysts Intl. Symposium, San Francisco, 12–14 September.
33. Ausbrooks, R. *et al.*: "Pore-Size Distributions in Vuggy Carbonates From Core Images, NMR, and Capillary Pressure," paper SPE 56506 presented at the 1999 SPE Annual Technical Conference and Exhibition, Houston, 3–6 October.
34. Pittman, E.D.: "Relationship of Porosity and Permeability to Various Parameters Derived From Mercury Injection-Capillary Pressure Curves for Sandstones," AAPG *Bull* (1992) **76,** No. 2, 191.
35. Nelson, P.H.: "Permeability-Porosity Relationships in Sedimentary Rocks," *The Log Analyst* (1994) **35,** No. 3, 38.
36. Mirotchnik, K., Allsopp, K., and Kantzas, A.: "Combination of NMR and Ultracentrifuge Techniques for Effective Carbonate Reservoir Characterization," paper SCA-9703 presented at the Soc. of Core Analysts 1997 Intl. Symposium, Calgary, 8–10 September.
37. Prammer, M.G. *et al.*: "Measurements of Clay-Bound Water and Total Porosity by Magnetic Resonance Logging," paper SPE 36522 presented at the 1996 Annual Technical Conference and Exhibition, Denver, 6–9 October.
38. Chitale, D.V., Day, P.I., and Coates, G.R.: "Petrophysical Implications of Laboratory NMR and Petrographical Investigation on a Shaly Sand Core," paper SPE 56765 presented at the 1999 SPE Annual Technical Conference and Exhibition, Houston, 3–6 October.
39. Brown, R.J.S. and Neuman, C.H.: "The Nuclear Magnetism Log—A Guide for Field Use," *The Log Analyst* (1982) **23,** No. 5, 4.
40. Cooper, R.K. and Jackson, J.A.: "Remote (Inside-Out) NMR: Part 1, Remote Production of a Region of Homogeneous Magnetic Field," *J. of Magnetic Resonance* (1980) **41,** 400.
41. Burnett, L.J. and Jackson, J.A.: "Remote (Inside-Out) NMR: Part 2, Sensitivity of Detection for External Samples," *J. of Magnetic Resonance* (1980) **41,** 406.
42. Jackson, J.A., Burnett, L.J., and Harmon, J.F.: "Remote (Inside-Out) NMR: Part 3, Detection of Nuclear Magnetic Resonance in a Remotely Produced Region of Homogeneous Magnetic Field," *J. of Magnetic Resonance* (1980) **41,** 411.

43. Jackson, J.A.: "Nuclear Magnetic Resonance Well Logging," *The Log Analyst* (1984) **25,** No. 5, 16.
44. Jackson, J.A.: "Los Alamos NMR Well Logging Project," *Concepts in Magnetic Resonance* (2001) **13,** No. 6, 368.
45. Chandler, R.N. *et al.*: "Improved Log Quality with a Dual-Frequency Pulsed NMR Tool," paper SPE 28365 presented at the 1994 SPE Annual Technical Conference and Exhibition, New Orleans, 25–28 September.
46. Prammer, M.G. *et al.*: "Measurements of Clay-Bound Water and Total Porosity by Magnetic Resonance Logging," *The Log Analyst* (1996) **37,** No. 6, 61.
47. Prammer, M. *et al.*: "Theory and Operation of a New, Multi-Volume NMR Logging System," paper DD presented at the 1999 Soc. of Professional Well Log Analysts Annual Logging Symposium, Oslo, Norway, 30 May–3 June.
48. Prammer, M.G. *et al.*: "A New Multiband Generation of NMR Logging Tools," *SPEREE* (February 2001) 59.
49. Kleinberg, R.L. *et al.*: "Novel NMR Apparatus for Investigating an External Sample," *J. of Magnetic Resonance* (1992) **97,** 466.
50. Morris, C.E. *et al.*: "Field Test of an Experimental Pulsed Nuclear Magnetism Tool," paper GGG presented at the 1993 Soc. of Professional Well Log Analysts Annual Logging Symposium, Calgary, 13–16 June.
51. Freedman, R. *et al.*: "Measurement of Total NMR Porosity Adds New Value to NMR Logging," paper OO presented at the 1997 Soc. of Professional Well Log Analysts Annual Logging Symposium, Houston, 15–18 June.
52. McKeon, D. *et al.*: "An Improved NMR Tool for Faster Logging," paper CC presented at the 1999 Soc. of Professional Well Log Analysts Annual Logging Symposium, Oslo, Norway, 30 May–3 June.
53. Heaton, N.J. *et al.*: "Applications of a New-Generation NMR Wireline Logging Tool," paper SPE 77400 presented at the 2002 Annual Technical Conference and Exhibition, San Antonio, Texas, 29 September–2 October.
54. DePavia, L. *et al.*: "A New Generation Wireline NMR Logging Tool," paper SPE 84482 presented at the 2003 SPE Annual Technical Conference and Exhibition, Denver, 5–8 October.
55. Minh, C.C. *et al.*: "Field Test Results of the New Combinable Magnetic Resonance Autotune Logging Tool," paper SPE 96759 presented at the 2005 SPE Annual Technical Conference and Exhibition, Dallas, 9–12 October.
56. Chen, S. *et al.*: "MR Explorer Log Acquisition Methods—Petrophysical-Objective-Oriented Approaches," paper ZZ presented at the 2003 Soc. of Professional Well Log Analysts Annual Logging Symposium, Galveston, Texas, 22–25 June.
57. Khamatdinov, R. *et al.*: "Field Test of a New Nuclear Magnetic Resonance Tool," paper AAA presented at the 2003 Soc. of Professional Well Log Analysts Annual Logging Symposium, Galveston, Texas, 22–25 June.
58. Bonnie, R.J.M. *et al.*: "Wireline T_1 logging," paper SPE 84483 presented at the 2003 SPE Annual Technical Conference and Exhibition, Denver, 5–8 October.
59. Heaton, N. *et al.*: "Saturation and Viscosity From Multidimensional Nuclear Magnetic Resonance Logging," paper SPE 90564 presented at the 2004 SPE Annual Technical Conference and Exhibition, Houston, 26–29 September.
60. Lippincott, B. *et al.*: "Advanced NMR Acquisition Has Improved the Assessment of Hydrocarbons in Place in the Gulf of Mexico," paper II presented at the 2004 Soc. of Petrophysicists and Well Log Analysts Annual Logging Symposium, Noordwijk, The Netherlands, 6–9 June.
61. Reppert, M. *et al.*: "Porosity and Water Saturation from LWD NMR in a North Sea Chalk Formation," paper D presented at the 2005 Soc. of Petrophysicists and Well Log Analysts Annual Logging Symposium, New Orleans, 26–29 June.
62. Prammer, M. *et al.*: "Directional Resonance: New Applications for MRIL," paper SPE 84479 presented at the 2003 SPE Annual Technical Conference and Exhibition, Denver, 5–8 October.
63. Speier, P. *et al.*: "Reducing Motion Effects on Magnetic Resonance Bound Fluid Estimates," paper II presented at the 1999 Soc. of Professional Well Log Analysts Annual Logging Symposium Oslo, Norway, 30 May–3 June.

64. Drack, E.D. *et al.*: "Advances in LWD Nuclear Magnetic Resonance," paper SPE 71730 presented at the 2001 SPE Annual Technical Conference and Exhibition, New Orleans, 30 September–3 October.
65. Prammer, M.G.: "NMR Logging-While-Drilling (1995–2000)," *Concepts in Magnetic Resonance* (2001) **13,** No. 6, 409.
66. Prammer, M.G. *et al.*: "A New Direction in Wireline and LWD NMR," paper DDD presented at the 2002 Soc. of Professional Well Log Analysts Annual Logging Symposium, Oiso, Japan, 2–5 June.
67. Prammer, M.G. *et al.*: "The Magnetic-Resonance-While-Drilling Tool: Theory and Operation," *SPEREE* (August 2001) 270.
68. Appel, M. *et al.*: "Nuclear Magnetic Resonance While Drilling in the Southern North Sea," *SPEREE* (October 2003) 351.
69. Morley, J. *et al.*: "Field Testing of a New Nuclear Magnetic Resonance Logging-While-Drilling Tool," paper SPE 77477 presented at the 2002 SPE Annual Technical Conference and Exhibition, San Antonio, Texas, 29 September–2 October.
70. Horkowitz, J. *et al.*: "Applications of a New Magnetic Resonance Logging-While-Drilling Tool in a Gulf of Mexico Deepwater Development Project," paper EEE presented at the 2002 Soc. of Professional Well Log Analysts Annual Logging Symposium, Oiso, Japan, 2–5 June.
71. Heidler, R., Morriss, C., and Hoshun, R.: "Design and Implementation of a New Magnetic Resonance Tool for the While Drilling Environment," paper BBB presented at the 2003 Soc. of Professional Well Log Analysts Annual Logging Symposium, Galveston, Texas, 22–25 June.
72. Borghi, M. *et al.*: "Magnetic Resonance While Drilling Streamlines Reservoir Evaluation," paper HHH presented at the 2005 Soc. of Professional Well Log Analysts Annual Logging Symposium, New Orleans, 26–29 June.
73. Kleinberg, R.L. and Vinegar, H.J.: "NMR Properties of Reservoir Fluids," *The Log Analyst* (1996) **37,** No. 6, 20.
74. Prammer, M.G., Bouton, J., and Masak, P.: "The Downhole NMR Fluid Analyzer," paper N presented at the 2001 Soc. of Professional Well Log Analysts Annual Logging Symposium, Houston, 16–20 June.
75. Bouton, J. *et al.*: "Assessment of Sample Contamination by Downhole NMR Fluid Analysis," paper SPE 71714 presented at the 2001 SPE Annual Technical Conference and Exhibition, New Orleans, 30 September–3 October.
76. Masak, P. *et al.*: "Field Test Results and Applications of the Downhole Magnetic Resonance Fluid Analyzer," paper GGG presented at the 2002 Soc. of Professional Well Log Analysts Annual Logging Symposium.
77. Akkurt, R. *et al.*: "Fluid Sampling and Interpretation with the Downhole NMR Fluid Analyzer," paper SPE 90971 presented at the 2004 SPE Annual Technical Conference and Exhibition, Houston, 26–29 September.
78. Stambaugh, B., Svor, R., and Globe, M.: "Quality Control of NMR Logs," paper SPE 63212 presented at the 2000 SPE Annual Technical Conference and Exhibition, Dallas, 1–4 October.
79. Marschall, D.M.: "HBVI—An NMR Method To Determine BVI as a Function of Reservoir Capillarity," paper KK presented at the 2000 Soc. of Professional Well Log Analysts Annual Logging Symposium, Dallas, 4–7 June.
80. Timur, A.: "Effective Porosity and Permeability of Sandstones Investigated Through Nuclear Magnetic Principles," *The Log Analyst* (1969) **10,** No. 1, 3.
81. Chen, S., Ostroff, G., and Georgi, D.T.: "Improving Estimation of NMR Log T_2 Cutoff Value With Core NMR and Capillary Pressure Measurements," paper SCA-9822 presented at the 1998 Soc. of Core Analysts Intl. Symposium, The Hague, 14–16 September.
82. Agut, R., Levallois, B., and Klopf, W.: "Integrating Core Measurements and NMR Logs in Complex Lithology," paper SPE 63211 presented at the 2000 SPE Annual Technical Conference and Exhibition, Dallas, 1–4 October.
83. Zhang, Q. *et al.*: "Some Exceptions to Default NMR Rock and Fluid Properties," paper FF presented at the 1998 Soc. of Professional Well Log Analysts Annual Logging Symposium, Keystone, Colorado, 26–29 May.
84. Coates, G.R. *et al.*: "A New Characterization of Bulk-Volume Irreducible Using Magnetic Resonance," *The Log Analyst* (1998) **39,** No. 1, 51.

85. Chen, S. *et al.:* "Methods for Computing SWI and BVI From NMR Logs," paper HH presented at the 1998 Soc. of Professional Well Log Analysts Annual Logging Symposium, Keystone, Colorado, 26–29 May.
86. Kleinberg, R.L. and Boyd, A.: "Tapered Cutoffs for Magnetic Resonance Bound Water Volume," paper SPE 38737 presented at the 1997 SPE Annual Technical Conference and Exhibition, San Antonio, Texas, 5–8 October.
87. Seevers, D.: "A Nuclear Magnetic Method for Determining the Permeability of Sandstone," paper L presented at the 1966 SPWLA Annual Logging Symposium, Tulsa, 8–11 May.
88. Timur, A.: "Pulsed Nuclear Magnetic Resonance Studies of Porosity, Movable Fluid, and Permeability of Sandstones," *JPT* (June 1969) 775.
89. Timur, A.: "An Investigation of Permeability, Porosity, and Residual Water Saturation Relationships for Sandstone Reservoirs," *The Log Analyst* (1968) **9,** No. 4, 8.
90. Ahmed, U., Crary, S.F., and Coates, G.R.: "Permeability Estimation: The Various Sources and Their Interrelationships," *JPT* (May 1991) 578.
91. Worthington, P.F.: "The Effect of Scale on the Petrophysical Estimation of Intergranular Permeability," paper A presented at the 2003 Soc. of Professional Well Log Analysts Annual Logging Symposium, Galveston, Texas, 22–25 June.
92. Bryant, S., Cade, C., and Mellor, D.: "*Permeability Prediction from Geologic Models,"* AAPG Bulletin (1993) **77**, No. 8, 1338.
93. Chang, D. *et al.:* "Effective Porosity, Producible Fluid and Permeability in Carbonates From NMR Logging," *The Log Analyst* (1997) **38,** No. 2, 60.
94. Babadagli, T. and Al-Salmi, S.: "Improvement of Permeability Prediction for Carbonate Reservoir Using Well Log Data," paper SPE 77889 presented at the 2002 SPE Asia Pacific Oil and Gas Conference and Exhibition, Melbourne, Australia, 8–10 October.
95. Marschall, D., Gardner, J., and Curby, F.M.: "MR Laboratory Measurements—Requirements to Assure Successful Measurements that Will Enhance MRI Log Interpretation," paper SCA-9704 presented at the 1997 Society of Core Analysts International Symposium, Calgary, 8–10 September.
96. Prammer, M.G.: "NMR Pore Size Distributions and Permeability at the Well Site," paper SPE 28368 presented at the 1994 SPE Annual Technical Conference and Exhibition, New Orleans, 25–28 September.
97. Winkler, M., Freeman, J., and Appel, M.: "The Limits Of Fluid Property Correlations Used in NMR Well Logging; an Experimental Study of Reservoir Fluids at Reservoir Conditions," paper DD presented at the 2004 Soc. of Petrophysicists and Log Analysts Annual Logging Symposium, Noordwijk, The Netherlands, 6–9 June.
98. Hirasaki, G.J., Lo, S.-W., and Zhang, Y.: "NMR Properties of Petroleum Reservoir Fluids," *Magnetic Resonance Imaging* (2003) **21**, No. 3–4, 269.
99. Chen, S. *et al.:* "Laboratory Investigation of NMR Crude Oils and Mud Filtrates Properties in Ambient and Reservoir Conditions," paper SPE 90553 presented at the 2004 SPE Annual Technical Conference and Exhibition, Houston, 26–29 September.
100. Prammer, M.G.: "Hydrocarbon Saturation Measurements by NMR," *Concepts in Magnetic Resonance* (2001) **13,** No. 6, 406.
101. Prammer, M.G. *et al.:* "Lithology-Independent Gas Detection by Gradient-NMR Logging," paper SPE 30562 presented at the 1995 SPE Annual Technical Conference and Exhibition, Dallas, 22–25 October.
102. Akkurt, R. *et al.:* "Enhanced Diffusion—Expanding the Range of NMR Direct Hydrocarbon-Typing Applications," paper GG presented at the 1998 Soc. of Professional Well Log Analysts Annual Logging Symposium, Keystone, Colorado, 26–27 May.
103. Timur, A.: "Nuclear Magnetic Resonance Study of Carbonate Rocks," *The Log Analyst* (1972) **13,** No. 5, 3.
104. Hurlimann, M.D.: "Diffusion and Relaxation Effects in General Stray Field NMR Experiments," *J. of Magnetic Resonance* (2001) **148,** 367.
105. Akkurt, R. *et al.:* "NMR Logging of Natural Gas Reservoirs," *The Log Analyst* (1996) **37,** No. 5, 33.

106. Coates, G.R., Gardner, J.S., and Miller, D.L.: "Applying Pulse-Echo NMR to Shaly Sand Formation Evaluation," paper B presented at the 1994 Soc. of Professional Well Log Analysts Annual Logging Symposium, Tulsa, 19–22 June.
107. Chitale, D.V., Day, P.I., and Coates, G.R.: "Petrophysical Implications of Laboratory NMR and Petrographical Investigation on a Shaly Sand Core," paper SPE 56765 presented at the 1999 SPE Annual Technical Conference and Exhibition, Houston, 3–6 October.
108. Moore, M.A. and Akkurt, R.: "Nuclear Magnetic Resonance Applied to Gas Detection in a Highly Laminated Gulf of Mexico Turbidite Invaded With Synthetic Oil Filtrate," paper SPE 36521 presented at the 1996 SPE Annual Technical Conference and Exhibition, Denver, 6–9 October.
109. Buller, D.: "Carbonate Evaluation Using NMR Time Domain Analysis," paper GGG presented at the 2000 Soc. of Professional Well Log Analysts Annual Logging Symposium, Dallas, 4–7 June.
110. Hou, B.L. and Miller, D.: "Determining Fluid Volume in Gas and/or Light Oil Reservoirs—Using a New Triple-Wait-Time NMR Logging Method," paper U presented at the 2000 Soc. of Professional Well Log Analysts Annual Logging Symposium, Dallas, 4–7 June.
111. Akkurt, R. *et al.*: "Determination of Residual Oil Saturation by Use of Enhanced Diffusion," *SPEREE* (June 1999) 303.
112. Horkowitz, J.P. *et al.*: "Residual Oil Saturation Measurements in Carbonates With Pulsed NMR Logs," *The Log Analyst* (1997) **38,** No. 2, 73.
113. Howard, J.J.: "Wettability and Fluid Saturations Determined From NMR T_1 Distributions," *Magnetic Resonance Imaging* (1994) **12,** No. 2, 197.
114. Flaum, C., Kleinberg, R., and Hurlimann, M.: "Identification of Gas With the Combinable Magnetic Resonance Tool (CMR)," paper L presented at the 1996 Soc. of Professional Well Log Analysts Annual Logging Symposium, New Orleans, 16–19 June.
115. Sun, B. *et al.*: "Two-Dimensional NMR Logging and Field Test Results," paper KK presented at the 2004 Soc. of Petrophysicists and Log Analysts Annual Logging Symposium, Noordwijk, The Netherlands, 6–9 June.
116. Slijkerman, W.F.J. *et al.*: "Processing of Multi-Acquisition NMR Data," *SPEREE* (2000) 492.
117. Chen, S. *et al*: "G-Te Correction for Processing Multigradient, Multiple-Te NMR Log Data," paper SPE 84481 presented at the 2003 SPE Annual Technical Conference and Exhibition, Denver, 5–8 October.
118. Freedman, R. *et al.*: "Wettability, Saturation, and Viscosity Using the Magnetic Resonance Fluid Characterization Method and New Diffusion-Editing Pulse Sequences," *SPEJ* (2003) 317.
119. Fang, S. *et al.*: "Quantification of Hydrocarbon Saturation in Carbonate Formations Using Simultaneous Inversion of Multiple NMR Echo Trains," paper SPE 90569 presented at the 2004 SPE Annual Technical Conference and Exhibition, Houston, 26–29 September.
120. Sun, B. and Dunn, K.J.: "A Global Inversion Method for Multi-Dimensional NMR Logging," *J. of Magnetic Resonance* (2005) **172,** 152.
121. Sun, B. and Dunn, K.J.: "Two-Dimensional Nuclear Magnetic Resonance Petrophysics," *Magnetic Resonance Imaging* (2005) **23,** No. 2, 259.
122. Bonnie, R.J.M. *et al.*: "Advanced Forward Modeling Helps Planning and Interpreting NMR Logs," paper SPE 71735 presented at the 2001 SPE Annual Technical Conference and Exhibition, New Orleans, 30 September–3 October.
123. Hurlimann, M.D. and Venkataramanan, L.: "Quantitative Measurement of Two-Dimensional Distribution Functions of Diffusion and Relaxation in Grossly Inhomogeneous Fields," *J. of Magnetic Resonance* (2002) **157,** 31.
124. Hurlimann, M.D. *et al.*: "Diffusion Editing: New NMR Measurement of Saturation and Pore Geometry," paper FFF presented at the 2002 Soc. of Professional Well Log Analysts Annual Logging Symposium, Oiso, Japan, 2–5 June.
125. Sun, B., and Dunn, K.: "Core Analysis With Two Dimensional NMR," paper SCA 2002-38 presented at the 2002 Soc. of Core Analysts Annual International Symposium, Monterey, California, 23–25 September.
126. Hurlimann, M.D. *et al.*: "Diffusion-Relaxation Distribution Functions of Sedimentary Rocks in Different Saturation States," *Magnetic Resonance Imaging* (2003) **21,** No. 3–4, 305.

127. Toumelin, E. *et al.*: "A Numerical Assessment of Modern Borehole NMR Interpretation Techniques," paper SPE 90539 presented at the 2004 SPE Annual Technical Conference and Exhibition, Houston, 26–29 September.
128. Hursan, G., Chen, S., and Murphy, E.: "New NMR Two-Dimensional Inversion of T_1/T_{2app} vs. T_{2app} Method for Gas Well Petrophysical Interpretation," paper GGG presented at the 2005 Soc. of Petrophysicists and Well Log Analysts Annual Logging Symposium, New Orleans, 26–29 June.
129. Flaum, M., Chen, C., and Hirasaki, G.: "NMR Diffusion Editing for D-T_2 Maps—Application to Recognition of Wettability Change," paper JJ presented at the 2004 Soc. of Petrophysicists and Log Analysts Annual Logging Symposium, Noordwijk, The Netherlands, 6–9 June.
130. Sun, B. *et al.*: "NMR Imaging with Diffusion and Relaxation," paper SCA2003-24 presented at the 2003 Soc. of Core Analysts Annual International Symposium, Pau, France, 21–24 September.
131. Leu, G. *et al.*: "NMR Identification of Fluids and Wettability In Situ in Preserved Cores," *Petrophysics* (2002) **43,** No. 1, 13.
132. Klimentos, T.: "NMR Applications in Petroleum Related Rock-Mechanics: Sand Control, Hydraulic Fracturing, Wellbore Stability," paper HHH presented at the 2003 Soc. of Petrophysicists and Well Log Analysts Annual Logging Symposium, Galveston, Texas, 22–25 June.
133. Van der Zwaag *et al.*: "Deformation and Failure of Rock Samples Probed by T_1 and T_2 Relaxation," *Magnetic Resonance Imaging* (2003) **21,** No. 3–4, 405.
134. Mullen, M. *et al.*: "Fluid Typing With T_1 NMR: Incorporating T_1 and T_2 Measurements for Improved Interpretation in Tight Gas Sands and Unconventional Reservoirs," paper III presented at the 2005 Soc. of Petrophysicists and Well Log Analysts Annual Logging Symposium, New Orleans, 26–29 June.
135. Longis, C., Vignau, S., and White, S.: "NMR and Capture Spectroscopy Help Resolve Producibility and Fluid Distribution in the North Alwyn Triassic," paper SPE 96609 presented at the 2005 SPE Annual Technical Conference and Exhibition, Dallas, 9–12 October.
136. LaTorraca, G. *et al.*: "Heavy Oil Viscosity Determination Using NMR Logs," paper PPP presented at the 1999 Soc. of Professional Well Log Analysts Annual Logging Symposium, Oslo, Norway, 30 May–3 June.
137. Galford, J.E. and Marschall, D.M.: "Combining NMR and Conventional Logs to Determine Fluid Volumes and Oil Viscosity in Heavy-Oil Reservoirs," paper SPE 63257 presented at the 2000 SPE Annual Technical Conference and Exhibition, Dallas, 1–4 October.
138. Seccombe, J. *et al.*: "Ranking Oil Viscosity in Heavy Oil Reservoirs," paper EE presented at the 2005 Soc. of Petrophysicists and Well Log Analysts Annual Logging Symposium, New Orleans, 26–29 June.
139. Nascimento, J. de D.S., and Gomes, R.M.R.: "Tar Mats Characterization from NMR and Conventional Logs, Case Studies in Deepwater Reservoirs, Offshore Brazil," paper FF presented at the 2004 Soc. of Petrophysicists and Log Analysts Annual Logging Symposium, Noordwijk, The Netherlands, 6–9 June.
140. Breda, E. *et al.*: "Detecting Restricted Diffusion Effect When Identifying Hydrocarbons with NMR Logs," paper SPE 90089 presented at the 2004 SPE Annual Technical Conference and Exhibition, Houston, 26–29 September.
141. Westphal, H. *et al.*: "NMR measurements in Carbonate Rocks: Problems and an Approach to a Solution," *Pure and Applied Geophysics* (2005) **162,** No. 3, 549.
142. Rose, D. *et al.*: "A Novel Approach to Real Time Detection of Facies Changes in Horizontal Carbonate Wells Using LWD NMR," paper CCC presented at the 2003 Soc. of Petrophysicists and Well Log Analysts Annual Logging Symposium, Galveston, Texas, 22–25 June.
143. Chen, S. *et al.*: "Simple, Robust NMR-Based Indicators for Detection of Hydrocarbon Gas or Oil, Borehole Contamination, and Vugs," paper SPE 96409 presented at the 2005 SPE Annual Technical Conference and Exhibition, Dallas, 9–12 October.
144. Fleury, M., Al-Nayadi, K., and Boyd, D.: "Water Saturation from NMR, Resistivity and Oil Base Core in a Heterogeneous Middle East Carbonate Reservoir," paper JJJ presented at the 2005 Soc. of Petrophysicists and Well Log Analysts Annual Logging Symposium, New Orleans, 26–29 June.
145. Volotikin, Y. *et al.*: "Constructing Capillary Pressure Curves from NMR Log Data in the Presence of Hydrocarbons," *Petrophysics* (2001) **42,** No. 4, 334.

146. Glorioso, J.C. *et al.*: "Deriving Capillary Pressure and Water Saturation from NMR Transversal Relaxation Times," paper SPE 81057 presented at the 2003 Latin American and Caribbean Petroleum Engineering Conference, Port-of-Spain, Trinidad, 27–30 April.
147. Grattoni, C.A. *et al.*: "An Improved Technique for Deriving Drainage Capillary Pressure from NMR T_2 Distributions," paper SCA2003-25 presented at the 2003 Soc. of Core Analysts Annual International Symposium, Pau, France, 21–24 September.
148. Dastidar, R., Rai, C., and Sondergeld, C.: "Integrating NMR with Other Petrophysical Information to Characterize a Turbidite Reservoir," paper SPE 89948 presented at the 2004 SPE Annual Technical Conference and Exhibition, Houston, 26–29 September.
149. Corbelleri, A. *et al.*: "Application of the Magnetic Resonance Logging in San Jorge Basin (Argentina)," paper VV presented at the 1996 Soc. of Professional Well Log Analysts Annual Logging Symposium, New Orleans, 16–19 June.
150. Davis, B., McDonald, T., and Aly, M.: "Practical Applications of NMR Technology Enhance Formation Evaluation, Testing and Completion Decisions," paper SPE 76717 presented at the 2002 SPE Western Regional/AAPG Pacific Section Joint Meeting, Anchorage, 20–22 May.
151. Soliman, M.Y. *et al.*: "Technology Integration of Conformance and Magnetic Resonance Imaging Enhances Stimulation Treatment," paper SPE 59530 presented at the 2000 SPE Permian Basin Petroleum Engineering Conference, Midland, Texas, 21–23 March.
152. Weiland, J., Teff, C., and Donovan, G.: "Integrating NMR, Rock Mechanics, and Production Evaluation to Optimize Hydrocarbon Production in Deepwater GOM," paper VVV presented at the 2004 Soc. of Petrophysicists and Log Analysts Annual Logging Symposium, Noordwijk, The Netherlands, 6–9 June.
153. Hodgkins, M.A. and Howard, J.J.: "Application of NMR Logging to Reservoir Characterization of Low-Resistivity Sands in the Gulf of Mexico," AAPG *Bulletin* (1999) **83,** No. 1, 114.
154. Ostroff, G.M. and Shorey, D.S.: "Integration of NMR and Conventional Log Data For Improved Petrophysical Evaluation of Shaly Sands," paper OOO presented at the 1999 Soc. of Professional Well Log Analysts Annual Logging Symposium, Oslo, Norway, 30 May–3 June.
155. Freedman, R. *et al.*: "Combining NMR and Density Logs for Petrophysical Analysis in Gas-Bearing Formations," paper II presented at the 1998 Soc. of Professional Well Log Analysts Annual Logging Symposium, Keystone, Colorado, 26–29 May.
156. Minh, C.C. *et al.*: "Sonic-Magnetic Resonance Method: A Sourceless Porosity Evaluation in Gas-Bearing Reservoirs," paper SPE 56767 presented at the 1999 SPE Annual Technical Conference and Exhibition, Houston, 3–6 October.
157. Chen, S. *et al.*: "Optimization of NMR Logging Acquisition and Processing," paper SPE 56766 presented at the 1999 SPE Annual Technical Conference and Exhibition, Houston, 3–6 October.
158. Akkurt, R., Prammer, M., and Moore, A.: "Selection of Optimal Acquisition Parameters for MRIL Logs," *The Log Analyst* (1996) **37,** No. 6, 43.
159. Tourmelin, E., Torres-Verdin, C., and Chen, S.: "Quantification of Multiphase Fluid Saturations in Complex Pore Geometries From Simulations of Nuclear Magnetic Resonance Measurements," paper SPE 77399 presented at the 2002 SPE Annual Technical Conference and Exhibition, San Antonio, Texas, 29 September–2 October.
160. Appel, M. *et al.*: "Fluid Identification in Heavy Oil Reservoirs by Nuclear Magnetic Resonance at Elevated Temperature," paper SCA-9917 presented at the 1999 Soc. of Core Analysts International Symposium, Golden, Colorado, 1–4 August.
161. Mirotchnik, K.D. *et al.*: "Low Field NMR Tool for Bitumen Sands Characterization: A New Approach," *SPEREE* (April 2001) 88.
162. Wyatt, D.F.: "Nuclear Magnetic Resonance Logging Aids in Defining Heavy Oil Types," paper SPE 54121 presented at the 1999 SPE International Thermal Operations and Heavy Oil Symposium, Bakersfield, California, 17–19 March.
163. Carmona, R. and Decoster, E.: "Assessing Production Potential of Heavy Oil Reservoirs From the Orinoco Belt with NMR Logs," paper ZZ presented at the 2001 Soc. of Professional Well Log Analysts Annual Logging Symposium, Houston, 17–20 June.
164. Padilha, S. and Baudot, P.: "Application of NMR Logging for Heavy Oil Identification and Quantification in OBM Environment," paper FF presented at the 2001 Soc. of Professional Well Log Analysts Annual Logging Symposium, Houston, 17–20 June.

165. Manalo, F.P., Bryan, J.L., and Kantzas, A.: "Addressing the Clay/Heavy Oil Interactions When Interpreting Low Field Nuclear Magnetic Resonance Logs," paper SPE/PS-CIM/CHOA 78971 presented at the 2002 SPE International Thermal Operations and Heavy Oil Symposium and International Horizontal Well Technology Conference, Calgary, 4–7 November.
166. Najia, W. *et al.:* "Nuclear Magnetic Resonance (NMR), A Valuable Tool for Tar Detection in a Carbonate Formation of Abu Dhabi," paper SPE 78485 presented at the 2002 SPE Abu Dhabi International Petroleum Exhibition and Conference, Abu Dhabi, UAE, 13–16 October.
167. Manalo, F. *et al.*: "Separating the Signals from Clay Bound Water and Heavy Oil in NMR Spectra of Unconsolidated Samples," paper SPE 84480 presented at the 2003 SPE Annual Technical Conference and Exhibition, Denver, 5–8 October.
168. Zhang, Q., Huang, C., and Hirasaki, G.: "Interpretation of Wettability in Sandstones with NMR Analysis," *Petrophysics* (2000) **41,** No. 3, 223.
169. Guan, H. *et al.:* "Wettablity Effects in a Sandstone Reservoir and Outcrop Cores from NMR Relaxation Time Distributions," *J. of Petroleum Science and Engineering* (2002) **34,** No. 1–4, 35.
170. Eidesmo, T., Relling, O., and Rueslatten, H.: "NMR Logging of Mixed-Wet North Sea Sandstone Reservoirs," paper PP presented at the 1996 Soc. of Professional Well Log Analysts Annual Logging Symposium, New Orleans, 16–19 June.
171. Akkurt, R. *et al.:* "Remaining Oil Saturation From NMR in a Mixed-Wet, Three-Phase Carbonate Reservoir," paper SPE 63216 presented at the 2000 SPE Annual Technical Conference and Exhibition, Dallas, 1–4 October.
172. Al-Mahrooqi, S.H. *et al.:* "An Investigation of the Effect of Wettability on NMR Characteristics of Sandstone Rock and Fluid Systems," *J. of Petroleum Science and Engineering* (2003) **39,** 389.
173. Fleury, M., Al-Nayadi, K., and Boyd, D.: "Water Saturation from NMR, Resistivity and Oil Base Core in a Heterogeneous Middle East Carbonate Reservoir," paper JJJ presented at the 2005 Soc. of Petrophysicists and Well Log Analysts Annual Logging Symposium, New Orleans, 26–29 June.
174. Looyestijn, W. and Hofman, J.: "Wettability Index Determination by Nuclear Magnetic Resonance," paper SPE 93624 presented at the 2005 SPE Middle East Oil and Gas Show and Conference, Bahrain, 12–15 March.
175. Howard, J.J.: "Quantitative Estimates of Porous Media Wettability From Proton NMR Measurements," *Magnetic Resonance Imaging* (1998) **16,** Nos. 5 and 6, 529.
176. Chen, J., Hirasaki, G., and Flaum, M.: "Effects of OBM Invasion on Irreducible Water Saturation —Mechanisms and Modifications of NMR Interpretation," paper SPE 90141 presented at the 2004 SPE Annual Technical Conference and Exhibition, Houston, 26–29 September.
177. Shafer, J. *et al.*: "Core and Log NMR Measurements Indicate Reservoir Rock is Altered by OBM Filtrate," paper CC presented at the 2004 Soc. of Petrophysicists and Log Analysts Annual Logging Symposium, Noordwijk, The Netherlands, 6–9 June.
178. Rodrigues, P., Alcocer, Y., and Matheus, J.: "NMR Signal Pattern Classification for Estimation of Petrophysical Properties," paper SPE 63258 presented at the 2000 SPE Annual Technical Conference and Exhibition, Dallas, 1–4 October.
179. Sigal, R.F. *et al.:* "A Method for Enhancing the Vertical Resolution of NMR Logs," paper SPE 63215 presented at the 2000 SPE Annual Technical Conference and Exhibition, Dallas, 1–4 October.
180. Akkurt, R. and Cherry, R.: "The Key to Improving the Vertical Resolution of Multi-Frequency Logging Tools," paper YY presented at the 2001 Soc. of Professional Well Log Analysts Annual Logging Symposium, Houston, 17–20 June.

SI Metric Conversion Factors

Unit		Factor	Exponent	SI
cm^2/sec	×	1.0*	E + 02	= mm^2/sec
cp	×	1.0*	E − 03	= Pa·s
°C		+ 273.15		= K
$dynes/cm^2$	×	1.0*	E − 01	= Pa
ft	×	3.048*	E − 01	= m
ft/min	×	5.080*	E − 03	= m/sec
°F		(°F + 459.67)/1.8		= K
gauss	×	1.0*	E − 04	= Tesla

in.	×	2.54*	E − 02 = m
ohm-m	×	1.0*	E − 02 = ohm-cm
psi	×	6.894 757	E + 00 = kPa

*Conversion factor is exact.

Chapter 3F
Mud Logging

Dennis E. Dria, Shell Intl. Exploration and Production Inc.

3F.1 Introduction

Mud logging, in its conventional implementation, involves the rig-site monitoring and assessment of information that comes to the surface while drilling, with the exclusion of data from downhole sensors. The term mud logging is thought, by some, to be outdated and not sufficiently descriptive. Because of the relatively broad range of services performed by the geologists, engineers, and technicians traditionally called mud loggers, the term "surface logging" is sometimes used, and the personnel performing the services may be called surface-logging specialists. Additional specialist designations may include pore pressure engineer, formation evaluation engineer, logging geologist, or logging engineer. For the sake of generality, the terms mud logging and mud logger are used here with the understanding that the hybrid discipline encompasses much more than monitoring the mud returns and that the trained specialists perform engineering and geological tasks that span several traditional disciplines.

There are several broad objectives targeted by mud logging: identify potentially productive hydrocarbon-bearing formations, identify marker or correlatable geological formations, and provide data to the driller that enables safe and economically optimized operations. The actions performed to accomplish these objectives include the following:

- Collecting drill cuttings.
- Describing the cuttings (type of minerals present).
- Interpreting the described cuttings (lithology).
- Estimating properties such as porosity and permeability of the drilled formation.
- Maintaining and monitoring drilling-related and safety-related sensing equipment.
- Estimating the pore pressure of the drilled formation.
- Collecting, monitoring, and evaluating hydrocarbons released from the drilled formations.
- Assessing the producibility of hydrocarbon-bearing formations.
- Maintaining a record of drilling parameters.

Mud logging service first focused on monitoring the drilling mud returns qualitatively for oil and gas content.[1,2] This included watching the mud returns for oil sheen, monitoring the gas evolving from the mud as it depressured at the surface, and examining the drill cuttings to determine the rock type that had been drilled, as well as looking for indication of oil on the cuttings. Detection of the onset of abnormal formation pressures using drilling parameters was proposed with the introduction of the *d* exponent.[3] Gas chromatography, which was developed early in the 20th century, saw its introduction in mud logging in the 1970s when electronics became sufficiently compact, rugged, and robust to be used at rig sites. The literature provides excellent reviews of the early history.[4,5]

Computerized data acquisition and the ability to routinely transfer continuously acquired data to the office data center enabled the broader application of more sophisticated interpretive techniques and the integration of data from different sources into the geological and reservoir model, in near real time. This, coupled with the blossoming of measurement-while-drilling (MWD) and logging-while-drilling (LWD) tools, moved the mud-logging unit into a new role as a hub for rig-site data gathering and transmission.[5] Starting in the 1980s, significant improvements to existing technologies, as well as major technical breakthroughs, have given the geologist and petroleum engineer a great number of powerful mud-logging tools to interpret and integrate geological, drilling, and geochemical data. These tools are discussed in subsequent sections of this chapter.

The traditional products delivered by a mud-logging vendor include geological evaluation, petrophysical/reservoir formation evaluation, and drilling engineering support services. In this overview, we consider that these products support three basic processes associated with drilling and evaluation of wells: formation evaluation (building or refining the geological and reservoir models), drilling engineering and operations (the planning and execution of the well construction process), and maintaining drilling and evaluation operations with appropriate health, safety, and environmental (HSE) consideration. The following section describes the path that drilling fluid follows during the drilling operation to explain where the logger obtains data.

3F.2 Mud Logging Data Acquisition

Fig. 3F.1 schematically shows the components of a drilling operation that have a part in mud logging. The most critical component is the drilling fluid (drilling mud), which, in addition to its role in drilling mechanics, carries most of the information from the formation up to the surface where it is acquired, decoded, or extracted from the mud stream by various techniques. Drilling liberates gas and liquid formation fluids, and circulation of the drilling fluid carries these to the surface (except during riserless drilling in a deepwater offshore environment, in which the drilling returns circulate only up to the sea floor). Cuttings, pieces of formation rock, are also carried in the circulated drilling fluid. MWD and LWD data are frequently encoded as pressure pulses and transmitted to the surface. Mud temperature is not a direct indicator of subsurface formation temperature, but monitoring the trend is important to understanding gas extraction efficiency and recycling. In deepwater drilling environments, the mud can be cooled significantly on the trip from the sea floor to the surface.

Drilling fluid is stored in the mud pit, drawn into the mud pumps, and pumped into the drillpipe via the kelly. Mud travels down the drillpipe, through any MWD tools and drill motors, and through the bit nozzles where its discharge aids drilling mechanics. At this point, the drilling fluid carries away rock fragments from drilled formation, along with any liberated reservoir fluids (water, oil, or gas). Cuttings and reservoir fluids are transported to the surface. Any gaseous components are dissolved in the base fluid of the drilling mud under most overbalanced drilling conditions. Drilling mud continues its flow up the wellbore drillstring annulus, through the casing-drillpipe annulus and the BOP stack, and, in the case of an offshore well, up the riser. At the bell nipple, the returning drill fluid is exposed to atmospheric pressure and flows down the mud return line. If an underbalanced drilling operation is being used, there is a rotating seal around the drillpipe, and the pressurized drilling returns stream moves through the "blooey line" to a separator and flair.

The return mud stream continues down the return line to the shaker box or "possum belly." This is the standard location for the "gas trap" gas extractor. Mud pours over the shaker screens, with cuttings getting discharged off the top of the screen, while the drilling fluid that

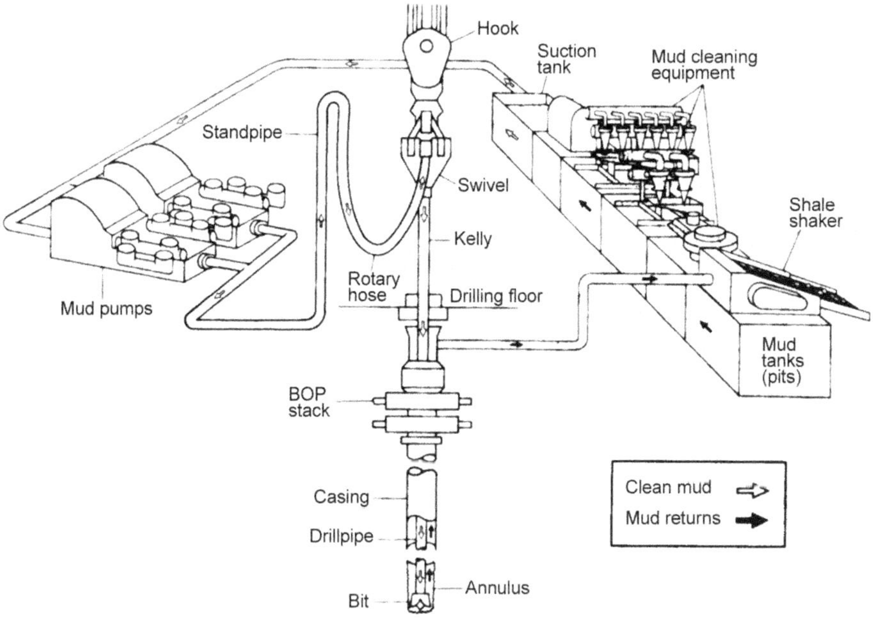

Fig. 3F.1—Drilling fluid flow path during drilling operations.

falls through the screens travels on through the degasser, desander, and desilter to the mud pits. The mud logger takes samples or acquires data at the following points in the process.

• Whole mud samples are taken at the mud suction pit and at the possum belly and are used to do whole-mud extraction using a steam still. They may be taken on an occasional basis during coring and wireline logging to assess the effects of mud filtrate and solids.

• Drill cuttings samples are taken off the shaker screen and off a "catch board" where cuttings fall from the screen to disposal. These are used for lithological and mineralogical description, paleo description, and sometimes "canned" for laboratory-based carbon-isotope analysis, detailed geological examinations such as thin-section preparation and analysis, chemostratigraphy, and source rock evaluation.

• Gas sampling is done through an extractor at the possum belly, in some cases at the bell nipple or off the mud return line to minimize losses to the atmosphere, and at the mud suction line or mud pit to monitor recycle gas content of the mud. After extraction, gas analysis may be performed at the sampling location,[6] or, more routinely, the gas is continuously transferred via a vacuum line to the logging unit where it is passes through the manifold of analytical instruments (total HC, GC, MS, H_2S, etc) and may be captured for laboratory-based analysis (carbon isotope, molecular composition).[7]

• Mud temperatures are monitored at the mud suction pit and mud return line.

The mud engineer collects mud samples for analyses that are used to determine any adjustments to mud properties needed for drilling.

Contamination is defined here as any material that does not come from the formation that has been drilled at the time that a specific volume element of mud exits the bit. Mud contamination has several potential sources:

• Air, which can enter the top of the drillpipe when the Kelly-drillpipe joint is broken during a connection.

• Pipe scale and pipe dope from inside the drillpipe (pipe dope fluoresces and may interfer with show identification or description).

• Rock sloughing or rubbing off formations further up hole.

• Cuttings that have bedded or built up because of improper hole cleaning dynamics that are mobilized by changes in mud viscosity, pumping rate, or drillpipe or collar rotation.
• Uphole fluids that flow or are swabbed into the annulus.
• Cuttings that have built up on the shaker screen or in the possum belly.

The logger should be watching for any change in cuttings or mud-conveyed hydrocarbon fluids that indicate contamination. Mud additives such as weighting agents and lost-circulation material are not considered contaminants, but must be monitored because some of these interfere with analytical observations and descriptions or give interfering instrument responses. Some base mud fluids, particularly some of the synthetic fluids, create challenges for the mud logger, as do some chemical additives (e.g., some sulfate or sulfonate wetting agents may give a false positive H_2S indication).

Samples of the drill cuttings are taken at the shale shaker. Wellsite geologists or engineers should specify the appropriate procedure for collecting samples, which may be done by the mud logger or the mud logger's sample catcher. Cuttings have a relatively short residence time on the shaker screen. Sampling protocol should include taking a composite sample with portions from different areas of the screen, combined with cuttings that have been retained on a "cuttings board." A cuttings board is a wooden board, steel angle iron, or other such device that is hung just below the base of the shaker screens to catch cuttings as they fall off the edge of the screen. Immediately after samples are collected, the screen and catch board should be washed down with clean drilling mud base fluid. The logger should mix this composite sample and take divided portions for cleaning, interpretation, and bagging. The wellsite planner should specify the sampling frequency (typically a composite over 10-, 30-, or 90-ft intervals or on a timed basis).

Gas sampling is traditionally done with a mechanical degasser, generically called a "gas trap." **Fig. 3F.2** shows an example.[8] Typically placed in the shaker box, the trap pulls in drilling mud through the centrifugal action of the stirrer. The mechanical action of the stirrer, combined with a slight vacuum pulled in the trap head space, allows the gas to partition between the liquid and gas phase. The head-space gas is pulled by vacuum through tubing, into the logging unit, and on through the gas analysis manifold.

Alternative methods for sampling gas may be accomplished by continuously operating controlled-volume mechanical or thermomechanical slip-stream gas extractors[9,10] and membrane-type extractors.[6] The mud logger may place the sampling point for these gas extraction devices at the bell nipple, the mud return line, or in the shaker box. Other methods require taking discrete samples, followed by thermal extraction techniques [such as the steam still, where samples of the whole mud are collected and portions heated in a steam-distillation apparatus (**Fig. 3F.3**)] and microwave heating methods.[8]

The gas manifold may include provisions for a portion of the gas stream to be pumped into sample containers, either laminated gas bags or stainless steel tubes* (see **Fig. 3F.4**). These gas samples are then shipped from the rig for laboratory analyses. There are new mass-spectrometer-based techniques that may not require a bulk extraction of the gas from the mud for analysis.[11]

Once gas is extracted from the drilling fluid, various analytical techniques determine properties of the gas at the rig site. The basic measurements include a determination of the "total" gas concentration and the composition and concentrations of the constituent components.

3F.3 Total Gas Analysis

The total gas analyzer (TGA), also referred to as the total hydrocarbon analyzer (THA), measures the total amount of gas, typically the total amount of combustible gas. The usual unit of TGA measurement is total methane equivalents (TME), which is essentially the BTU content

* Personal communication with D. Coleman, IsoTech Laboratories, Champaign, Illinois (2002).

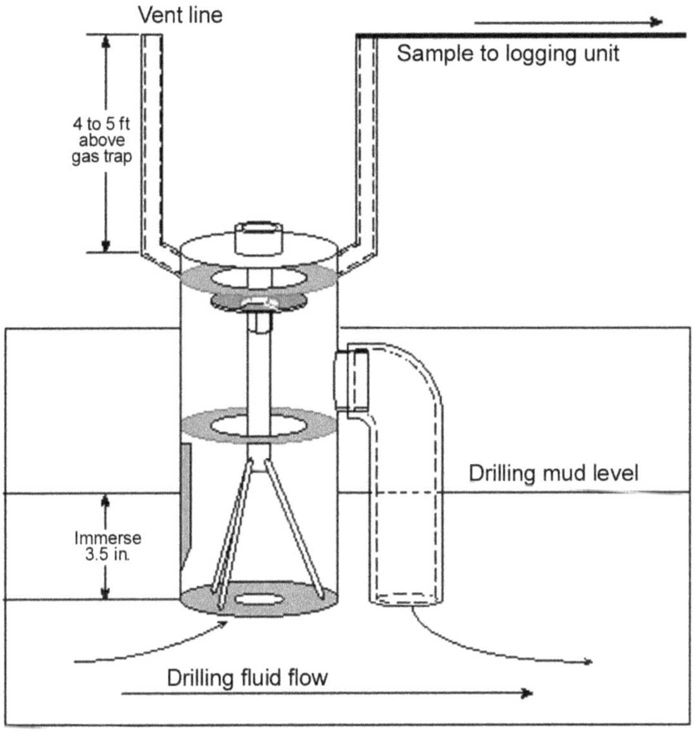

Fig. 3F.2—Schematic of a gas trap type gas extractor.

of the gas extracted from the drilling fluid, expressed as that which would be obtained from an equivalent concentration of pure methane in air. The TGA, while giving an undifferentiated indication of gas entrained in the drilling fluid, has the advantage of operating in a continuous mode. Today, most TGAs use either a thermal conductivity detector (TCD) or a flame ionization detector (FID).

3F.4 Component Analysis

There are several analytical techniques used at the rig site to measure the molecular composition and component concentrations of the reservoir fluids entrained to the surface in the drilling fluid. These techniques are the gas chromatograph (GC), the mass spectrometer (MS), and the infrared spectrometer.

3F.4.1 Gas Chromatograph. The most widely used technique, and among the more accurate, is the GC (**Fig. 3F.5**). This instrument separates components from a mixture by selectively adsorbing and desorbing each compound at different rates, either on the surface of a granular packing contained in a column made of small-diameter tubing or on the internal surface of the tubing itself. A small, usually submilliliter, volume of the mixture (i.e., the gas extracted from the drilling fluid) is injected into a carrier gas stream, which carries the sample into the GC column. Components start to separate, depending on their affinity for the active surface of the packing or the column tubing. At the end of the column, components elute, each with a unique retention time, and pass into a detector. The different detectors used to analyze the compounds eluting from the GC column include the FID, the TCD, the catalytic combustion detector (CCD), the MS detector, and the infrared (IR) absorption spectrometer.

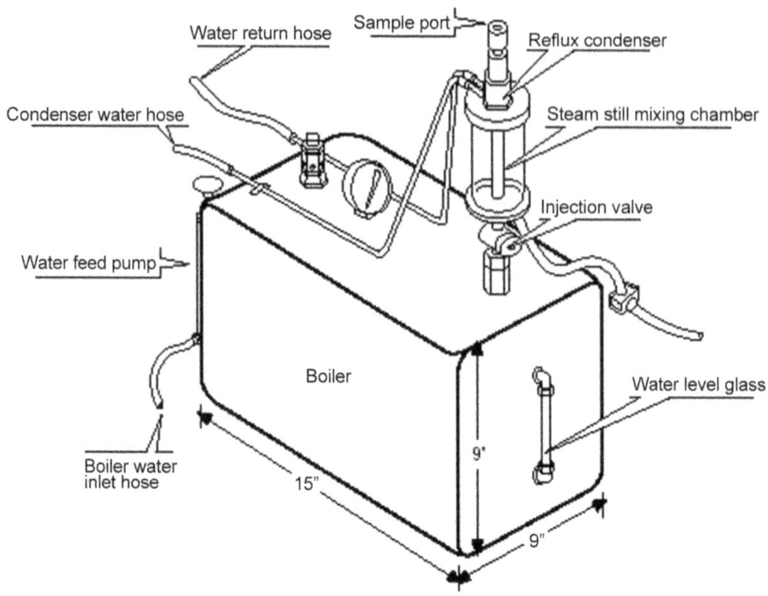

Fig. 3F.3—Schematic of a steam still-gas extractor.

Flame Ionization Detector. The FID uses a flame of burning hydrogen to combust the eluting compound. As the compounds undergo combustion, ionized single-carbon intermediates are captured by the collector electrode. Very sensitive circuitry (the electrometer) measures the extremely small, microamp-level current generated by this flow of ions, which is proportional to the total number of carbon atoms in the combusting gas mixture. **Fig. 3F.6** shows the FID device schematically. The hydrocarbon species present are known because the operator, by design, knows which hydrocarbon component is eluting; therefore, concentration of that component may be calculated.

Thermal Conductivity Detector. The TCD (or hot wire detector) measures the thermal conductivity of the gas mixture passing over a filament. The typical TCD uses a Wheatstone Bridge circuit to measure the resistance of a platinum filament. The filament is heated to a constant temperature. As a gas passes over the filament, the temperature is reduced by an amount that is related to the thermal conductivity of the gas and the gas-flow rate. The resistance of the filament changes with temperature. At a constant flow rate through the sensing cell, gases of different composition and, hence, different thermal conductivity cool the filament by different amounts. In most TCD devices, there are two cells: a reference cell through which the pure carrier gas (in these cases, usually air) flows and a sample cell through which the gas mixture extracted from the drilling fluid flows. The filaments from the cells form two of the four arms of the Wheatstone Bridge (**Fig. 3F.7**).

Catalytic Combustion Detector. The CCD measures the temperature change in a platinum filament when the gas mixture is combusted. Rather than using a flame, the CCD device has the filament imbedded in a catalytic bead, which catalyzes the combustion chemical reaction, allowing it to occur at temperatures lower than normal flames. Heat is released according to the "heat of combustion," and this released heat increases the filament temperature and, hence, its resistance. The resistance of the sample filament is compared with the resistance of a reference filament, in a fashion similar to the TCD, and provides an output that is a function of the composition of the gas mixture. This device provides an output signal that is more or less related to the BTU content of the gas mixture it is measuring.

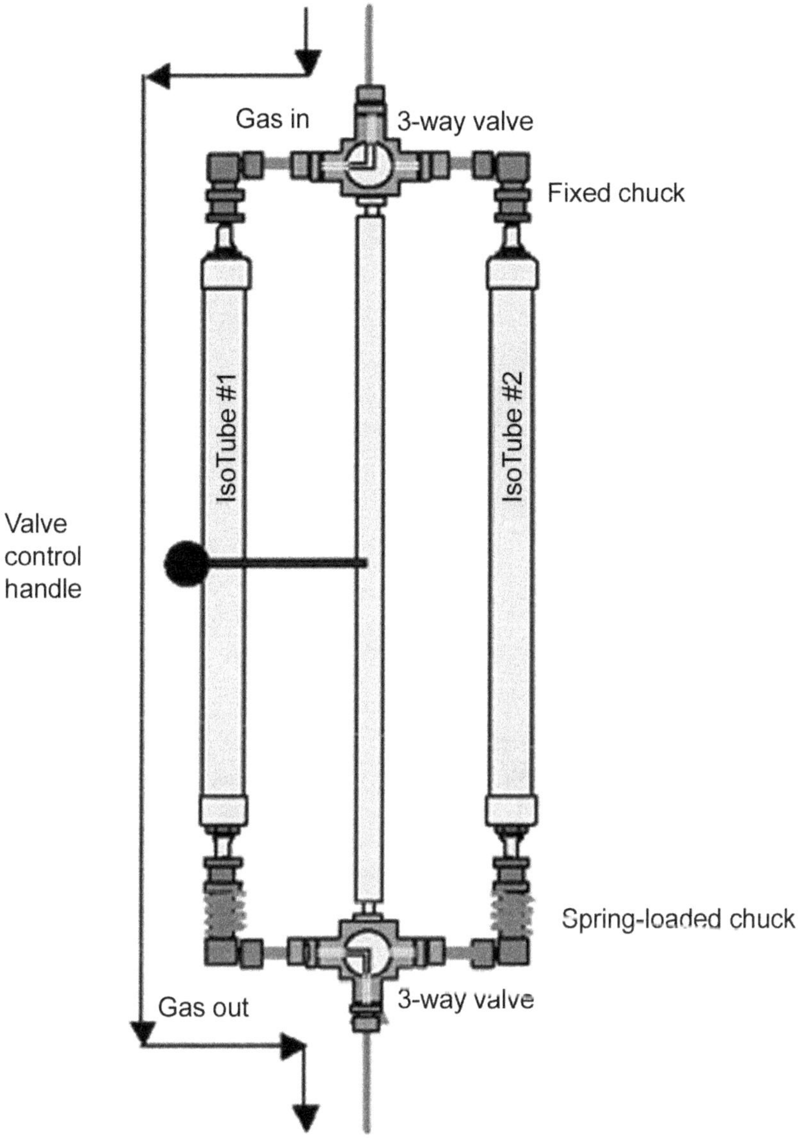

Fig. 3F.4—Low-pressure gas sampling tubes mounted two to a rack for continuous, sequential gas collection.

Mass Spectrometer Detector. The MS detector measures the quantity of a particular compound by determining the amount of material with a specific molecular weight. (**Fig. 3F.8** shows the magnetic-sector MS.) This is done by ionizing a small quantity of material, during which multiple fragments of the molecule may be produced. Each fragment has a specific mass-to-charge ratio. Ionized fragments are separated in space or time, on the basis of their respective mass-to-charge ratio, and are then quantified by an ion detector. These devices have been used only in laboratory-based analyses until recently, when compact, robust MS detectors have been introduced for rig-site use.[10] The MS has commonly been used to quantify components eluting from a GS (the GC-MS system) and has the advantage of being able to differentiate mixes of several components that may coelute, as long as the individual molecular species dif-

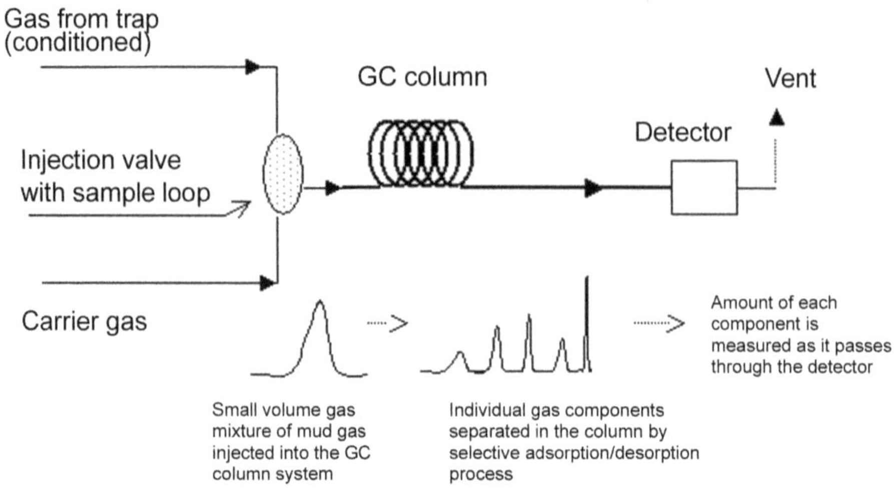

Fig. 3F.5—Schematic of a single-column gas chromatograph with generic detector.

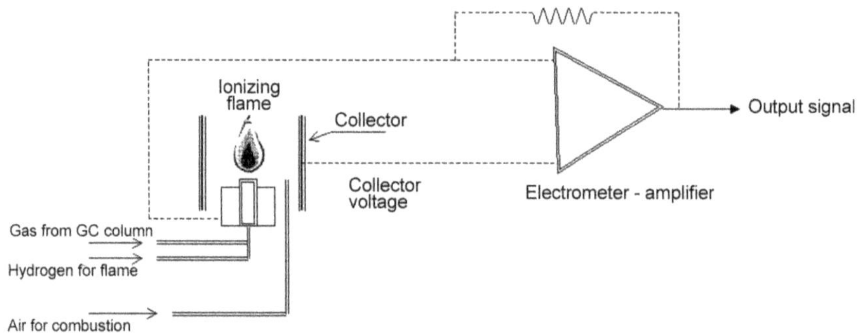

Fig. 3F.6—Flame ionization gas detector.

fer sufficiently in molecular weight. By principle of detection, this advantage allows simultaneous analysis of saturated and aromatic hydrocarbons, as well as the acid gases CO_2 and H_2S and inert gases such as nitrogen.

Recent advances in MS instrumentation and methods may allow its use at the rig site to measure directly the amounts of specific paraffinic and aromatic hydrocarbon compounds, CO_2, and H_2S without the need for chromatographically separating the components. This would replace the function of the GC for HC component analysis, along with in-mud hydrogen sulfide detectors. These specific methods[11] are in relatively early stages of development and are not described here in detail.

Infrared Absorption Spectrometer. On input of IR energy to a molecule, its bonds will vibrate in any of a variety of modes, such as stretching, bending, rotating, and twisting. The mode of vibration will vary depending on the specific bond and the wavelength of the exciting radiation. IR radiation passes through a cell filled with the gas mixture. Portions of the incident IR spectrum are absorbed by the molecules in the gas, according to the characteristic absorption wavelengths of the specific molecules comprising the gas. The amount of IR radiation absorbed is a function of the path length and the concentration.

For a variety of physical reasons, the absorption spectrum is not manifested as a series of discrete wavelengths, but rather as a series of broader bands that, particularly in hydrocarbon

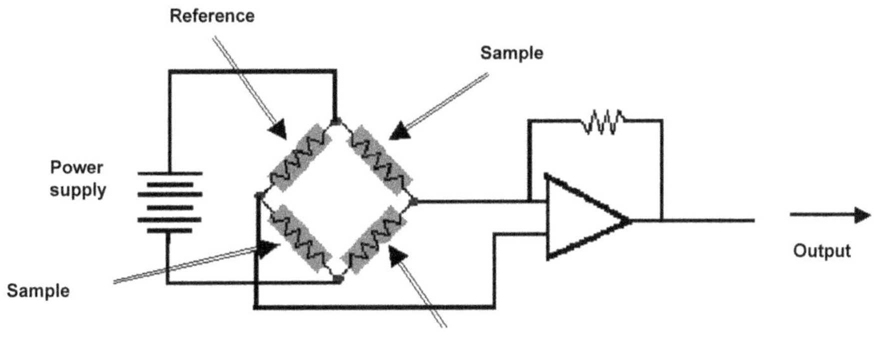

Fig. 3F.7—Thermal conductivity gas detector.

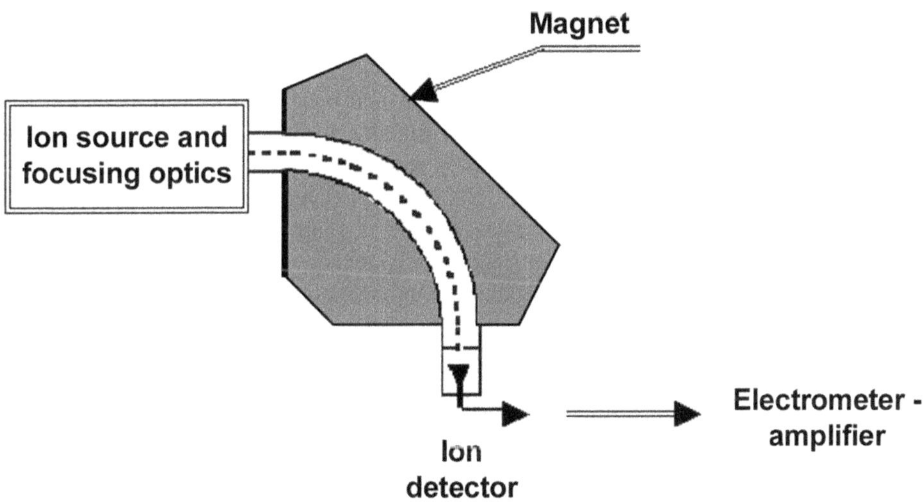

Fig. 3F.8—Magnetic sector mass spectrometer.

systems, overlap to the point of being difficult to explicitly deconvolve. Because IR sensors also tend to have poor resolution compared with band. When comparing differences in principles of operation, relative accuracy over a wide dynamic range of gas concentrations and operational robustness under rig site, for most applications the FID is the preferred detector system for both the GC and the THA.

3F.5 Cuttings Analysis

Cuttings provide the first opportunity and, in some wells, the only opportunity to actually look at the rock that has been drilled. Cuttings give the geologist information about the formation lithology needed for geologic correlation, the mineral composition for marker beds, input for the petrophysicist or log analyst, and, in some cases, enough hydrocarbon to allow some oil-quality measurements to be performed. Cuttings are also a source for microfossils used in biostratigraphy.

3F.5.1 Sample Lagging. Proper sample collection and preparation is of paramount importance. The first step is to know where the cuttings are from, which is done by performing a lag calculation. "Lagging" is performed by any of several methods. The most accurate method is to inject a tracer of some sort into the drilling fluid stream at the surface and time its exit as it is circulated out. The volume of the mudlines and drillpipe, from the point of injection to the bit, can be calculated from the pipe measurements (inside diameter and length). The total tracer residence time in the drill fluid is

$$t_t = t_s + t_{dp} + t_{an}, \qquad (3F.1)$$

$$\text{and } t_s + t_{dp} = q \cdot (V_s + V_{dp}), \qquad (3F.2)$$

so that

$$t_{an} = t_t - q \cdot (V_s + V_{dp}), \qquad (3F.3)$$

where t_{an} = the lag time from the bit to the point at the surface where the tracer is measured (usually the shaker box), t_t = the total elapsed pumping time from injection to measurement, q = the mud pump flow rate, V_{dp} = the internal volume of the drillpipe and drill collars, and V_s = the internal volume of any surface lines from the point of tracer injection to the kelly. Commonly used tracers include calcium carbide, air, and injected reference gas.

Solid carbide is put into the drillpipe when making a connection and pumped down with the circulating mud. Calcium carbide chemically reacts with water in the drilling fluid to form acetylene, which is then detected with the THA as it circulates out. Ease of handling and injection into the circulating mud system makes carbide lagging a popular choice. Disadvantages include the need for a manual operation, particularly the need for an additional person on the drill floor during the busy operation of making a connection, and the requirement that there be ample free water in the drilling fluid, which is not always the case with oil-based and synthetic mud systems.

Other materials may be injected into the drillpipe while making a connection. Virtually any benign pulverized solid that can be seen in the mud returns, such as ground-up bricks. If a high-viscosity sweep is being pumped, that can also be used to measure lag. These methods do require that logging personnel carefully watch the shaker screens to catch the appearance of the lag tracer.

When the kelly reaches the rotary table, the drillpipe is raised slightly so that the current connection between the kelly and the top pipe can be broken. This action reduces the mud pressure at the bit and can "swab" some formation fluids into the wellbore. The gas dissolved in this fluid is carried to the surface with circulation, appearing as "connection gas," and may be used as a lag indicator. Swabbing at connections may also bring gas in from locations above the bit, depending on pore pressure gradient, drilling fluid rheology, and annular geometry. If this occurs, an erroneous lag results.

A less frequently used method for lag determination involves the injection of a reference gas into the mud pump intake.[12] Reference gas injection, by its original design, serves primarily as an internal standard for quantifying mud gas concentration. A noninteracting gas is injected at low rates into the mud pump intake line, which is controlled to maintain a constant concentration of the referencing gas in the drilling fluid. Momentarily increasing or decreasing the injection rate results in a subsequent increase or decrease in gas concentration in the drilling fluid. This concentration perturbation is noted while monitoring the mud gas using the logger's gas analyzer, and its residence time in the system is used in the lag calculation. Advantages of this measurement include that it can be done on demand without waiting for a connection, it is independent of drilling fluid composition (no need for water in the case of nonaqueous drilling fluids), and that it can easily be automated.

3F.5.2 Cuttings Collection. Samples of drilled cuttings are normally taken at the shaker screens, although some have proposed and tested devices for diverting a small stream of cuttings-laden mud from the return line.[13] At predetermined depth or time intervals, the logger or sample catcher collects a composite sample that contains cuttings representative of the entire interval drilled since the previous collection. Very typically, cuttings samples will be taken every 30 ft (10 m), until target bed boundaries are approached, such as thin marker beds, anticipated reservoir sections, casing points, or coring points.

By placing a board at the base of the shaker screen in line to catch cuttings as they fall off the end of the screens, the sample catcher assures that a composite sample accumulates. At the desired sampling time, the sampler scrapes cuttings off the board into a sampling bucket and adds additional cuttings from several locations across the screen (the latter, to catch the very latest drilled material circulated up). This collection yields a composite sample that is fairly representative of the formations drilled since the last collection. Immediately after sampling, the sampler should hose down the screen and catching board with water for aqueous drilling fluids or with the appropriate base fluid for nonaqueous drilling fluids. Proper cleaning of the composited cuttings calls for placing the cuttings on a fine mesh screen and flushing with the base drilling fluid.

3F.5.3 Cuttings Sample Examination and Description. Proper handling, examination, and description guidelines should be specified when contracting mud-logging services. This section is not meant to replace an experienced professional geologist's interpretation of cuttings, but serves as a guide to the important "high points" that the logger should remember in applying procedures presented in appropriate geological guidelines and manuals. Several excellent references are available.[14–16] Swanson's thorough summary[17] is presented as a detailed manual.

Scanning a series of samples laid out on the counter allows the logger to get a good overview of trends, to highlight sample differences that may be the result of quality variation or contamination, and to help identify bed boundaries. Individual samples are then examined under a low-power stereomicroscope (10 to 50X) with either ample natural light or a lamp with a "blue" light or blue filter. Proper illumination is required so that the true colors of the sample constituent minerals are not distorted. Digital image capture of select samples adds significantly to the end-of-well documentation. A quick examination should identify all material present in significant quantity, including contaminants, metal, drilling additives, lost-circulation material, and suspected caved material, in addition to the drilled-formation rock cuttings. This examination should include an assessment of sample quality.

The well databases containing the cuttings log should include an estimate of the percentage of each rock type, which is an assessment of what is actually seen in each individual sample, as well as an interpretation of the lithology, which is based on all the data available to the logger. The logger's manual should call for a standardized description protocol containing rock type (with classification), color, texture (grain size, roundness, sorting), cement and/or matrix material, fossils, sedimentary structures, and porosity and oil shows. Standard abbreviations should be used in the description, as well as standard symbols as the log is being drawn.[17]

Many special tests are run on rock samples to make on-the-spot determination of specific minerals. These tests vary from such standard chemicals as alzarin red for calcite detection to calcimetry for quantitative determination of carbonate content. A qualified logging geologist will have experience in suggesting and implementing them.

The sample should be viewed under ultraviolet (UV) light, and any fluorescence noted (mineral or hydrocarbon). Certain types of hydrocarbons, in the rock pores or "stained" on the grain surface, may not fluoresce. To test for these, the rock samples may then be treated with an appropriate organic solvent while being viewed under UV light. "Streaming" fluorescence may be noted (streamers or wisps of hydrocarbon) as it moves from the rock cutting into the solvent surrounding the cutting. As the solvent dries, "residual cut" may be observed as a fluorescing ring or residue in the examination dish. These are examples of "cut fluorescence."

Sometimes there will be more extraneous material present in the cuttings sample than actual drilled formation fragments. Commonly seen nonformation solids include the following:

• Cavings from uphole, previously penetrated beds. These will be recognized as formation material logged earlier and individual cuttings may have shapes that distinguish them from bit cuttings.

• Recirculated cuttings. These are usually smaller, individual grains or microfossils.

• Lost circulation material. Extraneous solids, which have a size to plug up lost circulation zones, are added to the drilling fluid. These may be as unusual as cotton seed hulls, ground-up pecan shells, and shredded rubber.

• Cement fragments. Close attention should be paid to the drilling record.

• Drilling mud. Washing samples using procedures and solvents that are appropriate for the specific mud system. Consideration for the health of the logger and the environment is critical.

• Pipe dope. This contaminant may fluoresce, giving a false show indication.

This list of contaminant examples is not all inclusive, and judgment, combined with close attention to the drilling and mud engineers' records, is critical.

3F.6 Maintaining Data Quality

Many different data are available, obtained through technologies that range from tried-and-true classical "wet chemistry" techniques through high-tech sensors that use procedures established after countless years of thoughtful research, development, and field testing. Even the best planned operations may, from time to time, provide data of poor quality or even totally miss data from important geologic intervals. Proper planning of surface data logging operations should include provisions for "whole-system" qualification before starting the operation, as well as a plan for the occasional quality audit. Details will vary widely depending on the location of the operations, availability of staff, project magnitude, and economics. An appropriate quality assurance program may be as simple as receiving a weekly e-mail or fax with GC calibration information or as intensive as scheduling rig-site audits, depending on the exact circumstances and well logging objectives.

3F.7 Formation Evaluation

Formation evaluation, for this discussion, is the process of describing a geologic formation, and any fluids contained within, in terms of their constituent properties and determining the properties of the rock to assess the total and recoverable volume, value, and producibility of the fluids and the placement, and the engineering design and economics of drilling and completing the wells that are needed to produce the fluids. Many data are interpreted to evaluate a petroleum-bearing formation, and we discuss the interpretations of data acquired through surface data logging in terms of the rock formation and fluid properties they help determine.

3F.7.1 Fluid Type. The logging engineer or geologist gets information about the formation fluids directly from fluids that are released into the wellbore while drilling and circulating out suspended immiscibly in the drill fluid or remaining in the pores of larger cuttings that may not have been flushed. They receive information indirectly from remnants of the fluid that remain in pores of rock cuttings, as stains on the grain surface, or in solution in the drilling fluid.

Oil may be identified as a sheen on the surface of water-based drilling fluid. If the circulating fluid density is sufficiently low as to render an underbalanced drilling condition, oil may be produced in large enough quantities that a sample may be skimmed off a whole mud sample. Similarly, the underbalanced penetration of a gas-bearing formation yielding only a small quantity of free gas in the mud system at the bit will expand according to the real gas law as it is circulated to the surface, where it may be detected and possibly sampled (although, in an uncontrolled situation, this results in hazardous safety and environmental conditions). These are all fairly obvious, direct indicators.

The ratio of various mud-gas components (e.g., the molar or volumetric ratio of methane to ethane) may be crossplotted or correlated with fluid type.[18–21] The modified Pixler method,[19] for example, plots ratios of methane (C_1) to, respectively, ethane (C_2), propane (C_3), butanes and heavier (C_{4+}), and pentanes and heavier (C_{5+}):

$$C_1 / C_2, \quad \text{(3F.4)}$$

$$C_1 / C_3, \quad \text{(3F.5)}$$

$$C_1 / C_{4+}, \quad \text{(3F.6)}$$

$$\text{and } C_1 / C_{5+} . \quad \text{(3F.7)}$$

Through a correlated analysis, ranges of these ratio values had been determined that were representative of productive and nonproductive gas and oil (see **Fig. 3F.9**).

The Haworth *et al.* "wetness" method[20] defines several correlatable ratios: wetness, W_h; balance, B_h; and character, C_h.

$$W_h = \frac{C_2 + C_3 + iC_4 + nC_4 + C_5}{C_1 + C_2 + \ldots + C_5} . \quad \text{(3F.8)}$$

$$B_h = \frac{C_1 + C_2}{C_3 + iC_4 + nC_4 + C_5} . \quad \text{(3F.9)}$$

$$C_h = \frac{iC_4 + nC_4 + C_5}{C_3} . \quad \text{(3F.10)}$$

The values of these ratios determine, through the interpretation shown in **Fig. 3F.10**, whether the hydrocarbon is gas or oil, very dry gas, and, with enough data, whether the oil is tending towards being a light oil, heavy oil, or residual oil.

These ratio methods tend to work robustly only within the basins in which they are calibrated, the number and quality of data from the basin within which they are calibrated, and the errors inherent in obtaining the calibration set as well as the unknown data set. Examples of these inherent errors include the unknown or varying extraction efficiency at the gas trap and the accuracy of the gas-measurement system. Normalization techniques,[22,23] either against an internal standard or other parameters, can improve the consistency and robustness of the ratios.

3F.7.2 Fluid Properties. If determining fluid type is difficult because of errors and unknowns in measuring the composition of formation hydrocarbons entrained in the drilling fluid, even more difficulty can be anticipated in estimating the fluid properties, such as viscosity or API gravity. The more robust methods involve either quantitatively measuring the methane-through-pentane composition of the drilling fluid returns and correlating these with the HC fluid properties[24] or involve spectroscopy techniques to get particular spectra and correlating these to oil properties or type.[25,26]

The QGM™ process[8] involves the use of a gas-trap extractor that was characterized and field tested extensively. This trap, when used in a controlled fashion, can give a quantitative

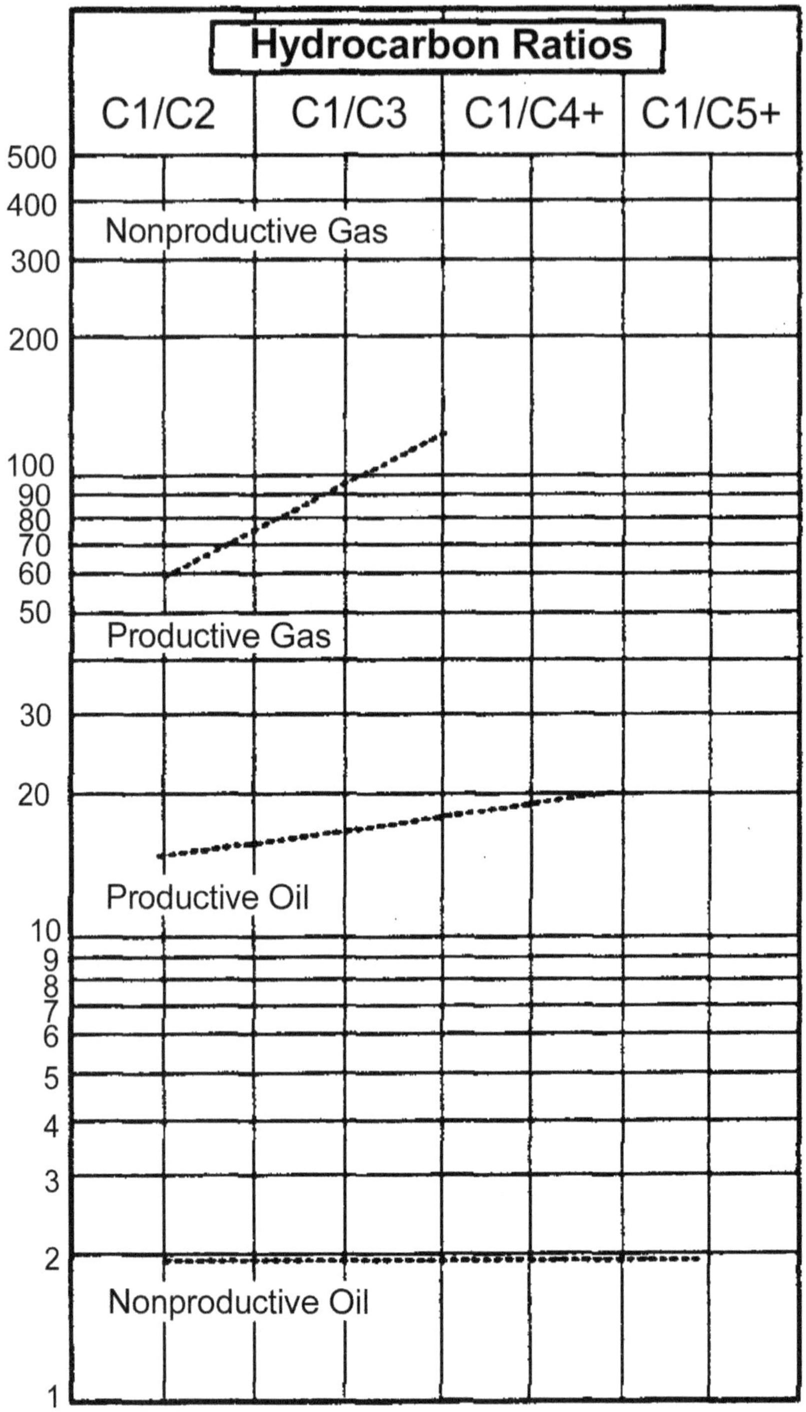

Fig. 3F.9—Template for plotting gas ratios with the modified Pixler method.

measurement of gas-trap extraction efficiency when calibrated with a measurement of thermally extracted mud-gas composition. With proper control and calibration, this process gives highly

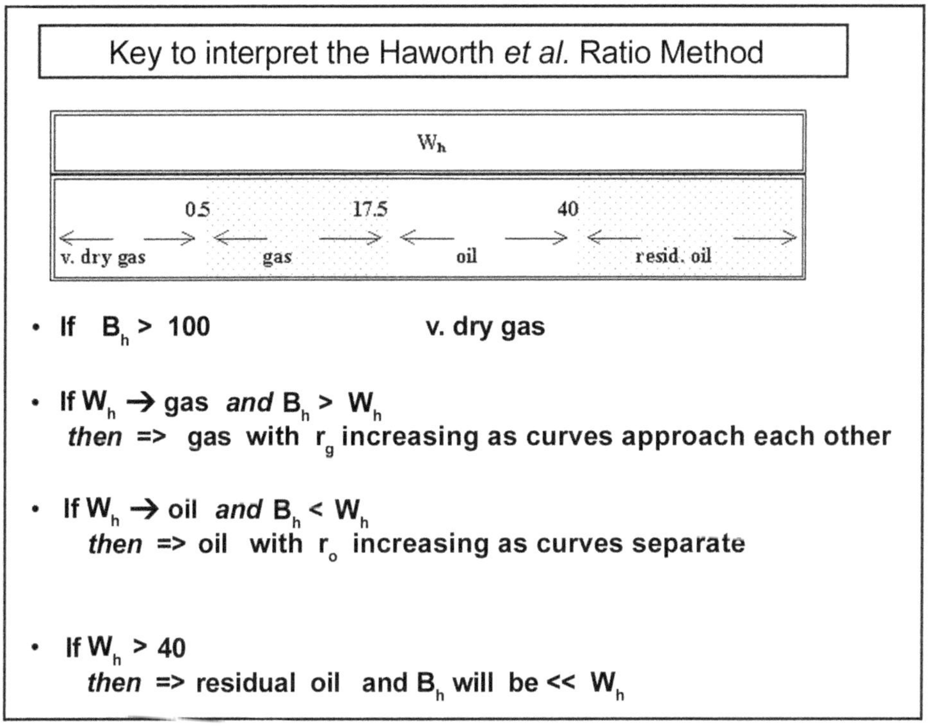

Fig. 3F.10—Interpretation legend for the Haworth *et al.* "wetness method."

accurate gas-in-mud compositions. With these, correlations have been developed that relate composition parameters to oil properties.

An enhanced quantitative UV-fluorescence technique[26] monitors the intensity of two emission wavelengths resulting from exciting an oil sample with UV radiation. The ratio of the intensities varies according to the oil composition or, more exactly, according to the concentrations of certain aromatic compounds in the oil sample.

3F.7.3 Formation Porosity. Formation porosity is estimated by visually inspecting rock cuttings, which requires a sufficient description and classification of the rocks. Several references supply discussion.[14–17] The requirement of rock fragments with intact pore systems eliminates the use of this method when drilling poorly consolidated or unconsolidated sands and when bit types, such as the PDC, reduce most of the potential reservoir cuttings to grain-sized fragments.

The method described by Boone[27] calculates the rock porosity that would be required to liberate the specific amount of gas measured when a set volume of formation is drilled. This assumes low fluid spurt loss while drilling (near-zero flushing of the rock ahead of the bit), which is frequently violated except in near-balance drilling and low-permeability formations.

3F.7.4 Formation Permeability. The formation permeability may be estimated visually by the logger from the grain size and sorting. At best, this may give order of magnitude estimates, but, frequently, the smaller, clay-sized particles, which may control the permeability if present in appreciable amounts, cannot be quantified with any confidence. Another limitation occurs because the cutting sample is acquired over a relatively large depth interval, and "smearing" of the data occurs in zones where grain size, sorting, and clay content vary significantly.

3F.7.5 Pore Pressure. Pore-pressure variations, particularly transitions from normal to geopressures, trend with several measurable parameters. These parameters are discussed more in the Drilling section.

3F.7.6 Geological or Petrophysical Information for Evaluation Away From the Wellbore. The correlation of geologic strata relies on determining certain characteristic properties or constituents of a particular geologic formation. Geologists use these properties in one of several techniques to correlate formations from well to well within a field as well as in broader, basin-wide studies. Particular formations may serve as "markers" by having one or a set of distinguishing characteristics. Mud-logging services provide rock-cutting samples for the determination of these characteristics and, in some cases, may acquire the data on site. The acquisition of cuttings samples for these tests may impact the cuttings sampling or preparation techniques and should be considered when specifying a mud-logging service.

Microscopic fossils identified and correlated by paleontologists yield "biomarkers," which become the working data for biostratigraphy. These biomarkers frequently appear in the cuttings in very low quantities because of the grinding and abrasion the cuttings see under the action of the drill bit and transport uphole in the returning mud stream. Their preparation calls for consistently applied cleaning with the use of a screen or sieve with openings small enough to retain the microfossils.

Specific mineral assemblages may be present in a particular geologic formation with compositions nearly constant or in relative abundances that are nearly constant over an extensive areal. The use of such mineral compositions as stratigraphically correlatable parameters is called chemostratigraphy. The ratio of concentrations of these certain minerals, as well as of other compounds associated with the deposition and diagenisis of the particular formation, must then be determined relatively quickly and cheaply. It has been shown that the ratios Ca/Al, Na/Al, and K/Al are related to clay mineral concentration ratios, and that relative amounts of such elements as Ti, Cr, and Zr are frequently associated with the clay minerals.[28–30] Chemostratigraphers determine the amounts of these elements using laboratory or rig-based instruments such as inductively coupled plasma spectrometry and laser-induced breakdown spectroscopy.

Pore fluids present during diagentic processes become crystallographically trapped in small volumes called fluid inclusions. These tiny drops of fluids, sealed in cavities typically 2 to 20 μm in size, are thought to be representative of the pore fluid composition that existed at the time of their formation.[31] After appropriately cleaning portions of the drill cuttings, geologists chose grain samples containing fluid inclusions for analysis, which includes heating to volatize the fluids and rupture the inclusion walls. Liberated vapor passes to an MS, which measures fluid composition for hydrocarbons, and other geochemical species such as H_2S, CO_2 acetic acid, and aromatics. Occasionally, a crude-oil fluid inclusion volume may be large enough to characterize the crude sufficiently to assess physical properties and the degree of biodegradation and thermal alteration.[32] Fluid inclusion stratigraphy, which uses included-fluid composition as the correlating parameter, has been proposed to assess the location of fluid contacts, identify migration pathways, characterize compartmentalization, and determine distance to pay zones.[31–33]

3F.8 Drilling Engineering and Operations

There is significant overlap between data gathered for geological, petrophysical, and reservoir engineering needs vs. data gathered for the driller. Information about pore pressure, formation gas, and rock type and strength are an integral part of well planning. Continuously tracking these parameters as a well is drilled and comparing the actual data with what was used in the well plan allows for quick response by the driller when trouble occurs. It also allows the driller to "fine tune" his operations to optimize drilling performance, which is measured by drill rate, trouble time and cost, and delivery of well specifications (e.g., in terms of being a producing asset, an exploration well, or an appraisal well). We will leave these items in the "evaluation" bin, with the acknowledgement that they could just as easily be lumped into the drilling engineering category. This section discusses the data types and processes that are used to a large extent, and in some cases exclusively, by the driller.

Any measurable parameter that gives an indication of pore pressure provides the driller with an estimate of the degree of overbalance, which directly affects the rate of penetration (ROP). Mud weight will be adjusted to be within the desired window for a well's particular set of drilling dynamics and rock strengths on the basis of modeled drill rate. The various measurements described in other sections of this chapter become important, to some degree, to the driller. These services may be provided by the mud logging contractor.

3F.8.1 Weight on Bit And Rate of Penetration. These data are collected to indicate drilling performance. The driller would like to know how to predict his drilling or penetration rate. Bourgoyne *et al.*[34] describes several models that have been developed and used. Jorden[3] proposed modifying the Bingham model and defined a normalized parameter called the drilling exponent (the *d*-exponent):

$$d_{\exp} = \frac{\log\left(\frac{R}{60N}\right)}{\log\left(\frac{12W}{1{,}000d_b}\right)}, \quad \text{(3F.11)}$$

where R = the penetration rate in ft/hr, N = the rotary speed in rpm, W = the weight on bit in Mlbf, and d_b = the bit diameter in inches. The *d*-exponent is sometimes corrected for mud density changes[35] by considering the effects of ρ_n, the mud density equivalent to a normal formation pore pressure, and ρ_e, the equivalent mud density at the bit while circulating:

$$d_c = d_{\exp}\frac{\rho_n}{\rho_e}. \quad \text{(3F.12)}$$

3F.8.2 Mud Pit Level. Indicators specify changes in the volume of mud in the pit. The total volume of mud changes continuously with depth as the hole volume increases. Rapid increases in pit volume may mean an influx of reservoir fluids, and well control measures may need to be implemented. A rapid decrease in volume indicates a downhole loss of mud, and lost-circulation material will probably be added to the drilling fluid.

3F.8.3 Mud Chloride Content. Mud chloride content is monitored in all systems, along with the water content in nonaqueous drilling fluid systems. Significant changes in content may indicate influx of formation water, which means that an underbalance condition may be close, and mud weight may need to be increased.

3F.8.4 Lithology and Mineralogy. Lithology and mineralogy may change as a fault is approached. Warmer water, with higher concentrations of dissolved salts, can flow along faults during some phases of their development. As the water moves into cooler zones, salts will precipitate, plugging pores and showing up in the cuttings. Indication of an approaching fault may warn of a potential jump across the fault, which, in some areas, is accompanied by a significant change in pore pressure. Advance knowledge of this allows the driller to adjust the mud weight before he encounters problems.

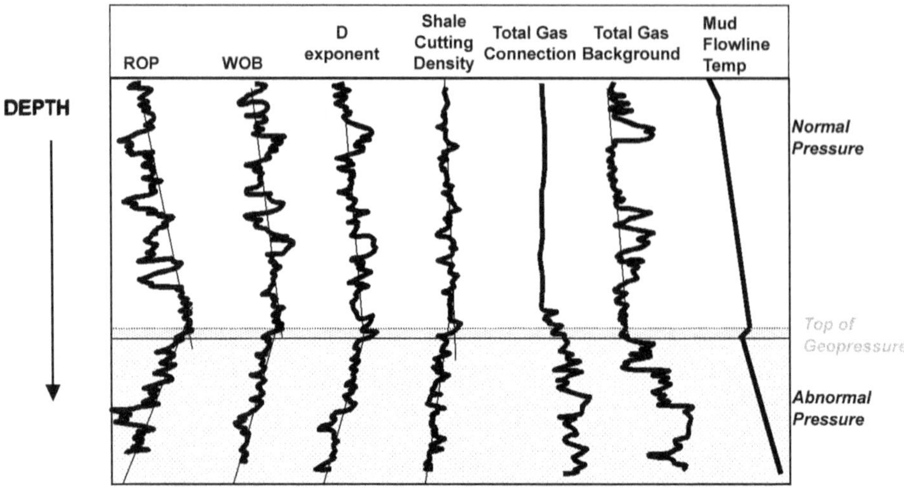

Fig. 3F.11—Schematic description of drilling and mud logging parameter changes with depth, normal pressure trends, and geopressure trends.

3F.8.5 Total Gas. Concentrations in the drilling fluid returns indicate the degree of overbalance or underbalance between the equivalent mud density and the formation pore pressure. The total gas concentration measured while drilling shales establishes a baseline or background level that is useful in tracking pore pressure, with the assumption that the shale pore fluids are in equilibrium with any neighboring permeable sands.

Fig. 3F.11 indicates how trends of several parameters vary with pore pressure and depth. Monitoring and plotting these can give indications of the transition from normally pressured zones to geopressures. The ROP, drilling exponent, shale cutting density, and background total gas all follow a normal trend with depth. Attempts to calibrate these measurements directly to pore pressure have been somewhat successful and usually are on the basis of establishing the trend for normally pressured formations. When a deviation from the normal trend occurs, correlations specific to the basin or geographic region are used to estimate the formation pore pressure. Most logging companies offer pore pressure service, which requires experienced pore pressure engineers who, frequently through experience, add subjective input to the model as well as the objective parametric inputs.

While the accuracy of these particular methods will vary from site to site, such plots are extremely useful in identifying the transition into geopressures (i.e., when passing from normally pressured to abnormally pressured zones). At the transition to geopressures, the trend lines change slope. Because some of the changes may be subtle, looking at all the available data helps pinpoint the transition.

3F.8.6 Connection Gas. As Sec. 3F.5.1 describes, connection gas is a good indicator of swabbing the wellbore at the bit (i.e., reducing the mud pressure at the bottom of the hole to below the pore pressure). If the pore pressure is less than the swabbed bottomhole pressure, little or no connection gas is seen. Some knowledge of the dynamic rheology of the drilling fluid is needed to perform input into a "swab model."

3F.8.7 Normal Geothermal Gradient. The normal geothermal gradient may shift on transition into geopressures. Other thermal nomalies, such as proximity to subsurface salt bodies, may interfer with this phenomenon. A more thorough discussion of these techniques and their application to detect overpressure may be found in several references.[3,4,34,35]

3F.8.8 Monitoring the Rate of Cuttings Return. As a formation is drilled, the cuttings should be circulated to the surface. Improper hole cleaning results in the downhole retention of cuttings, frequently as a cuttings bed on the low side of the hole in inclined wells. This causes an increased drag on the drillpipe and, if buildup is severe, may pack off the drillpipe, causing it to stick. Monitoring the rate of cuttings production from the drilling fluid returns indicates an approaching problem and warns the driller, which allows for remedial action before the pipe sticks. Watching for an increase in cuttings return rate can flag sloughing or extruding shale conditions, which call for an adjustment of mud density, as well as extreme washout conditions. Naegel *et al.*[36] describe a device for continuously weighing the cuttings as they come off the shaker screens and comparing this with what would be expected for a given ROP and mud pump rate.

3F.9 Health, Safety, and Environmental Considerations

Various parameters measured for formation evaluation and to monitor drilling operations and equipment are also indicators of conditions that could pose health, safety, and environmental concerns. Pore pressure changes that result in loss of well control pose obvious safety concerns. Any loss of control that results in a hydrocarbon release also poses serious environmental issues. Ambient monitoring for natural gas is done for health and fire safety. Monitoring hydrogen sulfide (H_2S) is essential in areas in which the potential has been shown historically to exist, as well as in rank wildcat wells in which the characteristics of the geological basin are poorly known.

Hydrogen sulfide is detectable by GC but can not be measured with an FID. Thermal conductivity, MS, and solid state sensors detect H_2S. The Delphian Mud Duck, which uses an electrochemical sensor, monitors dissolved H_2S, HS, and S^{2-} ionic concentrations to give total sulfide content of the drilling fluid. This tool continuously follows sulfide trends before its concentration increases to the point that gaseous H_2S is released from the mud. Draeger tubes are used for spot measurement of hydrogen sulfide, as a backup or as a check on other sensing equipment.

Nomenclature

B_h = balance ratio
C_1 = methane
C_2 = ethane
C_3 = propane
C_{4+} = butanes and heavier
C_{5+} = pentanes and heavier
C_h = character ratio
d_b = bit diameter, L, in.
d_c = corrected drilling exponent, dimensionless
d_{exp} = drilling exponent, dimensionless
N = rotary speed, rpm
q = mud pump flow rate, B/D
r_g = residual gas
r_o = residual oil
R = penetration rate, ft/hr
t_{an} = lag time from the bit to surface, min
t_{dp} = lag time in the drill pipe, min
t_s = lag time in surface lines, min
t_t = total elapsed pumping time from injection to measurement, min

V_{dp} = internal volume of the drillpipe and drill collars, bbl
V_s = internal volume of any surface lines tracer injection point to kelly, bbl
W = weight on bit, k-lbf
W_h = wetness ratio
ρ_e = equivalent mud density at the bit while circulating, m/L^3, g/cm^3
ρ_n = mud density equivalent to a normal formation pore pressure, m/L^3, g/cm^3

Acknowledgments
The author thanks Shell Intl. E&P Inc. for permission to publish this chapter and is very grateful to Geoservices, Datalog, and Halliburton for the help and advice he has had in gathering gas-sensing information. The reference list provides excellent sources for general information.[37–42]

References

1. Haywood, J.T.: "Continuous Logging at Rotary Drilling Wells," presented at the 1940 Drilling Practices Annual Meeting, Chicago, November.
2. Wilson, R.W.: "Mud Analysis Logging and Its Use in Formation Evaluation," AIME technical paper 587-G presented at the U. of Houston, October 1955.
3. Jorden, J.R.: "Application of Drilling Performance Data to Over Pressure Detection," *JPT* (November 1966) 1387.
4. Jorden, J.R. and Campell, F.L.: *Well Logging I–Rock Properties, Borehole Environment, Mud and Temperature Logging,* Monograph Series, SPE, Richardson, Texas (1984) **9.**
5. Whittaker, A.H.: "Mud Logging," *Petroleum Engineering Handbook*, SPE, Richardson, Texas (1992) Chap. 52.
6. Brumboiu, A.O. *et al.*: "Application of Semipermeable Membrane Technology in the Measurement of Hydrocarbon Gases in Drilling Fluids," paper SPE 62525 presented at the 2000 SPE/AAPG Western Regional Meeting, Long Beach, California, 19–23 June.
7. Ellis, L, Brown, A., Schoell, M., and Haught, M.: "Mudgas Isotope Logging While Drilling: A New Field Technique for Exploration and Production," presented at the 19th International Meeting on Organic Geochemistry, 6–10 September, 1999, Istanbul, Turkey, *Abstracts Part 1*, 67–68.
8. Wright, A.C., Hanson, S.A., and Delaune, P.L.: "A New Quantitative Technique for Surface Gas Measurements," paper A presented at the 1993 SPWLA Annual Logging Symposium, 13–16 June.
9. De Pazzis, L.L. *et al.*: "New Gas Logging System Improves Gas Shows Analysis and Interpretation," paper SPE 19605 presented at the 1989 SPE Annual Technical Conference and Exhibition, San Antonio, Texas, 8–11 October.
10. Brevier, J., Herzaft, B. and Mueller, N.: "Gas Chromatography—Mass Spectrometry (GCMS)—A New Wellsite Tool for Continuous C1-C8 Gas Measurement in Drilling Mud—Including First Original Gas Extrator and Gas Line Concepts. First Results and Potential," paper J presented at the 2002 SPWLA Annual Logging Symposium, Oiso, Japan, 2–5 June.
11. Sterner, M.: "Internal Marketing Literature," Fluid Inclusion Technology Inc., Tulsa (2002).
12. Amen, R.: "Quantifying Hydrocarbon Shows Using On-Line Gas Referencing," paper HH presented at the 1994 SPWLA Annual Logging Symposium, 19–22 June.
13. Williams, R.D. and Ewing, S.P. Jr.: "Improved Methods for Sampling Gas and Drill Cuttings," *SPEFE* (June 1989) 167.
14. Low, J.W.: "Examination of Well Cuttings," *Quarterly*, Colorado School of Mines, Golden, Colorado (October 1951) **46**, 1.
15. Archie, G.E.: "Classification of Carbonate Reservoir Rocks and Petrophysical Considerations," *Bull. AAPG* (February 1952) **36,** 278.
16. Dunham, R.J.: "Classification of Carbonate Rocks According to Depositional Texture," *Classification of Carbonate Rocks*, W.E. Ham (ed.), Memoir 1, AAPG, Tulsa (1962) 108–121.
17. Swanson, R.G.: *Sample Examination Manual*, American Assn. of Petroleum Geologists, Tulsa (1981).
18. Pixler, B.O.: "Formation Evaluation by Analysis of Hydrocarbon Ratios," *JPT* (June 1969) 665.

19. Ferrie, G.H., Pixler, B.O., and Allen, S.: "Well-Site Formation Evaluation by Analysis of Hydrocarbon Ratios," presented at the 1981 Annual Technical Meeting of the Petroleum Society of CIM, Calgary, 3–6 May.
20. Haworth, J.H., Sellens, M., and Whittaker, A.: "Interpretation of Hydrocarbon Shows Using Light (C1-C5) Hydrocarbon Gases from Mud-Log Data," *AAPG Bulletin* (August 1985) **69,** 1305.
21. Whittaker, A. and Sellens, M.: "Analysis Uses Alkane Ratios from Chromatography," *Oil & Gas J.* (18 May 1987) 42.
22. Whittaker, A. and Sellens, G.: "Normalization of Gas Shows Improves Evaluation," *Oil & Gas J.* (20 April 1987) 53.
23. Amen, R.M.: "Technique Uses Reference Gas to Assess HC Shows," *The American Oil & Gas Reporter* (April 1995) 57.
24. Wright, A.C.: "Estimation of Gas/Oil Ratios and Detection of Unusual Formation Fluids from Mud Logging Gas Data," paper CC presented at the 1996 SPWLA Annual Logging Symposium, 16–19 June.
25. Delaune, P.L.: "Surface Techniques to Measure Oil Concentrations While Drilling," paper KK presented at the 1992 SPWLA Annual Logging Symposium, June.
26. Delaune, P.L. *et al.:* "Enhanced Wellsite Technique for Oil Detection and Characterization," paper SPE 56802 presented at the 1999 SPE Annual Technical Conference and Exhibition, Houston, 3–6 October.
27. Boone, D.E.: "Gas Volumetric Analysis," *Petroleum Engineer Intl.* (August 1985).
28. Beeunas, M.A. *et al.*: "Application of Gas Geochemistry for Reservoir Continuity Assessment, Gulf of Mexico," presented at the 1996 AAPG/EAGE Research Symposium on Compartmentalized Reservoirs: Their Detection, Characterization and Management, 20–23 October.
29. Ehrenberg, S.N. and Siring, E.: "Use of Bulk Chemical Analyses in Stratigraphic Correlations of Sandstones: An example from the Statfjord Nord Field, Norwegian Continental Shelf," *J. of Sedimentary Petrology* (1992) **62,** 318.
30. Pearce, T.J. and Jarvis, I.: "Applications of Geochemical Data to Modeling Sediment Dispersal Patterns in Distal Turbidites: Late Quaternary of the Madeira Abyssal Plain," *J. of Sedimentary Petrology* (1992) **62,** 1112.
31. Hall, D., Shentwu, W., and Sterner, M.: "Using Fluid Inclusions to Explore for Oil and Gas," *Hart's Petroleum Engineer Intl.* (1997) No. 11, 29.
32. Hall, D.L., Bigge, M.A., and Jarvie, D.M.: "Fluid Inclusion Evidence for Alteration of Crude Oils," presented at the 2002 AAPG Annual Meeting, Houston, 10–13 March.
33. Barclay, S.A. *et al.:* "Assessment of Fluid Contacts and Compartmentalization in Sandstone Reservoirs Using Fluid Inclusions, An Example from the Magnus Oil Field, North Sea," *AAPG Bull.* (April 2000) 489.
34. Bourgoyne, A.T. *et al.: Applied Drilling Engineering,* second edition, SPE, Richardson, Texas (1991) Chap. 6.
35. Rehm, W.A. and McClendon, R.: "Measurement of Formation Pressure From Drilling Data," paper SPE 3601 presented at the 1971 SPE Annual Meeting, New Orleans, 3–6 October.
36. Naegel, M. *et al.:* "Cuttings Flow Meters Monitor Hole Cleaning in Extended Reach Wells," paper SPE 50677 presented at the 1998 SPE European Petroleum Conference, The Hague, 20–22 October.
37. *Gas Detection: Theory and Practice*, training manual, Geoservices Inc. (1995).
38. Mercer, R.F. and McAdams, J.B.: "Standards for Hydrocarbon Well Logging," presented at the 1982 SPWLA Annual Logging Symposium, 6–9 June.
39. *Mud Logging: Principles and Interpretation,* Exploration Logging Inc. (1979)
40. "Mud Logging: Methods and Interpretation Workshop," workshop notes, 1989 SPWLA Annual Logging Symposium, 11 June.
41. Helander, D.P.: *Fundamentals of Formation Evaluation*, OGCI Publications, Tulsa (1983).
42. "Advances in Mud Logging," presented at the 1995 SPWLA/CWLS Symposium, 17 June.

SI Metric Conversion Factors

ft × 3.048* E − 01 = m

*Conversion factor is exact.

Chapter 3G
Specialized Well-Logging Topics
Paul F. Worthington, Gaffney, Cline, and Assocs.

3G.1 Introduction

This chapter describes three categories of specialized well logging: downhole measurements that are concerned with the geometry and integrity of the wellbore; acoustic and electrical imaging of the geological architecture and fabric of the rock system penetrated by the well; and downhole measurements that use the Earth's gravitational and magnetic fields to infer large-scale changes through density and magnetization, respectively. Although some of these technologies are applied beyond the petroleum industry (e.g., geotechnical studies and hydrogeology), this overview concentrates on their hydrocarbon applications. Each topic has a brief introduction, a description of the principles of each method, a discussion of its practical application, and, in selected cases, an illustrative case history. **Fig. 3G.1** summarizes how these specialized logging methods relate to reservoir characteristics and the techniques for measuring them, as presented in Chaps. 3B through 3F and Chap. 4 in this section and Chap. 15 in the Drilling Engineering section of this *Handbook*.

3G.2 Geometry and Integrity of the Wellbore

3G.2.1 Directional Surveys. ***Principles.*** With the growth in drilling deviated, extended-reach, and horizontal wells, the location of the wellbore is increasingly a 3D problem. It is encountered in one of two situations: to direct and define the trajectory of the well during the drilling process (geosteering) or to characterize the well path after drilling. The former has contributed to huge increases in well productivity. The latter is a vital element of integrated reservoir studies in which the aim is to generate a 3D model of the reservoir based on correct well locations. This discussion is set within the context of the latter. Geosteering is a benefit of logging-while-drilling technology (see Chap. 15 in the Drilling Engineering section of this *Handbook*).

In directional-survey terminology, the azimuth is the orientation of the wellbore to "north," which has been variously defined as magnetic north, geographic north (latitude and longitude), or grid north [e.g., the Universal Transverse Mercator (UTM) geographic coordinate system]. Azimuth is typically measured clockwise from north. The inclination is the deviation of the wellbore from the vertical. The azimuth and inclination define the coordinates of the wellbore along its length, usually relative to the location of the wellhead. Tool face quantifies the direction in which the tool is pointing. It is the angle between a reference direction on a downhole

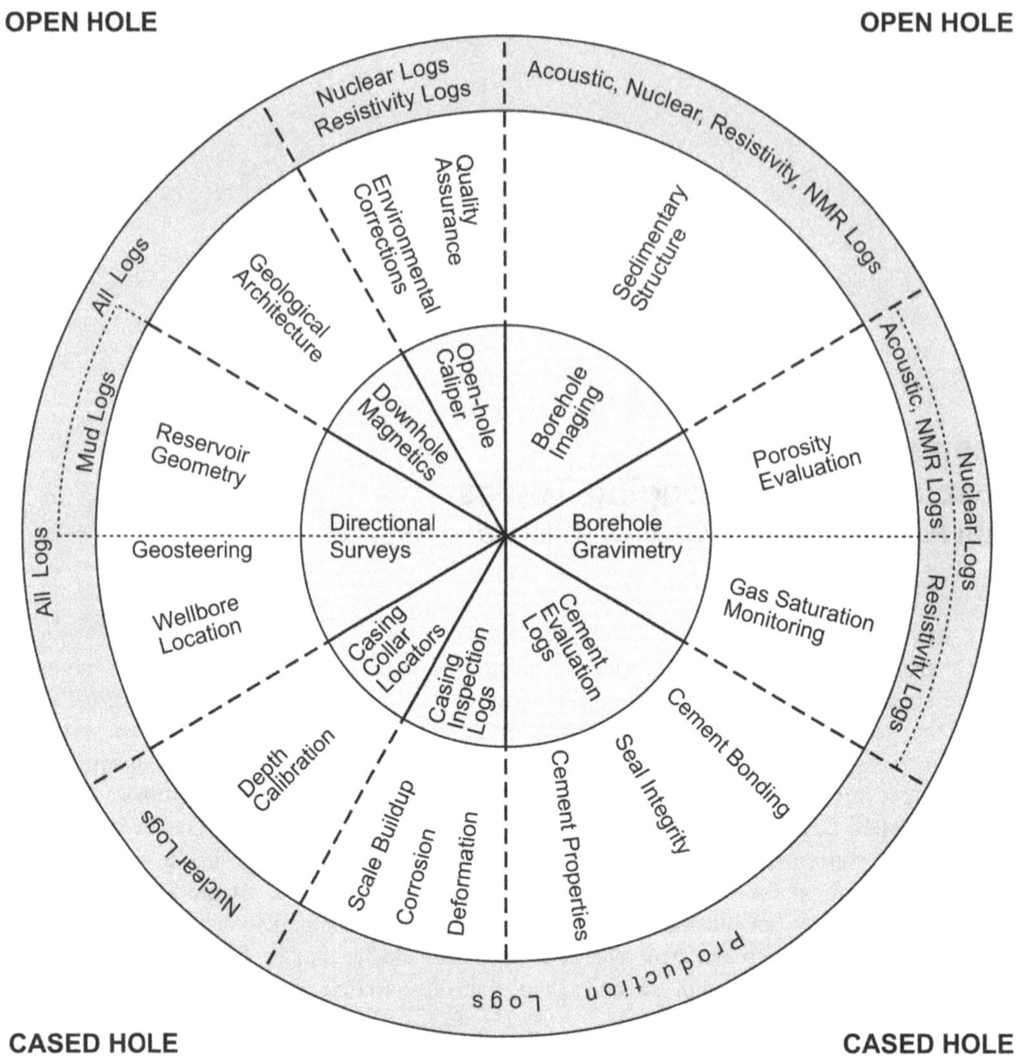

Fig. 3G.1—This chart is separated into three concentric areas: the middle annular area indicates the subsurface properties to be evaluated, the innermost area indicates the specialized logging tools discussed here, and the outermost area indicates the logging tools discussed in other subchapters of this *Handbook*. The corresponding innermost and outermost areas show how the different tools complement each other in the investigation of particular subsurface properties.

tool and a fixed reference in space. For near-vertical wells, the fixed reference is magnetic north and the (magnetic) tool face is the angle between magnetic north and the projection of the tool's reference direction onto a horizontal plane. For more deviated wells, the fixed reference is the top of the hole and the (gravity or high side) tool face is the tool orientation with respect to the top of the hole. Directional surveys allow azimuth and inclination to be determined so that the coordinates of well location can be computed at each survey station.

Measurement. There are two types of directional surveys: magnetic and gyroscope surveys. Traditionally, each of these has been run using either single- or multishot tools. The tools are self-contained and can be powered from the surface or with downhole batteries. For a detailed description, see Bourgoyne *et al.*[1]

Magnetic surveys can be run on the drillstring while tripping or on a wireline (conducting cable) after drilling. They are openhole measurements. In the simplest form, a magnetic-survey tool comprises a downhole inclinometer and a compass unit. More advanced tools comprise arrays of three-axis accelerometers (inclinometers) and magnetometers. A three-axis accelerometer measures three orthogonal components of gravity, which combine vectorially to give the direction of the Earth's gravitational field relative to the axis of the tool and a reference position on its circumference. A three-axis magnetometer measures the direction of the Earth's magnetic field relative to the tool axis. As **Fig. 3G.2** illustrates, by combining the accelerometer and magnetometer data, it is possible to calculate the inclination, I, and azimuth, A, of the tool. Through the tool face, it is then possible to calculate the inclination, α, and azimuth, β, of the wellbore itself. The magnetic measurements are impacted by short-term variations in the Earth's magnetic field and by proximity to magnetic materials such as drillpipe and casing or to magnetic minerals such as pyrite and/or hematite, the latter also occurring as a mud additive. These problems can be mitigated through the use of nonmagnetic drill collars and repeat measurements with different alignments of the tool in the wellbore. It should be noted that even where magnetic-survey tools are run in long lengths of nonmagnetic drill collars, there can still be a significant effect from steel drillstring tools.[3]

One of the problems with gyroscope surveys is that the instrumentation is very sensitive and cannot withstand downhole vibrations and stresses. For this reason, gyroscope surveys have to be run on wireline, usually while entering the hole with a few additional checkshots on coming out. They interrupt the drilling process, and this adds to cost. However, they can be run in cased hole. In the simplest form, a gyro tool comprises a downhole inclinometer and gyroscope unit. Gyroscope surveys are usually mechanically driven, but other methods, which use a rotating beam of light, have been developed on the basis of fiber optics and laser technology. Mechanical gyroscopes are grouped in terms of their freedom of movement and the number of flywheels (between one and three). Three-axis systems (i.e., three orthogonal accelerometers and gyroscopes mounted on an inertial platform) have proved superior, especially in highly deviated and horizontal wells, where, for example, the axial component of the Earth's gravity field is small. Here, the tool position is calculated from the accelerometer and gyroscope data through the tool's inertial navigation system.

Fig. 3G.3 shows the reporting format of a typical directional survey. A synopsis of the errors associated with these measurements and their application has been provided by Theys.[2]

Processing. The results of directional surveying usually take the form of inclination, α, and azimuth, β, of a borehole at a sequence of survey stations. The only other information available is the difference in measured depths for two adjacent stations, but this does not describe the shape of the well path. Starting with the coordinates of the surface reference point, the actual distance between adjacent survey stations needs to be calculated so that the coordinates of any station can be found by addition and those of any intermediate point can be found by interpolation. There are several ways of doing this interpolation. The following account is based on Inglis' work.[4]

The balanced tangential method assumes that the actual wellpath between two adjacent measurement stations can be approximated by two straight lines of equal length, $L/2$, shown as AX and BX in **Fig. 3G.4**. This leads to the following expressions for the incremental distances between adjacent survey stations in the vertical direction (ΔV), in the direction of the northing (ΔN), and in the direction of the easting (ΔE):

$$\Delta V = \Delta V_1 + \Delta V_2 = L/2\left(\cos \alpha_1 + \cos \alpha_2 \right), \quad \text{(3G.1)}$$

$$\Delta N = \Delta N_1 + \Delta N_2 = L/2\left(\sin \alpha_1 \cos \beta_1 + \sin \alpha_2 \cos \beta_2 \right), \quad \text{(3G.2)}$$

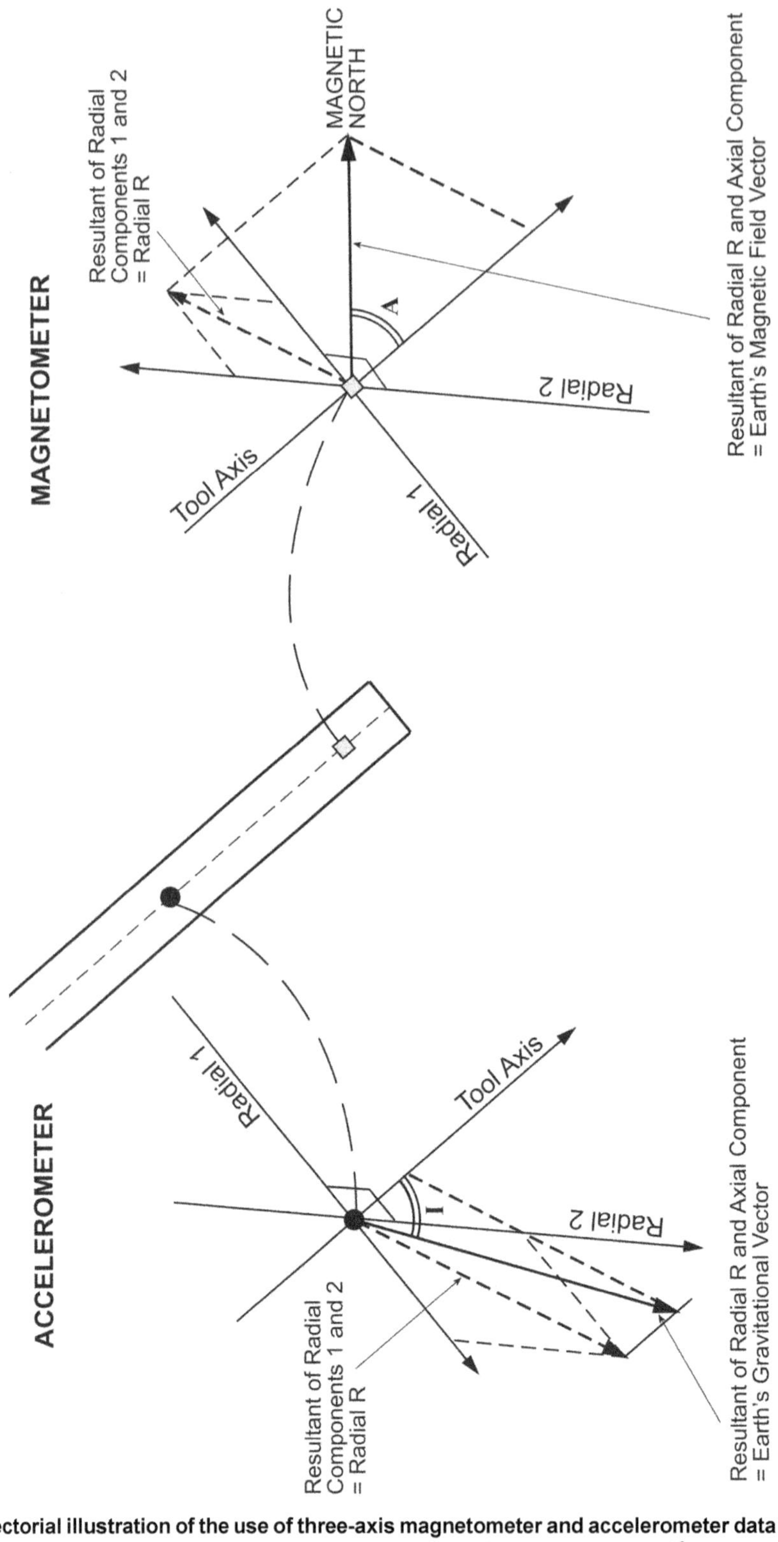

Fig. 3G.2—Vectorial illustration of the use of three-axis magnetometer and accelerometer data to calculate the inclination and azimuth of the directional-survey tool and of the wellbore itself.[2]

Well Path Plot

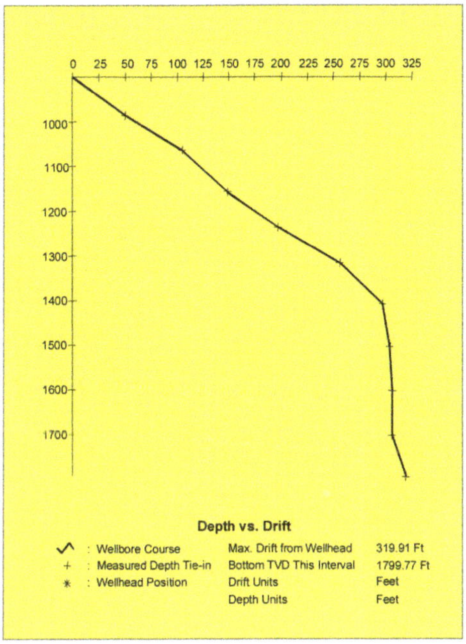

Plan View

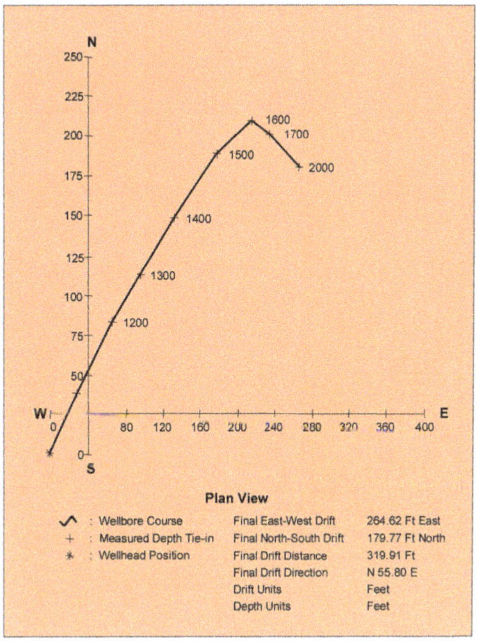

Tabular Listing

COMPANY PAGE 1
WELL

	DEVIATION			DRIFT COORDINATES STATION		DRIFT COORDINATES TOTAL	
DEPTH	ANGLE	AZIMUTH	TVD	N-S	E-W	N-S	E-W
1000.00	9.0	35.0	900.00	0.00N	0.00E	0.00N	0.00E
1100.00	50.0	37.0	985.19	38.66N	28.78E	38.66N	28.78E
1200.00	20.0	45.0	1066.17	40.70N	35.99E	82.37N	64.78E
1300.00	30.0	44.0	1156.69	30.15N	29.53E	112.52N	94.30E
1400.00	30.0	47.0	1243.29	35.04N	35.65E	147.55N	129.95E
1500.00	45.0	50.0	1322.40	40.01N	45.63E	187.56N	175.58E
1600.00	10.0	110.0	1409.73	20.63N	36.81E	208.19N	212.39E
1700.00	10.0	120.0	1508.21	7.31S	15.68E	200.88N	228.06E
1800.00	0.0	120.0	1607.70	4.35S	7.54E	196.53N	235.60E
1900.00	0.0	120.0	1707.70	0.00N	0.00E	196.53N	235.60E
2000.00	40.0	120.0	1799.78	16.76S	29.02E	179.77N	264.63E

THE FIRST POINT INFORMATION:

1000.00	9.0	35.0	900.00	0.00N	0.00E	0.00N	0.00E

THE LAST POINT INFORMATION:

2000.00	40.0	120.0	1799.78	16.76S	29.02E	179.77N	264.63E

FINAL EAST-WEST DRIFT = 264.63 E
FINAL NORTH-SOUTH DRIFT = 179.77 N
FINAL DRIFT DISTANCE = 319.91 FT
FINAL DRIFT DIRECTION = N 55.81 E

Fig. 3G.3—Example of the reporting format of a directional survey with depths in feet. Vertical scale on the Well Path Plot is true vertical depth (TVD). Depth markers on the Plan View trace are measured depths. The Tabular Listing links the two depth scales at measurement stations and contains the wellbore deviation, azimuth, and coordinates at the points of measurement. (Courtesy of Baker Atlas.)

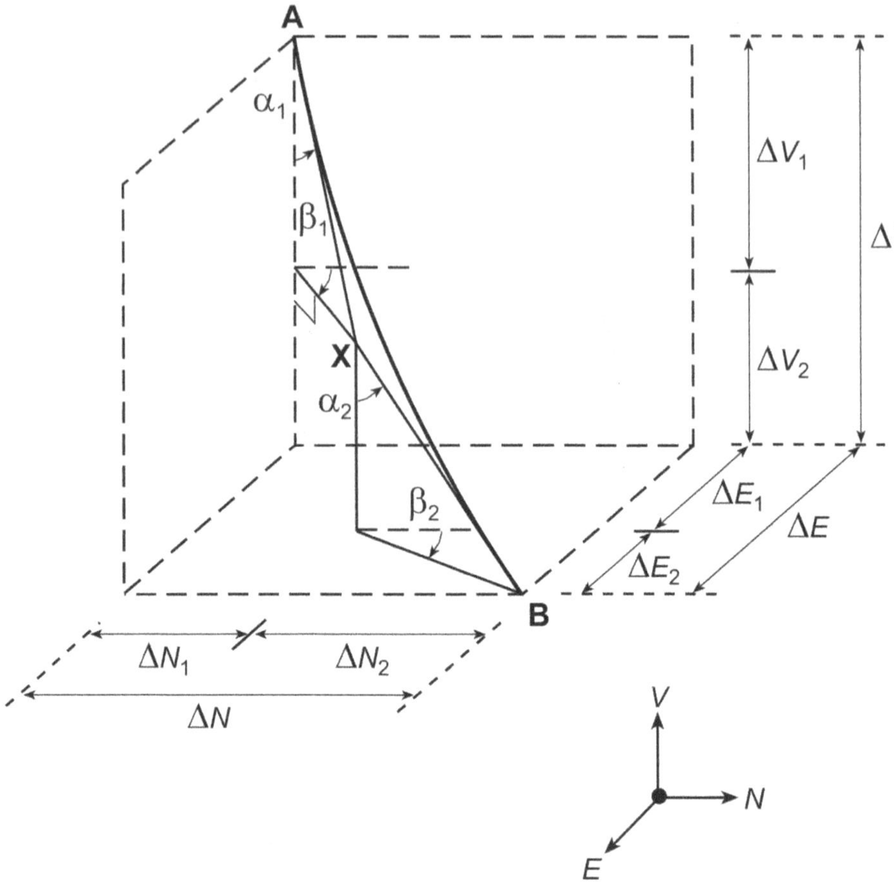

Fig. 3G.4—Principles of the balanced tangential method for modeling the well path between directional survey stations A and B.[4]

$$\text{and } \Delta E = \Delta E_1 + \Delta E_2 = L/2\left(\sin\alpha_1 \sin\beta_1 + \sin\alpha_2 \sin\beta_2 \right), \quad \text{(3G.3)}$$

where the subscripts 1 and 2 denote the upper and lower survey stations, respectively.

An improvement on the balanced tangential method is the minimum curvature method, which replaces the two straight lines by an arc. The position of the arc is based on the amount of bending in the wellpath between the two survey stations. The amount of bending is described by a ratio factor, F_r, and quantified by a dogleg angle, ψ (**Fig. 3G.5**), so that:

$$F_r = (360/\pi\psi)\tan(\psi/2), \quad \text{(3G.4)}$$

where

$$\cos\psi = \cos\alpha_1 \cos\alpha_2 + \sin\alpha_1 \sin\alpha_2 \cos\left(\beta_2 - \beta_1\right). \quad \text{(3G.5)}$$

The ratio factor is applied to each of the quantities ΔV, ΔN, and ΔE as calculated by the balanced tangential method (Eqs. 3G.1 through 3G.3). Thus, for example, Eq. 3G.1 becomes

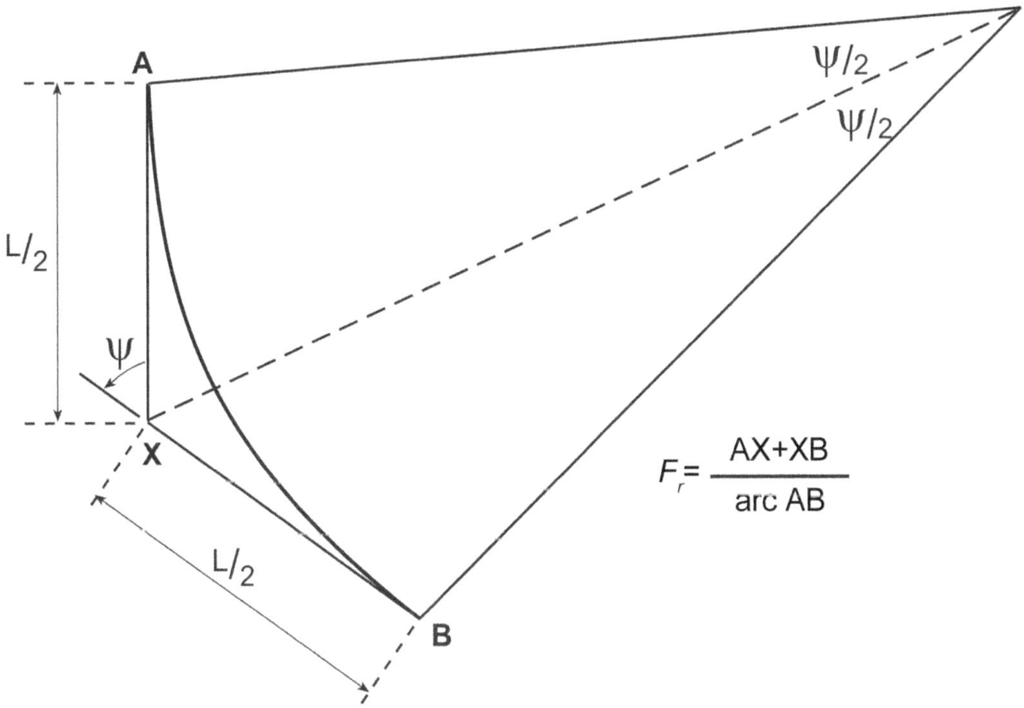

Fig. 3G.5—Principles of the minimum curvature method for modeling the well path between directional survey stations A and B, drawn in the plane of the wellbore.

$$\Delta V = F_r L/2(\cos\alpha_1 + \cos\alpha_2)\,. \quad \text{(3G.6)}$$

The minimum curvature method is the most widely used for computing the coordinate deliverables of directional surveys. Inglis[4] provided a detailed description of the calculations.

Applications. The uses of directional surveys include monitoring the actual wellpath to ensure that the drilling target has been reached, defining the X–Y coordinates of points in the wellbore, and determining the true vertical depth (TVD) for geological mapping. Other important operational objectives are to ensure that well paths do not collide (a noteworthy risk given that several production wells are drilled from the same platform) and that there are no changes in angularity (e.g., doglegs) that might impede tool deployment or production efficiency.

An important exercise for integrated reservoir studies is to use the TVD to evaluate the true stratigraphic thickness, h_{ts}, and the true vertical thickness, h_{tv}, of constituent reservoir beds. **Fig. 3G.6a** shows a well with inclination or deviation, α, penetrating a bed with dip, φ, over a measured bed thickness, h_m, for the particular case in which the azimuths of the well and the dip are the same. Here, the values of h_{ts} and h_{tv} can be calculated as

$$h_{ts} = h_m \cos(\alpha + \varphi) \quad \text{(3G.7)}$$

and $$h_{tv} = h_m \cos(\alpha + \varphi)/\cos\varphi\,. \quad \text{(3G.8)}$$

In the more general case, for which the azimuth of the well, β, is not the same as the azimuth of the dip, β_d (Fig. 3G.6b), the equations are more complex:

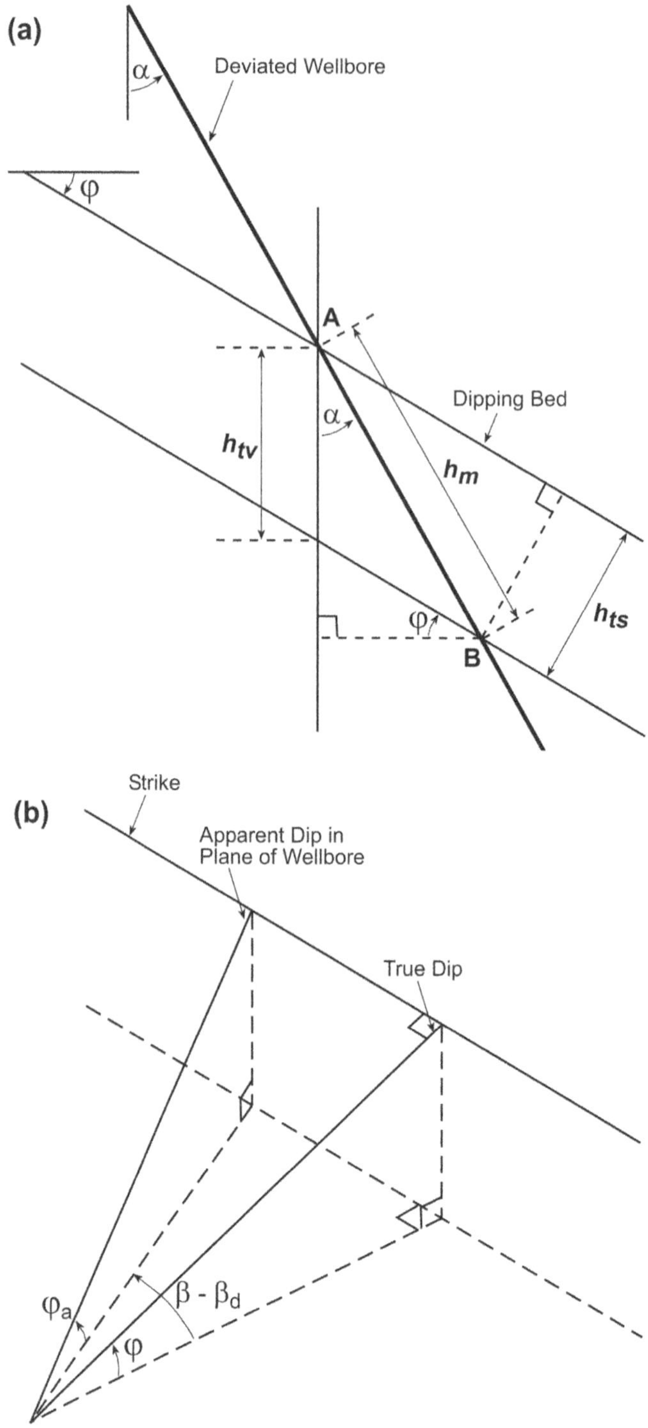

Fig. 3G.6—Basis for the evaluation of true stratigraphic thickness, h_{ts}, and true vertical thickness, h_{tv}, from processed directional-survey data using Eqs. 3G.9 and 3G.10, respectively. (a) The azimuths of the well and true dip are the same. (b) The azimuths of the well and true dip are not the same: the true dip is projected into the plane of the wellbore to become an apparent dip, so that the azimuths of the well and apparent dip are the same.

$$h_{ts} = h_m\left[\cos\alpha\cos\varphi - \sin\alpha\sin\varphi\cos\left(\beta-\beta_d\right)\right] \quad \text{(3G.9)}$$

$$\text{and } h_{tv} = h_m\left[\cos\alpha - \sin\alpha\tan\varphi\cos\left(\beta-\beta_d\right)\right]. \quad \text{(3G.10)}$$

As an illustration of the derivation of these expressions, Eq. 3G.10 can be derived from Eq. 3G.8 by projecting the true angle of dip, φ, onto the vertical plane that contains the wellbore azimuth within the layer of interest (Fig. 3G.6b). This introduces an apparent dip, φ_a, which is related to the true dip by the relationship:

$$\tan\varphi_a = \tan\varphi\cos\left(\beta-\beta_d\right). \quad \text{(3G.11)}$$

If the apparent dip is used instead of the true dip in Eq. 3G.8, we have

$$h_{tv} = h_m\cos\left(\alpha+\varphi_a\right)/\cos\varphi_a. \quad \text{(3G.12)}$$

Eq. 3G.12 will give the correct answer. If we now use Eq. 3G.11 to substitute for φ_a in Eq. 3G.12, we obtain Eq. 3G.10. Other derivations of Eq. 3G.10 are available (e.g., those that rotate the dip or deviation to zero to simplify the equations). Eqs. 3G.9 and 3G.10 can also be written in terms of wellbore coordinates.[5]

3G.2.2 Openhole Caliper Logs. Openhole calipers comprise up to six arms attached to the body of a sonde and held against the borehole wall by spring action. They provide a continuous measurement of borehole diameter. In general, one- and two-arm calipers measure only the maximum diameter where a hole is not circular. Four- and six-arm tools define the hole size and shape, and this is especially important in deviated wells and elliptically shaped wellbores. However, the size and pressure of the contacting arms also affect the measured data. This means that a caliper run with the density-log tool string (Chap. 3D in this section of the *Handbook*) might show a larger hole diameter than one run with the induction log (Chap. 3B in this section of the *Handbook*). This is because the density tool is a pad device, and the pad cuts through the mudcake to sense a larger diameter, whereas the induction tool is a mandrel device (i.e., it is essentially contained within a cylindrical housing). For these reasons, different open hole caliper logs should not be expected to show precise repeatability.

The movement of the caliper arms must be converted to something that is measurable at the surface. Most modern calipers use a potentiometer circuit connected to the caliper arms using transducers. The circuitry can use direct current or pulses. In the former case, the displacement of the arms translates directly to a voltage within the measuring circuit. Pulsed caliper tools use the potentiometer to deliver a variable voltage to a voltage-frequency circuit. The frequency of the pulse train is proportional to the extension of the caliper arm. A basic problem with armed calipers is that the extension of an arm is not directly proportional to the displacement of the transducer. This gives rise to a nonlinear response, which is linearized through data processing based on detailed calibration data. A caliper tool is designed to operate over a specified range of hole diameter. The design sets out to minimize nonlinearity of tool response over this range.

Openhole caliper data are used to estimate the volumes of gravel and cement needed for well-completion planning. Other openhole uses include providing information on the buildup of mudcake over permeable intervals and locating seats for packers (hydraulic seals used to isolate sections of the wellbore for flow-test purposes). They also indicate where boreholes are washed out or penetrate swelling clays as a result of rock/filtrate interaction. Yet again, calipers can be used to center or eccenter logging-tool strings. However, in a logging context, the greatest application of caliper data is in environmental corrections to other logs such as natural gamma ray, density, and neutron logs (Chap. 3D in this section of the *Handbook*). It is this application that makes the calibration and depth matching of caliper data especially important. Calibration is carried out using rings or sleeves of known diameter. Depth matching is usually done by applying to caliper data those same depth shifts that were generated by comparing the gamma log on the same tool string with the depth-reference gamma log where this has been measured on a different logging run.

3G.2.3 Casing-Collar Locators. The casing-collar locator (CCL) is an important tool because it is used for depth control. When combined with a gamma ray log, it allows depth correlation of a cased-hole logging run with the openhole logs and, therefore, reservoir units or zones. This is essential for subsequent downhole operations such as perforating. Because it constitutes the primary depth control, the CCL is run on almost every cased-hole tool string. The tool comprises a coil-and-magnet arrangement with a downhole amplifier. The most sensitive of these arrangements is two like-facing magnetic poles positioned on either side of a central coil. The magnetic lines of flux are distorted when the tool passes a location at which the metallic casing is enlarged by a collar. This distortion gives rise to a classical change in the magnetic field around the conducting coil, within which current is induced. The signal is amplified and recorded at the surface in the form of a voltage spike known as a collar "kick."

CCLs can be run in standard wireline logging mode or on a slickline (i.e., a nonconducting line).[6] Pure-memory slickline CCLs record their data simultaneously with, for example, a full production logging suite, but these CCL data are not available until the memory section has been retrieved and downloaded at the surface. Real-time slickline tools convert the voltage spike to a tension spike by using spring-loaded electromagnets that increase the apparent drag through the greater attraction between the electromagnets and the casing at collar locations. The tension spike can be detected at the surface. CCLs have had to be modified for coiled-tubing applications. The primary difference arose because the heaviness of a coiled-tubing string did not allow relatively small tension spikes to be detected with confidence. For this reason, a solenoid/piston/valve arrangement is used to transmit pressure spikes through the fluid within the coiled tubing to the surface, where they can readily be detected. These tools have recently been improved for high-pressure/high-temperature applications.[7] Some types of downhole tractor that are used to deploy tool strings in deviated wells also have the ability to produce a CCL during the tractor operation and thereby provide the same depth control.

3G.2.4 Casing Inspection Logs. There are four commonly used techniques for the inspection of casing: cased-hole calipers, flux-leakage tools, electromagnetic phase-shift tools, and ultrasonic tools.

Cased-Hole Calipers. Multifinger calipers are used to identify changes in casing diameter as indicators of wear and corrosion. They are also used to monitor casing deformation.[8] They can have up to 80 spring-loaded feelers or fingers, depending on the nominal casing diameter (**Fig. 3G.7**). Different multifinger caliper tools can log casing sizes from 4 to 20 in. [100 to 500 mm]. Smaller tools are available for tubing inspection. Each hardened finger can measure the internal casing diameter with a radial resolution of a few thousandths of an inch and a vertical resolution of a few hundredths of an inch at a typical logging speed of 1800 ft/hr [550 m/h]. Measurements are taken many times per second for each finger, giving a typical spatial-sampling interval of approximately 0.15 in. [4 mm] as the tool travels up the borehole. A finger extends where it encounters a pit or hole and retracts where there is scale present or there has been partial collapse. A potential disadvantage is that the fingers can damage the casing, although modern electronic tools have a very low finger pressure to avoid this. The tool also indicates which finger is the one on the highest side of the well. Moreover, fingers can be grouped azimuthally. All these data can be combined with the measurements of diameter to produce a 3D picture of the casing, including cross-sectional distortions and changes in the trajectory of the well axis as small as 0.01°. The data can be either transmitted to the surface where the tool is run on a wireline or stored downhole where the tool is deployed on a slickline.

There are two types of multifinger calipers, mechanical and electronic, although the distinction is misleading because all such calipers are mechanical in their deployment. The difference is in the way in which data are recorded. Older calipers were truly mechanical in that they were operated on a slickline and used a scribe chart for downhole data recording. These mechanical calipers have high temperature ratings because they are not limited by the ratings of downhole electronics [e.g., 600°F (315°C) for the Kinley caliper offered by the Expro Group]. Modern tools convert the mechanical data into electronic information for downhole memory storage or for transmittal uphole for real-time data display. Operating temperatures for these electronic tools are typically up to 350°F [177°C].

Multifinger tools contain an inclinometer so that tool deviation and orientation can be recorded. If these parameters are known, the high-quality output from modern multifinger calipers allows several image-based products to be generated. Deliverables include digital "maps" of the ovality of the casing and its internal diameter. The logs can be run and displayed in time-lapse mode to quantify the rates of corrosion or scale buildup. A digital image of variations in the inner diameter of the casing is the principal tool for identifying corrosion. In its basic form, this is an electronic version of what one might see using a downhole video camera; however, the electronic image can be rotated and inspected from any angle. Artificial colors are used to bring out anomalies.

Another processed product is the 3D shape of downhole tubulars to map the trajectory of the wellbore and to quantify casing deformation. An interesting example of the use of multifinger-caliper data to evaluate casing deformation in primary heavy-oil production in northeastern Alberta has been described by Wagg *et al.*[9] (**Fig. 3G.8**). Several postulates for formation movement were modeled and compared with the observed casing deformations. In the end, it was concluded that sand production from an elongated disturbed zone caused reservoir shortening to an extent that could account for the wellbore observations. The use of casing-deformation logs as a tool in reservoir geomechanics leads to an improved knowledge base for well design.

Although they are intended for cased-hole application, it is possible to use multifingered calipers in open hole. The results are much more detailed than with a standard openhole caliper, and the output can be displayed as images similar to those obtainable with ultrasonic imaging tools (see the "Ultrasonic Tools" section below).

Flux-Leakage Tools. Flux leakage is a semiquantitative method that uses a strong magnetic field to identify and, to a certain extent, quantify localized corrosion on both the inner and the outer surfaces of the casing. A downhole electromagnet that fits snugly within the casing creates a low-frequency or a direct-current magnetic field. This can be a permanent magnet so it is possible to use this tool on a memory string for which battery power is at a premium. Magnetic flux is concentrated within the casing, which is close to magnetic saturation. The tool contains spring-loaded, coil-type, pad-mounted sensors that are pushed close to the casing during logging. Where casing corrosion is encountered, the lines of flux "bulge out" from the casing as though they were leaking from it. The primary sensors pass through this excluded flux and measure the induced voltage. The amplitude and spatial extent of the sensor response is related to the volume and shape of the corrosion metal loss, thereby allowing an estimate of the size of the defect. Because the primary measurement cannot distinguish between internal and external casing defects, many tools use an additional higher-frequency eddy-current measurement. This is a shallower measurement that responds only to casing flaws on the inner wall. It uses a separate transmitter coil. The flux-leakage and eddy-current signals are distinguished using frequency filters.

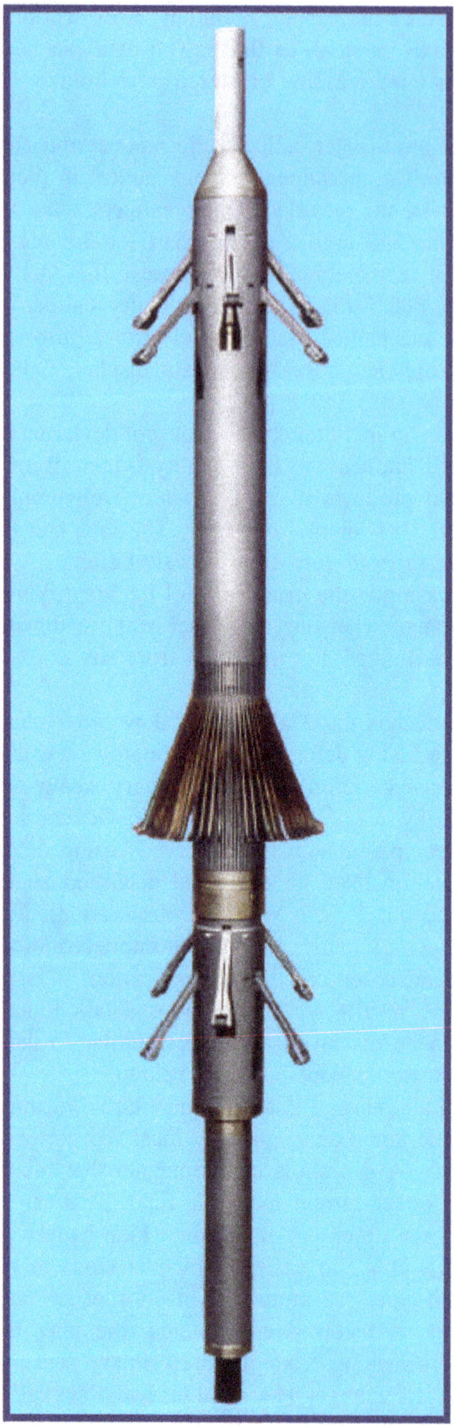

Fig. 3G.7—Multifingered caliper tool for deployment as a memory tool on slickline or as a surface-readout tool on monoconductor cable. This tool has 60 fingers, a 4-in. [100-mm] diameter, and a measurement range of 4.4 to 9.625 in. [114 to 245 mm]. It has a radial resolution of 0.005 in. [0.13 mm], a radial accuracy of ± 0.03 in. [0.75 mm], and a vertical resolution of 0.23 in. [5.84 mm] at a logging speed of 3,000 ft/hr [914 m/hr]. Pressure and temperature ratings are 15,000 psi [103 MPa] and 350°F [177°C], respectively. Note the tool centralizers. (Courtesy of Sondex.)

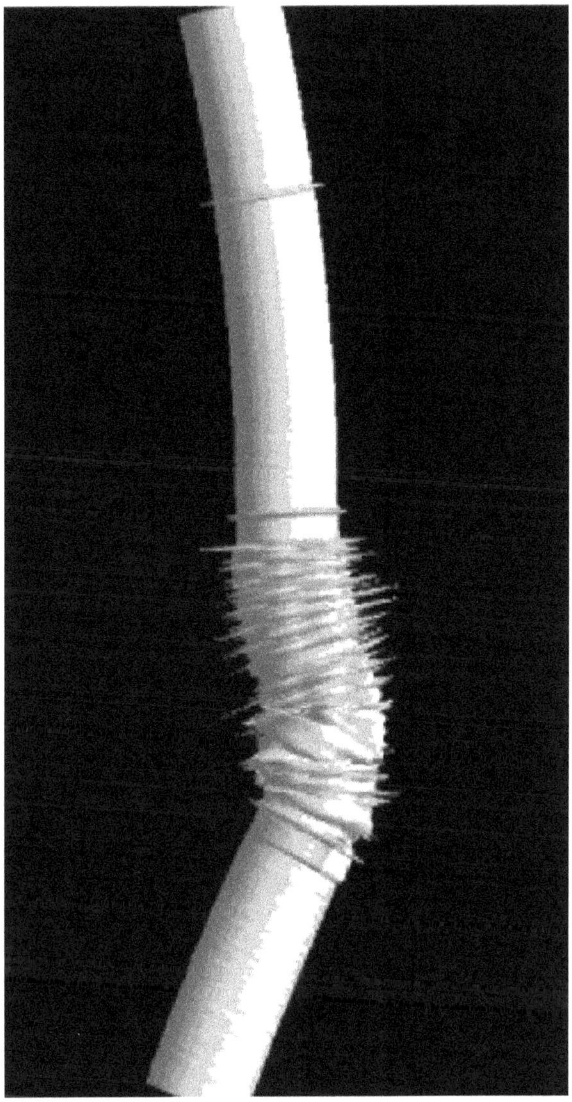

Fig. 3G.8—Digital image of casing deformation based on multifingered caliper data processed with C-FER Technologies' CalTran™ software. The "spikes" are indications of casing connections or perforations.[9]

The major advantage of flux-leakage tools is that they can identify localized casing defects such as corrosion patches, pits, and holes as small as 0.2 in. [5 mm] on both the inside and the outside of the pipe. A disadvantage is that the tool does not detect large areas of corrosion. It does not see nonmagnetic scale, which can degrade the sensor response. The tool is affected by changes in the electromagnetic properties of the casing. It is of limited accuracy, coverage, and resolution. The coil-sensor response is sensitive to logging speed, and this sensitivity makes quantitative interpretation more difficult.

Electromagnetic Phase-Shift Tools. The electromagnetic phase-shift technique provides an estimate of casing thickness across approximately 1 ft [300 mm] of casing length, so its spatial resolution is weaker than that of the first two methods. Electromagnetic phase-shift tools make measurements that are averages around the circumference of the pipe. They lack the localized investigative capability of flux-leakage tools and are best used to investigate larger-scale corrosion. Essentially, a transmitter coil generates a low-frequency alternating magnetic field, which couples to a receiver coil. It also induces eddy currents in the surrounding casing and formation. These eddy currents generate their own magnetic field, which is phase-shifted by the presence of casing. The phase-shifted field is superimposed on the transmitted field. This total field is detected by a receiver coil. The phase shift between the transmitted and received signals is related to the thickness, electrical conductivity, and magnetic permeability of the casing. If the last two are known, the casing thickness can be determined. Higher phase shifts indicate a higher casing thickness, all other things being equal. In practice, the electromagnetic properties of the casing can vary with composition, aging, and stress. To overcome this problem, modern tools comprise multiple sensor coils, which allow variations in the electromagnetic properties of the casing to be factored into the computation of casing thickness. Advantages are that the method is sensitive to large areas of corrosion and to gradual thinning of the casing. The sensors do not need to be in close proximity to the casing, so a single tool can examine a range of casing sizes. Disadvantages are the low spatial resolution and the lack of response to nonmagnetic scale. Moreover, the alternating-current magnet requires a relatively high power, which makes the tool difficult to deploy in memory mode.

Ultrasonic Tools. The ultrasonic method provides a full quantitative record of casing radius and thickness. The first ultrasonic casing-inspection tools were the borehole televiewers, but these only "saw" the inner casing surface and their use is now mainly in open hole (see Sec. 3G.3.2). Later tools had fixed ultrasonic transducers, but these were principally directed at cement evaluation, and they provided an incomplete coverage of casing-thickness measurements. This problem was overcome by a rotating ultrasonic transducer that was initially directed at cement evaluation (see Sec. 3G.2.5).

More recently, tools have been designed for a better spatial resolution.[10] Schlumberger's Ultrasonic Corrosion Imager (UCI™) was designed with a short-pulse 2-MHz transducer, 0.5 in. [12.5 mm] in diameter, focused at a distance of 2 in. [50 mm] from its front face. The higher-frequency measurement sharpened the spatial resolution so that internal pits of diameter 0.16 in. [4 mm] could be defined quantitatively. The velocity of sound in the borehole fluid is measured using a built-in reflector at a known offset while running into the hole. The wellsite computer calculates the internal radius from internal echo time and the measured fluid velocity. Downhole processing extracts the time difference between the internal and external echoes for an improved determination of casing thickness using the velocity of sound in steel. This information allows external casing defects to be identified. Azimuthal sampling interval is 2°. Vertical sampling interval in high-resolution mode is 0.2 in. [5 mm] at a logging speed of 425 ft/hr [130 m/hr]. The signal is attenuated by the borehole fluid. Best results are achieved with brine, oil, or very light drilling muds. **Fig. 3G.9** shows UCI images of 2D percentage metal loss and 3D views of casing integrity in a 5.5-in. [140-mm] saltwater-injection casing in Canada.[10]

Frisch and Mandal[11] described a "new generation of ultrasonic tools" for use in large-diameter casings. Their (Halliburton) tool uses two ultrasonic transducers, one of which rotates while the other is fixed for real-time measurements of borehole-fluid velocity. The tool operates in image mode or cased-hole mode. In image mode, the tool can be operated in open hole or in cased hole, where it examines only the inner casing surface. In cased-hole mode, it determines the inner radius and the casing thickness, so that defects on the outer casing can be discerned. Waveform processing allows the evaluation of cement bonding from the same logging run.

3G.2.5 Cement-Evaluation Logs. Conventional cement-bond logs (CBLs) comprise a pulsed transmitter and several receivers of acoustic energy positioned as a vertical array of transducers. The acoustic signal travels through borehole fluid, casing, cement, and the formation itself. The signal is received, processed, and displayed as a microseismogram. The recorded wave-

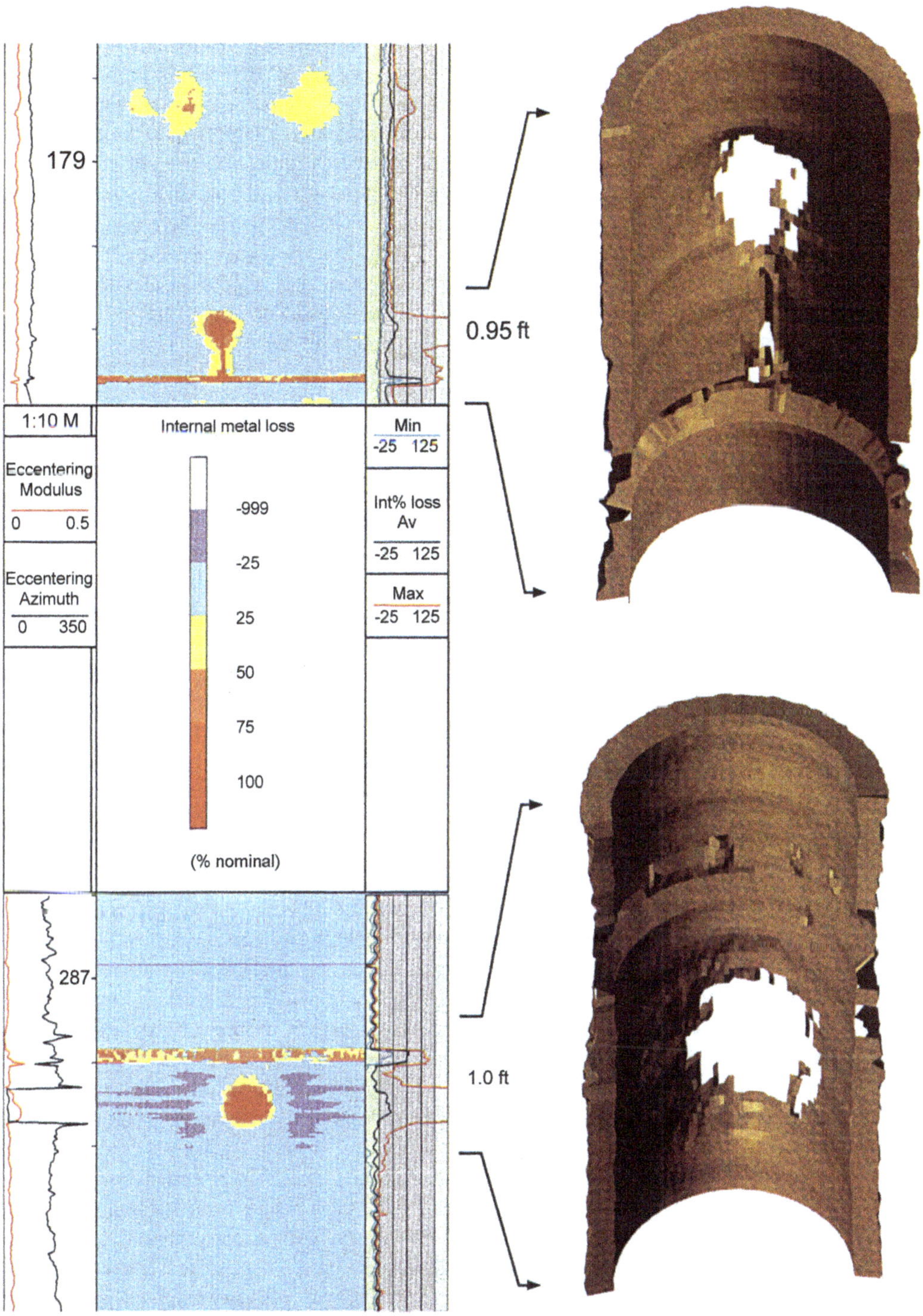

Fig. 3G.9—Example of casing inspection using the Ultrasonic Corrosion Imager (UCI™). The presentation includes digital 2D images of percentage metal loss, with good casing shown in light blue and holes indicated in red, together with 3D views of casing integrity. There are two holes in the 5.5-in. [140-mm] casing, each with a diameter of approximately 2 in. [50 mm]. In the upper image, note the deep groove from the casing hole down to the casing collar.[10] [Courtesy of the Soc. of Petrophysicists and Well Log Analysts (SPWLA)].

forms are presented together with the travel time and a casing-amplitude curve, which displays the amplitude of the acoustic signal that has traveled through the casing but not through the cement and formation. The waveform and amplitude data allow two bonds to be investigated. These are the bond between casing and cement and, to a lesser extent, that between cement and formation. A "straight" waveform display is traditionally interpreted to mean no cement bonding. Variations in the acoustic display are interpreted as indicating the presence of bonded cement.[12] These displays have been enhanced by the application of statistical variance processing to ultrasonic data.[13] CBLs clearly indicate the top of cement, where there is unbonded pipe, and they indicate where the pipe is well cemented (**Fig. 3G.10**). However, they are not reliable as indicators of hydraulic sealing by the cement, because they cannot detect small channels therein. Part of the problem is that conventional CBL transducer arrays are vertical, whereas bonding problems need to be investigated circumferentially.

Baker Atlas' Segmented Bond Tool (SBT™) uses six pads, on each of which there is a transducer arrangement of receivers and transmitters of acoustic energy.[14] The pads are in contact with the casing. Energy is transmitted at one pad and is received at an adjacent pad. The pad spacing is such that the first arrival is the wave that has passed through the casing. The rate of attenuation can be computed across each 60° segment of the casing circumference. A high rate of attenuation is indicative of a good cement bonding to the casing and an absence of channels within the cement. The method allows localized zones of good hydraulic seal to be identified in a way that is independent of borehole-fluid type. The bonding between cement and formation is investigated through a CBL-type receiver array for wave-train presentation (**Fig. 3G.11**).

Ultrasonic tools are superior to the acoustic CBLs, although they remain adversely affected by highly attenuating muds. They are often grouped as "cement evaluation tools." One of the earlier ultrasonic tools was actually called the Cement Evaluation Tool (CET™). This Schlumberger tool comprised an array of eight ultrasonic transducers that allowed a limited radial inspection of the casing and its annulus. The most recent tools have a single rotating transducer that incorporates both the source and receiver of ultrasonic energy. The tool has to be centered. The data for circumferential inspection of the casing, as described above, and for the evaluation of cement bonding are obtained on the same logging pass. Acoustic energy is reflected at interfaces that correspond to changes in acoustic impedance (the product of acoustic velocity and density). The first reflection is at the casing itself. The second reflection may be at the outside of the casing. If cement is bonded to the casing, there will be a strong reflection. If there is unset cement or water behind the casing, there will be a weak reflection. The received waveform is the sum of the reflected waveform from the original burst and the exponentially decaying waveform from the resonant energy that is trapped between the inner and outer edges of the casing. By analyzing the entire waveform, an acoustic-impedance map of the cement can be constructed. This map can indicate the presence of channels and their orientations.

Schlumberger's Ultrasonic Imager (USI™) is one such tool.[15] It operates from 200 to 700 Hz and provides a full high-resolution coverage of the casing and cement integrity. Channels as narrow as 1.2 in. [30 mm] can be detected. It is used with a conventional CBL tool. An interesting example of the complementary nature of these data has been presented by De Souza Padilha and Da Silva Araujo.[16] It deals with the problem of gas-contaminated cement, which has been a longstanding interpretation problem in the industry. Essentially, the CBL reads low-amplitude values in gas-contaminated cements. The USI cannot distinguish between gas-filled cement and fluids, but it can quantify the acoustic impedance of the cement. Therefore, the presence of gas-contaminated cement is indicated where the CBL reads low and the USI indicates fluids. If there is only gas behind the casing, the CBL reads high and the USI shows gas. The CBL and USI were used conjunctively to distinguish these cases. The application of statis-

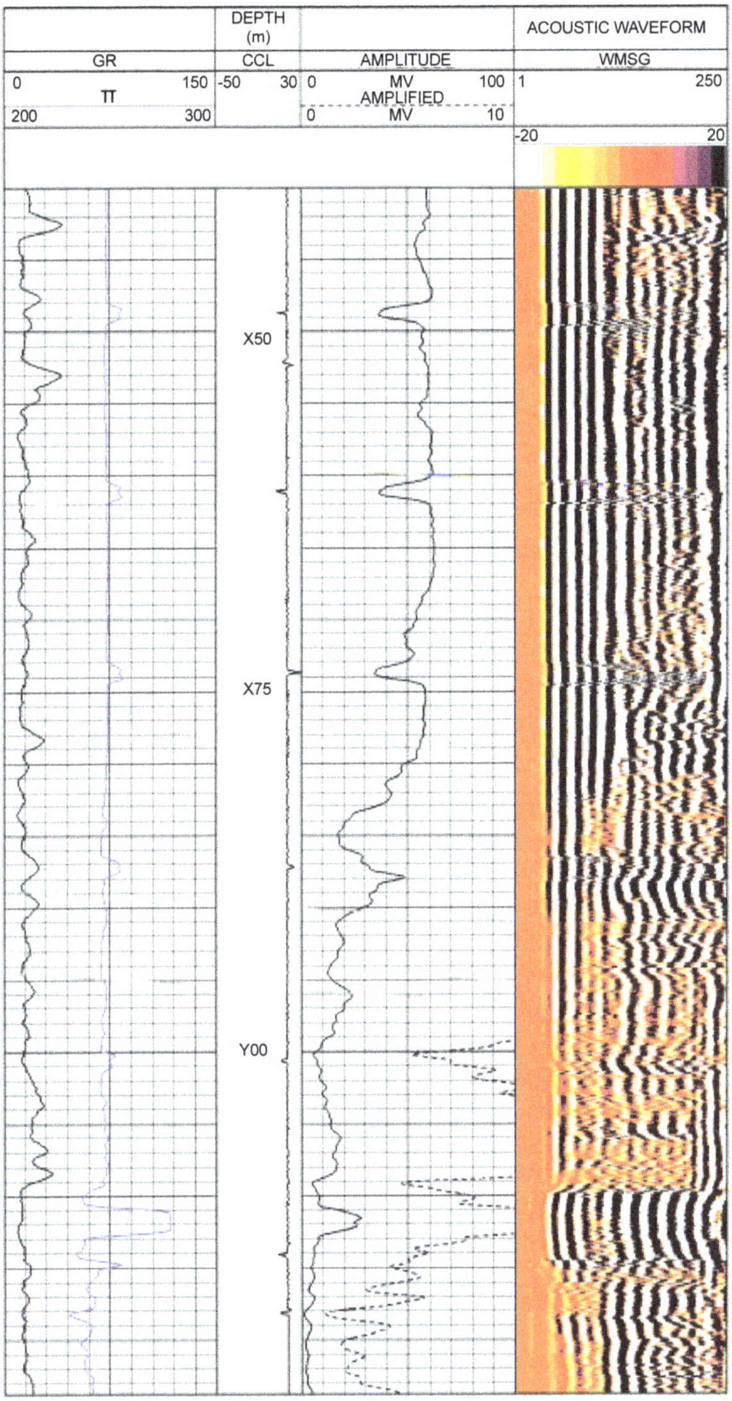

Fig. 3G.10—Example of a CBL. Track 1 contains the gamma ray (for correlation) and acoustic travel time (for quality control). Track 2 contains the amplitude curve and amplified amplitude, which indicates cement-to-casing bond. Track 3 contains the CBL waveform, which indicates cement-to-casing bond as well as cement-to-formation bond. Straight lines in the CBL waveform, along with high amplitude readings, indicate poor cement-to-casing bond. There is nearly free pipe above an apparent top of cement at a depth of approximately X80 depth units. At greater depths, the pipe is well bonded. (Courtesy of Halliburton.)

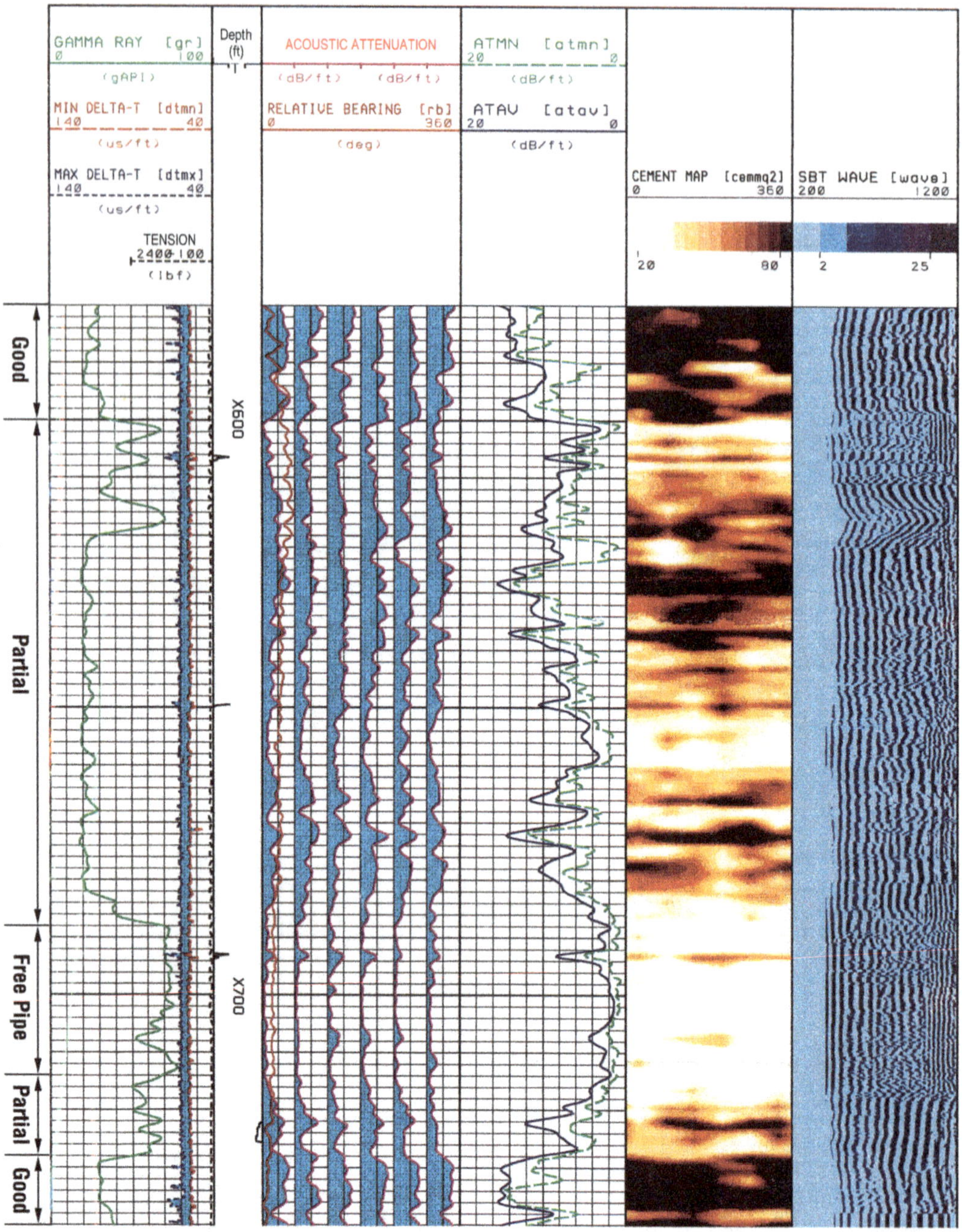

Fig. 3G.11—Example of cement evaluation using the Segmented Bond Tool (SBT™). Track 1 contains the gamma ray and two quality curves for pad contact with the borehole wall and for centralization, both of which are of high quality in this example. Track 2 contains the acoustic attenuation logs for the six pads. Track 3 shows the average and minimum attenuation at each sampling level. Track 4 presents a variable-attenuation log or cement map of the casing periphery vs. depth. Dark zones are the most strongly bonded. Track 5 is a CBL-type display. In this example, the partial bonding is sufficient to provide hydraulic isolation. There is poor cement condition between X688 and X714 depth units. Attempts to rectify this problem will be impeded by the hydraulic isolation above and below this interval. (Courtesy of Baker Atlas.)

tical variance processing to the conjunctive use of CBL and ultrasonic impedance data has led to an improved cement evaluation.[17] The CBL is also discussed in the chapter on Acoustic

Logging in this volume of the *Handbook*.

3G.2.6 Simultaneous Casing Inspection and Cement Evaluation. As indicated above, the ultrasonic tools can be operated to address two objectives concurrently: casing integrity and cement evaluation. A further example is Halliburton's Circumferential Acoustic Scanning Tool —Visualization version (CAST-V™), which allows separate or simultaneous casing inspection and cement evaluation.[18] The tool can operate in two modes: an image mode, whereby the scanner evaluates only the inner surface of the casing, or a cased-hole mode, whereby circumferential maps of casing thickness and acoustic impedance are used to assure casing integrity and to distinguish between fluids and cement in the annulus. **Figs. 3G.12 and 3G.13** show examples of CAST-V data displays. This tool can also operate in open hole as a formation imager (see Sec. 3G.3).

3G.3 Borehole Imaging

As introduced here, the term "borehole imaging" refers to those logging and data-processing methods that are used to produce centimeter-scale images of the borehole wall and the rocks that make it up. The context is, therefore, that of open hole, but some of the tools are closely related to their cased-hole equivalents. Borehole imaging has been one of the most rapidly advancing technologies in wireline well logging. The applications range from detailed reservoir description through reservoir performance to enhanced hydrocarbon recovery. Specific applications are fracture identification, analysis of small-scale sedimentological features, evaluation of net pay in thinly bedded formations, and the identification of breakouts (irregularities in the borehole wall that are aligned with the minimum horizontal stress and appear where stresses around the wellbore exceed the compressive strength of the rock).

The subject area can be classified into four parts: optical imaging, acoustic imaging, electrical imaging, and methods that draw on both acoustic and electrical imaging techniques using the same logging tool. Prensky[19] has provided an excellent review of this important subject.

3G.3.1 Optical Imaging. Downhole cameras were the first borehole-imaging devices. Today they furnish a true high-resolution color image of the wellbore. The principal drawback is that they require a transparent fluid in liquid-filled holes. Unless transparent fluid can be injected ahead of the lens, the method fails. This requirement has limited the application of downhole cameras. The other major historic limitation, the need to wait until the camera is recovered before the images can be seen, has fallen away with the introduction of digital systems.

The principal application of downhole video has been in air-filled holes in which acoustic and contact electrical images cannot be obtained. Most applications described in the literature are directed at fracture identification or casing inspection.

3G.3.2 Acoustic Imaging. Acoustic borehole-imaging devices are known as "borehole televiewers." They are mandrel tools and provide 100% coverage of the borehole wall. The first borehole televiewer, operating at a relatively high ultrasonic frequency of 1.35 MHz, was developed by Mobil Corp. in the late 1960s.[20,21] Since then, a succession of improvements have been made, principally through advances in digital instrumentation and computer-image enhancement. Modern tools contain a magnetometer to provide azimuthal information.

The borehole televiewer operates with pulsed acoustic energy so that it can image the borehole wall in the presence of opaque drilling muds. Short bursts of acoustic energy are emitted by a rotating transducer in pulse-echo mode. These travel through the drilling mud and undergo partial reflection at the borehole wall. Reflected pulses are received by the transducer. The amplitudes of the reflected pulses form the basis of the acoustic image of the borehole wall. These amplitudes are governed by several factors. The first is the shape of the borehole wall itself: irregularities cause the reflected energy to scatter so that a weaker reflected signal is

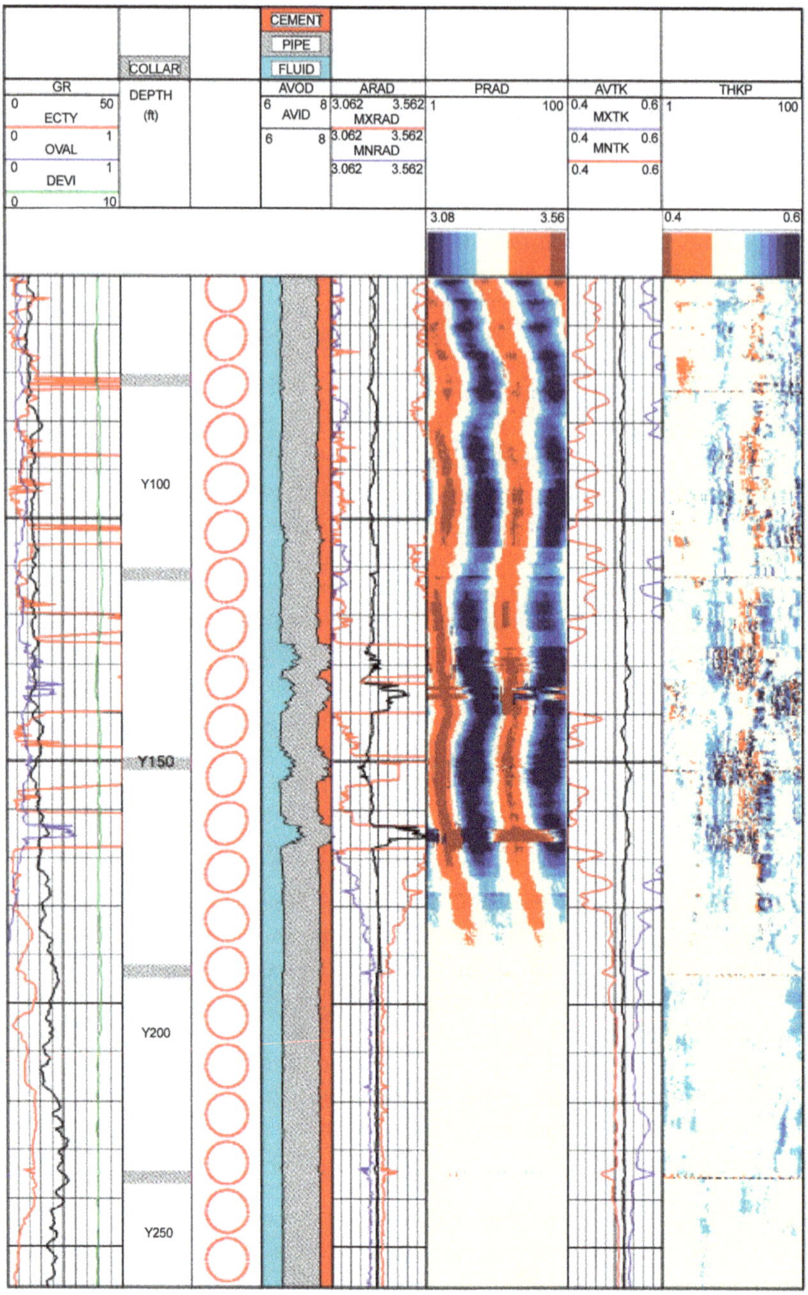

Fig. 3G.12—Example of casing inspection using the visualization version of the Circumferential Acoustic Scanning Tool (CAST-V™). The casing-evaluation presentation includes casing ovality, eccentricity, hole deviation, and gamma ray in Track 1. In this case, the eccentricity comprises both tool and casing eccentricity resulting from formation movement (salt flow). Track 2 shows a cross-sectional presentation of the pipe shape. Track 3 shows a cross section of the pipe wall. Track 4 provides the average, minimum, and maximum values of the pipe radius that is shown in Track 5. Track 6 provides the average, minimum, and maximum values of the pipe thickness that is the image shown in Track 7, where red indicates pipe thinning and blue indicates pipe thickening. (Courtesy of Halliburton.)

received by the transducer. Examples of these irregularities are fractures, vugs, and breakouts. Moreover, the reflected signal is degraded in elliptical and oval wellbores because of non-nor-

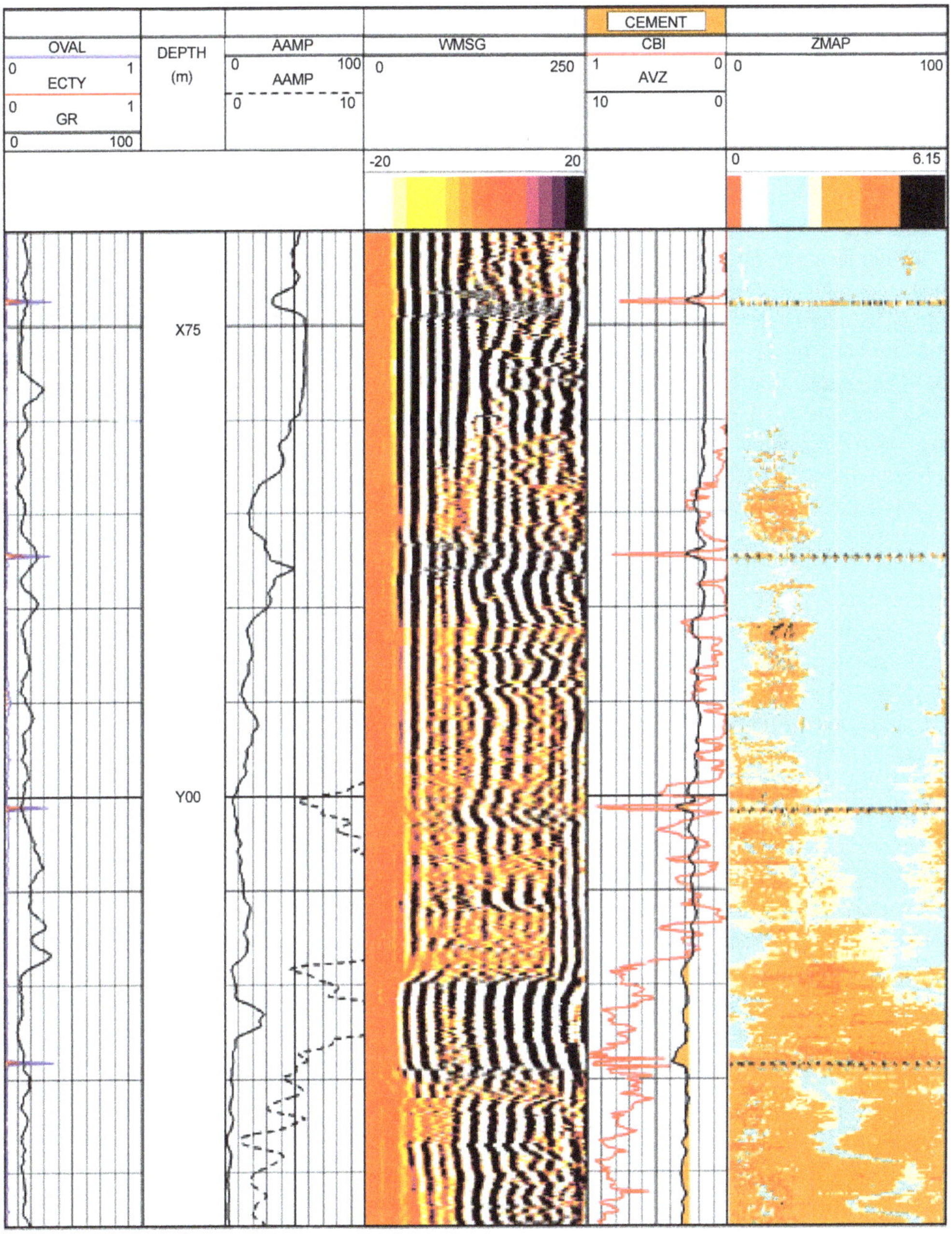

Fig. 3G.13—Example of cement evaluation using the visualization version of the Circumferential Acoustic Scanning Tool (CAST-V™). The data relate to an interval that overlaps with the conventional CBL in Fig. 3G.10. The cement-evaluation presentation includes casing ovality and tool eccentricity in Track 1. The conventional CBL output is shown in Tracks 2 and 3 as per Fig. 3G.10. Data from CAST-V are shown in Tracks 4 and 5. The image in Track 5 is an acoustic-impedance map from 0 to 360° (left to right) with 0° representing the high side of the hole. Track 4 contains the average impedance of the image in Track 5 and a cement-bond index (CBI) as a quick indication of the degree of bonding. Tracks 4 and 5 impart clarity to the interpretation of Fig. 3G.10 by more clearly showing no cement above X80 depth units, good cement below Y20 depth units and questionable bonding in between. (Courtesy of Halliburton.)

mal incidence. The second factor is the contrast in acoustic impedance between the drilling mud and the material that makes up the borehole wall. Acoustic impedance provides an acoustic measure of the relative firmness of the formations penetrated by the wellbore material and, thus, it has the capability of discriminating between different lithologies, with high acoustic impedance giving rise to high reflected amplitudes. Borehole televiewers work best where the borehole walls are smooth and the contrast in acoustic impedance is high. The third factor is the scattering or absorption of acoustic energy by particles in the drilling mud. This problem is more serious in heavily weighted muds, which are the most opaque acoustically, and it gives rise to a loss of image resolution.

The borehole televiewer can provide a 360° image in open or cased holes. It can operate in all downhole environments other than gas-filled holes. The travel time for the acoustic pulse depends on the distance between the transducer and the borehole wall, as well as the mud velocity. Modern televiewers allow some independent method of measuring the mud velocity. Thus, the borehole televiewer also operates as an acoustic caliper log. For best results, the tool should be centered, although correction algorithms have been developed for eccentered surveys.

An example of a modern ultrasonic imaging tool is Schlumberger's Ultrasonic Borehole Imager (UBI™), which is based on the cased-hole USI (see Sec. 3G.2.5) with two hardware modifications: a focused transducer was fitted for improved resolution, and an openhole centralizer was added.[22] The tool incorporates a rotating transducer within a subassembly. The size of the subassembly is selected on the basis of the diameter of the hole that is to be logged. The direction of rotation of the subassembly governs the orientation of the transducer. There are two positional modes: the standard measurement mode with the transducer facing the borehole wall (**Fig. 3G.14**) and the fluid-property mode with the transducer facing a target within the tool. In standard mode, the tool measures both amplitude and transit time at one of two frequencies, 250 or 500 kHz, with recommended logging speeds of 800 ft/hr [244 m/hr] and 400 ft/hr [122 m/hr], respectively, where logging speed is primarily determined by vertical sampling density and the rate of transducer rotation. The higher frequency allows a sharper image resolution of 0.2 in. [5 mm], but it is less effective in highly dispersive muds where the lower frequency should be used. The tool can also be used for investigating the geometry of the inner surface of casing where it is not desired to measure resonant ringing as an indicator of cement integrity.

Baker Atlas' Circumferential Borehole Imaging Log (CBIL™) has a similar range of capability but is rated to 20,000 psi [138 MPa] and 400°F [204°C]. Halliburton's Circumferential Acoustic Scanning Tool (CAST™) additionally offers simultaneous casing inspection and cement evaluation. It is rated to 20,000 psi [138 MPa] and 350°F [177°C], as is Schlumberger's UBI. Both of these tools predated the UBI.[23,24]

Data are usually presented as depth plots of enhanced images of amplitude and borehole radius. Applications include fracture detection, analysis of borehole stability, and identification of breakouts.

Fig. 3G.15 shows an example of breakout identification using an ultrasonic borehole televiewer.[25] The data presented are from the Cajon Pass scientific borehole in southeastern California. The aim was to investigate the orientation and magnitudes of in-situ stresses using borehole-image data. The televiewer has superseded multiarm dipmeter calipers for these applications. Although the caliper can reveal the orientation of breakouts, the tool provides little information about their size and, more generally, about the overall shape of the borehole wall. The ultrasonic televiewer can detect much smaller features than the multiarm caliper and can distinguish between features that are stress induced and those that are drilling artifacts.

3G.3.3 Electrical Imaging. Microresistivity imaging devices were developed as an advancement on dipmeter technology, which they have mostly superseded. Traditionally, they have required a conductive borehole fluid, but it will be seen later that this requirement has been obviated by oil-based-mud imaging tools. Originally, in the mid-1980s, they comprised two high-resolution pads with 27 button electrodes distributed azimuthally on each. This arrangement

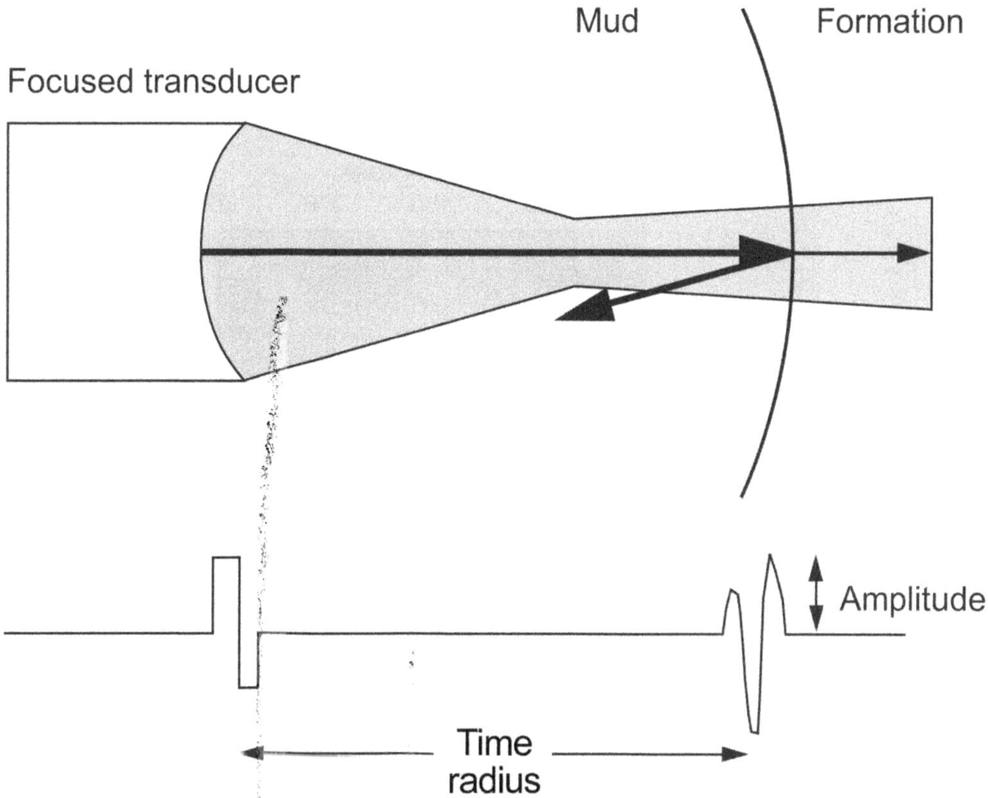

Fig. 3G.14—Principle of the Ultrasonic Borehole Imager (UBI™). The UBI measures reflection amplitude and radial distance using a direct measurement of mud velocity.[22]

provided a coverage of 20% of an 8.5-in. [216-mm] wellbore in a single pass.[26] This coverage was doubled by the development of a four-pad microresistivity imaging tool, each with 16 button electrodes arranged azimuthally in two rows of eight. The number of electrodes was limited by tool-transmission electronics.[27] Coverage was increased still further through the use of pads with flaps that opened to give a borehole-wall coverage of 80% in an 8-in. [203-mm] hole.[28] In another approach, the six-arm dipmeter evolved into a six-pad microresistivity imager,[29] Halliburton's Electrical Micro-Imaging tool (EMI™). Each pad contained 25 button electrodes also arranged azimuthally in two rows. With this arrangement, a 60% coverage was achieved in an 8-in. [203-mm] hole.

The measurement principle of the microresistivity imaging devices is straightforward. The pads and flaps contain an array of button electrodes at constant potential (**Fig. 3G.16**). An applied voltage causes an alternating current to flow from each electrode into the formation and then to be received at a return electrode on the upper part of the tool. The microelectrodes respond to current density, which is related to localized formation resistivity. The tool, therefore, has a high-resolution capability in measuring variations from button to button. The resistivity of the interval between the button-electrode array and the return electrode gives rise to a low-resolution capability in the form of a background signal. The tool does not provide an absolute measurement of formation resistivity but rather a record of changes in resistivity. The resolution of electrical microimaging tools is governed by the size of the buttons, usually a fraction of an inch. In theory, any feature that is as large as the buttons will be resolved. If it is smaller, it might still be detected. The tools can be run as dipmeters.

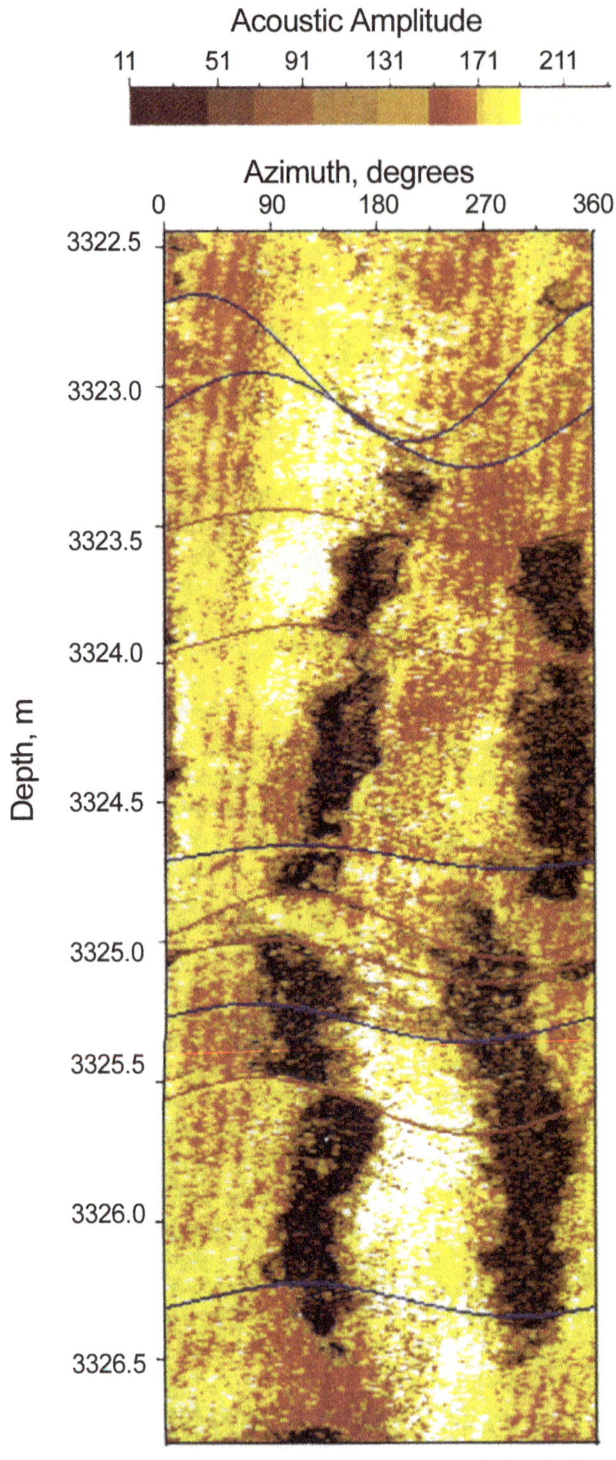

Fig. 3G.15—Example of breakout detection using an ultrasonic borehole televiewer. Breakouts are indicated by the low acoustic amplitude of the reflected signal, shown here as darker areas. The breakouts are rotated because of a drilling-induced slippage of localized faults.[25] (Courtesy of SPWLA.)

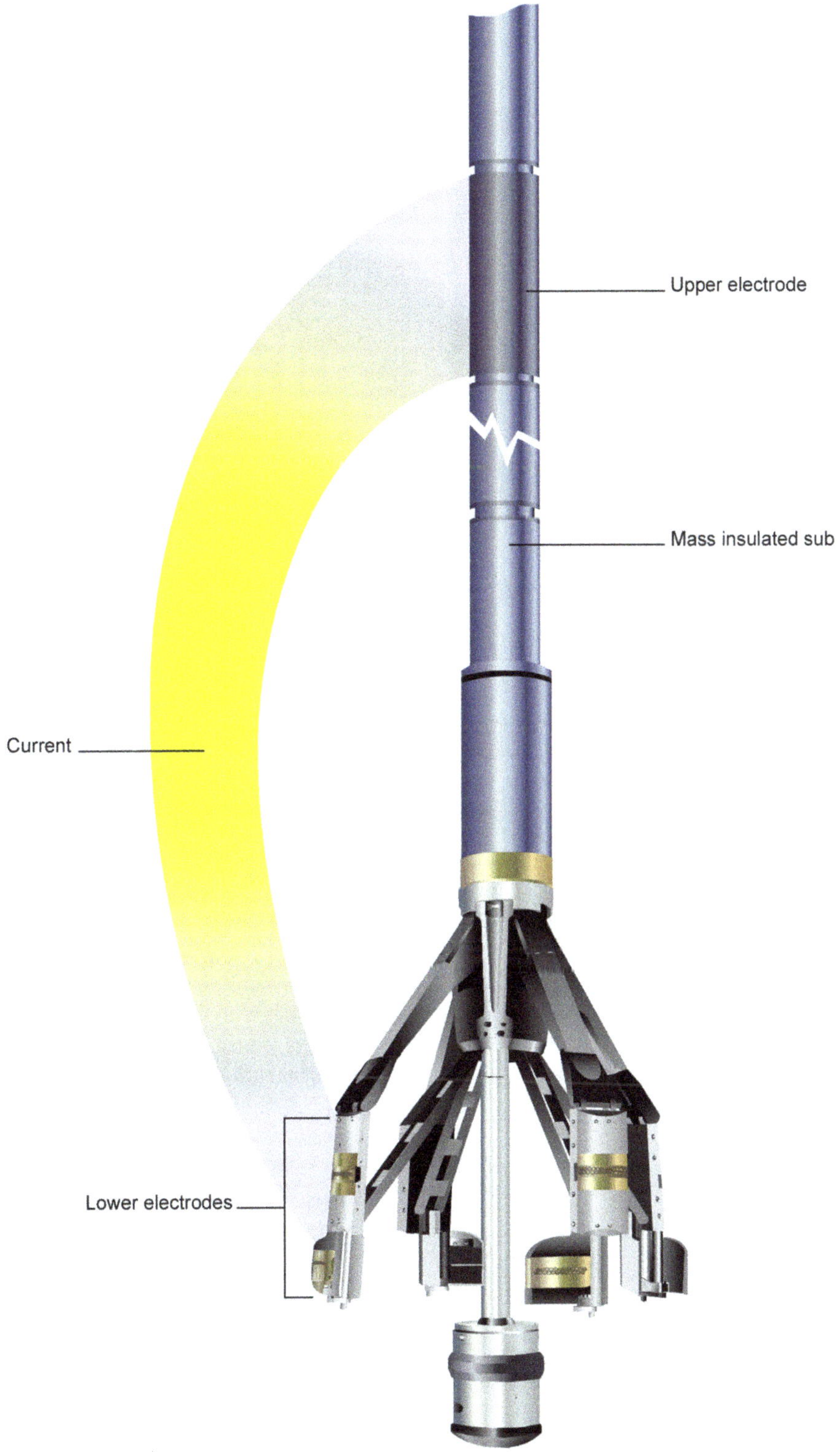

Fig. 3G.16—Measurement principle of microresistivity imaging devices illustrated by Schlumberger's Formation MicroImager (FMI™). (Courtesy of Schlumberger.)

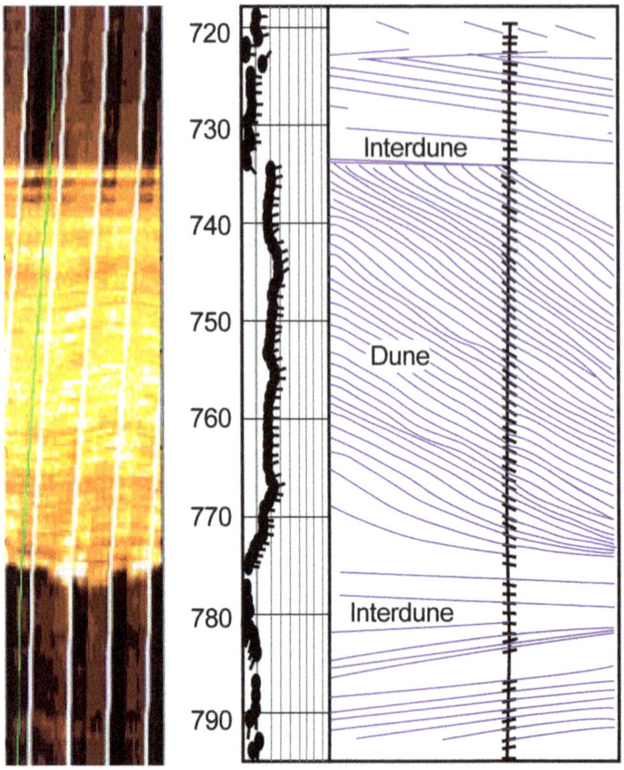

Fig. 3G.17—Recognition of sedimentary and structural features in microresistivity images. These Formation MicroImager (FMI™) images have been used to generate the dip information in Track 2. The combination of FMI images and dip data clearly differentiates the eolian and interdune sands in this 8.5-in. [216-mm] borehole. (Courtesy of Schlumberger.)

Data are presented as orientated, juxtaposed pad outputs whereby the cylindrical surface of the borehole wall is flattened out. This has the effect of distorting quasiplanar features such as dipping layers or fractures, which appear as sinusoidal in the data display. **Fig. 3G.17** shows a typical data display and identifies some of the key features.

Electrical microimaging tools have proved superior to the ultrasonic televiewers in the identification of sedimentary characteristics and some structural features such as natural fractures in sedimentary rocks. They are especially useful for net-sand definition in thinly laminated fluvial and turbidite depositional environments.

There are several microimaging tools available, each with similar capability. For example, Schlumberger's fullbore FMI™ has two horizontally offset rows of 24 button electrodes on each pad and each flap, making a total of 192 electrodes (**Fig. 3G.18**). The buttons have a diameter of 0.2 in. [5 mm], which determines the intrinsic spatial resolution of the tool. However, features as small as 50-μ fluid-filled fractures can be detected (but not fully resolved). Current is focused into the formation, where a depth of investigation of several tens of centimeters is claimed. However, the image probably relates to a depth of investigation of no more than 0.8 in. [20 mm]. The high-resolution image is normalized with respect to the low-resolution part of the signal or to another resistivity logging tool.

The conventional microresistivity imaging devices require a conductive mud in which to function. However, drilling with oil-based or synthetic muds has increased because of the improved drilling efficiency and greater borehole stability relative to water-based muds. Rather than have to change out the mud specifically for a microresistivity imaging survey, two other

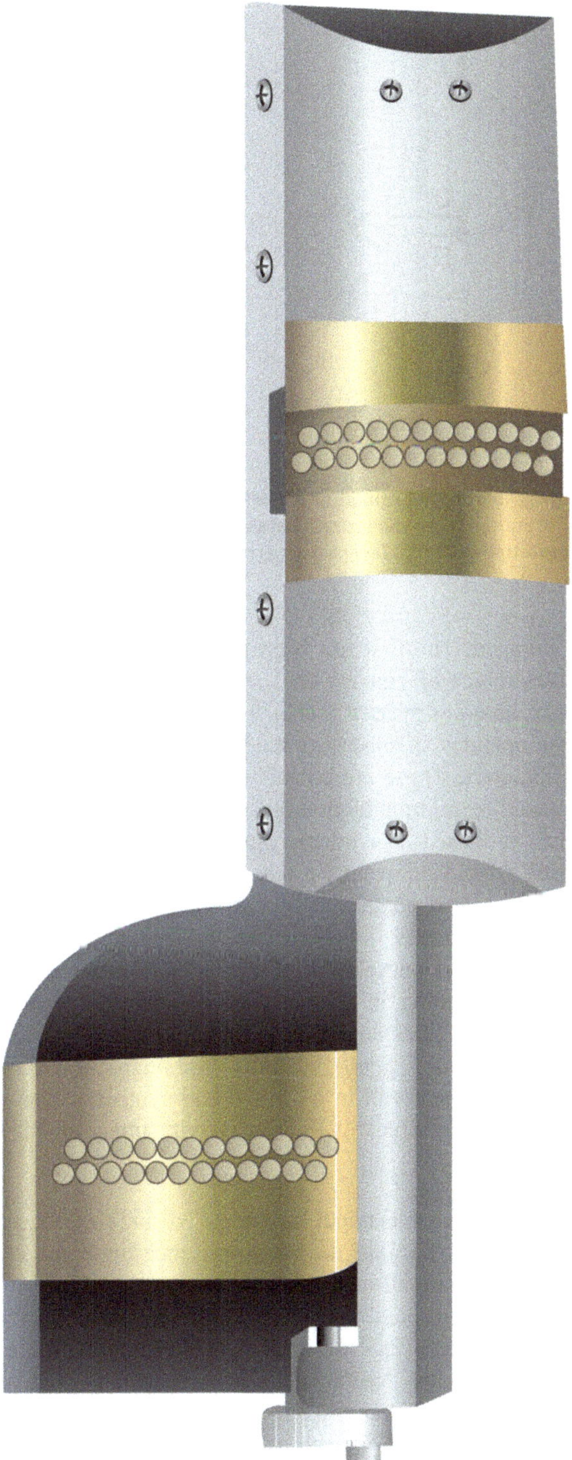

Fig. 3G.18—The Formation MicroImager (FMI™) pad and flap assembly with horizontally offset rows of electrode buttons. (Courtesy of Schlumberger.)

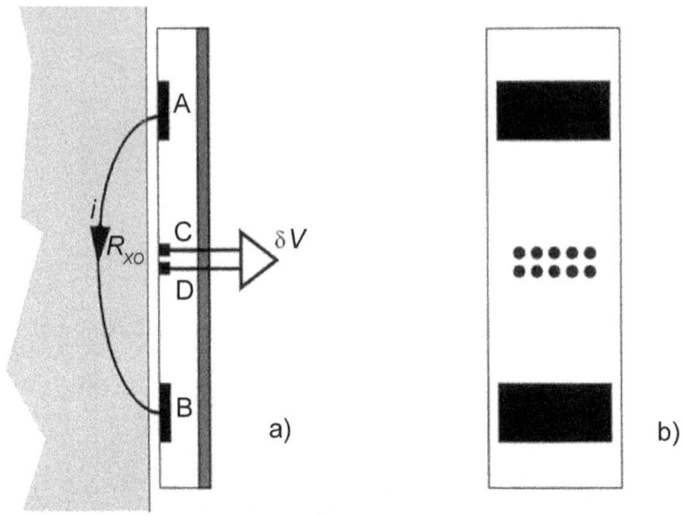

Fig. 3G.19—Principle of the Oil-Base MicroImager (OBMI™). A current, *i*, is applied between electrodes A and B. The potential difference, *δV*, is measured between electrodes C and D. An apparent formation resistivity, R_{xo}, is calculated using Ohm's law and an array geometry factor.[32] (Courtesy of SPWLA.)

approaches have been pursued. The first has been to develop a new synthetic mud that retains all the stabilizing characteristics of conventional synthetic muds but is sufficiently conductive to permit microresistivity imaging measurements.[30] The second has been to develop an electrical imaging device that operates in oil-based muds. This problem was addressed by the so-called oil-based mud dipmeters. These are conventional four-arm dipmeters for which the four microelectrodes are replaced by microinduction sensors.[31] More recently, contact resistivity methods have been applied in oil-based or synthetic muds.[32] Schlumberger's Oil-Base MicroImager (OBMI™) uses four pads positioned at 90° to one another to achieve a 32% coverage of an 8-in. [203-mm] wellbore. Each pad contains two current electrodes and a set of five pairs of closely spaced potential electrodes positioned centrally between the current electrodes (**Fig. 3G.19**). The arrangement is reminiscent of the Schlumberger electrode array that is still used for surface resistivity sounding in geoelectrical prospecting. However, in this downhole case, the aperture of the sensor gives an intrinsic spatial resolution of 0.4 in. [10 mm] with a nominal depth of investigation of 3.5 in. [90 mm]. Although the OBMI tool is sensitive to borehole rugosity, it has performed well in oil-based and synthetic muds for which water content lies between 1 and 30%. Examples of microresistivity image displays are shown in **Figs. 3G.20 and 3G.21**.

3G.3.4 Conjunctive Acoustic and Electrical Imaging. To some extent, the ultrasonic and electrical images are complementary because the ultrasonic measurements are influenced more by rock properties, whereas the electrical measurements respond primarily to fluid properties. Another difference is that the ultrasonic image covers 360°, whereas the electrical image is somewhat less than 80% of the surface of an 8-in. [203-mm] wellbore. Ultrasonic measurements can be made using the same tool in all types of drilling mud, and this can facilitate interwell comparisons. On the other hand, most microresistivity imaging devices require a water-based mud; otherwise, an alternative tool, such as the OBMI, has to be used.

These differences can be accommodated through the combined use of electrical and acoustic imaging. As an example, Baker Atlas' Simultaneous Acoustic and Resistivity Imager (STAR™) uses a combination of a CBIL and a six-pad resistivity imager with 12 electrodes per pad. The tool delivers a more complete data set than is achievable using either of the com-

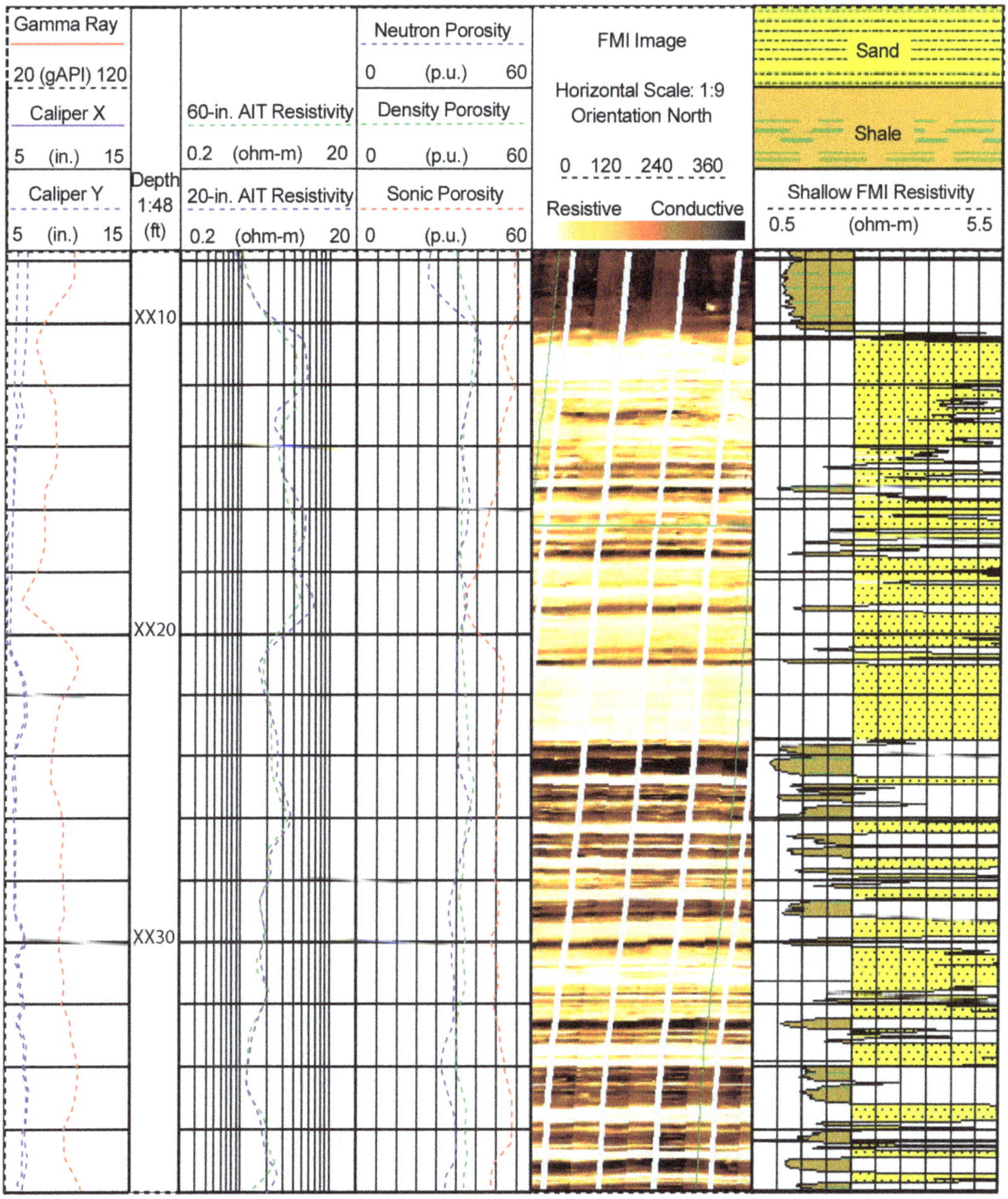

Fig. 3G.20—Example of the Formation MicroImager (FMI™) run in highly laminated sediments. The FMI tool is able to detect laminations as thin as 0.2 in. [5 mm]. In contrast, note the undiagnostic smoothed form of the conventional array induction logs around depth XX30 ft in Track 2. Microresistivity tools are able to detect pay in places where conventional log analysis might overlook it. Note the more complete description of borehole geometry afforded by the X and Y calipers in Track 1 (see Sec. 3G.2.2). (Courtesy of Schlumberger.)

ponents separately. The combined tool is 86 ft [26.2 m] in length with a diameter of 5.70 in. [145 mm]. It is rated to 20,000 psi [138 MPa] and 350°F [177°C].

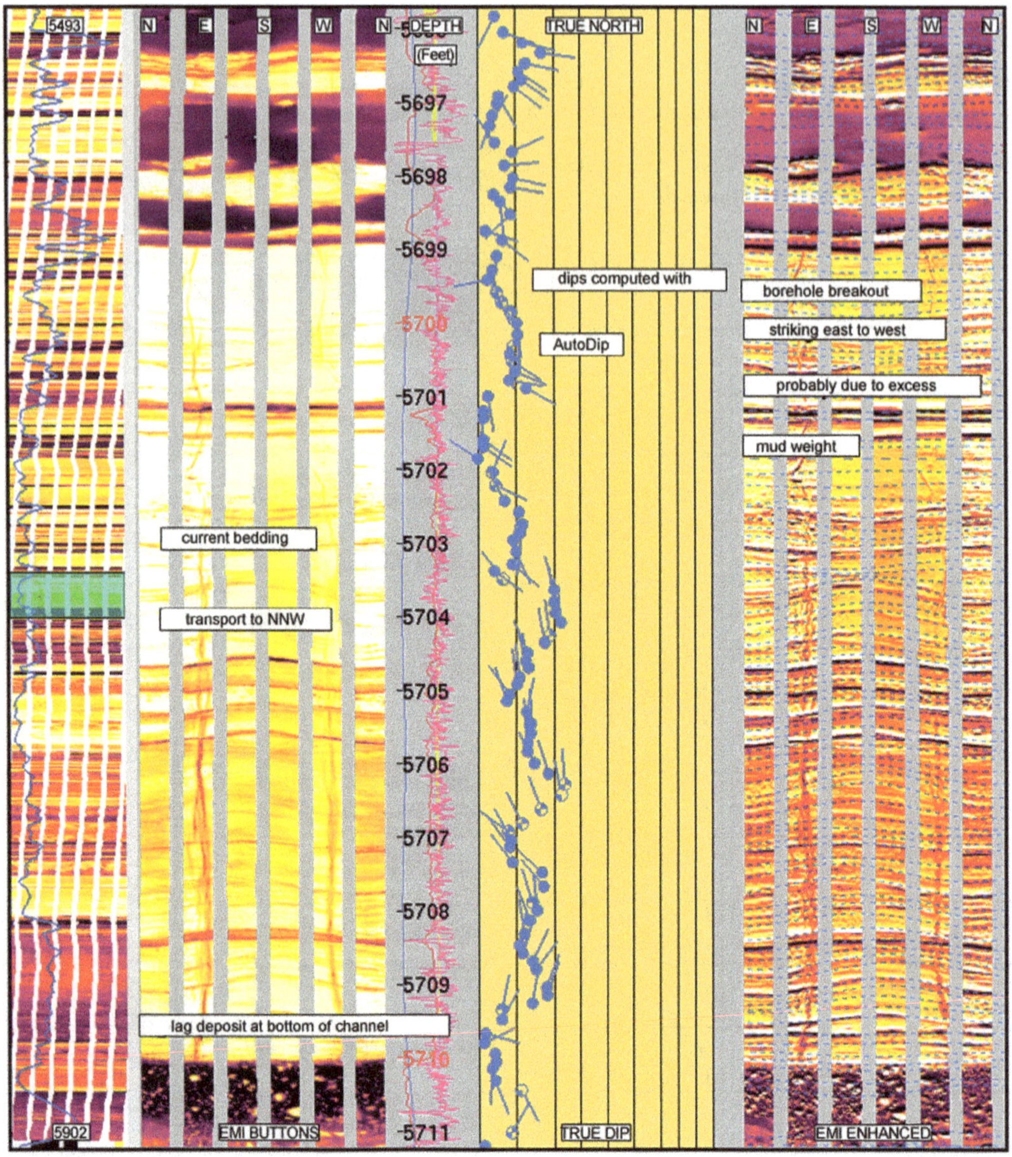

Fig. 3G.21—Example of an electrical microimage using the six-arm Electrical Micro-Imaging Tool (EMI™). Static (Track 2) and dynamic (Track 5) image enhancement has revealed a laminated sand/shale sequence and delivered computed dips (Track 4) of the sedimentary strata. The enhanced images also reveal drilling-induced fractures, which cut vertically across the bedding as sensed by Pads 2 and 5. (Courtesy of Halliburton.)

3G.4 Natural Field Methods

3G.4.1 Borehole Gravimetry. Borehole gravity was pioneered by Smith[33] and then applied to problems of reservoir evaluation by McCulloh *et al.*[34] The borehole gravity meter or gravimeter responds to variations in density. Modern instruments sense a rock volume that is approximately the same as that investigated by deep resistivity tools. Unlike the shallower-sensing density log, the borehole gravimeter is insensitive to wellbore conditions such as rugosity and the presence of casing. Its principal applications are (through-casing) time-lapse monitoring of saturations/fluid contacts in gas reservoirs and downhole calibration of surface geophysical mapping of geological structures.

Theory. Any two masses, m_i and m_j, separated by a distance, r, experience a gravitational force of attraction, f, which is expressed as

$$f = Gm_i m_j / r^2, \qquad (3G.13)$$

where G is the universal gravitational constant (6.6726×10^{-8} cgs units). More specifically, a mass, m, on the surface of the Earth would experience a gravitational force given by

$$f = GMm / R^2, \qquad (3G.14)$$

where M is the mass of the Earth, R is its radius, and an acceleration due to gravity, g, is given by

$$g = GM / R^2 . \qquad (3G.15)$$

Because the Earth is a rotating oblate spheroid, the quantity g at mean sea level varies with latitude, and it must be corrected for tidal effects. The unit of g is the Gal [1 cm/s^2]. Surface gravity surveys use the milliGal as the preferred unit. Borehole gravity surveys often use the microGal. The acceleration due to gravity, or just "gravity," is measured with a gravimeter.

Measurement. A borehole gravimeter follows the same principles of operation as a surface gravity meter. It is essentially a very sensitive spring balance. The weight of a horizontal hinged beam with a small mass attached to its free end is balanced by a combination of the tension in a compensating spring and an electrostatic force (**Fig. 3G.22**). When the acceleration due to gravity increases, the weight of the beam increases, and there is a greater tension in the spring. The spring tension is directly related to the acceleration due to gravity. It is controlled by an adjusting screw for which the number of turns is calibrated in gravitational units. The electrostatic force brings the beam to a horizontal position for reading purposes. It, too, is calibrated in gravitational units. The gravity reading is the difference between the spring tension and the electrostatic force: a tidal gravity correction has to be applied. In this way, differences in gravity can be measured between two places (e.g., between two depth locations in a borehole). Downhole measurements are made by occupying selected gravity stations. They are not continuous measurements with depth.

It can be shown[35] that the difference in gravity, Δg (mGal), between two locations at the top and bottom of an infinite horizontal reservoir layer penetrated by a vertical well is related to the density, ρ (g/cm^3), and thickness, h_m (m), of that layer by the expression:

$$\Delta g = \left[F - \left(4\pi\rho G \times 10^5\right)\right]h_m, \qquad (3G.16)$$

where F is the vertical gradient of gravity (mGal/m), and G is in cgs units. Eq. 3G.16 can be solved for the layer density so that

$$\rho = \frac{\left[F - \left(\Delta g / h_m\right)\right]}{4\pi G \times 10^5} . \qquad (3G.17)$$

The gradient, F (mGal/m), is a function of latitude, λ (degrees), and elevation, h (m), as per the Intl. Gravity Formula of 1967, as follows:

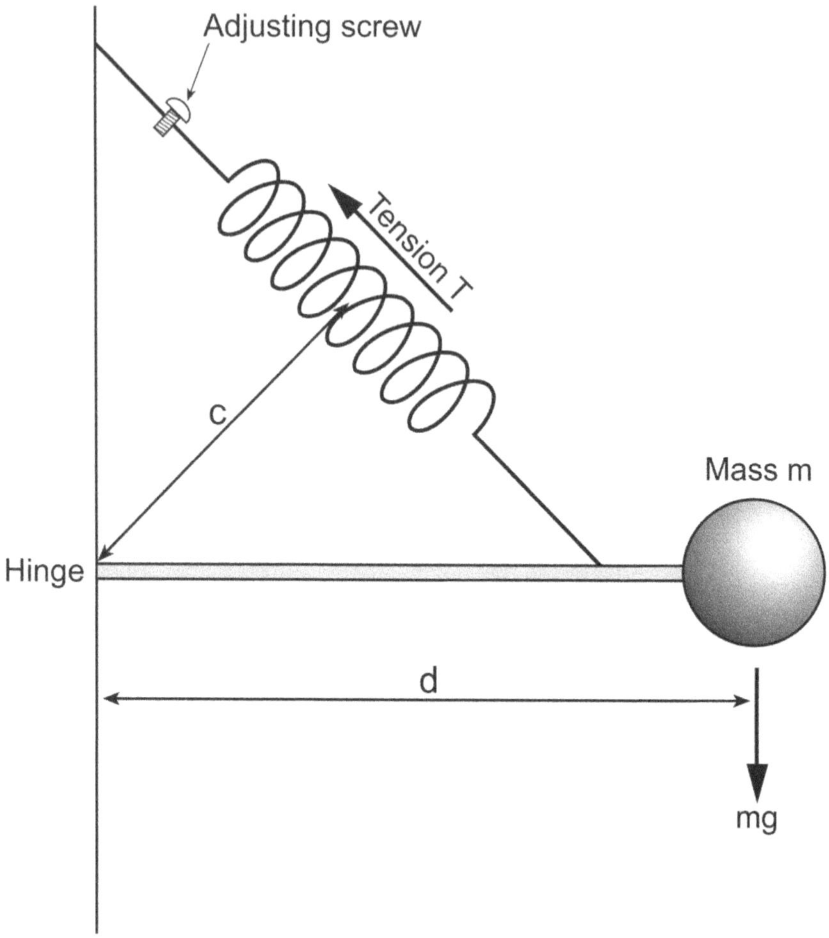

Fig. 3G.22—Principle of measurement of the borehole gravimeter. At equilibrium, the moments of force about the hinge are equal, so that *cT = mgd*. Because *d, m,* and *c* are design constants, *g* can be calculated if *T* is known.

$$F = 0.308768 - 0.000440\sin^2\lambda - 0.0000001442h \; . \quad (3G.18)$$

By substituting for F in Eq. 3G.17, we have

$$\rho = 3.6827 - 0.00524\sin^2\lambda - 0.00000172\bar{h} - 11.926(\Delta g / h_m), \quad (3G.19)$$

where $\bar{h}$ is the mean elevation of the layer (m), and $\Delta g/h_m$ is in mGal/m. Eq. 3G.19 is that most commonly used for deriving density from borehole gravity measurements. If the borehole is deviated at an angle, α, the measured depth interval has to be converted to a true stratigraphic thickness using a specific form of Eq. 3G.7 for zero dip. Corrections are needed where the model of an infinite layer breaks down because of the presence of structural discontinuities away from the wellbore. Modern borehole gravimeters can detect gravity differences of a few microGals.

The borehole gravity meter delivers an interval density. It is the only tool that can furnish through-casing density. Where the layer is heterogeneous, the computed density is an average or apparent density. The error in density is a function of the layer thickness. With a LaCoste-Romberg borehole gravimeter, a single measurement of gravity above and below a layer of thickness 6.6 ft [2 m] should result in an error in apparent density of approximately ± 0.025 g/cm^3. This expected error can be reduced through repeat measurements and by selecting a larger depth interval. Turning this around, the spatial resolution of a borehole gravimeter is governed by the accuracy to which density is required. For example, an accuracy of ±0.01 g/cm^3 would be achieved through three measurements of gravity at the top and at the base of the target layer, provided that the latter is at least 9 ft [2.7 m] thick.

Borehole gravity tools have different sizes for different hole conditions. For example, the EDCON tools range in diameter from 3.875 in. [98 mm] for low-temperature (110°C), low-pressure (8,000 psi) applications to 5.25 in. [133 mm] for high-temperature (204°C), high-pressure (20,000 psi) applications. The temperature range can be extended to 260°C with special ring seals. Because of the tool size, there are limits on the deviation of boreholes in which it can be deployed. The measurement stations are located relative to other logging runs by using the gamma log and the CCL (see Sec. 3G.2.3). The depth of investigation within a homogeneous layer is governed by the contrast between the mud filtrate and formation fluids. It is typically more than 23 ft [7 m]. A larger station spacing, h_m, will not increase this range. It will merely reduce the ability of the tool to see near-well density anomalies. Like surface gravity meters, the tool suffers from drift (of the spring tension), which makes accurate calibration difficult.

Application. Key thrusts in reservoir evaluation have been the sensing of vuggy, fractured, and heterogeneous reservoirs, in which a deep-sensing porosity measurement is needed to complement the volumetric-sensing capability of deep resistivity logs.[36,37] More recent applications have been directed at through-casing monitoring of gas saturations.[38,39] In this respect, it is noteworthy that the larger volumes sensed by the borehole gravity meter, relative to conventional density logs, are more closely associated with the simulator grid scale. Moreover, time-lapse gravity measurements are not degraded by structural anomalies.

As an example, time-lapse borehole gravimetry has been used to determine the residual oil saturation to gas within the oil rim of the onshore Rabi field in Gabon.[35] The reservoir comprises clean, coarse-grained sands with high-salinity formation water. The required accuracy for residual oil saturation was ± 10 saturation units. A baseline gravity survey was run over an undepleted oil-bearing interval near the gas/oil contact (GOC) (Run 1). As the reservoir is depleted, the GOC moves down and the oil saturation decreases toward its residual value. A second gravity survey (Run 2) allowed the change in gas saturation, ΔS_g, to be calculated from the change in measured density, $\Delta\rho_b$, porosity, ϕ, and the densities of oil, ρ_o, and gas, ρ_g.

$$\Delta S_g = \Delta\rho_b / \phi\left(\rho_g - \rho_o\right) . \qquad (3G.20)$$

Once ΔS_g was known, the oil saturation could be calculated, assuming no change in connate-water saturation.

Three surveys were undertaken twelve months apart (**Fig. 3G.23**). All measurements were made in a data-dedicated borehole. Gravity was measured four to six times at each station, with station intervals as low as 3.3 ft [1 m]. Stations were reoccupied with a shuttle-based system for enhanced depth control. The overall accuracy of the density difference in Eq. 3G.20 was 0.015 g/cm^3. This accuracy corresponds to an accuracy of 0.7 μGal on the station-specific readings and an accuracy of 1.0 μGal on the gravity difference. The residual oil saturation was determined as 15±10 saturation units. Of this uncertainty, eight saturation units could be ascribed to the borehole-gravity measurements and two saturation units to uncertainties in porosity and connate-water saturation. This study set new objectives and standards for borehole gravimetry.

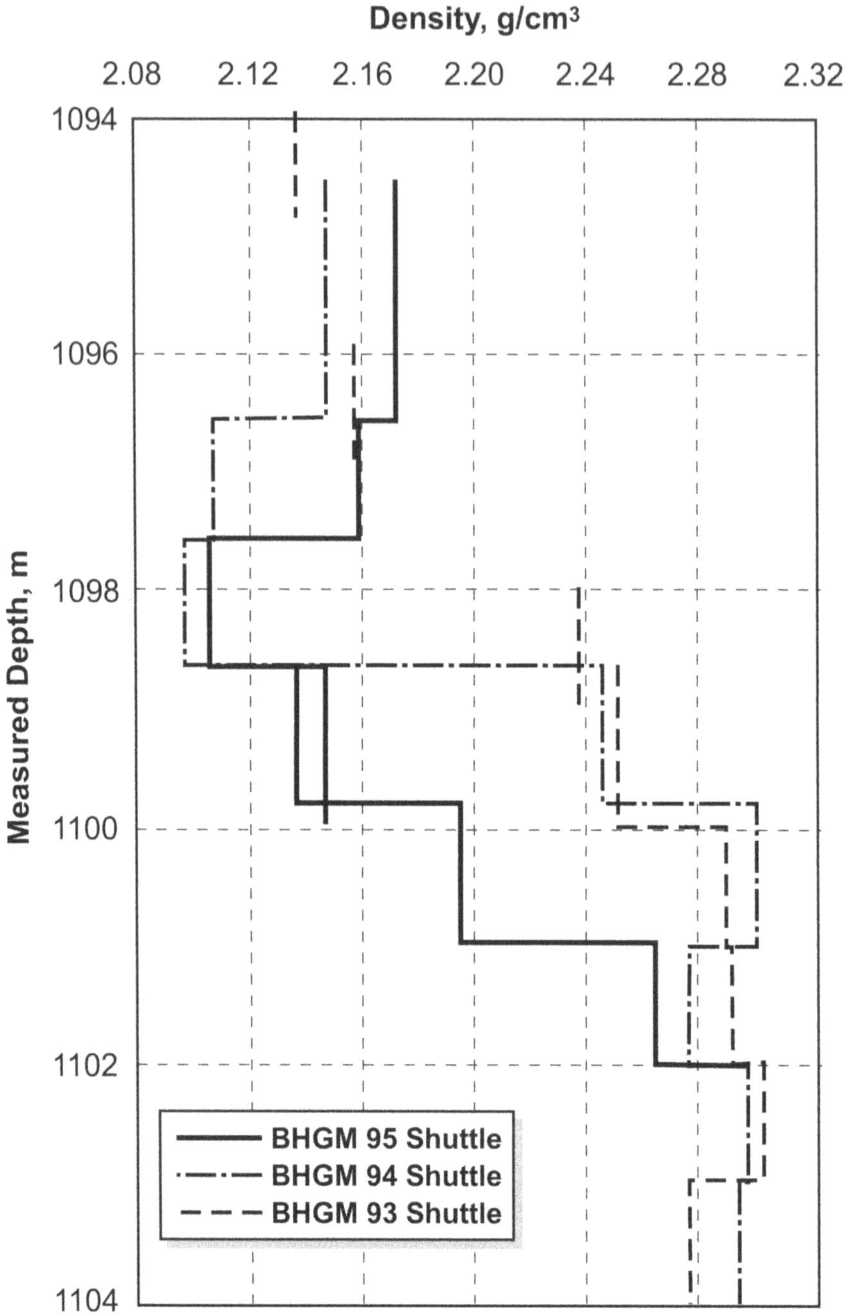

Fig. 3G.23—Time-lapse borehole gravimeter data around the GOC in the Rabi field, Gabon. Note the excellent repeatability of the 1995 shuttle data at approximately 1099 m measured depth. The density reduction caused by GOC movement is apparent between 1097 and 1101 m measured depth. The data recorded between 1098.5 and 1101 m measured depth were used to assess the remaining oil saturation to gas.[35]

3G.4.2 Downhole Magnetics. Downhole magnetic surveys have been most commonly applied in highly magnetized igneous rocks, which have usually been studied within pure geoscience, especially beneath the ocean floor. These rocks preserve the direction of the Earth's field at the time of their formation (i.e., the prevailing magnetic field is "frozen" in the rocks as they solidify, giving them a strong natural remnant magnetization). A primary application has been to identify points in time at which the Earth's magnetic field has undergone a polarity reversal. These reversals have been dated globally (e.g., isotopically in the case of volcanic series or by correlation with biostratigraphy in the case of volcaniclastics) and have given rise to a geomagnetic polarity time scale (GPTS) that is based on laboratory measurements. This, in turn, has allowed dates to be assigned to a given magnetozone that is bounded by reversal phenomena. It has been possible to recognize these reversals through downhole measurements and, therefore, to date the rocks accordingly.

Sedimentary rocks have much weaker remnant magnetizations than igneous sequences, and it has been much more difficult to investigate their magnetic character. However, recent advances in instrumentation have led to progress in downhole magnetic measurements of sedimentary strata.[40]

Theory. The following magnetic theory is extracted from Lalanne *et al.*[41] The magnetic field measured downhole has three parts: the Earth's magnetic field of the present day; the field that is induced in the rocks by the prevailing Earth's field; and the remnant magnetic field, which is the preservation in the rocks of a paleomagnetic field. The effect of the Earth's magnetic field can be accommodated during logging by extrapolating downhole the measurements made by a surface magnetometer that records diurnal variations in the Earth's field and allows the downhole data to be corrected for these variations where they are significant. The induced field is proportional to the magnetic susceptibility of the rock, which is governed by (ferro-magnetic) mineralogy and fluid composition. The remnant magnetic field adopts the direction of the Earth's field at the time that the rock was forming. For sediments, it is most pronounced in clays.

A measurement of magnetic induction or field strength, B_T, can be written as

$$B_T = B_o + B_i + B_r, \quad \text{(3G.21)}$$

where B_o is the magnetic induction associated with the present Earth's field, B_i is the magnetic induction caused by the field induced in the rock, and B_r is the magnetic induction caused by the remnant field. Magnetic induction is measured in units of nanoTesla (nT). It is a measure of field strength expressed in terms of the field's ability to induce magnetization. Typically, B_i and B_r are no more than a few tens of nanoTesla, and they have to be measured against a prevailing Earth's field that is a thousand times greater. Therefore, the exercise becomes very much one of analyzing residuals. For this reason, the prevailing Earth's field, B_o, is removed from the value of B_T, which then becomes a "net" field B_t.

$$B_t = B_i + B_r \,. \quad \text{(3G.22)}$$

The induced magnetic field, B_i (nT), is proportional to the magnetic susceptibility, χ (units SI), of the rock:

$$B_i = \mu_o H_T \chi \times 10^6, \quad \text{(3G.23)}$$

where μ_o is the magnetic permeability of the void ($4\pi \times 10^{-1}$ μH/m), and H_T is the Earth's magnetic field (A/mm). Therefore, B_i can be evaluated if susceptibility can be measured.

If B_t and B_i can be determined, B_r can be quantified,

$$B_r = B_t - B_i \,. \quad \text{(3G.24)}$$

If B_r is positive, the remnant magnetization and, therefore, the paleomagnetic field that caused it are aligned as per the present-day Earth's field, and B_r is described as "normal." Otherwise, its polarity is "reversed." The polarity of remnant magnetization is evaluated through a numerical comparison of B_t and B_i.

Measurement. A magnetic logging sonde developed for sedimentary rocks is Schlumberger's Geological High-Resolution Magnetic Tool (GHMT™). It actually comprises two tools: one to measure the total magnetic field and one to measure magnetic susceptibility. The tool housings are nonmagnetic and electrically insulating with a diameter of 4 in. [100 mm].

Because sedimentary rocks have a very low magnetization, a very high precision magnetometer is required. This requirement is satisfied by Schlumberger's Nuclear Magnetic Resonance Tool (NMRT™), which uses the principles of nuclear magnetic resonance whereby the frequency of precession of relaxing protons is proportional to B_T (see Chap. 3E in this section of the *Handbook*). The problem, therefore, reduces to a very precise measurement of frequency. The NMRT measurement has a sensitivity of 10^{-2} nT. Data are recorded at a logging speed of 1970 ft/hr [600 m/hr] with a sampling interval of 4 in. [100 mm].

A second tool measures the magnetic susceptibility of the rock, which is proportional to B_i (Eq. 3G.23). This tool, Schlumberger's Susceptibility Measurement Tool (SUMT™), uses the principles of electromagnetic induction. The voltage induced in the receiver coil increases with susceptibility, which is determined from the complex character of the induced signal. Susceptibility is dimensionless. The downhole measurement of susceptibility has a sensitivity of approximately 10^{-6} units SI. Data measured in sedimentary rocks are typically in the range 10^{-5} to 10^{-4} units SI. Data can be recorded at a logging speed of up to 3940 ft/hr [1200 m/hr] with a sampling interval of 6 in. [150 mm]. The tools are rated to a temperature of 257°F [125°C] and a pressure of 15,000 psi [103 MPa].

The key measurement deliverable from the combined use of these two integral tools is a depth record of the polarity of remnant magnetization based on the sign of B_r in Eq. 3G.24. This display is called the well magnetic stratigraphy (WMS).

Application. **Fig. 3G.24** illustrates the application of the measured data. It shows how a magnetostratigraphic sequence was established for a well in the Paris basin using GHMT log data.[40] For dating purposes, this sequence has to be tied to the GPTS.

Fig. 3G.25 illustrates another example, this time from the Ocean Drilling Program (ODP). This example reveals some geomagnetic features that are not yet part of the global standard. Notwithstanding these disparities, this chronal benchmarking allows the absolute dating of much of the sedimentary succession.

Future applications will examine the direction of remnant magnetization to investigate the movement of fault blocks and enhancing the fieldwide correlation of the sedimentary column.

3G.5 Discussion

There are two principal drivers for the further advancement of the technologies that have been described here.

The first is the need for improved reservoir characterization to help us deal with problematic reservoirs that have low-permeability characteristics, thin beds, laminations, low-resistivity-contrast pay, and fracture networks. Fracture networks lead us to the question of carbonates and their petrophysical differences from clastic rocks. One might ask why it is that with so much technology available, the industry still perceives a shortfall in its interpretative capability. The reason is that recent attention has been directed at data acquisition and management rather than methods of interpreting the data themselves. Thus, for example, we have not yet succeeded in reconciling petrophysical data measured at different scales. The gap between our ability to measure and our ability to interpret the measurements widened still further during the 1990s, the decade of the horizontal well. This drove the analysis of downhole measurements further

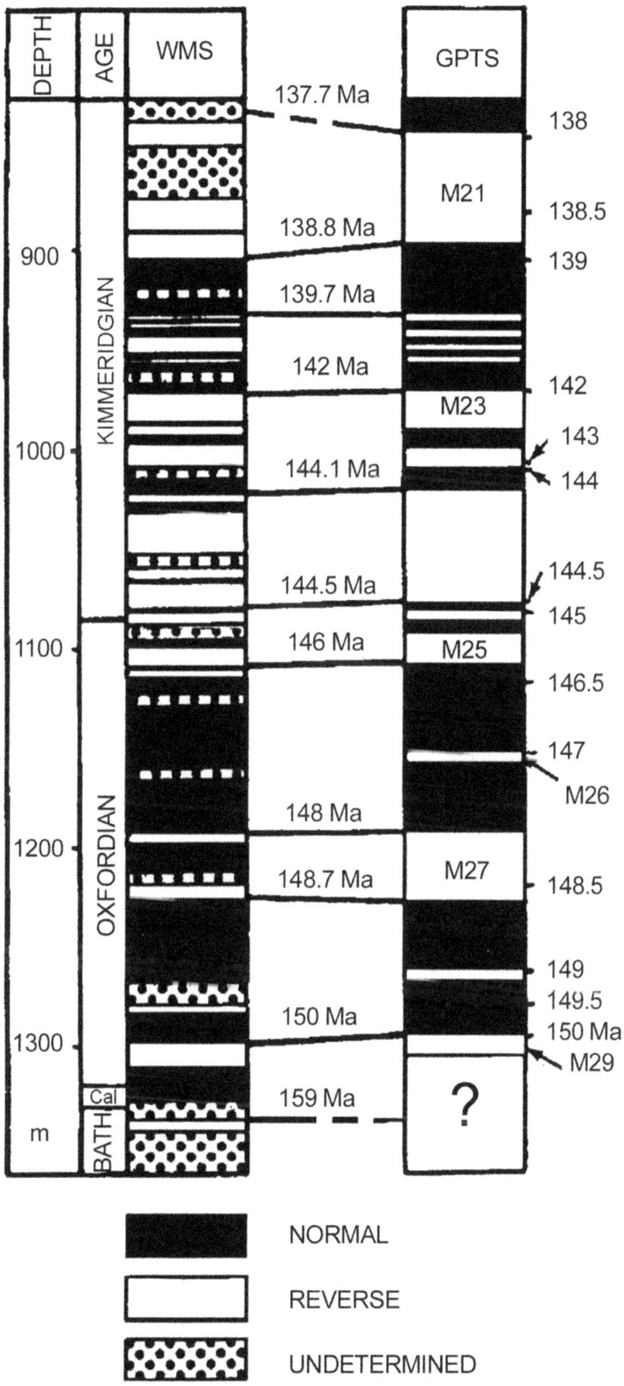

Fig. 3G.24—Comparison between the Geomagnetic Polarity Time Scale (GPTS) and the Well Magnetic Stratigraphy (WMS) for a borehole in the Paris basin. Normal polarity is black; reversed polarity is white. The GPTS and WMS are similar. This allows polarity to be correlated and absolute determinations of age to be made along the borehole column. In some places (e.g., at approximately 950 m depth), the detail of the GPTS is not reflected in the WMS, presumably because of low rock magnetization or, possibly, localized erosion. On the other hand, at approximately 1060 m, there are indications of detail that has yet to be accommodated within the GPTS.[40] (Courtesy of SPWLA.)

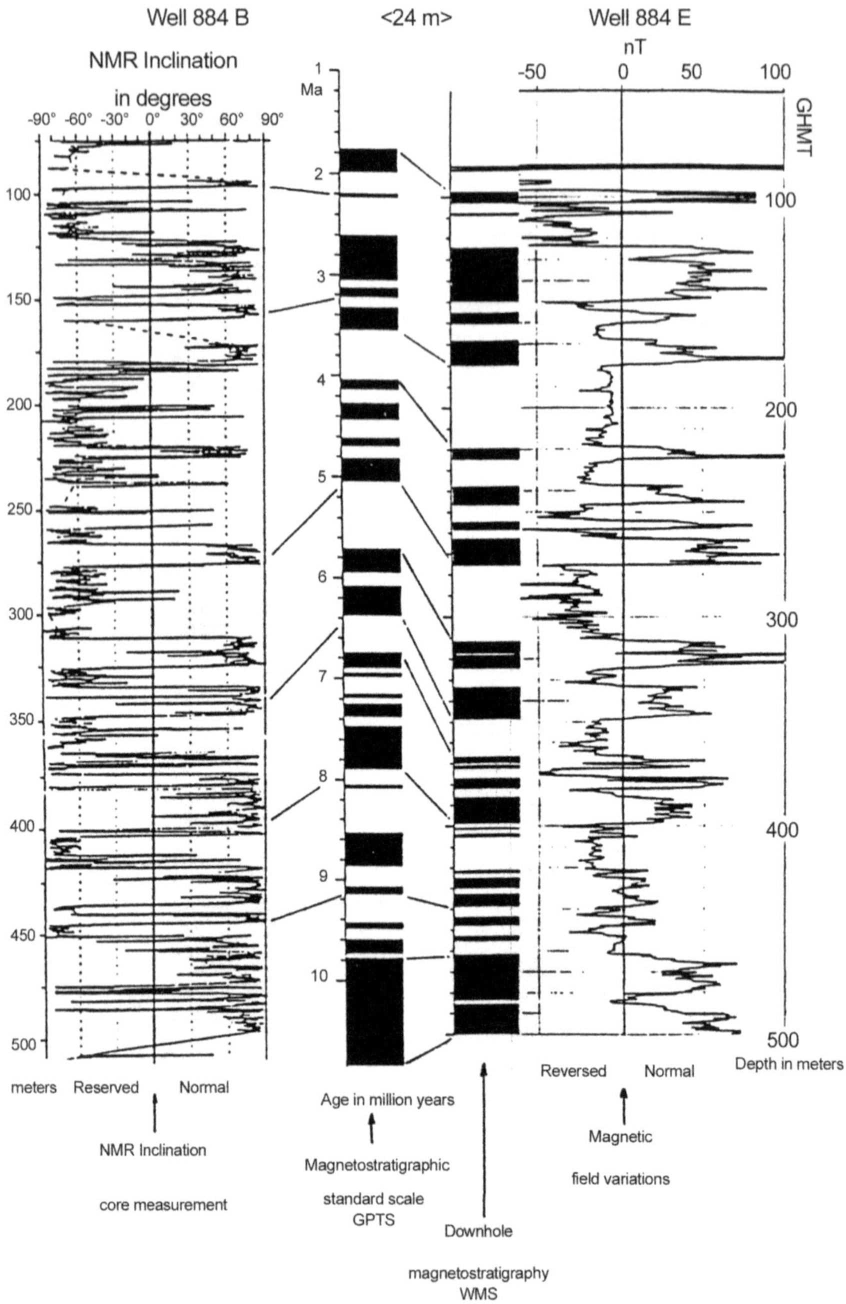

Fig. 3G.25—Geomagnetic data from Ocean Drilling Program (ODP) Leg 145. Wells 884B and 884E are 24 m apart. Normal polarity is black; reversed polarity is white. Well 884B was cored; core data have furnished the inclination of the remnant magnetization. These data have been correlated with the Geomagnetic Polarity Time Scale (GPTS). Well 884E was logged geomagnetically; log data have furnished the Well Magnetic Stratigraphy (WMS). The WMS shows detail that is not part of the GPTS. This could be highly significant because in these wells the susceptibility was very low indeed, so that the measured "net" field B_t was directly related to the remnant magnetization without being impacted by possible errors in susceptibility.[40] (Courtesy of SPWLA.)

into three dimensions and emphasized the need for us to get more out of our data if our reservoir models are to deliver the greatest benefit.

The second technology driver is the cost-effectiveness of multiwell platforms from which deviated, extended-reach, horizontal, and multilateral wells can be drilled to target hydrocarbon accumulations that have been identified in a reservoir model. This heralds a further thrust in the need to drill more difficult subsurface environments in a way that allows full control of the wellbore trajectory. This, in turn, will require a full casing- and cement-evaluation service, especially with regard to the monitoring of casing deformation. Only in this way can one be assured of an absence of flow constrictions or impediments to tool deployment.

Both reservoir characterization and the cost-effectiveness of multiwell platforms will continue to benefit from further developments in data recording, transmittal, processing, and visualization, which have underpinned the technical progress made to date.

3G.6 Summary

This chapter has addressed several specialized logging tools and the information they can provide in assessing borehole trajectory, wellbore conditions, and reservoir characteristics. Directional, caliper, and cement-bond surveys can be used to determine well location, the quality and condition of the open hole, the condition of the tubing, the presence of cement, the quality of the bond between tubing and cement, and, to a lesser extent, the degree of bonding between cement and formation. Borehole imaging can be used in open hole to picture the different strata encountered, and it is of particular use in detailing thinly bedded (sand/shale) intervals and identifying both natural and induced fractures. Finally, borehole gravimetry and downhole magnetics can be used to measure formation properties at a larger scale. Each of these specialized tools is experiencing a stronger application base as better sensor technology delivers the information needed for improved reservoir characterization and reservoir management.

Acknowledgments

The author is obliged to those oilfield service companies and petroleum technical societies who have granted permission for subject matter to be included here. Special mention is made of Baker Atlas, EDCON, Halliburton Energy Services, Schlumberger, and Sondex. This material has been included to give a balanced overview of the state of technology in this important subject area rather than to endorse any particular commercial service. In this rapidly changing world, the material is current as of June 2003.

Nomenclature

A = tool azimuth relative to magnetic north, –, degrees
B_o = magnetic induction associated with the present Earth's field, m/qt, nT
B_i = magnetic induction due to the field induced in the rock, m/qt, nT
B_r = magnetic induction due to the remnant field, m/qt, nT
B_t = "net" field, m/qt, nT
c = gravimeter design constant, L, m
d = gravimeter design constant, L, m
f = gravitational force of attraction, mL/t^2, N
F_r = ratio factor for the minimum curvature method
F = vertical gradient of gravity, 1/t^2, mGal/m
g = acceleration due to gravity, L/t^2, mGal
G = universal gravitational constant, L^3/mt^2, dyne cm^2/g^2
h = elevation, L, m
h_m = measured bed thickness, L, m
h_{ts} = true stratigraphic thickness, L, m
h_{tv} = true vertical thickness, L, m
H_T = Earth's magnetic field, q/tL, A/mm

i = current, q/t, A
I = tool inclination relative to gravitational vector, –, degrees
L = length of equal straight lines representing dogleg in wellbore, L, m
m_i, m_j = any two gravitationally attracting masses, m, kg
m = mass, m, kg
M = mass of the Earth, m, kg
r = distance between two gravitationally attracting masses, L, m
R = Earth's radius, L, km
R_{xo} = apparent formation resistivity, mL^3/tq^2, Ωm
T = tension, mL/t^2, N
$\bar{h}$ = mean elevation of layer, L, m
α = inclination or deviation of wellbore, –, degrees
φ = dip, –, degrees
φ_a = apparent dip, –, degrees
β = azimuth of wellbore, –, degrees
β_d = azimuth of dip, –, degrees
δV = potential difference, mL^2/qt^2, V
ΔE = incremental distances between adjacent survey stations in the direction of the easting, L, m
Δg = difference in gravity between two locations at the top and bottom of an infinite horizontal reservoir layer penetrated by a vertical well, L/t^2, Gal
ΔN = incremental distances between adjacent survey stations in the direction of the northing, L, m
ΔS_g = time-lapse change in gas saturation
ΔV = incremental distances between adjacent survey stations in the vertical direction, L, m
$\Delta\rho_b$ = change in measured density, m/L^3, g/cm^3
λ = latitude, –, degrees
μ_o = magnetic permeability of the void, mL/q^2, μH/m
ρ = density, m/L^3, g/cm^3
ρ_g = gas density, m/L^3, g/cm^3
ρ_o = oil density, m/L^3, g/cm^3
ψ = dogleg angle, –, degrees
ϕ = porosity
χ = magnetic susceptibility of the rock, –

Subscripts

1 = upper directional survey station
2 = lower directional survey station

References

1. Bourgoyne, A.T. Jr. *et al.*: *Applied Drilling Engineering,* Textbook Series, SPE, Richardson, Texas (1991) **2.**
2. Theys, P.: *Log Data Acquisition and Quality Control,* Editions Technip, Paris (1999).
3. Thorogood, J.L.: "Directional Survey Operations Management," *JPT* (December 1989) 1250.
4. Inglis, T.: *Directional Drilling,* Graham & Trotman, London (1987).

5. Boak, J.M.: "Conversion of Well Log Data to Subsurface Stratigraphic and Structural Information," AAPG Methods in Exploration Series, No. 10, American Assn. of Petroleum Geologists, Tulsa (1992) 289–293.
6. Larimore, D.R., Fehrmann, G.Z., and White, S.: "Field Cases of Cost Efficient Well Interventions Performed With Advanced Slickline Technology," paper SPE 38097 presented at the 1997 SPE Asia Pacific Oil and Gas Conference, Kuala Lumpur, 14–16 April.
7. Connell, M.L. *et al.*: "High-Pressure/High-Temperature Coiled Tubing Casing Collar Locator Provides Accurate Depth Control for Single-Trip Perforating," paper SPE 60698 presented at the 2000 SPE/ICoTA Coiled Tubing Roundtable, Houston, 5–6 April.
8. Wagg, B. and Matthews, C.: "Evaluating Casing Deformation," *World Oil* (2000) **221,** No. 12, 44.
9. Wagg, B. *et al.*: "Evaluating Casing Deformation Mechanisms in Primary Heavy Oil Production," paper SPE 54116 presented at the 1999 SPE International Thermal Operations and Heavy Oil Symposium, Bakersfield, California, 17–19 March.
10. Hayman, A.J. *et al.*: "Developments in Corrosion Logging Using Ultrasonic Imaging," *Trans.*, Soc. of Professional Well Log Analysts 36th Annual Logging Symposium, Paris (1995) paper W.
11. Frisch, G.J. and Mandal, B.: "Advanced Ultrasonic Scanning Tool and Evaluation Methods Improve and Standardize Casing Inspection," *Trans.*, Soc. of Professional Well Log Analysts 42nd Annual Logging Symposium, Houston (2001) paper X.
12. Frisch, G.J., Graham, L., and Wyatt, D.: "Economic Evaluation of the Use of Well Logs for Diagnosing Conformance Problems," paper SPE 40036 presented at the 1998 SPE Gas Technology Symposium, Calgary, 15–18 March.
13. Harness, P.E., Sabins, F.L., and Griffith, J.E.: "New Technique Provides Better Low-Density-Cement Evaluation," paper SPE 24050 presented at the 1992 SPE Western Regional Meeting, Bakersfield, California, 30 March–1 April.
14. Bigelow, E.L., Domangue, E.J., and Lester, R.A.: "A New and Innovative Technology for Cement Evaluation," paper SPE 20585 presented at the 1990 SPE Annual Technical Conference and Exhibition, New Orleans, 23–26 September.
15. Hayman, A.J., Hutin, R., and Wright, P.V.: "High Resolution Cementation and Corrosion Imaging by Ultrasound," *Trans.*, Soc. of Professional Well Log Analysts 32nd Annual Logging Symposium, Midland, Texas (1991) paper KK.
16. De Souza Padilha, S.T.C. and Da Silva Araujo, R.G.: "New Approach on Cement Evaluation for Oil and Gas Reservoirs Using Ultrasonic Images," paper SPE 38981 presented at the 1997 SPE Latin American and Caribbean Petroleum Engineering Conference, Rio de Janeiro, 30 August–3 September.
17. Frisch, G., Graham, L., and Griffith, J.: "A Novel and Economical Processing Technique Using Conventional Bond Logs and Ultrasonic Tools for Enhanced Cement Evaluation," *Trans.*, Soc. of Professional Well Log Analysts 41st Annual Logging Symposium, Dallas (2000) paper EE.
18. Graham, W.L. *et al.*: "Cement Evaluation and Casing Inspection With Advanced Ultrasonic Scanning," paper SPE 38651 presented at the 1997 SPE Annual Technical Conference and Exhibition, San Antonio, Texas, 5–8 October.
19. Prensky, S.: "Advances in Borehole Imaging Technology and Applications," *Borehole Imaging: Applications and Case Histories,* M. Lovell, G. Williamson, and P. Harvey (eds.) Geological Soc., London, Special Publications (1999) **159**, 1–43.
20. Zemanek, J. *et al.*: "The Borehole Televiewer—A New Logging Concept for Fracture Location and Other Types of Borehole Inspection," *JPT* (June 1969) 762.
21. Zemanek, J. *et al.*: "Formation Evaluation by Inspection With the Borehole Televiewer," *Geophysics* (1970) **35,** 254.
22. Hayman, A.J. *et al.*: "Improved Borehole Imaging by Ultrasonics," paper SPE 28440 presented at the 1994 SPE Annual Technical Conference and Exhibition, New Orleans, 25–28 September.
23. Faraguna, J.K., Chace, D.M., and Schmidt, M.G.: "An Improved Borehole Televiewer System: Image Acquisition, Analysis and Integration," *Trans.,* Soc. of Professional Well Log Analysts 30th Annual Logging Symposium, Denver (1989) paper UU.
24. Seiler, D. *et al.*: "Field Performance of a New Borehole Televiewer Tool and Associated Field Processing Techniques," *Trans.,* Soc. of Professional Well Log Analysts 31st Annual Logging Symposium, Lafayette, Louisiana (1990) paper H.

25. Barton, C.A. *et al.:* "Utilizing Wellbore Image Data To Determine the Complete Stress Tensor: Application to Permeability, Anisotropy and Well Bore Stability," *The Log Analyst* (1997) **38,** No. 6, 21.
26. Ekstrom, M.P. *et al.:* "Formation Imaging With Microelectrical Scanning Arrays," *The Log Analyst* (1987) **28,** No. 3, 294.
27. Boyeldieu, C. and Jeffreys, P.: "Formation Microscanner: New Developments," *Trans.,* Soc. of Professional Well Log Analysts 11th European Formation Evaluation Symposium, Paris (1988) paper X.
28. Safinya, K.A. *et al.:* "Improved Formation Imaging With Extended Microelectrical Arrays," paper SPE 22726 presented at the 1991 SPE Annual Technical Conference and Exhibition, Dallas, 6–9 October.
29. Seiler, D., King, G., and Eubanks, D.: "Field Test Results of a Six Arm Microresistivity Borehole Imaging Tool," *Trans.,* Soc. of Professional Well Log Analysts 35th Annual Logging Symposium, Tulsa (1994) paper W.
30. Laastad, H. *et al.:* "Water-Based Formation Imaging and Resistivity Logging in Oil-Based Drilling Fluids—Today's Reality," paper SPE 62977 presented at the 2000 SPE Annual Technical Conference and Exhibition, Dallas, 1–4 October.
31. Dumont, A., Kubacsi, M., and Chardac, J.L.: "The Oil-Base Mud Dipmeter Tool," *Trans.,* Soc. of Professional Well Log Analysts 28th Annual Logging Symposium, London (1987) paper LL.
32. Cheung, P. *et al.:* 2001. "Field Test Results of a New Oil-Base Mud Formation Imager Tool," *Trans.,* Soc. of Professional Well Log Analysts 42nd Annual Logging Symposium, Houston (2001) paper XX.
33. Smith, N.L.: "The Case for Gravity Data From Boreholes," *Geophysics* (1950) **15,** 605.
34. McCulloh, T.H., Kandle, G.R., and Schoellhamer, J.E.: "Application of Gravity Measurements in Wells to Problems of Reservoir Evaluation," *Trans.,* Soc. of Professional Well Log Analysts 9th Annual Logging Symposium, New Orleans (1968) paper O.
35. Alixant, J.-L. and Mann, E.: "In-Situ Residual Oil Saturation to Gas From Time-Lapse Borehole Gravity," paper SPE 30609 presented at the 1995 SPE Annual Technical Conference and Exhibition, Dallas, 22–25 October.
36. Jageler, A.H.: "Improved Hydrocarbon Reservoir Evaluation Through Use of Borehole-Gravimeter Data," *JPT* (1976) 709.
37. Gournay, L.S. and Lyle, W.D.: "Determination of Hydrocarbon Saturation and Porosity Using a Combination Borehole Gravimeter (BHGM) and Deep Investigating Electric Log," *Trans.,* Soc. of Professional Well Log Analysts 25th Annual Logging Symposium, New Orleans (1984) paper WW.
38. Popta, J.V. *et al.:* "Use of Borehole Gravimetry for Reservoir Characterization and Fluid Saturation Monitoring," paper SPE 20896 presented at the 1990 SPE European Petroleum Conference, The Hague, 22–24 October.
39. Brady, J.L., Wolcott, D.S., and Aiken, C.L.V.: "Gravity Methods: Useful Techniques for Reservoir Surveillance," paper SPE 26095 presented at the 1993 SPE Western Regional Meeting, Anchorage, 26–28 May.
40. Pages, G. *et al.:* "Wireline Magnetostratigraphy: Principles and Field Results," *Trans.,* Soc. of Professional Well Log Analysts 35th Annual Logging Symposium, Tulsa (1994) paper XX.
41. Lalanne, B. *et al.:* "Magnetic Logging: Borehole Magnetostratigraphy and Absolute Datation in Sedimentary Rocks," paper SPE 21437 presented at the 1991 SPE Middle East Oil Show, Bahrain, 16–19 November.

SI Metric Conversion Factors

degree ×	1.745329	E − 02	= rad
dyne ×	1.0*	E − 05	= N
ft ×	3.048*	E − 01	= m
°F	(°F − 32)/1.8		= °C
Gal ×	1.0*	E − 02	= m/s^2
in. ×	2.54*	E + 00	= cm
psi ×	6.894 757	E + 00	= kPa

*Conversion factor is exact.

Chapter 3H
Petrophysical Applications

H.R. (Hal) Warner Jr., Warner Consulting Services and **Richard Woodhouse,** Independent Consultant

3H.1 Introduction

This chapter discusses the determination of lithology, net pay, porosity, water saturation, and permeability from wellbore core and log data. The chapter deals with "Development Petrophysics" and emphasizes the integration of core data with log data; the adjustment of core data, when required, to reservoir conditions; and the calibration and regression line-fitting of log data to core data. The goal of the calculations is to use all available data, calibrated to the best standard, to arrive at the most accurate quantitative values of the petrophysical parameters (i.e., lithology, net pay, porosity, water saturation, and permeability). Log analysis, cased-hole formation evaluation, and production logging are not covered here.

The following topics are covered in this chapter: petrophysical data sources and databases, lithology determination, net-pay (or pay/nonpay) determination, porosity determination, fluid-contacts identification, water-saturation determination, permeability calculations, case studies, other considerations in petrophysical calculations, and summary and conclusions. This chapter concerns the foot-by-foot calculations of the five petrophysical parameters at the wellbore. It does not cover the propagation of the wellbore values, or "populating" of static or dynamic reservoir models, vertically and areally over the whole of the reservoir volume. Because other chapters in this section of the *Handbook* discuss the technical details of various well logs and coring, these details are generally assumed as the basis for this chapter and are referenced accordingly.

In practical terms, petrophysics is used for two types of calculations: determination of original hydrocarbons in place [original oil in place (OOIP) or original gas in place (OGIP)] and their distribution, and reservoir-engineering dynamic flow calculations. For the development geoscientists (geologists, geophysicists, and geostatisticians), petrophysics means developing the detailed stratigraphic, depositional, and diagenetic descriptions of the reservoir, both vertically and areally. To make accurate calculations of OOIP or OGIP and the various flow calculations, accurate foot-by-foot calculations of lithology, net pay, porosity, water saturation, and permeability are necessary. These calculations need to be made not only as overall calculations, but also so that the variation and distribution of these parameters are determined appropriately.

Some of the petrophysical calculations can be made in several ways, particularly for porosity and water saturation. One key to arriving at an accurate petrophysical calculation is to obtain the same quantitative result with a variety of techniques. An important consideration is the acquisition and handling of the various types of petrophysical data and, for each reservoir, the preparation of its unique petrophysical database. Petrophysical data take many forms and, for many reservoirs, may not be as comprehensive as desired. The technical personnel working with these data have to review what data are available, their quality, and what additional data might be acquired from the existing wellbores and from preserved and unpreserved cores. Finally, if there are sufficient financial stakes, new wells might be drilled, cores cut, various additional sample measurements made, and both conventional and special logs run to obtain other desired petrophysical information.

As mentioned previously, this chapter does not deal with log analysis, the petrophysical evaluations made when no core, mud log, or tester data are available. Log analysis is used universally and is generally successful in the identification of oil and gas reservoirs and in the preliminary estimation of their volumes.[1–7] However, log analysis augmented and calibrated with core[8] and other data provides the most accurate quantification of oil and gas volumes present in a well and best represents the practice of petrophysics.

This chapter is focused on petrophysical calculations at the reservoir level, where there are several to hundreds of wells with logs and significant amounts of core data that need to be integrated to develop the most accurate overall values for the petrophysical parameters over the whole of the reservoir. The techniques discussed also apply to single wellbores, but many of the complications are not a concern in single-well evaluation.

This chapter does not cover some special cases, such as oil shales, tight gas-sand reservoirs, or coalbed-methane reservoirs. Petrophysical calculations for tight gas-sand reservoirs and coalbed-methane reservoirs are discussed in separate chapters on these topics in the Emerging and Peripheral Technologies volume of this *Handbook*.

3H.2 Petrophysical Data Sources and Databases

In making the petrophysical calculations of lithology, net pay, porosity, water saturation, and permeability at the reservoir level, the development of a complete petrophysical database is the critical first step. This section describes the requirements for creating such a database before making any of these calculations. The topic is divided into four parts: inventory of existing petrophysical data; evaluation of the quality of existing data; conditioning the data for reservoir parameter calculations; and acquisition of additional petrophysical data, where needed. The overall goal of developing the petrophysical database is to use as much valid data as possible to develop the best standard from which to make the calculations of the petrophysical parameters.

3H.2.1 Inventory of Existing Petrophysical Data. To start the petrophysical calculations, the data that have been gathered previously from various wellbores throughout the reservoir must be identified, organized, and put into electronic form for future calculations.

In a typical reservoir, several "generations" of wells have been drilled. The exploration wells that discover and delineate the reservoir constitute the first generation of wells. These wells are usually drilled with scant knowledge of formation pressure, which results in deep mud-filtrate invasion in the reservoir interval for which there may be significant hole-washout problems. For this series of wells, the muds used may vary from one well to the next, and this phase may take from 1 to 10 years to complete. The second generation is the group of wells drilled during initial reservoir development. These wells are likely to be drilled with a common mud system, which might be either water-based or oil-based and will be tailored to help minimize the near-wellbore damage caused by detrimental mud-filtrate/reservoir-rock interactions. Third-generation wells may also be available. These wells would be those from later field-development activities and may have been drilled 5 to 15 years after the initial development wells were drilled.

The logs from these various generations of wells can vary in several regards. First, the logs may have been run by various service companies and may not be directly comparable to each other, even if they were from the same time period. Secondly, if the logs were run by the same service company, they may still have significant differences because different generations of logging tools were used for each set of wells. An additional difference may be between logging-while-drilling (LWD) logs and those run on wireline. LWD logging is often associated with high-angle wells. This geometry can lead to significantly different responses from zones previously seen in vertical wells and may lead to petrophysical mapping issues for high-angle-well evaluations, if not correctly accounted for.

Many of these same caveats hold for core data. The early wells may or may not have been completely cored through the reservoir interval. Later wells are more likely to be fully cored, although some zones of particular interest may have a greater concentration of cores. The routine data acquisition from cores also may vary because of different laboratories performing the core analysis on each well and because of changes in laboratory procedures and equipment over the intervening years. Also, different portions of the reservoir interval may have been analyzed with different techniques because they differ in degree of consolidation or rock heterogeneity. Special-core-analysis (SCAL) data are likely to be a variety of information because the SCAL programs for each well's cores will be unique to the perceived data needs when each well was drilled. With respect to the geologists' core descriptions, there may be differences between wells because different individuals prepared the various core descriptions with different techniques and emphasis. The number of petrographic measurements on cores [e.g., thin sections, scanning electron microscope (SEM), and X-ray diffraction (XRD)] is likely to vary widely from well to well.

First, the technical team must prepare several spreadsheets tabulating the basic information about each of the drilled wells. **Tables 3H.1 through 3H.3** show templates for a spreadsheet for the log data and two for the core data, respectively. These spreadsheets provide quick access by the technical team to see what data are available, what form the data are in, how much of each type of data is available, and where gaps exist in the database. Separate sets of spreadsheets should be prepared for each of the reservoirs, if there are several separate reservoirs in a particular oil or gas field.

Next, as much as possible of these detailed log and core data should be obtained in electronic form. There are a number of commercially available software packages that are useful for this purpose and can be used to access the log data and the routine-core-analysis data. However, the routine-core-analysis data may also be entered into spreadsheet form for easy use by the engineers and geoscientists. To the extent possible, the geologists' core descriptions and petrographic measurements should be converted to electronic form for use with the other types of petrophysical data. The SCAL data will require special spreadsheet formats because each of these types of data is unique.

Core and log databases should be considered as a part of the overall reservoir database. In establishing and maintaining the overall reservoir database, controls should be in place to ensure high quality of the data and the timely inclusion of all data that are obtained.

3H.2.2 Evaluation of the Quality of Existing Data. The second step in working with the petrophysical data is to evaluate the quality of each of these types of data. This step requires that the data inventory and database preparation steps are completed first so that this second step can occur as a systematic and complete process. The evaluation process is a "compare and contrast" exercise.

TABLE 3H.1—LOG DATABASE FOR EXPLORATION AND DEVELOPMENT WELLS

Well	Resistivity	Sonic	Density	Neutron	Gamma	Tester	Dip/Image	Other
E1	DIL, DLL, IES	LSS, BHC	FDC	CNL	GR	RFT	HDT	VSP, CST
E2	DIL	DDBHC	LDL	CNL	NGS, GR		HDT	CST
E3	DIL, DLL	LSS, DDBHC	LDL	CNL	GR			BGT
E4	DIL	LSS	LDL	CNL	GR	RFT	SHDT	VSP, CST
E6	DIL	LSS	LDL	CNL	GR	RFT	SHDT	VSP, CST
E8	DIT	BHC	LDL	CNL	GR	RFT	SHDT	VSP, CST
D1	AIT	DSI	LDL	DNL	NGS	RFT	UBI, OBDT	VSP
D2	AIT	DSI	LDL	DNL	NGS	RFT	UBI	VSP
D3	AIT	DSI	LDL	DNL	NGS	RFT	UBI	CS
D4	AIT	DSI	LDL	DNL	NGS	RFT	UBI	
D5	AIT	DSI	LDL	DNL	NGS	RFT	UBI	
D6	AIT	DSI	LDL	DNL	NGS	RFT	UBI	CS
D7	AIT	DSI	LDL	CNL, APS	NGS	MDT		
D8	AIT, CDR	DSI	ADN	ADN	GR	MDT		
D9	AIT	DSI	LDL	APS	NGS	MDT	UBI	

Note: ADN=azimuthal density neutron (logging while drilling); AIT=array induction tool; APS=accelerator porosity sonde; CDR=compensated dual resistivity (logging while drilling); CNL=compensated neutron log; CS=checkshot survey; CST=chronological sample taker; DIL=dual induction log; DIT-E=phasor dual induction log; DLL=dual laterolog; DNL=dual detector (thermal-epithermal) compensated neutron log; DSI=dipole sonic imager; GR=gamma ray; HDT=high-resolution dipmeter tool; IES=induction electrical survey (deep induction only); LDL=lithodensity log; LSS=long-spaced sonic; MDT=multidynamic tester; NGS=natural gamma ray spectroscopy tool; OBDT=oil-based dipmeter tool; RFT=repeat formation tester; SHDT=stratigraphic high-resolution dipmeter tool; UBI=ultrasonic borehole imager; and VSP=vertical seismic profile.

Log Data. The evaluation of log-data quality has many aspects.[7] First, the drilling-mud and hole-condition effects may lead to no valid readings being recorded on the logs. This should be noted in the petrophysical database. "Flags" of various types should be stored, for example, to denote intervals where the hole size exceeds some limit, or where there is cycle-skipping on the sonic logs. Logging tools sometimes become temporarily stuck as a log is being run. This results in constant readings on each of the several detectors on the tool string. When the tool is stationary, each detector on it becomes stuck at a different depth, so the interval of "stuck" log will vary for each log curve. For example, the neutron log typically sticks over an interval approximately 10 ft above the stuck interval on a density log. It may be possible to "splice" in a replacement section of log from a repeated log section, or the invalid readings may simply be deleted.

Second, each log is formally calibrated before the start of each logging run by various calibration standards. The logs are also checked again after the run. Calibration records may assist in determining the quality of the logs. Perhaps of equal importance are the written comments on the log heading made immediately after the job by the logging engineer.

Third, systematic influences on the quality of log readings should be corrected. For example, if some of the wells are drilled with water-based mud (WBM), the effect of WBM-filtrate invasion on various resistivity logs can be quantified. This is done by computations made using the various resistivity logs in the same wellbore; however, where deep invasion of WBM filtrate occurs, offsetting wells drilled with oil-based mud (OBM) give a good comparison. The induction logs in OBM wells can provide accurate true reservoir resistivity values in thick hydrocarbon zones. See the chapter on resistivity and SP logging in this volume of the *Handbook*

TABLE 3H.2—ROUTINE-CORE-ANALYSIS PROCEDURE BY WELL

Well	Plug Type	Laboratory No.	Net Confining Pressure (psi)	Porosity	Permeability	S_w
E2	H	1	350	Yes	Yes	No
E3	H,V	2	250	Yes	Yes	No
E3	H	1	350	Yes	Yes	No
E4	H,V	2	250	Yes	Yes	No
E5	H	2	250	Yes	Yes	No
D1	H	3	800 (CMS)	Yes	Yes	Yes
D3	H	3	800 (CMS)	Yes	Yes	Yes
D8	H	1	800 (CMS)	Yes	Yes	Yes
D9	H	1	800 (CMS)	Yes	Yes	Yes
D9	H	3	800 (CMS)	Yes	Yes	Yes

Note: H=horizontal core plugs, V=vertical core plugs, and CMS is an abbreviation for Core Measurement Systems.[9]

TABLE 3H.3—SUMMARY OF THE NUMBER OF ROUTINE CORE ANALYSES BY WELL AND BY ZONE

Well	E2	E3	E4	E5	D1	D3	D8	D9	Total
Reservoir No. 1 (by zone)									
1	0	47	18	35	14	117	0	0	231
2	0	0	0	49	91	108	0	136	384
3	70	0	0	52	99	96	0	0	317
4	53	0	121	56	123	136	0	0	489
5	0	0	0	88	94	108	0	0	290
6	0	59	116	12	43	92	0	0	322
Total	123	106	255	292	464	657	0	136	2033
Reservoir No. 2	0	0	0	0	0	0	177	0	177
Reservoir No. 3	0	0	44	0	0	42	0	0	86

Note: Tabulation is based on the number of horizontal core plugs and the number of porosity measurements.

for more information on how invasion effects can be handled. Boreholes are not always right cylinders. Holes sometimes become spiral shaped during the drilling process, and their logs show sinusoidal responses.

Routine-Core-Analysis Data. The evaluation of the quality of routine-core-analysis[9–11] data starts with crossplot comparisons of various wells' porosity, permeability, grain density, and saturation data on a reservoir or zone basis. With various wells' data presented on the same plot (permeability vs. porosity, grain density vs. porosity, S_w vs. permeability, or S_w vs. porosity), one can determine if there are significantly different trends from one well to the next. Differences may exist, and there may be good geologic reasons for such differences; however, some laboratory data may be of suspect quality and may require further review and inquiry.

With respect to core's fluid-saturation data, the evaluation process must be an end-to-end review process. This evaluation begins with knowledge of the drilling mud used and whether the cores were specially cut and preserved to try to obtain undisturbed connate S_w. Second, review the core sampling and routine-core-analysis laboratory procedures to understand how those steps impacted the final fluid-saturation data reported. Well-preserved OBM cores that are analyzed using the Dean-Stark water-extraction procedure typically provide the most valid values for connate S_w in the hydrocarbon column above the mobile gas/water or oil/water transition zone. In aquifers and other mobile-water intervals, OBM-filtrate invasion displaces mobile water during core cutting, and fluid "bleeding" occurs during core surfacing. As a result, OBM-core water saturations are too low and not representative of true resident saturations. They may also be more uncertain in very poor quality, high-S_w rock intervals in which it is difficult to make accurate porosity and water-volume determinations.

WBM cores can provide a qualitative measure of residual-oil saturations in the oil column after accounting for oil shrinkage and bleeding.[12] WBM cores in oil reservoirs may also provide information about oil/water contacts (OWCs) and the S_w of low-quality rock intervals above the OWC into which oil has never entered because the entry capillary pressure, P_{ce}, has not been exceeded.

The OBM and WBM routine-core-analysis data should be compared and contrasted to identify how each alone, and in combination, can be used to answer certain petrophysical issues. Also, the trends of these sets of data (e.g., permeability vs. porosity) should be compared as one of the evaluation tools.

Coring in friable, uncemented, and unconsolidated sands demands special coring, handling, and analysis techniques so that the grain structure is not altered. Modern practice is to use rapid drilling rates and fiberglass or aluminum inner core barrels to minimize the friction as the core enters the barrel. The core is tripped to surface slowly and smoothly to allow dissolved gas to exit the core without disruption. The last two stands of drillpipe are the most critical because during this time the volume of gas doubles, then quadruples. During the laying down of the inner core barrel, precautions are needed to prevent bending and core deformation. For transporting the core, the inner core barrel can be cut into short (1-m/3-ft) lengths, the ends sealed with caps, then the voids between the core and the sections of the inner core barrel filled with resin. Freezing the cores in their segments before transportation is sometimes used to prevent damage, but is not very effective when there is little formation water. The costs are also higher. Cross-section X-raying of the tubes reveals which cores are damaged and which are suitable for measurements in the laboratory. See the American Petroleum Institute's (API) *RP 40 Recommended Practices for Core Analysis* for details about the various types of core analysis and details of the laboratory procedures.[13]

Special Core Analysis Laboratory Data. SCAL-data evaluation begins with a comparison of the same type of data from different laboratories and whether data from each laboratory are internally consistent. SCAL data are much more difficult to measure, and the procedures often differ from laboratory to laboratory. The challenge is to determine which of these data are more correct and should be used to make various petrophysical-parameter calculations. With SCAL data, the best approach is to have those individuals who are expert in taking and evaluating these types of data review the procedures of the various laboratories and the reported data and provide an opinion about which of these data should be used and which should be discarded.

Capillary pressure (P_c /S_w) data can be susceptible to not being taken to fully equilibrated conditions because it occasionally takes longer for equilibrium to occur than typical laboratory procedures require. This is because the relative permeability of the wetting phase becomes so low that equilibrium is very slowly reached. Additionally, the porous-plate method is susceptible to loss of capillary contact between the core plug and the porous plate. In both situations at higher capillary pressure, P_c, reported S_w values will be too high.[14]

For the rock electrical-property laboratory measurements and how they are reported, the raw laboratory data should be reviewed very carefully to ensure that the data are of high quality and are properly reported for later S_w calculations. These measurements, as a function of brine saturation, again have the potential problem of nonequilibrium saturation distributions. Sometimes the saturation exponent, *n,* is a function of brine saturation, but this nonlinear behavior is typically not reported as such by the reporting laboratory. Restoration of the in-situ brine-saturation distribution is absolutely required for making laboratory rock-electrical-property and P_c measurements that lead to accurate reservoir S_w calculations, so it is best if any restored-state core-plug measurements agree with similar measurements made on native-state core plugs. Finally, the resistivity index (I_R) vs. S_w data should be taken over a range of S_w values equivalent to those found in the particular reservoir. Sometimes these data are taken only down to 30% pore volume (PV) S_w, yet some of the in-situ S_w values may be in the 5 to 20% PV saturation range. If this is the case, laboratory electrical-property measurements may not lead to accurate in-situ S_w calculations from resistivity logs for the low S_w values.

Internal consistency in a laboratory's reported results is a very good "first test" to determine if some of the data are immediately suspect. For example, if the measurements of the reduction in porosity from surface to reservoir stress vary from one set of measurements to another for a particular laboratory, then those measurements must be discarded or used very carefully. As another example, with respect to P_c /saturation measurements, there is an immediate concern if the air/water and air/oil P_c /saturation measurements do not reasonably overlay after accounting for the interfacial-tension (IFT) and contact angle difference between these fluid pairs. There would be a similar concern when mercury-injection P_c data are available. Again, experts in taking and using these types of data should evaluate the quality of the various sets of laboratory measurements.

3H.2.3 Conditioning the Data for Reservoir Parameter Calculations. Conditioning the log and core data for calculations of the various petrophysical parameters includes adjustments from surface-to-reservoir conditions, normalization, and environmental-correction factors. The emphasis here is on obtaining, at reservoir conditions, a set of reliable petrophysical data that involve many wells with many sets of log data and a variety of routine and SCAL core measurements.

Log Data. Each type of log data may need to be conditioned in unique ways for the subsequent petrophysical-parameter calculations. The purpose of each of these preliminary sets of calculations is to put that portion of the overall petrophysical database on a common basis.

The log data, as a whole for each well, must be depth aligned so that the various log measurements at each depth refer, as closely as possible, to the same rock volume. Differences in log response characteristics and in each tool's path in the hole make this a complex task. Lining up the bed boundaries is particularly important. Some commercial software programs have automatic routines that perform the depth-alignment process, but, more often, shifts are made manually. Borehole-size corrections are required for most logs, while the readings of some tools require correction for other factors, such as temperature, mud density, and mud resistivity.

Histograms are often used to compare the log values of one well with the typical field values of the same log type. This process can identify logs that are miscalibrated, and it may indicate by how much they should be corrected. Bad logs that cannot be corrected by reasonable-sized shifts should be discarded from the database.

Gamma Ray Logs. For reservoir-wide petrophysical calculations, gamma ray (GR) logs are often normalized to reduce the variation in their values from one well to the next in intervals that are considered to have the same rock properties, usually clean sands and specific thick shale intervals.

Density and Neutron Logs. Log calibrations[3–7] can be checked in situ because they read nearly constant values in certain formations. Anhydrite has a density value of 2.98 g/cm^3 and close to zero neutron porosity, while salt has a density value of 2.04 g/cm^3 and zero neutron porosity. There are few "marker" beds in which higher neutron-log readings can be verified. (See the chapter on Nuclear Logging in this volume of the *Handbook.*)

Sonic Logs. Sonic logs read near-constant velocity values inside steel casing (57 μsec/ft) and in evaporate formations such as anhydrite (50 μsec/ft) and salt (67 μsec/ft). Where they occur, these constant values are used to check the correct operation of the compressional-wave travel-time tool.

Resistivity (Laterolog and Induction) Logs. One of the main goals in developing the reservoirwide petrophysical database is to provide the most accurate true resistivity values for subsequent S_w calculations. (See the chapter on resistivity and SP logging in this section of the *Handbook* for information on electric logging.) Often there are several generations of laterolog and induction-logging devices that were run in the various wellbores, and this leads to a variety of data sources with different depths of investigation. Also, over the past two decades, various calculation techniques have been published to deconvolve the reported foot-by-foot resistivity values to obtain more-accurate estimates of true resistivity, R_t.[15,16] Some of the more modern induction tools incorporate deconvolution into the wellsite processing.[16]

Routine-Core-Analysis and SCAL Data. To prepare the routine-core-analysis data for use in reservoir petrophysical calculations, most of these data require adjustment from surface to reservoir conditions.

One of the first steps is aligning core data to depth-aligned log data. Frequently, a GR log of the core is measured in the laboratory, and this is used to depth match the core to the in-situ GR log. Also, in a sandstone with occasional shale intervals of low permeability, the core-analysis data must be aligned with those obvious from the downhole GR log. The log data are often digitized on a half-foot basis, but the core data are typically on a one-foot basis. As these different types of data are included in the same electronic database, care must be taken to ensure that some of the individual data points are not lost. This is likely to require a significant degree of user intervention. The core data for a particular well are really several subsets of data, each of which comes from an individual core-barrel run. These subsets of data must be kept together, and each may need to be individually depth-shifted to the log data.

Porosity Data. For the routine-core-analysis porosity data, SCAL measurements of core samples' porosity at various confining-stress levels are used as the basis for making compaction corrections. Historically, routine-core-analysis porosity measurements were taken at low stress conditions, and SCAL measurements on a small set (10 to 30) of core plugs were made to determine the relationship of porosity to overburden-stress level. Stressed-porosity values are plotted vs. laboratory stressed-porosity values to determine the relationship between the two. Regression of the porosity difference vs. surface porosity gives the same result as regressing stressed porosity vs. surface porosity. Two factors are important to consider in analyzing these plots: whether or not there is a systematic "baseline" laboratory effect related to the equipment's coreholder tightening against the core plug at the start of the test procedure, and whether there is a systematic relationship. Theoretically, the change in porosity is a function of porosity level; however, many sets of SCAL experimental data indicate the "baseline" effect can dominate the second effect.

In the past decade, an unsteady-state style of equipment has come into fairly common usage in which routine porosity measurements can be made at both low-stress and high-stress conditions.[9] With this equipment, reservoir-stress-level porosity measurements can be made on each and every core plug during the routine-core-analysis testing; however, typically, high-stress porosity measurements are made only on every fifth or tenth sample for later use in making the porosity adjustment from surface to reservoir conditions.

For some oil reservoirs, or OBM cores, there is a second adjustment that may need to be applied. This effect is that the routine-core-analysis procedures did not clean all of the heavy hydrocarbons from the pore space, and, hence, the measured porosity values are understated. This effect can be evaluated two ways: first, the measured grain densities may appear to be lower than expected for the particular rock type (i.e., grain densities of less than 2.65 g/cm^3 for clean sandstones); and second, if some of the routine core plugs are retested later, the cleaning solvent is found to discolor, and subsequent porosity values are found to be systematically higher than the original values. This second effect [0.5 to 1.0% bulk volume (BV)] can be as large as the stress-related porosity-reduction effect discussed in the previous paragraphs.

Permeability Data. For the routine-core-analysis permeability data, SCAL measurements of core samples' permeability at various stress levels are used as the basis for making compaction corrections. Historically, routine-core-analysis permeability measurements were taken at low-stress conditions, and SCAL measurements on a small set (10 to 30) of core plugs were made to determine the relationship of permeability to stress level. The "permeability ratio" values (stressed permeability divided by surface permeability) need to be plotted vs. surface-permeability values to determine the relationship, likely to be nonlinear, between the two. Determination of permeability at reservoir conditions is especially important in rocks with air permeabilities of less than 20 md. For low-permeability samples (< 2 md), reductions in permeability of 1 to 2 orders of magnitude have been observed between values at ambient conditions and those at reservoir stress levels.

OBM S_w Data. For routine-core-analysis S_w data from OBM cores, the adjustment from the surface values to those at reservoir conditions requires the application of several factors. First, the pore-volume reduction, as a result of the porosity adjustment discussed previously and because of the change in size of the core plug at stressed conditions, must be applied to the S_w data. Second, the water volume and S_w must be increased because of the effects of reservoir temperature and pressure, salinity, and gas in solution.

Other SCAL Measurements. For other SCAL measurements, some conditioning of these data may be required. For example, for the P_c/S_w data, the data must be converted from surface to reservoir conditions and a height-above-the-OWC (H_{owc}) basis by accounting for the oil/water or gas/water density difference at reservoir conditions and the change in IFT and contact angle between surface and reservoir conditions.

The SCAL electrical-property measurements of a, m, and n (and possibly Q_v, a^*, m^*, and n^*) will need to be considered in the light of the theoretical model that will be used to make S_w calculations from resistivity-log data. Many shaly-sand relationships for estimating S_w from R_t have been proposed.[3,8] These parameters are sometimes measured at overburden conditions.

Other Relevant Data. There are various other types of wellbore data that may need to be inventoried, organized, reviewed, and considered when making the various petrophysical calculations. Other wellbore data that can be particularly important include mud-log data, formation-pressure surveys, formation-tester fluid samples, drillstem-test fluid samples, and 3D-seismic data. Sections 3H.6 and 3H.7 discuss the uses of these data for the fluids-contact identification and for the S_w calculations.

3H.2.4 Acquisition of Additional Petrophysical Data, Where Critically Needed. Often when a new petrophysical evaluation of a reservoir is undertaken, there are significant gaps in the overall database after the existing data have been inventoried and evaluated. It is possible that an acceptable petrophysical evaluation can be completed within the constraints and limitations of the available data, but sometimes additional data are needed. These new data, typically additional SCAL data or possibly routine-core-analysis data, can be obtained from two sources: additional experimental measurements on core plugs taken from existing cores; and drilling, logging, and coring new wells to obtain the needed data. The second approach is used only if such expensive data gathering is required and economically justified.

To obtain additional data from existing cores, geologists can redescribe existing cores as needed. Additional porosity and permeability measurements can be made on newly cut core plugs from any of these cores. Additional valid fluid-saturation measurements might be made on well-preserved core samples; however, such measurements have to be checked carefully because such core samples often dry out over the years. Samples can be cut for more SCAL measurements if the experts consider that the rock samples can be restored for such testing.

In some cases, the need for additional data can justify the drilling of one or more new wells in which cores are cut and routine and special logs are run. In certain equity-redetermination situations, additional wells at specific locations have been drilled to gather additional data about the reservoir interval to more accurately calculate net pay, porosity, and water saturation. For some proposed reservoir-development projects, expensive new data, often of a special nature, can be economically justified because a new well can reduce risk and improve the likelihood of project success.

3H.3 Lithology Determination

Understanding reservoir lithology is the foundation from which all other petrophysical calculations are made. To make accurate petrophysical calculations of porosity, S_w, and permeability, the various lithologies of the reservoir interval must be identified and their implications understood. Lithology means "the composition or type of rock such as sandstone or limestone."[17] These few words belie a host of details about reservoir rocks, their depositional and diagenetic history, pore structure, and mineralogy. Geologists are trained to describe rocks, based on outcrops, cuttings, cores, and more-detailed mineralogical measurements, and they identify certain log-curve characteristics related to particular depositional environments (i.e., coarsening upwards, fining upwards, massive bedding, and the scales of interbedding). For reservoir petrophysical evaluations, a geologist must be on the technical team, and there must be cooperation between the geologists and engineers.

The lithology of a new oil or gas reservoir is understood on a preliminary basis by the wellsite geologist's description of mud-shaker cuttings and possibly by a few cores that are cut if the reservoir interval is sufficiently long. Lithology is also determined from the logs, because each main reservoir lithology has characteristic responses. Frequently, lithologies are derived by pattern recognition of the GR-, density-, and neutron-log responses.

Some subsequent delineation wells are likely to be cored over the entire of the reservoir interval. The geologist will make detailed descriptions of these cores and order a number of thin-section, SEM, XRD, and mercury-injection capillary pressure (MICP) or pore-size measurements on various rock types that have been identified. From these data and the routine-core-analysis data, geologists can construct their interpretation of the reservoir's depositional environment and insights into the nature of its pore system and mineralogy. Geologists typically prepare reservoir cross sections with seismic traces, well logs, and core descriptions to illustrate the depositional environments, rock types, and internal geometries. See the chapter on Reservoir Geology in this volume of the *Handbook* for additional discussion of geologic aspects of oil and gas reservoirs.

The lithology of a reservoir impacts the petrophysical calculations in numerous ways. The depositional environment and sediments being deposited will define the grain size, its sorting, and its distribution within the reservoir interval. In most sandstone reservoirs, the depositional environment controls the porosity/permeability relationship (see the chapter on permeability in the General Engineering volume of this *Handbook* for additional details about the controls on the absolute permeability because of grain size and rock sorting).

The mineralogy of a reservoir results from a combination of its depositional and diagenetic histories. For a sandstone reservoir, the depositional environment controls the percentages of quartz, chert, feldspar, and detrital clay-mineral grains and the other matrix material. These materials must be measured and their variations within the reservoir interval quantified. The diagenetic history determines the extent to which portions of the grains have been leached away; cements such as calcite, siderite, or pyrite have been deposited; and authigenetic clay minerals formed. The diagenetic history can be complicated and can be impacted by differences in burial history from one part of a reservoir to other parts, or by aspects of the hydrocarbon-filling history.

For carbonate reservoirs, most of the same factors as discussed for sandstones come into play, but the mineralogical considerations are different. For carbonate formations, the rock formations typically consist of interbedded sequences of carbonates, dolomites, anhydrite, salt, and shale layers. The keys to reservoir development within the carbonate layers are the original grain size and how it has been altered by chemical diagenetic processes. As these chemical reactions take place, the pore-size distribution and porosity level will change (e.g., by dolomitization). Carbonate-reservoir porosity is also greatly enhanced by weathering, dissolution, and fracturing.

As well as the basic mineralogy, clay-mineral properties are also of particular importance to the petrophysical calculations in the reservoir intervals. There are many types of clay minerals, and their impacts on well logs are quite different. Particularly, there may be differences between clay minerals that form shale layers and claystones and those that occur within sandstone intervals. While the detrital clay minerals in sandstones will likely be the same as the clay minerals in the shales, the authigenetic clay minerals in the sandstone pore system can be quite different. The types and amounts of the various clay minerals impact the rock pore system by affecting its electrical properties and permeability characteristics (see Sec. 3H.3.1). A few of the aspects of mineralogy that impact the petrophysical calculations are the extent to which various heavy minerals (e.g., anhydrite, calcite, dolomite, granites, pyrite, siderite) impact the density log; various light minerals (e.g., coal, halite salt) impact the density, neutron, and sonic logs; various radioactive minerals (e.g., uranium, thorium, and potassium salts such as K-feldspar) impact the GR log; and various electrically conductive minerals (e.g., clay minerals and pyrite) impact the resistivity logs.

3H.3.1 Clay-Mineral Properties. Clay minerals are, in general, composed of layered alumina and silicate molecules,[8,18,19] and the properties of the various clay minerals vary widely. Some swell when wet, are plastic, and can easily deform, while others are hard and dense. Clay minerals are extremely fine-grained, and those with the smallest grain size have a very high surface-area-to-volume ratio. Clay minerals (e.g., chlorite, illite, kaolinite, smectites, and mixed-layer clays) generally impair the permeability and porosity of the pores in which they reside; the permeability is sometimes impacted by an order of magnitude or more. However, it is the smectites (one of which is montmorillonite) that often cause very significant effects on the petrophysical measurements of porosity and water saturation.

In smectites, exchange cations and adsorbed water molecules are loosely bound between the silicate layers. Dehydration occurs whenever there is low humidity or an elevated temperature (e.g., in a dried-up lakebed or in a brick kiln). Loss of this adsorbed water is even more rapid at temperatures above the boiling point of water. For smectites, this is a problem during core analysis because the extraction and drying of the core samples is an essential step in the measurement procedures.

For resistivity logging, smectite clay minerals pose a further problem. The exchange cations and adsorbed water molecules lead to smectite exhibiting excess electrical conductivity. This occurs as exchange cations (e.g., sodium, calcium) migrate from site to site on the clay surfaces. The clays thus exhibit a lower resistivity and, in most cases, depress the bulk resistivity of reservoir rocks in which they reside. Cation exchange also occurs with the other clay minerals, but to a lesser extent.

Rock formations of pure clay minerals are rare. More typically, several species of clay mineral are associated together with clay-sized and silt-sized quartz, mica, and other rock grains. This association is known widely as "shale." In sandstone reservoirs, clay-mineral content typically ranges from 0 to approximately 10% BV. In shales, clay minerals occupy approximately 20 to 40% BV, and the remainder is often very-fine-grained quartz, volcanic minerals, carbonates, and organic matter. Besides the adsorbed water on the clay minerals, shales and authigenetic clay minerals also include additional formation water held in their micron-sized pore system by capillary retention. This water cannot be produced from the formation and is referred to as "capillary water."

During the burial of the sediments over geological time, the overburden stress and the pore-fluid pressures increase. The net result is that water is expelled from the shale beds into surrounding permeable beds. Young clay-mineral-bearing sediments at shallow burial depths are likely to be smectites (e.g., gumbo shale in the submerged Mississippi delta of Louisiana, U.S.A.). Clay minerals in deeply buried shales become less hydrated, and their forms can be altered by higher temperatures and pressures. Low-salinity water is sometimes observed in reservoir-rock pores adjacent to the shales and where clay-mineral-expelled water cannot escape from the permeable bed. This formation may become overpressured and can cause severe problems during drilling if not predicted or detected in time to alter the drilling program.

Although shale formations are not usually of commercial importance, the measurement and evaluation of core and log data in partly shaly reservoirs presents many difficulties that are not present in clay-mineral-free (clean) formations. A large body of technical literature addresses shaly-formation analysis because the shale, to one degree or another, affects all log and core measurements. Because these shaly rocks are so variable, a single model usually cannot fully describe all of their behaviors.

Reservoirs with a fractional shale content, V_{sh}, are common, and the clay minerals/shales take several physical forms, including laminated, structural, and pore-filling.[3,8] Laminated shales are thin detrital shale layers interbedded within a reservoir interval. Each represents short periods of deposition where the suspended finest sediments could settle out of the original sediment-rich river, lake, or seawater. Laminated shales may range from approximately one hundredth of a centimeter to 1 m thick. Shale deposits can be broken up and reworked after their original deposition and become "grains" in the same manner as quartz grains. Structural, or detrital, shale grains become a part of the grain composition of sandstone. As well as the clay minerals that are deposited directly as solids from lakes and marine environments, they also may be deposited from in-situ formation-water solutions that are rich in dissolved minerals. These clay minerals are called "authigenetic." By this mechanism, the pores of sandstone may become partly filled with various clay minerals and other minerals. Of this type, illite, kaolinite, chlorite, and smectite clays are most common. Each can take several physical and chemical forms within a pore. Several generations of pore-filling clay minerals may be present, representing different periods of geological time when changes occurred in formation-water composition or depth of burial.

3H.3.2 Evaluation of Shale Volume. Geological techniques, like XRD, are available to identify clay-mineral species and to quantify rock-component volumes in physical specimens. Such analyses can help calibrate the log-based methods for estimating V_{sh}, the bulk-volume fraction of shale. V_{sh} from the GR log is frequently used to determine nonpay.

Shale content can be estimated from well logs by many techniques, because shale affects the readings of most logs. The task for a particular field is to identify an evaluation technique that is reasonably accurate and as simple as possible. A method using a combination of the neutron and density logs is often applied for practical log analysis. Modern optimized simultaneous-equations log solutions[20] attempt to identify individual clay species. Nevertheless, for petrophysical studies for field development, it is the GR log that is probably used most frequently to evaluate V_{sh}. The GR-log readings are normalized to reduce hole-size variations and mud effects and differences among the tools. Normalization is achieved by finding typical GR values in the 100% "clean" sand and 100% shale formations for each well. These different endpoint values in each well are then equalized. The GR values in between the sand and shale levels are scaled to give V_{sh} values. The scaling is often linear, but nonlinear alternatives are available, if appropriate. Occasionally the GR log is affected by radioactive components that are not shale, and these need to be identified and assigned a revised V_{sh}. Water-based drilling muds sometimes contain high concentrations of potassium salts, and these may also lead to GR interpretation problems, such as invasion of potassium salts into the near-wellbore region.

The uncertainty of V_{sh} log evaluations is moderate to high. At low values, less than 30% BV, the authors estimate that V_{sh} may typically be accurate to approximately ± 10% BV at one standard deviation (SD). At values greater than 30% BV, the uncertainty increases. The uncertainties affect nonpay bed-boundary evaluations; however, the V_{sh} uncertainty also seriously impacts the accuracy of the effective-porosity and S_w estimates when smectites, with their high clay-mineral adsorbed-water fraction, are present. If the clay species is different (illite or chlorite with little or no adsorbed water), then the V_{sh} uncertainty has a much-reduced impact on porosity, as is discussed in Sec. 3H.5.3.

3H.3.3 Reservoir Zonation or Layering. An important conclusion from the geologists' technical studies is a definition of the extent to which the reservoir needs to be subdivided either vertically or areally. Besides all of the information developed by the geologists from the detailed core descriptions, the routine-core-analysis and SCAL data need to be analyzed for such effects. This is accomplished by preparing, for different possible layering within the reservoir interval, a variety of crossplots such as $\log_{10}$ (permeability) vs. porosity and OBM-core S_w vs. porosity or $\log_{10}$ (permeability) and comparing the data clouds and trends of those plots from one possible layer to the next. Significant differences should be expected between reservoir intervals with different depositional environments and differences in grain size and sorting. Areal variations may occur across a given vertical zone within the reservoir because of varying distances from the sediment source, differences in the depositional environment in various areas, or varying diagenetic effects.

For accurate petrophysical calculations, most large reservoirs will probably require a number of vertical subdivisions, usually termed zones or layers. Typical zones for a reservoir are 50 to 150 ft (15 to 45 m) thick. Areally, several square miles of reservoir can usually be included together; however, if the reservoir covers tens of square miles, it is likely to require some areal subdivision for accurate petrophysical calculations.

A second consideration is the amount of available data of various types. If a reservoir has a large database of log and core data from many wells, then the number of these subdivisions is not impacted by the quantity of data. However, if some types of data are very limited, then this consideration may control the degree of vertical and areal subdivision that can be used for various petrophysical calculations. The same degree of subdivision may not be required for net-pay calculations relative to porosity calculations or for porosity calculations relative to S_w calculations. In summary, more-accurate vertical and areal petrophysical calculations are made if the reservoir is appropriately subdivided.[21,22] **Fig. 3H.1** shows the vertical zonation used for the Prudhoe Bay field's Sadlerochit reservoir. **Figs. 3H.2 through 3H.4** are example plots based on real reservoir data that show how rock properties within the same field can vary from one reservoir to another and from one vertical portion of a reservoir interval to other parts.

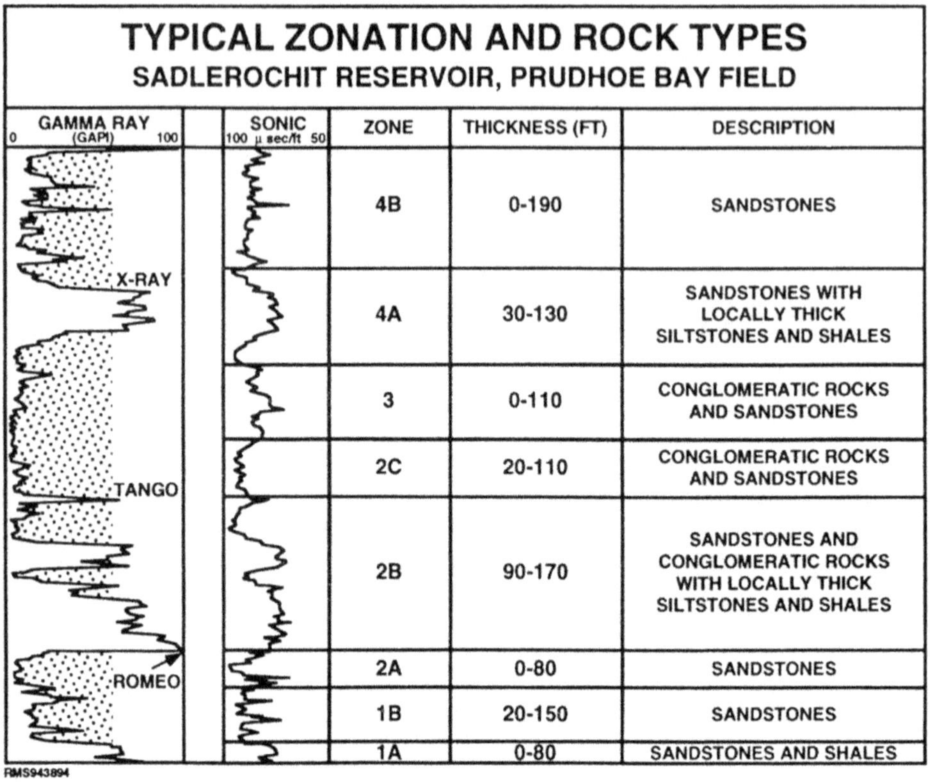

Fig. 3H.1—Reservoir zonation of the Sadlerochit reservoir, Prudhoe Bay field, Alaska, U.S.A.[21]

3H.4 Net-Pay (or Pay/Nonpay) Determination

The goal of the net-pay calculations is to eliminate nonproductive rock intervals and, from these calculations at the various wellbores, provide a solid basis for a quality 3D reservoir description and quantitative hydrocarbons-in-place and flow calculations. The determination of net pay is a required input to calculate the hydrocarbon pore feet, F_{HCP}, at a wellbore and its input to the overall reservoir OOIP or OGIP calculations. The total F_{HCP} at a well is the point-by-point summation over the reservoir interval with Eq. 3H.1. The top and base of the reservoir interval are defined by geologists on the basis of core descriptions and log characteristics.

$$F_{HCP} = \sum_{i=1}^{i=m} h_{ni} \phi_i \left(1 - S_w\right)_i \qquad (3H.1)$$

In the F_{HCP} calculation, net pay, h_{ni}, at each data point has a value of either 1 (pay) or 0 (nonpay). The "net-to-gross ratio" or "net/gross" (N/G) is the total amount of pay footage divided by the total thickness of the reservoir interval (for simplicity, the well is assumed here to be vertical). A N/G of 1.0 means that the whole of the reservoir interval is pay footage. In this formula, any foot (or half foot) that is defined as nonpay contributes absolutely nothing to the subsequent reservoir-engineering OOIP (or OGIP) and reserves calculations, even if it contains some amount of hydrocarbons. The net-pay determination should be performed in a reasonable practical manner, but it should be recognized that when any cutoff is used, the result will, to

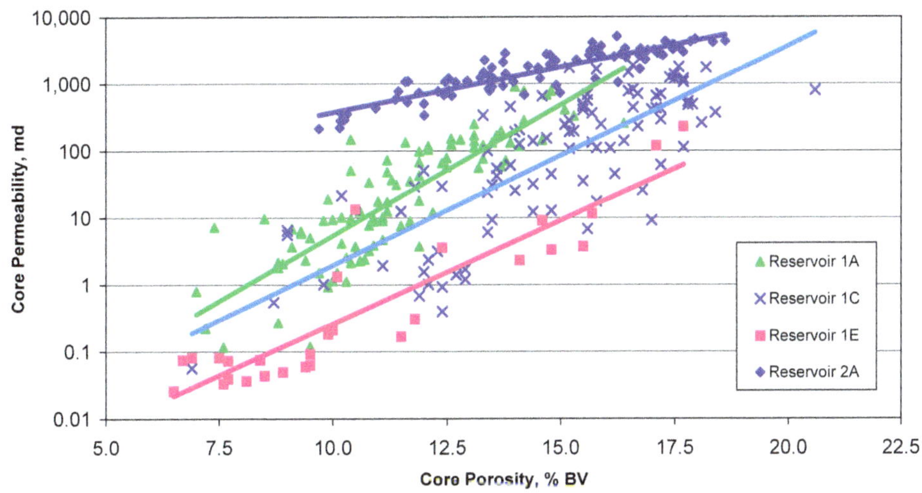

Fig. 3H.2—Typical reservoir permeability vs. porosity crossplot; data from an Asian gas field.

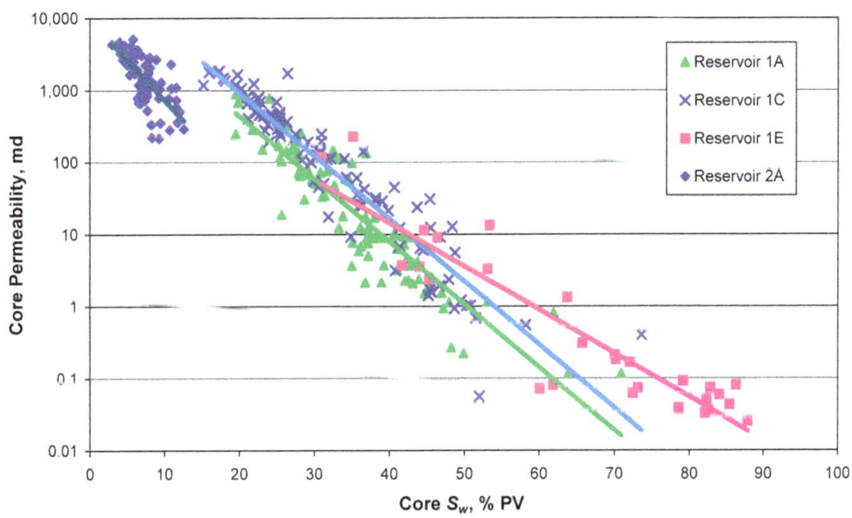

Fig. 3H.3—Typical reservoir permeability vs. S_w crossplot; data from an Asian gas field.

some extent, be arbitrary (see the chapter on reserves determination in this volume of the *Handbook*).

3H.4.1 Conceptual Bases for Net-Pay Calculations. Several conceptual bases for the petrophysical calculations of net pay are described here. At one limit, the whole of the reservoir interval can be treated as net pay (i.e., N/G equals 1.0). Another reasonable engineering approach is to define some lower limit on flow, below which each foot or half foot of the reservoir interval is deemed to be nonpay. A third approach is to use one or more log cutoffs that have been used historically within the petroleum industry. The advantages and disadvantages of these various approaches will be discussed in this section.

N/G = 1.0. One approach is to calculate the OOIP or OGIP assuming that all the reservoir interval is pay to determine the total volume of hydrocarbons present within the reservoir inter-

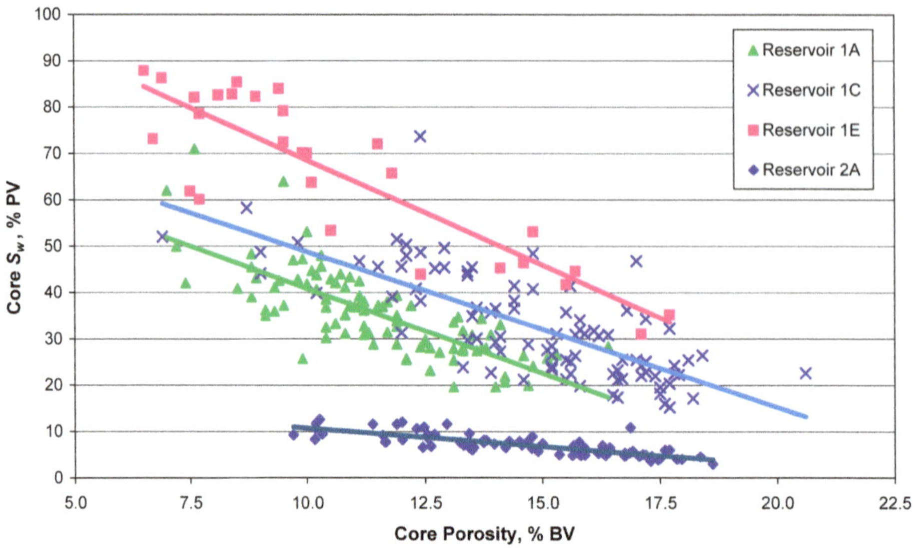

Fig. 3H.4—Typical reservoir S_w vs. porosity crossplot; data from an Asian gas field.

val. When using an N/G = 1 approach, the technical team needs to ensure that the calculations of porosity, permeability, and S_w are quantitatively reasonable over the whole range of values for each of these parameters. This calculation could be called a determination of the "total hydrocarbon resource" within the reservoir interval, and it provides a value for the total hydrocarbon potential of the reservoir. Some of these hydrocarbons will have low mobilities and will contribute little or nothing to hydrocarbon recovery. But, with this value available, the engineer has a measure of how well the reservoir is producing overall and what resources should be considered for improved-recovery-project evaluations. This value can be viewed as the ultimate "prize."

Another reason for setting N/G to 1.0 is that, with modern reservoir-engineering tools, it is technically feasible to treat the entire reservoir interval as pay. For example, with modern reservoir-engineering tools, a million (or more) cell reservoir-simulation model can be constructed in which a very detailed description of the vertical and horizontal variations in the reservoir-rock properties are incorporated. In this approach, the very-poor-quality portions of the reservoir are assigned low porosities, low permeabilities, and higher water-saturation values. Then, in the OOIP or OGIP calculations, these portions contain only small volumes of hydrocarbons and will contribute their appropriate, albeit small, share to pressure maintenance and recoverable hydrocarbons. This is in contrast to defining these poor-quality intervals as nonpay and defining *a priori* that they contribute nothing to OOIP or OGIP or reserves.

Mobility or Permeability Cutoff Approach. From first-principle calculations using Darcy's law, a reservoir engineer can define net pay by applying a fluid-flow cutoff. The choice of this cutoff would be related directly to the hydrocarbon mobility (rock permeability divided by hydrocarbon viscosity) in the different portions of the reservoir interval. With this approach, the net-pay permeability cutoff used in the point-by-point log calculations would be quite different between that for a gas reservoir (very low gas viscosity of approximately 0.02 cp), that for a light-oil reservoir (oil viscosity of 1 to 10 cp), and that for a heavy oil reservoir (oil viscosity of 10,000 cp or more).[23] Any portion of the reservoir interval that has a permeability at reservoir conditions below the cutoff would be defined to be nonpay. In the next section, the gas-reservoir situation is discussed separately from that for oil reservoirs.

The arbitrary nature of any net-pay cutoff is apparent when one notes the flow implications of using a permeability cutoff. If a rock interval has a permeability 1% greater than the cutoff value, it is included as net pay. However, if another rock interval has a permeability 1% less than the cutoff value, it is excluded as nonpay. The difference between the fluid-flow contributions from these two rock intervals is only 2%, yet one is allowed to contribute to the subsequent OOIP or OGIP and reserves calculations, while the other is not.

If a permeability cutoff is chosen, its application to the various wellbore data (cores and logs) generally takes three steps.[24] The first step is to apply the permeability cutoff to the routine-core-analysis permeability data. In this step, there are two checks that need to be made. First, a permeability/porosity plot needs to be prepared and outlier points identified. The outlier points need to be individually checked for validity. For example, a very-low-porosity shale sample may have dried out and a parting developed between the shale layers. This may lead to a very high permeability value that is not consistent with the rest of the rock characteristics. Bad routine-core-analysis data points should be excluded from the database. A second consideration is that often during the routine core analysis, the shale intervals are not sampled at the same frequency of core plugs as the other lithologies. This likelihood must be kept in mind in reviewing the routine-core-analysis database and in comparing the results of pay/nonpay calculations between cores and logs. The subsequent steps—the conversion of a permeability cutoff to a porosity or V_{sh} cutoff including the calibration of logs to the core standard, and the calculations from the logs of net pay for all wells over the reservoir interval—are discussed later in this section.

3H.4.2 Gas Reservoirs. For a gas reservoir being produced under pressure-depletion drive, any permeability cutoff applied should be very low. This is quite evident by the successful development of tight gas-sandstone reservoirs producing nearly 10 Bcf/D from 85,000 wells in the United States, some with average permeabilities in the microdarcy range (see the chapter on tight gas sands in the Emerging and Peripheral Technologies volume of this *Handbook* for additional details). In conventional gas reservoirs in which higher-quality rock intervals are interbedded with the poorer-quality ones, gas in the poorer-quality rocks will flow to the higher-quality rock intervals if there is any permeability between the two. An example calculation for gas flow from a 1-microdarcy layer with a pressure difference of 2,000 psi, over a thickness of 10 ft for an area of 10 acres, and for a period of 1 year shows that this layer would contribute 1 Bcf per year.

Because pressure-depletion-drive gas reservoirs are produced for decades and, if found at significant depths, have abandonment pressures less than 10% of their initial pressures, there are both long times and large pressure differentials to cause gas to flow from very-low-permeability and low-porosity rock intervals into higher-permeability conduits and on to the production wellbores. In many instances, the distance traveled to reach a higher-permeability layer is just a few feet vertically.

3H.4.3 Oil Reservoirs. For oil reservoirs, any permeability cutoff will be significantly higher than that for a gas reservoir, generally by a factor of 10 or 100 or more. A second aspect of oil reservoirs is that typically, only 10 to 20% of the OOIP will be produced by pressure-depletion drive (without assistance from gravity drainage) in which the pressure differential will affect all portions of the reservoir. However, during waterflooding, overall oil/water displacement efficiency will depend, in part, on how much of this displacement process occurs in poorer-quality oil-bearing rock intervals. Hence, the choice of oil-reservoir permeability cutoff needs to account for the oil/water relative permeability effects. Interwell injector/producer connectivity (or "floodability") is not a topic of this chapter. Connectivity will affect recovery but is considered a separate issue apart from individual-wellbore calculations of net pay.

Any permeability cutoff cannot be directly applied to foot-by-foot log calculations of net pay because there is no log that quantitatively measures permeability. A permeability cutoff typically is converted to a porosity cutoff and is subsequently applied to the logs through log porosity, bulk density, GR, or V_{sh} cutoffs. The procedures for applying a permeability cutoff to the logs are discussed later in this section.

3H.4.4 Other Historical Net Pay Cutoffs. The technical literature shows that a number of net-pay-cutoff approaches have been used over the years by petrophysicists, geologists, and reservoir engineers.[23–26] These cutoffs are often simply stated as a particular value for porosity, V_{sh}, and/or S_w. The justification for those cutoffs is rarely stated. Geologists often provide their foot-by-foot pay/nonpay identifications ("picks") with their detailed core descriptions.

In all cases, chosen cutoffs generally exclude poorer-quality rock intervals. The key issues become whether the cutoff is applied in an accurate, consistent, and systematic manner through net-pay calculations using log data and the volume of hydrocarbons that are excluded in the net-pay calculations. For example, geologists often define shales, and possibly siltstones and very shaly sandstones, as nonpay intervals; then the GR logs can be calibrated to this nonpay standard. A value of the geologists' core descriptions in the net-pay determination is that core descriptions are generally more continuous than are the routine-core-analysis data.

Like the geologist's criteria, porosity and V_{sh} cutoffs also exclude poorer-quality rocks because lower-porosity rocks also generally have lower permeability, higher water saturations, and little, if any, mobile oil. Higher V_{sh} values are generally indicative of more clay-mineral-rich rocks that also tend to have lower porosities, lower permeabilities, higher water saturations, and little, if any, mobile oil. If these two cutoffs are applied without consideration for flow implications, then some rock intervals containing significant volumes of hydrocarbons that can contribute significantly to production may be excluded before OOIP, OGIP, or reserves calculations.

An S_w cutoff is sometimes also used as a nonpay cutoff, often in addition to a porosity and/or V_{sh} cutoff. An S_w cutoff is typically justified on the basis that, at high S_w, gas or oil is immobile on the basis of relative permeability considerations. This approach does not account for, at original reservoir conditions, high-S_w rock intervals that contain lower hydrocarbon saturations but with those hydrocarbons in the larger pores of the rock. These hydrocarbons will have mobility and contribute to production, particularly for gas reservoirs in which, as the pressure declines, the gas phase expands (and gas saturation increases) and results in gas flow toward the production-well pressure sinks.

The difficulty with the use of porosity, V_{sh}, or S_w cutoffs, without reference to flow considerations, is that rock intervals evaluated to contain hydrocarbons may be excluded from the other reservoir-engineering calculations. Each of these approaches, when applied to the logs, requires that underlying physical relationships between log readings and these cutoffs be understood. Also, complications on logs (e.g., intervals of heavy minerals, radioactive minerals other than clay minerals, or hole washouts) need to be quantified and treated appropriately in net-pay calculations.

3H.4.5 Geologic Considerations in Net-Pay Determination. The primary geological considerations in determining pay and nonpay in the reservoir interval are depositional environment and hydrocarbon and structural history. The depositional environment provides a picture of whether the overall reservoir interval is sand rich (high N/G) or shale rich (low N/G) and the nature of the interbedding of high-quality rock with poor-quality rock. If the reservoir interval is quite interbedded with high-quality rock intimately layered with poor-quality rock on the scale of a few inches to a few feet, the poor-quality rock intervals, if they contain hydrocarbons, will likely contribute to production. However, if the layering is on a much larger scale with thick high-quality rock intervals separated from thick low-quality rock intervals, then the poor-quality rock intervals are much less likely to contribute significantly to production.

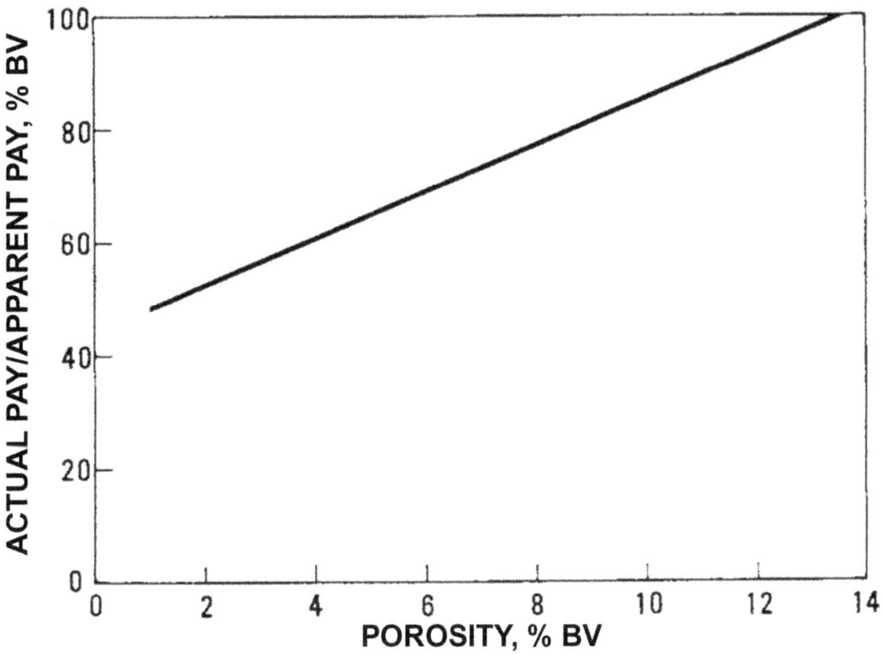

Fig. 3H.5—Apparent and actual pay for Robertson Upper Clearfork unit, west Texas, U.S.A.[26]

Regarding hydrocarbon and structural history considerations in net pay calculations, several fields have relict-oil intervals below the current OWC (e.g., Prudhoe Bay, Alaska North Slope, U.S.A.; San Andres carbonate reservoirs, west Texas, U.S.A.)[27–29] or relict-gas intervals below the current gas/water contact (GWC) (e.g., North Morecambe field, Irish Sea, U.K.).[30] These relict-oil columns would generally be considered to be nonpay intervals because of their high mobile-water saturations and lack of oil mobility. This is true for either primary production or waterflooding; however, for CO_2 enhanced oil recovery, the west Texas San Andres relict-oil intervals have been considered for development. A more significant situation is that of relict-gas saturations below the current GWC. This gas does not have immediate mobility; but if the aquifer is not strong, this gas will expand and can contribute to production as the reservoir pressure declines. Hence, a relict-gas interval should not necessarily be excluded in net-pay calculations.

George and Stiles[26] published an excellent example of the complications of net-pay calculations concerning the heterogeneous Clearfork carbonate oil reservoirs in west Texas, U.S.A. Their approach was to develop an empirical relationship between "actual pay" and "apparent pay" as a function of porosity in order to redetermine net pay to improve OOIP calculations and to obtain a reasonable distribution of net pay. They defined two net-pay cutoffs. The "actual pay" was defined as the net thickness of core samples with permeabilities greater than 0.1 md, and an "apparent pay" was defined as the net thickness of core samples with porosity greater than a specific cutoff. **Fig. 3H.5** shows the relationship of actual pay to apparent pay as a function of porosity. On the basis of this analysis, at a porosity level of 8% BV, 75% of the rock samples would be pay, while at a porosity level of 1% BV, 50% of the rock samples would be pay. By this methodology, wells with low porosity levels will not be all nonpay, but will be given a limited amount of pay. The purpose of their method was "to achieve a better distribution of porosity-feet" and "both total original oil in place and distribution of PV throughout the field will be realistic."[26]

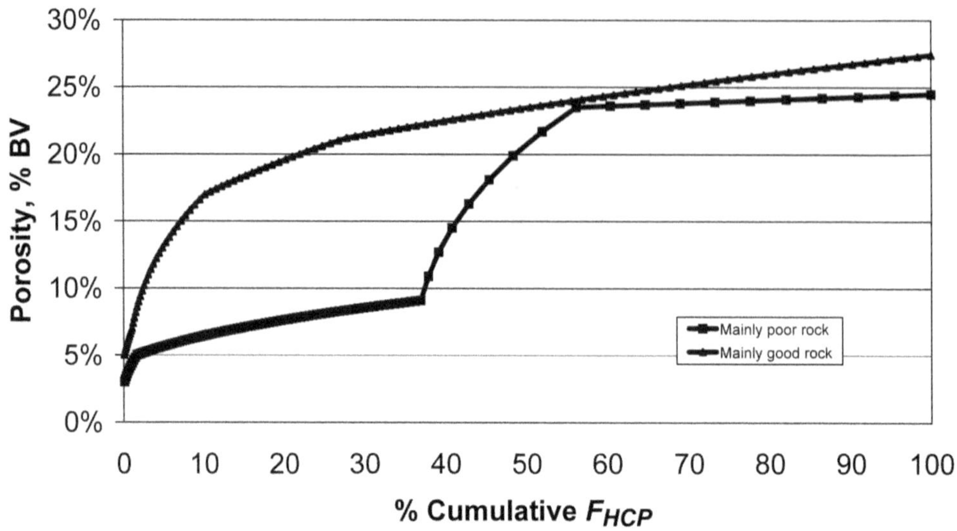

Fig. 3H.6—Illustrative plot of cumulative F_{HCP} vs. porosity for two types of reservoirs (one mainly good-quality rock, the other mainly poor-quality rock). The total F_{HCP} of the "good" reservoir is three times the total F_{HCP} of the "poor" reservoir.

Finally, the technical team needs to determine the implications of any net-pay cutoff. This is best done by plotting the cumulative hydrocarbon pore feet (F_{HCP}) percentage as a function of porosity and as a function of permeability (see **Figs. 3H.6 and 3H.7** for respective examples of these two types of plots). In this way, it is possible to determine what percentage of the hydrocarbons within the reservoir interval would be excluded by any particular net-pay cutoff. While this net-pay sensitivity method is a logical approach, the evaluation of porosity and water saturation is more uncertain in low-porosity rocks. Log calculations may indicate hydrocarbon saturations in rocks where no hydrocarbon actually exists.

3H.4.6 Application of Net-Pay Cutoffs to Well Logs. The four main steps in the application of a net-pay cutoff to a particular reservoir interval are to establish a standard, calibrate one or more logs to the chosen standard, confirm that the calibration step produces results consistent with the standard, and apply the calibrated model to all wells.

Establish a Standard. As discussed previously, the choice of the standard for the net-pay calculations should be reasonable but is, to some degree, arbitrary. The choice should be a single concept, such as a permeability cutoff, a porosity cutoff, or geologists' calls of pay/nonpay from core descriptions. The use of multiple cutoffs will lead to a very conservative result that eliminates rock intervals that are likely to contribute to production, particularly for gas reservoirs. This underestimation occurs because each of the individual cutoffs will, to some extent, define different datapoints as nonpay. Even after the best possible depth matching of the logs involved, remaining depth mismatches always occur, resulting in the double counting of nonpay at bed boundaries. The following discussion assumes that an air-permeability cutoff of 0.1 md has been chosen.

Calibrate One or More Logs to the Chosen Standard. Once the 0.1-md air-permeability cutoff has been chosen, it needs to be converted into a methodology that can be applied to foot-by-foot log calculations. Typically, this is done by converting the permeability-cutoff value into a porosity-cutoff value by a permeability-vs.-porosity semilog crossplot of routine-core-analysis data converted to reservoir conditions. Also, plots are made of the core permeability data vs. the various available log parameters to determine if there is a strong correlation that can be

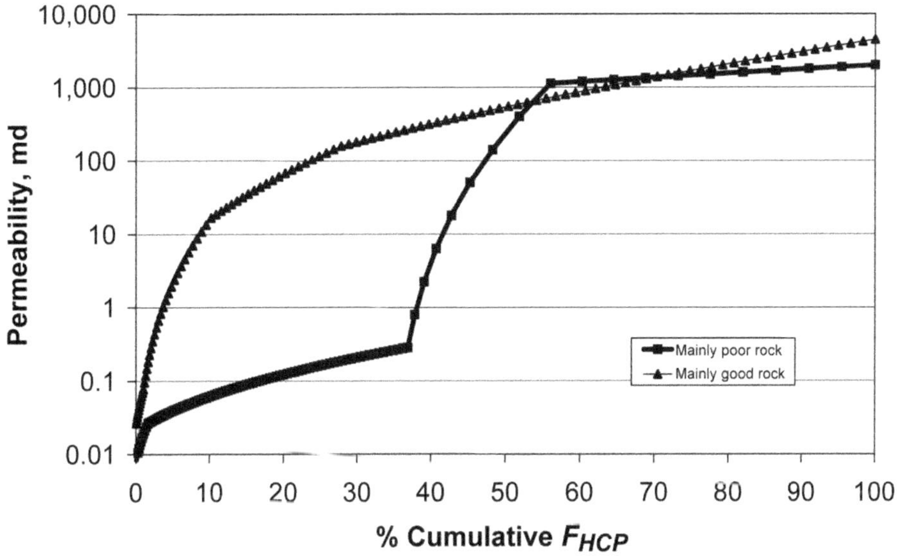

Fig. 3H.7—Illustrative plot of cumulative F_{HCP} vs. permeability for two types of reservoirs (one mainly good-quality rock, the other mainly poor-quality rock). The total F_{HCP} of the "good" reservoir is three times the total F_{HCP} of the "poor" reservoir.

used. Alternatively, a multivariate regression technique might be used to calibrate multiple logs to permeability. If a porosity cutoff is developed from the permeability cutoff, then it needs to be defined as a log-related cutoff, such as a log-derived porosity or density log cutoff, or a V_{sh} or GR-log cutoff. There are several variations on how this calibration step can be undertaken. The alternatives are not discussed here because each reservoir situation has unique characteristics.

Confirm That the Calibration Step Produces Results Consistent With the Standard. After the calibration step is completed, the resulting log calculations of pay/nonpay need to be checked against the core standard in the cored wells. This is needed to determine that the log calculations and their cutoffs do not overstate or understate the calibration standard of net pay of the reservoir interval. The goal is to develop the "best estimate" values in the reservoir-engineering calculations, not the "low estimate" or the "high estimate."

Apply the Calibrated Model to All Wells Over the Reservoir Interval. After the first steps have been successfully completed, the finalized net-pay log model can be applied to all wells' valid log data in the reservoir interval to develop point-by-point pay/nonpay determinations. For optimal results, it may be necessary to have different models in different areas of the reservoir. The results for each zone over the reservoir should be quality controlled. Maps should be examined looking for "bulls-eyes" that may represent either real geological effects, artifacts in the database, or bad calculations.

In the subsequent steps of calculating porosity, S_w, and permeability, those calculations will be made only for pay intervals. The nonpay intervals will be excluded from the core and SCAL database and the log database. In cases in which the depth matching of cores and logs presents difficulties, it is prudent to retain both core- and log-defined nonpay in the database. This will enable appropriate samples to be selected for various analyses, such as the evaluation of SCAL petrophysical properties.

3H.5 Porosity Determination

The accurate calculation of porosity at the wellbore is essential for an accurate calculation of OOIP or OGIP throughout the reservoir. The porosity and its distribution also need to be calculated as accurately as possible because they are almost always directly used in the S_w and permeability calculations and, possibly, in the net pay calculations. In most OOIP and OGIP studies, only the gross-rock-volume uncertainties have a greater influence on the result than porosity does. Occasionally, where porosity estimates are difficult, porosity is the leading uncertainty. Fractured and clay-mineral-rich reservoirs remain a challenge.

This section describes the methods that can be used to make porosity calculations at the wellbore with the available core and log data. For this discussion, it is assumed that the core data have been properly adjusted to reservoir conditions, that the data from various logs have been reviewed and validated as needed, and that all of the required depth-alignment work has been completed. Sec. 3H.2 discusses the specifics of these topics. This section discusses the use of core porosity data, total and effective porosity, core-log calculation approaches, consistency of calculations, and uncertainty.

3H.5.1 Use of Core Porosity Data. There are a few preliminary steps in the use of routine core porosity data over the reservoir interval. First, this data set needs to be restricted to those porosity measurements made in pay intervals; the nonpay porosity measurements should be excluded from the porosity calculations. Second, if more than one type of porosity measurements are made, then a hierarchy of these measurements needs to be developed for use in subsequent log/core porosity calculations. For example, it is possible that whole-core Boyles-law porosity measurements, core-plug Boyles-law porosity measurements, and sum-of-fluids porosity measurements are made with several of these measurements made on the same feet of core.[13] (See the chapter on petrophysical measurements in the General Engineering volume of this *Handbook.*) Also, if enough of two data types are taken for the same footage of cores, then these data need to be crossplotted to determine if there are any systematic differences between the various types of core porosity data. If more than one porosity measurement are available for a given depth, then only the highest priority in the hierarchy should be included in the log/core porosity calculations; otherwise, there will be unequal weighting in the statistical calculations.

The core porosity measured in shaly sands may include some volume that is associated with the dehydration of certain types of clay minerals (see Sec. 3H.3.1).[10,11,18,19] When smectite (montmorillonite) clay mineral is present as a significant fraction [e.g., a V_{cl} (not V_{sh}) of 40% BV], the core porosity may be increased, by approximately 12% BV, solely because of the smectite present. However, when other species of clay are present that have much less clay-mineral physically bound water (e.g., chlorite, illite, kaolinite), the clay water will add little to the core porosity. The effects of the presence of shale and clay minerals must be understood to yield the correct evaluation of the hydrocarbon and water content of a reservoir's pore network.

3H.5.2 Total and Effective Porosity. The estimation of porosity and water saturation in shaly formations is, where possible, based on various types of laboratory core data. Where core measurements are not available, estimates based wholly on log measurements and selected interpretation models are widely used. Rock models are based on "total" or "effective" porosity and "total" or "effective" S_w definitions. Both definitions account for the usual grain volume and hydrocarbon and capillary-water volumes seen in the porosity of nonshaly sands, and both models include volumes for the clay-mineral physically adsorbed water (sometimes known as clay-bound water) and the volume of dry clay minerals. **Fig. 3H.8** is a schematic of a shaly-sand-reservoir model. It indicates the various solid and fluid volumes and pore networks to which core measurements, density, and neutron logs correspond.

Core Analysis. This discussion is restricted to siliciclastic rocks; carbonate rocks are not discussed. The porosity measured on core plugs containing clay minerals is dependent on the methods used to clean and dry the sample before it is measured. Cleaning removes oil from the pores. It is widely, if not universally, accepted that drying the core in a vacuum oven at temperatures just above the boiling point of water (110°C) will remove most, or all, of the clay-

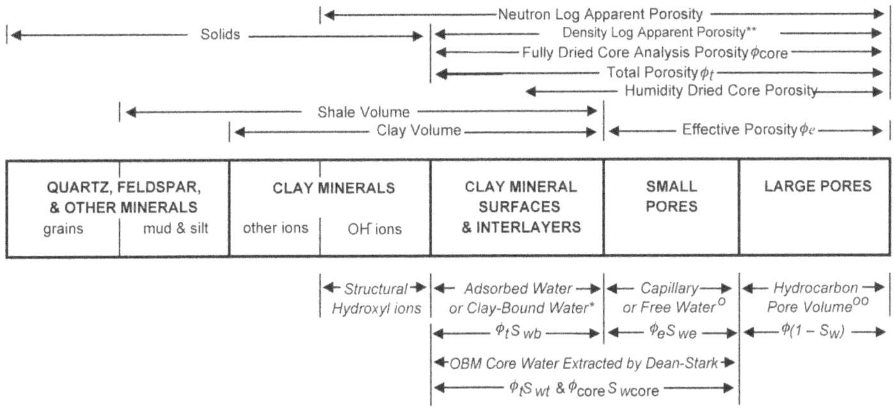

** Density log apparent porosity using the grain density of dried core and correct invaded-zone fluid density.
* Smectite clays have a large capacity to adsorb water, whereas most other clay minerals have little or no adsorbed (or bound) water.
°° This model has no mobile water, and the model sandstone will produce water-free hydrocarbons during its initial production phase.
° As height above the hydrocarbon-water contact increases, capillary water volume decreases and hydrocarbons displace capillary water.
Humidity dried core analysis porosity is shown as including some of the clay-bound water.
Box sizes are schematic.

Fig. 3H.8—Total, effective, and core porosities and their associated water saturations in shaly sands (after Woodhouse and Warner[31]).

mineral adsorbed water and the capillary water, but not the chemically bonded hydroxyl groups within the clay minerals. The standard reported core porosity, ϕ_c, is, therefore, a total porosity including the effective porosity, ϕ_e, and the clay-mineral adsorbed water volume, $V_{cl}\phi_{cl}$ or $V_{sh}\phi_{sh}$.

$$\phi_c = \phi_e + V_{cl}\phi_{cl} = \phi_e + V_{sh}\phi_{sh} \text{ . ... (3H.2)}$$

OBM-core S_w determined using Dean-Stark water-volume extraction is also accomplished with boiling toluene at approximately 110°C. Subsequent preparation of the OBM-core sample to measure porosity uses the same maximum temperature.

The core total porosity, ϕ_c, and core water saturation, S_{wc}, are, therefore, fully compatible with each other.[31] They are used together to accurately quantify hydrocarbon pore volume (V_{HCP}) [the "core" $V_{HCP} = \phi_c \times (1 - S_{wc})$]. When smectite clays are present, the core total porosity will be higher than the effective porosity, ϕ_e, as defined in the previous paragraph. However, for the same sample, the core water saturation, S_{wt} (capillary water plus clay-mineral adsorbed water volumes as a fraction of ϕ_t), will also be greater than S_{we} (capillary water volume as a fraction of the effective porosity) and will fully compensate the V_{HCP} for the increased porosity.

V_{HCP} is used instead of F_{HCP} in the following discussion because the equalities apply at all scales.

$$V_{HCP} = \phi_t \times (1 - S_{wt}) = \phi_e \times (1 - S_{we}) = \phi_c \times (1 - S_{wc}) \text{ . ... (3H.3)}$$

For porosity,

$$\phi_t \geq \phi_c \geq \phi_e \text{ . ... (3H.4)}$$

For water saturation,

$$S_{wt} \geq S_{wc} \geq S_{we} \text{ . } \quad (3H.5)$$

"Humidity-dried" cores, extracted at temperatures lower than boiling water and partially dried, retain some, or all, of the clay-mineral physically bound water and, therefore, approximate the effective porosity. There is, however, no generally accepted way to measure accurately the values of effective porosity on shaly-formation cores. In smectite-rich shaly sands, humidity-dried porosity is incompatible with the Dean-Stark S_w, and, if combined, the result will understate V_{HCP}.

Cores extracted at much higher temperatures may give compatible porosity and S_w results, but other problems can sometimes occur. The mineral gypsum dehydrates between 110 and 120°C and must be corrected for in the retort method of water-volume measurements. At 110°C, the structural, chemically bound hydroxyl groups that are part of the clay-mineral lattice are not liberated.[18] However, at very high temperature (500°C or more), these hydroxyl groups react to form water and condense as water in the collecting tubes of a high-temperature retort apparatus. This "structural water" is not captured during standard core analysis but is the hydrogen component of clay minerals that is detected by the neutron porosity log even though it is a part of the solid mineral (Fig. 3H.8).

Log Analysis. Log-interpretation methods for porosity and S_w all seek the same end result in terms of porosity and hydrocarbon volume. They are divided into methods that model the effective porosity and the clay-mineral adsorbed water separately and methods that model the total porosity containing both clay-mineral adsorbed water and capillary water. The total porosity is often calculated from the density log and the core grain density, as Fig. 3H.8 shows. Estimates of interconnected porosity (effective porosity) at reservoir conditions come from combinations of many different logs, but all of them attempt to quantify the clay-water volume fraction and subtract it from the total porosity.

3H.5.3 Core/Log Calculation Approaches. In calculating porosity values from the core and log data, the first step is to create depth plots and crossplots of the core data against the various log data, like those in **Figs. 3H.9 and 3H.10**, respectively. These crossplots visually show which of the logs has the strongest correlation with the core porosity measurements. For example, the density log readings vs. core porosity data may have a less-scattered data cloud than the sonic log vs. core porosity data, or, if there are heavy-mineral complications, the opposite may be the case. These crossplots show, and the correlation coefficient, r, of each correlation indicates which of the log/core combinations should be used for the log/core calibration step discussed next. Where possible, it is best to use a single-log porosity estimator because multiple-log estimators will have problems at bed boundaries because of imperfect depth matching. The volumes sampled by different logs also vary, with the neutron log "seeing" a larger volume than the density log.

The variance of core-plug data is always larger than the equivalent variance of a log because of the small plug volume compared with the larger volume seen by the running-average log reading. One way to reduce the core-plug variance is to create a modified core property curve that is the running average of the core data (a 1-2-1 filter may be appropriate). Core data modified in this way are considered by some to be a superior calibration standard.

Calibration Line-Fitting. The generally recommended method for obtaining a line-fit for porosity prediction is the "y-on-x" ordinary least-squares regression method.[32,33] The recommendation presumes that the calibration data set has accurate depth adjustments and is fully representative in all respects of the environment of the equation's future use. The dependent-variable calibration data, y, the values wanted in the future (e.g., core porosity), are regressed against the selected independent x-variable data [the values available to make the future prediction (e.g., the density log values)]. The same x-variable must be used when the calibration line

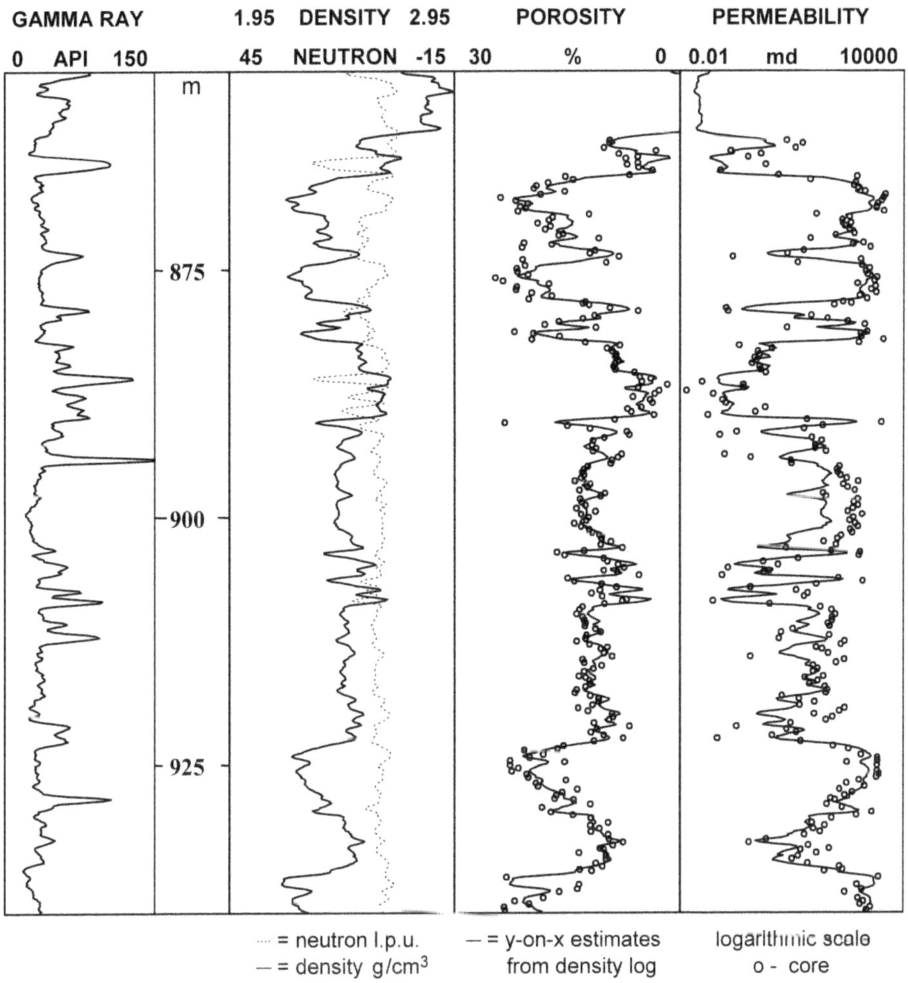

Fig. 3H.9—Well-log and core data from gas-bearing sandstone with predicted porosity and permeability (South Morecambe gas field, offshore U.K.). After Woodhouse.[32]

is applied in the uncored wells. Multiple regression uses more than one independent variable (e.g., the density and GR logs).

For *y*-on-*x* regression lines, removal of *x*-*y* data outliers, far from the general trend, must be considered. The *y*-on-*x* method assumes that the *y* and *x* measurements apply to the same rock sample, so data pairs that are not likely to represent similar rocks must be edited from the data set.

Although straight (linear) regression lines are often created with the data in their original form, the regression method applies equally to curved relationships. These are achieved by transforming one or more of the variables. For example, it is common to transform permeability to a logarithm, thereby creating a $\log_{10}$ (permeability) and linear porosity relationship. This transform preserves the geometric averages of the permeability. As an alternative, a permeability and exponential porosity relationship should be considered because this will preserve the arithmetic averages of permeability, instead of the geometric averages.

To some observers, the *y*-on-*x* line-fit initially is awkward and less central than some other line-fits. The reduced-major-axis (RMA) line-fit, for example, follows the intuitive middle

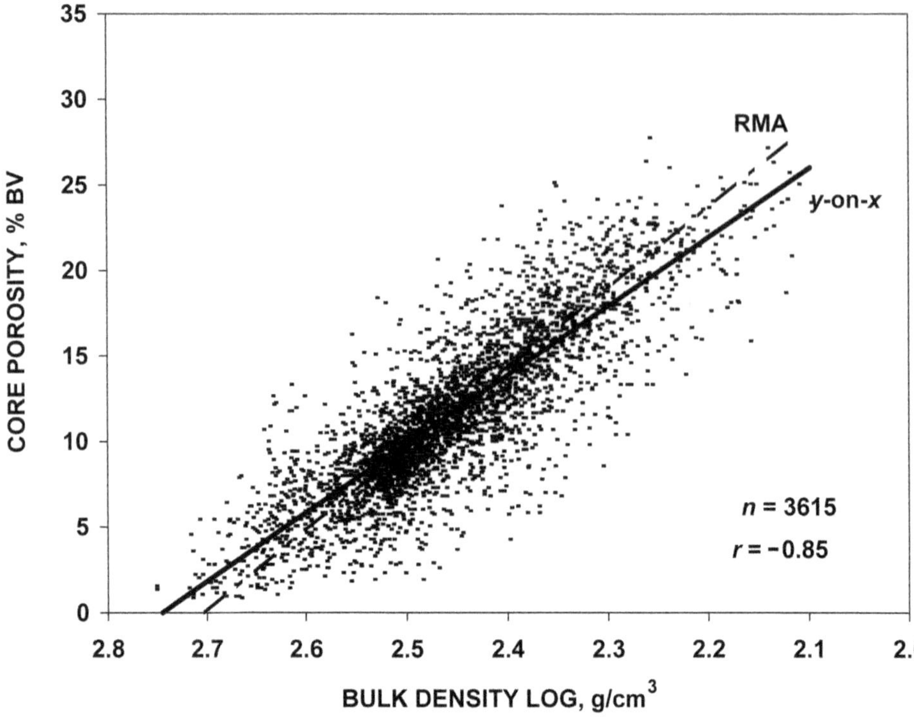

Fig. 3H.10—Core-plug porosity vs. well-log bulk density with *y*-on-*x* and RMA line fits (South Morecambe gas field, offshore U.K.) After Woodhouse.[32]

ground along the major axis of the data cloud (see Fig. 3H.10). The "structural" line-fits estimate the relationship that would be observed if both the y and x variables were error-free. The RMA line provides this relationship for the particular case when y and x have equal fractional errors. These structural line-fits are not generally used in practice because, for their future use, they apply to error-free x-input data, not the real measured data. The RMA slope is defined by the ratio of the SDs of y and x together with the sign of r. The y-on-x slope is equal to the RMA slope multiplied by the correlation coefficient r.

The initial impression of y-on-x does not weaken its status as the method with the lowest overall residual error in the required y estimates. When viewed from any position on the x-axis, the y-on-x line is central within the y-data values near that x-position (see **Fig. 3H.11**). Providing average y-estimates from measured x-data is the main feature of y-on-x lines; however, the y-variance of the calibration data is not preserved by y-on-x predictions and the extremes of the y-range are averaged. The RMA line-fit does honor the y-variance; but if y and x are only moderately correlated, high porosity values are overestimated and low porosity values are underestimated (see **Fig. 3H.12**). The depths of the RMA-predicted high and low y-values will not be at the same depths as the core high and low values.

Cores are not always regularly sampled (e.g., at one per foot) and are typically sampled at a lower frequency in the shale intervals. In these cases, the plotted data can sometimes have no trend, for example, in a high-porosity reservoir. External information, not in the standard log/core variables, can be used to provide a useful line-fit. The zero-porosity end of the line might be derived from the core grain-density data. Calibration lines may be calculated by joining this grain density to the means (arithmetic averages) of the x-y data, or a fixed-point regression might be used.

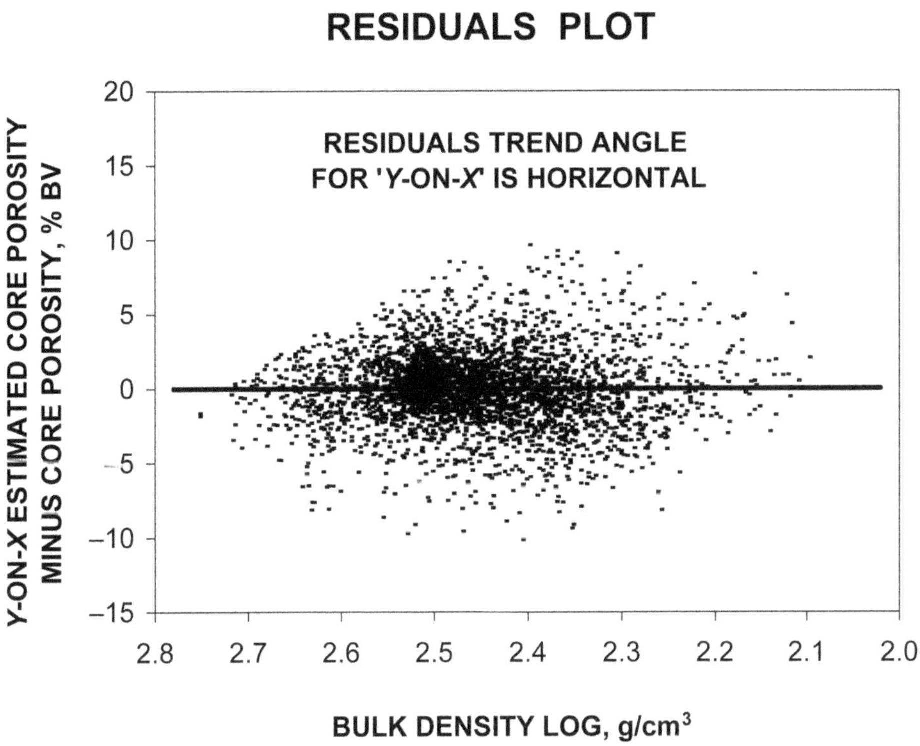

Fig. 3H.11—Residuals plot for *y*-on-*x* predicted porosity vs. well-log bulk density. Residuals are evenly distributed around zero (South Morecambe gas field, offshore U.K.). After Woodhouse.[32]

There are circumstances when line-fits such as RMA should be considered. Some reservoirs (e.g., carbonates and finely laminated sandstones) are so heterogeneous that it is difficult, if not impossible, to make accurate core-to-log depth adjustments. In these cases, where the data pairs do not reliably sample similar rocks, core porosity vs. log plots have a poor correlation, and the *y*-on-*x* line-fit has a low slope. Here, the RMA line can be a better practical approximation of the underlying core/log relationship. If the core/log correlation is very poor, deterministic[3] or simultaneous-equations log analysis[20]—without using the core—remain useful options.

Calibration lines should be determined for a single population, not a mixture of two or more populations. For example, when density logs are used to predict porosity, if two zones in a well have significantly different lithologies or grain densities, they should, where possible, be separated into different population groups. Likewise, significant grain-density trends across the reservoir area should be honored; however, this process must not be taken to an extreme. Calibration lines with excellent apparent correlations can be achieved with very fine subdivisions of the calibration sample data. Unfortunately, when they are applied in prediction mode to the uncored wells, these "overfitted" calibrations will not yield robust and accurate porosity estimates.

Density Log. The density log is often the best log for making porosity estimates.[34] In their simplest form, the density-log readings are considered to be a linear relationship between the zero-porosity limit where the density log reads the rock-matrix density and the 100% BV porosity limit where the density log reads the fluid density.

$$\phi = \frac{\rho_{ma} - \rho_b}{\rho_{ma} - \rho_{fl}}, \qquad \text{(3H.6)}$$

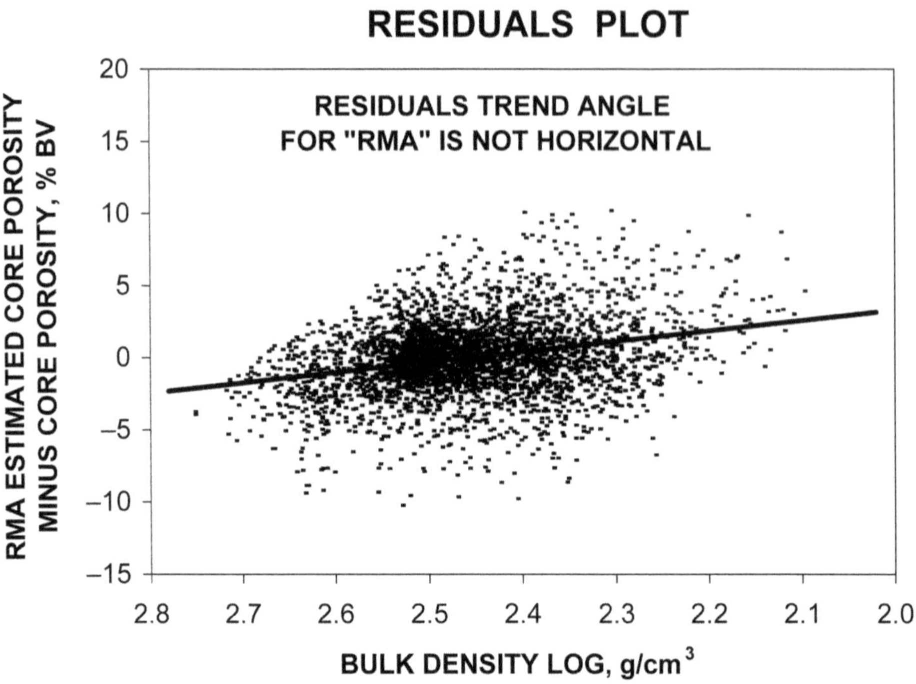

Fig. 3H.12—Residuals plot for RMA-predicted porosity. Residuals trend from negative at high densities to positive at low-density values (South Morecambe gas field, offshore U.K.). After Woodhouse.[32]

where ϕ = porosity, ρ_{ma} = matrix density, ρ_b = formation bulk density, and ρ_{fl} = fluid density.

This physical relationship assumes constant matrix density and insignificant variation in the fluid saturation and fluid density within the pore system. These are not necessarily the case in real reservoirs. Where core measurements and regression analysis are used to quantify the relationship, the regression coefficients (slope and intercept) do not represent true matrix and fluid-density properties. The regression coefficients are "catch-all" fitted parameters without a physical meaning.

When using the density log for porosity calculations, it should be expected that a different log/core relationship will be found for the aquifer, the oil column, and the gas column because of the different fluid densities in these various fluid environments. Also, there may be curvature of the relationship if the near-wellbore gas saturations increase as porosity increases. This curvature does not always occur, and gas saturations "seen" by the density tool (up to 4 in. into the borehole wall) may be fairly constant. Differences in the core/log relationship are also expected for WBM vs. OBM wells.

In a density-log vs. core-porosity crossplot, the low-porosity portion of the data cloud needs to be handled carefully. Typically, these rocks have lower porosity because of either a much higher clay-mineral content or various cements filling some of the pore system, either of which is likely to alter the average matrix density. Also, the lower-porosity rocks within the hydrocarbon column will have significantly higher S_w. Hence, either of these effects can cause curvature of this crossplot that will need to be accounted for in the correlation of log and core data.

For reservoirs that are buried sufficiently deep and in which no smectite clay mineral is present, the sandstone and shale core grain densities are often similar, and the core porosity of the shale is low, less than approximately 5% BV. In these particular conditions, there are very small volumes of adsorbed water in the clay molecules, ϕ_{cl} and ϕ_{sh} in Eq. 3H.2 are low, and it is possible to use the density log alone to estimate effective porosity. Neutron logs usually do not indicate such low apparent shale porosities because of chemically bound hydroxyl groups in clay-mineral structures and neutron-adsorbing elements. Sonic logs in shales usually read higher than quartz travel-time. The sonic and neutron are, therefore, not satisfactory single-log effective-porosity predictors in shaly sands.

Other Approaches. Evaluations based on the sonic log follow a logic similar to that of the density-log methods described previously. The sonic-derived porosities are particularly useful when conditions are adverse for the density log, such as in caved holes or when heavy minerals are present. See the chapter on acoustic logging in this volume of the *Handbook*.

Two-log combination solutions, such as density/neutron or density/GR, are useful in carbonate and siliciclastic reservoirs, including shaly sands. Gas-bearing sands may require multiple-log methods. Multiple-log and multivariate regression methods can be used but are often difficult to apply in practice. All multilog methods will have problems where one of the input logs reads incorrectly (e.g., hole washouts, tool sticking, cycle skipping, GR statistical variations, and poor depth alignment).

Some prefer to use core data for nonquantitative, visual comparison with log-analysis porosity developed from a variety of methods. When the volume of core data is low, making simple qualitative presentations of the measured core may be satisfactory. When there is sufficient core data to provide a representative sample of the formations in one or more wells, quantitative use will lead to more accurate OOIP and OGIP. It is not possible to provide strong guidance on the amount of core required, because the geological and engineering issues of each reservoir differ greatly. However, as a starting point, one might consider coring one well in approximately every 10 if this fits sensibly with the unique parameters of the reservoir under study.

3H.5.4 Consistency of Calculations and Uncertainty. Calculations of porosity and S_w must be compatible with each other. They must both be evaluated within the "total ϕ/S_w system," or both be evaluated within the "effective ϕ/S_w system" (see the discussion in the following S_w section). When preserved OBM ("native") core measurements of both porosity and water saturation are available, the core sample F_{HCP} [$= \phi \times (1 - S_w)$] is, after a few small standard adjustments, usually the best estimate of the V_{HCP} at each cored depth. This core-based V_{HCP} can be used to validate, or calibrate, the V_{HCP} given by either the "effective" or the "total" ϕ/S_w evaluation systems, as calculated from the log data (see Eq. 3H.3).

The uncertainty of porosity evaluations varies from case to case. The porosity of a single cleaned core plug can be repeated to within approximately ± 0.2% BV, where this uncertainty refers to one SD.[13] This very small instrument-repeatability uncertainty does not, however, include the many other noninstrumental variables that affect the systematic uncertainty and overall accuracy. Before the measurement is made, there may be core-plug cleaning problems from native salt in the pore space and incomplete oil removal. The drying time of water-adsorbing clay minerals adds further uncertainty. Surface roughness causes the plug volume to be uncertain especially when there are large grains and vugs. The uncertainty of the average core porosity will be improved when many plugs are selected at approximately one per foot without regard for the rock quality (i.e., randomly). However, because of commercial pressures and common sense, plug samples are not always selected at random, so care must be exercised, especially concerning the porosity values predicted at depths where core is not available. It must also be remembered that 1-in. core-plug samples, taken from each foot of whole core, sample only approximately 2% of the whole core volume.

Log readings are also uncertain, and, for example, the bulk-density-log random uncertainty may be approximately ± 0.015 g/cm^3 or approximately ± 1% BV.[7] Systematic errors, such as poor density-tool pad contact with the borehole wall, increase the uncertainty in some wells.

Further uncertainty in the final calculated porosity arises from the grain- and fluid-density values (or the related regression coefficients) and from mixed-mineral and shaly-sand effects.

The evaluations of zone-average porosity in the net-pay intervals in a single well that has relevant core control might have an accuracy of approximately ± 1.0% BV. This is largely the result of systematic uncertainty because the random uncertainties will be very small for zone-average values. In other words, an average porosity of 20% BV has an uncertainty range of 19 to 21% BV. This is a one SD estimate. In 32% of cases, zone-average uncertainties of greater than ± 1.0% BV are considered likely. Where core control is not available, these accuracy estimates should probably be doubled. Effective-porosity accuracy in very shaly sands is also more uncertain because of the associated V_{sh} estimates.

For fractured reservoirs, imaging tools now provide better visualizations of the borehole wall, but quantification of the open-fracture porosity, which may be approximately 0.1 to 1.0% BV, is highly uncertain. Production testing and test analysis are recommended to determine the nature and extent of any fracture system within the reservoir interval.

3H.6 Fluid-Contacts Identification

Defining the depths of the fluid contacts, GWC, OWC, and gas/oil contact (GOC), or defining both of the latter in some reservoir situations, is essential for volumetrics calculations and important for detailed petrophysical calculations. For example, for more-accurate porosity calculations, the reservoir's vertical interval needs to be subdivided by fluid type to account for differences in the average fluid saturation and, hence, differences in fluid density or sonic travel time in the various fluid intervals: gas cap, oil column, or aquifer. For the S_w calculations, the depth of the OWC or GWC, or more particularly the related free water level (FWL), is a required input for any S_w calculations using capillary pressure, P_c, data. These depths need to be defined in every wellbore, to the extent that they occur. This section addresses the methods used to make the most accurate determination of the GOC, OWC, and/or GWC depths at the wellbores. This section does not address the larger topic of how these fluid contacts may vary over the whole of the reservoir either because of faults, rock-quality variations, isolated sands, a reservoir's hydrocarbon-filling history, or hydrodynamics of the reservoir-aquifer system.

There are four types of data that can be used to define the fluid-contact depths in a wellbore: mud logs, cores (geologists' descriptions and routine-core-analysis data), resistivity and neutron logs, and formation-tester pressure surveys. These are listed approximately in the order in which these data are gathered from a wellbore. Each has its own strengths and weaknesses. Each is an independent source of information; therefore, the most accurate fluid contact is obtained by using all of the data available for a particular well. The first step in using any and all of these data sources is to align their depths as accurately as possible.

3H.6.1 Mud Logs. Mud logs record mud gas compositions and quantities and descriptions and analyses of drill cuttings. These provide information about the fluid content and lithology of the rock as it is drilled. These data have some depth uncertainty because of the lag time between a rock interval being drilled and the time the cuttings are recovered at the surface. The field personnel work to minimize this uncertainty by periodically dropping a bit of "carbide" into the drilling mud and then determining how long it takes for it to show up on the mud-returns gas-chromatograph output. The mud log gas-analysis data plotted vs. drilled depth, and adjusted for drilling rate, is a semiquantitative measure of the gas content over the reservoir interval. It can be used to determine a GWC or OWC because the background gas content per unit volume of aquifer brine is so low compared with that of free gas or gas dissolved in oil. The methane-concentration log is most useful for defining the top of the reservoir and the GWC for a gas reservoir, and the detailed gas analyses can also identify a GOC from increasing ratios of the heavier hydrocarbon components compared with methane as depth increases.

Direct observation of oil staining, and yellow or brown ultraviolet (UV) fluorescence on drill cuttings, identifies oil. When drill cuttings are crushed in solvents, mobile oil migrates to the solvent, which then fluoresces. Gas condensate has a bright white fluorescence. These characteristics assist in identifying oil and gas reservoirs, the top of the reservoir, GOC and OWC, and, possibly, the base of the reservoir.

3H.6.2 Water-Based-Mud Cores. WBM cores can provide direct observations of the OWCs and GOCs. Because of differences in the colors of the oil staining, the depths of gas, oil, tar, and relict-oil intervals can often be determined visually from the cores, especially when they are cut at a high rate of penetration. It is common practice to photograph cores in both white and UV light to provide an accessible, permanent visual record. These visual observations are typically complemented by the routine-core-analysis S_o data over the same depth ranges.[8] Gas and aquifer intervals have low core oil saturations, and tar intervals often have high core oil saturations.

3H.6.3 Log-Based Methods. The use of resistivity-log data is another method of determining OWC and GWC depth in a wellbore.[3] The resistivity logs are used to calculate S_w, and where there is a significant decrease in the S_w values (decreasing from near 100% PV as one moves up through the reservoir interval), that depth is defined as the fluid-contact depth. Also, the invasion profile of the shallow- vs. deeper-reading resistivity tools can be used to help define the depth interval in which the fluid contact occurs. This is true of either WBM- or OBM-drilled wells. Also in WBM-drilled wells, tar intervals can be defined by those depths over which the shallow- and deep-reading resistivity tools show a lack of oil-saturation change, which indicates that the hydrocarbons in the pore space are too viscous to be displaced by the WBM filtrate. If the reservoir is not too shaly, neutron logs can be one of the keys to identifying gas-bearing intervals.[3] A GOC, or a GWC, can be defined at the depth at which the neutron porosity significantly decreases and the density and sonic porosities slightly increase as one moves up through the reservoir interval.

3H.6.4 Formation-Pressure Surveys. The best data from which to determine FWL fluid contacts are given by the formation-pressure-testing tools (e.g., RFT, formation multitester, MDT, and RCI) that measure pressure surveys over reservoir intervals.[35] In going vertically from the gas cap into the oil column, or from the hydrocarbon column into the aquifer, there will be breaks in the formation-pressure vs. depth trends as one moves from a very low gas pressure gradient (0.10 psi/ft or less) to the higher oil pressure gradient (typically 0.25 to 0.35 psi/ft) and then to the water pressure gradient (0.40 to 0.55 psi/ft). When adequate data can be collected, the fluid contacts can be determined very accurately by identifying the depths at which the characteristic pressure gradients change. See the chapter on reservoir pressure and temperature in this volume of the *Handbook* for additional information on fluid-contact depth determination using pressure information.

Overall, the formation-pressure-survey data should be the primary source of data for defining the FWL fluid contacts. The other data should be used to complement these pressure data, or should be used together to define the fluid contacts if no pressure-survey data are available. In reservoirs with sand/shale sequences, sometimes the fluid contact is determined to be within a shale interval, even if that interval is only 10 ft thick or less. If this is the case, the best estimate depth of the fluid contact is at the mid-depth of the shale interval (unless the pressure-survey data indicate otherwise). The fluid contact may be different from the FWL, and it is the FWL depth that is important commercially and important when making S_w calculations from P_c/S_w data.

Since the introduction of 3D-seismic surveys, acoustic-impedance contrasts between gas-, oil- and water-bearing formations are increasingly used. The impedance is dependent on the density and acoustic velocity of each fluid. Many reservoirs exhibit a significant change of acoustic impedance at the fluid contacts, allowing the contacts defined at the wells to be propagated, with data control, into the undrilled areas of the reservoir. These impedance maps show visual "haloes" around a GWC or OWC. These same impedance changes may also be seen in vertical seismic sections and assist in identifying the top of the reservoir, GOC, GWC, and OWC.

3H.7 Water-Saturation Determination

S_w determination is the most challenging of petrophysical calculations and is used to quantify its more important complement, the hydrocarbon saturation $(1 - S_w)$. Complexities arise because there are a number of independent approaches that can be used to calculate S_w. The complication is that often, if not typically, these different approaches lead to somewhat different S_w values that may equate to considerable differences in the OOIP or OGIP volumes. The challenge to the technical team is to resolve and to understand the differences among the S_w values obtained using the different procedures, and to arrive at the best calculation of S_w and its distribution throughout the reservoir vertically and areally. In OOIP and OGIP calculations, it is important to remember the relative importance of porosity and S_w. A 10% PV change in S_w has the same impact as a 2% BV change in porosity (in a 20% BV porosity reservoir).

3H.7.1 Techniques for Calculating S_w. S_w in wellbores can be determined by the following primary methods:

- S_w calculations from resistivity well logs by application of a model relating S_w to porosity, connate-water resistivity, and various rock electrical properties.
- S_w calculations from laboratory capillary pressure/saturation (P_c/S_w) measurements by application of a model relating S_w to various rock and fluid properties and height above the free-water level.
- S_w calculations using OBM-core-plug Dean-Stark water-volume determinations.
- Combinations of these methods.

This listing is the chronological order in which data are likely to become available, not in a ranked order based on the accuracy of the various methods. In fact, the use of S_w values from properly handled and preserved OBM cores is superior to the other techniques in the oil (or gas) column above the mobile-water part of the oil/water (or gas/water) transition zone. The OBM cores are superior because the water-volume measurements on these core plugs are a direct determination of S_w. The extracted water is the reservoir's connate water. The resistivity-log data are inferential measurements that have to be converted to S_w values by a conceptual, theoretical, or empirical model. The use of P_c/S_w data requires other models and a number of input parameters whose values at reservoir conditions cannot be directly determined and, hence, must be estimated.

The choice of which S_w-calculation approach to use is often controlled by the availability of the various types of data. If no OBM cores have been cut, then this technique cannot be used unless funds are spent to acquire such data from one or more newly drilled wells. This is not a high incremental cost when OBM use is planned for other purposes. Resistivity logs are run in all wells, so these data are available for making standard-log-analysis S_w calculations. A key consideration when making calibrated S_w calculations is the availability of SCAL data on core samples from the particular reservoir; that is, the number of laboratory electrical-property and P_c/S_w core-plug measurements that have been made.

The technique chosen to calculate S_w is often a hybrid that combines the use of two of these basic data sources. For example, the OBM-core S_w data can be used in combination with the resistivity logs to expand the data set used to include all wells and the whole of the hydrocarbon column. Alternatively, the OBM-core S_w data can be used in combination with the $P_c/$

S_w data. In this way, the OBM-core S_w data define the S_w values for the majority of the reservoir, whereas the P_c/S_w data define the S_w values in the interval just above the fluid contact and perhaps in areas of the field where P_c data are available but OBM-core data are not.

3H.7.2 Data Availability and Data Quality. This section discusses the input-data availability and data-quality issues for each S_w technique. Details of the S_w physical-models equations are given in the next section. These considerations often control the initial choice of methodology to calculate S_w and need to be addressed at the start of the project to determine whether it is practically possible to fill gaps in the database in order to use a more accurate S_w-calculation approach. This discussion assumes that accurate porosity values are available from the routine-core-analysis database and that porosity is calculated point by point from the well logs. The discussion focuses on particular aspects that affect the choice of S_w methodology. Many of the database considerations were discussed in Sec. 3H.2.

Resistivity Logs. Wells generally have one variety or another of laterolog or induction resistivity log because they are broadly useful and because government regulations typically demand that they be recorded. This generally provides point-by-point data from the top of the hydrocarbon column down through any aquifer intervals that are present. However, in many fields, the early wells are spread thinly over the reservoir area, but the later development wells are drilled only in areas chosen to maximize rate and recovery while minimizing costs. This means that, often, few wells are drilled downdip where the hydrocarbon column thins because of an underlying aquifer, or in the potentially thin updip limits of the reservoir. In such areas, there may, therefore, be few resistivity logs.

Laterologs are preferred to induction logs when the drilling mud has moderate to high salinity. This limitation of induction tools arises because of the excessive conductivity signal from the borehole and the mud-filtrate-invaded zone. Deep laterolog tools read too high when measuring immediately beneath anhydrite and salt,[3] and alternative resistivity curves should be selected. When formation resistivity, R_t, is very high, previous generations of induction tools had limited accuracy, but current tools are much improved. Although the deep induction measurement is a running average over many vertical feet, modern tools include systems to deconvolve the raw log and provide a final log with a good vertical resolution.

Deep invasion of WBM filtrate affects all resistivity logs, and, in the extreme, the available resistivity log may be used only qualitatively. At the opposite extreme, when OBM filtrate invades a hydrocarbon reservoir, the invading OBM filtrate generally displaces only the reservoir oil and gas, leaving the S_w unchanged. Here, invasion of OBM does not usually change the deep-formation or the invaded-zone resistivity. For moderate invasion depths, the logging company charts are sometimes used to correct the deep-reading log to provide a better estimate of R_t.

Pad-mounted shallow-reading microresistivity logs measure R_{xo}, the resistivity of the mud-filtrate-invaded zone. When used together with the deeper-reading tools, these logs provide valuable information about the mobility of the reservoir fluids, including the presence of tar. In WBM wells, they also provide an estimate of the residual-hydrocarbon saturation, S_{orw}.

Connate-Brine Resistivity Data. An accurate value of connate-brine resistivity, R_w, or its values and distribution throughout the reservoir, are required for accurate S_w calculations using resistivity logs. Temperature estimates are also required.

A first check on the R_w of the aquifer is to back-calculate the apparent R_w with the Archie equation using the invasion-corrected resistivity logs and the best estimates of a and m parameters. Because S_w is typically 100% PV in the aquifer interval, the n value is not relevant here.

The spontaneous-potential (SP) log provides a second method to calculate R_w in wells drilled with WBM. Information on the mud-filtrate composition and temperature is used with the SP deflection to calculate R_w.[3,5] The moderately accurate calculation process is valid in the aquifer but is also valid in the hydrocarbon column if high resistivity does not suppress the SP response. When OBM-core salinity measurements are not available, the SP log provides the only evidence of possible R_w variations in the hydrocarbon column.

A third estimate of aquifer-water composition and R_w is often taken from samples recovered during flow tests of the aquifer interval; however, the R_w of the oil and/or gas column is not always the same as that of the aquifer interval.[36–38] Aquifer-interval flow tests must be validated and checked for contamination from mud-filtrate invasion.

For the oil or gas column, the determination of the R_w value or values is far more of a challenge because the reservoir water will not flow. The typical, but not necessarily correct, first assumption is that the hydrocarbon-column R_w is the same as that of the underlying aquifer. If wells have been cored with OBM, core plugs from the hydrocarbon- and water-bearing intervals can be analyzed for both their water volume and their salt content, particularly the chloride ion that in almost all cases dominates the anion side of the salinity determination.[36]

Fig. 3H.13 shows variation of chloride concentration with depth for a reservoir in Ecuador.[37] The chloride value can generally be used to quantify the reservoir-water salinity, from which the R_w at reservoir conditions can be calculated using standard water-resistivity vs. chloride charts or algorithms. For reservoirs in which there is a considerable CO_2 content (3+ mol%), the ion distribution at surface conditions will differ from that at reservoir temperature and pressure. Equilibrium ion-distribution calculations need to be made when adjusting the surface-salinity measurement to reservoir conditions.

Formation temperature affects the S_w estimates because, for constant formation-water composition, R_w varies with temperature.[3] Maximum downhole temperature is measured with most log runs and drillstem-tool (DST) tests, and these are widely used to estimate a temperature vs. depth profile. It can be argued that the temperature required for resistivity-derived S_w estimates is the prevailing temperature in the rock volume seen by the tool at the time of logging. At this time, the relevant rock is likely to be cooler than the original formation temperature. The error induced by the usual maximum temperature simplification is not large, and the cooling issue is generally ignored.

Electrical-Property SCAL Data. The third aspect of making these S_w calculations is the choice of the model for the "electrical network" within the rock. These models relate S_w to several formation variables including the bulk-formation resistivity and the formation-water resistivity. A number of models have been published [e.g., Archie, Waxman-Smits-Thomas (WST), dual-water (DW), and Indonesia].[16,31] Laboratory measurements of two or more types of electrical properties are taken. All of these models assume a homogeneous rock sample.

Archie Exponents.[39] First, a set of cleaned core plugs with a range of porosities are fully saturated with brine of known resistivity, and the bulk resistivity of each core plug is measured. For this simplest model, the slope of a line fitted to a log-log plot of the data set gives the cementation exponent, *m,* and the intercept is the cementation constant,[39] *a* (see **Fig. 3H.14**, where $a = 1$ and $m = 1.77$). These parameters are used to predict point-by-point F from porosity; leading to predictions of R_0 and S_w.

$$F = \frac{R_0}{R_w}, \qquad \text{(3H.7)}$$

where F = formation factor, R_w = brine-water resistivity, and R_0 = rock resistivity with zero oil and gas saturation (100% PV S_w). The plotted logarithmic data ($\log_{10}F$ and $\log_{10}\phi$) are fitted with a linear model of the form,

$$F = \frac{a}{\phi^m}, \qquad \text{(3H.8)}$$

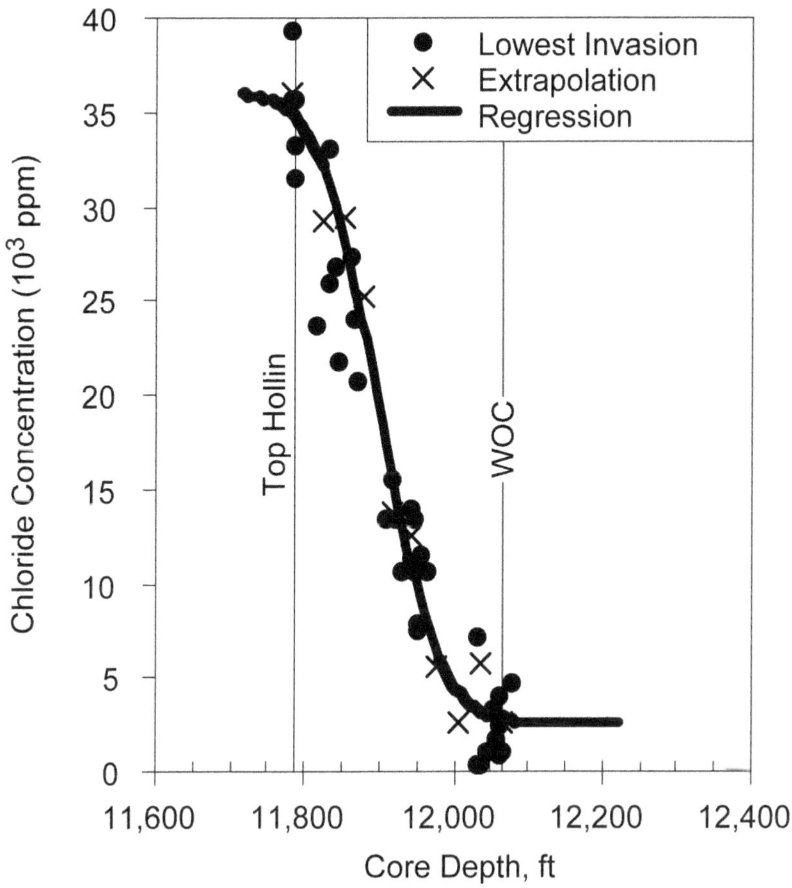

Fig. 3H.13—Formation-water salinity variation within the hydrocarbon column (Villano oil field, Ecuador).[37] Hollin is the name of the oil-reservoir formation, and WOC is the water/oil contact. Chloride concentration in the reservoir brine varies from approximately 2,000 to 35,000 ppm. Lowest invasion refers to limited OBM-filtrate invasion into the cores.

where ϕ= porosity, a = cementation constant, and m = cementation exponent. Therefore, m = − change in $\log_{10}F$/change in $\log_{10}\phi$ (the slope of the line-fit) and a = F at 100% BV porosity (the line-fit intercept).

This model was developed by Archie,[39] who proposed a = 1.0 and m = 1.8 to 2.0 for his data set. Subsequent work by Exxon researchers for several sandstone rocks recommended a = 0.61 and m = 2.15 (the Humble formula).[40] Carbonates also have been studied and yielded a recommendation to use $m = 1.87 + 0.019/\phi$ below 9% BV (the Shell formula).[3] However, carbonate pore and fracture networks vary greatly, and m values from 1.0 to 3.0 may be required. Clearly, m is not a constant, but varies with rock type.

When plotting these formation-factor data, it is typically assumed that the rock samples have similar pore geometry, but with differing levels of porosity and diagenesis. Reservoir-specific exponent values are likely to provide more-accurate S_w results than worldwide correlations. However, before reservoir-specific values are determined, descriptive and experimental data need to be studied to determine whether they need to be subdivided into various groupings that relate to distinct differences in lithological properties (grain size, sorting, or clay-mineral content).

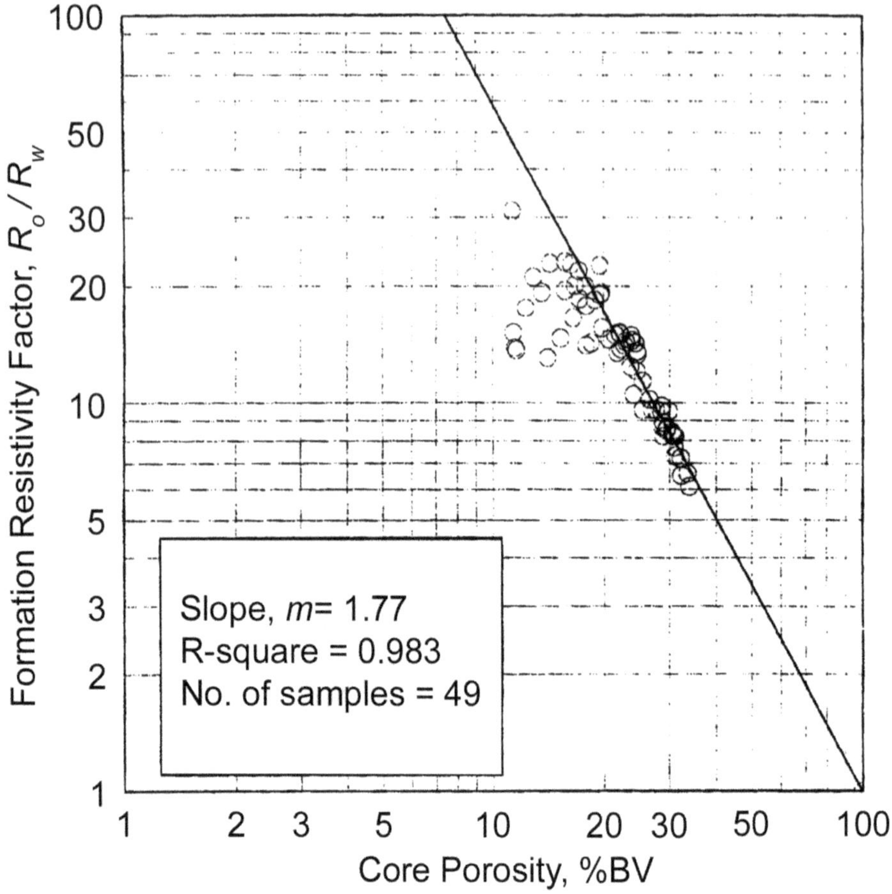

Fig. 3H.14—Laboratory-measured formation resistivity factor (*F*) vs. porosity for core plugs from Dulang field, Malaysia.[52]

In partially brine-saturated rocks, a related experimental study involves measuring electrical properties as a function of water saturation. In these experiments, the resistivity index (I_R), the ratio of the desaturated-rock resistivity to the 100% PV brine-saturated rock resistivity (R_t/R_0), is measured as a function of brine saturation. For example, in a porous-plate apparatus, S_w is changed by increasing gas pressure, and therefore capillary pressure, at the gas/water interface in the pores. Brine flows from the base of the plug via a porous-plate. From the measurements on each core plug, a log-log plot of I_R vs. S_w is made (see **Fig. 3H.15**, where n = 1.64). The slope of the line (almost always forced through I_R = 1.0 at S_w = 100% PV) is the Archie saturation exponent n (see Eqs 3H.9 and 3H.10). On the basis of experimental data, Archie[39] recommended that n = 2.0, and this value is still widely used when no experimental data are available. Although cementation exponents can be determined from log analysis, saturation exponents cannot and, therefore, require external information from core data.

$$n = \frac{-\Delta \log_{10} I_R}{\Delta \log_{10} S_w}, \quad \text{(3H.9)}$$

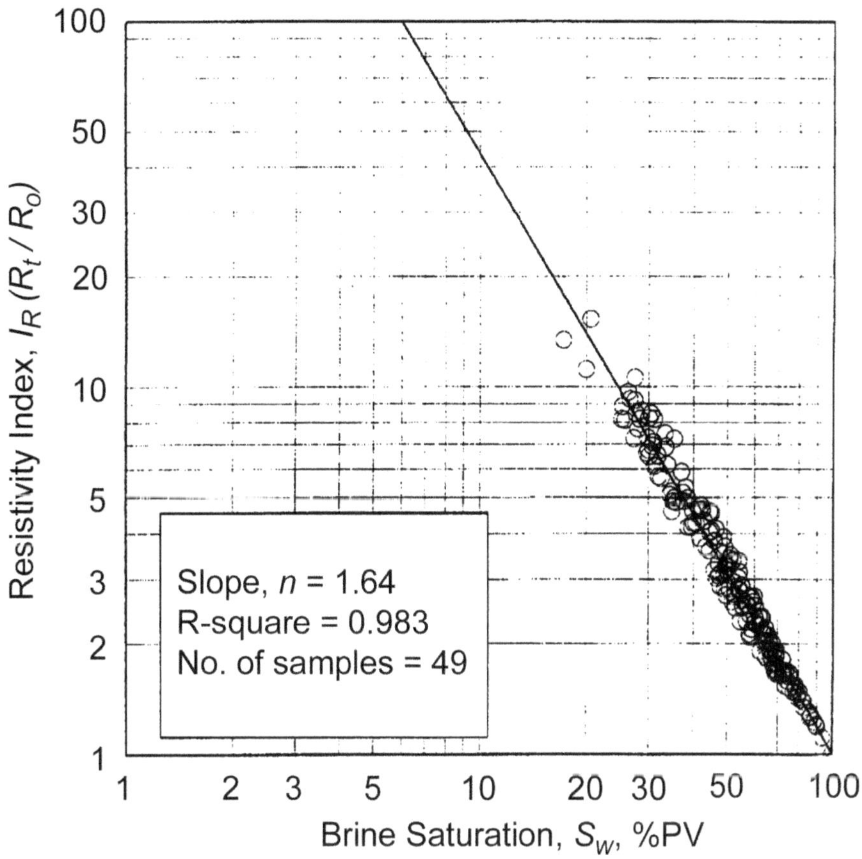

Fig. 3H.15—Laboratory-measured formation resistivity index (I_R) vs. brine saturation for core plugs from Dulang field, Malaysia.[52]

$$\text{and } S_w = \left(\frac{1}{I_R}\right)^{1/n} = \left(\frac{R_0}{R_t}\right)^{1/n}, \quad \text{(3H.10)}$$

where n = saturation exponent, the slope from the origin of a line-fit of several data points; I_R = resistivity index; and S_w = fractional brine water saturation.

A straight line-fit is usually used, but curved line-fits can be considered where necessary. Curvature is often the result of the clay-mineral content but may also result from an inhomogeneous water distribution at the pore scale (e.g., when microporous rock grains are present). When significant amounts of clay minerals are present in the rocks, other models are required to extend the Archie relationships. The WST model, discussed next, is based on laboratory SCAL measurements including cation-exchange capacity (CEC).

Waxman-Smits-Thomas Exponents and Cation-Exchange Capacity. WST cementation and saturation exponents (m^* and n^*) are required to apply the WST shaly-sand-model equation discussed in Sec. 3H.7.3. The quantity of cation-exchange sites per gram of rock sample (CEC) may be measured in the laboratory by several methods and, after converting to CEC per unit PV, is used as the model parameter Q_V.[41,42] The most reliable measurement of Q_V involves carrying out bulk-rock resistivity, R_0, tests at several brine resistivities and, therefore, is time consuming. The rock conductivity values ($1/R_0$) are plotted vs. brine conductivity ($1/R_w$) to identify the excess conductivity resulting from the shales and clay minerals. The slope of the fitted

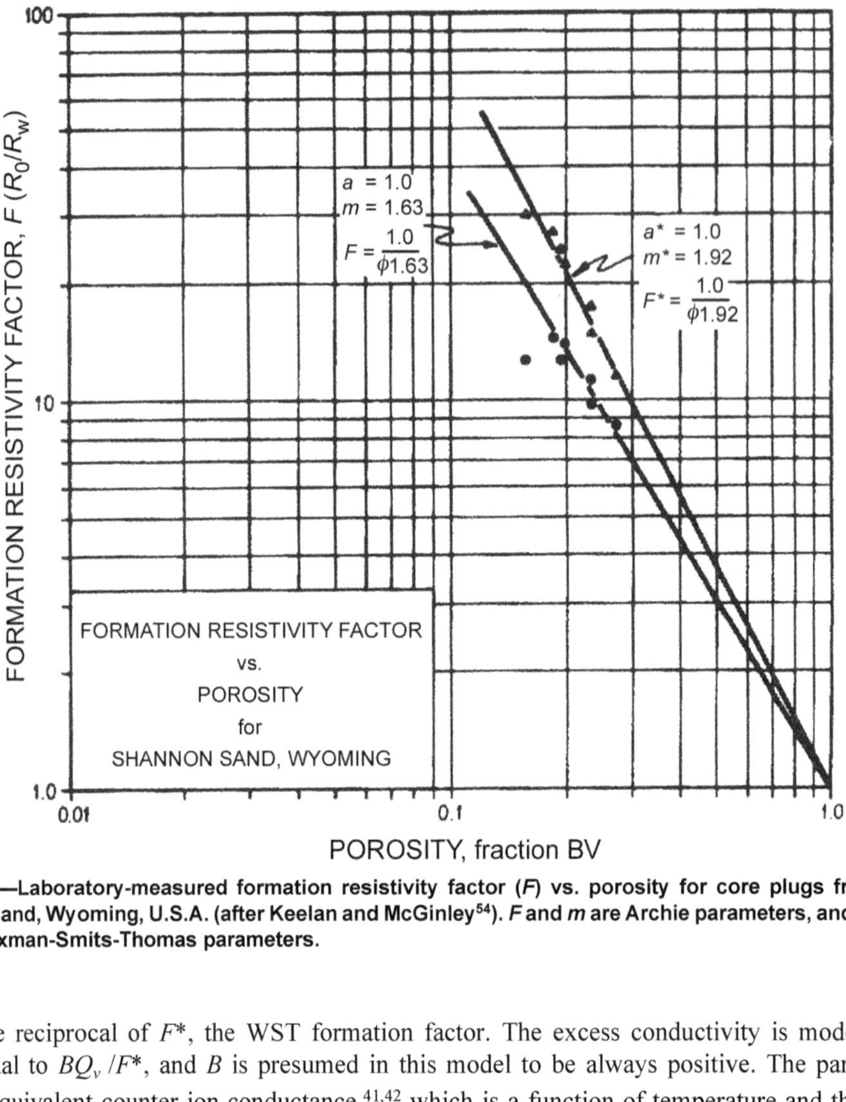

Fig. 3H.16—Laboratory-measured formation resistivity factor (*F*) vs. porosity for core plugs from the Shannon sand, Wyoming, U.S.A. (after Keelan and McGinley[54]). *F* and *m* are Archie parameters, and *F and *m** are Waxman-Smits-Thomas parameters.**

line is the reciprocal of F^*, the WST formation factor. The excess conductivity is modeled as being equal to BQ_v/F^*, and B is presumed in this model to be always positive. The parameter B is the equivalent counter-ion conductance,[41,42] which is a function of temperature and the free-water resistivity. Q_v is estimated from the values of F^* and B. Core resistivities are also measured when S_w is less than 100% PV and both the WST exponents m^* and n^* are derived (see **Fig. 3H.16 and Fig. 3H.17**, respectively). It should be noted that $m^* > m$ and $n^* > n$, except in "clean" sands.

Other CEC methods require the breaking up, disaggregation, and consequent partial loss of the real geometry of the rock's electrical network. These simpler methods, such as the ammonia method, use analytical-chemistry methods to measure CEC. After measuring porosity and grain density, this practical laboratory unit is converted to the required Q_v parameter.[43] These simpler CEC measurements are often made on sidewall cores and are used together with exponent values measured on cores from neighboring wells.

Numerous other shaly-sand models have been developed, and, unlike WST, many are calculated from effective porosity. These types of models are generally applied using Archie exponents. When using SCAL electrical-property data, there must be consistency between the electrical-network model used to derive the laboratory parameters and the model used in the

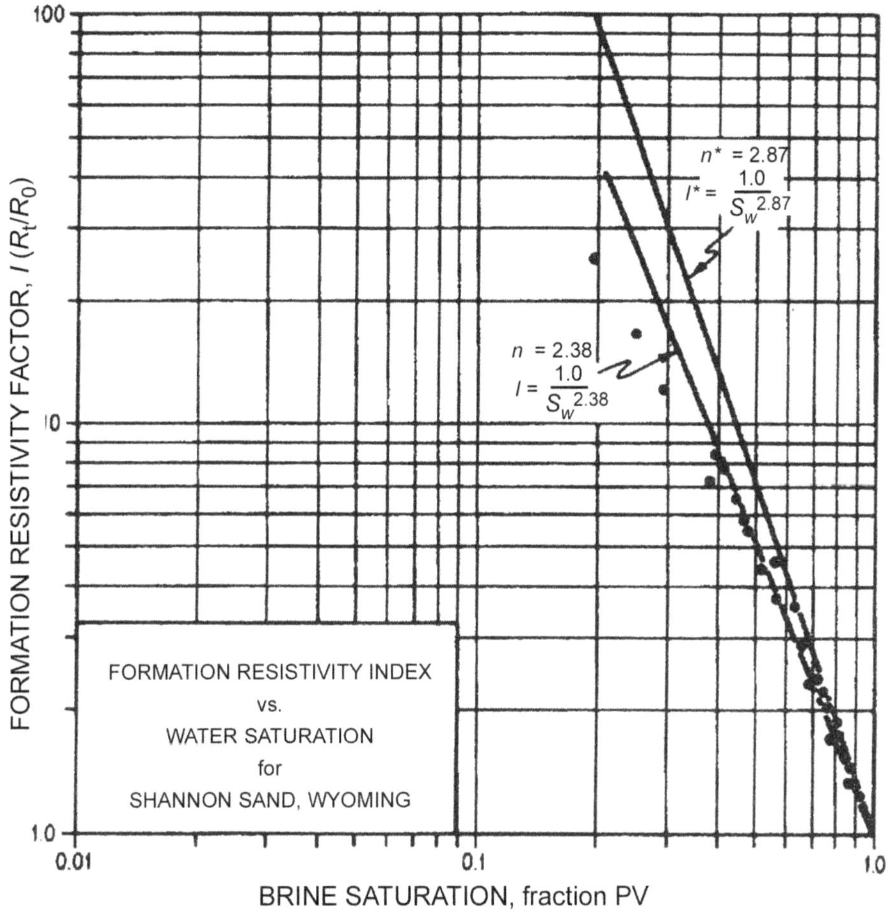

Fig. 3H.17—Laboratory-measured formation resistivity index (I_R) vs. brine saturation for core plugs from the Shannon sand, Wyoming, U.S.A. (after Keelan and McGinley[51]). n is the Archie saturation exponent, and n^* the Waxman-Smits-Thomas saturation exponent.

final S_w calculations from the porosity and the resistivity logs (e.g., if the laboratory provides standard Archie n values, these are not appropriate input to the WST equation).

Capillary Pressure SCAL Data. P_c data are a different type of SCAL data that can be taken experimentally in several ways. All P_c saturation tests respond to the pore-size distribution of the rock and the interfacial properties of the various solid/fluid systems. These data are obtained by desaturating core plugs, either using a centrifuge or a porous-plate apparatus. Initially, cleaned and dry plugs are saturated with either water or oil. The liquid is then displaced by air or nitrogen. Because air is very nonwetting compared with either water or oil, using these fluid pairs (air/water or air/oil) means that, as the P_c increases, the air will first occupy the largest pores. As the P_c and air saturation increase, the air will occupy smaller and smaller pores. The core plug begins the experiment saturated with the wetting phase, so the desaturation process provides data for the drainage P_c curve. After completing the drainage process, the core plug can be spun under the liquid in a centrifuge experiment, the liquid saturation will increase, and the imbibition P_c curve will be generated. Usually, only drainage P_c/S_w data are

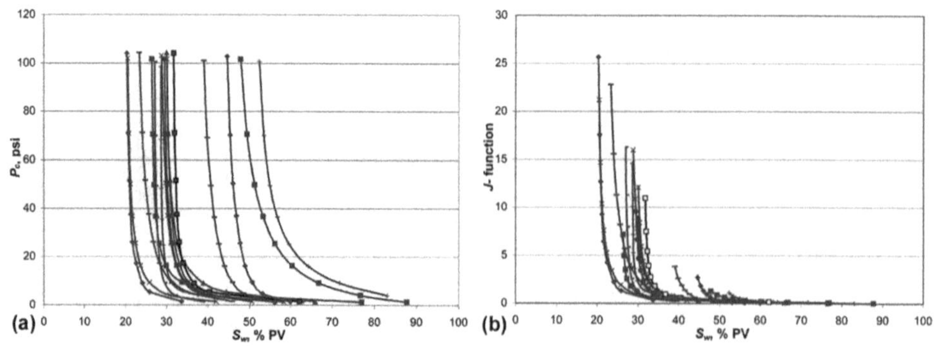

Fig. 3H.18—Example plots of centrifuge air/brine capillary pressure (P_c /S_w) data and its conversion to a *J*-function basis; data from an Asian gas field. The air/brine P_c value of 100 psi is equivalent to a height above free-water level of 200 to 350 ft, depending on the reservoir fluid's properties and the temperature and pressure conditions.

taken, and for most reservoir situations, these are the relevant data because they correspond to the original oil (or gas) trap-filling process.

MICP data are taken on cleaned and dried irregular core pieces. The core pieces are evacuated to a low vacuum, and mercury is injected with increasing pressure, up to 20,000 psi and sometimes higher. Corrections for clay-mineral adsorbed-water removed during drying can be made with the Hill-Shirley-Klein method.[44] The MICP experiment has the advantage of being run rapidly but is not a true wetting/nonwetting system. The sample cannot be used for subsequent SCAL tests because some mercury is retained within the core pieces at the end of the testing sequence. MICP data are widely used to measure pore-size distribution, but, when considering whether they should be used for accurate S_w calculations, MICP should be compared with air/water or air/oil P_c /S_w data.

The P_c /S_w data are usually compared first on a Leverett "*J*-function" basis.[45] The P_c data are converted to the *J*-function basis by multiplying each P_c value by the square root of its permeability divided by porosity and then dividing by the fluid-pair IFT multiplied by the contact angle (see Eq. 3H.11). *J*-function values differ depending on whether they are calculated in oilfield or metric units. The *J*-function approach assumes similar pore-size distribution in all cores tested. In this way, the various P_c /S_w data tend to converge when the underlying assumptions are met; however, there may still be enough scatter to suggest that the data need to be divided into two or more groupings (see **Fig. 3H.18**).

$$J(S_w) = \frac{P_c}{\sigma \times \cos\theta}\sqrt{\frac{k}{\phi}} \quad \text{.. (3H.11)}$$

From a *J*-function vs. S_w plot, the technical team can determine whether enough data have been obtained, whether new data need to be gathered to fill in portions of the data ranges, and whether the data indicate that subgroupings are appropriate and needed. Also, this plot indicates whether there are significant outliers that should be excluded or examined in more detail. A drawback to this averaging method is the introduction into S_w determination of four measured parameters and their associated errors (i.e., porosity, permeability, IFT, and contact angle).

Capillary pressure data may also be averaged by various models.[46,47] The relationship of S_w with permeability, and then porosity, is examined and is followed by examination of the height dependency.

OBM-Core S_w Data. The last type of S_w data discussed here is that obtained from routine core analysis of core plugs cut from OBM cores, either preserved as whole cores or else with core plugs cut at the wellsite and preserved individually. These data are taken foot-by-foot and are direct measurements of reservoir S_w values.[37,38,48–51] Many fields may never have had any wells cored with OBM; others may have only one or two OBM-cored wells. Even a single OBM coring of the full reservoir interval offers significant data that may impact the technical team's methodology for making the S_w calculations. It is better to have at least two wells fully cored with OBM from different areas of the reservoir.

To evaluate the OBM-core S_w data, they should be plotted as S_w vs. $\log_{10}$ (permeability) or vs. porosity to identify outliers and trends in the data. Particularly, the low-porosity/low-permeability data range should be examined for potential measurement problems. Sometimes, the raw laboratory measurements of water volume and PV data need to be reviewed for problem points and recalculations made where appropriate. Finally, the data should be divided into various possible interval groupings so that any needed zonation can be identified.

If reservoir connate water has flowed out of the core plug at any stage before the laboratory measurement, the OBM-core S_w data are clearly not representative of the in-situ reservoir S_w. This certainly occurs in water-bearing formations and can also occur in the lowest intervals of the oil/water or gas/water transition zones. These lowest intervals, which may be a few feet to approximately 30-ft thick, are precisely the same intervals in which a water cut is expected with the initial oil production. The mobile-water intervals can be identified in OBM wells where the shallow-reading induction-log resistivity is higher than the deep-reading induction-log resistivity. This pattern indicates higher oil saturations in the invaded zone compared with the original oil saturations. Where mobile water is observed, the OBM-core S_w measurements do not represent in-situ S_w and are too low.

3H.7.3 Application of Each S_w Technique and Its Strengths and Weaknesses. Methodologies for quantifying S_w at the wellbore are discussed here. The main features of each approach are described; however, in some cases, there are variations that are not addressed. For each technique, its strengths and weaknesses are discussed.

Calculating S_w From Resistivity Well Logs by Applying a Model Relating S_w to Porosity, Connate-Water Resistivity, and Various Rock Electrical Properties. The most common technique for calculating S_w is the use of resistivity logs with a model (empirical or theoretical) that relates S_w to R_t, R_w, and porosity. As mentioned previously, a large number of R_t/S_w models have been published. The models are applied at every data point in the reservoir where deep resistivity, porosity, and shale-volume estimates, if required, are available. The evaluation of all other necessary parameters (constant or variable R_w values, a, m, n, Q_V, V_{sh}, $R_0 = F \cdot R_w$, etc.) has also been discussed previously. Several commercial software packages are available that perform these S_w calculations for a variety of log models.

Clean-Sand (Archie) Model.

$$S_w = \left(\frac{aR_w}{R_t\phi^m}\right)^{1/n}, \quad \text{(3H.12a)}$$

and, alternatively,

$$S_w = \left(\frac{R_0}{R_t}\right)^{1/n}. \quad \text{(3H.12b)}$$

This model[39] is used for field studies in the many sandstone and carbonate reservoirs in which the clay-mineral content is low. This decision is strengthened after SCAL data have demonstrated that the simplest solution is satisfactory. When a significant fraction of smectite (montmorillonite) is present and where finely laminated sand and shale sequences occur, one of the shaly-sand models is very likely to be required. Low-resistivity pay is an issue in several oil-producing areas, such as the U.S. Gulf Coast, Egypt, and Indonesia, and hydrocarbon reserves can be missed and left undiscovered as a result of the resistivity suppression by clay minerals and shales.

Shaly-Sand Models. In the clean-sand model, the formation water is the only electrically conductive medium. In shaly rocks, R_t is suppressed and Archie S_w calculations are too high. As clay-mineral-rich rocks were studied and experimentally tested, more-complicated electrical models were developed to account for the effects of the geometries of conductive clay minerals and shale on rock resistivity. The primary goal of the shaly-sand models is to determine a working relationship between S_w using parameters similar to the Archie model, but also incorporating the quantity and specific electrical properties of the clay-mineral/shale. All of the shaly-sand models reduce to the Archie equation when the shale component is zero. For simplicity, in all of the shaly-sand models, the cementation constant, *a,* is taken to be 1.0 but, if required, can be easily associated again with the R_w term.

Laminated Sand/Shale Model. A parallel resistor model might be used for laminated sands, with multiple thin parallel layers of 100% shale interbedded with clean-sand layers. Thin, in this context, means that there are several beds within the vertical resolution of the resistivity-logging tool.

$$\frac{1}{R_t} = \frac{V_{sh}}{R_{sh}} + \frac{1 - V_{sh}}{R_{sd}}, \qquad \text{(3H.13)}$$

where the clean-sand resistivity $R_{sd} = R_w / \left(\phi^m_{sd} S^n_{wsd}\right)$. For this laminated shale/sand model, effective porosity depends simply on the sand fraction of the bulk volume:

$$\phi_e = \phi_{sd}\left(1 - V_{sh}\right), \qquad \text{(3H.14)}$$

$$\text{and } S_{we} = S_{wsd}\,. \qquad \text{(3H.15)}$$

The value of ϕ_{sd} may be assumed from neighboring thick sands, and all of the parameters, except the S_w of the sand, S_{wsd}, can be estimated.

Poupon-Leveaux (Indonesia) Model. The Indonesia model was developed by field observation in Indonesia rather than by laboratory experimental measurement support.[52] It remains useful because it is based on readily available standard log-analysis parameters and gives reasonably reliable results. The formula was empirically modeled with field data in water-bearing shaly sands, but the detailed functionality for hydrocarbon-bearing sands is unsupported, except by common sense and long-standing use. S_w results from the formula are comparatively easy to calculate and, because it is not a quadratic equation, it gives results that are always greater than zero. Several of the other quadratic and iterative-solution models can calculate unreasonable negative S_w results.

$$\frac{1}{R_t} = S^n_w \left[\left(\frac{V_{sh}^{\,2 - V_{sh}}}{R_{sh}} \right)^{1/2} + \left(\frac{\phi^m_e}{R_w} \right)^{1/2} \right]^2 . \qquad \text{(3H.16)}$$

$$S_w = \left\{ \left[\left(\frac{V_{sh}^{2-V_{sh}}}{R_{sh}} \right)^{1/2} + \left(\frac{\phi_e^m}{R_w} \right)^{1/2} \right]^2 R_t \right\}^{-1/n} . \qquad (3H.17)$$

The Indonesia model,[52] and other similar models, are often used when field-specific SCAL rock electrical-properties data are unavailable but are also sometimes used where the SCAL exponents do not measure the full range of shale volumes. Although it was initially modeled on the basis of Indonesian data, the Indonesia model can be applied everywhere. The inputs are the effective porosity, ϕ_e, shale volume and resistivity (V_{sh} and R_{sh}), and water and deep resistivities (R_w and R_t). The S_w output is usually taken to be the water saturation of the effective porosity, but it has been recently suggested that the output is likely to estimate S_{wt}.[31] Many other log-based shaly-sand models have been proposed[53] but, for brevity, are not discussed here.

Waxman-Smits-Thomas and Dual-Water Models. S_{wt}, the water saturation of the total porosity, is calculated at each reservoir data point by iterative solution of the complex multiparameter WST and DW equations (Eqs. 3H.18 and 3H.19). For brevity, the details[3,41–43,54–56] of the solution methods are not presented here. The WST and DW models are total-porosity/S_w system models.

The WST model is based on laboratory measurements of resistivity, porosity, and saturation of real rocks.[41–43,54] Q_v is the CEC per unit PV.

$$\frac{1}{R_t} = \phi_t^{m^*} S_{wt}^{n^*} \left(\frac{1}{R_w} + \frac{BQ_V}{S_{wt}} \right), \qquad (3H.18)$$

where S_{wt} = water saturation of the total porosity as shown schematically in Fig. 3H.8, B = specific cation conductance in (1/ohm•m)/(meq/mL), and Q_V = CEC in meq/mL of total PV. The exponents m^* and n^* apply to the total PV.

The DW model[31,55,56] is also based on the WST data. It uses clay-bound-water conductivity instead of WST's BQ_v factor (see Eqs. 3H.18 and 3H.19) and an alternative shale-volume descriptor, S_{wb}, the saturation of physically bound water in the total PV (see Fig. 3H.8).[3,43] When V_{sh} is zero, S_{wb} is zero; and when V_{sh} is 100% BV, S_{wb} and S_{wt} are also 100% PV.

$$\frac{1}{R_t} = \phi_t^{m_o} S_{wt}^{n_o} \left[\frac{1}{R_{wf}} + \frac{S_{wb}}{S_{wt}} \left(\frac{1}{R_{wb}} - \frac{1}{R_{wf}} \right) \right], \qquad (3H.19)$$

where R_{wb} = resistivity of clay-bound water (= $R_t\phi_t^m$) in the shales, and R_{wf} = resistivity of free formation water (= $R_t\phi_t^m$) in the shale-free water zones. Because of the different model assumptions, DW exponents m_o and n_o must always be smaller than the WST exponents[55] and may be values similar to "clean" sand exponents. Where the WST and DW models have been properly applied, the V_{HCP} results should be equal. All S_{wt} calculations from the WST and DW methods must be checked to ensure that they are greater than S_{wb}. After this check, they are used with ϕ_t to obtain the V_{HCP}. For the DW model, when the outputs require conversion to effective porosity, ϕ_e, and effective water saturation, S_{we}, the properties are converted with Eqs. 3H.20 and 3H.21, respectively.

$$\phi_e = \phi_t(1 - S_{wb}), \qquad (3H.20)$$

$$\text{and } S_{we} = \frac{S_{wt} - S_{wb}}{1 - S_{wb}} . \qquad (3H.21)$$

Strengths and Weaknesses of R_t-Based S_w Calculations. The greatest strength of S_w calculations from the R_t logs is that these calculations can be made at each net-pay depth with valid data for all wells within the log database. The calculations can account for any subsets of input parameters related to the individual zones.

The weaknesses of the R_t-based S_w calculations are that one has to select a model to describe the relationship of S_w to R_t, R_w, and a variety of other input parameters. Any model is an approximation to the real nature of the reservoir pore system and, typically, has limitations such as how the clay-mineral conductivity is modeled. Log-analysis estimates of V_{sh} are rather uncertain, so sands that are substantially free of clay minerals can easily, and incorrectly, be assigned significant clay volumes. In these circumstances, complex shaly-sand models may have been applied when it is more appropriate to model the sand as clean sand. Effective porosity is also impacted by the uncertain V_{sh} estimates. The R_w is often assumed to be constant within the hydrocarbon column, and usually there is little data regarding R_w other than from aquifer samples. In several cases in which the R_w distribution has been studied in depth, it was found to vary in systematic ways within the hydrocarbon column and not necessarily be the same as in the underlying aquifer.[36–38]

CEC can be measured in the laboratory, but in the reservoir it must be estimated by correlations with porosity or V_{sh}. For the laboratory CEC measurements, there are fundamental uncertainties such as the degree to which the clay-mineral geometry is altered by the disaggregation of the core. The total surface area and CEC may be enhanced by comminution (i.e., grinding to grain-size particles).[57]

The other input parameters for the S_w /R_t models are either based on "worldwide experience" (such as default exponent parameters in commercial software packages) or developed from SCAL rock-electrical-property measurements on a relatively small number of core plugs from the reservoir interval. Hence, there are relatively few data determining the parameters that are used for the log point-by-point S_w calculations. It has to be assumed that the manner in which the water saturation is distributed in the core plugs during these laboratory experiments is like that of the real reservoir. Because water is present during the laboratory measurements, clay minerals are rehydrated at the time of the tests.

Calculating S_w From Laboratory Capillary-Pressure/Saturation Measurements. A second S_w method that is totally independent of the resistivity logs uses laboratory-measured P_c /S_w data. The underlying concept of the use of capillary pressure data is that the reservoir has come to capillary equilibrium over geologic time (the millions of years since hydrocarbons have entered and filled the reservoir trap). This equilibrium is reproduced in laboratory experiments using the centrifuge, porous-plate, and MICP methods. The P_c /S_w data are measured on a selected set of reservoir core plugs representing a range of porosity and permeability values (and possibly also lithologies).

Centrifuge experiments are typically made on 1-in. core plugs over a period of several days in the intense gravitational field (up to 1000 G) of the centrifuge and are assumed to be equivalent to what occurs in a hydrocarbon reservoir over millions of years in a 1-G gravitational field and over lengths of 10 to hundreds of feet. These assumptions are broadly accepted as being reasonable, provided that the samples are not damaged during testing in the centrifuge. The reported P_c /S_w values are not the raw laboratory data. In the laboratory, the average saturation is determined at each centrifuge speed, and those raw data are input to a mathematical model to convert them to a tabulation of endface saturations and P_c values.

Porous-plate P_c tests are made on core plugs at several different gas pressures and are generally carried out at the same time as the resistivity experiments. After reaching equilibrium with no further brine flow at each pressure, the S_w is constant along each plug and is calculated from its weight loss.

TABLE 3H.4—TYPICAL IFT AND CONTACT ANGLES AND APPROXIMATE CONVERSION FACTORS FROM P_c TO H_{hwc} (after *Fundamentals of Rock Properties*[8])

Fluid-Pair Nonwetting/Wetting	Contact Angle (°)	IFT (dynes/cm)	Approximate Equivalent Vertical Height in Reservoir per psi of Laboratory P_c (ft per psi)
Laboratory P_c Pair			
Air/Brine	0	72	2 to 3.5
Oil/Brine	30	48	3.5 to 5
Mercury/Air	140	480	0.4 to 0.7
Air/Oil	0	24	6 to 11
Reservoir P_c Pair			
Oil/Brine	30*	30	
Gas/Brine	0*	50**	

*Typical contact angle for mainly water-wet system. **Pressure and temperature dependent. Approximate IFT value to depth of 5,000 ft. Note: These conversion-factor ranges apply in many gas and oil reservoirs, but conversion factors are larger for heavy-oil reservoirs. Assumes fluid density difference of 0.25 g/cm^3 for oil/water and 0.75 g/cm^3 for gas/water reservoirs. Eq. 3H.22 used to calculate height conversion factors.

MICP tests are made on dried core pieces and the volume of injected mercury, the nonwetting phase, is converted to an S_w value. This is considered to be total S_w if, at high enough pressures, mercury enters both the microporosity and dry clay-mineral porosity. Conversely, for centrifuge or porous-plate tests, where brine is present as the wetting phase, clay minerals probably hydrate, and their physically bound water is unlikely to be displaced during the test. Brine-related P_c/S_w measurements may give total or effective S_w, depending on the specific porosity measurement method used (i.e., whether the porosity occupied by the clay-mineral physically-bound water is included or excluded from the porosity calculation). Effective S_w values are always lower than total S_w values and should be very low at high capillary pressures if there is little nonclay-mineral-related microporosity.[58]

The conversion of the laboratory P_c/S_w data to reservoir conditions requires knowledge of the IFT and contact angle of the fluid pair used in the laboratory and properties of the brine and hydrocarbon fluids at reservoir conditions. These are needed to calculate the density of each phase and to estimate the IFT between the fluid pair at reservoir conditions. The P_c values (in psi) are converted to vertical height above the hydrocarbon/water contact, H_{hwc} (in feet), with the following formula:

$$H_{hwc} = \frac{\left[(\sigma \times \cos\theta)_r \times P_c\right]}{\left[0.433 \times (\rho_w - \rho_h) \times (\sigma \times \cos\theta)_s\right]}, \quad \text{(3H.22)}$$

where fluid densities (ρ) are in g/cm^3, and the subscripts are r = reservoir, s = surface, h = hydrocarbon, and w = water. **Table 3H.4** lists some typical values[8] for IFT, σ, and contact angle, θ, used in Eq. 3H.22 and provides approximate ranges for the factors for converting P_c-laboratory data to height above a reservoir free-water level. Height-P_c conversion factors are similar for many oil and gas reservoirs; the footnotes in Table 3H.4 describe the values that were assumed to calculate these ranges. More details of correlations for brine/hydrocarbon IFT as a function of oil or gas gravity have been published.[59] The reservoir-condition contact angle, θ, is usually taken as 0 for gas reservoirs and 0 or 30° for oil reservoirs because, generally, data are not available at reservoir conditions.

The suite of P_c/S_w data is typically converted to a mathematical relationship between S_w as the dependent variable and the independent variables—porosity, permeability, and H_{owc} or H_{gwc}.[46,47,60] Because permeability is usually determined as a function of porosity, it is often not included as an independent variable. Two of the mathematical forms that have been used are

$$S_w = A + B \times \phi + C \times \phi^2 + \frac{D}{H_{owc}} + \frac{E}{H_{owc}^2}, \qquad (3H.23)$$

$$\text{and } \log_{10} S_w = A \times \log_{10}\left(H_{owc} + B\right) + C \times \phi + D, \qquad (3H.24)$$

where A, B, C, D, and E are curve-fit constants. In Eq. 3H.24, B permits the removal of singularities at zero height.

In developing the coefficients for these relationships, any zonation of the reservoir intervals needs to be applied, and then separate sets of coefficients developed for each zone. The zonation can be based on geological interpretation of the reservoir depositional and diagenetic history and/or variation in the P_c/S_w curves for different parts of the reservoir interval.

The depth of the reservoir's OWC or GWC must be known in order to make S_w calculations using the P_c/S_w methodology. The calculations of S_w are made only above this depth. In reality, the H_{owc} or H_{gwc} is referenced to the FWL (i.e., the depth at which P_c=0 and which is deeper than the observed OWC or GWC). For a gas reservoir consisting of good-quality rocks, the difference between the FWL and the GWC is typically 1 ft or less. However, for an oil reservoir containing a heavier oil, this difference can be 10 feet or more, and, given four-way closure on an anticlinal structure, the impact on the OOIP volume between using the FWL vs. the observed OWC as the H_{owc} = 0 depth can equal a few percent of OOIP.

Once the various sets of coefficients have been developed and the P_c to H_{owc} (or H_{gwc}) conversion made, an S_w value can be calculated at each data point within the log database that has a valid porosity value and is above the OWC or GWC. Hence, there will be the same number of, or more, S_w values available from this S_w methodology as when using the R_t logs.

Strengths and Weaknesses of P_c/S_w-Based S_w Calculations. The strength of S_w calculations from P_c/S_w data is that, after making a correlation with porosity and height, a unique S_w value is available for all wells at all net-pay depths with valid porosity values in the log database. This also applies to the whole hydrocarbon column anywhere in the reservoir once the wellbore porosity values have been propagated into the full geocellular model grid. These calculations can account for any zonation and subsets of input parameters related to the individual zones.

A potential weakness in the P_c approach to S_w calculations is whether the laboratory measurements have been allowed sufficient time to reach equilibrium. If not, the S_w values, particularly at high P_c values, will be too high. Another potential weakness is the accuracy of the IFT value used in converting from surface to reservoir conditions; fortunately, these values vary over a limited range for most hydrocarbon/brine pairs. A third potential weakness is the definition of the FWL depth compared with the observed OWC or GWC. A fourth potential weakness is whether enough data have been taken to be representative, both vertically and areally, of the zones in the reservoir.[14]

The fifth potential weakness concerns the complexity of the reservoir's hydrocarbon-filling and structural history. In simple oil-reservoir situations and most gas-reservoir situations, this is not an issue. However, for oil reservoirs with tar mats and heavy-oil zones there is a complication because of the varying oil density near the OWC, including the possibility that the tar mat has a hydrocarbon density very close to that of the connate brine. Another aspect may be whether all or portions of the hydrocarbon column are on the imbibition cycle where imbibition P_c/S_w data are needed for the S_w calculations, not the typical drainage P_c/S_w data.[14]

Calculating S_w With OBM-Core-Plug Dean-Stark Water-Volume Determinations. The third method for determining the S_w in a reservoir's hydrocarbon column is to cut OBM cores and perform Dean-Stark water-volume determinations on the routine core plugs. Foot-by-foot S_w values can be calculated from these water volumes and the associated core-plug PVs. OBM cores are typically cut only in a few wells in a particular field. These S_w data can be applied to other, uncored wells in the reservoir if strong correlations between these values and porosity and/or permeability are identified. These data are not valid in the oil/water or gas/water transition zone or in the aquifer, intervals in which the connate brine is mobile. OBM-core S_w values may be found to be either higher or lower than those from the other two methods described previously.

Strengths and Weaknesses of OBM-Core S_w Values. The strength of S_w values from routine-core-analysis Dean-Stark S_w data is that these data are the most direct measure of reservoir connate S_w values above the oil/water or gas/water transition zone. Relative to the two methods discussed previously and the variations of these methods, the OBM-core S_w approach is a direct S_w determination and the other methods are indirect S_w-calculation approaches that require many more assumptions and inferences.

The weaknesses of the OBM S_w method are that it does not apply to the lowest parts of the oil/water or gas/oil transition zone where the brine phase has mobility and that, generally, the amount of OBM core S_w data is limited because the operator cuts cores with OBM only in a limited number of wells because of the expense. The first of these weaknesses can be overcome if the OBM S_w data are used in combination with either the resistivity logs or with P_c/S_w data.

Another consideration is that the whole project, from the mud formulation to the core-handling and -preservation procedures on through the routine-core-analysis measurements, needs to be monitored and reviewed in detail to ensure all steps were executed properly. This demands that considerable time and effort be spent by the technical team to ensure success; however, to some extent the same comment applies to the P_c/S_w and resistivity-log/S_w calculation approaches discussed previously.

3H.7.4 Integration of S_w Data From the Different Methods. Depending on the data availability in a particular reservoir situation, a combination of the various S_w approaches may prove superior to the use of a single type of data. The first step in going to a combination approach is to review the reservoir's database to identify any significant gap in vertical, or areal, coverage. The most obvious gap often occurs near the fluid contact, because there is little reason to drill wells in downdip locations, particularly during a reservoir's development phase. Three examples of combination approaches are described here.

Resistivity-Log Data With P_c/S_w Data. Resistivity-log-derived S_w results may not be available throughout the hydrocarbon column of a reservoir. To fill gaps and average the point-by-point data set, it is common practice to plot S_w as a function of height, to omit nonpay points, and to identify various porosity ranges by coding the data points. Resistivity-log-derived S_w data frequently shows V- or U-shaped patterns on these plots because of the shoulder/bed effects near nonpay sections (shales). The most accurate S_w values in such patterns are usually at the lowest S_w values where the thin-bed correction is minimized. In a manner similar to that described in the previous P_c/S_w section, height/saturation curves are often fitted to these resistivity-log-derived S_w data to enable reservoir hydrocarbons-in-place volumes to be calculated. The function forms are similar to or are the same as those described in the P_c/S_w section.[60]

Routine OBM Core S_w Data With P_c/S_w Data. Because there is a need to define the S_w characteristics of the oil/water or gas/water transition zone and because the OBM-core S_w data can be incorrect and too low in this interval, one approach is to use P_c/S_w data in combination with the routine OBM-core S_w data. This can be done by first correlating the OBM-core S_w data to porosity and assuming that this relationship is valid above the oil/water or gas/water transition zone. The functional form of this first relationship might be

$$S_w = A + B \times \phi + C \times \phi^2 \; . \qquad (3H.25)$$

The second step is to create a tabular data set in which the S_w/porosity correlation is used to calculate an array of S_w values for large H_{owc} or H_{gwc} values and a range of porosity values. For this part of the data set, S_w is assumed to be independent of the H_{owc} or H_{gwc} values. The P_c/S_w data converted to reservoir conditions is used to provide data points for low H_{owc} or H_{gwc} values and various porosity values. Statistical calculations are applied to the whole of this data set. The functional form of this second relationship might be

$$S_w + A' + B' \times \phi + C' \times \phi^2 + \frac{D}{H_{owc}} + \frac{E}{H_{owc}^2} \; . \qquad (3H.26)$$

With this functional form, the boundary conditions of the first step are automatically met in the second step.

Routine OBM-Core S_w Data With Resistivity-Log Data. To address the lack of valid OBM-core S_w data in the oil/water or gas/water transition zone discussed previously, it is also possible to combine OBM-core S_w data with the resistivity-log data to develop an overall S_w methodology. This approach assumes that a number of wells have been drilled through the OWC or GWC so that there are log resistivity values through the oil/water or gas/water transition zone. In this approach, the OBM-core S_w data are used to back-calculate the saturation-exponent, *n,* values over each zone so that the core-based V_{HCP} value equals that calculated from the resistivity logs (see Eq. 3H.26). Then the core-based saturation-exponent, *n,* values are applied to the noncored well's resistivity logs to calculate S_w point-by-point throughout the reservoir interval in all wells.[22,61] This approach assumes that the R_w, *a,* and *m* values have been determined from other experimental and fluid-sample data so that R_0 can be calculated.

$$n = \frac{\log_{10} R_0 - \log_{10} R_t}{\log_{10} S_{wc}}, \qquad (3H.27)$$

where R_0 is the bulk resistivity at S_w = 100% PV and is calculated with Eqs. 3H.7 and 3H.8. R_t is the deep-reading resistivity-log reading, and S_{wc} is the OBM-core S_w above the mobile-water transition zone. The resulting back-calculated *n* values at the core-plug depths are averaged for the zone. In some instances, *n* may be found to have an areal variation within a zone that should be taken into account in subsequent calculations.

Adjustments to S_w Data From Different Methods. In the previous section, a number of methodologies for S_w calculations have been described. These are basically three independent methods; hence, they can be used together to determine the accuracy of the S_w calculations throughout the hydrocarbon column. Because the methods are based on very different technical approaches and assumptions, if the different methods give essentially the same S_w answer, then it is highly likely that this is the correct S_w.

However, the challenge comes when, as is often the case, the different methods result in different S_w values and distributions. The OBM-core S_w values might be either higher or lower than those from the other two methods. The common misunderstanding that OBM-core S_w is likely to be too low is unsubstantiated. In a very large reservoir, it could go both ways depending on where one is in the reservoir.[22,61] If the values are quite different, two aspects of the calculations need to be reviewed in depth. First, the quality of the input laboratory data needs to be checked and how it was converted from raw data into the input values to the S_w calculations needs to be reviewed. Second, the assumptions and models used for the S_w calculations need to be checked. For example, with the P_c/S_w data, the assumed oil/water density difference may be considerably in error, or the shaly-sand S_w model may be inappropriate for the particular reservoir. As well as the S_w averages, the zone-average V_{HCP} values from the various methods should be compared, which includes porosity in the comparison calculations.

Core, Total, and Effective Systems Compatibility. The Archie R_t-based S_w equation models "clean" sands. Various other shaly-sand models use either the effective or the total-porosity systems. It is well known that these basic models, if applied properly to the same formation, must produce the same final V_{HCP} from their different calculation procedures (see Fig. 3H.8 and Eqs. 3H.3 through 3H.5 in Sec. 3H.5).[3] ϕ_t is greater than or equal to ϕ_e; however, at the same time, S_{wt} is greater than or equal to S_{we} and, when used together, the appropriate combinations must give the same V_{HCP} result. For the total-porosity system, $V_{HCP} = \phi_t \times (1 - S_{wt})$, whereas for the effective-porosity system, $V_{HCP} = \phi_e \times (1 - S_{we})$.

The V_{HCP} can also be estimated from a combination of core porosity and Dean-Stark S_w measured on preserved OBM cores. The several systems—core, total, and effective—must all give the same fundamental results, and the most accurate of them (the OBM-core method) can be used to calibrate and test the less accurate methods. When properly adjusted and applied (e.g., by improving the V_{sh} estimates or IFT values), all three methods give the same final V_{HCP}. If they do not agree, the likely sources of uncertainty and error must be examined.

It is clearly inconsistent and incorrect to mix the systems by, for example, reporting an effective porosity with a total S_w, a total porosity with an effective S_w, or a standard-core porosity with an effective S_w. System compatibility must also be maintained by correct use of the SCAL measurements and log-analysis formulae, when these are used to calibrate the resistivity logs and P_c/S_w methods. The differences should be resolved as much as possible. To the extent that they are not, the differences can be considered to be a measure of the uncertainty in the S_w calculations.

3H.7.5 Uncertainties. It is the uncertainty of the hydrocarbon saturation $(1 - S_w)$ that is economically important, not the absolute uncertainty in S_w. When uncertainties in S_w are evaluated, their importance in terms of S_o and S_g should be accounted for. The uncertainties of the several S_w-evaluation methods vary widely.

OBM-Core S_w Data. The water volume extracted from a single core plug may have a random and known systematic uncertainty of ± 0.05 cm^3, where each uncertainty refers to one SD. The PV of a typical 1-in. core plug is 4.0 cm^3 if the porosity is 20% BV. The water-volume uncertainty alone equates to an S_w uncertainty of ± 1% PV (0.05/4.0). The uncertainties in porosity have a further effect on this calculation.[48] An OBM-core S_w of 20% PV, therefore, has a combined 1-SD range from approximately 18 to 22% PV. At lower porosities and higher S_w values, the water-volume uncertainty may be ± 0.1 cm^3, leading to an S_w uncertainty of ± 3% PV, when the porosity is 15% BV. As porosity decreases, the uncertainty grows. Before the measurements are made, any water in the toluene and the Dean-Stark apparatus must be removed, or the S_w values will be overstated. The extraction time required to recover the water adsorbed on the clay minerals adds to the uncertainty.

The uncertainty of the average core S_w will be improved when plugs are selected at one or two per foot with equal spacing and without regard for the rock quality. However, as discussed previously, plug samples are not always selected at random, so care must be exercised, especially regarding the S_w values predicted at depths where core is not available. From a broader perspective, it must also be remembered that 1-in. core plugs only sample approximately 2% of the full-core volume. Because of these many factors, the authors estimate that uncertainties similar to those given concerning porosity also propagate to the zone-average OBM-core S_w values. Measurements in which larger core plugs are analyzed will reduce several of the uncertainties.

Resistivity-Log-Derived S_w Values. The log readings, typical SCAL-derived Archie exponents, and all of the other associated parameters are uncertain. For example, the resistivity-log uncertainty may be ± 50% when R_t is 500 ohm•m. The most important uncertainty contributors at low S_w values are likely to be R_t and n. S_w uncertainty in this circumstance is estimated at ± 5% PV (i.e., if S_w is calculated as 10% PV, the 1-SD range is 5 to 15% PV).[48] At lower porosity values and higher water saturations, similar methods led to uncertainty estimates of ± 9% PV. Given that further uncertainty in the final calculated S_w may arise from shaly-sand effects and many other sources, the authors believe that the ranges given apply equally to the overall systematic uncertainty of the S_w zone-average values. These estimates are all 1 SD; therefore, in 32% of cases, zone-average uncertainties are considered likely to be greater than the ranges given.

P_c-Derived S_w Values. The uncertainty estimates are the sum of several factors. Most of these factors have their greatest impact on the S_w calculations in the first 100 to 200 ft of the hydrocarbon column above the fluid contact. Therefore, because the transition zone is considerably longer in many oil reservoirs than in a gas reservoir, their impacts will be greater in most oil reservoirs. Above 200 ft, the S_w values are usually only changing slowly; hence, the primary consideration above the transition zone is whether the laboratory measurements are taken at equilibrium conditions.

The first factor in the uncertainty analysis is the fundamental assumption as to whether the drainage or imbibition P_c/S_w data should be used. In most cases, the drainage curves should be used, but, in a few situations, the reservoir may be on the imbibition cycle. In these situations, the improper choice of using the drainage P_c curve can lead to a +5 to 20% PV S_w error in the first 100 to 200 ft above the OWC.[28,29]

The second factor concerns the laboratory P_c/S_w measurements. If the measurements are not taken to equilibrium, then the S_w values at a particular P_c value will be too high. This can be +1 to 10% PV effect for the large H_{owc} or H_{gwc} range. The other key aspects for reported centrifuge laboratory results are how the raw laboratory measurements of water volumes were determined and how these data have been converted to the reported endface saturations. The water-volume measurements have the same-size potential error as discussed for the OBM Dean-Stark S_w measurements (± 1 to 3% PV). Differences in the laboratory calculation procedures can result in further variations of ± 1 to 3% PV in reported P_c/S_w results when using the same raw laboratory data. For porous-plate tests and others, the repeated handling of poorly cemented or uncemented core plugs can cause grain loss, which, after the final calculations, translates into small errors in S_w.

The third factor is how the suite of raw laboratory data for a particular reservoir interval are curve-fitted and presented in the final laboratory report as tabulations of P_c/S_w values for each core plug. Uncertainty in the application arises from how these reported values are averaged for use in the S_w calculations over the full range of reservoir porosity and permeability values. This uncertainty includes how the data are weighted and whether some potential outlier data from one or two core plugs distorts the averaged P_c/S_w curves. These uncertainties primarily affect the first 100 to 200 ft above the H_{owc} or H_{gwc} so that their impact depends on how thick the hydrocarbon column is and its distribution as a function of H_{owc} or H_{gwc}.

The final factor is the conversion of the averaged P_c/S_w curves (or equation) from surface to reservoir conditions, all of which affect the conversion of P_c values to H_{owc} or H_{gwc} values. This includes a number of subfactors, each with its own uncertainty level: IFT at surface and reservoir conditions, fluid-pair density difference at reservoir conditions, contact angles, and depth of the actual in-situ FWL compared with the OWC or GWC. The contact angles at surface and reservoir conditions are generally taken to be the same because no data are available to proceed otherwise. For these other factors, the uncertainty is considerably greater for an oil reservoir than for a gas reservoir; because the IFT values can be low and compared with those for a gas reservoir, the density differences are significantly less particularly if there is a vertical oil-gravity variation that results in a heavy-oil interval just above the OWC. All of these factors affect the H_{owc} or H_{gwc} values; therefore, their impact on the S_w calculations is predominantly in the first 100 to 200 ft above the fluid contact.

In summary, the use of P_c/S_w data can result in S_w uncertainty of ± 5 to 15% PV in the oil/water or gas/water transition zone. Above that transition zone, the uncertainty is related to whether the laboratory data were taken at equilibrium conditions and how the various P_c/S_w curves have been averaged together. In this range, the uncertainty is likely to be 3 to 10% PV.

3H.8 Permeability Determination

Point-by-point permeability values are needed over the reservoir interval at the wellbores for several purposes. First, the distribution and variation of the permeabilities are needed by the engineers to develop completion strategies. Second, this same information is needed as input to the geocellular model and dynamic-flow calculations (e.g., numerical reservoir-simulation models). For both of these, the first consideration is the location of shales and other low-permeability layers that can act as barriers or baffles to vertical flow. A second consideration is the nature of the permeability variation (i.e., whether the high-permeability rock intervals occur in specific layers and the low-permeability intervals occur in other layers, or that there is so much heterogeneity that the high- and low-permeability intervals are intimately interbedded with each other).

When good-quality core data are not available, estimates of permeability can be made from empirical equations. Permeability is controlled by such factors as pore size and pore-throat geometry, as well as porosity. To take some account of these factors, the widely used Timur equation[62] relates permeability to irreducible S_w and porosity, and therefore can be applied only in hydrocarbon-bearing zones. This form of his equation applies to a medium-gravity oil zone:

$$k = \frac{10^4 \phi_e^{4.5}}{S_w^2}, \quad \text{(3H.28)}$$

where k = absolute permeability in millidarcies, ϕ_e = effective (not total) porosity as a bulk volume fraction, and S_w = effective water saturation above the transition zone as a fraction of PV. Estimates that are based only on porosity are likely to have large prediction errors, especially in carbonate reservoirs. Equations of the following form, or a logarithmic-linear form, are useful particularly in sandstones:

$$\log_{10} k = C \log_{10} \phi_e + D, \quad \text{(3H.29)}$$

where parameters C and D are very approximate and equal to about 7, and k and ϕ_e are as defined following Eq. 3H.28. They should be adjusted according to local knowledge.

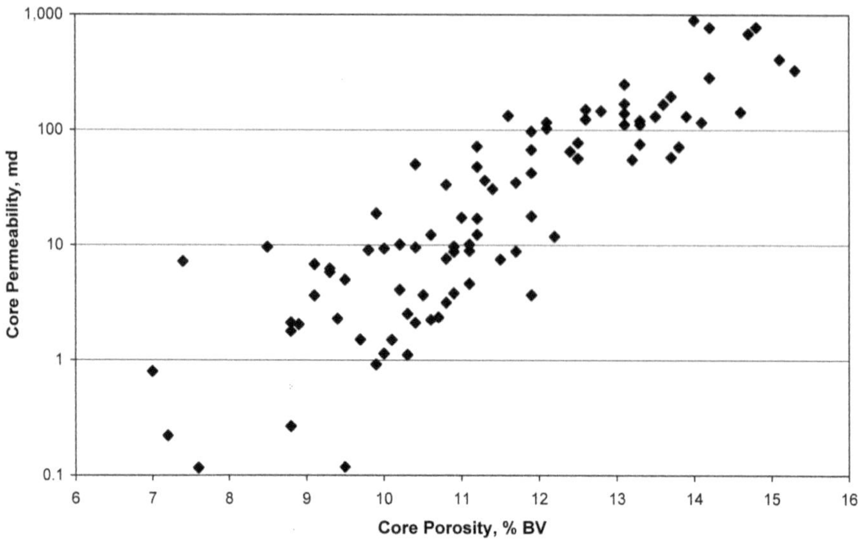

Fig. 3H.19—Core permeability vs. core porosity crossplot; data from an Asian gas field.

In field evaluation, the starting point for calculations of permeability is the routine-core-analysis data. These data, and the associated SCAL measurements of permeability and porosity as a function of overburden stress, are input to calculations to develop permeability values at reservoir conditions and the permeability vs. porosity correlation. The permeability vs. porosity correlation is often taken as semilogarithmic but usually with a steeper slope at low-porosity values. **Figs. 3H.19 and 3H.20** demonstrate the characteristics of these relationships. Fig. 3H.19 presents a typical permeability vs. porosity relationship from routine-core-analysis data (the scatter in these data increases at the lower-porosity levels). Fig. 3H.20 shows the permeability ratio (stressed permeability divided by unstressed permeability) vs. unstressed permeability. This ratio is much smaller for low-permeability values and approaches a value of 1.0 for the high-permeability values.

In developing the permeability vs. porosity relationships, the technical team needs to identify the extent to which the reservoir interval needs to be subdivided into zones or layers. The subdividing of the core data over the reservoir interval should be into logical subdivisions that are strongly influenced by the geologists' understanding of the depositional environment. This will naturally account for major differences in grain size, sorting, and key mineralogical factors. Alternatively, a sufficiently thick reservoir interval can be subdivided into layers of 50 to 100 ft each. A superior petrophysical methodology will be developed if a thick reservoir is appropriately subdivided, compared with treating the full reservoir interval with a single permeability vs. porosity correlation. A single permeability vs. porosity correlation for a reservoir interval with different depositional environments can lead to underprediction of permeability by an order of magnitude in an interval of better-sorted rocks compared with poorly sorted rocks (see Fig. 3H.3). Identifying the location and correct values of highest-permeability rocks is very important for reservoir flow modeling.

The result of modeling the relationship with the least-squares regression method is that the range of predicted permeability values is smaller than that of the original routine-core permeability data. This loss of range is made worse when the logarithm of permeability is used as the *y*-variable because the logarithmic model is a predictor of the geometric-average permeability.[32] While the permeability vs. porosity relationship is developed from the routine and SCAL core-analysis data, the application to the point-by-point well-log database requires the use of porosity values calculated from the logs. It is preferable to model the prediction

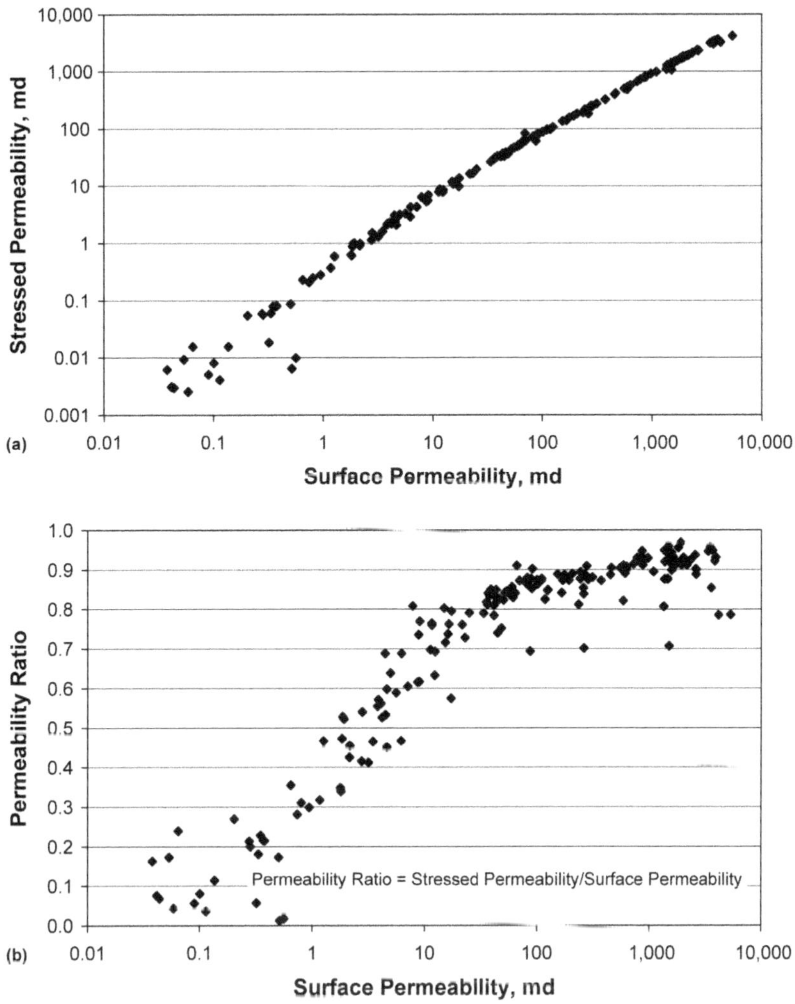

Fig. 3H.20—Crossplots of core permeability at stressed vs. surface conditions and core permeability ratio vs. core permeability at surface conditions; data from an Asian gas field. "Stressed" refers to the rock being subjected to simulated overburden pressure of approximately 4,500 psia. The permeability correction is larger at low permeabilities.

equation directly with core permeability and the basic log values[32] (see **Fig. 3H.21** and the calibration line-fitting in Sec. 3H.5.3). The *y*-on-*x* (dashed) line-fit in Fig. 3H.21 follows a curved trend on the logarithmic-linear plot and uses an arctangent function as the transformation. The solid line gives arithmetic average permeabilities at various bulk-density values. The arithmetic averages, which may be more appropriate in some reservoirs, are 2 to 3 times larger than the geometric averages. Alternative predictions of permeability may also be estimated using two-log or multiple regression analysis methods.

After the permeability values have been calculated point-by-point over the reservoir interval from the various wells' logs, these permeability values need to be compared with those derived at each well from the pressure-transient analysis (PTA) of the pressure-buildup (PBU) or falloff data. The PBU permeability values are average values for the interval open to flow into the wellbore. The type of average (arithmetic, geometric, harmonic, or somewhere in between) to use with the point-by-point permeability values depends on the nature of the depositional envi-

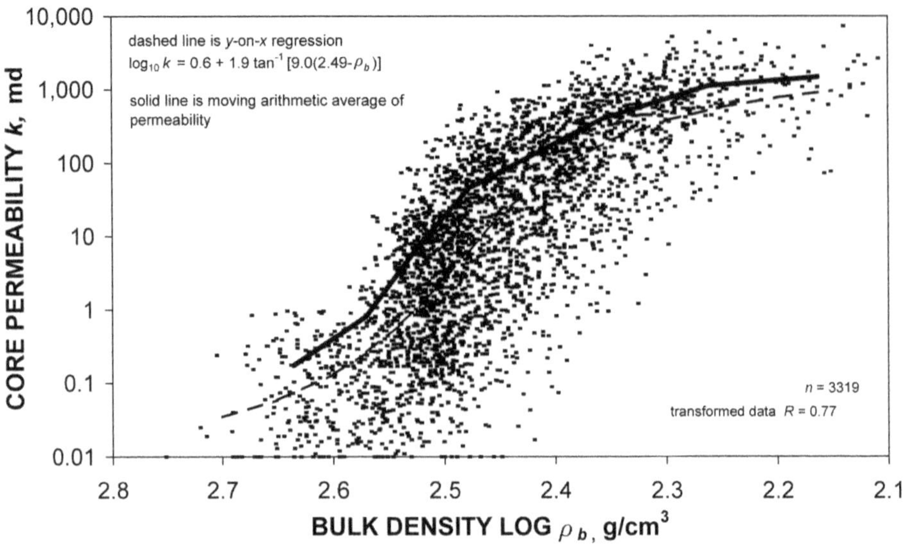

Fig. 3H.21—Nonlinear regression relationships for core permeability and bulk density log (South Morecambe gas field, offshore U.K.). After Woodhouse.[32] The two lines illustrate the significant difference between geometric and arithmetic averages.

ronment and whether the perforated intervals are a small fraction of the full reservoir interval. If there are significant differences between the two sets of average permeability values, then the technical team needs to determine the likely cause of the differences—small-scale fractures, relative permeability effects, or some other geological factors. The point-by-point permeability values may need to be adjusted on the basis of the technical teams' conclusions.

3H.9 Petrophysical Case Studies

This section presents brief summaries of detailed petrophysical evaluations of several fields that have been described in the SPE and Soc. of Professional Well Log Analysts (SPWLA) technical literature. These case studies cover some of the complications that occur when making net-pay, porosity, and S_w calculations.

3H.9.1 Prudhoe Bay Field. Prudhoe Bay is the largest oil and gas field in North America with more than 20 billion bbl of OOIP and an overlying 30 Tscf gas cap. In the early 1980s, the unit operating agreement required that a final equity determination be undertaken. In the course of this determination, an extensive field coring program was conducted, which resulted in more than 25 OBM cores being cut in all areas of the field and some conventional WBM and bland-mud cores in other wells. Also, several major laboratory programs were run to address various technical issues regarding the correct approach to calculate porosity and water saturation. The background geologic understanding of the major reservoir, the Ivishak or Sadlerochit, and various technical studies have been presented in a number of technical papers.[14,15,21,22,27,36,48,49,61,63,64]

Geologically, the Sadlerochit reservoir is a combination structural/stratigraphic trap consisting of a 500-ft reservoir interval covering an area of 15×35 square miles at a depth of 8,000 to 9,000 ft. This reservoir is mainly very high quality deltaic Permo-Triassic sandstones deposited in a braided stream environment, ranging from fine-grained to conglomeratic, with some limited intervals of shales found in various areas of the reservoir. The reservoir has been divided vertically into eight zones on the basis of differences in rock types (see Fig. 3H.1). The grains are primarily quartz, quartzite, and chert. Over geologic time, there have been significant leach-

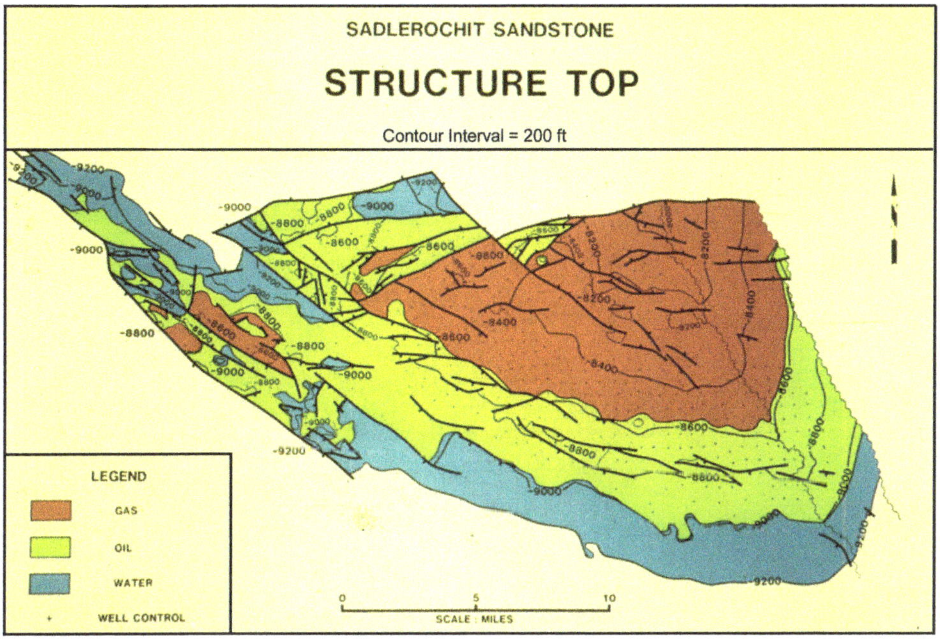

Fig. 3H.22—Areal structure map showing gas-cap and oil-column distribution in the Sadlerochit reservoir, Prudhoe Bay field, Alaska, U.S.A.[27]

ing and cementation. The chert grains have been leached to varying degrees, resulting in a significant intragranular component (10 to 60%) of the pore system. There are considerable amounts of siderite and pyrite cementation, together with quartz overgrowth and kaolinite cementation.[21]

The structural and hydrocarbon-filling histories of Prudhoe Bay are very complicated; this is clearly evident from a visual examination of the cores, the routine-core-analysis fluid-saturation data, and an interpretation of the seismic data.[27] Oil initially filled this trap 40 million years ago. Since that time, because of differential burial, there has been a change in the configuration of the trap. Where there was previously a 2,000-ft closure, currently there is less than 1,000-ft maximum closure. The reservoir "tilting" resulted in a relict-oil zone systematically varying in thickness, which underlies the current oil column particularly to the southeast. At its base, the relict-oil zone had a flat OWC at the end of the original oil filling. Later, gas migrated into the trap creating a large gas cap over the main area of the field, but also smaller ones in the west-end area of the field. At discovery, a 40- to 60-ft heavy-oil-tar (HOT) layer was found above the current OWC with the oil gravity decreasing upwards from there. No relict-oil interval and essentially no HOT interval are found in the western part of the reservoir (see **Figs. 3H.22 and 3H.23**).

The net pay was based on defining the shale and siltstone intervals as nonpay.[21] Except for the inclusion of a small number of highly cemented sandstone intervals, this is effectively equivalent to a permeability cutoff of 0.6 md on the basis of routine-core-analysis permeability data, unadjusted to reservoir conditions. The net pay was determined from geologists' core descriptions. A GR-log model was used to define the pay/nonpay intervals within the Sadlerochit reservoir interval using the more than 450 logged wells' normalized GR logs. Radioactive sandstone intervals had previously been edited out of the core-log database. This GR-log model had to account for both thick and thin shale intervals, which it did by using three parameters—GR-

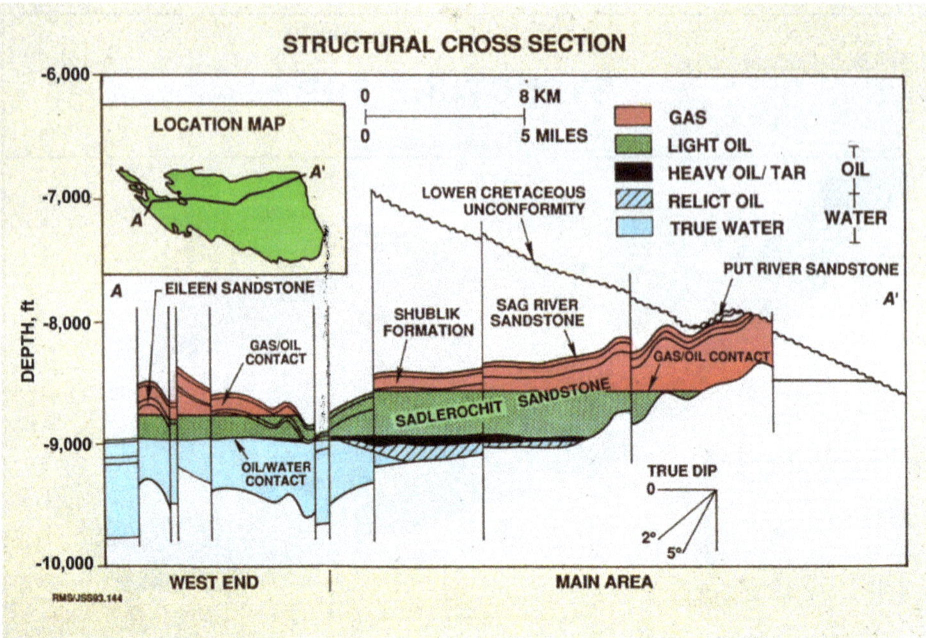

Fig. 3H.23—Cross section showing hydrocarbon distribution at discovery in the Sadlerochit reservoir Prudhoe Bay field, Alaska, U.S.A.[27]

sand, GR-thin, and GR-thick. The reservoir area was subdivided to account for areal variations when calibrating the GR-log model to the geologists' core picks of shale and siltstone intervals.

Porosity in the Sadlerochit interval was based on use of routine-core-analysis porosity data and sonic logs. The density log was not used because of the heavy-mineral effects discussed previously. The porosity data were not adjusted for overburden stress at reservoir conditions because it was found that the highly asphaltic crude oil had not been thoroughly cleaned from the core plugs during the routine laboratory procedures. Because there were generally both full-diameter Boyles-law and sum-of-fluids porosity measurements on the pebbly-sandstone and conglomeratic intervals (and both core-plug Boyles-law and sum-of-fluids porosity measurements for the other intervals), a hierarchy of these data was used when preparing the core standard for use with the sonic logs. The sonic log was calibrated linearly to the core porosity data. The calculations were performed individually for each of the eight zones and the different fluid intervals (gas cap, oil column, and aquifer). Finally, the reservoirwide porosity solution for each of these zone/fluid combinations was arealized to account for remaining systematic differences.

The OBM cores indicated that there is a wide range of S_w values, from less than 5% PV in the updip oil column and its overlying gas cap to as high as 50 to 70% PV in the west-end and southwestern portions of the Sadlerochit reservoir. Conventional log-analysis and capillary pressure methods suggested a much narrower range of S_w values. For these reasons, major studies were undertaken to identify the reasons for these differences and to determine if the OBM-core S_w data were valid. The primary conclusions were as follows:

• The OBM-core S_w data are valid.[22,48,49,61,63,64] This was confirmed by detailed observations and measurements at the wellsite at the time of coring some of the wells and at the commercial laboratory where the routine-core-analysis measurements were made.

• The centrifuge P_c data were found to agree with the OBM-core S_w data after the following effects were accounted for: the laboratory P_c/S_w measurements were run sufficiently long to reach equilibrium, it was recognized that various portions of the Sadlerochit reservoir were on different drainage and imbibition cycles on the basis of either the ancient or the current OWC, and the vertical variation of oil densities and the effect of the HOT interval were accounted for.[12]

• The S_w calculations from the SCAL measurements of rock electrical properties agreed with the OBM-core S_w values after the following effects were included: native-state rock electrical properties were measured at the wellsite and restored-state rock electrical properties were measured in the same S_w range as found from the OBM cores; and the variations in the R_w in the oil column and gas cap were included in these S_w calculations.[59] The n parameter was found to vary significantly as a function of S_w because of the complicated intergranular/intragranular (inhomogeneous) nature of the Sadlerochit pore system.[21]

• An independent in-situ measurement of S_w using the single-well chemical-tracer test produced results consistent with the OBM-core S_w at two updip well locations.[64]

The S_w methodology for the Sadlerochit reservoir was a combined use of the OBM-core S_w data with R_t values from the deconvolved [15] deep induction logs. Because of the low clay-mineral content of the Sadlerochit reservoir, an Archie S_w equation was used with n values derived for each zone at the OBM-cored wells and then trended over the reservoir. This results in foot-by-foot S_w calculations for the whole hydrocarbon column in the more than 450 logged wells. This very large database of S_w values was converted into a relationship involving porosity and H_{owc} and subdivided to account for systematic variations from one area to the next, before making the OOIP and OGIP calculations.

In summary, the Prudhoe Bay field shows that accurate petrophysical calculations can be complex, particularly in large reservoirs. If a typical approach had been used to make the various petrophysical calculations, the OOIP, OGIP, and the distribution of the hydrocarbons would have been significantly in error. Only through detailed core review by the geologists and by working through a number of complicated laboratory studies was closure reached. The same S_w values were then calculated using the various independent S_w methods.

3H.9.2 West Howard-Glasscock Unit Oil Field. This west Texas San Andres oil reservoir is an example of the petrophysical evaluation of a carbonate reservoir.[65] It is one of many San Andres oil reservoirs found in the Permian Basin of west Texas, U.S.A., and is presented to show that good-quality core/log petrophysical calculations can be undertaken even when a reservoir covers only a few square miles.

This study was performed in the 1970s after this unit was formed, and the waterflood of the San Andres reservoir was expanded. A total of 40 new wells was drilled with WBM and logged. Ten wells were extensively cored. The new wells were primarily for water injection and added to the existing 80 wells. This coring program was instituted because of the reservoir's complex and heterogeneous lithology and to aid log interpretation, geological mapping, and injection-well planning.

The routine-core-analysis measurements were primarily performed on whole core samples. **Fig. 3H.24** displays, through a permeability vs. porosity plot, the very heterogeneous nature of this reservoir. A number of plug samples (1.5 in. in diameter and 2 in. long) were cut for SCAL studies. **Figs. 3H.25 and 3H.26** show a comparison of the SCAL samples' permeability and porosity data with those from the adjacent whole-core data and show how heterogeneous this reservoir is. Previously published San Andres P_c/S_w correlations were used to calculate the irreducible S_w in the log calculations.[66]

The reported SCAL measurements focused on lithology, grain-density, and electrical-property measurements. The grain densities of the vast majority of the tested samples (82%) were in the 2.84- to 2.86-g/cm^3 range. The average m from tests on 32 core plugs was 2.1 (assuming a = 1.0). The average for n was 2.2; however, the data were quite scattered (see **Fig. 3H.27**).

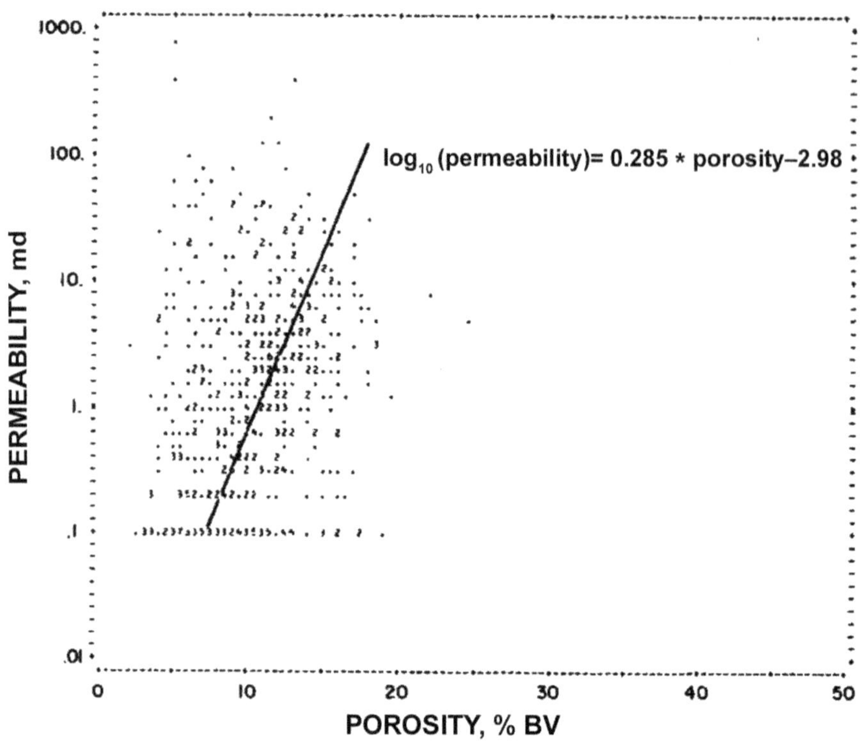

Fig. 3H.24—Core-permeability/core-porosity data for Zone D, West Howard-Glasscock Unit, west Texas, U.S.A.[65] The line fit is manually estimated.

Log computation software was used to calculate lithology and porosity. The volume of shale present was computed from the sonic and density logs because the GR log was not a good shale indicator because of the presence of many radioactive zones containing little or no shale. Variables calculated were primary porosity, secondary-porosity index, dolomite, sand, anhydrite, and shale. **Fig. 3H.28** presents a comparison of the computed and the core porosity values. The permeability, S_w, and S_{wi} were calculated from the logs on the basis of relationships derived from core data.

3H.9.3 Oil Fields—Offshore Peninsular Malaysia. Several technical papers have been written about the combined use of core and log data to increase the reliability of the petrophysical calculations for the various oil reservoirs found in this basin.[50,51] These papers particularly address how the calculations of fluid saturations were improved, and the uncertainty reduced, by a variety of coring and core-analysis studies. The bottom line to the technical work regarding the Dulang oil field was that the best estimate of OOIP was increased by approximately 30%, which led to management being more confident regarding further development.

The oil fields of the Malay basin consist of stacked sequences of laminated sands and shales. Dispersed clay minerals are present in the sands in the form of kaolinite, illite, and mixed-layer clays. The connate waters are generally of low salinity and impacted by the presence of meteoric (surface) waters. The resistivity-log readings in the sand intervals are suppressed because of the thin-bed effects.

For the reasons described in the previous paragraph, OBM- and bland-mud-coring programs were undertaken in the stacked oil reservoirs of several of these fields. These coring programs' design and execution paid great attention to all of the details before, during, and after the cor-

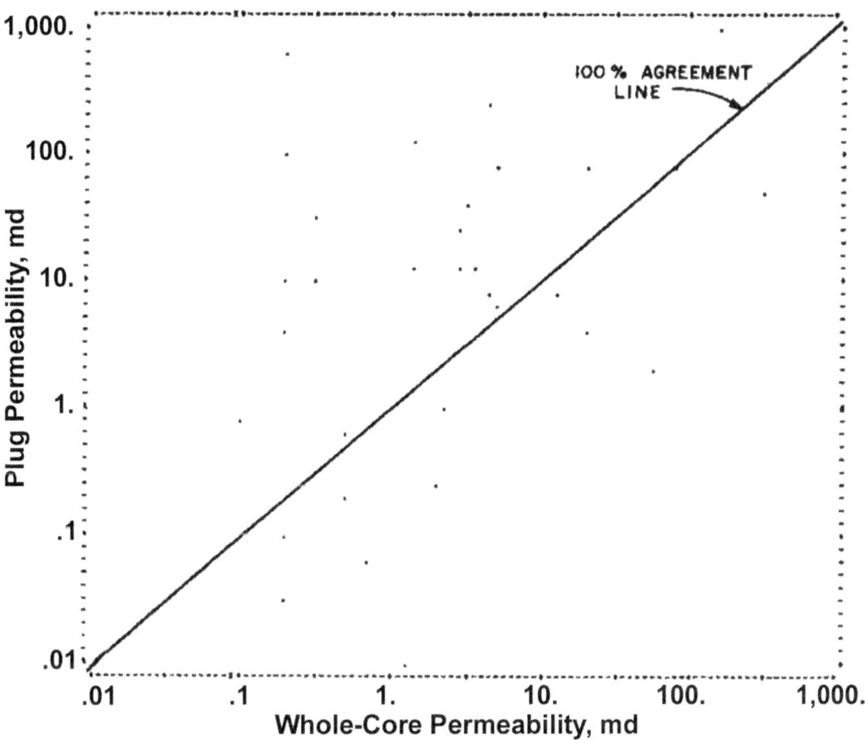

Fig. 3H.25—Comparison of whole-core permeability with core-plug permeability (West Howard-Glasscock unit, west Texas, U.S.A.).[65]

ing operations. The cores were generally cut at high rates of penetration (60 to 90 ft/hr) and with low overbalance pressure (maximum of 200 psi). Because of the friable and unconsolidated nature of some of the sands, full-diameter cores were frozen in dry ice before plug cutting with liquid nitrogen. Special testing, including the use of mud-system tracers, was conducted to determine that OBM filtrate had not impacted the connate-water saturations. Because some unpublished studies suggested that some of the connate water might be displaced as far as 100 to 150 ft above the OWC, core-layering tests were run to determine the depth and concentration of mud tracers within the cross section of the core. The layering tests found that OBM-filtrate invasion was insignificant.

The routine-core-analysis measurements of OBM cores included Dean-Stark S_w determinations and water-salinity studies. Sponge cores were analyzed for residual-oil saturations. SCAL core samples were taken for rock electrical-property measurements. **Fig. 3H.29** presents an example plot showing a comparison of the core results and the results calculated from the logs.[50]

The results from the analysis of the Dulang field's various studies were as follows:[51]

- The Dulang reservoirs' connate S_w was found, from OBM cores, to range from 20.1 to 43.1% PV with an average of 33.6% PV, compared with the previous range of 19 to 61% PV with an average of 39% PV. Also, the S_w varies from reservoir to reservoir, as controlled by the reservoir quality and depositional environments.
- The measurements of water salinity and R_w indicated that R_w varied with reservoir depth and was not constant as previously assumed (see **Table 3H.5**).
- The rock electrical-property measurements indicated that m and n have values of 1.77 and 1.64, respectively (assuming a = 1.0). n was significantly lower than the previously used value of 1.89 (see Table 3H.5).

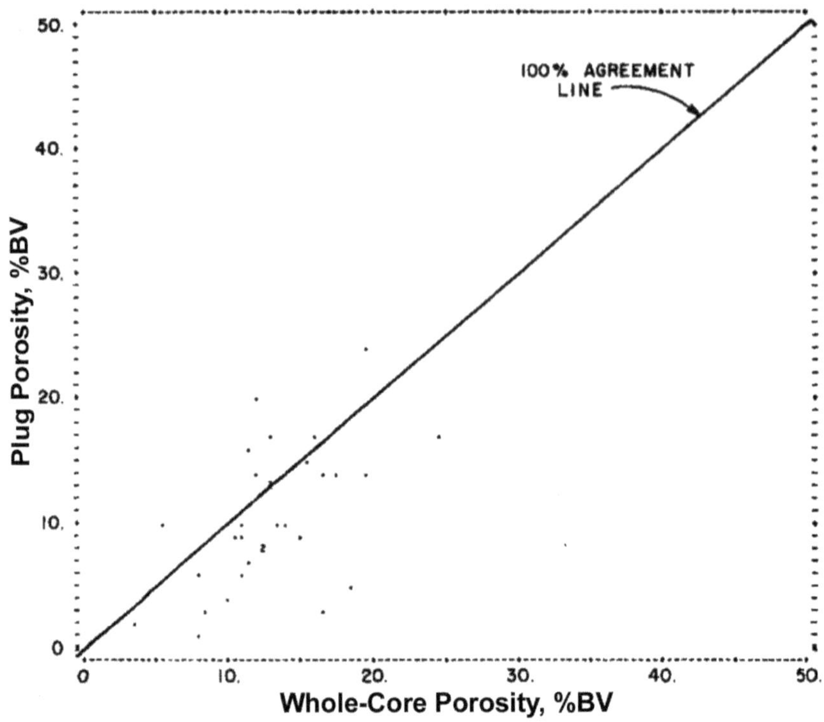

Fig. 3H.26—Comparison of whole-core porosity with core-plug porosity (West Howard-Glasscock unit, west Texas, U.S.A.).[65]

- The authors used these core data to calibrate the DW model.
- The residual-oil saturation data indicated that S_{orw} ranged from 25.8 to 29% PV with an average of 27.5% PV. Recovery factors ranged from 37.8 to 49.4% for the various reservoirs.

The effect of these changes in the petrophysical calculations for the various reservoirs was that S_w had been previously overestimated. The impact of the revised calculations and new core data for the Dulang field's OOIP was that it increased from 550 to 685 million STB.

3H.9.4 Block A-18, Gas/Condensate Field, Malaysia-Thailand Joint Development Area. This example used core P_c and log data to increase the reliability of the petrophysical calculations for gas/condensate reservoirs found offshore near the Thailand-Malaysia border.[67] It addresses how the S_w calculations were improved by the integrated use of resistivity logs in a shaly-sand model and P_c data measured on core plugs. The calibration method led to the resistivity model reproducing the S_w given by the core-measured P_c data. The end result of the work was a well-supported increase in OGIP estimates for the A-18 gas field, leading to a greater confidence in the development viability. The best estimate of gas reserves was increased by more than 20%.

In the first part of the calibration method, the shaly-sand sequence is classified into several facies according to the clay distribution. Initially, 10 facies are identified in the cores, but they are reduced to four petrophysical facies consisting of clean sands, sands with discontinuous clay, bioturbated and heterolithics, and poor reservoir mainly consisting of mudstones and coals. The electrical continuity of the shale components increases with each facies. Where core is not available, the facies type is estimated, preferably from resistivity imaging logs. P_c data were acquired on 1.5-in.-diameter plugs using the centrifuge method and spanned the full range

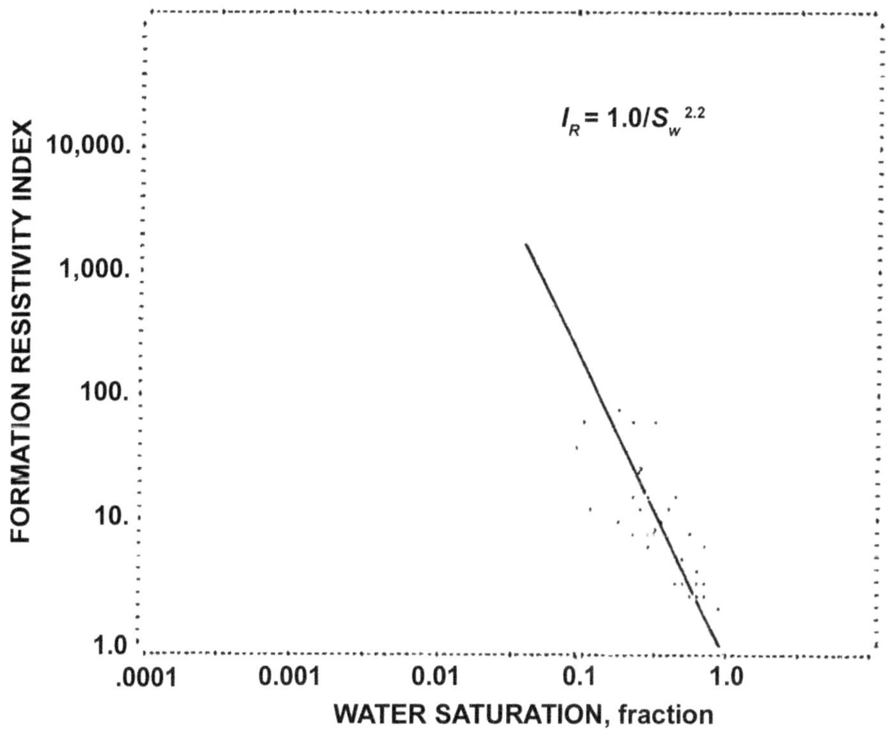

Fig. 3H.27—Laboratory-measured formation resistivity index (I_R) vs. brine saturation for core plugs from West Howard-Glasscock unit, west Texas, U.S.A.[65]

of reservoir parameters. A multiple-linear-regression prediction of S_w from P_c, porosity, and permeability was used to estimate S_w at every routine-core-plug depth.

In the second part of the calibration, the P_c-predicted S_w values were correlated with the Waxman-Smits shaly-sand model and the deep-resistivity log values at each foot of the core. The Waxman-Smits model and the resistivity log were forced to match the P_c-based S_w by back-calculation of the n^* value at each foot in a way similar to that described for OBM cores in the main S_w section of this chapter (Sec. 3H.7). A mean back-calculated n^* is then determined for each of the four facies. **Table 3H.6** gives the n^* mean values, along with the usual SCAL electrical-properties n^* values acquired by measuring core-plug resistivities in the laboratory. The P_c-calibrated n^* values are significantly lower than the SCAL electric-property n^* values, leading to significant increases in effective hydrocarbon column, especially in the bioturbated and heterolithic sands.

Total porosity at every logged foot was calculated from the density-log force fitting the line-fit through the core grain density. **Fig. 3H.30** shows mainly raw data with the S_w results at the right side.

3H.9.5 Whitney Canyon—Carter Creek Gas Field. This Wyoming, U.S.A., gas field produces from a 1,000-ft-thick complex carbonate with a wide range of minerals including dolomite, anhydrite, limestone, and quartz.[68,69] The study hoped to examine significant differences in well performance and to provide recommendations to increase recovery.

The main reservoir is in dolomitic rocks where there is a full continuum from slightly to fully dolomitized limestones. There is little or no relationship between permeability and porosity. Four rock types were identified by high-pressure mercury injection (HPMI) and, because

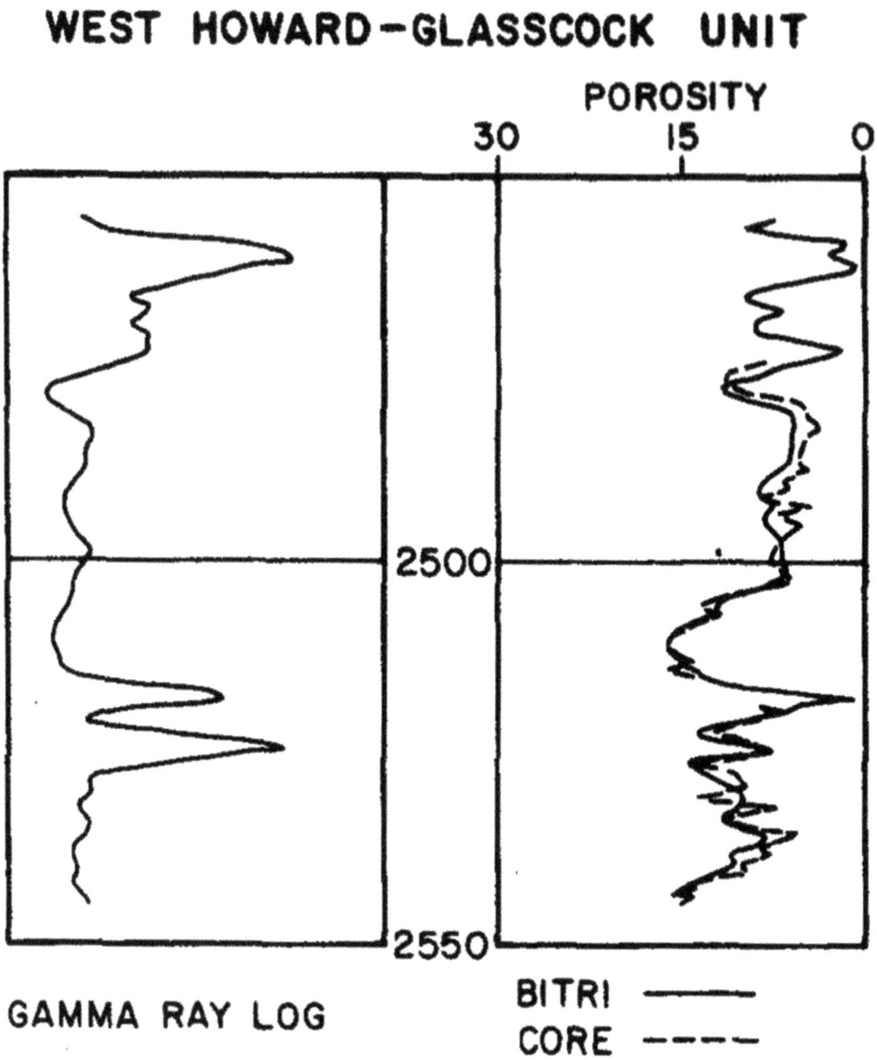

Fig. 3H.28—Agreement of core-analysis and Bitri-computed porosity for West Howard-Glasscock unit, west Texas, U.S.A.[65] "Bitri" refers to the log-response simultaneous-equations solution used to obtain porosity and rock-component volumes.

this reservoir has very high P_c (deriving from the 1,000-ft-thick gas column), even the poorest dolomitic rock has some gas saturation in the small pores. **Table 3H.7** gives a summary of the petrophysical properties of the four rock types.

Porosity and mineralogy were evaluated with a probabilistic simultaneous-equation method.[20] Gas corrections were required and had a significant effect on the results because, while the density-log reading is shallow, the neutron log is much deeper. Boron, a neutron absorber, is present in the dolomites and required a correction. S_w was calibrated to Dean-Stark initial S_w measurements and to the HPMI P_c measurements (**Fig. 3H.31**) because the laboratory Archie n measurements (1.1 to 1.5) were unreliable in the "tight" rocks. Force-fitted Archie exponents of m = 2.18 and n = 2.0 were used in the evaluations, and provided similar results to the OBM core S_w and HPMI P_c data.

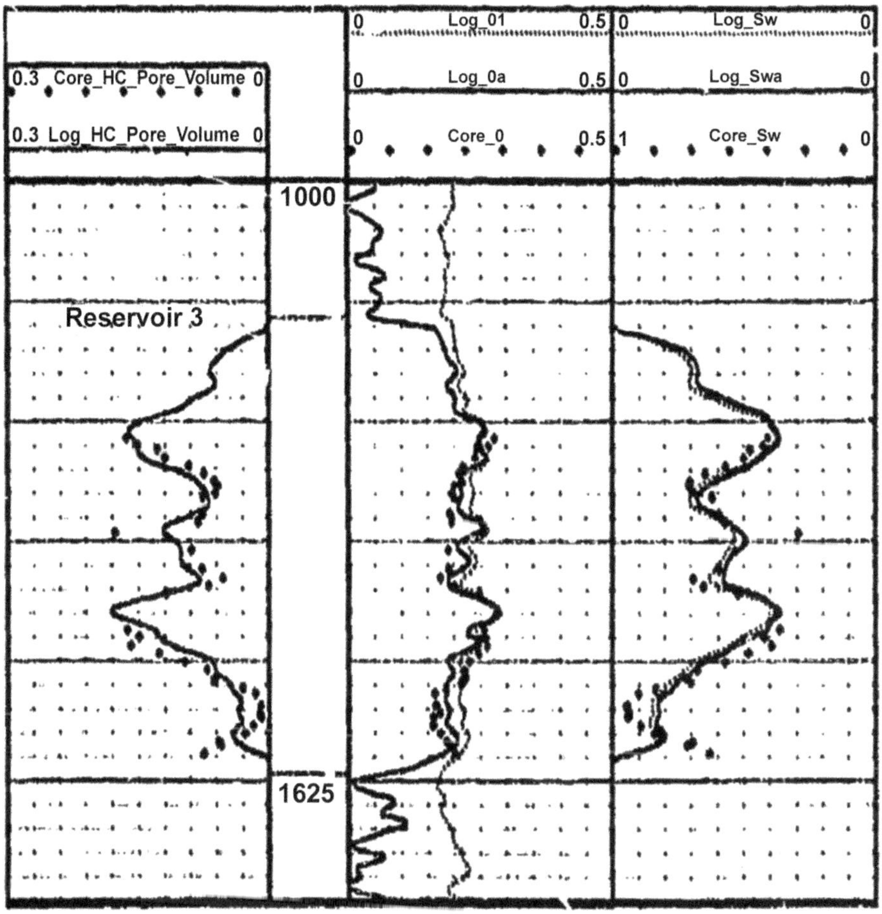

Fig. 3H.29—Plots of core- and log-calculated hydrocarbon pore volume, porosities, and S_w vs. depth for a Malaysian shaly-sandstone reservoir—Core #3.[51]

Gas saturations up to 75% PV were calculated for the poorest rock type and represented a significant prize for the production team. Positive steps were taken to ensure that this gas was accessed by production wells. The study determined that, as well as the usual acid stimulation of the higher-quality rock, the poorer rocks should also be separately stimulated with acid. Acid diversion to lower-permeability zones was used in new infill wells.

3H.10 Other Considerations in Petrophysical Calculations

While this chapter focuses on the calculations of various petrophysical properties using core and log data, there are other types of data obtained from oil and gas reservoirs that need to be considered when making petrophysical calculations. This section briefly discusses some of those data and how they can be used to improve the petrophysical evaluation. Sec. 3H.6 discussed some of these data sources, as used for determining fluid-contact depths.

3H.10.1 Mud-Log Gas and Oil Shows. The information gathered by the mud logger regarding lithology, oil and gas shows, and gas composition should be integrated with the petrophysical calculations. In a gas reservoir, the mud log gas shows can be used to reasonably locate the GWC. In an oil reservoir, this same information can be used to identify the depth of the OWC.

TABLE 3H.5—EFFECT OF PETROPHYSICAL PARAMETERS R_w, m, and n on S_w[66]

Petrophysical Parameters	1982 Values	1992 Values	Error in Parameter	ΔS_w (%)
Cementation exponent, m	1.72	1.77	0.05	3.4
Saturation exponent, n	1.89	1.64	–0.25	–13.9
Formation-water resistivity, R_w	0.185	0.150 (LD1)	–0.035	–10.0
		0.155 (LD2)	–0.030	–8.6
		0.155 (LD3)	–0.030	–8.6
		0.140 (E1)	–0.045	–12.9
		0.145 (E2)	–0.040	–11.4
		0.135 (E6)	–0.050	–14.3
		0.182 (E7)	–0.003	–0.9
		0.182 (E8)	–0.003	–0.9
		0.180 (E10/11)	–0.005	–1.4
		0.160 (E12/13)	–0.025	–7.2
		0.195 (E14)	0.010	2.9
		0.220 (E20/22)	0.035	10.0
		0.278 (E23)	0.093	26.6
		0.211 (E32*)	0.026	7.4
		0.211 (E34)	0.026	7.4
		0.205 (E36)	0.020	5.7
		0.220 (E40/42)	0.035	10.0
		0.220 (E48*)	0.035	10.0
		0.220 (E50*)	0.035	10.0
		0.220 (E51*)	0.035	10.0

*Estimated from nearby sands.

TABLE 3H.6—CHANGES IN EFFECTIVE HYDROCARBON COLUMN BY FACIES GROUP AS DETERMINED FROM SCAL ELECTRICAL-PROPERTY n^* VS. n^* BACK-CALCULATED FROM P_c and LOG DATA[67]

Facies Group	SCAL n^*	Mean Core/Log n^*	% Change in EHC
1. Clean sands	1.76	1.53	9
2. Sand with discontinuous clay	1.93	1.55	16
3. Bioturbated and thin heterolithic sands	2.03	1.40	31
4. Poor reservoir	1.54	1.23	20

Note: EHC is equivalent to F_{HCP}.

These data are useful to determine whether there is a relict-gas or -oil interval below the current GWC or OWC.

3H.10.2 Pressure Measurements and Fluid Samples From Formation-Tester Logging Runs.[35] At the time wells are drilled, a suite of pressure measurements are often made over the reservoir interval, and sometimes, a few fluid samples are taken. These pressure data can aid the petrophysical interpretation by indicating whether the vertical pressures are in equilibrium throughout the reservoir interval. If not, then the reservoir may actually be several different

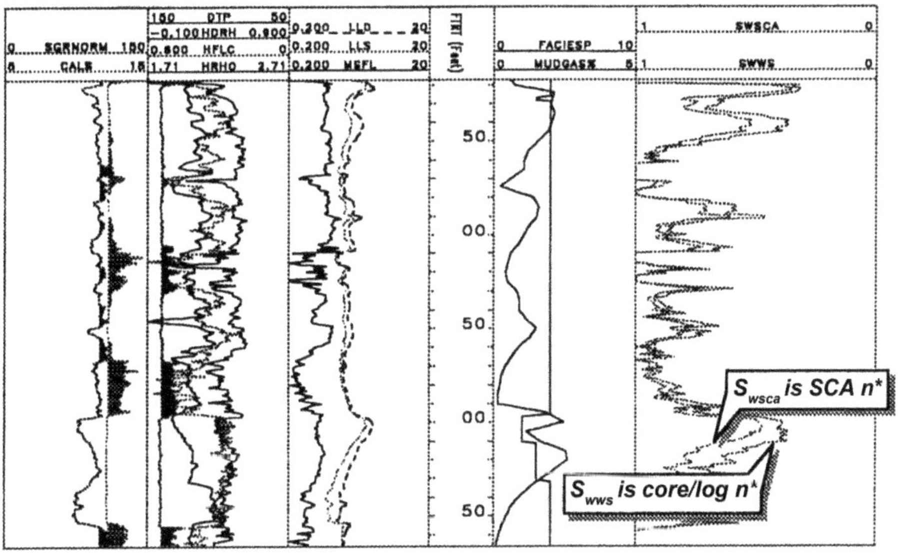

Fig. 3H.30—Water saturation S_{wws} calibrated to capillary pressure P_c core data and S_{wsca} calibrated to standard core-electrical-properties n^*.[67] The P_c-based solution provides lower S_w results in this offshore gas field in the Malaysia-Thailand Joint Development Area. From left to right: Track 1 = GR log and caliper showing washouts; Track 2 = high resolution density log and other porosity logs; Track 3 = laterolog deep resistivity, shallow resistivity, and microspherically focused log; Track 5 = facies identifier and mud-gas log; and Track 6 = S_{wws}, the P_c-calibrated core/log n^*-derived S_w and S_{wsca}, and the standard rock-electrical-properties n^*-derived S_w.

TABLE 3H.7—ROCK TYPE SUMMARY (after Gunter *et al.*[69])

Rock Type	Porosity Range (% BV)	Permeability Range (md air at 1,000 psi)	Irreducible S_w (% PV)	Equivalent P_{ce} (ft)	Pore-Throat Radius (μm)	Pore-Throat Category
1	2 to 20	0.2 to 20	5 to 15	4 to 15	>1	meso/macro
2	2 to 18	0.1 to 4	10 to 20	10 to 30	0.5 to 1	meso
3	2 to 16	0.01 to 1	15 to 25	30 to 100	0.1 to 0.5	micro
4	0 to 12	0.001 to 0.01	25 to 50	100 to 300	<0.1	nano

compartments with different OWCs, GWCs, or GOCs. The fluid samples are useful in defining which fluid is flowing in each of the tested intervals and the extent of compositional differences. In addition to pressure data, modern wireline formation-tester tools can determine hydrocarbon type, GOR, and API gravity from spectroscopic measurements before taking fluid samples.[70]

3H.10.3 Drillstem-Test Data. The pressure and flow data from DST tests must also be considered when performing a reservoir petrophysical evaluation. These tests provide information about the flowing fluids in different portions of the reservoir. Also, the pressure-transient-analysis calculations from the flow-rate and pressure data are used to calculate an average value for reservoir permeability. This needs to be compared with that calculated from the routine-core-analysis permeability data and the well logs. The permeability values determined from these two sources of data often disagree, with the value from the DST testing almost always taken to

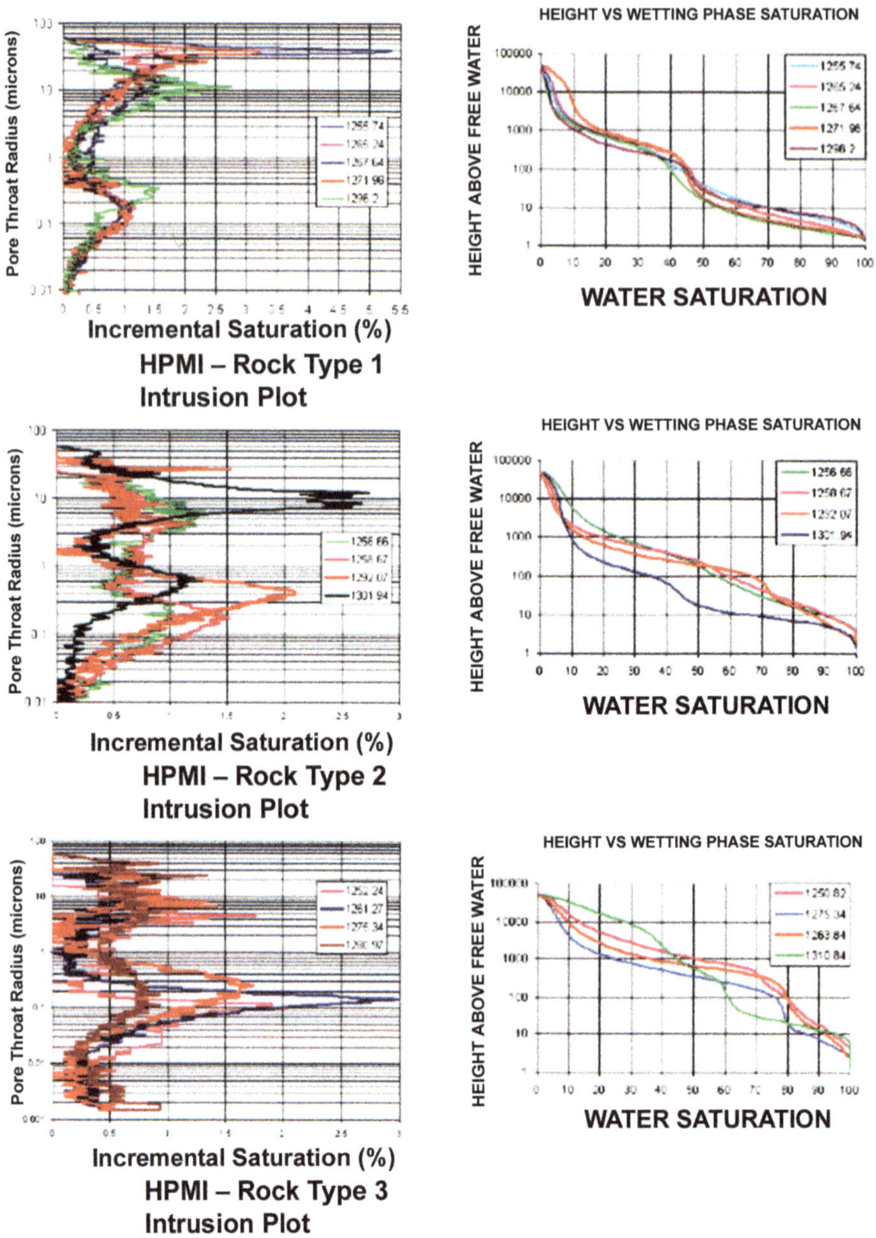

Fig. 3H.31—High pressure mercury injection (HPMI) plots for Whitney Canyon-Carter Creek field, Wyoming, U.S.A. for Rock Types 1, 2, and 3 (see Table 3H.6): (a) mercury intrusion as pore-throat radius changes and (b) wetting-phase saturation as height above free-water level (H_{GWC}) changes (after Gunter *et al.*[69]). Each rock type is represented by 4 or 5 core samples. Plots for Rock Type 4 are not included because those are the poorest rocks with nano-sized pore throats.

be more correct. The DST results can be used to develop adjustments to the methodology for calculating permeability from core and log data, but also to determine the underlying technical reason for the need for the adjustments. If a DST is run long enough, it may yield information on the drainage limits of a reservoir.

3H.10.4 Three-Dimensional Seismic Data. The interpretation of the 3D-seismic data provides complementary data to that gathered at the wellbores and, with the rapidly improving data quality and interpretation techniques, can add significantly to reservoir-characterization calculations. The depositional, diagenetic, structural, and hydrocarbon-filling histories of a reservoir can be better understood by including the broad picture available from the 3D-seismic data. Locations of major faults that may compartmentalize the reservoir can be identified and the extent and location of channel-sand deposits is possible. Seismic attributes may be correlated with rock properties and fluid contents, allowing prediction of properties at proposed well locations.

3H.11 Summary and Conclusions

The purpose of this chapter is to acquaint the reader with the various aspects of the quantitative petrophysical determination of lithology, net pay, porosity, fluid contacts, water saturation, and permeability. To make these calculations as accurately as possible, core and log data need to be integrated. The routine-core-analysis data adjusted to reservoir conditions should be used to calibrate the logs for more-accurate calculations at the various wells. The conclusions of this chapter are as follows:

- Lithology is determined by geologists working with cores and rock cuttings. This information can be combined with log characteristics to identify depositional environments and characterize how these change vertically and areally throughout the reservoir.
- The clay minerals present in the shales and the sandstone intervals, both as detrital and authigenic components, must be identified and quantified so that their effects on the logs and routine-core-analysis data can be adequately understood. Radioactive components present in the reservoir rocks must be identified and quantified so that the clay-mineral volumes derived from the GR log are not overstated.
- Net pay calculations determine how much of the reservoir interval contributes to the technical calculations of in-place hydrocarbon volumes and fluid flow. With modern reservoir engineering tools, it is possible to set N/G to 1.0 and work the various engineering calculations from that basis. If some portion of the reservoir interval is to be excluded as nonpay, the choice of cutoff should be based on flow considerations with a systematic and consistent approach. Whatever nonpay cutoff is used, that cutoff will be somewhat arbitrary.
- Porosity can be computed from a variety of well logs (density, sonic, or neutron) in combination with routine-core data adjusted to reservoir conditions. In sandstones in which the mineralogy and the hole conditions permit, foot-by-foot porosity calculations from the density log, calibrated to core, are likely to be the most accurate. Correct fluid values are an integral part of the log evaluation. Porosity needs to be calculated accurately because, as well as its primary use, these values are also required for S_w and permeability estimates used directly in the volumetrics and flow calculations. Minerals that affect the porosity calculations, such as clay and heavy minerals, need to be identified as part of the lithology determination.
- Water saturation can be computed by a number of independent methods using routine-core-analysis OBM-core Dean-Stark S_w data, SCAL capillary pressure data, resistivity logs used in combination with SCAL rock electrical-property measurements, or some combination of these three datasets. Adjusted OBM-core Dean-Stark S_w data are likely to be the most accurate method. Integrated use of these various technical approaches will result in the most accurate S_w solution overall. The relevant uncertainty here is not in S_w itself; it is the uncertainty in the complement, the hydrocarbon saturation $(1 - S_w)$, that is important.
- In the water-saturation calculation using resistivity logs, the connate-brine salinity and its resistivity, R_w, can vary within the hydrocarbon column, but the extent of this variation is often not measured. Also, the rock electrical properties may be a function of S_w. In most conventional S_w calculations using well logs, these are both assumed to be constant, and those assumptions can lead to significant errors in the calculated S_w values. The S_w calculations from the

resistivity logs and the various Archie parameters can be partially checked in aquifer intervals where S_w is known to be 100% PV.

• In the S_w calculation using P_c measurements, the laboratory tests are measurements of fluid volumes associated with cleaned and restored core plugs. For application, the reservoir values of the IFT, contact angle, and the wetting state of the reservoir generally must be estimated, along with several other factors. P_c laboratory tests do not always achieve the equilibrium water saturation, or the same water distribution within the pore network as is present in the real reservoir.

• Permeability is typically calculated from porosity logs through a permeability/porosity transform. Permeability values need to be adjusted to reservoir conditions. This adjustment is nonlinear, with poor-quality rocks having larger adjustments compared with those applied to high-quality rocks. Also, the calculated permeabilities at the wells should be compared with those obtained from PBU analysis of flow tests.

• The statistical correlation and calibration of core and log data requires that these data are properly depth aligned, have outliers deleted, and, if required, are mathematically transformed. A variety of line-fitting techniques are available, but the "*y*-on-*x*" approach generally results in the most accurate predictor, except in highly variable, heterogeneous rocks.

• In any reservoir with a thick hydrocarbon column and large areal extent, more accurate-petrophysical calculations are made if the reservoir is vertically zoned or layered. Different parameters for different areas of the reservoir may also be required for the most accurate solution.

Nomenclature

a = Archie cementation constant
a^* = Waxman-Smits cementation constant
A = Coefficient in various equations of this chapter
A' = Coefficient of term of Eq. 3H.26
B = Specific cation conductance, [(1/ohm•m) / (meq/mL)]
B' = Coefficient of term of Eq. 3H.26
C = Coefficient in various equations
C' = Coefficient of term in Eq. 3H.26
D = Coefficient in various equations
E = Coefficient in various equations
F = Archie formation factor
F^* = Waxman-Smits-Thomas formation factor
F_{HCP} = hydrocarbon pore feet, L, ft [m]
H_{gwc} = height above the gas/water contact, L, ft [m]
H_{hwc} = height above the hydrocarbon/water contact, L, ft [m]
H_{owc} = height above the oil/water contact, L, ft [m]
I_R = resistivity index
$J(S_w)$ = Leverett *J*-function
k = permeability, L^2, md [μm^2]
m = Archie cementation exponent
m^* = Waxman-Smits-Thomas cementation exponent
m_o = dual-water cementation exponent
n = Archie saturation exponent
n^* = Waxman-Smits-Thomas saturation exponent
n_o = dual-water saturation exponent
P_c = capillary pressure, m/Lt^2, psi
P_{ce} = entry capillary pressure, m/Lt^2, psi
Q_v = cation-exchange capacity of total PV, meq/mL

r = correlation coefficient
R_0 = rock resistivity with 100% PV water saturation, ohm•m
R_{sd} = clean-sand resistivity, ohm•m
R_{sh} = shale resistivity, ohm•m
R_t = true resistivity of uninvaded, deep formation, ohm•m
R_w = connate-brine resistivity, ohm•m
R_{wb} = clay-bound water resistivity, ohm•m
R_{wf} = free-formation-water resistivity, ohm•m
R_{xo} = shallow-reading invaded-zone microresistivity, ohm•m
S_g = gas saturation, %PV
S_o = oil saturation, %PV
S_{orw} = residual-oil saturation to water displacement, %PV
S_w = water saturation, %PV
S_{wb} = saturation of clay-bound water in the total porosity, %PV
S_{wc} = connate water saturation, %PV
S_{wc} = core water saturation, %PV
S_{we} = water saturation of the effective porosity, %PV
S_{wsd} = sand water saturation, %PV
S_{wt} = water saturation of the total porosity, %PV
V_{cl} = clay content, %BV
V_{HCP} = hydrocarbon pore volume, L^3, ft^3 [m^3]
V_{sh} = shale content, %BV
θ = contact angle, degrees
ρ_b = formation bulk density, m/L^3, g/cm^3
ρ_{fl} = fluid density, m/L^3, g/cm^3
ρ_h = hydrocarbon density, m/L^3, g/cm^3
ρ_{ma} = matrix or grain density, m/L^3, g/cm^3
ρ_w = water density, m/L^3, g/cm^3
σ = interfacial tension, m/t^2, dynes/cm
ϕ = porosity, %BV
ϕ_c = core porosity, %BV
ϕ_{cl} = clay porosity, %BV
ϕ_e = effective porosity, %BV
ϕ_{sd} = sand porosity, %BV
ϕ_{sh} = shale porosity, %BV
ϕ_t = total porosity, %BV

Subscripts

c = connate
c = core
e = effective
h = hydrocarbon
i = initial
r = reservoir
s = surface
t = total
w = water

References

1. Crain, E.R.: *Log Analysis Handbook,* PennWell, Tulsa (1986).
2. Dewan, J.T.: *Essentials of Modern Openhole Log Interpretation,* PennWell, Tulsa (1983).
3. *Log Interpretation Principles/Applications,* Schlumberger, Houston (1989).
4. *Introduction to Wireline Log Analysis,* Baker Atlas, Baker Hughes Inc., Houston (2002).
5. *Log Interpretation Charts,* Schlumberger, Sugar Land, Texas (2000).
6. *Log Interpretation Charts,* Baker Atlas, originally published by Western Atlas, Houston (1995).
7. Theys, P.P.: *Log Data Acquisition and Quality Control,* second edition, Editions Technip, Paris (1999).
8. *Fundamentals of Rock Properties,* Core Laboratories UK Ltd., Aberdeen (2002).
9. Keelan, D.K.: "Automated Core Measurement System for Enhanced Core Data at Overburden Conditions," paper SPE 15185 presented at the 1986 SPE Rocky Mountain Regional Meeting, Billings, Montana, 19–21 May.
10. Keelan, D.K.: "Core Analysis Techniques and Applications," paper SPE 4160 presented at the 1972 SPE Eastern Regional Meeting, Columbus, Ohio, 8–9 November.
11. Keelan, D.K.: "Critical Review of Core Analysis Techniques," *J. Cdn. Pet. Tech.* (April 1972) 42.
12. Rathmell, J.J., Braum, P.H., and Perkins, T.K.: "Reservoir Waterflood Residual Oil Saturation From Laboratory Tests," *JPT* (February 1973) 175.
13. *RP 40 Recommended Practices for Core Analysis,* second edition, API, Washington, DC (February 1998).
14. Richardson, J.G. and Holstein, E.D.: "Comparison of Water Saturations from Capillary Pressure Measurements With Oil-Base-Mud Core Data Ivishak (Sadlerochit) Reservoir, Prudhoe Bay Field," paper SPE 28593 presented at the 1994 SPE Annual Technical Conference and Exhibition, New Orleans, 25–28 September.
15. Woodhouse, R., Greet, D.N., and Mohundro, C.R.: "Induction Log Vertical Resolution Improvement in Vertical and Deviated Wells Using a Practical Deconvolution Filter," *JPT* (June 1984) 993.
16. Barber, T.D.: "Introduction to the Phasor Dual Induction Tool," *JPT* (September 1986) 1699.
17. Hynes, N.J.: *Dictionary of Petroleum Exploration, Drilling, and Production,* PennWell, Tulsa (1991).
18. Hensel, W.M. Jr.: "An Improved Summation-of-Fluids Porosity Technique," *SPEJ* (April 1982) 193.
19. Brown, G.: *The X-Ray Identification and Crystal Structures of Clay Minerals,* Mineralogical Soc. Clay Minerals Group, London (1961).
20. Mezzatesta, A., Rodriguez, E., and Frost, E.: "Optima: A Statistical Approach to Well Log Analysis," *Geobyte* (August 1988).
21. Sneider, R.M. and Erickson, J.W.: "Rock Types, Depositional History, and Diagenetic Effects: Sadlerochit Reservoir, Prudhoe Bay Field," paper SPE 28575 presented at the 1994 SPE Annual Technical Conference and Exhibition, New Orleans, 25–28 September.
22. Holstein, E.D. and Warner, H.R. Jr.: "Overview of Water Saturation Determination for the Ivishak (Sadlerochit) Reservoir, Prudhoe Bay Field," paper SPE 28573 presented at the 1994 SPE Annual Technical Conference and Exhibition, New Orleans, 25–28 September.
23. Cobb, W.M. and Marek, F.J.: "Net Pay Determination for Primary and Waterflood Depletion Mechanisms," paper SPE 48952 prepared for presentation at the 1998 SPE Annual Technical Conference and Exhibition, New Orleans, 27–30 September.
24. *Oil Reservoir Engineering,* S.J. Pirson (ed.), McGraw-Hill Book Co. Inc., New York City (1958) 443–445.
25. *Determination of Oil and Gas Reserves,* Petroleum Soc. of the Canadian Inst. of Mining, Metallurgy, and Petroleum, Petroleum Soc. Monograph 1 (1994) 45–46.
26. George, C.J. and Stiles, L.H.: "Improved Techniques for Evaluating Carbonate Waterfloods in West Texas," *JPT* (November 1978) 1547.

27. Erickson, J.W. and Sneider, R.M.: "Structural and Hydrocarbon Histories of the Ivishak (Sadlerochit) Reservoir, Prudhoe Bay Field," paper SPE 28574 presented at the 1994 SPE Annual Technical Conference and Exhibition, New Orleans, 25–28 September.
28. Lucia, F.J.: "San Andres and Grayburg Imbibition Reservoirs," paper SPE 59691 presented at the 2000 SPE Permian Basin Oil and Gas Recovery Conference, Midland, Texas, 21–23 March.
29. Thai, B.N. *et al.:* "Denver Unit Infill Drilling and Pattern Reconfiguration Program," paper SPE 59548 presented at the 2000 SPE Permian Basin Oil and Gas Recovery Conference, Midland, Texas, 21–23 March.
30. Cowan, G. and Boycott-Brown, T.: "The North Morecambe Gas Field," *United Kingdom Oil and Gas Fields, Commemorative Millennium Volume,* Geological Soc., London (2003).
31. Woodhouse, R. and Warner, H.R. Jr.: "Improved Log Analysis in Shaly Sandstones—Based on S_w and HCPV Routine Measurements on Preserved Cores Cut in Oil-based Mud," *Petrophysics* (May 2004) **45**, No. 3, 281. [An improved version of this paper is SPE 96618, presented at the 2005 SPE Annual Technical Conference and Exhibition, Dallas, 9–12 October.]
32. Woodhouse, R.: *Statistical Regression Line-Fitting in the Oil and Gas Industry,* PennWell, Tulsa (2003).
33. Jensen, J.L. *et al.: Statistics for Petroleum Engineers and Geoscientists,* second edition, Elsevier (2000).
34. Patchett, J.G. and Coalson, E.B.: "The Determination of Porosity in Sandstones and Shaly Sandstones, Part One–Quality Control," *The Log Analyst* (Nov 1979) **20,** 6.
35. *Formation Multi-Tester (FMT) Principles, Theory and Interpretation,* Baker Atlas, originally published by Western Atlas, Houston (1987).
36. McCoy, D.D., Warner, H.R. Jr., and Fisher, T.E.: "Water-Salinity Variations in the Ivishak and Sag River Reservoirs at Prudhoe Bay," *SPERE* (February 1997) 37.
37. Rathmell, J.J., Atkins, L.K., and Kralik, J.G.: "Application of Low Invasion Coring and Outcrop Studies to Reservoir Development Planning for the Villano Field," paper SPE 53718 presented at the 1999 SPE Latin American and Caribbean Petroleum Engineering Conference, Caracas, 21–23 April.
38. Rathmell, J.J. *et al.:* "Low Invasion, Synthetic Oil-Base Mud Coring in the Yacheng 13-1 Gas Reservoir for Gas-in-Place Calculation," paper SPE 29985 presented at the 1995 SPE International Meeting on Petroleum Engineering, Beijing, 14–17 November.
39. Archie, G.: "The Electrical Resistivity Log as an Aid in Determining Some Reservoir Characteristics," *Trans.,* AIME (1942) **146,** 54.
40. Winsauer, W.O., Shearin H.M., Masson P.H., and Williams M:. "Resistivity of Brine Saturated Sands in Relation to Pore Geometry," *AAPG Bull.* (February 1952) **36,** 2.
41. Waxman, M.H. and Smits, L.J.M.: "Electrical Conductivities in Oil-Bearing Shaly Sands," *SPEJ* (June 1968) 107.
42. Waxman, M.H. and Thomas, E.C.: "Electrical Conductivities in Sands—I. The Relation Between Hydrocarbon Saturation and Resistivity Index; II. The Temperature Coefficient of Electrical Conductivity," *SPEJ* (February 1974) 213.
43. Juhasz, I.: "The Central Role of Q_v and Formation Water Salinity in the Evaluation of Shaly Formations," *The Log Analyst* (June 1979) **20,** 4.
44. Hill, H.J., Shirley, O.J., and Klein, G.E.: "Bound Water in Shaly Sands—Its Relation to Q_v and Other Formation Properties," *The Log Analyst* (1979) **20,** No. 3, 3.
45. Leverett, M.C.: "Capillary Behavior in Porous Media," *Trans.,* AIME (1941) **142,** 152.
46. Heseldin, G.M.: "A Method of Averaging Capillary Pressure Curves," *The Log Analyst* (1974) **4,** 3.
47. Johnson, A.: "Permeability Averaged Capillary Data: A Supplement to Log Analysis in Field Studies," paper EE presented at the 1987 SPWLA Annual Symposium, London, June.
48. Woodhouse, R.: "Accurate Reservoir Water Saturations from Oil-Mud Cores: Questions and Answers from Prudhoe Bay and Beyond," *The Log Analyst* (May–June 1998) 23.
49. Richardson, J.G. *et al.:* "Validation of As-Received Oil-Based-Core Water Saturations From Prudhoe Bay," *SPERE* (February 1997) 31.
50. Dawe, B.A. and Murdock, D.M.: "Laminated Sands: An Assessment of Log Interpretation Accuracy by an Oil-Base Coring Program," paper SPE 20542 presented at the 1990 SPE Annual Technical Conference and Exhibition, New Orleans, 23–26 September.

51. Egbogah, E.O. and Amar, Z.H.B.T.: "Accurate Initial/Residual Saturation Determination Reduces Uncertainty in Further Development and Reservoir Management of the Dulang Field, Offshore Peninsular Malaysia," paper SPE 38024 presented at the 1997 SPE Asia Pacific Oil and Gas Conference, Kuala Lumpur, 14–16 April.
52. Poupon, A. and Leveaux, J.: "Evaluation of Water Saturations in Shaly Formations," *The Log Analyst* (July 1971) 3.
53. Worthington, P.F.: "The Evolution of Shaly-Sand Concepts in Reservoir Evaluation," *The Log Analyst* (January 1985) **23**.
54. Keelan, D.K. and McGinley, D.C.: "Application of Cation Exchange Capacity in a Study of the Shannon Sand of Wyoming," paper KK presented at the 1979 SPWLA Annual Symposium, June.
55. Clavier, C., Coates, G., and Dumanoir, J.L.: "Theoretical and Experimental Bases for the Dual-Water Model for Interpretation of Shaly Sands," *JPT* (April 1984) 153.
56. Best, D.L., Gardner, J.S., and Dumanoir, J.L.: "A Computer-Processed Wellsite Log Computation," paper Z presented at the 1979 SPWLA Annual Symposium.
57. Huff, G.F.: "A Correction for the Effect of Comminution on the Cation Exchange Capacity of Clay-Poor Sandstones," *SPEFE* (September 1987) 338.
58. Bryant, W.T. and Truman, R.B. II: "Proper Core-Based Petrophysical Analysis Doubles Size of Ha'py Field," paper SPE 77638 presented at the 2002 SPE Annual Technical Conference and Exhibition, San Antonio, Texas, 29 September–2 October.
59. Katz, D.L. and Firoozabadi, A.: "Predicting Phase Behavior of Condensate/Crude-Oil Systems Using Methane Interaction Coefficients," *JPT* (November 1978) 1649.
60. Harrison, B. and Jing, X.D.: "Saturation-Height Methods and Their Impact on Volumetric Hydrocarbon-in-Place Estimates," paper SPE 71326 presented at the 2001 SPE Annual Technical Conference and Exhibition, New Orleans, 30 September–3 October.
61. McCoy, D.D. and Grieves, W.A.: "Use of Resistivity Logs to Calculate Water Saturations in the Prudhoe Bay Field," *SPERE* (February 1997) 45.
62. Timur, A.: "An Investigation of Permeability, Porosity and Residual Water Saturation Relationships for Sandstone Reservoirs," *The Log Analyst* (July 1968) **9**, 4.
63. Charlson, G.S. *et al.:* "Application of Oil-Based-Mud Pressure Coring to the Determination of In-Situ Water Saturations," *SPEDC* (June 1997) 105.
64. Deans, H.A. and Mut, A.D.: "Chemical Tracer Studies To Determine Water Saturation at Prudhoe Bay," *SPERE* (February 1997) 52.
65. Wilson, D.A. and Hensel, W.M. Jr.: "Computer Log Analysis Plus Core Analysis Equals Improved Formation Evaluation in West Howard-Glasscock Unit," *JPT* (January 1978) 43.
66. Osborn, C.K. and Hogan, C.A.: "Corecom—A Practical Application of Core Analysis," Southwestern Petroleum Short Course, Texas Tech U., Lubbock, Texas (April 1972) 75.
67. Deakin, M. and Manan, W.: "The Integration of Petrophysical Data for the Evaluation of Low Contrast Pay," paper SPE 39761 presented at the 1998 SPE Asia Pacific Conference on Integrated Modelling for Asset Management, Kuala Lumpur, 23–24 March.
68. Logan, P.A. *et al.:* "Using Petrophysics to Improve Recovery: Whitney Canyon—Carter Creek Field, Western Wyoming Trust Belt, USA," paper W presented at the 1999 SPWLA Annual Symposium, Oslo, Norway, May.
69. Gunter, G.W. *et al.:* "Saturation Modeling at the Well Log Scale Using Petrophysical Rock Types and a Classic Non-Resistivity Based Method," paper Z presented at the 1999 SPWLA Annual Symposium, Oslo, Norway, May.
70. Hashem, M.N. *et al.:* "Determination of Producible Hydrocarbon Type and Oil Quality in Wells Drilled with Synthetic Oil-Based Muds," *SPEREE* (April 1999) 125.

SI Metric Conversion Factors

acre	×	4.046 856	E + 03	= m^2
bbl	×	1.589 873	E − 01	= m^3
Btu	×	1.055 056	E + 00	= kJ
cp	×	1.0*	E − 03	= Pa·s
dyne	×	1.0*	E − 02	= mN
ft	×	3.048*	E − 01	= m

°F		(°F − 32)/1.8		= °C
in.	×	2.54*	E + 00	= cm
mile	×	1.609 344*	E + 00	= km
psi	×	6.894 757	E + 00	= kPa

*Conversion factor is exact.

Chapter 4
Production Logging

R.M. McKinley, SPE, Consultant and **Norman Carlson,** Consultant

4.1 Introduction

This chapter consists of a general discussion of production logs, some misconceptions about what can be determined with these logs, and the requisite skills to obtain good results. Also included are discussions of practical matters such as required safety and environmental tips, sinker-bar weight, maximum tool length to pass through a bend, depth control, pricing, and record keeping. Attached as an Appendix is a set of tables prepared for readers that are consulting this text to find out quickly what type of production-logging tools are appropriate to a particular problem. These tables indicate what tools to use, how to use them, and what results (resolution) to expect. These tables were designed to be independent of the general discussion and can be used by themselves. The indexing scheme used in the tabulation is explained in the Appendix.

Production-logging tools find many applications from the time a well is drilled until abandonment and, occasionally, beyond. An appropriate categorization of production logs is by usage. This approach leads to the five distinct categories listed below that also represent a rough chronological order of tool evolution. Effective interpretation of the data from each type of log requires significant education and experience.

1. Diagnose production problems and allocate production.
2. Monitor cement placement.
3. Monitor corrosion.
4. Monitor reservoir fluid contacts.
5. Select zones for recompletion.

These are discussed as categories below.

4.1.1 Category One. Includes tools used to track movement of fluid either inside or immediately outside the casing of a well. The logs frequently used for such flow diagnosis and allocation include temperature surveys, mechanical flowmeter surveys, and borehole fluid-density or fluid-capacitance surveys. Each of these tools responds to fluid velocity or fluid type. The logs are run to determine if a production problem, such as excessive water or gas production, is the result of a completion problem or a reservoir problem. Their value thus resides in the guidance they give for continued expenditure on a well that is performing poorly. This type of application is largely responsible for the growth and evolution of modern production logging.

Also belonging in Category One are evaluations of the placement of acids or hydraulic-fracturing material and diagnoses of premature flow or lost circulation in a drilling well.

4.1.2 Category Two. There are two different objectives of cement-placement monitoring: to determine where the cement went (cement top) and to determine whether the cement provides zonal isolation. The logs used to locate the cement top include the temperature log, which responds to hydration heating; the unfocused gamma ray log, which responds to behind-pipe density; and the regular bond log, which measures the acoustical deadening of pipe.

Zonal isolation should be addressed when pressure imbalance causes crossflow through poorly cemented sections, leading to excessive production of unwanted fluids. The tools most often used for this purpose include cement-bond logs, temperature, noise, radioactive tracer, and neutron-activation logs. The temperature log detects alterations caused by flow, the noise log measures turbulent sound caused by flow, and the tracer log tracks tagged fluid behind casing. The neutron-activation log creates tracer in behind-pipe water.

4.1.3 Category Three. Corrosion-monitoring tools are specialized in nature and include mechanical caliper tools and electromagnetic casing-inspection tools. The mechanical caliper tools are used to assess corrosion internal to the casing and to measure the shape of casing as well as the amount of rod and drillpipe wear inside tubing or casing. The electromagnetic devices respond to changes in metal thickness either inside or outside the pipe containing the tool. These logging tools are either of the eddy-current type or of the flux-leakage type, or a combination of the two. The eddy-current devices measure the load on a coil resulting from eddy currents induced into the wall of the casing. This load increases with increases in wall thickness. The driven frequency of the coil determines the depth of penetration of the field into the casing wall. The flux-leakage devices measure, by means of pad-conveyed coils in contact with the pipe wall, the induced currents that result from magnetic field lines that escape at abrupt changes in metal-wall thickness. Both types of tools make indirect measurements that are then related to metal loss through calibration.

4.1.4 Categories Four and Five. The last two categories, monitoring of fluid contacts in formations and selection of recompletion zones, use the cased-hole nuclear logs such as the neutron, the pulsed-neutron, and the various spectral logs. Please refer to the chapter on nuclear logging in the Reservoir Engineering volume of this *Handbook* for information on these logs.

4.2 Misconceptions About Production Logging

There are three pervasive myths about production logging:

4.2.1 Anyone Can Run a Production Log. Quality control is paramount, and careful attention must be focused upon three parts of the logging operation: (1) procedure (this will usually determine the value of the resulting logs), (2) tool calibration, and (3) depth control.

4.2.2 Only One Logging Tool Is Needed. Just like openhole-logging tools, production-logging tools should be run in complementing suites so that one log can be compared with another. Seldom does a single log identify a problem sufficiently to prescribe a remedial action.

Table 4.1 lists the more common tool combinations used to diagnose problems and allocate flow. The Class A tools respond to flow either inside or outside the pipe containing the sonde and are usually employed for initial evaluation of a production problem. Class B tools respond to flow past the sensor and are used for detailed flow allocation from multiple entries into or exits from the pipe containing the sonde. The resolution of some of the Class A tools is actually better than some of the Class B members.

TABLE 4.1—LOGGING TOOLS FOR DIAGNOSING FLOW PROBLEMS AND ALLOCATING FLOW

Class A Tools	Class B Tools
Behind/Within-Pipe Detectors	Within-Pipe Detectors
Flowmeters	Flowmeters
Temperature	Continuous spinners
Noise	Diverting spinners
Radioactive tracer	Fluid Identifiers
Neutron activators (water)	Focused-gamma
Fluid density (unfocused)	Ray-density
Gamma-ray (gravel-pack logs)	Pressure-gradient
Uncompensated neutron (gas holdup)	Density
	Fluid-capacitance
	Fluid-resistance

4.2.3 The Answer (Anomaly) Will Jump Out From a Casual Scan of the Log. This myth is responsible almost entirely for the lack of adequate training in this area.

Production logs should be interpreted in a consistent fashion that first identifies normal or expected features. The abnormal portions can then be examined to determine which parts are pertinent to the problem and which parts are irrelevant. It is these irrelevant features that so often confound novices to the point that they delay or forego appropriate remedial action.

Once these three myths are set aside, the requisite skills, listed next, can be developed for use of production logs. In the authors' experience, a collaborative effort by service provider and client is needed to yield the most meaningful results. The effective user must be able to

1. Select the proper combination of tools.
2. Establish operating procedure.
3. Monitor data quality.
4. Interpret results.

4.3 Origin of Flow Problems

The more common types of flow problems associated with production-logging efforts are summarized in **Fig. 4.1**. Side A shows a drilling well that has crossflow or an underground blowout. Such flow can start for two different reasons. First, the well may "take a kick"; that is, entry into an abnormally pressured zone may result in drilling mud being blown out of the hole. Even though the well may be successfully shut in at the surface, the culprit zone is still free to flow into lower-pressured strata. Second, the well may "lose returns"; that is, a zone may fracture and allow drilling mud to leave the wellbore faster than it is pumped down the drillpipe. The loss in mud pressure may allow hydrocarbon crossflow as shown. Such a flow must be stopped before drilling can be resumed to prevent an uncontrolled blowout with the potential for injuries and equipment loss, as illustrated in **Fig. 4.2**.

Side B of Fig. 4.1 illustrates why a well completed for production may flow excessive amounts of water or gas. Any combination of three reasons may account for the unwanted flow:

1. Casing failure by stress or corrosion.
2. Lack of cement integrity behind the casing.
3. Mobile gas or water in the completed zone.

The first two situations result from completion failures that can usually be remedied. The third situation is caused by reservoir performance and typically remains as long as the zone is produced.

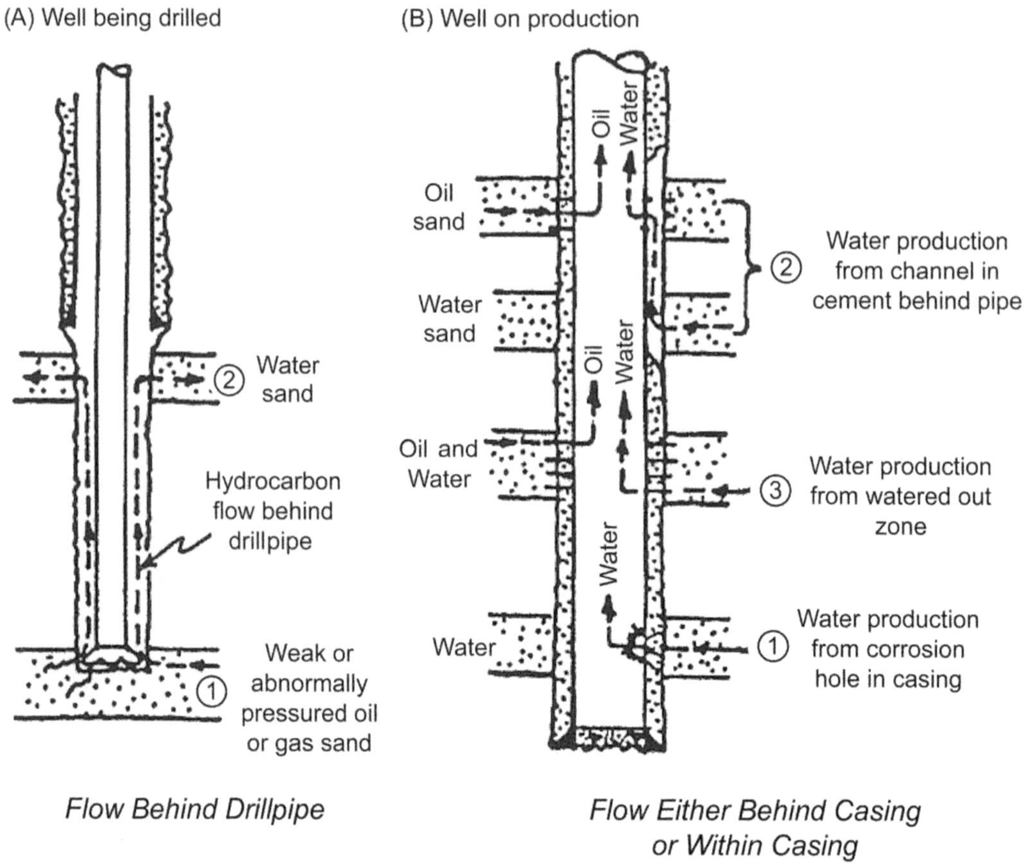

Fig. 4.1—Origin of flow problems.

All the problems illustrated in Fig. 4.1 have a common feature, namely the presence of flow either inside or outside the pipe or both. Those devices listed under Class A tools in Table 4.1 were selected because of their ability to respond to whatever situation exists in a particular well. The Class B tools were then selected as discriminators or as quantifiers for known flow conditions. The structure of Table 4.1 is a direct consequence of the situations illustrated in Fig. 4.1.

4.4 Safety and Environmental Tips

The following safety and environmental tips should be considered on every production-logging job:

1. All wireline units should be equipped with at least one 30-lbm dry chemical fire extinguisher.
2. Engines should be equipped with emergency shutoff devices on air intake (butterfly valve) and with spark-arrester-type mufflers.
3. Drip pans should be installed under all engines, pumps, oil reservoirs, and reels of skid-mounted units. Drip pans should be provided for trucks on barges.
4. Oil from the lubricator control head and oil from the line should not be permitted to drip into canals and streams.
5. Pieces of cut-off wireline or cable should be carried off by the wireline crew for disposal.

A Penrod Drilling Co. rig in the Gulf of Mexico sinks after the Exxon U.S.A. natural gas well it was drilling blew out, destroying the 147-ft-tall derrick and breaking one of the rig's three legs.

Fig. 4.2—Surface blowout of a gas well while drilling.

6. Bleeding of gas and oil into metal buckets should not be done. Use of a plastic bucket rather than a metal one will minimize the danger of a static-electricity spark.

7. Truck-mounted units should not be positioned over tall grass in such a way that mufflers or power-generator engines could ignite the grass.

When a potential risk is identified, the purchaser of the logging should make all required operational changes. Because service companies perform most of the production logging, it is also the purchaser's responsibility to ensure that a system is in place for (1) contractor selection and evaluation; (2) defining, communicating, and stewarding performance requirements; (3) managing interfaces between the purchaser and the contractor; and (4) monitoring performance and correcting deficiencies.

4.5 Sinker Bar Weight

A dead weight (sinker bar) is necessary to overcome the force of the wellhead pressure acting on the cross-sectional area of the logging cable. **Fig. 4.3** shows the required sinker bar weight in relation to the shut-in wellhead pressure. The weight shown is just enough to balance the force of the well pressure acting on the wireline.

Additional weight above that which is indicated on the graph is needed to realize downward movement of the logging string. As the inclination angle of the wellbore increases, it becomes especially important to increase the sinker bar weight over the value specified by the vertical axis of the figure. When an inclination angle requires unreasonably long sinker bars, roller centralizers are required.

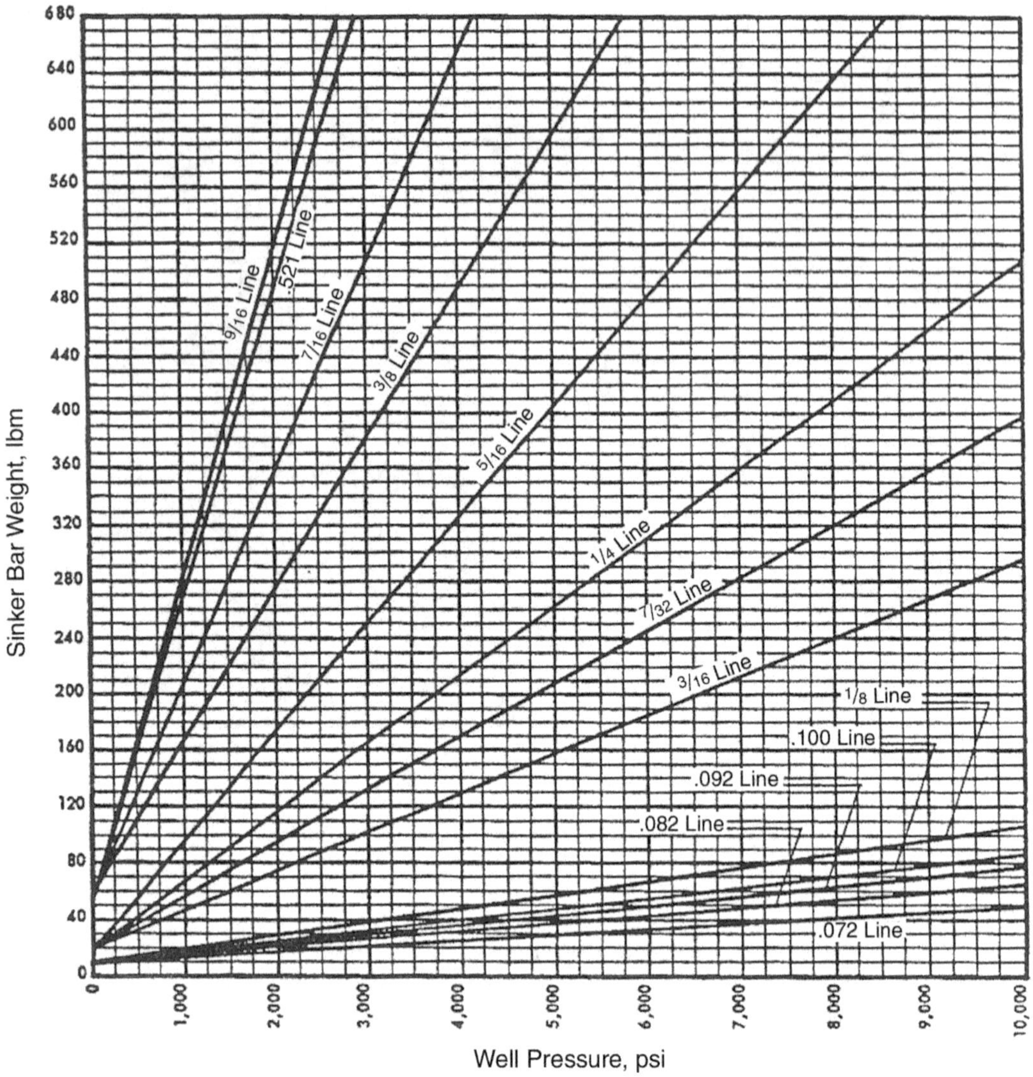

Note: Sinker bar weight given is at balance point.
Add weight as desired to obtain downward movement

Fig. 4.3—Sinker bar weight to balance wellhead pressure (Courtesy Bowen Tool Co.).

Typical slickline diameters are represented by the group of lines at the bottom of Fig. 4.3. The low sinker-bar weight needed to carry out a slickline survey, even at high wellhead pressures, requires only a short lubricator. As a result, slickline services are enjoying a rebirth. New versions of these tools contain sufficient downhole memory to record what is essentially a continuous log.

4.5.1 Maximum Tool Length To Negotiate Bend. Fig. 4.4a illustrates the maximum length of a tool that can pass a bend. The ends of the tool contact the bottom of the borehole, and its middle touches the top.

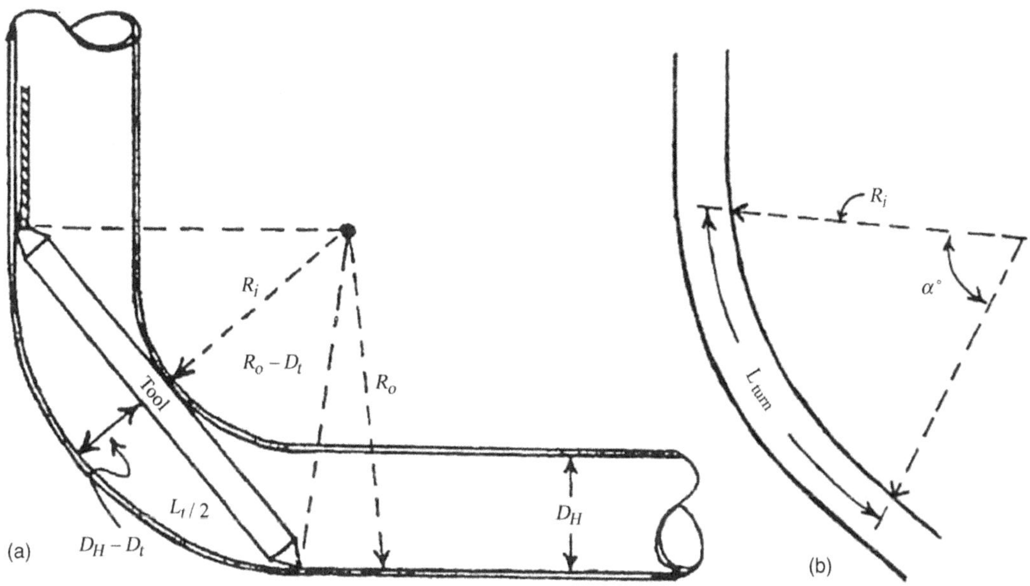

Fig. 4.4—(a) Tool limitations through a bend; (b) inside radius, bend angle, and distance to turn.

With the following equation, the maximum tool length that can pass through a bend can be calculated. While Fig. 4.4a shows a bend into a horizontal wellbore, the expression is valid for any bend and is independent of whether the final wellbore is horizontal.

$$L_t = 2(D_h - D_t)\left[1 + 2R_i / (D_h - D_t)\right]^{1/2} . \quad (4.1)$$

The expression for L_t involves the hole and tool diameters, as well as the inside radius of the bend. The inside radius can be expressed in terms of the angle of the bend, and the distance to bend through the specified angle (see Fig. 4.4b):

$$R_i = L_{\text{turn}} / [(\alpha\pi) / 180] . \quad (4.2)$$

An example of the use of the maximum-length equation:

$$D_h = 6 \text{ in.} = 0.5 \text{ ft};\ D_t = 1.5 \text{ in.} = 0.125 \text{ ft};\ \alpha = 90°;\ L_{\text{turn}} = 30 \text{ ft};\ (D_h - D_t) = 0.375 \text{ ft};$$

$$\text{and } R_i = 2 \times 30 / \pi = 19.098 \text{ ft} .$$

$$L_t = 2(0.375)(1 + 2x19.098 / 0.375)^{1/2};\ \text{and } L_t = 7.61 \text{ ft} .$$

Any 1.5-in. OD tool 7.61 ft in length can pass through the bend. If longer tools are required, then the tool string must be segmented with "knuckle" joints.

4.6 Depth Control

The counter wheels on a production-logging unit measure the length of cable in the well to an accuracy of 5 out of 15,000 ft, provided that a great deal of back and forth travel (yo-yoing) is not required to work the tool string down the well. Better depth control is obtained by placing a casing collar locator (CCL) sub at the top of any production-logging tool string. See logging chapters in the Reservoir Engineering volume of this *Handbook* for more on this tool. This sub generates a voltage spike as it moves past a change in metal thickness, particularly as it passes through the connection between joints of pipe. The resulting record of collars is the source of depth control.

Wells are perforated from a perforating depth control (PDC) log, a combination of a collar log and a cased-hole nuclear log such as a gamma ray log. The nuclear log is then depth correlated to a similar log run before the well was cased. This procedure ties the collar record into the depth scale on the openhole logs. Accuracy in this latter depth scale is maintained by means of magnetic flags placed at precise intervals—customarily, 100-ft intervals—along the openhole logging cable. The PDC log is a part of the file on a given well and provides the collar record that serves as the depth reference for subsequent production logging. A short joint of casing called a "pup" joint is often placed in the casing string as a depth marker. Otherwise, normal variations in length are used to correlate collar records.

Sometimes a radioactive collar that emits gamma rays is used as a marker for depth control. For flush-joint casing, collars are available that are strapped around the casing before it is run into the hole. Occasionally, radioactive bullets are fired into the formation before casing the well. Finally, it may become necessary to run a section of a nuclear log and to flag (mark) the logging cable on a particular operation to achieve the necessary depth control.

4.7 Pricing and Record Keeping

4.7.1 Pricing. An understanding of the price structure to perform production logging can help optimize data-acquisition costs. The total cost of a job is the sum of four separate charges:

1. ***Set-Up Charge.*** This is the amount the service company charges to bring their equipment to your well and "rig up" on it.
2. ***Pressure Charge.*** This charge is based on the surface pressure on the well and represents a rental fee on the equipment necessary to safely log the well.
3. ***Depth Charge.*** This is a charge based on the maximum depth reached in the well. It reflects the fact that at least three people are needed to rig up, to spool so much wireline off the drum of the logging unit, to reel this same amount of cable back onto the drum, and to rig down. It is the substitute for the familiar hourly charge.
4. ***Logging Charge.*** This is a per-foot charge for each tool run that reflects the length of interval actually logged by the tool.

From this structure, one can immediately see the potential savings to be had from being present during the logging operation. On-site changes in procedure to ensure the desired diagnostic logs can prevent having to start another day, resulting in incurring the first three charges again.

The footage-logging charge is often cited as the reason for running only one tool. This, however, is false economics because this charge, even for a three-tool suite, is usually no more than 40% of the total.

4.7.2 Record Keeping. Forward planning will ensure maximum long term use of log data. The log headings used by most service companies, while recording some specifics about the run itself, have little information on why and how the logs were run. Consequently, most professional loggers have their own forms to bridge this documentation gap. While the organization of these forms is a matter of personal preference, a good rule is to prepare a preliminary summary that specifies at least the following:

- Why the logging is undertaken.
- Previous production-logging summary.
- Current well-completion data with a wellbore sketch.
- Collars used for perforation.
- Depth reference point.

- Most recent well-test data.
- Anticipated total depth, bottomhole pressure, and temperature.

A second form completed at the time of logging is a chronology that lists

- Logs run and their order.
- Run number.
- String logged and its status for each run.
- Status of other strings or annuli.
- Logging direction and speed.
- Tool calibration checks.
- Intervals relogged.
- A summary of conclusions.

4.8 Operating Principles and Performance of Production-Logging Tools

The following sections describe operating principles for each of the tools listed in Table 4.1. The text will indicate applications for which a tool is best suited, those for which it is only partially suited and, when possible, those for which a tool is not suitable.

Some interpretive principles and recommended logging procedures will be presented in examples. However, the reader should refer to the Appendix for detailed information of this type. Oxygen-activation, cement-bond, and casing-inspection tools are not treated. These tools are, however, included in the application tables of the Appendix.

4.8.1 Temperature-Logging Tool. The tool includes a cage, which is open to the wellbore fluid, at the tool's bottom end. Inside the cage is a thermistor that senses the surrounding fluid temperature. The preferred sensor is a platinum element because the electrical resistance of the sensor varies linearly with temperature over a wide range and is stable over time. The circuitry of the tool is designed so that the voltage across the sensor is proportional to the sensor's electrical resistance.

In analog recording, the transmitted spikes per minute are converted to a voltage by a counting circuit. This voltage is recorded on a pen-and-ink strip chart recorded as the temperature (or gradient) trace. This is Trace 1 of **Fig. 4.5**. The scale of this trace is in °F. A recording sensitivity of 1°F/in. across the chart is strongly recommended for production wells.

In analog recording, the voltage output of the counting circuit is also input to a differentiating amplifier. The output of the amplifier is recorded as Trace 2 (the differential trace or the derivative of the temperature), and is proportional to the depth-rate-of-change of the temperature curve. Although no absolute scale is associated with the differential trace, it is useful for highlighting important changes of the slope of the temperature curve.

On the log, the temperature trace warms abruptly below Depth B. Because the tool is logging down and the temperature is increasing, the depth-rate-of-change of temperature is positive. Consequently the differential trace shows a strong positive excursion at B, highlighting the change of the slope of the temperature trace at this depth. The differential trace, when properly amplified as on Fig. 4.5, is easily worth the additional logging charge.

The flowing temperature Trace 1 in Fig. 4.5 also provides information on the production profile. Production commences at Depth A, where the flowing trace "stands up" to separate to the warm side of static. This location is at the bottom of the bottom set of perforations. The middle set of perforations, on the other hand, contributes nothing to the production. The largest contribution comes from near the bottom of the top set of perforations at Depth B, where a large mixing signature is evident. The volumetric rate of entry here is so large that the mixture temperature is "pulled" almost back to static temperature (i.e., to the entry temperature for the stream). There is one additional smaller entry at Depth C, whose mixing signature is hardly recognizable on the flowing Trace 1 but is clearly evident on the differential Trace 2. The location of this entry suggests that it is composed primarily of oil. The temperature traces there-

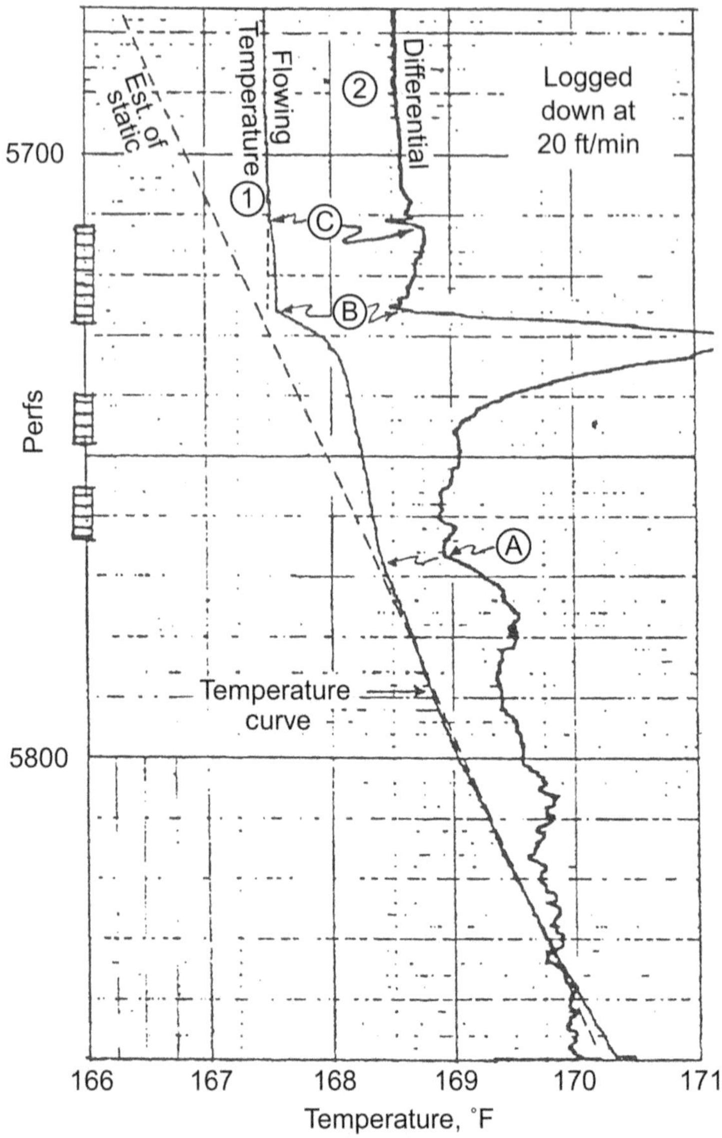

Fig. 4.5—Section of a temperature survey from a well flowing mainly water at 1,800 B/D.

fore show that the top set of perforations is responsible for the majority of both water (the major stream) and oil (the minor stream). The use of these mixing signatures to profile both single-phase and multiphase flow is described in detail in the Appendix under production-well profiling. The important point is that the size of a mixing signature relative to static temperature is dependent upon the thermal content (product of density, specific heat, and volumetric rate) both of the entry stream and of the stream in the casing immediately below the entry.

In digital recording, the spikes-per-minute from the logging cable are counted digitally at the surface, and the resulting count rate is converted to a temperature trace by the computer's program. Again, the temperature trace should be recorded at a sensitivity of 1°F across the chart. Digital recording degrades the sensitivity of the differential trace from that available with analog recording. Thus, the digitally determined differential trace is not as useful for highlighting important changes of the temperature curve's slope.

Depending on how carefully (or recently) a particular temperature tool was calibrated, there may be several °F difference between the recorded temperature and the true downhole temperature. However, the difference does not degrade the sensitivity of the differential trace. Provided that the temperature curve is recorded at the recommended sensitivity of 1°F/in. across the chart, and that the temperature log is carefully depth-correlated, the resulting temperature curve has more vertical resolution than does a curve from any other production-logging tools.

The temperature tool is most effective when located at the bottom of a tool string. In a production well, the tool should always be logged downward so as to enter undisturbed fluid. The log should be recorded at a constant logging speed not to exceed 30 ft/min. With digital recording, the maximum logging speed should be reduced to 20 ft/min.

Fig. 4.6 shows temperature logs from an exploration well that was perforated and acidized before the flow test. During the flow test, the surface rates were 2 MMscf/D gas and 500 B/D spent acid and formation water. The formation is gas-saturated limestone.

The temperature profile from the flow test is the solid trace; the line labeled "static" is a temperature log after a 6-day shut-in following the flow test. In the sump below the bottom perforation, there is no flow. The temperature here is approximately 6°F cooler than static. During drilling, mud circulation takes heat from the near-wellbore, leaving the near-wellbore below static temperature. Over time, heat flows from the formation farther away from the wellbore, where the temperature still is static. This heat flow, given enough time, restores the near-wellbore to static. In a no-flow interval, this heat flow is the only process for restoring the near-wellbore to static temperature.

At the time of the flow test, the near-wellbore temperature below the bottom perforation was still below static. The line labeled "cool temp" coincides with the flowing temperature profile below the bottom perforation and is parallel to the static temperature line determined subsequently. The cool temp line shows that, at the time of the flow test, the near-wellbore is below static throughout the perforated interval.

During the flow test, spent acid and formation water enter the wellbore at Depths 1, 2, 3, and 4. Each entry flow begins in the formation at static temperature. At Depth 1, the entry rate is small enough that the flow, after being cooled during passage through the cooler near-wellbore, enters the pipe at a flowing temperature that is considerably below static. Still, the entry causes an abrupt warming at Depth 1. Because the entry at this depth does not mix with a flow from below, the "entry temperature" (i.e., the temperature of the entry as it emerges into the wellbore) is the flowing temperature at the entry Location 1.

At Depth 2, additional liquid enters and mixes in the wellbore with the upward flow from Depth 1; the mixing of the two flows results in the abrupt warming at this depth. The flowing temperature at Location 2 is the final temperature after the mixing. The entry at Depth 2 comes into the wellbore at an entry temperature (before mixing) that exceeds the final temperature after mixing but is no greater than the static. Notice that the final temperature after mixing at Depth 2 is less reduced from static than is the entry temperature at Depth 1. Thus, the entry rate at Depth 2 exceeds that at Depth 1. The entry at 2 is cooled less by passage through the near-wellbore region than is the entry at 1, leaving the entry temperature at 2 higher than that at 1.

At Depth 3, liquid enters and mixes with the flow arriving from below. The mixing of the flows causes the abrupt warming at Depth 3. The flowing temperature at this depth is the final temperature after the mixing. The entry at Depth 3 comes into the wellbore at an entry temperature (before mixing) that exceeds the final temperature after mixing. However, this temperature is less reduced from static than is the final temperature at Depth 2. Thus, the entry rate at Depth 3 exceeds that at Depth 2. The entry at 3 is cooled less by the near-wellbore than the entry at 2, leaving the entry temperature at 3 higher than that at 2.

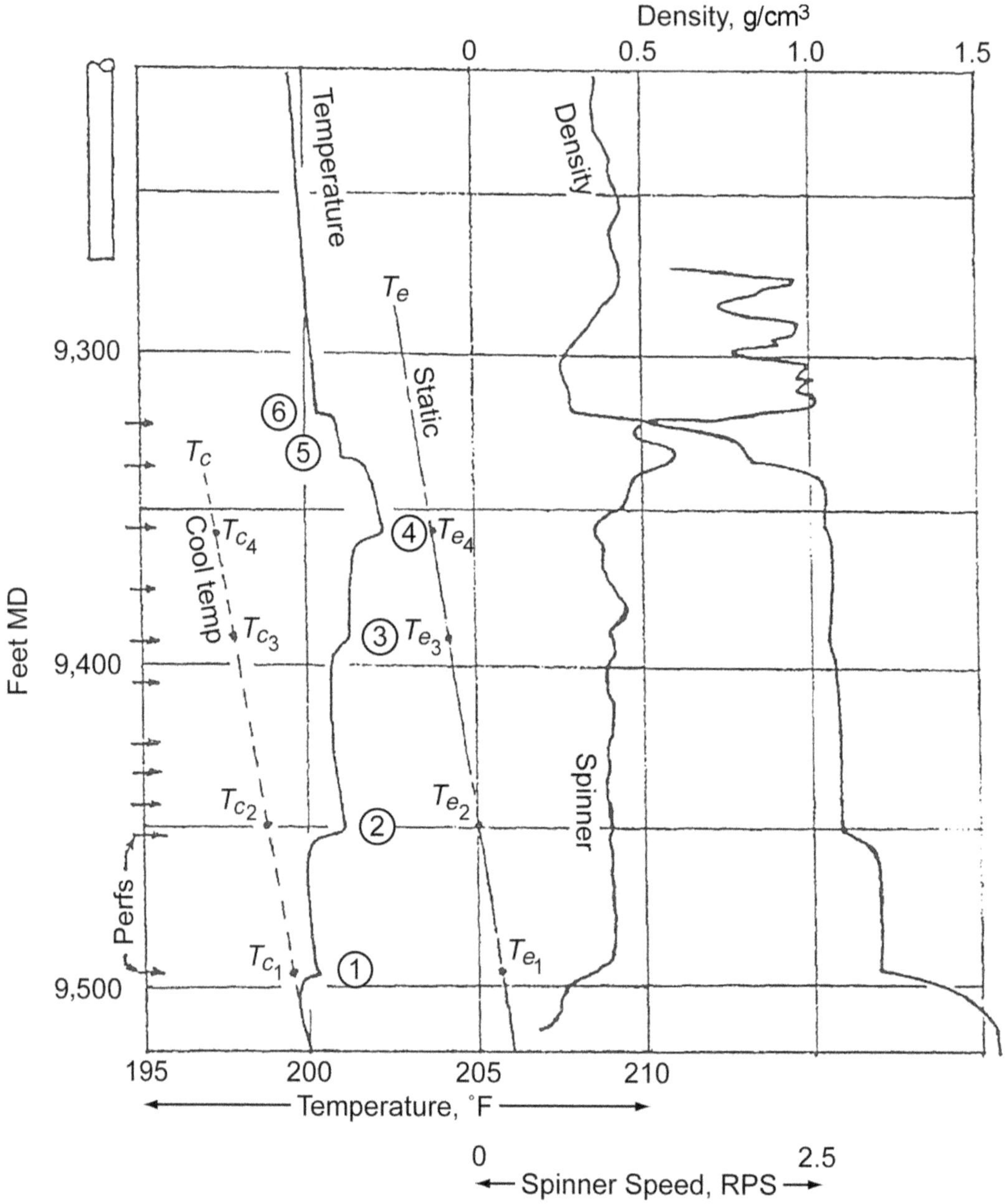

Fig. 4.6—Flowing logs from a new gas well flowing 2 MMcf/D.

By similar reasoning, the entry rate at Depth 4 is greater than the entry rate at Depth 3. In a recently drilled well or in an old gas well, provided that the near-wellbore is below static, the deepest entry, if it is a liquid, results in a warming signature. The warming is caused by the liquid's bringing of its warmer temperature into the wellbore itself. The contrast between the entry temperature and the cooler temperature of the wellbore below the entry is responsible for the warming signature. A common but mistaken view holds that the warming is caused by frictional heating of the liquid as it moves through the near-wellbore region and into the well. The fact that a water rate of only a few B/D is sufficient to produce a warm signature clearly refutes this "frictional" hypothesis. In a gas well, pressure loss in the near-wellbore and completion results in expansion of the gas during its passage into the wellbore. This expansion cools (or warms) the gas to a degree that depends on the flowing bottomhole pressure, temperature, and the amount of pressure loss. If the gas is cooled, it absorbs heat as it passes through the near-wellbore region before entry. Over time, the heat spreads up and down the near-wellbore, resulting in a below-static temperature in the near-wellbore region. Any subsequent liquid production produces the same warm signatures as described for a new well. In actual fact, the expanding gas warms at bottomhole pressures in excess of approximately 10,000 psi and cools at pressures lower than 5,000 or 6,000 psi. At intermediate pressures, the amount of cooling or warming is generally not significant.

In reference to the example of Fig. 4.6, the flowing bottomhole pressure and temperature are 1,325 psia and 200°F, respectively. The well flows at a drawdown in excess of 3,000 psia. At Depths 5 and 6, each temperature mixing signature is caused by expansion of a dry gas as it passes through the near-wellbore and completion. At Depth 5, the colder gas mixes in the wellbore with the warmer liquid arriving from below, giving the cooling signature at this depth. At Depth 6, the colder gas mixes with the warmer flow of gas and liquid arriving from below, giving the cooling signature at this depth. The next calculations focus again on the water entries.

Certain conventions apply to the volumetric rates q_i in the relationships to be discussed next. If water is produced, then the rates for all other phases are "water equivalent" rates that represent the multiplication of each actual rate by a ratio. The numerator in this ratio is the product of density and specific heat for the particular phase, whereas the denominator is the same product for water. If only gas and oil are produced, then the gas rate is presumed to be an "equivalent oil" rate obtained with a similar ratio for gas/oil. Once profiling is done by the following expressions, then the equivalent rates are converted to phase rates according to the equivalence ratios. Typically, 3 volumes of oil are the equivalent of 1 volume of water, whereas ~ 10 volumes of gas are the same equivalent.

The entry rates of the four liquid entries can be determined quantitatively from two relationships, a warming relationship where q_i ($i \geq 2$) refers to a given stream entering the wellbore at temperature T_i to mix with the total stream from below at temperature T_{i-1} and produce a mixture with temperature T_{mix}.

$$\left(\frac{q_i}{q_{i-1}}\right) = \left(\frac{T_{i-1} - T_{\text{mix}}}{T_{\text{mix}} - T_i}\right) \bullet \left(1 + \frac{q_2}{q_1} + \cdots + \frac{q_{i-1}}{q_1}\right), \quad (4.3)$$

and a warming relationship,

$$\frac{q_i}{q_1} = \left(\frac{\ln R_1}{\ln R_i}\right), \quad (4.4)$$

where

$$R_i = \left(\frac{T_{\text{geo}} - T_{\text{cool}}}{T_i - T_{\text{cool}}}\right)_i \quad (4.5)$$

and q_i ($i \geq 2$) denotes any stream entering the bottom stream – 1. In the ratio for R_i, the temperature T_i refers to the temperature at which stream-i arrives at the wellbore, whereas T_{geo} indicates normal geothermal temperature at the depth of the entry, and T_{cool} is the cooler temperature of the stream coming from below entry-i. Usually, the entry temperature T_i for a given liquid stream will be the same as geothermal temperature T_{geo} and only the mixing relationship

TABLE 4.2—PROCESS OF USING TRIAL VALUES

Trial Value T_{in2}	Rate Ratio, q_2/q_1	
	Mixing	Warming
201.5	2.75	2.35
201.6	2.20	2.56
201.55	2.44	2.50
201.54	2.50	2.49

(Eq. 4.3) is required to determine the relative size of the rates. The warming relationship (Eq. 4.4) is required when the wellbore has been cooled below geothermal, and liquid entries arrive at the wellbore at some unknown temperature cooler than geothermal. In this situation, a succession of liquid entries moves the flowing temperature profile successively closer to the geothermal temperature line without bringing the two into coincidence. This is the temperature behavior at Depths 1 through 4 of Fig. 4.6. The left side of the relationship in Eq. 4.5 is the ratio q_i/q_1 (i = 1 to 4 in the present example), which is the ratio of the entry rate q_i to the entry rate at Depth 1. The right side of the relationship in Eq. 4.5 incorporates numerical values that are available from the "cool temp" and "static" lines in the figure. The remaining variable is T_i, which is the temperature of the ith entry as it emerges into the wellbore, before mixing with the flow from below. This is the same as the "entry temperature" used in the text above. T_1 is the temperature at the warming signature at Depth 1, because there is no flow below this depth. The value of T_i at higher depths must be determined along with the rate ratios.

On the left side of the mixing equation appears the ratio q_i/q_1; on the right side all numerical values are available from the temperature profiles of Fig. 4.6 with the exception of T_i. Thus, there are two unknowns: q_i/q_1 and T_i. Various trial values of T_i are entered into the warming and mixing relationships to give two values for q_i/q_1. A solution for T_i is the trial value for which the two values of q_i/q_1 are substantially in agreement. A trial value for T_{ini} should exceed the final temperature after mixing at Depth i, but should be less than the static temperature at Depth i.

The process of using trial values is shown by **Table 4.2**, which relates to finding the value of q_2/q_1 and T_2 . As can be seen, the two values of the rate ratio are in close agreement at a trial temperature of 201.54°F.

From the same technique illustrated previously, the ratio of the third entry rate to the first is 4.65, and the ratio of the fourth entry rate to the first is 14.85. These ratios, along with a ratio of 1.00 for Entry 1, sum to 23.0 and provide the fractional value of each entry in the total liquid stream of 500 RB/D. Thus, Entry 1 amounts to 500 × (1/23) = 22 RB/D. Likewise, the entry rate at Depth 2 is 54 RB/D, at Depth 3, 101 RB/D, and at Depth 4, 323 RB/D. In Fig. 4.6, notice that the entry at Depth 4, with by far the largest rate, moves the temperature after mixing much closer to static than does any other entry, as expected. The density log of Fig. 4.6 shows that all four streams are composed primarily of water so that no phase conversion is needed.

4.8.2 Radioactive Tracer-Logging Tool. The tool has a reservoir to hold radioactive material and a pump section at the top. For injection-well logging, two gamma ray detectors below the reservoir and pump are preferable. Some tools employ only one detector, but this is less desirable. The tool includes the circuitry to amplify and transmit the detector counts to the surface, for recording.

Most natural radioactivity underground is from the decay of isotopes of potassium, thorium, and uranium. These materials concentrate in the shales, where they register approximately 100

API units on the gamma-ray log. Once downhole, a "slug" of tracer is ejected by the pump, under surface control. The activity owing to the ejected slug is much greater than the natural background activity. By tracking the progress of the slug down the wellbore, the exits of injected flow from the wellbore can be determined, as well as whether any of the injection, after exiting, passes through a channel close to the pipe.

There are two modes of logging: slug tracking and velocity shot. For slug tracking, the logging operator ejects a slug of tracer from the tool. After ejection, the tool is run up and down through the slug to ensure that the slug is uniformly mixed across the wellbore cross section. Then, the tool is lowered quickly and an upward logging pass is made at constant logging speed until the slug is detected. The time of detection of the peak and the depth of the peak are recorded. Then the tool is quickly lowered again, and another upward logging pass is made at the constant logging speed until the slug is detected again. Again, the time of detection of the peak and the depth of the peak are recorded. This process is repeated several times, resulting in a succession of detections of the same slug (see **Fig. 4.7**). As long as the peak progresses downward, there is flow in or near the wellbore. Once the peak stops, there is no flow in or near the wellbore below the stopping depth. For each detection, the area under the trace and above the common baseline of the traces is proportional to the percentage of injection still in or near the wellbore.

By visual inspection of the area under the traces in Fig. 4.7, nearly all of the injection reaches Depth D, and the injection leaves the wellbore between D and the bottom of the perforation set. A convenient measure proportional to slug area is the product of the slug's height, above a common baseline, and the slug's width at half-height. Numbers from the areas agree with those shown on the left side of Fig. 4.7 as derived from the travel times. Slug A is not yet mixed in the flow. Slug area has the advantage of being insensitive to variations in fluid velocity, allowing the approach to be extended to traced slugs moving behind casing. Notice in Fig. 4.7 that activity also is detected below the perforations. A slug was ejected below the perforations. Upward logging passes showed that the peak of this slug was stationary. Therefore, there is no wellbore flow below the perforations, and the activity below the perforations is attributable to tracer channeling downward behind the pipe.

Generally, only one gamma-ray detector is used for slug tracking. Slug tracking gives the best overview of where injection leaves the wellbore and whether, after exiting, any injection travels in a channel close to the pipe. The upward logging passes are made at high line speed. A constant logging speed should be used, and the same speed should be used for all passes.

Provided the ejected slug is uniformly mixed in the flow by movement of the tool up and down through the slug after ejection, the vertical distance (ft) between two successive peaks in total flow divided by the time (minutes) between detection of the peaks provides an accurate estimate of the average flow velocity of total injection. Such velocities are listed on the left of Fig. 4.7. The most frequently used tagging material for water is an aqueous solution of sodium iodide, which contains the isotope of iodine, I-131. The 8-day half-life is ideal. In solution, the iodine does not stick to rock surfaces; instead, with continued injection, the iodine is washed from the rock surfaces and carried away from the near-wellbore, beyond detection by the logging tool.

From inspection of slug spacing in Fig. 4.7, it is evident that slug tracking has limited vertical resolution. Furthermore, because 90% of the detected gamma rays originate within 1 ft of the detector, the tracer tool's depth of investigation is also limited and is much less than that of the temperature tool. Because of the limited depth of investigation, tracer that is channeling after exiting the wellbore must be close to the pipe to be detectable. Not all channels can be detected by the tracer tool. The same is true for fractures.

A velocity-shot survey is used in intervals where greater vertical resolution is desired. To perform a velocity shot, the logging operator stations the tool so that the detectors are at cho-

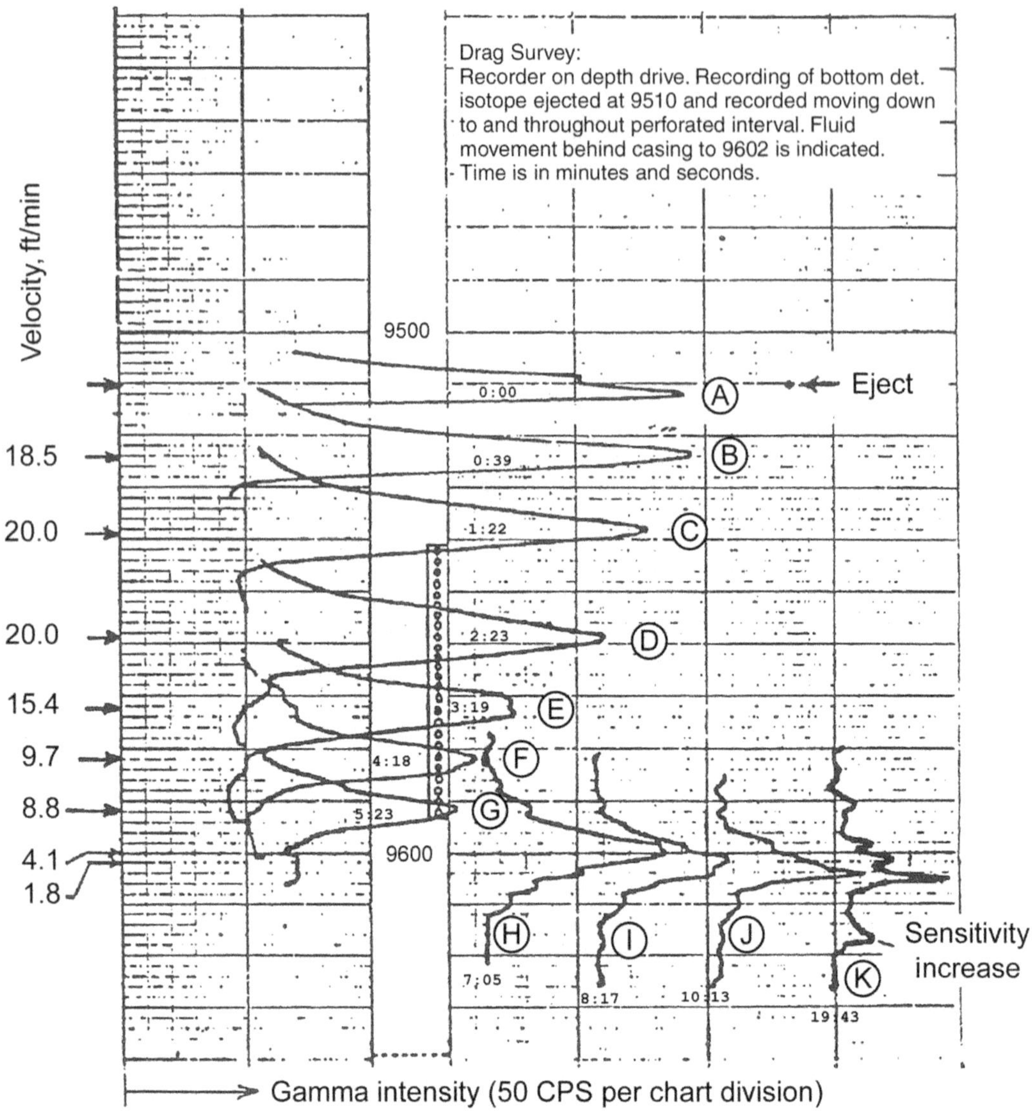

Fig. 4.7—A radioactive slug-tracking record from a water-injection well taking 600 BWPD.

sen locations. Then, with the tool stationary, a slug of tracer is ejected into the injection flow. As it passes downward, the slug is first detected by the top detector and then by the bottom detector, resulting in two traces on the log (see **Fig. 4.8**). On the chart, time increases from the bottom upward. Thus, the top detector gives the lower trace. The time interval between the two peaks (travel time) is inversely related to the velocity of the injection flow. The chart speed should be such that there are 5 to 10 vertical chart divisions between the peaks in total flow, although such is not the case in the figure.

The ratio of the travel time in total flow to the travel time at a selected position is the fraction of injection still in the wellbore at the selected position. However, dividing the separation between the detectors (ft) by the travel time (minutes) does not produce the average velocity of flow, as the slug cannot be uniformly mixed in the flow before it passes the detectors.

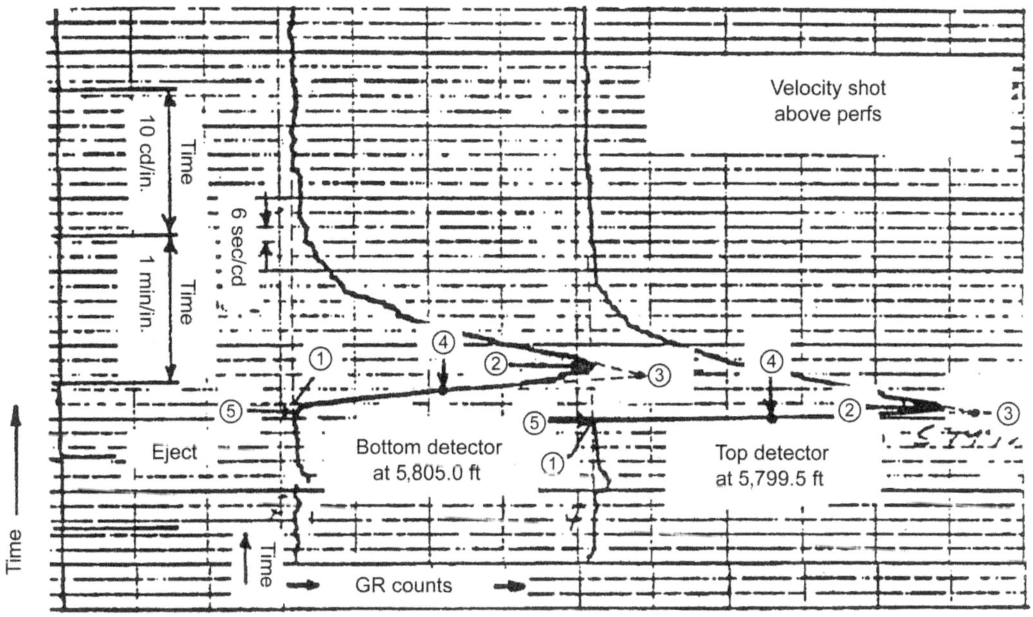

Fig. 4.8—Example of a radioactive tracer velocity-shot survey.

Two detectors are preferred for velocity shots. If there is only a single detector, there can be timing errors between initiating ejection of a slug and actual ejection downhole. These timing errors contaminate the measured travel times.

For detection of flow behind pipe, many logging operators prefer velocity shots. One detector is stationed within the perforations, while the other is stationed above or below the perforations to see if any flow channels up or down after exiting the wellbore. One difficulty with velocity shots is that they investigate only a very limited part of the total injection interval. In some circumstances, the results from velocity shots indicate the presence of a channel when in fact there is none. Whenever the results from velocity shots indicate an integrity problem, it is better to switch over to slug tracking, which investigates the overall injection interval as well as the wellbore above the interval.

Fewer applications of tracer logging occur in production wells. In a true single-phase flow, there is an appropriate tracer, whether the flow is water, oil, or gas. Because a slug is tracked for a while and then disappears uphole, multiple slugs are used, one for each producing interval under investigation. Usually, a well is logged from a bottom, no-flow interval up to an interval of total flow. Because of the unusual circulation patterns that can occur in multiphase flows, tracer results in these flows can be misleading.

4.8.3 Noise-Logging Tool. This tool "listens" passively to downhole noise, for example, from gas bubbling up through liquid in the wellbore. Behind pipe, a channeling flow often passes through cramped spaces and constrictions. These "tight spots" cause high velocities, sudden pressure reductions, and significant flow turbulence. The noise tool listens to the noise associated with the turbulence. Consequently, noise logging is an inexpensive way to investigate whether there is channeling in injection or production wells.

The tool includes piezoelectric crystals (transducers) which convert the oscillating pressure of wellbore sound to a corresponding oscillating voltage. At the surface, the oscillating voltage is applied to a speaker, permitting the logging operator to hear the downhole sound. The oscillating voltage also is applied to each of four high pass filters. A high pass filter is unresponsive to frequency components of the oscillating voltage that are below the filter's "cutoff" frequency, while it generates an oscillating output signal whose amplitude represents the content of the oscillating voltage of frequencies in excess of the cutoff. The cutoff frequencies most often used are 200, 600, 1,000, and 2,000 Hz.

For quality measurements, the operator stops the tool at a selected depth and waits for extraneous noise, such as the noise generated by tool and logging-cable movement, to subside. Hurrying contaminates the measurements with extraneous noise. Then, a recording system averages and records the amplitude of each filter's oscillating output signal. Subsequently, the tool is moved to a new measurement depth; after waiting for the extraneous noise to subside, the operator initiates the averaging and recording of the amplitudes at the new depth. Thus, four values are recorded for each measurement depth. These are plotted in millivolts across the chart at a measurement depth. The amplitude of the 200-Hz filter output is the greatest; as the cutoff frequency increases, the amplitude correspondingly decreases, with the amplitude of the 2,000-Hz filter output being the least. To form a log, the plotted 200-Hz amplitudes are joined by line segments between successive measurement depths, as are the amplitudes for 600, 1000, and 2,000 Hz.

Because a filter's output signal is oscillatory, it consists of positive excursions from neutral alternating with negative excursions. There are two ways of measuring the amplitude of the output. One measures the amplitude from the peak of a positive excursion to the trough of the following negative excursion. The resulting amplitude is known as "peak to peak," and the recording is characterized as "standard gain" or "standard sensitivity." The other amplitude measurement is from the peak of a positive excursion to neutral, with the amplitude being characterized as "peak" and the recording sensitivity as "one-half standard gain." Other measures are sometimes used.

A single-station measurement typically lasts 3 to 4 minutes. Relocating the tool usually requires 1 minute. Thus, the logging rate is approximately 15 stations per hour. A 4-hour logging run accommodates 60 measurements. Thirty of these are used for a coarse-measurement grid, with successive measurements separated by one-thirtieth of the total survey interval. The remaining 30 measurements are used for detailing areas of interest, such as a noise peak identified by the coarse grid, a gas/liquid interface in the wellbore, a plug, or a packer. For detailing, the separation between successive measurements can be as little as 1 to 3 ft.

Fig. 4.9 shows the "dead-well" response recorded at one-half standard gain in a shut-in injection well. As expected, the 200-Hz amplitude is greatest, and the amplitudes progressively decrease as the frequency increases. Notice that the trace for one frequency never crosses the trace for another frequency. If such crossings occur, the log is faulty. If a dead-well response does not appear at all, the log is suspect. The figure is illustrative, and the amplitudes of the dead-well response in other wells may vary from those shown.

The cutoff frequencies were carefully determined from laboratory measurements. The 200-Hz cutoff rejects most logging-truck vibrations transmitted through the logging cable, while it is low enough to identify gas bubbling up through liquid. The 1,000-Hz cutoff permits identification of single-phase flow, the strongest frequencies of which occur at 1,000 Hz or higher. The determination of whether behind-pipe flow is single-phase or gas-through-liquid is facilitated by the 600- and 2,000-Hz cutoffs.

One service company uses software processing that implements four different *high*-cutoff frequencies with the same *low*-frequency cutoff of 100 Hz in all four instances. Thus, the computed amplitudes pertain to the frequency "bands" between 100 Hz and each of the high-cutoff frequencies. One difficulty is that the lower-frequency components of the sound, whatever their origin, have amplitudes more variable with time than those of the higher frequencies. Also, extraneous noise sources tend to be more intense at the lower frequencies. The result is that

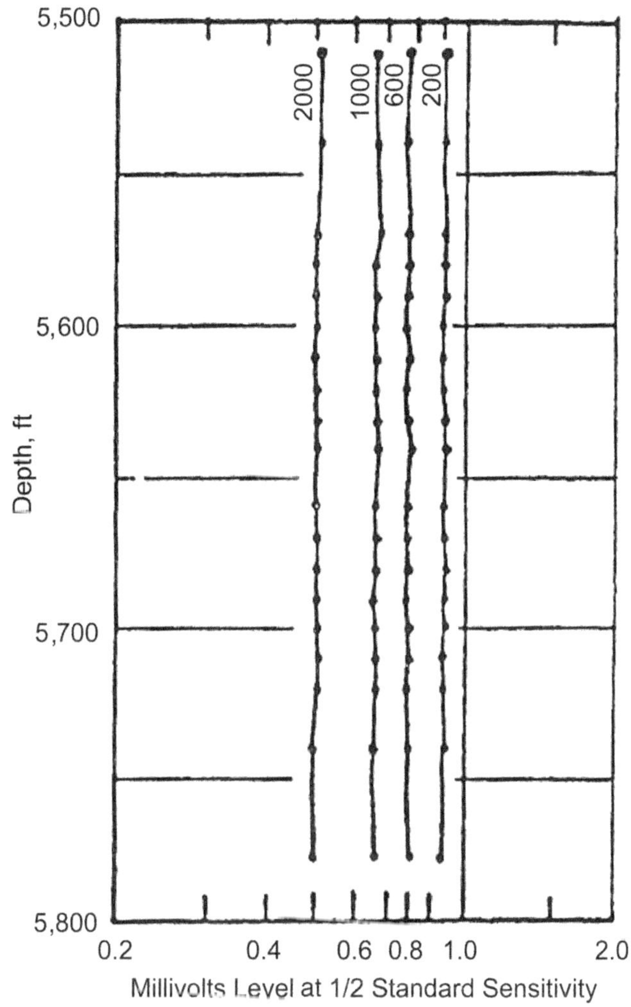

Fig. 4.9—Shut-in injection well.

each of the bandpass amplitudes can be contaminated by the variable amplitudes and the extraneous sources.

Noise can travel appreciable distances to the noise tool, both vertically and horizontally. The user must be sure that all extraneous noise from surface operations (with the exception of logging-truck vibrations) and flow in nearby wells is eliminated, so that the only noise recorded is that associated with the suspected well problem.

Fig. 4.10 pertains to a 20-B/D water flow behind pipe into a depleted gas zone. Only one noise peak occurs, corresponding to a single "tight spot" between the water sand and the gas sand. Almost 30 measurement stations were used. Farthest away from the peak, the spacing between measurements was greatest. Closer to the peak, the spacing was reduced. The measurements were densest in the locality of the peak for best detailing of the peak itself. The single-phase nature of the behind-pipe flow is evident from the "bunching" of the 200-, 600-, and 1,000-Hz traces at the peak. Gas-through-liquid flow would separate the 200- and 600-Hz traces, in addition to the separation of the 1,000- and 2,000-Hz traces.

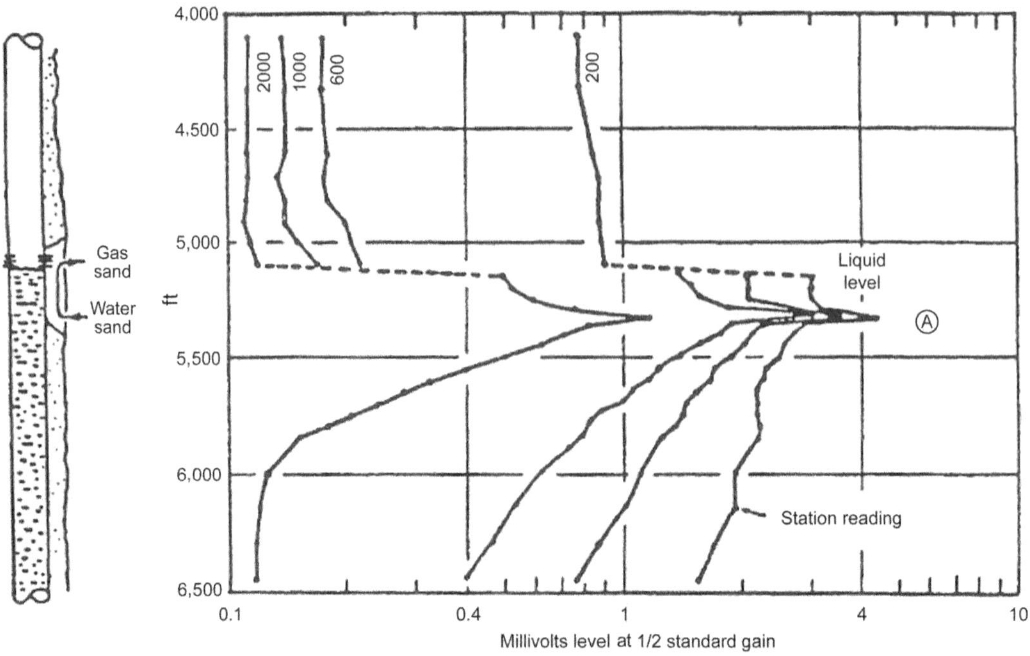

Fig. 4.10—Noise log format with a 20-B/D water flow behind pipe into a gas zone depleted by 250 psi.

Because there is no noise peak at the water or gas sand, the entry and exit of flow to and from the channel occur in absence of a tight spot at either sand. With the known 250-psi pressure drop between the sands confined to a single constriction, it is possible to estimate the flow rate of the channeling water from the pressure drop and the level of the 1,000-Hz trace at the peak (twice the level shown on the log, which was recorded at half-standard gain), yielding the 20-B/D rate. The rate equations are discussed in the appendix. If there were multiple tight spots, the estimate could still be made based upon total pressure drop and upon the sum of all the peak 1,000-Hz noise levels.

Above the gas/water interface in the wellbore, all four traces in Fig. 4.10 decline rapidly to dead-well response because of two attenuating factors. First, much of the sound reflects downward at the interface itself. Second, the tool sensor is constructed for coupling to liquid rather than to gas. In the water below the interface, only the 2,000-Hz trace, which attenuates the most rapidly with distance, reaches dead-well level. Because the water transmits sound at much less attenuation with distance than the gas, the 2,000-Hz trace requires almost 700 ft to reach the dead-well level in the water. In the gas, the dead-well level is reached in considerably less distance. In the water, the other traces show typical noise behavior; the attenuation with distance is least for the 200-Hz trace and increases with increasing frequency. In a production or injection well, it is not unusual to find a gas/liquid interface.

4.8.4 Focused Gamma-Ray Density-Logging Tool. The tool incorporates a compacted slug of Cesium-137 at the bottom of an open cage. The Cesium emits gamma rays, and a lead lens focuses gamma rays in a narrow beam parallel to the axis of the cage. Because the cage is open, wellbore fluid is present inside the cage, and the fluid is in the path of the focused beam. The gamma rays have an energy level low enough that the rays are deflected by the electron cloud surrounding the nucleus of any atom. Furthermore, the amount of backscatter (or absorption) is directly related to the density of the electron cloud and, therefore, to the density of the wellbore fluid. Those gamma rays that are backscattered or absorbed do not emerge from the wellbore fluid at the top of the cage.

At the upper end of the tool cage, a gamma detector responds to the gamma rays that still remain when the beam emerges from the wellbore fluid. A counter determines the counts/min (intensity) of the gamma rays; this information is transmitted through the logging cable to the surface.

The preferred gamma-ray detector is a scintillation crystal because of its high sensitivity and short dead time (the interval required between detection of different gamma rays). Many tools use Geiger tubes as the gamma-ray detector. A Geiger tube has a much longer dead time than a scintillation crystal. It is necessary to use multiple Geiger tubes to ensure detection of all the gamma rays. Preferably, there are eight tubes, but tools of lesser quality may have as few as three.

Count rate is inversely related to the density of the fluid in the wellbore and is recorded in g/cm^3 vs. depth. The count-rate/density relationship is calibrated first with air in the open cage, yielding a count rate corresponding to a density of 0 g/cm^3 and then with tap water in the open cage, yielding a count rate corresponding to a density of 1 g/cm^3. For some tools, there is an aluminum block of known density that can be placed in the open cage. There is enough nonlinearity in the relationship to justify a calibration with three known density values, but typical practice is to use only two. Often, the recording system is adjusted so that the range from 0 to 1 g/cm^3 spans the right-side track of a typical log. With 20 chart divisions in this track, the recorded log is shown at a sensitivity of 0.05 g/cm^3. Attempts to increase the recording sensitivity appreciably over that used for the typical display result in a trace with many irregularities because of the statistical nature of nuclear events. The irregularities can make log interpretation difficult.

The advantage of the focused tool is that it measures only the density of the wellbore fluid. If an unfocused (gravel-pack) tool is used, the gamma rays investigate not only the density of the wellbore fluid, but also that of the pipe wall as well as the material and fluids that are close to the outside wall. The gamma tool can distinguish between the phases in a two-phase mixture, but it cannot distinguish among the phases of a three-phase mixture.

The tool should be logged downward at logging speeds between 15 and 30 ft/min in a flowing production well. A constant logging speed should be used, and the same speed should be used for all runs. It is best to log downward from a starting depth above all the perforations, and to log all the way to the deepest depth that can be surveyed. In a slugging or churning multiphase flow, the log may show variable behavior even in intervals that are not perforated. In that case, another logging run is advisable to establish the degree of repeatability. If the result is less than desirable, a stationary measurement can be time-averaged for each selected location. Usually a logging run is also made with the well shut-in. This log should be run after at least two or three hours of shut-in, to be sure that the fluid distribution is stable. If the well has been shut in overnight, a shut-in log can be recorded, but the well must flow for 2 or 3 hours before the first flowing log is recorded. If the tool is centralized, there is a tendency for the recorded density to be somewhat low in a multiphase flow relative to the average density across the wellbore cross section. This is because the lighter phase tends to rise through the center of the cross section, leaving the remainder of the cross section for the heavier phase.

Although the health, safety, and environmental risks are generally low, radiation safety procedures should be strictly followed when calibrating or running a gamma-ray density logger. Needless exposure to the radiation from the tool when it is at the surface should be avoided. Logging-company personnel should have current radiation training and certification. Because of the long half-life of Cesium-137, the legal restrictions on the use of the tool vary from state to state and country to country. If the tool is dropped in the well or becomes stuck, it must be retrieved or cemented over.

Another common density tool is the pressure-gradient instrument. As the name implies, this device determines density from a differential; pressure measurement across a spacing of a few feet along the wellbore. These instruments are often called "gradiomanometers" or "differential manometers." Owing to their linearity, a two-point, air/water calibration is sufficient for such instruments. Furthermore, the resolution of the tools is higher than that of the gamma-ray densiometers. However, at high fluid velocities, the apparent density provided by these tools is corrupted by frictional losses in pressure and requires correction. Likewise, in wellbore intervals containing intense fluid turbulence, the apparent values are again corrupted and are uncorrectable. Finally, the apparent density must be corrected for deviation of the wellbore from the vertical.

A pressure-gradient density log is shown in Fig. 4.6. Near the top of the figure, about halfway across the chart, is a trace labeled "Density." This trace is scaled from 0 to 1.5 g/cm^3. Below the perforations, there is a stagnant mud column in the wellbore; the mud's density is somewhat in excess of 1.5 g/cm^3.

Spent acid and formation water enter the wellbore at the bottom perforation, Depth 1, where a density of approximately 1.2 g/cm^3 is measured. This density continues to the perforations at Depth 2. At Depth 2, spent acid, water, and a small amount of gas enter the wellbore and mix with the flow from below. After mixing, the gas contributed at Depth 2 causes a density decrease from approximately 1.2 g/cm^3 to approximately 1.1 g/cm^3. The density remains at that level to Depth 3.

Additional entries of spent acid and gas at Depths 3 and 4 result in further, slight reductions in density.

Major gas entries at Depths 5 and 6 reduce the density to less than 0.3 g/cm^3.

Redistribution of phases occurring in the 20 ft above 9,300 ft results in a density averaging approximately 0.4 g/cm^3 below and within the tubing. The actual density of the gas/liquid mixture decreases significantly in the tubing stream relative to the mixture density in the casing just below the end of tubing. This happens because the gas holdup in the stream in the tubing increases in response to the increase in fluid velocity. This significant decrease is hidden on the density trace by the increased frictional loss associated with the increased velocity.

The four liquid entries and the two major gas entries correlate with the temperature behavior. Please refer to the discussion of the temperature-logging tool for a detailed interpretation of the temperature data.

4.8.5 Fluid-Capacitance-Logging Tool. The tool includes an inside dielectric probe located on the tool's axis. The probe is surrounded by an outside housing that is open to the wellbore fluid. Together, the probe, the housing, and the fluid constitute an electrical capacitor, the capacitance level of which depends on the particular fluid, or fluids, within the capacitor.

Circuitry within the tool is connected to the electrical capacitor, with the result that the circuitry generates an oscillating signal that varies inversely with the capacitance level. Water has the greatest capacitive effect, resulting in the lowest frequency. Gas has the least capacitive effect, resulting in the highest frequency. The frequency with oil is intermediate to those of water and gas. However, the oil frequency is much closer to the gas frequency than to the water frequency. Consequently, the tool distinguishes principally between water and hydrocarbons.

Preferably, the tool is calibrated at the surface in produced water from the well, establishing the trace for water. Normally, the recording system is adjusted so that the water trace is at the left edge of the track. Air customarily establishes the trace for gas. Normally, the recording system is adjusted so that the air trace is at the right edge of the track. If the well produces any oil, the tool can be calibrated in produced oil, establishing the trace for oil. Sometimes tap water is used to establish the water trace.

Obviously, the tool poses no hazard to personnel who are exposed to it at the surface. If the tool is dropped into the well and it must be left there, it is not necessary to cement it over, as with a nuclear tool. Furthermore, the recording sensitivity can be greatly increased above normal sensitivity because the tool produces a signal that is "clean" (free of statistical events), unlike a nuclear tool. At such an increased sensitivity, the tool can detect the slightest "whiff" of hydrocarbon that passes close enough to its sensor. With the sensitivity increased, the tool also can detect very small amounts of water dispersed in oil.

A small gas entry into water looks to the fluid-capacitance tool just about the same as a small oil entry. Whereas the small oil entry, because of its low density contrast with water, changes the fluid density only slightly, the small gas entry, because of its low density relative to water, changes the density log significantly. Thus, by a comparison of the two log types, an analyst can fathom the nature of a hydrocarbon entry.

From the prior discussion, it is obvious that the fluid-capacitance tool can distinguish between water and the two hydrocarbons, but it cannot distinguish one hydrocarbon from the other. Also, the tool has a very nonlinear response over the range from water to hydrocarbon. During use downhole, there can be a calibration drift because of filming of the housing or the dielectric probe, or both. If the drift is severe, the film possibly can be removed with the tool pulled into the tubing, where the velocity of flow may be high enough to remove the film.

In a production well, the tool should be logged down at a logging speed between 20 and 30 ft/min. Maintain a constant logging speed, and use the same speed for all passes. The log should begin at a location above the perforations and end at the deepest depth that can be reached. In a slugging or churning multiphase flow, the log may show variable behavior, even in intervals that are not perforated. In that case, another logging run is advisable to establish the degree of repeatability. If results are less than desirable, a stationary measurement can be time-averaged at each selected location. Usually, a log is run with the well shut-in after flow. If the well has been shut in before logging, a shut-in log can be recorded, but the well must flow for 2 or 3 hours before the first flowing log.

Fig. 4.11 pertains to a well producing 3,520 RB/D at 68% oil and 32% water. Notice the shut-in log (left trace); at the bottom, below the perforations, the water response is near the left edge of the track. At 8,250 ft, the log shows a water/oil interface in the wellbore. In the oil above the interface, the response appears near the right edge of the track. In gas, the response would be approximately 2,350 Hz.

Below the perforations, the flowing log shows a water response indicating stagnant water. Across the bottom perforations, the log shifts somewhat to the right, indicating some contribution to the oil production.

At 8,420 ft, there is a spike in the oil direction caused by perforations which jet oil directly at the tool's sensor. Just above the spike, the log is somewhat farther in the oil direction than it is just below, identifying the additional oil in the wellbore.

Near the top of the upper perforations (8,400 ft), there is a major shift in the oil direction. Moreover, the log response persists from this location to the end of tubing. This means that the major contribution to the oil production is from the top part of the upper perforations. Above 8,350 ft, the elevated fluid velocity within the tubing results in the oil being more effective at sweeping the water out of the pipe's cross section than it is in the casing. The reduced presence of water across the tubing cross section results in a shift of the log in the oil direction. The presence of the water production is indicated because the log never shifts as far right as the oil response identified by the shut-in log.

Note that in Fig. 4.11, the flowing trace in the tubing crosses the oil/water contact on the shut-in trace at approximately 62% of the total deflection from water to oil. If the tool's response was completely linear in holdup, then the flowing trace would cross at 68% of the total deflection (i.e., at a point slightly closer to the oil frequency). Unfortunately, the "calibration"

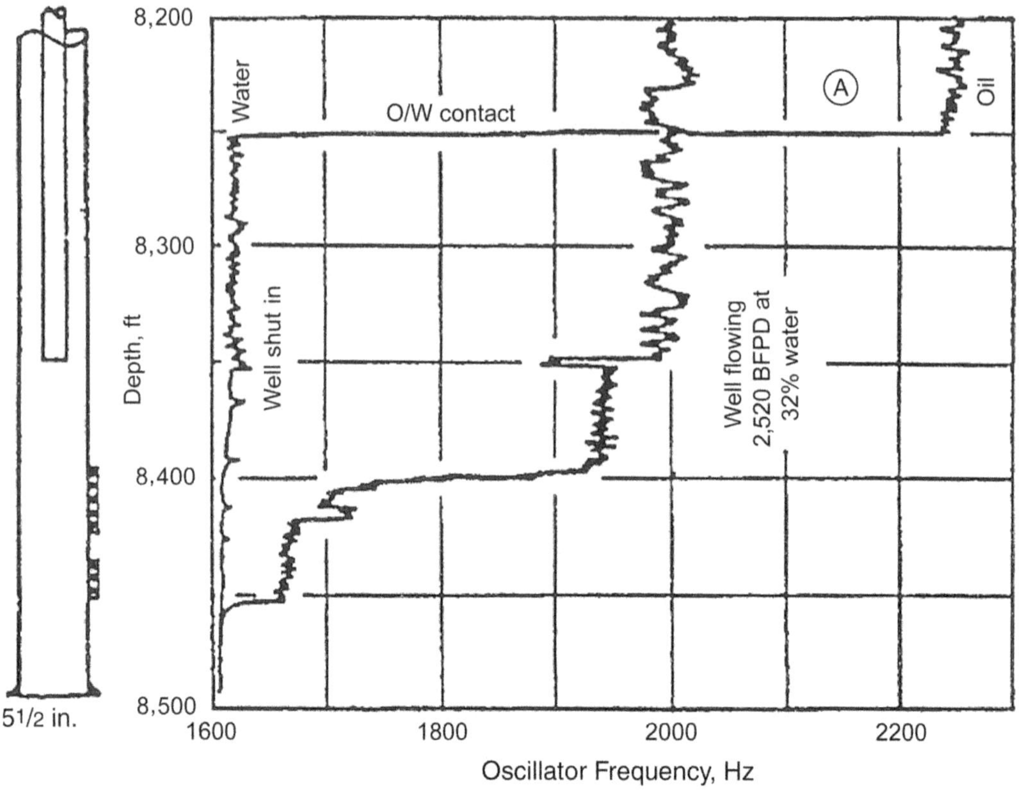

Fig. 4.11—Fluid-capacitance surveys from a well that produces 2,520 RB/D, 68% oil, 32% water.

for these instruments depends upon the viscosity of the oil owing to the filming of oil on the electrode. The smaller the diameter of the electrode, the larger this effect. In gas/water flows, water tends to film the electrode instead, which biases the "calibration" toward water.

4.8.6 Unfocused Gamma Ray (Gravel-Pack) Density Logger. The tool incorporates a compacted slug of Cesium-137 near the bottom of the device. A gamma-ray detector, located approximately 20 in. above the slug, responds to incident gamma radiation. A counter determines the counts/min (intensity) of the gamma rays; this information is transmitted through the logging cable to the surface, where the count rate is plotted against depth.

From the Cesium-137 source, some of the gamma rays are transmitted to the detector through the tool body ("direct" transmission), some by the wellbore fluid between the tool and the casing ("indirect" transmission), and some through the material outside the casing (also "indirect" transmission). Not all of the gamma rays from the source reach the detector because of backscattering by the wellbore fluid, but the majority of the detector's response is attributable to this transmission.

As in the focused detectors, a scintillation crystal is preferable, but Geiger tubes are used in many tools. Preferably there are eight tubes, but tools of lesser quality may have as few as three.

When the tool is at the surface, radiation safety procedures should be strictly followed. Needless exposure to the radiation from the tool should be avoided. Logging-company personnel should have current radiation training and certification. Because of the long half-life of Cesium-137, the legal restrictions on the use of the tool vary from state to state and country to country. If the tool is dropped in the well or becomes stuck, it must be retrieved or cemented over.

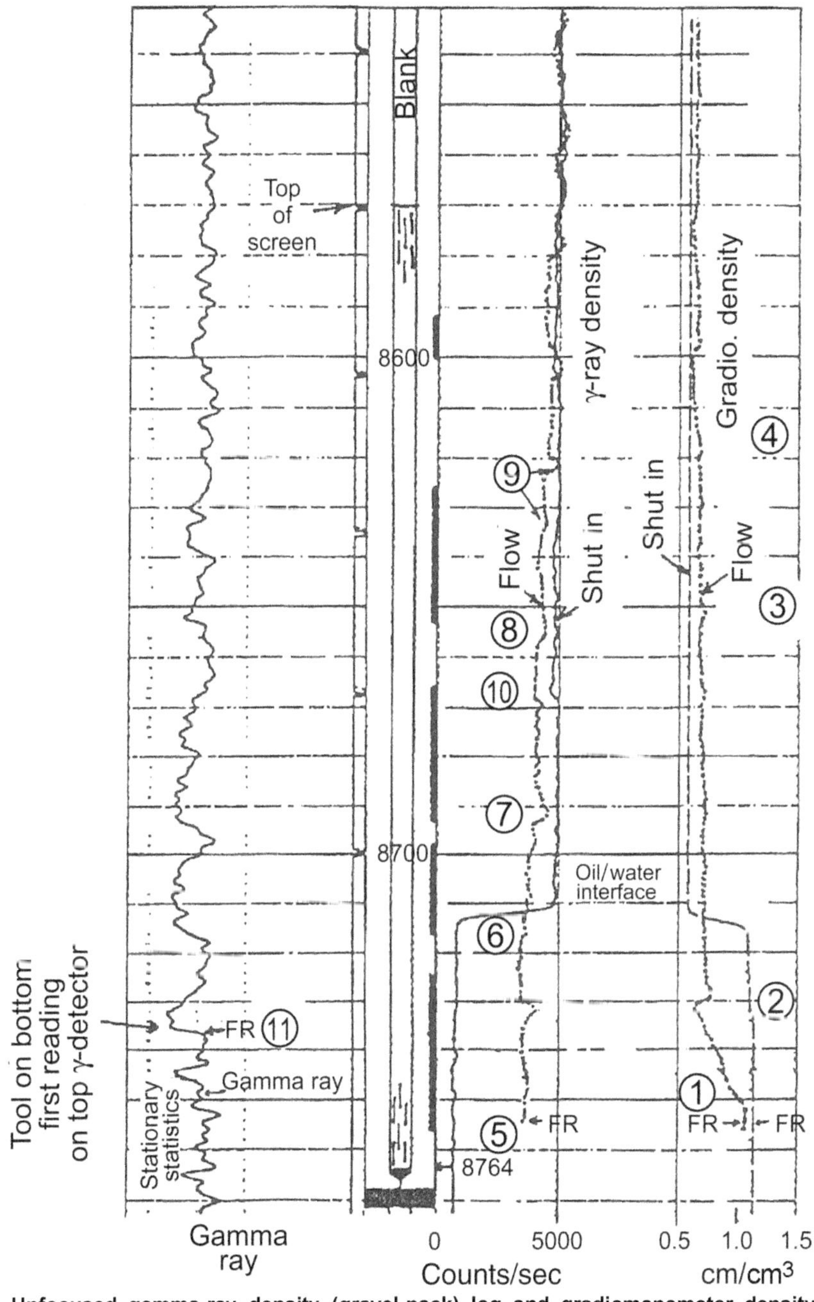

Fig. 4.12—Unfocused gamma-ray density (gravel-pack) log and gradiomanometer density log on a 2,264-BOPD cased gravel pack (well deviation 38°).

Logging-speed and shut-in times for a shut-in survey are the same as recommended for the focused tools.

An example of an unfocused gamma log appears in **Fig. 4.12**. A gradiomanometer (pressure-gradient) density log also is shown. Logs were recorded with the well flowing and shut-in.

During shut-in, both tools identify an oil/water interface at Depth 6. The gradio shows that the density of water in the screen below the interface is approximately 1.1 g/cm^3. Although the gradio responds only to fluid inside the screen, 1.1 g/cm^3 is also the density of the fluid in the porosity of the pack below the interface, because this porosity is water-filled during shut-in. Above the interface, the oil density in the screen is approximately 0.6 g/cm^3; although the gradio surveys only within the screen, the density of the fluid in the porosity of the pack is also 0.6 g/cm^3, because this porosity is oil-filled during shut-in.

Because gamma-ray count rate is inversely related to density, the count rate on the shut-in unfocused log is lowest in the water below the interface. Whether the transmission is direct or indirect, the transmission is lower and the count rate is lower when the density of the fluid in the transmission path is higher.

Above the interface, the unfocused gamma ray shows its greatest count rate in the oil in the screen and pack.

During flow, the gradio shows essentially water in the screen below Depth 1. Farther down, at Depth 5, however, the unfocused gamma ray shows a much higher count rate than during shut-in. Because the screen at this depth is water-filled, as during shut-in, the increased response implies more transmission of gamma rays through the pack than during shut-in. More transmission through the pack is attributable to a lower density fluid in the pack. Thus, oil is present in the pack on the high side of the casing at Depth 5, but it is not in the screen. From Depths 1 to 2, the gradio shows decreasing density in the screen. This means that oil enters the stagnant water in the screen between Depths 1 and 2. Consequently, the oil in the pack at Depth 5 flows up the pack and enters the screen between Depths 1 and 2.

There is an oil-jet entry in the screen at Depth 2. Correspondingly, the gradio shows a slight spike toward lower density. The unfocused gamma ray shows a spike toward higher count rate because the oil jet lowers the density around the tool, causing more transmission of gamma rays to the detector.

At Depths 6, 7, 8, and 9, the unfocused gamma ray shifts somewhat toward higher count rate. From Depths 6 to 7, the gradio, however, shows no change of the density of the fluid in the screen. The same is true for Depths 10 to 9. Consequently, the shifts toward higher count rate of the unfocused log are attributable to increased transmission of gamma rays through the pack. Each shift, then, implies an increased presence of oil in the pack, and thus, an entry of oil to the pack at each of the four depths. At Depth 4, a slight decrease of the gradio response indicates that oil from the four entries flows up the pack and enters the screen over the 10 ft immediately above this depth.

Both logs show little or no contribution from the top, short interval at 8,600 ft. The gradio shows no change of the density of the fluid in the screen, and the unfocused gamma ray also shows no change, implying that no change occurs in the pack.

At Depth 10 (8,670 ft), the unfocused gamma ray response decreases on both the flowing and shut-in logs. During shut-in, the screen and the pack are both oil-filled above and below Depth 10; thus, the decrease during shut-in cannot be explained by a change of the density of the fluid, whether in the screen or in the pack. Also, the flowing gradio response is unchanged at Depth 10; thus, the decrease during flow cannot be attributed to a change of the density of the fluid within the screen. Instead, the decrease is attributable to a change in the porosity of the pack, with the porosity decreasing at Depth 10, resulting in an increased density of the pack above Depth 10, relative to the pack density below this depth. Above Depth 10, the increased pack density results in lower transmission of gamma rays through the pack and, thus, a lower count rate. Consequently, the decrease appears on both the flowing and shut-in logs.

In this example, it is important to note that the unfocused gamma ray, by itself, cannot distinguish between entries to the screen and entries to the pack. In addition to the unfocused gamma ray, this distinction requires a gradio log, as in this case, or a focused gamma-ray log.

The comparison of the two density logs has shown that, at least in the lower parts of the hole, oil moves upward as a separate layer on the high side of the hole. The discussion in the

Appendix shows that in such situations, the count rate from the unfocused density tool is linear in the fraction of water occupying the casing's cross-sectional area. If this concept is applied to the data in Fig. 4.12, then the two unfocused-density trace shows a constant water holdup of 28% below Depth 6 during flow. This quantitative feature of the tool can be exploited in the high-angle holes discussed in the next section of the text.

4.8.7 Special Logging Tools for Fluid Identification in High-Angle Wells. Layered flow often occurs in high-angle wells. (i.e., a water layer in the lower part of the wellbore cross-section, an oil layer above the water, and a gas layer at the upper part of the cross-section). While the tools used in vertical wells have proven effective in high-angle wells on most occasions, special tools have been developed for studying two- and three-phase flow. These tools make use of arms to position electrodes across the casing diameter. Consequently, they are "blind" to flow outside a screen or perforated liner. The brief descriptions of these tools that follow are based on the limited published information and personal discussions with suppliers.

Halliburton Gamma-Ray Backscatter Gas-Holdup Tool.*** The purpose of this instrument is to distinguish liquid holdup (the percentage of the wellbore cross section occupied by liquid) from gas holdup (the percentage of the cross section occupied by gas) and provide a cross-sectional average for the mixture.

The gamma-ray source is the same Cesium-137 isotope used in the gamma-ray densiometers. A shielded scintillation crystal detector receives its signal from radiation backscattered by the density of material around the tool. The tool is used primarily to differentiate between gas and liquids. The contrast in density between gas and liquid causes a large change in signal level, while the small difference between oil and water densities causes a correspondingly small change in signal level. The value used for liquid density in the holdup calculation does not make much difference as long as it is in the range common to liquids. Thus, the corresponding error in holdup is at worst 10 to 15% and probably better because only backscattered radiation is involved. To compensate for pipe size, variable spacing between source and detector and/or shield should be used to eliminate backscatter from the formation outside the pipe.

For a limited range of gas-holdup values, flow-loop data for the backscatter tool show a linear response over the range of the data. Sensitivity to small gas concentrations is sufficient. These data do not shed any light on the matter of compensation for variable pipe size, however. If formation backscatter is high, then the gas point will show a much higher count rate and the small difference in liquid signal at zero gas holdup will become a relatively large difference. In-situ calibration in a shut-in well is therefore recommended for this tool if it is to function quantitatively.

Baker Atlas Multi-Capacitance Flowmeter. This centralized tool includes a spinner flowmeter at its bottom end. The wellbore cross section is considered as being divided into eight levels, and a positive orientation section of the tool ensures that the levels are perpendicular to true vertical.

Twenty-eight capacitance sensors are deployed on "wings" from the tool in such a way that there are capacitance sensors spanning each of the eight levels. An array of capacitance sensors spans levels at a first position along the tool's axis. Another array of sensors spans levels at a second axial position.

During logging, the various capacitance measurements for each level are recorded and converted to values for the gas, oil, and water holdups. An across-the-wellbore, bidirectional velocity profile is constructed from transit-time measurements of the capacitance sensors on Levels 1 and 2 (bottom levels) and 7 and 8 (top levels). The construction involves cross-corre-

[*] Personal communication, Halliburton (1993).

lation of some of the sensor responses. The spinner flowmeter provides velocity information related to the wellbore centerline. Stationary measurements can be made as well.

The holdups, the velocity profile, and the cross-sectional area of the wellbore are combined to determine the flow rate of each fluid as a function of the axial position along the wellbore.

A possible limitation is that capacitance sensors sometimes film in heavier oils; in turn, an oil film biases the capacitance measurements toward oil and gas and away from water. Another possible limitation is that the distinction between oil and gas may not be nearly as great as the distinction between oil and water. Finally, the concept of velocity determination from two measurements at the same radius along the axis supposes perfectly layered flow free of circulatory velocities. This condition, however, is at odds with the assumption of capacitive-event creation for cross-correlation.

Schlumberger Flowview. In a horizontal well, five types of segregated flow are usually defined. These are: stratified with a flat interface, stratified with a wavy interface, stratified with a bubbly interface, lighter phase slugging over the heavy phase, and one phase existing as bubbles in the other phase.

Because of segregated flow, the new tool string consists of both traditional and nontraditional sensors. Some of these sensors average over the entire cross section of the casing, whereas others make measurements at different locations in the cross section. The various sensor sections are referred to as "items" in the description that follows. The tool string described next is for mainly oil/water flow. For three-phase flow, a larger tool string, called the Flagship (Schlumberger), is available.

Item 1 is a full-bore spinner. This item gives information about composite fluid velocity, which in multiphase flow is difficult to relate to fluid-flow rates even after an image of the layers in the well has been generated by an imaging sensor such as the Flowview Plus (Schlumberger).

Item 2 is the Flowview Plus. The main results from this tool are eight-electrode measurements of water holdup to provide an approximate image of how the fluids are segregated in the cross section. The fluid image greatly aids in the interpretation of the spinner response.

This item consists of two Flowview tools, combined so that one is rotated 45° to the other and separated from it by at least 3 ft. Each Flowview makes four independent measurements of borehole fluid holdup and bubble counts, distributed in different quadrants of the pipe cross section. The tool is self-centralized and uses matchstick-sized electrical probes to measure the resistivity of the wellbore fluid—high for hydrocarbons and low for water. The probes are located inside the tool's four centralizer blades to protect them from damage. The opening of the blades positions each probe at midradius in the casing. The tool can run up to $9^5/_8$-in. casing.

Each probe is sensitive to the resistivity of the fluid that impinges upon its sharp leading edge. If the fluids are distinct and not in a fine droplet emulsion form, and the bubble size is larger than the tip of the probe (less than 1 mm), then both water-holdup and bubble-count measurements may be obtained from the output of the probe. Local water holdup is equated to the fraction of the time that the probe is conductive, whereas bubble count comes from the average frequency of the output. The local water holdup from each of the eight probes is used to generate the water/hydrocarbon distribution in the well's cross section.

Item 3 is the Reservoir Saturation Tool. This tool is a pulsed-neutron tool that can be operated in lifetime mode or spectral carbon/oxygen mode. Its main applications are for estimation of oil, gas, and water holdups and determination of water-phase velocity by oxygen activation.

Item 4 consists of pressure, temperature, and deviation sensors. In addition to the spinner, these are the sensors found on traditional production-logging tool strings. The new string, however, locates the temperature sensor nearly 15 ft from the end of the string, making the measurement subject to interferences from fluid mixing by tool movement.

From a combination of the holdups, the cross-sectional area of the wellbore, and the fluid velocities, the rates of the individual phases are estimated as a function of position along the wellbore's axis.

The small resistivity probes tend to film with oil when in use in heavier oils and with water in gas/water flow. In these cases, the measurements of the affected probes are biased toward the filming phase.

4.8.8 Diverting-Spinner Flowmeter. These are the most accurate of the spinner devices when low total rates and multiphase flows occur. The stream is diverted through the tool's barrel, thereby raising the velocity of flow and increasing the sensitivity to the point that diverting spinners can detect rates as low as 10 to 15 B/D. A typical $1^{11}/_{16}$-in. tool has a barrel ID of approximately 1.45 in. A flow of 10 B/D results in a velocity of 3.4 ft/min inside the barrel. Because of the limited clearance between the spinner and the barrel, this velocity is enough to overcome friction and turn the spinner.

Furthermore, a flow of 100 B/D passes through the barrel at 34 ft/min, which is sufficient to start the homogenization of the flow, which eventually eliminates phase influence. In casing, a rate of 2,000 B/D is needed to have the same effect around a continuous spinner. Another benefit to multiphase-flow application is that the tool can be calibrated directly for such flow. As long as the diversion is effective, casing size is not a parameter in the calibration, and the tool can be calibrated in the low-rate range where phase bias is still important even inside the tool.

One type of diverting flowmeter is the hem-packer. The lower end of the diverter contains a coated fabric "skirt" with a 1-in. hem at the bottom. The skirt is opened by the metal struts to which it is attached, so that the momentum of the stream can effect a reasonable seal. A downhole pump inflates the hem with oil contained in the tool body, resulting in a seal that diverts practically the full stream through the barrel and past the spinner. A downhole motor expands and retracts the struts. The response of the tool to single-phase flow is nearly linear. Two interchangeable spinner elements are available—one with a small pitch for high sensitivity at low rates and a second with a larger pitch for rates up to approximately 2,400 RB/D. Because of its fragility, the tool never found widespread use, but is still available at some locations.

Another type of diverting flowmeter is the basket (metal-petal) flowmeter. On the lower end, the basket is opened by motorized compression of its several struts, each of which is tack-welded on its underside to a corresponding petal cut from sheet metal. As the struts are compressed to open the basket, the petals slip over each other so as to maintain some overlap even when the basket is fully opened. As the basket opens more to accommodate larger casing diameters, overlap between the petals decreases and leakage through the basket increases. Strut travel is limited to prevent excessive opening and leakage. Standard $1^{11}/_{16}$-in. tools accommodate casings as large as 5.5 in. nominal. Some $1^{11}/_{16}$-in. tools can accommodate 7-in. casing. This is an operational feature that should be checked carefully when ordering this service. Also, it is prudent to inquire whether the tool has been calibrated in a section of pipe of diameter close to the intended application. If not, the service company may not know the effect of leakage on spinner response.

At the point where the outside of the struts meets the casing's inside wall, the metal petals (which are attached to the inside of the struts) cannot deform enough to effect a complete seal. Because of leakage around the struts the low flow-rate response of the tool is nonlinear. Attaching the petals to the outside of the struts would improve the seal, but would wear the petals quickly during use, destroying the tool's main advantage, its ruggedness.

There is very small clearance between the spinner and the ID of the barrel. This feature assures almost no diversion of flow around the spinner. On the other hand, small particles of debris can plug the tool.

As the spinner turns, it generates a specific number of voltage pulses per revolution. Thus, the pulse rate from the tool can be transmitted through the logging cable for surface recording and determination of the corresponding revolutions per second. The number of pulses per revolution varies considerably from manufacturer to manufacturer. When ordering the service, it is recommended to inquire as to the number of pulses per revolution (more is better).

Typical basket flowmeters are rated for temperatures in the range of 320 to 350°F and for pressures in the range of 15,000 to 20,000 psia. The 1.70-in. tool typically accommodates 3,000 B/D (maximum); the 2.25-in. tool, 5,000 B/D; and the 3-in. tool, 8,000 B/D. In each case, the tool length is approximately 60 in.

Measurements are made with the tool stationary. In a production well, the tool is lowered to the deepest measurement depth and then opened. After recording the measurement at this depth, the tool is pulled up (while open) to the next measurement depth, and so on. When opened in a production well, the tool can be damaged considerably by movement downwards. When considering a diverting flowmeter for an injection well, the user should inquire of the service company whether its tool is capable of such operation.

The chance of having a diverting flowmeter stick in the hole is greater than with a continuous flowmeter. In a sandy flow, for example, the basket recesses may plug, in which case the basket cannot close downhole. If the tool is stuck, the cable can almost always be pulled loose from the cable head and retrieved. If the flowmeter is stuck in the casing, the least expensive approach may be to jar the tool to the bottom of the hole and leave it. If stuck in the tubing, it may be necessary to pull the tubing. Horizontal wells typically have dirt along the bottom of the wellbore; the flow carries the dirt into the diverting flowmeter, which usually plugs.

Diverting flowmeters are easily calibrated in a flow loop, in both single- and multiphase flows. While continuous spinner flow-loop calibrations require many correction factors for their use, such is not the case for the diverting flowmeter.

Only a few companies have extensive flow-loop facilities. If flow loops are not used for calibration, the customary approach is to apply the same calibration to all types of flow or, even worse, to use the same sensitivity coefficient for all situations. Flow-loop facilities at various universities do offer an alternative source of equipment.

Flow loops generally cannot be pressurized or heated, so the calibrations they produce for diverting flowmeters are not exactly correct for multiphase flows, especially gas/liquid. The error, however, is much less than with a corresponding calibration for a continuous flowmeter. Further, some degree of correction is possible for a diverting flowmeter calibration, because diversion of flow around the spinner element is not a problem.

A single-phase calibration of a packer flowmeter in liquid flow is linear, showing constant sensitivity (slope), with no change of the response because of casing size. The corresponding calibration of a basket flowmeter is nonlinear because of leakage around the struts and through the basket. Moreover, the degree of leakage increases with the casing size. For a packer tool in water flow, the RPS-flow rate relation involves a threshold rate (below which the spinner does not turn); for rates above the threshold spin rate, the spin rate increases linearly with rate. For a basket tool in water flow, the initial response (at lower rates) is nonlinear because of leakage around the struts. The final response (at rates approaching maximum) is nonlinear because of leakage through the overlap between petals (i.e., flow through the basket itself). Because of leakage around the struts at lower rates, the basket also has a threshold. Unlike the packer calibration, there is no single sensitivity to the basket flowmeter calibration. The sensitivity is lower at lower rates, then increases with rate until it reaches a maximum at midrange, after which it decreases with increasing rate.

For a packer flowmeter in air at atmospheric pressure, the earlier portion of the tool's response is nonlinear in gas rate. The lower the gas rate, the less the sensitivity because of the tendency of the gas to "spiral" through the spinner element rather than turn it. With increasing rate, the response becomes linear (constant sensitivity). The value of this sensitivity, however, is approximately one-half the sensitivity in water because of the presence of spiral-velocity components in the low-pressure stream. At normal wellbore pressure, gas density is sufficient to largely eliminate these distortions; however, tool sensitivity remains slightly lower than that for liquid.

In a multiphase flow, the diverting flowmeter should be calibrated at light-phase cuts of 0, 20, 40, 60, 80, and 100%. For each cut, the total flow rate should vary from the minimum the tool can sense up to the maximum the tool can accommodate. This process generates a family of calibration curves, with the total flow rate on the horizontal axis, the light-phase cut as the parameter of the curves, and the RPS response on the vertical axis. The total flow rate should begin at minimum and then include 100, 200, 300, 400, 500, 700, and 1,000 B/D. Above 1,000 B/D, the total flow rate should increase in 500-B/D increments until the maximum is reached. The calibrations should be performed in a pipe size close to the user's intended application. The deviation angle is unimportant.

When requesting the diverting flowmeter, the user should verify that the service company has a calibration of the tool in the same phases as those that flow downhole in the user's well, and in a pipe size close to that in the user's intended application. Unfortunately, very few companies have such calibrations. If the calibration is at a different deviation angle than in the user's well, this difference is not important.

An example of the use of a diverting flowmeter appears in **Fig. 4.13**, which also shows two temperature traces. A 24-hour shut-in temperature log, recorded before the flowing log, is dashed. A flowing temperature log, after 4 hours of flow, is solid. An open set of perforations is shown near the bottom of the figure. Six squeezed perforation sets are above the open set. The stationary measurements from the diverting flowmeter are shown at locations labeled A through D, top-to-bottom.

Above Depth F, the flowing temperature log is initially constant. A water flow of sufficient rate that the flow carries its temperature up with it originates at Depth F. At Depth D, the diverting flowmeter response is 7.1 RPS, showing that the water from F enters the pipe through the open perforations.

At Depth E, the flowing temperature log cools over a short interval, indicating that the water from F mixes with colder water entry at E. The size of the mixing signature at Depth E indicates that the entry accounts for 43% of the total flow, with the remaining 57% produced from the open perforations at Depth F. The flowmeter response at Depth C is 11.7 RPS, also showing that the colder water enters the pipe through squeezed perforations.

Above C, the flowmeter response is 11.5 RPS at Depth B and 11.6 RPS at Depth A. Thus, no more water enters the pipe above C. For a linear response from the diverting flowmeter, 7.1 RPS is 61% of 11.7 RPS; therefore, 61% of the water (approximately 610 BWPD) is from F, with the remaining 39% (approximately 390 BWPD) entering through the squeezed set. This is all the information the diverting flowmeter can yield, because the flowmeter responds only to flow in the pipe. On the basis of this information, one would assume that the 390 BWPD flow comes from the formation at E.

The flowing temperature log, which responds to flow within and behind the pipe, shows that the assumption just made is fallacious. Above depth G, the flowing temperature log is more constant (less slope) than it is below G. Above G the flowing temperature log responds to the full 1,000 B/D water flow inside the casing. At this rate, the stream cools slowly as it moves upward. Immediately below G, the temperature increases more rapidly with depth. This behavior is consistent with an "entry" at depth G that "pulls" the temperature toward its shut-in value. However, the spinner fails to show such an entry at this depth. Instead, the 390 B/D water flow channels down from G and enters the pipe through the squeezed set at Depth E, creating the mixing signature where it mixes with the flow from F. Therefore, the source of the

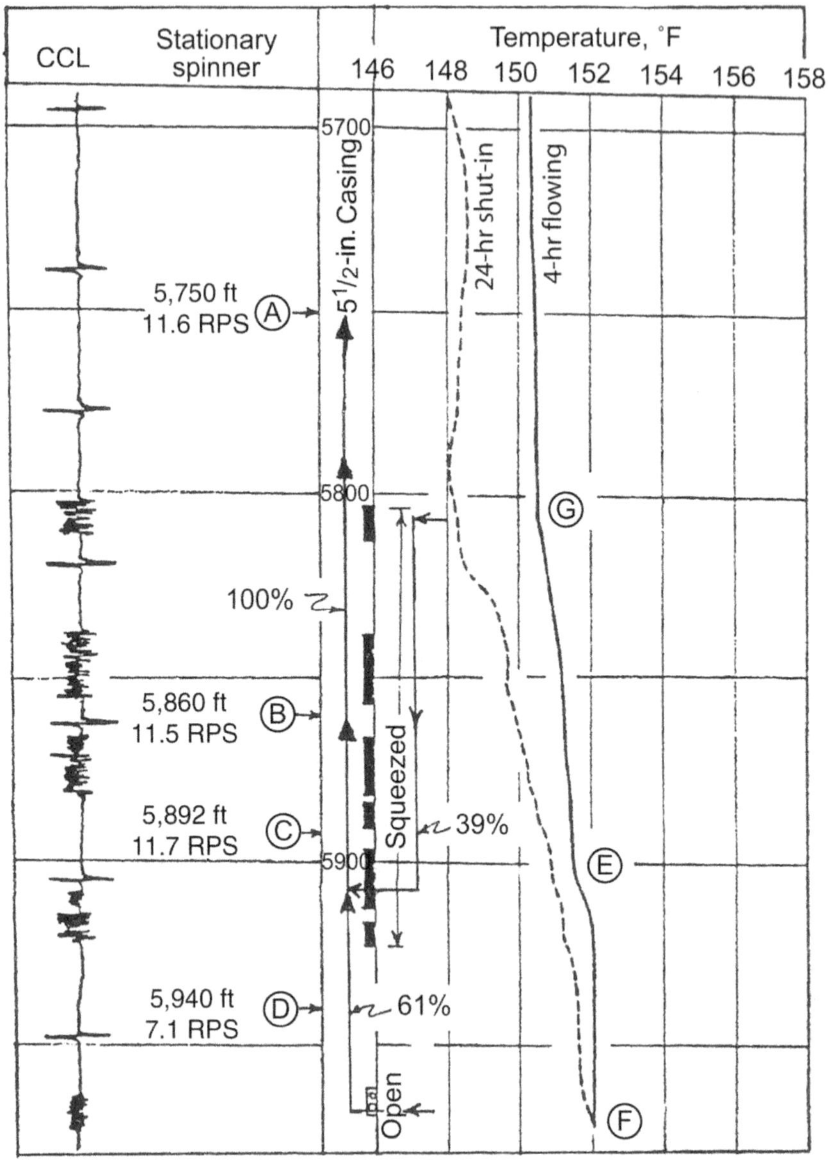

Fig. 4.13—Spinner and temperature survey in well producing 1,000 BWPD with minor oil cut.

colder water entering through the squeezed set is the formation at G, and not the formation at E. Between Depths E and G, water flows up inside the pipe at 1,000 B/D, while the 390 B/D water flow from G travels down behind the pipe. Because the net upward flow is less than 1,000 B/D and the temperature tool responds to the flow within and behind the pipe, the temperature trace between E and G cools more rapidly with decreasing depth than above G, where the net upward flow is 1,000 B/D.

4.8.9 Continuous and Fullbore Spinner Flowmeters. There is no generic difference between a "continuous" spinner and a "fullbore" spinner. In the case of the fullbore, the spinner element folds into a diameter no greater than that of the tool when in the tubing, but expands into a larger diameter for surveying in the casing. The continuous spinner does not have this capability. The difference between the two is too small to justify a separate discussion of each.

The continuous meter derives its name from the need to move the tool fast enough to overcome frictional torque and start the spinner element rotating. It also derives its name from the in-situ calibration procedure that uses logging runs at several different cable speeds with the well shut-in at the surface. Neither continuous nor fullbore, however, can provide a log that is quantitative whenever the fluid velocity is sporadic, that is, changing in the logged interval.

The continuous and fullbore tools share three features. First, the spinner element on each is at the very bottom of the tool string. If the temperature tool is run in combination with the spinner, then the thermometer's sensor will usually be located in the tool string above the spinner. Thus, the sensor will be 4 to 6 ft above the bottom of the string. Moreover, this 4 to 6 ft will include at least one centralizer. The mixing caused by the passage of the centralizer in front of the temperature sensor will decrease the vertical resolution of the temperature log by a few ft. Even so, the vertical resolution of the temperature tool to localized entries is still far superior to that of the spinner tool, and the temperature tool should be included in the tool string.

The second common feature is the presence of at least two centralizers in any tool string containing the spinner flowmeter. The centralizers ensure that the spinner element samples the same location in the wellbore's cross section at each depth. This consistency is necessary for the same tool calibration to apply at each depth and for the relative profile to be representative of flow rates.

The third shared feature is that both tools use a four-blade propeller-type (or turbine-type) spinner element. Although the design of the spinner element may vary, the four-blade feature is retained.

The spinner element can rotate either clockwise or counterclockwise (as viewed down the tool barrel). The direction of rotation depends upon the movement of the fluid relative to the barrel of the tool, that is, upon the direction of fluid movement as seen by "rider" on the tool. Usually, the pitch of the spinner is such that relative movement of fluid up the barrel causes the spinner to turn clockwise, whereas relative movement of fluid down the barrel causes a counterclockwise rotation of the spinner. Consequently, movement of the tool downward in stagnant fluid causes relative movement of fluid up the barrel and rotates the spinner clockwise. Movement upward in a stagnant fluid causes relative fluid movement down the tool barrel and spins the element in a counterclockwise direction. Movement of the tool in a direction opposite to the direction of flow causes the spinner to turn in the sam direction, clockwise or counterclockwise, throughout the logged interval including those intervals with stagnant fluid. Passes made downward in a production well therefore cause a clockwise rotation over the interval from 100% to 0% flow. Passes made upward in an injection well cause a counterclockwise rotation. On the other hand, movement of the tool in the direction of flow at a line speed less than full-steam velocity causes a reversal in the direction of spinner rotation at some depth within the flow interval. The spinner first loses speed as flow velocity approaches tool speed. It then stops when fluid velocity reaches tool speed less the frictional threshold speed for the tool. The spinner remains stopped until fluid velocity changes by an amount that equals two threshold values, at which point the spinner begins to turn again, but in a direction opposite the previous one. As fluid velocity changes further, the spinner maintains its opposite rotational direction. Such a pass should not be used for percentage flow profiling, because two threshold values are "lost" in the record of spin rate.

Many spinners, however, do not record the direction of rotation. Even worse, some spinners have lower sensitivities when rotating counterclockwise. In event the direction of rotation changes and the spinner does not record it, the change can be recognized from its signature: a drop of the RPS to zero, followed by a resumption of the RPS to above-zero values.

Both continuous and fullbore tools are intended for quantitative use only in flow streams having a single component of velocity directed along the axis of the tool. Furthermore, the stream should be a single-phase or a high-rate multiphase (10,000 B/D or more). In the multiphase case, the well should have low deviation angle. The meters are designed to function quantitatively in environments such as that indicated in **Fig. 4.14**. This spinner pass downward is from a cased, vertical well producing single-phase gas from three isolated perforation sets, the deepest of which is a short interval at 11,700 ft. In the left track, the cable speed trace shows a logging speed of 56 ft/min. This speed produces the 1.2-RPS rate of spin below the deepest perforations. This part of the RPS trace is labeled "zero-flow reference" because the RPS response is caused by moving the tool downward through the stagnant fluid below the deepest set.

The completion in the figure has at least 50 ft of separation between successive perforation intervals. This is sufficient distance for any tangential velocity components associated with an entry at a given interval to die out before additional gas enters at the next interval above. Thus, the spinner attains a constant rotational rate above a given entry that reflects the axial velocity produced by the combined flow from that entry and all others below. In the figure, this constant rate is attained within approximately 10 ft above the entry. The amount of deflection to the right of the no-flow reference is proportional to the flow rate in the wellbore at the respective depth. Under these conditions, flow profiling is done simply from a determination of the fraction that a given stable deflection represents of the total stable deflection above all entries. Thus, the top set contributes approximately two-thirds of the total gas while the bottom set contributes only approximately 15% of the total.

The survey of the figure appears to show simple and direct a spinner record is to interpret. The validity of the interpretation, however, cannot be judged on the basis of the spinner log itself. Most wells that produce only oil or gas will have a stagnant column of mud or workover fluid standing to some depth in the wellbore unless the deepest entry is at a high rate. A wellbore-fluid density log for the present example could show a stagnant water column standing in the wellbore, at least to the bottom of the middle perforation set. If that is the case, then the spinner record above the bottom set is in response to lifting of water by the buoyant rise of gas through it and not a response to the single-phase velocity of gas flow upward. Then, if the response is assumed to be from single-phase gas flow, the contribution from the bottom set will be greatly exaggerated.

Furthermore, the type of completion in Fig. 4.14 is seldom encountered. The type of completion more likely to be associated with spinner logs is shown in **Fig. 4.15**. Here, there is a long bottom interval perforated over 65 ft and a short perforated interval only 5 ft above the bottom one. The figure shows a fullbore-spinner log on the right track and a wellbore-fluid density log on the left track. Both logs were run downward at a cable speed of 200 ft/min with the well producing at a combined rate of 10,000 B/D of oil and water at a 50% water cut. The objective of the logs was to profile the relative oil and water production within the long interval and to determine the relative contributions of each from the short interval.

Below the bottom of the long interval, the wellbore-fluid density is 1.0 g/cm^3, identifying water. Above the bottom of the long interval, the fluid density decreases up to approximately Depth D because of the entry of oil. The increasing presence of oil in the water shows that the bottom interval is productive up to D. Above D, there is very little change of the fluid density, suggesting that the upper part of the long interval and the short interval are not significantly productive.

Below D, the RPS trace in the 7-in. casing is ever-changing. The trace never attains a stable RPS value, which is necessary for quantitative analysis of the flow velocity. This fact alone means that the production from the long part of the bottom interval cannot be profiled in the straightforward manner of Fig. 4.14. Possibly, the absence of a stable RPS response is because

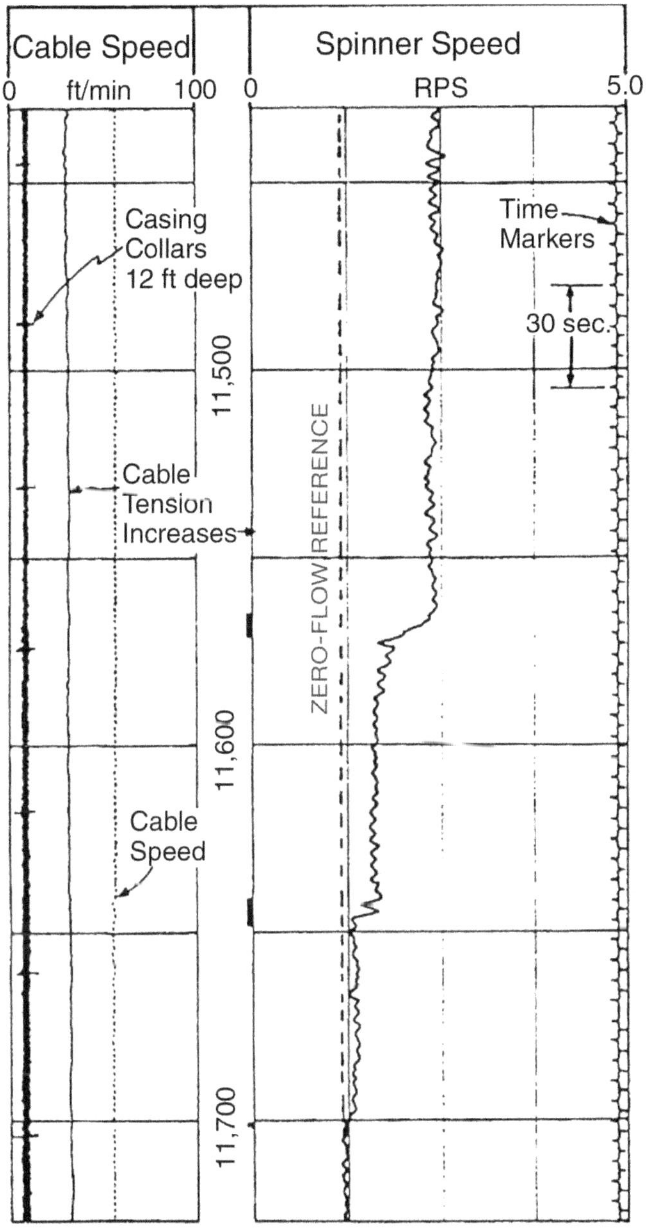

Fig. 4.14—Ideal completion for continuous flowmeter application (courtesy Schlumberger Wireline).

of the high logging speed, which raises the rotational inertia of the spinner element. Quantitative analysis also requires that the flow velocity be entirely axial. But such is not the case in this example. At Depth A, there is a jetting entry, the tangential velocity of which causes a spiked increase in the RPS response. The fluid-density log at this depth decreases for a short interval, identifying oil. At Depth C, the fluid-density trace spikes in response to a fluid jet that appears to put water between the sensor ports spaced 2 ft apart. If this is the case, then the rate is insignificant because the density is the same on either side of the spike. At D, there is a spiked increase in the RPS response resulting from the tangential velocity of a jet entry. This

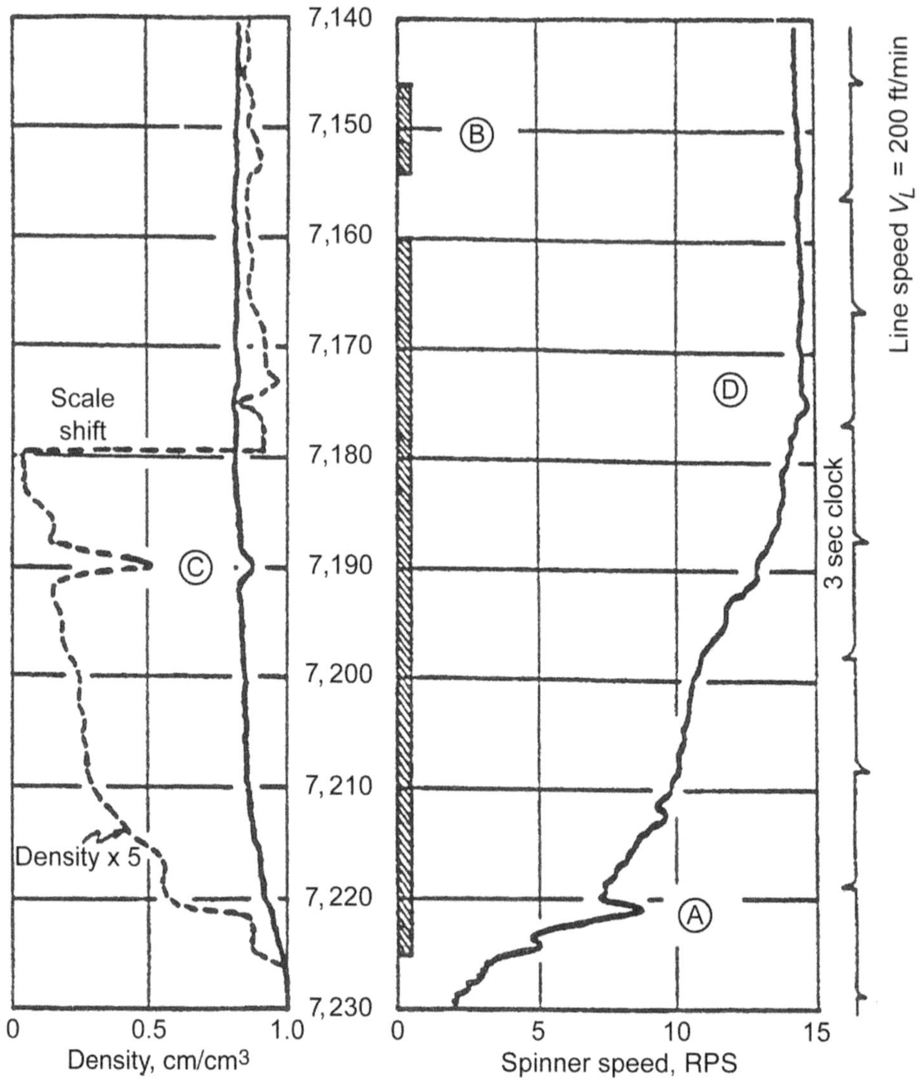

Fig. 4.15—A "fullbore" spinner and "gradiomanometer" density survey from a well flowing 10,000 B/D at 50% water cut.

is not to say that all perforation jets cause an increase of the RPS response. In other cases, a jet entry may have a tangential velocity that decreases the RPS trace. This happens at 7,211 ft, for example.

The density trace shows that essentially all the oil has entered the wellbore by a depth of 7,180 ft. The spinner shows that essentially all the flow is in the wellbore by this depth. Consequently, both oil and water enter below this depth. The reader can apply similar reasoning to conclude that approximately 70% of the flow enters below 7,200 ft, bringing with it a major part of the oil and water production.

Above D, the RPS trace slowly diminishes. This means that the flow at D has a swirling (tangential) component that dies away as the flow moves up. Swirls can require several hundred feet of travel to decay completely. At D, the swirl is contributing to the RPS response; as the swirl dies away, the contribution diminishes and the RPS decreases. In other cases, the swirl may detract from the RPS response; as the swirl decays, the RPS increases. Even if the

RPS trace were to show some evidence of production from the short interval, the presence of the swirl would preclude a quantitative analysis of the flow velocity.

The logging operator used the high logging speed in this example to minimize the distortion of the RPS trace by jet entries, as at Depth A. Actually, this procedure maximizes the distortion by biasing all fluctuations to the high side of their excursions rather than to their average.

The previous two examples should remind the reader that an apparently simple record from a direct measurement can have subtle meanings and may fail to present a complete accounting of the situation.

Continuous tools are available in a wide range of configurations, more so than the fullbore tools. The diameters range from 1⅜-in. to 2⅛-in. The 1⅜-in. versions, with centralizers, should pass through 2-in. tubing. If the tubing includes landing nipples, such as the 1.82-in. size, it is difficult to force the centralizers through the nipples. Safety considerations preclude even an attempt.

The centralizers also come in a variety of configurations. Powered centralizers offer the least problem to entry through tubing. These are closed by strong springs when in the tubing. In the casing, a downhole motor opens the centralizer against the spring force. In event of a failure of the motor downhole, the powered centralizer has a shearing mechanism so that the constriction at the tubing's end can be used to close the centralizer when re-entering the tubing. The chance of sticking a centralized string is greater than that for an uncentralized one.

Some continuous tools have the spinner element inside a bow-spring cage with no additional protection from damage. In others, the element is inside a rigid cage having the same diameter as the tool. Still others place the element inside a short section of tubing having the same diameter as the barrel. The latter are immune to the tangential velocity of a jet-entry; they are not immune to swirl, which has both axial and tangential components. If the flow is multiphase, the trace from such a tool is noisier because of the light phase's tendency to pass through the "chimney."

To measure the RPS of the spinner element, the most common means is a magnet and pickup coil. A narrow magnet is attached lengthwise to a section of the spinner shaft. The magnet rotates under the pickup coil, which is divided into independent segments so that the coil generates a number of inductive current spikes per revolution of the shaft. The resolution, however, does not approach that from tools utilizing an optical sensing assembly, which also detects the direction of rotation. Another variation uses a single sector of pickup coil with three bar magnets embedded in the rotor at unequal azimuth angles. While reducing the tool's resolution, this approach does detect the direction of rotation.

Magnets should be located inside the tool barrel so that iron particles in the wellbore fluid attach themselves to the outside of the tool and do not interfere with rotation of the spinner. Magnets located on the spinner shaft before it enters the barrel can attract iron particles and eventually make rotation impossible.

Most continuous spinners are rated for pressures in the range of 15,000 to 20,000 psi and temperatures of 350 to 400°F. Some tools can accommodate 500°F, but they employ vacuum flasks and thus have a diameter of at least 2.5 in.

Practically all fullbore spinners are copies of the original Schlumberger version with slight modifications here and there. Fullbore tool diameters range from 1½ to $1^{11}/_{16}$ in.

These tools use specific diameter spinner elements to accommodate specific casing sizes. Typically, three different diameters are used to cover casings in the range of 4 to $9^{5}/_{8}$ in. Pressure and temperature ratings are the same as for the continuous spinners, which were stated previously.

The difference in resolution between continuous and fullbore spinners is small; the following comments apply to both types. Neither is very effective for quantitative resolution of low-

TABLE 4.3—TYPICAL RATE RESOLUTION UNDER DIFFERENT FLOW CONDITIONS IN VERTICAL WELLBORES

Type of Fluid in Wellbore	Typical Rate Resolution
Single-phase liquid	60 to 100 B/D
Single-phase gas	200 Mcf/D
Oil and water	300 B/D
Gas and liquid	600 B/D

rate, multiphase production. In the U.S., the spinners' greatest quantitative application is injection profiling. **Table 4.3** summarizes typical rate resolution under different flow conditions in vertical wellbores, although actual resolutions are quite dependent on flow conditions.

These numbers should be viewed in the sense that the given amount of flow is lost, even qualitatively, for entries above the deepest entry. The qualitative resolution of the deepest entry is better than the numbers in the table.

Refer to **Fig. 4.16**, which pertains to the record of a high-quality, high-resolution continuous tool that detects the spinner element's direction of rotation. The well, with three perforation sets, produces 1,800 B/D at 28% oil, 72% water, and no gas. The survey shows two logging runs, one up and one down, with each at 22 ft/min cable speed (left track). There are two respective RPS records (right track). Zero RPS is at the fifth chart division from the left, and 1 RPS is spread over four chart divisions (a sensitivity of 0.25 RPS per chart division), showing the tool's high resolution.

On the up run (dashed trace) in the stagnant fluid below the bottom perforation set, an imaginary observer riding on the tool would perceive the fluid velocity as down the barrel so that this trace records counterclockwise rotation (CCW) below the bottom set. On the down run through the stagnant fluid below the bottom set, the imaginary observer would perceive the fluid velocity as up the barrel; therefore, the rotation is clockwise (CW). Below the bottom set, the RPS of the up run is more irregular than on the down run, because the tool, when jerked off the bottom, requires some distance to reach a steady speed. Thus, the comments below pertain to the down run.

On the down run in the stagnant fluid (which is single-phase formation brine or workover fluid) below the bottom set of perforations, the RPS shows a very steady value of three chart divisions; that is, 0.75 RPS. The tool used for this survey has a threshold velocity of 5 ft/min. This amount of line speed in the stagnant fluid is required to overcome frictional torque and start the spinner rotating. Therefore, the velocity driving the spinner on the down run through the stagnant fluid is 22 – 5 = 17 ft/min. The sensitivity of the tool is 17/0.75 = 22.67 ft/min/RPS. In the stagnant fluid, a sustained defelection of 0.2 chart division would be recognizable on the RPS trace. The single-phase sensitivity of this tool is

$$0.2(\text{chart div}) \times 0.25(\text{RPS}/\text{chart div}) \times 22.67\ \text{ft}/\text{min}/\text{RPS} = 1.13\ \text{ft}/\text{min}\ .$$

The 7-in., 23-lbm/ft casing has a capacity of .0393 bbl/ft; thus, this velocity corresponds to a flow rate of

$$1.13(\text{ft}/\text{min}) \times 0.0393(\text{bbl}/\text{ft}) \times 1{,}440(\text{min}/\text{D}) = 64\ \text{B}/\text{D}\ .$$

This value is at the low end of the range listed in Table 4.3 because the flowmeter in this case has very high resolution. On the down run, the flow from the bottom two sets causes an average deflection of 0.8 chart divisions to the right of the steady response in the stagnant fluid

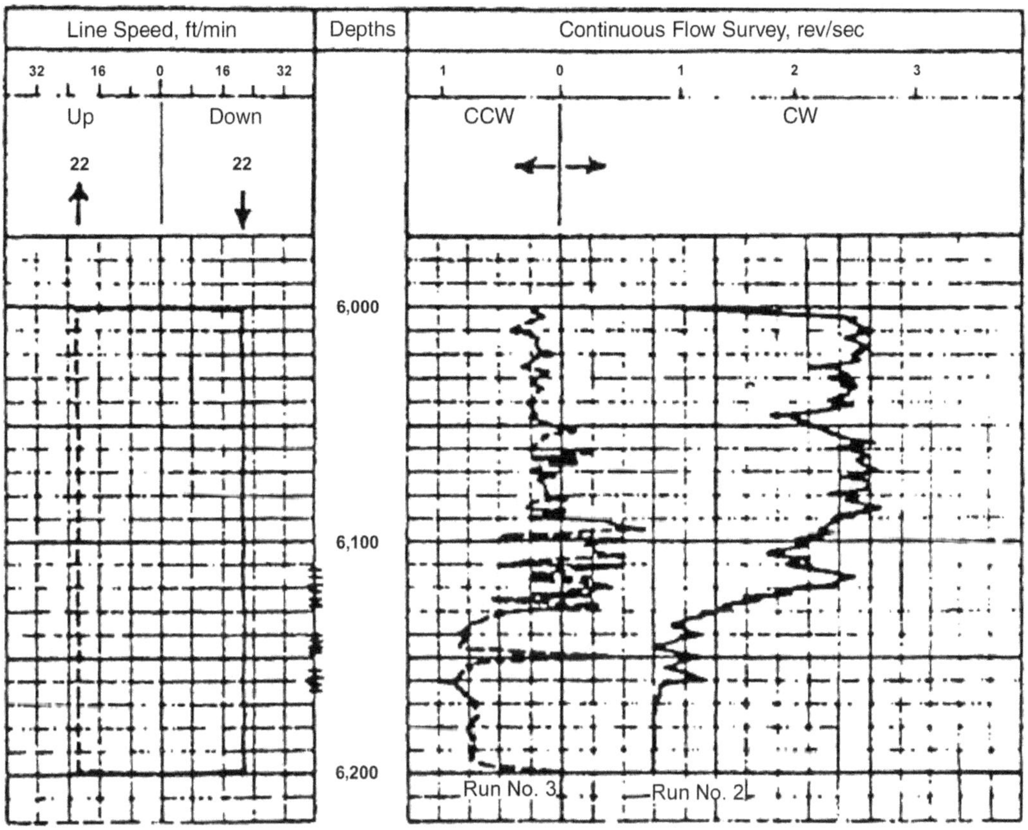

Fig. 4.16—High-resolution, continuous-spinner record from a production well.

below the bottom set (the no-flow reference). In full flow, the average deflection is 6.5 chart divisions to the right of reference. Thus, the relative contribution of the bottom two sets is (0.8/6.5) × 1,800 = 222 B/D. A flowing temperature survey in the same well (not shown), analyzed separately, establishes a more reliable estimate of the contribution of the bottom two sets: 430 B/D. The flowmeter lost 200 B/D, which is consistent with the resolution stated for oil and water flows (see Table 4.3). As oil, because of its buoyancy, rises through water in the wellbore interval defined by the bottom two sets, the oil churns and circulates the water even if the water is flowing; in turn, this action in the heavy phase can either slow down or speed up the spinner element, depending upon what component of circulation in the heavy phase most affects the tool. On the down run, the spinner is slowed, diminishing the spinner-estimated relative contribution of these sets. On the other hand, the spinner element speeds up slightly on the up run in response to a circulation downward.

The calculation immediately preceding is an example of relative profiling; that is, using the flowmeter log to establish relative contributions, with the total flow rate known independently. When an absolute flow rate is needed, the method of downhole calibration can be used. Downhole calibration is appropriate for injection flows or single-phase production except for, perhaps, the deepest entry, which may be submerged in captive completion fluid.

To perform a downhole calibration, the well is shut in at the surface. Both up and down runs are made through the static fluid at various cable speeds. On the calibration plot, cable velocities appear on the horizontal axis, with downward logging speeds as positive velocities and upward speeds as negative velocities. Values of RPS appear on the vertical axis, with values positive for down runs and negative for up runs. For each run, a point is plotted that corresponds to the cable velocity and its respective RPS value. A "best-fit" straight line is constructed for the plotted points of the down runs; this line corresponds to clockwise rotation and is located in the first quadrant of the plot. A second best-fit straight line is constructed for the plotted points of the up runs; this line corresponds to counterclockwise rotation and appears in the third quadrant.

The line for the down runs intersects the cable velocity axis to the right of the origin, and the velocity value at this intersection is the ideal or extrapolated threshold velocity for down runs. This value is slightly less than the actual speed needed to overcome frictional torque and start the spinner element rotating. The slope of this line is the sensitivity of the tool during down runs. The line for the up runs intersects the cable velocity axis to the left of the origin, and the absolute velocity value at the intersection is the threshold velocity for up runs. The value of the slope of the line is the sensitivity of the tool during up runs. For a quality tool, the two threshold velocities are nearly the same, as are the two sensitivities.

As an example of the use of the downhole calibration, consider a down run with the well producing. The RPS value at a depth of interest is taken from the log and on the plot is projected horizontally onto the straight line for the down runs. The corresponding value on the velocity axis is the velocity that drives the spinner element at this location. From this velocity, subtract the cable speed to obtain the net velocity. This velocity value is the apparent flow velocity. Multiplication of the net velocity (ft/min) by the pipe capacity (bbl/ft), by 1,440 (min/D) and by an independently determined calibration factor yields the desired flow rate. The calibration factor is needed because the apparent velocity as measured by a spinner is larger than the average or superficial velocity of the stream.

As another example, consider an up run with the well producing and with the spinner turning counterclockwise. The RPS value at a depth of interest is taken from the log and projected onto the line for the up runs. The corresponding value on the velocity axis is the velocity that drives the spinner element. From the absolute value of this velocity, subtract the positive value of the cable speed to obtain the net velocity. The rate is calculated from this net velocity in the same way. Had the spinner reversed to spin clockwise, the speed would be projected onto the calibration line for down runs and the cable speed added to the result to obtain the net velocity. An algebraic formulation of spinner speed as a function of cable and fluid velocities is given in the Appendix under the category of injection profiling.

Some additional comments are in order relative to the spinner traces in Fig. 4.16. Note first the pass upward at a line speed of 22 ft/min (the dashed-line trace). This line speed is close to the value of the upward superficial velocity for the full-flow stream (18 ft/min). Consequently, the spinner deflection is small approximately 0.25 rps counterclockwise above 6,050 ft. Below this depth, this low speed accentuates the fluid turbulence associated with the lifting and fallback of water as oil moves through it. This churning and circulatory action causes the spinner speed to flip-flop across the zero-rps axis with a reversal in the direction of spin on each crossing. On the downward pass at 22 ft/min (the solid-line trace), the higher-frequency oscillations are much smaller than on the upward pass. With the relative sped of the fluid to the tool being approximately 10 times larger than on the up pass, the sample time is too small for the spinner response to mirror the full extent of the high-frequency fluctuations seen on the up pass. As a result, one sees on the downward pass primarily the occasional lower-frequency events associated with gross slugging or "heading" in the flow.

The reduced effect of turbulence on the pass downward illustrates an adage that one still finds repeated in the literature to the effect that a high logging speed should be used to "minimize" the effect of turbulence. Although the claim is true, the implication that the response is more accurate is not. The reduced sample time caused by high relative speed between fluid and tool introduces a "hidden" bias toward higher spinner speed in flows affected by multiphase turbulence. The positive bias results from the greater influence of the life surges relative to the fallback flow along the casing wall. Such a bias is evident on the record from the upward pass, the dashed line in Fig. 4.16. An average value for the spinner speeds on the section of record in the interval of 6,090 to 6,130 ft is clearly to the clockwise side of the zero axis by approximately 0.25 rps, whereas the value above 6,050 ft is approximately 0.25 rps counterclockwise. The gross bias of approximately 0.5 rps represents a flow of 160 B/D in this example. The bias increases as the relative speed increases. Some spinner tools rectify the output pulses before conversions to rps values. On records from such tools, turbulence of any type produces a bias to higher speeds, owing to the interaction between sample time and spinner inertia.

4.9 Production-Logging Suite Examples

The earlier discussion emphasized the importance of using suites of production logs, rather than relying on a single log. In addition, the discussion indicated the applicability of production logs at various stages of a well's life, from drilling to abandonment, and even beyond. To further emphasize these points, four examples of production-logging suites (during drilling, during primary recovery, during tertiary recovery, and after abandonment) are described. These discussions include the objective of running a particular production-logging suite, stating why each log was used, and indicating what each log contributed toward achieving the objective of an understanding of the subject well's behavior.

4.9.1 Gas Kick While Drilling. In this development well, 10¾-in. pipe is set to 3,500 ft, and $7^5/_8$-in. casing is set to slightly below 14,000 ft. Behind the $7^5/_8$-in. casing, the top of the cement is just below 6,000 ft. During the coring of a gas sand at 15,000 ft, a pressure kick occurred, gas pressure was then lost at the surface, and mud was added periodically to keep the drillpipe full.

On the day after the gas kick, noise and temperature logs were recorded during the same run inside the drillpipe with the well static. These logs were run to identify the flow path of a likely underground blowout. The temperature log is basic to understanding a well's behavior, and the noise log is particularly responsive to the movement of gas through liquid.

At 15,000 ft, the noise log (**Fig. 4.17a**) exhibits a nearly "dead-well" noise level, indicating no activity (i.e., no fluid movement). Above 15,000 ft, however, the noise log departs from dead-well response. This departure shows that the gas, which enters the wellbore just above 15,000 ft, moves uphole. Above 14,000 ft, the noise log decreases very rapidly to dead-well response, showing that the gas moving uphole enters a formation at 14,000 ft, approximately 100 ft above the $7^5/_8$-in. casing shoe. The log thus shows a crossflow of gas (single-phase) from approximately 14,800 to 14,000 ft. The absence of gas pressure at the surface on the annulus between the drillpipe and the 7.625-in. casing means that the annulus is plugged by a bridge at some unknown depth above 14,000 ft.

A temperature log is quite sensitive to liquid flow and will reveal details regarding the flow path of any mud returns, which should move uphole in the annulus between the drillpipe and the $7^5/_8$-in. casing, owing to the periodic addition of mud to the drillpipe at the surface. Each time mud is added at the surface, some hotter mud is pushed upward in the annulus. For a given length of wellbore, the volume of hotter mud in the annulus greatly exceeds the volume of cooler mud in the drillpipe. During the time between additions, the two volumes equate in temperature to a value warmer than static. The periodic addition of mud at the surface should therefore produce what appears to be a "production" profile on a temperature survey. The temperature log in Fig. 4.17b does indeed show such a profile below Depth C. From this depth downward to 14,000 ft, the temperature is warmer than static, indicating displacement of annular mud upward to Depth C. This depth happens to be the location of a zone that was fractured and used during drilling for disposal of sour mud from the mud pit. The log therefore shows the movement of mud from deep in the well upward and into the disposal zone at Depth C.

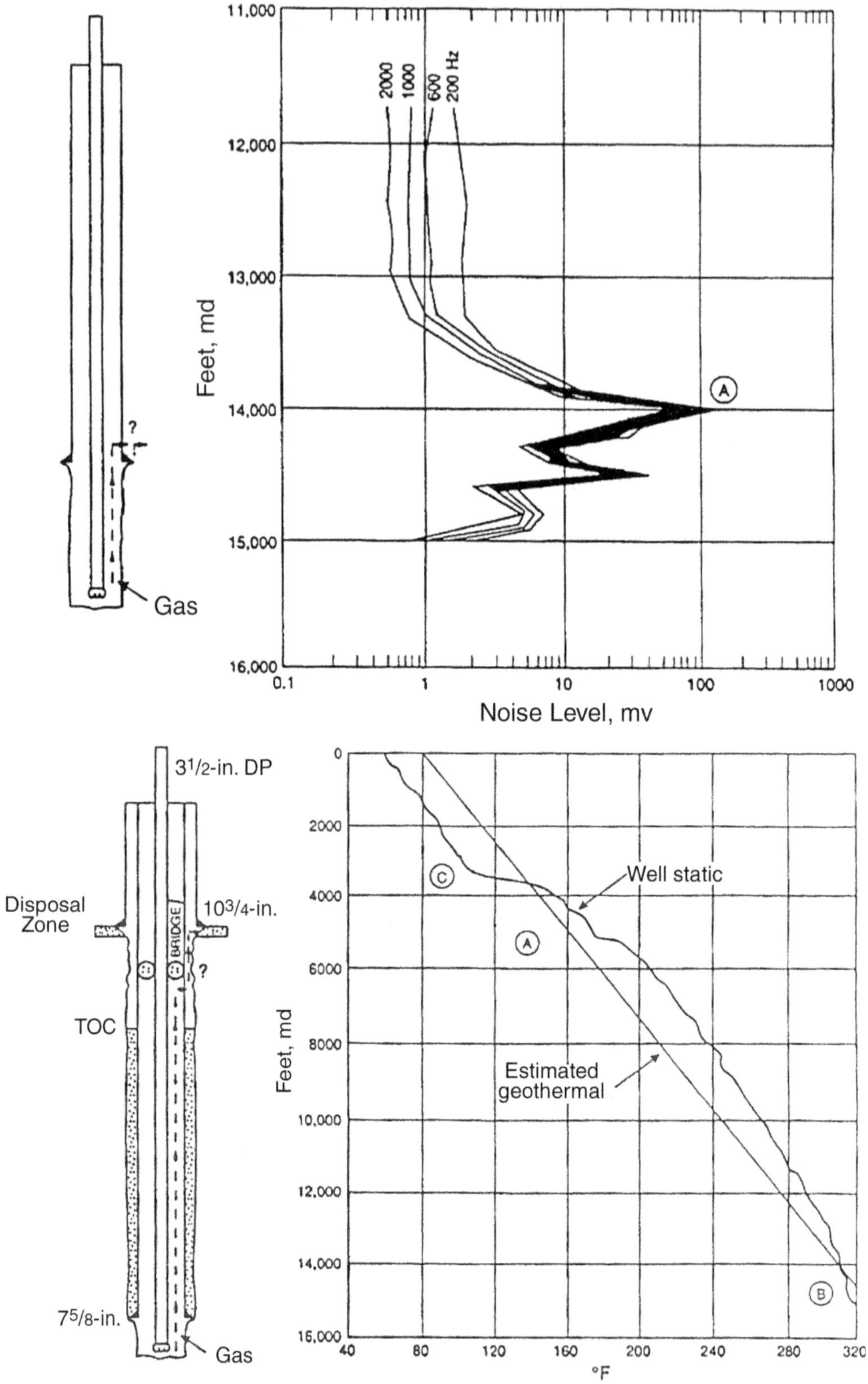

Fig. 4.17—(a) Noise log with well shut in at surface; (b) temperature log with well shut in at surface.

Above Depth C, the temperature trace lies to the cool side of static, reflecting the addition of cool mud to the drillpipe without displacement of warmer mud upward in the annulus.

For this flow path to exist, the 7.625-in. casing must also have failed at some depth. The temperature trace shows the location of this failure at Depth A just below 5,000 ft. Above this depth, the displaced mud moves behind the casing and loses heat to the formation at an increased rate. The corresponding mud temperature, although still warmer than static, is not as warm as the mud still in the annulus below Depth A. A mud path is therefore established as shown by the dashed line on the wellbore schematic on the left side of Fig. 4.17b. The only remaining task is the location of the depth of the failure in the drillpipe.

To investigate the mechanical integrity of the drillpipe, a radioactive tracer log was run, with the tracer tool stationary in the drillpipe at various depths while mud was pumped into the drillpipe continuously. The tracer traveled down the drillpipe with the mud flow, and the time lapse between its ejection and its subsequent detection by a detector below the ejector was measured. Time-lapse measurements made at various depths indicated a significant mud leak from the drillpipe in the interval between 12,800 and 12,900 ft. As further confirmation of the mud leak, tracer was ejected at 12,800 ft, and the tracer tool was quickly repositioned to 12,500 ft. While the logging tool was at 12,500 ft, the detector responded to the ejected tracer as it moved upward in the drillpipe-casing annulus. As a result of this survey, one can further conclude that the annulus is plugged to gas flow at some depth below the leaking joint of drillpipe at approximately 12,800 ft. The location of this bridge was determined from a free-point survey (see appendix), allowing a joint of split drillpipe to be replaced. The underground gas crossflow was then "killed" with mud.

Once the underground blowout was eliminated, the drillpipe was removed and the $7^5/_8$-in. casing string was examined with a casing collar locator to complete the investigation of mechanical integrity. The collar-locator log showed a 4-ft vertical separation in the casing string at Depth A because of a casing joint that was unscrewed from the collar above.

In this example, a careful analysis of a suite of inexpensive production logs yields definitive information of a diagnostic nature. Please note that the first suggestion of mechanical integrity problems resulted from a thorough examination of the temperature profile. Because the temperature profile was analyzed during the logging, the tracer and collar-locator logs were run while at the wellsite, avoiding the additional setup, pressure, and depth charges that would occur if the service company were to return for the tracer and collar-locator logs at a later date.

4.9.2 Profiling Commingled-Gas Production. This well produces gas from four perforated intervals; however, the bottom two intervals are in such close proximity that they are considered as one perforation set. Therefore, the perforations are referred to as the "top set" (A), the "middle set" (B), and the "bottom set" (C). The well produces 175 to 320 Mscf/D, with negligible water production. The objective of the production logs was to determine the fraction of total production contributed by each perforation set. With the well flowing, fluid density, temperature, and continuous-spinner flowmeter profiles were recorded. With the well flowing, a diverting flowmeter log was also recorded with the flowmeter stationary below and above each perforation set.

Viewed upward from the bottom, the fluid-density profile (**Fig. 4.18**) responds to a column of water below the bottom perforation set. Then it decreases above each of the three perforation sets, indicating that each set produces gas. Above the top perforation set, the measured density still exceeds the density of the gas, showing that water is present in the wellbore throughout the survey interval. Even though the water production is negligible, the gas rate is insufficient to lift water out of the wellbore.

Below the bottom perforation set, the temperature profile (Fig. 4.18) coincides with the estimated geothermal temperature profile. At each perforation set, the temperature profile shows cooling, which occurs because of the expansion of the produced gas as it passes through the near-

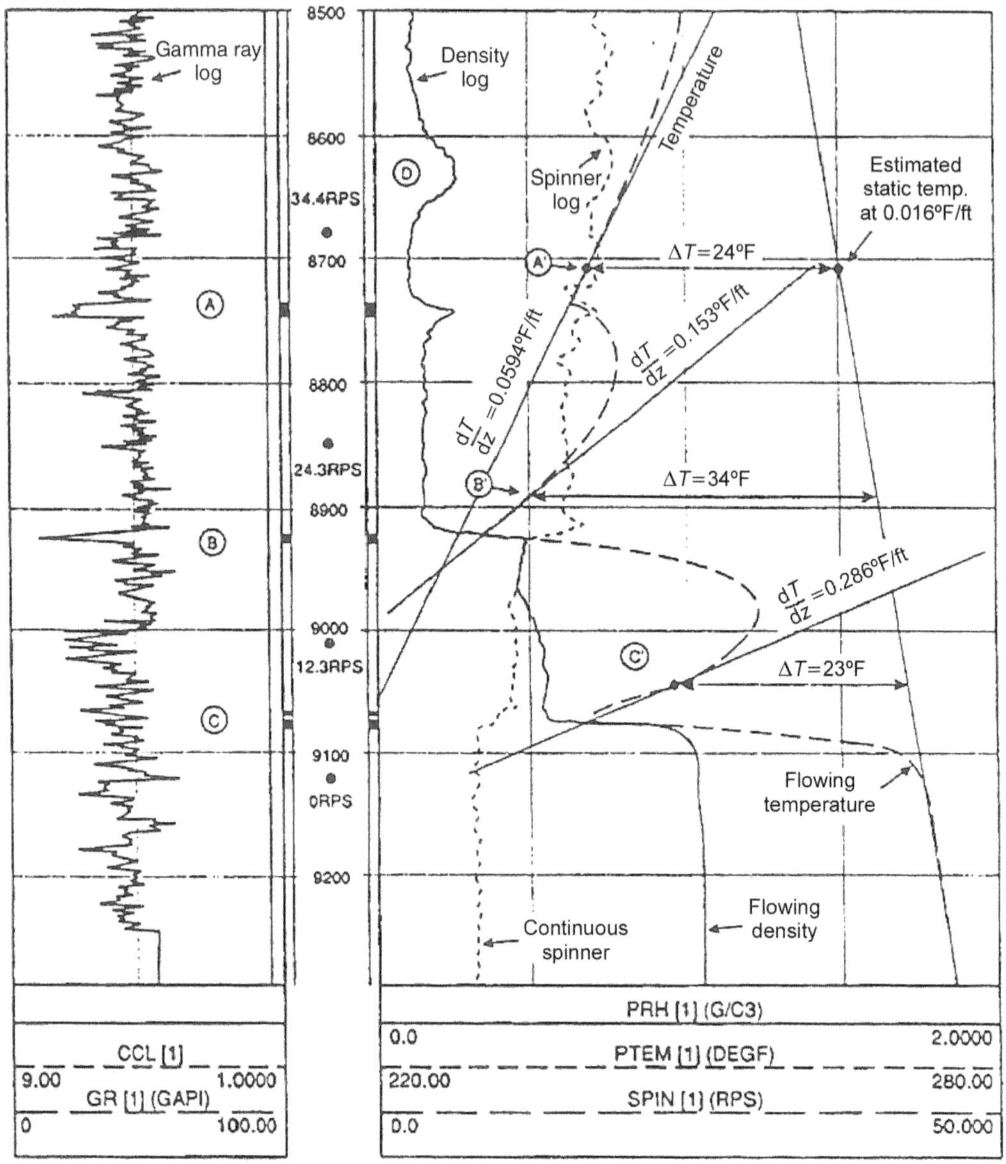

Fig. 4.18—Log suite in flowing gas well with commingled production from three zones.

wellbore region and the perforated completion. These cooling effects corroborate the gas production at each set indicated by the fluid-density profile. Between perforation sets, the temperature profile warms toward geothermal and then cools as it approaches the perforation set above. The tangent lines with their slopes illustrate the relative rates of warming of the stream above each perforation set.

Although there are two phases present in the wellbore (the produced gas and the captive water), the diverting flowmeter can respond to the gas flow rate at each measurement depth. This flowmeter was calibrated in a flow loop by measuring its response to a variety of air flows as they were passed through a column of stationary water. A "percolation calibration" was generated by plotting the diverting flowmeter's responses to the various air-flow rates. With this percolation calibration, the flowmeter's downhole measurements, shown in the depth

TABLE 4.4—PRODUCTION PROFILE FROM CALIBRATED DIVERTING FLOWMETER

Perforation Set	Fraction of Production
Top	0.44
Middle	0.39
Bottom	0.17

TABLE 4.5—PRODUCTION PROFILE FROM TEMPERATURE LOG

Perforation Set	Fraction of Production
Top	0.45
Middle	0.35
Bottom	0.20

track, were interpreted to yield the profile in **Table 4.4**, which shows the fraction of total gas production from each perforation set.

An independent analysis of the production profile was made from the temperature log by a profiling technique described in the Appendix under injection-well profiling. At a fixed distance above each perforation set, a tangent line was constructed to the temperature profile at a depth on its recovery toward geothermal. For each tangent, a determination was made of the slope of the tangent and the difference between the profile temperature at the depth of tangency and the corresponding geothermal temperature. By appropriately combining these data, the fractional contributions of the three perforation sets were obtained, as shown by **Table 4.5**. The two analyses are in good agreement.

The fractional contributions of the three sets also were determined by a linear analysis of both the diverting-flowmeter log and the continuous-spinner-flowmeter log. With this approach, a perforation set's fractional contribution to production is determined as the ratio of the change of the flowmeter's response across the perforation set to the total change of response from below the bottom perforation set to above the top set. The linear analyses of the two flowmeter logs both erroneously indicate that only one-quarter of the total production is from the top set, as opposed to the correct contribution of nearly one-half. A linear analysis is appropriate when only one phase is present in the wellbore but not when there are two.

In this example, the production profiles based on the temperature log and a proper analysis of the diverting flowmeter log corroborate each other, showing the advantage of more than one log in the suite. Also, this example demonstrates that when reliable data are analyzed with an inappropriate method (the linear-flowmeter technique), appreciable errors can result.

4.9.3 Profiling Oil Production Under WAG Recovery. In this well, 5½-in. casing is set to 4,463 ft. Below the casing, oil is produced in the open hole under WAG (water-alternating-gas) recovery.

The well produces 1381 RB/D of water, 119 RB/D of oil, and 245 RB/D of CO_2. Carbon dioxide, CO_2, dissolves primarily in the oil and secondarily in the water. Water is the continuous phase in the wellbore. The produced oil, with CO_2 in solution, bubbles (or "percolates") up through the flowing water. The logs in this suite were run with the objective of identifying the following:

1. Intervals in which water enters the wellbore.

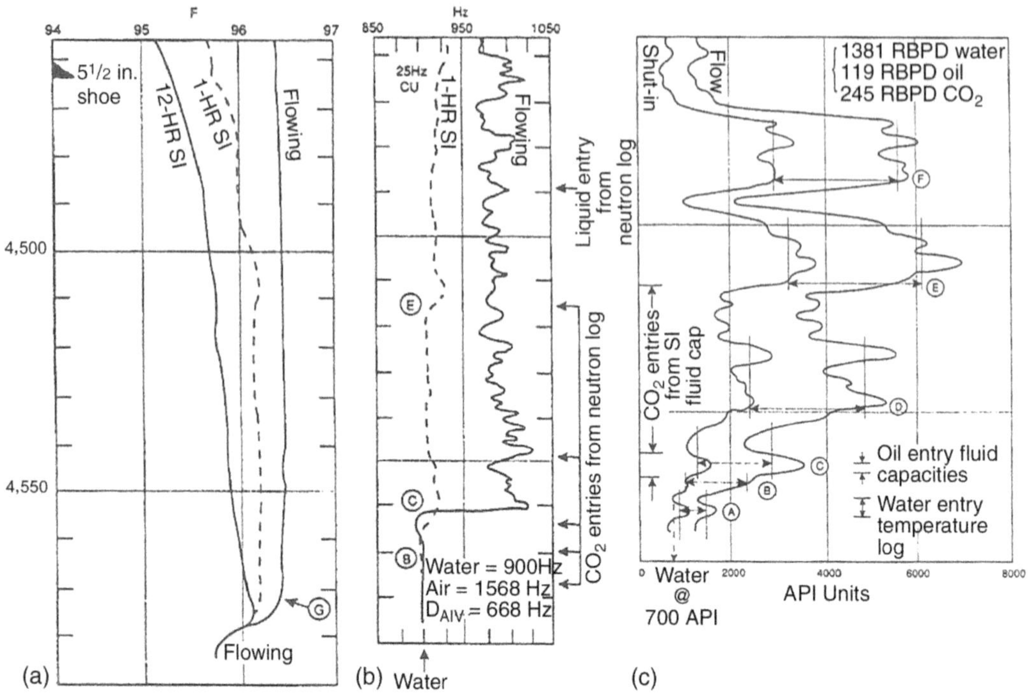

Fig. 4.19—(a) temperature logs; (b) fluid-capacitance logs; and (c) neutron logs.

2. Whether CO_2 is dissolved in the water.
3. Intervals in which oil (with CO_2 in solution) enters the wellbore.

The well was logged flowing, then shut in overnight. Shut-in logs were recorded the next morning. The following logs were run: fluid capacitance, fluid density, neutron, and temperature.

A comparison of flowing and shut-in temperature profiles (**Fig. 4.19a**) shows that the major production originates from a thin interval just above 4,575 ft (Depth G). Because the thermal content of the stream is essentially water, a thin interval at Depth G is therefore the source of water production. Injected water travels along a permeable streak known to exist at the based of the porous interval. This is not a good profile for oil recovery. The water is warm because warmer brine is being injected into a formation cooled by years of waterflooding.

The fluid-capacitance log (Fig. 4.19b) (well flowing) responds to the deepest oil entry at Depth C on an up run. However, fouling of the capacitance probe by the heavy oil renders the remainder of the up run useless for detecting additional oil entries. The probe was cleaned by stationing it in the tubing, where the elevated flow velocity removed the heavy oil film; however, it again fouled upon exit from the tubing. The shut-in capacitance profile, recorded later, reveals an additional oil entry at Depth E.

One usually depends on the response of the flowing fluid-capacitance log to determine whether an entering fluid is water or oil. In this example, the failure of the flowing capacitance log to respond to all oil entries with the exception of the deepest one (C) was anticipated in advance of logging and is not a problem because the water production is localized, and a comparison of flowing and shut-in neutron logs, which respond to a change of CO_2 concentration in the wellbore, can detect oil entries above Depth C.

The comparison of the separations between flowing and shut-in neutron logs (Fig. 4.19c) reveals the following:

1. CO_2 is dissolved in the water entering at (A).

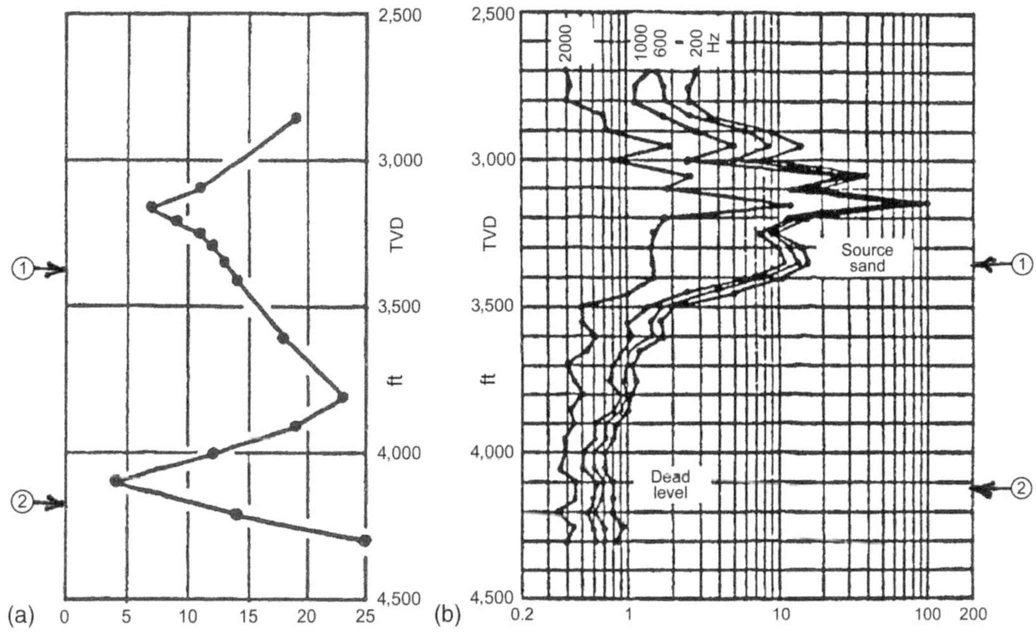

Fig. 4.20—(a) Magnetometer survey in relief well; (b) noise log in relief well.

2. CO_2 is dissolved in the water entering below (B), but at higher concentration than in the entry below (A).

3. CO_2 is dissolved in the oil entry at (C), which is the deepest oil entry according to the flowing capacitance log.

4. Oil with dissolved CO_2 enters just below (D).

5. CO_2 is dissolved in the fluid entering at (E), which is oil according to the shut-in capacitance log.

6. There is a fluid entry at (F), but with no CO_2 dissolved in the entering fluid. Because the water is produced below (C), the entering fluid is probably oil. The injected CO_2 is not reaching as far up the formation as Depth F. Still, the gravity migration of carbon dioxide upward in the formation above the bottom permeability streak is much better than one might suspect from the water production profile alone.

In this example, none of the three log types (temperature, fluid capacitance, and neutron) is capable of accomplishing the logging objective by itself. Moreover, no combination of two of the logs is capable of fulfilling the objective. Only a thorough analysis of the three logs taken together can accomplish the objective, showing the importance of a carefully selected, comprehensive suite of logs.

4.9.4 Gas Blowout After Abandonment. A well was drilled through two gas zones on the way to test a deeper oil zone. The well was abandoned and the wellhead cut at the seafloor.

Six months after abandonment, there was a gas blowout to the surface, causing the sea to churn. A relief well was drilled in order to flood the blowing zone. A magnetometer survey (**Fig. 4.20a**) shows the distance of the relief well from the original wellbore. The separation of the two wellbores "corkscrews" between approximately 5 and 25 ft, an effect caused by changes in the direction of the original wellbore. The two gas-zone locations are marked on the figure.

A noise log (Fig. 4.20b) was run in the relief well to identify which of the two gas zones was the source of the gas blowout. The noise log is applicable for this purpose because of its ability to "listen" to the sound in the original wellbore.

The dead-well noise levels in the interval of the relief well at the location of Gas Zone (2) shows it is not the source of the blowout. The noise-log response is elevated above the dead-well level in an interval of the relief well corresponding to the location of Gas Zone (1) showing that this gas zone is the source of the blowout. Above this location, the noise log exhibits several peak responses, two of which are larger than the response at the source of the gas blowout. The location of the largest corresponds to the location of least separation above Zone (1) in Fig. 4.20a. The other spikes are likely at locations of tight spots in the flow path.

It is also possible to estimate the distance between a blowing well and a relief well a point source of noise in the blowing well. This is done from the rate at which the noise level attenuates with distance along the relief wellbore from the peak noise location, and is possible because the noise level is inversely proportional to distance from the source. For example, the 200-Hz noise level at the location of zone (1) in Fig. 4.20b is 17 millivolts. By 3,500 ft, a distance of 140 ft in the relief well, this level has dropped to 2.0 millivolts. The separation between the wells at zone (1) is therefore estimated by the expression:

$$L_{sep} = \frac{140}{\sqrt{\left(\frac{17}{2}\right)^2 - 1}} = 16.6 \text{ ft.}$$

This value compares to the 13–14 distance indicated by the magnetometer survey. Likewise, the separation at the large peak at 3,150 ft is estimated from the decay at a value of 6.5 ft, which is the same distance as that shown on the magnetometer survey.

In this example, only two production logs were used. In considering this point, it is important to remember that the logging objective is quite limited: to determine which of the two known gas zones is the source of the blowout. Given that the logging is in the relief well, the noise log is better suited to this purpose than any other production log.

In aggregate, the previous four examples give the reader an appreciation of what can be accomplished with production logs during the entire life of a well.

4.10 Summary

This chapter describes the various tools used in production logging and provides insight on how to interpret results. In addition, the Appendix provides a "road map" of what logs to use for specific circumstances and what analysis methods apply. In particular, logging tools for making temperature surveys, several types of gamma-ray measurements, and spinner surveys are described. Examples are included for each tool and for four examples that illustrate why logging suites are needed to pinpoint what is happening in a well. Specific information is included on logging suites and skills needed to make successful surveys and meaningful interpretations. There are two specific recommendations that merit repeating: logging suites can be used to reduce ambiguous interpretations, and careful planning and documentation will ensure the current and future usability of production-logging results.

Nomenclature

a = coefficient
A = coefficient; also area, L^2
B = coefficient
C = casing capacity, bbl/ft
C = count rate, counts/sec
C_p = specific heat

D_h = casing or hole diameter, ft
D_t = tool diameter, ft
f = frequency
L_t = maximum tool length to pass through bend, ft
L_{turn} = distance to bend, ft
N = noise frequency
N_p = number of pipe strings
q = flow rate, B/D–scf/D
R = ratio
R_i = inside bend radius (or turning radius), ft
S = spinner turn rate, RPS
T = temperature, °F
V = linear velocity
y = holdup of a given phase
z = distance in z vertical direction
α = angle of bend, degrees
Δ = difference in two values
ρ = density, g/cm^3

Subscripts

1 = reference to bottom set of perforations
cool = area cooled below geothermal gradient
flow = flowing condition
g = gas
geo = geothermal
hc = hydrocarbon
i = counter
in = property of fluid flowing into wellbore from a set of perforations
mix = mixture of flowing fluids
rel = relative
S = slippage
shut-in = shut-in condition

Acknowledgments
The authors wish to express their appreciation to the ExxonMobil Co. for permission to use the information contained in this chapter. The information was extracted largely from an internal document that required many hours of preparation by the authors.

References

1. "Polaris-Production Optimization Log and Reservoir Information Solutions," Baker-Hughes Brochure, Houston (1999).
2. Hupp, D. and Schnorr, D.R.: "Evaluating High-Angle Wells With Advanced Production-Logging Technology," paper SPE 57690 presented at the 1999 SPE Annual Technical Conference and Exhibition, Houston, 3–6 October.

General References
The following brief collection of references to production-logging technology is definitely not intended to be comprehensive. Instead, the entries are chosen as being representative of the available literature within the constraint that each reference contains information of general in-

terest. The reader should be warned that some of the articles are strongly biased toward the tools offered by a particular company to the exclusion of other equally effective but usually older devices. The inclusion of such articles does not constitute an endorsement by us. Furthermore, the reader should know that log interpretations offered in some of the papers differ from those that would follow from the application of the interpretative guidelines presented in the Appendix to this chapter.

The references are organized into categories for ease of use. Within each category, articles are listed in alphabetical order according to the surname of the lead author without regard to the relative significance of the article.

Books

Bateman, R.M.: *Cased-Hole Log Analysis and Reservoir Performance Monitoring,* Intl. Human Resources Development Corp., Boston, Massachusetts (1985).

Carlson, N.R.: *Interpretative Methods for Production Well Logs,* Publication 9411, Dresser Atlas (Baker Hughes) Co., Houston (1982).

Hill, A.D.: *Production Logging—Theoretical and Interpretative Elements,* Monograph Series, SPE, Richardson, Texas (1990) **14.**

McKinley, R.M.: *Temperature, Radioactive Tracer and Noise Logging for Injection Well Integrity,* U.S. Environmental Protection Agency, Publication EPA/600/R-94/124, R.S. Kerr Research Laboratory, Ada, Oklahoma (1994).

Smolen, J.J.: *Production Log Interpretation,* Publication C-11811, Schlumberger Ltd., Houston (1973).

Smolen, J.J.: *Production Logging,* Reprint Series No. 19, SPE, Richardson, Texas (1985).

Smolen, J.J.: *Cased Hole and Production Log Evaluation,* PennWell Publishing Co., Tulsa (1996).

Survey Papers

Connolly, E.T.: "Resume and Current Status of the Use of Logs in Production," paper presented at the 1965 SPWLA Annual Well Logging Symposium, Dallas, 4–7 May.

Hill, A.D. and Oolman, T: "Production Logging Tool Behavior in Two-Phase Inclined Flow," *JPT* (October 1982) 2432.

McKinley, R.M.: "Production Logging," paper SPE 10035 presented at the 1982 SPE International Petroleum Exhibition and Technical Symposium, Beijing, 17–26 March.

Petevello, B.G.: "Evaluation of Well Performance Through Production Logging," paper presented at the 1975 Formation Evaluation Symposium, Calgary, 5–7 May.

Wade, R.T. *et al.:* "Production Logging—The Key to Optimum Well Performance," *JPT* (February 1965) 137.

Operational Topics

Garwood, G.L.: "Equipment Selection for Sour Gas Condensate Wells," *The Drilling Contractor* (March–April 1974) 40.

Hammack, G.W., Myers, B.D., and Barcenas, G.H.: "Production Logging Through the Annulus of Rod-Pumped Wells to Obtain Flow Profiles," paper SPE 6042 presented at the 1976 SPE Annual Technical Conference and Exhibition, New Orleans, 3–6 October.

Howell, E.P., Smith, L.J., and Blount, C.G.: "Coiled-Tubing Logging System," *SPEFE* (March 1988) 37; *Trans.,* AIME, **285.**

Instrument Evolution

Anderson, R.A. *et al.:* "A Production Logging Tool With Simultaneous Measurements," *JPT* (February 1980) 191.

Brown, G.A., Kennedy, B., and Meling, T.: "Using Fiber-Optic Distributed Temperature Measurements to Provide Real-Time Reservoir Surveillance Data on Wytch Farm Field Horizontal Extended-Reach Wells," paper SPE 62952 presented at the 2000 SPE Annual Technical Conference and Exhibition, Dallas, 1–4 October.

Dennis, B.R. *et al.:* "High-Temperature Borehole Instrumentation," Report LA-10558-HDR, Los Alamos Natl. Laboratory, Los Alamos, New Mexico (1985).

Hupp, D. *et al.:* "Polaris—Production Optimization Log and Reservoir Information Solutions," Brochure *A*99118, Baker-Atlas Div. of Baker Hughes, Houston (1999).

Hupp, D. *et al.:*: "Evaluating High-Angle Wells With Advanced Production-Logging Technology," paper SPE 56790 presented at the 1999 SPE Annual Technical Conference and Exhibition, Houston, 3–6 October.

Hupp, D. *et al.:* "Memory Pressure and Production Logging Tools," Brochure, Sondex Ltd., Bramshill, U.K.

Tello, L.N. *et al.:* "A Dipole Array Sonic Tool for Vertical and Deviated Wells," paper SPE 56790 presented at the 1999 SPE Annual Technical Conference and Exhibition, Houston, 1–4 October.

Papers Aimed at Computerized Interpretation

Catala, G.N., Torre, A.J., and Theron, B.E.: "An Integrated Approach to Production Log Interpretation," paper SPE 25654 presented at the 1993 SPE Middle East Oil Show, Manama, Bahrain, 3–6 April.

Holm, W.H. *et al.:* "A Cased Hole Well Log Computer System," paper presented at the 1983 SPWLA Annual Well Logging Symposium, Calgary, 27–30 June.

Nerby, G.: "A New Approach to Production Log Analysis," *Graham and Trotman's Proceedings*, 1989 North Sea Oil and Gas Reservoirs Conference II, Trondheim, Norway, May.

Nerby, G.: "Plato—The Answer to Your Production Logging Analysis," Brochure, IPS Interpretative Software Products Co., Houston (2000).

Temperature Logging

Agnew, B.G.: "Evaluation of Fracture Treatments with Temperature Surveys," *JPT* (July 1966) 892.

Carlson, N.R. *et al.:* "The Significance of the Temperature Log in Multiphase Flow," paper presented at the 1986 SPWLA Annual Well Logging Symposium, Houston, 9–13 June.

Cooke, C.E. Jr.: "Radial Differential Temperature (RDT) Logging—A New Tool for Detecting and Treating Flow Behind Casing," *JPT* (June 1979) 676.

Dobkins, T.A.: "Improved Methods to Determine Fracture Height," *JPT* (April 1981) 719.

Jameson, L.R.: "Some Applications of Differential Temperature Logging," paper SPE 1977 presented at the 1967 SPE Regional Secondary Recovery Symposium, Pampa, Texas, 26–27 October.

Millikan, C.V.: "Temperature Surveys in Oil Wells," *Trans.,* AIME (1941) **142,** 15.

Pierce, A.E. *et al.:* "Temperature surveys spot well ills, solve operating problems," *Oil & Gas J.* (June 1966) 96.

Ramey, H.J. Jr.: "Wellbore Heat Transmission," *JPT* (April 1962) 427; *Trans.,* AIME, **225.**

Smith, R.C.: "Temperature Log Interpretation," Publication L-36, The Welex (Halliburton) Co.

Smith, R.C. and Steffensen, R.J.: "Interpretation of Temperature Profiles in Water-Injection Wells," *JPT* (June 1975) 777; *Trans.,* AIME, **259.**

Noise Logging

Enright, R.J.: "Sleuth for Down-Hole Leaks," *Oil & Gas J.* (1955) **38,** 78.

Koerner, H.B. Jr. and Carroll, J.C.: "Use of the Noise Log as a Downhole Diagnostic Tool," paper SPE 7774 presented at the 1979 SPE Middle East Technical Conference and Exhibition, Manama, Bahrain, 25–29 March.

Korotaev, Y.P.: "Acoustic Method of Delineating Operating Intervals in Gas-Bearing Formations," *Gazovaya Prom.* (1970) No. 11, 14.

McKinley, R.M., Bower, F.M., and Rumble, R.C.: "The Structure and Interpretation of Noise From Flow Behind Cemented Casing," *JPT* (March 1973) 329; *Trans.,* AIME, **255.**

Myung, J.L.: "Fracture Investigation of the Devonian Shale Using Geophysical Well Logging Techniques," paper SPE 6366 presented at the 1976 SPE Eastern Regional Meeting, Columbus, Ohio, 18–19 May.

Pennebaker, E.S. Jr. and Woody, R.T.: "The Temperature-Sound Log and Borehole Channel Scans for Problem Wells," paper SPE 6782 presented at the 1977 SPE Annual Technical Conference and Exhibition, Denver, 9–12 October.

Rambow, F.H.K.: "Active Listening: An Alternative Method for Detecting Flow and Measuring Flow Velocity Behind Casing," *The Log Analyst* (November–December 1991) 645.

Robinson, W.S.: "Recent Application of the Noise Log," paper presented at the 1976 SPWLA Annual Well Logging Symposium, Dallas, June 9–12.

Shuck, L.V. *et al.:* "Noise Characteristics of Oil Wells and Reservoirs," paper SPE 5147-B presented at the 1974 SPE Annual Meeting, Houston, 6–9 October.

Stein, N. *et al.:* "Sand Production Determined From Noise Measurements," paper SPE 3498 presented at the 1971 SPE Annual Meeting, New Orleans, 3–6 October.

Radioactive-Tracer Logging

Akers, T.J. *et al.:* "Radioactive Tracer Logging in Laminar Flow," paper presented at the 1985 CWLS Formation Evaluation Symposium, Calgary, 29 September–2 October.

Anthony, J.L. and Hill, A.D.: "An Extended Analysis Method for Two-Pulse Tracer Logging," *SPEPE* (March 1986) 117; *Trans.,* AIME, **281.**

Bearden, W.G. *et al.:* "Interpretation of Injectivity Profiles in Irregular Bore Holes," *JPT* (September 1970) 1089.

Gadeken, L.L.: "The Interpretation of Radioactive-Tracer Logs Using Gamma-Ray Spectroscopy Measurements," *The Log Analyst* (January–February 1991) 24.

Hill, A.D. and Solares, J.R.: "Improved Analysis Methods for Radioactive Tracer Injection Profile Logging," *JPT* (March 1985) 511.

Hill, A.D., Boehm, K.E., and Akers, T.J.: "Tracer Placement Techniques for Improved Radioactive Tracer Logging." *JPT* (November 1988) 1484.

Kelldorf, W.F.N.: "Radioactive Tracer Surveying—A Comprehensive Report," *JPT* (June 1970) 661.

Killian, H.W.: "Fluid Migration Behind Casing Revealed by Gamma Ray Logs," *The Log Analyst* (January–March 1966) 46.

Lichtenberger, G.J.: "A Primer on Radioactive Tracer Injection Profiling," Southwestern Petroleum Short Course, Lubbock, Texas (1981) 251–263.

Roesner, R.E. *et al.:* "New Logging Instruments for Polymer and Water Injection Wells," paper presented at the 1983 SPWLA Annual Well Logging Symposium, Calgary, 27–30 June.

Self, C. and Dillingham, M.: "A New Fluid Flow Analysis Technique for Determining Bore Hole Conditions," paper SPE 1752 presented at the 1967 SPE Mechanical Engineering Aspects of Drilling and Production Symposium, Fort Worth, Texas, 5–7 March.

Simpson, G.A. and Gadeken, L.L.: "Interpretation of Directional Gamma Ray Logging Data for Hydraulic Fracture Orientation," paper SPE 25851 presented at the 1993 SPE Low Permeability Reservoirs Symposium, Denver, 26–28 April.

Small, G.P.: "Steam-Injection Profile Control Using Limited-Entry Perforations," *SPEPE* (September 1986) 388.

Wiley, R. and Cocanower, R.D.: "A Quantitative Technique for Determining Injectivity Profiles Using Radioactive Tracers," paper SPE 5513 presented at the 1975 SPE Annual Technical Conference and Exhibition, Dallas, 28 September–1 October.

Oxygen-Activation Logging

Arnold, D.M. and Paap, H.J.: "Quantitative Monitoring of Water Flow Behind and in Wellbore Casing," *JPT* (January 1979) 121; *Trans.,* AIME, **267.**

Hill, F.L. *et al.:* "New Instrumentation and Interpretive Methods for Identifying Shielded Water Flow Using Pulsed Neutron Technology," paper presented at the 1989 CWLS Formation Evaluation Symposium, Calgary, September.

Ierubino, J.V. and Ginest, N.H.: "Use of Pulsed Neutron Logging Techniques to Prove Protection of Underground Sources of Drinking Water," paper SPE 19616 presented at the 1989 SPE Annual Technical Conference and Exhibition, San Antonio, Texas, 8–11 October.

Lamb, G. *et al.:* "Measurement of Water Flow in Deviated Production Wells by Oxygen Activation Logging," paper presented at the 1983 SPWLA Annual Well Logging Symposium, Calgary, 27–30 June.

Scott, H.D. *et al.:* "Applications of Oxygen Activation for Injection and Production Profiling in the Kuparuk River Field," *SPEFE* (June 1993) 103.

Thornhill, J.T. *et al.:* "Injection Well Mechanical Integrity," Publication EPA/625/9-87/007, U.S. Environmental Protection Agency, R.S. Kerr Research Laboratory, Ada, Oklahoma (1987).

Wydrinsky, R. and Katahara, K.W.: "Effects of Thermal Convection on Oxygen Activation Logs," paper SPE 24739 presented at the 1992 SPE Annual Technical Conference and Exhibition, Washington, DC, 4–7 October.

Flowmeter Logging

Dale, C.R.: "Bottom Hole Flow Surveys for Determination of Fluid and Gas Movements in Wells," *Trans.,* AIME (1949) **186,** 205.

Hess, A.E.: "A Heat-Pulse Flowmeter for Measuring Low Velocities in Boreholes," Report 82-669 (Open File), U.S. Dept. of the Interior Geological Survey, Denver, Colorado (1982).

Hess, A.E.: "Identifying hydraulically conductive fractures with a slow-velocity borehole flowmeter," *Canadian Geotechnical J.* (1986) **23,** No. 1, 69.

Kading, H.W.: "Horizontal Spinner, A New Production Logging Technique," Southwestern Petroleum Short Course, Lubbock, Texas (1975).

Kragas, T.K. *et al.:* "Downhole Fiber-Optic Flowmeter: Design, Operating Principle and Testing," paper SPE 77655 presented at the 2002 SPE Annual Technical Conference and Exhibition, San Antonio, Texas, 29 September–2 October.

Leach, B.C. *et al.:* "The Full Bore Flowmeter," paper SPE 5089 presented at the 1974 SPE Annual Meeting, Houston, 6–9 October.

McBane, R.A., Campbell, R.L. Jr., and DiBello, E.G.: "Acoustic Flowmeter Field Test Results," *SPEPE* (February 1991) 49.

Piers, G.E., Perkins, J., and Escott, D.: "A New Flowmeter for Production Logging and Well Testing," paper SPE 16819 presented at the 1987 SPE Annual Technical Conference and Exhibition, Dallas, 27–30 September.

Rumble, R.C.: "A Subsurface Flowmeter," *Trans.,* AIME (1955) **204,** 258.

Fluid-Capacitance Logging

Carlson, N.R. *et al.:* "Applications of the Fluid Capacitance Log in Multiphase Flows," paper B presented at the 1986 Symposium of the Soc. of Aberdeen Well-Log Analysts, Aberdeen.

Carlson, N.R. *et al.:* "Water-Oil Flow Surveys with Basket Fluid Capacitance Tool," paper F presented at the 1982 SPWLA Annual Well Logging Symposium, Corpus Christi, Texas, 6–9 July.

Carlson, N.R. *et al.:* "New Method for Accurate Determination of Water Cuts in Oil-Water Flows," paper presented at the 1989 SPWLA Annual Well Logging Symposium, 22–25 June.

Guo, H. *et al.:* "The Design and Development of Microwave Holdup Meter and Application in Production Logging Interpretation of Multiphase Flows," paper SPE 26451 presented at the 1993 SPE Annual Technical Conference and Exhibition, Houston, 3–6 October.

Knight, B.L.: "Flow-Loop Evaluation of Production Logging Holdup Meter," *The Log Analyst* (July–August 1992) 412.

Multiphase Logging in Vertical Wellbores

Carlson, N.R. *et al.:* "Importance of Production Logging Suites in Multiphase Flows," paper P presented at the 1983 SPWLA Annual Well Logging Symposium, Calgary, 27–30 June.

Cmelik, H.R.M.: "A Controlled Environment for Measurements in Multiphase Vertical Flow," paper R presented at the 1979 SPWLA Annual Well Logging Symposium, Tulsa, 3–6 June.

Cmelik, H.R.M. and Sarabian, R.A.: "Quantitative Analysis of Production Logs in Two-Phase Gas-Liquid Systems," paper SPE 8761 presented at the 1979 SPE Production Technology Symposium, Lubbock, Texas, 5–6 November.

Darvarzani, J.D.: "Investigation of the Flow of Oil and Water Mixtures in Large Diameter Vertical Pipes," paper Q presented at the 1983 SPWLA Annual Well Logging Symposium, Calgary, 27–30 June.

Nicolas, Y. and Witterholt, E.J.: "Measurements of Multiphase Fluid Flow," paper SPE 4023 presented at the 1972 SPE Annual Meeting, San Antonio, Texas, 8–11 October.

Multiphase Logging in Deviated and Horizontal Wellbores

Barnette, J.C., Copoulos, A.E., and Biswas, P.B.: "Acquiring Production Logging Data with Pulsed Neutron Logs in Highly Deviated Wells with Multiphase Flow in Prudhoe Bay, Alaska," paper SPE 24089 presented at the 1992 SPE Western Regional Meeting, Bakersfield, California, 30 March–1 April.

Carlson, N.R. and Davarzani, M.J.: "Profiling Horizontal Oil-Water Production," *JPT* (July 1991) 780.

Carnegie, A., Roberts, N., and Clyne, I.: "Application of New Generation Technology to Horizontal Well Production Logging—Examples from the North West Shelf of Australia," paper SPE 50178 presented at the 1998 SPE Asia Pacific Oil and Gas Conference and Exhibition, Perth, Australia, 12–14 October.

Chauvel, Y. and Clayton, F.: "Quantitative Three-Phase Profiling and Flow Regime Characterization in a Horizontal Well," paper SPE 26520 presented at the 1993 SPE Annual Technical Conference and Exhibition, Houston, 3–6 October.

Maher, T. and Trcka, D.: "Inflow Fluid Typing in Screened Horizontal Completions Using a Pulsed Neutron Holdup Imager," paper SPE 56646 presented at the 1999 SPE Annual Technical Conference and Exhibition, Houston, 3–6 October.

Nice, S.B.: "Production Logging in Horizontal Wellbores," paper presented at the 1992 *World Oil* Horizontal Well Conference, Houston, 9–11 November.

Roscoe, B.A.: "Three-Phase Holdup Determination in Horizontal Wells using a Pulsed-Neutron Source," paper SPE 37147 presented at the 1996 SPE International Conference on Horizontal Well Technology, Calgary, 18–20 November.

SI Metric Conversion Factors

bbl ×	1.589 873	E − 01	= m³
ft ×	3.048*	E − 01	= m
°F	(°F − 32)/1.8		= °C
psi ×	6.894 757	E + 00	= kPa

*Conversion factor is exact.

Appendix—Production-Logging Application Tables: Tool Selection, Operating Procedures, and Interpretation Aids

Introduction. This section of the chapter provides an extensive set of tables intended to aid in the practical application of production-logging technology. For a given problem, the reader is guided first in the selection of the set of logging tools most appropriate. Next, suggestions are given on the proper procedure for each tool's use. This is an important part of the guidance, because the way logging records are obtained is often the most important part of the operation. Finally, the user is provided with comments regarding what the records should show relative to the problem. Recognition of expected results is equally important because irrelevant features on a log can easily prevent its proper interpretation. The tables are independent of the preceding body of the text and presume that the user lacks detailed knowledge of the subject.

The tables are unique to the literature in their ability to provide the user with detailed guidance without the investment of extensive search time. This feature is a consequence of the indexing approach to the information in the tables. The first level in the classification system is the nature of the well and the type of completion in which the tools are to be run. This environmental factor is called the *Well Category*. The tools that are best for one type of completion may be completely inappropriate for another type. For a given *Well Category*, the second level is the type of problem of interest to the user. This level is called *Problem Type*. Its identification leads to the final or third level in the classification—namely, to a table of tools appropriate to the specific well category and problem type. This final table provides the information listed above. Such an approach leads to much duplication of material, but is the one most beneficial to the user. Available publications attempt the classification on the basis of problem type alone, but this approach leads to so many disclaimers "for this situation and for that situation" that the result is confusion rather than guidance.

To make use of the application tables, one therefore enters the compilation sequentially through two indexing tables, identified as **Table 4.A1**: *Well Category* and **Table 4A2**: *Problem Type*. Having selected a category number from Table 4.A1 (Level 1), the user than locates this number in Table 4.A2 (Level 2), where a listing of general problem types is associated with each category number. Each problem type is identified by its index number, made up of a category number followed by a capital letter. For example, the designation 1A identifies a well still being drilled (well category) but experiencing problems of pipe sticking (pipe manipulation) or cross-sectional constriction (problem type). Having selected an index number from Table 4.A2, the reader next locates the third-level table having this number/letter designation. This final table subdivides the general problem into more specific problems (where appropriate) and provides a listing of recommended logging tools along with suggestions concerning their proper use and comments on what one should expect to see in the records from each tool. The tools are listed in the order of their likelihood to resolve the reasons for a particular problem; consequently, tool exclusion for a particular job should start at the bottom of the list and work upward.

The well category, Table 4.A1, cannot include all possible combinations of tubing placement relative to casing. The user may therefore not find a single category that describes completely an unusual completion. Instead, the completion may have at certain locations features common to one particular listing, whereas, at other locations, the features may coincide with a different listing. In such a case, the user will need to digest the contents of several tool-selection tables to devise a tool string appropriate to the completion.

A general comment is in order relative to tool selection. Slickline or downhole-memory logging tools have evolved to the point that multiple traces can be recorded on a single pass. This gives the ability to record traces from depth-control sensors, such as a collar locator and a gamma-ray tool, simultaneously with traces from production-logging sensors. The previously limited quality in depth control has therefore been eliminated. Furthermore, sample rates can be

TABLE 4.A1—WELL CATEGORY (FUNCTION AND COMPLETION)

Category No.	Well Type
1	Well still being drilled
2	Newly cased, unperforated well
3	Well being stimulated, fractured, or gravel-packed
4	Injection well—cased or openhole without screen or liner (any deviation)
5	Injection well with screen or liner (any deviation)
6	Injection well with tubing across perforations (dual completions)
7	Production well—vertical, cased, or openhole without screen or liner
8	Production well—vertical with screen, liner, or dual completion
9	Production well—highly deviated without screen or liner
10	Production well—highly deviated with screen, liner, or dual completion
11	Production well shut in owing to annular pressure or to suspected crossflow behind casing
12	Well down for tubing and/or casing inspection

set sufficiently high (up to 5 samples per second) so that the tools produce traces with a quality equal to that from electric-line surveys. In fact, the newer sensors on slickline strings often provide better traces. A corresponding improvement in surface interrogation and processing hardware now makes it possible to prepare a log from a slickline string within a few hours after tool retrieval. Memory tools now have practically all the advantages of electric-line tools with the added potential for safer and less-expensive operation.

Despite the extensive nature of the tables, their primary function is to provide realistic information on the selection of logging tools for a particular environment and a specific problem. The tables cannot make the user proficient in the very important task of interpretation. Moreover, computer software cannot do this either; intervention by an expert analyst is still required to ensure correct interpretation. The user should not hesitate to seek such intervention, preferably from experts other than employees of the logging company involved.

Classification Tables. The two classification Tables 4.A1 and 4.A2 are provided here. These tables are used sequentially to navigate the extensive set of tables devoted to tool selection. Table 4.A1 is used to select a category number best describing the type of well to be logged. This numeral is then entered in Table 4.A2 and associated with the capital letter best describing the type of problem to be resolved. The resulting *Index Number* identifies one of the *Tool-Selection Tables.* This final table provides the user with the information necessary to plan and conduct the logging operation.

Tool-Selection Tables. These guidance tables list recommended logging tools in order of general effectiveness, giving comments regarding procedures for tool use and indicating the normal features that should appear on the log traces. Each table is identified by an index code made up of a number and a capital letter, such as 2C. A specific index code is obtained through the use of Tables 4.A1 and 4.A2. Each tool-selection table deals with a particular problem area for a particular well configuration.

TABLE 4.A2—PROBLEM TYPE

Category No.	Tool-Selection Table Number According to Problem
1	Well Still Being Drilled
	1A: Pipe manipulation or constriction difficulties
	1B: Lost returns
	1C: Underground flow (blowout)
2	Newly Cased, Unperforated Well
	2A: Cement placement
	2B: Casing inspection for wear, collapse, and bends
	2C: Annular pressure, crossflow behind casing, and well certification for hazardous injection
3	Well Being Stimulated, Fractured, or Gravel-Packed
	3: Acid, fracture proppant, or gravel placement
4	Injection Well—Cased or Openhole Without Screen or Liner
	4A: Demonstration of confinement
	4B: Injection profile determination
5	Injection Well With Screen or Liner
	5A: Demonstration of confinement
	5B: Injection profile determination
6	Injection Well With Tubing Across Perforations
	6A: Demonstration of confinement
	6B: Injection profile determination
7	Production Well—Vertical, Cased, or Openhole Without Screen or Liner
	7A: Tracking single-phase production
	7B: Tracking two-phase production
	7C: Tracking three-phase production
8	Production Well—Vertical With Screen, Liner, or Dual Completion
	8A: Tracking single-phase production
	8B: Tracking two-phase production
	8C: Tracking three-phase production
9	Production Well—Highly Deviated Without Screen or Liner
	9A: Tracking single-phase production
	9B: Tracking two-phase production
	9C: Tracking three-phase production
10	Production Well—Highly Deviated With Screen, Liner, or Dual Completion
	10A: Tracking single-phase production
	10B: Tracking two-phase production
	10C: Tracking three-phase production
11	Production Well Shut In Owing to Annular Pressure or to Suspected Crossflow Behind Casing
	11A: Location of source of annular pressure
	11B: Detection of crossflow behind pipe
12	Well Down for Tubing or Casing Inspection
	12A: Detection of corrosion or wear to string containing tool
	12B: Detection of holes or parting in casing external to one containing tool

TABLE 4.A3—1A: DRILLPIPE MANIPULATION AND CONSTRICTION PROBLEMS (WELL STILL BEING DRILLED)

Problem	Recommended Logging Tools	Operational Tips and Expected Results
Stuck pipe	Stuck-pipe log (axial cement-bond log)	This log is the trace of the transmitted amplitude (millivolts) from an acoustical device having a transmitter and a receiver spaced approximately 3 ft apart. A large, nearly constant amplitude denotes free pipe. Intervals where pipe is stuck show small, irregular amplitude. Log is only qualitative in the degree of pipe immobility; consequently, a backoff operation should not be done on the basis of this log. Its advantage is that a long interval of pipe can be logged relatively quickly.
	Gamma-ray gravel-pack log (unfocused gamma-density log, not neutron gravel-pack log)	This device uses gamma-ray transmission to respond to the density of material within a radius of approximately 6 in. and should be run centralized. Count rate is related inversely and exponentially to density. Larger count rates correspond to lower density, whereas higher density produces lower count rates. Free pipe gives the highest and most constant count rate. Log is qualitative in degree of pipe mobility and should not be the basis of a backoff procedure. Its advantage is that a long interval of pipe can be logged quickly. Its disadvantage is that it contains radioactive material that must be cemented in place if the tool cannot be retrieved.
Free pipe location	Free-point indicator log	This device locates the depth above which pipe is free and should be used to plan a backoff job after a stuck-pipe survey. It uses an articulated, two-section tool with load cells in each part to make stationary measurements joint-by-joint. If the bottom section is anchored in a location where the pipe is stuck, whereas the top section is in free pipe, then the top section will stretch or rotate relative to the bottom in response to a pull or a torque on the pipe string at the surface. Free pipe will show little or no relative movement.
Pipe bend or kink	"Dog-leg" or kink-meter log	These tools make a nonoriented measurement of the angle of deviation of the pipe from the vertical. The typical sensor is a pendulum-type element. The device therefore responds only to bends in the pipe and not to pipe collapse.
Pipe collapse	X-Y caliper log	This is a simple caliper device that measures two pipe diameters oriented 90° to each other. The measurement is sufficient to detect changes from the round, but is not sufficient to define the shape of the pipe's cross section for collapse extent.
Casing external to drillpipe (dropped casing)	Gravel-pack log (see previous)	The same operational principles outlined previously also apply here. Interpretation of the log is greatly aided if it is possible to log across one end of the parted casing external to the drillpipe.
	Eddy-current wall-thickness log (an induction tool normally used for pipe inspection)	This is actually the best tool for this problem, but versions small enough to run in drillpipe are scarce. The instrument is a two-coil device having an active coil driven at a very low frequency (~1Hz) and a passive coil to detect the induced field. The phase lag of the induced signal behind the driven signal is the measured trace. This lag is directly proportional to the thickness of metal within the induced field. The interpretation is aided by the ability to log across one end of the parted casing. A multi-frequency tool should be operated at its lowest frequency for greatest depth of field penetration. The tool, of course, will not function in non-magnetic metal.

TABLE 4.A4—1B: LOST RETURNS (WELL STILL BEING DRILLED)		
Problem	Recommended Logging Tools	Operational Tips and Expected Results
Losing returns while circulating	Temperature survey while pumping down drillpipe	At least 60 bbl of mud should be pumped prior to logging. One display should be at a relatively high sensitivity of 1°F/in. At the lost-returns zone, the temperature trace will show a step increase to a warmer value that is maintained below the zone if all or most of the injection enters the zone. Only an increase to larger gradient (slope) at and below the zone may be evident for partial loss of returns (small loss to zone). The temperature behavior is the result of the warmer temperature brought upward from the bottom of the hole by flow up the annulus on the outside of the drillpipe.
	Temperature survey with pumping down annulus (backside)	This approach to the problem requires the least amount of mud for its execution, but is often not feasible for reasons of well control. Circulation should be stopped for approximately 3 hours before this operation is commenced to allow some recovery in wellbore temperature. A base survey should then be recorded prior to the start of injection. One display should be at a sensitivity of at least 2°F/in. of chart. A logging pass downward can be commenced after approximately 10 minutes of injection. Below the injection zone, the injection temperature trace will show a rapid warming back to or slightly above the base temperature. If multiple injection zones are present, then the trace will show only partial recovery at each zone above the deepest.
	Radioactive tracer survey while pumping down the drillpipe	Tracer slug should be ejected in the drillpipe near the bit and subsequently followed as it travels up the annulus behind the pipe. Slug size should be tailored for the expected travel distance up the annulus. A 10-microcurie slug will survive for approximately 3,000 ft before dispersion causes the activity to be lost in background. Tracer adsorption on the borehole wall is generally not a problem for an Iodine-131 tracer. If loss is only partial, pumping should cease before the tracer is returned to the surface.
	Activated oxygen survey with tool configured for upflow for injection down drillpipe and for downflow for annular injection.	Water-based mud must be used for this survey. The pulsed-neutron tool creates its own tracer through the activation of the oxygen in the water. The velocity of movement of the activated oxygen relative to the tool is then measured. Stationary measurements are therefore the easiest to interpret. The resolution of these tools is determined by fluid velocity and is generally in the range of 4 ft/min. Typical resolution rates for annular flow are therefore in the range of 300 B/D. The survey should not be attempted from vessels experiencing significant vertical motion. This survey is usually considerably more expensive than the other surveys, but is listed because radioactive tracer service offshore may not be available.
Location of lost-returns zone after occurrence	Temperature survey	Cool "anomalies" identify those zones into which mud has been injected relative to inactive parts of the borehole. One display should be at a sensitivity of at least 1°F/in. of chart. Survey should be run within days of a mud loss of 100 bbl or less and within a week or two of larger losses. Owing to the character normally present on the temperature trace, expert interpretation is usually required to identify the lost-returns zones. Two surveys run at least 18 hours apart aids in the interpretation because injection zones warm slower than do inactive parts of the wellbore.

TABLE 4.A5—1C: UNDERGROUND FLOW (BLOWOUT) (WELL STILL BEING DRILLED)

Problem	Recommended Logging Tools	Operational Tips and Expected Results
Detection of flow underground	Temperature survey	Because a base survey is generally unavailable under these circumstances, an absolute interpretation of the traces is required. One display at a sensitivity of at least 2°F per in. of chart is advisable. Two surveys recorded at least 8 hours apart aids this interpretation because any crossflow interval will change its temperature less than will inactive sections of hole. Liquid entry into the cooled wellbore will bring its warmer temperature with it to give a warm anomaly at the entry location. This temperature is carried away in the direction of flow in proportion to the rate. A rough rule of thumb for water is that the trace remains vertical for 20 ft for each 100 B/D rate (0.20 ft/B/D). The entire crossflow interval will either "stand up" more or parallel the gradient at inactive (static) locations. The warming makes this survey particular sensitive to liquid and able to detect rates of a few B/D. Annular liquid rates behind drillpipe can be estimated from the slope dT/dz, °F/ft, and amount of warming $\Delta T = T_{flow} - T_{static}$, °F, at a given location in the crossflow path from the expression: $$q\ (\text{B/D}) = A \times \frac{\Delta T}{\left(\frac{dT}{dz}\right)},$$ where A = 1.0 for water flow and A = 2.5 for oil flow. The result is an approximation only. Gas entry will generally produce a cool anomaly at the point of entry. For dry gas, this cooling can amount to 1°F for each 50-psi drop in pressure experienced by the gas. If the flow path is short, then the entire section will be cooled. If, however, the path is over a long section of hole, then the trace will cross to the warm side of "static" at some point. The previously mentioned rate expression applies to downhole B/D of dry gas if A = 10.

TABLE 4.A5—1C: UNDERGROUND FLOW (BLOWOUT) (WELL STILL BEING DRILLED) (Continued)

Problem	Recommended Logging Tools	Operational Tips and Expected Results
Detection of flow underground	Noise-log survey	This device is a microphone that measures the amount of turbulent noise produced by the movement of fluids. It makes stationary measurements of the sound level, millivolts, above four high-pass cutoff frequencies, typically set at 200, 600, 1,000, and 2,000 Hz. Each reading includes levels at all frequencies above the cutoff values; consequently, the readings decrease with increasing cutoff frequency. The cutoff values are chosen so as to distinguish two flow conditions: single phase and gas movement through liquid. The movement of gas through liquid produces a high level of noise in the low-frequency range below 600 Hz. This fact makes the tool particularly sensitive to small gas flows through liquid, the application for which it was devised. Gas percolation at downhole rates as low as 10 ft^3/d can be resolved. Below 400 ft^3/d, gas rate behind drillpipe can be estimated from the millivolts of noise in the band N_{200}–N_{600} from the relationship $$q_g(\text{ft}^3/\text{d}) = 0.35 \times (N_{200} - N_{600}) .$$ Higher rates are underestimated by this expression. Single-phase flow, by contrast, produces noise primarily in response to acceleration past tight spots in the flow path. The resulting sound is of relatively weak amplitude and has a frequency in the range of 1,000 Hz or higher. If a restriction holds a pressure drop Δp psi in response to a rate q B/D, then the single-phase rate of either liquid or gas behind drillpipe can be estimated from the millivolt value of the noise cut at 1,000 Hz by the expression $$q \times \Delta p = 700 \times N_{1000} .$$ If the total pressure drop over the entire flow path is used in this expression, then the summed noise levels at each constriction should also be used. Both of these flow expressions suppose that the tool is calibrated to read standard peak-to-peak millivolts.
	Activated oxygen survey for water with tool configured for suspected direction of flow	This tool activates the oxygen in water to create its own radioactive tracer, whose movement is then timed to provide a velocity of water with a resolution of approximately 4 ft/min. This device also takes stationary measurements. If the direction of flow is unknown, then two surveys may be necessary. The tool must be retrieved or cemented in place.

TABLE 4.A6—2A: CEMENT PLACEMENT (NEWLY CASED, UNPERFORATED WELL)

Problem	Recommended Logging Tools	Operational Tips and Expected Results
Cement top location	Temperature survey	Must be recorded within limited time frame, typically 12 to 24 hours, after cement is circulated into place. The cement company can usually provide the proper time for logging. Owing to the heat of hydration, temperature will increase at the top of the cement column and will remain warmer throughout the column. Below the cement top, temperature will be more "noisy" than above because the variations in caliper are accentuated by the heat of hydration. Display sensitivity should be such as to avoid "wraps" in the record for a few hundred feet on either side of the cement top so that the increase in both temperature and "noise" is evident. Highly retarded cements usually produce very little heat of hydration for such a survey. This approach has the advantage that, unlike cement-bond logs, it does not depend upon intimate contact between cement and the outside of the casing at the top of the column.
	Axial (regular) cement-bond log	This is an acoustical device having one transmitter and two receivers. The latter are 3 and 6 ft from the transmitter. The 3-ft receiver measures the signal amplitude at the start of the wave train and thus gives a millivolt reading of the transmission along the casing wall. Free pipe produces the highest and nearly constant amplitude on this "pipe signal" trace. Typical values are in the range of 5 to 100 millivolts. This constancy above the top of cement is the "calibration" feature to the log. At the top of cement, signal amplitude will usually decrease. The signal will drop to a few millivolts if the casing is in intimate contact with cement and will usually become more erratic with depth otherwise. The 6-ft receiver measures the entire wave train as a function of time since the transmitter was pulsed. It thus captures signal transmitted along all paths linking transmitter and receiver. It is best displayed on a black-and-white variable density format (VDL record) so that lines vertical with depth denote arrivals at a constant sonic velocity. Such lines at the front of the record denote transmission along the casing wall (Pipe Signal), whereas lines at the back of the record result from transmission through the liquid inside the casing itself (Borehole Fluid Signal). If nothing but mud is present in the annulus behind casing, then the VDL display with consist of little more than pipe signal over the first two-thirds of record merging into borehole fluid signal over the final third. Cement in the annulus will pass enough signal to the formation and on to the receiver to interfere with and to degrade the borehole fluid signal. Consequently, the VDL trace often locates the top of cement even in the absence of contact with the casing.

TABLE 4.A6—2A: CEMENT PLACEMENT (NEWLY CASED, UNPERFORATED WELL) (Continued)

Problem	Recommended Logging Tools	Operational Tips and Expected Results
Cement top location	Axial (regular) cement-bond log	The application of pressure to the casing at the surface generally improves the quality of a bond log through improved contact between casing and cement. Any pressure at all is beneficial, but values of 500 to 2,000 psi are typical. Pressure may not offer any improvement whenever pipe and cement are separated by a viscous film on the outside of the casing. This is a problem with oil-based mud and with polymer mud. For cement to provide support to the casing, it must be rigid (i.e., must have some compressive strength). Usually, a 3-day waiting period prior to logging is sufficient for the cement to cure to approximately 75% of its ultimate strength. This time should be verified with the cement company. Foamed and other light-weight (extended) cements never develop much rigidity and thus fail to dampen pipe signal significantly. Cement-top location is therefore a subtle determination made from the VDL display. The same approach must also be used for bond logs recorded in fiberglass casing, but for a different reason. Fiberglass does not transmit much pipe signal, whether free or in contact with cement. Compensated versions of the tool with dual transmitters should be used in wells that are intentionally deviated because of their tolerance to variations in offset between tool and casing. On the other hand, the use of radial (circumferential) tools is not necessary.
Evaluation of cement placement	Axial (regular) cement-bond log in combination with radial (circumferential) -bond log	The objective is to determine whether cement has displaced mud in the rugose annulus behind casing. The principal record is the one from the axial tool, whereas the radial record is primarily of value in the determination of why poor contact may exist between casing and cement. The operational principles of the axial tool are described previously. The logging should be done under surface pressure and the wave-train record should be of the black-and-white, VDL type.

TABLE 4.A6—2A: CEMENT PLACEMENT (NEWLY CASED, UNPERFORATED WELL) (Continued)

Problem	Recommended Logging Tools	Operational Tips and Expected Results
Evaluation of cement placement	Axial (regular) cement-bond log in combination with radial (circumferential) -bond log	Three types of radial tools are available. The simplest is an axial-transmission tool with a set of 8 button receivers arrayed at 45° increments around the circumference of the tool. The array is spaced approximately 1 to 3 ft from an omnidirectional transmitter. The display is a variable-density representation of the eight values for pipe-signal amplitude. White generally represents large pipe signal (poor contact), whereas black denotes small pipe signal and very good contact. The second type of radial tool contains six pads that are held in contact with the inside casing wall by bow springs. Each pad contains a transmitter crystal and a receiver crystal. The pads are aligned so as to make six measurements of the pipe-signal attenuation coefficient over a transmission of approximately 6 in. forming a saw-tooth pattern around the circumference of the casing. The six coefficients are displayed both as six curves and as a variable-density display. This is the radial tool of choice. The third type of radial tool is a reflection-type device operating in the ultrasonic range. It contains a single crystal that rotates as it is pulsed. Between pulses, the crystal records the wave train reflected from both the inside and the outside walls of the casing. These data are processed so as to measure the amplitude of the signal reflected from the outside wall under the premise that a large reflection means little solid support to the casing, whereas a small amplitude implies good support. As one would suspect, the signal range from "good-to-bad" is quite small. This amplitude signal is presented on a yellow-to-dark-brown-colored, variable-density display called a "cement map." The reflection from each pulse is affected by conditions immediately outside such a small segment of the circumference that the record indicates anomalies that often have little or no significance to effective placement. If the well is deviated, then the radial tools should be run in the "oriented" mode so as to distinguish the high side of the hole from the low side. The VDL display from the axial tool is used to determine whether the annulus is essentially filled with cement. For casing 5 in. in diameter or larger, a full annlus will allow sufficient signal to travel through the formation to interfere with and to usually obliterate the borehole fluid signals otherwise appearing at the end of the record. In 4.5-in. casing, formation signal may end soon enough to leave approximately 1 stripe of borehole signal at the end. For smaller casing, formation signal may cease before the last third of the record, leaving this part free for borehole fluid signal. The annulus may be filled even though the pipe-signal trace indicates free pipe. In those locations showing poor contact on the pipe-amplitude trace, the radial record is used to determine circumferential extent of the lost contact. This information along with porosity data from an included gamma log can help in distinguishing among the many conditions that can cause loss of contact.

TABLE 4.A6—2A: CEMENT PLACEMENT (NEWLY CASED, UNPERFORATED WELL) (Continued)

Problem	Recommended Logging Tools	Operational Tips and Expected Results
Evaluation of cement placement	Axial (regular) cement-bond log in combination with radial (circumferential) -bond log	Bond logs measure acoustical coupling, a quantity only marginally related to the ability of the cement to seal against the movement of fluid behind casing. Many conditions can destroy coupling without allowing the transmission of fluid over any significant distance. The most that can be concluded from a bond log is that it shows no reasons to suspect that the cement should not seal. This does not mean that the cement will seal.
	Combination of openhole-caliper log and temperature survey after cement placement	This combination is actually the best way to determine how effectively cement has displaced mud from a rugose borehole. The temperature trace should be run according to the same time guidelines used for the location of the top of cement and should be displayed at a sensitivity sufficiently high to accentuate the "noise" arising from caliper variations below the top. If the placement is effective, then locations showing hole enlargements on the caliper trace will show as warm spots on the temperature trace owing to the increased volume of hydrating cement. Likewise, sections of hole with reduced caliper will appear as cooler spots on the temperature log. If such a calibration exists between the surveys, then the placement should be sufficient to seal the annulus.
	Gamma-ray gravel-pack log (unfocused gamma-ray) before and after cement is circulated in place	This device uses the transmission of gamma rays to respond to density within roughly a 6-in. radius around it. Higher count rate means lower density, whereas lower count rate signifies higher density. The dependence is exponential in density, giving it a resolution of approximately 0.03 g/cm^3. For the approach to be effective, the difference in density between cement and displaced mud should be at least twice this amount, or 0.06 g/cm^3. The advantage is that no waiting time is required after the cement slurry is circulated.

TABLE 4.A7—2B: CASING INSPECTION FOR WEAR, COLLAPSE, AND BENDS (NEWLY CASED, UNPERFORATED WELL)

Problem	Recommended Logging Tools	Operational Tips and Expected Results
Casing wear	Multifeeler, mechanical caliper survey with display for each feeler or for cluster of feelers	These devices record the movement of anywhere from 16 to 84 individual feelers along the inside wall of the casing. The resolution to a change in offset between the centralized tool and the pipe wall is approximately 0.005 in. Earlier versions recorded only the minimum and maximum deflections. Modern versions record either the movement of each feeler or the average movement for each of several clusters of feelers along with traces for minimum and maximum excursions. Even a 16-arm log provides a reasonable depiction of the shape of the casing's cross section, thereby distinguishing wear from collapse. Drillpipe wear (increased diameter) is generally restricted to one side of the casing over long stretches of vertical intervals, whereas casing collapse produces either simple ovality in shape or ovality with flattening on one side of the casing. These devices have no orientation detectors; consequently, it may be difficult to track gradual wear along the casing length. Rotation of the tool as it moves upwards places first one set of feelers in the worn area and then another set. The traces from tools with 64-plus feelers are so closely spaced on the display that a "wave" of change in deflection will sweep across the record as the tool rotates. These waves identify locations where the deflections of individual feelers can be examined for circumferential detail. These devices also respond to bends in the casing, provided the radius of the bend is small enough to displace the tool body off the centerline of the casing as the instrument moves through the bend. The signature is a diameter with one radius smaller than normal and the other larger. Collapse and bending often occur in concert, so that a detailed picture of the cross-section is needed if the two affects are to be separated.
	Reflection-type radial-bond log	This device has a rotating transmitter crystal that is pulsed at ultrasonic frequency as it rotates. Between pulses, the crystal records the echo wave train from the casing wall. When operated in the casing-inspection mode, the device records first the transit time for the echo from the inside wall of the casing as well as the transit time through the wellbore liquid over a fixed distance. The two measurements provide a pipe radius corresponding to each pulse. These radii are presented as a colored, variable-density map over the circumference of the inside wall. In deviated wells, the tool provides an oriented display to remove the effects of tool rotation. Each reflected wave train is also processed to identify the resonant frequency, if present, in the portion following the echo from the inside wall. This frequency is inversely proportional to wall thickness. This information is displayed on a second colored, variable-density map of wall thickness over the circumference of the casing. This display can also be oriented to remove the effect of rotation. These displays of caliper and wall thickness are easy to interpret and are more reliable than results from operation in the cement-bond mode. Still, the accuracy is less than that provided by mechanical calipers. The thickness measurements are highly sensitive to small displacements of the tool off pipe center. In fact, at shallow depths, sufficient centralization sometimes cannot be maintained for the tool to even operate.

TABLE 4.A7—2B: CASING INSPECTION FOR WEAR, COLLAPSE, AND BENDS (NEWLY CASED, UNPERFORATED WELL) (Continued)

Problem	Recommended Logging Tools	Operational Tips and Expected Results
Casing wear	Eddy-current wall-thickness log	This is an electromagnetic induction device having very high resolution to changes in wall thickness. The normal taper in a joint of seamless casing (~0.125 in. over 40 ft) can be spread easily over three or more chart divisions. The tool uses a coil driven at a very low frequency (~ 1 Hz) to induce a field throughout the thickness of the wall. A passive coil is used to measure the lag angle of the induced field behind the driven field. Ideally, this angle is directly proportional to wall thickness. Although included on the log, the "caliper" trace from a coil driven at high frequency is responsive primarily to sudden changes in internal radius. Normal taper is difficult to discern on this trace. Furthermore, the casing versions of this tool are too long and of too large a diameter to negotiate bends.
Casing collapse (or split or part)	Tight-spot survey	These are minimum caliper tools containing one set of centralizer arms hinged at center and whose maximum compression is recorded. The devices tend to remain on pipe center even in sections of collapsed casing. The tools are short and small in diameter, and are less likely to stick in the well while providing some idea of the severity of collapse.
	Multifeeler mechanical-caliper survey	As indicated previously, this is the best device available for defining the shape of the cross section of casing. In collapsed casing, the record will reveal an oval shape with perhaps flattening along one side of the casing. The tools are long and strongly centralized. Safety precautions may preclude their use until the severity of the problem is known.
	Acoustical caliper survey (radial, reflection-type bond log)	The "caliper" map provided by this "pulse-echo" tool is described in this table under Casing Wear. Because of its strong centralization and relatively large diameter (2 or more in.), the tool should only be run where severity of collapse is known.
	Casing-collar log	A simple collar record is usually sufficient to detect parted casing. The tool should be run with a decentralizing magnet on the bottom. Relative to a normal casing signal, a large deflection is induced as the tool passes across each end of the parted casing. The amount of part is determined by the spacing between the centers of the signal.

TABLE 4.A7—2B: CASING INSPECTION FOR WEAR, COLLAPSE, AND BENDS (NEWLY CASED, UNPERFORATED WELL) (Continued)

Problem	Recommended Logging Tools	Operational Tips and Expected Results
Casing bends (dog legs)	Kink-meter or dog-leg survey	These are relatively compact devices that measure the inclination of the tool's axis from the vertical. This measurement is typically made with a pendulum free to rotate around the tool axis. Passes upward and downward or successive passes in the same direction may produce different dog-leg angles because of the way the tool is displaced off casing center on a given pass. Despite this sensitivity to offset, the tool provides an estimate of radius of curvature in a relatively safe procedure. These devices should thus be the first ones run. However, they cannot detect the presence of collapsed casing in conjunction with a bend.
	Multifeeler mechanical-caliper survey	As described previously under Casing Wear, these devices respond to bends because of the displacement of the tool off pipe center as it navigates a bend. The classic signature of a pure bend is therefore a diameter with one radius too short and the other too long. A bend with collapse is therefore easily distinguished once the existence of the bend is established. The radius of the bend must be larger than ideal if the casing is also out of round.
Condition of external casing strings	Eddy-current wall-thickness survey	These tools are electromagnetic field induction devices. They utilize a coil driven at very low frequency (~1 Hz) to spread a magnetic field within a radius of 8 to 10 in. about the coil. A second sensing coil is used to detect the lag angle between the driven field and the induced field. This lag is proportional to the amount of metal within the induced field. The angle is therefore influenced by the first two casing strings external to the one containing the tool. Collars on the first string external to the string containing the device are clearly evident on the trace of phase lag. Accordingly, the trace is sensitive to parts in the casing external to the one containing the tool. Likewise, the trace will detect significant holes in the external string. However, it will not detect splits along the axis of the external string unless these have expanded into holes. Finally, the trace will usually detect an end to the second external string.

TABLE 4.A8—2C: ANNULAR PRESSURE, CROSSFLOW BEHIND CASING, AND CERTIFICATION FOR HAZARDOUS WASTE INJECTION (NEWLY CASED, UNPERFORATED WELL)

Problem	Recommended Logging Tools	Operational Tips and Expected Results
Location of source of annular pressure	Temperature surveys before and during (or after) annulus bleed	The before and after traces should be merged on a display with a sensitivity of 1°F per in. of chart. A second display should be at a reduced sensitivity sufficient to prevent "wrapping" over the logged interval. Because the wellbore is new and cooler than normal geothermal temperature, any movement of liquid to the well will produce a warm signature at the source depth. This makes the survey quite sensitive to low rates. The warmer temperature will be carried upward from the source in proportion to the flow rate. For annular flow behind the string containing the tool, liquid rated can be estimated from the slope *dT/dz*, °F/ft, and amount of warming $\Delta T = T_{\text{flow}} - T_{\text{static}}$, °F, at a given location in the crossflow path from the Romero-Juarez expression: $q\ (\text{B/D}) = A \times \dfrac{\Delta T}{\left(\dfrac{dT}{dz}\right)}$, where A = 1.0 for water flow and A = 2.5 for oil flow. The result is an approximation that supposes no cooling to the liquid as a result of the dissolution of gas from water or the flashing of gas from oil. The expression underestimates rates for short-time flow behind multiple strings. Expansion at a gas source usually produces a cool signature. For dry gas, this cooling amounts to approximately 1°F for each 50-psi drop in pressure. This cooler temperature is carried upward in proportion to the gas rate. This rate can be approximated from the previous expression with A = 10.

TABLE 4.A8—2C: ANNULAR PRESSURE, CROSSFLOW BEHIND CASING, AND CERTIFICATION FOR HAZARDOUS WASTE INJECTION (NEWLY CASED, UNPERFORATED WELL) (Continued)

Problem	Recommended Logging Tools	Operational Tips and Expected Results
Location of source of annular pressure	Noise-logging surveys run in same way as temperature surveys	This survey is particularly sensitive to the percolation of a small rate of gas into liquid, a condition where the temperature survey loses sensitivity. The tool is a microphone that measures the amount of turbulent noise produced by the movement of fluids. It makes stationary measurements of the sound level, in millivolts, above four high-pass cutoff frequencies, typically set at 200, 600, 1,000, and 2,000 Hz. Each reading includes levels at all frequencies above the cutoff values; consequently, the readings decrease with increasing cutoff frequency. The cutoff values are chosen so as to distinguish two flow conditions: single-phase and gas movement through liquid. The movement of gas through liquid produces a high level of noise in the low-frequency range below 600 Hz. This fact makes the tool particularly sensitive to small gas flows through liquid, the application for which it was devised. Gas percolation at downhole rates as low as 10 ft^3/d can be resolved. Below 400 ft^3/d, gas rate behind drillpipe can be estimated from the millivolts of noise in the band N_{200}–N_{600} from the relationship $$q_g(\text{ft}^3/\text{d}) = 0.35 \times N_p \times (N_{200} - N_{600}),$$ where N_p is the number of pipe strings between the source and the tool. Higher rates are underestimated by this expression. Single-phase flow, by contrast, produces noise primarily in response to acceleration past tight spots in the flow path. The resulting sound is of relatively weak amplitude and has a frequency in the range of 1,000 Hz or higher. If a restriction holds a pressure drop Δp psi in response to a rate q B/D, then the single-phase rate of either liquid or gas behind drillpipe can be estimated from the millivolt value of the noise cut at 1,000 Hz by the expression $$q \times \Delta p = 700 \times N_p \times N_{1000},$$ where the pressure drop Δp is in psi, the rate q is in B/D and N_p is the number of pipe strings between source and tool. If the total pressure drop over the entire flow path is used in this expression, then the summed noise levels at each constriction should also be used. Both of the previous flow expressions suppose that the tool is calibrated to read standard peak-to-peak millivolts. The tool makes stationary measurements, meaning that approximately 60 to 80 locations can be surveyed on a typical run. Sample density can be compressed at locations such as suspect zones, cement tops, and casing shoes. Otherwise, 50-ft increments are typical.

TABLE 4.A8—2C: ANNULAR PRESSURE, CROSSFLOW BEHIND CASING, AND CERTIFICATION FOR HAZARDOUS WASTE INJECTION (NEWLY CASED, UNPERFORATED WELL) (Continued)

Problem	Recommended Logging Tools	Operational Tips and Expected Results
Location of source of annular pressure	Activated-oxygen survey for water movement	The operating principles for this device are described previously in Table 1B under Lost Returns. Because the direction of flow is generally unknown, the tool must be configured for water movement either upward or downward. For some service companies, this may require two separate runs into the hole. The "calibration" feature is a measurement, preferably in a shale zone, in a location where flow behind casing is absent. This calibration is essential to the reliable detection of small crossflow rates near the tool's resolution (2 to 3 ft/min water velocity).
Certification for annular disposal of mud and cuttings	Temperature surveys before and during the injection of mud or brine down the annulus	This procedure is sometimes called a "pump-in test" to determine the base of injection. The two surveys should be merged on a common display at a sensitivity of approximately 2°F per in. of chart. The injection trace will separate, usually to the cool side, from the "before" trace down to the base of injection, where it will begin an exponential approach to the "before" trace. The coincidence of the two traces does not occur immediately below the base of injection owing to the combination of tool response time and thermal conduction along the wall of the casing containing the tool. Still, base detection is easy.
	Radioactive tracer surveys during and after injection of tracer into annulus at surface	Some states require this test for certification, although it is both expensive and often misleading unless conducted carefully. The cost arises from the use of two service companies, one to pump the tracer at the surface and another to do the logging. The potential for erroneous interpretation arises from the tendency for pockets of tracer to be trapped in pockets in the annular space and thereby move down the hole behind the main slug. To prevent this stringing out of tracer from clouding the interpretation, pumping should be continued after the main slug has been pumped away until all trailing slugs are also pumped away. A common failing is to stop the pumping too soon and then to detect tracer above the target zone, a condition often attributed to seal failure on the part of the external casing shoe above the zone. The injection of 5 to 10 millicuries of tracer at the surface is generally sufficient to track injection for approximately 5,000 ft down the annulus. Before injection commences, a section of gamma-ray log should be recorded past the target zone so as to define the normal gamma activity in the hole and the noise statistics for the tool.

TABLE 4.A8—2C: ANNULAR PRESSURE, CROSSFLOW BEHIND CASING, AND CERTIFICATION FOR HAZARDOUS WASTE INJECTION (NEWLY CASED, UNPERFORATED WELL) (Continued)

Problem	Recommended Logging Tools	Operational Tips and Expected Results
Certification for the injection of hazardous waste	Activated-oxygen survey configured for flow upward	Federal regulations require this survey in a well penetrating drinking-water sands even before the well is perforated for hazardous injection. This requirement is to insure that the wellbore does not allow crossflow of brine into depleted-water sands (so-called "Unrelated Flow"). Contemporary practice, however, is to delay the survey until the well is perforated and to make the measurements before and with the well on injection. The procedure is the same either way. Operational principles are described previously under Annular Pressure. "Calibration" measurements are first made in a deeper shale where crossflow is not expected. The tool is then positioned for measurements in first shale below the deepest drinking-water sand. The detection of water moving upward behind casing at a velocity in excess of 3 ft/min prevents certification. Precautions should be taken to prevent water circulation inside the logged string. For example, heavy mud should not be circulated out of the hole with brine before the test. This prevents the slumping of viscous layers of mud that have adhered to the casing wall.
Detection of crossflow behind casing	See Table 4.A5 (1C): Detection of flow underground	The tool preference carries over intact to this environment. Likewise, the features to expect are unchanged. Because the new wellbore is cooler than geothermal, the temperature surveys are the most sensitive to water crossflow because of water's ability to produce warm signatures at low rates. Likewise, the noise logger remains the most sensitive to small flows of gas upward through water behind casing. If the rate expressions are used, then the noise levels must be multiplied by N_p = number of pipe strings separating sensor from source. Table 4.A5 (1C) assumes this value is 1.

TABLE 4.A9—3: ACID, FRACTURE-PROPPANT, OR GRAVEL-PLACEMENT (WELL BEING STIMULATED, FRACTURED, OR GRAVEL-PACKED)

Problem	Recommended Logging Tools	Operational Tips and Expected Results
Single-stage acid placement	Temperature surveys before and after acid placement	This is listed as the prime survey because the injection of acid often fractures into locations not intended. The two surveys should be merged on a display scaled at least at 2°F per in. of chart. The "after" temperature survey will show cool signatures at locations into which the flush pad was injected. Locations where acid remains close to the wellbore will show warm signatures if acid reacts with formation or mud solids. Otherwise, cool anomalies will appear. The base of injection is located by the convergence of the "after" trace with the 'before" trace. If ball diverters are used, the amount of acid injection is limited and may produce only small signatures.
	Radioactive tracer surveys before and after treatment	Both the acid and the flush liquid are tagged at the surface, typically with Iridium-192. The operation usually requires two contractors—one to pump tag at the surface, and one to do the logging. Inerpretation of the records is simple but subject to certain peculiarities. First, if the tagged fluid is displaced more than approximately 2 ft away from the wellbore (in a fracture, for example), then tag activity approaches background level. Second, tagged fluid tends to hang in collars along the wellbore and in eddies formed at the end of the treatment string.
Multistage acid placement	Temperature surveys before and after treatment	Ideally, a survey would be recorded after each stage, but such a practice is usually impractical. Before and After surveys are subject to conditions listed previously.
	Radioactive-tracer spectral surveys before and after treatment	For all stages except the last, only the diverting fluid is usually tagged. A different tracer is used for each stage so that a spectral survey (gamma-ray energy survey) can identify subsequently the location of each tracer. Metal tracers such as antimony (Sb), scandium (Sc), silver (Ag), and gold (Au) are used commonly in the form of powders with neutral buoyancy. The final acid stage and its flush are tagged with Iridium-192. The use of a spectral log has an advantage over a normal gamma-ray log. The energy levels can be used to make an estimate of the distance from the tool to the tracer. Tracer hang-up in the wellbore can thus be distinguished from displaced tracer. Two contractors are again usually require—one to tag at the surface and one to log.
Single-stage fracture placement	Temperature surveys including a "before" log and at least one "after" log recorded 12 to 48 hours later	This is the best procedure for the determination of the vertical extent of a fracture so far as fluid communication is concerned. The large volumes of pumped fluid produce large cool signatures on the "after" log. However, the fracture wings continue to close on the proppant for typically 12 hours or longer. Unequal rates of closure pumps liquid from one wing to the other during this time. This return produces warm signatures at locations where the liquid passes through the wellbore. For surveys run shortly after a fracturing operation, a warm "nose" may mark the top of a fracture. This activity is usually over after 24 to 48 hours so that the size of the cold signatures can be used to assess relative degrees of treatment.

TABLE 4.A9—3: ACID, FRACTURE-PROPPANT, OR GRAVEL-PLACEMENT (WELL BEING STIMULATED, FRACTURED, OR GRAVEL-PACKED) (Continued)

Problem	Recommended Logging Tools	Operational Tips and Expected Results
Single-stage fracture placement	Radioactive-tracer survey to locate tagged proppant	Special solids are available that can be made radioactive at location and added to the fracture proppant continuously. This approach avoids the transport of large amounts of precoated solids. If this is not a problem, then solids coated with resin containing Iridium-192 are available. The same interpretative restrictions mentioned previously for tagged acid apply here, particularly the problem of hang-up in collars. An additional problem is the inevitable settling of tagged sand to the bottom of the wellbore.
Multistage fracture placement	Temperature surveys before and 24 to 48 hours after treatment	Remarks concerning the long shut-in time for the "after" survey are given above for single-stage fracturing. The time is more critical here because the size of the cool signatures can be used to judge the relative effectiveness of each stage. The larger the cooling anomaly on the survey, the greater the displacement away from the wellbore.
	Spectral gamma-ray survey to locate radioactive tracers particular to each stage	The same metal tracers are used to tag solids as described previously for multistage acidization. The same contamination problems arise as for single-stage fracturing.
Gravel placement	Gravel-pack survey before and after placement	Most companies use an unfocused, gamma-ray tool for this survey. This device measures the gamma-ray transmission over 17 to 23 in. to determine the density of material in an 8- to 12-in. radius about the tool. Gravel behind the screen causes a large reduction in count rate relative to liquid. A uniformly low count rate is desired over the packed interval.
	Gravel-pack survey before and after placement	Some companies use a compensated neutron-porosity tool for this survey. Gravel behind the screen reduces the liquid porosity. Effective gravel placement will produce a relatively uniform porosity of approximately 40%. Whatever the type, the tool contains a radioactive source. If lost in the hole, it must be either recovered or cemented in place. Modern practice has tended towards the use of prepacked screens, making these types of surveys unnecessary. However, the unfocused gamma tool has remained in widespread use as an "interface" device used to locate the hydrocarbon/brine interface in storage caverns.
	Gamma-ray survey to detect tagged gravel	Tagged solids are added continuously at the surface to the gravel during its placement. The sole advantage to this approach is that some companies have gamma detectors that rotate to give an indication of the placement around the circumference of the screen. This type of information is useful for packs in holes having high deviation angles (horizontal wells).

TABLE 4.A10—4A: DEMONSTRATION OF CONFINEMENT (INJECTION WELL—CASED OR OPENHOLE WITHOUT SCREEN OR LINER)

Problem	Recommended Logging Tools	Operational Tips and Expected Results
Determination of extent of fluid injection	Three temperature surveys: One with well on injection and two, spaced approximately 12 hours apart, with the well shut in	Temperature surveys are essential to this problem because the majority of wells on continuous injection fracture at some time in their history. Flow in these fractures, either above or below the target interval, is detected by the temperature surveys but may be missed by radioactive-tracer surveys. The three surveys should be merged and have one display at a sensitivity of 1°F per in. of chart and one display free of "wraps." The survey recorded with the well on injection shows those locations where fluid is lost from the vicinity of the wellbore. Below each such location, the injection trace will shift toward the shut-in traces. Depending upon the amount of loss, the shift may be a step change in temperature or simply an increase in gradient. The base of injection is at the location where the injection trace begins an uninterrupted, exponential-type return to shut-in temperature. Movement upward behind casing to a thief zone will produce on the injection trace at the zone the same type of shift as described previously but of less magnitude. This behavior is caused by the slightly warmer temperature carried upward by the stream. A shut-in survey recorded after a few hours of shut-in time will show cool signatures at all the porosity taking injection, provided the trace is not corrupted by backflow from the closure of unpropped fractures. It is not a good survey from which to judge relative amounts of injection, however. A survey after 12 to 48 hours is much better suited to this purpose. Those zones having received larger volumes of injection (farther displacement of fluid away from the wellbore) will remain cooler than those zones taking lesser volumes. This distinction persists after years of injection, but the contrast shrinks to a fraction of a degree after approximately 6 months. If the well is on gas injection, then expansion at each injection zone may produce a temperature at that zone that is cooler than the temperature above. Dry gas expanding adiabatically cools approximately 1°F for each 50-psi drop in pressure. Other than this localized cooling, the injection trace behaves in the same manner described previously. Whenever gas moves behind casing, however, the shut-in traces may also show localized cooling at tight spots in the path as well as at receptor porosity.

TABLE 4.A10—4A: DEMONSTRATION OF CONFINEMENT (INJECTION WELL—CASED OR OPENHOLE WITHOUT SCREEN OR LINER) (Continued)

Problem	Recommended Logging Tools	Operational Tips and Expected Results
Determination of extent of water injection	Radioactive-tracer survey with the well on injection	These surveys make use of a tool consisting of two of more gamma detectors along with a reservoir filled with 50 to 100 cm^3 of field brine tagged with 5 to 10 microcuries of Iodine-131. A piston-type pump is operated at the surface to discharge discrete slugs of this tagged brine into the wellbore. The amount of tracer ejected on each "shot" is determined by the length of time the pump is on. For this particular problem, the best procedure is to eject a slug of tracer at least 50 ft above the target zone with the well on full injection. Successive logging runs are then made upward with one of the detectors active so as to locate the position of the slug at the time of each pass. These "chased-slug" passes are continued until the slug of tagged brine is pumped beyond recognition depth (~1 ft from tool). When merged, these successive records present a graphic picture of what happens to the slug as it moves down the wellbore. So long as the slug remains in the casing or open hole, it produces on the record a bell-shaped pattern of activity. However, as the leading edge of the slug moves into a confined space, such as the annulus behind casing, it begins to string out or unspool like a ribbon from the main slug. The "ribbon" will continue to unwind until the main slug passes the entry location. The signature of a "channel" (annular flow) is therefore unmistakably unique. The sensitivity of this survey to liquid movement immediately behind pipe is unsurpassed. A rate of only a few B/D can be detected in the manner described. The same sensitivity also prevails for the detection of crossflow in a shut-in wellbore. On the other hand, the tool may not detect flow in a fracture behind pipe unless the liquid remains in the immediate vicinity. The method suffers from two disadvantages. First, the steam can only be tagged while it is inside the wellbore. Second, the service is becoming difficult to obtain. Very few companies now run radioactive tracer surveys on offshore locations. Furthermore, tools for general use in gas wells are nonexistent except for certain special applications such as CO_2 injection (carbon-12) and steam injection (Xenon tag).

TABLE 4.A10—4A: DEMONSTRATION OF CONFINEMENT (INJECTION WELL—CASED OR OPENHOLE WITHOUT SCREEN OR LINER) (Continued)

Problem	Recommended Logging Tools	Operational Tips and Expected Results
Demonstration of injection-packer seal	Radioactive-tracer survey with well on injection	The same tracer tool as described above is best suited to this problem. However, the "chased-slug" approach described above is unsuitable because of the danger associated with the operation of pulling the tool into a tubing string at high logging speed. Instead, a stationary or "velocity-shot" approach is utilized. The tool is positioned according to the length of tailpipe extending below the packer. If the pipe is short, then the tool is placed in the tubing so that the top detector (usually located above the ejection sub) is approximately 3 ft above the packer-seal elements. A bottom detector (located below the ejector) should also be active for this procedure. Whether it is positioned in the tubing or below it is not important. With the tool stationary and the well on injection, a slug of tracer is ejected and records of activity made for each detector with the recorder on time drive. The record from the bottom detector should show the passage of tracer on its way down the hole. To satisfy regulatory constraints for hazardous liquids, the record should be continued for a time long enough to allow tracer to reach the top detector at a velocity of 1% of injection velocity. Typically, a 30-minute record satisfies this constraint.
	Activated-oxygen survey with well on injection	This is a pulsed-neutron tool that creates its own radioactive tracer (activated oxygen by bombardment of water) with neutrons and then times the passage of the tracer between two detectors. Stationary measurements give the best resolution. The device has two detectors spaced approximately 1 ft apart and 2 to 3 ft above the neutron source. On this test for flow upward, the detectors should be located above the source. Both detectors as well as the source should be inside the tubing string. With a tracer half-life of only 7.4 seconds, velocity resolution is poor compared to that for the chemical-tracer tool. However, some regulations allow the activation tool as a substitute for the latter.
"Channel" check for annular seal above and below perforations	Radioactive-tracer surveys with well on injection	If the chased-slug surveys described above under Injection Extent have been done, then this check has already been made. Otherwise, stationary velocity-shot records can be taken with the bottom detector positioned first above the perforations and then below (ejector in perforations). If tracer movement is detected below the perforations, then a slug should be ejected in the sump and logged on at least three passes.

TABLE 4.A11—4B: INJECTION PROFILE DETERMINATION (INJECTION WELL—CASED OR OPENHOLE WITHOUT SCREEN OR LINER)

Problem	Recommended Logging Tools	Operational Tips and Expected Results
Determine injection profile	Three temperature surveys: One with well on injection and two, spaced approximately 8 hours apart, with well shut in	Most wells that have been on continuous injection for any length of time are fractured. The distribution of the unpropped fractures along the borehole determines the locations where fluid exits the well. These locations, however, may not correspond to the locations of the porosity that ultimately stores the fluid. Temperature traces can detect this storage porosity. The trace with the well on injection shows the locations where fluid leaves the vicinity of the wellbore to enter porosity. At each location where fluid enters porosity or circulates in a fracture, the injection trace will exhibit a shift toward the shut-in traces. Depending upon the relative size of the loss, the shift may take the form of a step change (large loss) or a bend toward the shut-in traces (gradient increase at small loss). Starting at the top, one can examine the slope of the injection trace to identify the exit locations down to the base of injection where the trace returns to the shut-in traces. The volumetric rate remaining in the wellbore vicinity at any location is determined from the injection trace from its slope *dT/dz* and its displacement from the longest shut-in trace $\Delta T = T_{\text{flow}} - T_{\text{shut-in}}$ by the Romero-Juarez relationship $q = A \cdot \frac{\Delta T}{\frac{dT}{dz}}$ The value of the constant A is not required for relative profiling. Values for the ratio on the right side of this expression are determined from the top in the full stream downward to a location above the last exit at the base of injection. The profile is then the ratio of each value to the value in the full stream. The slopes should be determined far enough below an exit to escape the effect of the associated shift. For many injection wells, the displacement is relatively constant, meaning that only values for reciprocal slopes are needed. The profile determined in this manner lacks the accuracy of one obtained from either tracer surveys or flowmeter survey, but is typically a better representation of the actual injection profile. In a fractured wellbore, the temperature-based profile is usually quite different from either a tracer or a flowmeter survey. The long-time shut-in trace defines true storage locations as either cool or warm anomalies depending upon which side of the shut-in traces the injection trace is located. The trace will allow the elimination of "exits" that cause only circulation patterns in fractures and "exits" that arise from slope changes because of variations in rate. This problem exists because multiple injection wells are often attached to a single supply header.

TABLE 4.A11—4B: INJECTION PROFILE DETERMINATION (INJECTION WELL—CASED OR OPENHOLE WITHOUT SCREEN OR LINER) (Continued)		
Problem	Recommended Logging Tools	Operational Tips and Expected Results
Determine injection profile	Three temperature surveys: One with well on injection and two, spaced approximately 8 hours apart, with well shut in	When compared to the shorter-time shut-in trace, the long shut-in trace also allows an assessment of the historical progression of the injection program. Those zones that have taken a large volume of fluid will show anomalies whose sizes change more slowly with time than do the anomalies at zones that have received lesser volumes. This condition persists after years of injection, but the differences shrink quickly to fractions of a degree.
	Radioactive tracer surveys with well on injection at rates no larger than approximately 5,000 B/D	This tool is one next most likely to detect flow in fractures as long as the stream remains within ~1 ft of the tool. The device is described in Table 4.A10 (4A), to which the reader should refer before continuing here. A merged record of chased-slug surveys should be obtained first to obtain a graphic picture of the tracer slug as it moves down the hole. An exit profile can be derived from this record in two ways. The most common approach is to record the time at which the peak activity is attained on each pass through the ejected slug. An injection velocity is then determined from the difference in peak location and peak time for two adjacent slugs. These velocities are then proportioned for the profile. A second, preferred approach is the slug-area method that equates the area under the activity trace for a given slug to the volumetric rate in the hole at that location. A ratio of areas to that in total flow then provides the profile. This approach is unaffected by variations in rate and retains its accuracy for annular flow. Because only a ratio of areas is involved, one can determine quickly a value that is proportional to area by taking the product of slug height (in millimeters above base activity, for example) and the slug width at a distance above the base of one-half the peak height. If the curve is Gaussian, then this number is 14% smaller than the actual area. At injection rates greater than a few hundred B/D, the chased-slug approach gives a vertical resolution of only 10 to 20 ft. To supplement this data, approximately 10 to 12 stationary velocity-shot records should be made at different locations in the hole. A slug of tracer is ejected above two detectors spaced 4 to 10 ft apart with the recorder on time drive. A travel time between two detectors is determined from each shot record. These values are ratioed inversely for profiling. Vertical resolution less than detector spacing is obtained from measurements on closer stations. Such travel times are quite sensitive to variations in rate, however.

TABLE 4.A11—4B: INJECTION PROFILE DETERMINATION (INJECTION WELL—CASED OR OPENHOLE WITHOUT SCREEN OR LINER) (Continued)

Problem	Recommended Logging Tools	Operational Tips and Expected Results
Determine injection profile	Diverting spinner-flowmeter surveys with well on injection at rates no larger than 2,000 B/D	These are the most sensitive of the spinner-type flowmeters with a resolution in the range of 10 to 15 B/D. They make stationary measurements using either a packer or extended metal vanes to divert the flow past a spinner element contained in the barrel of the tool. Because of the limited clearance between tool wall and spinner element, these devices plug easily. They should not be run in highly deviated wells where debris tends to settle to the low side of the casing everywhere. Profiling from these stationary measurements is straightforward. The service company can also usually provide a calibration curve for single-phase flow giving rate as a function of spinner speed. The metal-petal devices have been known to stick in the end of tubing strings owing to a failure of the petals to close into the tool body. Furthermore, all spinner devices have the disadvantage of identifying only locations where fluid exits the casing or open hole.
	Continuous or fullbore spinner surveys both with well on injection and shut-in	These devices produce continuous traces of spinner speed, RPS, as a function of depth in the wellbore. Other than at very low spin rates, the speed is linearly related to fluid velocity relative to the tool. Profiling is therefore a simple matter of proportioning the changes in spinner speed from that speed induced by line speed. Continuous flowmeters have spinner elements whose diameter is slightly smaller than the diameter of the tool itself, whereas fullbore flowmeters have large-diameter elements hinged so as to open only when the tools emerge from the end of tubing. Both versions have poorer rate resolution than do the diverting types. Depending upon casing size, the resolution for either type is in the range of 70 to 200 B/D for liquid and 100 to 500 Mcf/D for gas.

TABLE 4.A11—4B: INJECTION PROFILE DETERMINATION (INJECTION WELL—CASED OR OPENHOLE WITHOUT SCREEN OR LINER) (Continued)

Problem	Recommended Logging Tools	Operational Tips and Expected Results
Determine injection profile	Continuous or fullbore spinner surveys both with well on injection and shut-in	The principal difference between the two types of spinners lies in the fluid velocity required to start the spinner turning with the tool stationary (i.e., in the threshold velocity). This value depends upon the density of the fluid, the frictional resistance in the tool bearings, and the amount of diversion of fluid around the element rather than past it. Fullbore spinners and high-sensitivity continuous spinners both have a threshold velocity in the range of 5 to 8 ft/min in liquid. In gas, however, the fullbore tools have a much smaller threshold velocity (10 to 15 ft/min) than do the continuous tools (25 to 50 ft/min). Both types typically use an element with a pitch of approximately 6 in. This is the distance fluid must move to turn the element one revolution. The corresponding theoretical sensitivity is 0.050 RPS per ft/min. Actual values are more like 0.044 RPS per ft/min. Values for sensitivity and threshold velocity are best determined by calibration in the well shut-in so as to avoid the influence of rate variations. At least three passes at sequentially doubled line speeds should be made for tool movement both upward and downward in the wellbore. The slowest line speed should be sufficient to turn the spinner at a speed of at least 1 RPS. Even after such a calibration, one must still have an independently determined tool factor to convert spinner speed into the superficial velocity required for rate calculation. This factor usually comes from flow-loop data. The most reliable injection profile is obtained from the pass made against flow (upward) at a line speed no greater than 30 ft/min in liquid and 50 to 70 ft/min in gas. Common practice, however, is to log in both directions at three line speeds.

TABLE 4.A12—5A: DEMONSTRATION OF CONFINEMENT [INJECTION WELL WITH SCREEN OR SLOTTED (PERFORATED) LINER]

Problem	Recommended Logging Tools	Operational Tips and Expected Results
Determination of injection extent	Three temperature surveys: one with well on injection and two, spaced approximately 8 hours apart, with well shut in	This completion is just as subject to formation fracturing as is the completion described in Table 4.A10 (4A). The user should refer to Table 4.A10 (4A) for details of this procedure. Briefly, on the flowing trace, each exit will produce the normal shift back toward the shut-in traces, even though the exit is behind the screen or liner. One new feature is introduced by this completion. Because of a partially plugged screen or widely spaced holes in the liner, fluid may exit the screen or liner and then move upward to a higher receptor zone. In this case, one will observe what appears to be a normal exit signature at both the receptor and exit locations, but both are muted and, in fact, may appear only as "kinks" to an increased gradient between the receptor zone and the exit.
	Radioactive-tracer surveys with well on injection	The use of this tool is also described in Table 4.A10 (4A). The chased-slug approach is recommended for completion because of the variation in fluid velocity caused by changes in gravel porosity or hole size around a liner. Slug area is essentially preserved when the tracer passes into the annulus behind the screen or liner even though the slug may lose its "bell" shape. The chased-slug records will also reveal those locations where fluid exits the screen or liner and moves upward to a higher receptor zone.
Demonstration of packer seal	Radioactive-tracer surveys with well on injection	As described in Table 4.A10 (4A), the stationary, velocity-shot procedure should be used for this problem.
	Activated-oxygen survey with well on injection	The procedure is the same as that described in Table 4.A10 (4A). The tool is a substitute for the chemical-tracer device in those cases where the latter is unavailable. It provides only stationary, velocity-shot type data. Its velocity resolution, 2 to 3 ft/min, is not nearly as good as that of the chemical-tracer device.
"Channel" check for annular seal above and below perforations	Radioactive-tracer survey (activated-oxygen survey as substitute)	Same procedure as described in Table 4.A10 (4A).

TABLE 4.A13—5B: INJECTION PROFILE DETERMINATION [INJECTION WELL WITH SCREEN OR SLOTTED (PERFORATED) LINER]		
Problem	Recommended Logging Tools	Operational Tips and Expected Results
Determine injection profile	Three temperature surveys: one with well on injection and two, spaced approximately 8 hours apart, with well shut in	The possibility of flow in fractures makes this the preferred initial approach. The procedure is the same as that described in Table 4.A11 (4B), to which the reader should refer. One new difficulty is introduced by this completion. Owing to plugged screens or to widely spaced holes in liner, flow may exit screen or liner and move back upwards to a higher receptor zone. Normal but muted exit signatures will appear both at the location of the receptor and at the location of the exit. However, the circulation gradient existing between these locations does not satisfy the rate-estimation expression given in Table 4.A11 (4B). Only qualitative assessments based on exit-signature size can be made for such rates from the flowing trace. The same sort of assessment can also be made from the trace made after the longest shut-in time.
	Radioactive-tracer surveys with well on injection at rates no larger than approximately 5,000 B/D	In the absence of flow in fractures, this is the most accurate method of profiling. The procedure is described in Table 4.A11 (4B). Because of the mixture of flow both inside and outside the screen/liner, the chased-slug approach should be used and the profiling should be based on slug area. This approach remains valid for slugs that exit the screen/liner and move "strung out" upward or downward in the annulus.
	Activated-oxygen surveys with well on injection	This tool is a pulsed-neutron device that creates its own tracer, activated oxygen, by neutron bombardment of water. It makes station measurements of the travel time of the tracer between detectors spaced approximately 1 ft apart. Hence, the device gives data corresponding to velocity-shot data from a chemical-tracer tool. However, the velocity resolution is much less at approximately 2-3 ft/min. The profile may also be only qualitative owing to distortion from variations in fluid velocity. The possibility for flow either upward or downward means that the tool must be configured for both cases. For some companies, this means separate trips into the hole. Furthermore, if lost in the hole, the tool must either be "fished" out or cemented in place.
	Continuous flowmeter surveys with well on injection	These devices are described in Table 4.A11 (4B). They respond only to flow inside the screen/liner. Consequently, the profile is qualitative at best.

TABLE 4.A14—6A: DEMONSTRATION OF CONFINEMENT [INJECTION WELL WITH TUBING ACROSS PERFORATIONS (DUAL COMPLETIONS, REGULATED INJECTION)]

Problem	Recommended Logging Tools	Operational Tips and Expected Results
Determine extent of injection	Three temperature surveys: One with well on injection and two, spaced about 8 hours apart, with well shut in	This type of completion is no less susceptible to formation fracturing than are the other types considered previously. Temperature surveys are therefore the preferred initial logging method. The procedure is described in Table 4.A10 (4A), to which the reader should refer. A new feature is introudced for wells with tailpipe extending across perforations and for wells injecting from a tubing string equipped with regulator valves. In such cases, fluid may exit the tubing and then flow back upward to receptor zones. These locations are recognized on the injection trace by the muted exit signatures (shifts in temperature or bending towards shut-in traces) that appear at both receptor and exit locations. The receptor zones are also evident on the shut-in traces.
	Radioactive tracer surveys with well on injection	This method is not suitable for dual completions for which logging must be done in the shut-in long string. The procedure is described in Table 4.A10 (4A). The chased-slug approach is recommended because of the likely presence of flow in both directions behind the tubing string.
	Activated-oxygen surveys with well on injection	This is the only tracer method possible for dual completions. The device creates its own tracer by the bombardment of water with pulsed neutrons. The passage of the tracer between detectors spaced approximately 1 ft apart is timed to give data comparable to velocity-shot data from a chemical-tracer survey. If flow upward as well as downward is possible, then the tool must be configured for both. For some companies, this requires separate trips into the hole. A lost tool must be fished out or cemented in place.
Demonstration of packer seals	Radioactive-tracer survey or activated-oxygen survey with the well on injection	Stationary velocity shots are used for this test. The procedure is described in Table 4.A10 (4A) for both tools. In dual completions, only the activated-oxygen survey is available. Its relatively low resolution of 2 to 3 ft/min velocity may not satisfy some regulations. A lost tool must be recovered or cemented in place.
"Channel" check for annular seal	Radioactive-tracer (or activated-oxygen surveys) on injection	See Table 4.A10 (4A) for procedure. In dual completions, only the activated-oxygen survey is available. Tool must be configured for annular flow both above and below perforations.

TABLE 4.A15—6B: INJECTION PROFILE DETERMINATION [INJECTION WELL WITH TUBING ACROSS PERFORATIONS (DUAL COMPLETIONS, REGULATED INJECTION)]

Problem	Recommended Logging Tools	Operational Tips and Expected Results
Determine injection profile	Three temperature surveys: one with well on injection and two, spaced approximately 8 hours apart, with well shut in	Flow in unpropped fractures is common to all injection wells; consequently, temperature surveys are the best approach initially. The procedure for flow profiling from the injection trace is described in Table 4.A11 (4B), to which the reader should refer. In a dual completion, the procedure is unchanged from that described in Table 4.A11 (4B). For injection through tubing-string regulator valves or into wells with tailpipe extending into the perforations or open hole, a new feature is present. Exits from the tubing may move upward, downward or in both directions as the stream splits. Those exits giving rise to flow upwards to a receptor zone cause, on the injection trace, normal but muted exit signatures at both exit location and receptor location. The circulation gradient existing between the two locations, however, does not satisfy the profiling expression given in Table 4.A11 (4B). The injection trace gives only a qualitative idea of the rates based on the relative size of the exit signatures. Similar information is available from the two shut-in surveys.
	Radioactive-tracer surveys with well on injection at rates no larger than approximately 5,000 B/D	This approach is not applicable to dual completions because the flow is entirely external to the long tubing string. In the other completions, the tool is the next one most likely to detect flow in fractures. The logging procedure is described in Table 4.A11 (4B), to which the reader should refer before going on to the special features unique to these completions. The chased-slug procedure is the appropriate approach because of flow both inside and outside the tubing string. Slug areas must be used for profiling. If the completion is one with a long tailpipe, then the slug should be injected inside the tubing string so that it can be captured just below its exit from the end of tubing. Such a record reveals any split into upward and downward-moving components and provides the area representative of the full stream. If the completion is one of regulator valves in a plugged tubing string, then the first slug should be ejected above the top valve and then chased past all deeper valves so as to determine the relative exits from each valve. Additional slugs should then be ejected above each regulator valve so as to capture the slug as it first passes into the annulus and on its subsequent travel to receptor zones. This approach provides a slug area that is compensated both for dispersion in the annulus and for the minor amount of attenuation offered by the tubing wall. The dispersion can affect the area significantly because it changes the distribution of tracer around the detector.

TABLE 4.A15—6B: INJECTION PROFILE DETERMINATION [INJECTION WELL WITH TUBING ACROSS PERFORATIONS (DUAL COMPLETIONS, REGULATED INJECTION)] (Continued)

Problem	Recommended Logging Tools	Operational Tips and Expected Results
Determine injection profile	Activated-oxygen survey with well on injection	This is the only tracer survey available for a dual completion. In this case, the tool need be configured only for flow downward. The best resolution is achieved from stationary measurements. This device creates its own tracer by the bombardment of water with pulsed neutrons. For one version of the tool, the passage of the slug past two detectors spaced 1 ft apart is timed directly. A second version of the tool determines velocity from the decay in count rate as measured simultaneously at two detectors spaced approximately the same distance apart. Either way, the tool functions as a velocity-shot device for annular flow behind the long string. Tool resolution is approximately 2 to 3 ft/min. In wells with extended tailpipe, the tool must be configured for flow both upward and downward. Area corrections must be made to the velocities of flow upward behind the tailpipe relative to velocities measured below the end of the pipe. The device is not suitable for use in wells on injection through regulator valves. Any flow downward behind tubing is obscured by the large signal from flow inside the tubing.
	Continuous-spinner surveys with well on injection	Continuous-spinner flowmeters can be used to determine rates inside the tubing string of a well on injection through regulator valves. Of course, no information is provided relative to flow behind the tubing string. The device is described in Table 4.A11 (4B) where its calibration is discussed. This aspect is not important for use in a small tubing string because the diversion around the element is insignificant. Furthermore, the distinction between spinner-derived fluid velocity and superficial velocity is moot. The tool should operate at a sensitivity equal to the theoretical value for the element's pitch.

TABLE 4.A16—7A: TRACKING SINGLE-PHASE PRODUCTION (PRODUCTION WELL—VERTICAL, CASED, OR OPENHOLE WITHOUT SCREEN OR PERFORATED LINER)

Problem	Recommended Logging Tools	Operational Tips and Expected Results
Profile production	Continuous or fullbore spinner flowmeter survey both with well on production and shut in	Either type of device is suitable for liquid production, but the fullbore type is preferable for gas production. The continuous types have either propeller or turbine (screw thread) elements having a diameter slightly less than the diameter of the tool barrel. The fullbore types have four longer vanes that are hinged so as to open only when the tool emerges from the production tubing. Both devices are run continuously to give a record of spinner turn rate in RPS as a function of depth. Common practice is to log both upward and downward at three different line speeds in each direction. Actually, the most reliable profile is usually obtained from the pass made against the flow (downward) at a line speed of approximately 30 ft/min. This pass is least biased by variations in fluid velocity and gives the best vertical resolution. Depending upon casing size, both types of spinners have rate resolutions in the range of 70 to 200 B/D for liquid or 100 to 500 Mcf/D for gas. The principal difference between the two types lies in the amount of fluid that diverts around the spinner element without turning it. This effect is incorporated into a threshold velocity V_{th} (i.e., a velocity required to turn the element with the tool stationary). In liquid, both fullbore and high-sensitivity continuous tools have a threshold velocity in the range of 5 to 8 ft/min. In gas, however, the fullbore type has a lower threshold (10 to 15 ft/min) than does the continuous type (25 to 50 ft/min). Once the elements are turning at rates in excess of approximately 0.5 RPS, both types produce signals that are related linearly to flow velocity relative to the tool. Thus, spinner turn rate, S RPS, is proportional to relative velocity, V_{rel}, $S = a \cdot (V_{rel} - V_{th}) = a \cdot (V_{line} + V_{fluid} - V_{th})$ A relative velocity directed up the tool barrel turns the spinner in a clockwise direction. The sign conventions in the expression assign a + sign to both spinner speed and threshold velocity when the spinner turns clockwise. Both are negative for counterclockwise spin. Line speed is positive for downward logging and negative for upward passes. Fluid velocity is then positive for flow upward and negative for flow downward. On passes upward with the flow, the spinner first turns counterclockwise until fluid velocity increases to within threshold value of line speed and the spinner stops turning. An additional increase in fluid velocity equal to twice the threshold value starts the spinner turning again, but in a clockwise direction. The constant a is a sensitivity value for the element expressed in units of RPS per ft/min relative velocity.

TABLE 4.A16—7A: TRACKING SINGLE-PHASE PRODUCTION (PRODUCTION WELL—VERTICAL, CASED, OR OPENHOLE WITHOUT SCREEN OR PERFORATED LINER) (Continued)

Problem	Recommended Logging Tools	Operational Tips and Expected Results
Profile production	Continuous or fullbore spinner flowmeter survey both with well on production and shut in	This value is related to the pitch of the element, P_{inches}, as a = 12/(60P). A common pitch is 4 in., corresponding to a sensitivity of 0.05 RPS/(ft/min). The actual value is approximately 0.044.
		Values for sensitivity and threshold velocity are best determined from calibration pass in a shut-in well. The well is logged both upward and downward, each at three progressively doubled line speeds, starting with a speed sufficient to turn the spinner no slower than 1 RPS. For these passes, the relative velocity is the line speed, which is given a + sign for travel down and a – sign for travel up. The threshold velocity is similarly signed.
		The number produced for fluid velocity by a spinner tool in casing is larger than the superficial velocity needed for volumetric rate calculation. The value of the reduction factor depends on the element size relative to casing size. It is usually determined for a given pipe size from measurements in a flow loop.
		Owing to tool inertia, both devices lack the vertical resolution needed to locate precisely entries within perforated intervals. A very localized entry can be smeared out over 10-20 ft. Within such intervals, the spin rate can either increase or decrease in response to lateral velocities from perforations. Furthermore, high rates can carry for over 100 ft up the well any swirl components induced by entries.
	Diverting-spinner surveys for rates below 2,000 B/D	The devices make station measurements using an inflated packer or extended metal vanes to divert flow past a turbine element inside the tool barrel. Consequently, their rate resolution, 10 to 15 B/D, is much better than that of the continuous devices. Packers usually provide complete diversion, so that spin rate is linear in B/D rate. A single calibration coefficient is therefore sufficient. The metal vanes, used more commonly, leak so that diversion is not complete. Calibration curves for specific casing sizes are therefore needed. The vanes sometimes jam open, sticking the tool as it enters the end of tubing.
	Density survey	Oil rates in excess of 400 B/D and gas rates in excess of 200 Mcf/D are required to remove completion liquid from a vertical wellbore. Productive locations below the depth at which such velocities are attained are therefore submerged in completion liquid. Hydrocarbon percolates upward through this liquid at "slippage" velocities in the range of 25 ft/min for oil in water and 45 ft/min for gas in water. In the process, the continuous-liquid phase is churned up and down at equivalent lift velocities. Spinner speeds tend to reflect these higher velocities and are thus not comparable to speeds measured higher in the hole. Instead, rates are typically determined from percolation theory. For a casing capacity C bbl/ft, a slippage velocity V_S ft/min and an hydrocarbon holdup y_{hc}, the B/D rate is $$q_{hc} = 1440\, y_{hc} V_s C$$

TABLE 4.A16—7A: TRACKING SINGLE-PHASE PRODUCTION (PRODUCTION WELL—VERTICAL, CASED, OR OPENHOLE WITHOUT SCREEN OR PERFORATED LINER) (Continued)

Problem	Recommended Logging Tools	Operational Tips and Expected Results
Profile production	Density survey	The density survey not only identifies the depth below which percolation occurs, but also provides a value for hydrocarbon holdup from the density of the mixture as $$y_{hc} = \frac{\rho_{\text{liquid}} - \rho_{\text{mixture}}}{\rho_{\text{liquid}} - \rho_{hc}}$$ Two types of density tools are available. One is a differential pressure device that measures pressure gradient over a distance of approximately 2 ft. These devices are linear and need only two calibration points, air and water. The second type is a gamma-ray transmission device that uses a tightly focused chemical gamma source at the bottom of a cage approximately 8 in. in length, having detectors placed at its top. Owing to gamma scatter out of the cage, these devices are usually nonlinear, requiring three calibration points. Either type is suitable for this problem. However, if the gamma device is lost in the hole, then it must be fished out or cemented in place.
	Temperature surveys with the well on production and at least one with the well shut in for 1 to 6 hours	The flowing and shut-in traces should be merged on a display of 1°F per in. of chart. Another display should be made at a reduced scale that eliminates "wraps." Even for single-phase production, temperature surveys are still valuable. In the first place, hydrocarbon can travel to perforations from locations behind casing. The base of production is at the location where the merged surveys begin to separate. For oil, the flowing trace "stands up" to the warm side of the shut-in trace(s). A gas source typically moves the flowing trace to the cool side of the shut-in owing to expansion cooling. If either oil or gas moves downward behind casing to perforations, then the flowing trace will show a "mixing" signature at the source in the form of a shift or a bend toward shut-in temperature. At the perforations, the temperature will warm upon shut in. Two shut-in traces best illustrate this direction. Second, the vertical resolution on a flowing temperature trace within a perforated interval is much superior to that of a trace from a spinner flowmeter. Over a productive interval, temperature traces invariably show a sequence of essentially "point" entries rather than the drawn-out change in spin rate typically exhibited on a spinner trace.

TABLE 4.A16—7A: TRACKING SINGLE-PHASE PRODUCTION (PRODUCTION WELL—VERTICAL, CASED, OR OPENHOLE WITHOUT SCREEN OR PERFORATED LINER) (Continued)

Problem	Recommended Logging Tools	Operational Tips and Expected Results
Profile production	Temperature surveys with the well on production and at least one with the well shut in for 1 to 6 hours	Finally, production can also be profiled from the temperature traces. Each entry above the bottom one normally produces, on the flowing trace, a mixing signature in the form of a step shift at the entry toward the shut-in traces. This signature is the result of the mixing of an entering stream at a temperature less than that of the stream arriving from below. This mixing is so rapid that it is essentially achieved under adiabatic conditions. The temperature of the mixture T_{mix} must conserve energy. If a stream enters at a volumetric rate q_{in} and temperature T_{in} and mixes with a stream from below at a rate q_b and temperature T_b, then energy conservation requires that $$\left(\rho C_p\right)_{in} q_{in}\left(T_{mix}-T_{in}\right)=\left(\rho C_p\right)_b q_b\left(T_b-T_{mix}\right),$$ where ρC_p is the product of density and specific heat capacity. Because this product is the same for both streams, the ratio of volumetric rates is given by $$\frac{q_{in}}{q_b}=\frac{T_b-T_{mix}}{T_{mix}-T_{in}}=\frac{\Delta T_b}{\Delta T_{in}}$$ This expression is applied to the second entry off bottom assuming a unit rate for bottom entry q_1, to establish a relative value for q_2. The process is continued for all higher entries. Oil-entry temperature for each stream can be taken as static temperature. A value for entry temperature at each depth can be found from the 6-hour shut-in trace. For gas, the entry temperature can be taken as the coolest temperature on either a 1-hour shut-in trace or on a 6-hour trace. Uncertainty in entry temperatures is greatly attenuated in the percentages for the production profile. As a result, a profile determined in this fashion is often more accurate within a perforated interval than is one derived from spinner deflections. If the completion is new, then there is a cooled zone arond the wellbore at temperature T_{cool}. The presence of this zone is evidenced by the temperature in the sump. Liquid entries therefore show as warm signaures on the temperature traces. On the flowing trace, one may see from the bottom upward several steps to warmer temperature indicative of entries of liquid at progressively increasing rates. Liquid arrives at the outer edge of the cool zone at normal geothermal temperature T_{geo} but in moving through the zone cools to arrive at the wellbore at a temperature T_{in} that is warmer than the cool temperature T_{cool} but is cooler than geothermal. The flowing temperature trace provides an entry temperature only for the bottom entry. For all higher entries T_{in} is unknown.

TABLE 4.A16—7A: TRACKING SINGLE-PHASE PRODUCTION (PRODUCTION WELL—VERTICAL, CASED, OR OPENHOLE WITHOUT SCREEN OR PERFORATED LINER) (Continued)

Problem	Recommended Logging Tools	Operational Tips and Expected Results
Profile production	Temperature surveys with the well on production and at least one with the well shut in for 1 to 6 hours	The mixing equation thus contains another unknown besides q_{in}, meaning that another equation is needed to profile. If one supposes that the entry is a "point" entry and that the formation in the cool zone changes temperature slowly with time, then one can arrive at a second expression for the ratio of any higher rate to the bottom rate q_1: $$\frac{q_k}{q_1}=\frac{(\rho C_p)_1}{(\rho C_p)_k}\cdot\frac{\ln(R_1)}{\ln(R_k)},$$ where $$R_i=\frac{T_{geo-i}-T_{cool-i}}{T_{in-i}-T_{cool-i}}$$ for i – 1, 2, etc. The ratio of density-specific heat products is set equal to 1 to provide the second equation. Extrapolation of sump temperature gives values for T_{cool}.
	Radioactive tracer surveys with well on production	This is an approach that should be used only as a last resort because of the production of radioactive fluid at the surface. Dilution in the surface facilities must be sufficient to meet exposure regulations. Tracer surveying is discussed in detail in Table 4.A11 (4B), to which the user should refer. A stationary, velocity-shot approach is used with a progression from the top of production down. The flowing temperature trace is used to select locations for shots. On the other hand, the chased-slug procedure offers two advantages unique to the tracer tool. The first relates to a slug ejected in the full stream and tracked upward. With the tracer well mixed over the casing cross section, the centroid of the slug moves at a velocity equal to the superficial velocity required for calculation of volumetric rate. The tool is thus the only absolute flowmeter available. The second feature is the ability of the tool to measure very small rates of zonal crossflow in a shut-in well.
Location of source of hydrocarbon behind casing	Temperature surveys both with well on production and shut in	The merged surveys should be displayed on one scale at 1°F per in. of chart and another scaled to prevent wraps. The features to expect for flow behind casing from either above or below the perforations are described here under Profiling Production.
	Noise-logging survey with well on production	Single-phase flow inside casing produces such a low level of noise that the well can be logged while on production. The tool is described in detail in Table 4.A5 (1C) under Detection of Flow Underground, to which the used should refer. The expression given in this table for estimation of single-phase rate q_{in} B/D past a constriction with pressure drop Δp psi from the 1,000-Hz noise level N_{1000} is modified to include the number of pipe strings between flow and tool N_p to give $$\Delta p\times q=700\times N_p\times N_{1000}$$ The flow can be either gas or liquid as long as the gas is not mixed with liquid.

TABLE 4.A17—7B: TRACKING TWO-PHASE PRODUCTION (PRODUCTION WELL—VERTICAL, CASED, OR OPENHOLE WITHOUT SCREEN OR PERFORATED LINER)

Problem	Recommended Logging Tools	Operational Tips and Expected Results
Profiling commingled oil/water production	Temperature survey with well on production followed by two shut-in surveys spaced 3 hours apart	The traces should be merged on one display at 1°F per in. of chart. Another display should be at a scale that prevents wraps. Trace behavior is the same as described in Table 4.A16 (7A) under Profile Production. The user should refer to this table. The distinction for the oil/water case relates to the effect of the individual phases on the temperature traces. A unit volume of water affects the temperature trace more than does a unit volume of oil owing to its higher value for the product of density and specific heat. This product is approximately 2.5 times larger for water than for oil. If a stream has a volumetric composition that is 50% oil and 50% water, then the "thermal" barrels in 100 bbl of mixture is 50 bbl of water and 50/2.5 = 20 bbl of oil as far as the temperature traces are concerned. The "thermal" composition is therefore approximately 29% oil and 71% water. Mixing signatures are used to profile by the method described in Table 4.A16 (7A) in detail for streams with equal values for the product of density and specific heat. This equality is retained here so that the relative rates so obtained are "thermal" rates rather than volumetric rates. The conversion to volumetric rates is obtained once the volumetric composition of the wellbore streams is determined from the fluid-identification traces described here. For a given stream, the volumetric composition is converted to a "thermal" composition as illustrated here. Each thermal rate is then decomposed into its oil and water thermal rates. These rates are then referenced to their corresponding thermal component in the full stream for volumetric percentages.
	Fluid-capacitance surveys both with well on production and shut in	This device is the best available for the detection of small oil entries into water or vice versa. The tool's electrode is exposed to wellbore fluid and acts as a capacitor that regulates the frequency output of an oscillatory circuit. In water, a polar compound, the effective capacitance is high and the output frequency is low. In oil, the capacitance is small and the output frequency is high. Likewise, the output frequency in gas is high at a value only slightly larger than that for oil. The device therefore basically distinguishes water from hydrocarbon. The spread in frequency between oil and water depends upon the tool's manufacturer, ranging from a low of 1,000 Hz to a high of 4,000 Hz. The spread measured at the surface between fresh water and any oil will be close to the downhole value. The effect of salinity is small, as the water readings downhole will show.

TABLE 4.A17—7B: TRACKING TWO-PHASE PRODUCTION (PRODUCTION WELL—VERTICAL, CASED, OR OPENHOLE WITHOUT SCREEN OR PERFORATED LINER) (Continued)

Problem	Recommended Logging Tools	Operational Tips and Expected Results
Profiling commingled oil/water production	Fluid-capacitance surveys both with well on production and shut in	The tool produces such a "clean" signal that, whatever the spread, the display sensitivity can be increased to the point that one chart division represents only 0.25% of the spread. This is the reason for the tool's ability to detect very small entries of oil into water. In theory, the output frequency should be inversely proportional to water holdup y_w in the wellbore. In fact, the relationship is quite nonlinear and is just as well represented as a function of frequency rather than its reciprocal. Flow-loop calibrations are needed to define the relationship for a particular tool. Furthermore, the electrode is subject to filming with oil in an oil/water environment and with water in a gas/water environment. One therefore needs calibrations both for light oil and for heavy oil. For oil viscosity greater than a few cp at downhole temperature, the first contact with oil films the electrode so severely that additional oil entries are hidden. On either end of the water holdup range, the tool is relatively linear. For y_w < 0.30, the water holdup can be represented as a linear function of frequency f as $$y_w = A\left(\frac{f_{\text{oil}} - f_{\text{mix}}}{f_{\text{oil}} - f_{\text{water}}}\right).$$ The coefficient A typically lies in the range of 0.5 to 1.0. At the other end, for y_w > 0.70, the relationship is $$y_w = 1 - B\times\left(1 - \frac{f_{\text{oil}} - f_{\text{mix}}}{f_{\text{oil}} - f_{\text{water}}}\right),$$ with B typically in the range of 1 to 3. If the full stream has water holdup in this range, then the tool response to this stream can be used as a calibration point to determine the value for the appropriate constant.
	Fluid density surveys both with well on production and shut in	Density tools are available in two versions as described in Table 4.A16 (7A) under Profile Production, to which the user should refer. The differential-pressure (gradient) types have a resolution of 0.01 g/cm^3, whereas the gamma-transmission types are limited by gamma statistics to a resolution of 0.025 g/cm^3 or less. Some companies "manufacture" their density traces from a pressure trace. This approach relies on two measurements taken at different times rather than simultaneously, making the result highly sensitive to tool flexure and to rate variations. The resolution for such traces is therefore much poorer than the previous values. Unlike the capacitance probe, a fluid density device produces a value that is linear in wellbore holdup. If a mixture density is measured at a value ρ_{mix}, then the oil holdup in the well is given by $$y_o = \frac{\rho_{\text{water}} - \rho_{\text{mix}}}{\rho_{\text{water}} - \rho_{\text{oil}}}.$$

TABLE 4.A17—7B: TRACKING TWO-PHASE PRODUCTION (PRODUCTION WELL—VERTICAL, CASED, OR OPENHOLE WITHOUT SCREEN OR PERFORATED LINER) (Continued)

Problem	Recommended Logging Tools	Operational Tips and Expected Results
Profiling commingled oil/water production	Fluid density surveys both with well on production and shut in	The density for the pure phases can be determined usually from the shut-in survey. For a density difference between oil and water of 0.20 g/cm³, the resolutions given here amount to oil holdups of 5% for the differential-pressure (gradient) tools and 12.5% for the gamma-transmission devices. For heavy oil with a density of 0.9 g/cm³, the resolution for even for the gradient tools drops to an oil holdup of 10%. Oil holdup values are needed to determine the oil "cut" (volumetric percentage) of the stream in the wellbore at a given depth. The so-called slippage velocity V_s is defined such that the oil cut C_o is given by an expression also involving the average or superficial velocity V_{ave}. $C_o = y_o \cdot \left[1-(1-y_o)\cdot\frac{V_s}{V_{ave}}\right]$. This value is the number referred to under temperature surveys as being needed in order to convert thermal rates into volumetric rates. The expression, however, contains two unknowns, the slippage velocity and the average flow velocity. Numerous correlations exist to refine the approximate value of 25 ft/min mentioned in Table 4.A16 (7A) according to the type of flow regime and density difference. An initial estimate for average velocity can be made from spinner data if available. Otherwise, a value consistent with thermal rate is assumed. In either case, an interative calculation must be made until the assumed average rate agrees with the value calculated from the thermal profile.
	Diverting-spinner survey with production no greater than 2,000 B/D	This device is described in Table 4.A16 (7A) under Profile Production, to which the user should refer. The tool has two distinct advantages for this problem. With the flow sped up, the oil and water phases tend to move together without slippage between the two. The tool can therefore be calibrated for two-phase flow in terms of total rate as a function of spinner speed with the light-phase cut (oil cut) as the parameter on a family of curves. Unfortunately, few service companies have such calibration data. The increased speed also means the oil holdup in the tool barrel is very nearly equal to the flowing cut. Most such devices are therefore also equipped with a fluid-capacitance probe in the barrel.
	Continuous or fullbore-spinner surveys with the well on production and shut in	These two types of spinner flowmeters are described in Table 4.A16 (7A) under Profile Production, to which the user should refer. Both devices are poorly suited to the quantification of multiphase flow of any type. The deflections tend to reflect the entries of the lightest phase, oil in this case, rather than average flow velocity. In fact, the traces can be used to obtain an approximate-percentage production profile for the light phase.

TABLE 4.A17—7B: TRACKING TWO-PHASE PRODUCTION (PRODUCTION WELL—VERTICAL, CASED, OR OPENHOLE WITHOUT SCREEN OR PERFORATED LINER) (Continued)

Problem	Recommended Logging Tools	Operational Tips and Expected Results
Profiling commingled gas and water production (or gas and oil)	Temperature surveys both with well on production and after shut-in periods of 1 and ~4 hours	The traces should be merged on one display scaled at 1°F per in. of chart and one scaled so as to avoid wraps. Gas wellbores are generally cooler than geothermal throughout. Consequently, water production warms the wellbore at the source. This makes the surveys particularly sensitive to low rates of water production. If warm signatures are not evident on the flowing trace, then the water enters commingled with a major part of the gas production and is hidden by expansion cooling. Even then, warm signatures are usually evident on the shut-in traces. Profiling is done from mixing signatures as described in Table 4.A16 (7A) under Profile Production, to which the user should refer. The distinction here is the effect that the individual phases have on the temperature traces. A unit volume of water has much more effect than does a unit volume of gas because the product of density and specific heat for water is much higher than that for gas. Typically, at least 10 volumes of gas are required to have the same effect as one volume of water, although the exact equivalency depends upon bottomhole temperature and pressure. The conversion of the thermal rates to volumetric rates for each phase is done as described above for oil/water.
	Density surveys both with the well on production and shut in	These surveys are identical to those of oil-water production described earlier. Gas holdup corresponding to a mixure density ρ_{mix} is given by $y_g = \dfrac{\rho_{water} - \rho_{mix}}{\rho_{mix} - \rho_g}$. A typical value for density difference between water and gas is 0.80 g/cm³. For gradient-type density tools, a resolution of 0.01 g/cm³ corresponds to a gas holdup of 1.2%, whereas a resolution of 0.025 g/cm³ for gamma-transmission tools corresponds to a gas holdup of 3.1%. Both types of tools are therefore sensitive to small gas entries. The same comments apply to gas cut (volumetric percentage in stream) as made above for oil/water flow. The gas cut C_g is related to holdup, gas slippage velocity V_s, and average fluid velocity V_{ave} by the expression $C_g = y_g \cdot \left[1 + (1 - y_g) \cdot \dfrac{V_s}{V_{ave}}\right]$. The typical value of 45 ft/min for slippage velocity can be refined by reference to correlations. As explained above for oil/water production, the gas cut is used to decompose thermal rates into gas and water volumetric rates.

TABLE 4.A17—7B: TRACKING TWO-PHASE PRODUCTION (PRODUCTION WELL—VERTICAL, CASED, OR OPENHOLE WITHOUT SCREEN OR PERFORATED LINER) (Continued)

Problem	Recommended Logging Tools	Operational Tips and Expected Results
Profiling commingled gas and water production (or gas and oil)	Back-scatter gamma-ray density surveys with well on production and shut in	These devices are referred to as gas holdup meters because of their ability to distinguish gas from liquid. They are relatively insensitive to the density of the surrounding liquid. Like the gravel-pack device, they contain an unfocused, chemical source of gamma rays. The detector, however, is shielded from direct transmission of gamma rays to the point that it responds primarily to rays back-scattered from the material around the tool. The back-scatter from gas is insignificant relative to that from liquid. Normally the logarithm of the tool's count rate should be inversely proportional to the gas holdup around the tool, at least in the absence of layered flow. Some limited calibration data suggest the gas holdup may be more linearly related to count rate directly as in layered flow [Table 4.A23 (9B)]. At this point, gas holdup y_g is taken as a linear function of the logarithm of count rate C. $$y_g = 1 - \frac{\ln(C_{\text{mix}}) - \ln(C_{\text{gas}})}{\ln(C_{\text{liquid}}) - \ln(C_{\text{gas}})}.$$ Calibration in the shut-in well is necessary because the pure-phase counts depend upon casing size. The devices retain their vitality in mixtures of water and completion fluid.
	Diverting-spinner flowmeter survey	Same comments as above for oil/water production. The distinction here is that gas leakage through the metal-vane type diverters is worse for gas/water production. Furthermore, calibration curves for this case are practically nonexistent.
	Continuous or fullbore flowmeter surveys both flowing and shut in	Same comments as above for oil/water production. Owing to the increased chaos in the wellbore, spinner speeds are even less related to average flow velocity than was the case for oil/water production. In fact, it is common to see the spinner turn backward above the deepest gas entry in response to circulation in the continuous water phase. This artifact is more pronounced for fullbore tools with their larger spinners. Generally, however, the spinner deflections indicate gas entries and can be used to profile relative gas rates approximately. Gas jetting from perforations into water tends to produce a relatively stable foam whenever a trace of condensate is present to act as a surfactant. The spinner sensitivity coefficient to foam flow is so small that the element can nearly cease to spin.
	Noise-logging survey with well on production	This tool is described in detail in Table 4.A5 (1C) under Detection of Underground Flow. The user should refer to this material.

TABLE 4.A17—7B: TRACKING TWO-PHASE PRODUCTION (PRODUCTION WELL—VERTICAL, CASED, OR OPENHOLE WITHOUT SCREEN OR PERFORATED LINER) (Continued)

Problem	Recommended Logging Tools	Operational Tips and Expected Results
Profiling commingled gas and water production (or gas and oil)	Noise-logging survey with well on production	The advantage of the device lies in its ability to detect very small gas entries into liquid, an ability that extends to the entry of occasional individual bubbles. For gas entry into a wellbore, the gas rate q_g in ft^3/d at downhole conditions can be estimated for rates less than 200 ft^3/d from the noise level in standard millivolts in the 200- to 600-Hz band from the expression $$q_g = 0.07 \times (N_{200} - N_{600}).$$ This device will also detect the creation of foam in the wellbore. Foam is an insulator to sound; consequently, the noise levels above a gas entry that foams are much less than the levels immediately below the entry.
Location of water source behind casing	Temperature surveys: one with well on production and two, spaced approximately 4 hours apart, with well shut in	The traces should be merged on one display scaled at 1°F per in. of chart. Another display should be scaled so as to avoid wraps. Temperature logging is recommended for this problem because the well must be flowed for the logging. This causes problems for other sensors of flow behind casing. Most oil wells produce a small amount of gas that, although insignificant volumetrically, creates a noisy environment for the noise logger. The production of multiple phases sets up a circulation pattern in the water phase that extends for great distances below the perforations. This circulation is detected by the oxygen-activation tools as water flow upward. A water source moving behind casing from below is detected as a base of production located below the deepest perforations. This base is the depth at which the flowing trace first "stands up" to separate to the warm side of the shut-in traces. The flowing trace will also show a normal mixing signature (a shift towards the shut-in traces) at the deepest entry from the perforated zone. At the base of perforations, temperature will decrease on the shut-in surveys. The two traces better illustrate this direction of change. A water source located above the top perforations will cause a normal mixing signature on the flowing trace at the source location and at the top of perforations, where the temperature will increase upon shut-in. Temperature surveys have the advantage that water, with its relatively large value for the product of density and specific, has a larger influence than does either oil or gas. Approximately 2.5 bbl of oil are required to give the same effect as 1 bbl of water. The disparity is even greater for gas. At least 10 bbl of gas are required to equal 1 bbl of water.
	Flowing fluid-capacitance survey for oil detection and flowing density survey for gas	These devices are described individually above. It is just as necessary to identify the source of hydrocarbon as it is to find the source of water. A workover that eliminates both water and hydrocarbon is certainly not desired. If the water is from a source external to the perforated zone, then it is highly unlikely that it will enter the wellbore commingled with hydrocarbon. The fluid-identifier traces are needed to establish qualitatively a separation between water and hydrocarbon sources.

TABLE 4.A18—7C: TRACKING THREE-PHASE PRODUCTION (PRODUCTION WELL—VERTICAL, CASED, OR OPENHOLE WITHOUT SCREEN OR PERFORATED LINER)

Problem	Recommended Logging Tools	Operational Tips and Expected Results
Profile commingled gas, oil, and water production	Temperature surveys: one with well on production and two, spaced approximately 4 hours apart, with well shut in	The traces should be merged with one display scaled at 1°F per in. of chart and another scaled so as to avoid wraps. Thermal profiling is done from mixing signatures as described in Table 4.A16 (7A) under Profile Production. The user should read this material. The distinction here is that water, with its relatively large product of density and specific heat, has the larger influence on the temperature traces. Approximately 2.5 bbl of oil are needed to equal the affect of 1 bbl of water. The equivalent for gas is at least 10 bbl. The thermal profile, assuming equal density-specific heat product for all phases, can be done quite accurately. To find individual phase rates, volumetric composition of the stream at various locations in the wellbore must be converted to "thermal" composition. For example, if a stream has a 50% water cut, a 25% oil cut, and a 25% gas cut, then 100 bbl of the mixture contains 50 "thermal" bbl of water, 25/2.5 = 10 thermal bbl of oil, and 25/10 = 2.5 "thermal" bbl of gas. The thermal composition is then 80% water, 16% oil, and 4% gas. The uncertainty is introduced in the determination of stream composition from the fluid-identifier traces. The determination of holdup values in the wellbore for a three-phase mixture is inaccurate. The conversion of these values to their corresponding stream compositions (cuts) introduces more error. Consequently, the resulting profiles tend to be only semiquantitative at best.
	Fluid-capacitance surveys with well flowing and shut in	These devices distinguish water from hydrocarbon. They are described in Table 4.A17 (7B) under Profiling. The user should refer to this material. The endpoint calibration expressions in this table are approximately valid for the determination of the water holdup in the liquid phase (i.e., in the mixture of oil and water). If this value is denoted by y_{wc}, then the water holdup in the wellbore is $y_w = y_{wc} \cdot (1 - y_g)$, where y_g is the gas holdup in the wellbore. The oil fraction $y_{oc} = 1 - y_{wc}$ is adjusted in the same manner to arrive at oil holdup in the wellbore. The chaos created in the wellbore by the flow of gas keeps the liquid phase mixed to the point that oil slippage relative to water is insignificant. The water and oil cuts (compositions) are the same as the holdup fractions in the wellbore.
	Back-scatter gamma-ray density surveys with well both flowing and shut in	This device is described in Table 4.A17 (7B) under Profiling. The user should read this material. It is the ideal tool for the determination of gas holdup y_g in the wellbore because of its insensitivity to the density of the surrounding liquid. Its output is typically scaled linearly in counts/seconds. For this application, the user should request that the output be calibrated in terms of gas holdup. Otherwise, a calibration equation is given in Table 4.A17 (7B).

TABLE 4.A18—7C: TRACKING THREE-PHASE PRODUCTION (PRODUCTION WELL—VERTICAL, CASED, OR OPENHOLE WITHOUT SCREEN OR PERFORATED LINER) (Continued)

Problem	Recommended Logging Tools	Operational Tips and Expected Results
Profile commingled gas, oil, and water production	Fluid density survey with well both flowing and shut in	These tools are described in Table 4.A17 (7B). Either type, differential-pressure or gamma-transmission, is suitable to this application. With a water holdup having been estimated from a fluid-capacitance survey, the measured value for mixture density can be converted to a value for gas holdup by the expression $$y_g = \frac{y_w \rho_w + (1-y_w)\rho_o - \rho_{\text{mix}}}{\rho_o - \rho_g},$$ where the subscripts *o, g,* and *w* refer to oil, gas, and water, repectively. Density for oil and water are determined from the shut-in traces, but gas density is usually estimated. Owing to uncertainly in the value for water holdup, this value may not agree with that obtained from the back-scatter density survey. Adjustments should be made to y_w to bring the two values into agreement. If the fluid-capacitance trace is not recorded, then the previous expression can be used to determine water holdup in the wellbore from the back-scatter gas holdup value $$y_w = \frac{\rho_{\text{mix}} - \rho_g y_g - (1-y_g)\rho_o}{\rho_w - \rho_o}$$ Hence, the water cut C_w in the well-mixed liquid phase as $C_w = y_w/(1-y_g)$. The corresponding oil cut is $C_o = 1 - C_w$. The chaos created by the gas flow keeps the oil and water sufficiently mixed that slippage between the two is minor. Only the gas phase need be corrected to obtain flowing gas cut C_g from the equation $$C_g = y_g \cdot \left[1 + (1-y_g)\cdot\frac{V_s}{V_{\text{ave}}}\right],$$ where the slippage velocity $V_S \cong 45$ ft/min is the value appropriate to the rise of gas through water. An estimate of the average or superficial velocity is selected consistent with the thermal rate as an initial estimate, and then iterated upon until value calculated from the volumetric rate equals the assumed value. The three cut values are converted into "thermal" cuts as explained above under Temperature. The thermal phase components are then determined from the total thermal rate at depth. Each of these values is ratioed to the "thermal" cut in the full stream as a volumetric percentage of each phase (and, hence, a volumetric rate).
	Diverting-spinner flowmeter surveys with the well flowing at less than 2,000 B/D	As described in Table 4.A17 (7B), these devices are ideal for multiphase flow because they can be calibrated in terms of B/D vs. spinner speed to give a family of curves with gas cut as the parameter. Unfortunately, such calibration data are lacking, so the principal advantage offered by the device lies in the values for total rate and superficial velocity.
	Continuous or fullbore spinner surveys with well both on production and shut in	As discussed in Table 4.A17 (7B), the principal contribution from these devices is an approximate profile for the light phase—gas, in this case. The entry of this phase produces the major deflections on the trace.
Location of water water behind casing		This is the same problem covered at the end of Table 4.A17 (7B), to which the user is referred.

TABLE 4.A19—8A: TRACKING SINGLE-PHASE PRODUCTION (PRODUCTION WELL—VERTICAL WITH SCREEN, PERFORATED LINER, OR DUAL COMPLETION)

Problem	Recommended Logging Tools	Operational Tips and Expected Results
Profiling production	Temperature surveys with the well on production and at least one after 1 to 6 hours shut in	This is exactly the same problem described in Table 4.A16 (7A). The presence of flow both inside and outside the screen/liner has an effect on temperature traces. The flow profiling from mixing signature proceeds in the manner described in the table. The principal distinction here is that the spinner surveys that occupied a top position in Table 4.A16 (7A) are relegated to secondary status for this completion, being superseded by temperature surveys.
	Density surveys flowing and shut in (not for dual completions)	As explained in Table 4.A16 (7A), this survey is run to determine the depth below which hydrocarbon-productive zones are submerged in residual completion fluid. Although this condition has no effect on the profiling procedure from temperature surveys, it does influence the output of the flowmeters as well as the activated-oxygen tracer tool.
	Gravel-pack (unfocused gamma transmission) survey in dual completion with well shut in	This tool has a chemical source of gamma rays positioned approximately 17 to 23 in. below a detector. The gamma-ray transmission is influenced by the density of the material within approximately a 6-in. radius about the tool. The count rate is attenuated exponentially by the density. A high count rate indicates a low density and vice versa. This tool can easily locate the interface between hydrocarbon and workover fluid behind the long string in a dual completion.
	Continuous spinner surveys with the well on production (not for dual completions)	These devices are described in Table 4.A11 (4B), to which the user may refer. Their contribution here is qualitative owing to the fact that fluid may enter at locations other than where it enters the screen/liner, and that fluid velocity will vary with conditions behind screen/liner.
	Activated-oxygen survey with well on production	This is the only tracer survey possible for a dual completion. This tool is described in Table 4.A13 (5B) under Injection Profiling, to which the used should refer. The tool is configured for flow upward. The apparent velocity is affected by swirl patterns in the annulus.
Source of hydrocarbon behind casing	Same as listed in Table 4.A16 (7A)	

TABLE 4.A20—8B: TRACKING TWO-PHASE PRODUCTION (PRODUCTION WELL—VERTICAL WITH SCREEN, PERFORATED LINER, OR DUAL COMPLETION)

Problem	Recommended Logging Tools	Operational Tips and Expected Results
Profiling commingled production of oil and water	Temperature surveys with well on production followed by surveys after 3 and 6 hours shut in	The profiling procedure is described in Table 4.A17 (7B), to which the reader should refer. As explained in that table, one must convert the "thermal" production profile, which is usually accurate, into a volumetric profile for each phase. This is done from phase "cuts" (compositions) as determined by fluid-identifier tools. These cuts are a source of error.
	Gravel-pack (unfocused gamma transmission) surveys both flowing and shut in	This device is described under Profiling in Table 4.A19 (8A). The distinction introduced for this problem is that any fluid identifier must give a cross-sectional average so as to sample fluid both inside and outside the liner/screen or in the annulus behind the long string in a dual completion. Furthermore, one should be able to estimate holdup values from the tool signal. In vertical wellbores, this tool's output, *C* counts/sec, is related exponentially to density (i.e., to holdup). Thus, water holdup y_w is approximated by the expression $$y_w = 1 - \frac{\ln\left(\frac{C_{\text{mix}}}{C_{\text{water}}}\right)}{\ln\left(\frac{C_{\text{oil}}}{C_{\text{water}}}\right)}$$ The counts for the pure phases are taken from the shut-in survey.
	Pulsed-neutron surveys both with well on production and shut in	These surveys include the neutron lifetime log and the carbon/oxygen log. The lifetime survey provides an estimate for the capture cross section or Sigma value for the borehole fluids. In theory, this value is a holdup-weighted average of the Sigma values for the two phases. Water holdup in terms of Sigma for a mixture Σ_{mix} is therefore $$y_w = \frac{\Sigma_{\text{mix}} - \Sigma_{\text{oil}}}{\Sigma_{\text{water}} - \Sigma_{\text{oil}}}$$ Values for pure-phase Sigma come from the shut-in surveys. The estimate for water holdup is less accurate than that from a gravel-pack survey owing to the higher noise level in the raw signal and to inaccuracy in processing the signal. A carbon/oxygen log is a spectral survey of induced gamma radiation. Energy levels are processed to give a volumetric ratio of carbon-to-oxygen in the material around the tool. This ratio is converted to a ratio of oil-to-water holdup, the sum of which is 1.0. The holdup estimates are even less accurate than are the ones from the lifetime log. The survey is therefore of qualitative value (comparison of flowing to shut-in traces.
	Fluid-capacitance surveys both flowing and shut in (not for dual completions)	This device is described under profiling in Table 4.A17 (7B). For this problem, its ability to detect very small oil entries is exploited in a qualitative manner. In a vertical wellbore, the light phase (oil) tends to move inside the screen/liner.

TABLE 4.A20—8B: TRACKING TWO-PHASE PRODUCTION (PRODUCTION WELL—VERTICAL WITH SCREEN, PERFORATED LINER, OR DUAL COMPLETION) (Continued)

Problem	Recommended Logging Tools	Operational Tips and Expected Results
Profiling commingled production of oil and water	Continuous spinner surveys with well on production (not for dual completions)	These tools are described in Table 4.A11 (4B) under profiling. Their value to this problem is purely qualitative. The chaos caused by an oil entry into water manifests itself inside the screen/liner. The survey therefore serves to identify locations where oil enters.
Profiling commingled gas and water (or gas and oil)	Temperature surveys with well on production followed by surveys after 1 and 6 hours shut in	The traces are used exactly as described under gas/water profiling in Table 4.A11 (4B). The thermal profile from mixing signatures is more accurate than are the stream compositions determined from the fluid-identifier surveys and used to convert to a volumetric profile.
	Gravel-pack (unfocused gamma-transmission) density survey both flowing and shut in	This device is described in Table 4.A19 (8A). In a gas/water (oil) environment, the water (oil) holdup is a nonlinear function of logarithm of the count rate *C*. The calibration equation is of the form $y_w = a R_{\text{mix}} + (1-a)(R_{\text{mix}})^2$, where a ≅ 0.45 and $R_{\text{mix}} = \dfrac{\ln(C_{\text{gas}}) - \ln(C_{\text{mix}})}{\ln(C_{\text{gas}}) - \ln(C_{\text{water}})}$
	Back-scatter gamma-ray survey with well on production and shut in	This is the "reflection" counterpart to a gravel-pack survey and is referred to as a gas-holdup survey. The tool is described in Table 4.A17 (7B) under Profiling Gas/Water. It functions in the same capacity as described in this table. However, the sensitivity to gas holdup outside the screen/liner or behind the long string of a dual completion is reduced by the presence of the intervening metal. Its advantage over the gravel-pack tool is that it may be more readily available and may be run in combination with other tools. The service company is unlikely to have a calibration for this environment.
	Gradient density surveys flowing and shut in (not for dual wells)	This tool is described under Profiling in Table 4.A16 (7A). Its value here is primarily as a qualitative trace that should corroborate the features evident on the gravel-pack log.
	Pulsed-neutron surveys both flowing and shut in	See previous discussion for oil/water profiling.

TABLE 4.A20—8B: TRACKING TWO-PHASE PRODUCTION (PRODUCTION WELL—VERTICAL WITH SCREEN, PERFORATED LINER, OR DUAL COMPLETION) (Continued)

Problem	Recommended Logging Tools	Operational Tips and Expected Results
Profiling commingled gas and water (or gas and oil)	Noise-logging survey with the well flowing	This device is described in Table 4.A5 (1C) under Detection of Underground Flow. It is an excellent survey for the detection of gas entry into liquid, either inside pipe or outside. The expression giving an estimate of gas rate behind the long string of a dual completion in terms of the noise level difference between the 200- and 600-Hz cut is $q_g = 0.35 \times (N_{200} - N_{600})$, where the rate is in downhole ft^3/d and the noise levels are in standard millivolts. The companion expression for gas entry into a screen or liner is $q_g = 0.07 \times (N_{200} - N_{600})$. Unfortunately, these linear expressions apply for relatively low rates no greater than approximately 200 ft^3/d. At higher rates, the expressions become nonlinear. In fact, at locations where gas jets from perforations are fast enough to foam water, noise levels drop to very low values.
	Neutron survey from a simple correlation tool with well flowing and shut in	This is an inexpensive way to detect gas entry into a screen or behind the long string in a dual completion. The tool consists of a chemical neutron source and a single detector spaced 17 to 23 in. above the source. It has a small diameter (1.373 in.) and is widely available. Because the tool responds to formation porosity, the shut-in trace is essential to interpretation. The tool has the advantage that a gas entry causes the same response in either water or oil. Because gas, with its low hydrogen content, is an ineffective neutron attenuator, a gas entry causes the flowing trace to shift away from the shut-in trace toward higher count rates. The size of the shift is exponentially related to the change in gas holdup in the liquid. However, the calibration is even more nonlinear than that for the gravel-pack correlation given previously. This survey can also be used to detect the entry of oil containing dissolved CO_2, a situation encountered in locations with active carbon-dioxide injection projects.
Location of a water source behind casing		This is the same problem discussed at the end of Table 4.A11 (4B). The user should refer to this table.

TABLE 4.A21—8C: TRACKING THREE-PHASE PRODUCTION (PRODUCTION WELL—VERTICAL WITH SCREEN, PERFORATED LINER, OR DUAL COMPLETION)

Problem	Recommended Logging Tools	Operational Tips and Expected Results
Profiling commingled gas, oil, and water production	Temperature survey with well flowing and two after shut-in times of 3 and 6 hours	The traces should be merged on one display scaled at 1°F per in. of chart and another scaled so as to avoid wraps. The procedure is the same as that of Table 4.A17 (7B). Water has the greatest effect—approximately 2.5 bbl of oil and 10 bbl or more of gas equal 1 bbl of water. Although the thermal profile is accurate, the volumetric profile is likely to be mainly qualitative in this situation. The reason is that stream compositions are required to convert the thermal profile to a phase-volumetric profile. These compositions are almost impossible to obtain with any accuracy.
	Back-scatter gamma density survey both flowing and shut in	This "gas-holdup" device is described in Table 4.A17 (7B) under Profiling. Its advantage here is that its response to gas is relatively insensitive to liquid density. If the device is calibrated for this environment, then it provides a value for gas holdup y_g directly. The corresponding value $y_{liq} = 1 - y_g$ represents an oil/water mixture created by the chaos associated with gas flow in liquid. It is the density of this mixture that determines the response from the gravel-pack tool listed next.
	Gravel-pack (unfocused gamma transmission) density survey both flowing and shut in	This device is described in Table 4.A20 (8B), where its nonlinear calibration equation is given for a flow containing liquid and gas. In the case of three-phase flow, water holdup y_w becomes y_{liq} for the oil/water mixture. Likewise, the count rate C_{water} becomes C^*_{liq} for the oil/water mixture. This value is no longer a constant, but varies with location in the hole. If y_{liq} is known from the back-scatter survey, then a value for C^*_{liq} can be calculated from the gravel-pack log at any location in the well. For these values, one uses the linear expression $$y^*_w = 1 - \frac{\ln\left(\frac{C^*_{liq}}{C_{water}}\right)}{\ln\left(\frac{C_{oil}}{C_{water}}\right)}$$ to obtain values of y^*_w and $y^*_o = 1 - y^*_w$. These are the compositions by which the flowing liquid cut must be multiplied by to arrive at the water and oil cuts, respectively. Under the assumption that the only slippage is between the gas and liquid phases, the gas cut is obtained from the expression $$C_g = y_g \cdot \left[1 + (1 - y_g) \cdot \frac{V_s}{V_{ave}}\right],$$ where $V_s \cong 45$ ft/min is the value for gas slippage through water and V_{ave} as estimated from the thermal rate. The corresponding liquid cut is $C_{liq} = 1 - C_g$. The corresponding oil and water cuts are obtained from this last value to give the three cuts needed to convert the thermal profile to a volumetric profile. A check is available on the calculations from the fact that the actual water holdup is given by $y_w = y^*_w y_{liq}$ and oil holdup is $y_o = y^*_o y_{liq}$. The three holdups must sum to 1.

TABLE 4.A21—8C: TRACKING THREE-PHASE PRODUCTION (PRODUCTION WELL—VERTICAL WITH SCREEN, PERFORATED LINER, OR DUAL COMPLETION) (Continued)

Problem	Recommended Logging Tools	Operational Tips and Expected Results
Profiling commingled gas, oil, and water production	Pulsed neutron surveys with well flowing and shut in	These surveys are described in Table 4.A9 (3) under oil/water profiling. Their use here is qualitative.
	Gradient-type density survey both flowing and shut in (not in duals)	Same comments as in Table 4.A20 (8B).
	Continuous-spinner surveys both flowing and shut in (not in duals)	These are described in Table 4.A17 (7B). Deflections are caused primarily by gas entries. Consequently, the trace can be used to profile gas production approximately.
	Noise-logging survey flowing	Same comments as in Table 4.A20 (8B).
Location of water source behind casing		This is the same problem discussed at the end of Table 4.A20 (8B).

TABLE 4.A22—9A: TRACKING SINGLE-PHASE PRODUCTION (PRODUCTION WELL—HIGHLY DEVIATED WITHOUT SCREEN OR LINER)

Problem	Recommended Logging Tools	Operational Tips and Expected Results
Profiling production	Elevation survey for wells deviated at more than 85°	Attempts to drill a wellbore nearly horizontal result in delayed corrections that cause variations in elevation from a few ft up to 20 ft. It is virtually impossible to produce a well at rates high enough to displace residual workover liquids and/or mud from the low spots along the length of the wellbore over the long section intended to produce. During production, hydrocarbon moves in waves along the top of this trapped liquid at locations where inclines upward in the direction of flow are encountered. Logging values measured at such locations are accordingly distorted.
	Density survey with well shut in	These devices are described in Table 4.A16 (7A). The type of tool is unimportant for this application because they are run simply to verify the presence of trapped completion fluid or mud.
	Continuous or fullbore spinner surveys both flowing and shut in	The procedure for the use of these tools is described under Profiling in Table 4.A16 (7A). The distinction here is that spinner responses used to profile should be taken in relatively level sections of the hole or, if this is not possible, at the top of downslope locations. The devices tend to plug because of debris on the low side of the hole. Diverting-spinner tools should be avoided for this reason.
	Temperature surveys both flowing and shut in for 1 to 6 hours	For wellbores with deviation angles less than 75°, a merged display scaled at 1°F per in. of chart is sufficient to clearly define mixing signatures for profiling according to the method described in Table 4.A16 (7A). For wells approaching horizontal, the merged display should be at 1°F per 2.5 in. of chart or even at 1°F per 5 in. of chart. At this magnification, variations in elevation may appear as corresponding variations in temperature on the shut-in trace. Mixing signatures are accordingly affected. In new wells, horizontals included, liquid flow to the wellbore will cause warm signatures because of cooling around the well residual from the drilling operations. Because of this effect, this is the time when profiles with the best accuracy will be obtained. The warm signatures have a size increasing with increasing rate in the manner described in Table 4.A16 (7A). At any time, a liquid rate of approximately 500 B/D is sufficient to remove the normal variations in temperature present on the shut-in traces. Active locations are therefore recognizable. Likewise, gas production usually creates cold signatures at entry locations with a size of approximately 0.02°F per psi drop in pressure. Temperature surveys remain a valuable source of information even in horizontal wells.
Location of hydrocarbon source behind casing	Temperature survey flowing and shut in	The traces should be treated in the manner described previously. The same features and behavior can be expected at the locations of the source. In particular, the comparison of flowing and shut-in traces is important.

TABLE 4.A23—9B: TRACKING TWO-PHASE PRODUCTION (PRODUCTION WELL—HIGHLY DEVIATED WITHOUT SCREEN OR LINER)

Problem	Recommended Logging Tools	Operational Tips and Expected Results
	Elevation survey for wells deviated 80° or more	In nearly horizontal wellbores, delayed corrections during drilling introduce changes in elevation ranging from a few to 20 ft. The resulting high and low spots appear all along the long interval in the productive zone. Water or completion fluid is trapped in the low spots and restricts the cross section of the hole. When the well is produced, the light phase moves on top of the heavy phase. In long, nearly horizontal sections of hole, the flow consists of well-defined layers, but in practice, these sections appear to be limited in number. When the stream encounters an upward-sloping section of hole, the heavy phase lags so that its layer thickness increases. The light phase then moves in waves or surges along the top of the hole. This action induces strong circulation velocities into the heavy phase to the extent that a device measuring the upward velocity of the heavy phase records an increase even though the cross-section occupied by the heavy phase increases. The behavior on a downward-sloping section is the reverse. The heavy phase races ahead and thins in the cross-sectional area. The resulting drag action induces circulation into the light phase.
Profiling commingled oil and water production	Temperature surveys with the well flowing and after shut-in times of 3 and 6 hours	One shut-in trace after approximately 3 hours suffices for horizontal completions. The merged surveys should be scaled and interpreted as described in Table 4.A22 (9A). The artifact added by this completion is related to the shut-in traces. Upon shut in, the heavy phase drops back down the wellbore into the productive interval. In wells deviated by approximately 10° or more, the heavy phase circulates on the low side of the hole, usually behind the liner or screen, bringing cooler temperature with it. Consequently, the shut-in traces in the sump may lie to the cool side of the flowing trace rather than coincide with it. The two traces, however, will parallel each other, indicating no net upward flow. The flowing chaos described above under elevation survey is an advantage to temperature surveys and a detriment to all others.
	Fluid-capacitance surveys both flowing and shut in	These devices are described in Table 4.A17 (7B). Any oil that enters the wellbore rises quickly to the high side of the hole. Unless entering on the high side, the oil still must pass through the well's cross section. Thus, very small entries are detectable. The sensor is essentially a "line-of-sight" element capable of revealing layered flow by pure-phase values.

TABLE 4.A23—9B: TRACKING TWO-PHASE PRODUCTION (PRODUCTION WELL—HIGHLY DEVIATED WITHOUT SCREEN OR LINER) (Continued)

Problem	Recommended Logging Tools	Operational Tips and Expected Results
Profiling commingled oil and water production	Differential-pressure (gradient-type) density surveys both flowing and shut in (for wells with 75° or less) (not for duals)	These devices are described in Table 4.A16 (7A). For deviations of 75° or less, they seem to function quite well. When corrected for deviation, they yield holdup values that often agree much better with expected values than do values from the more elaborate multiarm devices described below. The reason for this is likely the result of circulatory chaos in the flow combined with a 2-ft sample interval. For this environment, the hydrocarbon holdup values seem to obey simple momentum-displacement relationships to either light-phase cut C_{hc} or light-phase rate q_{hc} so that $C_{hc} = \left(y_{hc}\right)^n$ or $q_{hc} = \left(y\right)^n_{hc}$, where the constant n=1-z is established from full-stream values.
	Gravel-pack (unfocused gamma-transmission) surveys both flowing and shut in	This device is described in Table 4.A20 (8B), where its calibration equation is given for vertical wellbores. In layered flow, however, the logarithmic relationship no longer applies. Data from horizontal flow loops show that water holdup is linear in count rate C instead. The calibration equation thus becomes $y_w = \dfrac{C_{\text{oil}} - C_{\text{mix}}}{C_{\text{oil}} - C_{\text{water}}}$. Pure-phase count rates are provided by the shut-in survey.
	Continuous spinner survey with well flowing (not for duals)	These devices are described in Table 4.A11 (4B). In horizontal wellbores, it is a struggle just to keep the element from plugging. Even if this does not happen, the trace shows activity primarily at locations where light phase enters. In lesser deviated wellbores, however, the spinner deflections can often provide an approximate production profile for the light phase.
	Pulsed-neutron lifetime or spectral surveys flowing and shut in	These logs are described in Table 4.A20 (8B). The behavior of neutron transmission in layered flows, however, is quite complicated. Certain relative locations between tool and interface can produce count rates higher than the highest pure-phase value. Consequently, these traces are at best qualitative in this environment.
	Flowing surveys with radioactive or chemical tracers specific to each phase (not for duals)	A variety of tracers have been proposed to track the movement of individual phases. These include radioactive tags for oil and for water, chemical tags that increase capture-cross-section of each phase and activation of oxygen in the water phase. All of this activity is motivated by the fact that a rate cannot be calculated from a holdup value without an associated velocity. The conclusion from these efforts is that tracers behave just as badly in this environment as they do in vertical wells. Circulation patterns cause the tracers to spread, split, disperse or move in the wrong direction. Most wellbores lack the long sections of purely horizontal hole required to complete the profile.

TABLE 4.A23—9B: TRACKING TWO-PHASE PRODUCTION (PRODUCTION WELL—HIGHLY DEVIATED WITHOUT SCREEN OR LINER) (Continued)

Problem	Recommended Logging Tools	Operational Tips and Expected Results
Profiling commingled oil and water production	Multiarm conductivity surveys both flowing and shut in (not for duals)	The one device of this type available has conductivity electrodes on each of the four bow-spring centralizers that open the fullbore-spinner element. When opened, the centralizers position each electrode at the midpoint of four radii spaced on 90° angles. Conductivity is not measured directly owing to its strong dependency on salinity. Instead, the fraction of time that each probe conducts in a time sample is measured. During this fraction, the electrode is in contact with water. The assumption is then made that this time distribution of water equals its space distribution; that is, the time fraction equals the water holdup. On the display from the tool, the four holdup values are interpolated to provide a colored display of water distribution over the casing's cross section. A trace of average holdup is also presented. An orientation angle is presented for the arm at left of the display. The small diameter of the electrodes make them quite sensitive to oil films to the point that they can become nonconductive even though they are surrounded by water. Unlike the fluid-capacitance probe, there is no way to calibrate out the effect of filming on the average holdup value.
	Multiarm fluid capacitance surveys both flowing and shut in (not for duals)	These devices contain two or more bow springs with arrays of button electrodes attached fore and aft along each spring. The one tool of this type that is widely available has two opposing bow springs and contains a positioning arm that points the same spring downward as the springs open. The result is a high-density measurement of the distribution of oil and water along a diameter perpendicular to the bottom of the hole. The relatively small electrodes are more subject to interference from oil filming than is the larger-diameter regular fluid-capacitance probe. This tool also makes a cross-correlation in time of the fluctuations at similarly positioned arrays on each end of both springs to establish a travel time. A velocity distribution over the diameter is then derived from these travel times. From these data, an average velocity is estimated for each phase. In perfectly layered flow, this is a valid approach. In practice, however, the tool is unable to distinguish an event created by circulatory movement upward from any other.
Profiling commingled gas and water production (or gas and oil)	Temperature surveys both flowing and shut in for periods of 1 and 4 hours	A single shut-in survey will usually suffice for a horizontal well. The same comments apply as listed in Table 4.A22 (9A). The distinction here is that even in horizontal holes, gas expansion generally cools approximately 0.02°F per psi drop in pressure at entry locations. These cooled locations are easily identified on the shut-in survey. The thermal profile derived from mixing signatures as described in Table 4.A17 (7B) is relatively accurate. The volumetric profile derived with the aid of flowing cuts is likely to be only qualitative owing to error in the determination of composition from fluid-identifier traces.

TABLE 4.A23—9B: TRACKING TWO-PHASE PRODUCTION (PRODUCTION WELL—HIGHLY DEVIATED WITHOUT SCREEN OR LINER) (Continued)

Problem	Recommended Logging Tools	Operational Tips and Expected Results
Profiling commingled gas and water production (or gas and oil)	Back-scatter gamma density surveys both flowing and shut in	This device is described in Table 4.A17 (7B), where a correlation equation is given for gas holdup independent of liquid density. In this environment, the tendency of the flow to layer means that the equation should be altered to replace logarithms by count rates directly $y_g = 1 - \dfrac{C_{\text{mix}} - C_{\text{gas}}}{C_{\text{liquid}} - C_{\text{gas}}}$. Shut-in traces give pure-phase counts.
	Gravel-pack (unfocused gamma-transmission) density survey flowing and shut in	This device is described in Table 4.A19 (8A). In layered flows with only one liquid phase present (water or oil), the water holdup becomes a nonlinear function of count rate according to $y_w = aR_{\text{mix}} + (1-a)(R_{\text{mix}})^2 \; ; \; R_{\text{mix}} = \dfrac{C_{\text{gas}} - C_{\text{mix}}}{C_{\text{gas}} - C_{\text{liquid}}}$, where the pure-phase counts come from the shut-in traces.
	Gradient-type density surveys both flowing and shut in (not in duals)	See previous under profiling oil and water. This device is sensitive to small gas entries and will corroborate the back-scatter survey.
	Pulsed-neutron surveys both flowing and shut in	These surveys are described in Table 4.A20 (8B). The expressions given in that table for holdup do not apply to layered flows, making these surveys qualitative only.
	Flowing spinner surveys	See comments under oil/water above. May provide an approximate gas-rate profile.
	Neutron survey with a simple correlation tool both flowing and shut in	This tool is described in Table 4.A17 (7B). Like the back-scatter gamma tool, it is also able to distinguish gas from liquid irrespective of the type of liquid. In a highly deviated wellbore, the formation component in the signal is relatively constant. Furthermore, this tool has been calibrated for layered flows when centralized (U.S. Patent #5,708,203, 13-Jan-1998). The count-rate ratio in this patent is defined as is R_{mix} above.
	Optical gas-detection surveys both flowing and shut in	One company runs an optical sensor that combines measurements of light reflection and transmission to arrive at a bubble "count rate" and an assumed bubble size to estimated gas holdup.
	Noise-logging survey with the well flowing	This survey is described in Table 4.A20 (8B). The same concepts apply to this environment at locations where gas enters liquid.
Water channeling		This is the same problem discussed at the end of Table 4.A17 (7B).

TABLE 4.A24—9C: TRACKING THREE-PHASE PRODUCTION (PRODUCTION WELL—HIGHLY DEVIATED WITHOUT SCREEN OR LINER)

Problem	Recommended Logging Tools	Operational Tips and Expected Results
	Elevation survey	See Table 4.A23 (9B). The distinction in this environment is that low spots in the wellbore may contain a layer of oil on top of water whereas high spots will contain gas at top.
Profiling production of gas, oil, and water	Temperature surveys flowing and after shut-in times of 3 and 6 hours	See Table 4.A23 (9B) under profiling oil and water and profiling gas and water.
	Back-scatter gamma-ray density survey both flowing and shut in	See Table 4.A23 (9B) under profiling gas and water. This tool distinguishes gas from liquid. In this case, the liquid is a combination of oil and water. With wellbore deviations of no more than 75°, the two phases will be mixed together by lifting action of the gas. The calibration equation in Table 4.A23 (9B) for gas holdup y_g contains a liquid count rate, C_{liq}, that in theory changes with composition along the productive length of the wellbore. The same is true for horizontal wells with the additional complexity of changes in form from well-mixed to combined layers of oil on water. Fortunately, the tool is relatively insensitive to liquid density.
	Gravel-pack (unfocused gamma transmission) density survey both flowing and shut in	See Table 4.A23 (9B) under profiling gas and water. The nonlinear calibration equation in this table also contains a count rate for the oil/water mixture. Unfortunately, this device is sensitive to liquid density as the shut-in surveys will show. If the flowing trace is to give a value for water holdup y_w, then the method outlined in Table 4.A21 (8C) will have to be used with the logarithm of count rates replaced by count rates.
	Fluid capacitance surveys flowing and shut in (not in duals)	See Table 4.A23 (9B) under profiling oil and water. The tool functions just as well in this environment for the detection of small oil entries and for detection of layered flow. Its response is relatively insensitive to gas in the presence of oil.
	Gradient-type density surveys flowing and shut in (not in duals)	See Table 4.A23 (9B) under profiling oil and water and profiling gas and water. This device will detect very small gas entries and in conjunction with the fluid-capacitance trace will help distinguish gas entries from oil entries.
	Flowing spinner survey	See comments in Table 4.A23 (9B). This survey may provide an approximate gas-rate profile.
	Noise-logging survey with well flowing	See Table 4.A20 (8B). This tool is sensitive to small entries of gas into liquid. The gas-rate estimations in the table are valid at entry locations before gas rises to the top of the hole.
	Neutron survey with simple correlation tool flowing and shut in	See Table 4.A23 (9B) under gas and water profiling. This tool detects gas holdup independent of liquid composition. It is one of the few tools that has been calibrated for layered flow.

TABLE 4.A24—9C: TRACKING THREE-PHASE PRODUCTION (PRODUCTION WELL—HIGHLY DEVIATED WITHOUT SCREEN OR LINER) (Continued)

Problem	Recommended Logging Tools	Operational Tips and Expected Results
Profiling production of gas, oil, and water	Multiarm fluid-conductivity surveys with well flowing (not for duals)	See Table 4.A23 (9B) under profiling oil and water. This tool can provide a value for water holdup provided oil does not film on the electrodes.
	Multiarm fluid-capacitance surveys with well both flowing and shut in (not for duals)	See Table 4.A23 (9B) under profiling oil and water. The tool provides a distribution profile of gas, water, and oil across a diameter perpendicular to the bottom of the casing in sections of the hole where flow is layered.
	Flowing surveys with radioactive or chemical tracers specific to individual oil and water phases (not for duals)	See Table 4.A23 (9B). These surveys are intended to provide velocities for the oil and the water phases. However, the chaos created by gas entries is so severe that the utility of the survey is restricted to horizontal sections of the hole free of gas entries.
	Pulsed-neutron surveys with well both flowing and shut in	See Table 4.A23 (9B). The chaos resulting from the presence of gas increases the noise level on these surveys greatly. Furthermore, their processing software was not intended to apply to layered flows.

TABLE 4.A25—10A: TRACKING SINGLE-PHASE PRODUCTION
(PRODUCTION WELL—HIGHLY DEVIATED WITH SCREEN, LINER, OR DUAL COMPLETION)

Problem	Recommended Logging Tools	Operational Tips and Expected Results
	Elevation survey	See Table 4.A22 (9A). This survey identifies low spots that may contain trapped completion fluid
	Density survey with well shut in	These devices are described in Table 4.A16 (7A). The survey verifies the presence of trapped fluids.
Profiling production	Temperature surveys both flowing and shut in for 1 to 6 hours	Table 4.A22 (9A) outlines the use of these devices even in horizontal holes. The mixing signature method of profiling is not affected by the presence of trapped liquids in low spots.
	Continuous spinner flowmeter survey both flowing and shut in (not for use in dual completions)	These devices are described in Table VIIA. Their use here, however, is purely qualitative owing to their response only to flow inside the screen/liner. The traces will identify the locations where velocity increases because of trapped completion fluids or changes in hole size.
Location of hydrocarbon source behind casing	Temperature surveys both flowing and shut in	These surveys should be scaled and interpreted in the manner described in Table 4.A22 (9A). A general discussion of the features to expect on the merged traces appears in Table 4.A16 (7A).
	Noise-logging survey with well shut in	The use of this device for detection of flow behind casing is described in Table 4.A8 (2C). A shut-in condition is chosen so as to detect continuing flow from above.

TABLE 4.A26—10B: TRACKING TWO-PHASE PRODUCTION (PRODUCTION WELL—HIGHLY DEVIATED WITH SCREEN, LINER, OR DUAL COMPLETION)

Problem	Recommended Logging Tools	Operational Tips and Expected Results
	Elevation survey for well deviated 80° or more	See Table 4.A23 (9B), where the behavior of multiphase flow is discussed. The distinction to be made in this environment is that the light phase tends to move along the top of the hole and not enter the screen/liner completely until the top of the screen or the last hole in the liner. Likewise, the heavy phase moves along the low side of the hole until the top of the screen/liner.
Profiling commingled oil and water production	Temperature surveys both flowing and after shut-in times of 3 and 6 hours	See comments in Table 4.A23 (9B) and general behavior in Table 4.A16 (7A). The distinction for this environment is related to the merged traces. When the well is shut in, the heavy phase tends to flow back down the well on the low side of the casing and circulate all the way to the sump. Consequently, the shut-in traces in the sump may be located slightly to the cool side of the flowing trace; however, they will parallel the flowing trace owing to the lack of production in the sump.
	Fluid-capacitance traces both flowing and shut in	These devices are described in Table 4.A17 (7B) and their use in deviated wells is discussed in Table 4.A23 (9B). This device will also identify locations where flow is layered.
	Gravel-pack (unfocused gamma-transmission) density surveys both flowing and shut in	This device is described in Table 4.A17 (7B), where its calibration equation for water holdup in sections without layered flow is presented. The corresponding calibration for water holdup in layered flow appears in Table 4.A23 (9B).
	Continuous-spinner survey with well flowing (not for use in duals)	See comments in Table 4.A23 (9B). The survey may provide an approximate production profile for the oil phase.
	Pulsed-neutron lifetime or spectral surveys both flowing and shut in	The presence of the screen/liner does not prevent these devices, described in Table 4.A20 (8B), from providing values called "borehole fluid" that capture the cross section or "borehole carbon/oxygen" ratio. The software, however, was not written for layered flow [see comments in Table 4.A23 (9B)].
Profiling commingled gas and water production (or gas/oil)	Temperature surveys both flowing and after shut-in periods of 1 and 4 hours	See comments in Table 4.A23 (9B). As indicated in that table, cooling signatures at gas-entry locations are the distinctive feature. The distinction for this environment is the tendency for the gas to move along the top of the hole outside the screen/liner until its last chance for entry at the top of screen/liner. This does not interfere with profiling, but may produce an additional mixing signature at the top of the screen/liner.
	Back-scatter gamma-ray surveys both flowing and shut in (so-called gas-holdup surveys)	See comments in Table 4.A23 (9B), where the calibration equation for layered flow is given. This device should provide a value for gas holdup in the present environment if it is calibrated with the well shut-in at locations where only gas is present behind the screen/liner and at locations where only water (or oil) is present.

TABLE 4.A26—10B: TRACKING TWO-PHASE PRODUCTION (PRODUCTION WELL—HIGHLY DEVIATED WITH SCREEN, LINER, OR DUAL COMPLETION) (Continued)

Problem	Recommended Logging Tools	Operational Tips and Expected Results
Profiling commingled oil and water production	Gravel-pack (unfocused gamma-transmission) both flowing and shut in	See comments in Table 4.A23 (9B), where the calibration equation is given for layered flow. The value of the survey in this environment lies in its ability to provide an independent value for water (oil) holdup and thereby check the accuracy of the gas holdup determined from the back-scatter survey. The two holdups should sum to 1.0.
	Gradient-type density surveys both flowing and shut in (not in duals)	Table 4.A23 (9B) explains why these devices remain useful in wells with no more than approximately 75° deviation. They function in this environment to detect small gas entries.
	Continuous-spinner survey with well flowing (not for use in duals)	See comments in Table 4.A23 (9B). In wells with no more than 75° deviation, the survey may provide an approximate rate profile for the gas.
	Neutron survey with simple correlation tool both flowing and shut in	This device is a backup for the back-scatter gamma-ray survey. It also provides a gas holdup that is independent of liquid type. Its calibration for layered flow is discussed in Table 4.A23 (9B).
	Optical gas-detection surveys	See comments in Table 4.A23 (9B). At gas entries other than at the top of the hole, the tool should provide gas-holdup data for comparison to back-scatter values.
	Noise-logging survey with well flowing	This survey is described in Table 4.A20 (8B). The same concepts and rate expressions apply to this environment at locations where gas enters liquid.
Water channeling		This is the same problem discussed at the end of Table 4.A17 (7B).

TABLE 4.A27—10C: TRACKING THREE-PHASE PRODUCTION (PRODUCTION WELL—HIGHLY DEVIATED WITH SCREEN, LINER, OR DUAL COMPLETION)

Problem	Recommended Logging Tools	Operational Tips and Expected Results
	Elevation survey	See Table 4.A26 (10B). The distinction for this environment is that the low spots may have a layer of oil on top of water, whereas gas will be present at the top of high spots.
Profiling production of gas, oil, and water	Temperature surveys both flowing and after shut-in times of 3 and 6 hours	See Table 4.A26 (10B) under profiling both oil and water and gas and water.
	Fluid-capacitance surveys both flowing and shut in (not in duals)	See Table 4.A26 (10B) under profiling oil and water production. The distinction here is that the liquid phase is an oil/water mixture produced by gas lift (except in layered flow).
	Gradient-type density surveys with well flowing (not in duals)	See Table 4.A26 (10B) under profiling gas and water. They complement the fluid-capacitance survey in this environment to distinguish small gas entries from large oil entries.
	Continuous flowmeter survey with well flowing (not in duals)	See Table 4.A26 (10B) under profiling gas and water
	Gravel-pack density surveys both flowing and shut in	See Table 4.A26 (10B) under profiling oil and water production. The distinction here is that the liquid phase is a mixture of oil and water carried by gas lift (except in layered flow)
	Back-scatter gamma-ray surveys both flowing and shut in	See Table 4.A26 (10B) under profiling gas and water production. These devices are called gas-holdup meters and are relatively insensitive to liquid density.
	Neutron survey with simple correlation tool both flowing and shut in	See Table 4.A26 (10B) under profiling gas and water production. The device is a backup to the back-scatter gamma survey. It is totally independent of liquid density.
	Optical gas-detection surveys	See Table 4.A26 (10B).
	Noise-logging survey flowing	See Table 4.A26 (10B).
Water channeling		Same problem discussed at end of Table 4.A17 (7B).

TABLE 4.A28—11A: LOCATION OF ANNULAR PRESSURE SOURCE (PRODUCTION WELL SHUT IN OWING TO ANNULAR PRESSURE OR CROSSFLOW BEHIND CASING)

Problem	Recommended Logging Tools	Operational Tips and Expected Results
Location of annular pressure source	See same problem in Table 4.A8 (2C)	The location of the source is determined by a comparison of the "Before" and "After" annular bleed surveys. The distinction between this table and Table 4.A8 (2C) lies entirely in their environments. In this case, the logging will most likely be carried out in a tubing string under pressure from its hydrocarbon content. The tubing must be sealed to flow at the surface, meaning that wing valves must be checked for their ability to seal completely. Furthermore, careful attention must be paid to the seal on the lubricator provided by the grease-injection unit during logging. Because the well is not at geothermal temperature, this survey remains the most sensitive to water sources because of water's ability to warm or cool (depending upon shut-in temperature) at relatively low rates. Likewise, the noise logger remains the most sensitive to small gas sources that must move through liquid behind or in casing. If the tubing string contains gas, then the noise levels must be multiplied by a coupling factor $F \cong 2$ before application of the rate expressions in Table 4.A8 (2C).

TABLE 4.A29—11B: DETECTION OF CROSSFLOW BEHIND PIPE (PRODUCTION WELL SHUT IN OWING TO ANNULAR PRESSURE OR CROSSFLOW BEHIND CASING)

Problem	Recommended Logging Tools	Operational Tips and Expected Results
Detection of crossflow behind casing	See Table 4.A5 (1C)—detection of flow underground	The environmental differences are the same ones listed in Table 4.A28 (11A). The procedures and expected features remain the same as those discussed in Table 4.A5 (1C). If the rate expressions given for the noise logger are used, then the noise levels must first be multiplied by N_p = Number of pipe strings between source and sensor and by a coupling factor $F = 2$ if the tool is in gas rather than liquid.

TABLE 4.A30—12A: DETECTION OF CORROSION OR WEAR TO STRING CONTAINING TOOL (WELL DOWN FOR TUBING OR CASING INSPECTION)

Problem	Recommended Logging Tools	Operational Tips and Expected Results
Corrosion detection	Multifeeler mechanical-caliper surveys for internal corrosion	These devices are available for both tubing and casing inspection. They are described in detail in Table 4.A7 (2B) (Casing Wear), to which the user should refer before continuing here. Because a direct measurement is made, these tools are the best for the detection of internal pitting, which appears as high-frequency spiking to larger radii. This kind of "hash" is easily recognizable on the display of individual feeler records. However, scale buildup on the inside surface of the tubing or casing can also produce "hash" in the records with the spikes being toward smaller radii. It is therefore very helpful to have "min-max" traces (minimum and maximum deflections) on the display as well. These traces reveal immediately the direction of the spikes. Furthermore, the min-max traces are essential to the recognition of extended regions of wear or relatively uniform loss of wall from corrosion. Because of tool rotation, such regions are difficult to spot otherwise.
	Eddy-current wall-thickness and caliper surveys for both internal and external corrosion	These electromagnetic-induction devices are also available for both tubing and casing. Their description appears in Table 4.A7 (2B), to which the user should refer. These tools are the most sensitive available for the detection of external metal loss. The trace from the wall-thickness coils can be displayed easily at a sensitivity of 0.02 in. per chart division (i.e., at 4% normal wall thickness per chart division). In seamless tubing or casing, the normal taper (approximately 0.125 in. per 40 ft) can be used as a calibration point. Corrosion pits also appear as high-frequency spiking to lesser values of wall thickness on the display. Internal or external pitting produces equal effects on the trace. The two are distinguished by the caliper trace. This signal is the variation in voltage of a field induced by a coil driven at a high frequency to prevent penetration into the metal wall. This trace is displayed so as to spike to larger radii at internal pits. Of course, the tool does not function in nonmagnetic metals.

TABLE 4.A30—12A: DETECTION OF CORROSION OR WEAR TO STRING CONTAINING TOOL (WELL DOWN FOR TUBING OR CASING INSPECTION) (Continued)

Problem	Recommended Logging Tools	Operational Tips and Expected Results
Corrosion detection	Magnetic flux-leakage surveys for internal and external corrosion	These are casing-inspection devices that use either a permanent magnet or an electromagnet to induce a magnetic field over the wall thickness of the casing. At locations where metal thickness changes abruptly, the magnetic field either expands out of (metal loss) or collapses into (metal gain) the wall. Either way, the field acquires a component perpendicular to the casing wall. Pads containing coils to cut this field are pressed tightly against the casing wall. Two rings, each containing between 6 and 12 pads, are used so that the pads can be staggered to sample more closely around the casing circumference. The polarity of the signal is displayed so that, with the tool moving upwards, the passage of a coil over a sudden decrease in wall thickness induces a spike to the right on the display from that coil. A sudden increase in wall thickness induces a "negative" spike (i.e., a deflection to the left). The polarity of a spike is used to distinguish, for example, the bands on centralizers or "scratchers" attached to the outside of the casing. The size of the induced spike is, of course, smaller for a change in thickness on the outside casing wall than for a change on the inside wall. On modern versions of this tool, the output from each of the 12 to 24 coils is displaced as a function of depth. Two additional traces display the maximum and the average spike size without regard to polarity. Pitting induces a swarm of spikes with positive polarity on the individual traces. The display of these traces, along with the average trace, shows the radial extent of the corrosion, whereas the maximum trace shows the severity of pitting. To be effective, the trace of maximum deflection is recorded at a sensitivity such that the "road noise" induced by tool motion corresponds to the signal produced by approximately 10% of internal thickness loss and approximately 30% external loss. The sparser the pitting density, the smaller the spikes. Casing strings external to the one being logged further decrease spike amplitude. Without calibration data, one can only use the response at collars (≈100% loss) as a relative guide.

TABLE 4.A30—12A: DETECTION OF CORROSION OR WEAR TO STRING CONTAINING TOOL (WELL DOWN FOR TUBING OR CASING INSPECTION) (Continued)

Problem	Recommended Logging Tools	Operational Tips and Expected Results
Corrosion detection	Magnetic flux-leakage surveys for internal and external corrosion	Each pad also contains a pair of coils, one driven at high frequency and one to sample induced voltage, to examine the inside wall of the casing. This "caliper" trace is used to discriminate internal corrosion from external.
		These devices respond only to abrupt changes in wall thickness and, consequently, can miss severe wear or the equivalent from overlapping pits. It should therefore be run in conjunction with a mechanical caliper. This procedure also allows one to judge the percentage of spurious events to be expected on the maximum trace.
	Casing differential voltage survey	The device uses two (or more) contact electrodes to measure the difference in voltage, caused by galvanic currents, in sections along the casing wall. A section with a voltage swing, top-to-bottom, in the positive direction identifies an area losing metal ions to the earth. A current is calculated for such sections and converted into a rate of metal loss.
	Radial-type (pulse-echo) bond-logging survey for caliper and wall thickness	This device is described in Table 4.A7 (2B). It provides colored-density displays for wall thickness and for internal radius. It is an excellent tool for the detection of wear and its equivalent from overlapping pit-corrosion. It is less effective for high-frequency pitting that has yet to coalesce.
Casing collapse, part or split		See Tool Selection in Table 4.A7 (2B).
Bends or "dog-legs"		See Tool Selection in Table 4.A7 (2B).

TABLE 4.A31—12B: DETECTION OF HOLES OR PARTING IN CASING EXTERNAL TO THE ONE CONTAINING THE TOOL (WELL DOWN FOR TUBING OR CASING INSPECTION)

Problem	Recommended Logging Tools	Operational Tips and Expected Results
Holes or parts in external casing	Eddy-current wall-thickness surveys	See Tables 4.A7 (2B) and 4.A28 (12A).
	Gravel-pack (unfocused gamma-transmission) survey	See Table 4.A19 (8A).

Chapter 5
The Single-Well Chemical Tracer Test—A Method For Measuring Reservoir Fluid Saturations In Situ

Harry Deans and **Charles Carlisle,** Chemical Tracers, Inc.

5.1 Introduction

The single-well chemical tracer (SWCT) test is an in-situ method for measuring fluid saturations in reservoirs. Most often, residual oil saturation (S_{or}) is measured; less frequently, connate water saturation (S_{wc}) is the objective. Either saturation is measured where one phase effectively is stationary in the pore space (i.e., is at residual saturation) and the other phase can flow to the wellbore. Recently, the SWCT method has been extended to measure oil/water fractional flow at measured fluid saturations in situations in which both oil and water phases are mobile.

The SWCT test is used primarily to quantify the target oil saturation before initiating improved oil recovery (IOR) operations, to measure the effectiveness of IOR agents in a single well pilot and to assess a field for bypassed oil targets. Secondarily, it is used to measure S_{wc} accurately for better evaluation of original oil in place (OOIP). Fractional flow measurement provides realistic input for simulator models used to calculate expected waterflood performance.

This chapter familiarizes the reader with the SWCT method, and offers guidelines for selecting suitable test wells and for planning and executing the field operations on the target well. Test interpretation is also discussed and illustrated with typical examples.

5.2 History

The first SWCT test for S_{or} was run in the East Texas Field in 1968.[1] Patent rights were issued in 1971.[2] Since then, numerous oil companies have used the SWCT method.[3–7] More than 400 SWCT tests have been carried out, mainly to measure S_{or} after waterflooding.

The SWCT method has gained considerable recognition over the past few years because of increasing interest in the quantitative measurement of S_{or}. Some experts[8,9] consider the SWCT test to be the method of choice because of its demonstrated accuracy and reasonable cost.

5.3 Measuring S_{or}

Accurate S_{or} measurement is important because of a combination of basic problems in oil recovery. The industry still produces less than half the oil in the reservoirs discovered, and nearly all that oil is produced using traditional primary and secondary recovery methods.[10] Furthermore, as the cost of finding new reserves continues to increase, especially in the U.S.A., the oil re-

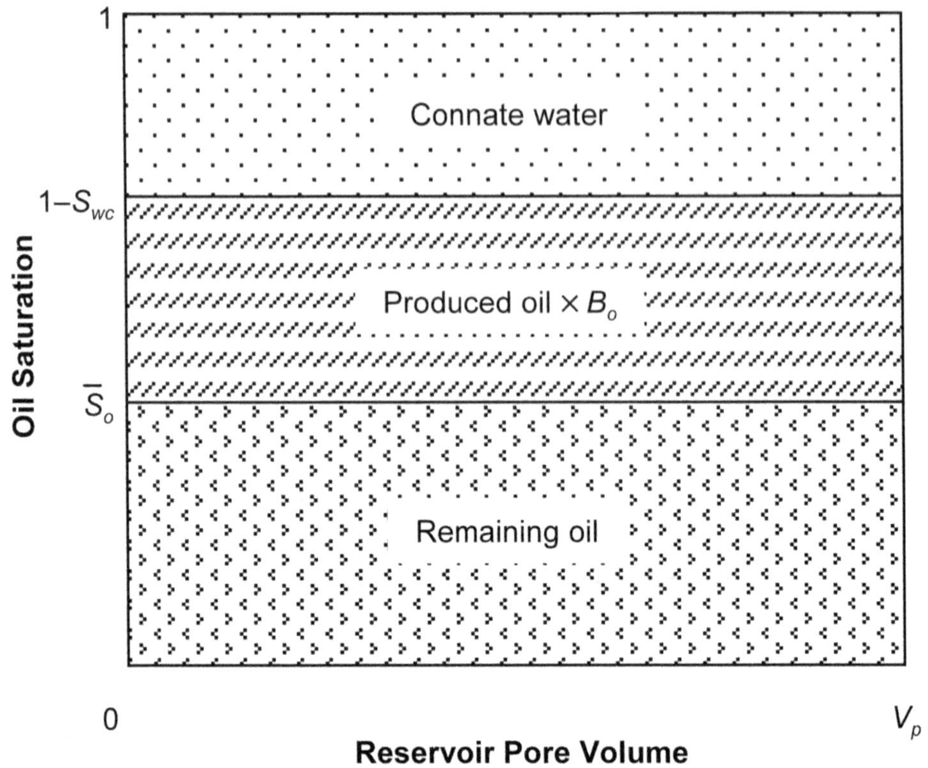

Fig. 5.1a—Reservoir evaluation by material balance alone in a waterflooded oil field. B_o = formation volume factor, oil RB/STB.

maining in old fields becomes a significant economic target for infill drilling and IOR projects. See Chaps. 12 through 17 in the Reservoir Engineering and Petrophysics volume of this *Handbook* for more detail on these methods.

In every target field, the quantity and location of the remaining oil must be determined. **Fig. 5.1a** illustrates the principle of material balance, as applied to an oil reservoir. The entire area of the graph represents the reservoir pore volume (V_p), which is known with varying degrees of uncertainty. The produced oil, corrected back to reservoir conditions, is the middle area; its accuracy, however, depends on how thoroughly the production records are kept.

The uppermost area is the connate water, which is known only as well as available methods and coverage of measurements allow. The lowermost area, the remaining oil, can be expressed as an average saturation of oil ($\bar{S}_o$), if we accept the total pore volume V_p, produced oil, and S_{wc} values as being accurate.

If a given field has been waterflooded, the fraction of the OOIP displaced by the water is a critical parameter. Testing for S_{or} in watered-out wells in the field can determine the maximum waterflood displacement efficiency. A significant difference between the material-balance S_o and the measured S_{or} would indicate the presence of bypassed oil. This would signify that parts of the reservoir had not been contacted by injected water, or had not received sufficient water throughput to reach S_{or}. This concept is shown in Fig. 5.1b.

A reliable in-situ measurement of S_{or} simultaneously defines the target for enhanced oil recovery (EOR) and allows estimation of the potential bypassed (mobile) oil in the field. This moveable oil is the target for infill drilling and/or flood sweep efficiency improvements.

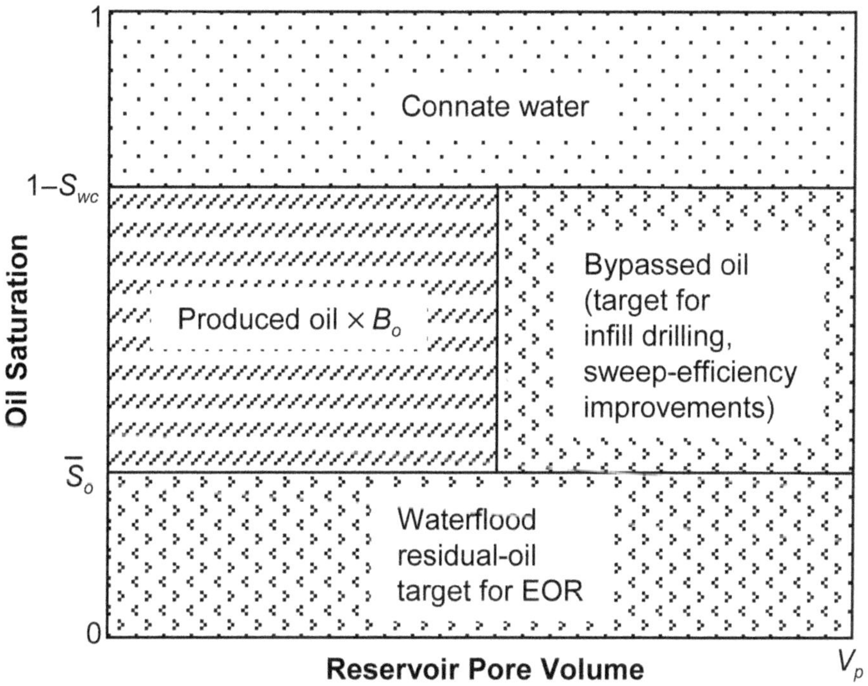

Fig. 5.1b—Reservoir evaluation by material balance with measured S_{or}.

Because S_{or} varies greatly with formation type, oil/water properties, and other variables that are not completely understood (e.g., wettability changes caused by water flood practices), S_{or} measurements range from < 10% to > 45%. There is no reliable way to predict S_{or} with acceptable accuracy for most reservoirs. Furthermore, measuring residual oil is not easy. Laboratory corefloods performed at other than native state wettability are unreliable.[11]

Well logs can give vertical profiles of S_{or} under optimal conditions, but their results are not absolute. Logs of all types require calibration by an independent method, which gives either a quantitative S_o at some point or an average S_o over some layer. Pressure cores or sponge cores can provide this calibration, but require a new well and are subject to saturation disturbances caused by mud filtrate invasion.

An advantage of the SWCT method is that it pushes tracers beyond damaged regions near the wellbore and into layers that are known to be at residual oil conditions. The tracers go where water is contacting the residual oil, as shown in Fig. 5.1b.

In an SWCT test, the formation volume sampled is large enough to be representative. A typical test quickly investigates hundreds of barrels of pore space in an existing watered-out well. The tracer-bearing fluids are produced back into the well without disturbing the formation, allowing further testing.

5.4 Measuring S_{wc}

Connate water saturation is more variable (and less predictable) than S_{or}, with S_{wc} measurements ranging from > 50% in certain rock types to < 5% in unusual reservoir situations. Large variations within the producing intervals of major reservoirs are well documented.[9]

As Figs. 5.1a and 5.1b show, estimates of recoverable oil depend fundamentally on knowing S_{wc}. Oil-based coring can provide reliable results and is an effective choice if its expense can be justified at the time a well is drilled. Electric logs can give vertical profiles, but as with S_{or}, require calibration for quantitative S_{wc} measurement.

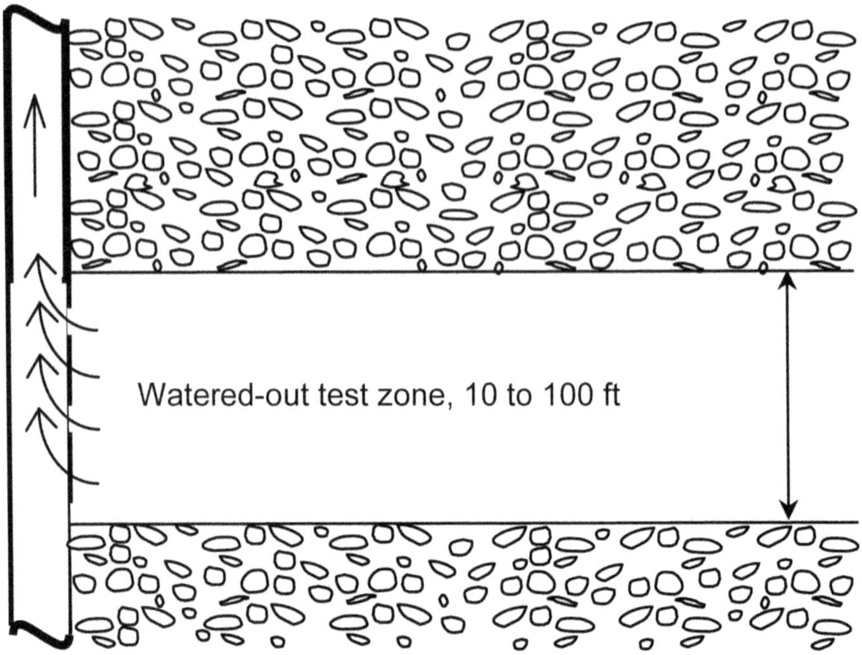

Fig. 5.2—Candidate SWCT test well.

Deans and Shallenberger[12] reported the first application of the SWCT method for measuring S_{wc}. As in the case when measuring S_{or}, the reservoir volume sampled is large and the test is nondestructive, which is especially important for a producing oil well. The tracers used in measuring S_{wc} are nonhazardous oxyhydrocarbons, so that no contaminated oil needs to be disposed of after the SWCT test. Most importantly, though, the test holds a high success rate—every known SWCT test for S_{wc} has yielded quantitative results.

5.5 SWCT S_{or} Tests

5.5.1 How the SWCT Test Works. The SWCT test for S_{or} uses only one well and involves the injection and back production of water carrying chemical tracers. A typical target interval for SWCT testing is shown in **Fig. 5.2**. The candidate well should be completed only to the watered-out zone of interest (zone at S_{or}). The water used normally is from the formation to be tested, and often is collected during the initial setup for the test.

The injected volume is divided into two parts: the partitioning tracer bank, which carries a small concentration of tracer (usually some type of ester) dissolved in water, and the push volume of water, which pushes the partitioning tracer bank away from the wellbore 10 to 20 ft. A material-balance tracer (normally a water-soluble alcohol) is added to the entire injected volume to differentiate it from the formation water being displaced. This injection step is shown in **Fig. 5.3**.

The primary tracer is an alkyl ester. The esters used in SWCT testing usually are more soluble in oil than in water. This solubility preference is expressed quantitatively by the ester's oil/water partition coefficient, K_e:

$$K_e = \frac{C_{eo}}{C_{ew}}, \quad (5.1)$$

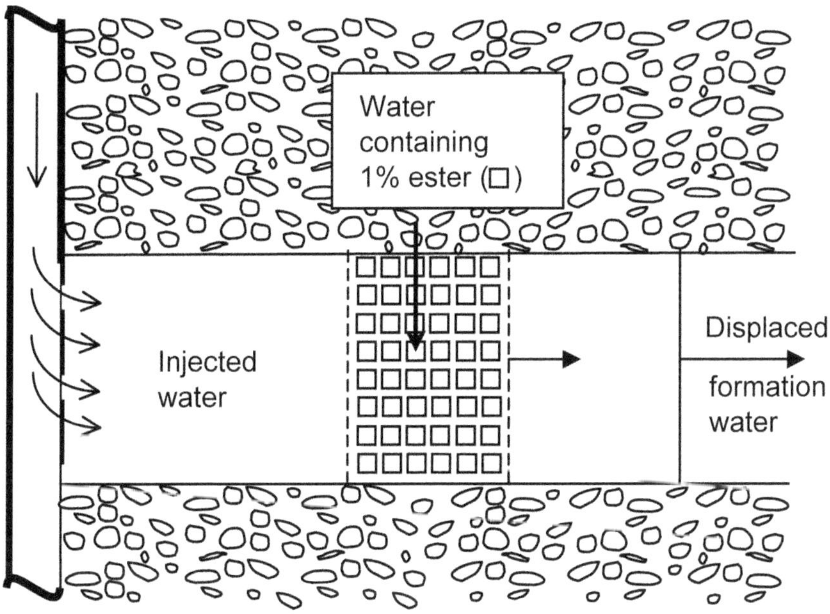

Fig. 5.3—Injection of ester tracer and push volume.

where C_{eo} = the concentration of ester in oil; C_{ew} = the concentration of ester in water; C_{eo} and C_{ew} are values at equilibrium.

For example, if the partition coefficient is four, the ester prefers the oil phase four times more than the water phase. For each tracer to be used in each test, the actual value of K_e must be measured in the laboratory at reservoir conditions. Oil and water samples are collected from the target formation for this purpose.

As the ester tracer enters the pore space containing the residual oil, it partitions between the oil and water phases. The ester maintains a local equilibrium concentration in the oil phase, controlled by the ester's partition coefficient, even though the water is flowing.

Because the oil is stationary and the water is moving, the ester tracer moves more slowly through the reservoir pore space than does the water with which it was injected. The ester's velocity thus is a function of the water velocity, the ester partition coefficient, and the S_{or}. Fig. 5.3 schematically shows the radial position of the injected ester and water. The material-balance tracer, normally an alcohol, is nearly insoluble in oil, so that it travels at approximately the same velocity as the water, reaching as far into the formation as the injected water.

For simplicity, Fig. 5.3 illustrates the no-dispersion case however, in all real reservoirs, the tracers disperse significantly, though this effect does not alter the basic mechanism of retardation of ester velocity by the residual oil.

After the ester and push injections are completed, the well is shut in for one to ten days, depending on the reactivity of the ester and the reservoir temperature. This shut-in period allows some of the ester to react with water in the reservoir, which forms a new tracer in situ, the secondary (or "product") tracer.

Reacting an alcohol and an organic acid makes an alkyl ester. At reservoir temperature, however, when dissolved in water, this ester slowly breaks down again into the alcohol and acid: Ester $+ H_2O$ = Alcohol + Acid. The shut-in period must be long enough for measurable alcohol to form in situ by this reaction (**Fig. 5.4**).

It is the alcohol formed that makes the S_{or} measurement possible. The acid formed during the reaction is not observed because it is neutralized by the natural base components of the

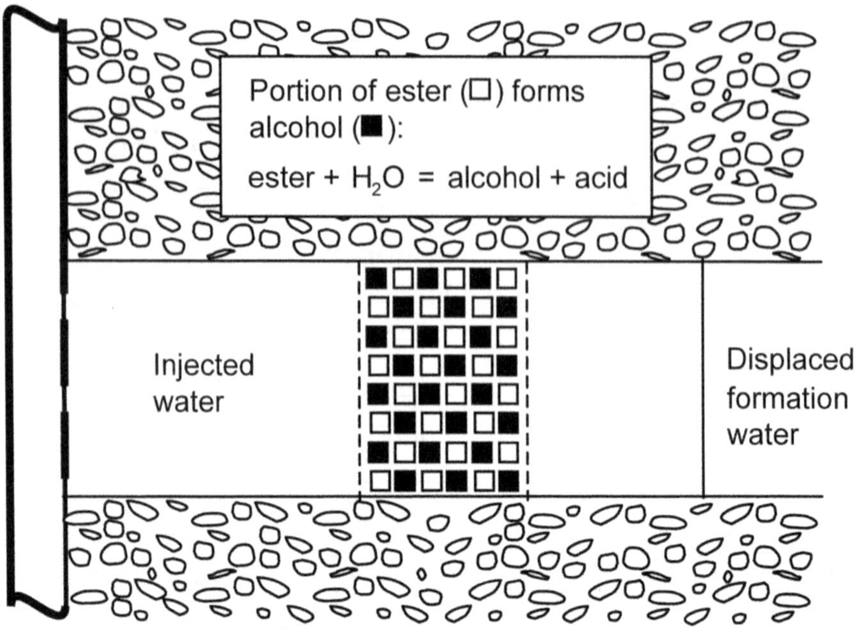

Fig. 5.4—Shut-in (reaction) period.

reservoir. The alcohol, however, is not in the original formation water, and can be detected at very low concentrations in the produced water, thus acting as a unique, secondary tracer.

At the end of the reaction period, the remaining ester and the product alcohol tracers are located together 10 to 20 ft from the wellbore. The tracers then are ready to be back produced to the wellbore and monitored at the surface in the produced water.

Fig. 5.5 shows the chromatographic separation of the product alcohol and ester tracers. This separation occurs because the product alcohol and water velocities are essentially the same, whereas the ester production velocity is slower because the ester must partition between the oil and water phases during production in the same manner described in the injection step. For an animation of the entire process, see http://www.pe.utexas.edu/~castillo/lwl/ [CD edition includes **video**].

Throughout the production period, samples of the produced water are collected frequently at the surface. Total produced volume is measured at the time each sample is taken. At a portable laboratory at the wellsite, the samples are analyzed immediately for product alcohol and remaining ester tracer concentrations. A plot of concentration of tracers vs. total volume produced is developed during the production, as the samples are analyzed.

Fig. 5.6 shows a hypothetical example of typical tracer concentration profiles. All three tracer profiles show the effects of dispersion, which always occurs during flow in porous formations. The unreacted ester, for example, was injected as a square wave and returns as a Gaussian peak.

5.5.2 Summary of SWCT Test Features. The important SWCT test features are summarized below:

• The S_{or} measurement is made in situ in the waterflooded layers of the target formation. The tracers can go only where the injected water goes.

• Compared to coring or logging method results, the S_{or} results are from a relatively large reservoir volume.

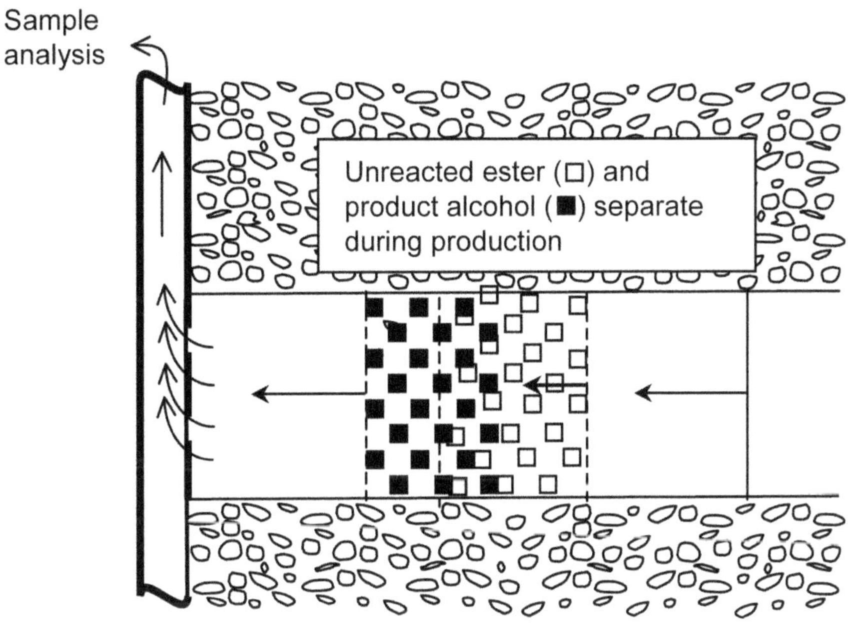

Fig. 5.5—Production period.

• The S_{or} measurement is carried out on an existing well, and usually in an existing completion, which can be perforated or openhole.

• Because the S_{or} measured actually is the volume fraction of oil in the pore space, the measurement is independent of porosity.

5.5.3 How the SWCT Test Works Quantitatively. Before discussing the design and interpretation of SWCT tests, we need to establish the quantitative relationship between tracer velocity, the tracer distribution coefficient, and S_{or}. **Fig. 5.7** schematically shows a local population of ester tracer molecules in a control volume (V_c) of a pore. The tracer is assumed to be locally in equilibrium, even though the water phase is moving and the oil phase is fixed (residual oil conditions).

The number of ester molecules in the water (n_{ew}) is given by

$$n_{ew} = C_{ew} S_w V_c, \quad (5.2)$$

and the number of ester molecules in the oil (n_{eo}) is given by

$$n_{eo} = C_{eo} S_{or} V_c, \quad (5.3)$$

where C_{ew} and C_{eo} = the concentration of ester (molecules/unit volume) in water and oil, respectively; and S_w = saturation of water, in fraction of PV.

We determine the retardation factor for ester (β_e) by dividing these two equations:

$$\beta_e = \frac{n_{eo}}{n_{ew}}. \quad (5.4)$$

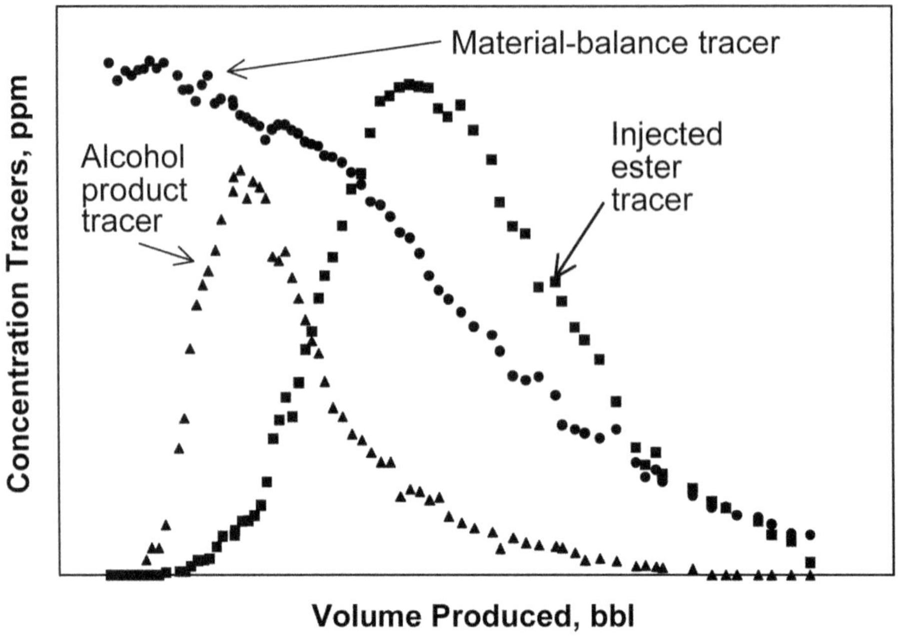

Fig. 5.6—Hypothetical SWCT test tracer-concentration profiles.

The larger β_e is, the more the ester tracer is retarded by the residual oil. Substituting in Eqs. 5.2 and 5.3 and canceling V_c yields:

$$\beta_e = \left(\frac{C_{eo}}{C_{ew}}\right)\left(\frac{S_{or}}{S_w}\right). \quad (5.5)$$

Because $C_{eo}/C_{ew} = K_e$ (see Eq. 5.1), the equilibrium distribution coefficient of ester between oil and water, and $S_w = 1 - S_{or}$, this becomes

$$\beta_e = K_e\left(\frac{S_{or}}{1 - S_{or}}\right). \quad (5.6)$$

The typical ester molecule spends a fraction of its time (f_t) in water and the rest of its time ($1 - f_t$) in oil. Elementary probability theory requires that

$$\frac{n_{eo}}{n_{ew}} = \frac{1 - f_t}{f_t}. \quad (5.7)$$

The probable behavior of each ester molecule is the same as the behavior of a large population of identical molecules. From Eq. 5.6, then:

$$\frac{1 - f_t}{f_t} = \beta_e. \quad (5.8)$$

Solving for f_t:

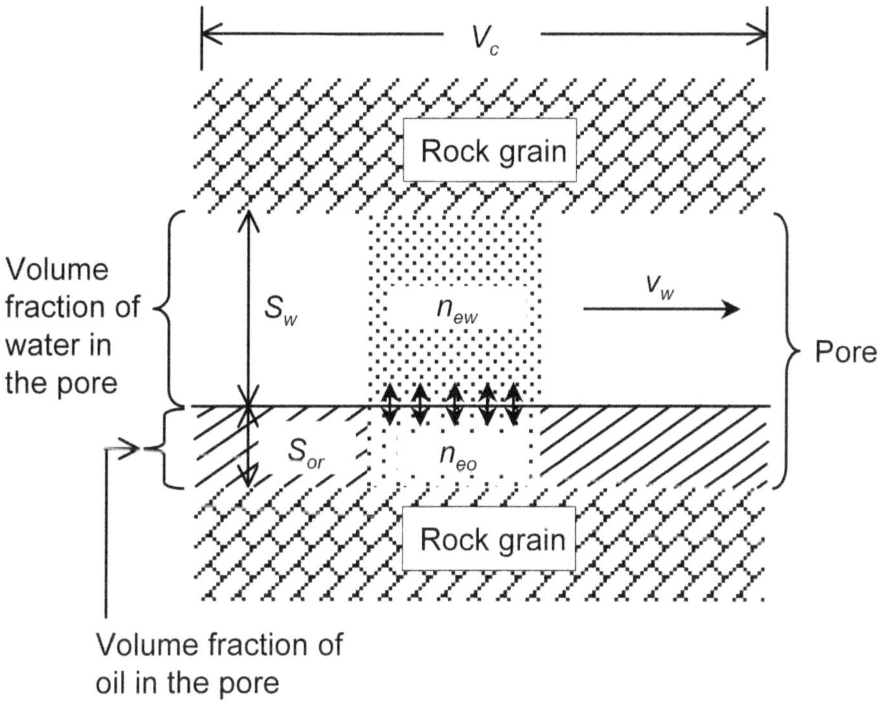

Fig. 5.7—Ester distribution between oil and water in a pore.

$$f_t = \frac{1}{1+\beta_e} \quad \text{(5.9)}$$

The typical ester molecule will travel at the time-weighted average velocity:

$$v_e = f_t v_w + (1 - f_t) v_o, \quad \text{(5.10)}$$

where v_e, v_w, and v_o = the time-weighted velocities of the tracer molecule, water, and oil, respectively. Because $v_o = 0$ if oil is at residual saturation, the last two equations combine to give

$$v_e = \frac{v_w}{1+\beta_e} \quad \text{(5.11)}$$

Eq. 5.11 is the fundamental equation for tracer chromatography in a porous medium. Solving Eq. 5.11 for β_e:

$$\beta_e = \frac{v_w}{v_e} - 1. \quad \text{(5.12)}$$

If we can develop a way to measure v_e and v_w using an in-situ test, then β_e can be evaluated. We then can measure K_e in the laboratory (at reservoir conditions) and substitute it into Eq. 5.6 to solve for S_{or}:

$$S_{or} = \frac{\beta_e}{\beta_e + K_e} \text{ . } \qquad (5.13)$$

Both the well-to-well flow of two tracers and the SWCT test are methods that have been used to measure v_e and v_w in the reservoir. Well-to-well was the first suggested of these[13] and in it, tracers A and B are injected (dissolved in water) into Well 1, and water is produced from nearby Well 2 until the tracers arrive. Tracer A is a partitioning tracer, such as an ester, that has a partition coefficient ranging from 2 to 10. Tracer B is assumed to have an equilibrium distribution coefficient (K_B) of zero (insoluble in oil), so that its time-weighted velocity (v_B) = v_w. Tracer B arrives first, after time t_B, and tracer A later, after time t_A. Because the two tracers travel the same distance,

$$\frac{v_w}{v_A} = \frac{v_B}{v_A} = \frac{t_A}{t_B}, \qquad (5.14)$$

where v_A = the time-weighted velocity of tracer A. β_A then is determined from Eq. 5.12. Once the equilibrium distribution coefficient of tracer A (K_A) is measured, S_{or} follows from Eq. 5.13.

Often, the well-to-well method is impractical for determining S_{or} because well spacing is too large. At the velocities that can be achieved between such wells, t_A is months, or even years. Also, where different permeability layers exist, the different travel time in each layer causes excessive dispersion of tracers traveling from Well 1 to Well 2.

Chap. 6 in this volume presents the technology of well-to-well tracer tests and discusses where they can be beneficial to the management of a reservoir.

5.5.4 The SWCT Test. Using the SWCT test avoids the problems of too-wide well spacing and excessive tracer dispersion caused by layering. In the SWCT test, the tracer-bearing fluid is injected into the formation through the test well and then produced back to the surface through the same well. The time required to produce the tracers back can be controlled by controlling the injected volume on the basis of available production flow rate from the test well.

In a single-well test, tracers injected into a higher-permeability layer will be pushed farther away from the well than those in a lower-permeability layer, as indicated in **Fig. 5.8a**; however, the tracers in the higher-permeability layer will have a longer distance to travel when flow is reversed. As the tracer profiles in Fig. 5.8b show, the tracers from different layers will return to the test well at the same time, assuming that the flow is reversible in the various layers.

Flow reversibility in a single-well test is desirable for the reason just explained, but it complicates S_{or} measurement. The simple strategy of injecting two tracers with different partition coefficients does not work in this case. As **Fig. 5.9** shows, tracers A and B separate during injection, just as in a well-to-well test, and at the end of injection, tracer B will be farther from the wellbore than tracer A because the presence of the residual oil retards A. During production, then, tracer B has farther to travel to return to the well. The separation achieved during injection is reversed during back production, so that A and B arrive back at the wellbore at the same time (if they were injected at the same time). No measurement of S_{or} is possible using Eqs. 5.12 and 5.13 because no information can be obtained about the ratio of velocities of A and B.

One possible way of avoiding this reversibility problem is to generate the second tracer in the formation instead of injecting it. The steps are: (1) Inject tracer A and push it into the target formation, as described above; (2) Stop flow to allow part of the injected A to react, forming tracer B in the same pore space where A is located, after the reaction time, A and B are together; (3) The fluid is then produced back into the test well; A and B must separate if

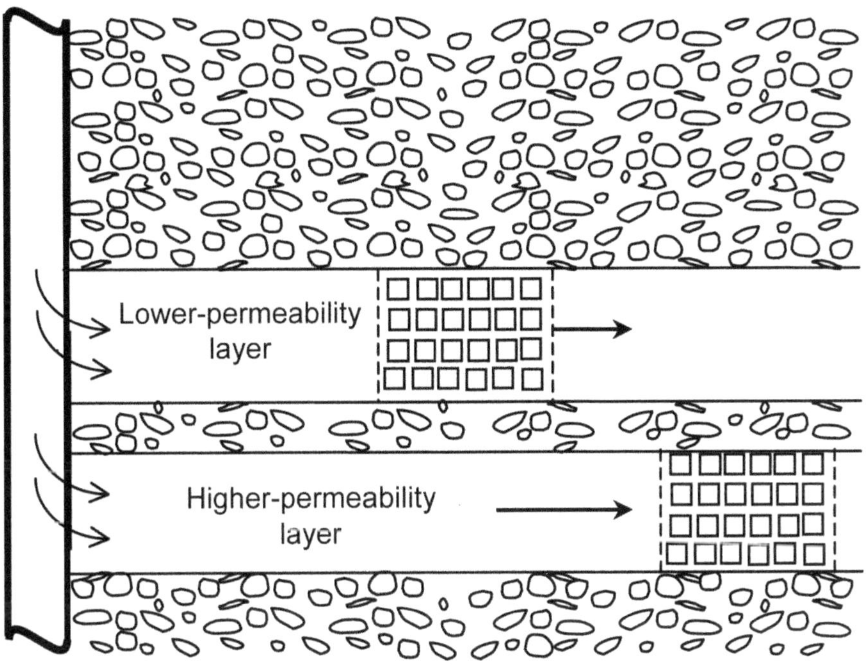

Fig. 5.8a—Injection of tracers into two layers of different permeability.

their equilibrium distribution coefficients (K_A and K_B) are different and if residual oil is present. This concept is the basis for the SWCT test patent.[2]

The practicability of this method depends on finding suitable tracers. The demands on partitioning tracer A are especially severe: inexpensive, available in reasonable quantities at high purity, nontoxic, not present in the reservoir fluids, and easily measured at low concentrations in water. It must have an appropriate K_A for the oil, water, and temperature of the target field, and, most importantly, it must react at a rate that allows formation of enough (but not too much) of a suitable tracer B. K_B must be different from K_A, and tracer B also must be measurable at low concentrations in the produced water, and not be present in the reservoir fluids.

A methyl, ethyl, or propyl ester of formic or acetic acid has proved suitable in every reservoir tested. These simple chemicals are sufficiently soluble in water, and have an appropriate range of K values and reaction rates. They are relatively inexpensive and nontoxic at the concentrations used, and they react with water to produce alcohols, which are not found in crude oils and can be detected readily in the produced fluid.

For best results,[3] choose an ester with a retardation factor (β_e) in the optimum range ($0.5 < \beta_e < 1.5$). This requires that K_e be in the range $0.5(1.0 - S_{or})/S_{or} < K_e < 1.5(1.0 - S_{or})/S_{or}$. Use the best available estimate of S_{or} to fix this range. Then choose the optimum ester using available correlations[14] for the dependence of K_e on temperature and water salinity.

In several past SWCT tests, two esters (e.g., methyl acetate and ethyl acetate) were used simultaneously to give two different depths of investigation for S_{or} in the same test. The multiple-ester test design has become increasingly popular in recent years.

5.5.5 Test Design. The design and implementation of a SWCT test for S_{or} is straightforward. Certain facts about the target formation are needed to begin test design. Some essential reservoir properties include oil cut of the test well; reservoir temperature; reservoir lithology; production rate; test interval size and average porosity; and formation water salinity.

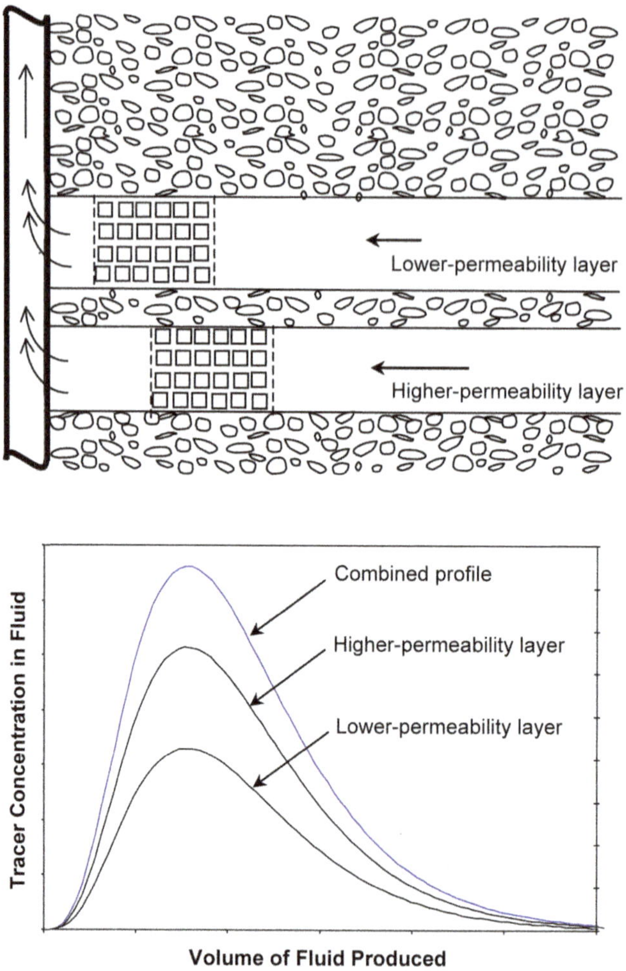

Fig. 5.8b—Production of tracers from two layers of different permeability.

Oil Cut of Test Well. Candidate wells for S_{or} measurement must be able to produce formation water to the surface. The produced fluid should be nearly all water to ensure that the test interval is at or near S_{or}. In cases where the test interval produces high oil cut, water can be injected into the test interval before testing to water-out the zone before tracers are injected.

Reservoir Temperature. Reservoir temperature dictates which esters are suitable for the SWCT test. The formate esters hydrolyze approximately 50 times faster than do the acetate esters, and are used in the reservoir temperature range of 70 to 135°F. The slower-reacting acetate esters generally are used in the 130 to 250°F range.

Reservoir Lithology. SWCT testing has been done in a variety of test conditions. In sandstone reservoirs, SWCT tests give satisfactory results for a wide range of test designs. Test timing, total injected volume, and the ester used can vary considerably for the same zone, with little effect on test interpretability.

However, SWCT tests in carbonate formations require much more precise design. In a given carbonate test zone, subtle changes in test design can cause significant variation in the tracer profile shapes. Each of our past carbonate test designs has required significant tailoring to overcome the dispersed nature of the production profiles generally present in carbonate test

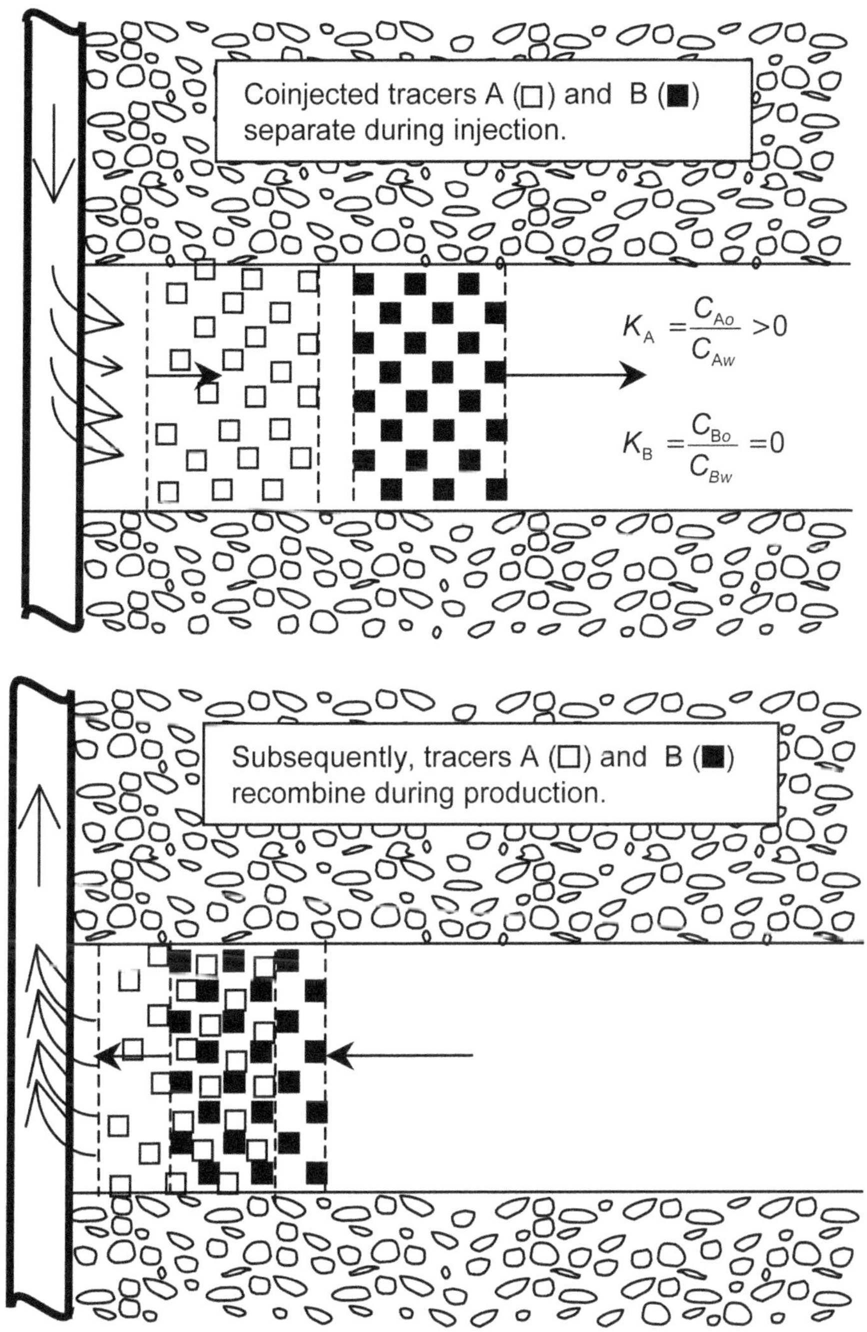

Fig. 5.9—Reversing composition profiles in a single-well injection/production test using tracers. C_{Ao} = concentration of tracer A in oil, C_{Aw} = concentration of tracer A in water, C_{Bo} = concentration of tracer B in oil, and C_{Bw} = concentration of tracer B in water.

results.[15] The reason for this dispersed nature is that the assumption of local equilibrium is not always valid for carbonate reservoir tests. This problem is illustrated in Case 4 below.

Production Rate. The production rate of the candidate well controls the test size or volume to be injected. The amount of water that can be produced in one day is a normal test volume;

two days' production is a practical upper limit. In normal productivity reservoirs, the injection is sized to give a 15- to 20-ft depth of investigation into the formation. The injection rate of the SWCT test usually is approximately the same as the well's production rate. Care must be exercised to avoid fracturing the formation during test injection.

Test-Interval Size and Average Porosity. SWCT test can be carried out on cased and open-hole completions. Estimated interval size and porosity are used to calculate a theoretical radius of investigation for the design injection volume, although S_{or} results are not dependent on either. Interval sizes of from 1 to 100 ft have been tested with satisfactory results. Generally, testing on 10- to 50-ft zones generates more definitive results.

Formation-Water Salinity. Because it normally avoids chemical incompatibility, water that is produced from the test well generally is used to carry the chemical tracers into the reservoir. Because K_e depends on it, water salinity also is a factor in choosing the best ester for the SWCT test. To avoid difficulties during injection of the tracers, closely examine the water quality (especially suspended solids) and any precipitation tendencies.

5.5.6 Field Procedures. After selecting the ester and sizing the test volume, determine the field-test location, production method, and safety requirements. Then, schedule and implement the test.

Four lift methods have been used in SWCT testing. They are free flow, electric submersible pump, rod pump, and gas lift. These lift mechanisms are listed roughly in order of desirability, but all are satisfactory.

Before the field test, the candidate well should be produced long enough to establish the oil cut, measure the stabilized production rate, and clean up the tubular goods in the completion. Then accumulate produced water for the upcoming test in clean tanks near the well. Position the portable laboratory/pumping system near the test well.

Pure tracer chemicals usually are delivered to the wellsite in 55-U.S.-gal drums. The tracers either can be batch-mixed with formation water before injection, or continuously metered into the water during injection. **Fig. 5.10** shows a schematic for a typical field setup for continuous metering of chemical in the injection water. For batch mixing, the tank can serve as the mixing vessels. With either batch-mixing or continuous metering, filter the water to one-micron or higher quality to prevent plugging when the fluids enter the reservoir.

After a short period for analytical equipment checkout, inject the chemical solution of ester and material-balance tracer and push it according to test design. Samples of the injection water should be analyzed periodically to verify tracer concentrations, and volume, rate, and pressure information should be monitored carefully throughout the injection. Be careful not to part the formation by exceeding the fracture gradient.

Once the injection is complete, the well is secured for the planned shut-in period. When the shut-in period is over, the well is placed on production. **Fig. 5.11** shows a general schematic for the production phase of a SWCT test. The produced water flows through a portable separator (if necessary) to the storage tanks on location, where its volume is carefully measured. Production volume also can be measured using a field production test separator, if one is available.

During production, water samples should be taken near the wellhead and analyzed on location for tracer concentrations. On-site chemical analysis is necessary to gather data that are accurate for the time of production, whereas sending the samples to a service laboratory for analysis would allow additional hydrolysis of the ester to take place during transport. At the time each sample is taken, the total production volume is recorded and plots of tracer concentration vs. volume produced are generated. These tracer concentration profiles are the essential field data for the SWCT test.

Because of dispersion, the total produced volume required normally is two to three times the injected test volume. Injected volume usually is one day's production, and two to three days normally are required for the back-production phase of an SWCT test.

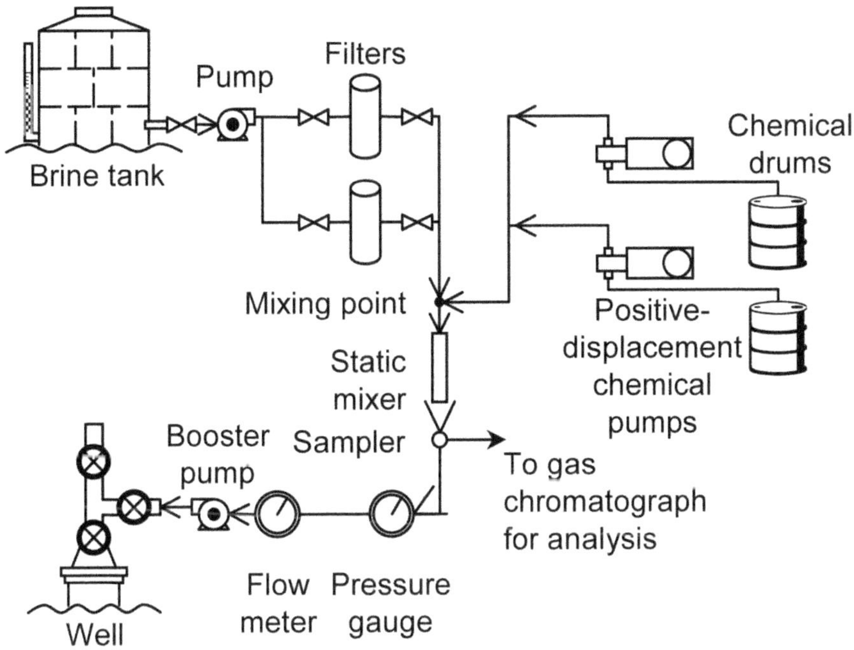

Fig. 5.10—Schematic of SWCT test injection step field setup.

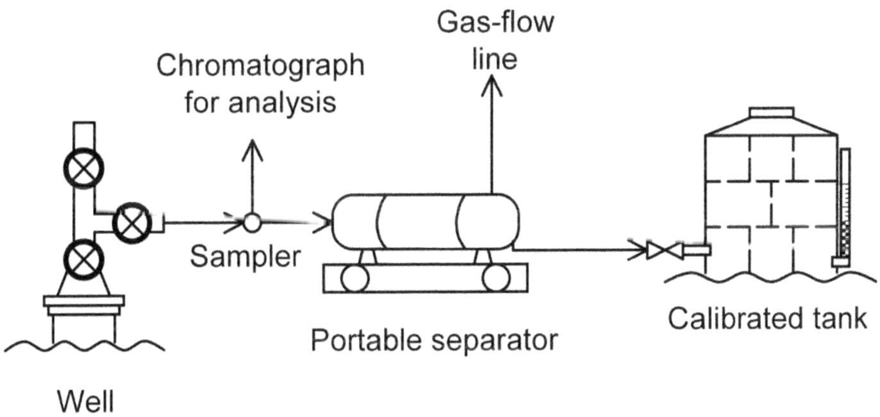

Fig. 5.11—Schematic of field setup for SWCT test production step.

5.5.7 Test Data Interpretation—Case 1. Tomich *et al.*[1] report one of the earliest SWCT tests, which was performed on a Frio Sandstone reservoir on the Texas Gulf Coast. The results of this test are used here to demonstrate the details of SWCT test interpretation for an ideal situation.

The test well in the Tomich *et al.*[1] report was in a fault block that had been depleted for several years. Because of the natural water drive and high permeability of the sand, the formation was believed to be near true S_{or}. When the well was returned to production (gas lift), it produced 100% water at a rate of 1,000 BWPD.

On the basis of observed reservoir temperature (160°F) and brine salinity [100,000 ppm total dissolved solids (TDS)], ethyl acetate was chosen as the primary tracer. Formation oil and

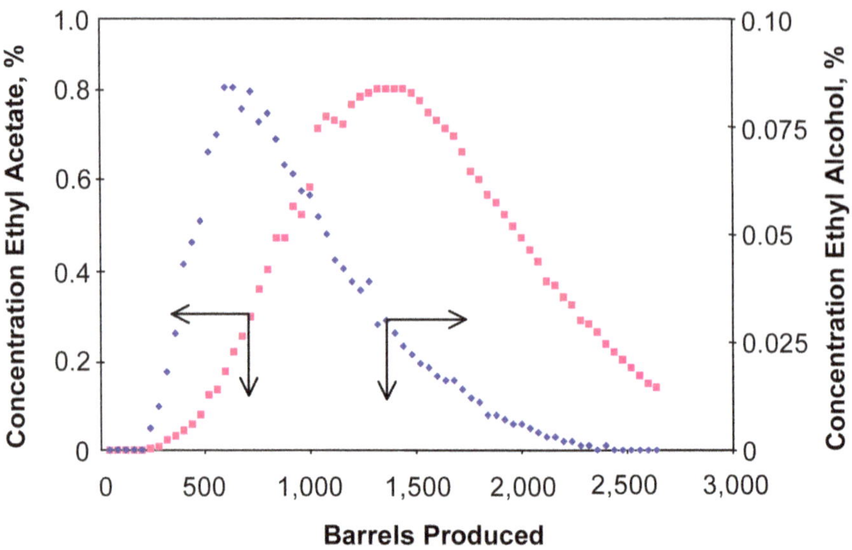

Fig. 5.12—SWCT test results from a Frio Sandstone well.

water samples were obtained for laboratory measurement of K at reservoir temperature conditions. The value of K_e measured for ethyl acetate was 6.5 at these conditions.

The test injection consisted of 1,000 bbl of formation water carrying ethyl acetate (13,000 ppm) and methyl alcohol (5,000 ppm), followed by a push bank of 1,000 bbl of formation water carrying methyl alcohol (5,000 ppm), injected over a period of two days. An eight-day shut-in period followed. During the production period, samples were collected regularly and analyzed on site using gas chromatography. The observed data are plotted as tracer concentration vs. produced volume in **Fig. 5.12**.

In ideal cases, when enough data have been gathered to define the tracer profiles, it is possible to use Eqs. 5.11 through 5.13 to approximate S_{or} in the field. If product tracer B and unreacted ester A begin together in the formation, the produced volume when A arrives back at the well (Q_{pA}) is related to the produced volume when B arrives (Q_{pB}) by the formula

$$Q_{pA} = Q_{pB}(1 + \beta_A), \quad (5.15)$$

where Q_{pA} and Q_{pB} are in bbl.

This suggests that if on the same graph we plot normalized concentration of A vs. volume produced (Q_p) and normalized concentration of B vs. $Q_p(1 + \beta_A)$, the two curves should coincide. Because we do not know β_A, this must be done by trial and error (i.e., β_A is adjusted until the best possible match of the two profiles is found).

Fig. 5.13 demonstrates this procedure. Profiles from the SWCT test (Fig. 5.12) first were normalized by dividing each observed concentration by the peak value measured for that tracer. β_A then was varied to obtain the plot shown. The best-fit value for β_A was 0.97.

Using Eq. 5.13, the S_{or} is approximated as

$$S_{or} = \frac{\beta_A}{\beta_A + K_A} = \frac{0.97}{0.97 + 6.5} = 0.13.$$

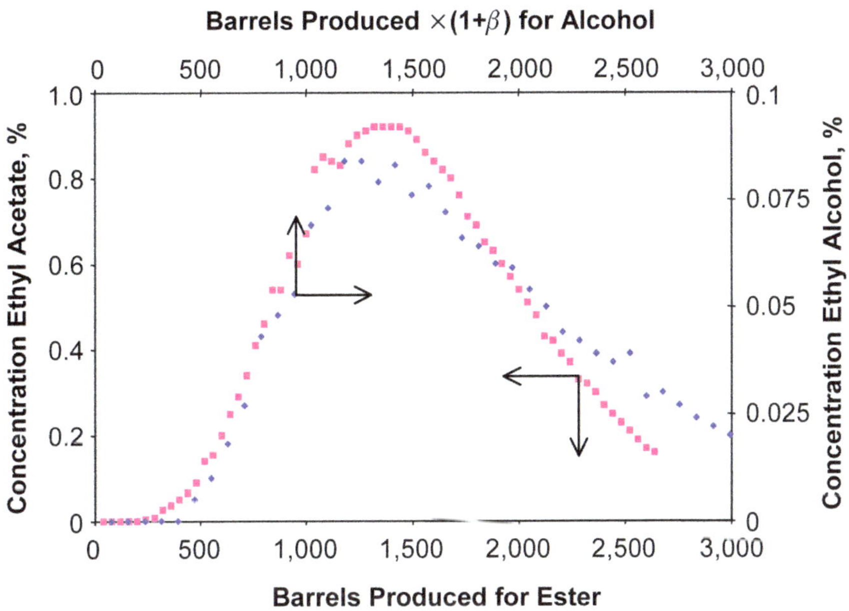

Fig. 5.13—Approximation of S_{or} using alcohol/ester superposition.

Agreement is only fair over the entire curve in Fig. 5.13; still, this result is quite close to the final interpreted result, obtained as described below under core/simulation. The approximated S_{or} depends on the validity of several idealized assumptions, which are rarely satisfied in practice:

• That no B is present in the injected A, and that most of the hydrolysis reaction occurs during shut-in (no flow), so that B and unreacted A are exactly together before backflow begins. In reality, some reaction takes place during injection and production, and there always is some B present in the A as purchased. These two sources explain the high ethyl alcohol "tail" observed in Fig. 5.13.

• That S_{or} is uniform throughout the formation tested. In practice, however, there often is evidence of several layers with different saturations in a given completion.

• That the injected tracer-carrying fluids are stationary in the reservoir throughout the shut-in period. Actually, fluids sometimes relocate during the shut-in period, because of pressure gradients across the field or pressure differences within subzones of the test completion.

• That the fluids inject uniformly into subzones and subsequently produce back to the test well reversibly. Total fluid flow reversibility is not observed in many tests.

Obtaining a more reliable answer for S_{or} requires a detailed simulation of the actual test procedure for a given SWCT test. Simulators are available to account for all the nonideal situations listed above. Simulation is time-consuming and is an added expense, but in our experience, it always is justifiable and its expense is small compared to that of field data acquisition.

Theory of Simulation. Mathematical modeling of the ideal SWCT test assumes that the carrier fluid flow is incompressible, pseudosteady state, single-phase, and radial only; that the formation is a homogeneous layer of thickness (h) and porosity (ϕ) extending from the wellbore radius (r_w) to an external boundary radius (r_e), where reservoir pressure is constant.

With these assumptions, the interstitial fluid velocity (v_{fi}) is given by

$$v_{fi} = \frac{q}{2\pi h \phi S_f r} \text{ for } r_w \le r \le r_e, \quad (5.16)$$

where r = radial position; q = fluid flow rate in the single well ($q > 0$ is injection and $q < 0$ is production); h = height of test zone; ϕ = porosity; and S_f = flowing fluid saturation, constant.

The additional assumptions regarding the tracers are:

• That K_A for tracer A is constant and that A is in local equilibrium between flowing fluid (saturation S_f) and residual fluid (saturation S_r). For Case 1, $S_r = S_{or}$ and $S_f = 1.0 - S_{or}$.

• That tracer A (primary) reacts in situ to form tracer B (product) at a rate given by

$$R_H = k_H C_A, \quad (5.17)$$

where R_H = hydrolysis reaction rate, in moles of A per vol-day; C_A = concentration of A, in mol/volume; and k_H = hydrolysis rate constant (days^{-1}) in the aqueous phase. Tracers are dispersed in the radial flow with an effective dispersion coefficient (D_a), given by

$$D_E = \alpha v_{fi}, \quad (5.18)$$

where α = dispersivity (ft).

The material balance for tracer A is the partial differential equation:

$$\left(1+\beta_A\right)\frac{\partial C_A}{\partial t} + \frac{q}{2\pi h\phi(1.0 - S_{or})}\frac{1}{r}\frac{\partial C_A}{\partial r} + k_H C_A = \frac{1}{r}\frac{\partial}{\partial r}\left(rD_E\frac{\partial C_A}{\partial r}\right). \quad (5.19)$$

(Accumulation of A in both phases) (Convection of A) (Reaction) (Dispersion)

Similar equations apply for product tracer B and the material-balance tracers.

SWCT test simulation requires the numerical solution of these equations for A, B, and the material-balance tracer. The equations first are converted to finite-differenced form, based on the perfectly mixed cell model.[16] The simulation program then solves the finite-differenced equations for the concentrations of the tracers in a radial series of cells over the test time interval.

Input to the simulation program consists of:

• Known parameters, which are: q as a function of time (injection, shut-in, production), injected concentration of tracer A (C_A) and concentration of tracer B (C_B) as a function of time; r_w, h, ϕ; K_A and K_B.

• Unknown (estimated) parameters, which are: k_H, S_{or}, and the radial dimension of the cells (ΔR). Note: according to the theory of the perfectly mixed cell model for small time intervals, $\Delta R = 2\alpha$.

5.5.8 Finding Best-Fit Parameters Using the Simulator—Case 1 Simulation. Simulation of a relatively ideal test now will be demonstrated using the tracer profiles of Case 1, shown in Fig. 5.12. The known parameters are input to the program, along with estimated values for ΔR, k_H, and S_{or}.

First, several runs are made with different ΔR values, keeping all other parameters constant. The simulation predicts values for tracer concentrations vs. volume produced. **Fig. 5.14** shows the C_A results superimposed on the field data, for different values of ΔR. The best-fit value appears to be approximately 0.50 ft. (Further refinement, using least-squares criteria, yields a value of 0.52 ft.) Note that ΔR affects the shapes of the concentration peaks, not their average positions on the produced volume axis. The shapes of C_A and C_B peaks are similarly affected.

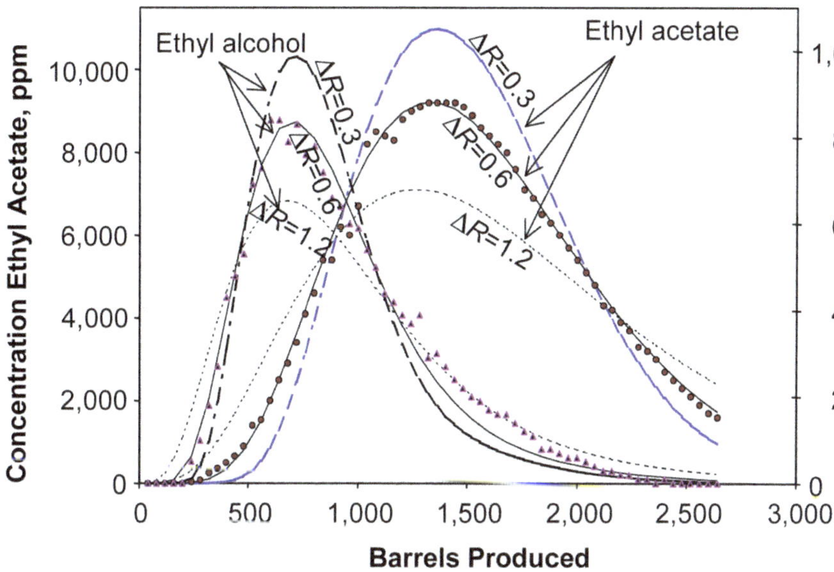

Fig. 5.14—Variation of Δ*R* shown for three cases.

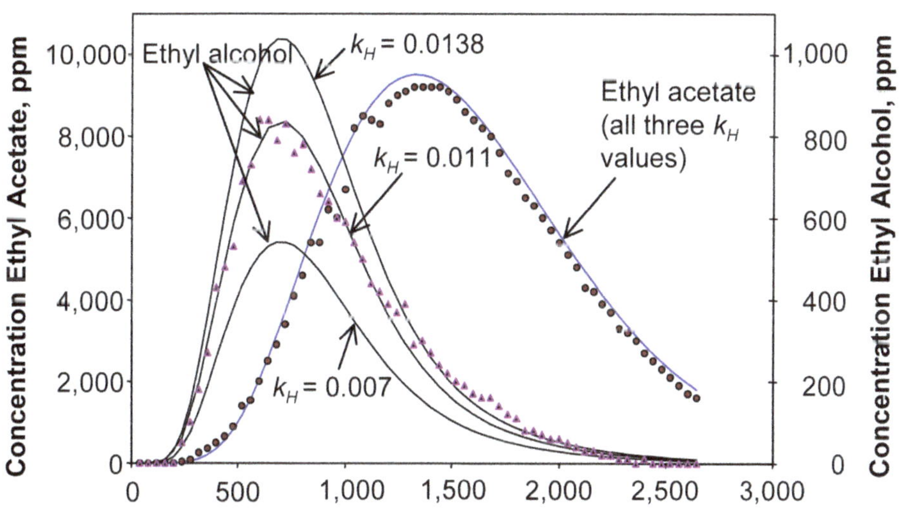

Fig. 5.15—Variation of k_H shown for three cases.

The next step is to find the best-fit value of k_H, which determines how much product tracer B will form. In **Fig. 5.15**, the simulated profiles for C_B are superimposed on the field data for C_B, for a range of values of k_H. Note that k_H affects the height of the C_B peak, but not its position. Because of the relatively small amount of hydrolysis in this case (less than 10% of the injected ester was hydrolyzed), the height of the simulated C_A peak changes very little when k_H is varied in this range of values.

Finally, the input value for S_{or} is varied. Minor adjustments are made in k_H to keep the height of the C_B peak constant. As **Fig. 5.16** shows, changing the S_{or} moves the *position* of the product tracer peak on the "produced volume" axis. Changing the S_{or} does not affect the pre-

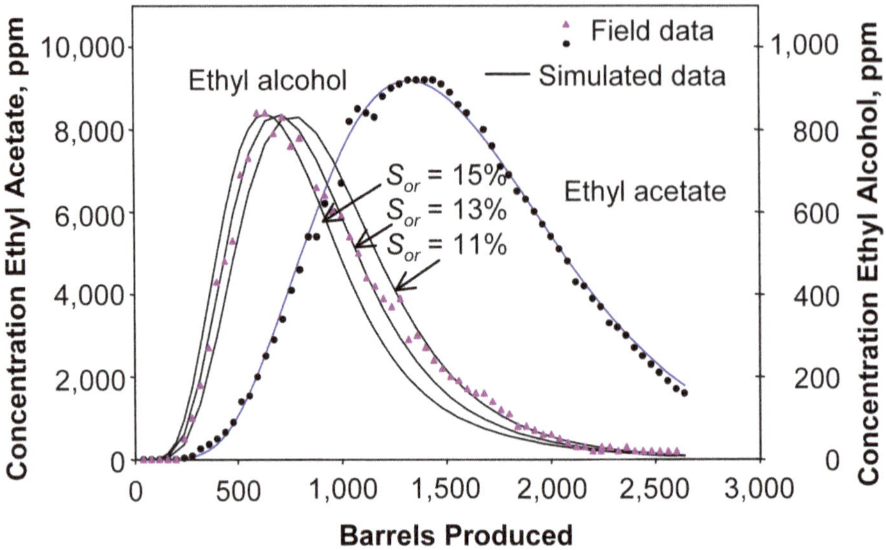

Fig. 5.16—Varying S_{or} to match the ethyl alcohol profile.

dicted position of the unreacted ester peak because of the reversibility effect that was discussed earlier.

The final best-fit simulation is shown with the field data in Fig. 5.16. The best-fit estimates of the unknown parameters for Case 1 are ΔR = 0.52 ft, k_H = 0.011 days^{-1}, and S_{or} = 0.13. To indicate the level of precision expected in the test, simulated C_B peaks also are shown for S_{or} = 0.11 and S_{or} = 0.15. S_{or} = 0.13 ± 0.02 is the best estimate for S_{or} for this reservoir, using the SWCT method.

5.5.9 How To Simulate Nonideal SWCT Tests. The 400+ SWCT tests performed to date have demonstrated that most cases are not ideal. Some reservoir factors cause the produced concentration profiles to have irregular shapes. The challenge to the test interpreter in those cases is to deduce what has caused the irregularities and to modify the simulation to obtain a credible match to the field profiles.

The three major nonideal conditions observed in SWCT tests, in order of their potential encounter, include fluid movement in the formation at the test site; in carbonates, a lengthy time required for local equilibrium to be achieved by diffusion in liquids, as compared to test duration; and in sandstones, nonreversing flow behavior in formation layers.

Fluid Movement. In active parts of a reservoir (i.e., when other producers or water injectors are close to the test well), there might be fluid movement in the formation at the test site. This is known as fluid drift. The tracers injected with the SWCT test fluids are subjected to a flow field that is not radial and reversible, as is assumed by the simulation theory above. Such was the situation during the first SWCT test.[1] A specialized simulator was developed to interpret that test.

The simulation theory assumes that a linear flow field with a fixed drift velocity, v_D, is superimposed on the radial flow at the test well. This requires the numerical solution of partial differential equations involving two space dimensions and time. Obtaining a best fit to field data involves varying drift velocity, v_D, in addition to the unknowns ΔR, k_H, and S_{or}. The drift velocity is caused by a regional pressure gradient in the formation, where as the radial component of velocity is caused by injection or production at the test well. The original simulator is furnished to licensed users of the SWCT method.

Local Equilibrium Time Length. The pore geometry of most carbonates is such that the basic assumption of local equilibrium of partitioning tracers is not valid. A significant fraction of the pore space is not directly in the flow paths, but is only in diffusional contact with these paths. Because the time required for local equilibrium to be achieved by diffusion in liquids can be long compared to the duration of the SWCT test, dual-porosity models must be used to simulate SWCT tests in carbonates. In addition to ΔR, k_H, and S_{or} discussed earlier, several new unknown parameters must be added to the list:

- The fraction of the total PV that appears to be poorly connected for flow (dead-end fraction).
- The effective diffusion parameter for each tracer in the poorly connected fraction.
- S_{or} in the poorly connected pores.

The dual-porosity simulator has been used to interpret more than 30 tests to date. Case 3 below uses the dual-porosity simulation model.

For a summary of SWCT test experience in carbonate formations, see Deans and Carlisle.[15]

Nonreversing Flow. Interpretation of SWCT tests in sandstone formations has revealed a very common nonideality. The formations tested appear to consist of two or more layers, as might be expected, but reversing behavior is not observed. One likely explanation is the existence of local pressure differences between the layers, caused by activity at other wells in the reservoir. Qualitatively, the effects of such pressure differences on the SWCT test results are:

1. A layer at higher pressure accepts less fluid during injection and produces more fluid during production than a parallel layer at lower pressure. This results in a nonideal profile because the tracer bank from the higher-pressure layer will return earlier than it should and the tracer from the lower-pressure layer will be late.
2. During the shut-in period for ester hydrolysis, fluid will flow through the wellbore from the higher-pressure layer to the lower-pressure layer. One tracer bank moves back toward the well and the other moves away during the shut-in period. This special flow condition, called crossflow, adds to the separation caused by effect 1.
3. As mentioned earlier, apparent differences in S_{or} sometimes are observed in different layers in the same formation. Along with nonreversing flow, these differences require the use of a multilayer simulator program to interpret certain SWCT tests.

Once the specific nonideal conditions are recognized, the SWCT test simulation proceeds as before. The unknown parameter set now contains an estimate of the number of layers present, the fraction of total fluid injected/produced from each layer, the crossflow between layers, and the S_{or} for each layer. Because of the large number of adjustable parameters, semi-automated optimum-seeking subprograms have been developed help find the best fit. Seetharam and Deans[17] demonstrate that an accurate flow-weighted S_{or} is obtainable using such a multilayer simulator to match field data, even though the layering parameters are not unique. Case 2 (below) presents an example of a SWCT test in a layered sandstone formation.

In all SWCT test simulations, the objective is to find the simplest simulator model that adequately matches the observed tracer profiles. The same model must match all tracers used in a given test (ester, product alcohol, and material-balance tracers). In some tests, one or more additional tracers may be used in different injection patterns to identify wellbore and/or formation flow irregularities. The more tracers that are used, the more difficult it is for the simulator to find a reasonable model to fit all field-measured tracer profiles. Our experience suggests, however, that several models can adequately describe a given set of profiles, and that the least-sensitive parameter to changing models is the main target, S_{or}.

5.5.10 Simulation of SWCT Tests in Layered Formations—Case 2. The second field test example is from a candidate test zone with the following characteristics:

- Produced water cut of 0% (100% oil).
- Reservoir temperature of 234°F.

- Lithology is sandstone.
- Produced by gas lift.
- Production rate of 500 B/D.
- Perforated interval of 45 ft.
- Average porosity of 22%.
- Brine salinity of 43,000 ppm.

No well was available in this field that produced water only. In this case, the test well initially produced 100% oil. To generate residual oil saturation near the well, 6,500 bbl of filtered water first was injected into the test zone. The waterflood injection took eight days to complete, after which the well produced 100% water during a short test.

Samples of produced oil and water were collected for ester K value measurement. Ethyl acetate was selected as a suitable partitioning tracer. Its K value was 3.65, based on laboratory measurements at reservoir conditions. With an assumed S_{or} of 25%, the corresponding value of β is 0.94, which is close to the optimum value of 1.0.

For the test that followed the waterflood, 135 bbl of formation water was injected containing ethyl acetate (7,000 ppm), normal propyl alcohol (NPA) (3,900 ppm), and isopropyl alcohol (IPA) (12,700 ppm). This was followed by 550 bbl of formation water carrying only isopropyl alcohol (12,700 ppm). The injection rate was 650 B/D and wellhead pressure was 1,200 psia during injection. Wellhead pressure was monitored carefully during injection to avoid fracturing the test zone.

The SWCT test injection required 1.15 days. The well then was shut in for five days to allow hydrolysis of a fraction of the ethyl acetate tracer. After hydrolysis, the gas lift system was turned on and the well was produced through a separator at an average rate of 500 B/D, with a stable gas lift ratio of 620 scf/bbl.

The production period was 2.7 days, throughout which samples of the produced water were taken at 5- to 15-minute intervals and analyzed immediately at the wellsite for tracer content. Total produced volume was carefully recorded at the time of each sampling. Total produced volume was 1,350 bbl. No oil cut was observed in the produced fluids at the end of the 1,350-bbl production period.

The field data from this test are plotted in **Figs. 5.17 and 5.18** as tracer concentration vs. total produced volume, along with simulation results. The best-fit simulation required four layers, with minor irreversible flow. The ethyl acetate contributions from the four layers are shown in Fig. 5.17, along with the composite concentration, which is the flow-weighted sum of the four layers. The two main layers accepted 73% of the injected tracer, and produced the same fraction. The early layer took 13% of the injection and gave back 19% of the production, and the late layer took 14% of the injected fluid, but only produced back 9%.

The predicted ethanol concentrations for the same four-layer model are shown in Fig. 5.18. To obtain the ethanol best fit, only the S_{or} was varied for the four layers. The two main layers have an S_{or} ranging from 22 to 28%. A radial gradient of oil saturation caused by the waterflood performed before the SWCT test would produce a similar effect.

On the basis of this model, the tracer-injection average S_{or} is calculated to be 26 ± 2%. This is a permeability-thickness-weighted average for the PV accessed by the ethyl acetate, which is roughly the volume of a cylinder 8 ft in radius and 45 ft high.

5.5.11 Simulation of SWCT Tests in Dual-Porosity Media. Tracer test results from many carbonate formations seem far from ideal when compared to those from sandstone formations. One of the fundamental assumptions of the SWCT test—local equilibrium of tracer in all the available fluid—is invalid for carbonates. This also is the case with fractured sandstones.

Fig. 5.19 shows the assumed situation in these cases. Tracer material being transported through the well-connected pores can diffuse into the fluid in the dead-end pores. Depending

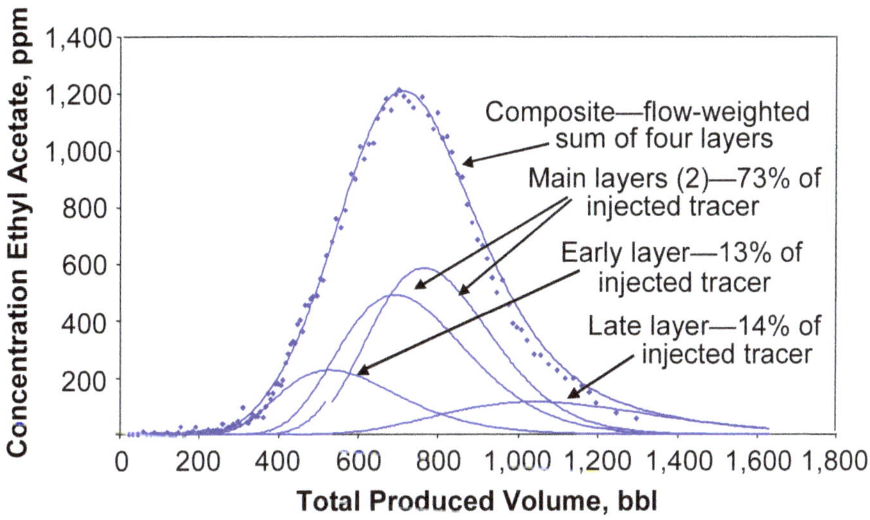

Fig. 5.17—Ethyl acetate field data and simulation, Case 2.

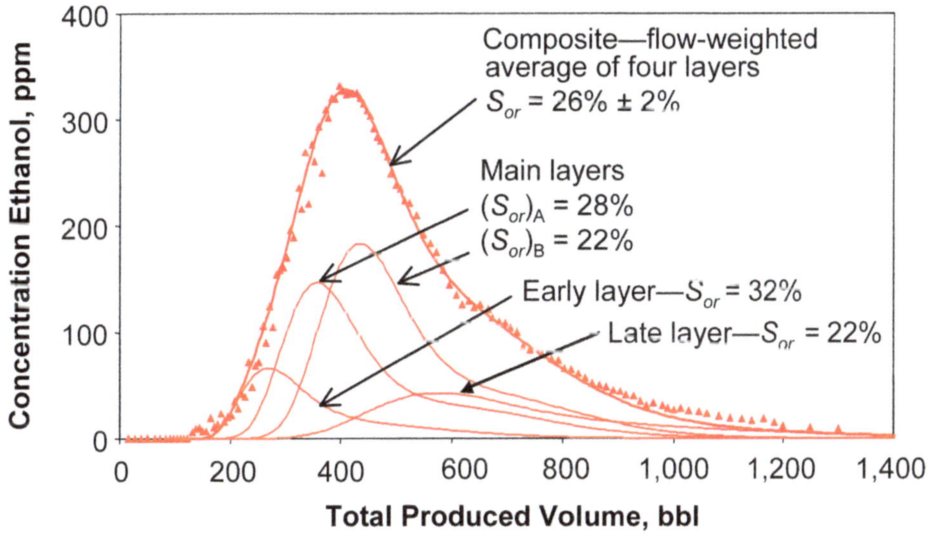

Fig. 5.18—Ethanol field data and simulation results, Case 2.

on the geometry of the pore system and the flow rates, however, the tracer might not have enough time to approach equilibrium by diffusion during the test.

A dual-porosity simulator accounts for this effect.[15] First, a "source" term is added to each tracer material-balance equation, which now describes the flow of tracer in the well-connected pores. New material-balance equations then are written to describe the diffusion of each tracer in the local dead-end pores. The diffusion equations are connected to the flowing pore equations through the source term in the original material balances.

The new model introduces three new parameters:

• The S_{or} in the dead-end pores, which is not necessarily the same as S_{or} for the well-connected pore space.

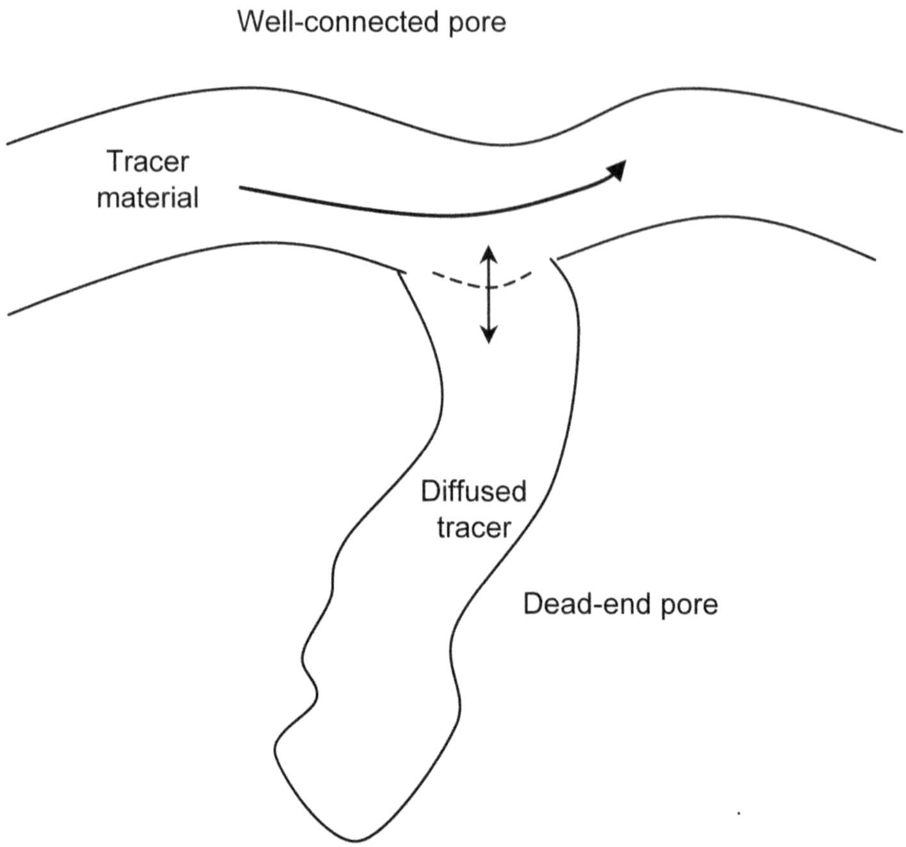

Fig. 5.19—Schematic of dual-porosity system.

- The fraction of total porosity that is in the dead-end pores.
- A diffusion parameter that controls the rate at which tracers diffuse in the dead-end pores (this parameter is smaller for long, thin pores and larger for short, fat pores).

For details on the numerical solution of the dual-porosity model equations, consult Deans and Carlisle.[15]

5.5.12 Simulation of SWCT Tests in Carbonate Formations—Case 3. The field test example for Case 3 demonstrates the extreme nonidealities that are possible in carbonate formations. The test well in this case was completed in a carbonate reef structure in Alberta, Canada. Test well characteristics were:

- Produced water cut of 99%.
- Reservoir temperature of 133°F.
- Lithology is vuggy carbonate.
- Production rate of 320 B/D.
- Perforated interval of 13 ft.
- Average porosity of 12.8%.
- Brine salinity of 113,000 ppm.
- S_{or} (anticipated) of 20%.

Because of the low reservoir temperature and high brine salinity, ethyl formate was chosen as the partitioning tracer for this test. This was designed to be a relatively small-volume test because previous large-volume tests in the same formation had not produced definitive results

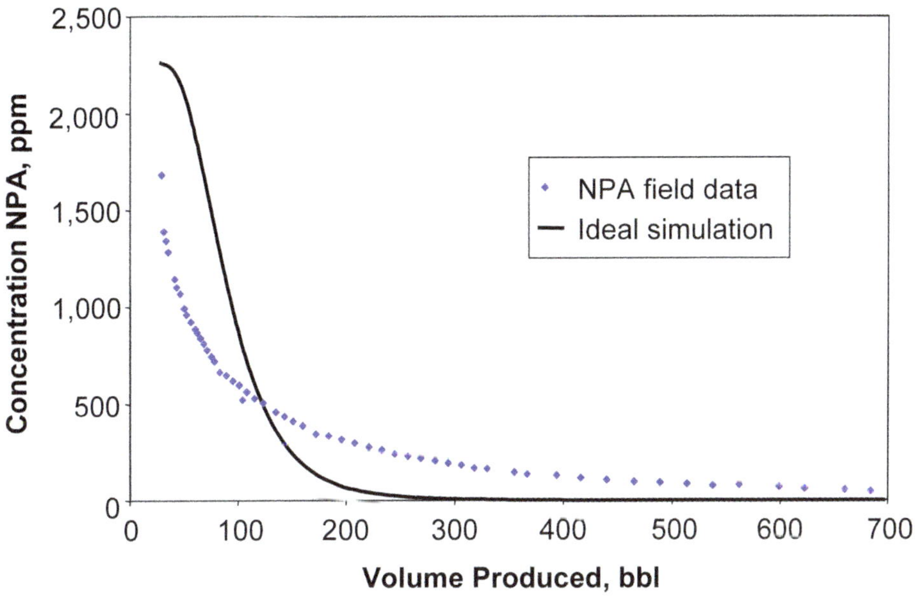

Fig. 5.20—NPA field data and ideal simulation, Case 3.

and had taken many days to complete production. Because ethyl formate is highly reactive, a decision was made to try to complete the test in less than 5 days. At an injection rate of approximately 300 B/D, 20 bbl of formation water was injected carrying ethyl formate (10,800 ppm), IPA (cover tracer) (4,700 ppm), and NPA (material-balance tracer) (2,300 ppm). This bank was followed with a net push injection of 50 bbl of water containing NPA (2,300 ppm).

The well was shut in for 1.3 days to allow part of the ethyl formate to hydrolyze, producing the secondary tracer, ethyl alcohol. The well then was produced at a rate of 300 B/D for 3.0 days. Samples were taken at regular intervals and analyzed at the wellsite for ethyl formate, ethyl alcohol, NPA, and IPA. These data are plotted as concentration vs. produced volume in **Figs. 5.20 through 5.24**.

Fig. 5.20 shows the simulation of the NPA material-balance tracer profiles with an ideal model that assumes local equilibrium. Several features are obvious:

- The ideal model is inadequate.
- Although all the injected water contained 2,300 ppm of NPA, the first returns from the formation contain significantly lower concentrations.
- The field data "tail" very badly—NPA is still being produced after 600 bbl, even though only 70 total bbl containing NPA were injected into the formation.

Figs. 5.21 through 5.24 show the dual-porosity test simulation results for each of the four tracers. The fraction of dead-end pores is 0.80 for all these simulated profiles. The S_{or} in both flowing and dead-end pores was 0.24. The dimensionless diffusion parameters ranged from 0.22 for ethyl formate to 0.28 for NPA and IPA.

5.5.13 Single-Well (One-Spot) Improved Oil Recovery (IOR) Pilot. The SWCT test can be used to evaluate an IOR process quickly and inexpensively. The one-spot procedure takes advantage of the nondestructive nature of the SWCT method.

The single-well (one-spot) pilot is carried out in three steps. First, S_{or} for the target interval is measured as described earlier. Then an appropriate volume of the IOR fluid is injected into the test interval and pushed away from the well with water. Finally, the SWCT test is repeated

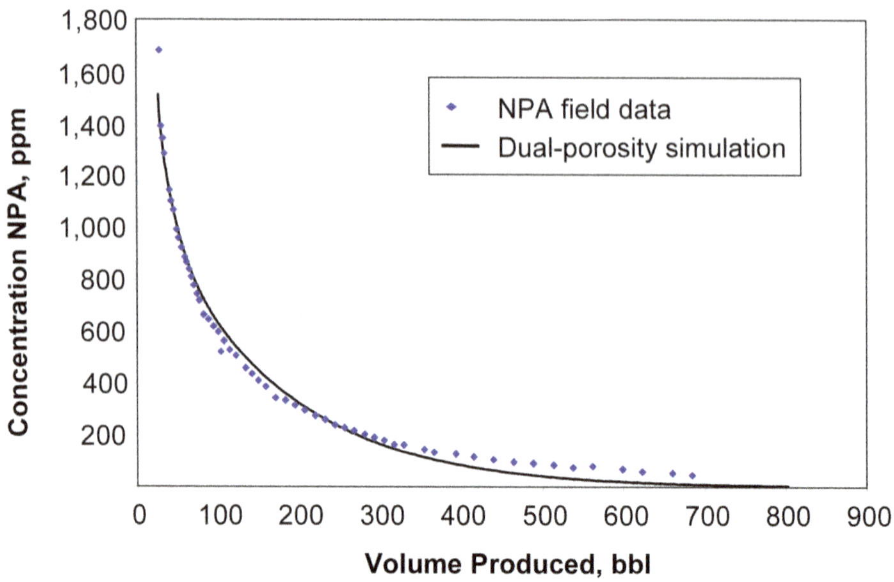

Fig. 5.21—NPA material-balance tracer simulation, dual-porosity model, Case 3.

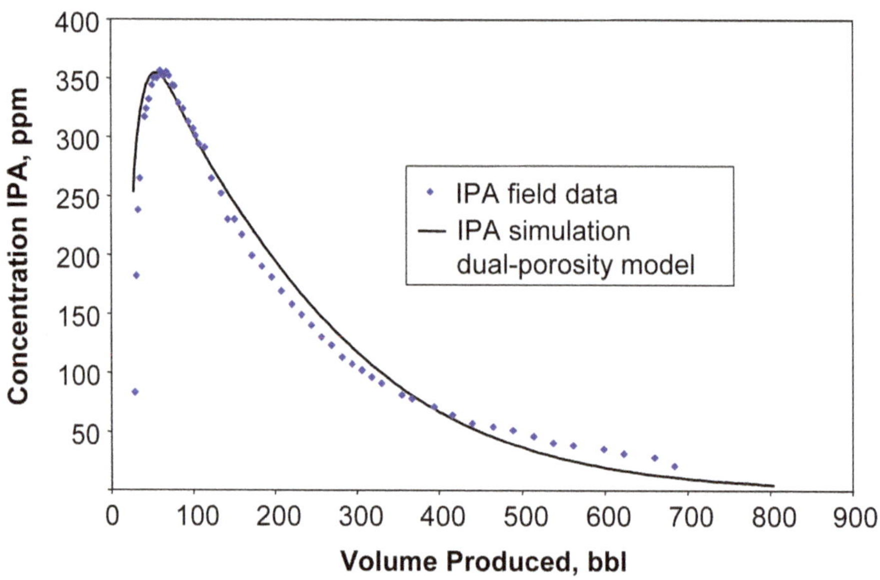

Fig. 5.22—IPA cover tracer simulation, dual-porosity model, Case 3.

within the treated region. Typically, the entire process is completed in a few weeks, as compared to the much longer duration of pilot programs, even small-patterned ones.

The reduction in S_{or} observed in the postflood SWCT test is a measure of the IOR flood performance. By using several esters with different partition coefficients, it is possible to measure S_{or} as a function of radial position from the test well. This procedure also can be used to evaluate the stability of the IOR process because tracer test volume is increased relative to the IOR displaced volume.

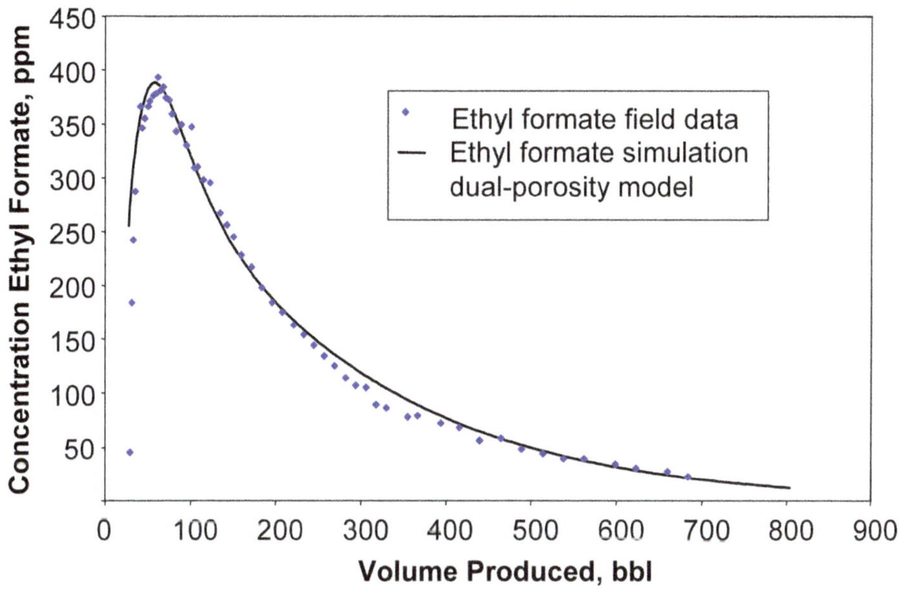

Fig. 5.23—Ethyl formate simulation, dual-porosity model, Case 3.

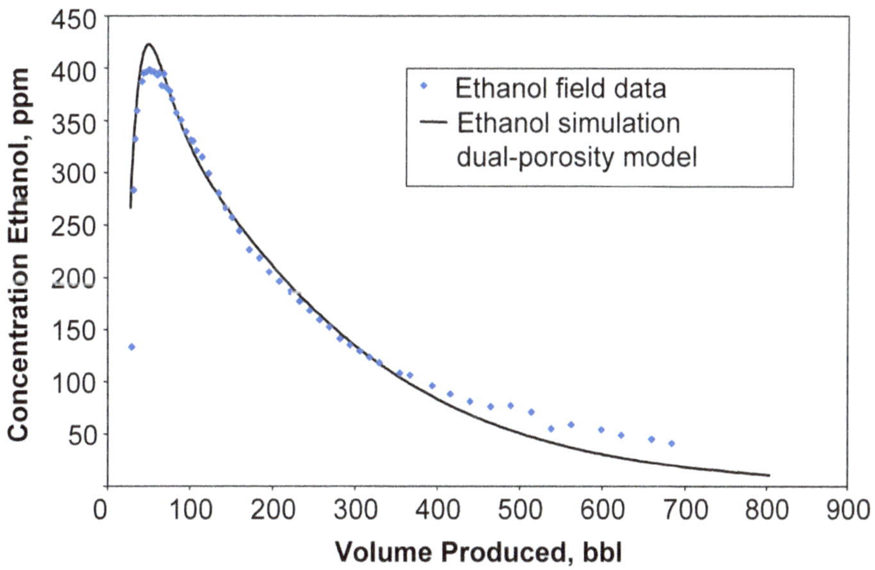

Fig. 5.24—Ethanol simulation, dual-porosity model, Case 3.

• The single-well IOR pilot concept has been applied in a number of different projects,[18–21] in the course of which the following IOR processes have been evaluated:

• Surfactant, surfactant-polymer, and alkaline-surfactant-polymer (ASP).
• CO_2 miscible.
• Caustic and caustic-polymer.
• Hydrocarbon-miscible.

In every case, the process was evaluated successfully using the SWCT test. In most of these sequences, the IOR fluid injection and push steps were carried out by the SWCT test

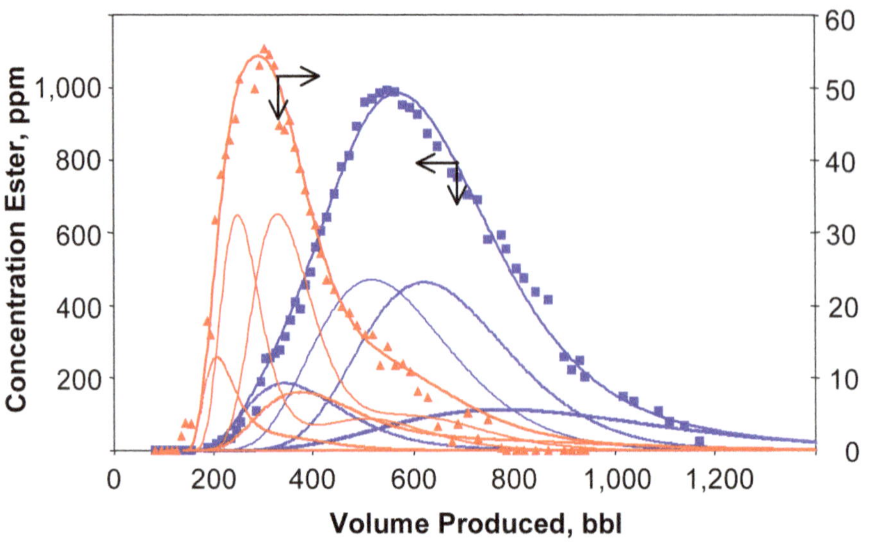

Fig. 5.25—One-spot pilot evaluation of hydrocarbon-miscible IOR before IOR bank.

field group as a part of the field test program. A typical single-well pilot evaluation of an IOR process requires four to six weeks to complete.

5.5.14 SWCT Tests for In-Situ Evaluation of Hydrocarbon-Miscible IOR—Case 4. A series of SWCT tests was performed in the Aurora Field on the Alaska North Slope,[22] U.S.A., to provide an in-situ formation evaluation and to evaluate hydrocarbon-miscible IOR. The target well produced 100% oil.

The steps in the test sequence were:

1. An SWCT test for S_{wc} (see below under *Other Field Measurements*).
2. A multiple tracer test to determine oil/water fractional flow vs. saturation.
3. A waterflood to reduce the near-wellbore oil saturation to near-residual conditions.
4. An SWCT test for S_{or}.
5. Injection of a bank of hydrocarbon-miscible solvent.
6. A waterflood to displace the IOR bank away from the test well.
7. A second SWCT test for S_{or}, to determine the post-IOR oil saturation in the flooded region.

Steps 4 through 7 constitute the "one-spot" pilot for the hydrocarbon-miscible IOR process.

The results of step 4 are shown in **Fig. 5.25**. The field data are plotted, along with the best-fit four-layer simulation model. Two nonideal effects are indicated by these data:

• Nonreversing flow is evidenced by the early hump and the extended tail on the tracer profiles.

• A radial gradient of oil saturation apparently was present in the formation because the waterflood of step 3 was not large enough to reduce the formation to true residual oil.

The four-layer model can account for both of these effects. The ester-weighted S_{or} measured in step 4 is 32 ± 3%.

The field data and best-fit simulation of the aftertest (step 7) are shown in **Fig. 5.26**. Again, two distinct nonideal effects are apparent:

• Nonreversing flow again is a factor. An early hump and a tail indicate that small layers are needed to improve the fit of the simulation model.

• In this case, the late layer, which produces the tail, also is anomalous regarding oil saturation. Whereas both of the other layers require an S_{or} of 6% to produce the best fit, the late layer appears to have an S_{or} of 26%. A likely explanation for this remarkable difference is

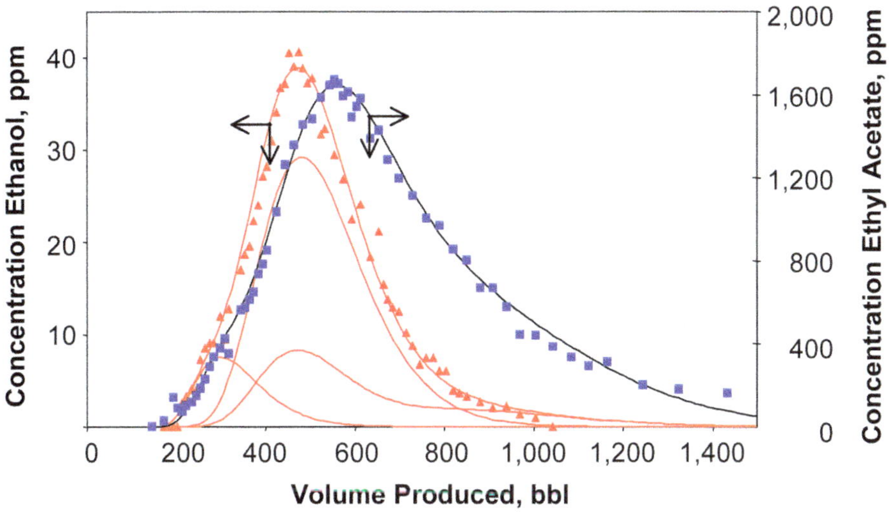

Fig. 5.26—One-spot pilot evaluation of hydrocarbon-miscible IOR after IOR bank.

gravity override by the miscible solvent bank. There appears to be a layer that was not contacted by the IOR bank.

5.6 Other Field Measurements

5.6.1 Connate Water Saturation Testing. In certain situations, it is necessary to obtain a reliable measurement for S_{wc} in an oil reservoir. The SWCT method has been used successfully for this purpose in six reservoirs.[23] The SWCT test for S_{wc} usually is carried out on wells that are essentially 100% oil producers. The procedure is analogous to the SWCT method for S_{or}, taking into account that oil is the mobile phase and water is stationary in the pore space.

Because oil is the mobile phase, it is used to carry the chemical tracers into and back out of the formation. The ester dissolved in oil is injected and then pushed away from the well. The first bank carries the ester plus material-balance tracer, as in the S_{or} test. The push bank contains only the material-balance tracer.

During the flow into the formation, the ester partitions between the oil and water phases. This partitioning slows the ester velocity only slightly because the ester is more soluble in the oil phase. After the push step, the ester bank is located 10 to 20 ft from the wellbore.

During the 2- to 10-day shut-in period, a portion of the ester hydrolyzes (reacts with the connate water) to form the alcohol product tracer. The product alcohol is much more soluble in the connate water than in the oil phase.

The well then is placed on production. Samples of the produced oil are taken frequently and analyzed for tracer content. The unreacted ester comes back first because it travels at nearly the same velocity as the carrier oil. The product alcohol is produced later because it travels more slowly than the oil. Because of its preferential solubility in the stationary water phase, the alcohol has a large β value compared to that of ester. The difference in arrival volumes between the ester and the delayed alcohol tracer thus is a function of S_{wc}.

As in the SWCT test for S_{or}, β can be calculated directly from the concentration vs. produced volume profiles. Using the partition coefficients measured in the laboratory, S_{wc} can then be calculated. However, the best-fit S_{wc} result actually is determined by matching the field tracer profiles using our multilayer simulator program for the connate water test.

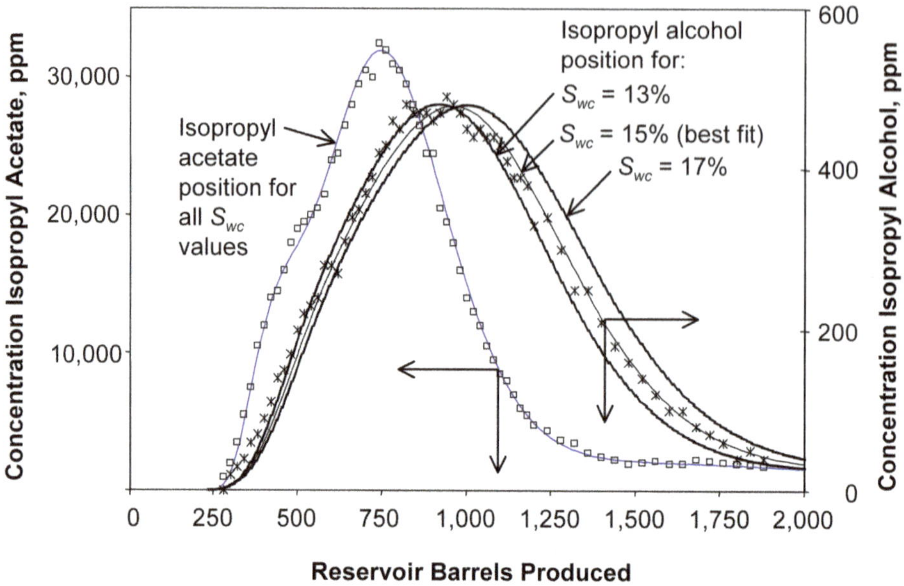

Fig. 5.27—Field data and simulated tracer profiles for an SWCT test for S_{wc}.

5.6.2 SWCT Test for Connate Water—Case 5. An example of a production profile from a SWCT test for connate water is shown in **Fig. 5.27.**[23] The target formation was in the Prudhoe Bay Field on the Alaska North Slope, U.S.A. The reactive ester injected was isopropyl acetate, which was dissolved in light diesel oil. The product tracer in this case was IPA.

The best-fit simulation model for this case used three layers, to account for the slightly nonreversing flow behavior. The injected ester-weighted average result was S_{wc} = 0.15 ± 2%. This is in close agreement with the average S_{wc} determined from oil-based cores on the same well.[23]

5.6.3 Gas-Saturation Testing. In formations where the pore space is occupied by a stationary gas phase and a mobile water phase, such as in a watered-out gas reservoir, the residual gas saturation (S_{gr} may need to be measured in situ. The S_{gr} also can be determined using a single-well injection/production test method.[24]

S_{gr} measurement involves injecting and immediately producing a suitable volume of water. The water used for injection typically is produced from the target well before the test and stored in tanks on the surface. During production, the amount of gas dissolved in the water (R_{sw}) that is produced from the formation is measured.

The injected water is essentially free of dissolved hydrocarbon gases (i.e., it is "dead"). As the injected water enters the formation that contains residual gas, the water dissolves the gas, becoming saturated at the temperature and pressure in the reservoir. A region of increasing radius around the wellbore is stripped of gas. Injection continues until a volume (V_I) has entered the formation. A material-balance tracer such as methanol is added to all injected water.

At the end of the injection, the stripped region near the well is a volume of pore space that is filled with dead water (V_d). Just beyond this dead-water zone, the pore space contains the original gas saturation and gas-saturated injected water.

Immediately after injection, water is produced back from the formation through the well to the surface. The first volume of water produced is the V_d from the leached region, and so contains no dissolved gas. This dead water is followed by gas-saturated water from the region outside the stripped zone. The gas-saturated water is able to move through the stripped region

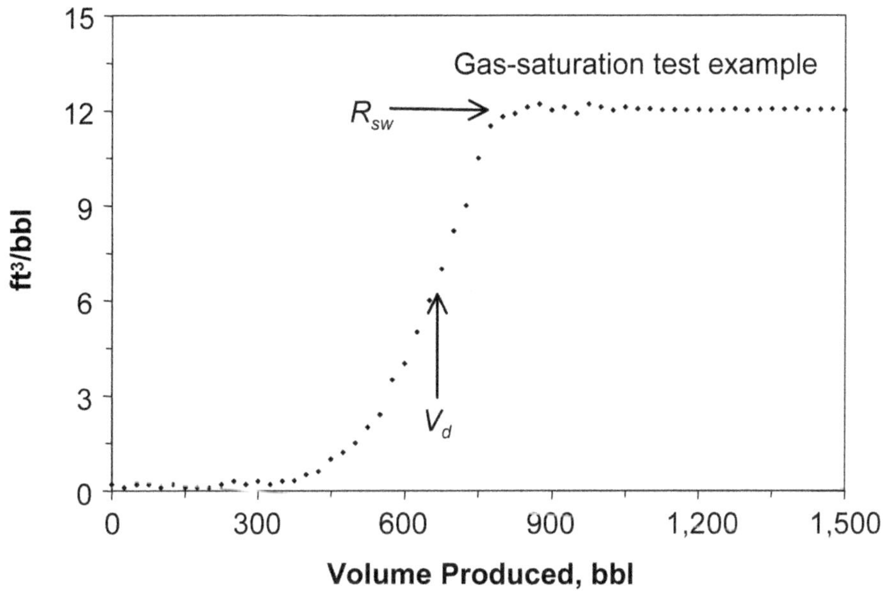

Fig. 5.28—Gas-saturation test example.

without giving up its dissolved gas to reform the original gas saturation. This nonreversing behavior is the reason the method works in a single well.

After the volume of dead water is produced, the observed gas content of the produced water rises to the solubility R_{sw} (scf/bbl) of gas at reservoir conditions. The volume of gas-saturated water ($V_I - V_d$) produced after the dead water contains the gas that was dissolved from the stripped zone. Thus, if B_g is the gas formation volume factor at reservoir conditions, the reservoir volume of gas dissolved in this fluid (V_g) is:

$$V_g = \left(V_I - V_d\right) R_{sw} B_g \text{ . } \qquad (5.20)$$

But this is just the volume of gas originally in the stripped zone, which is:

$$V_g = V_d S_{gr} \text{ . } \qquad (5.21)$$

Equating the two and solving for S_{gr}:

$$S_{gr} = \left(V_I - V_d\right) R_{sw} \left(\frac{B_g}{V_d}\right) \text{ . } \qquad (5.22)$$

Fig. 5.28 is a typical profile for gas content vs. produced water volume from a single-well injection test for S_{gr}. The volume required to produce the inflection point of the gas content profile is V_d. The final level of gas content is R_{sw}, the solubility of gas at reservoir conditions. These two values, along with the total injected volume V_I, are the only field data required to calculate S_{gr}. The procedure for measuring S_{gr} is described in detail by Bragg and Shallenberger.[24]

Performing the S_{gr} test is relatively simple. The only operational requirements are that a method must be available to produce water from the formation in the test well and that there must be a way to measure the gas content of the water leaving the formation, with which the fluid lifting mechanism does not interfere. Because of the nonreversing feature, the test cannot be repeated in a given completion. The gas that is dissolved from the near-well pore space is not replaced during the production step.

This method can be used to measure either hydrocarbon gas or carbon dioxide (CO_2) saturation after displacement by water, and in some cases where residual oil, residual gas, and mobile water all share the pore space (three-phase system), this test has been conducted in tandem with the SWCT test. The ester solution and push volume are added to the end of the dead-water injection. A shut-in period also is added. The S_{or} value measured by the SWCT test is for the two-phase (oil and dead water), gas-leached zone near the well. The S_{gr} obtained from the gas content profile represents the fraction of the near-well pore space filled with gas before the test injection.

5.7 Summary

This *Handbook* chapter has familiarized the reader with the SWCT method by presenting the essential details of its theory to establish its credibility. It also has covered the planning and execution of the field operations to show the practicality of the method, and has outlined the process of test data interpretation and applied it to real test data to demonstrate the SWCT method's sensitivity and accuracy. High-quality fluid-saturation data always have been essential to the reservoir engineer, and in today's challenging oil production operations, the information provided by SWCT testing methods is more important than ever before. In the coming years, this unique and intriguing method is likely to be used more widely in reservoir evaluation.

Nomenclature

B_o = formation volume factor, oil, RB/STB
B_g = formation volume factor, gas, RB/Mscf
C_A = concentration of tracer A, mol/vol
C_{Ao} = concentration of tracer A in oil
C_{Aw} = concentration of tracer A in water
C_B = concentration of tracer B, mol/vol
C_{Bo} = concentration of tracer B in oil, mol/vol
C_{Bw} = concentration of tracer B in water, mol/vol
C_{eo} = concentration of ester in oil, mol/vol
C_{ew} = concentration of ester in water, mol/vol
D_E = effective dispersion coefficient, ft^2/D
f_t = fraction of time
h = thickness, ft
k_H = hydrolysis rate constant (days^{-1}) in the aqueous phase
K = equilibrium partition coefficient
K_A = equilibrium partition coefficient for tracer A
K_B = equilibrium partition coefficient for tracer B
K_e = oil/water partition coefficient for ester
n_{eo} = number of ester molecules in oil
n_{ew} = number of ester molecules in water
q = fluid-flow rate in a single well, B/D
Q_p = produced volume, bbl
Q_{pA} = volume required to produce tracer A, bbl
Q_{pB} = volume required to produce tracer B, bbl
r = radial position, ft
r_e = external boundary radius, ft

r_w = wellbore radius, ft
R_H = hydrolysis reaction rate, mol/vol-day
R_{sw} = gas solubility in produced water at reservoir conditions, scf/bbl
S_f = flowing fluid saturation, fraction of PV
S_{gr} = residual gas saturation, fraction of PV
S_o = oil saturation, fraction of PV
$\bar{S}_o$ = average oil saturation in the reservoir, fraction of PV
S_{or} = residual oil saturation, fraction of PV
S_r = residual fluid saturation fraction of PV
S_w = saturation of water, fraction of PV
S_{wc} = connate water saturation, fraction of PV
t = time, day
t_A = arrival time of tracer A, day
t_B = arrival time of tracer B, day
v_A = average velocity of tracer chemical A, ft/D
v_B = average velocity of tracer chemical B, ft/D
v_D = fluid drift velocity in the formation, ft/D
v_e = time-weighted velocity of ester
v_{fi} = interstitial fluid velocity, ft/D
v_o = time-weighted velocity of oil
v_w = time-weighted velocity of water
V_c = control volume of a pore, bbl
V_d = volume of dead (gas-free) water produced, bbl
V_g = the reservoir volume of gas dissolved in gas-saturated water, scf/bbl
V_I = volume of water injected, bbl
V_p = total pore volume, bbl
V_w = velocity of water
α = dispersivity
β = retardation factor
β_A = retardation factor for tracer A
β_e = retardation factor for ester
ΔR = radial dimension (ft) of a cell
ϕ = porosity

References

1. Tomich, J.F. *et al.*: "Single-Well Tracer Method to Measure Residual Oil Saturation," *JPT* (February 1973) 211.
2. Deans, H.A.: "Method of Determining Fluid Saturations in Reservoirs," U.S. Patent No. 3,623,842 (1971).
3. Deans, H.A. and Majoros, S.: "The Single-Well Chemical Tracer Method for Measuring Residual Oil Saturation," Final report, Contract No. DOE/BC20006-18, U.S. DOE, Washington, DC (October 1980).
4. O'Brien, L.J., Cooke, R.S., and Willis, H.R.: "Oil Saturation Measurements at Brown and East Voss Tannehill Fields," *JPT* (January 1978) 17.
5. Sheely, C.Q.: "Description of Field Tests to Determine Residual Oil Saturation by Single-Well Tracer Method," *JPT* (February 1978) 194.

6. Thomas, E.C. and Ausburn, B.E.: "Determining Swept-Zone Residual Oil Saturation in a Slightly Consolidated Gulf Coast Sandstone Reservoir," *JPT* (April 1979) 513.
7. Nute, A.J.: "Design and Evaluation of a Gravity-Stable, Miscible CO_2-Solvent Flood, Bay St. Elaine Field," paper SPE 11506 presented at the 1983 SPE Middle East Oil Technical Conference, Manama, Bahrain, 14–17 March.
8. Chang, M.M. *et al.*: "Evaluation and Comparison of Residual Oil Saturation Determination Techniques," *SPEFE* (March 1988) 251.
9. Donaldson, E.C. and Staub, H.L.: "Comparison of Methods for Measurement of Oil Saturation," paper SPE 10298 presented at the 1981 SPE Annual Technical Conference and Exhibition, San Antonio, Texas, 5–7 October.
10. Energy Information Administration (EIA). 1980. Annual Report to Congress, Volume 3, DOE/EIA-0173(79)/3. Washington, DC (July 1979).
11. Salathiel, R.A.: "Oil Recovery by Surface Film Drainage in Mixed-Wettability Rocks," *JPT* (October 1973) 1216; *Trans.*, AIME, **255**.
12. Deans, H.A. and Shallenberger, L.K.: "Single-Well Chemical Tracer Method to Measure Connate Water Saturation," paper SPE 4755 presented at the 1974 SPE Improved Oil Recovery Symposium, Tulsa, 22–24 April.
13. Cooke, C.E. Jr.: "Method of determining Residual Oil Saturation in Reservoirs," U.S. Patent No. 3,590,923 (1971).
14. Carlisle, C.T. *et al.*: "Development of a Rapid and Accurate Method for Determining Partition Coefficients of Chemical Tracers Between Oils and Brines (for Single-Well Tracer Tests)," Contract No. DOE/BC/10100-4, U.S. DOE, Washington DC (December 1982).
15. Deans, H.A. and Carlisle, C.T.: "Single-Well Tracer Test in Complex Pore Systems," paper SPE/DOE 14886 presented at the 1986 SPE/DOE Symposium on Enhanced Oil Recovery, Tulsa, 20–23 April.
16. Deans, H.A.: "A Mathematical Model for Dispersion in the Direction of Flow in Porous Media," *SPEJ* (March 1963) 49; *Trans.*, AIME, **228**.
17. Seetharam, R.V. and Deans, H.A.: "CASTEM-A New Automated Parameter-Estimation Algorithm for Single-Well Tracer Tests," *SPERE* (February 1989) 35.
18. Sheely, C.Q. Jr. and Baldwin, D.E.: "Single-Well Tracer Test for Evaluating Chemical Enhanced Oil Recovery Processes," *JPT* (August 1982) 1887.
19. Holland, K.M. and Porter, L.T.: "Single-Well Evaluation Program for Micellar/Polymer Recovery, Main and 99 West Pools, West Coyote Field, California," paper SPE 11990 presented at the 1983 SPE Annual Technical Conference and Exhibition, San Francisco, 5–8 October.
20. Edinga, K.J., McCaffery, F.G., and Wytrychowski, I.M.: "Cessford Basal Colorado A Reservoir —Caustic Flood Evaluation," *JPT* (December 1980) 2103.
21. Cockin, A.P. *et al.*: "Design, Implementation and Simulation Analysis of a Single-Well Chemical Tracer Test To Measure the Residual Oil Saturation to a Hydrocarbon Miscible Gas at Prudhoe Bay," paper 48951 prepared for presentation at the 1998 SPE Annual Technical Conference and Exhibition, New Orleans, 27–30 September.
22. Carlisle, C.T. *et al.*: "Aurora Single Well Tracer Test (SWTT) Results," poster presented at the 2002 SPE Western North American Region/Pacific Section AAPG Joint Meeting, Anchorage, 20–22 May.
23. Deans, H.A. and Mut, A.D.: "Chemical Tracer Studies To Determine Water Saturation at Prudhoe Bay," *SPERE* (February 1997) 52.
24. Bragg, J.R. and Shallenberger, L.K.: "In-Situ Determination of Residual Gas Saturation by Injection and Production of Brine," paper SPE 6047 presented at the 1976 SPE Annual Technical Conference and Exhibition, New Orleans, 3–6 October.

SI Metric Conversion Factors

bbl	×	1.589 873	E − 01	= m^3
°F		(°F − 32)/1.8		= °C
ft	×	3.048*	E − 01	= m
ft^3	×	2.831 685	E − 02	= m^3
day	×	1.0*	E − 00	= d

lbm mol	×	4.535 924	E − 01	= kmol
psia	×	6.894 757	E − 00	= kPa
res bbl	×	1.689 873	E − 01	= res m^3

*Conversion factor is exact.

Chapter 6
Well-To-Well Tracer Tests

Øyvind Dugstad, Inst. for Energy Technology

6.1 Introduction

Well-to-well tracer tests contribute significantly to the reservoir description, which is essential in determining the best choice of production strategy. Direct dynamic information from a reservoir may be obtained, in principle, from three sources: production history, pressure testing, and tracer testing.

The value and importance of tracer tests are broadly recognized. Tracer testing has become a mature technology, and improved knowledge about tracer behavior in the reservoir, improved tracer analysis, and reduction of pitfalls have made tracer tests reliable. Many tracer compounds exist; however, the number of suitable compounds for well-to-well tracers is reduced considerably because of the harsh environment that exists in many reservoirs and the long testing period. Radioactive tracers with a half-life of less than one year are mentioned only briefly in this chapter because of their limited applicability in long-term tests.

Tracers may be roughly classified as passive or active. In principle, a passive tracer blindly follows the fluid phase in which it is injected. Active tracers interact with the other fluids in the system or with the rock surface. Interpretation of tracer-production curves must account for this. The results from the application of active tracers may give information about fluid saturation and rock surface properties. This information is especially important when enhanced-oil-recovery techniques that use expensive fluids such as surfactants, micellar fluids, or polymers are considered.

In the last 50 years, many tracer studies have been reported and even more have been carried out without being published in the open literature. Wagner[1] pointed out six areas in which tracers could be used as a tool to improve the reservoir description.

• Volumetric sweep. The volume of fluid injected at an injection well until breakthrough of the traced fluid at an offset producer is a measure of the volumetric sweep efficiency between that pair of wells. Very small volumes injected before breakthrough [relative to the interwell pore volume (PV)] would indicate the existence of an interwell open fracture (or a very thin high-permeability stringer) and would give an idea of the volume of that channel. Knowledge of channels is important to the sizing of remedial treatment.

• Identification of offending injectors. Problem injection wells can be identified by associating the breakthrough of a specific tracer to its point of injection. At this well, a remedial treatment to seal a channel normally would be applied.

• Directional flow trends. When fluids are injected in a regular pattern (five-spot, nine-spot, line drive, etc.) and the fluids injected at each well tagged with a different tracer, directional flow trends will be obvious from the repeated early tracer breakthrough at producers in a preferential direction from the injectors. Where directional flow trends are prevalent, the interwell sweep efficiency often can be improved by altering the injection pattern and/or the injection and withdrawal rates at selected wells.

• Delineation of flow barriers. Faults with large displacement along the fault plane and permeability pinchouts can represent barriers to the flow of fluids perpendicular to their axis. Normally, such barriers are detected by bottomhole pressure buildup surveys run in nearby wells. However, the course of these barriers can be delineated further from the production well's response (or lack of it) to traced water injection at an array of wells surrounding the producer.

• Relative velocities of injected fluids. When different fluids are injected simultaneously, alternately, or sequentially in the same well with each fluid tagged with a different nonadsorbing tracer, the relative velocities of these fluids can be measured from the individual-tracer arrival time at offset producers. For example, assume that traced solvents and traced water are injected alternately in the same well. The early arrival of one of the traced fluids at the producing well would indicate that the early arriving fluid had contacted less of the reservoir than the slower fluid. This shows a need to alter one of the fluid injection cycles to achieve more uniform sweep of the reservoir. Similarly, in a micellar flood in which water is injected sequentially, the overrunning or fingering of one injected fluid through another points out the need for better fluid-mobility control to achieve more uniform sweep by the various injected fluids.

• Evaluations of sweep-improvement treatments. Remedial treatments to correct sweep problems can be evaluated by comparing the before-and-after-treatment interwell volumetric sweep as determined by tracing.

Many companies apply tracer on a routine basis. The reservoir engineer's problem generally is a lack of adequate information about fluid flow in the reservoir. The information obtained from tracer tests is unique, and tracer tests are a relatively cheap method to obtain this information. The information is an addendum to the general field production history and is used to reduce uncertainties in the reservoir model.

Tracer tests provide tracer-response curves that may be evaluated further to obtain relevant additional information. Primarily, the information gained from tracer testing is obtained simply by observing breakthrough and interwell communication. Adequate data presentation and simple hand calculation can give further knowledge about the flow behavior in the reservoir. More quantitative information can be obtained by fitting response curves obtained from numerical simulation to the observed response curves. Additional information also can be obtained by applying analytical procedures on the basis of generic or simplified reservoir models.

6.2 Types of Tracers Available

A passive tracer that labels gas or water in a well-to-well tracer test must fulfill the following criteria. It must have a very low detection limit, must be stable under reservoir conditions, must follow the phase that is being tagged and have a minimal partitioning into other phases, must have no adsorption to rock material, and must have minimal environmental consequences. The tracers discussed in the following sections have properties that make them suitable for application in well-to-well test in which dilution volumes are large. For small fields in which the requirement with respect to dilution is less important, other tracers can be applied.

6.2.1 Radioactive Water Tracers. In most field studies, the tracer is expected to behave exactly as the water it is going to trace. Very few compounds will behave as passive tracers in all situations, but near-passive tracers will, in many applications, work satisfactorily. If the objec-

TABLE 6.1—RADIOACTIVE WATER TRACERS USED IN OIL RESERVOIR EXAMINATIONS (after Bjørnstad and Maggio[2])

Water Tracer Compound	Half-Life (year)	Main Radiation Characteristics (keV)	Comments
HTO	12.32	β^- (18)	Generally applicable
CH_2TOH [1)]	12.32	β^- (18)	Caution at T > 100°C (partition into gas phase), some biodegradation below 70 to 80°C
$CH_3CTOHCH_3$	12.32	β^- (18)	Reasonably general application
C_3H_7CHTOH	12.32	β^- (18)	Reasonably general application, partition somewhat into oil
$S^{14}CN^-$	5730	β^- (156)	For T < 90 to 100°C, long-term experiments
$^{36}Cl^-$	3.10^5	β^- (709)	High-temperature reservoirs, long term, EMS-analysis
$^{22}Na^+$	2.6	β^+ (545), γ (1274.5)	High-temperature tracer in saline reservoirs, long term, slight reversible sorption

tive is to measure fluid communication exclusively, a near-passive tracer may be as good as a true passive tracer. **Table 6.1** lists the most frequently applied radioactive tracers.

The compound that best fulfills the passive tracer requirements is tritiated water (HTO). The passive water tracer mimics all movements and interactions that the water molecules do in the traced water volume. For instance, HTO can track the free movement in and out of dead-end pores because it is insensitive to coulombic forces set up by negatively charged rock surfaces. Other interactions can exchange with connate water in the rock pores or exchange with crystal water molecules. Thus, HTO sometimes seems to lag behind the injection water breakthrough as measured, for instance, by a salt balance (ionic logging). This has sometimes been interpreted that HTO is unstable at reservoir conditions and that it may be subject to isotope-exchange reactions of tritium with hydrogen in neighboring hydrogen-containing compounds, some of which are stationary.[2] Isotopic exchange, however, is expected to be negligible.

Other noncharged tracers are methanol, CH_2TOH, and the other light alcohols. These tracers will behave qualitatively similar to HTO with respect to transport but have different interactions. Other organic molecules may be applied as radioactive water tracers and can be labeled either with tritium or with ^{14}C. Larger alcohols, however, may have a partition coefficient that may cause a considerable retention.

Zemel[3] measured partition coefficients of some alcohols. Over a limited range, the effect of temperature on the partition coefficient, K, can be represented by a semiempirical equation:

$$\ln K = A + \frac{B}{T}, \qquad (6.1)$$

where T is the temperature in Kelvin, A is a constant related to the enthalpy, and B is a constant related to the entropy change. **Table 6.2** gives B and A values for some alcohols.

Anions are the more applicable electrically charged tracers; however, in laboratory experiments, clear evidence of ion exclusion can be seen (i.e., negatively charged species tend to be repelled from the negatively charged rock surfaces). As a result, the tracers tend to flow in the middle of the fluid-conducting pores and will not easily enter into dead-end pores or through

TABLE 6.2—PARTITION COEFFICIENT PARAMETERS FOR SOME WATER TRACERS (after Zemel[3])

Tracer	*K* at 194°F	*A*	*B*
Isopropyl alcohol	0.26	2.15	-992.6
n-propyl alcohol	0.28	1.72	-883.1
n-butyl alcohol	1.23	2.52	-882.2

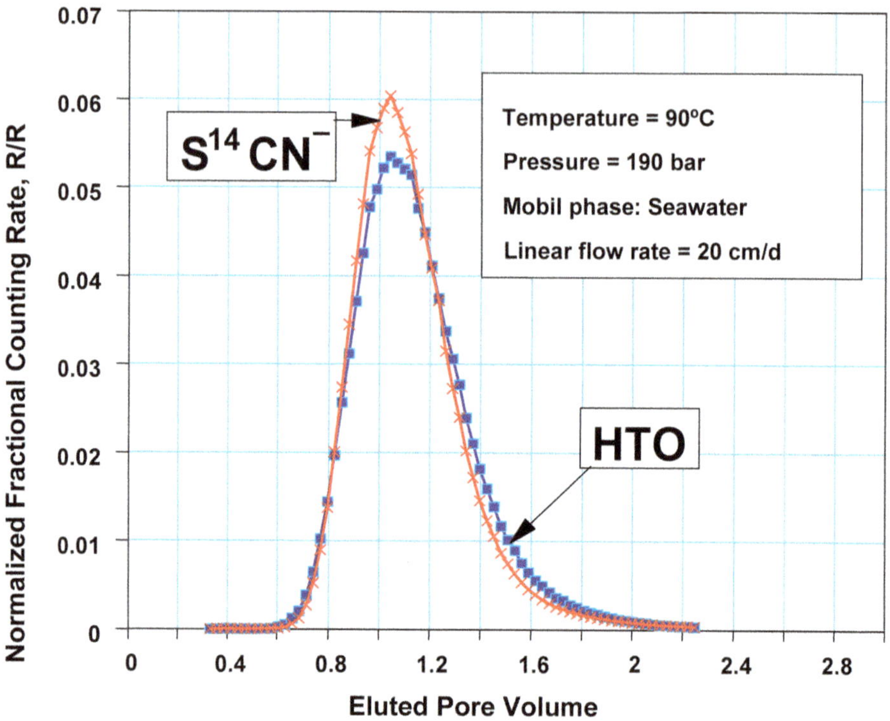

Fig. 6.1—Production curve of $S^{14}CN$ compared with the production curve of HTO in a dynamic flooding laboratory test (carbonate rock) (after Bjørnstad and Maggio[2]).

narrow pore throats, which results in a smaller available PV for anions than for noncharged species. The production profile differs in reproducible ways from that of HTO.

Anionic tracers are represented here by thiocyanate or $S^{14}CN^-$. **Fig. 6.1** gives a typical production profile from flooding experiments. This profile is compared with the production profile of the simultaneously injected HTO. It is evident from the curve that the breakthrough of HTO occurs before that of $S^{14}CN^-$ and that the tail of the HTO profile is more pronounced. This profile difference is qualitatively the same for all near-passive anionic water tracers. The retention volume may be represented by the peak maximum value or the mass middle point (first moment, μ_1) for nonsymmetrical profiles. These values are found best by fitting the profile with an analytical function consisting of polynomials.

^{35}SCN is applicable only to small reservoirs because the half-life of ^{35}S is only 87 days. $^{36}Cl^-$ has shown to be an excellent tracer. There are no possibilities for thermal degradation, and it follows the water closely. The $^{36}Cl^-$ is a long-lived nuclide (3×10^5years), and the detec-

TABLE 6.3—MOST FREQUENTLY APPLIED CHEMICAL WATER TRACERS (after Bjørnstad *et al.*[6])

Nonradioactive Water Tracers	Normal Abbreviation	Chemical Formula
Thiocyanate	SCN^-	SCN^-
2-fluorobenzoic acid	2-FBA	C_6H_4FCOOH
4-fluorobenzoic acid	4-FBA	C_6H_4FCOOH
2,6-fluorobenzoic acid (FBA)	2,6-DFBA	$C_6H_3F_2COOH$
15-20 other FBAs	Other FBAs	$C_6H_{5-x}F_x$ COOH, $C_6H_{5-n}(CF_3)_nCOOH$

tion method is atomic mass spectroscopy rather than radiation measurements. The disadvantage is that the analysis demands very sophisticated equipment and is relatively time consuming.

For mono-valent anions, the retention factors (see Eq. 6.2) are in the range of 0 to -0.03, which means that such tracers pass faster through the reservoir rock than the water itself (represented by HTO). A compound such as $^{35}SO_4^{2-}$ may be applied in some very specific cases but should be avoided normally because of absorption.

Some anionic tracers may show complex behavior. Radioactive iodine ($^{125}I^-$ and $^{131}I^-$) breaks through before water but has a substantially longer tail than HTO. Both a reversible sorption and ion exclusion seem to play a role here. $^{125}I^-$ and $^{131}I^-$ have half-lives of 60 and 8 days, respectively, which makes the compounds less attractive as tracers in large reservoir segments.

Cationic tracers are, in general, not applicable; however, experiments have qualified $^{22}Na^+$ as an applicable water tracer in highly saline (total dissolved solids concentration > seawater salinity) waters. In such waters, the nonradioactive sodium will operate as a molecular carrier for the tracer molecule. Retention factor has been measured in the range of 0.07 (see Eq. 6.2) at reservoir conditions in carbonate rock (chalk).[4] Accordingly, the tracer is delayed somewhat by sorption and ion exchange to reservoir rock but in a reversible manner.

Wood[5] reported the use of ^{134}Cs, ^{137}Cs, ^{57}Co, and ^{60}Co cations as tracers. The same cations also were injected as ethylene diamine tetra-acetic acid (EDTA) complex in a carbonate reservoir. The EDTA complexes were recovered completely in a 3-day push-and-pull test. For the cations, only the Cs^+ were produced while the Co^{3+} never appeared in the producer; however, the Cs^+ cations generally cannot be used. It will adsorb strongly on clay-containing rock.

$^{56}Co(CN)_6^3$ is a stable complex that has been tried as tracer. Not all trials have been successful, and the compound is not normally applied. Especially at temperatures greater than 90°C, they should be avoided.[2] The complex can be labeled with several isotopes of cobalt (^{56}Co, ^{57}Co, ^{58}Co, ^{60}Co) in addition to ^{14}C.

6.2.2 Chemical Water Tracers. Application of several nonradioactive chemical tracers has been reported in the literature. **Table 6.3** lists the most frequently used chemical water tracers. The most-applied nonradioactive anion is the thiocyanate anion. It has a low natural background in the reservoir, and a detection limit in the range of 1 μg/L (1 ppb) can be obtained by electrochemical detection after separation on a high-pressure liquid chromatograph. In the reservoir, this tracer will behave as the radioactive labeled $S^{14}CN^-$ or as the $^{35}SCN^-$.

Nitrate (NO_3^-), bromide (Br^-), iodide (I^-), and hydrogen borate (HBO_3^-) also have been applied as tracers.[3] These can be analyzed by ion chromatography or by high-pressure liquid chromatograph. A minimum detection limit in the range of 25 to 1 ppb is reported for the different compounds. This detection limit will not be sufficient in many situations. It is also a problem for several of these compounds that the background concentration normally will be

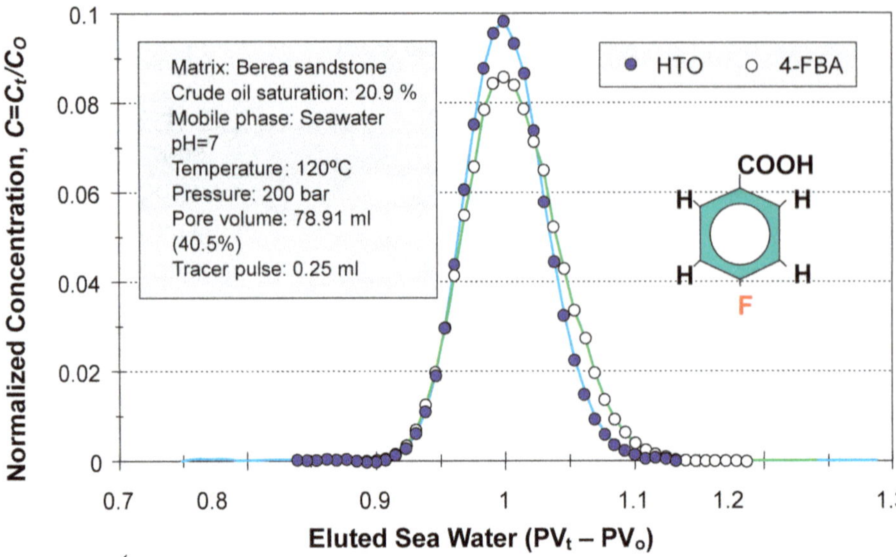

Fig. 6.2—Response curve of 4-FBA compared with the response curve of tritiated water in a laboratory flood (after Bjørnstad and Maggio[2]).

too high. Nitrate is the cheapest of the tracers, but its background may be in the range of 500 ppb, which means that an excessive amount of tracer is needed. Examples of organic molecules are fluorecien, rhodamine B, methanol, ethanol, and isopropanol.

The most-applied chemical water tracers are fluorinated benzoic acids. A large suite of mono-, di- and tri-fluorinated benzoic acids have been qualified as tracers. Because their thermal stability is variable, the compounds must be selected with care, especially for high-temperature reservoirs. Trifluoromethylbenzoic acids also may work as tracers; however, these compounds interact with the oil phase to a larger degree and retention of these compounds is observed.

The partitioning of acids to the oil phase is generally low. The partitioning will depend on the pH and at lower pH, when a larger portion of the compounds is in the undissociated form, higher partitioning can be expected. **Fig. 6.2** shows laboratory test results in which the 4-FBA is compared with tritiated water in a packed-column flooding test. The two tracers are injected simultaneously and the residence time distribution measured. Both tracers are produced at the same time, and only a marginal retention of the 4-FBA is recorded. This experiment is performed by an oil saturation of approximately 20%. In most practical applications, this retention can be neglected.

There is active research to identify new tracers. Among the new tracers are deuterated compounds. Deuterated fatty acids probably will work well as tracers but, because of production costs, their applicability is limited. A new group of potential tracers are shorter DNA fragments. These compounds have two major advantages. They have an extremely low detection limit and exist in an almost unlimited number of distinguishable variations. It is, however, uncertain if it is possible to produce modifications that can be qualified with respect to flow behavior and stability.

6.2.3 Radioactive Gas Tracers. Several authors report the use of radioactive gas tracers in oilfield applications. The tracers most frequently applied have been tritiated methane, tritiated ethane, and ^{85}Kr.[7,8,9] In addition, the use of tritiated butane, ^{127}Xe, ^{133}Xe, and tritiated hydrogen gas (HT) has been reported. **Table 6.4** lists the most widely applied radioactive tracers.

TABLE 6.4—MOST USED RADIOACTIVE GAS TRACERS IN OIL RESERVOIR EXAMINATIONS (after Bjørnstad and Maggio[2])

Gas Tracer Compound	Half-Life	Main Radiation Characteristics (keV)
CH_3T*	12.32 years	β^-(18)
CH_2TCH_3*	12.32 years	β^-(18)
CH_3CHTCH_3*	12.32 years	β^-(18)
^{85}Kr	10.76 years	β^-(687)
$^{14}CO_2$	5,730 years	β^-(156)
^{133}Xe	5.25 days	β^-(346) # (81)
^{127}Xe	36.4 days	γ (202.9, 172.1, 375.0)

*These molecules may also be labelled by ^{14}C.

The tritium-labeled compounds may be measured directly in the gas phase by proportional counter techniques. To obtain the low detection limit normally required in well-to-well tracer studies, it is normal to convert the gas to water by combustion. The produced water then is counted by a liquid scintillation technique. The detection limit of the water produced from this process depends on the instrumentation but can be less than 1 Bq/L. Typically, approximately 5 to 10 mL of water must be produced to obtain the low detection limits. The water, mixed with a liquid scintillator, is counted for a few hours. The same technique can be applied for the different tritium-labeled hydrocarbon gases. If several tritium-labeled compounds are present in the same sample, it will be necessary to perform a chemical separation of the compounds before oxidation to produce the water. This separation complicates the analysis considerably; therefore, it is not common to use several of these tritiated compounds in the same reservoir segment.

The chemical properties of tritium-labeled hydrocarbon compounds are equal to their nonlabeled homologues in almost all practical situations, which means that their behavior in the reservoir will be the same as the nonlabeled hydrocarbons. The partition coefficients of these compounds, therefore, can be found by ordinary pressure-volume-temperature (PVT) packages like those applied in phase-equilibria calculations. The partition coefficient will influence the residence time in the reservoir and the concentration of the tracer in the production line. It is important to take into account partitioning both in the evaluation of flow behavior in the reservoir and in the understanding of sampling quality. Because of partitioning to the oil phase, some of the tracers will be in the oil phase at the sampling point. Calculation of total recovery of the tracer then requires partition coefficient and gas/oil ratio (GOR) at sampling conditions. This problem will be more significant for heavier hydrocarbons than for the lighter ones.

All the hydrocarbons that are labeled with tritium can also be labeled with ^{14}C. These molecules are, in general, more expensive and their analyses are more complicated. From a general demand for reducing the application of long-lived radioactive tracers, the ^{14}C-labeled compounds are less attractive than the tritium-labeled compound.

Other alternatives are the noble gases. The noble gases are virtually inert against chemical reaction. ^{85}Kr has a half-life of 10.76 years and is a beta emitter with an energy of 687 keV. The two xenon isotopes, ^{127}Xe and ^{133}Xe, are also inert noble gases that may be applied in special situations in which rapid response is expected. The half-lives are 36.4 and 5.25 days, respectively.

TABLE 6.5—APPLICABLE CHEMICAL GAS TRACERS (after Bjørnstad *et al.*[6])

Tracer Type	Common Abbreviation	Detection Method	Disadvantage
Perfluorocarbons	PDMCH, PMCP, PMCH, 1,2-PDMCH, 1,3-PDMCH	GC/MS or GC/ECD	High GWP
Sulfur hexafluoride	SF_6	GC/MS or GC/ECD	High GWP
Freons	Freon-11, Freon-12, Freon-113	GC/MS or GC/ECD	Ozone depleting
Deuterated hydrocarbons	CD_4, C_2D_6, C_3D_8	GC/MS	High cost

TABLE 6.6—PROPERTIES OF THE MOST FREQUENTLY APPLIED PERFLUOROCARBON GAS TRACERS (after Dugstad[11])

Perfluorocarbons Tracer	Abbreviation	Chemical Formula	Molecular Weight	Boiling Point
Perfluoromethylecyclopentane	PMCP	C_6F_{12}	300	48
Perfluorodimethylcyclobutane	PDMCB	C_6F_{12}	300	45
Perfluoromethylecyclohexane	PMCH	C_7F_{14}	350	76
1,3 Perfluorodimethylcyclohexane	1,3-PDMCH	C_8F_{16}	400	102

For CO_2 as injection gas, $^{14}CO_2$ will be an ideal tracer. CO_2 will interact with the formation water, which must be considered when $^{14}CO_2$ is used as a tracer. A few papers have reported application of HT.9 Hydrogen gas is reactive, and the tracer is lost in the reservoir.

6.2.4 Chemical Gas Tracers. As early as 1946, Frost[10] reported the use of helium as a tracer under gas injection. The background of the noble gases in the reservoir is, however, generally too high, which makes nonradioactive noble gases unattractive as tracers. These gases can be applied only when the dilution volume is very small. **Table 6.5** lists the most frequently applied chemical gas tracers.

One group of tracers that has found wide application and has become the most widely applied of the chemical gas tracers today is the perfluorocarbon (PFC) group of molecules. PFCs have excellent tracer properties such as high stability, chemical inertness, and high detectability. The most frequently applied compounds are perfluorodimethylcyclobutane (PDMCB), perfluoromethylcyclopentane (PMCP), perfluoromethylcyclohexane (PMCH), and 1,2- and 1,3-perfluorodimethylcyclohexane (1,2-/1,3-PDMCH). **Table 6.6** lists some of their properties. PFCs are liquids with a density of approximately 1.8 g/mL at standard conditions; therefore, they can be injected into the reservoir with high-pressure liquid pumps.[11]

One technique for analyzing PFCs is gas chromatography (GC) in combination with an electron-capture detector. The electron-capture detector is extremely sensitive to perfluorinated hydrocarbons and especially the cyclic compounds.

Senum[12] reported a method to analyze the PFC content in a hydrocarbon gas from a production stream. The gas contained in pressure bombs is flushed through a capillary absorption tube sampler (CATS) filled with activated carbon that absorbs the PFCs. The PFC-containing pellets are desorbed thermally, and the gas is directed through a combustion system composed of a precolumn, catalysts, and traps to remove the hydrocarbons before the PFCs enter the main separation column on the GC/electron-capture detector system to determine the amount of each tracer.

Another applicable technique is GC in combination with mass spectroscopy.[13,14] This technique distinguishes tracers from other compounds not only by chromatographic separation but also by molecular mass. This reduces background noise, which is essential to obtaining a low limit of detection. The technique employing CATS has made the collection and logistics of gas samples much more efficient. Shipment and handling of high-pressure sampling cylinders are expensive and complicated. Sampling on CATS made a large improvement, making gas tracing on remote locations easier to operate. This technique allows analysis of tracer quantities in low 10^{-13} L/L concentrations. Because of the very sensitive analytical techniques, the amount of tracer needed even in large reservoirs is only a few kilograms. The CATS technology is applicable only to PFC tracers and has not been developed for the other gas tracers.

In addition to the PFC tracers, SF_6 is a frequently applied chemical tracer. SF_6 can be measured in very low concentration on electron-capture and mass-spectroscopy detectors. The tracer is stable at reservoir conditions, and it is relatively cheap. This compound is a gas at standard conditions and must be injected with gas booster pumps. The PFCs and SF_6 have a very high global-warming potential; therefore, finding an alternative tracer is needed.

Another group of compounds that have good tracer properties is freons; however, because of their ozone-depleting character, these compounds are rarely applied. Hydrocarbon gases, in which hydrogen is substituted with the deuterium isotope, will work as a tracer, but because of high production costs, these compounds are not in common use.

6.2.5 Health, Safety, and Environment Constraints. The regulations for the use of radioactive and chemical substances vary from country to country. The application of radioactivity is generally restricted, and it is important to plan the tracer test with the actual regulations in mind.

Radioactive tracers are either gamma emitters, beta emitters, or both. Tritium is a low-energy beta emitter. A sheet of paper stops this beta radiation, and it will not cause any harm to humans as long as it is kept outside the body. If spillage occurs, operators can be exposed to radiation because of direct contact with skin or because of evaporation and inhalation. Procedures must be followed carefully to ensure a safe operation.

Radioactive tracers such as ^{22}Na and ^{131}I emit gamma radiation. This radiation penetrates steel walls, which means that an operator can be exposed to this radiation. The tracers are transported in lead containers and special handling procedures are needed to reduce the radiation exposure to a minimum. The activity, measured in becquerels, is normally 2 to 3 orders of magnitude less than the activity applied for tritium-labeled compounds.

Gas tracers like the PFCs have a high global-warming potential. Most of the back-produced gas will be burned in the end, and the tracers will decompose; therefore, the amount of PFCs entering the atmosphere is low. The tracers, however, may be banned in the future, and research activities are ongoing to identify new gas tracers that are more environmentally friendly.

6.3 Tracer Flow in Porous Reservoir Rock

6.3.1 Retention Caused by Partitioning Between Phases. When tracers are flowing in the reservoirs, it is normally a requirement that the compounds follow the phase they are going to trace. The best example of a passive water tracer is HTO. The HTO will, in all practical aspects, follow the water phase.

For gas tracers, there are no known passive tracers. All gas compounds will, to a certain degree, partition between the phases. The most ideal gas tracer is tritiated methane. This gas tracer follows the methane component in the gas phase closely, and the PVT properties of this gas tracer can be found with ordinary PVT calculations. The properties of the other radioactive hydrocarbon gas tracers may be found with the same PVT calculations by examining their respective nonradioactive homologues.

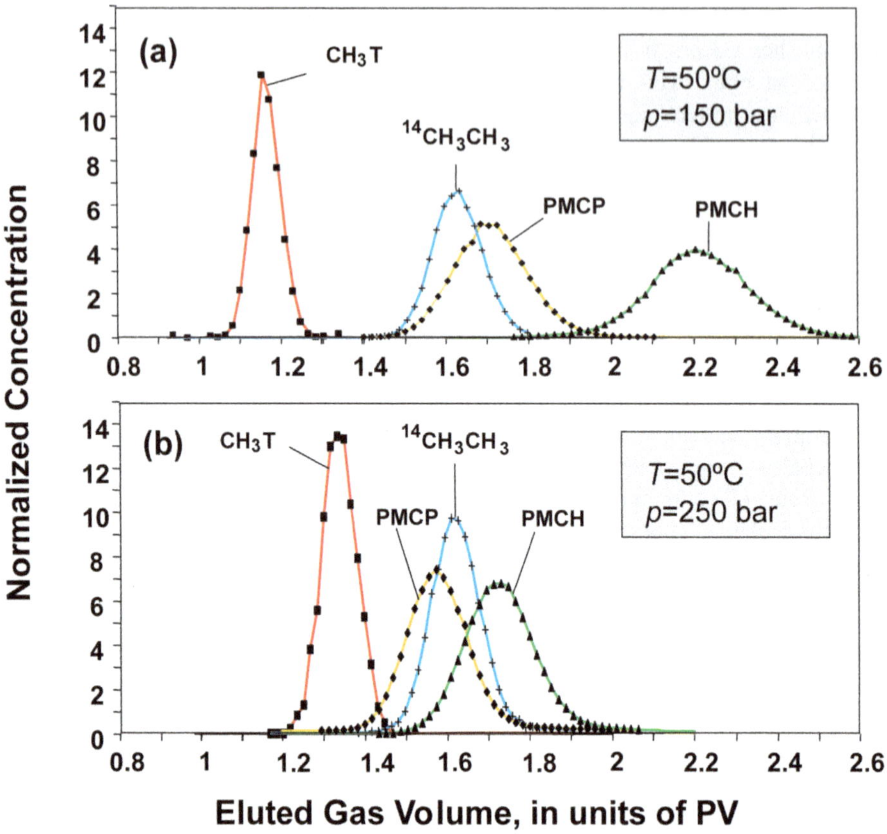

Fig. 6.3—Tracer-retention experiments in a packed column of Ottawa sand. Gas tracers are compared at two different conditions (after Dugstad[11]).

In **Fig. 6.3**, the flowing properties of a selection of some gas tracers are compared. These tracers were tested in a 12-m-long slimtube with an inner diameter of 0.5 cm. The filling material was crushed Ottawa sand. Tritiated methane, CH_3T, was the reference tracer. The oil saturation in the experiments is 30% and the production curve, when a small slug of tracer is flowing through a porous medium, is measured. All tracers, even the tritiated methane, are delayed with respect to the average flow rate in the reservoir. A nonpartitioning tracer would have been produced after one PV (available gas volume) had been flushed through the system.

Normally, the K_C values (see Eq. 6.3) decrease and, thereby, the variations in retention time are reduced when the pressure is increased. The K_P value (see Eq. 6.4) converges to 1 for all compounds when miscible conditions are obtained. The separation of the tracer production curves is reduced considerably in Fig. 6.3b when the pressure increases from 150 to 250 bar. It also is observed that the ethane tracer, $^{14}CH_3CH_3$, is produced before that of the PMCP at 150 bar while, at 250 bar, PMCP is in front of ethane. The partitioning properties of the PFC have different characteristics than the partitioning properties of the hydrocarbons.

Water tracers, like gas tracers, may partition to the oil phase. Many water tracers exist that behave almost as ideal tracers. **Fig. 6.4** shows laboratory tests in which some benzoic acid tracers are compared with the HTO. These experiments are carried out in a packed column of 2-m length. The packing material was crushed Berea sandstone. Other types of equipment also are used frequently. It is common to use cores of consolidated reservoir or reservoir-like rock (i.e., sandstones and/or carbonate material).

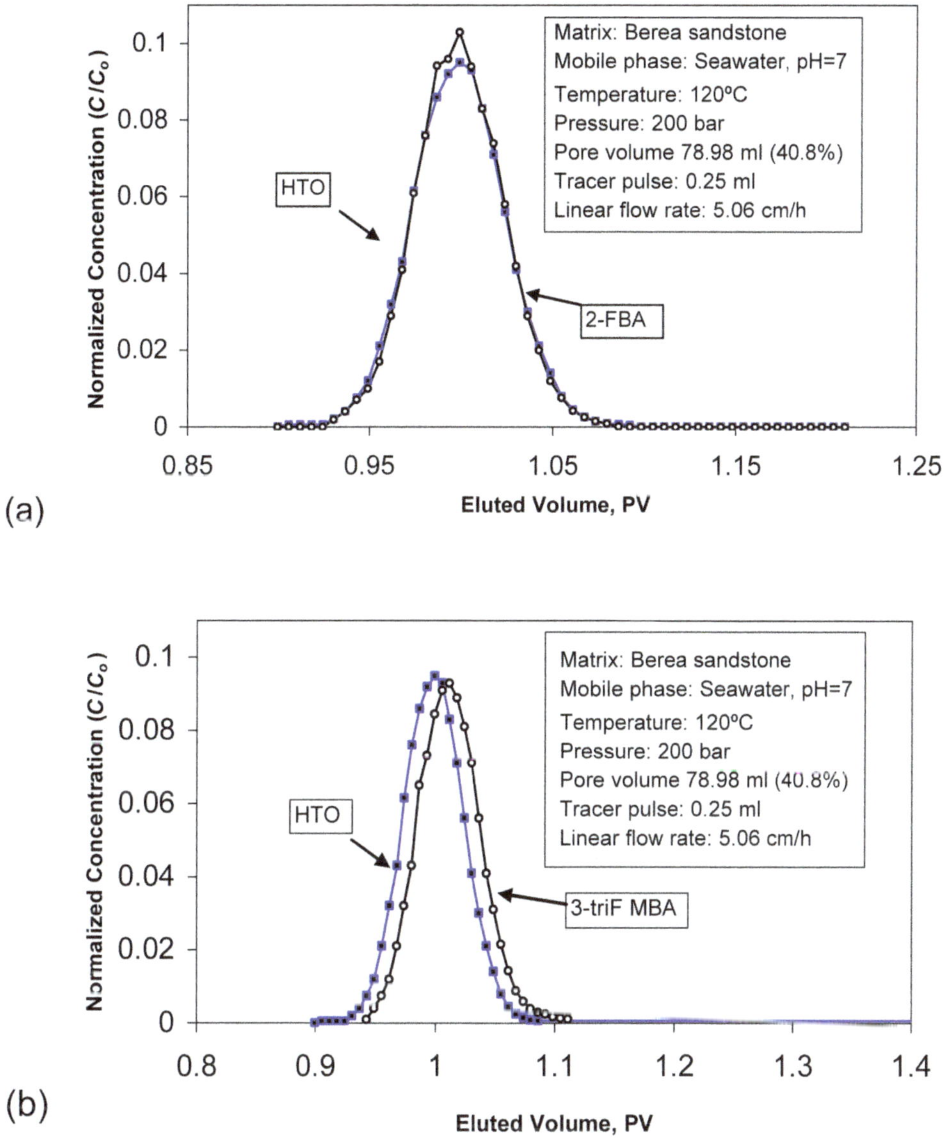

Fig. 6.4—Tracer-dispersion curves of (a) 2-fluorobenzoic acid and (b) 3-trifluoromethyl-benzoic acid compared with tritiated water from flooding experiments in sandstone (after Bjørnstad *et al.*[6]).

Retention factors, β, may be derived from Eq. 6.2 on the basis of the production profiles found by such experiments.

$$1 + \beta = \frac{V_T}{V_S}, \qquad (6.2)$$

where V_T is the retention volume for the tracer candidate and V_S is the retention volume for the standard reference tracer. The retention volume for the standard reference tracer (nonpartitioning) may further be regarded as the volume of the mobile phase, V_M, in this system. If other retention effects can be excluded, the retention factor is an expression for the delay caused by partitioning the tracer between the mobile and the stationary phase.

Partition Coefficients. The partition coefficient depends on the temperature and pressure as well as the composition of the system. Unlike the radioactively labeled hydrocarbons, the partition coefficient of the PFC gas tracers are not easily obtained. The equation of state (EOS) used in PVT calculations is not optimized for these types of compounds, and lack of interaction parameters makes partition coefficient estimation on the basis of the EOS difficult. Normally, laboratory measurements are necessary.

When the composition, temperature, and pressure are given, each component will have a given partitioning coefficient at equilibrium. This component may be defined in different ways. Two different definitions are used frequently for gas/oil systems:

$$K_C \equiv \frac{C_O}{C_G} \quad \text{(6.3)}$$

$$\text{and } K_P \equiv \frac{y_i}{x_i} \,. \quad \text{(6.4)}$$

C_O = the tracer concentration in the oil phase, and C_G = the tracer concentration in the gas phase. The partition coefficient, K_P, is calculated by dividing the mole fraction of the tracer in the gas phase, y_i, by the mole fraction of the tracer in the oil phase, x_i .

When the molar composition and the densities of the two phases are known, the relation between the two values can be expressed as

$$K_C = \frac{\delta_O \sum y_i M_W}{\delta_G \sum x_i M_W} \cdot \frac{1}{K_P} \,. \quad \text{(6.5)}$$

For simple systems, partition coefficients can be found when the vapor pressure, p_v, of the tracer compound is known. At low pressure, the partitioning between the phases will obey Henry's law, and the Henry's law constant, H_C, is

$$H_C = \frac{p_i}{x_i} \,. \quad \text{(6.6)}$$

In the case in which Raoult's law is valid (ideal behavior), the partial pressure, p_i, of the component in the gas phase is given as

$$p_i = x_i p_v \quad \text{(6.7)}$$

$$\text{and } y_i = \frac{p_i}{p}, \quad \text{(6.8)}$$

where p is the total pressure. Combining these two equations gives

$$\frac{y_i}{x_i} = \frac{p_v}{p} = K_P \,. \quad \text{(6.9)}$$

The vapor pressure of the pure component divided by the total pressure gives the K_P value. This relation will normally be too simple to be accurate but may have sufficient validity to carry out a rough estimate.

Flow Equations When One Phase Is Mobile. The partition coefficient may be measured directly by measuring the concentration of the component in each of the two phases in an equilibrium system.

In a dynamic situation, the K_C value can be found by measuring the retention time of the actual component. K_C values can be calculated when the saturation and the retention time are known. In a dynamic column experiment in which the column contains only two phases (one stationary), the fraction of time, t_f, the tracer stays in the gas phase may be expressed as the number of tracer molecules in the gas phase in a reference block of the column divided by the total number of tracer molecules in the same reference block.

$$t_f = \frac{C_M V_M}{C_M V_M + C_S V_S} \quad \text{.......... (6.10)}$$

Dividing by $C_M V_M$ above and below the fraction time and inserting Eq. 6.3 leads to

$$t_f = \frac{1}{1 + K_C \left(\frac{V_S}{V_M} \right)} \quad \text{.......... (6.11)}$$

C_M and C_S are tracer concentrations in mobile and stationary phases, respectively. The tracer is produced when one retention volume, V_T, has been injected. This volume multiplied by the fraction of the time the tracer stays in the mobile phase is equal to the volume of the mobile phase, V_M, in the system.

$$t_f = \frac{V_M}{V_T} \quad \text{.......... (6.12)}$$

A combination of Eqs. 6.11 and 6.12 gives

$$K_C = \frac{(V_T - V_M)}{V_S} \quad \text{.......... (6.13)}$$

Assuming steady-state flow, V_T / V_S may be replaced by the t_T / t_S, where t_T and t_S are the retention time of the partitioning tracer and the nonpartitioning reference tracer, respectively. Applying Eqs. 6.14 and 6.15 and rearranging, the oil saturation in the flooded porous medium can be estimated (with retention factor β found in Eq. 6.2) by Eq. 6.16.

$$V_p = V_S + V_M, \quad \text{.......... (6.14)}$$

$$S_o = \frac{V_S}{V_p}, \quad \text{.......... (6.15)}$$

$$\text{and } S_o = \frac{1}{K_C / \beta + 1} \quad \text{.......... (6.16)}$$

In Eq. 6.14, V_S is assumed to be the volume of the oil phase.

Flow Equations When Two Phases Are Mobile. Eqs. 6.10 through 6.16 assume one mobile and one stationary phase. For a generic two-phase case, the tracer flow equation may be expressed in a dimensionless form where S = saturation, C = concentration, K = partition coefficient, and f = fractional flow and when dispersion is neglected.

$$\frac{\partial}{\partial t^*}\{[S+(1-S)K]C\}+\frac{\partial}{\partial x^*}\{[f+(1-f)K]C\}=0. \qquad (6.17)$$

$x^* = x/L$ and $t^* = (q\,t)\,/\,(a\,L\,\phi)$ to obtain the dimensionless form of the system.

Assuming a two-phase gas/oil system in which the tracers are flowing partly in the gas phase and partly in the oil phase, the linear velocity of the gas and oil phases are given as[15]

$$\overline{U}_o=\left(\frac{f_o}{S_o}\right)\left(\frac{q}{a\phi}\right) \qquad (6.18)$$

$$\text{and } \overline{U}_g=\left(\frac{f_g}{S_g}\right)\left(\frac{q}{a\phi}\right), \qquad (6.19)$$

where f_o and f_g are the fractional flow of oil and gas, respectively, and q = volumetric flow rate, ϕ = porosity, and a = flow cross section. As previously shown, the fraction of time the tracer stays in the gas may be expressed by Eq. 6.10. The mean velocity of the tracer in a two-phase flow may be expressed by Eq. 6.20 when accepting that V_S/V_M is equal to S_o/S_g.

$$\overline{U}_{tr}=\left\{\left(\frac{f_g}{S_g}\right)\frac{1}{1+K_C\left(\frac{S_o}{S_g}\right)}+\left(\frac{f_o}{S_o}\right)\left[1-\frac{1}{1+K_C\left(\frac{S_o}{S_g}\right)}\right]\right\}\left(\frac{q}{a\phi}\right). \qquad (6.20)$$

Rearrangement gives

$$\overline{U}_{tr}=\left(\frac{f_g+K_Cf_o}{S_g+K_CS_o}\right)\left(\frac{q}{a\phi}\right)=f_t\left(\frac{q}{a\phi}\right). \qquad (6.21)$$

This tracer-flow-velocity model may be applied when the K_C values are known. **Fig. 6.5** gives a graphical outline of f_t (based on Eq. 6.21) for five different partitioning coefficients. The fractional flow curve for the gas, f_g, is arbitrarily chosen. The f_t curves = 1 at the saturation where $f_o/S_o = f_g/S_g$ (i.e., when both phases are flowing with the same linear velocity). When the K_c value = 1, the tracer flow rate is independent of saturation and shape of the fractional flow curve. The f_t curves plotted in Fig. 6.5 show that the relative flow of the tracers depends strongly on the saturation. Below the gas saturation at which the two phases flow with the same linear velocity, the lower-partitioning tracers will be in front while above that saturation decreasing K_c value will give increasing flow rate.

Residual-Oil Measurements. Eqs. 6.10 through 6.21 give the basis for estimating remaining oil in the reservoir. In the most simple system, in which the oil saturation is stationary, S_o may be calculated from the knowledge of the partition coefficient of two tracers and the peak of the tracer-response curve by

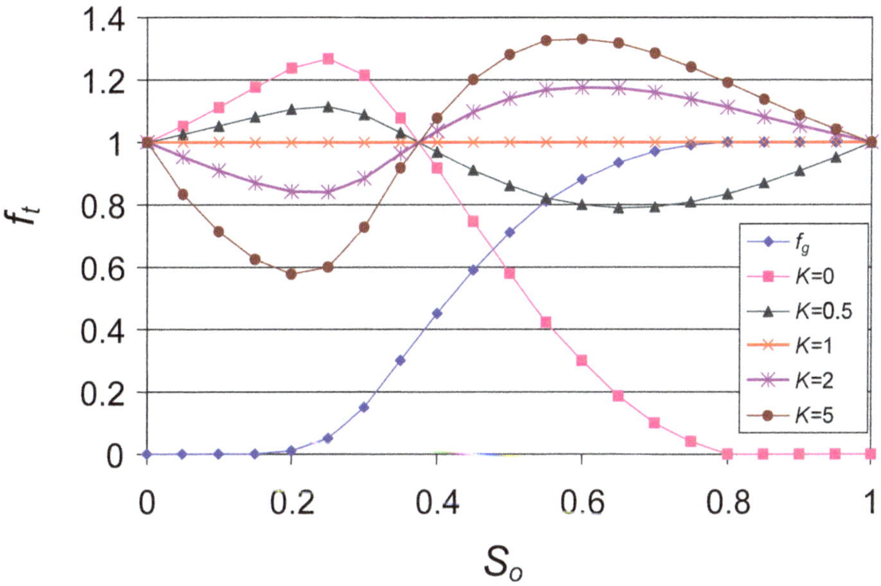

Fig. 6.5—Relative tracer flow rates in a two-phase system for a certain fractional flow curve (after Dugstad, Bjørnstad, and Hundere[15]).

$$S_o = \left[\left(\frac{V_{Ti}K_{ci} - V_{Tj}K_{cj}}{V_{Tj} - V_{Ti}}\right) + 1\right]^{-1} . \quad (6.22)$$

Under ideal conditions, V_T may be replaced by the retention time of the two tracers. There is always a question about which retention time to use in the calculations. Alternatives are breakthrough, production peak, moment (mass middle point), or other specific landmarks. The production curve is a superposition of contribution from individual streamtubes in the reservoir; therefore, the choice of retention time will reflect either saturations in certain streamtubes or an average value. A sensitivity evaluation of Eq. 6.22 shows that it is preferable to apply two tracers with K_C values far apart from each other. The most efficient would be to include one tracer that has no partitioning to the liquid phase. Eq. 6.22 then is simplified to Eq. 6.16.

One of the questions that may be raised against the method is the possibility for obtaining equilibrium between the phases. The calculations are based on a real equilibrium between the phases, which depends on saturation, diffusion rates, flow rates, pore structure, and partition coefficient.

To obtain reliable results, it is crucial to understand the flow situation. If the oil phase is stationary, Eq. 6.22 may give a satisfactorily result. If two phases are flowing, it is important to know the fractional flow curves (i.e., relative permeability), and a reservoir simulator will be necessary to obtain a reliable result.

6.3.2 Biodegradation. Microbial stability of water tracers may be a problem. This problem will be less important at higher temperatures at which the microbial activity is lower. The problem, however, must be addressed in sample handling and storage. Some tracers, in special situations, may biodegrade after sampling. To avoid such degradation, a biocide may be added to the sample immediately after collection. Adding 0.1 ppm NaN_3 to the stock solution can prevent bacterial growth.

Fluorobenzoic acid tracers have been reported to biodegrade when exposed to seawater. The biodegradation has been measured for monofluorobenzoic acid and di- and tri-fluorobenzoic acid. No degradation was observed for the di- and tri-fluor compounds. Alcohols also may biodegrade under certain conditions. In general, the odd-carbon-number alcohols are more resistant to bacterial attack than the even-number ones.

6.3.3 Ion Exchange. Ions adsorbed on the reservoir surface are free to exchange with ions in the water. This ion exchange is a reversible process and tends to obtain equilibrium between surface concentration and the concentration in the water. Different adsorption isotherms may be used to describe this situation. In most cases, this ion exchange can be described by a Langmuir isotherm (Eq. 6.24).

The flow of the tracer through the reservoir may be influenced by adsorption to the grain surface. To come into contact with the grain surface, the gas tracer may need to diffuse through a film of oil, water, or both. The adsorption depends on the partitioning of the tracer into the two liquid phases in the reservoir. Different models are applied to describe the adsorption on the grains. The simplest form is the linear relation in which the adsorbed amount, a_c, is proportional to the concentration in the contacting phase C_c.

$$a_c = K \cdot C_c \text{ . } \qquad (6.23)$$

A tracer in a dynamic system will be retained independently of concentration. In many cases, however, the active adsorption sites on the surface will be saturated and the amount adsorbed will not increase linearly with the concentration. In some cases, a more accurate expression is the Langmuir isotherm,

$$a_c = \frac{U \cdot c}{1 + V \cdot c} \text{ . } \qquad (6.24)$$

U and V are two parameters that decide the shape of the curve for the actual system at a specific temperature. c = the concentration in the liquid phase caused by partitioning between the gas and liquid phase, and a_c = the amount of tracer adsorbed to the grain. At low concentration, which is the case when tracers are considered, this equation will be a straight line with slope U. Other isotherms also may be considered but are less likely to be needed. Different adsorption isotherms will influence the produced tracer peak.

If the tracer is positively charged, it can be exchanged with cations adsorbed to the reservoir surface. The affinity to the surface may vary, and some cations will be more tightly adsorbed than others. The effect on the tracer flow will be retention of the tracer. According to the most likely adsorption isotherm, the adsorption will be linear with the tracer concentration as long as the concentration is very low.

^{22}Na has shown to move through porous media with only minor delay caused by sorption and ion exchange to reservoir rock but in a reversible manner. Most other cations (Cs^+, Co^{3+}) have shown strong adsorption and cannot be used as tracers.

6.4 Planning and Design of Tracer Tests

6.4.1 Timing of Tracer Programs. The timing for tracer injection depends on the information that is requested. Normally, it is desirable to inject the tracer early in the injection process to obtain information as soon as possible and be able to take the necessary actions to optimize the production strategy. However, it has been seen that water tracer is lost in the reservoir because of imbibition of water. Therefore, if there is large imbibition potential in the reservoir and water is trapped in volumes that do not contribute to the normal flow path, tracers can be trapped and only minor tracer concentrations leak out in the flowing part of the reservoir. This information can be valuable on its own, but if the problem is not considered, it may contribute to wrong interpretation of the results. In this situation, water may break through from an injector that is traced without observing any positive detection of tracer. In general, injecting the tracer in the very front of an injection program is not recommended.

Tracers should not be injected until the pattern area has been pressured up with the injected fluid. If the traced injected fluid is spent in collapsing a gas phase or otherwise pressuring the reservoir, the volume injected to tracer breakthrough, which indicates the volumetric sweep to breakthrough, may be significantly larger than that determined after the pattern area has been pressured.

Well-to-well tracer tests may be performed, in principle, either as a pulse injection or as a continuous injection. The most often applied method is pulse injection. The continuous injection may be useful especially where unsaturated water-wet rock may absorb short tracer pulses by water imbibition from the injected waterfront edge. To obtain a constant concentration in the injected water, a constant recording of the water-injection rate is necessary. In general, continuous injection is more complicated and a tracer engineer must follow the injection for a long period. For pulse injection, however, the whole program can be carried out within a few hours.

6.4.2 Field-Test Design. The design of a field tracer test has two components: a tracer part and an analytical part. The tracer component includes choosing the tracer for each well, estimating the required amount of each tracer, dealing with the regulations, and planning for acquisition and injection of the tracer into the ground. The analytical part includes selecting an analytical strategy, setting up a sampling program, and determining the background realistic detection limits in the samples collected from the particular field.

Traditionally, two methods have been applied to determine how much tracer must be added to obtain a tracer-production response significantly above background. In general, it is desirable to inject as small an amount as possible to reduce environmental problems, contamination, and costs. Both radioactive- and chemical-tracer compounds are potentially harmful to the environment; therefore, the amount applied must be kept to a minimum. In some cases, the maximum permissible concentration of the tracer that may be released to the environment is limited by requirement from the authorities.

In many situations, produced gas containing tracer is reinjected into the reservoir. This has, in some cases, contaminated the whole reservoir and destroyed the possibility of applying the same tracers in another segment of the reservoir. If not considered, flow conditions can be misinterpreted because of positive identification of tracer originated from reinjection of tracer in uncontrolled conditions.

The third reason for the use of low concentrations is to reduce costs. In small reservoirs, the tracer cost is a minor part of the total cost connected to the test but, in larger reservoirs, the tracer costs may be significant, especially if exotic tracers are required.

One method to estimate the tracer concentration is to assume that the injected tracer is diluted uniformly by the entire swept volume when it is produced. Sufficient tracer is added to ensure detection at this dilution volume. The peak tracer concentration is presumed to be well above the average.

The first part of the calculation is to estimate a dilution volume, which is obtained by calculating the water- or gas-filled PV between the injector and the production wells. The first approximation is to assume radial flow from injector, but it usually is modified by any known reservoir conditions, such as known flow channels or barriers, large permeability variations, or other constraints causing nonradial pressure gradients. It is important to know porosity, net pay zone, and distances between wells for the calculation. Optimal design can be obtained by performing full-field simulations. The problem with this method is that many of the parameters are unknown, which again is the main reason for performing the tracer test.

The smallest injection pulse required is usually the amount of tracer needed to obtain an average concentration of 10 times the minimum limit of detection in the dilution volume. This is expected to give a peak production on the order of 100 times the detection limit. To be able to follow production curves reflecting the contribution from different layers and zones, it is important to have at least this amount of tracer injected.

In the ideal situation, the dilution volume, V_d, can be calculated with a radial approximation.

$$V_d = \pi \times r^2 \times h \times S_w \times \phi \times F, \qquad (6.25)$$

where r = distance between the injector and the producer, h = height of the reservoir zone, ϕ = porosity, and S_w = water saturation. For a gas tracer, S_g should substitute for S_w. In addition, the expansion of the gas has to be taken into account. F is a correction factor that accounts for nonsymmetry caused by barriers, well location, and other restrictions causing changes in the drainage area.

An analytical method to determine tracer production curves in a layered system exists.[16] The method assumes that the tracer pulse moves radially from the injector to the producers through homogeneous, noncommunicating layers with longitudinal dispersion in the direction of flow. The number of layers, their thickness, and their permeability are used to represent the reservoir heterogeneity. The tracer pulse moving through each layer is diluted at the producers by untagged water from other streamlines in the same layer. This dilution effect is a consequence of pattern geometry. Production of the combined tracer responses from all these layers makes up the response curve of tracer concentration as a function of the cumulative volume of water injected.

The model predicts peak height, breakthrough time, and shape of the produced tracer-response curve from the amount of tracer injected and the pattern geometry. The fundamental equation (Eq. 6.26) used to calculate tracer-pulse flow in a streamtube is further applied to give analytical expressions for pattern-breakthrough curves.[17]

$$\frac{C(\psi)}{C_o} = \frac{1}{2}\operatorname{erfc}\left(\frac{L - s_1}{\sqrt{2\sigma_1^2}}\right) - \frac{1}{2}\operatorname{erfc}\left(\frac{L - s_2}{\sqrt{2\sigma_2^2}}\right). \qquad (6.26)$$

The concentration of a tracer at any location within the streamtube, ψ, is the difference between two terms, given by Eq. 6.26. L = the distance along a streamline, and s_1 and s_2 = the front and end location of the tracer pulse in the streamline, respectively. σ = the standard deviation that, in a radial system, can be found from

$$\sigma^2 = 2\alpha\frac{r}{3}, \qquad (6.27)$$

where α = dispersivity and r = the radius at the front defined at a location corresponding to the 50%-concentration point. On the basis of these predictions, tracer amounts can be injected to ensure a peak concentration in the production well that is well above detection limits. In practice, however, such calculations are often based on so many unknown or estimated parameters that it is advisable to inject more tracer than these calculations predict.

A reservoir simulator probably can provide the best estimate for tracer amounts; however, these simulations are based on detailed reservoir description. At the stage of tracer injection, the reservoir model is generally uncertain and, again, it is advisable to inject more than these calculations predict. Sufficient tracer must be injected to enable measurement of unexpected flow behavior.

6.4.3 Collection of Samples for Tracer Analysis. Sampling frequencies will depend strongly on the field considered. Tracer tests are, in general, a method to verify proposed flow scenarios. To cover unexpected behavior, sampling should be started long before expected breakthrough. Sampling frequency should be highest at the start of the flood to avoid missing early breakthrough.

Water samples may be stored in bottles that normally are collected from the separator. Water sampling is cheap, and frequent sampling is advised. Initially, only some of the collected samples need be analyzed; intermediate samples can be discarded if no tracer is found. Once tracer is found, samples are analyzed backward until the tracer breakthrough is found. In certain situations, some tracers may biodegrade after sampling. To avoid this, a biocide may be added to the sample immediately after collection. Adding 0.1 ppm NaN_3 to the stock solution can prevent bacterial growth.

Samples can be collected at the wells or from a test separator shared by a number of wells, provided the stream is sampled only near the end of the test when the water is representative of the currently sampled well. Well sampling usually involves a water-separation problem, which can be avoided by sampling at the separator. The optimal situation is to have a dedicated separator for the particular well. New technology involving multibranched wells, horizontal wells with perforation in several reservoir compartments, subsea manifolds, or even subsea separators cause additional problems in providing relevant samples.

Gas tracers normally have been collected on pressure cylinders. The gas may be collected directly on the flowline or from a separator. The tracer content in the gas will depend on where the sample is collected. The partition coefficient of the tracer depends strongly on the pressure. A sample collected on the flowline at 100 bars and one collected at a separator at 30 bars will give different results. Further estimation of the total amount of tracer produced also will depend on the GOR at the sampling point. To calculate the produced-tracer amount in one particular well, it is necessary to know the pressure, temperature, and GOR at the sampling point. In addition, it is necessary to establish the partition properties of the tracer at these particular conditions. This can be done by measurement or by applying PVT models; however, existing PVT models will not treat all tracer types with the same accuracy.

Collection of gas tracers is more expensive than the collection of water tracers. When gas is collected on pressure cylinders, the cost of the cylinders will add considerably to the analysis cost. A new technology[14] has been developed in which the tracers are absorbed by an activated carbon trap. These CATS are used for collection of PFC gas tracers. The tracer then is collected in these tubes in the field, and only small tubes, without any surplus pressure, are shipped to the laboratory. Special sampling equipment is available that ensures a reliable and reproducible sampling. The CATS method is applicable only for a limited number of tracers, and in many situations, it is still necessary to collect samples with pressure cylinders. The cylinder size and the amount of gas collected will depend on the tracers and the concentration expected in the produced gas. Normally, a 200-mL cylinder will be sufficient.

The laboratory applies a large variety of techniques to measure the concentrations of the tracers. The different techniques will have degrees of uncertainty and the differences between detection limit and quantification limit should be distinguished. At concentrations close to the detection limit, it may be very difficult to obtain accurate quantification of the tracer; therefore, some laboratories report only "detected" without quantification when the concentration is low.

The detection limit obtained by an analytical procedure will be influenced by the quality of the sample. In formation water, the detection limit may be different from the detection limit in production water that also contains emulsion breakers, scale inhibitors, corrosion inhibitors, and other additives. It is, therefore, important to have a good cooperation between the field operators and the laboratory to obtain the best quality on the field samples.

Because of the very sensitive analytical techniques needed, it is essential to avoid cross contamination. It is important to plan tracer operations carefully to avoid any possibilities of close contact between injection equipment and sampling equipment. For example, it might be a source of contamination if injection pumps or tracer containers are transported after injection in the same van or stored in the same building as sample bottles or sample equipment.

6.5 Interpretation of Field Data

6.5.1 Different Levels of Interpretation. To obtain high-quality tracer-response curves that are the basis for the further interpretation, a well-designed sampling program is needed. In general, more samples will give the potential for extraction of more information from field tests. Too often, interpretation is difficult because of limited tracer data.

The final objective of a well-to-well study is the interpretation of the response curves. A good analysis of the information given by the tracers, in combination with other available data, gives a better understanding of the flow in the reservoir, not just verification of communication between injector and producer.

The response curve can be analyzed from three different points of view or complexity levels. The simplest interpretation is the qualitative one. By observing the curves, the following pattern characteristics can be derived: injection-water arriving time (breakthrough); existence of high-permeability channels, barriers, and fractures between wells; communication between different layers; stratification in the same layer; and preferential flow directions in the reservoir. Furthermore, the cumulative response can be obtained by integrating the concentration vs. time curve if the production flow rate is known. From this new curve, the fraction of injection water reaching each producer is easily calculated. A common spreadsheet is the best way to make these calculations. This type of interpretation can be carried out without any advanced simulation. It is important to integrate the data obtained from the geological model, primary production data, 4D seismic data (if available), and tracer data in a systematic way.

The second level of analysis uses basic mathematical models to fit simple response curves by means of theoretical expressions and to decompose complex responses in several simpler functions. In this way, partial residence times, as well as other parameters, can be determined for each function. Mathematical models also allow for the evaluation of important parameters, such as permeability, and make it possible to predict the behavior of unknown patterns.

Finally, complex mathematical models like numerical finite-element simulators or stream-tube modeling can be used to achieve a deeper analysis. Most of the commercially available simulators have tracer options with varying degrees of complexity. Not all simulators include important physical effects like partitioning, dispersion, and adsorption. A tracer simulator that can be coupled to full-field reservoir simulators has been developed recently. Simulation is a tool to improve the existing reservoir model; therefore, it is crucial to have access to the best available model to enable an efficient optimization based on the tracer-production data.

6.5.2 Evaluation Based on Hand Calculation and Adequate Data Presentation. Qualitative interpretation of field data is illustrated by examples from the Snorre field in the North Sea. **Fig. 6.6** shows the segment of the field that was tested and the location of injectors (triangles) and producers (circles).[18] A water-alternating-gas (WAG) injection program is being used in the field.

Fig. 6.7 shows a way to present field data for a gas-tracer program. The data are plotted as a time response. Alternatively, the data can be plotted as a volumetric response. To obtain a better understanding, it is important to plot gas-tracer-production data with GOR. The tracer data are plotted relative to the amount of tracer injected, which means that the concentrations presented for each individual tracer can be compared directly.

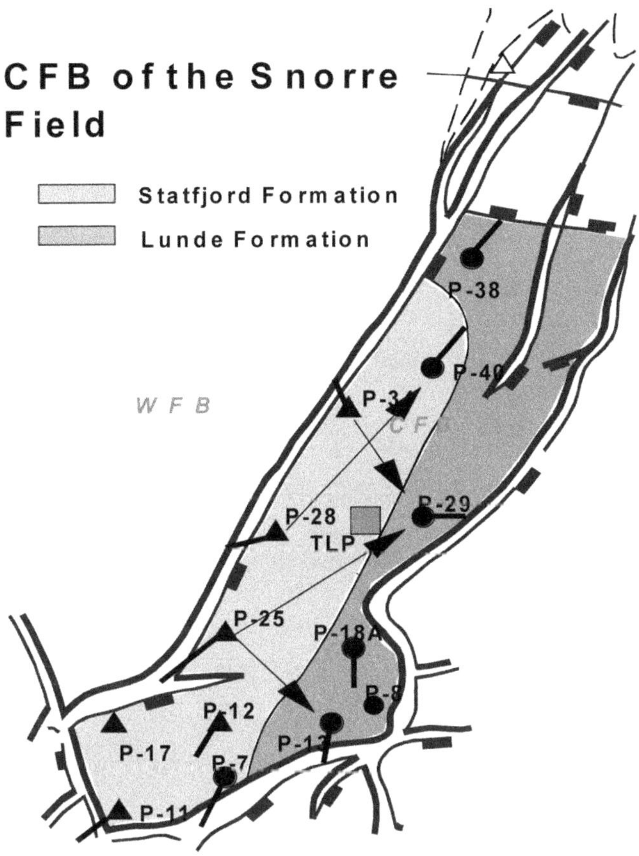

Fig. 6.6—Well location in the central fault block (CFB) of the Snorre field. Arrows indicate the main producers of injected tracers.[18]

From Fig. 6.7, it is possible not only to find communication but also to extract information about the prevailing flow regime. The gas tracer injected into Well P25 breaks through without any significant increase in GOR, which means that the gas has not followed a free gas layer but has been dissolved in the oil. Three different gas tracers were injected in Well P28 at three different times. These tracers were produced only after a significant GOR was recorded. All the tracers were produced at almost the same time. This shows that gas has been trapped and that the gas in the individual gas slugs has subsequently filled this trap. At a certain time, the gas trap has been filled and the gas moves on to the producers. When reaching the producer, the gas contains all three different tracers injected over a 3-year period.

From the data in Fig. 6.7, it is possible to carry out a spreadsheet calculation to estimate the recovery of the injected tracer in each individual production well. This shows that a large percentage of the injected gas has moved in different directions. Breakthrough time also may be estimated, either as the first sample that contains tracer above the detection limit or as an extrapolation backward of the initial phase of the production curve. However, the accuracy of this calculation is limited by the sampling frequency.

To fully understand the flow behavior, it is important to add all information available. The picture may be modified by the knowledge of injection rates and production rates in the neighboring wells.

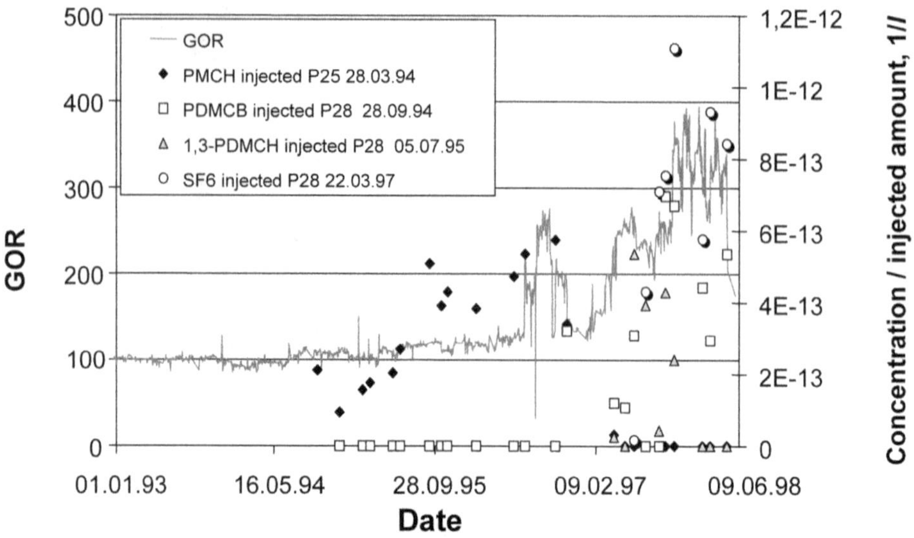

Fig. 6.7—Gas tracer data from the Snorre field in the North Sea.[18]

Because of maintenance and operational problems, production wells may be shut in for some periods. Because of fluid drift in the reservoir during shut-in, samples collected immediately after such periods may give unexpected results. This additional information can be exploited to better understand the pressure distribution in the field. Somaruga *et al.*[19] used well shut-in in a systematic way to obtain additional information. If the changes that may occur during shut-in are not considered, it may cause misinterpretation of the tracer-response curves.

The water-tracer-response curves should be presented together with the water cut. **Fig. 6.8** shows water-tracer-response curves in the same well as discussed previously. Initially, a small water-cut increase is recorded. The tracer injected in Well P34 follows this increase. When the water cut starts to increase more rapidly, one of the other tracers starts production while the concentration of the first tracer vanishes. This method of presentation increases the understanding of the flow behavior and the water contribution from the different injectors.

Response Curves. The time response is the graphic representation of the concentration of activity (after background subtraction and decay correction) as a function of time. Preprocessing the experimental data can smooth the response. From this curve, the cumulative response (recovered activity vs. time) is derived by a simple numeric integration.

Concerning the cumulative response, the Eq. 6.28 gives the cumulative tracer recovered up to an instant, t_i.

$$m_{t_i} = \int_0^{t_i} Q_{p_t} C_t \mathrm{d}t, \qquad (6.28)$$

where m_{t_i} = the total tracer recuperation up to t_i, Q_{p_t} = the production engineering water flow rate as a function of time, C_t = tracer concentration as a function of time, and t_i = the elapsed time (days after the injection). Q_{p_t} often will be available as a discrete value representing each day. C_t will be available only as an individual measurement according to the sampling program. The accuracy of the total recovery will depend on the sampling frequency.

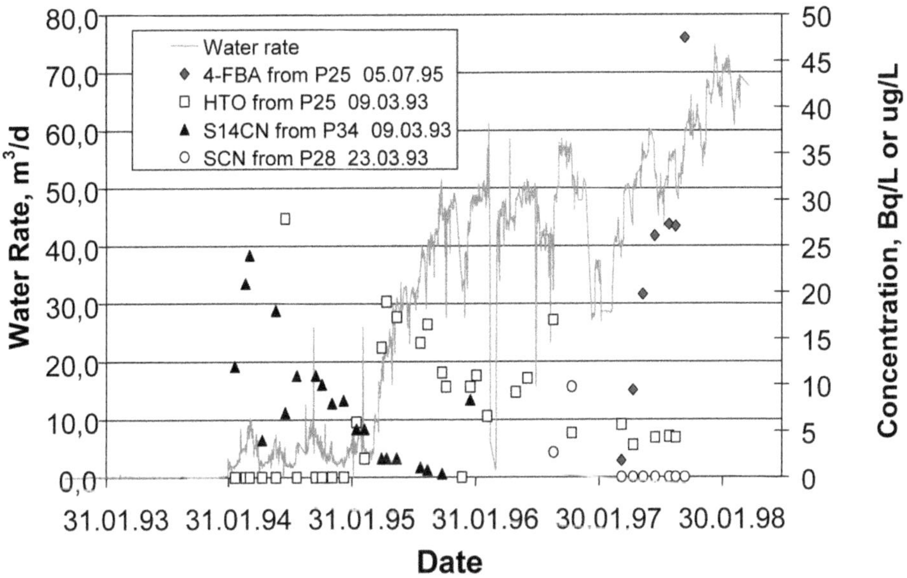

Fig. 6.8—Water-tracer-response curve from the Snorre field in the North Sea.[18]

Sweep Volume. Instead of the time-response curves, the data are often presented as volumetric-response curves. The presentation is especially convenient for estimating the swept volume between the particular injector and producer pair. Swept volume, V_S, can be estimated from the product of the mean produced volume, $\bar{V}$, and the ratio of the mean volumetric water rate between the injector and the producer, Q_m, to the mean volumetric water production rate, Q_p.[3]

$$V_s = \frac{Q_m}{Q_p}\bar{V} \quad (6.29)$$

$\bar{V}$ is calculated from the first moment of the produced-tracer concentration, C.

$$\bar{V} = \frac{\int_o^\infty CV \, dV}{\int_o^\infty C \, dV} \quad (6.30)$$

Q_m, the flow rate between injector and producer, is estimated from the fraction of injected tracer produced at the well at the mean injection rate, Q_i.

$$Q_m = \frac{m}{M} Q_i, \quad (6.31)$$

where m = the amount of tracer produced at a given well and M = the amount of tracer originally injected. The swept volume can be expressed in terms of the injected and produced flow rates and the fraction, m/M, of water going from the injector to the producer.

$$V_s = \frac{m}{M}\frac{Q_i}{Q_p}\bar{V} \text{ . ... (6.32)}$$

Because of limited data, the tracer-production curve is composed of a discontinuous set of points and, in practice, the integral in Eq. 6.30 is approximated by

$$\bar{V} = \frac{\sum_0^\infty CV\Delta V}{\sum_0^\infty C\Delta V} \text{ . ... (6.33)}$$

The calculation of swept volume often is complicated by a lack of data. The mean produced volume may need to be estimated on the basis of extrapolated tracer-production curves. In many cases, the sampling program is either not finished or it has been stopped without following the tail of the production curve. To fit the missing data in the tail of the production curve, an exponential decline approximation can be applied.

$$C = C_e \exp -\left(\frac{V - V_e}{a}\right), \text{ .... (6.34)}$$

where C_e = the measured tracer concentration at the value of V_e from which the exponential fit starts. The slope of the line is represented by $1/a$. If Eq. 6.34 is substituted for C in Eq. 6.30 and the integral replaced by a finite sum, the following expression is valid:

$$\bar{V} = \frac{\sum_0^{V_e} CV\mathrm{d}V + aC_e\left(a + V_e\right)}{\sum_0^{V_e} C\mathrm{d}V + aC_e} \text{ . ... (6.35)}$$

To find the swept volume in an incomplete data set, this equation is used for $\bar{V}$ in Eq. 6.32.

6.5.3 Evaluation Based on Analytical Solutions. Eq. 6.26 gives the fundamental equation for tracer breakthrough in a streamtube. A correlation into a single breakthrough curve for various repeated homogeneous-flooding patterns has been found.[16] In a layered system, the overall tracer-response curve is a combination of responses from individual layers. The individual-layer responses are predictable from the correlated pattern-breakthrough curve; however, the tracer arrival time at the production well and the concentration contribution from each layer are functions of the porosity, permeability, and thickness of that layer. Conversely, the decomposition of an actual tracer-production curve from a multilayered system into the individual-layer responses can yield individual-layer parameters. Computer algorithms exist that deconvolve the overall tracer profile into the individual-layer responses and compute ϕh and $kh/\Sigma kh$ of the individual layers.

6.5.4 Evaluation of Tracer-Response Curves With Numerical Simulation. The most thorough analyses of tracer data are carried out in combination with a reservoir simulator applied for that particular field. Many of the applied reservoir simulators, based on numerical solution of flow equations applying finite-difference methods, have options for handling tracers.[20] In addition, streamline models[21,22] exist that handle tracer flow. However, normal simulators have limitations. One problem is the limited possibilities for including physical properties such as partitioning between phases and adsorption to grain surfaces. This problem is of special importance for gas tracers. There is also a problem connected to handling of dispersion because the tracer pulse is, in its initial phase, much smaller than a typical gridblock size.

To overcome some of these problems, Sagen *et al.*[23] developed a simulation module to improve the accuracy of the calculations. To resolve the influence of reservoir heterogeneities on the measured tracer response, an accurate numerical treatment of the tracer equation is needed. This is especially important when narrow tracer slugs are injected in a reservoir. This tracer module calculates tracer flow with an explicit method for integration of the convection part of the tracer equation. To reduce numerical smearing of the tracer pulses, the timestep for the tracer calculation is selected as large as possible but may still be much smaller than the timesteps of the host reservoir simulator, which most often uses an implicit formulation. In the tracer module, the main tool for reducing numerical dispersion is the use of a second-order numerical scheme for integrating the tracer equation. A separate grid-refinement option for tracer calculation is available. In combination, these methods lead to a good resolution of narrow slugs propagating through the reservoir. The method of separate grid refinement is far less time consuming than performing the whole reservoir simulation on a refined grid. This tracer simulation module has been coupled to some of the standard reservoir simulation tools.

6.6 Field Experience

Interwell tracer tests are widely used. This chapter reviews some of the studies reported in open literature. The selection introduces different problems to be addressed, but the original papers should be studied to obtain a more detailed description of the programs.

6.6.1 Tracers in WAG Programs. The Snorre field is a giant oil reservoir (sandstone) in the Norwegian sector of the North Sea. Injection water and gas were monitored with tracers,[18] and the resulting tracer measurements are discussed in Sec. 6.5.2.

The same tracers used in the Snorre field have been injected in the Gullfaks field[24] in the North Sea. The tracers identified unexpected communication paths between layers. The results contributed to methods for improving the WAG recovery performance.

6.6.2 Water and Gas Tracer Injection in Fractured Reservoir. A tracer test was carried out in the Spraberry trend in west Texas.[25] In this field, the oil was produced primarily from fractures and not from the matrix porous sandstone. To carry out the most efficient waterflooding strategy, the knowledge of fracture direction was imperative. Instead of a costly water-injection pilot project, natural gas with tracer was injected into a central well for 16 weeks to define the fracture pattern. The ^{85}Kr tracer applied was injected continuously with a concentration 28 times minimum detectability, which was 13,000 Bq/m^3 (10^{-8} Ci/ft^3) gas. The detector system applied was a thin-wall, beta-sensitive Geiger-Muller tube that records the tracer amount in the effluent gas continuously.

At the beginning of the thirteenth week, a 2-Ci slug was injected, and at the beginning of 14th week, a 1-Ci slug was injected. Radioactive gas was recorded only in two wells. No radioactivity increases were detected at any of the other 10 wells monitored, either continuously or intermittently, during the tracer survey. The cyclic nature of tracer appearance was attributed to the necessity for overcoming the varying hydrostatic head of oil in the well and to the appreciable difference in effective permeability of the principal and cross-fracture systems. Apparently, neither of the two wells in which tracer was recorded intersected the same fracture plane as the injection well. The tracer survey confirmed predictions about the general line of fracture orientation in the lower formation of the Sprabarry trend.

6.6.3 Gas and Water Tracers in the El Furrial Field. Viela *et al.*[26] reported the application of both gas and water tracers in the El Furrial field. The radioactive water tracers applied were isopropyl alcohol, HTO, ^{22}Na, and thiocyanate. Breakthrough time recorded was between 1 and 3 years. In the same field, gas tracers also were added. The tracers applied were PMCP, PDMCB, and SF_6. The conclusion was that the tracer survey was important to verify expected flow behavior and to identify unexpected communication paths.

6.6.4 Tracers for Gas Injection. Welge[27] described one of the pioneering works of tracer application to follow injected gas. The communication between the wells in a small part of the Cromwell pool in central Oklahoma was investigated. The well spacing in the area was approximately 65 m. Three radioactive tracers were used: HT (only tritiated hydrogen is mentioned which may be either HT or T_2), CH_3T, and ^{85}Kr. One Ci of the tritiated compounds and 280 mCi of ^{85}Kr were injected into the same well. In the nearest well, the tracers were produced after 7 days, which indicates an average flow rate of 10 m/day or more than 1 ft/hr. The tracers were not injected simultaneously; therefore, a comparison of the response curves should be done with care. The curves show small variation with the methane peak produced a day or two later than tritium and ^{85}Kr. It also seems that the ^{85}Kr data are a bit more spread out than the other data. The test results were used to improve the knowledge about the sweep area and the volumetric flow in the different directions.

Calhoun[28] reported the use of several gas tracers in the Fairway field. The program consisted of three phases of injection. In the first phase, 10 Ci of tritiated hydrogen and ^{85}Kr were injected. Four months later, 10 Ci of tritiated methane, tritiated hydrogen, and ^{85}Kr were injected in a new set of injectors. In the third phase, 10 Ci of tritiated methane and ^{85}Kr were injected again in several wells. The extensive tracer program showed the source of gas breakthrough of 25 production wells. Continuous sampling of these wells after gas breakthrough has indicated a change in front configurations and fluid-migration pattern. Tracer responses proved the need for controlled injection and withdrawal to even out sweep configurations. Furthermore, tracers have indicated those areas in which injection rates and gas/water cycles should be changed to reduce fingering of the injected gas. This knowledge has been useful in alternating injection cycles and rates to control GORs at a reasonable level. Radioactive-tracer results also indicate that high-pressure gas injection at the Fairway is yielding additional oil recovery, either through swelling of the residual oil or by partial miscibility.

Tinker[29] reported the use of tritiated hydrogen and ^{85}Kr as tracers under methane injection in the Trembler zone II reservoir of the East Coalinga field, California. The tracers revealed that desaturated intervals were often continuous over wide areas of the field. Those intervals could greatly influence an injection project by acting as a thief zone for injection water. The tracer study suggested generally better reservoir-sand continuity than could be inferred from a related outcrop study.

6.6.5 Tracers in Enriched-Gas Injection. In the Prudhoe Bay field,[30] ^{85}Kr, tritiated methane, ethane, and propane have been used as tracers for an enriched-gas flooding. In the South Swan Hills unit, Alberta, Canada,[31] tracers were applied to follow enriched gas in a water/solvent cyclic-injection program. This is a limestone reservoir, and the tracers applied were tritiated hydrogen gas, tritiated ethane, and ^{85}Kr.

To gain information about the interwell reservoir characteristics, relative fluid velocities, and volumetric sweep efficiency in the early life of this project, it was desirable to trace the interwell movement of both the injected solvent and water. Tracers were added in as many as 14 solvent and 14 water injectors. The tracer responses in the wells were applied to redesign the injection program to achieve better sweep efficiency. The results proved to be of value as a qualitative indication of sweep, showing water and solvent flowing together to the majority of

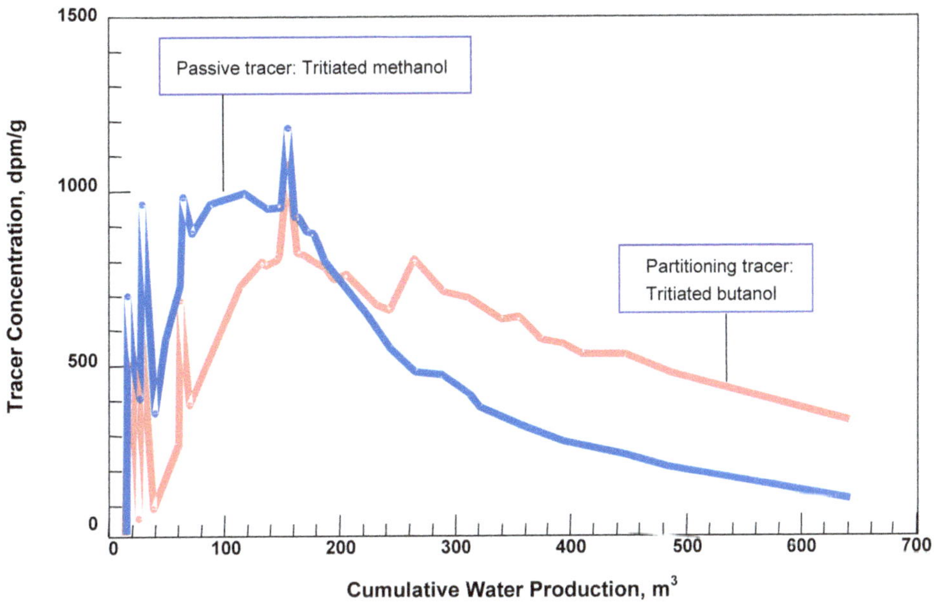

Fig. 6.9—Production curve of tritiated methanol and tritiated butanol in the Leduc field.[35]

the offset wells. In some cases, unexpected flow paths were identified. No quantitative interpretation and information on project performance, however, have been derived from these data.

6.6.6 High-Pressure N_2 Miscible Injection. In the Jay/Little Escambia Creek field, five radioactive tracers and one chemical tracer were used to tag injected N_2.[32] The radioactive tracers used were ^{85}Kr, tritiated hydrogen, methane, ethane, and propane. Sulfur hexafluoride was the lone chemical tracer. The N_2 was injected at a pressure of up to 7,600 psig for a period of 1 to 2 weeks before the well was switched over to water injection. The tracers produced are ^{85}Kr and tritiated propane. The main purpose of tracer injection was to determine the source of N_2 breakthrough. This knowledge enabled adjustment of injection rates and volumes to improve area convergence on the production wells.

Also in the Fordyce field,[33] ^{85}Kr, HT, CH_3T, and C_2H_5T were applied in a high-pressure (> 7,000 psig) miscible injection program. The gas injected was dry natural gas with an N_2 content of approximately 5%. The tracer-production data were used in a gas-sweep model to predict gas movement and to localize unswept areas.

6.6.7 Labeling of CO_2 Injection. Craig[34] reported the use of halogen compounds to trace injected CO_2. The compounds applied were freon-11, freon-12, freon-113, and sulfur hexafluoride. The halocarbons were reported to be detectable in concentrations down to 0.5 ppb in laboratory tests on produced fluids. The detection was carried out by separation through a GC column and registration by an electron-capture detector. These four tracers were injected in nine injection wells, and registration of the tracer content in the production was done from 23 production wells.

6.6.8 Residual-Oil Saturation in a Leduc Miscible Pilot.[35] Because of the large remaining oil in place, Leduc Woodbend D-2A had the potential to be an ideal miscible-flood candidate. Before the miscible injection, it was important to quantify the remaining oil. An injector/producer pair on 64-m spacing was chosen as a pilot for the tracer test. The partitioning tracer applied was tritiated butanol. **Fig. 6.9** shows the production curves of the tracers. On the basis of the retention of the butanol and methanol and the partition coefficient of these two tracers at the actual conditions, the residual oil between the two wells was measured. The tritiated methanol was regarded as a nonpartitioning tracer. The residual oil, S_{or}, was calculated on the basis of different arrival times of the two tracers, as shown in the production curve. The "peak," "half-peak height," and "breakthrough" gave the S_{or} at 34, 35, and 38%, respectively. The results were compared with the oil saturation obtained from sponge coring and single-well push-and-pull tracer tests. The sponge core gave an S_{or} at 33%, while the single-well tracer test gave a result from 35 to 40%, depending on the porosity model applied.

In summary, all field tracers, whether chemical or radioactive, are currently the only feasible, direct means of tracking the movement of injected fluids in a reservoir. In many fields, this information has been crucial for improving injection and production programs. Investments in new wells and equipment for injection programs are large and decisions should be based on the best possible data. A tracer test is a cost-efficient method to obtain important data that allow the analysis of injection and production options.

Nomenclature

a = cross section, L^2
a_c = adsorbed amount
A = constant related to the enthalpy
B = constant related to the entropy change
c = concentration in the liquid phase caused by partitioning between the gas and liquid phase
C = concentration
C_c = concentration in the contacting phase
C_e = measured tracer concentration
C_G = concentration of tracer in gas phase
C_M = concentration of tracer in mobile phase
C_O = concentration of tracer in oil phase
C_S = concentration of tracer in stationary phase
C_t = tracer concentration as a function of time
f = fractional flow
f_g = fractional flow for gas
f_o = fractional flow for oil
f_t = flow of tracer (see Eq. 6.21)
F = correction factor
h = height of reservoir zone, L
H_C = Henry's law constant
k = permeability, L^2
K = partition coefficient
K_C = partition coefficient based on concentration
K_{ci} = partition coefficient of component i
K_{cj} = partition coefficient of component j
K_P = partition coefficient based on mol fraction
L = distance, L
m = amount of tracer produced in a given well
m_{t_i} = total tracer recuperation up to t_i
M = amount of tracer originally injected

M_W = molecular weight, m
p = pressure, m/Lt^2
p_i = partial pressure, m/Lt^2
p_v = vapor pressure, m/Lt^2
q = volume flux, L^3/t
q_T = total volume flux, L^3/t
Q_i = mean volumetric injection rate, L^3/t
Q_m = mean volumetric water rate between injector and producer, L^3/t
Q_p = mean volumetric water production rate, L^3/t
Q_{p_t} = production-water flow rate as a function of time, L^3/t
r = radius of tracer front in radial system, L
s = frontal location in streamtube, L
s_1 = front location of tracer pulse in the streamline, L
s_2 = end location of tracer pulse in the streamline, L
S = saturation
S_g = gas saturation
S_o = oil saturation
S_{or} = residual oil
S_w = water saturation
t = time, t
t^* = dimensionless time
t_f = fraction of time, t
t_i = elapsed time for tracer *i,* t
t_S = retention time of nonpartitioning reference tracer, t
t_T = retention time of partitioning tracer, t
T = temperature, T
U = parameter that decides the shape of the curve for the actual system at a specific temperature
$\overline{U}_g$ = average linear flow rate of gas phase, L/t
$\overline{U}_L$ = average linear flow rate of liquid phase, L/t
$\overline{U}_{tr}$ = average velocity of tracer, L/t
V = parameter that decides the shape of the curve for the actual system at a specific temperature
$\overline{V}$ = mean produced volume, L^3
V_d = dilution volume, L^3
V_e = produced volume from which the exponential fit starts, L^3
V_M = volume of mobile phase, L^3
V_p = pore volume, L^3
Vs = retention volume of standard reference tracer, L^3
V_T = retention volume of tracer candidate, L^3
V_{Ti} = retention volume of tracer candidate *i,* L^3
V_{Tj} = retention volume of tracer candidate *j,* L^3
x^* = dimensionless distance
x_i = mole fraction of tracer *i* in liquid phase
y_i = mole fraction of tracer *i* in gas phase
α = dispersivity

β = retention factor
δ_G = density of gas, m/L^3
δ_O = density of oil, m/L^3
μ = viscosity, m/Lt
μ_1 = first momentum
σ = standard deviation
ϕ = porosity
ψ = streamtube

References

1. Wagner, O.R.: "The Use of Tracers in Diagnosing Interwell Reservoir Heterogeneities—Field Results," *JPT* (November 1977) 1410.
2. Bjørnstad, T. and Maggio, G.M.: "Radiotracer Applications in Industrial Processing, Oil & Geothermal Reservoirs," *IAEA Technical Report Series,* Intl. Atomic Energy Agency, Vienna, Austria (2002).
3. Reprinted from *Developments in Petroleum Science*, **43,** Zemel, B., "Tracers in the Oil Fields," 199 (1995) with permission from Elsevier.
4. Bjørnstad, T., Haugen, O.B, and Hundere, I.A.: "Dynamic Behaviour of Radio-Labelled Water Tracer Candidates for Chalk Reservoirs," *J. Petroleum Science Engineering* (1994) **10.**
5. Wood, K.N., Lai, F.S., and Heacock, D.W.: "Water Tracing Enhances Miscible Pilot," *SPEFE* (March 1993) 65.
6. Bjørnstad, T. *et al.:* "Interwell Tracer Technology in Oil Reservoirs: State-of-the-Art," *Tracing and Tracing Methods, Nancy, France, 2001, Recent Progress en Génie des Procèdes* (May 2001) **15,** No. 79.
7. McIntyre, F.J. *et al.:* "Radioactive Tracers Monitor Solvent Spreading in Rainbow Vertical Hydrocarbon Miscible Flood," *SPERE* (February 1988) 273.
8. Howell, W.D., Armstrong, F.E., and Watkins, W.: "Radioactive Gas Tracer Survey Aids Waterflood Planning," *World Oil* (1961) **41,** No. 2, 152.
9. Calhoun, T.G. and Tittle, R.M.: "Use of Radioactive Isotopes in Gas Injection," paper SPE 2277 presented at the 1968 SPE Annual Meeting, Houston, 29 September–2 October.
10. Frost, E.M.: "Helium Tracer Studies in the Elk Hills, California Field," BM-R1-3897, U.S. Bureau of Mines (June 1946).
11. Dugstad, Ø.: "An Experimental Study of Tracers for Labelling of Injection Gas in Oil Reservoirs," PhD dissertation, U. of Bergen, Norway (1992).
12. Senum, G.I. *et al.:* "Hydrocarbon Precombusting Catalyst Survey and Optimization for Perfluorocarbon Tracer Analysis in Subsurface Tracer Application," BNL 42769, Brookhaven Natl. Laboratories Report Series (May 1989) 34.
13. Galdiga, C.U. and Greibrokk, T.: "Ultra Trace Detection of Perfluorocarbon Tracers in Reservoir Gases by Absorption/Thermal Desorption in Combination with NICI-GC/MS," *Fresenius J. Analytical Chemistry* (May 2000) **367,** No. 1.
14. Galdiga, C.U. *et al.:* "Experience With the Perfluorocarbon Tracer Analysis by GC/NICI-MS in Combination With Thermal Desorption," *Tracing and Tracing Methods, Nancy, France, 2001, Recent Progress en Génie des Procèdes* (May 2001) **15,** No. 79.
15. Dugstad, Ø., Bjørnstad, T., and Hundere, I.A.: "Measurement and Application of Partition Coefficients of Compounds Suitable for Tracing Gas Injected into Oil Reservoirs," *Revue de L'Institut Français du Pétrole* (March–April 1992) **15,** No. 79.
16. Abbazadeh-Dehghani, M. and Brigham, W.E.: "Analysis of Well-to-Well Tracer Flow to Determine Reservoir Layering," *JPT* (October 1984).
17. Abbazadeh-Dehghani, M. and Brigham, W.E.: "Computation of Tracer Production Curves from Various Flooding patterns," *Proc.,* Annual Heavy Oil/EOR Contractor Reports, U.S. DOE, San Francisco (1981) L-1,L-15.

18. Dugstad, Ø. *et al.*: "Application of Tracer to Monitor Fluid Flow in the Snorre Field: A Field Study," paper SPE 56427 presented at the 1999 SPE Annual Technical Conference and Exhibition, Houston, 3–6 October.
19. Somaruga, C., Gazzera, C., and Wouterland, C.: "Application of 'BIPI' Tests Employing Natural and Artificial Tracers in Waterflooded Oil Fields of the Neuqina Basin (Argentina)," *Tracing and Tracing Methods, Nancy, France, 2001, Recent Progress en Génie des Procèdes* (May 2001) **15,** No. 79.
20. Agca, C., Pope, G.A., and Sepehrnoori, K.: "Modelling and analysis of tracer flow in oil reservoirs," *J. Petroleum Science and Engineering* (1990) **4,** 3.
21. Yoon, S. *et al.*: "In-Situ Characterization of Residual NAPL Distribution Using Streamline-Based Inversion of Partitioning Tracer Tests," paper SPE 52729 presented at the 1999 SPE/EPA Exploration and Production Environmental Conference, Austin, Texas, 28 February–3 March.
22. King, M.J. and Datta-Gupta, A.: "Streamline Simulation: A Current Perspective," *In-Situ* (1998) **22,** No. 1, 91.
23. Sagen, J. *et al.*: "Reservoir Chemical-Thermal Simulation with Tracers," paper SPE 36921 presented at the 1996 SPE European Petroleum Conference, Milan, Italy, 22–24 October.
24. Kleven, R.: "Non-Radioactive Tracing of Injection Gas in Reservoirs," paper SPE 35651 presented at the 1996 SPE Gas Technology Conference, Calgary, 28 April–1 May.
25. Armstrong, F.E.: "Field use of Radioactive Gas Tracers," *Petroleum Engineer* (December 1960).
26. Vilela, M.A., Zerpta, L.B., and Mengual, R.: "Water and Gas at El Furrial Field," paper SPE 53737 presented at the 1999 SPE Latin American and Caribbean Petroleum Engineering Conference, Caracas, 21–23 April.
27. Welge, H.J.: "Super Sleuths Tracer Flow of Injected Gas," *Oil & Gas J.* (August 1955) 77.
28. Calhoun, T.G. and Hurford, G.T.: "Case History of Radioactive Tracers and Techniques in Fairway Field," *JPT* (October 1970) 1217.
29. Tinker, G.E.: "Gas Injection with Radioactive Tracers to Determine Reservoir Continuity, East Coaling Field, California," *JPT* (November 1973) 1251.
30. Rupp, K.A. *et al.*: "Design and Implementation of a Miscible Water-Alternating-Gas Flood at Prudhoe Bay," paper SPE 13272 presented at the 1984 SPE Annual Technical Conference and Exhibition, Houston, 16–19 September.
31. Davis, J.A., Blair, R.K., and Wagner, O.R.: "Monitoring and Control Program for a Large Scale Miscible Flood," paper SPE 6097 presented at the 1976 SPE Annual Technical Conference and Exhibition, New Orleans, 3–6 October.
32. Langston, E.P. and Shirer, J.A.: "Performance of Jay/LEC Fields Unit Under Mature Waterflood and Early Tertiary Operation," *JPT* (February 1985) 261.
33. Mayne, C.J. and Pendleton, R.W.: "Fordoche: An Enhanced Oil Recovery Project Utilizing High-Pressure Methane and Nitrogen Injection," paper SPE 14058 presented at the 1986 SPE International Meeting on Petroleum Engineering, Beijing, 17–20 March.
34. Craig, F.F.: "Field Use of Halogen Compounds To Trace Injected CO_2," paper SPE 14309 presented at the 1985 SPE Annual Technical Conference and Exhibition, Las Vegas, Nevada, 22–25 September.
35. Wood, K.N., Tang, J.S., and Luckasavich, R.J.: "Interwell Residual Oil Saturation at Leduc Miscible Pilot," paper SPE 20543 presented at the 1990 SPE Annual Technical Conference and Exhibition, New Orleans, 23–26 September.

SI Metric Conversion Factors

bar	×	1.0*	E + 05	= Pa
Ci	×	3.7*	E + 10	= Bq
ft	×	3.048*	E − 01	= m
°F		(°F − 32)/1.8		= °C
°F		(°F + 459.67)/1.8		= K
eV	×	1.602 19	E − 19	= J
mL	×	1.0*	E + 00	= cm^3
psi	×	6.894 757	E + 00	= kPa

*Conversion factor is exact.

Chapter 7
Reservoir Pressure and Temperature

David Harrison, Schlumberger, and **Yves Chauvel,** Gamma Experts

7.1 Introduction

The practice of using bottomhole pressure measurements to improve oil and gas production and solve problems of reservoir engineering began around 1930. Initially, pressures were calculated using fluid levels; a later method was to inject gas into the tubing until the pressure became constant. The earliest bottomhole pressure measurements were made with one-time-reading pressure bombs and maximum-indicating or maximum-recording pressure gauges that lacked the accuracy, reliability, or durability of present-day technology.

The varied uses of bottomhole pressure and temperature measurements have increased in scope during the past two decades as instrumentation technologies have produced more reliable and accurate tools. These advances have made more applications possible, including use in multilayer reservoirs, horizontal wells, interference testing, and drawdown test interpretation.

This chapter is focused mainly on the types of measurements made and the tools available. Some information is included on interpretation techniques to connect the data acquisition with its use in characterizing a reservoir and its contents. Detailed explanations of these interpretation techniques can be found in other chapters in this *Handbook*.

7.2 Reservoir Pressure

The measurement commonly referred to as "bottomhole pressure" is a measurement of the fluid pressure in a porous reservoir. The reservoir pore-fluid pressure is a fraction of the overburden pressure that is supported by the fluid system. The other portion is supported by the rock and generates the in-situ rock stress. The overburden pressure is created by the weight of the rocks composing the lithostatic column at the point of observation. Hence, the difference between the overburden pressure and the vertical rock stress can approximate the pore pressure.

At original, or virgin, conditions, the vertical pressure profile reflects the distribution of fluids in the reservoir and may also indicate compartmentalization resulting from fluid flow barriers. Within any reservoir compartment, the pressure gradient reflects the density of the continuous fluid phase in the reservoir, hence the position of fluid contacts. **Fig. 7.1** illustrates a typical pressure profile showing gas-, oil-, and water-bearing intervals of a reservoir at initial conditions.

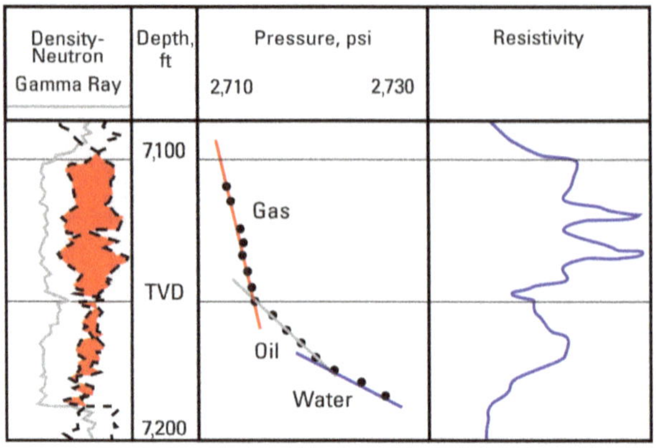

Fig. 7.1—Pressure gradients in a well drilled in a virgin reservoir.

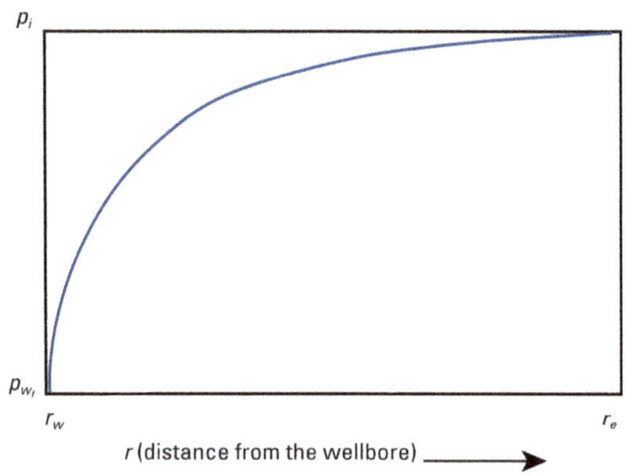

Fig. 7.2—Pressure profile away from the wellbore.

In a developed reservoir, differential depletion of lithostatic layers with various permeabilities and the movement of fluid contacts can change the pressure profile. Monitoring the static pressures vs. time in developed reservoirs is a crucial tool for reservoir management.

7.2.1 Pressure Distribution in the Reservoir During Fluid Flow. The Fluid Flow chapter in this volume of the *Handbook* explains the factors that govern the flow of fluids through porous media for steady-state, pseudosteady-state, and transient flow conditions.

Steady-State Flow. Steady-state flow is characterized by simultaneous constant pressure and flow rate. From the equation for steady-state radial flow,[1]

$$p_i - p_{wf} = \frac{141.2 q B \mu}{kh} \ln\left(\frac{r_e}{r_w}\right), \qquad (7.1)$$

the pressure profile away from a producing well can be calculated. A typical result is shown in **Fig. 7.2**.

Pseudosteady-State Flow. Pseudosteady-state flow behavior is observed when a well reaches stabilized production from a limited drainage volume. For constant-rate production under

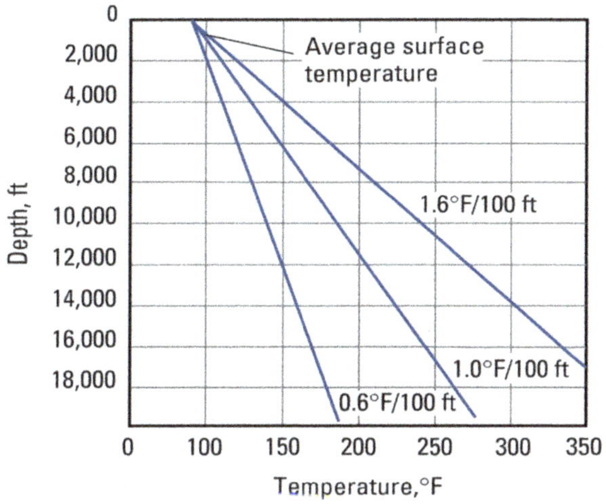

Fig. 7.3—Definition of an average geothermal gradient.

pseudosteady-state conditions, the difference between the flowing wellbore pressure and the average reservoir pressure in the drainage volume is constant, and the pressure drawdown is a linear function of time. The late-time buildup pressure will level off to the average reservoir pressure if the buildup duration is sufficiently long. Pressure depletion occurs with continued pseudosteady-state production.

Transient Flow. Transient flow is most often modeled with the radial diffusivity equation, which allows modeling pressure vs. time and pressure vs. distance from an observation point (typically, a well).

At a sufficiently large time, the pressure disturbance anywhere in the reservoir is proportional to the logarithm of the inverse square of the radius away from the origin of the disturbance. Thus, the magnitude of the disturbance is maximum near its origin (the wellbore) and rapidly terminates away from the wellbore. Because the pressure wave is affected by the reservoir fluid transmissibility, kh/μ, higher transmissibility results in smaller pressure differentials and vice versa. This effect explains why high-resolution pressure gauges are necessary to measure meaningful pressure differentials in reservoirs with high transmissibility. The radius of influence of a pressure disturbance is proportional to the square root of time. This is why the well testing time necessary to observe distant boundaries becomes prohibitively expensive, particularly in low-productivity reservoirs.

The variations of pressure as a function of time, which can be formulated by solving the radial diffusivity equation for specific cases, have given rise to well-testing applications.

7.3 Reservoir Temperature

Reservoir temperature is governed primarily by the reservoir's proximity to the earth's mantle, and by the relative heat exchange capacities and thermal conductivities of the formations forming the lithostatic sequence that includes the reservoir.

The geothermal gradient resulting from the heat-exchange process varies from basin to basin, but within a specific area the variations are small. In most hydrocarbon-producing areas, the gradient is usually in the range of 0.6 to 1.6°F per 100 ft of depth increase (**Fig. 7.3**). Areas where the earth's crust is thinner than average, such as volcanic and geothermal areas, have much higher gradients. In thin-crust areas the gradient change averages 4°F per 100 ft of depth increase. Local temperature gradients at depth have been reported as high as 10°F per 100 ft approaching singularities (e.g., major faults, areas of tectonic movement) in the earth's crust in geothermal areas.

To determine a precise geothermal gradient, the selected well must be shut in, without disturbance, for a period of time sufficient to let conduction effects equilibrate the temperatures. The temperature survey should be conducted from surface to bottom on the first descent into the well and at a slow speed (ideally, no more than 30 ft/min to accommodate the thermal inertia of the sensor). This procedure is necessary because the passage of the thermometer alters the static gradient. Even if a precise gradient is not required, following the procedure is still recommended for running temperature surveys in wells (shut-in, injecting, or flowing). Anomalies present during the first descent—whether observed or not—may disappear on subsequent surveys after disruption of the initial thermal equilibrium.

The virgin reservoir temperature may be determined when drilling initial exploration and appraisal wells by using the maximum temperatures recorded on successive logging runs or wireline sampling operations. The technique[1] calls for plotting T_{bh} vs. $(t_k + \Delta t_c)/\Delta t_c$ on a linear scale.

These data are interpreted in Horner analysis fashion by drawing a straight line through the data points and extrapolating to the reservoir temperature at $(t_k + \Delta t_c)/\Delta t_c = 1$, which corresponds to infinite shut-in time. Even though this approach is not mathematically rigorous, it provides reliable estimates of the static temperature when short circulation times are assumed. This technique is especially applicable to regions with high geothermal gradients, where the temperatures recorded at the time of logging runs can be significantly lower than the static temperature.

7.4 Metrology of Bottomhole Pressure and Temperature Gauges

Metrology is the science and process of ensuring that a measurement meets specified degrees of accuracy and precision. Bottomhole pressure-gauge and temperature-gauge performance depends on the static and dynamic metrological parameters described here. The pressure measurement equipment consists of the pressure transducer, associated electronics, and telemetry. Each component uniquely influences the measurement quality.

7.4.1 Static Metrological Parameters. The main static metrological parameters affecting gauge performance are accuracy, resolution, stability, and sensitivity.

Accuracy. Accuracy is the maximum pressure error exhibited by the pressure transducer under the following applied conditions: fitting error, pressure hysteresis, and repeatability. The fitting error, also called the mean quadratic deviation (MQD), is a measure of the quality of the mathematical fit of the sensor response at a constant temperature. Pressure hysteresis is the maximum discrepancy of the transducer signal output between increasing and decreasing pressure excursions. Repeatability is defined as the discrepancy between two consecutive measurements of a given pressure at the same temperature.

Resolution. Resolution is the minimum pressure change detected by the sensor. When referring to the resolution of a bottomhole pressure gauge, it is important to account for the associated electronics, because the gauge is always used in series with the electronics. Thus, the resolution of the measurement is the lower of the resolution of the gauge and its electronics. Another important consideration is that the resolution must be evaluated with respect to a specific sampling rate, because an increase of the sampling rate worsens the resolution. The electronic noise of strain-gauge transducers is often the major factor affecting resolution. Mechanically induced noise may further limit gauge resolution because some gauges behave like microphones or accelerometers. This effect may be significant during tests when there is fluid or tool movement downhole.

Stability. A pressure sensor is stable if it can retain its performance characteristics for a relatively long time period. Stability is quantified by the sensor mean drift (psi/D) at a given pressure and temperature. Three levels of stability can be defined: short-term stability for the first day of a test, medium-term stability for the following 6 days, and long-term stability for a minimum of one month.

Sensitivity. Sensitivity is the ratio of the transducer output variation induced by a change in pressure to that change in pressure. The ratio represents the slope of the line produced by a plot of the transducer output vs. pressure input. The plotted sensitivity should be, but is not always, linear with respect to pressure.

7.4.2 Dynamic Metrological Parameters. Four aspects are used to evaluate the dynamic metrology of pressure gauges.

Transient Response During Temperature Variation. The sensor response is monitored under dynamic temperature conditions while holding the applied pressure constant. The transient response characterizes the time required to get a reliable pressure measurement for a given temperature variation. Peak error and stabilization time are calculated.

Transient Response During Pressure Variation. The sensor response is recorded before and after a pressure variation at a constant temperature. Peak error and stabilization time are calculated.

Dynamic Response During Pressure and Temperature Shock. The sensor response is recorded before and after a temperature shock.

Dynamic Temperature Correction on the Pressure Measurement. The most advanced quartz-gauge transducers are based on a single-crystal sensor design. The crystal is activated on two distinct resonating modes, which are sensitive to both pressure and temperature but with different sensitivities. An advantage of this design is that the measured temperature is the temperature of the crystal. "Dynamic temperature correction" is used to adjust the pressure measurement of single-crystal sensors in real time for any remnant temperature effect. The nonuniform temperature of the crystal, especially while undergoing strong pressure or temperature variations or both, may induce such effects.

7.5 Calibration and Standard Evaluation Tests for Pressure Gauges

Calibration is essential for obtaining good temperature and pressure data. To ensure that a pressure gauge gives a pressure as close as possible to the real pressure over its entire operating range, it must be calibrated. Calibration involves establishing transfer functions to convert raw output from the pressure and temperature data channels into scaled pressure and temperature readings. These transfer functions are 2D (in pressure and temperature) polynomial models—the degree of which is a function of the accuracy required for the measurement.

The calibration process consists of applying known pressures and temperatures to the desired operational ranges. The raw pressure and temperature output signals are received and entered into a polynomial optimization routine. Input pressures are applied with a dead-weight tester, and input temperatures are generated by an oil bath. The following steps are required for a complete master calibration.

7.5.1 Choosing the Pressure-Temperature Calibration Points. Clearly, more calibration points yield a more accurate calibration; however, the inertia in temperature equilibration is a limiting factor. The best practice is to never use fewer than 100 pressure-temperature calibration points and distribute the points in a scheduled time routine, such as shown in **Fig. 7.4**.

7.5.2 Deriving the Pressure Calibration Function. The pressure calibration function is a polynomial of order N in pressures and order M in temperatures:

$$p_c = \sum_{i=j=0}^{i=N,\, j=M} A_{ij}\left(S_p - S_{po}\right)^i \left(S_t - S_{to}\right)^j, \quad (7.2)$$

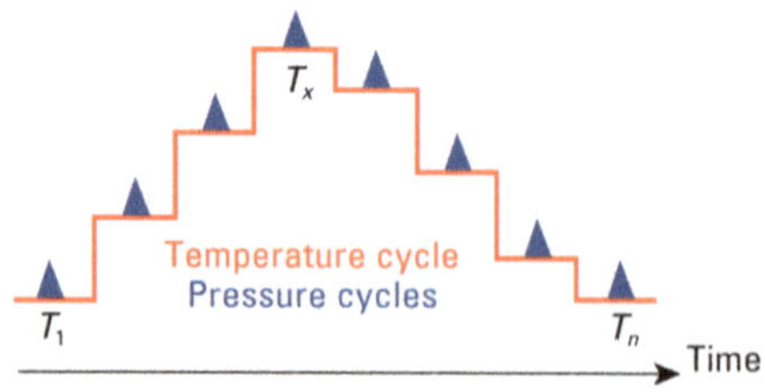

Fig. 7.4—Scheduled time routine for a pressure gauge calibration test.

in which the A_{ij} calibration coefficients are determined by a linear regression providing a least-squares minimization; S_p and S_t are the pressure and temperature outputs, respectively; and S_{po} and S_{to} are the corresponding offsets. Usually the number of A_{ij} coefficients can be limited to approximately 15. During this step, the peak error and MQD are determined.

7.5.3 Temperature Calibration Function. It is often useful, though not always necessary, to calibrate the transducer to output a scaled temperature measurement. The temperature calibration function is a polynomial of order N in temperatures and order M in pressures:

$$T_c = \sum_{i=j=0}^{i=N,j=M} A'_{ij}(S_t - S_{to})^i (S_p - S_{po})^j, \quad \text{(7.3)}$$

in which the A_{ij}' calibration coefficients are determined by a linear regression providing a least-squares minimization; S_p, S_t, S_{po}, and S_{to} are as described previously.

7.5.4 Determining Nonlinearity in Pressures and Temperatures. Several other tests supplement the master calibration.

Pressure Thermal Sensitivity. The pressure thermal sensitivity represents the error (psi) that results if the temperature measurement is in error 1°C.

Maximum Hysteresis During the Calibration Cycle. This test is determined from calibration data.

Calibration Check. A calibration check verifies the consistency of the sensor readings when the applied pressures and temperatures are different from those used during the calibration cycle. The calibration check is performed in the laboratory at the time the sensor is evaluated and is essentially a rerun of a master calibration.

Other Procedures and Tests. Standard procedures are typically used in evaluating pressure transducers to compare different technologies and certify the calibration parameters. The most commonly used standard procedures are as follows:

- Complete master calibration.
- Calibration check.
- Stability tests: middle term and long term.
- Repeatability test.
- Resolution test.
- Noise or short-term stability test.
- Dynamic tests: temperature shock, temperature transient, temperature response time, and pressure shock.

7.6 Metrology of Temperature Gauges

When the temperature is used to correct pressure gauge readings, it must come from the pressure-sensing element, not from the wellbore fluid. On the other hand, bottomhole-fluid temperature measurements are performed with sensors that are in immediate contact with the wellbore fluid and have a minimum thermal inertia (1 or 2 seconds) to follow the variations of the fluid temperature as closely as possible. For this reason, temperature measurements available from pressure-gauge technology are rarely valid for traditional wellbore temperature profiling, which uses the wellbore fluid temperature as a diagnostic tool to detect anomalies in the expected flow patterns in and around the wellbore.

Typical wellbore-fluid temperature measurements have a resolution in the range of 0.05°F and accuracy in the range of 1°F. Accuracy in thermometry is not always a prerequisite because temperature measurements are often normalized between themselves—e.g., from pass to pass or from flowing run to shut-in run in production-logging applications. Accuracy is necessary, however, to compare absolute bottomhole temperatures to draw geothermal maps; to design temperature-sensitive operations, such as stimulation or drilling operations using temperature-sensitive chemicals; or to operate close to the limitations of equipment, such as in geothermal or other high-temperature oil and gas fields.

Resolution is of paramount importance for applications such as the diagnosis of flowing wells or when the measured temperature is the temperature of a pressure-sensing element—the reading of which is affected by minute changes in sensor temperature. High-resolution wellbore-fluid thermometry is also used in extremely slanted and horizontal wells, in which true vertical depth (TVD) variations, and therefore geothermal temperature variations, are small.

7.7 Pressure Transducer Technology

All pressure transducers operate on the principle of converting a pressure change into a mechanical displacement, or deformation. Deformation of the sensing element is then converted into an electrical signal that is processed by the measuring system. Types of pressure transducers available in the field, either individually or in combination, are mechanical, capacitance, strain gauge, and quartz gauge.

7.7.1 Mechanical Pressure Transducers. The first pressure transducers had mechanical-force-summing elements that converted energy into mechanical displacement, or deformation, and then coupled the generated force to a recording device. In the Amerada gauge, a popular mechanical pressure transducer, the pressure-sensing element is a helical Bourdon tube. The tube is of sufficient length to rotate a clock-driven stylus a full circumference inside the cylindrical chart holder. The chart, usually made of coated metal, is recovered at the end of the test, unfolded until flat, and read on a high-precision optical machine. The transducer also incorporates a vapor-type recording thermometer to make temperature corrections on the pressure measurements.

Mechanical transducers have largely been abandoned because of their obsolete metrological characteristics and lack of surface readout (SRO). They are still used occasionally for basic applications at the lower end of the economic spectrum, for some very high-temperature applications, or as backup for an electronic pressure gauge.

7.7.2 Capacitance Pressure Transducers. Capacitance transducers have a variable-gap capacitor in which the sensing element is formed by two metallic or quartz plates. As the external pressure increases, the deflection of the sensing plate creates a change in the capacitance that can be mathematically related to the applied pressure. Capacitance transducers have the advantages of good frequency response, low hysteresis, good linearity, and excellent stability and repeatability. The disadvantages are high sensitivity to temperature, acceleration, orientation, and mechanical noise.

Fused quartz has excellent elastic behavior (low hysteresis) and is chemically inert. These properties make it an almost ideal material for manufacturing small capacitor modules with a high temperature rating.

7.7.3 Strain Pressure Transducers. Many types of strain-gauge transducers are in use. Strain gauges have become very popular because of their ruggedness, low cost, and good dynamic behavior. Their metrological characteristics have greatly improved in recent years; gauges with an accuracy of a few psi and resolution as low as 0.05 psi are available. The primary limitation of strain gauges is their tendency to drift, although that aspect of the measurement has improved.

A strain gauge has a strain-sensitive resistor directly attached to a measuring sensor; when the sensor is subjected to pressure, it deforms. The resulting displacement changes the resistor length, hence its resistance. The applied pressure is calculated from a calibrated relationship to the change in resistance at a given temperature.

Bonded Wire Transducers. In this design, introduced by the Paine Corporation in the 1970s, two sets of wire, called the "active" windings, are wrapped around a cylindrically shaped tube-sensing member. As pressure increases, the tube bore is stretched, causing a change in the wire resistance. Another two sets of wire—the reference, or "passive," windings—are wrapped on the upper part of the tube, which is not exposed to pressure. These four sets of wire form a Wheatstone bridge that allows the electrical output to be reduced to a pressure reading.

Thin-Film Transducers. The thin-film sensor consists of a resistor pattern that is vaporized or sputter-deposited onto the force-summing element (the measuring diaphragm). In some transducers the resistors are not directly mounted on the diaphragm but are on a beam linked to the diaphragm by a push rod.

Sapphire Transducer. In the Schlumberger improvement of the diaphragm-type thin-film transducer, the sensing resistors are mounted on a miniature substrate of industrial sapphire. The Sapphire* pressure-gauge system is vacuum-filled and the resistor pattern forms a Wheatstone bridge. This system benefits from the elastic performance of the sapphire and its stable deformation properties. The result is a sensor with good repeatability, good stability, low hysteresis, and low drift. A high-gauge factor improves the resolution over traditional designs. The main disadvantages are low output level and high cost.

7.7.4 Quartz Pressure Transducers. Quartz-crystal pressure transducers vibrate at their resonating frequency when excited by a suitable external energy source. The resonating frequency is affected by both the pressure and temperature to which the crystal is exposed. Because of the excellent gauge factor yielded by this physical process, quartz-crystal pressure transducers have exceptional accuracy, resolution, and long-term stability. The disadvantages are high cost and high sensitivity to temperature, although the most recent designs are much less temperature-sensitive.

Hewlett-Packard Design. The Hewlett-Packard (HP) design has been in use since the early 1970s. It features a two-crystal arrangement of a measure and a reference crystal. The measure crystal is exposed to both pressure and temperature. The reference crystal is exposed only to temperature and is used to compensate for temperature effects on the measure crystal. Both crystals are factory-matched so that their frequency characteristics in temperature are approximately the same. The measure crystal senses the pressure directly rather than through a mechanical linkage or other force-summing device. This has the effect of optimizing the metrology of the measurement.

The output from the crystal pair and associated electronics is calibrated to yield the measured pressure by means of a 2D cubic polynomial including 16 coefficients. The values of these coefficients are determined at least annually during the gauge master calibration.

* Mark of Schlumberger.

Quartzdyne Design. The Quartzdyne design features three resonating crystals: the measure crystal, which is exposed to both pressure and temperature, and the temperature and reference crystals, which are exposed only to temperature. The measure crystal is a thick-walled, hollow quartz cylinder closed at both ends. The resonating element is a disk placed in the center, which divides the cylinder into halves. Separate conductive plates are located on the front and back of the resonator disk. Fluid pressure on the exterior walls hydrostatically compresses the quartz cylinder, producing internal compressive stresses in the resonator. The oscillating frequency of the resonator changes in response to these internal stresses.

The reference crystal oscillates at a fixed high frequency, which is subtracted from both the measure crystal and temperature crystal resonating frequencies. The temperature compensation is performed based on these low-frequency signals. The calibration procedure involves a fourth-order polynomial.

Because of its small size, the Quartzdyne design provides good thermal performance and low cost, although somewhat at the expense of accuracy.

Schlumberger Design. In the Schlumberger CQG* Crystal Quartz Gauge design, the transducer features a single quartz crystal structure in which a resonator is coupled with a dual-mode oscillator. The resonant frequency of the first mode is highly sensitive to pressure, and that of the second mode is more sensitive to temperature. The sensor consists of a cylindrical quartz body fitted with two end caps. The end caps maintain a vacuum inside the sensor. The resonator is a plate etched out of the quartz cylinder that features shaped surfaces acting as vibrating lenses. The resonating frequency of the plate varies with changes in pressure and temperature.

From a static point of view, the main advantage of this design is that pressure and temperature are measured at the same location, which minimizes time and space delays for thermal corrections. From a dynamic point of view, this design leads to very small peak transient errors in the thermal response that can be further minimized by using real-time dynamic compensation. The calibration involves a fourth-order polynomial.

The main disadvantages of the CQG design are fragility and high cost.

Paroscientific Design. The Paroscientific design uses a quartz crystal operating in flexure mode to measure force. To derive a pressure output, a force-summing device such as a Bourdon tube or bellows must be used. Thus, the transducer senses pressure through the force-summing device and is not in direct contact with the wellbore fluid. This design tends to improve temperature characteristics but dampens the response and downgrades the measurement metrology. A temperature sensor comprising a quartz torsional tuning fork provides temperature compensation.

Quartztronics Design. The Quartztronics design is a modified HP design, with a specially cut resonator and a noncylindrical cell geometry. The result is a smaller, lower cost pressure transducer with a higher pressure range.

The transducer features a temperature-sensing crystal and a reference crystal, both located close to the measure crystal. This configuration provides improved pressure- and temperature-transient responses in comparison with the HP design. The two crystals are thermally matched to the measure sensor and within a pressure-proof package bonded to one of the end caps of the measure sensor.

7.8 Temperature Sensors

7.8.1 Mechanical Transducers. The first bottomhole thermometers were mechanical. They were identical to bottomhole mechanical pressure gauges, except that a thermometer sensor was substituted for the pressure sensor. This type of thermometer has been mostly replaced by sensors and recording elements that are easier to use and have higher precision and accuracy.

7.8.2 Thermistors. Thermistors are temperature-sensitive resistive elements made of semiconductor material with a negative coefficient of resistance. The physical effect governing a thermistor's change of resistance is the increased number of conducting electrons for a corresponding increase in temperature. Thermistors can be built up to 100 times more sensitive to temperature change, for the same resistivity change, than resistance temperature detectors (RTDs), which are described next. The main drawback of thermistors is their operating temperature limitation of approximately 300°F.

7.8.3 Resistance Temperature Detectors. RTDs rely on the increase in resistance of metals in response to increasing temperature. The resistor consists of a coil of fine metal wire or a film of pure metal deposited on a nonconductive surface. Different metals with different resistances are used, but platinum has become the most popular because of its excellent accuracy, large linear range of operation, and wide temperature range (higher than 1000°F is possible). The RTD is usually encased in a probe that is directly exposed to the well fluids. The RTD is commonly the active leg of a Wheatstone bridge.

7.9 Optical Fiber Measurement of Pressure and Temperature

Several systems are being developed to provide pressure and temperature measurements distributed over the length of an optical fiber that is permanently deployed in the completion. An advantage of fiber optic technology is that the sensors have no electronic components at depth, so they tend to be more reliable. Furthermore, optical sensors are immune to shock, not prone to electromagnetic interference, and operable at high temperatures.

Fiber optic technology is based on exposing the fiber to periodic ultraviolet (UV) light patterns that induce a "grating" on it. Pressure and temperature variations change the reflection wavelength of the gratings and can be decoded with respect to the fixed, incipient operating wavelength. The system is self-referencing.

Every point distributed along the length of the fiber has the potential to generate a different temperature measurement. The advantages are measurement of a permanent temperature gradient over the length of the fiber and the ability to select specific measurement points. Single-point and distributed temperature sensors have been reported to operate successfully in steamflood wells up to 575°F. In one reported case, temperature measurements taken along a horizontal wellbore at different times showed steamchests, water breakthrough, crossflow, and flow behind pipe.

Pressure is measured by sensors located at discrete, fixed points along the fiber. At the sensors, the fiber is cut, and its ends are placed face-to-face in a proximal arrangement. The face-to-face spacing is measured by successive reflections of the light wave. Changes in the value of the spacing reflect the environmental pressure around the fiber at that point. The self-referencing technique uses the distributed temperature measurement for suitable corrections.

7.10 Acquiring Bottomhole Pressure and Temperature Data

The acquisition of bottomhole pressure and temperature data can be planned and executed in a cost-effective manner with a minimum disruption to normal operating routines. In many cases, early on-site interpretation is useful in guiding decisions about continuing the acquisition program. Several questions should be answered at the design stage.

- What are the objectives of measurement: static pressure, reservoir dynamics, fluid characterization, vertical pressure and temperature profile, well flow characterization, or other?
- Is the environment openhole or cased hole? An exploration or development well?
- Is there a need for real-time SRO measurements or can the data be recorded downhole and reviewed later?
- What metrology is needed for the measurements (e.g., maximum temperature and pressure, measurement resolution and accuracy)?[2]

• How will the gauges be conveyed to the bottomhole measuring points?

• Is there a need to perform continuous or repeated measurements over months or years?

• What economics apply, and do they imply a possible tradeoff with the quality or quantity of the measurements?

This section addresses these issues within the categories of SRO vs. downhole recording (DHR), surface vs. downhole shut-in, mode of conveyance, drillstem tests (DSTs), openhole wireline formation testing, production logging, measurement while perforating, and permanent installations in "smart" completions.

7.10.1 Surface Readout vs. Downhole Recording. Measurements can be transmitted to the surface, usually via an electric cable, or recorded in downhole memory powered by batteries.

SRO has the obvious advantage of providing data in real time. Real-time readouts are especially beneficial for transient measurements that require time for the pressure to stabilize and radial flow to develop. Because stabilization times depend on reservoir and fluid properties and because the determination of these parameters is often the purpose of pressure measurements in well tests, predicting the duration of stabilization periods is often difficult. SRO is preferred in these cases.

Some applications, usually those in the lower economic tier, can be conducted without the need for SRO. The benefits include lower operating costs and a fixed operations schedule. The drawback is the difficulty of guaranteeing the quality of the acquired data, including the potential for significant operating losses if the bottomhole recording equipment malfunctions. For these reasons, DHR should be chosen only when the measurement target does not necessarily depend on stabilization times or when stabilization times are already known (e.g., to measure the average reservoir pressure in a reservoir of known mobility).

Many industry tools provide both SRO and DHR. The measuring section of these tools is common to both options. In the SRO option, the sensor electronics are coupled to a telemetry system for uphole transmission, and the cable supplies downhole power. In the DHR option, downhole batteries supply power, and the data are stored in memory boards for future readout or downloading to suitable computer systems.

7.10.2 Surface Shut-In vs. Downhole Shut-In. The practice of downhole shut-in during a buildup test as opposed to surface shut-in is discussed in "Pressure Transient Testing." The advantages of downhole shut-in include control of the wellbore volume (afterflow), reduction of the duration of buildup tests, and choice on the recording mode (SRO or DHR).

Downhole shut-in can be performed during conventional DSTs or during tests performed on production wells. In a DST, the downhole shut-in valve is usually the main test valve. Shut-in is performed at the end of a flow period by actuating the test valve. Traditionally buildup pressures are recorded in DHR mode. Another procedure is to use the DST valve as the main shut-in valve while the DataLatch* electrical wireline downhole recorder/transmitter is used to observe pressures in real time during the buildup. DataLatch technology is described in "Drillstem Testing."

In producing wells, downhole shut-in is performed by setting a valve assembly in the tubing before performing the test. The tubing must have been previously equipped with locator nipples so that the valve can be anchored at the appropriate depth. The valve is run on either a slickline carrying a DHR pressure gauge or an electric line equipped with an SRO recording gauge. The shut-in valve is actuated by a sequence of pulls and releases on the slickline or cable. Commonly operated shut-in valves can perform in the range of up to 12 open-close cycles, after which the valve assembly is released from the setting nipples by an appropriate pull on the line. Other versions of shut-in valves can be operated by a clock, small explosive squib, or battery.

7.10.3 Bottomhole Conveyance of Gauges. Pressure and temperature gauges can be conveyed to the downhole environment by a number of methods, possibly in tandem with other measurements. Several possibilities are discussed here.

Drillstem Test String. Refer to the DST subsection 7.10.4.

Electric Line. Electric line operations provide surface readout and can be conducted any time during the life of the well. In openhole, wireline pressure testing offers a unique opportunity to efficiently collect distributed pressure data on the entire stratigraphic sequence penetrated by the well. In cased holes, pressure and temperature measurements are taken repeatedly along with other production logging measurements to monitor well performance and diagnose flow and completion problems. In addition, pressure buildups and other transient tests are frequently conducted in producing wells, using production logging tools or with "hanging" gauges, to check for variations in the productivity index and for skin development and to monitor multilayer producing systems. In the exploration environment, however, electric lines are not typically used because of the risks associated with having cable in the well while flowing. These risks include difficulty closing safety valves or subsea trees with cable in the borehole and sticking of the cable because of sanding from unconsolidated formations or hydrate formation in subsea wells. The DataLatch electrical wireline downhole recorder/transmitter, introduced in the late 1980s, has largely mitigated this risk. The DataLatch system is briefly described in "Drillstem Testing."

Slickline. Slickline pressure and temperature surveys are performed with hanging gauges in situations that do not require SRO. Slickline operations are more cost-effective than electric line operations; however, the data quality usually does not match that of SRO data. Depth control is one of the critical factors affecting data accuracy. On the other hand, surface pressure control is easier in slickline operations because of their smaller diameter, typically in the range of 0.1 in. in diameter. A promising development is the "slick conductor line," a thin, hollow-core, 100% steel cable laid around an electric conductor to provide limited SRO capabilities.

Coiled Tubing. A popular alternative to drillpipe, coiled tubing is used to convey downhole gauges and other equipment in deviated holes when gravity is insufficient to pull the tools to the bottom of the well. Insufficient gravity occurs where well deviation exceeds values in the range of 60 to 70°, depending on tool weight and length, friction coefficients, pipe roughness, and the presence and type of completion components. In horizontal wells, the coiled tubing may not reach the toe of the completion because of a helical lockup of the coil inside the completion. Coiled tubing may be equipped with an internal electric cable running the length of the coil to support SRO operations.

Tractors. Tractors are an emerging technology that complements the use of coiled tubing in difficult, deviated completions. Tractors are self-powered and operated by electric line. They can negotiate bends, crawl up or down, and overcome the limitations of coiled tubing in long horizontal wells. Their main limitation is the large amount of cable power required for operation.

Wireless Transmission. Wireless transmission is a technique that has been in use since the late 1980s. It attempts to provide the advantages of SRO without using an electric line. The downhole tool, a sub that is part of a DST string, features a pressure gauge, battery pack, telemetry, recorder board, and antenna. The antenna sends the signals collected from the pressure recorder at a frequency suitable for transmission through the formation strata. At the surface, the signals are picked up by an array of suitably deployed stake antennae. This technique is limited to land operations and depths of approximately 8,000 ft.

7.10.4 Drillstem Testing. A DST string is a complex array of downhole hardware used for the temporary completion of a well. DSTs provide a safe and efficient method to control flow while gathering essential reservoir data in the exploration, appraisal, and development phases of a reservoir or to perform preconditioning or treatment services before permanent well completion. **Fig. 7.5** shows a typical DST string with its essential components. In exploration well testing in particular, the DST string usually includes tubing-conveyed perforating (TCP) guns, which are shot underbalanced (i.e., wellbore pressure is less than reservoir pressure) at the initial well completion.

DST strings include gauge carriers, which are collars that normally contain up to four pressure gauges bundled together, affording redundancy in long tests in which one or more gauges are likely to fail. These gauges perform only DHR measurements.

Most modern DST strings are fullbore strings, which means they have a flush opening running completely through the string of tools. The opening enables running pressure gauges and other slim tools (typically $1^{11}/_{16}$ in.) in SRO mode.

The DataLatch system is a DST string component that combines the advantages of DHR (typically during flow periods) with the advantages of SRO (typically during shut-in periods). Pressure data are recorded in the tool's memory boards. In suitable conditions, a LINC* Latched Inductive Coupling tool is run with an electric cable and latched into the DataLatch mandrel. This combination is used to read out the memories, reprogram the gauge acquisition schedule if necessary, and monitor the test in real time. The DataLatch system is unique for bottomhole pressure measurement applications. It allows simultaneous and continuous acquisition of three different measurements during the course of a DST test: rathole or reservoir pressure, cushion or tubing pressure, and annulus pressure.

7.10.5 Openhole Wireline Pressure Testing. Wireline pressure testing is conducted using tools lowered on an electric cable or coiled tubing in deviated wells. The tools consist of function-specific modules selected for a specific operation. **Fig. 7.6** shows the modular arrangement of a modern wireline pressure tester.[3] In a complete configuration, the tools may include a single-probe module for basic pressure testing and sampling, a dual-probe module for permeability applications, a flow-control module for flexible schedule testing, a fluid analyzer module for optical fluid properties monitoring in real time, multisample module for representative fluid sampling at reservoir pressure-volume-temperature (PVT) conditions, sample modules for large-volume fluid sampling, a pumpout module that can recirculate mud filtrate and other unrepresentative fluids out of the tool flow system before representative sampling, and a dual-packer module for very large area sampling, interference testing, and DST emulation.

Wireline pressure testing encompasses several applications described in "Specific Applications and Interpretations." The applications include measuring static reservoir pressure, collecting representative fluid samples, determining anisotropic permeability, identifying reservoir permeability barriers, determining reservoir fluid gradients and densities, and determining rock stress components.

As in conventional well testing, a wireline sampling operation must be carefully designed beforehand to achieve the desired objectives. Key parameters that require special attention include the test type, fit of the probe type to the reservoir's mechanical and hydraulic deliverability characteristics, volumes to be withdrawn, type of pressure gauges for the expected reservoir permeability, test sequence, number of pressure points to be taken, and interpretation objectives. The actual downhole tool configuration and the test sequence directly reflect the design study that precedes the operation.

7.10.6 Production Logging. The uses, benefits, and interpretation techniques of production logging are discussed in another chapter in this section of the *Handbook*.

7.10.7 Measurements While Perforating. Pressure and temperature measurements may be performed concurrently with shaped-charge perforating. This technique, called measurement while perforating (MWP),[4] includes the following applications:

• Before perforating, MWP verifies the completion fluid density, directly measures the wellbore pressure, and adjusts the perforating underbalance.

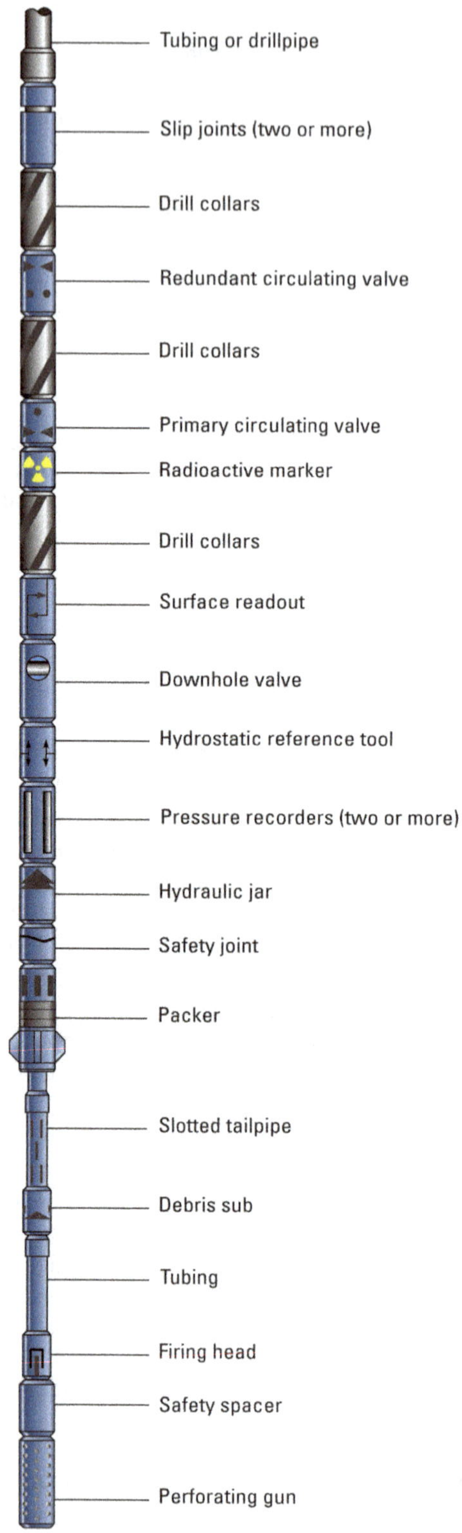

Fig. 7.5—Pressure-controlled DST string.

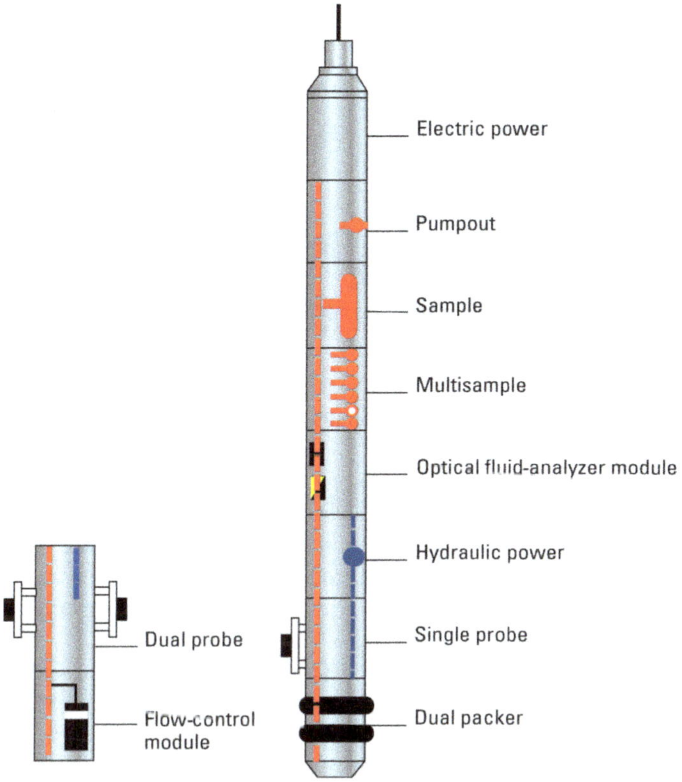

Fig. 7.6—Wireline tool components.

• During perforating, MWP positively detects the detonation of the perforating gun.

• After perforating, MWP observes the pressure responses and interprets them as a transient test, and it monitors the fluids produced by the perforation.

The MWP system is especially adapted to low-flow-rate or short-duration tests such as IMPULSE* measurement while perforating tests, closed-chamber tests, and slug tests. Flow tests can also be analyzed.

MWP must be performed in SRO mode; otherwise, the technique offers no benefit. An SRO tool such as the DataLatch recorder/transmitter or an electric line tool must be used. The production well environment is preferred for MWP, with the perforating guns conveyed on an electric line with the MWP sub above the guns. A typical MWP sub for pressure and temperature measurements incorporates a gamma ray detector and casing collar locator for depth control and suitable shock absorbers to mechanically decouple the guns from the measurement system. An optional bottom electric adapter can fire the guns electrically, which is often the procedure in non-TCP applications.

Because MWP also measures the wellbore fluid temperature, it is used for the same applications as production logging.

7.10.8 Permanent Pressure Measurement Installations. Permanent monitoring systems are placed downhole with the completion string near the depth of the reservoir to be monitored. They are connected to the surface by a cable that runs the full length of the completion string and exits the wellbore through suitable connectors crossing any subsurface safety systems and the wellhead. Advanced telemetry allows querying these sensors at any time throughout the life of the reservoir. Most systems in operation today record bottomhole pressure and temperature.

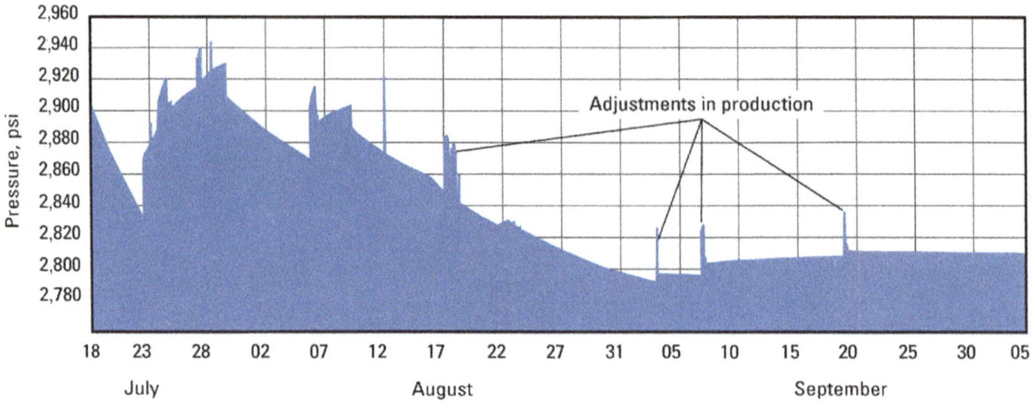

Fig. 7.7—Readout of a recording over an 80-day period from a permanent gauge.

Permanent systems[5] are engineered specifically for monitoring applications and have a life expectancy of several years. The digital electronics within the gauges are designed for extended exposure to high temperature without required maintenance. The metrology characteristics emphasize long-term stability rather than fast dynamic response. Quartz crystals as well as sapphire-based sensors can be used.

The cables for permanent installations are designed to withstand pressure, temperature, and exposure to highly corrosive fluids. They must also be mechanically rugged to prevent damage during installation. Usually single conductor cables are used.

The connections are similarly designed for durability. They include bottomhole connectors (power and pressure) to the permanent gauge mandrel and uphole connectors that cross through the wellhead. The complexity of the surface installation varies, depending on whether the wellhead is located at the surface—as on a land well or wellhead exposed above the sea on a platform —or subsea. For subsea wells, the acquisition system is typically through existing data-gathering systems through umbilicals. On platforms, several permanent gauges may be connected to an autonomous surface unit that records the measurements and communicates with shore facilities through standard or advanced (satellite) transmission links.

Fig. 7.7 shows a continuous data stream from an 80-day recording in which the pressure measurement was used to optimize production. The surface production rate was repeatedly adjusted to yield an acceptable bottomhole flowing pressure.

7.11 Wellhead Environments

Although the focus of this chapter is bottomhole measurements, it is worthwhile to mention a few interesting points about the environments of surface and subsea measurements.

7.11.1 Surface Acquisition of Downhole Data. Current specifications of surface acquisition systems, sensors, umbilicals, and piping commonly used in the industry are 0 psi and –40°F for the lower range of pressures and temperatures, respectively. The temperature specification in particular presents no obstacle to testing operations in extremely cold areas, such as the Arctic and similar cold-weather territories.

7.11.2 Subsea Acquisition of Pressure and Temperature. Pressure and temperature measurements are sometimes required at the subsea tree level. The measurements are mainly used to monitor the operating conditions of the landing string near the ocean floor. Applications include ensuring that the maximum temperature rating of the elastomers in the blowout preventer

(BOP) is not exceeded, and providing data to help prevent hydrate formation during deep-sea cleanup and well testing operations.

7.12 Specific Applications and Interpretation

Bottomhole pressure data are vital for understanding reservoir performance and predicting future behavior. Applications include volumetric calculations (e.g., reserves), reservoir dynamic properties (e.g., permeability), drainage volumes (e.g., compartmentalization and flow barriers), fluid properties (e.g., density, phase behavior), well tubular and artificial lift design (e.g., size selection and lifting systems), evaluation of reservoir energy and fluid contacts with time, and input to numerical reservoir simulation models.

7.12.1 Depth Datum of Pressure. Static pressures should be corrected to a fixed depth datum to eliminate the influence of the fluid pressure gradient for building isobaric maps, using bottomhole pressure to calculate inflow performance relationship (IPR) diagrams for multilayer pressure data sets, or interpreting vertical permeability barriers from a pressure differential between two reservoir layers.

Pressures are adjusted to a fixed datum by calculating the hydrostatic potential (also called the datum-corrected pressure) as follows:

$$\Phi = p - 0.00135\rho_{fl}g(D - D_0)\cos\delta\,. \quad (7.4)$$

The potentials (adjusted pressures) reflect the dynamics of fluid movement in the reservoir better than the raw pressure data can. Reservoir layers with different potentials flow into one another if put in communication (e.g., if they are completed in the same wellbore). Isobaric maps built on datum-depth-corrected pressures reveal flow within a specific reservoir layer if this layer shows different potentials in different regions of the reservoir. In addition, vertical permeability barriers are qualified in terms of potential differences between the two adjacent reservoir units separated by the barrier.

D_0, the datum depth, can be arbitrary and has no influence on the interpretation of the hydrostatic potential, assuming a constant reservoir fluid density across the different wells or layers. Typically, a datum at the midpoint of the hydrocarbon column is selected to study pressure trends across the reservoir and phase behavior effects. A datum within a well may prove more useful for analyzing potential differences between multiple reservoir layers crossed by the well.

7.12.2 Static Pressure. Static pressure measurements always result from some form of transient test, in which a large or small amount of fluid is withdrawn from the well before the pressures are allowed to stabilize. Static pressures are acquired during wireline testing at the rate of approximately one measurement every few minutes because only very small amounts of fluid samples are withdrawn. Conversely, static pressures take much longer to stabilize in conventional well testing because the much larger fluid samples withdrawn create much greater pressure disturbances.

Static Pressure From Buildup Tests. The static pressure of a reservoir is one of the interpretation outputs of pressure transient tests. Many short-duration buildup tests (including wireline pressure "pretests") are designed solely for measuring the static reservoir pressure. The interpretation of buildup tests to determine the static reservoir pressure is discussed in another chapter in this section of the *Handbook.*

Average Reservoir Pressure. The average reservoir pressure can be determined arithmetically by averaging the datum-corrected pressures of a given layer in all wells, with each pressure weighted by the net thickness of the reservoir at the well. A better average pressure is determined by recording the pressures, either actual or weighted, on a map of the area and drawing

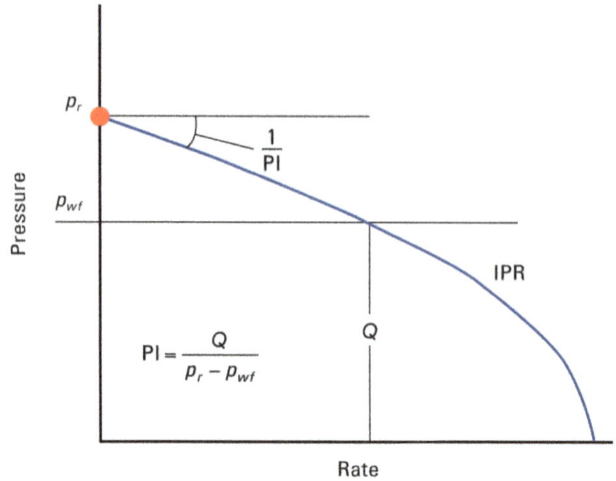

Fig. 7.8—PI of a reservoir with a single producing layer.

isobars from which the average pressure weighted for an area is determined by planimetery (or gridding) of the isobars.

Static Pressure Determined From the Productivity Index. The productivity index (PI) of a producing layer, J, is defined as the ratio of the downhole production rate of the layer to the pressure drawdown under which the layer produces:

$$J = \frac{Q}{\left(p_r - p_{wf}\right)} \quad \text{.......... (7.5)}$$

On a plot of bottomhole flowing pressure vs. downhole flow rate, the PI is represented by the inverse of the slope of the IPR line describing the pressure-rate characteristics of the producing layer (**Fig. 7.8**).

The static reservoir pressure and PI of a reservoir with a single producing layer can be determined with production logging measurements without the need to shut in the well. The well must be flowed at several different flow rates (typically three or four) and allowed to stabilize between successive rate changes. Bottomhole pressure and flow-rate measurements are performed for each value of the surface flow rate. The IPR is drawn through the data points on a pressure vs. rate plot, and extrapolation of the IPR line to a zero-flow condition gives the static pressure.

In most gas wells and in oil wells drawn below the bubblepoint pressure, the IPR may not be linear. Although the same procedure can be used, the IPR shape should fit the curved nature of the data (e.g., a quadratic fit if turbulence is the cause of the nonlinearity of the IPR). Once properly fitted, the y-axis intercept (zero flow) of the modeled IPR gives the static pressure, and the x-axis intercept (atmospheric pressure) gives the absolute open flow (AOF) potential. The AOF potential of a gas well is generally a better indicator of its performance than its PI because PI is not constant and the IPR is represented by a curved line. The AOF of a gas well is determined by plotting the gas potential, $m(p)$, as a function of the flow rate of each flow period of an isochronal test. In some cases of low reservoir pressure, Δp^2 can be used instead of $m(p)$. More information on determining gas flow potential and reserves is in another chapter in this *Handbook.*

The same procedure applies to multilayer completions—plotting bottomhole pressure vs. rate for each layer of the system. To interpret the pressure data of comparable layers, however,

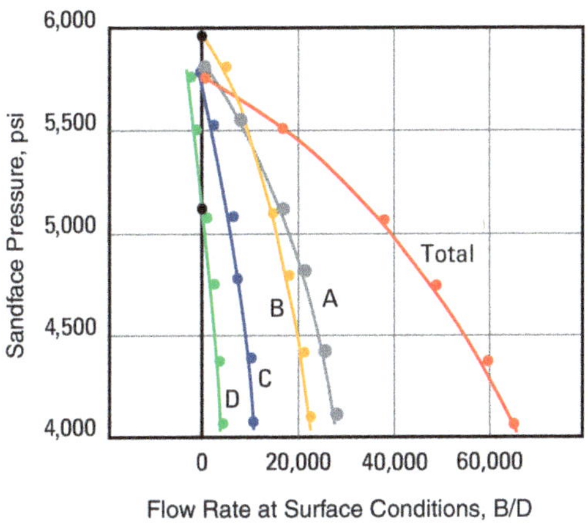

Fig. 7.9—SIP analysis of a four-layer commingled reservoir with crossflow.

the pressures must be corrected to a common arbitrary datum depth to readily differentiate whether the layers belong to the same hydraulic system. The results of the procedure, called selective inflow performance (SIP), include the static pressure and the PI per layer. The SIP procedure has become very popular for commingled producing systems, especially in gas wells because of the shorter stabilization times involved. SIP overcomes a fundamental limitation of commingled producing systems where the layer static pressures are not available by direct measurement, not even by shutting in the well, unless all the reservoir layers are in a strict hydraulic equilibrium.

Fig. 7.9 shows this technique applied to a multilayer reservoir comprising four layers: A, B, C, and D. The "Total" curve represents the global performance of the whole system, intersecting the pressure axis at a value that represents the wellbore pressure when shutting in the well. Obviously, this shut-in pressure differs from the pressure of each of the individual wells because the whole system is not at hydraulic equilibrium. Crossflows develop when shutting in this well, and high-pressure Layers A and B flow into depleted Layers C and D.

7.12.3 Pressure-Depth Plots. Vertically distributed wellbore and formation pressures, such as those measured by a wireline pressure tester, can be used to build mud and reservoir pressure profiles. If the measured interval is sufficiently thick, accurate pressure gradients may be established. As already mentioned, the gradients can in turn be used to spot permeability barriers and reservoir fluid contacts and to determine the reservoir fluid density.

Thick beds have a greater pressure change from top to bottom than thin beds. Therefore, the resolution of the pressure gauge becomes increasingly important the thinner the beds are. Another important factor is the number of pressure measurements taken within the bed of interest. **Fig. 7.10** shows that increasing the number of pressure points greatly reduces the statistical error in determining the true gradient.

In some plots, the recorded pressures may not fall on a linear gradient. One example of this condition is when pressure points are not taken in a uniform depth-increasing or depth-decreasing sequence. This situation favors dispersion of the pressure measurements because of gauge hysteresis and lack of temperature stabilization. A procedure to help determine the reservoir fluid density consists of comparing the fluid density with the mud density over a set of tests taken with a wireline tester. As shown in **Fig. 7.11**, if the fluid pressures vary by Δp_{fl} and the mud pressures vary by Δp_m over the depth interval ΔD and a vertical well is assumed, then the following can be written:

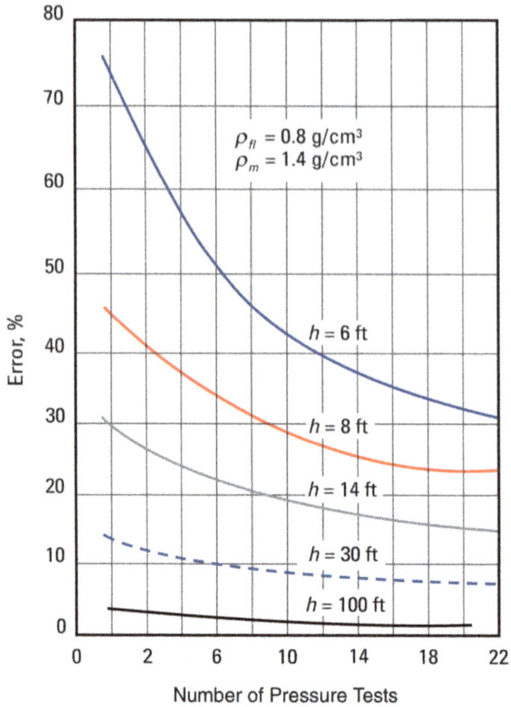

Fig. 7.10—Gradient accuracy vs. bed thickness.

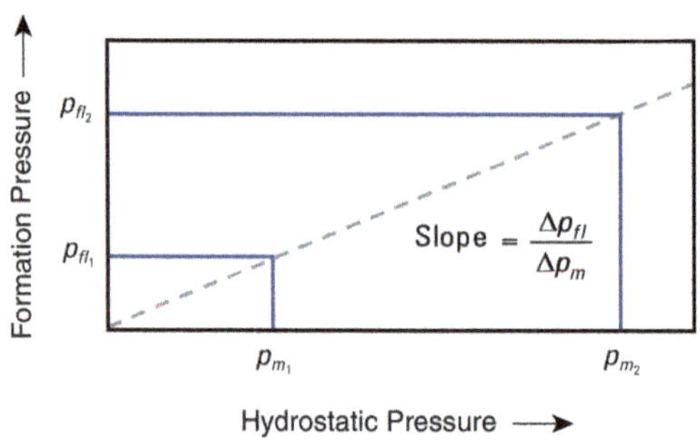

Fig. 7.11—Plot of reservoir (formation) pressure vs. mud (hydrostatic) pressure.

$$\Delta p_m = 0.00135 \rho_m g \Delta D, \quad (7.6)$$

$$\Delta p_{fl} = 0.00135 \rho_{fl} g \Delta D, \quad (7.7)$$

then by elimination:

$$\rho_{fl} = \rho_m \left(\frac{\Delta p_{fl}}{\Delta p_m} \right). \quad (7.8)$$

Because mud pressures are consistent over greater depth intervals, ρ_m is usually known. Eq. 7.8 then can be used to improve the reservoir fluid density determination.

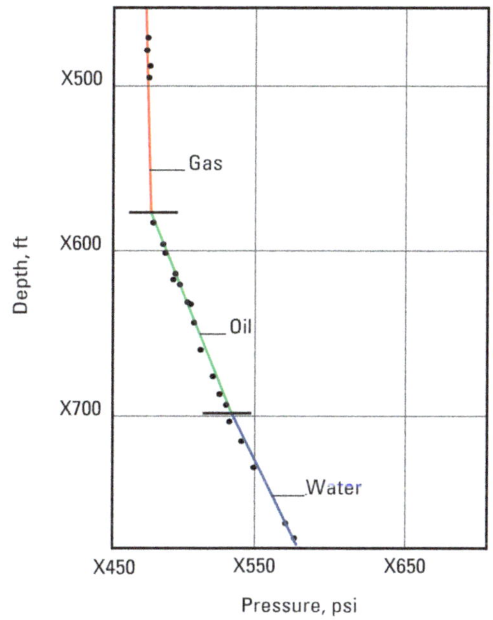

Fig. 7.12—Fluid types and contacts in a virgin reservoir.

Virgin Reservoirs. In virgin reservoirs, the static reservoir pressures are unaffected by fluid withdrawal and the observed gradients therefore reflect the density of the original fluids. The "breaks," where the slope changes in the gradient, reflect the original fluid contacts as illustrated in **Fig. 7.12**.

Permeability barriers can also be identified as illustrated in **Fig. 7.13**. The barrier is indicated in Fig. 7.13a by the hydrostatic potential difference between the layers above and below the detected permeability barrier of approximately 20 psi. The line with a gradient of 0.497 psi/ft represents the mud pressure, which was measured in the same trip in the well while acquiring the formation pressure. In Fig. 7.13b, the reservoir fluid gradients differ across the permeability barrier. Nevertheless, a potential difference of approximately 140 psi across the barrier is interpreted as indicating a no-flow barrier. Zero permeability is implied. Otherwise, the pressure would have equilibrated on both sides of the barrier over geologic time.

Sometimes the gradients must be extrapolated to confirm fluid contacts. The gas/water contacts in **Fig. 7.14** cannot be identified by the pressure profile of Well 1 or Well 2. By extrapolating the water gradient of Well 1 and the gas gradients of Well 2, however, it is possible to determine the position of the gas/water contacts in three zones. This extrapolation shows that pressure readings taken near the wellbore in this case reflect pressures that exist deep within the formation. From the gradient interpretation, the fluid in the upper formation is water, and there are two gas/water contacts in the lower formation.

It is important to note when extrapolating gradients from reservoir pressures in low-permeability reservoirs that the pressures may be affected by supercharging. Supercharging is caused by the nonzero, small value of the mudcake permeability. This permeability allows a finite continuous flow of filtrate across the mudcake. In a low-permeability formation, the resistance to fluid flow created by the mudcake can be on the same order of magnitude as the resistance of the formation to accepting fluid. A standard wireline pressure measurement is therefore insufficient to measure the pressure of the virgin formation because a residual finite pressure difference remains between the formation at the mudcake interface and the virgin formation some distance away. Supercharged points plot to the right of a normal fluid gradient line.

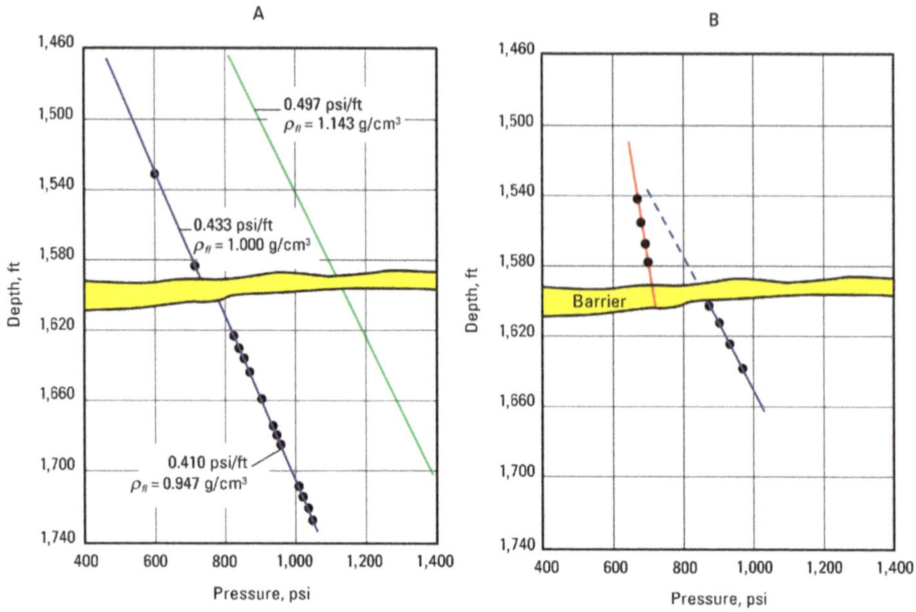

Fig. 7.13—Detecting permeability barriers.

Developed Reservoirs. Differential depletion is most likely to occur in developed reservoirs, destroying the original gradients. In addition, differential depletion generates vertical flow in the reservoir. Vertical flow may also result from partial completion effects that superimpose the corresponding pressure gradients on the fluid density gradients. A typical pressure-depth profile in a well drilled in a field under production is in **Fig. 7.15**. The well was completed in an interval in Zone 1. The pressure profile taken some time after initial completion clearly shows that the pressure in Zone 1 has been drawn down by fluid withdrawal. The pressure in Zones 2, 3, 4, and 5, which were not perforated, has also been affected by vertical flow through the reservoir. The measured gradients reflect the pressure drop created by vertical flow. The sharp pressure drop across Zone 2 reflects the very low permeability of this zone.

In spite of the blurring of fluid gradients in developed reservoirs, vertically distributed reservoir pressures are still useful for correlating formations hydraulically from well to well. The initial correlation made on the basis of openhole logs (left side) in **Fig. 7.16** had to be modified because of the reservoir pressure data. Although time equivalent and present in both Wells 1 and 2, Zones A and B show different pressure regimes in the two wells and are not in hydraulic communication (right side).

7.12.4 Pressure Probes in Duplex or Triplex. Taking pressure points with a multiple-probe wireline tester eliminates the uncertainty of the depth measurement for the set of points taken at a tool station. Modern wireline testers include a multiple-probe system that can measure pressure at a sink, or flowing, probe, at the same depth at a "horizontal probe" opposite the sink probe, and at a "vertical probe" at some vertical distance on a generatrix (i.e., parallel to the tool axis) with the sink probe. Both data density and data consistency increase greatly when this probe arrangement is used.

7.12.5 Effect of Capillary Pressure. Several studies have shown that a wireline formation tester actually measures the pressure of the continuous phase in the invaded region around a wellbore; typically this is the drilling fluid filtrate. The measured tester pressure is thus differ-

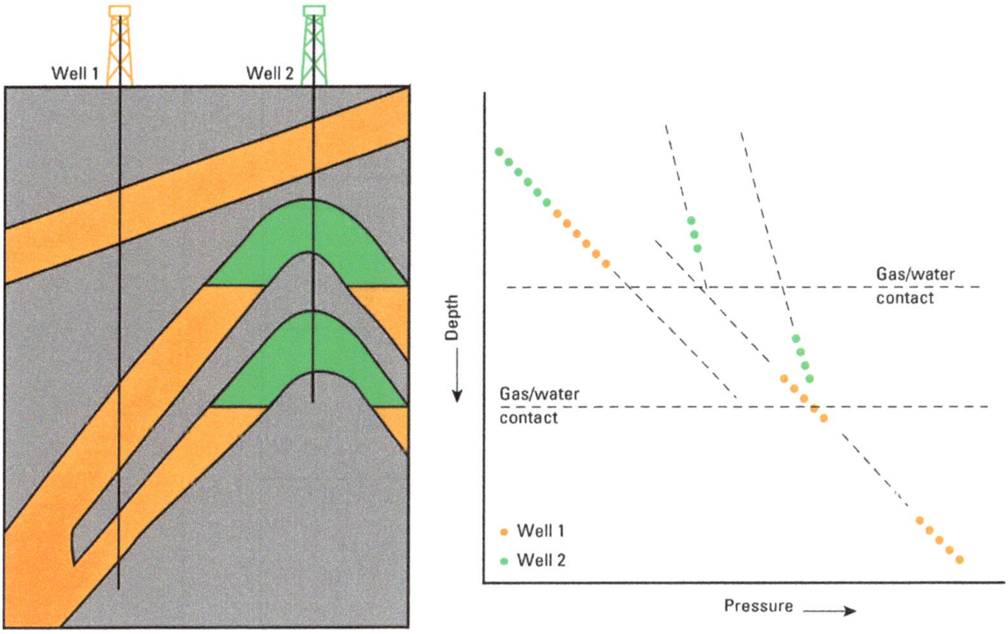

Fig. 7.14—Extrapolation of pressure gradients to fluid contact levels.

ent from the reservoir pressure by the amount of capillary pressure. The capillary pressure affects the saturation of the wetting phase in the reservoir. The combined effects of rock wettability and capillary pressure can be reflected as changes in the pressure gradient, fluid contact level, or both on pressure-depth profiles, especially those recorded with oil-base mud in the borehole.[6,7]

The principles, measurement techniques, and applications of capillary pressure measurements are discussed in another chapter in this *Handbook*. Refer to that chapter for information on the variation of water saturation with height, entry or threshold pressure, free-water level (FWL), and oil/water contact (OWC).

The first and most conspicuous effect of capillary pressure on wireline tester pressure gradient profiles is the creation of a break in the gradient at the FWL that may not coincide with the OWC interpreted from other measurements such as resistivity logs. The depth difference increases as the displacement pressure (a function of pore-throat diameter) increases. Fine-grained reservoirs with small pore throats are most likely to exhibit potentially large depth differences between the OWC and FWL.

The second, potentially more deleterious effect of capillary pressure on wireline tester pressure gradient profiles is that the measured pressure may differ from the true formation pressure. The difference results in an unrecognized shift of the gradient to the right or left of the true reservoir fluid gradient. The effect of the gradient shift, which is equal to the amount of capillary pressure, is to displace the observed break in the gradient. Interpreting this break as the FWL yields an erroneous depth, located either above or below the true FWL, depending on the conditions described next.

Fig. 7.17 presents pressure-depth plots and capillary pressure profiles in a water-wet reservoir drilled with water-base mud and oil-base mud. Fig. 7.17c shows the capillary pressure profile expected in a water-wet oil-bearing section of the reservoir. The capillary pressure is greater in the virgin zone because oil is the nonwetting fluid. The wireline tester measures a filtrate pressure in the invaded zone that is lower because of the absence of capillary pressure.

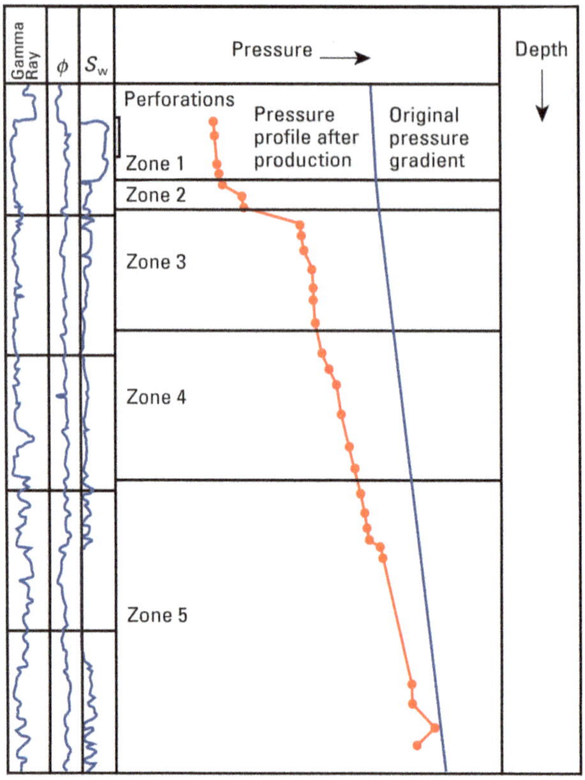

Fig. 7.15—Pressure profile of a well in a developed field.

The result is a gradient shift to lower pressure values and the FWL is interpreted above its true location. There is no shift in the water-bearing section of the reservoir because the capillary pressure is zero in both the virgin zone and invaded zone.

When oil-base mud is used in a water-wet reservoir, the effect of the capillary pressure causes the measured pressures to differ from the true formation pressures only in the water-bearing section of the reservoir (Fig. 7.17b). Fig. 7.17d shows the capillary profile in the water-bearing section of the reservoir for this case.

Similar data and effects for water- and oil-base muds used in oil-wet reservoirs are shown in **Fig. 7.18**.

One possible method to correct for wettability and capillary pressure effects on wireline formation tester pressures is the Leverett J-function[8]:

$$J(S_w) = p_c \frac{\sqrt{k/\phi}}{\sigma \cos \delta} \text{ . } \qquad (7.9)$$

Laboratory measurements of p_c, k, and Φ are used to develop a relationship for a reservoir. The amount of capillary pressure determined by the J-function is added to the measured pressure:

$$p_{\text{corrected}} = p_{\text{measured}} + p_c(S_{xo}), \qquad (7.10)$$

where $p_c(S_{xo})$ is the capillary pressure in the filtrate-invaded zone, for which the water saturation is traditionally called S_{xo}.

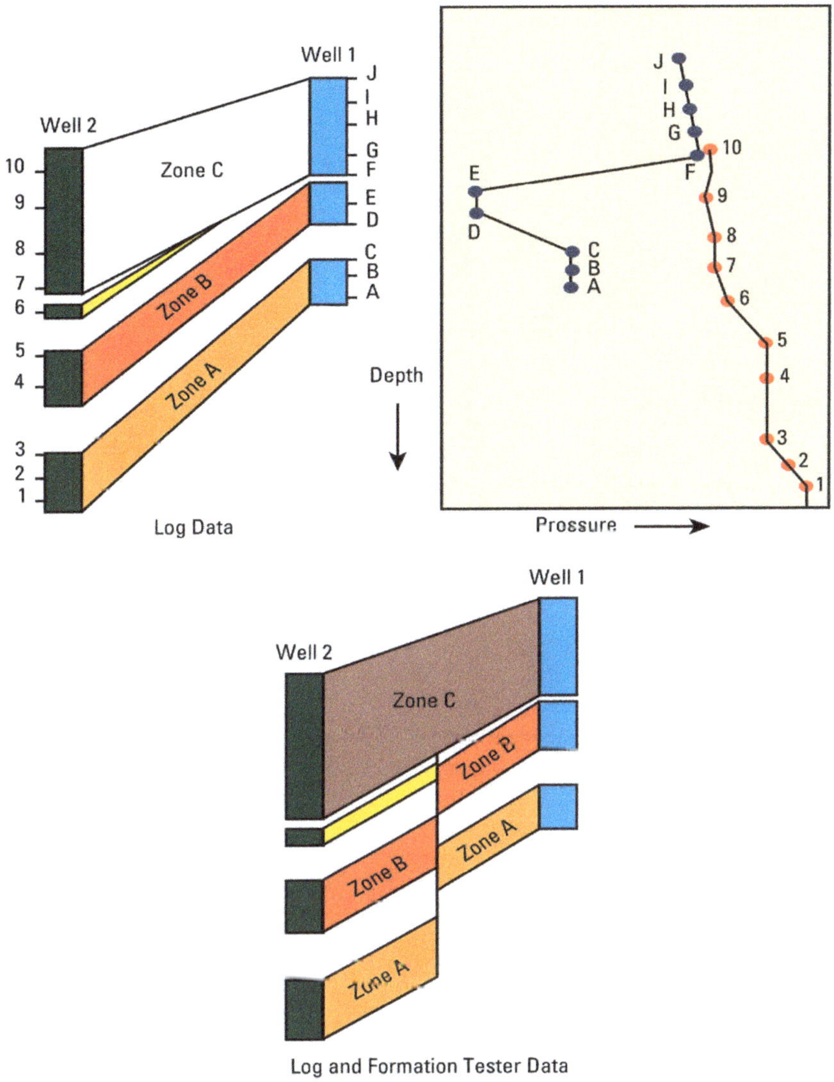

Fig. 7.16—Hydraulic correlation between two wells.

Alternatively, if a nuclear magnetic resonance log (NMR) is available, the in-situ capillary pressure correction can be performed directly. NMR logs have the capability to model the pore-size distribution. The method also makes use of laboratory experiments on cores to calibrate the correction.[9,10]

7.13 Pressure Transient Testing

The interpreted pressure transient test is a primary source of dynamic reservoir data. Tests on oil and gas wells are performed at various stages of drilling, completion, and production. The test objectives at any stage range from simple measurement of reservoir pressure to complex characterization of reservoir features. Most pressure transient tests can be classified as either single-well productivity tests or descriptive reservoir tests.

Productivity tests are conducted to determine well deliverability, characterize formation damage and other sources of skin effect, identify produced fluids and determine their respective

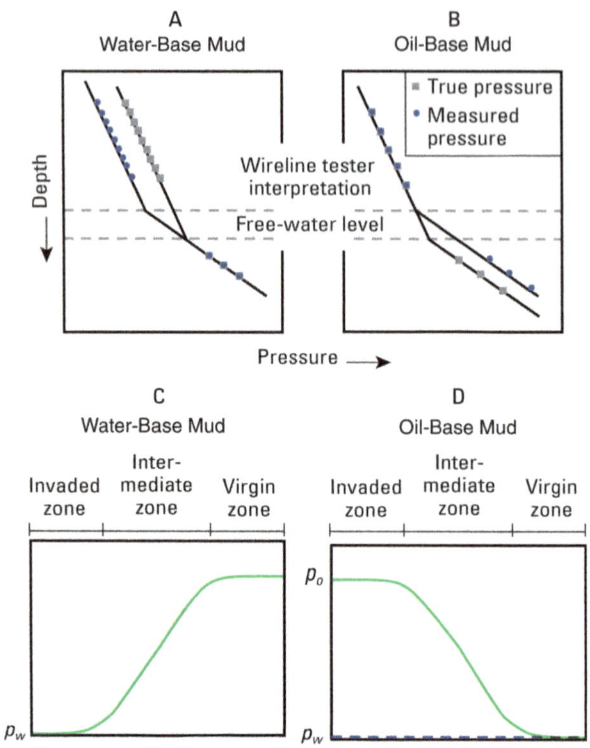

Fig. 7.17—Pressure-depth plots and capillary pressure profiles—water-wet reservoirs.

volume ratios, measure reservoir pressure and temperature, obtain representative fluid samples suitable for PVT analysis, evaluate completion efficiency, and evaluate workover or stimulation treatments.

Descriptive reservoir tests are conducted to assess reservoir extent and geometry, determine hydraulic communication between wells, characterize reservoir heterogeneities, and evaluate reservoir parameters.

Detailed information on the use and interpretation of transient pressure data is presented in another chapter in this *Handbook.* Some special analytical techniques follow.

7.13.1 Pressure Flow Convolution and Deconvolution. The pressure-flow convolution involves simultaneous bottomhole flow rate and pressure measurements to correct for the variations of bottomhole pressure caused by flow rate fluctuations during drawdown tests.[11]

The bottomhole pressure and flow rate are mathematically convolved (coupled) as follows:

$$p_D(t_D) = \int_0^{t_D} q_D(\tau) p'_D(t_D - \tau) d\tau, \quad (7.11)$$

where p_D, the pressure function equivalent to a constant flow rate situation, is obtained by mathematical deconvolution of the pressure from the flow-rate fluctuations. When software deconvolution operators are used, trial and error is required to convolve a flow-rate schedule with a pressure function that approximates the true constant rate-equivalent pressure function, thus reproducing the measured pressures. The process can be made to converge rather rapidly for a

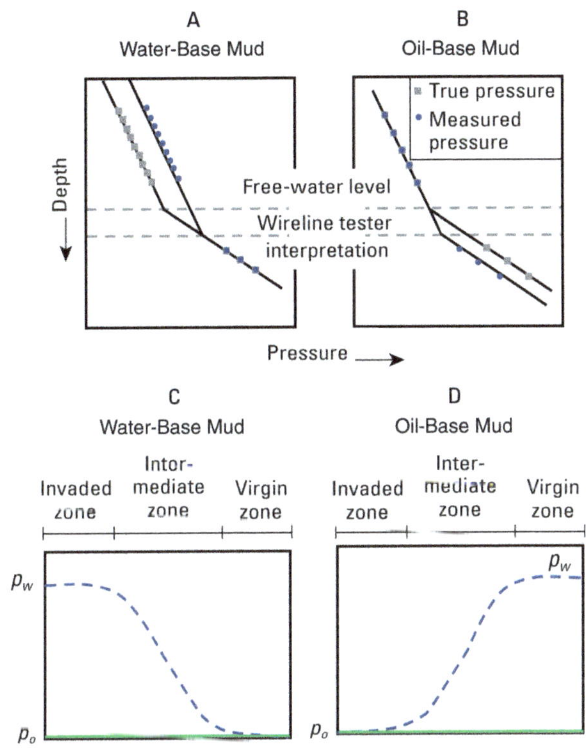

Fig. 7.18—Pressure-depth plots and capillary pressure profiles—oil-wet reservoirs.

pressure measurement of a given resolution, as long as the results allow for an acceptable margin of error.

Fig. 7.19 shows an example in which the transient consists of a step-rate change from a high value with a downhole spinner flowmeter rotation rate of approximately 17 revolutions per second (rps) to a lower value with a flowmeter response of approximately 7 rps. Clearly, the pressure and flow-rate data mirror each other, which is precisely the effect of the convolution. A constant flow-rate function was sought to interpret this test. The technique used here makes use of semilog analysis, in which rate-normalized pressures are plotted vs. the "sandface convolution time" (a time function akin to a generalized superposition function). The result (**Fig. 7.20**) is a straight line on the semilog plot, which in turn can be interpreted to yield the test objectives of the permeability and skin effect.

7.13.2 Benefits of Downhole Shut-In. **Fig. 7.21** shows superimposed log-log plots for two buildup tests run on the same well. The surface shut-in test barely reaches radial flow after 200 hours. However, a large fraction of the wellbore volume is eliminated in the downhole shut-in test, and consequently radial flow is detected almost as early as the first minute after shut-in, and confirmed after 1 hour. This test, which lasts 100 hours, could well have been aborted after a maximum of 5 hours without any loss of information.

7.13.3 Multilayer Tests. To interpret tests when several layers are producing in a commingled environment, a generalization of the pressure-flow convolution is used. Conventional well tests performed on commingled multilayer reservoir systems normally do not yield interpretable data. The different dynamic reservoir parameters (i.e., *kh*, skin effect, static pressure, boundary condition, heterogeneity) of each layer induce off-phase flow rate events in the layer that do

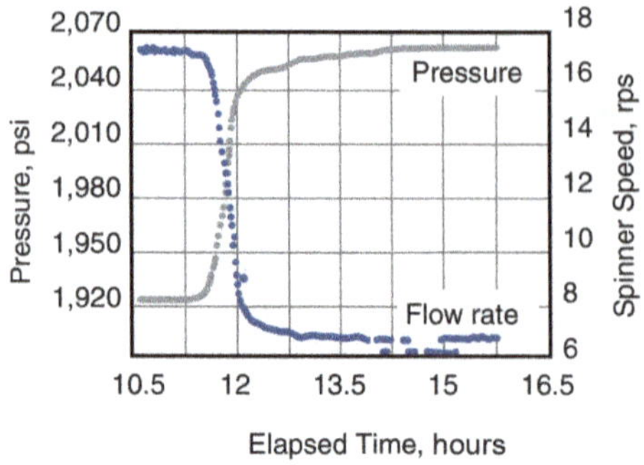

Fig. 7.19—Pressure and spinner data in a step-rate change test.

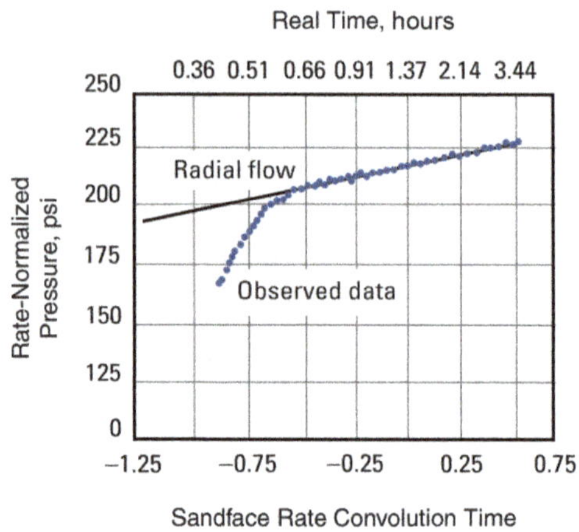

Fig. 7.20—Interpretation of test data in Fig. 7.19.

not superimpose themselves to yield a predictable sandface pressure response. By using simultaneous bottomhole pressure and flow rate measurements and designing the drawdown test as a succession of step-rate tests (**Fig. 7.22**), a rigorous solution to deriving the dynamic reservoir parameters can be obtained for each layer.[12,13]

The following steps describe a typical design for a three-layer multirate test:

1. The well is shut in, and the pressure and flow sensors (typically conveyed by a production logging tool) are positioned above the top of the uppermost layer. The well is opened to the smallest choke opening, and the ensuing transients of rate and pressure are recorded until stabilization occurs. Finally, a continuous flow profile is recorded across the set of producing layers.

2. The pressure and flow sensors are repositioned above the top of the middle layer. The well is opened to the intermediate choke opening, and the ensuing transients of rate and pressure are recorded until stabilization occurs. A second continuous flow profile is recorded across the set of producing layers.

3. The pressure and flow sensors are repositioned above the top of the lowermost layer. The well is opened to the largest choke opening, and the ensuing transients of rate and pressure are recorded until stabilization occurs. A third continuous flow profile is recorded across the set of producing layers.

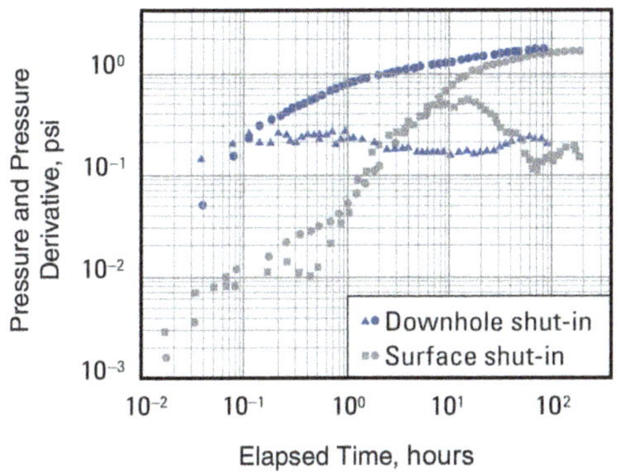

Fig. 7.21—Compared pressures of a surface shut-in and downhole shut-in test.

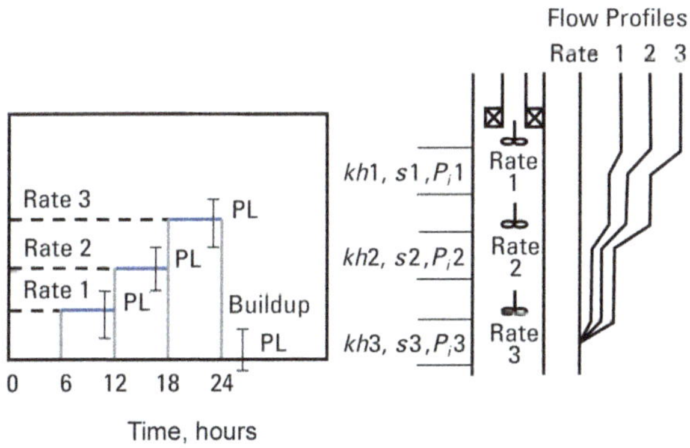

Fig. 7.22—Typical design for a three-layer multirate test (PL = production logging).

4. In a last, optional step, the pressure and flow sensors are repositioned above the top of the uppermost layer and the well is shut in again. The observed transients of rate and pressure are recorded as in a traditional buildup test.

The interpretation of this data set (which includes SIP data) makes extensive use of the pressure-flow convolution to extract the individual layer parameters. After the results are obtained, it is advised to verify their quality by forward-simulating the commingled pressure and flow response of the layered system and by matching the simulated responses to the measured data. A single-well numerical simulator is used with the layered system described by the interpreted values of permeability and skin effect for each layer. The surface flow rate schedule is input, and the simulator predicts the commingled pressure response of the system as well as the individual layer flow history for the entire test, which must match the measured downhole pressure and flow rate records.

7.13.4 Wireline Pressure Transient Tests. Some interpretation techniques are unique to wireline testers because of the specific hardware used to perform the tests. Wireline testers investigate a smaller region around the wellbore because of the smaller volumes flowed. Wireline pressure testing offers unique advantages over drillstem testing, however, because of the variety of options available in the downhole hardware configuration, multiprobe arrangements, and packer devices. Stewart and Wittmann first described some salient techniques specific to wireline pressure testing in 1979.[14]

In wireline pressure testing, the static pressure is measured by shutting in the sampling system after retrieving a small sample, typically 5 to 20 cm^3. The subsequent buildup duration is short, and the stabilized static pressure is typically obtained within a few seconds to a maximum of approximately 30 minutes.

In low-permeability situations, the buildup may take much longer. Continued testing with the tool hanging stationary at the same depth, firmly seated against the formation, may be impractical. In addition, pressure measurements may be affected by the supercharging phenomenon as described previously, resulting in understated pressures.

Packer Probe Tests: Small-Scale Drillstem Testing. A packer probe fitted into the string of a modern wireline tester increases the area of the formation open to the flow during formation sampling, typically by a factor of up to several thousand. This increase multiplies flow rates by the same factor, which in turn greatly increases the depth of investigation. In some cases, a packer probe test has a depth of investigation similar to that of a small-scale DST.

Packer and Multiple-Probe Tests for Vertical Interference Testing. A packer probe can be used in tandem with a vertical probe to test for vertical permeability.[15] The vertical probe is located on a generatrix (parallel to the tool axis) with the sink (packer) probe of the downhole tool. The distance of the vertical probe from the sink probe is adjustable. Whereas the pressure response at the sink probe depends on the local values of the permeability tensor, called k_x, k_y, and k_z, which are the permeabilities along arbitrary axes x, y, and z, respectively, the pressure response at the vertical probe (which is considered an observation probe) is a function of both the horizontal permeability at the vertical probe and the vertical permeability being measured. Thus, both pressure responses must be modeled simultaneously by a numerical parameter estimator.

Fig. 7.23 depicts the results of a tandem test. The dots are pressure measurements and the dashed curves are the pressures reconstructed from probe responses calculated from the interpretation results. Fig. 7.23a shows the response at the sink (packer) probe, and Fig. 7.23b shows the response at the vertical probe, which was set approximately 1 hour after the packer was set. The test sequence included a number of open-close cycles generated at the sink probe by the use of a flow-control module. A sample was also taken between times 2,800 and 3,800 seconds. The vertical probe response clearly shows the delayed interference response that occurred after that probe was set. From this data set, the horizontal mobility k_h/μ was calculated as 1.0 md/cp, and the vertical mobility k_v/μ was calculated as 0.3 md/cp. **Fig. 7.24** shows the log-log plot of the buildup between times 3,800 and 4,700 seconds that occurred after the sample was taken.

7.13.5 NODAL Analysis. The techniques and uses of NODAL* production system analysis are explained in another chapter in this *Handbook*. The objective of NODAL analysis is to predict well-producibility characteristics, also referred to as vertical lift properties (VLP), for various tubular and pressure configurations. If the pressure data are limited to sandface and wellhead measurements, the normal procedure is to generate several sets of VLP characteristics and select the one that best represents the measured pressure data. There may not be a unique solution. Recording a continuous profile of pressure vs. depth can alleviate nonuniqueness because the profile constitutes a precise measurement of the multiphase pressure losses that take place in a well. Using a continuous profile for input leads to better optimization of production rates with NODAL analysis.

NODAL analysis, aided by distributed pressure measurements, is the best way to design gas-lift systems. Gas-lift valve placement involves matching the pressure drop in the valves with the amount of pressure available in the well above the valve opening pressure. The pressure drop in the tubing, in turn, depends on the location and flow capacity of the valves.

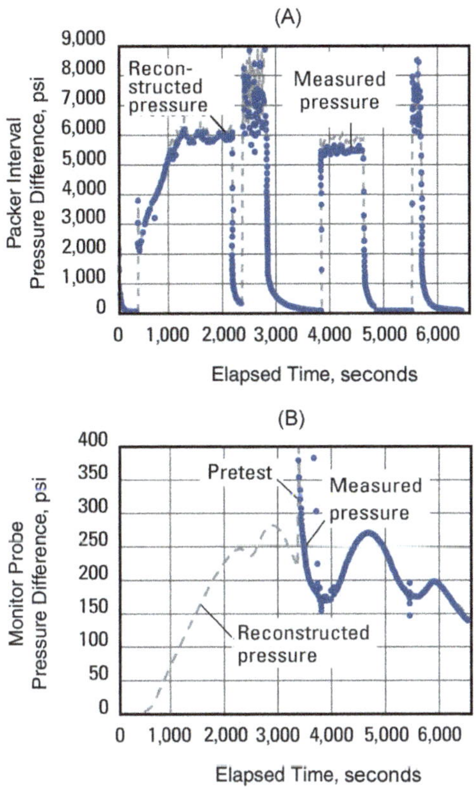

Fig. 7.23—Sink probe (a) and vertical probe (b) pressure responses during a vertical interference test conducted with wireline tester.

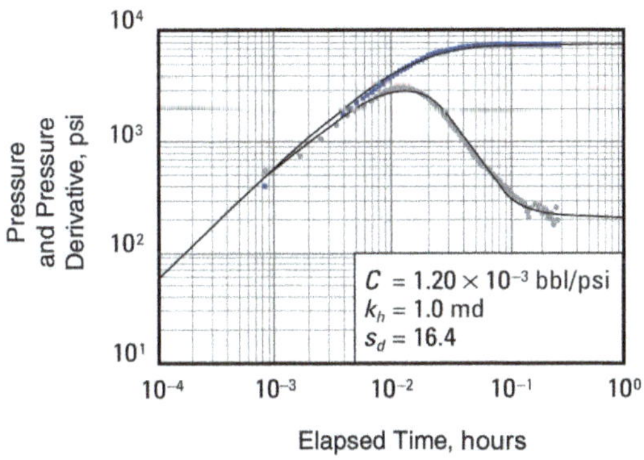

Fig. 7.24—Log-log plot of post-sampling buildup from the vertical interference test of Fig. 7.23.

7.13.6 Using Pressure To Characterize Reservoir Fluids. Pressure and temperature provide important information about the phase behavior and calibration of the equation-of-state for a fluid and average fluid density in flowing wells.

The average fluid density can be calculated by differentiating the pressure measurement vs. depth. In the absence of fluid friction on pipes, the acceleration and kinetic terms can be written as follows:

$$\frac{dp}{dD} = 0.00135\rho_{fl} g \cos \delta \ . \qquad (7.12)$$

In a well flowing above the bubblepoint, the bubblepoint pressure can be inferred from a plot of the fluid density in the tubing. At the depth where the pressure reaches bubblepoint pressure, gas starts evolving from solution, and the density of the fluid shows a break to lower values. Density can be measured by differential pressure measurements.

Similarly, the pressure gradient in wet gas wells shows a break when the dewpoint is reached and condensate forms.

7.13.7 Temperature Profiles in Production and Injection Wells. All the mass-transfer processes taking place in and around a wellbore produce changes in the wellbore temperature. Measuring the wellbore temperature is a good diagnostic tool for applications such as identifying fluid entries into and exits from the wellbore, monitoring exothermic reactions such as cement hydration, determining the effects of temperature change on compression or decompression (Joule-Thompson effects), detecting the movement of fluids behind the casing, and identifying nongeothermal fluid entries into the wellbore. Another chapter in this section of the *Handbook* explains the use and interpretation of these data.

Recommendations for Temperature Profiling. To obtain good-quality temperature profile data, the following procedures are recommended.

1. Record a complete profile from surface to total depth (bottom of the well) on the first descent into the well. If the well is shut in, the thermal equilibrium becomes disrupted after the first passage of the temperature sensor, and unrecorded temperature anomalies may be lost forever. If the well is flowing, the first descent is a unique opportunity to diagnose leaks, spurious flow, or loss of completion integrity.
2. It may be possible to record a representative geothermal gradient if the well is shut in.
3. Record shut-in profiles if possible. Always compare shut-in profiles with the flowing profiles.
4. Repeat all runs.
5. In dual completions, run the temperature log in both tubing strings because the two logs are not identical.
6. Use short depth scales for presentation. They highlight temperature anomalies better than large depth scales.
7. Always interpret temperature logs together with flowmeter data.

Detecting Cement Tops. Cement hydration is an exothermic reaction that generates sufficient heat for determining the presence of cement behind a casing string by a temperature survey up to several days after cementing. The character of the anomaly above the cement top may be a large, sharp increase, in some cases up to 50°F, or a very slight increase in gradient.

The principal influence on the survey is the time elapsed between placing the cement and running the survey. Other influential conditions include cement texture, chemical composition, rate of hydration, mass of cement in place, and the thermal conductivity of the adjacent formation. The maximum temperature usually occurs 4 to 9 hours after cementing, but reliable data can be determined in most areas after 48 hours. The rate of hydration affects temperature change more than the total amount of heat liberated. Although hydration continues indefinitely, the rate decreases rapidly from the peak. A washed-out section of hole may be responsible for a large, sharp increase in temperature that falsely indicates a cement top. A small temperature change or slight change in gradient could be caused by a small annular area or dilution of the cement with drilling mud. These factors, which influence the size of the temperature anomaly at the top of the cement in a well, vary widely in their effect. Even for an unfavorable combination of factors, however, sufficient heat is typically generated to determine the cement top.

Vertical Extent of Fracturing and Detecting Lost Circulation. The temperature of fluids and solids injected during a frac job is low relative to that of the formation which causes anomalies in the geothermal profile. This effect also applies to lost circulation zones that receive excessive amounts of drilling mud. Diagnosis of these anomalies with temperature surveys can supply quantitative data on the fracture size and amount of mud lost.

Nomenclature

A_{ij} = pressure calibration coefficient of orders *i* and *j*, dimensionless
A_{ij}' = temperature calibration coefficient of order *i* and *j*, dimensionless
B = formation volume factor, RB/STB
C = wellbore storage coefficient, bbl/psi
D = depth along the wellbore, ft
D_0 = depth along wellbore of the reference datum, ft
g = acceleration of earth's gravity, ft/sec^2
h = reservoir thickness, ft
J = productivity index, B/D/psi
$J(S_w)$ = Leverett J-function, dimensionless
k = reservoir permeability, md
k_h = reservoir horizontal permeability, md
k_x = permeability along arbitrary axis x, md
k_y = permeability along arbitrary axis y, md
k_z = permeability along arbitrary axis z, md
$m(p)$ = real gas pseudopressure, psi^2/cp
p = pressure, psi
p_c = pressure calibration function, dimensionless
$p_c(S_{xo})$ = capillary pressure in filtrate invaded zone, psi
p_D = observed pressure response at the wellbore, dimensionless
p_D' = derivative of the sought constant rate pressure response, dimensionless
p_{fl} = fluid pressure, psi
p_i = original reservoir pressure, psi
p_m = mud pressure, psi
p_o = oil-phase pressure, psi
p_r = reservoir pressure, psi
p_w = water-phase pressure, psi
p_{wf} = bottomhole flowing pressure, psi
q = flow rate, STB/D
$q_D(\tau)$ = variable sandface flow rate, dimensionless
Q = downhole flow rate, B/D
r = distance from wellbore axis, ft
r_e = external boundary radius of the well, ft
r_w = wellbore radius, ft
s = skin effect, dimensionless
s_d = skin effect due to damage, dimensionless
S_p = pressure output of a pressure gauge, psi
S_{po} = pressure output offset of a pressure gauge, psi
S_t = temperature output of a pressure gauge, °F
S_{to} = temperature output offset of a pressure gauge, °F
S_w = water saturation, fraction

t_D = dimensionless time
t_k = circulation time, hr
T_{bh} = observed temperature in a well, °F
T_c = temperature calibration function, dimensionless
δ = deviation of the well, assumed constant between D and D_0, deg
ΔD = vertical depth differential, ft
Δp = pressure change, psi
Δp_{fl} = fluid pressure differential, psi
Δp_m = mud pressure differential, psi
Δt_c = time after circulation, hr
μ = fluid viscosity, cp
ρ = reservoir fluid density, g/cm^3
ρ_{fl} = reservoir fluid density, g/cm^3
ρ_m = mud density, g/cm^3
σ = surface tension, psi.ft
τ = integration variable, dimensionless
ϕ = reservoir porosity, fraction
Φ = hydrostatic potential, psi

References

1. Dowdle, W.L. and Cobbs, W.M.: "Static Formation Pressure From Well Logs—An Empirical Method," *JPT* (November 1975) 1326.
2. Veneruso, A., Ehlig-Economides, C.A., and Petitjean, L.: "Pressure Gauge Specification Considerations in Practical Well Testing," paper SPE 22752 presented at the 1991 SPE Annual Conference and Exhibition, Dallas, 6–9 October.
3. Colley, N. *et al.:* "The MDT Tool: A Wireline Testing Breakthrough," *Schlumberger Oilfield Review* (April 1992) 58.
4. Davies, J. *et al.:* "Spinners Run While Perforating," paper SPE 38549 presented at the 1997 SPE Offshore Europe Conference, Aberdeen, 9–12 September.
5. Eck, J. *et al.:* "Downhole Monitoring: The Story So Far," *Schlumberger Oilfield Review* (Winter 1999) 20.
6. Elshahawi, H., Fathy, K., and Hiekal, S.: "Capillary Pressure and Rock Wettability Effects on Wireline Formation Tester Measurements," paper SPE 56712 presented at the 1999 SPE Annual Technical Conference and Exhibition, Houston, 3–6 October.
7. Elshahawi, H., Samir, M., and Fathy, K.: "Correcting for Wettability and Capillary Pressure Effects on Formation Tester Measurements," paper SPE 63075 presented at the 2000 SPE Annual Technical Conference and Exhibition, Dallas, 1–4 October.
8. Leverett, M.C.: "Capillary Behavior in Porous Solids," *Trans.,* AIME (1941) **142,** 152.
9. Lowden, B.: "Some Simple Methods for Refining Permeability Estimates From NMR and Generating Capillary Pressure Curves," *DiaLog, The On-Line Newsletter of the London Petrophysical Society* (March 2000) **8**, No. 1, http://www.lps.org.uk/site_pages/dialog.htm#top.
10. Marschall, D. *et al.*: "Method for Correlating NMR Relaxometry and Mercury Injection Data," *Trans.,* Intl. Symposium of the SCA, San Francisco, 12–15 September.
11. Joseph, J., Ehlig-Economides, C.A., and Kuchuk, F.: "The Role of Downhole Flow and Pressure Measurements in Reservoir Testing," paper SPE 18379 presented at the 1988 SPE European Petroleum Conference, London, 16–18 October.
12. Kuchuk, F., Karakas, M., and Ayestaran, L.: "Well Test Analysis of Commingled Zones Without Crossflow," paper SPE 13081 presented at the 1984 SPE Annual Technical Conference and Exhibition, Houston, 16–19 September.

13. Ayestaran, L. *et al.*: "Layered Reservoir Testing," *Schlumberger Technical Review* (October 1987) **35,** No. 4, 4.
14. Stewart, G. and Wittmann, M.: "Interpretation of the Pressure Response of the Repeat Formation Tester," paper SPE 8362 presented at the 1979 SPE Annual Technical Conference and Exhibition, Las Vegas, Nevada, 23–26 September.
15. Head, E.L. and Bettis, F.E.: "Reservoir Anisotropy Determination With Multiple Probe Pressures," *JPT* (December 1993) 1177.

SI Metric Conversion Factors

bbl	×	1.589 873	E − 01	= m^3
cp	×	1.0*	E − 03	= Pa·s
ft	×	3.048*	E − 01	= m
°F		(°F − 32)/1.8		= °C
in.^{3}	×	1.638 706	E + 01	= cm^3
psi	×	6.894 757	E + 00	= kPa
psi^2	×	4.753 8	E + 01	= kPa^2

*Conversion factor is exact.

Chapter 8
Fluid Flow Through Permeable Media

John Lee, Texas A&M U.

8.1 Overview

This chapter discusses fluid flow in petroleum reservoirs. Basic concepts, which include flow equations for unsteady-state, pseudosteady-state, and steady-state flow of fluids, are discussed first. Various flow geometries are treated, including radial, linear, and spherical flow. The pseudosteady-state equations provide the basis for a brief discussion of oil well productivity, and the unsteady-state equations provide the basis for a lengthy discussion of pressure-transient test analysis. For pressure-transient test analysis, semilog techniques, type curves, damage and stimulation, modifications for gases and multiphase flow, the diagnostic plot, bounded reservoirs, average pressure in the drainage area, hydraulically fractured wells, and naturally fractured reservoirs are included. The chapter also discusses transient and stabilized flow in horizontal wells and gas-well deliverability tests. It concludes with considerations of coning in vertical and horizontal wells.

8.2 Basic Concepts

8.2.1 The Ideal Reservoir Model. Many important applications of fluid flow in permeable media involve 1D, radial flow. These applications are based on a model that includes many simplifying assumptions about the well and reservoir. These assumptions are introduced as needed to combine the law of conservation of mass, Darcy's law, and equations of state to achieve our objectives.

Consider radial flow toward a well in a circular reservoir. Combining the law of conservation of mass and Darcy's law for the isothermal flow of fluids of small and constant compressibility yields the radial diffusivity equation,[1]

$$\frac{\partial^2 p}{\partial r^2}+\frac{1}{r}\frac{\partial p}{\partial r}=\frac{\phi \mu c}{0.0002637k}\frac{\partial p}{\partial t} \quad \text{(8.1)}$$

In the derivation of this equation, it is assumed that compressibility of the total system, c_t, is small and independent of pressure; permeability, k, is constant and isotropic; viscosity, μ, is independent of pressure; porosity, ϕ, is constant; and that certain terms in the basic differential equation (involving pressure gradients squared) are negligible. The grouping $0.0002637k/\phi\mu c_t$ is called the hydraulic diffusivity and is given the symbol η.

8.2.2 Line-Source Solution to the Diffusivity Equation. Assume that a well produces at constant reservoir rate, qB; the well has zero radius; the reservoir is at uniform pressure, p_i, before production begins; and the well drains an infinite area (i.e., that $p \rightarrow p_i$ as $r \rightarrow \infty$). Under these conditions, the solution to Eq. 8.1 is[1]

$$p = p_i + 70.6\frac{qB\mu}{kh}Ei\left(\frac{-948\phi\mu c_t r^2}{kt}\right), \quad (8.2)$$

where p is the pressure at distance r from the well at time t, and

$$Ei(-x) = -\int_x^\infty \left(\frac{e^{-u}}{u}\right)du, \quad (8.3)$$

the Ei function or exponential integral.

The Ei-function solution is an accurate approximation to more exact solutions to the diffusivity equation (solutions with finite wellbore radius and finite drainage radius) for $3.79 \times 10^5 \phi\mu c_t r_w^2/k < t < 948\phi\mu c_t r_e^2/k$. For smaller times, the assumption of zero well size (line source or sink) limits the accuracy of the equation; for larger times, the reservoir's boundaries affect the pressure distribution in the reservoir, so that the reservoir is no longer infinite acting.

For the argument, x, of the Ei function less than 0.01, the Ei function can be approximated with negligible error by

$$Ei(-x) = \ln(1.781x). \quad (8.4)$$

For $x > 10$, the Ei function is zero for practical applications in flow through porous media. For $0.01 < x < 10$, Ei functions are determined from tables or subroutines available in appropriate software.[4]

8.2.3 Altered Zone and Skin Factor. In practice, most wells have reduced permeability (damage) near the wellbore resulting from drilling or completion operations. Many other wells are stimulated by acidization or hydraulic fracturing. Eq. 8.2 fails to model such wells properly. Its derivation includes the explicit assumption of uniform permeability throughout the drainage area of the well up to the wellbore. Hawkins[2] pointed out that if the damaged or stimulated zone is considered equivalent to an altered zone of uniform permeability. k_s, and outer radius, r_s, the additional pressure drop, Δp_s, across this zone can be modeled by the steady-state radial flow equation

$$\Delta p_s = 141.2\frac{qB\mu}{k_s h}\ln\left(\frac{r_s}{r_w}\right) - 141.2\frac{qB\mu}{kh}\ln\left(\frac{r_s}{r_w}\right)$$

$$= 141.2\frac{qB\mu}{kh}\left(\frac{k}{k_s} - 1\right)\ln\left(\frac{r_s}{r_w}\right). \quad (8.5)$$

Eq. 8.5 states that the pressure drop in the altered zone is inversely proportional to k_s rather than to k and that a correction to the pressure drop in this region must be made. Combining Eqs. 8.2 and 8.5, we find that the total pressure drop at the wellbore is

$$p_i - p_{wf} = -70.6\frac{qB\mu}{kh}Ei\left(-\frac{948\phi\mu c_t r_w^2}{kt}\right) + \Delta p_s$$

$$= -70.6\frac{qB\mu}{kh}\left[Ei\left(-\frac{948\phi\mu c_t r_w^2}{kt}\right) - 2\left(\frac{k}{k_s} - 1\right)\ln\left(\frac{r_s}{r_w}\right)\right]. \quad \text{(8.6)}$$

For $r = r_w$, the argument of the Ei function is sufficiently small after a short time that we can use the logarithmic approximation; thus, the drawdown is

$$p_i - p_{wf} = -70.6\frac{qB\mu}{kh}\left[\ln\left(\frac{1688\phi\mu c_t r_w^2}{kt}\right) - 2\left(\frac{k}{k_s} - 1\right)\ln\left(\frac{r_s}{r_w}\right)\right]. \quad \text{(8.7)}$$

We can conveniently define a dimensionless skin factor, s, in terms of the properties of the equivalent altered zone:

$$s = \left(\frac{k}{k_s} - 1\right)\ln\left(\frac{r_s}{r_w}\right) = \frac{kh\Delta p_s}{141.2qB\mu}. \quad \text{(8.8)}$$

Thus, the drawdown is

$$p_i - p_{wf} = -70.6\frac{qB\mu}{kh}\left[\ln\left(\frac{1688\phi\mu c_t r_w^2}{kt}\right) - 2s\right]. \quad \text{(8.9)}$$

Eq. 8.9 provides some insight into the physical significance of the algebraic sign of the skin factor. If a well is damaged ($k_s < k$), s will be positive, and the greater the contrast between k_s and k and the deeper into the formation the damage extends, the larger the numerical value of s, which has no upper limit. Some newly drilled wells will not flow before stimulation; for these wells, $k_s = 0$ and $s \to \infty$. If a well is stimulated ($k_s > k$), s will be negative, and the deeper the stimulation, the greater the numerical value of s. Rarely does a stimulated well have a skin less than –7, and such skin factors arise only for wells with deeply penetrating, highly conductive hydraulic fractures. If a well is neither damaged nor stimulated ($k = k_s$), $s = 0$.

The altered zone near a well affects only the pressure near that well; that is, the pressure in the unaltered formation away from the well is not affected by the existence of the altered zone. Thus, use Eq. 8.9 to calculate pressures at the sandface of a well with an altered zone, and Eq. 8.2 to calculate pressures beyond the altered zone in the formation surrounding the well. See Sec. 8.4 for more information on damage and stimulation.

8.2.4 Inertial-Turbulent Flow and Rate-Dependent Skin. The diffusivity equation, Eq. 8.1, assumes that Darcy's law represents the relationship between flow velocity and pressure gradients in the reservoir, an assumption that is adequate for low-velocity or laminar flow. However, at higher flow velocities, deviations from Darcy's law are observed as a result of inertial effects or even turbulent flow effects. In 1D radial flow, these inertial/turbulent effects (often

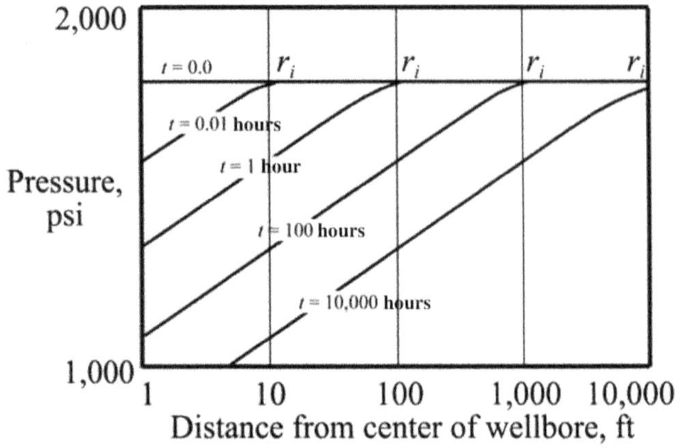

Fig. 8.1—Pressure distributions at increasing times.

called non-Darcy flow effects) are confined to the region near the wellbore in which flow velocities are largest. This results in an additional pressure drop similar to that caused by skin, but the additional pressure drop is proportional to flow rate. The apparent skin, s', for a well with non-Darcy flow near the wellbore is given by[3]

$$s' = s + Dq, \quad (8.10)$$

where D is the non-Darcy flow factor for the system. D can be regarded as constant, although, in theory, it depends slightly on near-well pressure. In practice, non-Darcy flow is ordinarily important only for gas wells, which have high-flow velocities near the wellbore, but it can be important for oil wells with high-velocity flow in some situations.

8.2.5 Radius of Investigation and Stabilization Time. Radius of investigation is the distance a pressure transient has moved into a formation following a rate change in a well. This distance is related to formation rock and fluid properties and time elapsed since a rate change in the well. Consider this concept by visualizing the pressure distributions at increasing times as **Fig. 8.1** shows for a well producing at constant rate from a reservoir initially at uniform pressure. (These pressure distributions were calculated using the *Ei*-function solution to the diffusivity equation.)

Important observations about this figure include the following:

• The pressure in the wellbore, at $r = r_w$, decreases steadily as flow time increases; likewise, pressures at other fixed values of r also decrease as flow time increases.

• The pressure drawdown (or pressure transient) caused by producing the well moves further into the reservoir as flow time increases. For the range of flow times shown, there is always a point beyond which the drawdown in pressure from the original value is negligible. This time-dependent point of "negligible drawdown" can be considered to be a radius of investigation.

Analysis shows that the time, t, at which a pressure disturbance reaches a distance, r_i, which is called the radius of investigation, is given by the equation[4]

$$r_i = \sqrt{\frac{kt}{948\phi\mu c_t}}. \quad (8.11)$$

Investigators differ on the numerical constant in Eq. 8.11, but this difference is of little practical importance if the radius of investigation is used as a semiquantitative indicator of the distance into the reservoir to which formation properties have influenced the response of a well in a pressure-transient test.

The radius of investigation has several applications in pressure-transient test analysis and design. A qualitative use is to help explain the shape of a pressure buildup or drawdown curve. For example, a buildup test plot may have a complex shape at early times when the radius of investigation is in the altered zone near the wellbore, where the permeability is different from formation permeability. Or a buildup test plot may change shape at long times when the radius of investigation reaches the general vicinity of a reservoir boundary.

The radius-of-investigation concept provides a guide for well-test design. For example, you may want to sample reservoir properties at least 1,000 ft from a test well. The radius of investigation concept allows you to estimate the time required to achieve the desired depth of investigation.

Eq. 8.11 also provides a means to estimate the time required to achieve "stabilized" flow; that is, the time required for a pressure transient to reach the boundaries of a tested reservoir. For example, if a well is centered in a cylindrical drainage area of radius r_e, then the time required for stabilization, t_s, is

$$t_s = \frac{948 \phi \mu c_t r_e^2}{k} \quad \text{(8.12)}$$

For other drainage shapes, the time to stabilization can be quite different, as discussed later.

8.2.6 Pseudosteady-State Flow. The *Ei*-function solution to the radial diffusivity equation is valid only while a reservoir is infinite-acting; that is, until boundaries begin to affect the pressure drawdown at the well. For the constant rate flow of a well centered in its drainage area of radius, r_e, and modeled by the *Ei*-function solution, these effects begin at $t = 948\ \phi \mu c_t r_e^2/k$. Before these boundary effects, the regime is called unsteady-state flow. After boundary effects are felt fully, the solution to the radial diffusivity equation for a well centered in a cylindrical drainage area and producing at constant rate is[4]

$$p_{wf} = p_i - 141.2 \frac{qB\mu}{kh} \left[\frac{0.000527kt}{\phi \mu c_t r_e^2} + \ln\left(\frac{r_e}{r_w}\right) - \frac{3}{4} \right] \quad \text{(8.13)}$$

This equation for calculating pressure in the wellbore becomes valid for $t > 948\ \phi \mu c_t r_e^2/k$ at the same time at which the *Ei*-function solution becomes invalid.

Another form of Eq. 8.13 is useful for some applications. It involves replacing original reservoir pressure, p_i, with average pressure, $\bar{p}$, within the drainage volume of the well. The volumetric average pressure within the drainage volume of the well can be found from material balance. The pressure decrease $(p_i - \bar{p})$ resulting from removal of qB RB/D of fluid for t hours (a total volume removed of $5.615qBt/24$ ft^3) is

$$p_i - \bar{p} = \frac{\Delta V}{c_t V} = \frac{5.615qB(t/24)}{c_t\left(\pi r_e^2 h \phi\right)} = \frac{0.0744qBt}{\phi c_t h r_e^2} \quad \text{(8.14)}$$

Substituting in Eq. 8.13, the time-dependent terms cancel, and the result is

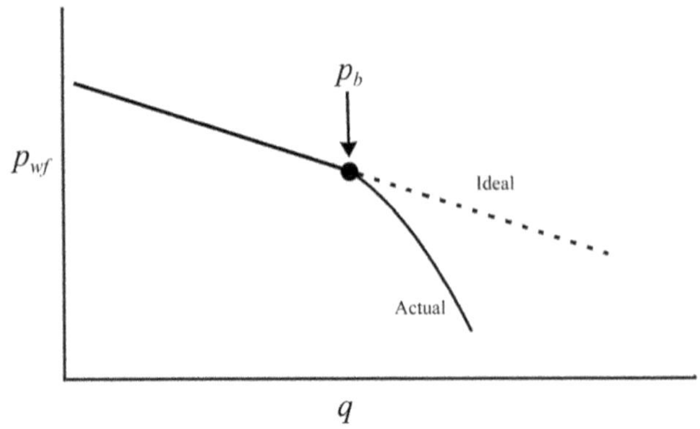

Fig. 8.2—Field data drop below idealized straight line at pressures below bubblepoint.

$$p_{wf} = \bar{p} - 141.2\frac{qB\mu}{kh}\left[\ln\left(\frac{r_e}{r_w}\right) - \frac{3}{4}\right]. \quad (8.15)$$

Eqs. 8.13 and 8.15 are more useful in practice if they include skin factors to account for damage or stimulation. In Eq. 8.15,

$$\bar{p} - p_{wf} = 141.2\frac{qB\mu}{kh}\left[\ln\left(\frac{r_e}{r_w}\right) - \frac{3}{4}\right] + (\Delta p)_s, \quad (8.16)$$

$$\bar{p} - p_{wf} = 141.2\frac{qB\mu}{kh}\left[\ln\left(\frac{r_e}{r_w}\right) - \frac{3}{4} + s\right], \quad (8.17)$$

and $$p_i - p_{wf} = 141.2\frac{qB\mu}{kh}\left[\frac{0.000527kt}{\phi\mu c_t r_e^2} + \ln\left(\frac{r_e}{r_w}\right) - \frac{3}{4} + s\right]. \quad (8.18)$$

8.2.7 Productivity Index. The productivity index, J, of an oil well is the ratio of the stabilized rate, q, to the pressure drawdown, ($\bar{p}$–p_{wf}), required to sustain that rate. For flow from a well centered in a circular drainage area, Eq. 8.17 allows us to relate productivity index to formation and fluid properties:

$$J \equiv \frac{q}{\bar{p} - p_{wf}} = \frac{kh}{141.2B\mu\left[\ln\left(\frac{r_e}{r_w}\right) - \frac{3}{4} + s\right]}. \quad (8.19)$$

Thus, if a well is tested at several different stabilized rates and the stabilized flowing bottomhole pressure (BHP), p_{wf}, is measured at each rate (that is, if pseudosteady-state is attained at each rate), Eq. 8.19 implies that a plot of test data should produce a straight line with slope J and intercepts $q = 0$ when $p_{wf} = \bar{p}$ and $q = J\bar{p}$ when $p_{wf} = 0$. (See **Fig. 8.2**.) In practice, actual field data will fall below the theoretical straight line for pressures below the bubblepoint pressure of the oil because of increasing gas saturations and oil viscosities that increase the resistance to flow.

8.2.8 Generalized Drainage Area Shapes. Eq. 8.17 is limited to a well centered in a circular drainage area. A similar equation models pseudosteady-state flow in more general reservoir shapes[4]:

$$\bar{p} - p_{wf} = 141.2\frac{qB\mu}{kh}\left[\frac{1}{2}\ln\left(\frac{10.06A}{C_A r_w^2}\right) - \frac{3}{4} + s\right], \quad (8.20)$$

where A is the drainage area in square feet, and C_A is the dimensionless shape factor for a specific drainage-area shape and configuration. **Table 8.A-1** (Appendix) gives values of C_A.

The productivity index, J, can be expressed for general drainage-area geometry as

$$J = \frac{q}{\bar{p} - p_{wf}} = \frac{0.00708kh}{B\mu\left[\frac{1}{2}\ln\left(\frac{10.06A}{C_A r_w^2}\right) - \frac{3}{4} + s\right]}. \quad (8.21)$$

Other numerical constants tabulated in Table 8.A-1 allow us to calculate the maximum elapsed time during which a reservoir is infinite-acting (so that the *Ei*-function solution can be used), the time required for the for the pseudosteady-state solution to predict pressure drawdown within 1% accuracy, and time required for the pseudosteady-state solution to be exact. For a given reservoir geometry, the maximum time a reservoir is infinite acting can be determined using the entry in the column "Use Infinite System Solution With Less Than 1% Error for t_{DA}<." This t_{DA} is defined as $0.0002637kt/\phi\mu c_t A$, so this means that the time in hours is calculated from

$$t = \frac{\phi\mu c_t A t_{DA}}{0.0002637k}. \quad (8.22)$$

Time required for the pseudosteady-state equation to be accurate within 1% can be found from the entry in the column titled "Less Than 1% Error for t_{DA}." Finally, the time required for the pseudosteady-state equation to be exact is found in the entry in the column "Exact for t_{DA}>."

Figs. 8.3 and 8.4 show the flow regimes that occur at various times. These figures show p_{wf} in a well flowing at constant rate, plotted as a function of time on both logarithmic and linear scales. In the transient region, the reservoir is infinite acting and is modeled by Eq. 8.9, which implies that p_{wf} is a linear function of log t. In the pseudosteady-state region, the reservoir is modeled by Eq. 8.20 in the general case or Eqs. 8.15 or 8.13 for the special case of a well centered in a cylindrical drainage area. Eq. 8.13 shows a linear relationship between p_{wf} and t during pseudosteady-state flow. This linear relationship also exists in generalized reservoir geometries.

At times between the end of the transient region and the beginning of the pseudosteady-state region, there is a transition region, sometimes called the late-transient region. This region is, for practical purposes, nonexistent for wells centered in circular, square, or hexagonal drainage areas, as Table 8.A-1 indicates. However, for a well off-center in its drainage area, the late-transient region can span a significant time region, as Table 8.A-1 also indicates.

8.2.9 Steady-State Flow. Pseudosteady-state flow describes production from a closed drainage area (one with no-flow outer boundaries, either permanent and caused by zero-permeability rock or "temporary" and caused by production from offset wells). In pseudosteady-state, reser-

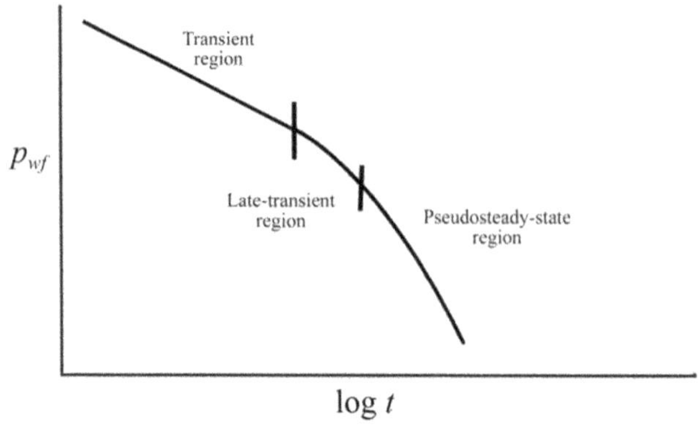

Fig. 8.3—Flow regions on semilogarithmic coordinates.

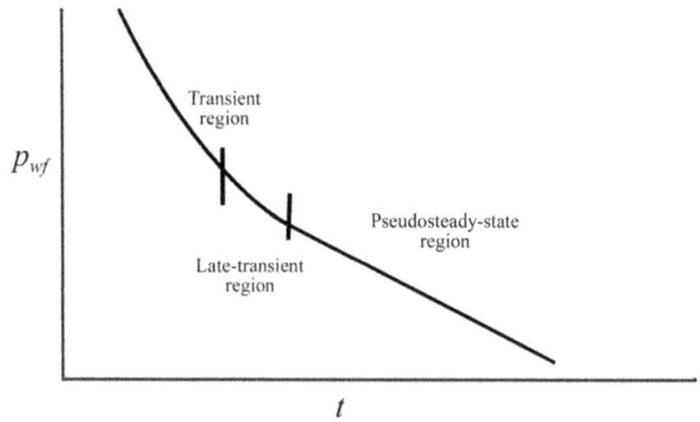

Fig. 8.4—Flow regions on Cartesian coordinates.

voir pressure drops at the same rate with time at all points in the reservoir, including at the reservoir boundaries. Ideally, true steady-state flow can occur in the drainage area of a well, but only if pressure at the drainage boundaries of the well can be maintained constant while the well is producing at constant rate. While unlikely, steady-state flow is conceivable for wells with edgewater drive or in repeated flood patterns in a reservoir. The solution to the radial diffusivity equation is based on a constant-pressure outer boundary condition, instead of a no-flow outer boundary condition. The steady-state solution, applicable after boundary effects have been felt, is

$$p_i - p_{wf} = 141.2\frac{qB\mu}{kh}\left[\ln\left(\frac{r_e}{r_w}\right) + s\right] \text{ . } \quad (8.23)$$

8.2.10 Constant Pressure in the Well. Both the steady-state solution (constant pressure at the outer boundaries) and the pseudosteady-state solution (no-flow at the outer boundaries) assume constant rate production in the well. A well is actually more likely to be produced at something close to constant flowing BHP than constant rate. When pressure transients reach no-flow drainage area boundaries, the flow regime is not pseudosteady state; instead, it is more correctly called boundary-dominated flow. If the drainage boundaries are maintained at constant pressure, however, steady-state flow is achieved when the pressure transient reaches the reservoir boundaries.

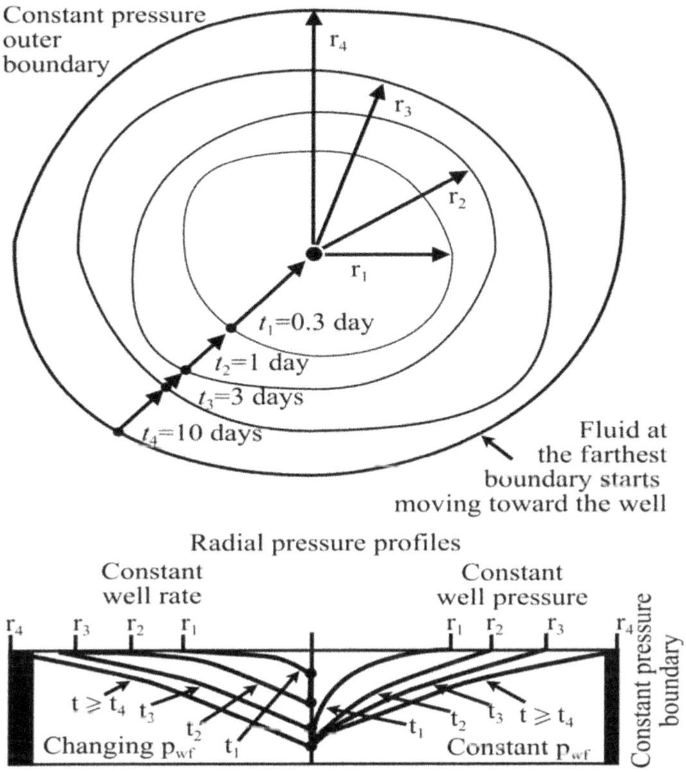

Fig. 8.5—Flow regimes with constant well flow rate or constant well pressure and constant-pressure outer boundaries.

These different flow regimes are clarified with figures showing pressure distributions in the drainage area of wells with constant flow rate and constant-pressure outer boundaries (**Fig. 8.5**); constant BHP and constant-pressure outer boundaries (also Fig. 8.5); constant flow rate and no-flow outer boundaries (**Fig. 8.6**); and constant BHP and no-flow outer boundaries (also Fig. 8.6).

8.2.11 Wellbore Storage. The *Ei*-function solution to the diffusivity equation assumes constant flow rate in the reservoir, starting at time zero. In practice, only the rate at the surface can be controlled. Under ideal conditions, a constant surface rate can be maintained, but the first fluid produced will be fluid that was stored in the wellbore, and, at first, the flow rate from the reservoir into the wellbore will be zero. As the wellbore is unloaded, the reservoir rate approaches the surface rate (**Fig. 8.7**). Only as the reservoir and surface rates become approximately equal does the *Ei*-function solution become valid. This wellbore unloading during flow tests is a special case of a general phenomenon called wellbore storage.

For a pressure buildup test, the surface rate is zero starting at the instant of shut-in. However, fluid continues to flow into the wellbore from the reservoir because of existing pressure gradients. Idealized models of pressure buildup tests assume a reservoir rate of zero starting at the time of shut-in for the test. This assumption is obviously violated because of the afterflow into the wellbore. As the afterflow rate diminishes, the downhole rate approaches the surface rate (zero), and only as the afterflow rate approaches zero closely can the idealized models closely approximate actual well behavior (**Fig. 8.8**). Afterflow during buildup tests is another special case of wellbore storage.

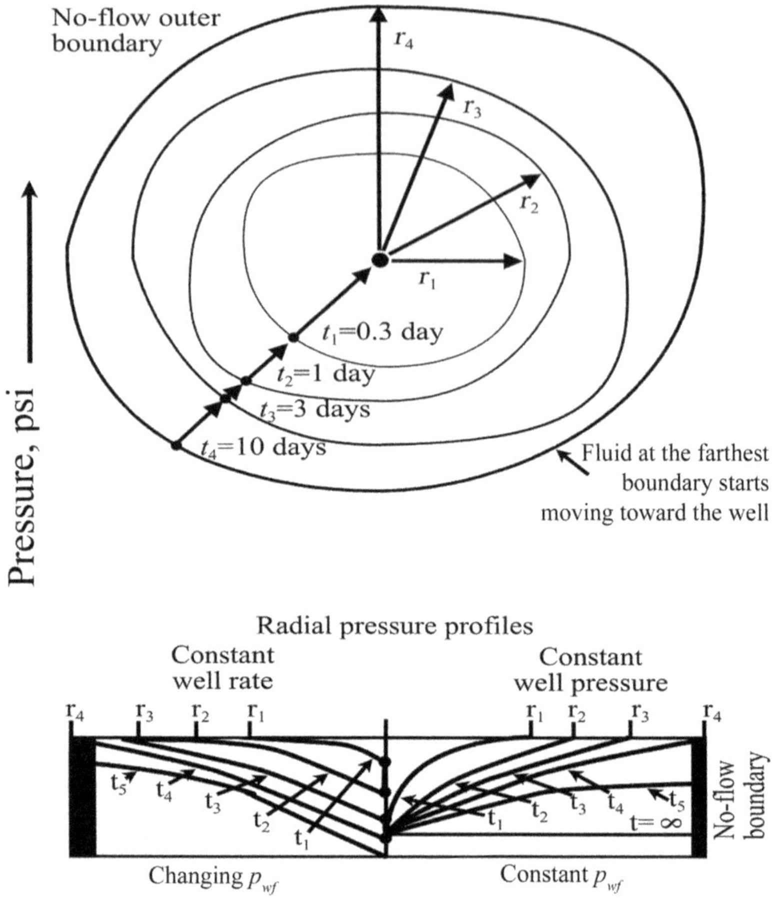

Fig. 8.6—Flow regimes with constant well flow rate and no-flow outer boundaries.

The relationship between changes in bottomhole pressure and wellbore unloading or afterflow rates can be modeled with mass balances on the wellbore. There are two special cases of interest: a wellbore completely filled with a single-phase fluid (**Fig. 8.9**, usually gas in practice) and a wellbore with a rising or falling liquid/gas interface in the well (**Fig. 8.10**).

For the wellbore filled with a single-phase fluid,[4]

$$q_{sf} = q + \frac{24 V_{wb} c_{wb}}{B} \frac{dp_w}{dt} \quad \text{(8.24)}$$

For a well with a rising or falling liquid/gas interface,[4]

$$\frac{(24)(144)}{5.615 \rho_{wb}} \frac{g_c}{g} A_{wb} \frac{d(p_w - p_t)}{dt} = \left(q_{sf} - q\right) B \quad \text{(8.25)}$$

In most applications, p_t is assumed to be constant, a convenient but frequently inaccurate simplification. Both equations can be written in the general form

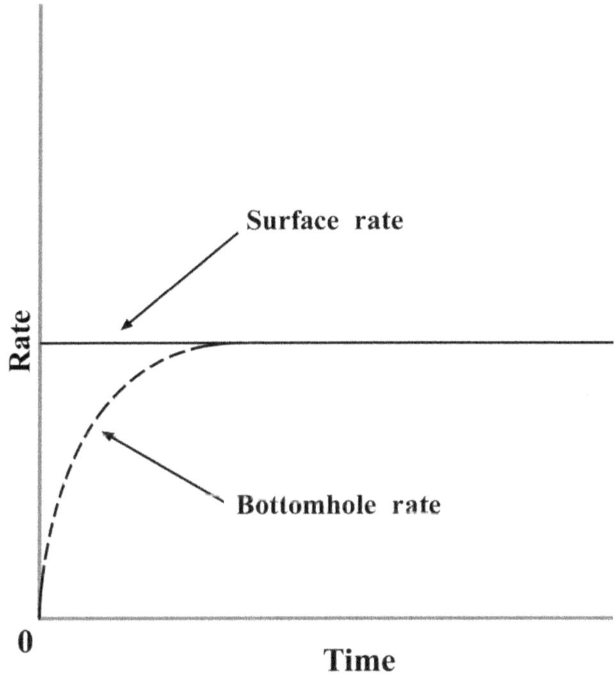

Fig. 8.7—Reservoir rate approaches surface rate as the wellbore is unloaded.

$$q_{sf} = q + \frac{24C}{B}\frac{dp_w}{dt}, \quad (8.26)$$

where, for a fluid-filled wellbore,

$$C = c_{wb}V_{wb}, \quad (8.27)$$

and, for a moving liquid/gas interface with unchanging surface pressure,

$$C = \frac{25.65A_{wb}}{\rho_{wb}}. \quad (8.28)$$

C is called the wellbore storage coefficient.

For special cases in which, at earliest times for a flowing well, all the production is coming from fluid stored in the wellbore and none is entering the wellbore from the formation (or, for a shut-in well, the rate of afterflow is equal to the rate before shut in), the integration of Eq. 8.26 yields

$$\Delta p = \frac{qB\Delta t}{24C}, \quad (8.29)$$

where Δp is the pressure change in the time because either the start of flow or shut in and Δt is the elapsed time. On a log-log plot of Δp vs. Δt during these early times, a straight line with a slope of unity will result. For any point on this unit slope line, the wellbore storage coeffi-

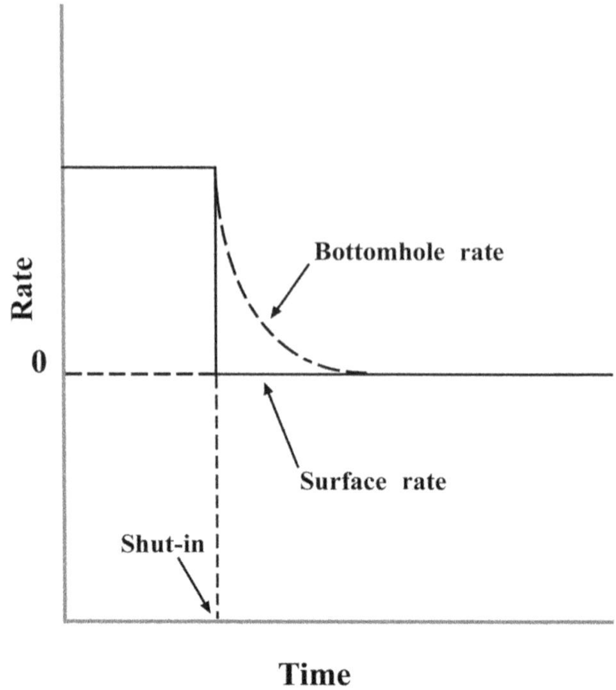

Fig. 8.8—Downhole rate approaches surface rate as afterflow rate diminishes.

cient, C, can be found from any point on the line (Δt, Δp) and Eq. 8.29 (**Fig. 8.11**). Alternatively, the slope ($qB/24C$) of a plot of Δp vs. Δt on Cartesian coordinates also leads to an estimate of the wellbore-storage coefficient.

8.2.12 Linear Flow. Linear flow occurs in some reservoirs with long, highly conductive vertical fractures; in relatively long, relatively narrow reservoirs (channels, such as ancient stream beds); and near horizontal wells during certain times. For unsteady-state linear flow in an unbounded (infinite-acting) reservoir,[4]

$$p_{wf} = p_i - 16.26\frac{qB\mu}{kA}\left(\frac{kt}{\phi\mu c_t}\right)^{\frac{1}{2}} - 70.6\frac{qB\mu}{kh}s_f \ . \quad (8.30)$$

8.2.13 Spherical Flow. Spherical flow occurs in wells with limited perforated intervals and into wireline formation test tools. The solution to the spherical/cylindrical, 1D form of the diffusivity equation, subject to the initial condition that pressure is uniform before production and the boundary conditions of constant flow rate and an infinitely large drainage area, is[5]

$$p_{wf} = p_i - \frac{70.6qB\mu}{k_s r_s} + \frac{2456\sqrt{\phi\mu c_t}qB\mu}{k_{sp}^{3/2}}\frac{1}{\sqrt{t}} - \frac{70.6qB\mu}{k_{sp}r_{sp}}s, \quad (8.31)$$

$$\text{where } k_{sp} = \left(k_r k_z^{1/2}\right)^{2/3}, \quad (8.32)$$

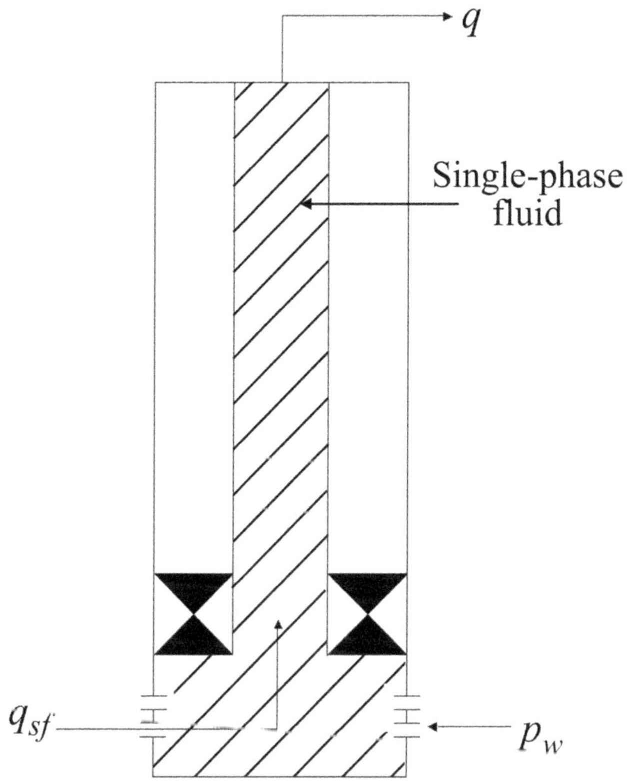

Fig. 8.9—Schematic of wellbore containing single-phase liquid or gas.

and r_{sp} = the radius of the sphere into which flow converges.

8.2.14 Superposition. The principle of superposition indicates that the total pressure at any point in a reservoir is the sum of the pressure drops at that point caused by flow in each of the wells in the reservoir. A simple illustration of this principle is the case of three wells in an infinite reservoir. Consider wells A, B, and C, that start to produce at times t_A, t_B, and t_C in an infinite-acting reservoir (**Fig. 8.12**). Application of the principle of superposition shows that[4]

$$(p_i - p_{wf})_{\text{total at Well A}} = (p_i - p)_{\text{due to A}} + (p_i - p)_{\text{due to B}} + (p_i - p)_{\text{due to C}} \cdot \quad (8.33)$$

For an infinite-acting reservoir, use the *Ei*-function solutions, including the logarithmic approximation at Well A:

$$\begin{aligned}&(p_i - p_{wf})_{\text{total at Well A}}\\ &= -70.6\frac{q_A B\mu}{kh}\left[\ln\left(\frac{1{,}688\phi\mu c_t r_{wA}^2}{k(t-t_A)}\right) - 2s_A\right]\\ &\quad -70.6\frac{q_B B\mu}{kh} Ei\left(\frac{-948\phi\mu c_t r_{AB}^2}{k(t-t_B)}\right)\end{aligned}$$

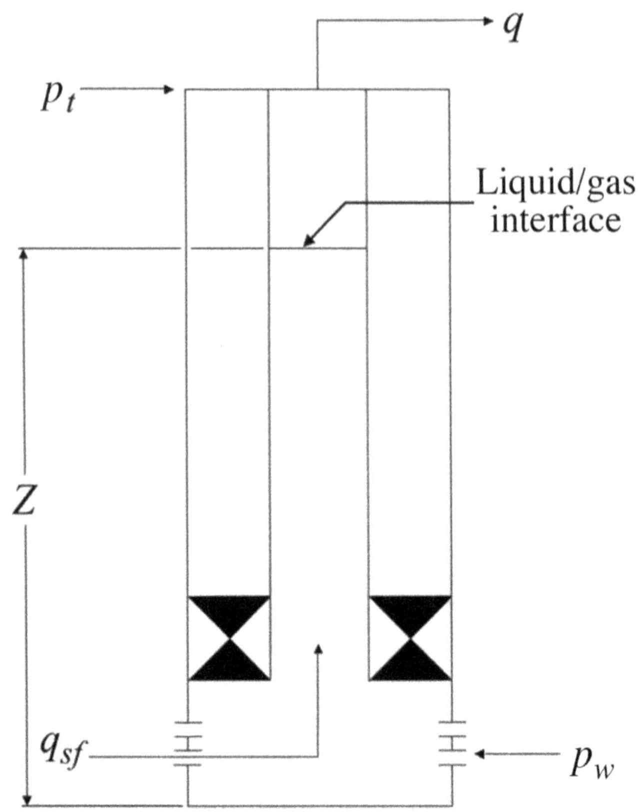

Fig. 8.10—Schematic of wellbore with moving liquid/gas interface.

$$-70.6\frac{q_C B\mu}{kh}Ei\left(\frac{-948\phi\mu c_t r_{AC}^2}{k\left(t-t_C\right)}\right), \quad (8.34)$$

where t_A, t_B, and t_C are times at which wells A, B, and C will begin to produce. The skin factor for Well A is included in Eq. 8.29. The skin factors for other wells are not, because skin factors for individual wells affect only pressures measured inside altered zones for those wells.

Next, consider the use of superposition to model the effects of boundaries in bounded reservoirs. Consider the well in **Fig. 8.13**, a distance L from a single no-flow boundary (such as a sealing fault). Mathematically, this problem is identical to the problem of a well at distance $2L$ from an "image" well; that is, a well that has the same production history as the actual well. The reason that the two-well system simulates the behavior of a well near a boundary is that a line equidistant between the two wells can be shown to be a no-flow boundary. That is, along this line the pressure gradient is zero, which means that there can be no flow. Thus, this problem is a simple problem of two wells in an infinite reservoir:

$$p_i - p_{wf} = -70.6\frac{qB\mu}{kh}\left(\ln\frac{1{,}688\phi\mu c_t r_w^2}{kt} - 2s\right) - 70.6\frac{qB\mu}{kh}Ei\left(\frac{-948\phi\mu c_t(2L)^2}{kt}\right). \quad (8.35)$$

The drawdown term of the image well does not include a skin factor.

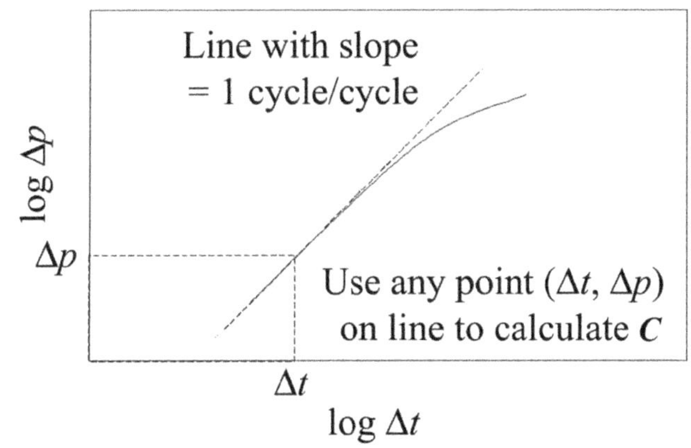

Fig. 8.11—Unit slope line on a log-log plot of Δ*p* vs. Δ*t*.

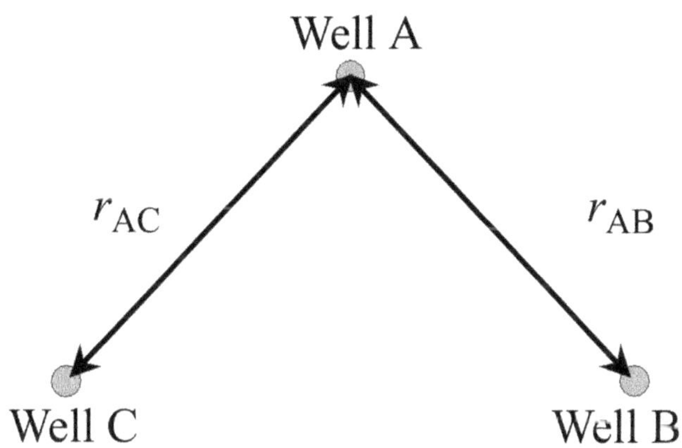

Fig. 8.12—Multiple-well system in infinite reservoir.

As examples, extend the imaging technique to model wells between boundaries intersecting at 90° (**Fig. 8.14**); wells between two parallel boundaries (**Fig. 8.15**); wells near single constant-pressure boundaries (**Fig. 8.16**); and wells at various locations in closed reservoirs (**Fig. 8.17**).

One of the most frequently used applications of superposition is to model variable-rate production. Consider **Fig. 8.18**, in which a well in an infinite-acting reservoir produces at rate q_1 from time 0 to time t_1; q_2 from t_1 to t_2, and q_3 for times greater than t_2. To model the total drawdown for $t > t_2$, add three drawdowns: the drawdown because of a well producing at rate q_1 starting at time zero and continuing to produce to time t; the drawdown because of a well producing at rate $(q_2 - q_1)$, starting at time t_1 and continuing to time t; and the drawdown because of a well producing at rate $(q_3 - q_2)$ starting at time t_2 and continuing to time t. The total drawdown is thus

$$p_i - p_{wf} = (\Delta p)_1 + (\Delta p)_2 + (\Delta p)_3 = -70.6\frac{\mu q_1 B}{kh}\left[\ln\left(\frac{1{,}688\phi\mu c_t r_w^2}{kt}\right) - 2s\right]$$

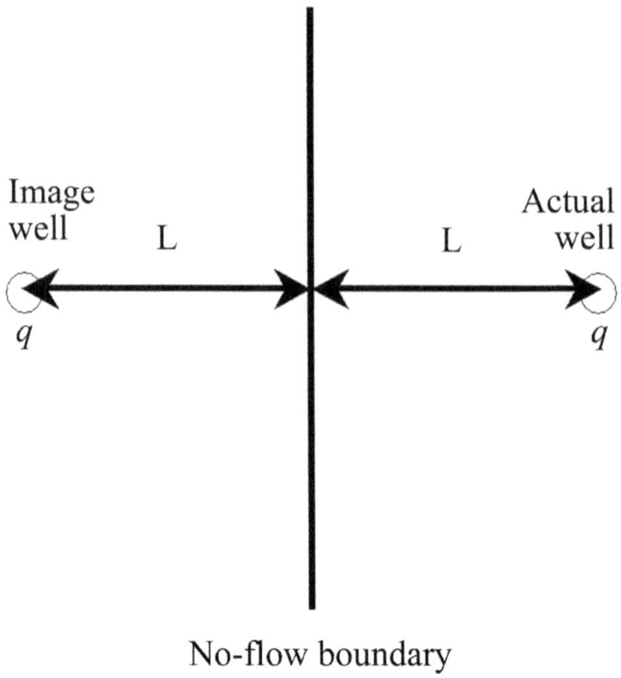

Fig. 8.13—Well near no-flow boundary illustrating use of imaging.

$$-70.6\frac{\mu(q_2-q_1)B}{kh}\cdot\left\{\ln\left[\frac{1{,}688\phi\mu c_t r_w^2}{k(t-t_1)}\right]-2s\right\}$$

$$-70.6\frac{\mu(q_3-q_2)B}{kh}\cdot\left\{\ln\left[\frac{1{,}688\phi\mu c_t r_w^2}{k(t-t_2)}\right]-2s\right\} \quad (8.36)$$

Horner[6] proposed a convenient alternative to superposition to model the many changes in rate in the history of a typical well. With this approximation, the sequence of *Ei* functions reflecting rate changes can be replaced with a single *Ei* function that contains a single producing time and a single producing rate. The single rate is the most recent nonzero rate at which the well has produced, q_n. The single producing time, called t_p, is the ratio of cumulative production, N_p, to q_n.

$$t_p=\frac{N_p}{q_n} \quad (8.37)$$

This approximation preserves the material balance in the drainage area of the well and properly gives greatest weight to most recent rate (as opposed to average rate), which dominates the pressure distribution near a well out to the radius of investigation achieved while the well was produced at rate q_n. The approximation is particularly useful for hand calculations. Given the widespread availability of computer software for analyzing flow and buildup tests on well, the use of more rigorous superposition to model variable-rate production histories is generally more appropriate.

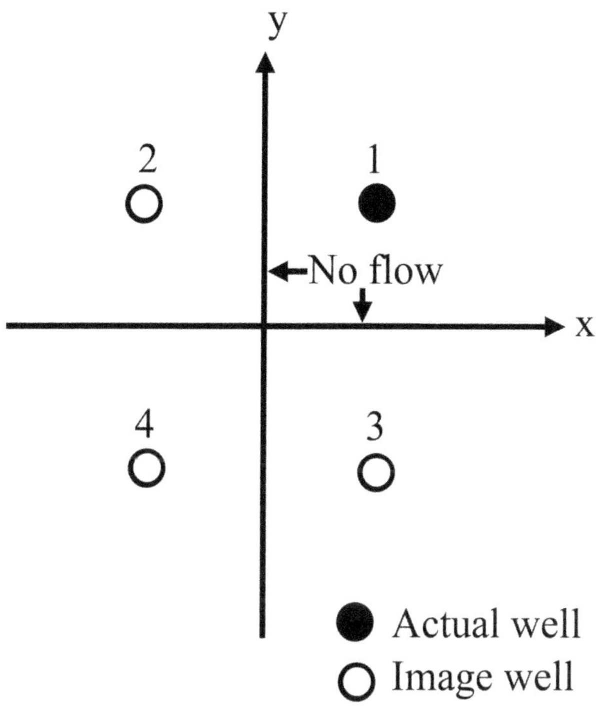

Fig. 8.14—Image-well locations for two intersecting no-flow boundaries.

8.2.15 Semilog Methods for Flow Tests. The logarithmic approximation to the *Ei*-function solution can be used as a basis for analysis of an ideal constant-rate flow test in a well. Written in terms of $\log_{10}$, this equation, which models the BHP for a well in a homogeneous-acting formation with an infinite-acting drainage area and, in absence of wellbore unloading, becomes

$$p_{wf} = p_i - 162.6\frac{qB\mu}{kh}\left[\log_{10}(t) + \log_{10}\left(\frac{k}{\phi\mu c_t r_w^2}\right) - 3.23 + 0.869s\right]. \quad (8.38)$$

This expression has the same form as the equation of a straight line, $y = mx + b$, with the analogies

$$p_{wf} \sim y, \quad (8.39)$$

$$\log_{10}(t) \sim x, \quad (8.40)$$

$$-162.6qB\mu / kh \sim m, \quad (8.41)$$

and $p_i - m\left[\log_{10}\left(\frac{k}{\phi\mu c_t r_w^2}\right) - 3.23 + 0.869s\right] \sim b$. (8.42)

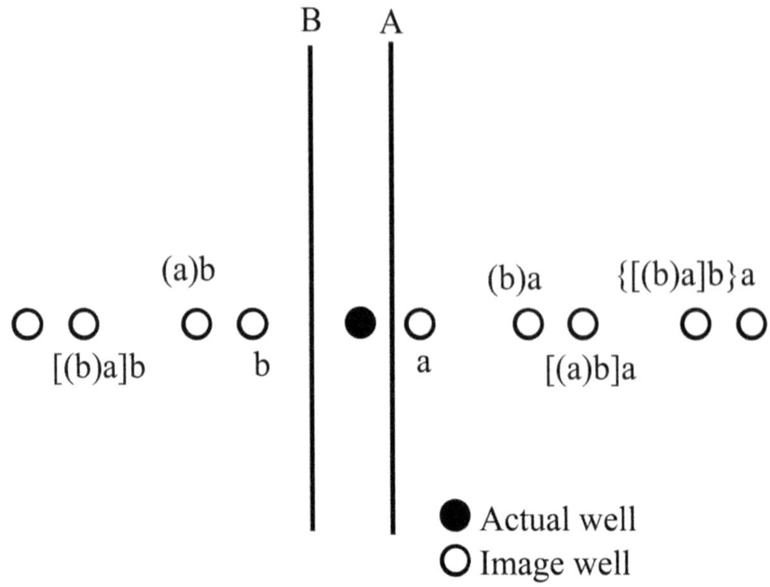

Fig. 8.15—Image-well locations for well between parallel faults.

These analogies suggest a graphical method of analysis. Eq. 8.38 indicates that a plot of p_{wf} vs. $\log_{10}(t)$ should be a straight line with slope m that will allow an estimate of effective permeability to the single liquid phase flowing. (See **Fig. 8.19**.)

$$k = \frac{162.6qB\mu}{(-m)h} \quad (8.43)$$

From the intercept, b, at t = 1 hr [$\log_{10}(1) = 0$], p_{1hr}, calculate the skin factor.

$$s = 1.151\left[\frac{p_i - p_{1hr}}{-m} - \log_{10}\left(\frac{k}{\phi\mu c_t r_w^2}\right) + 3.23\right] \quad (8.44)$$

In these equations, the slope, m, is given by

$$m = \frac{(p_{wf2} - p_{wf1})}{\left[\log_{10}(t_2) - \log_{10}(t_1)\right]} = \frac{(p_{wf2} - p_{wf1})}{\left[\log_{10}(t_2/t_1)\right]} \quad (8.45)$$

Eq. 8.45 indicates that m is most easily determined by choosing values of times t_1 and t_2 that differ by powers of 10 and is especially easy if t_1 and t_2 differ by one log cycle. The intercept, p_{1hr}, is the pressure at a time of 1 hour on the best straight line through the data. It may be necessary to extrapolate the straight line to a time of one hour to read the intercept.

8.2.16 Semilog Methods for Pressure Buildup Test. Consider the rate history for an idealized pressure test shown in **Fig. 8.20**. A well is produced at constant rate q for a time t_p, and then the well shut in (q = 0) for a pressure buildup test. The rate history is modeled as the sum of two constant flow rate periods, one at rate q, beginning at t = 0, and the other at rate $-q$,

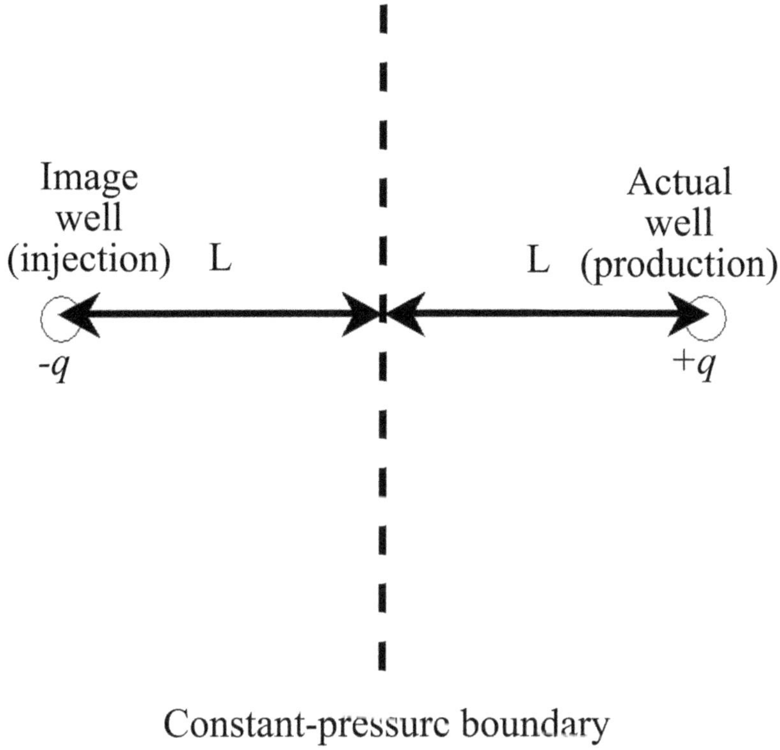

Fig. 8.16—Well near single constant-pressure boundary.

beginning at $t = t_p$, at which the time elapsed since shut-in, Δt, is zero. Use the log approximation to the *Ei*-function solution to model the drawdown, and sum them as **Fig. 8.21** shows. Represented mathematically, the superposition process is

$$p_{ws} = p_i - 162.6\frac{qB\mu}{kh}\left[\log_{10}(t_p + \Delta t) + \log_{10}\left(\frac{k}{\phi\mu c_t r_w^2}\right) - 3.23 + 0.869s\right]$$

$$-162.6\frac{(-q)B\mu}{kh}\left[\log_{10}(\Delta t) + \log_{10}\left(\frac{k}{\phi\mu c_t r_w^2}\right) - 3.23 + 0.869s\right] \quad \text{(8.46)}$$

This can be simplified to

$$p_{ws} = p_i - 162.6\frac{qB\mu}{kh}\log_{10}\left(\frac{t_p + \Delta t}{\Delta t}\right) \quad \text{(8.47)}$$

Like the drawdown equation, Eq. 8.43 can be interpreted as the equation of a straight line. The analogies are

$$p_{ws} \sim y, \quad \text{(8.48)}$$

$$\log_{10}(t) \sim x, \quad \text{(8.49)}$$

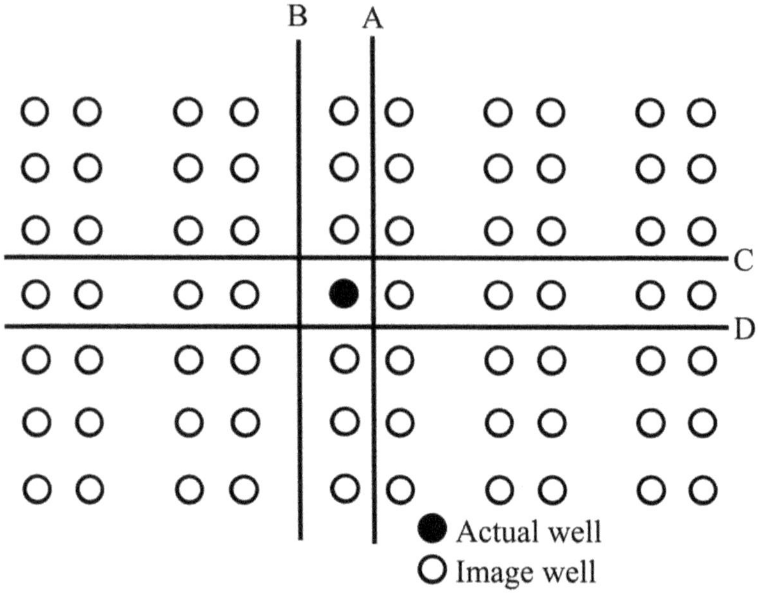

Fig. 8.17—Image-well locations for well between parallel boundaries.

$$-162.6qB\mu / kh \sim m, \quad (8.41)$$

$$\text{and } p_i \sim b \,. \quad (8.50)$$

The group $[(t_p + \Delta t)/\Delta t]$ is called the Horner time ratio (HTR) or sometimes simply the Horner time. Our simple model, which describes a buildup test in a homogeneous, infinite-acting reservoir, a well with one constant rate before shut in and without afterflow (wellbore storage), indicates that a graph of p_{ws} vs. the HTR should fall on a straight line. From the slope, m, of this line, the permeability to the single-phase liquid flowing into the wellbore can be estimated. The intercept, b, at $\log_{10}[(t_p + \Delta t)/\Delta t] = 0$ or $[(t_p + \Delta t)/\Delta t] = 1$ provides an estimate of original drainage area pressure, p_i.

Obtain the slope, m, from

$$\frac{p_{ws2} - p_{ws1}}{\log_{10}\left(\frac{t_p + \Delta t}{\Delta t}\right)_2 - \log_{10}\left(\frac{t_p + \Delta t}{\Delta t}\right)_1}, \quad (8.51)$$

where $\{[(t_p+\Delta t)/\Delta t]_1, p_{ws1}\}$ and $\{[(t_p+\Delta t)/\Delta t]_2, p_{ws2}\}$ are any two points on the straight-line (**Fig. 8.22**). Normally, choose $[t_p + \Delta t)/\Delta t]_1$ and $[(t_p + \Delta t)/\Delta t]_2$ to be powers of 10.

In Fig. 8.22, which is called a Horner plot, the HTR on the horizontal axis decreases from left to right, so that shut-in time increases from left to right. In some Horner plots, the HTR increases from left to right; in that case shut-in time increases from right to left.

Skin factor can be estimated from a pressure buildup test, even though the skin factor does not appear in the buildup equation, Eq. 8.47. Simultaneously solve the equation modeling the drawdown at the instant of shut in (at time t_p) with Eq. 8.47, discard terms that are ordinarily negligible, and arrive at the result

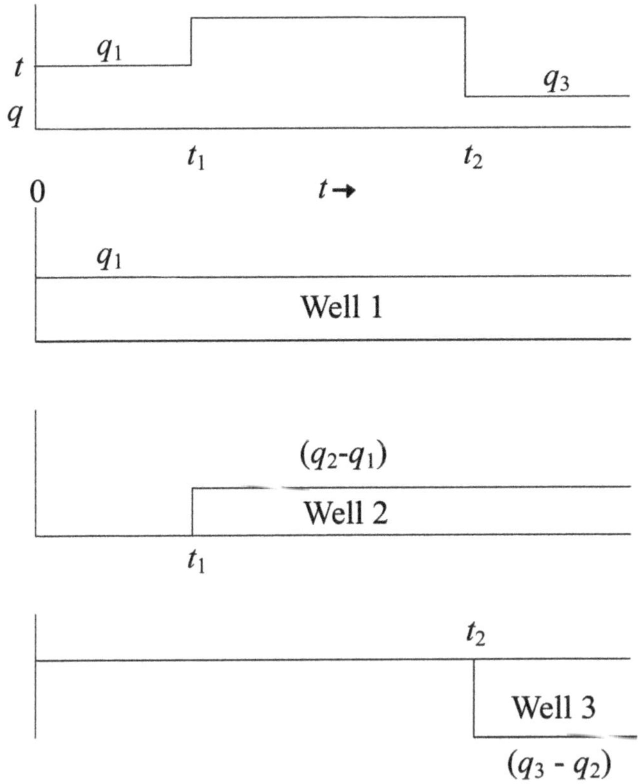

Fig. 8.18—Production schedule for variable-rate well.

$$s = 1.151\left[\frac{p_{1\text{hr}} - p_{wf}}{-m} - \log_{10}\left(\frac{k}{\phi\mu c_t r_w^2}\right) + 3.23\right] . \qquad (8.52)$$

The radius-of-investigation concept is also useful for pressure buildup tests, as **Fig. 8.23** illustrates. The approximate position of the point at which the pressure has built up to a uniform level intersects the region in which the pressure is little affected by the shut-in is given by Eq. 8.11, with elapsed time, t, interpreted as shut-in time, Δt.

8.3 Type Curves

Type curves provide a powerful method for analyzing pressure drawdown (flow) and buildup tests. Fundamentally, type curves are preplotted solutions to the flow equations, such as the diffusivity equation, for selected types of formations and selected initial and boundary conditions. Because of the way they are plotted (usually on logarithmic coordinates), it is convenient to compare actual field data plotted on the same coordinates to the type curves. The results of this comparison frequently include qualitative and quantitative descriptions of the formation and completion properties of the tested well.

8.3.1 Dimensionless Variables. The solutions plotted on type curves are usually presented in terms of dimensionless variables. To review dimensionless variables, consider the *Ei*-function solution to the flow equation, Eq. 8.2, presented in terms of dimensional variables:

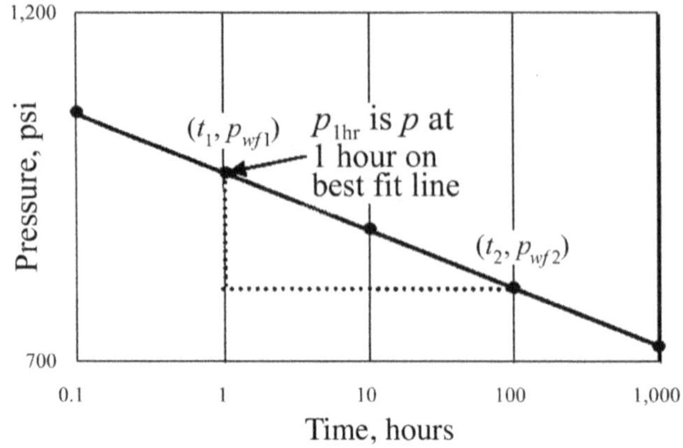

Fig. 8.19—Straight line with slope *m* allows us to estimate effective permeability to the single liquid phase flowing.

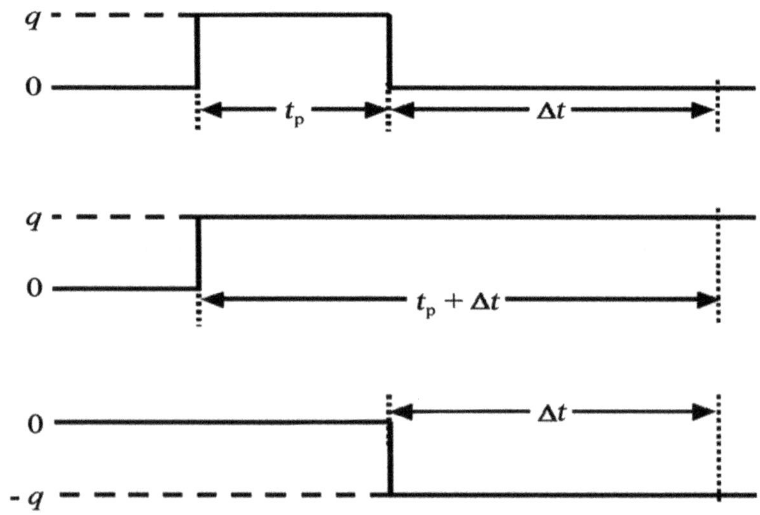

Fig. 8.20—Rate history for an idealized pressure buildup test.

$$p = p_i + 70.6\frac{qB\mu}{kh}Ei\left(\frac{-948\phi\mu c_t r^2}{kt}\right). \qquad (8.2)$$

Eq. 8.2 can be rewritten in terms of conventional definitions of dimensionless variables. (Variables that when the parameters are expressed in terms of the fundamental units of mass, length, and time, have no dimensions are sometimes said to have dimensions of zero.)

$$p_D = -\frac{1}{2}Ei\left(-\frac{r_D^2}{4t_D}\right). \qquad (8.53)$$

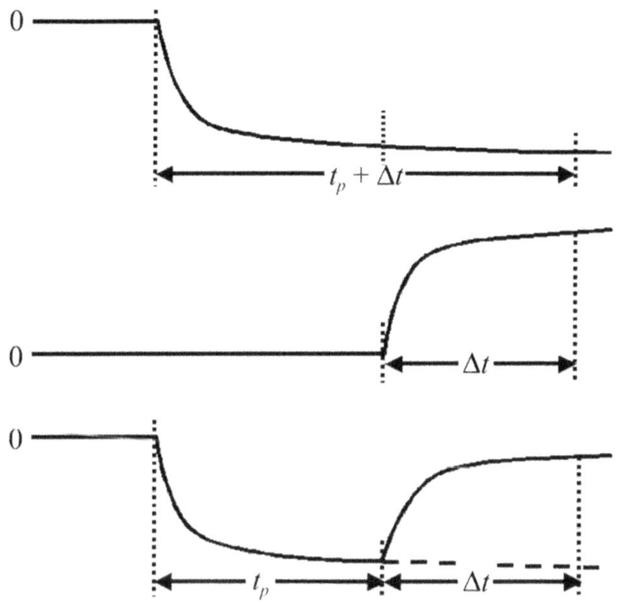

Fig. 8.21—Sum of *Ei*-function solutions to model pressure buildup test.

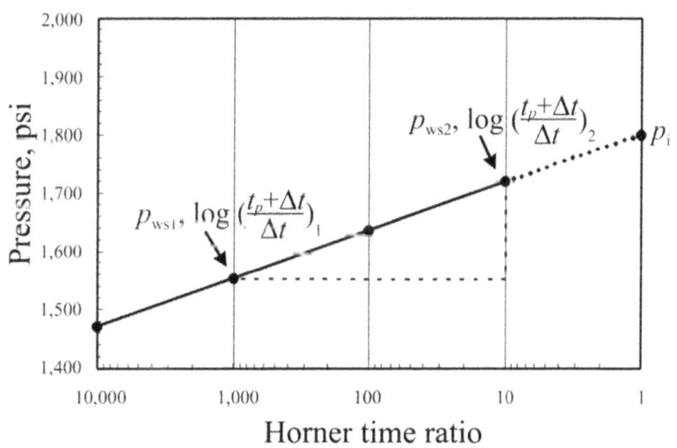

Fig. 8.22—Horner plot.

In Eq. 8.53, the definitions of the dimensionless variables are

$$p_D \equiv \frac{kh(p_i - p)}{141.2qB\mu}, \quad (8.54)$$

$$t_D \equiv \frac{0.0002637kt}{\phi \mu c_t r_w^2}, \quad (8.55)$$

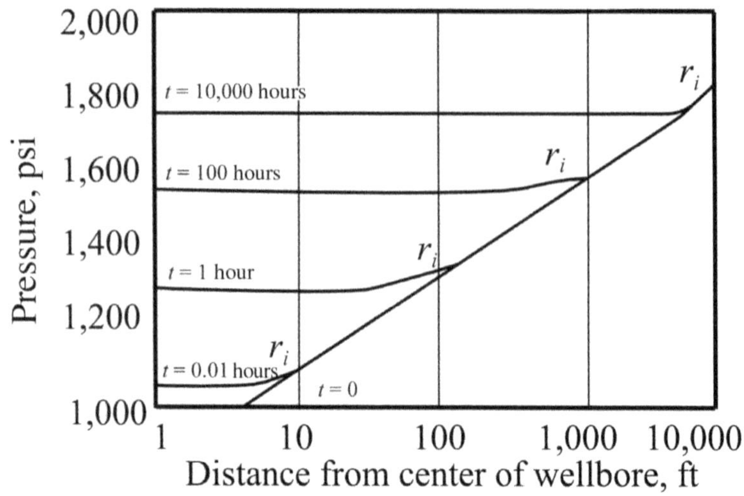

Fig. 8.23—Radius of investigation in pressure buildup tests.

$$\text{and } r_D \equiv \frac{r}{r_w} \quad \text{.......... (8.56)}$$

The dimensionless form of Eq. 8.2 has the advantage that this solution, p_D, to the diffusivity equation can be expressed in terms of a single variable, t_D, and single parameter, r_D. This leads to much simpler graphical or tabular presentation of the solution than would direct use of Eq. 8.2. Solutions to the diffusivity equation for more realistic reservoir models also include the dimensionless skin factor, s, and wellbore storage coefficient, C_D, where

$$C_D = \frac{0.8936C}{\phi c_t h r_w^2} \quad \text{.......... (8.57)}$$

8.3.2 Gringarten Type Curve. Gringarten *et al.*[7] presented a type curve, commonly called the Gringarten type curve, that achieved widespread use. It is based on a solution to the radial diffusivity equation and the following assumptions: vertical well with constant production rate; infinite-acting, homogeneous-acting reservoir; single-phase, slightly compressible liquid flowing; infinitesimal skin factor (thin "membrane" at production face); and constant wellbore-storage coefficient. These assumptions indicate that the type curve was developed specifically for drawdown tests in undersaturated oil reservoirs. The type curve is also useful to analyze pressure buildup tests and for gas wells.

In the Gringarten type curve, p_D is plotted vs. the time function t_D/C_D, with a parameter C_De^{2s} (**Fig. 8.24**). Each different value of C_De^{2s} describes a pressure response with a shape different (in theory) from the responses for other values of the parameter. However, adjacent pairs of curves can be quite similar, and this fact can cause uncertainty when trying to match test data to the "uniquely correct" curve.

8.3.3 Derivative Type Curve. The derivative type curve proposed by Bourdet *et al.*[8] eliminates the ambiguity in the Gringarten type curve. The "derivative" referred to in this type curve is the logarithmic derivative of the solution to the radial diffusivity equation presented on the Gringarten type curve. Two limiting forms of this solution help illustrate the nature of

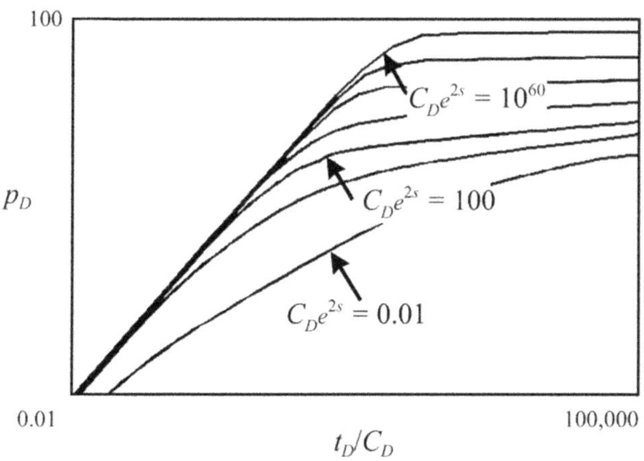

Fig. 8.24—Gringarten type curve with parameter C_De^{2s}.

the derivative type curve. First, consider that part of a test response where the distorting effects of wellbore storage have vanished. This portion of the test is described by the logarithmic approximation to *Ei*-function solution, Eq. 8.9:

$$p_i - p_{wf} = -70.6\frac{qB\mu}{kh}\left[\ln\left(\frac{1688\phi\mu c_t r_w^2}{kt}\right) - 2s\right]. \qquad (8.9)$$

The derivative of $(p_i - p_{wf})$ with respect to $\ln(t)$, expressed more simply as $t\partial\Delta p/\partial t$, is $70.6qB\mu/kh$, a constant. In terms of dimensionless variables, $t_D(\partial p_D/\partial t_D) = 0.5$. Thus, when the distorting effects of wellbore storage have disappeared, the pressure derivative will become constant in an infinite-acting reservoir, and, in terms of dimensionless variables, will have a value of 0.5.

When wellbore storage completely dominates the pressure response (all produced fluid comes from the wellbore, none from the formation),

$$\Delta p = \frac{qB\Delta t}{24C}. \qquad (8.29)$$

The derivative, $t\partial\Delta p/\partial t$, is $qBt/24C$, the same as the pressure change itself. In terms of dimensionless variables, the derivative becomes

$$p_D' = t_D(\partial p_D/\partial t_D) = t_D/C_D. \qquad (8.58)$$

The implication of Eq. 8.58 is that, on logarithmic coordinates, graphs of p_D and $t_D(\partial p_D/\partial t_D)$ vs. t_D/C_D will coincide and will have slopes of unity.

For values of $t_D(\partial p_D/\partial t_D)$ between the end of complete wellbore storage distortion and the start of infinite-acting radial flow, no simple solutions are available to guide us, but **Fig. 8.25** shows the derivatives, including those times. Note the unit slope lines at earliest times and the horizontal derivative at later times. The shapes of the derivative stems are much more distinctive than those for the pressure-change type curve.

For test analysis, we plot pressure change, p_D, and pressure derivative $[t_D(\partial p_D/\partial t_D)]$ on the same graph (**Fig. 8.26**). On this graph, a specific value of the parameter C_De^{2s} refers to a pair

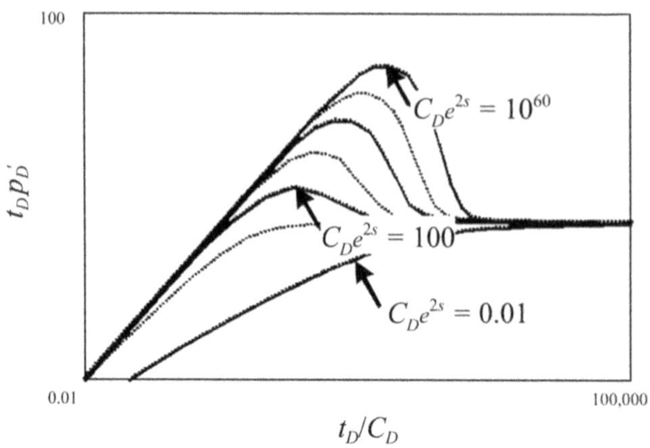

Fig. 8.25—Bourdet's derivative type curves.

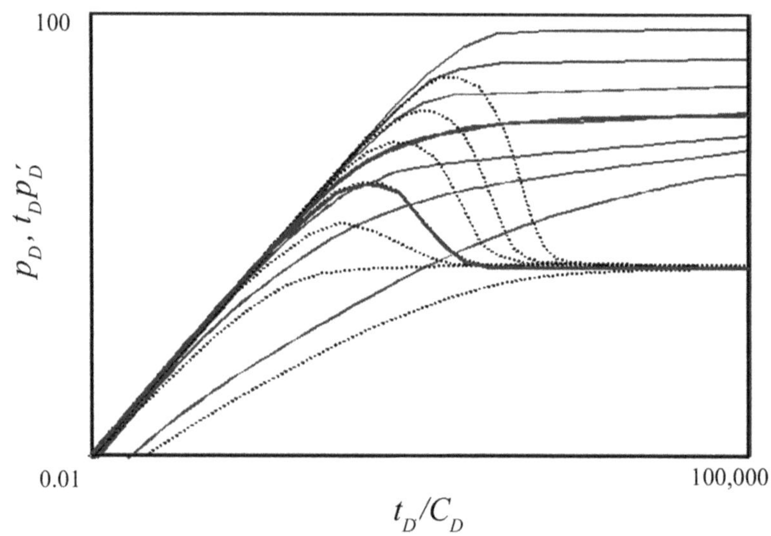

Fig. 8.26—Combined pressure change, pressure derivative type curves.

of curves—one pressure-change curve and one pressure-derivative curve. Time regions can be defined conveniently on the basis of the combined pressure (**Fig. 8.27**) and pressure derivative type curves.

The shape of the pressure- and pressure-derivative type curves provides a qualitative estimate of skin factor (**Fig. 8.28**). For a well with a large skin factor, the derivative rises to a maximum and then falls sharply before flattening out for the middle-time region (MTR). The pressure change curve rises along the unit-slope line and then flattens quickly. The pressure-change and pressure-derivative curves are separated by approximately two log cycles when wellbore storage (WBS) ends.

When the skin is near zero, the pressure derivative rises to a maximum and then falls only slightly before flattening for the MTR. The pressure change and pressure derivative are separated by approximately one log cycle when WBS ends. When the skin factor is negative, the pressure derivative approaches a horizontal line from below. The pressure change and pressure

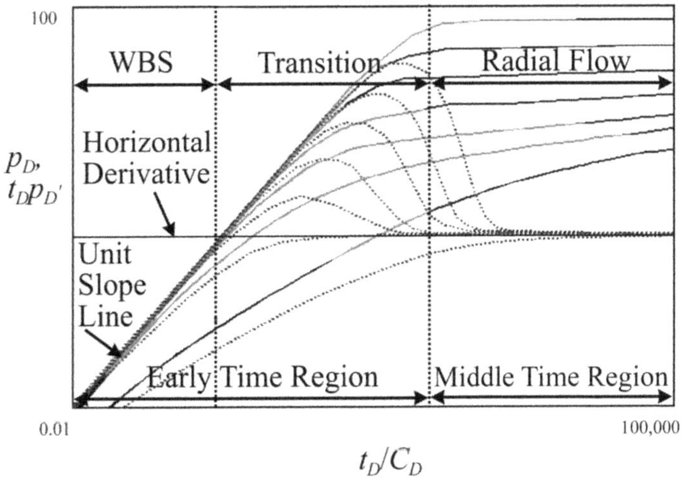

Fig. 8.27—Combined pressure-change/pressure-derivative type curves define time regions conveniently.

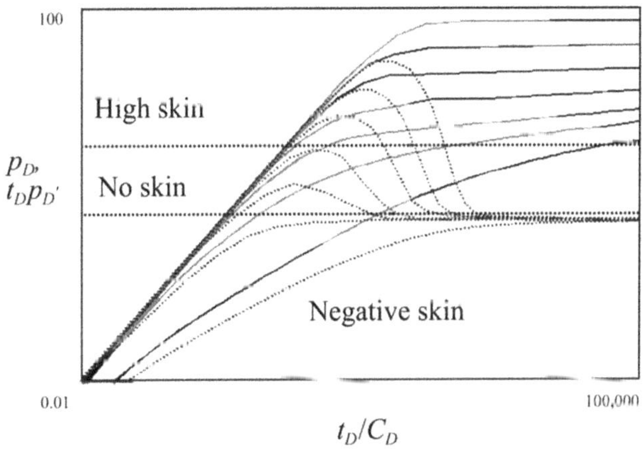

Fig. 8.28—Shape of the type curves provides a qualitative estimate of skin factor.

derivative curves leave the unit slope line at relatively early times and take a relatively long time to reach the MTR.

8.3.4 Differences in Drawdown and Buildup Test Type Curves. The shapes of drawdown and buildup type curves are different, as **Fig. 8.29** illustrates. In this simplified case, in which wellbore storage distortion is absent, a well has produced for a dimensionless producing time, t_{pD}, of 10^5, before shut-in. In the figure, note that, on a plot of p_D and p_D' (the derivative) vs. t_D (dimensionless time since each test began), the shapes of the buildup and drawdown curves for infinite-acting radial flow coincide up to $t_D = 10^4$ and then begin to deviate. The buildup pressure-change curve is "flatter" than the drawdown curve at later times in an infinite-acting reservoir, and thus the slope of the buildup curve (the derivative) tends to deviate from the drawdown derivative. For many years, test analysts used a rule of thumb that buildup tests could be analyzed on a drawdown type curve only up to a maximum time of one-tenth the producing time before shut-in. That rule of thumb is appropriate for the conditions in Fig. 8.29.

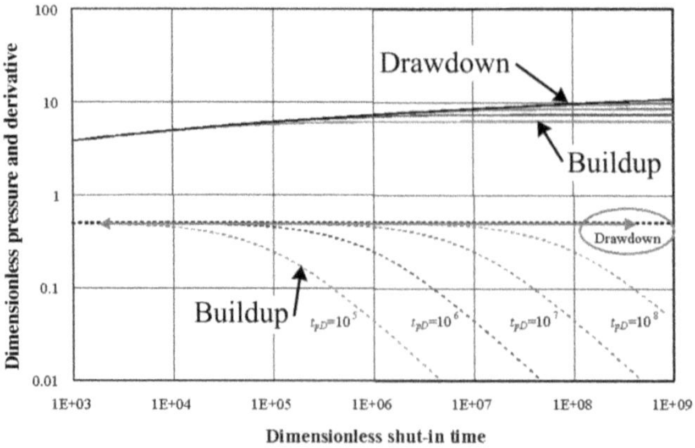

Fig. 8.29—Buildup and drawdown type curves have different shapes.

8.3.5 Equivalent Drawdown Time. Agarwal[9] suggested a method of plotting pressure change data from a buildup test on a logarithmic graph that alters the shape so that it corresponds to that of a constant rate flow test during infinite-acting radial flow. The basis for Agarwal's "equivalent time" is a combination of logarithmic approximations to *Ei*-function solutions to the diffusivity equation. The equation modeling the drawdown at the instant of shut-in is

$$p_i - p_{wf} = 162.6\frac{qB\mu}{kh}\left[\log_{10}(t_p) + \log\left(\frac{k}{\phi\mu c_t r_w^2}\right) - 3.23 + 0.869s\right]. \quad (8.59)$$

We model a buildup test with

$$p_i - p_{wf} = 162.6\frac{qB\mu}{kh}\left[\log_{10}(t_p + \Delta t) + \log\left(\frac{k}{\phi\mu c_t r_w^2}\right) - 3.23 + 0.869s\right]$$

$$-162.6\frac{qB\mu}{kh}\left[\log_{10}(\Delta t) + \log\left(\frac{k}{\phi\mu c_t r_w^2}\right) - 3.23 + 0.869s\right]. \quad (8.60)$$

Combining Eqs. 8.59 and 8.60 and simplifying,

$$p_i - p_{wf} = +162.6\frac{qB\mu}{kh}\left[\log_{10}(t_p) + \log\left(\frac{k}{\phi\mu c_t r_w^2}\right) - 3.23 + 0.869s\right]$$

$$-162.6\frac{qB\mu}{kh}\left[\log_{10}(t_p + \Delta t) + \log\left(\frac{k}{\phi\mu c_t r_w^2}\right) - 3.23 + 0.869s\right]$$

$$+162.6\frac{qB\mu}{kh}\left[\log_{10}(\Delta t) + \log\left(\frac{k}{\phi\mu c_t r_w^2}\right) - 3.23 + 0.869s\right], \quad (8.61)$$

which can be rewritten as

$$p_{ws} - p_{wf} = 162.6\frac{qB\mu}{kh}\left[\log_{10}\left(\frac{t_p\Delta t}{t_p + \Delta t}\right) + \log\left(\frac{k}{\phi\mu c_t r_w^2}\right) - 3.23 + 0.869s\right]. \quad (8.62)$$

The forms of Eqs. 8.62 and 8.59 are the same; thus Eq. 8.62 is an "equivalent" drawdown equation, with the equivalent pressure change, $(p_{ws} - p_{wf})$, a function of equivalent time, $\Delta t_e = t_p\Delta t/(t_p + \Delta t)$. The analogies between these equations suggest that, just as $\Delta p = p_i - p_{wf}$ vs. t were plotted for drawdown tests, $\Delta p = p_{ws} - p_{wf}$ vs. Δt_e can be plotted for buildup tests and achieve the same shapes on logarithmic graphs. However, the theoretical basis for this radial-equivalent time indicates that the equivalence exists only for infinite-acting radial flow and not for data influenced by wellbore storage or by effects of boundaries or other conditions that cause the flow pattern to deviate from radial. In practice, buildup test data for infinite-acting radial flow, including data distorted by wellbore storage, are transformed to the same shape as drawdown test data. However, data affected by boundaries or by linear flow (as in wells with hydraulic fractures) may not be transformed accurately.

Radial equivalent time has the properties

$$\Delta t_e = \frac{t_p\Delta t}{t_p + \Delta t}$$

$$= \frac{t_p}{t_p + \Delta t}\Delta t \approx \Delta t,\ \Delta t << t_p$$

$$= \frac{\Delta t}{t_p + \Delta t}t_p \approx t_p,\ \Delta t >> t_p. \quad (8.63)$$

8.3.6 Type-Curve Matching. The steps in type-curve matching for wells with infinite-acting radial flow are outlined here. Details vary for more complex reservoirs, but the general procedure is similar to that for infinite-acting reservoirs.

- Plot field data on log-log coordinates with the same size log cycles as the type curve.
- Align the horizontal sections of the field data and the type curve.
- Align unit slope regions on the field data and the type curve.
- Select the value of C_De^{2s} that best matches the field data.
- Select pressure and time match points (corresponding values of real and dimensionless variables from field data and type curve plots) from anywhere on the plot.
- Calculate permeability from the pressure match-point ratio,

$$(\Delta p/\Delta p_D)_{MP} = kh/141.2qB\mu, \quad (8.64)$$

$$\text{or } k = 141.2(qB\mu/h)(\Delta p/\Delta p_D)_{MP}. \quad (8.65)$$

- Calculate C_D from the time match-point ratio,

$$[t/(t_D/C_D)]_{MP} = (\phi\mu c_t r_w^2/0.0002637k)C_D, \quad (8.66)$$

$$\text{or } C_D = (0.0002637k/\phi\mu c_t r_w^2)[t/(t_D/C_D)]_{MP}. \quad (8.67)$$

- Calculate s from the matching stem value, C_De^{2s}:

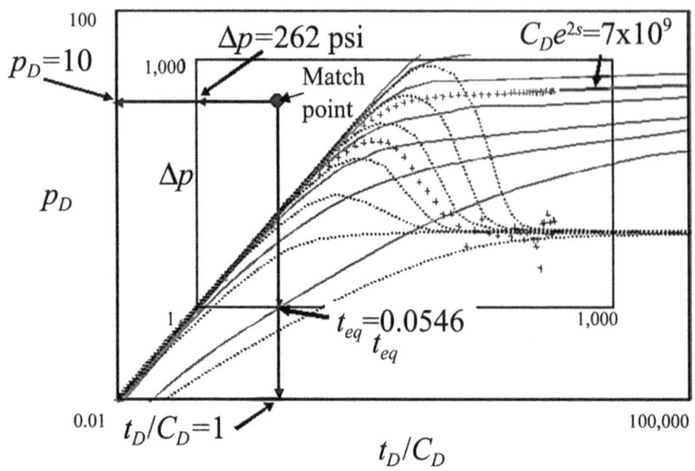

Fig. 8.30—Example interpretation of match points.

$$s = 0.5 \ln \left(C_D e^{2s} / C_D \right). \quad (8.68)$$

Fig. 8.30 shows an example interpretation of match points. In practice, this matching and match-point interpretation procedure is done on the computer and monitor, and much of the process is transparent to the analyst.

8.4 Damage and Stimulation

8.4.1 Causes of Formation Damage. The causes of formation damage that lead to positive skin factors include damage caused by drilling-fluid invasion, production, or injection.

When mud filtrate invades the formation surrounding a borehole, it will generally remain in the formation even after the well is cased and perforated. This mud filtrate in the formation reduces the effective permeability to hydrocarbons near the wellbore. It may also cause clays in the formation to swell, reducing the absolute permeability of the formation. In addition, solid particles from the mud may enter the formation and reduce permeability at the formation face.

The production process may also reduce permeability and introduce a positive skin factor. For example, in an otherwise undersaturated oil reservoir, pressure near the well may be below the bubblepoint pressure, causing a free-gas saturation and reducing the effective permeability to oil. In a retrograde gas reservoir, the pressure near the wellbore may drop below the dewpoint and an immobile liquid phase may form and reduce the effective permeability to gas near the wellbore.

Injection can also cause damage. The water injected may be dirty; that is, it may contain fines that may plug the formation and reduce permeability. In other cases, the injected water may be incompatible with the formation water, causing solids to precipitate and plug the formation. In still other cases, the injected water may be incompatible with clays in the formation (e.g., fresh water can destabilize some clays, causing fines to migrate and plug the formation).

8.4.2 Altered Zone and Skin Effect. A two-region reservoir model (**Fig. 8.31**) is a convenient representation of a damaged well (and some stimulated wells with radially symmetric permeability alteration around the wellbore). In this model, the altered zone around the wellbore is assumed to have uniform permeability k_s out to a radius r_s, beyond which the formation permeability, k, is unaltered.

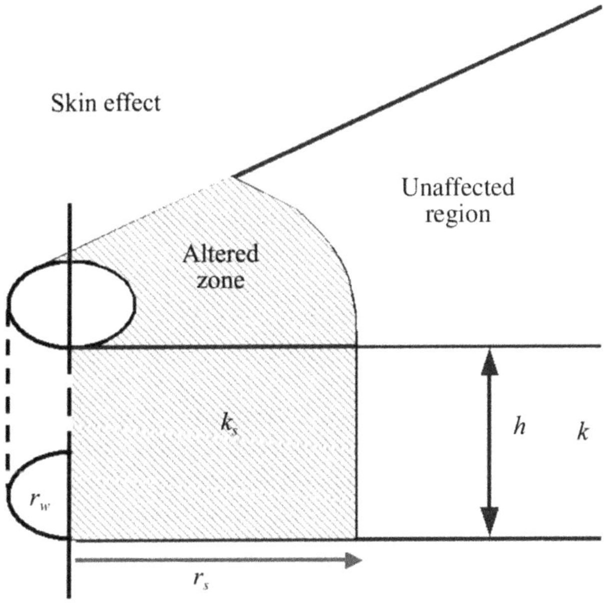

Fig. 8.31—Two-region reservoir model with altered zone around wellbore and unaltered formation permeability beyond.

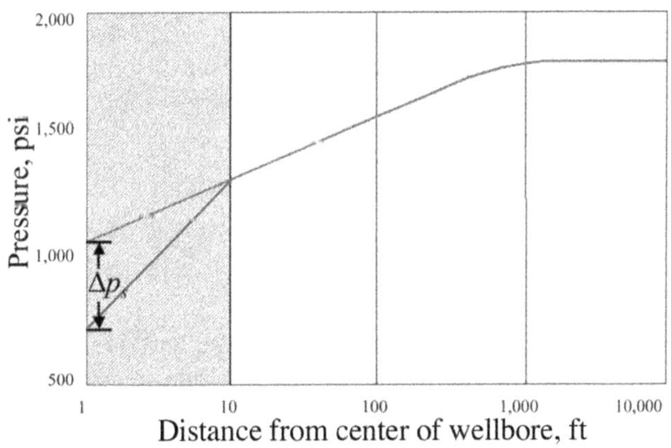

Fig. 8.32—Additional pressure drop in damaged well is caused by reduced permeability in altered zone.

For a damaged well, the reduced permeability in the altered zone causes an additional pressure drop, Δp_s (**Fig. 8.32**). The dimensionless skin factor, s, and the additional pressure drop across the altered zone are related by

$$\Delta p_s = \frac{141.2 q B \mu}{kh} s \text{ . } \qquad (8.69)$$

For a well with a known skin factor, s, Eq. 8.69 provides a method of translating the somewhat abstract dimensionless skin factor into a more concrete characterization of the practical effect of damage or stimulation.

In a two-region reservoir model, the skin factor, s, is related to the properties of the altered zone:

$$s = \left(\frac{k}{k_s} - 1\right) \ln\left(\frac{r_s}{r_w}\right) . \qquad (8.70)$$

Rearrange Eq. 8.70 and solve for the permeability of the altered zone:

$$k_s = \frac{k}{1 + \dfrac{s}{\ln\left(r_s / r_w\right)}} . \qquad (8.71)$$

Rearrangements of Eq. 8.70 provide a second method of translating skin into a more concrete characterization of a well with altered permeability near the wellbore. If the depth of damage can be estimated for a well with a known skin factor, s, the permeability of the altered zone can be estimated. Even if the depth of permeability alteration, r_s, is estimated Eq. 8.71 can still provide a reasonable estimate of altered zone permeability because r_s appears in a logarithmic term. Alternatively, an estimate of the permeability reduction ratio (for example, from laboratory tests on cores) can produce an estimate of the depth of damage from another rearrangement of Eq. 8.70,

$$r_s = r_w \exp\left[s k_s / \left(k - k_s\right)\right] . \qquad (8.72)$$

8.4.3 Apparent Wellbore Radius. A third method of translating skin to a more concrete characterization of near-well conditions is to calculate apparent or effective wellbore radius, r_{wa}. Apparent wellbore radius is defined as

$$r_{wa} = r_w e^{-s} , \qquad (8.73)$$

$$\text{or } s = -\ln\left(r_{wa} / r_w\right) . \qquad (8.74)$$

For a stimulated well, the pressure drawdown at the wellbore is the same as it would be in a formation with unaltered permeability but with wellbore radius equal to the apparent wellbore radius. This concept has value in some simulation applications. Note that r_{wa} can be calculated from the actual wellbore radius and skin factor.

Eqs. 8.73 and 8.74 are also useful to illustrate the minimum (i.e., the most-negative possible) skin factor. This minimum skin, s_{min}, occurs when the apparent wellbore radius is equal to the drainage radius of the well:

$$s_{min} = -\ln\left(r_e / r_w\right) . \qquad (8.75)$$

For a well with a circular drainage area of 40 acres for which r_e is 745 ft and a wellbore radius of 0.3 ft, the minimum skin (maximum stimulation) is $s_{min} = -\ln\left(r_e / r_w\right) = -(745/0.3) = -7.82$. Such a skin implies increasing the permeability throughout the entire altered zone to infinity—clearly an idealistic “upper limit.” More realistically, research[10] has shown that the half-length, L_f, of a highly conductive vertical fracture is related to r_{wa} by

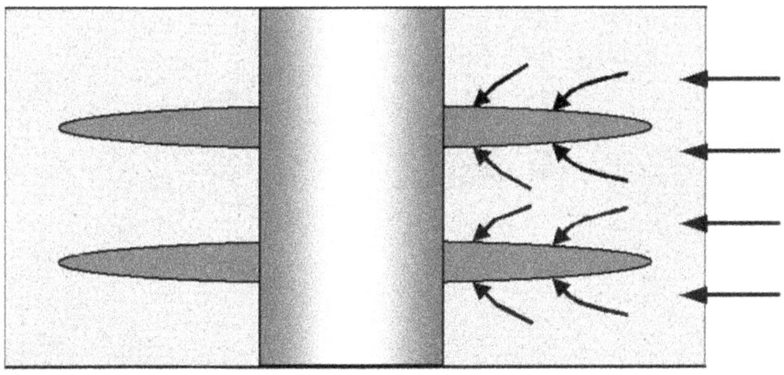

Fig. 8.33—When perforation spacing is large, converging flow results in positive skin factor.

$$r_{wa} = \frac{L_f}{2} = r_w e^{-s}, \quad (8.76)$$

$$\text{or } s = -\ln\left(\frac{L_f}{2r_w}\right). \quad (8.77)$$

Thus, for $L_f = r_e = 745$ ft, $s = -7.12$ is a more realistic minimum (for the given drainage radius and wellbore radius).

8.4.4 Flow Efficiency. A fourth way to characterize a well with nonzero skin is to calculate the flow efficiency of the well. Flow efficiency, E_f, is defined as the ratio of the actual productivity index of the well (including skin) to the ideal productivity index if the skin factor were zero. Because the productivity index is the ratio of stabilized flow rate to pressure drop required to sustain that stabilized rate,

$$J_{\text{actual}} = \frac{q}{\left(\overline{p} - p_{wf}\right)}, \quad (8.78)$$

$$J_{\text{ideal}} = \frac{q}{\left(\overline{p} - p_{wf} - \Delta p_s\right)}, \quad (8.79)$$

$$\text{and } E_f \equiv \frac{J_{\text{actual}}}{J_{\text{ideal}}} = \frac{\overline{p} - p_{wf} - \Delta p_s}{\overline{p} - p_{wf}}. \quad (8.80)$$

For a well with neither damage nor stimulation, $E_f = 1$; for a damaged well, $E_f < 1$; and for a stimulated well, $E_f > 1$.

8.4.5 Geometric Skin. When the area open to flow decreases, the pressure drop is greater than when the area is unchanged all the way to the formation face. Examples include flow converging to perforations (**Fig. 8.33**), partial penetration (**Fig. 8.34**), and an incompletely perforated interval (**Fig. 8.35**).

Fig. 8.33 illustrates flow converging into perforations. When the perforation spacing is too large, this converging flow results in a positive skin factor. The skin increases as vertical permeability decreases and increases as shot density decreases.

Partial Penetration. Fig. 8.34 illustrates flow converging into an interval that is only partly penetrated by perforations. When a well is completed in only a fraction of the productive inter-

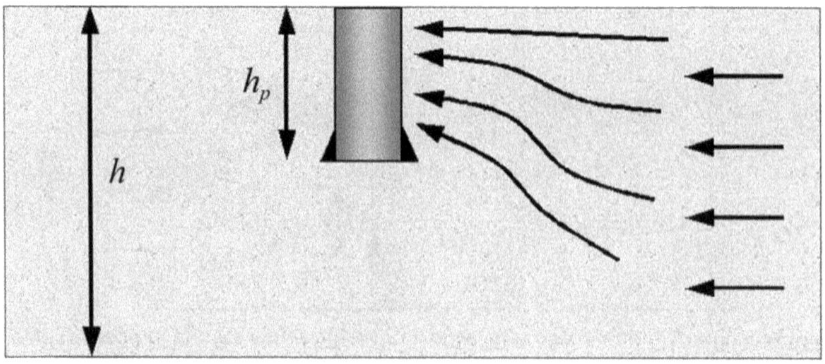

Fig. 8.34—When interval is only partially penetrated by perforations, flow converges through small area, increasing pressure drop near the well.

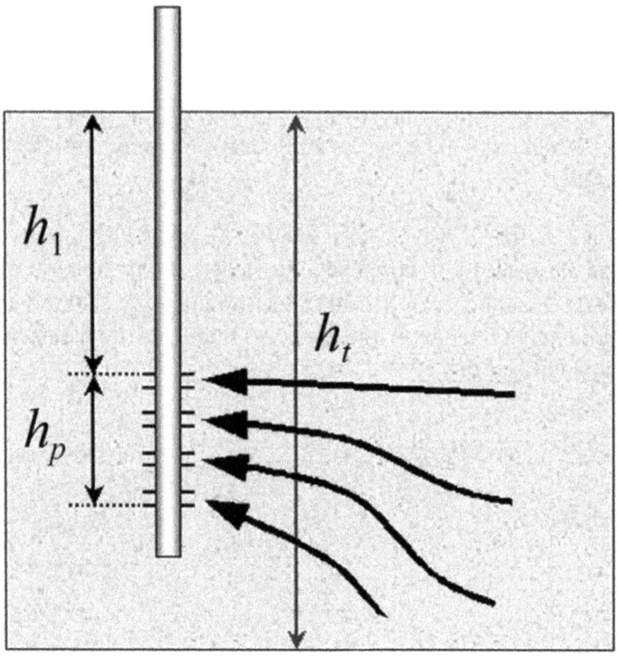

Fig. 8.35—In an incompletely penetrated well, only a portion of the total productive interval (*h*) is perforated, increasing pressure drop and skin.

val, the flow must converge through a smaller area, increasing the pressure drop near the well (compared to a fully completed interval). The additional pressure drop near the well results in a more positive skin. It increases as the vertical permeability decreases and as the perforated interval as a fraction of the total interval decreases. Formation damage (reduced permeability) near the completion face can significantly increase the additional pressure drop and thus the calculated skin factor.

Incompletely Perforated Interval. Partial penetration is a special case of an incompletely perforated interval (Fig. 8.35). In the general case, the well is perforated starting at a distance h_1 from the top of the productive interval and has perforations extending over a distance, h_p, in an interval of total thickness, h. The total skin for the well in this general situation is

$$s = (h/h_p)s_d + s_p \;. \quad (8.81)$$

In Eq. 8.81, s_d is the skin caused by formation damage, and s_p is the skin resulting from an incompletely perforated interval. This equation is not valid for a stimulated well.

The skin factor for an incompletely perforated interval, s_p, can be quantified by[11]

$$s_p = \left(\frac{1}{h_{pD}} - 1\right)\ln\frac{\pi}{2r_D} + \frac{1}{h_{pD}}\ln\left[\frac{h_{pD}}{2+h_{pD}}\left(\frac{A-1}{B-1}\right)^{1/2}\right], \quad (8.82)$$

$$\text{where } h_{1D} = h_1/h, \quad (8.83)$$

$$h_{pD} = h_p/h, \quad (8.84)$$

$$A = \frac{1}{h_{1D} + h_{pD}/4}, \quad (8.85)$$

$$r_D = \frac{r_w}{h}\left(\frac{k_v}{k_h}\right)^{1/2}, \quad (8.86)$$

$$\text{and } B = \frac{1}{h_{1D} + 3h_{pD}/4} \;. \quad (8.87)$$

The most significant limitation in applying Eq. 8.82 in practice is the difficulty in estimating accurately the vertical-to-horizontal-permeability ratio, k_v/k_h. Fortunately, this ratio appears only in a logarithmic term in Eq. 8.82, so errors will not seriously distort the calculated value of s_p.

Deviated Well. For a deviated well (**Fig. 8.36**), which penetrates the formation at an angle other than 90°, more surface is in contact with the formation. This introduces a negative skin factor, s_θ, which makes the total skin factor, s, more negative.

$$s = s_d + s_\theta \;. \quad (8.88)$$

The effect increases as the vertical permeability increases and increases as the angle from the vertical, θ_w, increases. The deviated well skin factor, s_θ, is given by a correlation of simulated results[12] (valid for $\theta_w < 75°$):

$$s_\theta = -\left(\frac{\theta'_w}{41}\right)^{2.06} - \left(\frac{\theta'_w}{56}\right)^{1.865}\log\left(\frac{h_D}{100}\right), \quad (8.89)$$

$$\text{where } \theta'_w = \tan^{-1}\left(\sqrt{\frac{k_v}{k_h}}\tan\theta_w\right), \quad (8.90)$$

$$\text{and } h_D = \frac{h}{r_w}\sqrt{\frac{k_h}{k_v}} \;. \quad (8.91)$$

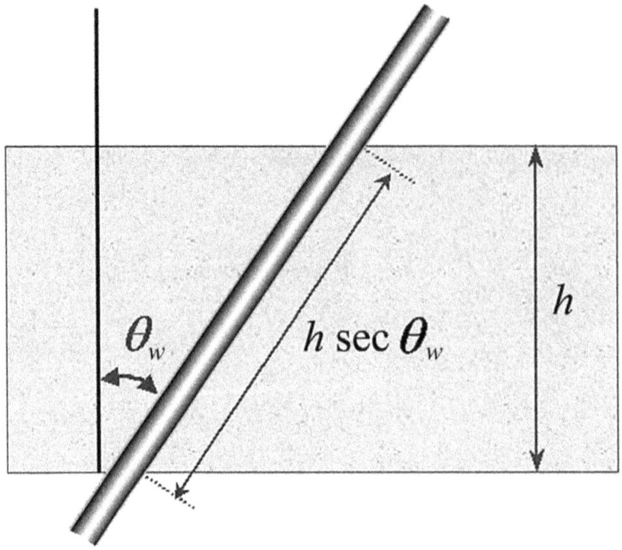

Fig. 8.36—More surface of a deviated well is in contact with formation, introducing a negative skin factor.

Gravel-Pack Skin. When a well is gravel packed (**Fig. 8.37**), there is a pressure drop through the gravel pack within the perforations, given by[13]

$$s_{gp} = \frac{khL_g}{2nk_{gp}r_p^2}, \quad (8.92)$$

where s_{gp} is the skin factor because of Darcy flow through the gravel pack; h, the net pay thickness, ft; k_{gp}, the permeability of the gravel in the gravel pack, md; k, the reservoir permeability, md; L_g, the length of the flow path through the gravel pack, ft; n, the number of perforations open; and r_p, the radius of the perforation tunnel, ft. Eq. 8.92 does not include the effects of non-Darcy flow, which may be extremely important in high-rate gas wells.

Completion Skin. For a perforated well, any reduced permeability, k_{dp}, in the zone surrounding the perforations (**Fig. 8.38**) introduces an additional pressure drop. The additional skin is[14]

$$s_{dp} = \left(\frac{h}{L_p n}\right)\left(\ln \frac{r_{dp}}{r_p}\right)\left(\frac{k}{k_{dp}} - \frac{k}{k_d}\right), \quad (8.93)$$

$$\text{and } s = s_p + s_d + s_{dp}, \quad (8.94)$$

where s_p is the geometric skin from flow converging to the perforations; s_d, the damage skin; s_{dp}, perforation damage skin; k_d, permeability of the damaged zone around the wellbore, md; k_{dp}, permeability of the damaged zone around perforation tunnels, md; k, reservoir permeability, md; L_p, length of perforation tunnel, ft; n, number of perforations; h, formation thickness, ft; r_d, radius of the damaged zone around the wellbore, ft; r_{dp}, radius of the damages zone around the perforation tunnel, ft; r_p, radius of the perforation tunnel, ft; and r_w, wellbore radius, ft. Eq. 8.94 does not include the effects of non-Darcy flow.

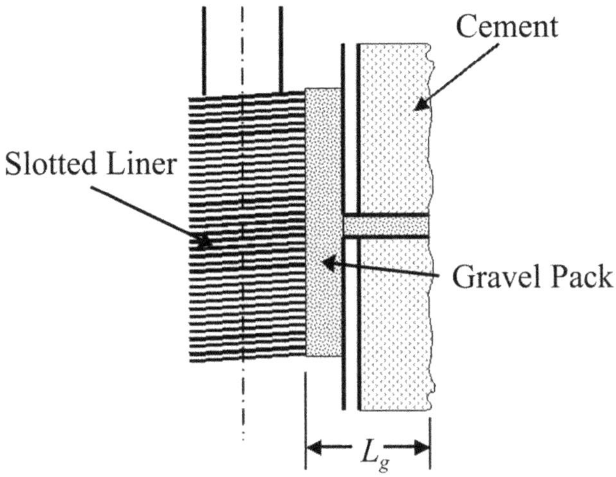

Fig. 8.37—Pressure drop in a gravel-packed well includes the skin effect in perforations.

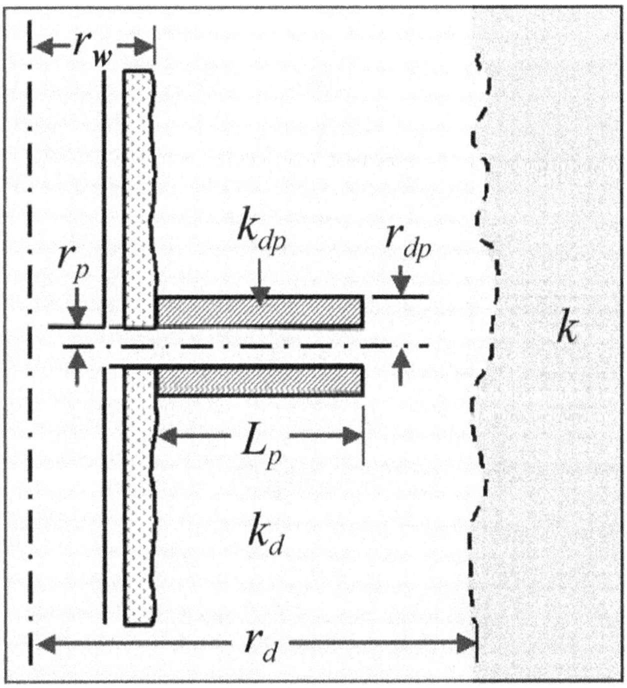

Fig. 8.38—Reduced permeability in damaged zone surrounding perforations introduces additional pressure drop and skin.

Hydraulically Fractured Wells. Wells are frequently fractured hydraulically to improve their productivity, especially in low-permeability formations where fractures increase the effective drained area and in high-permeability formations where they penetrate near-well damage or promote sand control. These fractures, almost always vertical (**Fig. 8.39**), are high-conductivity paths between the reservoir and the wellbore. If the fracture conductivity is large enough relative to the formation permeability and fracture length, the pressure drop within the fracture

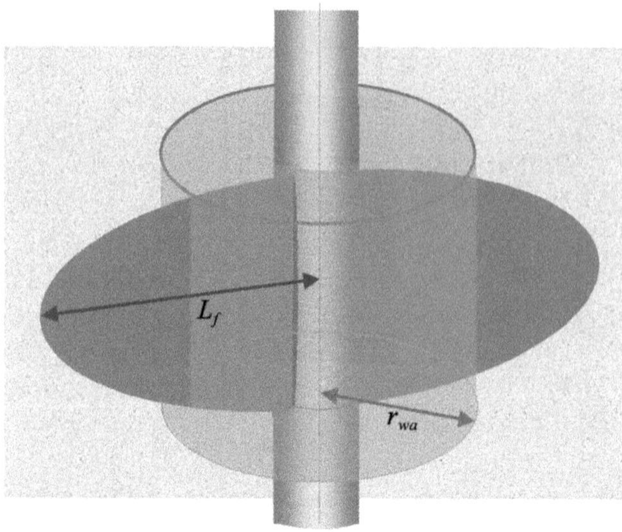

Fig. 8.39—Hydraulic fractures change the flow pattern in a reservoir and introduce a negative geometric skin effect.

will be negligible. This distributes the pressure drop caused by fluid influx into the wellbore over a much larger area, resulting in a negative skin factor, which is interpreted as a geometric skin.

Dimensionless fracture conductivity, C_r, is defined by

$$C_r = \frac{w_f k_f \pi}{k L_f}, \quad (8.95)$$

where w_f is the fracture length, ft; k_f, the permeability of the proppant in the fracture; k, the formation permeability, md; and L_f, the fracture half-length, ft. Pressure drop in the fracture is negligible for $C_r > 100$.

8.5 Modifications for Gases and Multiphase Flow

8.5.1 Diffusivity Equation for Gas Flow. The diffusivity equation for liquids, Eq. 8.1,

$$\frac{1}{r}\frac{\partial}{\partial r}\left(r\frac{\partial p}{\partial r}\right) = \frac{\phi \mu c_t}{0.0002637k}\frac{\partial p}{\partial t}, \quad (8.1)$$

was derived from three principles: conservation of mass, the equation of state for slightly compressible liquids, and Darcy's law. This form of the diffusivity equation is linear, which makes solutions (such as the *Ei*-function solution) much easier to find and which allows us to use superposition in time and space to develop solutions for complex flow geometries and for variable rate histories from simple, single-well solutions.

8.5.2 Pseudopressure. Other forms of the equation for flow of gases must be developed because the equation of state for a slightly compressible liquid will not be applicable. First, introducing the real gas law,

$$pV = znRT, \quad (8.96)$$

to replace the slightly compressible equation of state results in a more complex, nonlinear partial differential equation. This equation can be partially linearized by introducing the pseudopressure transformation,[15]

$$p_p(p) = 2\int_{p_o}^{p} p\,dp/\mu z, \quad (8.97)$$

where p_0 is an arbitrary "base" pressure, frequently chosen to be zero psia. The resulting form of the diffusivity equation is

$$\frac{1}{r}\frac{\partial}{\partial r}\left(r\frac{\partial p_p}{\partial r}\right) = \frac{\phi\mu c_t}{0.0002637k}\frac{\partial p_p}{\partial t}. \quad (8.98)$$

Eq. 8.98 has the same form as the diffusivity equation for slightly compressible liquids, with pressure replaced by pseudopressure, p_p. However, this equation is nonlinear because the product μc_t is a strong function of pressure. Fortunately, research has shown that the equation can be treated as linear, and the *Ei*-function is valid for gases if μc_t is evaluated at the pressure at the beginning of a flow period until the time when boundaries begin to have a significant influence on the pressure drop at the well; that is, as long as the reservoir is infinite-acting.

8.5.3 Pressure-Squared and Pressure Approximations. By assuming that the product μz is constant, then, from Eq. 8.97, pseudopressure becomes

$$p_p(p) = \frac{1}{\mu z}\left(p^2 - p_0^2\right), \quad (8.99)$$

and the diffusivity equation becomes

$$\frac{1}{r}\frac{\partial}{\partial r}\left(r\frac{\partial p^2}{\partial r}\right) = \frac{\phi\mu c_t}{0.0002637k}\frac{\partial p^2}{\partial t}. \quad (8.100)$$

The independent variable has become p^2, and, in terms of this variable, the *Ei*-function solution is valid when the assumption that μz is constant is valid. This is true (based on empirical evidence) even though Eq. 8.100 is nonlinear (pressure-dependent μc_t), but it is valid only for an infinite-acting reservoir.

Fig. 8.40 shows the range of validity of this assumption for a reservoir temperature of 200°F and several different gas gravities. The μz product is fairly constant at pressures below approximately 2,000 psia (the shaded area in the figure). Conclusions are similar at other temperatures from 100 to 300°F.

By assuming that the group $p/\mu z$ is constant, from Eq. 8.97, pseudopressure becomes

$$p_p(p) = (p/\mu z)(p - p_0), \quad (8.101)$$

and the diffusivity equation becomes

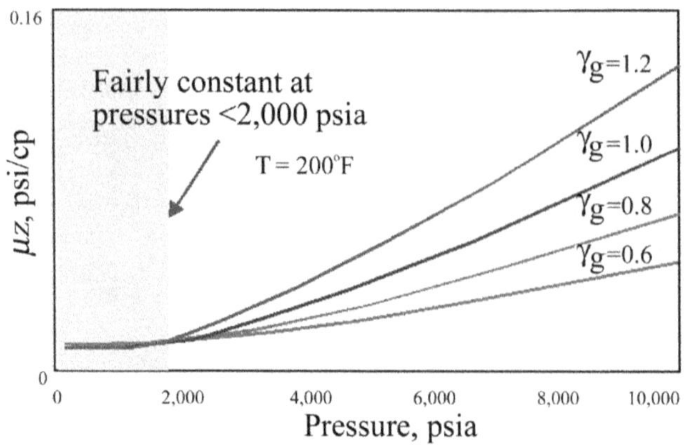

Fig. 8.40—*μz* product is fairly constant at low pressures. Shaded area indicates pressure range in which *μz* is constant.

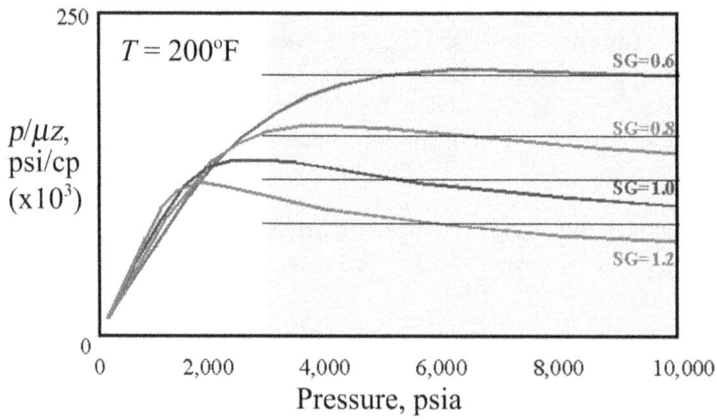

Fig. 8.41—*p/μz* product is fairly constant for high pressures (*p* > 3,000 psia). Shaded area indicates pressure range in which *μz* is constant.

$$\frac{1}{r}\frac{\partial}{\partial r}\left(r\frac{\partial p}{\partial r}\right)=\frac{\phi\mu c_t}{0.0002637k}\frac{\partial p}{\partial t}\ . \qquad (8.102)$$

The independent variable has become p, and, in terms of pressure, the Ei-function is valid (from empirical evidence) when the assumption that $p/\mu z$ is constant is valid. This is true even though Eq. 8.102 is nonlinear (pressure-dependent μc_t), but is valid only for an infinite-acting reservoir.

Fig. 8.41 shows the range of validity of this assumption (shaded area in the figure) for a reservoir temperature of 200°F and several different gas gravities. The group $p/\mu z$ is fairly constant at pressures above approximately 3,000 psia as it is at other temperatures from 100 to 300°F.

The implication of these results is that the choice of variable for gas well-flow equations depends on the situation. The pressure-squared approximation is valid only for low pressures (p < 2,000 psia), the pressure approximation is valid only for high pressures (p > 3,000 psia), and the pseudopressure transformation is valid for all pressure ranges. For pressure transient test analysis using software, the pseudopressure is almost always the optimal variable to use. For hand analysis, only pressure or pressure-squared approaches are feasible.

8.5.4 Pseudotime. Although the diffusivity equation written for gas flow has the same form as the diffusivity equation for slightly compressible liquids, with pressure replaced by pseudopressure, it is a nonlinear equation because the product, μc_t, is strongly pressure dependent. In some cases, the remaining nonlinearity cannot be ignored. To solve this problem, Agarwal[16] introduced the pseudotime transformation to further linearize the diffusivity equation for gas. (The linearization is not rigorous, but is adequate for many practical purposes.[17]) The definition of pseudotime is

$$t_{ap} \equiv \int_0^{\Delta t} \frac{dt}{\mu(p)c_t(p)} \; . \qquad (8.103)$$

In terms of pseudotime, t_{ap}, the diffusivity equation becomes

$$\frac{1}{r}\frac{\partial}{\partial r}\left(r\frac{\partial p_p}{\partial r}\right) = \frac{\phi}{0.0002637k}\frac{\partial p_p}{\partial t_{ap}} \; . \qquad (8.104)$$

Subsequent studies[18] have shown that the pseudotime transformation is particularly useful for analysis of flow and buildup tests distorted by wellbore storage when using type curves designed to model flow of slightly compressible liquids.

Because the pressure in the integrand of Eq. 8.103 is a function of position in the reservoir, it is not obvious where the pressure is to be evaluated. Empirical observations[18] indicate that the pressure should be evaluated at BHP during wellbore storage distortion for both buildup and flow tests. During the middle time region for buildup tests, it should be evaluated at BHP, and, for flow tests, at the average reservoir pressure at the start of the test. For flow tests in infinite-acting reservoirs, this is equivalent to using ordinary time as the independent variable.

8.5.5 Normalized Transformed Variables. The pseudopressure and pseudotime transformations provide excellent results when used as part of the analysis procedure for gas well tests. However, they are inconvenient for two reasons: the values of both variables will often be in the range of 10^5 to 10^9, and they do not have units of actual pressure and time. Thus, the intuitive "feel" for the transformed variables is lost, and they may tend to be regarded as "black box" output—never helpful in test analysis. The use of pseudopressure and pseudotime require different test interpretation equations for oil wells than for gas wells.

These difficulties are overcome by normalizing pseudopressure and pseudotime by multiplying them by constants[19]:

$$p_a(p) \equiv \left(\frac{\mu z}{p}\right)_i \int_{p_0}^{p} \frac{p\,dp}{\mu z} = \left(\frac{\mu z}{2p}\right)_i p_p(p), \qquad (8.105)$$

$$\text{and } \Delta t_a \equiv \left(\mu c_t\right)_i \int_0^{\Delta t} \frac{dt}{\mu(p)c_t(p)} = \left(\mu c_t\right)_i \Delta t_{ap} \; . \qquad (8.106)$$

TABLE 8.1—PLOTTING METHODS AND INTERPRETATION EQUATIONS FOR WELL TESTS

	Flow Test	Buildup Test
Oil		
Semilog graph variables	p_{wf} vs. t	p_{ws} vs. $(t_p + \Delta t)/\Delta t$
Permeability	$k = \dfrac{162.6 q_o B_o \mu_o}{-mh}$	$k = \dfrac{162.6 q_o B_o \mu_o}{-mh}$
Skin factor	$s = 1.151\left[\dfrac{p_i - p_{1hr}}{-m} - \log\left(\dfrac{k}{\phi \mu_o c_t r_w^2}\right) + 3.23\right]$	$s = 1.151\left[\dfrac{p_{1hr} - p_{wf}}{-m} - \log\left(\dfrac{k}{\phi \mu_o c_t r_w^2}\right) + 3.23\right]$
Definition of $p_{MBH,D}$	NA	$\dfrac{kh(p^* - \bar{p})}{70.6 q_o B_o \mu_o}$
Gas, Using Adjusted Variables		
Semilog graph variables	$p_{a,wf}$ vs. t	$p_{a,ws}$ vs. $(t_p + \Delta t_a)/\Delta t_a$
Permeability	$k = \dfrac{162.6 q_g B_{gi} \mu_i}{-mh}$	$k = \dfrac{162.6 q_g B_{gi} \mu_i}{-mh}$
Skin factor	$s' = 1.151\left[\dfrac{p_{a,i} - p_{a,1hr}}{-m} - \log\left(\dfrac{k}{\phi \mu_i c_{ti} r_w^2}\right) + 3.23\right]$	$s' = 1.151\left[\dfrac{p_{a,1hr} - p_{a,wf}}{-m} - \log\left(\dfrac{k}{\phi \mu_i c_{ti} r_w^2}\right) + 3.23\right]$
Definition of $p_{MBH,D}$	NA	$\dfrac{kh(p_a^* - \bar{p}_a)}{70.6 q_g B_{gi} \mu_i}$
Gas, Using Pressure and Time		
Semilog graph variables	p_{wf} vs. t	p_{ws} vs. $(t_p + \Delta t)/\Delta t$
Permeability	$k = \dfrac{162.6 q_g \bar{B}_g \bar{\mu}}{-mh}$	$k = \dfrac{162.6 q_g \bar{B}_g \bar{\mu}}{-mh}$
Skin factor	$s' = 1.151\left[\dfrac{p_i - p_{1hr}}{-m} - \log\left(\dfrac{k}{\phi \bar{\mu} \bar{c}_t r_w^2}\right) + 3.23\right]$	$s' = 1.151\left[\dfrac{p_{1hr} - p_{wf}}{-m} - \log\left(\dfrac{k}{\phi \bar{\mu} \bar{c}_t r_w^2}\right) + 3.23\right]$
Definition of $p_{MBH,D}$	NA	$\dfrac{kh(p^* - \bar{p})}{70.6 q_g \bar{B}_g \bar{\mu}}$

This normalization, or multiplication by appropriate constants, gives the new variables the same units—and similar ranges—as pressure and time, respectively. With these transformations, the equations for analysis of gas wells in terms of normalized pseudopressure and pseudotime, which are called adjusted pressure and adjusted time, are obtained from the equations for analysis of oil well tests by simple substitution. Of course, the transformations require the computer. Commercial well-test analysis software often provides these transformations.

Table 8.1 summarizes plotting methods and interpretation equations for oil well tests. It also presents information for gas well tests analyzed with ordinary pressure and time, adjusted pressure and time, pressure squared and time, and, finally, pseudopressure and time. The table includes a definition of p_{DMBH}, a dimensionless pressure defined by Matthews, Brons, and Hazebroek[20] that is useful in estimating current average drainage pressure. See this topic in Section 8.8.

TABLE 8.1—PLOTTING METHODS AND INTERPRETATION EQUATIONS FOR WELL TESTS (continued)		
	Gas, Using Pressure Squared and Time	
	Flow Test	Buildup Test
Semilog graph variables	p_{wf}^2 vs. t	p_{ws}^2 vs. $(t_p + \Delta t)/\Delta t$
Permeability	$k = \dfrac{1637 q_g T \bar{z}\bar{\mu}}{-mh}$	$k = \dfrac{1637 q_g T \bar{z}\bar{\mu}}{mh}$
Skin factor	$s' = 1.151\left[\dfrac{p_i^2 - p_{1hr}^2}{-m} - \log\left(\dfrac{k}{\phi\bar{\mu}\bar{c}_t r_w^2}\right) + 3.23\right]$	$s' = 1.151\left[\dfrac{p_{1hr}^2 - p_{wf}^2}{-m} - \log\left(\dfrac{k}{\phi\bar{\mu}\bar{c}_t r_w^2}\right) + 3.23\right]$
Definition of $p_{MBH,D}$	NA	$\dfrac{kh\left(p^{*2} - \bar{p}^2\right)}{711 q_g T \bar{z}\bar{\mu}}$
	Gas, Using Pseudopressure and Time	
	Flow Test	Buildup Test
Semilog graph variables	p_p vs. t	p_p vs. $(t_p + \Delta t)/\Delta t$
Permeability	$k = \dfrac{1637 q_g T}{-mh}$	$k = \dfrac{1637 q_g T}{-mh}$
Skin factor	$s' = 1.151\left[\dfrac{p_{p,i} - p_{p,1hr}}{-m} - \log\left(\dfrac{k}{\phi\bar{\mu}\bar{c}_t r_w^2}\right) + 3.23\right]$	$s' = 1.151\left[\dfrac{p_{p,1hr} - p_{p,wf}}{-m} - \log\left(\dfrac{k}{\phi\bar{\mu}\bar{c}_t r_w^2}\right) + 3.23\right]$
Definition of $p_{MBH,D}$	NA	$\dfrac{kh\left(p_p^* - \bar{p}_p\right)}{711 q_g T}$

In Table 8.1, the HTR for gas well buildup tests is best estimated to be simply $(t_p + \Delta t_a)/\Delta t_a$. This conclusion is based on the findings of Spivey and Lee.[18] Thus, when using adjusted pressure and time, the HTR is calculated using the actual producing time, t_p.

8.5.6 Non-Darcy Flow. The flow equations shown to this point assume that Darcy's law is an appropriate model for gas flow into wells. However, as the flow velocity and Reynolds number near the well increase, the result is a transition from laminar and turbulent flow and then to turbulent flow. This transitional (and possibly turbulent) flow is called non-Darcy (non-laminar) flow. The high velocities at which the flow is transitional occur in the immediate vicinity of the well, and the additional pressure drop caused by this transitional flow is similar to a zone of altered permeability that is characterized with a skin factor. In the case of non-Darcy flow, however, the additional "skin effect" caused by the deviations from Darcy's law is rate dependent.

An adequate model for the apparent skin factor, s′, determined from a flow or buildup test is

$$s' = s + D\left|q_g\right| \quad \text{.......... (8.107)}$$

In Eq. 8.107, s is the "true" skin because of damage or stimulation; D is a non-Darcy flow coefficient (assumed constant), with units of D/Mscf; and q_g is the gas flow rate with units of Mscf/D. The absolute value of the gas rate is used because the contribution to the skin is positive regardless of whether the gas well is a producer or an injector.

The true skin for a gas well cannot be obtained from information in a single test conducted at constant rate (including a buildup test following constant-rate production). However, skin calculated from tests conducted at several different rates (for example, associated with a multi-

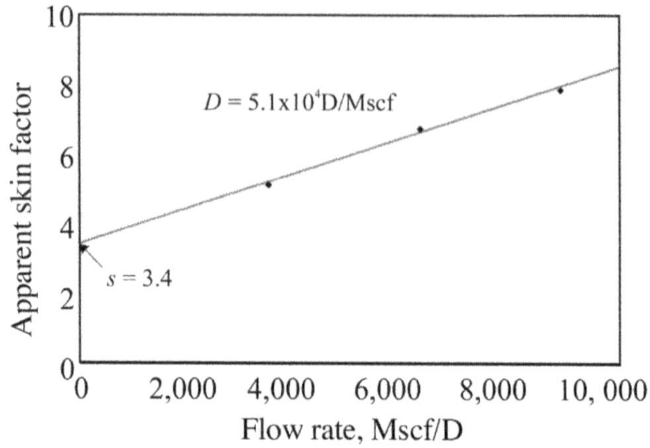

Fig. 8.42—Process to obtain true skin from rate-dependent skin factor.

point deliverability test on a well) can be used to determine the true skin and the non-Darcy flow coefficient. **Fig. 8.42** illustrates the process for a well tested at three different rates, with an apparent skin factor determined at each rate.

The apparent skin factor extrapolated to zero rate is the true skin (in this case, 3.4), and the slope of the curve is the non-Darcy flow coefficient, D (in this case, 5.1×10^{-4} D/Mscf). When this method is used, take care to ensure that the permeabilities obtained from the different tests are the same; otherwise, the skin factors will be inconsistent and erroneous.

Often, only one test is available. In this case, the non-Darcy flow coefficient, D, can be estimated from[3]

$$D = \frac{2.715 \times 10^{-15} \beta k_g M p_{sc}}{h r_w T_{sc} \mu_{gwf}} \quad \text{(8.108)}$$

The turbulence parameter, β, can be estimated from[21]

$$\beta \approx 1.88 \times 10^{10} k^{-1.47} \phi^{-0.53} \quad \text{(8.109)}$$

The correlation represented by Eq. 8.109 will provide only a crude estimate of the turbulence parameter, β. Further, the correlation assumes that the non-Darcy flow occurs in the formation near the wellbore rather than through the perforations. In a gravel-packed well, the most significant additional pressure drop caused by non-Darcy flow may occur in the perforation channels through the casing.

8.5.7 Multiphase Flow. The equations modeling flow in reservoirs can be modified to include multiphase flow. Perrine[22] suggested simple and easily applied modifications and Martin[23] gave them a theoretical basis. These modifications are based on the simplifying assumption that the saturation gradients in the drainage area of the tested well are small. Thus, as examples, the modifications may lead to reasonable approximations for solution-gas drive reservoirs and are inappropriate for water-drive reservoirs with a water bank (and saturation discontinuity) in the drainage area of the tested well. The Perrine-Martin modification for constant-rate flow in an infinite-acting reservoir is

$$p_{wf} = p_i + 162.6\frac{q_{Rt}}{\lambda_t h}\left[\log\left(\frac{1{,}688\phi c_t r_w^2}{\lambda_t t}\right) - 0.868s\right], \quad (8.110)$$

and the Horner equation modeling a buildup test in an infinite-acting reservoir becomes

$$p_{ws} = p_i - 162.6\frac{q_{Rt}}{\lambda_t h}\log\frac{t_p + \Delta t}{\Delta t}. \quad (8.111)$$

In Eqs. 8.110 and 8.111, q_{Rt} represents the total reservoir flow rate (RB/D) and is given by

$$q_{Rt} = q_0 B_0 + \left(q_g - \frac{q_o R_s}{1{,}000}\right)B_g + q_w B_w, \quad (8.112)$$

and λ_t represents the total mobility, given by

$$\lambda_t = \frac{k_o}{\mu_o} + \frac{k_w}{\mu_w} + \frac{k_g}{\mu_g}. \quad (8.113)$$

The total mobility, λ_t, can be determined from a pressure buildup test run on a well that produces two or three phases simultaneously. Because Eq. 8.111 implies that λ_t is related to the slope, m, of a Horner plot of p_{ws} vs. $\log\left(t_p + \Delta t\right)/\Delta t$ by

$$\lambda_t = \frac{-162.6 q_{Rt}}{mh}. \quad (8.114)$$

The slope, m, of a plot of p_{wf} vs. log (t) data from a constant-rate flow test has the same interpretation. Perrine[22] also showed that the permeability to each phase flowing can be estimated from the relations

$$k_o = -162.6\frac{q_o B_o \mu_o}{mh}, \quad (8.115)$$

$$k_g = -162.6\frac{\left(q_g - \frac{q_o R_s}{1{,}000}\right)B_g \mu_g}{mh}, \quad (8.116)$$

$$\text{and } k_w = -162.6\frac{q_w B_w \mu_w}{mh}. \quad (8.117)$$

The quantity $(q_g - q_o R_s/1{,}000)B_g$ in Eqs. 8.112 and 8.116 is the free-gas flow rate in the reservoir; that is, the difference in the total gas rate, q_g, and the dissolved gas rate, $q_o R_s/1{,}000$. Skin factor for multiphase flow test analysis using semilog plots is calculated from

$$s = 1.151\left[\frac{\Delta p_{1\,\mathrm{hr}}}{-m} - \log\left(\frac{\lambda_t}{\phi c_t r_w^2}\right) + 3.23\right]. \quad (8.118)$$

For analysis of tests using type curves, note that the pressure match point on a type curve is related to total and individual phase mobilities and rates by

$$\frac{\lambda_t}{q_{Rt}} = \frac{k_o}{\mu_o q_o B_o} = \frac{k_w}{\mu_w q_w B_w}$$

$$= \frac{k_g}{\mu_g\left[\left(q_g - q_o R_s\right)/1{,}000\right]B_g} = \left(\frac{141.2}{h}\right)\left(\frac{p_D}{\Delta p}\right)_{\text{MP}}, \quad \text{(8.119)}$$

and the time match point is related to the dimensionless storage coefficient by

$$C_D = \left[0.0002637\phi c_t (tC_D/t_D)_{\text{MP}}/r_w^2\right]\lambda_t$$

$$= \left[0.0002637(tC_D/t_D)_{\text{MP}}/r_w^2\right](k_o/\mu_o)\left(q_{Rt}/q_o B_o\right)$$

$$= \left[0.0002637(tC_D/t_D)_{\text{MP}}/r_w^2\right](k_w/\mu_w)\left(q_{Rt}/q_w B_w\right)$$

$$= \left[0.0002637(tC_D/t_D)_{\text{MP}}/r_w^2\right](k_g/\mu_g)\left[q_{Rt}\Big/\left(q_g - \frac{q_o R_s}{1{,}000}\right)B_g\right]. \quad \text{(8.120)}$$

The practical implication of Eqs. 8.119 and 8.120 is that total mobility and individual phase permeability are determined from the pressure-match point on a type-curve match. The dimensionless storage coefficient is determined from the time-match point resulting in the calculation of skin factor from

$$s = 0.5 \ln\left(C_D e^{2s}/C_D\right), \quad \text{(8.121)}$$

just as for single-phase flow. When the conditions for applicability of the Perrine-Martin approximations (small saturation gradients in the drainage area of the tested well) are not satisfied, use of a reservoir simulator for test analysis is an appropriate alternative.

8.6 Diagnostic Plot

8.6.1 Introduction. The diagnostic plot is a log-log plot of the pressure change and pressure derivative (vertical axis) from a pressure transient test vs. elapsed time (horizontal axis). **Fig. 8.43** shows an example. The diagnostic plot can be divided into three time regions: early, middle, and late. At the earliest times on a plot (the early-time region), wellbore and near-wellbore effects dominate. These effects include wellbore storage, formation damage, partial penetration, phase redistribution, and stimulation (hydraulic fractures or acidization). At intermediate times (the middle-time region), a reservoir will ordinarily be infinite acting. For a homogeneous reservoir, the pressure derivative will be horizontal during this time region. Data in this region lead to the most accurate estimates of formation permeability. At the latest times in a test (the late-time region), boundary effects dominate curve shapes. The types of boundaries that may affect the pressure response include sealing faults, closed reservoirs, and gas/water, gas/oil, and oil/water contacts. Several common flow regimes and the diagnostic plots associated with these flow regimes are discussed in the remainder of Section 8.6.

8.6.2 Volumetric Behavior. Volumetric behavior is defined as that pressure response time dominated by the wellbore, reservoir, or part of the reservoir acting like a uniform-pressure "tank" with fluid entering or leaving the tank. The most common example of volumetric behav-

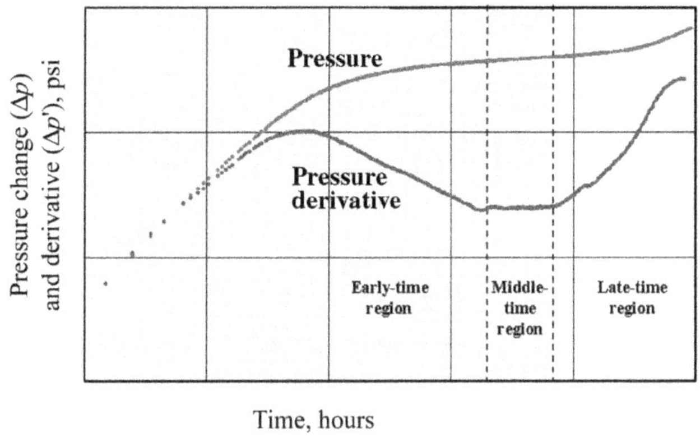

Fig. 8.43—Diagnostic plot showing time regions.

ior is wellbore storage, which dominates during the early-time region. The "tank" is the wellbore, in which the pressure is uniform. Fluid either leaves this tank (earliest times in a flow test, before the reservoir begins to respond) or enters the tank (earliest times in a buildup test). Another example is pseudosteady-state (boundary-dominated) flow in a closed reservoir during constant-rate production. In this case, the reservoir is the tank; pressure is changing at the same rate throughout (although it is not the same at all points), and fluid is leaving the reservoir through the producing well. As a final example, in a test the reservoir may behave like a tank with recharge (fluid influx) entering from a secondary source of pressure support, such as a large supply of hydrocarbons in a lower-permeability medium in pressure communication with the reservoir being tested.

The equation modeling wellbore storage (derived from a mass balance on the wellbore) is

$$\Delta p = \frac{qBt}{24C} \quad \text{.......... (8.29)}$$

The equation modeling pseudosteady-state flow in a cylindrical drainage area is

$$p_i - p_{wf} = \frac{0.0744qBt}{\phi c_t h r_e^2} + \frac{141.2qB\mu}{kh}\left[\ln\left(\frac{r_e}{r_w}\right) - \frac{3}{4} + s\right] \quad \text{.......... (8.18)}$$

The general form is

$$\Delta p = m_V t + b_V \quad \text{.......... (8.122)}$$

The derivative of the general form is

$$t\frac{\partial \Delta p}{\partial t} = t\frac{\partial\left(m_V t + b_V\right)}{\partial t} = m_V t \quad \text{.......... (8.123)}$$

The implication is that the derivative plot will have unit slope (up one log cycle as it moves over one log cycle) on log-log coordinates, and the pressure change plot will approach unity at long times when b_v is not equal to zero (**Fig. 8.44**). In wellbore storage, b_v is zero, and

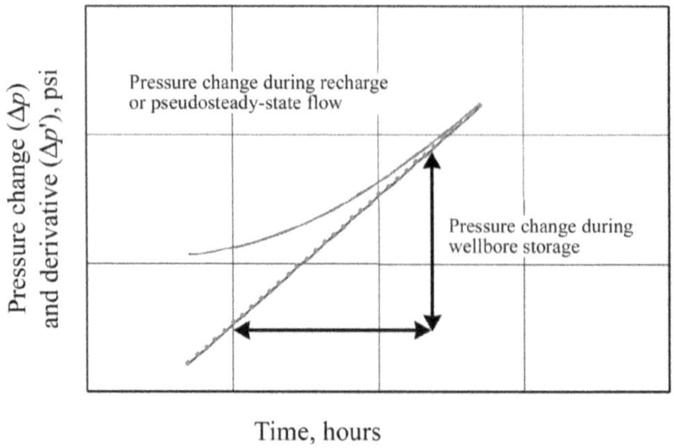

Fig. 8.44—Volumetric flow produces derivative with unit slope.

the derivative and pressure change plots will lie on top of one another. During pseudosteady-state flow or recharge, the pressure change and pressure derivative plots will not coincide.

8.6.3 Radial Flow. Infinite-acting radial flow is common in reservoirs, and data in the radial flow regime can be used to estimate formation permeability and skin factor. Common situations in which radial flow occurs include flow into vertical wells after wellbore storage distortion has ceased and before boundary effects, hydraulically fractured wells after the transient has moved well beyond the tips of the fracture, horizontal wells before the transient has reached the top and bottom of the productive interval, and horizontal wells after the transient has moved beyond the ends of the wellbore.

The equation used to model radial flow for a well producing at constant rate is the familiar logarithmic approximation to the line-source solution,

$$\Delta p = \frac{162.6 q B \mu}{kh}\left[\log\left(\frac{kt}{\phi \mu c_t r_w^2}\right) - 3.23 + 0.869 s\right] . \qquad (8.124)$$

Equations modeling radial flow have the general form

$$\Delta p = m \log(t) + b, \qquad (8.125)$$

with derivative

$$t\frac{\partial \Delta p}{\partial t} = \frac{m}{2.303} . \qquad (8.126)$$

On the diagnostic plot (**Fig. 8.45**), radial flow is indicated by a horizontal derivative.

8.6.4 Linear Flow. Linear flow is also common and occurs in channel reservoirs, hydraulically fractured wells, and horizontal wells. Data from linear flow regimes can be used to estimate channel width or fracture half-length if an estimate of permeability is available. In horizontal wells, an estimate of permeability perpendicular to the well can be made if the productive well length open to flow is known.

An equation that models linear flow in a channel reservoir of width w is

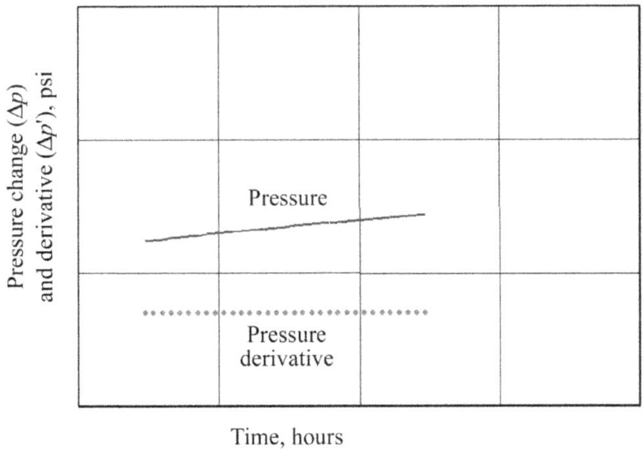

Fig. 8.45—Radial flow appears as a horizontal derivative on the diagnostic plot.

$$\Delta p = \frac{16.26 q B \mu}{k h w} \left(\frac{k t}{\phi \mu c_t} \right)^{1/2} + \Delta p_s \ . \quad \text{(8.127)}$$

For a hydraulically fractured well with fracture half-length L_f,

$$\Delta p = \frac{4.064 q B \mu}{k h L_f} \left(\frac{k t}{\phi \mu c_t} \right)^{1/2} + \Delta p_s \ . \quad \text{(8.128)}$$

The general form is

$$\Delta p = m_L t^{1/2} + b_L \ . \quad \text{(8.129)}$$

The derivative is

$$t \frac{\partial \Delta p}{\partial t} = \frac{1}{2} m_L t^{1/2} \ . \quad \text{(8.130)}$$

Linear flow on the diagnostic plot is indicated when a derivative follows a half-slope line—that is, a line that moves up vertically by one log cycle for each two cycles of horizontal movement (**Fig. 8.46**). The pressure change may or may not also follow a half-slope line. In a hydraulically fractured well, the pressure change will follow a half-slope line unless the fracture is damaged. In a channel reservoir, a hydraulically fractured well with damage, or a horizontal well, the pressure change will approach the half-slope line from above.

8.6.5 Bilinear Flow. Bilinear flow occurs primarily in wells with low-conductivity hydraulic fractures. Flow is linear within the fracture to the well, and also linear (normal to fracture flow) from the formation into the fracture. Estimates of fracture conductivity, $w_f k_f$, can be made with data from this flow regime when estimates of formation permeability are available.

For a hydraulically fractured well, an equation that models bilinear flow is

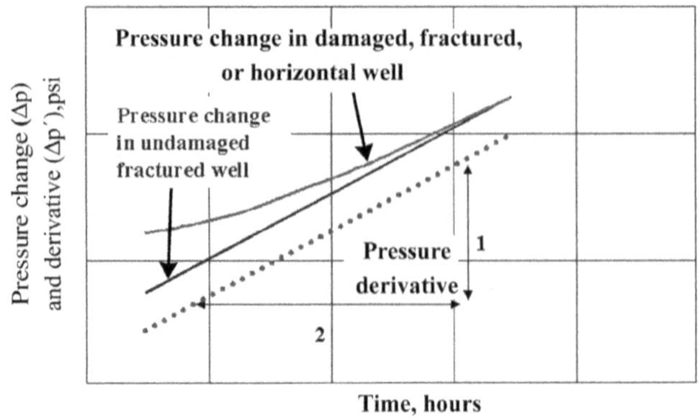

Fig. 8.46—Linear flow derivative follows a half-slope line on a diagnostic plot.

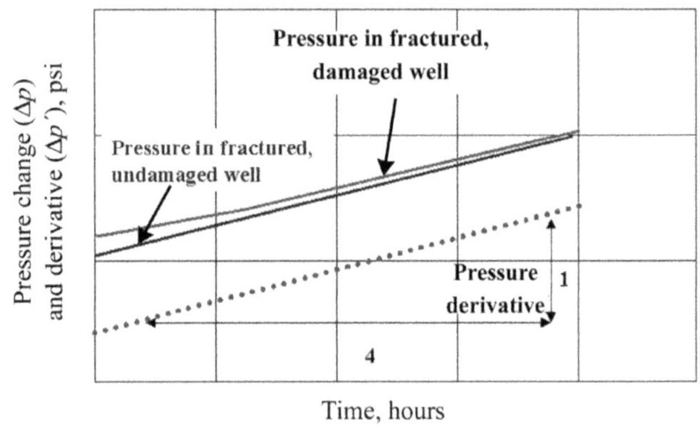

Fig. 8.47—Bilinear flow derivative follows a quarter-slope line on the diagnostic plot.

$$\Delta p = \frac{44.1 qB\mu}{h}\left(\frac{1}{w_f k_f}\right)^{1/2}\left(\frac{t}{\phi\mu c_t k}\right)^{1/4} + \Delta p_s \, . \qquad (8.131)$$

The general form is

$$\Delta p = m_B t^{1/4} + b_B \, . \qquad (8.132)$$

The derivative is

$$t\frac{\partial \Delta p}{\partial t} = \frac{1}{4} m_B t^{1/4} \, . \qquad (8.133)$$

Bilinear flow derivatives plot as a quarter-slope line on the diagnostic plot (**Fig. 8.47**). The quarter-slope line moves up one log cycle as it moves over four log cycles. The pressure change does not necessarily follow a quarter-slope line. In a damaged, hydraulically fractured well, the pressure change curve will approach the quarter-slope line from above; in an undamaged hydraulically fractured well ($\Delta p_s = 0$), the pressure change will typically follow the quarter-slope line when the effects of wellbore storage have ended.

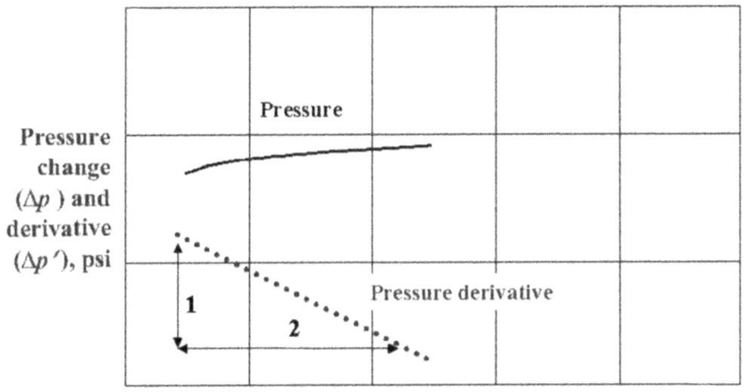

Fig. 8.48—Spherical flow derivative has slope of –½.

8.6.6 Spherical Flow. The flow pattern is spherical when the pressure transient can propagate freely in three dimensions and converge into a "point." This can occur for wells that penetrate only a short distance into the formation (actually hemispherical flow), wells that have only a limited number of perforations open to flow, horizontal wells with inflow over only short intervals, and during wireline formation tests. Data in the spherical-flow regime can be used to estimate the mean permeability,

$$k_s = \left(k_h k_z^{1/2}\right)^{2/3} \text{ . ... (8.134)}$$

An equation that models spherical flow is

$$p_{wf} = p_i - \frac{70.6 q B \mu}{k_s r_s} + \frac{2456\sqrt{\phi \mu c_t} q B \mu}{k_{sp}^{3/2}} \frac{1}{\sqrt{t}} - \frac{70.6 q B \mu}{k_{sp} r_{sp}} s, \text{ .... (8.31)}$$

$$\text{where } k_{sp} = \left[k_r k_z^{1/2}\right]^{2/3}, \text{ .... (8.32)}$$

and r_{sp} is the radius of the sphere into which flow converges. The general form is

$$\Delta p = -m_s t^{-1/2} + b_s, \text{ .... (8.135)}$$

and the derivative is

$$\frac{\partial \Delta p}{\partial t} = \frac{1}{2} m_s t^{-1/2} \text{ . ... (8.136)}$$

Spherical flow on the diagnostic plot produces a derivative line with a slope of –½. The pressure change during spherical flow approaches a horizontal line from below, and never exhibits a straight line with the same slope as the derivative (**Fig. 8.48**). Spherical flow can occur during either buildup or drawdown tests.

8.6.7 Flow Regimes on the Diagnostic Plot. A major application of the diagnostic plot is the potential that it provides in identifying the flow regimes that appear in a logical sequence during a buildup or flow test. For example, consider **Fig. 8.49**. At early times, the unit slope line

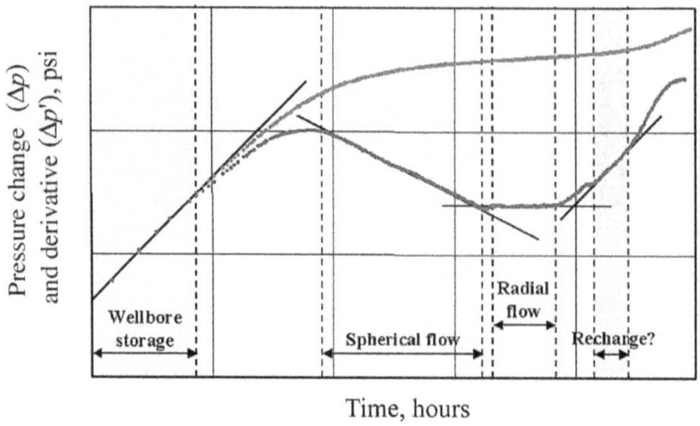

Fig. 8.49—Well test diagnostic plot indicating several flow regimes.

on both derivative and pressure change, indicating wellbore storage. Later, a derivative with a slope of −½, indicating possible spherical flow, followed by a horizontal derivative, indicating infinite-acting radial flow. Boundary effects, including a unit-slope line, follow, indicating possible recharge of the reservoir pressure.

8.7 Behavior of Bounded Reservoirs

8.7.1 Introduction. Reservoir boundaries have significant influences on the shape of the diagnostic plot. The effects of boundaries appear following the middle-time region (infinite-acting radial flow) in a test. Recognizing the influence of boundaries can be as important as analyzing the test quantitatively. However, a problem in recognition is that many reservoir models may produce similar pressure responses. The model selected to interpret the test quantitatively must be consistent with geological and geophysical interpretations. Once the proper reservoir model has been determined, test analysis may be relatively straight-forward type-curve matching or regression analysis using modern well-test analysis software.

The shapes of the diagnostic plots for a buildup test and a drawdown test are essentially identical during the early- and middle-time regions for most tests. However, boundary effects can cause quite different shapes for a given reservoir model at late times in buildup and drawdown tests. This problem is augmented by the common use of "equivalent time" functions to analyze buildup tests on drawdown type curves. (There are different equivalent time functions for radial flow, linear flow, and bilinear flow, as discussed in more detail in the section on analysis of hydraulically fractured wells.)

Basically, equivalent time functions apply rigorously only to situations where either the producing time and the shut-in time both lie within the middle-time region or, as is commonly the case, the shut-in time is much less than the producing time before shut in.

To further complicate matters for buildup test analysis, the shape of the derivative curve depends on how the derivative is calculated and plotted. The derivative of pressure change may be taken with respect to the logarithm of either shut-in time or equivalent time. The derivative may then be plotted vs. either of these time functions, and the shape differs for each plotting function. Some pressure transient test analysis software allows the user a choice in the time function used in taking the derivative and another choice in time plotting function; for other software, the time functions used are "hard-wired." The results can be bewildering.

8.7.2 Well in an Infinite-Acting Reservoir. Infinite-acting, radial flow reservoirs were described in the previous section. **Figs. 8.50 and 8.51** show their diagnostic curves. For these plots, the derivative was taken with respect to shut-in time and derivative and pressure change

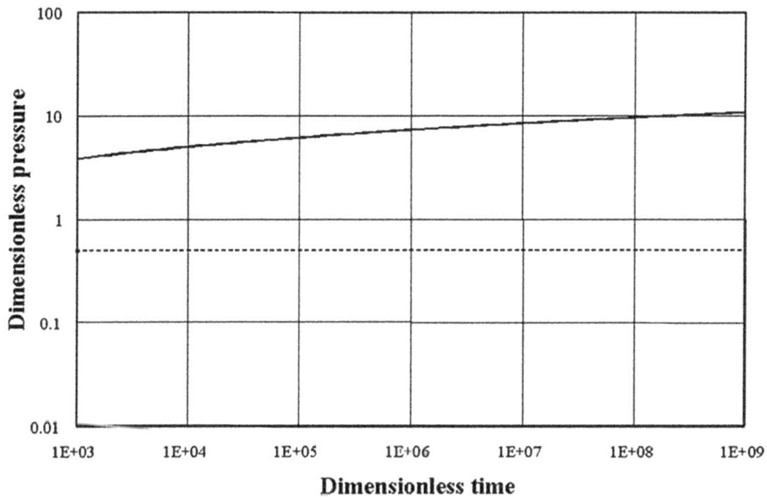

Fig. 8.50—Diagnostic plot for constant-rate drawdown test in an infinite-acting reservoir.

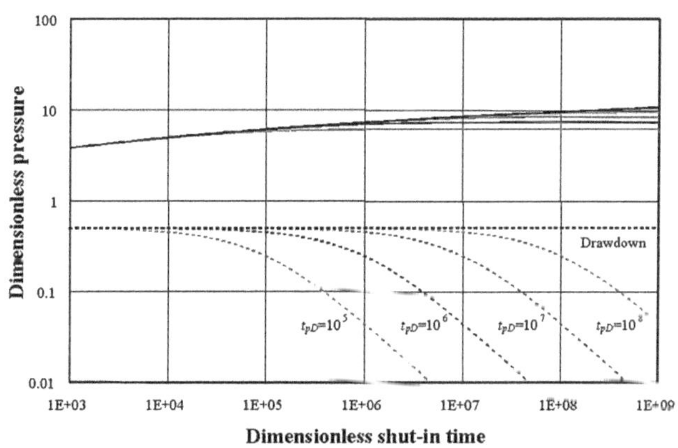

Fig. 8.51—Diagnostic plot for buildup test.

curves are plotted vs. shut-in time. Both pressure and time are in terms of dimensionless variables. Wellbore storage distortion is not included in any of the diagnostic plots in this section.

Notice the significant difference in the shapes of both the derivative and pressure change curves for buildup and drawdown tests, with the pressure change curves flattening for buildup tests and the derivatives moving downward with an ultimate slope of –1. The time at which the flattening of the pressure change curve (and corresponding downward movement of the derivative) becomes apparent is a function of the producing time before shut-in. The longer the producing time, the longer the flattening is delayed and the longer the time the buildup diagnostic plot is essentially identical to the drawdown diagnostic plot.

Fig. 8.52 is the diagnostic plot that results when the derivative is taken with respect to radial equivalent time and the time-plotting function is radial equivalent time. The drawdown and buildup curves appear to be identical for all times. However, the radial equivalent time has a maximum value of the producing time before shut-in, so, for the buildup plots, the curves terminate at these maximum values of the time plotting function, and all points "stack up" at these limiting values of the plotting function. Our conclusion is that radial equivalent time is

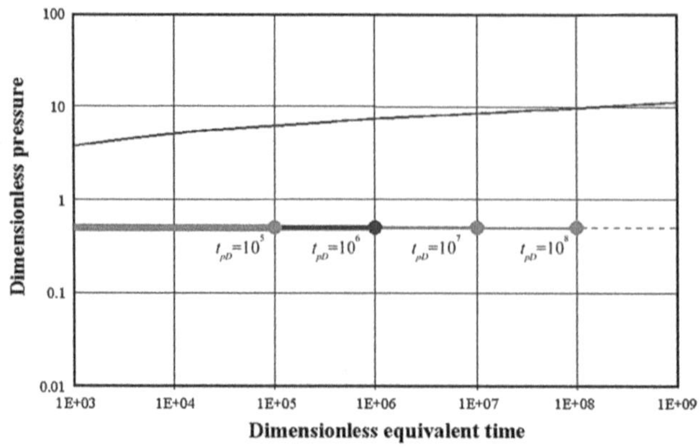

Fig. 8.52—Diagnostic plot when derivative is taken with respect to and plotted against radial equivalent time.

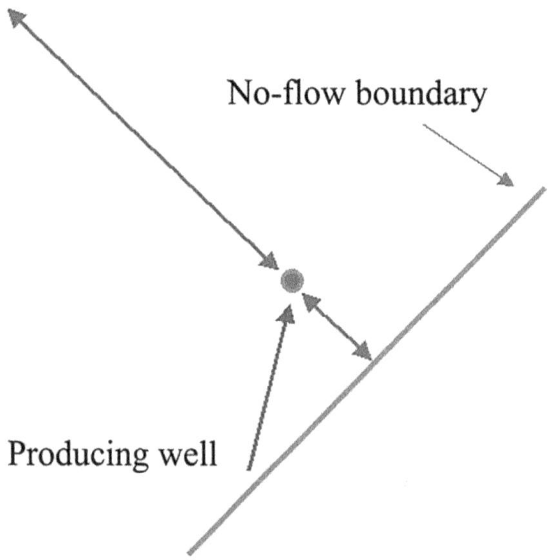

Fig. 8.53—Well near a single no-flow boundary.

more satisfactory as a variable for taking the derivative and as a plotting function for an infinite-acting reservoir because the shape of the diagnostic plot is the same as for a constant-rate drawdown test.

8.7.3 Linear No-Flow Boundary. When a well is near a single no-flow boundary (**Fig. 8.53**) or, as a practical matter, much closer to one boundary than to any other, and when sufficient time has elapsed for the boundary to have an influence on the pressure response during the test, the characteristic diagnostic plot, as **Fig. 8.54** shows, results for a constant-rate drawdown test. (Wellbore storage may distort some of the earlier data on this diagnostic plot.) The derivative will double in value over approximately 1⅔ log cycles (from 0.5 to 1.0 on a plot of dimensionless variables). Similar responses occur in naturally fractured reservoirs with transient flow from the matrix to the fractures.

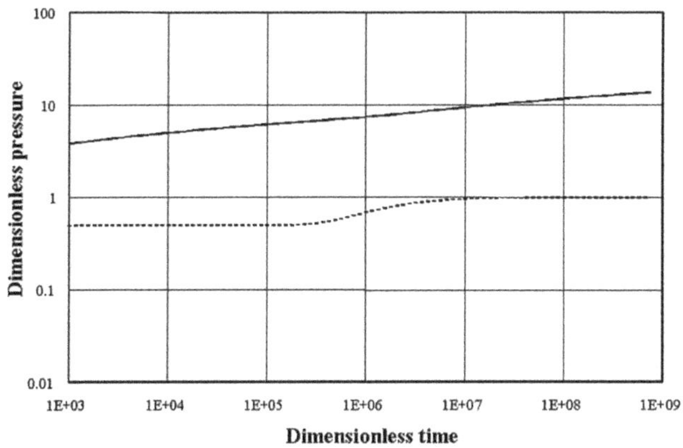

Fig. 8.54—Diagnostic plot for constant-rate drawdown test influenced by a single no-flow boundary.

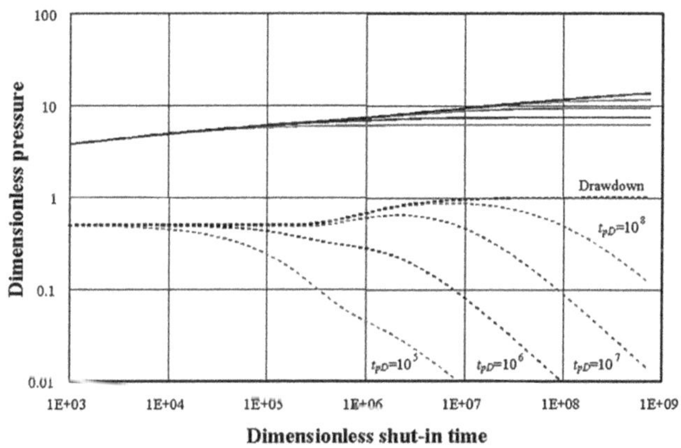

Fig. 8.55—Diagnostic plot for buildup test with derivative taken with respect to shut-in time. Long producing time before shut-in produces curve resembling plot for drawdown test.

Fig. 8.55 is the diagnostic plot for a buildup test with the derivative taken with respect to shut-in time and plotted vs. shut-in time. (Wellbore storage may distort some of the earlier data on this plot.) The longer the producing time before shut-in, the more nearly the shape of the diagnostic plot for a buildup test resembles the diagnostic plot for a drawdown test. The derivative has a slope of –1 for shut-in times much longer than producing time before shut-in.

Fig. 8.56 is the diagnostic plot for a buildup test with derivative taken with respect to radial equivalent time and plotted vs. equivalent time. Derivatives double over a small fraction of a log cycle for short producing times and, in general, the shapes of the diagnostic plots for buildup tests are similar to drawdown diagnostic plots only for longer producing times before shut-in.

Fig. 8.57 is the diagnostic plot for a buildup test, with derivative taken with respect to radial equivalent time and plotted vs. shut-in time. In this case, the diagnostic plot is similar to the drawdown response, but the plots are not identical. Notice that the derivative doubles over approximately 1⅔ log cycle. This procedure for taking the derivative and preparing the diagnostic plot is the most satisfactory of the alternatives considered.

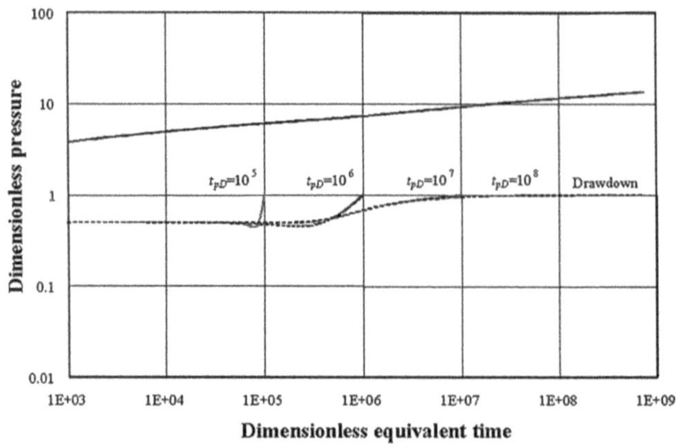

Fig. 8.56—Diagnostic plot for buildup test with derivative taken with respect to and plotted against equivalent time shows that derivative doubles over small fraction of log cycle for short producing times.

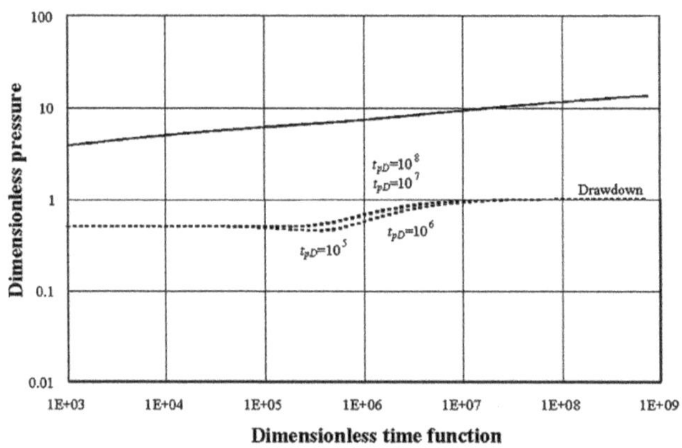

Fig. 8.57—Diagnostic plot for buildup test with derivative taken with respect to radial equivalent time and plotted vs. shut-in time resembles drawdown response.

8.7.4 Linear Constant-Pressure Boundary. When a well is much nearer a single boundary (similar to Fig. 8.53) but with a constant-pressure at that boundary and boundary effects are encountered during the test, the diagnostic plot shown in **Fig. 8.58** will result in a constant-rate drawdown test. (Wellbore storage effects could also occur early in the test.) The derivative has a slope of –1 at late times on the diagnostic plot.

Fig. 8.59 is the diagnostic plot for a buildup test, with derivative taken with respect to shut-in time and plotted vs. shut-in time. This diagnostic plot is identical to the drawdown plot if steady state was achieved during the flow period preceding the buildup test. For other cases, with shorter producing times, the derivative has a slope steeper than the drawdown slope of –1.

Fig. 8.60 is the diagnostic plot for a buildup test, with derivative taken with respect to radial equivalent time and plotted vs. equivalent time. For short producing times, the derivative falls precipitously.

Fig. 8.61 is the diagnostic plot for a buildup test, with derivative taken with respect to radial equivalent time and plotted vs. shut-in time. The shapes of the diagnostic plots are similar to, but not identical to, the drawdown diagnostic plot for all producing times before shut-in. The diagnostic plot prepared in this way is the most satisfactory of the alternatives considered.

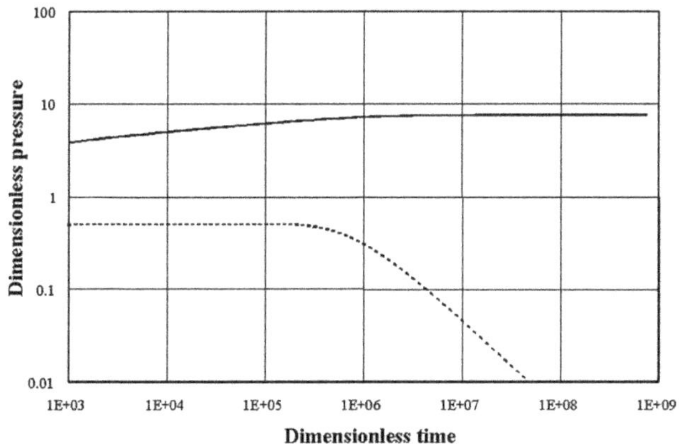

Fig. 8.58—Derivative has a slope of –1 for well located near a single, constant-pressure boundary.

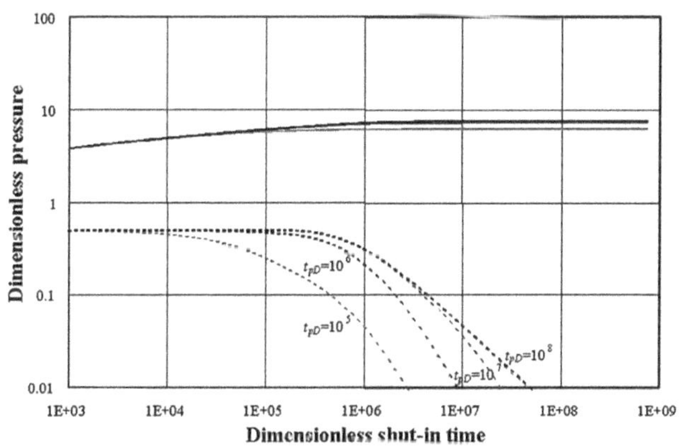

Fig. 8.59—Diagnostic plot for buildup test with derivative taken with respect to and plotted vs. shut-In time is similar to drawdown plot if steady state achieved during flow period preceding buildup test.

8.7.5 Well in a Channel. When a well is between two parallel no-flow boundaries and the pressure transient encounters both during a test long before the ends of the reservoir influence the test data, the diagnostic plot in **Fig. 8.62** results for a constant-rate drawdown test. Before the boundary effects, with characteristic derivative slope of ½, wellbore storage, radial flow (or hemiradial flow if the well is much nearer one boundary than the other) will usually appear on the diagnostic plot. Diagnostic plots with similar shapes occur for a well between two sealing faults, a hydraulically fractured well with a high-conductivity fracture, and a horizontal well during early linear flow.

Fig. 8.63 is the diagnostic plot for a buildup test, with derivative taken with respect to shut-in time and plotted vs. shut-in time. The longer the producing time before shut-in, the more similar the curve shape is to the drawdown-test diagnostic plot. The derivative has a slope of –½ when shut-in time is much larger than producing time.

Fig. 8.64 is the diagnostic plot for a buildup test with derivative taken with respect to radial equivalent time and plotted vs. equivalent time. This plot is not particularly useful for test analysis. However, linear equivalent time produces a more useful diagnostic plot as long as channel ends do not affect the pressure response.

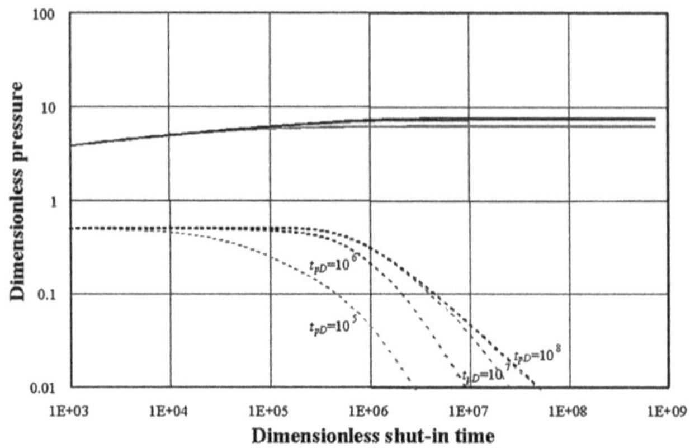

Fig. 8.60—Diagnostic plot for buildup test with derivative taken with respect to and plotted against equivalent time shows that derivative falls precipitously for short producing times.

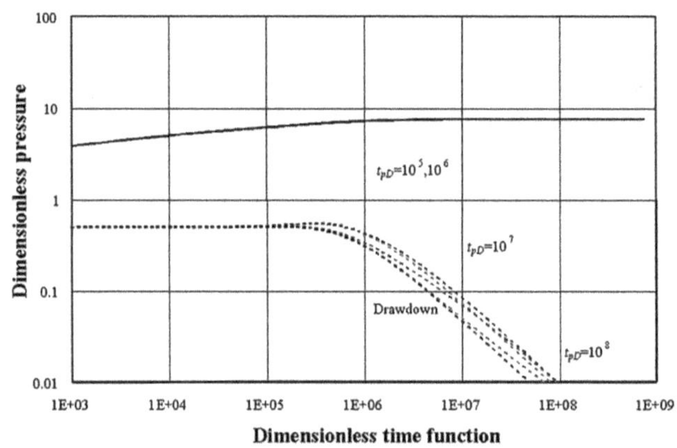

Fig. 8.61—Diagnostic plot for buildup test with derivative taken with respect to and plotted against shut-in time resembles drawdown diagnostic plot of all producing times before shut-in.

Fig. 8.65 is the diagnostic plot for a buildup test with derivative taken with respect to radial equivalent time and plotted vs. shut-in time. The derivative is similar to, but not identical to, the drawdown response. This method is the most useful for test analysis among the alternatives discussed.

8.8 Estimating Average Reservoir Pressure

Two different method types, one using data from the middle-time region and the second using data from the late-time region (LTR), are commonly applied in estimating average reservoir pressure. The middle-time region methods are the Matthews-Brons-Hazebroek (MBH) method[20] and the Ramey-Cobb method.[24] The LTR methods are the modified Muskat method[25] and the Arps-Smith method.[26]

8.8.1 Middle-Time Region Methods. The MTR methods are based on extrapolation of the middle-time region and the correction of the extrapolated pressure. The advantage of these methods is that they use pressure data only from the middle-time region, which means they require relatively short tests. The disadvantages are the need for accurate fluid property estimates, a known drainage area shape and size, and the location of the well within the drainage area.

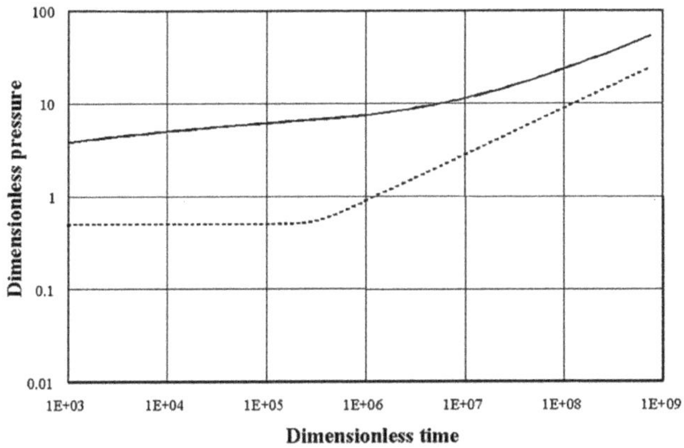

Fig. 8.62—Constant-rate drawdown diagnostic plot, well between two no-flow boundaries, characteristic derivative slope of ½.

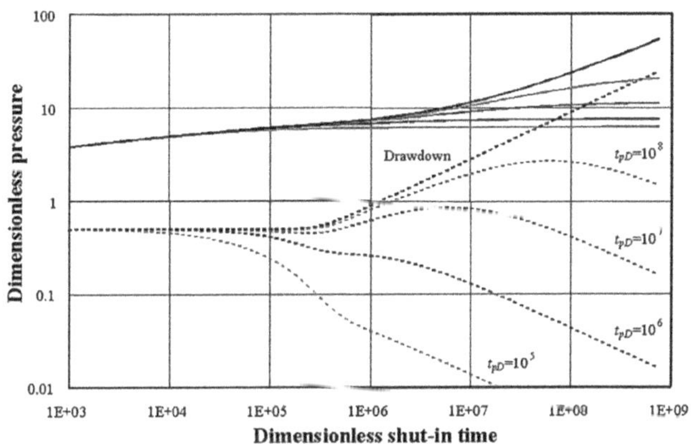

Fig. 8.63—Diagnostic plot for buildup test in well between two boundaries with derivative taken with respect to shut-in time. Longer producing time results in curve more similar to drawdown-test diagnostic plot. (Derivative has a slope of –½ when shut-in time is much longer than producing time.)

Drainage Area Shapes. The MTR methods depend on the shape of the drainage area. Matthews-Brons-Hazebroek[20] developed a series of curves that model buildup tests in many shapes. As a matter of interest, these graphs were generated using image wells to simulate boundaries.

Figs. 8.66 through 8.68 illustrate representative dimensionless pressures as calculated by the MBH method. Fig. 8.66 is a plot of dimensionless pressure as defined by the MBH method plotted against dimensionless producing time calculated using the drainage area. Dimensionless pressure is defined as

$$p_{D\text{MBH}} = \frac{kh}{70.6qB\mu}(p^* - \overline{p}) = \frac{2.303}{m}(p^* - \overline{p}), \quad \text{(8.137)}$$

and dimensionless time is

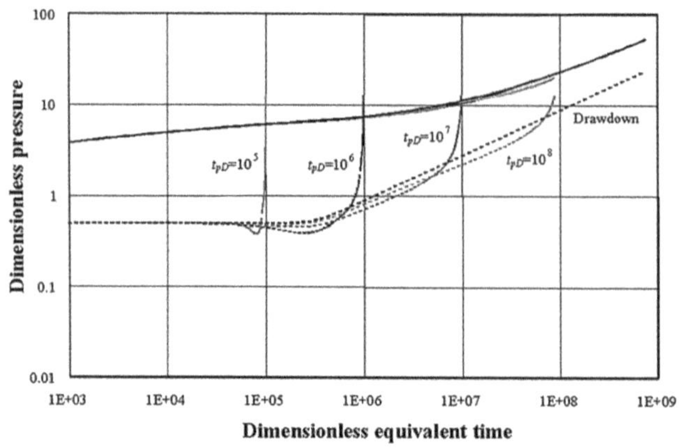

Fig. 8.64—Diagnostic plot for buildup test with derivative taken with respect to and plotted against equivalent time is not particularly useful for test analysis.

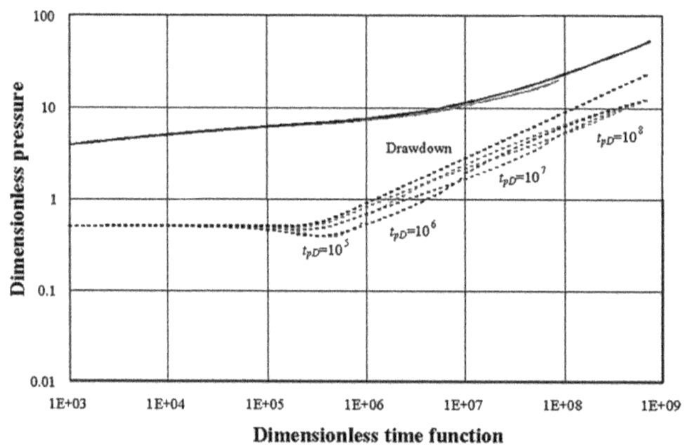

Fig. 8.65—Diagnostic plot for buildup test in well between two flow barriers. Derivative taken with respect to equivalent time and plotted against shut-in time resembles drawdown response.

$$t_{pAD} = \frac{0.0002637 k t_p}{\phi \mu c_t A} . \qquad (8.138)$$

In Eq. 8.137, p^* = the extrapolated pressure at a HTR of unity, $\bar{p}$ = the current average drainage area pressure, and m = the slope of the MTR straight line on a Horner plot. In Eq. 8.138, t_p = the producing time before shut-in, and A = the well's drainage area expressed in square feet.

The four different curves in Fig. 8.66 represent four different locations of a well within a square drainage area. On this plot of dimensionless pressure on a linear scale vs. dimensionless time on a logarithmic scale, these curves eventually become straight lines. For example, for a well centered in a square drainage area, the line becomes straight at a dimensionless time of approximately 0.2. The time at which the line becomes straight is an indication that a well has reached pseudosteady-state flow at that dimensionless time.

Fig. 8.67 shows the Matthews-Brons-Hazebroek correlations for a well in the various positions in a 2×1 rectangle. The wells eventually reach pseudosteady state and the lines become

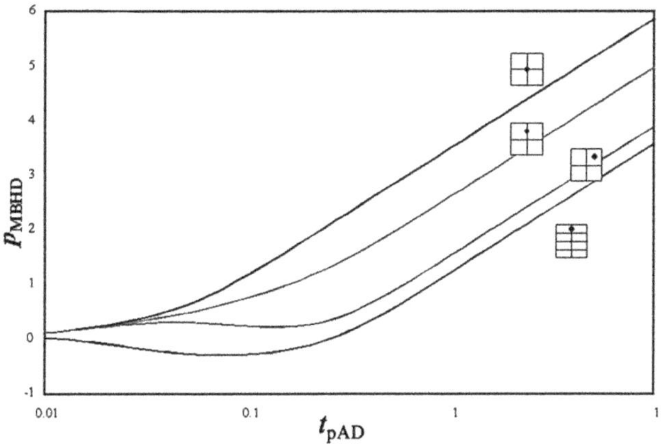

Fig. 8.66—MBH pressures for wells in square drainage area.

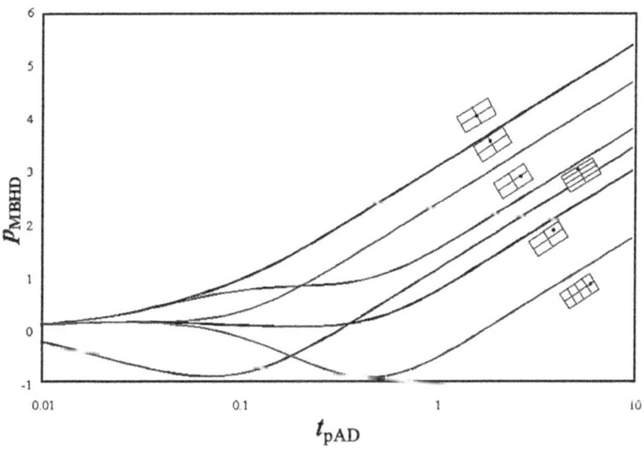

Fig. 8.67—MBH pressures for wells located in a 2×1 rectangular reservoir.

straight, but, in general, the time to reach pseudosteady state is longer for the 2×1 rectangle than it was for the square rectangle. Furthermore, the farther the well is off center within the drainage area, the longer the time required to reach pseudosteady state. The difference is on the order of one full log cycle between the case in which the well is centered in the drainage area and that for a well most off-centered in the drainage area, which is the lowest curve on this plot. Fig. 8.68 shows the MBH pressures for wells in various positions in a 4×1 rectangle. Matthews-Brons-Hazebroek generated many similar graphs for other drainage-area shapes.

Example of the Matthews-Brons-Hazebroek Method. This method will be applied to a well in a reservoir with the following properties: t_p = 482 hours, ϕ = 0.15, μ = 0.25 cp, c_t = 1.615 × 10^{-5}, and A = 1,500 × 3,000 ft (a 2 × 1 reservoir, well centered).

First, plot well shut-in pressure against the HTR on semilog coordinates. In **Fig. 8.69**, which is an ordinary HTR plot, the wellbore storage affects the data at large values of HTR, followed by the straight-line middle-time region, in turn followed by a deviation of the curve as it begins moving toward a fully built-up pressure.

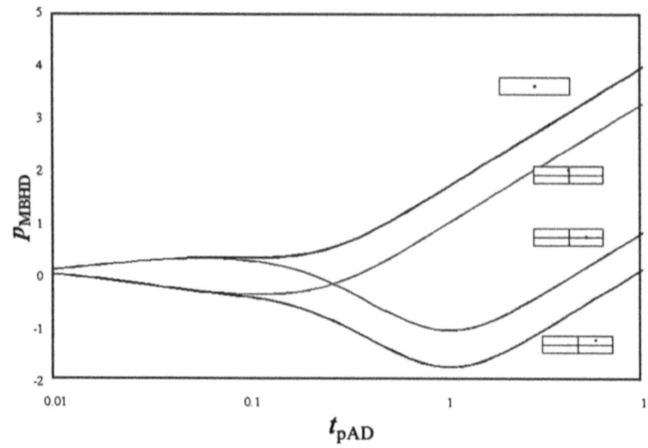

Fig. 8.68—MBH pressures for wells located in a 4×1 rectangular reservoir.

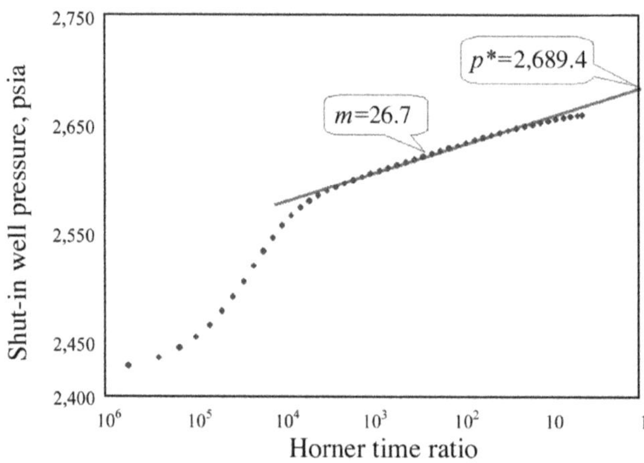

Fig. 8.69—Horner plot yields slope, Horner time ratio, and extrapolated pressure.

The MTR straight line on this Horner graph is extrapolated to a HTR of 1 to determine p^*. In this case, p^* = 2,689.4 psi. From the slope of the semilog straight line, 26.7 psi/cycle, we calculate k = 7.5 md.

Next, calculate the dimensionless producing time, t_{pAD}, with Eq. 8.138.

$$t_{pAD} = \frac{0.0002637 k t_p}{\phi \mu c_t A}$$

$$= \frac{(0.0002637)(7.5)(482)}{(0.15)(0.25)\left(1.615 \times 10^{-5}\right)(1{,}500)(3{,}000)} = 0.35.$$

To calculate dimensionless production time, use the same producing time used in preparing the Horner graph. If the actual producing time is quite long, replace it with the time required to reach pseudosteady state, but remember to use the same producing time in the HTR and in calculating the dimensionless time for the MBH function. The time to reach pseudosteady state

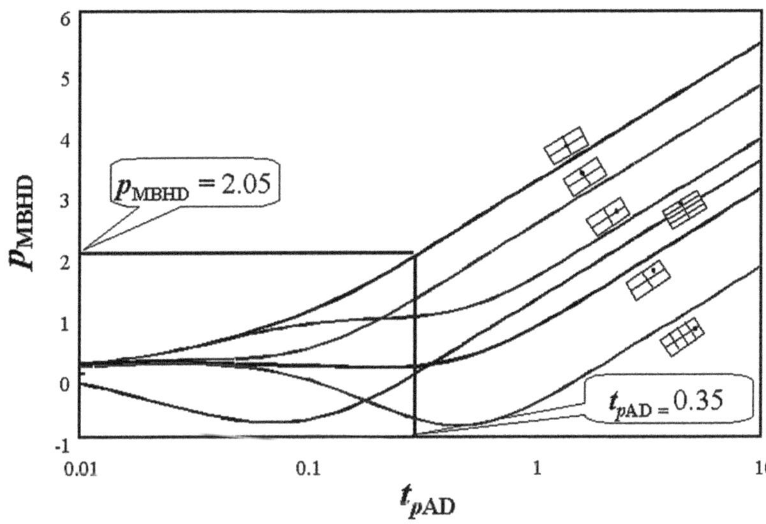

Fig. 8.70—Entering chart at producing time yields MBH pressure.

is determined by observing the appropriate MBH graph and finding when the dimensionless pressure vs. time becomes a straight line.

The next step is to select the appropriate MBH chart for the drainage area shape and well location being evaluated. Because the example well is centered in a 2×1 rectangle, choose Fig. 8.67. On this chart, enter the graph at a dimensionless producing time of 0.35, as illustrated in **Fig. 8.70**, and read across to find the dimensionless pressure, p_{MBHD}, which has a value of 2.05.

The next step is to calculate the average reservoir pressure, $\bar{p}$. From rearrangement of Eq. 8.137,

$$\bar{p} = p^* - \frac{m}{2.303} p_{MBHD} \cdot \quad (8.139)$$

In this case, the extrapolated p^* = 2,689.4 psi, the slope of the MTR = 26.7, and the dimensionless pressure = 2.05. Thus,

$$\bar{p} = 2689.4 - \frac{26.7}{2.303}(2.05) = 2665.6 \text{ psi.}$$

Ramey-Cobb Method. The Ramey-Cobb method[24] also uses information from a Horner plot of buildup test data. After determining permeability from the Horner plot, dimensionless producing time, $t_{pAD,}$ can be calculated.

The third step differs from the MBH method in that the Dietz shape factors, C_A, from Table 8.A-1 for the drainage-area shape and well location that best describes the tested well are used. (For the physical significance of the shape factor, see Ramey and Cobb.[24]) For the example well, the drainage area is a 2 × 1 rectangle, and the shape factor is 21.8369. Ramey and Cobb found a relationship between shape factor and the HTR at which the pressure on the MTR is current average drainage area pressure, $\bar{p}$. The relationship is

$$\left(\frac{t_p + \Delta t}{\Delta t}\right)_{\bar{p}} = C_A t_{pAD} \cdot \quad (8.140)$$

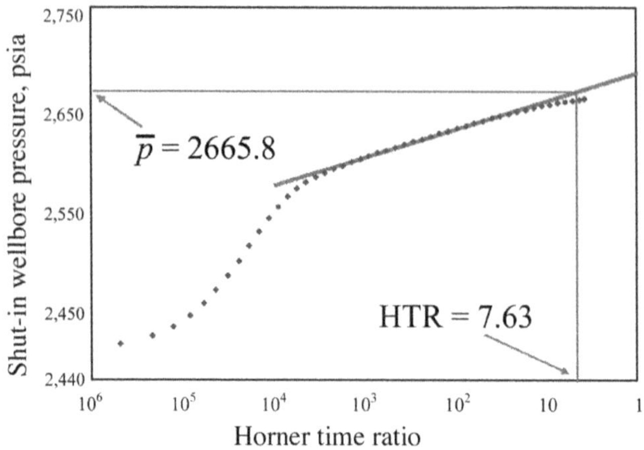

Fig. 8.71—Ramey-Cobb results replicate those of MBH method.

In the example test, the dimensionless producing time is 0.35, so the HTR that corresponds to the average reservoir pressure is 7.63.

$$\left(\frac{t_p + \Delta t}{\Delta t}\right)_{\bar{p}} = C_A t_{pAD} = (21.8)(0.35) = 7.63.$$

Enter the Horner plot at a HTR of 7.63, read up to the extrapolated MTR straight line, then read across to the vertical axis. The resulting average reservoir pressure is 2,665.8 (**Fig. 8.71**). The result, for practical purposes, is identical to the result obtained using the MBH method.

The MBH and Ramey-Cobb methods use only data in the MTR. Once enough data is available to identify the MTR, the test can be terminated, which reduces test costs. The disadvantages of these methods are the need to know the drainage area size, shape, location of the well within that drainage area, and an accurate measurement of fluid properties. In the MBH method, the well can be in transient flow at the time of shut-in, but in the Ramey-Cobb method, the well must have reached pseudosteady state before shut-in. Results for the two methods should be identical, because they are based on the same theory. When it is applicable (pseudosteady state before shut-in), the Ramey-Cobb method is preferred because it is easier to apply.

8.8.2 Late-Time Region Methods. Methods using LTR data are based on extrapolation of the post-middle-time region data trend. The advantages of these methods are that they require neither accurate fluid property estimates nor the drainage area size and shape. They do require that the well be reasonably centered within its drainage area. The disadvantage is that they require the post-middle-time region transient data. Thus, they require longer and more expensive shut-in tests to provide the data required for analysis.

Modified Muskat Method. The modified Muskat method is based on the theoretical observation first published by Larsen[25] that, for late-time data (after boundary effects have appeared), the difference between current average reservoir pressure, $\bar{p}$, and shut-in BHP, p_{ws}, declines exponentially. In equation form,

$$\bar{p} - p_{ws} = Ae^{bt}, \quad (8.141)$$

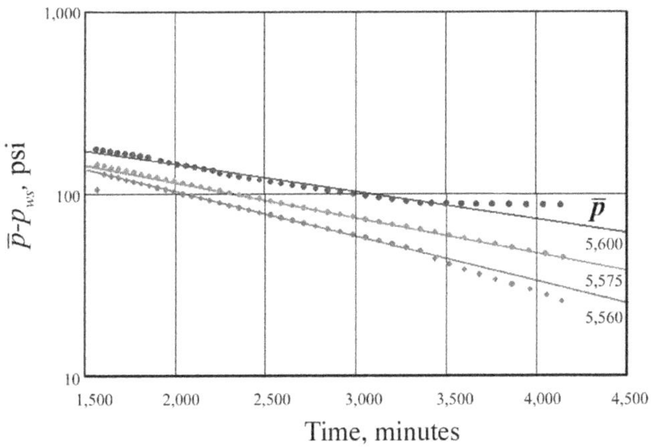

Fig. 8.72—Trial-and-error plot identifies best value of average pressure.

$$\text{or } \ln\left(\overline{p} - p_{ws}\right) = \ln(A) - bt \quad (8.142)$$

Eq. 8.142 leads to a procedure for estimating average drainage-area pressure, $\overline{p}$. This method requires a trial-and-error approach. To select data suitable for analysis with this method, use the diagnostic plot to determine the start of boundary effects. Then assume a value for $\overline{p}$, and plot $\log\left(\overline{p} - p_{ws}\right)$ vs. time. If the curve is concave downward, the assumed pressure is too low; if the curve is concave upward, the assumed pressure is too high. Try different values for $\overline{p}$ until the graph is a straight line, as predicted by theory.

On the example (**Fig. 8.72**), once the data begin to fall on a straight line, they tend to remain on that straight line. Shown are curves for assumed values of $\overline{p}$ = 5,600; 5,575; and 5,560. On the first curve, for $\overline{p}$ = 5,600, the final data points are trending above the straight line. For the lower curve, with $\overline{p}$ = 5,560, the last few data points are trending below the straight line. For the assumed value $\overline{p}$ = 5,575, all of the data points fall on a straight line making this assumption the right estimate of $\overline{p}$. The advantage to this method is that it is very easy to apply. It works best with a well reasonably centered within a drainage area.

The weaknesses of this method are that it is more sensitive to estimates that are too low rather than to estimates that are too high and that it is not easily automated and, therefore, not as widely incorporated into well-test analysis software as some other methods.

Arps-Smith Method. This is an alternative method for analyzing LTR data.[26] The theoretical basis for this originally empirical method is also Eq. 8.141. Differentiating Eq. 8.141 with respect to time,

$$\frac{dp_{ws}}{dt} = Abe^{-bt} = b\left(\overline{p} - p_{ws}\right). \quad (8.143)$$

To apply this method, plot the change in BHP with time, dp_{ws}/dt vs. p_{ws}, on Cartesian coordinates. On such a plot, data for the LTR should fall on a straight line, and extrapolation of that line to $dp_{ws}/dt = 0$ provides an estimate of the average drainage area pressure, $\overline{p}$.

In **Fig. 8.73**, the final points from an example test fall on a straight line. Extrapolating the straight line to the horizontal axis gives the average pressure at the intercept. For this example, which is the same test illustrated with the modified Muskat method, the average pressure is 5,575 psi, which is the same value found with the Muskat method.

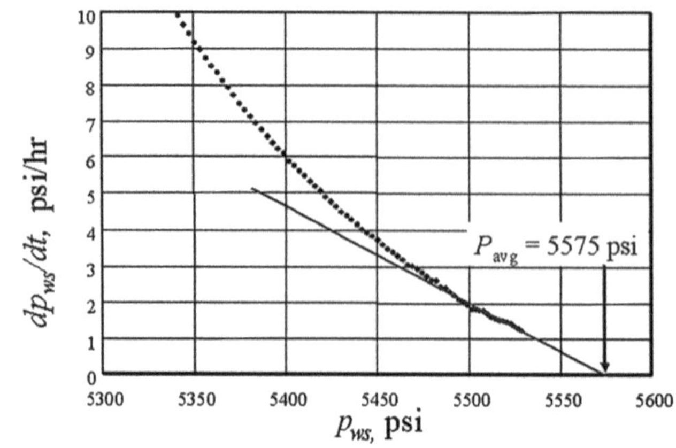

Fig. 8.73—Extrapolation of straight line to *x*-axis yields average pressure.

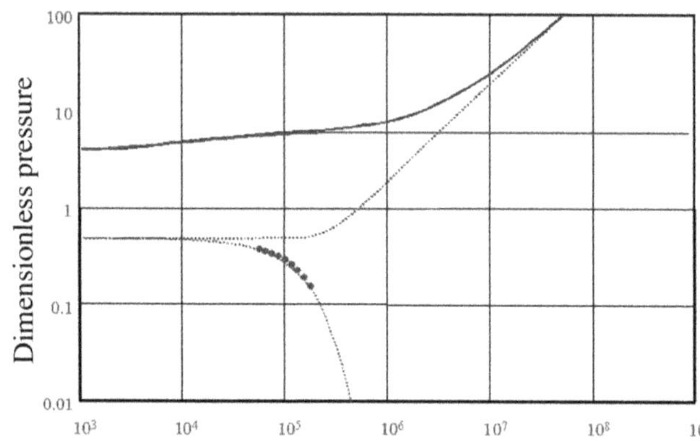

Fig. 8.74—Darker dots represent data in range of Muskat, Arps-Smith methods.

The advantages of the Arps-Smith method are that it is simple to apply and easily automated (which means that it is easily implemented into well-test analysis software or into spreadsheets). The disadvantages are that it requires data in the LTR, which means that it requires longer, more expensive tests. It assumes that shut-in pressure approaches average pressure exponentially, which is most nearly true for wells centered in the drainage area, and it requires numerical differentiation of pressure with respect to time, which tends to magnify any noise that may be present in the data.

The modified Muskat and Arps-Smith methods actually apply for shut-in times in the range,

$$\frac{250\phi\mu c_t r_e^2}{k} \leq \Delta t \leq \frac{750\phi\mu c_t r_e^2}{k} \quad \text{.......................................} \quad (8.144)$$

In **Fig. 8.74**, the data points with darker dots are on the type curve for the derivative. These are the data in the range for which the modified Muskat and Arps-Smith methods work.

Fig. 8.74 illustrates one of the disadvantages of these two methods. Many other reservoir models will exhibit similar diagnostic plots, but data like that shown with the dark dots in this figure will not extrapolate to the correct average reservoir drainage area pressure. Examples of these other cases include dual-porosity reservoirs during the early transition from fracture flow to total system flow, layered reservoirs, and composite reservoirs with an inner zone mobility much lower than the outer zone mobility.

8.9 Hydraulically Fractured Wells

Many wells—particularly gas wells in low-permeability formations—require hydraulic fracturing to be commercially viable. Interpretation of pressure-transient data in hydraulically fractured wells is important for evaluating the success of fracture treatments and predicting the future performance of fractured wells. This section includes graphical techniques for analyzing post-fracture pressure transient tests after identifying several flow patterns that are characteristic of hydraulically fractured wells. Often, identification of specific flow patterns can aid in well test analysis.

8.9.1 Flow Patterns in Hydraulically Fractured Wells. Five distinct flow patterns (**Fig. 8.75**) occur in the fracture and formation around a hydraulically fractured well.[27] Successive flow patterns, which often are separated by transition periods, include fracture linear, bilinear, formation linear, elliptical, and pseudoradial flow. Fracture linear flow (Fig. 8.75a) is very short-lived and may be masked by wellbore-storage effects. During this flow period, most of the fluid entering the wellbore comes from fluid expansion in the fracture, and the flow pattern is essentially linear.

Because of its extremely short duration, the fracture linear flow period often is of no practical use in well test analysis. The duration of the fracture linear flow period is estimated by[27]

$$t_{L_fD} \cong \frac{0.1C_r^2}{\eta_{fD}^2}, \quad \text{(8.145)}$$

where t_{L_fD} is dimensionless time in terms of fracture half-length,

$$t_{L_fD} \cong \frac{0.0002637k_t}{\phi\mu c_t L_f^2} \quad \text{(8.146)}$$

The dimensionless fracture conductivity, C_r, is

$$C_r = \frac{w_f k_f}{\pi k L_f}, \quad \text{(8.147)}$$

and η_{fD} is dimensionless hydraulic diffusivity defined by

$$\eta_{fD} = \frac{k_f \phi c_t}{k\phi_f c_{ft}} \quad \text{(8.148)}$$

Bilinear flow (Fig. 8.75b) evolves only in finite-conductivity fractures as fluid in the surrounding formation flows linearly into the fracture and before fracture tip effects begin to influence well behavior. Fractures are considered to be finite conductivity when $C_r < 100$. Most of the fluid entering the wellbore during this flow period comes from the formation. During the bilinear flow period, BHP, p_{wf}, is a linear function of $t^{1/4}$ on Cartesian coordinates.

A log-log plot of $(p_i - p_{wf})$ as a function of time exhibits a slope of ¼ unless the fracture is damaged. The pressure derivative also has a slope of ¼ during this same time period. The

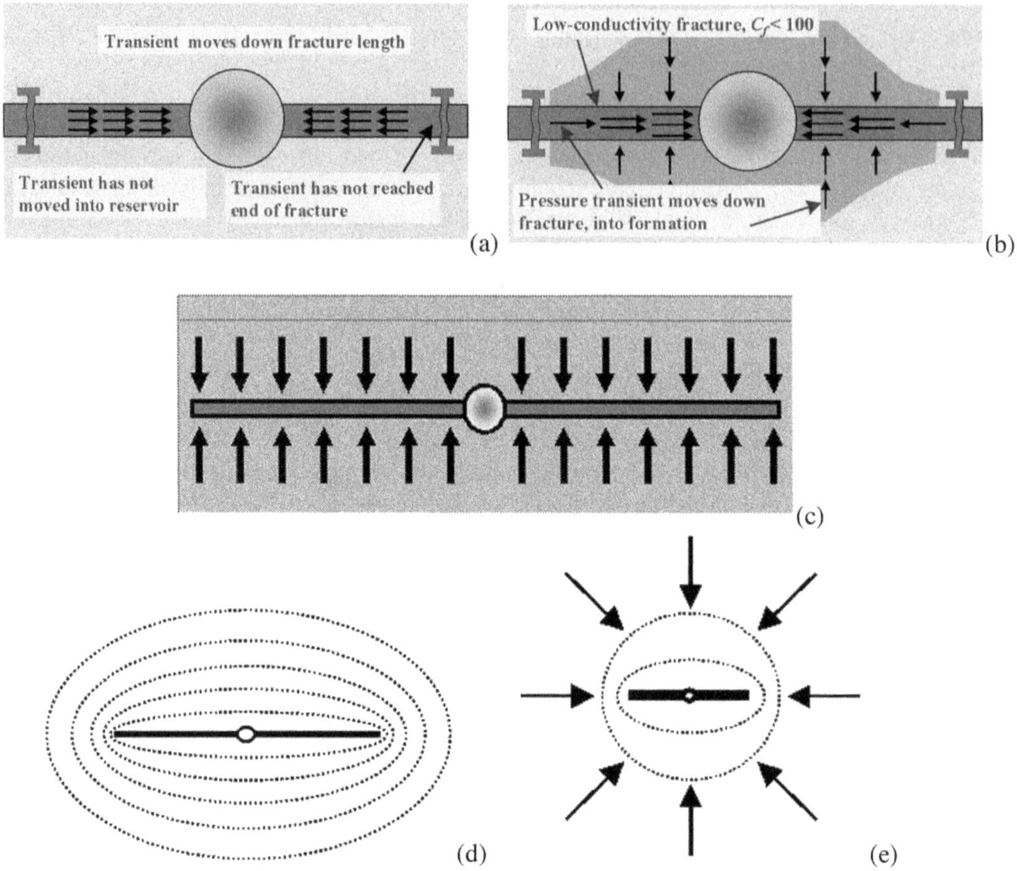

Fig. 8.75—Flow periods in a vertically fractured well.[27]

duration of bilinear flow depends on dimensionless fracture conductivity and is given by Eqs. 8.149a through 8.149c[27] for a range of dimensionless times and fracture conductivities:

$$t_{L_f D} \cong \frac{0.01}{C_r^2} \text{ for } C_r > 3, \qquad (8.149a)$$

$$t_{L_f D} \cong 0.0205\left(C_r - 1.5\right)^{-1.53} \text{ for } 1.6 \le C_r \le 3, \qquad (8.149b)$$

$$\text{and } t_{L_f D} \cong \left(\frac{4.55}{\sqrt{C_r}}\right)^{-4} \text{ for } C_r < 1.6 \qquad (8.149c)$$

Formation linear flow (Fig. 8.75c) occurs only in high-conductivity ($C_r \geq 100$) fractures. This period continues to a dimensionless time of $t_{L_f D} \cong 0.016$. The transition from fracture linear flow to formation linear flow is complete by a time of $t_{L_f D} = 10^{-4}$. On Cartesian coordinates, p_{wf} is a linear function of $t^{1/2}$, and a log-log plot of $(p_i - p_{wf})$ has a slope of ½ unless the fracture is damaged. The pressure derivative plot exhibits a slope of ½. Elliptical flow (Fig.

TABLE 8.2—DATA FOR EXAMPLE 8.1

Case No.	L_f (ft)	k (md)	Time to End of Linear Flow (hours)	Time to Start of Pseudoradial Flow (hours)
1	100	1.00	0.273	51.2
2	100	0.01	27.3	5,120 (213 days)
3	1,000	0.01	2,730 (114 days)	512,000 (58 years)
Other data: ϕ = 0.15, μ = 0.03 cp, C_r = 100, c_t = 1×10^{-4} psi^{-1}				

8.75d) is a transitional flow period that occurs between a linear or near-linear flow pattern at early times and a radial or near—radial flow pattern at late times.

Pseudoradial flow (Fig. 8.75e) occurs with fractures of all conductivities. After a sufficiently long flow period, the fracture appears to the reservoir as an expanded wellbore (consistent with the effective wellbore radius concept suggested by Prats *et al.*[10]). At this time, the drainage pattern can be considered as a circle for practical purposes. (The larger the fracture conductivity, the later the development of an essentially radial drainage pattern.) If the fracture length is large relative to the drainage area, then boundary effects distort or entirely mask the pseudoradial flow regime. Pseudoradial flow begins at $t_{L_fD} \cong 3$ for high-conductivity fractures ($C_r \geq 100$) and at slightly smaller values of t_{L_fD} for lower values of C_r.

These flow patterns also appear in pressure-buildup tests and occur at approximately the same dimensionless times as in flow tests. The physical interpretation is that the pressure has built up to an essentially uniform value throughout a particular region at a given time during a buildup test. For example, at a given time during bilinear or formation linear flow, pressure has built up to a uniform level throughout an approximately rectangular region around the fracture. At a later time during elliptical flow, pressure has built up to a uniform level throughout an approximately elliptical region centered at the wellbore. At a given time during pseudoradial flow, pressure has built up to a uniform level throughout an approximately circular region centered at the wellbore. The area of the region and the pressure level within that area increase with increasing shut-in time. Example 8.1 illustrates how to estimate the duration of flow periods for hydraulically fractured wells.

Example 8.1: Estimating Duration of Flow Periods in a Hydraulically Fractured Well. For each case, estimate the end of the linear flow period and the time at which pseudoradial flow period begins. Assume that pseudoradial flow begins when $t_{L_fD} = 3$. **Table 8.2** gives the data for each case.

Solution. The end of the linear flow regime occurs at a dimensionless time of $t_{L_fD} \cong 0.016$ or, using Eq. 8.146,

$$t = \frac{\phi\mu c_t L_f^2 t_{L_fD}}{0.0002637k} = \frac{(0.15)(0.03)\left(1\times10^{-4}\right)L_f^2(0.016)}{0.0002637k} = \frac{2.73\times10^{-5}L_f^2}{k}.$$

Similarly, the time to reach pseudoradial flow is $t_{L_fD} \cong 3$, or

$$t = \frac{\phi\mu c_t L_f^2 t_{L_fD}}{0.0002637k} = \frac{(0.15)(0.03)\left(1\times10^{-4}\right)L_f^2(3)}{0.0002637k} = \frac{5.12\times10^{-3}L_f^2}{k}.$$

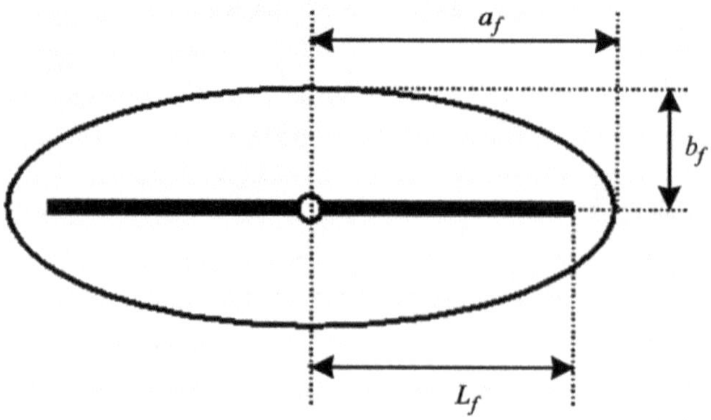

Fig. 8.76—Elliptical flow pattern around a vertically fractured well.

Table 8.2 summarizes the results.

8.9.2 Flow Geometry and Depth of Investigation of a Vertically Fractured Well. Fluid flow in a vertically fractured well has been described using elliptical geometry.[28] The equation for an ellipse with its major axis along the x-axis and minor axis along the y-axis is

$$\frac{x^2}{a_f^2} + \frac{y^2}{b_f^2} = 1, \quad (8.150)$$

where the endpoints of the major and minor axes are ($\pm a_f$, 0) and (0, $\pm b_f$), respectively. The foci of the ellipse are $\pm c_f$ where $c_f^2 = a_f^2 - b_f^2$. In terms of a well with a single vertical fracture with two wings of equal length, L_f, the relation becomes $L_f^2 = a_f^2 - b_f^2$, where L_f is the focal length of the ellipse. **Fig. 8.76** shows the elliptical geometry of a vertically fractured well.

Hale and Evers[28] defined a depth of investigation for a vertically fractured well. Their definition is based on a definition of dimensionless time at a distance b_f, the length of the minor axis:

$$t_{bD} = \frac{0.0002637kt}{\phi \mu c_t b_f^2}. \quad (8.151)$$

Solving for the length of the minor axis,

$$b_f = \left(\frac{0.0002637kt}{\phi \mu c t_{bD}}\right)^{1/2}. \quad (8.152)$$

Assuming that pseudosteady-state flow exists out to distance, b_f, at dimensionless time t_{bD} = $1/\pi$ as in linear systems, Eq. 8.152 becomes

$$b_f = 0.0002878\left(\frac{kt}{\phi \mu c_t}\right)^{1/2}, \quad (8.153)$$

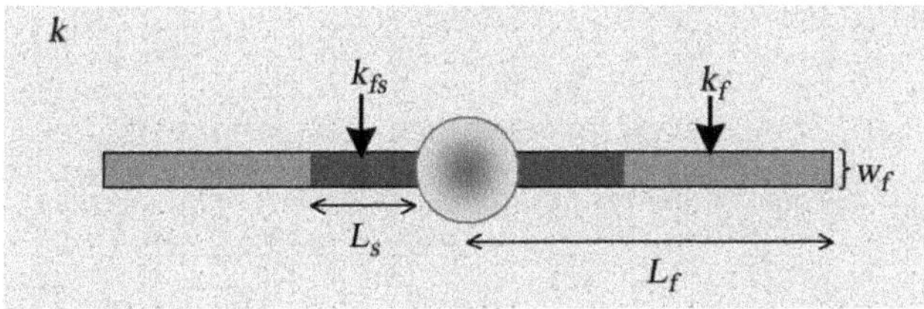

Fig. 8.77—Permeability differs between propped and near-wellbore portions of fracture.

which represents the depth of investigation in a direction perpendicular to the fracture at time, t, for a vertically fractured well. In gas wells, the terms μ and c_t should be $\bar{\mu}$ and $\bar{c}_t$, evaluated at average drainage-area pressure, $\bar{p}$.

The elliptical pattern of the propagating pressure transient can be fully described in terms of the lengths of the major axis, a_f, the minor axis, b_f, and the focus, L_f. Using the estimate of b_f from Eq. 8.153 and an estimate of L_f obtained by one of the methods described in sections that follow, the length of the major axis can be estimated from

$$a_f = \sqrt{L_f^2 + b_f^2} \; . \qquad (8.154)$$

Given values of a_f and b_f, the depth of investigation at a particular time, t, in any direction from the fracture can be calculated using Eq. 8.150. Furthermore, the area, A, enclosed by the ellipse at time, t (the area of the reservoir sampled by the pressure transient), is given by

$$A_f = \pi a_f b_f \; . \qquad (8.155)$$

The coefficient 0.0002878 in Eq. 8.153 is strictly correct only for highly conductive fractures ($C_r \geq 100$). As C_r becomes smaller, the ratio a_f/b_f also becomes smaller. The lower bound of a_f/b_f is 1 (a circle) as C_r approaches 0.

8.9.3 Fracture Damage. Two major types of fracture damage are frequent: choked fracture damage and fracture-face damage. The choked-fracture damage means that the fracture has a reduced permeability in the immediate vicinity of the wellbore (**Fig. 8.77**). In this case, k_f is used for the permeability in the propped portion of the fracture farther along the wellbore, and k_{fs} for reduced permeability near the wellbore, out to a length, L_s, in the fracture.

The choked-fracture skin factor, s_f, is[29]

$$s_f = \frac{\pi k L_s}{k_{fs} w_f} \; . \qquad (8.156)$$

Fracture face damage in a hydraulically fractured well (**Fig. 8.78**) is a permeability reduction around the edges of the fracture, usually caused by invasion of the fracture fluid into the formation or an adverse reaction with the fracturing fluid. The equation for fracture face skin is[29]

$$s_f = \frac{\pi w_s}{2L_f}\left(\frac{k}{k_s} - 1\right) . \qquad (8.157)$$

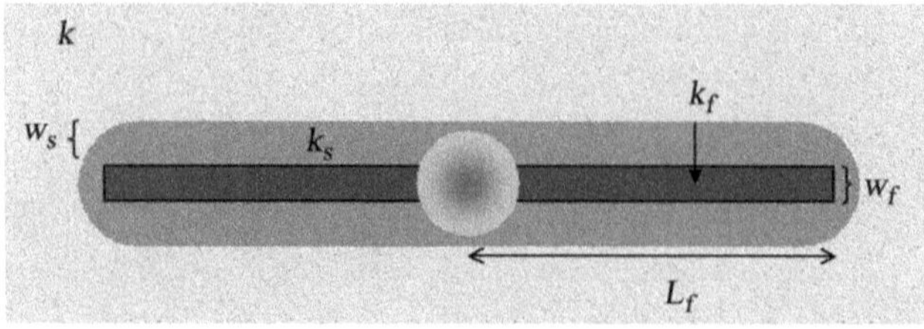

Fig. 8.78—Permeability reduction around edges of fracture represents fracture face damage.

8.9.4 Specialized Methods for Post-Fracture Well-Test Analysis. Generally, the objectives of post-fracture pressure-transient test analysis are to assess the success of the fracture treatment and to estimate the fracture half-length, fracture conductivity, and formation permeability. Three specialized methods of analyzing these post-fracture transient tests are included in this section: pseudoradial flow, bilinear flow, and linear flow.

Bilinear Flow Method. The bilinear flow method[30] applies to test data obtained during the bilinear flow regime in wells with finite-conductivity vertical fractures. Bilinear flow is indicated by a quarter-slope line on a log-log graph of pressure derivative vs. t or Δt_e.

During bilinear flow,

$$p_D = \frac{1.38}{\sqrt{C_r}} t_{L_f D}^{1/4} + s_f, \quad (8.158)$$

$$\text{and } t_{L_f D} \frac{dp_D}{dt_{L_f D}} = \frac{0.345}{\sqrt{C_r}} t_{L_f D}^{1/4}. \quad (8.159)$$

The following procedure is recommended for analyzing test data obtained in the bilinear flow regime (that is, data in the time range with quarter slope on the diagnostic plot). In Step 1, note the use of "bilinear equivalent time," Δt_{Be}. Radial equivalent time is rigorously correct as a plotting function only for infinite-acting radial flow.

1. For a constant-rate flow test, plot p_{wf} vs. $t^{1/4}$ on Cartesian coordinates. For a buildup test, plot p_{ws} vs. $\Delta t_{Be}{}^{1/4}$, where

$$\Delta t_{Be}^{1/4} = \left[t_p^{1/4} + \Delta t^{1/4} - \left(t_p + \Delta t\right)^{1/4} \right]. \quad (8.160)$$

2. Determine the slope, m_B, of the straight line region of the plot.
3. Determine the pressure extrapolated to time zero, p_o, and the fracture skin, s_f, from

$$s_f = \frac{0.00708kh}{qB\mu}\left(p_i - p_o\right) \text{ or } \frac{0.00708\text{ kh}}{qB\mu}\left(p_o - p_{wf}\right) \quad (8.161)$$

for drawdown and buildup tests, respectively.

4. From independent knowledge of k (for example, from a prefracture well test), estimate the fracture conductivity, $w_f k_f$, using m_B and the relationship

$$w_f k_f = \left(\frac{44.1 qB\mu}{hm_B}\right)^2 \left(\frac{1}{\phi \mu c_t k}\right)^{0.5}, \quad (8.162)$$

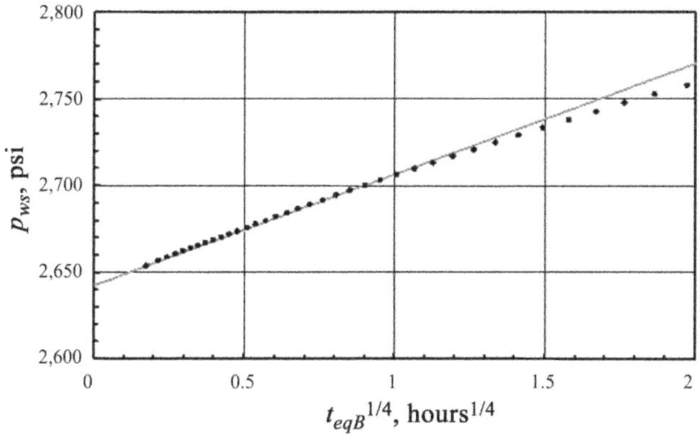

Fig. 8.79—Bilinear flow analysis.

where $\bar{\mu}$ and $\bar{c}_t$, evaluated at $\bar{p}$, are used for a gas well test.

Fig. 8.79 is an example of bilinear flow analysis. The bilinear flow analysis method has the following important limitations.

- No estimate of fracture half-length, L_f.
- In wells with low-conductivity fractures, wellbore storage frequently distorts early test data for a sufficient length of time so that the quarter-slope line characteristic of bilinear flow may not appear on a log-log plot of test data.
- An independent estimate of k is required. This suggests that prefracture well tests should be conducted before fracturing the well, thus obtaining independent estimates of formation properties.

Linear Flow Method. The linear flow method[30] applies to test data obtained during formation linear flow in wells with high-conductivity fractures ($C_r \geq 100$). After wellbore storage effects have ended, formation linear flow occurs up to a dimensionless time of $t_{L_fD} = 0.016$, which means that a log-log plot of pressure derivative against time will have a slope of one-half. The plot of pressure change vs. time, however, will have a half-slope only if the fracture skin is zero. The pressure and pressure derivative are

$$p_D = \left(\pi L_{fD}\right)^{1/2} + s_f, \quad (8.163)$$

$$\text{and } t_{L_fD}\frac{dp_D}{dt_{L_fD}} = \frac{1}{2}\left(\pi t_{L_fD}\right)^{1/2}, \quad (8.164)$$

so that

$$\log\left(t_{L_fD}\frac{dp_D}{dt_{L_fD}}\right) = \frac{1}{2}\log\left(t_{L_fD}\right) + \frac{1}{2}\log(\pi), \quad (8.165)$$

which indicates that a log-log plot of the derivative against time will have a slope of one-half. Radial equivalent time applies rigorously only for radial flow in an infinite-acting reservoir. When linear flow is the flow pattern occurring at both times ($t_p + \Delta t$) and Δt, a more useful equivalent time function is the linear equivalent time, Δt_{eL}.

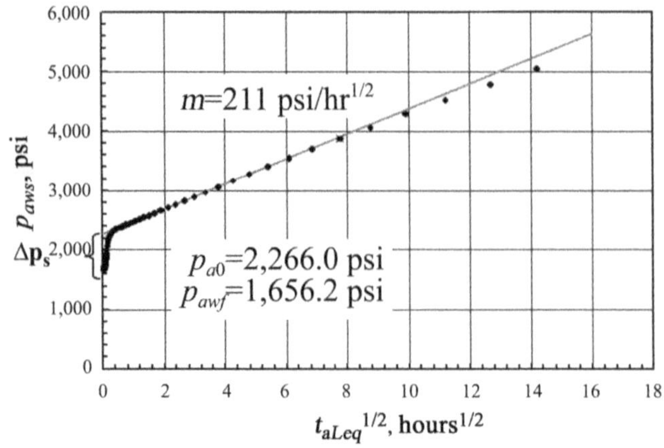

Fig. 8.80—Plot used in linear flow analysis.

$$\sqrt{\Delta t_{eL}} = \sqrt{t_p} + \sqrt{\Delta t} - \sqrt{t_p + \Delta t} \; . \quad (8.166)$$

Test conditions in which linear flow occurs at both (t_p + Δt) and Δt are rare, and, consequently, Eq. 8.166 is not necessarily rigorously correct for well-test analysis. Fortunately, when $t_p >> \Delta t_{max}$, $\Delta t_{eL} \approx \Delta t$. **Fig. 8.80** is an example of a plot used in linear flow analysis.

The linear flow analysis method also has limitations.

- The method applies only for fractures with high conductivities. Strictly speaking, linear flow occurs for the condition of uniform flux into a fracture (same flow rate from the formation per unit cross-sectional area of the fracture at all points along the fracture) rather than for infinite fracture conductivity. Therefore, only very early test data ($t_{L_f D} \leq 0.016$) exhibit linear flow in a high-conductivity fracture.
- Some or all of these early data may be distorted by wellbore storage, further limiting the amount of linear-flow data available for analysis.
- Estimating fracture half-length requires an independent estimate of permeability, k, which suggests the need for a prefracture well test.

Pseudoradial Flow Method. The pseudoradial flow method applies when a short, highly conductive fracture is created in a high-permeability formation, so that pseudoradial flow develops in a short time. The time required to achieve pseudoradial flow for an infinitely conductive fracture ($C_r \geq 100$) in either a flow test or a pressure buildup test is estimated by

$$t_{L_f D} = \frac{0.0002637 k t}{\phi \mu c_t L_f^2} \cong 3. \quad (8.167)$$

The beginning of pseudoradial flow is characterized by the flattening of the pressure derivative on a log-log plot and by the start of a straight line on a semilog plot. Hence, when the pseudoradial flow regime is reached, conventional semilog analysis can be used to calculate permeability and skin factor. For a highly conductive fracture, skin factor is related to fracture half-length by[10]

$$L_f \cong 2 r_w e^{-s} \; . \quad (8.168)$$

Fig. 8.81 shows an example.

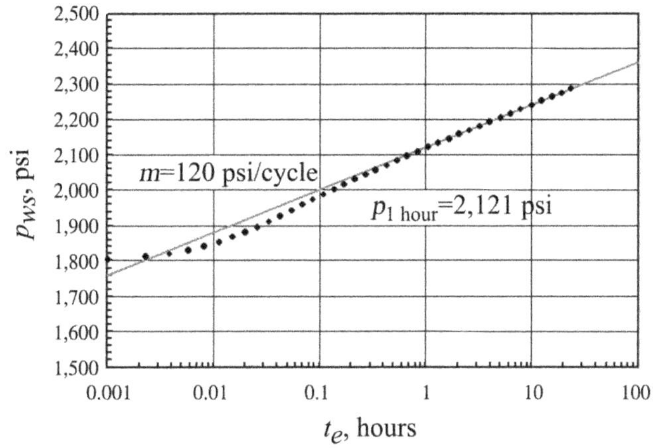

Fig. 8.81—Pseudoradial flow analysis.

A recommended procedure for analyzing test data from the pseudoradial flow regime is as follows.

- For a drawdown test, plot p_{wf} vs. log t. For a buildup test, plot p_{ws} vs. the HTR.
- Determine the position and slope, m, of the semilog straight line and the intercept, p_{1hr} on the line.
- Using m, calculate values of k and s (or s' for a gas well).
- Calculate the fracture half-length, L_f, using Eq. 8.168.

The pseudoradial flow method has the following limitations that seldom make it applicable in practice.[30]

- The conditions that are most favorable for the occurrence of pseudoradial flow are short, highly conductive fractures in high-permeability formations. These formations, however, are rarely fractured. The most common application of hydraulic fractures—wells with long fractures in low-permeability formations—require impractically long test times to reach pseudoradial flow.
- For gas wells, the apparent skin factor, s', calculated from test data is often affected by non-Darcy flow.
- The pseudoradial method applies only to highly conductive ($C_r \geq 100$) fractures. For lower conductivity fractures, fracture lengths calculated using the skin factor (Eq. 8.168) will be too low.

8.9.5 Using Type Curves for Hydraulically Fractured Wells. Type curves are the most common method of analyzing hydraulically fractured wells. The independent variable for most type curves for analyzing hydraulically fractured wells is the dimensionless time based on hydraulic fracture half-length, $t_{L_f D}$. The dependent variable is usually the dimensionless pressure, p_D.

For type curves used for manual type-curve matching, most vary only one parameter. The Cinco type curve[27] is obtained for zero $C_{L_f D}$ and s_f; the only parameter is dimensionless fracture conductivity, C_r or F_{cD} (where $F_{cD} = \pi c_r$). The choked-fracture skin is analyzed by assuming $C_{L_f D}$ and infinite C_r with single parameter s_f. The wellbore-storage type curve[31] sets s_f to 0 and C_r (F_{cD}) to infinity and varies the coefficient $C_{L_f D}$.

When using type curves in commercial software, the computer can set any two of the three parameters to fixed values (other than their limiting values) and vary the third parameter to obtain the matching stems.

Procedures for Analyzing Fractured Wells With Type Curves. The following steps outline the procedure for analyzing fractured wells with type curves.

- Graph field data pressure change and pressure derivatives.
- Match field data to the appropriate type curve.
- Find the match point and matching stem.
- Calculate the formation permeability from the pressure match point.
- Calculate L_f from the time match point.
- Interpret the matching stem value appropriate for a given type curve. For one type curve, this can be $w_f k_f$, which will provide an estimate of fracture conductivity. For another, it can be s_f, the choked-fracture skin, or, for a third, it can be C, the wellbore-storage coefficient.

To interpret the match points for a test with unknown permeability, use Eqs. 8.169 and 8.170. The formation permeability, k, is determined from the pressure match point; that is, the relationship between the pressure derivative and pressure change found at a match point given by

$$k = \frac{141.2qB\mu}{h}\left(\frac{p_D}{\Delta p}\right)_{\text{MP}} . \qquad (8.169)$$

From the time match point, calculate the fracture half-length:

$$L_f = \sqrt{\frac{0.0002637k}{\phi\mu c_t}\left(\frac{\Delta t}{t_{L_f D}}\right)_{\text{MP}}} . \qquad (8.170)$$

Matching can be ambiguous for hydraulically fractured wells; the data can appear to match equally well in several different positions. The ambiguity can be reduced or eliminated if a prefracture permeability is determined, and the post-fracture test data forced to match the permeability.

Type Curves Used for Analysis in Fractured Wells. *Cinco Type Curve.* The Cinco type curve (**Fig. 8.82**),[27] assumes that $C_{L_f D} = 0$ and $s_f = 0$. The type-curve stems on this curve are obtained by varying values of C_r or F_{cD}. With the Cinco type curve, the fracture conductivity, $w_f k_f$, can be determined from the matching parameter:

$$w_f k_f = \pi k L_f C_r . \qquad (8.171)$$

Choked-Fracture Type Curve. **Fig. 8.83** shows the choked-fracture type curve.[29] The choked-fracture type curve is generated with wellbore-storage coefficient, $C_{L_f D}$, of zero and infinite fracture conductivity, C_r. On this type curve, the stems represent different values of the fracture skin, s_f. The fracture skin, s_f, can be used to find the additional pressure drop from

$$\Delta p_s = \frac{141.2qB\mu}{kh} s_f . \qquad (8.172)$$

Wellbore-Storage Type Curve. The wellbore-storage type curve (**Fig. 8.84**) takes into account the possibility of wellbore storage. The wellbore-storage type assumes $s_f = 0$ and $C_r = \infty$. To interpret a best-fitting stem for this type curve, use the following:

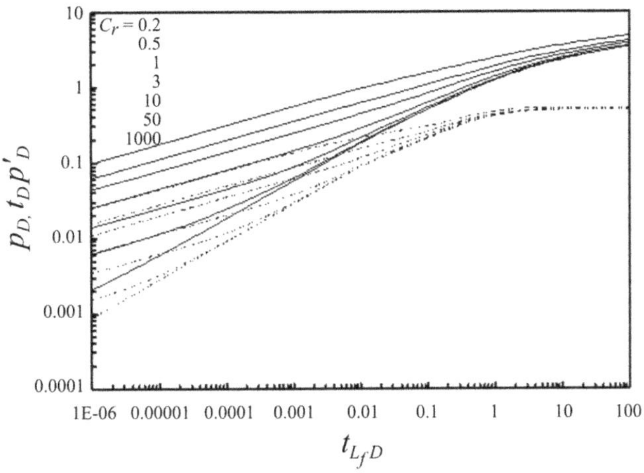

Fig. 8.82—Cinco type curve.

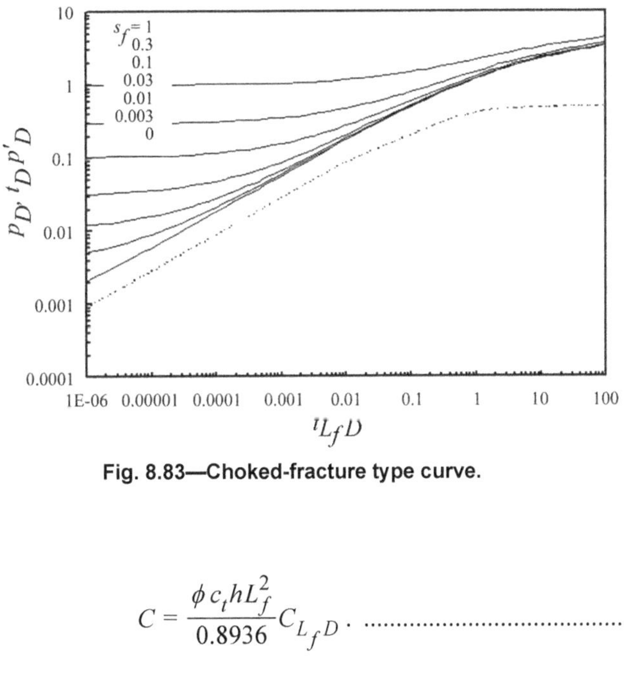

Fig. 8.83—Choked-fracture type curve.

$$C = \frac{\phi c_t h L_f^2}{0.8936} C_{L_f D} . \quad (8.173)$$

8.9.6 Limitations of Type-Curve Analysis in Hydraulically Fractured Wells. Although it is the most common methodology for analyzing hydraulically fractured well, type-curve analysis still has some limitations.

First, type-curves for analysis of hydraulically fractured wells are usually based on solutions for constant-rate drawdown tests. For buildup tests, shut-in time itself may possibly be used as a plotting function in those cases in which producing time is much greater than the shut-in time. Equivalent time can be used in some cases, but equivalent time has different definitions depending on the flow regime: radial, linear, and bilinear flow. Another possibility is to use a "superposition" type curve, which depends on the specific durations of flow and buildup periods. Superposition type curves can be readily generated with computer software.

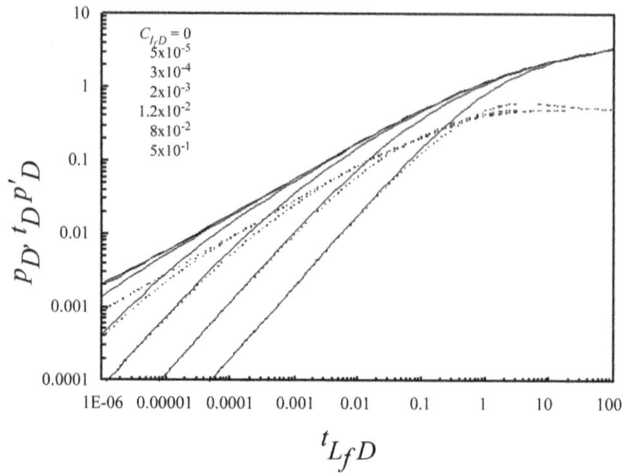

Fig. 8.84—Wellbore-storage type curve.

Another problem with type curves is that they may ignore important behavior. The type curve that takes into account wellbore storage does not consider a variable wellbore storage coefficient. This can be caused by phase redistribution in the wellbore, for example. The widely available type curves that have been discussed do not include boundary effects. With gas wells, the probability of non-Darcy flow is high, but available type curves don't take this into account.

An independent estimate of permeability may also be needed. A number of different type curves or a variety of stems on a given type curve may seem to match test data equally well. To remove this ambiguity, the best solution is to have an independent estimate of permeability.

8.10 Naturally Fractured Reservoirs

This section focuses on interpretation of well test data from wells completed in naturally fractured reservoirs. Because of the presence of two distinct types of porous media, the assumption of homogeneous behavior (discussed in previous sections) is no longer valid in naturally fractured reservoirs. This section includes two naturally fractured reservoir models, the physics governing fluid flow in these reservoirs and semilog and type-curve analysis techniques for well tests in these reservoirs.

8.10.1 Naturally Fractured Reservoir Models. Naturally fractured reservoirs are characterized by the presence of two distinct types of porous media: matrix and fracture. Because of the different fluid storage and conductivity characteristics of the matrix and fractures, these reservoirs often are called dual-porosity reservoirs. **Fig. 8.85** illustrates a naturally fractured reservoir composed of a rock matrix surrounded by an irregular system of vugs and natural fractures. Fortunately, it has been observed that a real, heterogeneous, naturally fractured reservoir has a characteristic behavior that can be interpreted using an equivalent, homogeneous dual-porosity model such as that shown in the idealized sketch.

Several models have been proposed to represent the pressure behavior in a naturally fractured reservoir. These models differ conceptually only in the assumptions made to describe fluid flow in the matrix. Most dual-porosity models assume that production from the naturally fractured system comes from the matrix, to the fracture, and then to the wellbore (i.e., that the matrix does not produce directly into the wellbore). Furthermore, the models assume that the matrix has low permeability but large storage capacity relative to the natural fracture system,

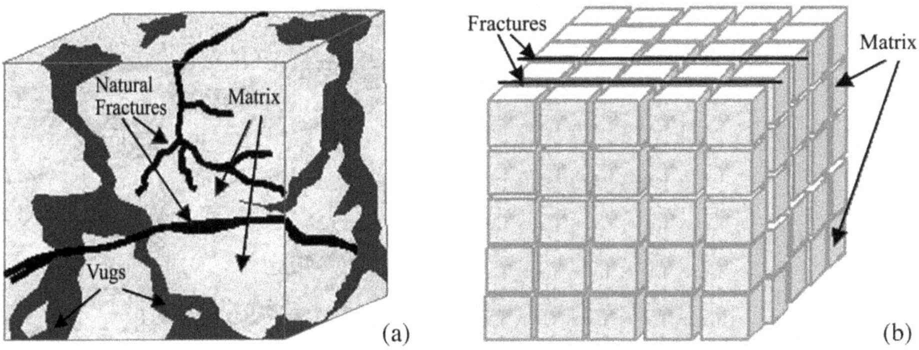

Fig. 8.85—Actual (a) and idealized (b) dual-porosity reservoir model.[32]

while the fractures have high permeability but low storage capacity relative to the natural fracture system. Warren and Root[32] introduced two dual-porosity parameters, in addition to the usual single-porosity parameters, which can be used to describe dual-porosity reservoirs.

Interporosity flow is the fluid exchange between the two media (the matrix and fractures) constituting a dual-porosity system. Warren and Root[32] defined the interporosity flow coefficient, λ, as

$$\lambda = \alpha r_w^2 \frac{k_m}{k_f}, \qquad (8.174)$$

where k_m is the permeability of the matrix, k_f is the permeability of the natural fractures, and α is the parameter characteristic of the system geometry.

The interporosity flow coefficient is a measure of how easily fluid flows from the matrix to the fractures. The parameter α is defined by[33]

$$\alpha = \frac{4j(j+2)}{L^2}, \qquad (8.175)$$

where L is a characteristic dimension of a matrix block and j is the number of normal sets of planes limiting the less-permeable medium (j = 1, 2, 3). For example, j = 3 in the idealized reservoir cube model in Fig. 8.85. On the other hand, for the multilayered or "slab" model shown in **Fig. 8.86**,[34] j = 1. For the slab model, letting $L = h_m$ (the thickness of an individual matrix block), λ becomes

$$\lambda = 12 r_w^2 \frac{k_m}{k_f h_m^2}. \qquad (8.176)$$

The storativity ratio,[33] ω, is defined by

$$\omega = \frac{(\phi V c_t)_f}{(\phi V c_t)_{f+m}} = \frac{(\phi V c_t)_f}{(\phi V c_t)_f + (\phi V c_t)_m}, \qquad (8.177)$$

where V is the ratio of the total volume of one medium to the bulk volume of the total system and ϕ is the ratio of the pore volume of one medium to the total volume of that medium.

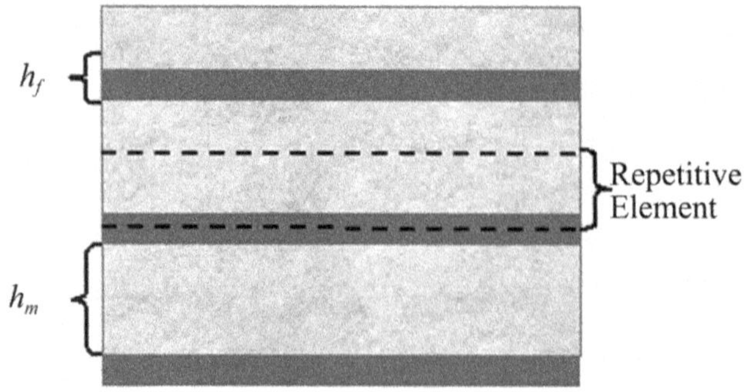

Fig. 8.86—Schematic reservoir with rectangular matrix elements.[34]

Subscripts f and $f + m$ refer to the fracture and to the total system (fractures plus matrix), respectively. Consequently, the storativity ratio is a measure of the relative fracture storage capacity in the reservoir.

Many models have been developed for naturally fractures reservoirs. Two common models, pseudosteady-state and transient flow, that describe flow in the less-permeable matrix are presented here. Pseudosteady-state flow was assumed by Warren and Root[32] and Barenblatt *et al.*[35]; others, notably deSwaan,[36] assumed transient flow in the matrix. Intuition suggests that, in a low-permeability matrix, very long times should be required to reach pseudosteady-state and that transient matrix flow should dominate; however, test analysis suggests that pseudosteady-state flow is quite common. A possible explanation of this apparent inconsistency is that matrix flow is almost always transient but can exhibit a behavior much like pseudosteady-state, if there is a significant impediment to flow from the less-permeable medium to the more-permeable one (such as low-permeability solution deposits on the faces of fractures).

8.10.2 Pseudosteady-State Matrix Flow Model. The pseudosteady-state flow model assumes that, at a given time, the pressure in the matrix is decreasing at the same rate at all points and, thus, flow from the matrix to the fracture is proportional to the difference between matrix pressure and pressure in the adjacent fracture. Specifically, this model, which does not allow unsteady-state pressure gradients within the matrix, assumes that pseudosteady-state flow conditions are present from the beginning of flow.

Because it assumes a pressure distribution in the matrix that would be reached only after what could be a considerable flow period, the pseudosteady-state flow model obviously is oversimplified. Again, this model seems to match a surprising number of field tests. One possible reason is that damage to the face of the matrix could cause the flow from matrix to fracture to be controlled by a sort of choke (the thin, low-permeability, damaged zone) and, therefore, is proportional to pressure differences upstream and downstream of the choke. In the next two sections, semilog and type-curve analysis techniques are presented for well tests in naturally fractured reservoirs exhibiting pseudosteady-state flow characteristics.

Semilog Analysis Technique. The pseudosteady-state matrix flow solution developed by Warren and Root[32] predicts that, on a semilog graph of test data, two parallel straight lines will develop. **Fig. 8.87** shows this characteristic pressure response.

The initial straight line reflects flow in the fracture system only. At this time, the formation is behaving like a homogeneous formation with fluid flow originating only from the fracture system with no contribution from the matrix. Consequently, the slope of the initial semilog straight line is proportional to the permeability-thickness product of the natural fracture system, just as it is for any homogeneous system. Following a discrete pressure drop in the fracture

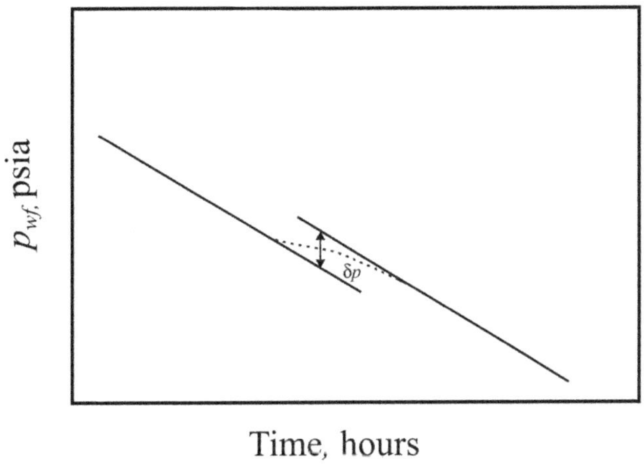

Fig. 8.87—Characteristic pressure response of flow test exhibiting pseudosteady-state matrix flow.

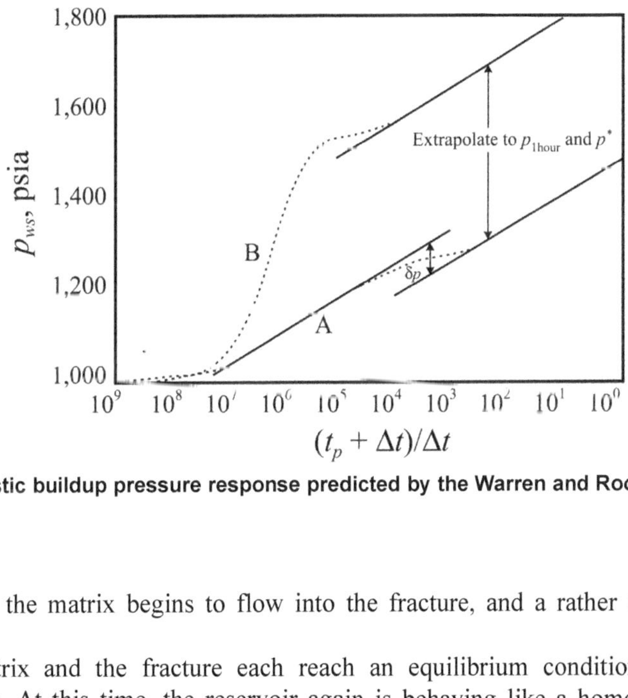

Fig. 8.88—Characteristic buildup pressure response predicted by the Warren and Root[32] pseudosteady-state model.

system, the fluid in the matrix begins to flow into the fracture, and a rather flat transition region appears.

Finally, the matrix and the fracture each reach an equilibrium condition, and a second straight line appears. At this time, the reservoir again is behaving like a homogeneous system, but now the system consists of both the matrix and the fractures. The slope of the second semilog straight line is proportional to the total permeability-thickness product of the matrix/fracture system. Because the permeability of the fractures is much greater than that of the matrix, the slope of the second line is almost identical to that of the initial line.

Similar shapes are predicted for pressure buildup tests (**Fig. 8.88**). The lower curve, *A*, represents the ideal buildup test plot predicted by Warren and Root.[32] The shape of a semilog plot of test data from a naturally fractured reservoir is almost never the same as that predicted by Warren and Root's model. Wellbore storage almost always obscures the initial straight line and often obscures part of the transition region between the straight lines. The upper curve, *B*, in Fig. 8.88 shows a more common pressure response.

The reservoir permeability-thickness product, kh [actually the kh of the fractures, or $(kh)_f$, because $(kh)_m$ is usually negligible], can be obtained from the slope, m, of the two semilog straight lines. Storativity, ω, can be determined from their vertical displacement, δp. The interporosity flow coefficient, λ, can be obtained from the time of intersection of a horizontal line, drawn through the middle of the transition curve, with either the first or second semilog straight line.[33]

When semilog analysis is possible (i.e., when the correct semilog straight line can be identified), the following procedure is recommended for semilog analysis of buildup or drawdown test data from wells completed in naturally fractured reservoirs. Although presented in variables for slightly compressible fluids (liquids), the same procedure is applicable to gas well tests when the appropriate variables are used.

• From the slope of the initial straight line (if present) or final straight line (more likely to be present), determine the permeability-thickness product, kh. In either case, the slope, m, is related to the total kh of the system, which is essentially all in the fractures. The permeability-thickness product is given by

$$(kh)_f = \bar{k}h = \frac{162.6 qB\mu}{|m|}, \qquad (8.178)$$

where $\bar{k}$ is equal to $(kh)_f/h$. Strictly speaking, the slope of the second straight line is related to $[(kh)_f + (kh)_m]$, but $(kh)_m$ ordinarily is negligible compared to $(kh)_f$.

• If both initial and final straight lines can be identified (or the position of the initial line can at least be approximated) and the pressure difference, δp, established, then the storativity ratio, ω is calculated from

$$\omega = 10^{-\delta p/m}. \qquad (8.179)$$

If the times of intersection of a horizontal line drawn through the midpoint of the transition data with the first and second semilog straight lines are denoted by t_1 and t_2, respectively, the storativity ratio may also be calculated from

$$\omega = \frac{t_1}{t_2}. \qquad (8.180)$$

For a buildup test, where the times of intersection of a horizontal line drawn through the midpoint of the transition data with the first and second semilog straight lines are denoted by $[(t_p + \Delta t)/\Delta t]_1$ and $[(t_p + \Delta t)/\Delta t]_2$, respectively, the storativity ratio may be calculated from

$$\omega = \frac{[(t_p + \Delta t)/\Delta t]_2}{[(t_p + \Delta t)/\Delta t]_1}. \qquad (8.181)$$

• The interporosity flow coefficient, λ, is calculated[33] for a drawdown test by

$$\lambda = \frac{(\phi V c_t)_f \mu r_w^2}{\gamma \bar{k} t_1} = \frac{(\phi V c_t)_{f+m} \mu r_w^2}{\gamma \bar{k} t_2}, \qquad (8.182)$$

or for a buildup test by

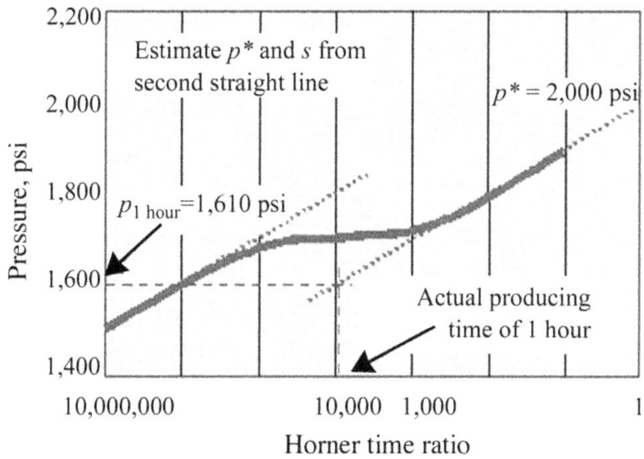

Fig. 8.89—Extrapolation of second semilog straight line to p^*.

$$\lambda = \frac{(\phi V c_t)_f \mu r_w^2}{\gamma \bar{k} t_p}\left(\frac{t_p + \Delta t_1}{\Delta t_1}\right) = \frac{(\phi V c_t)_{f+m} \mu r_w^2}{\gamma \bar{k} t_p}\left(\frac{t_p + \Delta t_2}{\Delta t_2}\right), \quad \text{(8.183)}$$

where $\gamma = 1.781$.

The terms $(\phi V)_m$ and $(c_t)_m$ in Eq. 8.183 are obtained by conventional methods. A porosity log usually reads only the matrix porosity (not the fracture porosity) and thus gives ϕ_m, while $(c_t)_m$ is the sum of c_oS_o, c_gS_g, c_wS_w, and c_f. V_m usually can be assumed to be essentially 1.0. From the definition of ω in Eq. 8.177,

$$(\phi V c_t)_f = (\phi V c_t)_m \left(\frac{\omega}{1-\omega}\right). \quad \text{(8.184)}$$

• The second semilog straight line should be extrapolated to p_{1hr}, and the skin factor is

$$s = 1.151\left[\frac{\Delta p_{1\,hr}}{|m|} - \log\left(\frac{\bar{k}}{\phi \mu c_t r_w^2}\right) + 3.23\right], \quad \text{(8.185)}$$

where $\Delta p_{1\,hr}$ is equal to $(p_i - p_{1\,hr})$ for a drawdown test or $\left[p_{1\,hr} - p_{wf}(\Delta t = 0)\right]$ for a buildup test.

• The second semilog straight line should be extrapolated to p^* (**Fig. 8.89**). From p^*, $\bar{p}$ can be found using conventional methods (such as the Matthew-Brons-Hazebroek p^* method).

Type Curve Analysis Technique. Particularly because of wellbore-storage distortion, type curves are quite useful for identifying and analyzing dual-porosity systems. **Fig. 8.90** shows an example of the Bourdet *et al.*[37] type curves developed for pseudosteady-state matrix flow. Initially, test data follow a curve for some value of C_De^{2s} where C_D is the dimensionless wellbore storage coefficient. In Fig. 8.90, the earliest data for the well follow the curve for $C_De^{2s} = 1$. The data then deviate from the early fit and follow a transition curve characterized by the parameter λe^{-2s}. In Fig. 8.90, the data follow the curve for $\lambda e^{-2s} = 3\times10^{-4}$. When equilibrium is reached between the matrix and fracture systems, the data then follow another C_De^2s curve. In the example, the later data follow the $C_De^{2s} = 0.1$ curve.

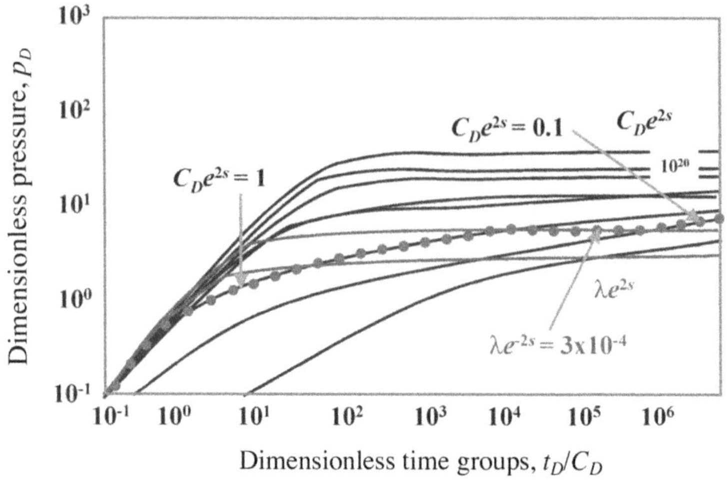

Fig. 8.90—Type curves for pseudosteady-state matrix flow. (After Bourdet *et al.*[37])

At earliest times, the reservoir is behaving like a homogeneous reservoir with all fluid originating from the fracture system. During intermediate times, there is a transition region as the matrix begins to produce into the fractures. At later times, the system again is behaving like a homogeneous system with both matrix and fractures contributing to fluid production.

Fig. 8.91 illustrates the derivative type curves for a formation with pseudosteady-state matrix flow.[37] The most notable feature, characteristic of naturally fractured reservoirs, is the dip below the homogeneous reservoir curve. The curves dipping downward are characterized by a parameter $\lambda C_D/\omega\ (1 - \omega)$, while the curves returning to the homogeneous reservoir curves are characterized by the parameter $\lambda C_D/(1 - \omega)$. Test data that follow this pattern on the derivative type curve can reasonably be interpreted as identifying a dual-porosity reservoir with pseudosteady-state matrix flow (a theory that needs to be confirmed with geological information and reservoir performance). Pressure and pressure derivative type curves can be used together for analysis of a dual-porosity reservoir. The pressure derivative data are especially useful for identifying the dual-porosity behavior. Manual type-curve analysis for well in naturally fractured reservoirs is tedious, and the interpretation involved is difficult. Most current analysis uses commercial software.

Transient Matrix Flow Model. The more probable flow regime in the matrix is unsteady-state or transient flow; that is, flow in which an increasing pressure drawdown starts at the matrix/fracture interface and moves further into the matrix with increasing time. Only at late times should pseudosteady-state flow be achieved, although a matrix with a thin, low-permeability damaged zone at the fracture face may behave as predicted by the pseudosteady-state matrix flow model even though the flow in the matrix is actually unsteady-state.

A semilog graph of test data for a formation with transient matrix flow has a characteristic shape different from that for pseudosteady-state flow in the matrix. Three distinct flow regimes have been identified that are characteristic of dual-porosity reservoir behavior with transient matrix flow. **Fig. 8.92** illustrates these flow regimes on a semilog graph as regimes 1, 2, and 3.

Flow regime 1 occurs at early times during which all production comes from the fractures. Flow regime 2 occurs when production from the matrix into the fracture begins and continues until the matrix-to-fracture transfer reaches equilibrium. This equilibrium point marks the beginning of flow regime 3, during which total system flow, from matrix to fracture to wellbore, is dominant. The same three flow regimes appear when there is pseudosteady-state matrix flow.

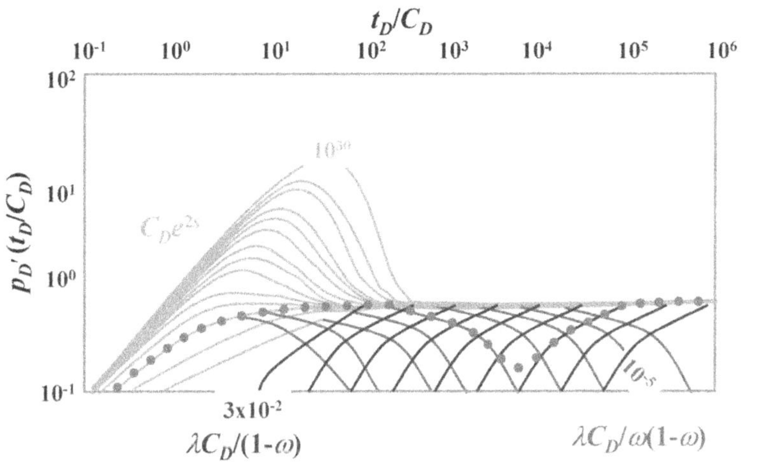

Fig. 8.91—Derivative type curves for pseudosteady-state matrix flow. (After Bourdet *et al.*[37])

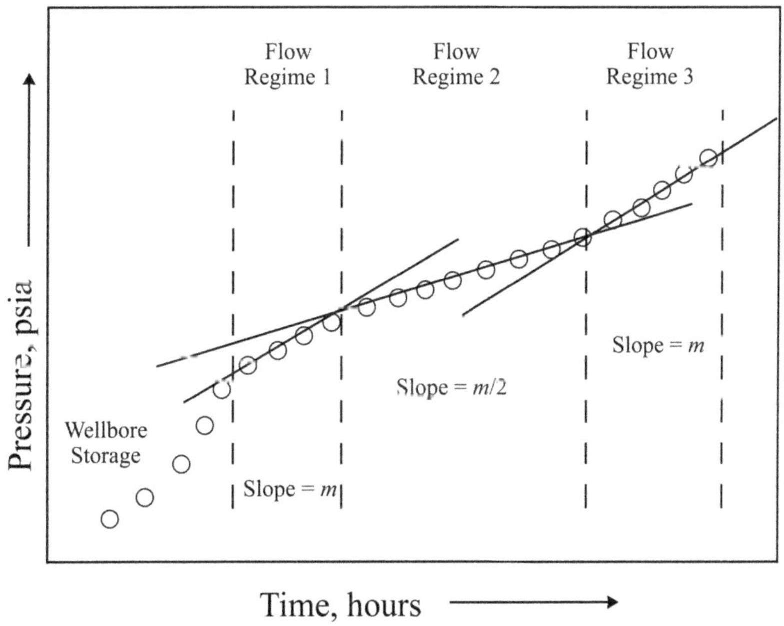

Fig. 8.92—Flow regimes in a dual-porosity reservoir with transient matrix flow.[34]

The duration and shape of the transition flow regimes, however, is considerably different for the two matrix flow models.

Serra *et al.*[34] observed that pressures from each of these flow regimes will plot as straight lines on conventional semilog graphs. Flow regimes 1 and 3, which correspond to the classical early- and late-time semilog straight-line periods, respectively, have the same slope. Flow regime 2 is an intermediate transitional period between the first and third flow regimes. The semilog straight line of flow regime 2 has a slope of approximately one-half that of flow regimes 1 and 3. If all or any two of these regimes can be identified, then a complete analysis is possible using semilog methods alone. Certain nonideal conditions, however, may make this analysis difficult to apply.

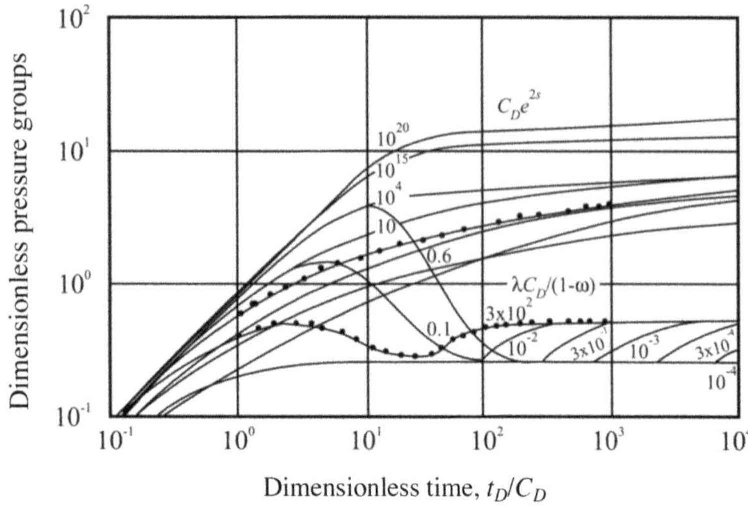

Fig. 8.93—Type curves for transient matrix flow.[37]

Flow regime 1 often is distorted or obscured by wellbore storage, which often makes this flow regime difficult to identify. Flow regime 2, the transition, also may be obscured by wellbore storage. Flow regime 3 sometimes requires a long flow period followed by a long shut-in time to be observed, especially in formations with low permeability. Furthermore, boundary effects may appear before flow regime 3 is fully developed.

Semilog Analysis Techniques. Serra *et al.*[34] presented a semilog method for analyzing well test data in dual-porosity reservoirs exhibiting transient matrix flow (Fig. 8.92). They found that the existence of the transition region, flow regime 2, and either flow regime 1 or flow regime 3 is sufficient to obtain a complete analysis of drawdown or buildup test data. Further, they assumed unsteady-state flow in the matrix, no wellbore storage, and rectangular matrix-block geometry, as Fig. 8.86 shows. The rectangular matrix-block geometry is adequate, although different assumed geometries can lead to slightly different interpretation results.

The major weakness of the Serra *et al.* method is that it assumes no wellbore storage. In many cases, flow regimes 1 and 2 are partially or even totally obscured by wellbore storage, making analysis by the Serra *et al.* method impossible or difficult. Despite this limitation, the Serra *et al.* method has great practical value when used in conjunction with type-curve methods. These calculations of the Serra *et al.* method apply to both buildup and drawdown test data and are applicable for well test analysis of slightly compressible liquids and gas well tests.

Type Curve Analysis Technique. Bourdet *et al.*[37] presented type curves for analyzing well tests in dual-porosity reservoirs including the effects of wellbore storage and unsteady-state flow in the matrix. The type curves are useful supplements to the Serra *et al.* semilog analysis. **Fig. 8.93** gives an example of the pressure and pressure derivative type curves for transient matrix flow. Early (fracture-dominated) data are fit by a C_De^{2s} value indicative of homogeneous behavior. Data in the transition region are fit by curves characterized by a parameter β'. Finally, data in the homogeneous-acting, fracture-plus-matrix flow regime are fit by another C_De^{2s} curve.

On the derivative type curve, early data also are fit by a derivative curve reflecting homogeneous behavior. **Fig. 8.94** shows an actual example. If wellbore-storage distortion ceases before the transition region begins (which did not happen in the example but is possible in other cases), the derivative data will be horizontal and should be aligned with the $(t_D/C_D)p_D' = 0.5$

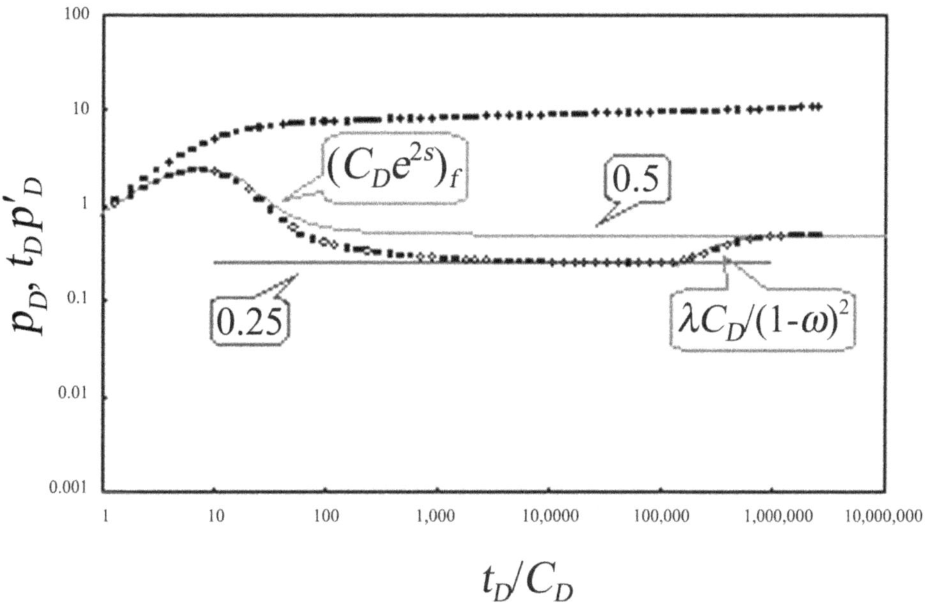

Fig. 8.94—Example of using type curves for transient matrix flow.[37]

curve. However, if the transition region is present (recall that its semilog slope is half that of the middle-time straight line), the derivative curve will flatten and should be aligned with the $(t_D/C_D)p_D' = 0.25$ curve as shown in this example. The homogeneous (fracture-plus-matrix) data should, after wellbore distortion has ceased and before boundary effects have appeared, be horizontal on the derivative type curve and should be aligned with the $(t_D/C_D)p_D' = 0.5$ curve as this example shows.

Manual type-curve matching is tedious and difficult, especially with the interpolation involved. Analysis ordinarily uses commercially available software to analyze these kinds of tests after the reservoir model has been identified.

8.11 Horizontal Well Analysis

Productivity estimates in horizontal wells are subject to more uncertainty than comparable estimates in vertical wells. Further, it is much more difficult to interpret well test data because of 3D flow geometry. The radial symmetry usually present in a vertical well does not exist. Several flow regimes can potentially occur and need to be considered in analyzing test data from horizontal wells. Wellbore storage effects can be much more significant and partial penetration and end effects commonly complicate interpretation.

In vertical wells, variables such as average permeability, net vertical thickness, and skin are used. Horizontal wells need more detail. Not only is vertical thickness important, but the horizontal dimensions of the reservoir, relative to the horizontal wellbore, need to be known.

8.11.1 Steps in Evaluating Horizontal Well-Test Data. There are three basic steps in evaluating pressure-transient data from a horizontal well. First, identify the specific flow regimes in the test data. Second, apply the proper analytical and graphical procedures to the data. Finally, evaluate the uniqueness and sensitivity of the results to properties derived from analysis or simply assumed.

Identify Flow Regimes. As discussed in previous sections, evaluation of data from a vertical wellbore will generally center on a single flow regime, such as infinite-acting radial flow,

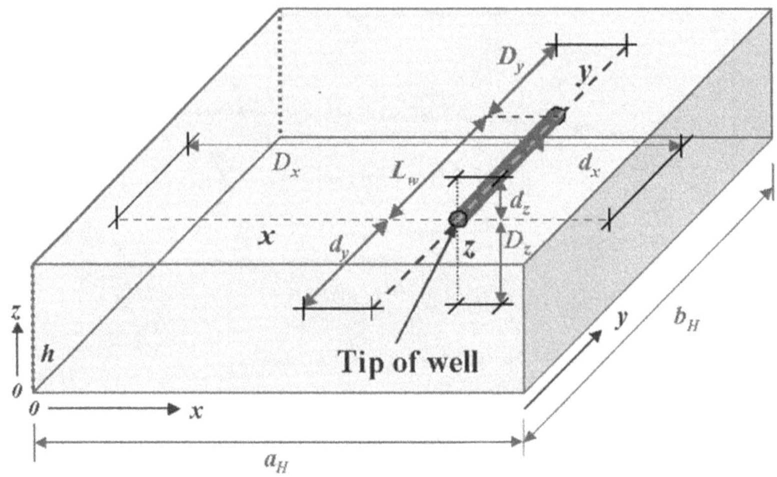

Fig. 8.95—Well and reservoir geometry and nomenclature for a horizontal well.

known as the MTR. However, a pressure-transient test in a horizontal well can involve as many as five major and distinct regimes that need to be identified. These regimes may or may not occur in a given test and may or may not be obscured by wellbore storage effects.

Apply the Proper Procedures. Each flow regime can be modeled by an equation that can be used to estimate important reservoir properties. At best, only groups of analytical parameters can be determined directly from equations. It is imperative that the proper analytical and graphical procedures be applied to the data. In many cases, when solving for specific parameters, the application of these analytical expressions may involve a complex iterative procedure.

Evaluate Uniqueness and Sensitivity. Experience indicates that results of horizontal well test analysis are seldom unique, so it is important that the uniqueness and sensitivity of the results to assumed properties be evaluated. Simulation of the test using properties that have been determined from the test can confirm that at least the analysis is consistent with the test data. A simulator can also determine whether other sets of formation properties will also lead to a fit of the data.

8.11.2 Horizontal Well Flow Regimes. Different formation properties can be calculated from the data in each of the five different flow regimes. Any flow regime may be absent from a plot of test data because of geometry, wellbore storage, or other factors. Nor does the fact that they can appear mean that they do appear. The five different flow regimes that can occur are early radial, hemi-radial, early linear, late pseudoradial, and late linear.

Fig. 8.95 shows a horizontal well with length, L_w, within a reservoir that is assumed to be a rectangular parallelepiped or a "box reservoir" drainage area. In this discussion, it is assumed that the axes of the coordinate system coincide with the direction of principal permeability and the well produces over its entire length, L_w.

The axes for this box are the usual x-, y-, and z-axes. Notice that the x-axis is measured along the bottom edge of the reservoir, going from left to right in the direction perpendicular to the well. The y-axis lies along the axis from front to back of the reservoir, parallel to the wellbore. The z-axis is oriented in the direction of reservoir thickness.

The total width of the reservoir perpendicular to the wellbore is a_H, the total length in a direction parallel to the wellbore is b_H, and the total height of the reservoir is the net pay thickness, h. Notice the parameters for the distance from the well to the various borders. Along the axis of the well, the shortest distance from the end of the well to a boundary is d_y, and the

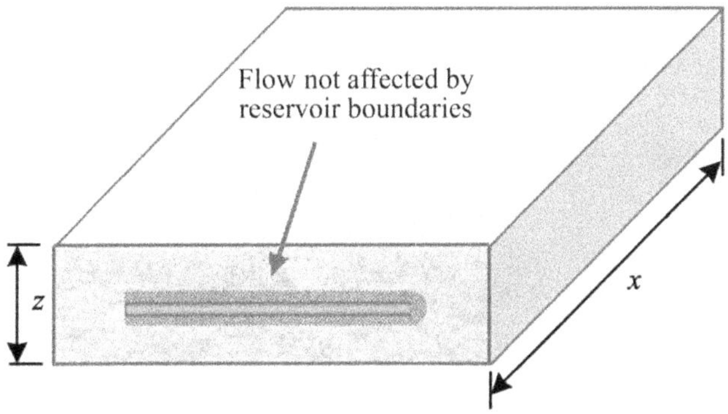

Fig. 8.96—Early-radial flow is not influenced by reservoir boundaries.

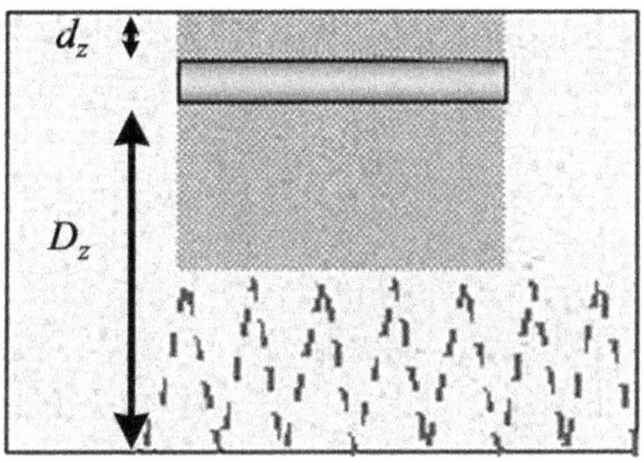

Fig. 8.97—Hemiradial flow occurs when effects of one vertical boundary are felt before the effects of the other.

longest distance from the other end of the well to the boundary is D_y. In the vertical direction, the shortest distance to a vertical boundary is d_z, and the longest distance to a vertical boundary is D_z.

Characteristics of Flow Regimes. Consider a well producing at a constant rate. The early radial flow regime occurs before the area drained or the pressure transient caused by this production encounters either of the vertical boundaries of the reservoir. **Fig. 8.96** shows a radial flow pattern penetrating out into the reservoir. Actually, however, this flow pattern is likely to be elliptical, moving further into the reservoir at a given time in the higher-permeability x-direction than in the lower-permeability z-direction. This phenomenon causes no significant complications in our analysis.

When the wellbore is much nearer one vertical boundary than the other, another flow regime, called half-radial or hemiradial flow (**Fig. 8.97**) may exist. Hemiradial flow can occur immediately following the early radial flow regime, if the well is much nearer one of the vertical boundaries than the other. Eventually, the area affected by the production will include the entire thickness of the reservoir. When that happens, a linear flow pattern may develop, as **Fig. 8.98** shows.

Eventually, flow will begin to come into the wellbore from beyond the ends of the well. Until these end effects become important, early linear flow continues. Once end effects become important a transition period is followed by a later pseudoradial flow regime, as **Fig. 8.99** illustrates.

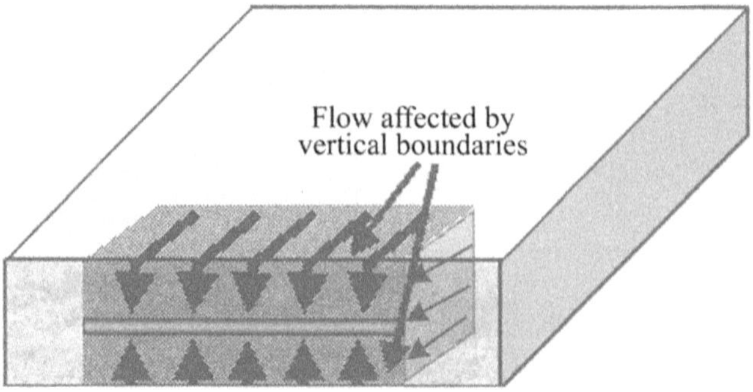

Fig. 8.98—Early-linear flow may develop after both vertical boundaries are encountered.

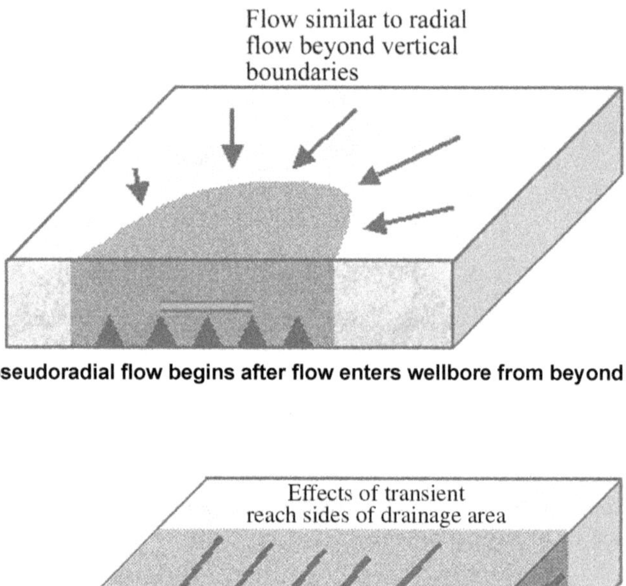

Fig. 8.99—Late pseudoradial flow begins after flow enters wellbore from beyond ends of well.

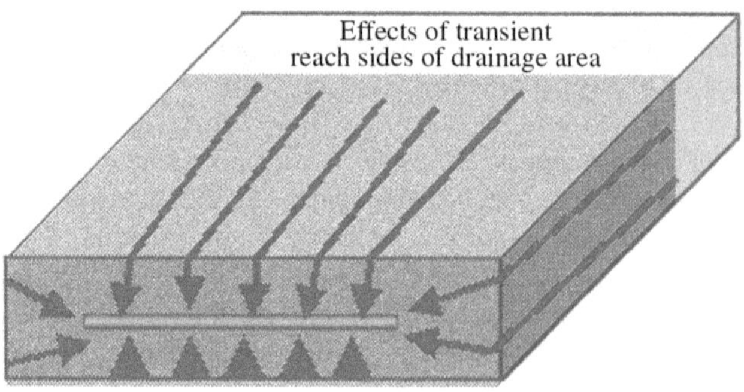

Fig. 8.100—Late-linear flow includes flow from the drainage volume perpendicular to the well.

This flow regime continues until the area affected by the production reaches one of the sides of the reservoir. Once the area affected is the entire width of the reservoir (that is, the pressure transient has reached both sides of the reservoir), then the late-linear flow regime begins (**Fig. 8.100**).

8.11.3 Identifying Flow Regimes in Horizontal Wells. All of these flow regimes in a test can be identified on a diagnostic log-log plot of the pressure change, Δp, and pressure derivative, p', against the logarithm of time (**Fig. 8.101**).

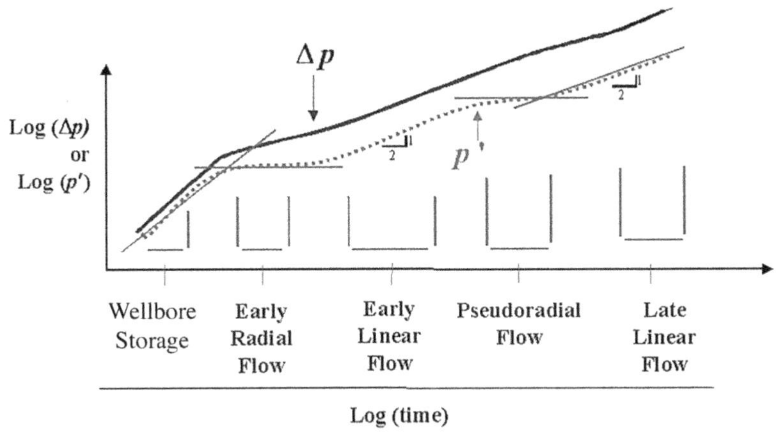

Fig. 8.101—Idealized diagnostic plot identifies all normally expected flow regimes.

A unit-slope line appears during wellbore storage; a horizontal derivative during early radial flow, and then, later, in pseudoradial flow; and a half-slope line in early-linear flow and then in late-linear flow. (These half-slope lines appear on the derivative but not on the pressure-change curves.) This does not imply that all these flow regimes will appear in any given test; in fact, that would be rare. But these are the shapes that identify the flow regimes that may appear in the test being analyzed.

The shapes that may appear in a drawdown test (which is the basis of Fig. 8.101) may not appear in a buildup test because of the complex superposition of flow regimes. For example, a test would have to be in linear flow both at time ($t_p + \Delta t$) and at time Δt to ensure appearance of a derivative with half-slope; this is highly unlikely. The best way to solve the problem is to ensure that a buildup test on a horizontal well is run with a producing time, t_p, much greater than the maximum shut-in time in the test (that is, $t_p > 10\ \Delta t_{max}$).

Table 8.A-2 (see Appendix) summarizes the working equations for permeability, skin, and start and end of each of the recognized flow regimes. Different investigators have found different equations for start and end of various flow regimes, especially linear flow regimes. This is partly because of a difference of assumptions about flow into the wellbore. Uniform flux or infinite conductivity models are common; neither is rigorously applicable in practice.[38] In this section, the equations for duration of flow regimes derived by Odeh and Babu[39] are used. This model assumes uniform flux into the wellbore.

Early-Radial Flow. Early-radial flow is similar to the radial flow period in a vertical well (Fig. 8.96). The governing equation for this flow regime is

$$p_i - p_{wf} = \frac{162.6 qB\mu}{\sqrt{k_z k_x} L_w}\left[\log_{10}\left(\frac{\sqrt{k_x k_z}\, t}{\phi \mu c_t r_w^2}\right) - 3.227 + 0.868 s_d\right]. \quad (8.186)$$

Data for this period may be masked by wellbore storage effects, but, when present, they may be analyzed on a semilog plot.

The early-radial flow regime may in theory start at time zero, in absence of wellbore storage effects. The end of the early-radial flow regime may occur when the transient reaches a vertical boundary or when flow comes from beyond the end of the wellbore. The end of the period is the smaller of these two values.

Eq. 8.187[39] says that the period must end when the transient reaches the nearest boundary, d_z, from the well. This equation includes the permeability in the vertical direction:

$$t_{\text{Eerf}} = \frac{1{,}800 d_z^2 \phi \mu c_t}{k_z} \quad \text{(8.187)}$$

The radial flow regime may also end when flow from beyond the end of the wellbore becomes important. Eq. 8.188 gives the time by

$$t_{\text{Eerf}} = \frac{125 L_w^2 \phi \mu c_t}{k_y} \quad \text{(8.188)}$$

L_w is the completed length of the well, and k_y is the permeability in the direction parallel to the wellbore. The actual end is the lesser of the two times calculated from Eqs. 8.187 and 8.188. It is helpful to check the expected duration of the early-radial flow regime after estimating the parameters necessary to make these calculations.

Eq. 8.186 suggests that possible radial flow on the diagnostic plot be identified and then bottomhole flowing pressure be plotted against time during the appropriate time range on semilog coordinates. The slope of the straight line that results is

$$|m_{\text{erf}}| = \frac{162.6 q B \mu}{L_w \sqrt{k_x k_z}} \quad \text{(8.189)}$$

The group $\sqrt{k_x k_z}$ can be found from the slope, m_{erf}:

$$\sqrt{k_x k_z} = \frac{162.6 q B \mu}{|m_{\text{erf}}| L_w} \quad \text{(8.190)}$$

Effective completed length of the well must be known to make this calculation. This is not necessarily the same as the perforated or completed length of the well. Some sections of the well may not produce at all.

The equation for calculating the altered permeability skin, s_d, for early-radial flow is

$$s_d = 1.151 \left[\frac{\Delta p_{1\text{hr}}}{|m_{\text{erf}}|} - \log \left(\frac{\sqrt{k_x k_z}}{\phi \mu c_t r_w^2} \right) + 3.23 \right] \quad \text{(8.191)}$$

When analyzing a buildup test rather than a constant-rate flow test, plot the HTR or equivalent time on the horizontal axis of the semilog plot, and then plot shut-in or equivalent time on the vertical axis. Note that this plotting is correct only if ($t_p + \Delta t$) and Δt appear in this time period simultaneously; that is, radial flow must exist at both time ($t_p + \Delta t$) and time Δt. This is unlikely because radial-flow regime may exist at time Δt, but a different flow regime is likely at time ($t_p + \Delta t$).

Example 8.2: Well Erf-1. For drawdown test data from Well Erf-1,[39] the diagnostic plot indicates the data from approximately 0.24 to 24 hours may be in early-radial flow. The following information is available for this well: q = 800 STB/D, μ = 1 cp, B = 1.25 RB/STB, r_w = 0.25 ft, ϕ = 0.2, c_t = 15×10^{-6} psi^{-1}, centered in box-shaped drainage area, h = 200 ft, b_H = 4,000 ft, and a_H = 2,000 ft, L_w=1,000 ft, and, from analysis of data in early linear flow regime, k_x = 200 md. **Table 8.3** shows the pressure change data for 0.24 to 24 hours.

Plot $(p_i - p_{wf}) = \Delta p$ vs. t on semilog coordinates (**Fig. 8.102**). The plot results in a straight line with a slope of 8 psi/cycle. In Fig. 8.102, at t = 2.4 hours, the points begin to deviate from the straight line, as expected from calculations for flow regime duration that follow. The pressure change at 1 hour is 39 psia. Using the slope of 8 and Eq. 8.190,

$$\sqrt{k_z k_x} = \frac{162 q B \mu}{|m_{\text{erf}}| L_w}$$

$$= \frac{162.6(800)(1.23)(1)}{|8|1000}$$

$$= 20.3 \text{ md}$$

$$\text{and } k_x k_z = 412 \text{ md} .$$

Thus, because k_x = 200 md, k_z = 2 md. Using the value of 39 for Δp_{1hr} from Fig. 8.102, skin from Eq. 8.191 is

$$s_d = 1.151\left[\left(\frac{\Delta p_{1\text{ hr}}}{|m_{\text{erf}}|}\right) - \log\left(\frac{\sqrt{k_x k_z}}{\phi \mu c_t r_w^2}\right) + 3.23\right]$$

$$= 1.151\left[\left(\frac{39}{8}\right) - \log\left(\frac{20.3}{(0.2)(1)\left(15 \times 10^{-6}\right)(0.25)^2}\right) + 3.23\right] = 0.0812 .$$

The start of the early-radial flow regime is controlled by wellbore storage, which appears to have vanished at times earlier than 0.24 hours in this example. The end of the early radial flow regime is expected at the lesser of the two values derived from Eqs. 8.187 and 8.188. For a centered well, $d_z = h/2$ = 100 ft, and Eq. 8.187 gives

$$t_{E\text{erf}} = 1{,}800 d_z^2 \phi \mu c_t / k_z$$

$$= (1{,}800)(100)^2(0.2)(1)\left(1.5 \times 10^{-5}\right)/2 = 27 \text{ hours} .$$

Assuming $k_y = k_x$ = 200 md, Eq. 8.188 gives

$$t_{E\text{erf}} = 125 L_w^2 \phi \mu c_t / k_y$$

$$= (125)(1000)^2(0.2)(1)\left(1.5 \times 10^{-5}\right)/200 = 1.875 \text{ hours.}$$

Thus, expect the early-radial flow regime to end at approximately 1.875 hours, which is the smaller value and is consistent with observed test data.

8.11.4 Hemiradial Flow. The hemiradial flow period (Fig. 8.97) will occur only when the well is close to one of the vertical boundaries (either the upper or the lower boundaries) and is analogous to a vertical well near a fault. The governing equation is[38]

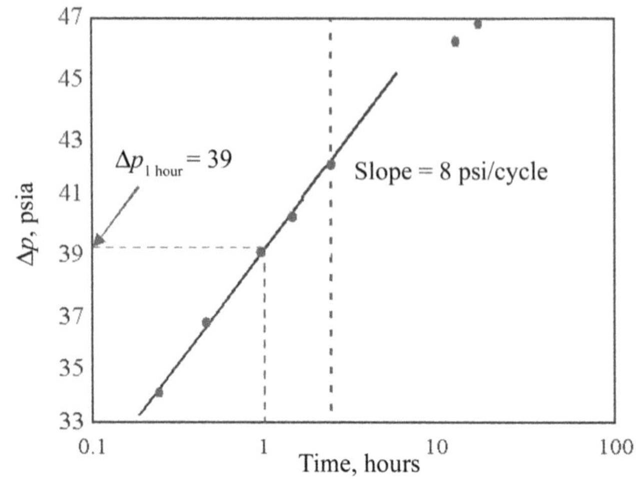

Fig. 8.102—Early-radial flow is indicated by semilog straight line for well Erf-1.

TABLE 8.3—PRESSURE CHANGE DATA FOR WELL Erf-1

Time (hours)	$\Delta p=(p_i-p_{wf})$ (psi)
0.24	33.96
0.48	36.41
0.96	38.85
1.44	40.26
2.40	41.96
12.50	46.23
16.80	46.91
24.00	47.65

$$p_i - p_{wf} = \frac{325.2 q B \mu}{\sqrt{k_z k_x} L_w} \left[\log_{10}\left(\frac{\sqrt{k_x k_z} t}{\phi \mu c_t r_w^2} \right) - 3.227 + 0.868 s_d \right] . \qquad (8.192)$$

A horizontal derivative on the diagnostic plot identifies hemiradial flow. If data appear to fall into this flow regime, a straight line on a semilog plot would provide more confidence that radial flow has been identified. Consistency checks in the analysis coupled with a well survey will be required to distinguish hemiradial flow from early radial flow.

The time range in which the analysis for hemiradial flow is valid begins after the closest vertical boundary, d_z, affects the data and before the farthest boundary, D_z, affects them. In the absence of wellbore storage, the start of hemiradial flow is given by

$$t_{\mathrm{Shrf}} \cong \frac{1{,}800 d_z^2 \phi \mu c_t}{k_z} . \qquad (8.193)$$

Note that the start of the hemiradial flow regime involves the shortest distance to a vertical boundary and the permeability in the vertical direction. However, wellbore storage will most likely determine the actual start of hemiradial flow.

The end of hemiradial flow occurs when pressure is affected by the farther vertical boundary or flow from beyond the ends of the wellbore, whichever occurs first. It is the smaller of the times calculated using Eqs. 8.194 and 8.195. If the hemiradial flow regime ends when pressure reaches the farthest vertical boundary, it depends on the distance, D_z, and the vertical permeability, k_z:

$$t_{E\mathrm{hrf}} = \frac{1800 D_z^2 \phi \mu c_t}{k_z} \quad \text{(8.194)}$$

When the appearance of end effects—flow from beyond the ends of the wellbore—causes the end of the hemiradial flow regime to appear, the end of the flow regime occurs when

$$t_{E\mathrm{hrf}} = \frac{125 L_w^2 \phi \mu c_t}{k_y} \quad \text{(8.195)}$$

The completed length of the well, L_w, and the permeability k_y, parallel to the wellbore appear in this equation. These parameters determine when enough flow has come from beyond the ends of the wellbore to distort the radial flow pattern that appeared earlier.

$$\left| m_{\mathrm{hrf}} \right| = \frac{325.2 q B \mu}{L_w \sqrt{k_x k_z}} \quad \text{(8.196)}$$

gives the slope of the semilog straight line for semiradial flow, m_{hrf}. The multiplier, 325.2, is twice the multiplier for early-radial flow. The equation to estimate the damage skin factor is also similar to that for radial flow but has a multiplier that differs by a factor of two.

$$s_d = 2.303 \left[\frac{\Delta p_{1\mathrm{hr}}}{\left| m_{\mathrm{hrf}} \right|} - \log \left(\frac{\sqrt{k_x k_z}}{\phi \mu c_t r_w^2} \right) + 3.23 \right] \quad \text{(8.197)}$$

The equations relating slope and permeability and the equation for skin are similar in a buildup test to those for a drawdown test. The pressure change in the equation for skin is $[p_{1\mathrm{hr}} - p_{wf}(\Delta t = 0)]$. Semilog plots of buildup test data from the hemiradial flow regime cannot be analyzed rigorously using data from a Horner plot unless the pressure data at $(t_p + \Delta t)$ and at time Δt are simultaneously in this flow regime. As a practical matter, the hemiradial flow regime is likely to appear clearly in the buildup test only when the producing time is much greater that the shut-in time.

8.11.5 Early Linear Flow. The governing equation for early-linear flow is[38]

$$p_i - p_{wf} = \frac{8.128 q B}{L_w h} \sqrt{\frac{\mu t}{k_x \phi c_t}} + \frac{141.2 q B \mu}{L_w \sqrt{k_x k_z}} \left(s_c + s_d \right) \quad \text{(8.198)}$$

The "convergence skin," s_c, is discussed later in this section. The start of the early-linear flow regime (Fig. 8.98) depends on the farthest distance to a vertical boundary, D_z, and the vertical permeability, k_z.[39]

$$t_{Self} = \frac{1{,}800 D_z^2 \phi \mu c_t}{k_z} \; . \qquad (8.199)$$

Not until flow reaches that farthest vertical boundary can a linear flow pattern begin toward the well. This flow period ends when fluids flow from beyond the ends of the wellbore. Thus,

$$t_{Eelf} = \frac{160 L_w^2 \phi \mu c_t}{k_y} \; . \qquad (8.200)$$

Notice that the end depends on the effective completed length of the well, L_w, and on the permeability in the direction parallel to the well. This is the time in which end effects—flow beyond the ends of the well—begin to significantly distort the linear flow pattern.

The early-linear flow regime is identified in a drawdown test with a half-slope for the derivative. (Because of the skin effect, the pressure change curve on the diagnostic plot will only approach a half-slope asymptotically.) For data identified as being in this flow regime, plot pressure against the square root of time.

The slope of the straight line on such a plot, m_{elf}, can be used to estimate the square-root of k_x, the horizontal permeability perpendicular to the well:

$$\sqrt{k_x} = \frac{8.128 qB}{|m_{elf}| L_w h} \sqrt{\frac{\mu}{\phi c_t}} \; . \qquad (8.201)$$

To calculate the damage skin,

$$s_d = \frac{L_w \sqrt{k_x k_z}}{141.2 qB\mu} \Delta p_{t=0} - s_c \; . \qquad (8.202)$$

This equation includes a convergence skin, s_c, which is[39]

$$s_c = \ln\left(\frac{h}{r_w}\right) + 0.25 \ln\left(\frac{k_x}{k_z}\right) - \ln\left[\sin\left(\frac{\pi d_z}{h}\right)\right] - 1.838 . \qquad (8.203)$$

This convergence skin is an additional pressure drop that acts like a skin effect caused by flow moving from throughout the entire formation until it converges down to the small wellbore in the middle of the formation (**Fig. 8.103**). This convergence skin is defined in terms of the ratio of the permeability in the *x*-direction, which is perpendicular to the wellbore, to the vertical permeability. It also involves the distance to the nearest vertical boundary, d_z, and the net pay thickness, h.

Kuchuk[40] derived a different equation for convergence skin. As a practical matter, the Odeh-Babu and Kuchuk equations lead to the same result. When there has been a single rate preceding shut-in during early-linear flow, the buildup pressure is plotted against $\sqrt{t_p + \Delta t} - \sqrt{\Delta t}$ on a Cartesian plot, which is sometimes called a tandem root plot. The permeability, k_x, is calculated from the slope, m_{elf}, of the plot and Eq. 8.201. k_x has the same relationship to the slope that existed in a drawdown test. Skin for this flow regime is calculated with Eqs. 8.202 and 8.203.

Plots of buildup data from the early-linear flow regime cannot be analyzed rigorously with a plot of p_{ws} vs. $\sqrt{t_p + \Delta t} - \sqrt{\Delta t}$ (that is, the tandem-root plot) unless data at $(t_p + \Delta t)$ and Δt

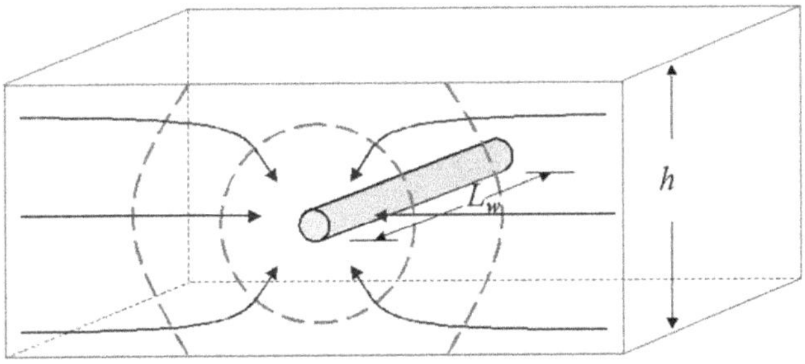

Fig. 8.103—Convergence skin accounts for increased pressure drop as flow deviates from the full vertical thickness of the reservoir into a radial pattern as it enters the well.

are simultaneously in this flow regime—highly unlikely—or unless t_p is much greater than Δt, in which case simply plot p_{ws} vs. $\sqrt{\Delta t}$. Little error results from ignoring the $(t_p + \Delta t)$ term, which is essentially constant.

Example 8.3: Well Elf-2. The diagnostic plot for a drawdown test from Well Elf-2[39] indicates data in the early-linear flow regime because the derivative has a half-slope. The following data apply to this well: q = 800 STB/D; μ = 1 cp; B = 1.25 RB/STB; r_w = 0.25 ft; ϕ = 0.2; $c_t = 15\times10^{-6}$ psi^{-1}; centered in box-shaped drainage area 100 ft thick, 4,000 ft long, 4,000 ft wide; L_w = 2,500 ft; and, from early radial-flow regime data, $k_x k_z$ = 8,000 md^2. In addition, **Table 8.4** shows the pressure-change data for this well.

A plot of pressure change vs. the square root of time (**Fig. 8.104**) indicates early linear flow. The final point on the straight line is at a time of approximately 24 hours, but there may have been some deviation from the straight line by this time. From the slope of the straight line, m_{elf} = 0.934 psi/hr$^{1/2}$ and Eq. 8.201, calculate the permeability in the horizontal plane perpendicular to the wellbore.

$$\sqrt{k_x} = \frac{8.13qB}{m_{\text{elf}} L_w h}\sqrt{\frac{\mu}{\phi c_t}}$$

$$= \frac{(8.13)(800)(1.25)}{(0.934)(2{,}500)(100)}\sqrt{\frac{1}{(0.2)(15)\left(10^{-6}\right)}}$$

$$= 20.10 \text{ md}^{1/2},$$

or $k_x = (20.1)^2$ = 404 md.

Analysis of data from the early radial flow regime indicated that $k_x k_z$ is 8,000 md^2; thus, k_z is approximately 19.8 md. To calculate s_d, use Eqs. 8.202 and 8.203, noting that the value for $(p_i - p_{wf})$ = 3.1 at t = 0.

$$s_d = \frac{L_w\sqrt{k_x k_z}}{141.2qB\mu}\Delta p_{t=0} - s_c$$

$$= \frac{(2{,}500)\sqrt{(8{,}000)}(3.10)}{(141.2)(800)(1.25)(1)} - s_c = 4.91 - s_c\,.$$

$$s_c = \ln\left(\frac{h}{r_w}\right) + 0.25\ln\left(\frac{k_x}{k_z}\right) - \ln\left[\sin\left(\frac{\pi d_z}{h}\right)\right] - 1.838$$

$$= \ln\left(\frac{100}{0.25}\right) + 0.25\ln\left(\frac{404}{19.8}\right) - \ln\left[\sin\left(\frac{\pi(50)}{100}\right)\right] - 1.838 = 4.91.$$

Then, $s_d = 4.91 - 4.91 = 0$.

Check the expected time range for the early-linear flow regime.

Using $D_z = h/2 = 50$ ft and $k_z = 20$ md, calculate the beginning of linear flow with Eq. 8.199:

$$t_{\text{Self}} = \frac{(1{,}800)D_z^2\phi\mu c_t}{k_z}$$

$$= \frac{(1{,}800)(50)^2(0.2)(1)\left(1.5\times10^{-5}\right)}{20} = 0.675 \text{ hours.}$$

Assuming $k_y = k_x$ at 400 md, use Eq. 8.200 to find the end of early linear flow.

$$t_{\text{Eelf}} = \frac{160L_w^2\phi\mu c_t}{k_y}$$

$$= \frac{(160)(2{,}500)^2(0.2)(1)\left(1.5\times10^{-5}\right)}{400} = 7.5 \text{ hours.}$$

These limits are reasonably consistent with the time range analyzed assuming early linear flow.

8.11.6 Late Pseudoradial. The governing equation for late-pseudoradial flow is[38]

$$p_i - p_{wf} = \frac{162.6qB\mu}{\sqrt{k_yk_x}h}\left[\log_{10}\left(\frac{k_yt}{\phi\mu c_tL_w^2}\right) - 2.303\right] + \frac{141.2qB\mu}{L_w\sqrt{k_xk_z}}\left(s_c + s_d\right). \quad (8.204)$$

The late-pseudoradial flow period occurs only if [39]

$$\frac{L_w}{b_H} < 0.45. \quad (8.205)$$

Here, b_H is the dimension of the reservoir parallel to the wellbore. As long as the completed length of the well is relatively short compared with the length of the drainage area late-pseudoradial flow can occur.

The start of this flow period occurs when fluid flows from well beyond the ends of the wellbore (Fig. 8.99). It is approximated with[39]

$$t_{\text{Sprf}} = \frac{1{,}480L_w^2\phi\mu c_t}{k_y}. \quad (8.206)$$

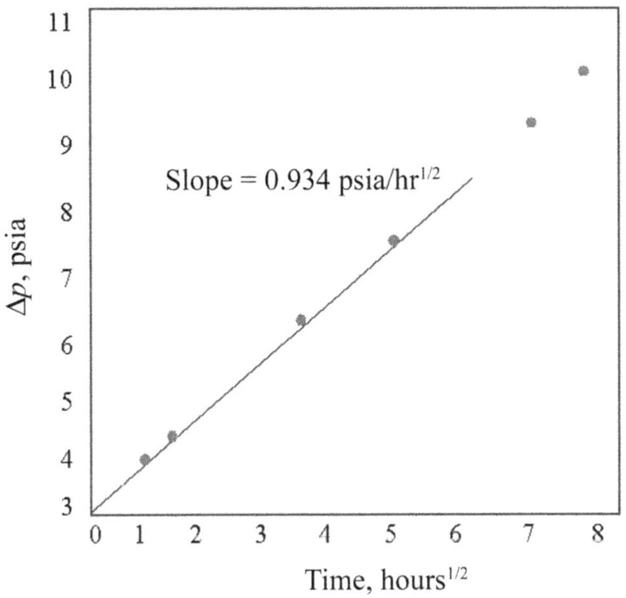

Fig. 8.104—Early-linear flow is indicated by straight line on square-root-of-time plot for well Elf-2.

TABLE 8.4—PRESSURE CHANGE DATA FOR WELL Elf-2

Time (hours)	$\Delta p = (p_i - p_{wf})$ (psi)
1.2	4.11
2.4	4.54
12.0	6.33
24.0	7.65
48.0	9.49
60.0	10.26

This starting time depends on the completed length of the well, L_w, and on the permeability in the direction of the well, k_y. The end of this period, like others in this section, is approximated by the minimum of the results of two calculations. The first,

$$t_{E\text{prf}} = \frac{2{,}000\phi \mu c_t \left(L_w/4 + d_y\right)^2}{k_y}, \quad \text{(8.207)}$$

depends on d_y and the length of the wellbore along with k_y, the permeability in the direction parallel to the wellbore. This is the time at which horizontal boundary effects first appear.

The other equation gives a time at which the radial-flow pattern begins to be distorted depending on the shortest distance, d_x, from the well to a boundary perpendicular to the wellbore and on k_x, the permeability in that direction.

$$t_{E\text{prf}} = \frac{1{,}650\phi\mu c_t d_x^2}{k_x} \quad \text{(8.208)}$$

Whenever boundary effects first appear, whether in a direction that is parallel to the well or perpendicular to the axis of the well, the late-pseudoradial flow period will end.

The diagnostic plot helps identify the late-pseudoradial flow regime with the characteristic horizontal derivative. For data in the appropriate time range, prepare a semilog plot of pressure against time for a drawdown test. The slope of this plot will be m_{prf} and the relationship between that slope and the square root of $k_x k_y$, or the permeabilities in the horizontal plane, is given by

$$\sqrt{k_x k_y} = \frac{162.6 qB\mu}{|m_{\text{prf}}|\, h} \quad \text{(8.209)}$$

The skin equation is similar in form to those seen before:

$$s_d = \left(1.151\sqrt{\frac{k_z}{k_y}}\frac{L_w}{h}\right)\left[\frac{\Delta p_{1\text{hr}}}{|m_{\text{prf}}|} - \log\left(\frac{k_y}{\phi\mu c_t L_w^2}\right) + 1.76\right] - s_c \quad \text{(8.210)}$$

Again, the total skin depends on $\Delta p_{1\text{hr}}$. The convergence skin (Eq. 8.203) is subtracted from the "total" skin to determine the damage skin.

For a buildup test preceded by production at a single rate, plot pressure against the HTR on a semilog graph. Permeability is calculated from Eq. 8.209, the same as for a drawdown test. The skin equation is basically the same as for a drawdown test, except that the $\Delta p_{1\text{hr}}$ is now $p_{1\text{hr}} - p_{wf}$. To obtain p^*, the extrapolated pressure, extrapolate the semilog straight line to a HTR of unity.

Semilog plots of buildup test data from the late-pseudoradial flow regime cannot be analyzed rigorously using a Horner plot unless pressures at $(t_p + \Delta t)$ and at time Δt are simultaneously in the pseudoradial flow regime, which is highly unlikely. However, little error appears if the producing time before shut-in is much greater than the maximum shut-in time achieved in the buildup test.

Example 8.4: Well Prf-3. The diagnostic plot suggests that a constant-rate drawdown test from Well Prf-3[39] includes data in the late-pseudoradial period. The following data are available from the test: q = 800 STB/D, μ = 1 cp, B = 1.25 RB/STB, r_w = 0.25 ft, ϕ = 0.2, c_t = 15×10^{-6} psi^{-1}, h = 150 ft, L_w = 900 ft, a_H = 5,280 ft, b_H = 5,280 ft, well centered in drainage volume, k_x = 100 md (from analysis of early linear flow), $k_x k_z$ = 1,000 md^2, and k_z = 10 md (from analysis of early radial flow). **Table 8.5** gives the pressure change, $\Delta p = p_i - p_{wf}$ vs. time.

A plot of pressure change vs. the logarithm of time (**Fig. 8.105**) confirms pseudoradial flow. A straight line fits all the data from 192 to 432 hours; the slope of the line, m_{prf}, is 15.3 psi/cycle, and $\Delta p_{1\text{ hr}}$ = 18.94 psi (extrapolated). Then, from Eq. 8.209,

$$\sqrt{k_x k_y} = \frac{162.6 qB\mu}{|m_{\text{prf}}|\, h} = \frac{(162.6)(800)(1.25)(1.0)}{(15.3)(150)} = 70.8 \text{ md.}$$

Thus, $k_y = \dfrac{\sqrt{k_x k_z}}{k_x} = \dfrac{(70.8)^2}{100} = 50.2$ md.

From Eqs. 8.210 and 8.203,

$$s_d = 1.151\sqrt{\frac{k_z}{k_y}}\frac{L_w}{h}\left[\frac{\Delta p_{1\,\mathrm{hr}}}{|m_{\mathrm{prf}}|} - \log\left(\frac{k_y}{\phi\mu c_t L_w^2}\right) + 1.76\right] - s_c,$$

$$\text{and } s_c = \ln\left(\frac{h}{r_w}\right) + 0.25\ln\left(\frac{k_x}{k_z}\right) - \ln\left[\sin\left(\frac{\pi d_z}{h}\right)\right] - 1.838.$$

Here,

$$s_c = \ln\left(\tfrac{150}{0.25}\right) + 0.25\ln\left(\tfrac{100}{10}\right) - \ln\left[\sin\left(\tfrac{\pi(75)}{150}\right)\right] - 1.838 = 5.13,$$

$$\text{and } s_d = 1.151\sqrt{\frac{10}{50.2}}\frac{(900)}{(150)}\left[\frac{18.94}{15.3} - \log\left(\frac{50.2}{(0.2)(1)\left(1.5\times10^{-5}\right)(900)^2}\right) + 1.76\right] - 5.13 = 0.057.$$

The pseudoradial flow regime should start at the time given by Eq. 8.206:

$$t_{S\mathrm{prf}} = \frac{1{,}480L_w^2\phi\mu c_t}{k_y} = \frac{(1{,}480)\left(900^2\right)(0.2)(1)\left(1.5\times10^{-5}\right)}{50.2} = 71.6\text{ hours}.$$

It should end at a time given by the lesser of values from Eqs. 8.207 and 8.208. From Eq. 8.207,

$$t_{E\mathrm{prf}} = \frac{2{,}000\phi\mu c_t\left(L_w/4 + d_y\right)^2}{k_y} = \frac{(2{,}000)(0.2)(1)\left(1.5\times10^{-5}\right)(900/4 + 2{,}190)^2}{50.2} = 697\text{ hours},$$

where $d_y = \frac{1}{2}(5{,}200 - 900) = 2{,}190$ ft for this centered well. From Eq. 8.208,

$$t_{E\mathrm{prf}} = \frac{1{,}650\phi\mu c_t d_x^2}{k_x} = \frac{(1{,}650)(0.2)(1)\left(1.5\times10^{-5}\right)(2{,}640)^2}{(100)} = 344\text{ hours}.$$

The smaller of these two values is 344 hours, which is thus the expected end of pseudoradial flow. The data on the figure that lie on the straight line show the time range from 192 to 432 hours, which is generally consistent with the expected duration of the flow regime.

8.11.7 Late-Linear Flow. The governing equation for late-linear flow is[38,39]

$$p_i - p_{wf} = \frac{8.128qB}{b_H h}\sqrt{\frac{\mu t}{k_x\phi c_t}} + \frac{141.2qB\mu}{b_H\sqrt{k_x k_z}}\left(s_p + s_c + \frac{b_H}{L_w}s_d\right). \quad \text{(8.211)}$$

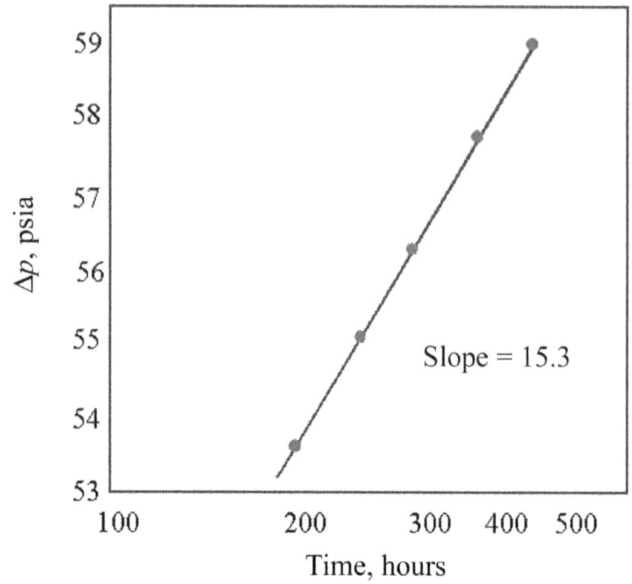

Fig. 8.105—Pseudoradial flow indicated by semilog straight line for well Prf-3.

TABLE 8.5—PRESSURE CHANGE DATA FOR WELL Prf-3	
Time (hours)	$\Delta p = (p_r - p_{wf})$ (psi)
192	53.69
240	55.13
288	56.32
360	57.81
432	59.05

The late-linear flow regime starts after the pressure transient has reached the boundaries in the z- and y-directions, and the flow behavior with regard to these directions has become pseudosteady state, as Fig. 8.100 shows.

The start of this time period is the maximum of two equations.[39] The first depends on the time to reach the boundary, D_y, beyond the end of the horizontal well. It also depends on the permeability, k_y, in the direction parallel to the wellbore.

$$t_{\text{SIIf}} = \frac{4{,}800\phi\mu c_t (D_y + L_w/4)^2}{k_y} \quad \text{(8.212)}$$

Another requirement for the start of the late-linear flow regime is the time to reach the maximum vertical distance, D_z, divided by the vertical permeability:

$$t_{\text{SIIf}} = \frac{1{,}800\phi\mu c_t D_z^2}{k_z} \quad \text{(8.213)}$$

Usually, the start of the late-linear flow regime is dictated by the time to reach the boundaries in the y-direction. The end of this period is given by the equation

$$t_{E\mathrm{llf}}=\frac{1{,}650\phi\mu c_t d_x^2}{k_x} \quad \text{(8.214)}$$

The end of the late-linear flow regime depends on reaching the nearest boundary in the direction perpendicular to the wellbore, which is the distance, d_x, away, and on the permeability k_x in that direction.

Identify the late-linear flow regime by a half-slope on the derivative in the diagnostic plot of drawdown test data. (The pressure change may approach a half-slope asymptotically.) For data that appear to be in this flow regime, plot pressure against the square root of time. From the slope m_{llf} of the plot, estimate permeability in the x-direction from

$$\sqrt{k_x}=\frac{8.128qB}{|m_{\mathrm{llf}}|\,b_H h}\sqrt{\frac{\mu}{\phi c_t}} \quad \text{(8.215)}$$

Alternatively, if k_x is known from an early-linear flow regime, estimate b_H, the length of the drainage area, from

$$b_H=\frac{8.128qB}{|m_{\mathrm{llf}}|\,h}\sqrt{\frac{\mu}{\phi c_t k_x}}, \quad \text{(8.216)}$$

This late-linear flow regime is the only period that provides the data to calculate the total skin, s, including the partial-penetration skin, s_p, and the convergence skin, s_c. To calculate the damage skin, s_d, use

$$s_d=\frac{L_w}{b_H}\left(\frac{b_H\sqrt{k_x k_z}(\Delta p_{t=0})}{141.2qB\mu}-s_p-s_c\right). \quad \text{(8.217)}$$

The total skin depends on $\Delta p_{t=0}$. Subtracting the partial penetration skin, s_p, and the convergence skin, s_c, from the total skin yields the damage skin.

The partial-penetration skin is a complex function that is calculated with Eqs. A-25 through A-35 in Table 8.A-2. For a buildup test, plot pressure against the HTR. From the slope, m_{llf}, calculate k_x with Eq. 8.215, exactly the same as for drawdown tests. Or, if k_x is known, estimate the length, b_H, of the drainage area with Eq. 8.216. Calculate the damage skin, s_d, from a pressure buildup test from Eq. 8.217, where $\Delta p_{t=0}=(p_{t=0})_{\mathrm{ext}}-p_{wf(t=0)}$.

Note that the same difficulty arises in using superposition to find plotting functions plots of buildup data from the late-linear flow regime as existed with the previous flow regimes. Pressures at both time $(t_p+\Delta t)$ and time Δt must be in the late linear flow regime for a tandem-root plot to be valid. However, if $t_p \gg \Delta t_{\max}$, there is little error.

Example 8.5: Well Llf-4. The diagnostic plot for a drawdown test from Well Llf-4[39] appears to include data in the late-linear flow regime (derivative with half-slope). The following data applies to this well: q = 800 STB/D, μ = 1 cp, B = 1.25 RB/STB, r_w = 0.25 ft, ϕ = 0.2, c_t = 15 × 10^{-6} psi^{-1}, h = 150 ft, L_w = 1,000 ft, b_H = 2,000 ft (well centered), a_H = 6,968 ft (well centered), D_z = 85 ft, d_z = 65 ft, $k_x k_z$ = 1,000 md^2 (from analysis of early-radial flow), and $k_x k_y$ = 5,000 md^2 (from analysis of pseudoradial flow). **Table 8.6** gives pressure change, $\Delta p = p_i - p_{wf}$, data vs. time.

Fig. 8.106 is a plot of pressure change vs. the square root of time. The straight line on this plot for the entire time range (60 to 240 hours) confirms late-linear flow for this range. The slope of the line is 1.56 psi/hr$^{1/2}$, and the intercept is $\Delta p_{t=0}$ =28.4 psi.

From Eq. 8.215,

$$\sqrt{k_x} = \frac{8.128 qB}{|m_{\text{llf}}|\, b_H h}\sqrt{\frac{\mu}{\phi c_t}} = \frac{(8.128)(800)(1.25)}{(1.56)(2{,}000)(150)}\sqrt{\frac{1}{(0.2)\left(1.5\times10^{-5}\right)}} = 10.0.$$

Then, k_x = 100 md. Because $k_k k_z$ = 100 md^2, k_z = 10 md. Also, because $k_x k_y$ = 5,000 md^2, k_y =50 md. From Eq. 8.217,

$$s_d = \frac{L_w}{b_H}\left(b_H \frac{\sqrt{k_x k_z}\,\Delta p_{t=0}}{141.2 qB\mu} - s_c - s_p\right)$$

$$= \frac{(1{,}000)\sqrt{1{,}000}(28.4)}{(141.2)(800)(1.25)(1)} - \frac{1{,}000}{2{,}000}\left(s_p + s_c\right) = 6.36 - 0.5\left(s_c + s_p\right).$$

From Eq. 8.203,

$$s_c = \ln\left(\frac{h}{r_w}\right) + 0.25\ln\left(\frac{k_x}{k_z}\right) - \ln\left[\sin\left(\frac{\pi d_z}{h}\right)\right] - 1.838$$

$$= \ln\left(\frac{150}{0.25}\right) + 0.25\ln\left(\frac{100}{40}\right) - \ln\left[\sin\left(\frac{65\pi}{150}\right)\right] - 1.838 = 5.16.$$

Calculate the partial penetration skin, s_p, using the appropriate equation from among Eqs. A-25 through A-35 in Table 8.A-2. First, calculate

$$\frac{a_H}{\sqrt{k_x}},\ \frac{b_H}{\sqrt{k_y}},\ \text{and}\ \frac{h}{\sqrt{k_z}}$$

to determine whether "Case 1" or "Case 2" applies:

$$\frac{a_H}{\sqrt{k_x}} = \frac{6{,}968}{\sqrt{100}} = 696.8,$$

$$\frac{b_H}{\sqrt{k_y}} = \frac{2{,}000}{\sqrt{50}} = 283;\ \frac{0.75 b_H}{\sqrt{k_y}} = 212,$$

$$\text{and}\ \frac{h}{\sqrt{k_z}} = \frac{150}{\sqrt{10}} = 47.4;\ \frac{0.75}{\sqrt{k_z}} = 35.6.$$

Because

$$\frac{a_H}{\sqrt{k_x}} > \frac{0.75 b_H}{\sqrt{k_y}} >> \frac{0.75 h}{\sqrt{k_z}},$$

this is Case 1 (Eq. A-26). Use Eqs. A-25 through A-31 from the table. From Eq. A-27, $s_p = p_{xyz} + p'_{xy}$. From Eq. A-25,

$$p_{xyz} = \left(\frac{b_H}{L_w} - 1\right)\left\{ \ln\left(\frac{h}{r_w}\right) + 0.25 \ln\left(\frac{k_x}{k_z}\right) - \ln\left[\sin\left(\frac{\pi d_z}{h}\right) - 1.838\right]\right\}$$

$$= \left(\frac{2{,}000}{1{,}000}\right)\left\{ \ln\left(\frac{150}{0.25}\right) + 0.25 \ln\left(\frac{100}{10}\right) - \ln\left[\sin\left(\frac{65\pi}{150}\right) - 1.838\right]\right\} = 5.16.$$

From Eq. A-28,

$$p'_{xy} = \frac{2b_H^2}{L_w h}\sqrt{k_z / k_y}\left\{F\left(\frac{L_w}{2b_H}\right) + 0.5\left[F\left(\frac{4y_m + L_w}{2b_H}\right) - F\left(\frac{4y_m - L_w}{2b_H}\right)\right]\right\}.$$

Here, from Eq. A-29,

$$y_m = d_y + \frac{L_w}{2} = 500 + \frac{1{,}000}{2} = 1{,}000.$$

The well is centered, so $d_y = D_y = 500$ ft.

$$\frac{L_w}{2b_H} = \frac{1{,}000}{(2)(2{,}000)} = 0.25.$$

From Eq. A-30,

$$F(u) = -u\left[0.145 + \ln(u) - 0.137(u)^2\right],\ u < 1.$$

$$F\left(\frac{L_w}{2b_h}\right) = -(0.25)\left[0.145 + \ln(0.25) - 0.137(0.25)^2\right] = 0.313.$$

Also,

$$\frac{4y_m + L_w}{2b_H} = \frac{(4)(1{,}000) + 1{,}000}{(2)(2{,}000)} = 1.25,$$

$$\text{and } \frac{4y_m - L_w}{2b_H} = 0.75.$$

From Eq. A-31,

$$F(u) = (2-u)\left[0.145 + \ln(2-u) - 0.137(2-u)^2\right],\ u > 1,$$

$$F\left(\frac{4y_m + L_w}{2b_H}\right) = (2-1.25)\left[0.145 + \ln(2-1.25) - 0.137(2-1.25)^2\right] = -0.165,$$

and

$$F(u) = -u\left[0.145 + \ln(u) - 0.137(u)^2\right],\ u < 1,$$

$$F\left(\frac{4y_m - L_w}{2b_H}\right) = -0.75\left[0.145 + \ln(0.75) - 0.137(0.75)^2\right] = 0.165.$$

Then,

$$p'_{xy} = \frac{(2)(2{,}000)^2}{(1{,}000)(150)}\sqrt{\frac{10}{100}}\{0.313 + 0.5[-0.165 - 0.165]\} = 2.50.$$

Then,

$$s_p = 5.16 + 2.50 = 7.66.$$

Thus, $s_d = 6.36 - 0.5(7.66 + 5.16) = -0.05$.

Now check the expected duration of the late-linear flow regime. The start is the larger of values from Eqs. 8.212 and 8.213. From Eq. 8.212,

$$t_{\text{SIlf}} = \frac{4{,}800\phi\mu c_t\left(D_y + L_w/4\right)^2}{k_y}$$

$$= \frac{4{,}800(0.2)(1)\left(1.5\times10^{-5}\right)\left(500 + \frac{1{,}000}{4}\right)^2}{50} = 162 \text{ hours.}$$

From Eq. 8.213,

$$t_{\text{SIlf}} = \frac{1{,}800\phi\mu c_t D_z^2}{k_z}$$

$$= \frac{1{,}800(0.2)(1)\left(1.5\times10^{-5}\right)(85)^2}{10} = 3.90 \text{ hours.}$$

Thus, the start is expected to be at approximately 162 hours. Eq. 8.214 gives the end of the flow regime.

$$t_{\text{EIlf}} = \frac{1{,}650\phi\mu c_t d_x^2}{k_x}$$

$$= \frac{1{,}650(0.2)(1)\left(1.5\times10^{-5}\right)(3{,}484)^2}{100} = 601 \text{ hours.}$$

The data in this example spanned the time range from 60 to 240 hours. Some of the data that fall on the straight line are, in theory, from times before the start of the late-linear flow regime, but they appear to cause no problem in determining the slope of the line.

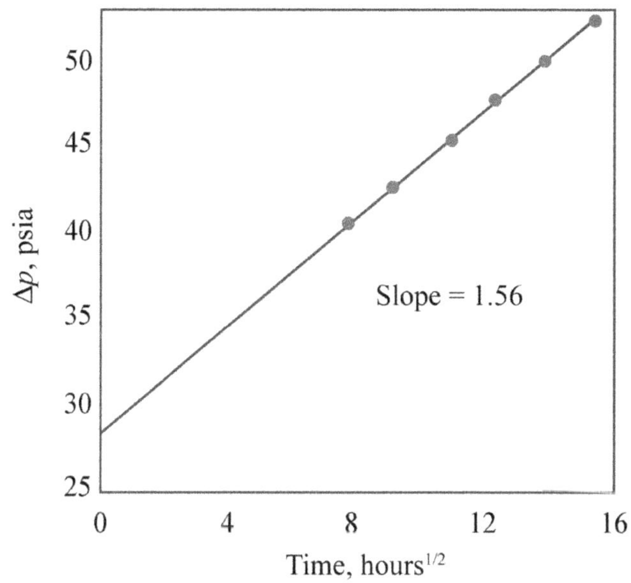

Fig. 8.106—Late linear flow indicated by straight line on square-root-of-time plot for well LIf-4.

TABLE 8.6—PRESSURE CHANGE DATA FOR WELL LIf-4

Time (hours)	$\Delta p = (p_r - p_{wf})$ (psi)
60	40.50
84	42.72
120	45.25
156	47.92
192	50.07
240	52.65

8.11.8 Field Examples.[41] The following field examples illustrate the procedures used in analyzing horizontal well-test data.

Field Example Well A. **Table 8.7** summarizes the reservoir and completion properties for Well A. The target for Well A, a horizontal exploration well, was vertical tectonic fracture development in a low-permeability shale. Because of the fractures, the permeability is assumed to be isotropic ($k_h = k_z$) and a result of the fractures. **Fig. 8.107** is a diagnostic plot for Well A and includes a history match using an analytical model.

During the early part of the test there is a unit-slope line representing wellbore storage. Following that wellbore storage, there is a transition to radial flow. The final few data points may be in radial flow. On the Horner plot (**Fig. 8.108**), the last few data points fall on a straight line. From the slope of the straight line, the apparent permeability is 0.011 md and the altered zone skin is 2.9. There is no evidence of boundary effects on this Horner plot. The existence of the semilog straight line is not assured, but the data are at least on the verge of reaching it.

Guided by the Horner analysis results, engineers simulated these data with an analytical horizontal well model. The initial match was poor. The match was improved considerably by

TABLE 8.7—PARAMETERS FOR EXAMPLE WELL A	
L_d, ft	2,470
L_w, ft	-
r_w, ft	0.25
ϕ, %	5
h, ft	150
q, STB/D	104
B_o, RB/STB	1.40
μ, cp	0.45
t_p, hours	238

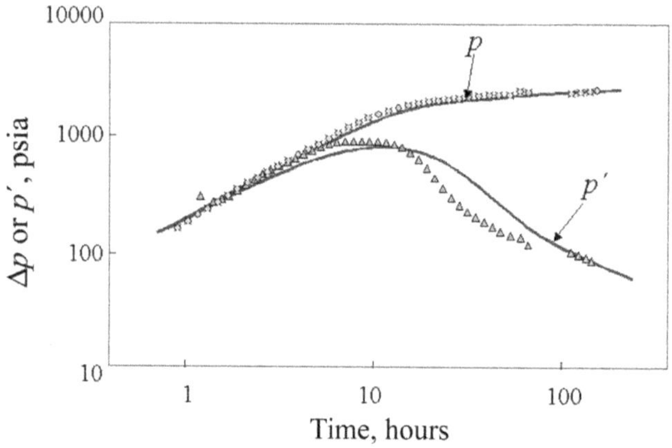

Fig. 8.107—Diagnostic plot for Well A.

introducing a no-flow boundary approximately 16 ft above the wellbore, which led to a permeability estimate of 0.027 md and an apparent skin of 11.5. It was concluded that the flow regime observed in the test was hemiradial flow.

The final match, shown on the type-curve plot in Fig. 8.107, is still not a good match at all times, but the author stated that the poor match in the transition region could be the result of phase-redistribution effects in the wellbore. The distance to the no-flow boundary that led to the best match compared favorably with well survey data, which indicated that the well was drilled approximately 20 ft below the upper limit of the productive horizon.

Field Example Well B. Well B is in a west Texas carbonate formation. It was expected to have isotropic permeability caused by fracturing and dissolution. **Table 8.8** gives the field data for this well.

Fig. 8.109 is the diagnostic plot for this well. After wellbore storage, a short period of radial flow appears to be followed by the onset of linear flow, because p' approaches a slope of 0.5. In the time region where the derivative is horizontal, a straight line on the Horner plot (**Fig. 8.110**) yields k = 0.14 md. Using these results in the analysis shows that the end of radial flow occurs at t_{Erf} = 165 hours.

A tandem-root plot (**Fig. 8.111**) indicates linear flow and also suggests a distance to the nearest boundary of 29 ft. This is in good agreement with geological observations and helps to verify the assumption of isotropic permeability. The history match with an analytical horizontal well model, shown in Fig. 8.109, confirms the results of the Horner and tandem-root plots.

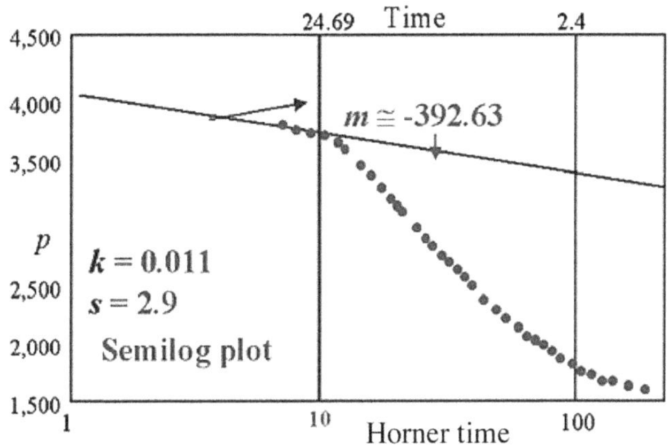

Fig. 8.108—Horner plot suggests beginning of middle-time region appears in test data.

TABLE 8.8—PARAMETERS FOR EXAMPLE WELL B	
L_d, ft	2,000
L_w, ft	-
r_w, ft	0.30
ϕ, %	17
h, ft	75
q, STB/D	200
B_o, RB/STB	1.60
μ, cp	1.80
t_p, hours	1,320

Field Example Well C. Well C data are from a buildup test of a horizontal well in a high-permeability sandstone where a 54-ft oil column overlies an extensive aquifer estimated to be approximately 180 ft thick. **Table 8.9** shows the available data.

The diagnostic plot (**Fig. 8.112**) shows essentially no wellbore storage and a constant derivative, indicating radial, hemiradial, or elliptical flow at early times. The rapid decline of the derivative at the end of the test is caused by the aquifer underlying the oil column, which is acting like a constant-pressure boundary. A history match with an analytical horizontal well model with one no-flow and one constant-pressure boundary (the lower boundary), yielded k_h = 313 md, k_z = 7.5 md, s_a = 1.5, and L_w = 356 ft. The no-flow boundary was estimated to be approximately 112 ft below the wellbore.

The time at which the early radial flow regime ends—the time where the derivative ceases to be flat on the diagnostic plot—is approximately 1.5 hours. For a wellbore with a volume of 130 bbl filled with a fluid of compressibility of 3.5×10^{-6} psi, the duration of wellbore storage (the unit-slope line) is estimated to be 0.0005 hours. With the gauge sampling rate set at 0.017 hours, the wellbore-storage unit slope simply could not be detected and does not appear at all on this plot. **Fig. 8.113** is the Horner plot for this test. A straight line appears in the same range as the flat derivative on a diagnostic plot. From the slope of the line, the permeability is estimated to be 53 md, close to the regression analysis match value of 48 md.

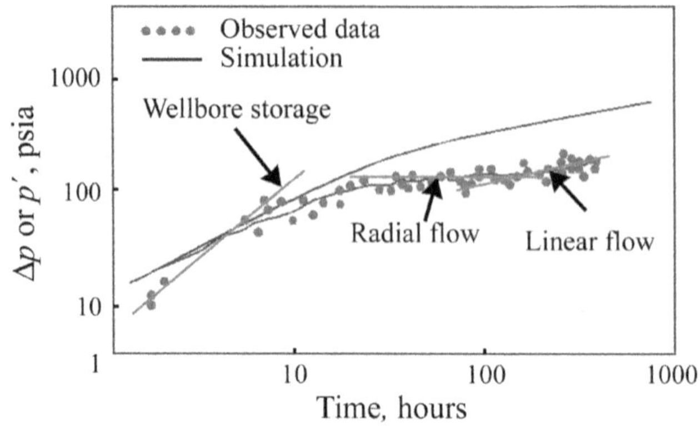

Fig. 8.109—Diagnostic plot for Well B indicates radial and linear flow regimes.

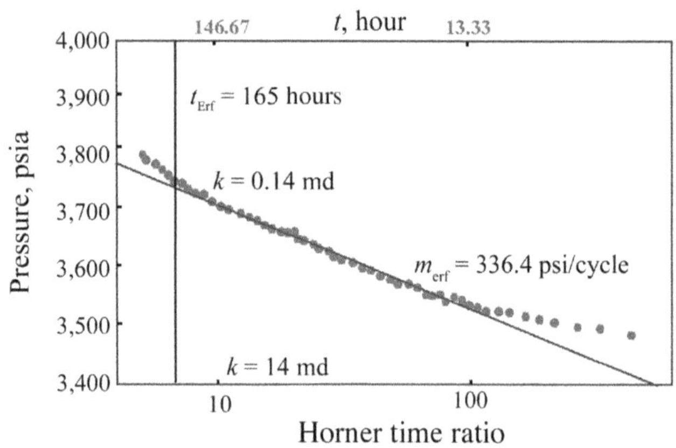

Fig. 8.110—Horner plot confirms early-radial flow.

8.11.9 Running Horizontal Well Tests. The measurements in horizontal wells are usually made above the wellbore with the pressure gauge still in the vertical section.

The test string may often be too rigid to pass through the wellbore. However, in most cases, conventional hardware can be used for horizontal well tests. With longer horizontal wellbores, wellbore storage is an inherent problem for testing, even for buildup tests with downhole shut-in. As mentioned previously, problems arise in conducting buildup tests with short-duration production periods because superposition is inappropriate; therefore, Horner plots and tandem-root plots, which depend on superposition being applicable, are often inappropriate.

Another problem in conducting buildup test following a short production period is that significant pressure gradients along the length of the wellbore may cause crossflow within the wellbore during shut-in, so fluid may flow from one region to another in the wellbore. These gradients can be removed and this crossflow eliminated by a longer-duration flow period preceding a buildup test.

Factors That Affect Transient Responses. A number of factors may affect the transient response of a horizontal well test: horizontal permeability (normal and parallel to well trajectory), vertical permeability, drilling damage, completion damage, producing interval that may be effectively much less than drilled length, and variations in standoff along length of well.

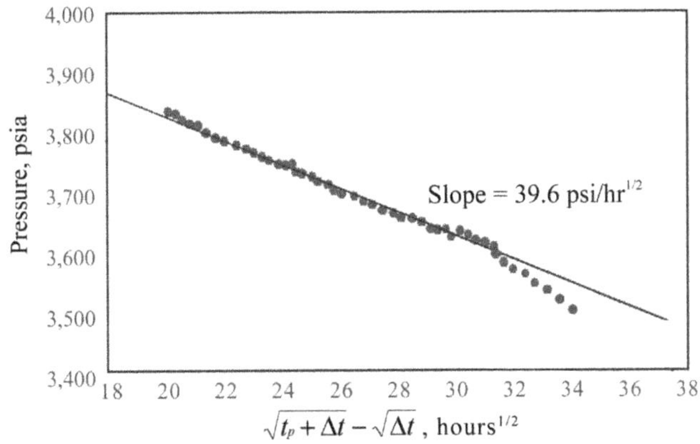

Fig. 8.111—Tandem-root plot confirms linear flow.

TABLE 8.9—WELL AND FORMATION PROPERTIES FOR EXAMPLE WELL C	
L_d, ft	1,400
L_w, ft	484
r_w, ft	0.41
ϕ, %	17
h, ft	54
q, STB/D	2,760
B_o, RB/STB	1.10
μ, cp	4.88
t_p, hours	36

In summary, seven or more factors may affect interpretation for horizontal wells in homogeneous reservoirs before the effects from boundaries. The problem is complex, so test results are frequently inconclusive. Furthermore, wellbore storage inhibits determination of properties associated with early-time transient behavior such as vertical permeability and damage from drilling and completion. Middle- and late-time behavior may require several hours, days, or months to appear in transient data.

Some practical steps will help ensure interpretable test data. First, it is helpful to run tests in the pilot hole before kicking off to drill the horizontal borehole section. From a test in the vertical section, it is possible to get usable estimates of horizontal and vertical permeabilities using modern wireline test tools. Second, a good directional drilling survey can frequently provide an adequate estimate of standoff. A production-log flow survey conducted with coiled tubing can determine what part of the wellbore is actually producing and, therefore, help provide an estimate of effective productive length. Wells in developed reservoirs should be flowed long enough to bring pressures along the wellbore to equilibrium and thus minimize crossflow. For high-rate wells, continuous borehole pressure and flow-rate measurements acquired during production can be used to interpret the pressure-drawdown transient response. If the downhole rates are not measured, the buildup test should be conducted with downhole shut-in to minimize wellbore storage distortion of test data.

8.11.10 Estimating Horizontal Well Productivity. Because of two fundamental problems, estimating the productivity of a horizontal well accurately is even more difficult than estimating

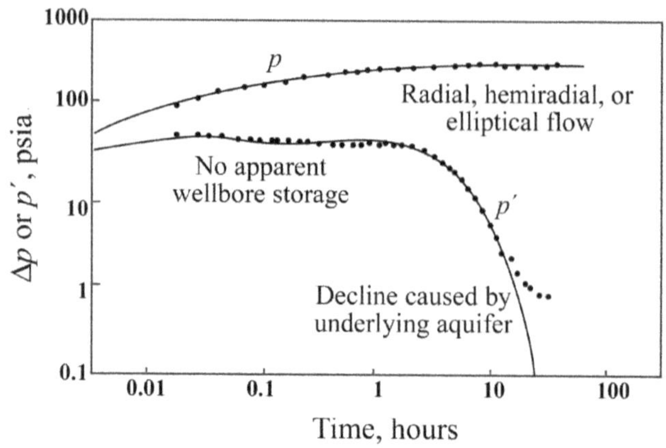

Fig. 8.112—Diagnostic plot for Well C shows no wellbore storage and constant derivative, with aquifer underlying oil column.

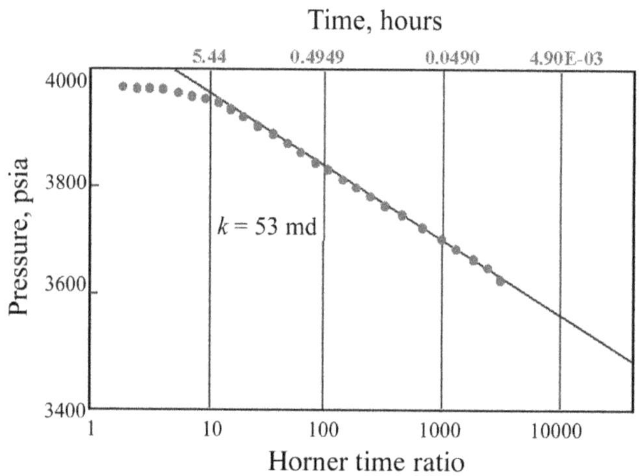

Fig. 8.113—Horner plot gives permeability of 5.3 md, close to that from regression analysis.

the productivity of a vertical well. The theoretical models available have a number of simplifying assumptions and the data required for even these simplified models are not likely to be available. Still, we must make estimates and decisions based on those estimates. In this section, two productivity models that have proved useful in practice are discussed. The first, published by Babu and Odeh[42] in 1989, is limited to single-horizontal wells. The second, published by Economides, Brand, and Frick[43] in 1996, is more general and is useful for multilateral wells.

Babu-Odeh Method. Babu and Odeh[42] obtained a rigorous solution to the diffusivity equation for a well in a box-shaped reservoir, subject to certain limiting assumptions. The assumptions include the following:

• Fluid flows to the well uniformly at all points along the wellbore (uniform flux) and the well is completed uniformly.

• The sides of the drainage volume are aligned with the principal permeability direction.

• The wellbore is parallel to the sides of the drainage area and is oriented parallel to one direction of principal permeability and perpendicular to the other two.

• The boundaries of the reservoir are all no-flow boundaries and the well reaches stabilized, pseudosteady-state flow.

• The formation damage around the wellbore is uniform at all points along the wellbore.

Fig. 8.95 introduces the nomenclature in the Babu and Odeh solution. The solution is quite complex but is approximated accurately with an equation written in the same form as the pseudosteady-state flow equation for a vertical oil well producing a single-phase, slightly incompressible liquid.

$$q = \frac{0.00708 b_H \sqrt{k_x k_z}\left(\bar{p} - p_{wf}\right)}{B\mu\left[\ln\left(\frac{A^{1/2}}{r_w}\right) + \ln C_H - 0.75 + s_p + \left(\frac{b_H}{L_w}\right)s_d\right]} \quad \text{(8.218)}$$

$$J = \frac{q}{\left(\bar{p} - p_{wf}\right)} = \frac{0.00708 b_H \sqrt{k_x k_z}}{B\mu\left[\ln\left(\frac{C_H A^{1/2}}{r_w}\right) - 0.75 + s_p + \left(\frac{b_H}{L_w}\right)s_d\right]} \quad \text{(8.219)}$$

Table 8.A-2 gives equations to estimate C_H and s_p. Two examples adapted from Babu and Odeh[42] illustrate the application of these equations.

Example 8.6. A horizontal well 1,000 ft long (L_w) is drilled in a box-shaped drainage volume 4,000 ft long (a_H), 2,000 ft wide (b_H), and 100 ft thick (h). The well is off-center in the y-direction (parallel to the well), so d_y = 250 ft and D_y = 750 ft. The well is also off-center in the x-direction so that d_x = 1,000 ft and D_x = 3,000 ft. Finally, the well is centered in the z-direction so that $d_z = D_z$ = 50 ft. The wellbore radius is 0.25 ft; $k_x = k_y$ = 200 md and k_z = 50 md. Fluid properties are B_o = 1.25 RB/STB and μ = 1 cp. Calculate the productivity index.

Solution.

From Eq. A-38,

$$\ln C_H = 6.28\frac{a_H}{h}\sqrt{\frac{k_z}{k_x}}\left[\frac{1}{3} - \frac{d_x}{a_H} + \left(\frac{d_x}{a_H}\right)^2\right] - \ln\left(\sin\frac{\pi d_z}{h}\right) - 0.5\ln\left[\left(\frac{a_H}{h}\right)\sqrt{\frac{k_z}{k_x}}\right] - 1.088$$

$$= \frac{(6.28)(4{,}000)}{(100)}\sqrt{\frac{50}{200}}\left[\frac{1}{3} - \frac{1{,}000}{4{,}000} + \left(\frac{1{,}000}{4{,}000}\right)^2\right] - \ln\left[\sin\left(\frac{50\pi}{100}\right)\right] - 0.5\ln\left[\left(\frac{4{,}000}{100}\right)\sqrt{\frac{50}{200}}\right] - 1.088$$

$$= 15.73$$

and $\frac{a_H}{\sqrt{k_x}} = \frac{4{,}000}{\sqrt{200}} = 283$; $\frac{0.75 b_H}{\sqrt{k_y}} = \frac{(0.75)(2{,}000)}{\sqrt{200}} = 106$; and $\frac{0.75h}{\sqrt{k_z}} = \frac{(0.75)(100)}{\sqrt{50}} = 10.6$. Thus, use Case 1 equations (Eqs. A-26 through A-31) to calculate s_p.

$$s_p = p_{xyz} + p'_{xy}.$$

$$p_{xyz} = \left(\frac{b_H}{L_w} - 1\right)\left\{ \ln\left(\frac{h}{r_w}\right) + 0.25 \ln\left(\frac{k_x}{k_z}\right) - \ln\left[\sin\left(\frac{\pi d_z}{h}\right)\right] - 1.838\right\}$$

$$= \left(\frac{2{,}000}{1{,}000} - 1\right)\left\{ \ln\left(\frac{100}{0.25}\right) + 0.25 \ln\left(\frac{200}{50}\right) - \ln\left[\sin\left(\frac{50\pi}{100}\right)\right] - 1.838\right\} = 4.50.$$

To calculate p'_{xy}, determine y_m, $L_w/2b_H$, $(4y_m - L_w)/2b_H$, and $(4y_m + L_w)/2b_H$.

$$y_m = d_y + L_w/2 = 2.50 + 1{,}000 = 750 \text{ ft},$$

$$L_w/2b_H = (1{,}000)/(2)(2{,}000) = 750 \text{ ft},$$

$$(4y_m - L_w)/2b_H = [(4)(750) - 1{,}000]/(2)(2{,}000) = 0.5,$$

$$\text{and } (4y_m + L_w)/2b_H = [(4)(750) + 1{,}000]/(2)(2{,}000) = 1.0.$$

Thus,

$$F(L_w/2b_H) = F(0.25) = -(0.25)\left[0.145 + \ln(0.25) - 0.137(0.25)^2\right] = 0.313,$$

$$F\left[(4y_m - L_w)/2b_H\right] = F(0.5) = -(0.5)\left[0.145 + \ln(0.5) - 0.137(0.5)^2\right] = 0.291,$$

$$\text{and } F\left[(4y_m + L_w)2b_H\right] = F(1.0) = -(1.0)\left[0.145 + \ln(1.0) - 0.137(1.0)^2\right] = -0.008.$$

Then,

$$p'_{xy} = \frac{2b_H^2}{L_w h}\sqrt{k_z/k_y}\left\{F\left(\frac{L_w}{2b_H}\right) + 0.5\left[F\left(\frac{4y_m + L_w}{2b_H}\right) - F\left(\frac{4y_m - L_w}{2b_H}\right)\right]\right\}$$

$$= \frac{(2)(2{,}000)^2}{(1{,}000)(100)}\sqrt{\frac{50}{200}}[0.313 + 0.5(-0.008 - 0.291)] = 6.54.$$

Then, $s_p = p_{xyz} + p'_{xy} = 4.50 + 6.54 = 11.0$, and

$$J = \frac{q}{(\bar{p} - p_{wf})} = \frac{0.00708 b_H\sqrt{k_x k_z}}{B\mu\left[\ln\left(\frac{C_H A^{1/2}}{r_w}\right) - 0.75 + s_p + \left(\frac{b_H}{L_w}\right)s_d\right]}$$

$$= \frac{(7.08 \times 10^{-3})(2{,}000)\sqrt{(50)(200)}}{(1.25)(1)\left\{\ln\left[\frac{[(4{,}000)(100)]^{1/2}}{0.25}\right] + 15.73 - 0.75 + 11.0\right\}} = 33.5 \text{ STB/D/psi}.$$

Example 8.7. A horizontal well is drilled in a box-shaped reservoir with the following characteristics: L_w = 1,000 ft, a_H = 2,000 ft long, b_H = 4,000 ft wide, and h_w = 2,000 ft thick. The well is off-center in the y-direction (d_y = 1,000 ft; D_y = 2,000 ft), centered in the x-direction (d_x = D_x = 1,000 ft), and off-center in the z-direction (d_z = 50 ft; D_z = 150 ft).

Permeabilities are $k_x = k_y = 100$ md and $k_z = 20$ md. Wellbore radius is 0.25 ft, $B_o = 1.25$ RB/STB, $\mu = 1$ cp, and $s_d = 0$. Find the productivity index, J.

Solution.

From Eq. A-38 (Table 8.A-2),

$$\ln C_H = 6.28\frac{a_H}{h}\sqrt{\frac{k_z}{k_x}}\left[\frac{1}{3} - \frac{d_x}{a_H} + \left(\frac{d_x}{a_H}\right)^2\right] - \ln\left(\sin\frac{\pi d_z}{h}\right) - 0.5\ln\left[\left(\frac{a_H}{h}\right)\sqrt{\frac{k_z}{k_x}}\right] - 1.088$$

$$= \frac{(6.28)(2{,}000)}{(200)}\sqrt{\frac{20}{100}}\left[\frac{1}{3} - \frac{1{,}000}{2{,}000} + \left(\frac{1{,}000}{2{,}000}\right)^2\right] - \ln\left[\sin\left(\frac{50\pi}{200}\right)\right] - 0.5\ln\left[\left(\frac{2{,}000}{200}\right)\sqrt{\frac{20}{100}}\right] - 1.088$$

$$= 0.850\ .$$

Note that

$$\frac{b_H}{\sqrt{k_y}} = \frac{4{,}000}{\sqrt{100}} = 400;\quad \frac{1.33a_H}{\sqrt{k_x}} = \frac{(1.33)(2{,}000)}{\sqrt{100}} = 266;\ \text{and}\ \frac{h}{\sqrt{k_z}} = \frac{(200)}{\sqrt{20}} = 44.7.$$

Thus, use Case 2 equations to calculate s_p.

$$s_p = p_{xyz} + p_y + p_{xy}.$$

$$p_{xyz} = \left(\frac{b_H}{L_w} - 1\right)\left\{\ln\left(\frac{h}{r_w}\right) + 0.25\ln\left(\frac{k_x}{k_z}\right) - \ln\left[\sin\left(\frac{\pi d_z}{h}\right)\right] - 1.838\right\}$$

$$= \left(\frac{4{,}000}{1{,}000}\quad 1\right)\left\{\ln\left(\frac{200}{0.25}\right) + 0.25\ln\left(\frac{100}{20}\right) - \ln\left[\sin\left(\frac{50\pi}{200}\right)\right] - 1.838\right\} = 16.79.$$

To calculate p_y, determine y_m. From Eq. A-29 (Table 8.A-2),

$$y_m = d_y + \frac{L_w}{2} = 1{,}000 + \frac{1{,}000}{2} = 1{,}500\ \text{ft}.$$

$$p_y = \frac{6.28b_H^2}{a_H h}\frac{\sqrt{k_x k_z}}{k_y}\left[\left(\frac{1}{3} - \frac{y_m}{b_H} + \frac{y_m^2}{b_H^2}\right) + \frac{L_w}{24b_H}\left(\frac{L_w}{b_H} - 3\right)\right]$$

$$= \frac{6.28(4{,}000)^2}{(2{,}000)(200)}\frac{\sqrt{(100)(20)}}{100}\left[\left(\frac{1}{3} - \frac{1{,}500}{4{,}00} + \left(\frac{1{,}500}{4{,}000}\right)^2\right) + \frac{1{,}000}{(24)(4{,}000)}\left(\frac{1{,}000}{4{,}000} - 3\right)\right] = 7.90.$$

From Eq. A-35 (Table 8.A-2),

$$p_{xy} = \left(\frac{b_H}{L_w} - 1\right)\left(\frac{6.28a_H}{h}\sqrt{\frac{k_z}{k_x}}\right)\left(\frac{1}{3} - \frac{d_x}{a_H} + \frac{d_x^2}{a_H^2}\right),\ d_x \geq 0.25a_H$$

$$= \left(\frac{4{,}000}{1{,}000} - 1\right)\left(\frac{(6.28)(2{,}000)}{200}\sqrt{\frac{20}{100}}\right)\left(\frac{1}{3} - \frac{1{,}000}{2{,}000} + \left(\frac{1{,}000}{2{,}000}\right)^2\right) = 7.02.$$

Thus, $s_p = 16.79 + 7.90 + 7.02 = 31.7$. Then, from Eq. A-37 (Table 8.A-2),

$$J = \frac{q}{\left(\bar{p} - p_{wf}\right)} = \frac{0.00708 b_H \sqrt{k_x k_z}}{B\mu \left[\ln\left(\frac{C_H A^{1/2}}{r_w}\right) - 0.75 + s_p + \left(\frac{b_H}{L_w}\right) s_d \right]}$$

$$= \frac{\left(7.08 \times 10^{-3}\right)(4{,}000)\sqrt{(100)(200)}}{(1.25)(1)\left[\ln\left(\frac{\sqrt{(200)(2{,}000)}}{0.25}\right) + 0.850 - 0.75 + 31.7 + 0 \right]} = 25.6 \text{ STB/D/psi.}$$

Economides* et al. *Method. Economides *et al.*[45] presented a more general method to estimate productivity index for a horizontal well. The method has the advantage that it is applicable to multilateral wells in the same plane and is not limited to wells aligned with principal permeabilities. It includes solutions for wells with no pressure drop in the wellbore (infinite conductivity, as opposed to wells with uniform flux). It has the disadvantage that it requires interpolation in a table in which only certain drainage area shapes are given.

The basic working equation for the productivity index in this method is

$$J = \frac{\bar{k} b_H}{887.22 B\mu \left(p_D + \frac{b_H}{2\pi L_w} \Sigma s \right)}, \quad \text{(8.220)}$$

where Σs refers to damage skin, turbulence, and other pseudoskin factors. In Eq. 8.220,

$$p_D = \frac{b_H C_H}{4\pi h} + \frac{b_H}{2\pi L_w} s_c, \quad \text{(8.221)}$$

where

$$s_c = \ln\left(\frac{h}{2\pi r_w}\right) - \frac{h}{6L_w} + s_e, \quad \text{(8.222)}$$

and s_e, describing eccentricity effects in the vertical direction, is

$$s_e = \frac{h}{L_w}\left[\frac{2d_z}{h} - \frac{1}{2}\left(\frac{2d_z}{h}\right)^2 - \frac{1}{2} \right] - \ln\left[\sin\left(\frac{\pi d_z}{h}\right) \right]. \quad \text{(8.223)}$$

$s_e = 0$ when a well is centered in the vertical plane. This convergence skin differs only slightly from that used by Babu and Odeh. The difference is $0.25 \ln\left(k_x / k_z\right) + h / L_w\left[2d_z / h - 1/2\left(2d_z / h\right)^2 - 2/3\right]$, which is usually small (< 0.5). **Table 8.10** gives values of C_H for several drainage areas and multilateral configurations. The equations as written are for isotropic reservoirs. Certain variable transformations are required before substituting into the working equation:

$$L' = L_w \alpha^{-1/3} \beta. \quad \text{(8.224)}$$

$$r_w' = r_w \frac{\alpha^{2/3}}{2}\left(\frac{1}{\alpha\beta} + 1\right), \quad \text{(8.225)}$$

where

$$\alpha = \sqrt{\frac{\left(k_x k_y\right)^{1/2}}{k_z}}, \quad \text{(8.226)}$$

$$\text{and } \beta = \left(\sqrt{\frac{k_x}{k_y}}\cos^2\phi + \sqrt{\frac{k_y}{k_x}}\sin^2\phi\right)^{1/2}. \quad \text{(8.227)}$$

ϕ is the azimuth of the well trajectory (relative to the y-axis). Reservoir dimensions:

$$x' = x\frac{\sqrt{k_y k_z}}{\bar{k}}, \quad \text{(8.228)}$$

$$y' = y\frac{\sqrt{k_x k_z}}{\bar{k}}, \quad \text{(8.229)}$$

$$z' = z\frac{\sqrt{k_x k_y}}{\bar{k}}, \quad \text{(8.230)}$$

$$\text{and } \bar{k} - \sqrt[3]{k_x k_y k_z}. \quad \text{(8.231)}$$

Two examples, one from an isotropic reservoir and one from an anisotropic reservoir, illustrate this method.

Example 8.8. Economides *et al.*[43] provide this example. Consider a horizontal well 1,500 ft long in a reservoir with b_H = 2,000 ft, a_H = 4,000 ft, h = 20 ft, r_w = 0.4, $k_x = k_y = k_z$ = 10 md, B_o = 1.25 RB/STB, and μ = 1 cp. Assume that the well is centered vertically so that s_e = 0. Also, assume Σs = 0.

Solution.

From Eq. 8.223,

$$s_c = \ln\left(\frac{h}{2\pi r_w}\right) - \frac{h}{6L_w} + s_e$$

$$= \ln\left(\frac{20}{2\pi(0.4)}\right) - \frac{20}{(6)(1{,}500)} + 0 = 2.07.$$

(As a matter of interest, the Babu and Odeh s_c for this case is also 2.07.) From Table 8.10, for $2b_H = a_H$ and L_w/b_H = 1,500/2,000 = 0.75, C_H = 2.53. From Eq. 8.221,

TABLE 8.10—HORIZONTAL PLANE SHAPE FACTORS

		L_w/b_H	C_H
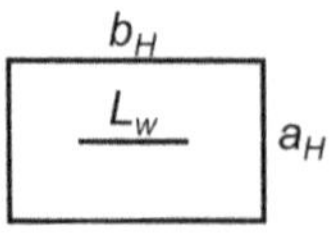	$b_H = 4a_H$	0.25	3.77
		0.5	2.09
		0.75	1.00
		1.	0.26
	$b_H = 2a_H$	0.25	3.19
		0.5	1.80
		0.75	1.02
		1	0.52
	$b_H = a_H$	0.25	3.55
		0.5	2.21
		0.75	1.49
		1.	1.04
	$2b_H = a_H$	0.25	4.59
		0.5	3.26
		0.75	2.53
		1	2.09
	$4b_H = a_H$	0.25	6.69
		0.5	5.35
		0.75	4.63
		1.	4.18
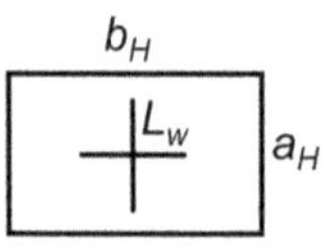	$b_H = a_H$	0.25	2.77
		0.5	1.47
		0.75	0.81
		1	0.46
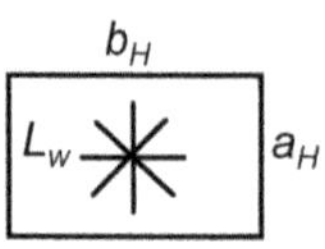	$b_H = a_H$	0.25	2.66
		0.5	1.36
		0.75	0.69
		1	0.32
		θ	
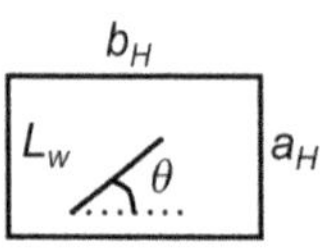	$b_H = a_H$	0	1.49
	$L_w/b_H = 0.75$	30	1.48
		45	1.48
		75	1.49
		90	1.49

$$p_D = \frac{b_H C_H}{4\pi h} + \frac{b_H}{2\pi L_w} s_c$$

$$= \frac{(2{,}000)(2.53)}{(4\pi)(20)} + \frac{(2{,}000)(2.07)}{(2\pi)(1{,}500)} = 20.58.$$

Then, from Eq. 8.220,

$$J = \frac{\bar{k} b_H}{887.22 B\mu \left(p_D + \frac{b_H}{2\pi L_w} \Sigma s \right)}$$

$$= \frac{(10)(2{,}000)}{(887.22)(1.25)(1)(20.58 + 0)} = 0.876 \text{ STB/D/psi}.$$

Example 8.9. Rework the Babu-Odeh Example 8.7 using the Economides *et al.* method.
Solution.
First transform the variables. From Eqs. 8.226 and 8.227,

$$\alpha = \sqrt{\frac{\left(k_x k_y\right)^{1/2}}{k_z}} = \sqrt{\frac{[(100)(100)]^{1/2}}{(20)}} = 2.236.$$

Because the well is parallel to the x-axis, $\phi = 0$, and

$$\beta = \left(\sqrt{\frac{k_x}{k_y}}\right)^{1/2} = 1.$$

From Eq. 8.231,

$$\bar{k} = \sqrt[3]{k_x k_y k_z} = 58.5 \text{ md}.$$

From Eq. 8.224,

$$L' = L_w \alpha^{-1/3} \beta = (1{,}000)(2.236)^{-1/3}(1) = 765 \text{ ft}.$$

From Eq. 8.225,

$$r_w' = r_w \frac{\alpha^{2/3}}{2}\left(\frac{1}{\alpha\beta} + 1\right) = (0.25)\frac{(2.236)^{2/3}}{2}\left[\frac{1}{(2.236)(1)} + 1\right] = 0.309 \text{ ft}.$$

From Eqs. 8.228 through 8.230,

$$a_H' = a_H \frac{\sqrt{k_y k_z}}{\bar{k}} = \frac{2{,}000\sqrt{(100)(20)}}{58.5} = 1{,}529 \text{ ft},$$

$$b_H' = b_H \frac{\sqrt{k_x k_z}}{\bar{k}} = \frac{4{,}000\sqrt{(100)(20)}}{58.5} = 3{,}058 \text{ ft},$$

$$h' = h \frac{\sqrt{k_x k_y}}{\bar{k}} = \frac{200\sqrt{(100)(100)}}{58.5} = 342 \text{ ft},$$

$$\text{and } d_z' = d_z \frac{\sqrt{k_x k_y}}{\bar{k}} = \frac{(50)\sqrt{(100)(100)}}{58.5} = 85.5 \text{ ft}.$$

Thus, the equivalent system is a rectangular-shaped drainage area twice as long parallel to the wellbore (3,050 ft) as perpendicular, with $L'/b_H' = 765/3{,}058 = 0.25$. In the original example, the well was off-center in the horizontal plane; here, assume that a centered well is an adequate approximation. From Table 8.10, $C_H = 3.19$.

From Eq. 8.223,

$$s_e = \frac{h'}{L_w'}\left[\frac{2d_z'}{h'} - \frac{1}{2}\left(\frac{2d_z'}{h'}\right)^2 - \frac{1}{2}\right] - \ln\left[\sin\left(\frac{\pi d_z'}{h'}\right)\right]$$

$$= \frac{342}{765}\left[\frac{(2)(85.5)}{342} - \frac{1}{2}\left(\frac{(2)(85.5)}{342}\right)^2 - \frac{1}{2}\right] - \ln\left[\sin\left(\frac{85.5\pi}{342}\right)\right] = 0.291.$$

Then,

$$s_c = \ln\left(\frac{h'}{2\pi r_w'}\right) - \frac{h'}{6L_w'} + s_c$$

$$= \ln\left(\frac{342}{2\pi(0.309)}\right) - \frac{342}{(6)(765)} + 0.291 = 5.39.$$

(The Babu-Odeh s_c is 5.60 for this case.) Then, from Eq. 8.221,

$$p_D = \frac{b_H' C_H}{4\pi h'} + \frac{b_H'}{2\pi L_w'} s_c$$

$$= \frac{(3{,}050)(3.14)}{4\pi(342)} + \frac{(3{,}050)(5.39)}{2\pi(765)} = 5.70.$$

Finally, from Eq. 8.220,

$$J = \frac{\bar{k} b_H'}{887.22 B\mu\left(p_D + \frac{b_H'}{2\pi L_w'}\Sigma s\right)}$$

$$= \frac{(58.5)(3{,}058)}{(887.22)(1.25)(1.0)(5.70+0)} = 28.3 \text{ STB/D/psi}.$$

The result is slightly larger than the result using the Babu-Odeh method (J = 25.6 STB/D/psi). At least part of the reason for the difference is that, in this example, it was necessary to assume that the well was centered in its drainage volume, which was not true in the original example. The optimal location of a horizontal well to maximize productivity is to center it within its drainage volume.

Comparison of Recent and Older Horizontal Well Models. Ozkan[44] compared "contemporary" (generally 1990s) and "conventional" horizontal well models in a paper published in 2001. He pointed out that the older models are used for both pressure-transient test analysis and for estimating well productivity. Ozkan stressed three limitations of the conventional models, which include the Babu-Odeh model and other pioneering work.

Conventional models usually assume that the horizontal well is parallel to one of the principal permeability directions (preferably the minimum permeability direction in the horizontal plane). In many cases, this is not true. In fact, in many cases the principal permeability directions are unknown. When the principal permeability directions are known, corrections to length are possible (as in the Economides *et al.* model); if they are not known, there is no way to correct the analysis. Contemporary models show that the error in permeability estimates approaches 50% when the deviation angle exceeds 50°. Unfortunately, the models also indicate that there is nothing in a well's response that provides any indication that the assumption that the well is parallel to a principal permeability direction is incorrect.

Ozkan pointed out that the damaged region around a horizontal well probably is nonuniform with distance (perhaps with the greatest damage near the heel of the well and the least near the toe, because filtrate invasion is of much longer duration near the heel). If there is variable permeability along the path of the well, the situation is even more complicated. Some contemporary models can take this variation into account; however, most conventional models cannot. Conventional models usually assume (implicitly) uniform skin effect along the wellbore. However, the contemporary models will not be helpful if the skin distribution along the length is unknown.

Ozkan notes that it is a common practice to complete horizontal wells selectively. Also, in other cases, some segment of the well may not be open to flow of reservoir fluids because of relatively low permeabilities or relatively large local skin effects. The absolute amount of the well that is open to flow and the location of the open intervals affect the pressure response in the well. Some contemporary well models can take these effects into account, but, again, the capabilities of the newer models may be limited if the location and length of the open intervals is unknown.

Many models assume negligible pressure drop in the wellbore (infinite conductivity). Others assume the same flow rate per unit length at all points along the well bore (uniform flux). In fact, there is likely to be finite pressure drop in the wellbore, resulting in neither uniform flow nor infinite conductivity. Contemporary models in which a reservoir model is coupled to a wellbore model can take these effects into account.

Unfortunately, contemporary horizontal well models have not led to simple, easily applied methods of well-test analysis or of predicting well productivity. Further, their full utility depends on availability of detailed well and reservoir description data. At present, the major use of such models may be to quantify the possible errors that arise from uncertainty and to be used to history-match observed information when sufficient data are available.

8.12 Deliverability Testing of Gas Wells

8.12.1 Introduction. This section discusses the implementation and analysis of the four most common types of gas-well deliverability tests: flow-after-flow, single-point, isochronal, and modified isochronal tests. A summary of the fundamental gas-flow equations, both theoretical and empirical, used to analyze deliverability tests in terms of pseudopressure is followed by a focus on specific tests and testing procedures, advantages and disadvantages of each testing method, and common analysis techniques. Examples illustrating deliverability tests analyses are included.

8.12.2 Types and Purposes of Deliverability Tests. Deliverability testing refers to the testing of a gas well to measure its production capabilities under specific conditions of reservoir and bottomhole flowing pressures (BHFPs). A common productivity indicator obtained from these tests is the absolute open flow (AOF) potential. The AOF is the maximum rate at which a well could flow against a theoretical atmospheric backpressure at the sandface. Although in practice the well cannot produce at this rate, regulatory agencies sometimes use the AOF to allocate allowable production among wells or to set maximum production rates for individual wells.

Another application of deliverability testing is to generate a reservoir inflow performance relationship (IPR) or gas backpressure curve. The IPR curve describes the relationship between surface production rate and BHFP for a specific value of reservoir pressure (that is, either the original pressure or the current average value). The IPR curve can be used to evaluate gas-well current deliverability potential under a variety of surface conditions, such as production against a fixed backpressure. In addition, the IPR can be used to forecast future production at any stage in the reservoir's life.

Several deliverability testing methods have been developed for gas wells. Flow-after-flow tests are conducted by producing the well at a series of different stabilized flow rates and measuring the stabilized BHP. Each flow rate is established in succession without an intermediate shut-in period. A single-point test is conducted by flowing the well at a single rate until the BHFP is stabilized. This type of test was developed to overcome the limitation of long testing times required to reach stabilization at each rate in the flow-after-flow test.

Isochronal and modified isochronal tests were developed to shorten tests times for wells that need long times to stabilize. An isochronal test consists of a series of single-point tests usually conducted by alternately producing at a slowly declining sandface rate without pressure stabilization and then shutting in and allowing the well to build to the average reservoir pressure before the next flow period. The modified isochronal test is conducted similarly, except the flow periods are of equal duration and the shut-in periods are of equal duration (but not necessarily the same as the flow periods).

8.12.3 Theory of Deliverability Test Analysis. This section summarizes the theoretical and empirical gas-flow equations used to analyze deliverability tests. The theoretical equations developed by Houpeurt[45] are exact solutions to the generalized radial-flow diffusivity equation, while the Rawlins and Schellhardt[46] equation was developed empirically. All basic equations presented here assume radial flow in a homogeneous, isotropic reservoir and therefore may not be applicable to the analysis of deliverability tests from reservoirs with heterogeneities, such as natural fractures or layered pay zones. These equations should not be used to analyze tests from hydraulically fractured wells during the fracture-dominated linear or bilinear flow periods. Finally, these equations assume that wellbore-storage effects have ceased. Unfortunately, wellbore-storage distortion may affect the entire test period in short tests, especially those conducted in low-permeability reservoirs.

Theoretical Deliverability Equations. The early-time transient solution to the diffusivity equation for gases for constant-rate production from a well in a reservoir with closed outer boundaries, written in terms of pseudopressure, p_p,[47] is

$$p_p(p_s) - p_p(p_{wf}) = \frac{1.422 \times 10^6 qT}{k_g h} \times \left[1.151 \log\left(\frac{k_g t}{1{,}688 \phi \bar{\mu}_g \bar{c}_t r_w^2} \right) + s + Dq \right], \quad (8.232)$$

where p_s is the stabilized shut-in BHP measured before the deliverability test. In new reservoirs with little or no pressure depletion, this shut-in pressure equals the initial reservoir pressure, $p_s = p_i$, while in developed reservoirs, $p_s < p_i$.

The late-time or pseudosteady-state solution is

$$p_p(\bar{p}) - p_p(p_{wf}) = \frac{1.422 \times 10^6 qT}{k_g h} \times \left[1.151 \log\left(\frac{10.064 A}{C_A r_w^2} \right) - \frac{3}{4} + s + Dq \right], \quad (8.233)$$

where $\bar{p}$ is current drainage-area pressure. Gas wells cannot reach true pseudosteady state because $\mu_g(p)c_t(p)$ changes as $\bar{p}$ decreases. Note that, unlike $\bar{p}$, which decreases during pseudosteady-state flow, p_s is a constant.

Eqs. 8.232 and 8.233 are quadratic in terms of the gas flow rate, q. For convenience, Houpeurt[45] wrote the transient flow equation as

$$\Delta p_p = p_p(p_s) - p_p(p_{wf}) = a_t q + b q^2, \quad (8.234)$$

and the pseudosteady-state flow equation as

$$\Delta p_p = p_p(\bar{p}) - p_p(p_{wf}) = a q + b q^2, \quad (8.235)$$

where

$$a_t = \frac{1.422 \times 10^6 T}{k_g h} \left[1.151 \log\left(\frac{k_g t}{1{,}688 \phi \bar{\mu}_g \bar{c}_t r_w^2} \right) + s \right], \quad (8.236)$$

$$a = \frac{1.422 \times 10^6 T}{k_g h} \left[1.151 \log\left(\frac{10.06 A}{C_A r_w^2} \right) - \frac{3}{4} + s \right], \quad (8.237)$$

$$\text{and } b = \frac{1.422 \times 10^6 TD}{k_g h}. \quad (8.238)$$

The coefficients of q (a_t for transient flow and a for pseudosteady-state flow) include the Darcy flow and skin effects and are measured in (psia^2/cp)/(MMscf/D) when q is in MMscf/D. The coefficient of q^2 represents the inertial and turbulent flow effects and is measured in (psia^2/cp)/(MMscf/D)^2 when q is in MMscf/D.

The Houpeurt equations also can be written in terms of pressure squared and are derived directly from the solutions to the gas-diffusivity equation, assuming that $\mu_g z$ is constant over the pressure range considered. For transient flow,

$$\Delta p^2 = p_s^2 - p_{wf}^2 = a_t q + b q^2, \quad (8.239)$$

and for pseudosteady-state flow,

$$\Delta p^2 = \bar{p}^2 - p_{wf}^2 = aq + bq^2 \; . \qquad (8.240)$$

The flow coefficients are

$$a_t = \frac{1.422 \times 10^6 \bar{\mu}_g \bar{z} T}{k_g h}\left[1.151 \log\left(\frac{k_g t}{1.688 \phi \bar{\mu}_g \bar{c}_t r_w^2}\right) + s\right], \qquad (8.241)$$

$$a = \frac{1.422 \times 10^6 \bar{\mu}_g \bar{z} T}{k_g h}\left[1.151 \log\left(\frac{10.06 A}{C_A r_w^2}\right) - \frac{3}{4} + s\right], \qquad (8.242)$$

$$\text{and } b = \frac{1.422 \times 10^6 \bar{\mu}_g \bar{z} T D}{k_g h} \; . \qquad (8.243)$$

When the Houpeurt equation is presented in terms of pressure squared, the coefficients of q are measured in psia^2/(MMscf/D) when q is in MMscf/D, while the coefficient of q^2 is measured in units of psia^2/(MMscf/D)^2 when q is in MMscf/D. For convenience, all equations and examples in this section are presented with q measured in MMScf/D.

The pressure-squared form of the equation should be used only for gas reservoirs at low pressures (less than 2,000 psia) and high temperatures. To eliminate doubt about which equations to choose, use of the pseudopressure equations, which are applicable at all pressures and temperatures, is recommended. Consequently, all the analysis procedures in this section are presented in terms of pseudopressure.

An advantage of the pseudopressure form of the theoretical deliverability equation is that the flow coefficients are independent of the average reservoir pressure and, therefore, do not change as $\bar{p}$ decreases during a flow test conducted under pseudosteady-state flow unless s, k, or A changes. Because the non-Darcy flow coefficient is a function of $\mu_g(p_{wf})$, the coefficient b will change slightly if the BHFP is changed. In contrast, because of the pressure dependency of the gas properties on average reservoir pressure, the flow coefficients for the pressure-squared form of the deliverability equation must be recalculated for every new $\bar{p}$ value. When s, k, or A changes with time, the only way to update the deliverability curve is to retest the well.

Empirical Deliverability Equations. In 1935, Rawlins and Schellhardt[46] presented an empirical relationship that is used frequently in deliverability test analysis. The original form of their relation, given by Eq. 8.244 in terms of pressure squared, is applicable only at low pressures:

$$q = C\left(\bar{p}^2 - p_{wf}^2\right)^n \; . \qquad (8.244)$$

In terms of pseudopressure, Eq. 8.244 becomes

$$q = C\left[p_p(\bar{p}) - p_p(p_{wf})\right]^n, \qquad (8.245)$$

which is applicable over all pressure ranges. In Eqs. 8.244 and 8.245, C is the stabilized performance coefficient and n is the inverse slope of the line on a log-log plot of the change in pressure squared or pseudopressure vs. gas flow rate. Depending on the flowing conditions, the theoretical value of n ranges from 0.5, indicating turbulent flow throughout a well's drainage area, to 1.0, indicating laminar flow behavior modeled by Darcy's law. The value of C changes depending on the units of flow rate and whether Eq. 8.244 or 8.245 is used. All equations and examples in this section are presented with q measured in MMscf/D.

Houpeurt proved that neither Eq. 8.244 nor Eq. 8.245 can be derived from the generalized diffusivity equation for radial flow of real gas through porous media. Although the Rawlins and Shellhardt equation is not theoretically rigorous, it is still widely used in deliverability analysis and has worked well over the years, especially when the test rates approach the AOF potential of the well and the extrapolation from test rates to AOF is minimal.

8.12.4 Stabilization Time. Unlike pressure-transient tests, the analysis techniques for conventional flow-after-flow and single-point tests require data obtained under stabilized flowing conditions. Although isochronal and modified isochronal tests were developed to circumvent the requirement of stabilized flow, they may still require a single, stabilized flow period at the end of the test. Consequently, there is a need to understand the meaning of stabilization time and have a method to estimate its value.

Stabilization time is defined as the time when the flowing pressure is no longer changing or is no longer changing significantly. Physically, stabilized flow can be interpreted as the time when the pressure transient is affected by the no-flow boundaries, either natural reservoir boundaries or an artificial boundaries created by active wells surrounding the tested well. Consider a graph of pressure as a function of radius for constant-rate flow at various times since the beginning of flow. As Fig. 8.1 shows, the pressure in the wellbore continues to decrease as flow time increases. Simultaneously, the area from which fluid is drained increases, and the pressure transient moves farther out into the reservoir.

The radius of investigation, the point in the formation beyond which the pressure drawdown is negligible, is a measure of how far a transient has moved into a formation following any rate change in a well. The approximate position of the radius of investigation at any time for a gas well is estimated by Eq. 8.246[48]:

$$r_i = \sqrt{\frac{k_g t}{948\phi\bar{\mu}_g\bar{c}_t}} \quad \text{(8.246)}$$

Stabilized flowing conditions occur when the calculated radius of investigation equals or exceeds the distance to the drainage boundaries of the well (i.e., $r_i \geq r_e$). Substituting r_e and rearranging Eq. 8.246, yields an equation for estimating the stabilization time, t_s, for a gas well centered in a circular drainage area:

$$t_s = \frac{948\phi\bar{\mu}_g\bar{c}_t r_e^2}{k_g} \quad \text{(8.247)}$$

As long as the radius of investigation is less than the distance to the no-flow boundary, stabilization has not been attained and the pressure behavior is transient. To illustrate the importance of stabilization times in deliverability testing, stabilization times were calculated as a function of permeability and drainage area for a well producing a gas with a specific gravity of 0.6 from a formation at 210°F and an average pressure of 3,500 psia ($\bar{\mu}_g$ = 0.02 cp and $\bar{c}_t$ = 2.468 × 10^{-4}psia^{-1}), with a porosity of 10%. **Table 8.11** shows that, for wells completed in low-permeability reservoirs, several days—or even years—are required to reach stabilized flow, while wells completed in high-permeability reservoirs stabilize in a short time.

A more general equation for calculating stabilization time is

TABLE 8.11—EFFECTS OF PERMEABILITY AND DRAINAGE AREA ON STABILIZATION TIME

	A = 40 acres		A = 640 acres	
k (md)	t_s (hours)	t_s (days or years)	t_s (hours)	t_s (days or years)
0.01	25,953	3 years	415,242	47 years
0.1	2,595	108 days	41,524	4.7 years
1.0	259.5	10.8 days	4,152.4	173 days
10.0	25.95	1.1 days	415.2	17.3 days
100.0	2.59	0.11 days	41.52	1.73 days
1000.0	0.259	0.011 days	4.15	0.173 days

TABLE 8.12—EFFECTS OF PERMEABILITY AND HYDRAULIC FRACTURE HALF-LENGTH ON TIME TO REACH PSEUDORADIAL FLOW

Case	k (md)	L_f (ft)	t_{prf} (hours)
1	1	100	51.2
2	0.01	100	5,120 (213 days)
3	0.01	1,000	512,000 (58 years)

$$t_s = \frac{\phi \bar{\mu}_g \bar{c}_t A t_{DA}}{0.0002637 k_g}, \qquad (8.248)$$

where t_{DA} is dimensionless time for the beginning of pseudosteady-state flow. Values for t_{DA} are given in Table 8.A-1 for various reservoir shapes and well locations.[49] The time required for the pseudosteady-state equation to be exact is found from the entry in the column "Exact for t_{DA} >."

The Rawlins-Schellhardt and Houpeurt deliverability equations assume radial flow. If pseudoradial flow has been achieved, however, these analysis techniques can be used for hydraulically fractured wells. The time to reach the pseudoradial flow regime, t_{prf}, occurs[30] at $t_{L_f D} = 3$ and is estimated with

$$t_{prf} = \frac{11{,}400 \phi \bar{\mu}_g \bar{c}_t L_f^2}{k_g}. \qquad (8.249)$$

To illustrate the importance of achieving pseudoradial flow during a deliverability test, values of t_{prf} were calculated for a hydraulically fractured well completed in a reservoir with ϕ = 0.15, $\bar{\mu}_g$ = 0.03 cp, and $\bar{c}_t = 1 \times 10^{-4}$ psia^{-1} and with the range of permeabilities and hydraulic fracture half-lengths in **Table 8.12**. The results illustrate that a well with a long fracture in a low-permeability formation will take far too long to stabilize for conventional deliverability testing.

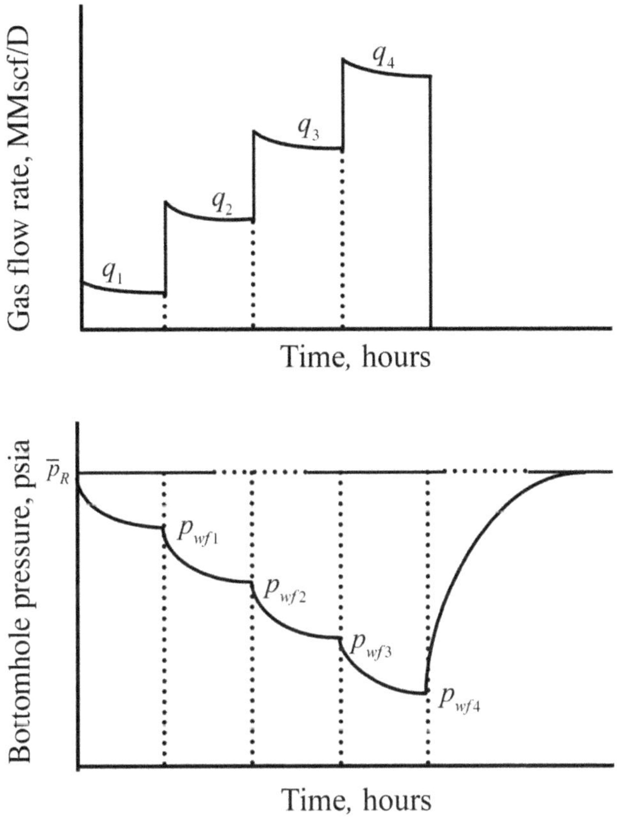

Fig. 8.114—Pressure and flow rate history of a typical flow-after-flow test.

8.12.5 Analysis of Deliverability Tests. This section discusses the implementation and analysis of the flow-after-flow, single-point, isochronal, and modified isochronal tests. Both the Rawlins and Schellhardt and Houpeurt analysis techniques are presented in terms of pseudopressures.

Flow-After-Flow Tests. Flow-after-flow tests, sometimes called gas backpressure or four-point tests, are conducted by producing the well at a series of different stabilized flow rates and measuring the stabilized BHFP at the sandface. Each different flow rate is established in succession either with or without a very short intermediate shut-in period. Conventional flow-after-flow tests often are conducted with a sequence of increasing flow rates; however, if stabilized flow rates are attained, the rate sequence does not affect the test.[46] The requirement that the flowing periods be continued until stabilization is a major limitation of the flow-after-flow test, especially in low-permeability formations that take long times to reach stabilized flowing conditions. **Fig 8.114** illustrates a flow-after-flow test.

Rawlins-Schellhardt Analysis Technique. Recall the empirical equation that forms the basis for the Rawlins-Schellhardt analysis technique:

$$q = C\left[p_p(\bar{p}) - p_p(p_{wf})\right]^n \text{ . ... (8.245)}$$

Taking the logarithm of both sides of Eq. 8.245 yields the equation that forms the basis for the Rawlins-Schellhardt analysis technique:

$$\log(q) = \log(C) + n\left\{\log\left[p_p(\bar{p}) - p_p(p_{wf})\right]\right\} . \quad (8.250)$$

The form of Eq. 8.250 suggests that a plot of log (Δp_p) vs. log (q) will yield a straight line of slope $1/n$ and an intercept of $\{-1/n[\log(C)]\}$. The AOF potential is estimated from the extrapolation of the straight line to Δp_p evaluated at a p_{wf} equal to atmospheric pressure (sometimes called base pressure). This analysis technique is illustrated with Example 8.10.

Houpeurt Analysis Technique. Flow-after-flow tests require stabilized data or data measured during pseudosteady-state flow. Houpeurt[45] gives the theoretical equation for pseudosteady-state flow, which was derived from the gas-diffusivity equation, as

$$\Delta p_p = p_p(\bar{p}) - p_p(p_{wf}) = aq + bq^2 . \quad (8.235)$$

The coefficients a and b have theoretical bases and can be estimated if reservoir properties are known or they can be determined from flow-after-flow test data. Dividing both sides of Eq. 8.235 by the flow rate, q, and rearranging yields the equation that is the basis for the Houpeurt analysis technique:

$$\frac{\Delta p_p}{q} = \frac{p_p(\bar{p}) - p_p(p_{wf})}{q} = a + bq . \quad (8.251)$$

The form of Eq. 8.251 suggests that a plot $\Delta p_p / q$ vs. q will yield a straight line with a slope b and an intercept a. The AOF is estimated in the Houpeurt deliverability analysis by solving Eq. 8.235 for $q = q_{AOF}$ at $p_{wf} = p_b$.

Example 8.10: Analysis of a Flow-After-Flow Test. Estimate the initial stabilized AOF potential of a well[50] with the well and reservoir properties listed. Use both the Rawlins-Schellhardt and the Houpeurt analysis techniques. In addition, estimate the AOF potential 10 years later when the static drainage area pressure has decreased to 350 psia. Evaluate the AOF potential at p_b = 14.65 psia. **Table 8.13** summarizes the flow-after-flow test data. L = 3,050 ft, r_w = 0.5 ft, M_a = 20.71 lbm/lbm-mole, T = 90°F = 555°R, A = 640 acres, ϕ = 0.25, C_A = 30.8828, and h =200 ft.

Current $\bar{p}$ = 407.6 psia, $p_p(\bar{p}$ = 407.6) = 1.617 × 10^7 psia^2/cp. $\bar{p}$ after 10 years = 350 psia, $p_p(\bar{p}$ = 350) = 1.2239 × 10^7 psia^2/cp. p_b = 14.65 psia, $p_p(p_b)$ = 2,674.8 psia^2/cp.

The pseudopressure in this example (and all others in this section) were calculated using the methods suggested by Al-Hussainy *et al.*[15] These methods, which involve numerical evaluation of the integral in Eq. 8.97 and which require computational routines to estimate gas viscosity, μ, and deviation factor, z, are widely available in basic reservoir fluid flow analysis software.

Solution.

Rawlins-Schellhardt Analysis. Plot Δp_p vs. q on log-log graph paper (**Fig. 8.115**). **Table 8.14** gives the plotting functions. Construct the best-fit line through the data points. All data points lie on the best-fit line and will be used for all subsequent calculations.

Next, determine the deliverability exponent using least-squares regression analysis. Alternatively, because Points 1 and 4 both lie on the perceived "best" straight line through the test data, the reciprocal slope is estimated to be

TABLE 8.13—FLOW-AFTER-FLOW TEST DATA, EXAMPLE 8.10

q (MMscf/D)	p_{tf} (psia)	p_{wf} (psia)	$p_p\,(p_{wf})$ (psia^2/cp)
0	375.2	407.60	1.6173×10^7
4.288	371.2	403.13	1.5817×10^7
9.265	361.3	393.03	1.5032×10^7
15.552	343.8	375.79	1.3736×10^7
20.177	327.1	359.87	1.2591×10^7

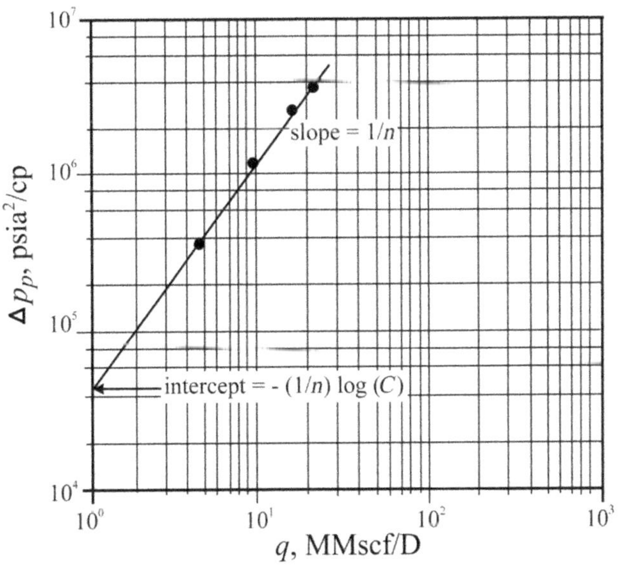

Fig 8.115—Rawlins-Schellhardt analysis, Example 8.10.

$$n = \frac{\log\left(q_4/q_1\right)}{\log\left(\Delta p_{p4}/\Delta p_{p1}\right)} = \frac{\log\,(20.177/4.28)}{\log\left(3.582\times10^6/3.560\times10^5\right)} = 0.671.$$

Now, calculate the AOF of the well. Because $0.5 \le n \le 1.0$, calculate C using either regression analysis or a point from the best-fit straight line through the test data. Estimating C with regression analysis results in $\log(C) = \alpha = -3.09$. Thus,

$$C = 10^{\alpha} = 10^{(-3.09)} = 8.13\times10^{-4}.$$

Estimating C using Point 4 from the best-fit line and Eq. 8.245:

$$C = \frac{q_4}{\left(\Delta p_{p4}\right)^n} = \frac{20.177}{\left(3.582\times10^6\right)^{0.671}} = 8.07\times10^{-4}.$$

Therefore, the AOF potential of this well is

TABLE 8.14—PLOTTING FUNCTIONS FOR THE RAWLINS-SCHELLHARDT ANALYSIS, EXAMPLE 8.10	
q (MMscf/D)	$\Delta p_p = p_p(p) - p_p(p_{wf})$ (psia²/cp)
4.288	3.560×10^5
9.265	1.1416×10^6
15.552	2.437×10^6
20.177	3.582×10^6

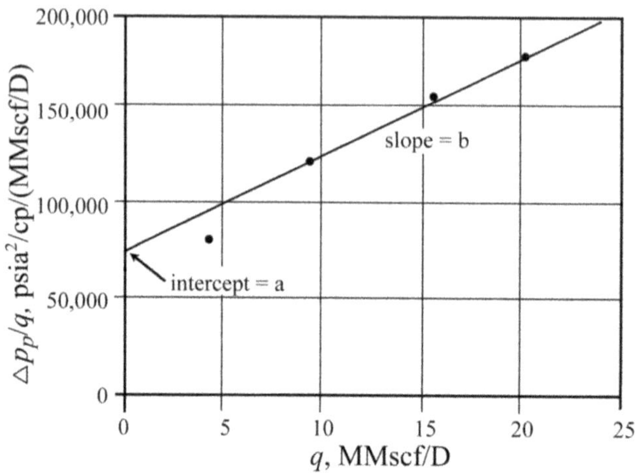

Fig. 8.116—Houpeurt analysis, Example 8.10.

$$q_{\text{AOF}} = C\left[p_p(\bar{p}) - p_p(14.65)\right]^n$$

$$= 8.13 \times 10^{-4}(1.6173 \times 10^7 - 2.6738 \times 10^3)^{0.67} = 55.0 \text{ MMscf/D at } \bar{p} = 407.6 \text{ psia.}$$

To update the AOF to a new average reservoir pressure, recall that for pseudopressure analysis, neither C nor n changes as drainage area pressure decreases. The AOF for the new drainage area pressure becomes

$$q_{\text{AOF}} = C\left[p_p(\bar{p}) - p_p(14.65)\right]^n$$

$$= 8.13 \times 10^{-4}(1.2239 \times 10^7 - 2.6748 \times 10^3)^{0.67} = 45.6 \text{ MMscf/D at } \bar{p} = 350 \text{ psia.}$$

Houpeurt Analysis. Plot $\Delta p_p/q$ vs. q on Cartesian graph paper (**Fig. 8.116**). **Table 8.15** gives the plotting functions. Construct the best-fit line through the last three data points. The first point, corresponding to the lowest flow rate, does not follow the trend and will be ignored in subsequent analyses.

Determine the deliverability coefficients, a and b, from a least-squares regression analysis, excluding the first point. The result is

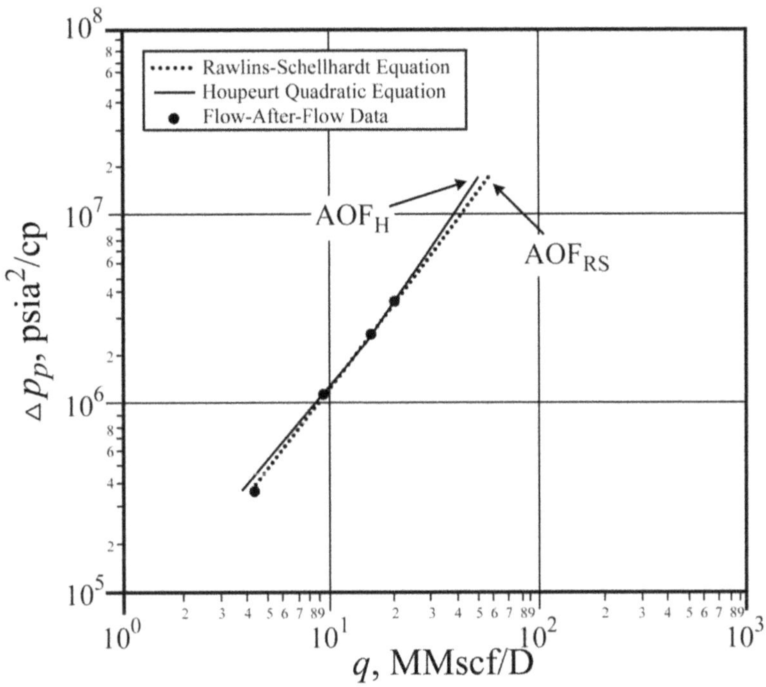

Fig. 8.117—Comparison of Rawlins-Schellhardt and Houpeurt methods.

TABLE 8.15—PLOTTING FUNCTIONS FOR THE HOUPEURT ANALYSIS, EXAMPLE 8.10

q (MMscf/D)	$\Delta p_p/q = [p_p(p) - p_p(p_{wf})]/q$, [(psia^2/cp)/(MMscf/D)]
4.288	0.8302×10^5
9.265	1.232×10^5
15.552	1.567×10^5
20.177	1.775×10^5

$$a = 7.75 \times 10^4 \text{psia}^2/\text{cp}/(\text{MMscf/D})$$

$$\text{and } b = 5.00 \times 10^3 \text{psia}^2/\text{cp}/(\text{MMscf/D})^2 .$$

Alternatively, use Points 2 and 4 from the line drawn through the test data to calculate a and b:

$$b = \frac{\left(\frac{\Delta p_p}{q}\right)_4 - \left(\frac{\Delta p_p}{q}\right)_2}{q_4 - q_2}$$

$$= \frac{1.775 \times 10^5 - 1.232 \times 10^5}{20.177 - 9.265} = 4.976 \times 10^3 \text{psia}^2/\text{cp}/(\text{MMscf/D})^2 .$$

$$a = \left(\frac{\Delta p_p}{q_1}\right)_1 - bq_1$$

$$= 1.775 \times 10^5 - (4{,}976)(20.177) = 7.71 \times 10^4 \text{psia}^2/\text{cp}/(\text{MMscf/D})^2 .$$

Then,

$$q_{AOF} = \frac{-a + \sqrt{a^2 + 4b\left[p_p(\bar{p}) - p_p(14.65)\right]}}{2b}$$

$$= -\left(7.75 \times 10^4\right).$$

$$+ \frac{\sqrt{\left(7.75 \times 10^4\right)^2 + 4\left(5.00 \times 10^3\right)\left[\left(1.62 \times 10^7\right) - 2{,}674.8\right]}}{2\left(5.00 \times 10^3\right)} = 49.7 \text{ MMscf/D at } \bar{p} = 407.6 \text{ psia.}$$

To update the AOF, note that for pseudopressure analysis neither a nor b changes as drainage area pressure changes. Therefore, the AOF for the new drainage area pressure is

$$q_{AOF} = \frac{-a + \sqrt{a^2 + 4b\left[p_p(\bar{p}) - p_p(14.65)\right]}}{2b}$$

$$= \frac{-(7.75 \times 10^4) + \sqrt{(7.75 \times 10^4)^2 + 4(5.00 \times 10^3)\left[(1.2239 \times 10^7) - 2{,}674.8\right]}}{2(5.00 \times 10^3)}$$

$$= 42.2 \text{ MMscf/D at } \bar{p} = 350 \text{ psia .}$$

A comparison (**Fig. 8.117**) of the results from the two parts of Example 8.10 shows that the Rawlins-Schellhardt equation appears to be valid for this range of test data; however, the line representing the Houpeurt equation deviates from the Rawlins-Schellhardt equation as BHFP decreases. Although the Rawlins-Schellhardt method is valid under many testing conditions, this deviation suggests that extrapolating the empirical equation over a large interval of pressure may not predict well behavior correctly.

Single-Point Tests. A single-point test is an attempt to overcome the limitation of long test times. A single-point test is conducted by flowing the well at a single rate until the sandface pressure is stabilized. One limitation of this test is that it requires prior knowledge of the well's deliverability behavior, either from previous well tests or possibly from correlations with other wells producing in the same field under similar conditions. Ensure that the well has flowed long enough to be out of wellbore storage and in the boundary-dominated or stabilized flow regime. Similarly, for hydraulically fractured wells, the well must be flowed long enough to be in the pseudoradial flow regime and then stabilized.

To analyze a single-point test with the Rawlins-Schellhardt method, n must be known or estimated. An estimate of n can be obtained either from a previous deliverability test on the well or from correlations with similar wells producing from the same formation under similar conditions. The calculation procedure is similar to that presented for flow-after-flow tests. The AOF can be estimated graphically by drawing a straight line through the single flow point with a slope of $1/n$ and extrapolating it to the flow rate at $\Delta p_p = p_p(\bar{p}) - p_p(p_b)$. The AOF can also be calculated with

$$q_{\text{AOF}} = C\left[p_p(\bar{p}) - p_p(p_b)\right]^n,$$

where C is estimated with

$$C = q/\left[p_p(\bar{p}) - p_p(p_{wf})\right]^n.$$

To use the Houpeurt analysis technique, the slope, b, of the line on a plot of

$$\Delta p_p/q = \left[p_p(\bar{p}) - p_p(p_{wf})\right]/q \text{ vs } q$$

must be known. If a value of b is unavailable, estimate b using Eq. 8.238. Note that estimates of the formation properties are necessary to use Eq. 8.238. The remaining analysis procedure is similar to that for flow-after-flow tests.

Isochronal Tests. The isochronal test[51] is a series of single-point tests developed to estimate stabilized deliverability characteristics without actually flowing the well for the time required to achieve stabilized conditions at each different rate. The isochronal test is conducted by alternately producing the well then shutting it in and allowing it to build to the average reservoir pressure before the beginning of the next production period. Pressures are measured at several time increments during each flow period. The times at which the pressures are measured should be the same relative to the beginning of each flow period. Because less time is required to build to essentially initial pressure after short flow periods than to reach stabilized flow at each rate in a flow-after-flow test, the isochronal test is more practical for low-permeability formations. A final stabilized flow point often is obtained at the end of the test. **Fig. 8.118** illustrates an isochronal test.

The isochronal test is based on the principle that the radius of drainage established during each flow period depends only on the length of time for which the well is flowed and not the flow rate. Consequently, the pressures measured at the same time periods during each different rate correspond to the same transient radius of drainage. Under these conditions, isochronal test data can be analyzed using the same theory as a flow-after-flow test, even though stabilized flow is not attained. In theory, a stabilized deliverability curve can be obtained from transient data if a single, stabilized rate and the corresponding BHP have been measured and are available.

The transient flow regime is modeled by

$$p_p(p_s) - p_p(p_{wf}) = \frac{1.422 \times 10^6 qT}{k_g h} \times \left[1.151 \log\left(\frac{k_g t}{1{,}688\phi\bar{\mu}_g\bar{c}_t r_w^2}\right) + s + Dq\right], \quad \text{(8.232)}$$

where p_s is the stabilized BHP measured before the test. The transient equation can be rewritten in a form similar to the stabilized equation for a circular drainage area. To start this process, write

$$p_p(p_s) - p_p(p_{wf}) = \frac{1.422 \times 10^6 qT}{k_g h} \times \left[\ln\left(\frac{1}{r_w}\right) + \ln\left(\frac{k_g t}{377\phi\bar{\mu}_g\bar{c}_t}\right)^{1/2} - \frac{3}{4} + s + Dq\right] \quad \text{(8.252)}$$

Further, a transient radius of drainage is defined as

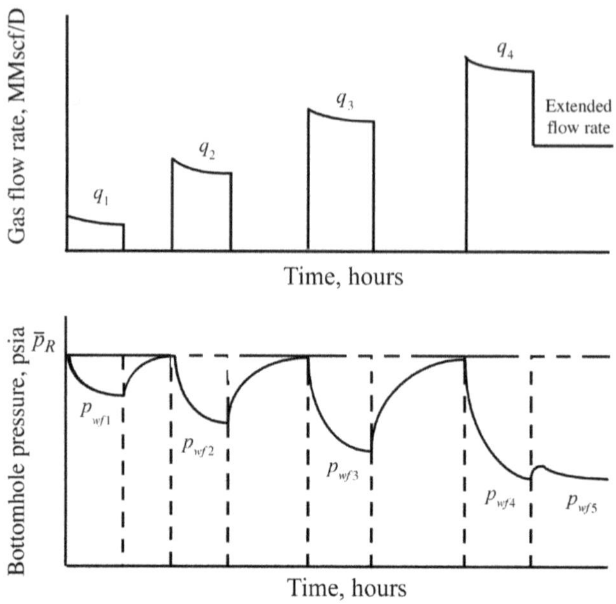

Fig. 8.118—Pressure and flow rate history of a typical isochronal test.

$$r_d = \sqrt{\frac{k_g t}{377\phi\bar{\mu}_g\bar{c}_t}} \quad \text{. .... (8.253)}$$

By substituting Eq. 8.253 into Eq. 8.252 and rearranging, the transient solution becomes

$$p_p(p_s) - p_p(p_{wf}) = \frac{1.422 \times 10^6 qT}{k_g h} \times \left[\ln\left(\frac{r_d}{r_w}\right) - \frac{3}{4} + s + Dq \right], \quad \text{......... (8.254)}$$

which is valid at any *fixed* time because r_d is a function of time and not of flow rate. r_d has no rigorous physical significance. It is simply the radius that forces the transient equation to resemble the pseudosteady-state equation. In addition, do not confuse r_d with r_i, which is the transient radius of investigation given by Eq. 8.246.

Similar to Houpeurt's equations, rewrite Eq. 8.254 as

$$p_p(p_s) - p_p(p_{wf}) = a_t q + bq^2, \quad \text{......... (8.255)}$$

where

$$a_t = \frac{1.422 \times 10^6 T}{k_g h}\left[\ln\left(\frac{r_d}{r_w}\right) - \frac{3}{4} + s \right] \quad \text{......... (8.256)}$$

$$\text{and } b = \frac{1.422 \times 10^6 DT}{k_g h} \quad \text{. .... (8.238)}$$

b is not a function of time and will remain constant. Similarly, the intercept a_t is constant for each fixed time line or isochron.

The theory of isochronal test analysis implies that the transient pressure drawdowns corresponding to the same elapsed time during each different flow period will plot as straight lines with the same slope b. The intercept a_t for each line will increase with increasing time. Therefore, draw a line with the same slope, b, through the final, stabilized data point, and use the coordinates of the stabilized point and the slope to calculate a stabilized intercept, a, independent of time, where (for radial flow) the stabilized flow coefficient is defined by

$$a = \frac{1.422 \times 10^6 T}{k_g h}\left[\ln\left(\frac{r_e}{r_w}\right) - \frac{3}{4} + s\right] . \quad (8.257)$$

Rawlins-Schellhardt Analysis. In logarithmic form, the empirical equation introduced by Rawlins and Schellhardt for analysis of flow-after-flow test data is

$$\log(q) = \log(C) + n\left\{\log\left[p_p(\bar{p}) - p_p(p_{wf})\right]\right\} . \quad (8.250)$$

For isochronal tests, plot transient data measured at different flow rates but taken at the same time increments relative to the beginning of each flow period. The lines drawn through data points corresponding to the same fixed flow time prove to be parallel, so the value of n is constant and independent of time. However, the intercept, log (C), is a function of time, so a different intercept must be calculated for each isochronal line. This "transient" intercept is log (C_t). In terms of this transient intercept, Eq. 8.248 becomes

$$\log(q) = \log(C_t) + n\left\{\log\left[p_p(p_s) - p_p(p_{wf})\right]\right\} . \quad (8.258)$$

$\bar{p}$ is replaced by p_s in the modified equation.

The conventional Rawlins-Schellhardt method of isochronal test analysis is to plot

$$\log\left[\Delta p_p = p_p(p_s) - p_p(p_{wf})\right] \text{ vs } \log(q)$$

for each time, giving a straight line of slope $1/n$ and an intercept of

$$\left\{-1/n\left[\log(C_t)\right]\right\} .$$

Houpeurt Analysis. Recall that the Houpeurt equation for analyzing flow-after-flow tests is

$$\frac{\Delta p_p}{q} = \frac{p_p(\bar{p}) - p_p(p_{wf})}{q} = a + bq . \quad (8.251)$$

Eq. 8.251 assumes stabilized flow conditions; however, in isochronal testing, measured transient data are being recorded. Consequently, for each isochronal (or fixed time) line, the equation for transient flow conditions is

$$\frac{\Delta p_p}{q} = \frac{p_p(p_s) - p_p(p_{wf})}{q} = a_t + bq, \quad (8.259)$$

where

$$a_t = \frac{1.422 \times 10^6 T}{k_g h}\left[\ln\left(\frac{r_d}{r_w}\right) - \frac{3}{4} + s\right], \quad \text{(8.256)}$$

$$\text{and } b = \frac{1.422 \times 10^6 DT}{k_g h}. \quad \text{(8.238)}$$

The form of Eq. 8.259 suggests that a plot of $\Delta p_p/q = [p_p(p_s) - p_p(p_{wf,s})]/q$ vs. q will yield a straight line with slope b and intercept a_t. This theory can then be extended to the stabilized point and calculate a stabilized intercept, a, using the coordinates of the stabilized point. The slope b remains the same.

Example 8.11: Analysis of Isochronal Tests. Estimate the AOF of this well[51] using both the Rawlins and Schellhardt and the Houpeurt analyses. **Table 8.16** summarizes the isochronal test data. Assume p_b= 14.65 psia.

$$\bar{p} \approx p_s = 352.4 \text{ psia}, \; p_p(\bar{p}) \approx p_p(p_s) = 9.9715 \times 10^6 \text{ psia}^2/\text{cp}.$$

$$p_b = 14.65 \text{ psia}, \; p_p(p_b) = 2{,}098.7 \text{ psia}^2/\text{cp}.$$

Solution.

Rawlins-Schellhardt Analysis Technique. First, plot $\Delta p_p = p_p(p_s) - p_p(p_{wf})$ vs. q on log-log coordinates (**Fig 8.119**) and include the single stabilized, extended flow point. **Table 8.17** gives the plotting functions.

Calculate the deliverability exponent, n, for each line or isochron using least-squares regression analysis. Note that, because the first data point for each isochron does not align with the data points at the last three flow rates (Fig. 8.119), the first data point is ignored in all subsequent calculations.

Table 8.18 summarizes the deliverability exponents determined with a least-squares regression analysis for each isochron. The arithmetic average of the n values in Table 8.18 is 0.89.

Because $0.5 \le \bar{n} \le 1.0$, AOF can be calculated or determined graphically using **Fig. 8.120**. AOF will be calculated in this example. First, determine the stabilized performance coefficient using the coordinates of the stabilized, extended flow point and $n = \bar{n}$:

$$C = \frac{q_s}{\left[p_p(p_s) - p_p(p_{wf,s})\right]^n} = \frac{1.156}{\left(2.443 \times 10^6\right)^{0.89}} = 2.39 \times 10^{-6}.$$

Then calculate the AOF potential:

$$\begin{aligned} q_{\text{AOF}} &= C\left[p_p(\bar{p}) - p_p(p_b)\right]^n \\ &= 2.39 \times 10^{-6}\left(9.9715 \times 10^6 - 2{,}098.7\right)^{0.89} = 4.04 \text{ MMscf/D}. \end{aligned}$$

To determine the AOF graphically, first calculate the pseudopressure at p_b and compute

$$\Delta p_p = p_p(p_s) - p_p(p_b) = 9.9715 \times 10^6 - 2{,}098.7 = 9.969 \times 10^6.$$

Then, draw a line of slope $1/\bar{n}$ through the stabilized flow point, extrapolate the line to the flow rate at $\Delta p_p = p_p(p_s) - p_p(p_b)$, and read the AOF directly from the graph. The result is $q_{\text{AOF}} = 4.04$ MMscf/D.

Houpeurt Analysis Technique. Plot $\Delta p_p/q = [p_p(p_s) - p_p(p_{wf})]/q$ vs. q on Cartesian graph paper (**Fig. 8.121**). **Table 8.19** gives the plotting functions. Construct best-fit lines through the isochronal data points for each time. Note that, for each flow time, the point corresponding to the lowest rate does fit on the same straight line, so all four data points will be used for the analysis of each isochron.

Next, determine the slope b of each line or isochron. Values of b from least-squares regression analysis are summarized in **Table 8.20**. The arithmetic average value of the slopes in Table 8.20 is 2.074×10^4 psia²/cp/(MMscf/D)².

Calculate the stabilized isochronal deliverability line intercept using $\Delta p_p/q = 2.113 \times 10^6$ psia²/cp/(MMscf/D) at the extended, stabilized point.

$$a - (\Delta p_p / q) - bq$$

$$= 2.113 \times 10^6 - (2.074 \times 10^4)(1.156) = 2.109 \times 10^6 \text{ psia}^2/\text{cp}/(\text{MMscf/D}) .$$

Calculate the AOF potential using the average value of b and the stabilized value of a.

$$q_{\text{AOF}} = \frac{-a + \sqrt{a^2 + 4b\left[p_p(p_s) - p_p(p_b)\right]}}{2b}$$

$$= \frac{-\left(2.109 \times 10^6\right) + \sqrt{\left(2.109 \times 10^6\right)^2 + 4\left(2.074 \times 10^4\right)\left(9.97 \times 10^6\right)}}{2\left(2.074 \times 10^4\right)} = 4.53 \text{ MMscf/D}.$$

Fig. 8.122 illustrates the results.

Modified Isochronal Tests. The time to build up to the average reservoir pressure before flowing for a certain period of time still may be impractical, even after short flow periods. Consequently, a modification of the isochronal test was developed[52] to shorten test times further. The objective of the modified isochronal test is to obtain the same data as in an isochronal test without using the sometimes lengthy shut-in periods required to reach the average reservoir pressure in the drainage area of the well.

The modified isochronal test (**Fig. 8.123**) is conducted like an isochronal test, except the shut-in periods are of equal duration. The shut-in periods should equal or exceed the length of the flow periods. Because the well does not build up to average reservoir pressure after each flow period, the shut-in sandface pressures recorded immediately before each flow period rather than the average reservoir pressure are used in the test analysis. As a result, the modified isochronal test is less accurate than the isochronal test. As the duration of the shut-in periods increases, the accuracy of the modified isochronal test also increases. Again, a final stabilized flow point usually is obtained at the end of the test but is not required for analyzing the test data.

The well does not build up to the average reservoir pressure during shut-in; the analysis techniques for the modified isochronal tests are derived intuitively. Recall the transient flow equation, expressed in terms of the reservoir pressure at the start of flow, on which isochronal testing is based:

TABLE 8.16—ISOCHRONAL TEST DATA, EXAMPLE 8.11

Time (hours)	q (MMscf/D)	p_{wf} (psia)	$p_p(p_{wf})$ (psia^2/cp)
0.5	0.983	344.7	9.6386×10^6
1.0	0.977	342.4	9.5406×10^6
2.0	0.970	339.5	9.4179×10^6
3.0	0.965	337.6	9.3381×10^6
0.5	2.631	329.5	9.0027×10^6
1.0	2.588	322.9	8.7351×10^6
2.0	2.533	315.4	8.4371×10^6
3.0	2.500	310.5	8.2458×10^6
0.5	3.654	318.7	8.5674×10^6
1.0	3.565	309.5	8.2071×10^6
2.0	3.453	298.6	7.7922×10^6
3.0	3.390	291.9	7.5435×10^6
0.5	4.782	305.5	8.0534×10^6
1.0	4.625	293.6	7.6136×10^6
2.0	4.438	279.6	7.0990×10^6
3.0	4.318	270.5	6.7797×10^6
214 (Extended Flow Point)	1.156	291.6	7.5285×10^6

TABLE 8.17—PLOTTING FUNCTIONS FOR RAWLINS-SCHELLHARDT ANALYSIS, EXAMPLE 8.11

t (hours)	q (MMscf/D)	Δp_p (psia^2/cp)	t (hours)	q (MMscf/D)	Δp_p (psia^2/cp)
0.5	0.983	0.3329×10^6	2.0	3.654	1.4041×10^6
	0.977	0.4309×10^6		3.565	1.7644×10^6
	0.970	0.5536×10^6		3.453	2.1793×10^6
	0.965	0.6334×10^6		3.390	2.4280×10^6
1.0	2.631	0.9688×10^6	3.0	4.782	1.9181×10^6
	2.588	1.2364×10^6		4.625	2.3579×10^6
	2.533	1.5344×10^6		4.438	2.8725×10^6
	2.500	1.7257×10^6		4.318	3.1918×10^6
			214	1.156	2.443×10^6

$$p_p(p_s) - p_p(p_{wf}) = \frac{1.422\times10^6 qT}{k_g h}\left[\ln\left(\frac{r_d}{r_w}\right) - \frac{3}{4} + s + Dq\right]. \quad \text{(8.254)}$$

In new reservoirs with little or no pressure depletion, p_s equals the initial reservoir pressure ($p_s = p_i$); in developed reservoirs, $p_s < p_i$. In addition, the transient drainage radius, r_d, in Eq. 8.254 is defined as

TABLE 8.18—DELIVERABILITY EXPONENTS FOR VARIOUS FLOWING TIMES, EXAMPLE 8.10

t (hours)	n
0.5	0.88
1.0	0.91
2.0	0.89
3.0	0.88

TABLE 8.19—PLOTTING FUNCTIONS FOR HOUPEURT ANALYSIS, EXAMPLE 8.11

t (hours)	q (MMscf/D)	$\Delta p_p/q$ (psia^2/cp/MMscf/D)	t (hours)	q (MMscf/D)	$\Delta p_p/q$ (psia^2/cp/MMscf/D)
0.5	0.983	3.387×10^5	2.0	0.970	5.707×10^5
	2.631	3.682×10^5		2.533	6.058×10^5
	3.654	3.843×10^5		3.453	6.311×10^5
	4.782	4.011×10^5		4.438	6.473×10^5
1.0	0.977	4.410×10^5	3.0	0.965	6.564×10^5
	2.588	4.777×10^5		2.500	6.903×10^5
	3.565	4.949×10^5		3.390	7.162×10^5
	4.625	5.098×10^5		4.318	7.392×10^5

TABLE 8.20—SLOPES, b, OF THE ISOCHRONS, EXAMPLE 8.11

t (hours)	b [psia^2/cp/(MMscf/D)^2]
0.5	1.644×10^4
1.0	1.904×10^4
1.5	2.255×10^4
2.0	2.492×10^4

$$r_d = \sqrt{\frac{k_g t}{377\phi \bar{\mu}_g \bar{c}_t}} \quad \text{. ......} \quad (8.253)$$

Because r_d is a function of time and not of flow rate, Eq. 8.254 is valid at any fixed time. For modified isochronal tests, use Eq. 8.254, in which the stabilized shut-in BHP, p_s, is replaced with shut-in BHP, p_{ws}, measured before each flow period, where $p_{ws} \le p_s$,

$$p_p(p_{ws}) - p_p(p_{wf}) = \frac{1.422\times10^6 qT}{k_g h}\left[\ln\left(\frac{r_d}{r_w}\right) - \frac{3}{4} + s + Dq\right] \quad \text{. ......} \quad (8.260)$$

Eq. 8.260 can be rewritten as

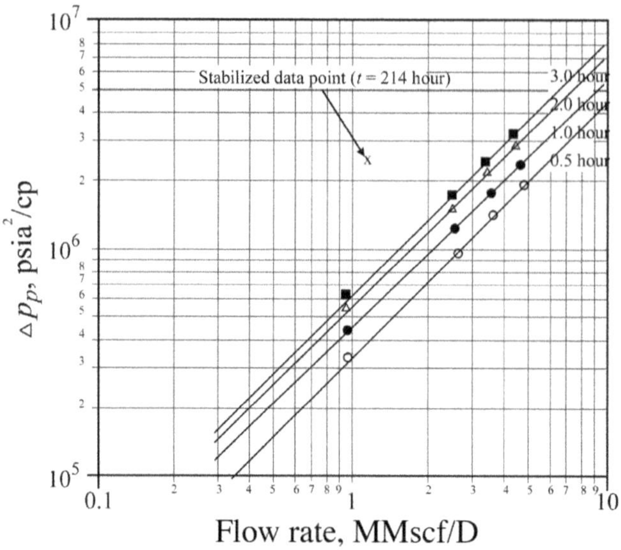

Fig. 8.119—Rawlins-Schellhardt analysis, Example 8.11.

$$p_p(p_{ws}) - p_p(p_{wf}) = a_t q + bq^2, \quad \text{(8.261)}$$

$$\text{where } a_t = \frac{1.422 \times 10^6 T}{k_g h}\left[\ln\left(\frac{r_d}{r_w}\right) - \frac{3}{4} + s\right] \quad \text{(8.256)}$$

$$\text{and } b = \frac{1.422 \times 10^6 DT}{k_g h}. \quad \text{(8.238)}$$

Eq. 8.238 indicates that b is independent of time and will remain constant during the test. Similarly, Eq. 8.256 indicates that a_t is constant for a fixed time. The similarity of Eqs. 8.254 and 8.260 for the isochronal and modified isochronal tests, respectively, suggests that the modified isochronal test data can be analyzed like those from an isochronal test.

The theory developed for the modified isochronal test implies that, if the intuitive approximation of using p_{ws} instead of p_s is valid, the transient data will plot as straight line for each time with the same slope, b. The intercept, a_t, will increase with increasing time. By drawing a line with slope b through the stabilized data point and using the coordinates of the stabilized point and the slope, a stabilized intercept, a, that is independent of time can be calculated, where

$$a = \frac{1.422 \times 10^6 T}{k_g h}\left[\ln\left(\frac{r_e}{r_w}\right) - \frac{3}{4} + s\right]. \quad \text{(8.257)}$$

To calculate the AOF of the well, use the average reservoir pressure, p_s, measured before the test instead of the p_{ws} value, or

$$q_{AOF} = \frac{-a + \sqrt{a^2 + 4b\left[p_p(p_s) - p_p(p_b)\right]}}{2b}. \quad \text{(8.262)}$$

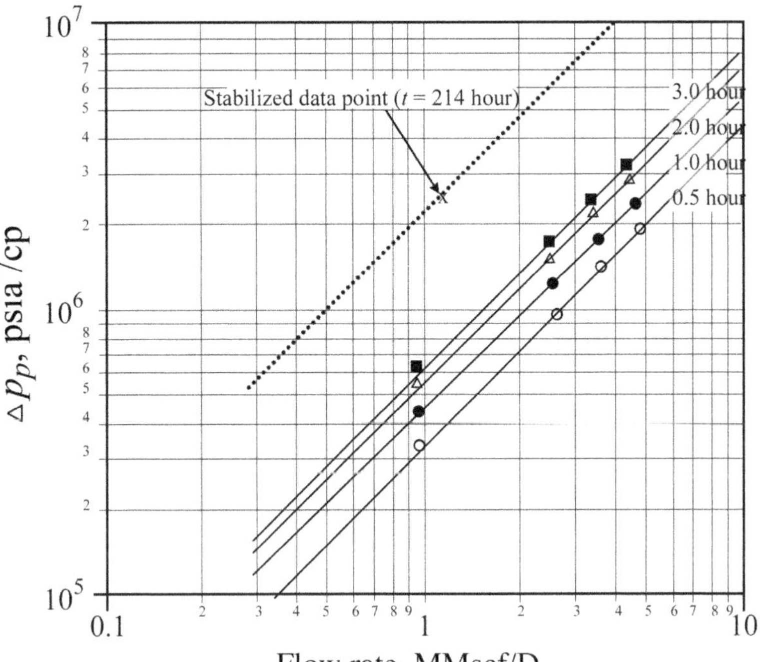

Fig. 8.120—Rawlins-Schellhardt analysis result, Example 8.11.

Two variations of the modified isochronal test are considered: tests with a stabilized flow point obtained at the end of the test and tests run without that final point.

Modified Isochronal Tests With a Stabilized Flow Point. Rawlins-Schellhardt Analysis. Recall the empirical Rawlins and Schellhardt equation in terms of transient isochronal test data:

$$\log(q) = \log(C_t) + n\left\{\log\left[p_p(\bar{p}_s) - p_p(p_{wf})\right]\right\}. \quad (8.258)$$

As in the graphical analysis techniques for isochronal tests, plot several trends of data taken at different times during a modified isochronal test. The slope n of each line through points at equal time values will be constant. However, the intercept, $\log(C_t)$, is a function of time but not flow rate. Therefore, a different intercept should be calculated for each isochronal test. Use $p_p(p_{ws})$ instead of $p_p(p_s)$ in Eq. 8.258, which gives

$$\log(q) = \log(C_t) + n\left\{\log\left[p_p(p_{ws}) - p_p(p_{wf})\right]\right\}. \quad (8.263)$$

The conventional analysis technique for modified isochronal test data is to plot log $[p_p(p_{ws}) - p_p(p_{wf})]$ vs. log (q) for each time, giving a straight line of slope $1/n$ and an intercept of $\{-1/n [\log(C_t)]\}$. The Rawlins-Schellhardt analysis procedure for modified isochronal tests with a stabilized flow point is similar to that presented for isochronal tests, except the plotting functions are developed in terms of the shut-in pressure measured immediately before the next flow period. Only the stabilized, extended flow point is plotted in terms of the average reservoir pressure measured before the test, p_s. Example 8.12 illustrates the procedure.

Houpeurt Analysis. As shown previously, the Houpeurt deliverability equation in terms of transient isochronal test data is

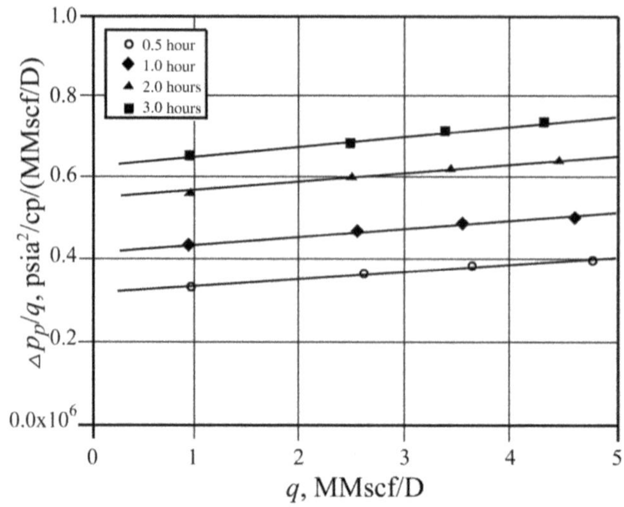

Fig. 8.121—Houpeurt analysis of isochronal test data, Example 8.11.

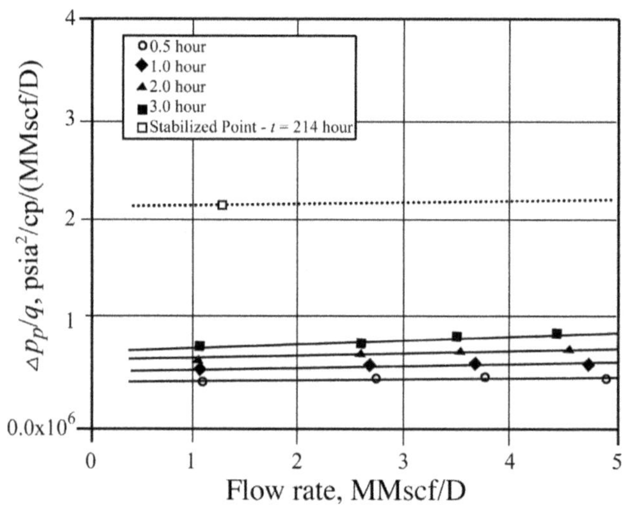

Fig. 8.122—Houpeurt analysis of isochronal test data result, Example 8.11.

$$\frac{\Delta p_p}{q} = \frac{p_p(p_s) - p_p(p_{wf})}{q} = a_t + bq \; . \qquad (8.259)$$

For modified isochronal test data, Eq. 8.259 should be modified with the assumption that $p_p(p_{ws})$ can be used instead of $p_p(p_s)$. With this assumption, Eq. 8.259 becomes

$$\frac{\Delta p_p}{q} = \frac{p_p(p_{ws}) - p_p(p_{wf})}{q} = a_t + bq, \qquad (8.264)$$

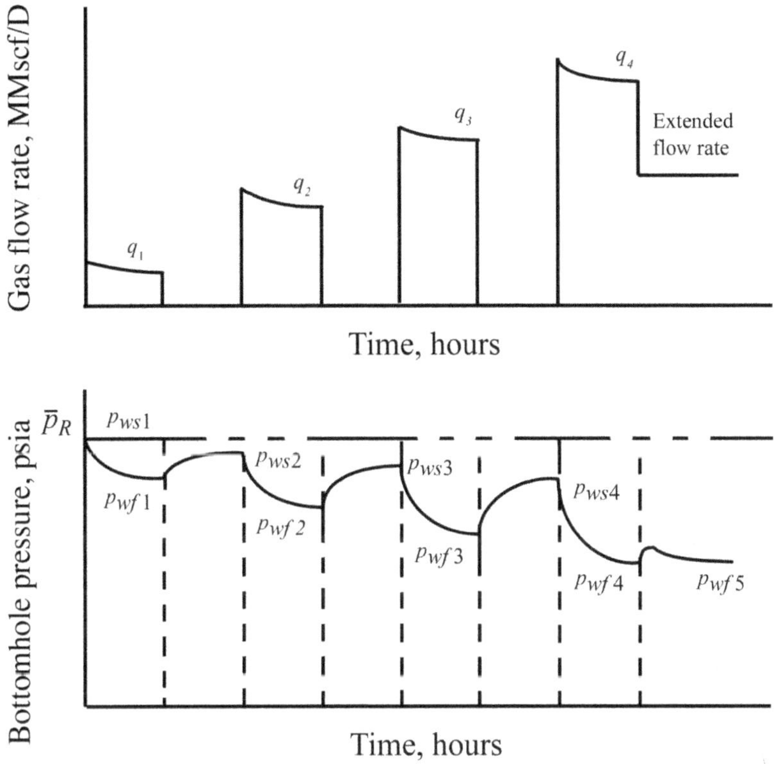

Fig. 8.123—Pressure and flow rate history of a typical modified isochronal test.

$$\text{where } a_t = \frac{1.422 \times 10^6 T}{k_g h}\left[\ln\left(\frac{r_d}{r_w}\right) - \frac{3}{4} + s\right], \quad \text{(8.256)}$$

$$\text{and } b = \frac{1.422 \times 10^6 DT}{k_g h}. \quad \text{(8.238)}$$

The form of Eq. 8.264 suggests that a plot of

$$\Delta p_p / q = \left[p_p(p_{ws}) - p_p(p_{wf})\right] / q \text{ vs. } q$$

will be a straight line with a slope b and intercept a_t. This theory can be extended to the stabilized point, and we can calculate a stabilized intercept, a, using the coordinates of the stabilized point, or

$$a = \frac{\left[p_p(p_s) - p_p(p_{wf,s})\right]}{q_s} - bq_s = \frac{\Delta p_p}{q_s} - bq_s. \quad \text{(8.265)}$$

The slope b of the line through the stabilized point should remain the same. In addition, the average reservoir pressure, which is measured before the test, must be used to evaluate the pseudopressure, $p_p(p_s)$ in Eq. 8.265. Example 8.12 illustrates the Houpeurt analysis procedure for modified isochronal tests with a stabilized flow point, which is similar to that presented for isochronal tests.

Example 8.12: Analysis of a Modified Isochronal Test With a Stabilized Flow Point. Using the following data taken from Well 4,[53] calculate the AOF using both Rawlins and Schellhardt and Houpeurt analysis techniques. Assume p_b = 14.65 psia, where $p_p(p_b)$ = 5.093×10^7 psia²/cp. **Table 8.21** gives the test data. h = 6 ft, r_w = 0.1875 ft, ϕ = 0.2714, T = 540°R (80°F), $\bar{p} \approx p_s = 706.6$ psia, $\bar{\mu}_g = 0.015$ cp, $\bar{z} = 0.97$, $\bar{c}_g = 1.5 \times 10^{-3}$ psia^{-1}, γ_g = 0.75, S_w = 0.30, $c_f = 3 \times 10^{-6}$ psia^{-1}, and A = 640 acres (assume that the well is centered in a square drainage area).

Solution.

Rawlins-Schellhardt Analysis. Plot

$$\Delta p_p = p_p(p_{ws}) - p_p(p_{wf}) \text{ vs. } q$$

on log-log graph paper (**Fig. 8.124**). **Table 8.22** gives the plotting functions. In addition, plot on the same graph the values of Δp_p that corresponds to the stabilized, extended flow point evaluated at p_s.

$$\Delta p_p = p_p(p_s) - p_p(p_{wf})$$
$$= 5.093 \times 10^7 - 3.276 \times 10^7 = 1.817 \times 10^7 \text{psia}^2/\text{cp} .$$

For each time, construct the best-fit line through the data points. Because the first data points for each isochron do not follow the trend of the higher rate points, they will be ignored for all subsequent calculations.

Calculate the deliverability exponent, n, for each line or isochron. For this example, use least-squares regression analysis. For example, at t = 0.5 hours, n_1 = 0.72. **Table 8.23** summarizes the deliverability exponents.

The arithmetic average of the values in Table 8.23 is

$$\bar{n} = \frac{n_1 + n_2 + n_3 + n_4}{4} = \frac{0.72 + 0.74 + 0.74 + 0.78}{4} = 0.74.$$

Because $0.5 \le \bar{n} \le 1.0$, determine the stabilized performance coefficient, C, using the coordinates of the stabilized, extended flow point and $n = \bar{n}$. Note that the pseudopressure used to calculate the stabilized C value is evaluated at p_s measured at the beginning of the test, rather than p_{ws}. From Eq. 8.245,

$$C = \frac{q}{\left[p_p(p_s) - p_p(p_{wf})\right]^n} = \frac{2.665}{(5.093 \times 10^7 - 3.276 \times 10^7)^{0.74}} = 1.132 \times 10^{-5} .$$

Then,

$$q_{\text{AOF}} = C\left[p_p(p_s) - p_p(p_b)\right]^n$$
$$= 1.132 \times 10^{-5}(5.0935 \times 10^7 - 2{,}766.6)^{0.74} = 5.7 \text{ MMscf/D}.$$

To determine the AOF graphically draw a line of slope $1/\bar{n}$ through the extended flow point, extrapolate the line to the flow rate at $\Delta p_p = p_p(p_s) - p_p(p_b)$, and read the AOF directly from the graph (**Fig. 8.125**).

Houpeurt Analysis. Plot

$$\Delta p_p / q = \left[p_p(p_{ws}) - p_p(p_{wf}) \right] / q \text{ vs. } q$$

on Cartesian coordinates (**Fig 8.126**). In addition, plot the $\Delta p_p / q$ value that corresponds to the stabilized, extended flow point. **Table 8.24** gives the plotting functions. Construct best-fit lines through the modified isochronal data points for each time. The first data point at the lowest rate for each isochron does not fit on the same straight line as the last three rate points and is ignored in subsequent calculations.

Determine the slopes of the lines, b, for each isochron by least-squares regression analysis of the best-fit lines through the data points. For example, at t = 0.5 hours, b_1 = 9.654 × 10^5 psia^2/cp/(MMscf/D)^2. **Table 8.25** summarizes the slopes of the isochrons. The average arithmetic values of the slopes in **Table 12.15** is

$$\bar{b} = \frac{b_2 + b_3 + b_4}{3} = \frac{(8.678 + 8.711 + 8.780) \times 10^5}{3} = 8.723 \times 10^5 \text{ psia}^2/\text{cp}/(\text{MMscf/D})^2 .$$

Calculate the stabilized isochronal deliverability line intercept, a:

$$a = \left[\frac{p_p(p_s) - p_p(p_{wf})}{q} \right] - bq$$

$$= \frac{1.817 \times 10^7}{2.665} - (8.732 \times 10^5)(2.665) = 4.493 \times 10^6 \text{ psia}^2/\text{cp}/(\text{MMscf/D}) .$$

Calculate the AOF potential using $\bar{b}$ and the stabilized a value:

$$q_{\text{AOF}} = \frac{-a + \sqrt{a^2 + 4b[p_p(\bar{p}) - p_p(14.65)}}{2b}$$

$$= \frac{-(4.493 \times 10^6) + \sqrt{(4.493 \times 10^6)^2 + 4(8.723 \times 10^5)\left[5.093 \times 10^7 - 2{,}766.6\right]}}{2(8.723 \times 10^5)} = 5.5 \text{ MMscf/D}.$$

Fig. 8.127 shows the data for this example.

Modified Isochronal Tests Without a Stabilized Flow Point. Because the well is not required to build up to the average reservoir pressure between the flow periods, the modified isochronal approximation shortens test times considerably. However, the test analysis relies on obtaining one stabilized flow point. Under some conditions, environmental or economic concerns prohibit flaring produced gas to the atmosphere during a long production period, thus preventing measurement of a stabilized flow point. These conditions often occur when new wells are tested before being connected to a pipeline.

TABLE 8.21—MODIFIED ISOCHRONAL TEST DATA, EXAMPLE 8.12

Time (hours)	p_{wf} (psia) q =1.520 MMscf/D	q =2.041 MMscf/D	q =2.688 MMscf/D	q =3.122 MMscf/D
0 (p_{ws})	706.6	706.6	703.5	701.2
0.5	655.6	724.5	578.5	541.7
1.0	653.6	620.7	573.9	537.8
1.5	652.1	619.9	572.3	536.3
2.0	651.3	619.1	570.8	534.7
	p_p (p_{wf}) (psia)			
0 [p_p (p_{ws})]	5.093×10^7	5.093×10^7	5.093×10^7	5.015×10^7
0.5	4.379×10^7	3.970×10^7	3.403×10^7	2.979×10^7
1.0	4.352×10^7	3.922×10^7	3.348×10^7	2.936×10^7
1.5	4.332×10^7	3.911×10^7	3.330×10^7	2.919×10^7
2.0	4.321×10^7	3.901×10^7	3.312×10^7	2.902×10^7

Extended Flow Point

p_{wf} = 567.7 psia
t= 24 hours

p_s = 706.6 psia
p_p (p_{wt}) = 3.276×10^8 psia^2/cp

q = 2.665 MMscf/D
p_p = (14.65) = 2,766.6 psia^2/cp

TABLE 8.22—PLOTTING FUNCTIONS FOR RAWLINS-SCHELLHARDT ANALYSIS, EXAMPLE 8.12

Time (hours)	Δp_p (psia^2/cp) q =1.520 (MMscf/D)	q =2.041 (MMscf/D)	q =2.688 (MMscf/D)	q =3.122 (MMscf/D)
0.5	0.714×10^7	1.123×10^7	1.645×10^7	2.039×10^7
1.0	0.714×10^7	1.171×10^7	1.700×10^7	2.082×10^7
1.5	0.761×10^7	1.182×10^7	1.718×10^7	2.099×10^7
2.0	0.772×10^7	1.192×10^7	1.736×10^7	2.113×10^7

Two methods have been developed to analyze modified isochronal tests without a stabilized flow point. The Brar and Aziz method[53] was developed for the Houpeurt analysis, while the stabilized C method[54] was developed for the Rawlins and Schellhardt analysis. The stabilized C method requires prior knowledge of permeability and skin factor or determination of these properties using the methods Brar and Aziz proposed for analyzing modified isochronal tests. Both methods require knowledge of the drainage area shape and size.

Brar and Aziz Method—Houpeurt Analysis. The Brar and Aziz method[53] is based on the transient Houpeurt deliverability Eqs. 8.234, 8.236, 8.238, and p_s, the stabilized BHP measured before the deliverability test.

Rewriting Eq. 8.236 as

$$a_t = m' \log(t) + c'_t \quad \text{(8.266)}$$

TABLE 8.23—SLOPES, *n*, OF ISOCHRONS, EXAMPLE 8.12

t (hours)	n
0.5	0.72
1.0	0.74
1.5	0.74
2.0	0.78

TABLE 8.24—PLOTTING FUNCTIONS FOR HOUPEURT ANALYSIS, EXAMPLE 8.12

	$\Delta p_p/q$ (psia²/cp/MMscf/D)			
Time (hours)	q =1.520 (MMscf/D)	q =2.041 (MMscf/D)	q =2.688 (MMscf/D)	q =3.122 (MMscf/D)
0.5	0.470×10^7	0.550×10^7	0.612×10^7	0.653×10^7
1.0	0.488×10^7	0.574×10^7	0.632×10^7	0.667×10^7
1.5	0.501×10^7	0.579×10^7	0.639×10^7	0.672×10^7
2.0	0.508×10^7	0.584×10^7	0.646×10^7	0.678×10^7

TABLE 8.25—SLOPES, *b*, OF ISOCHRONS, EXAMPLE 8.12

t (hours)	b (psia²/cp/(MMscf/D)²)
0.5	8.654×10^5
1.0	8.678×10^5
1.5	8.711×10^5
2.0	8.780×10^5

$$\text{where } m' = \frac{1.632\times10^6 T}{k_g h} \quad \text{(8.267)}$$

$$\text{and } c' = \frac{1.422\times10^6 T}{k_g h}\times\left[1.151\log\left(\frac{k_g}{1{,}688\phi\bar{\mu}_g\bar{c}_t r_w^2}\right)+s\right]. \quad \text{(8.268)}$$

m' and c' can be calculated using regression analysis of Eq. 8.266. Alternatively, these variables can be computed directly from the slope and the intercept of a plot of a_t vs. log t. Then calculate the permeability from the slope,

$$k_g = \frac{1.632\times10^6 T}{m'h}. \quad \text{(8.269)}$$

Combining Eqs. 8.267 and 8.268 yields an equation for the skin factor,

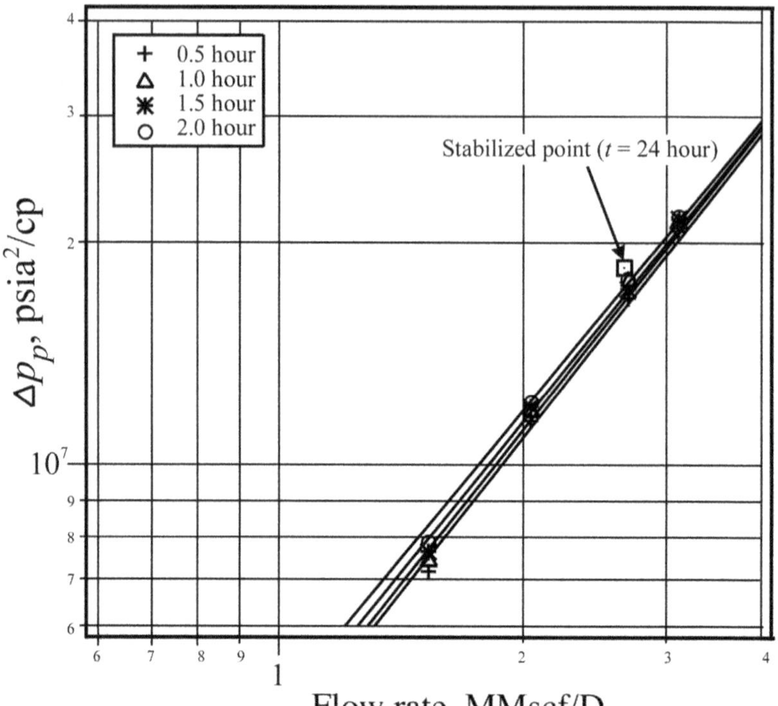

Fig. 8.124—Rawlings-Schellhardt plot of modified isochronal test data, Example 8.12.

$$s = 1.151\left[\left(\frac{c'}{m'}\right) - \log\left(\frac{k_g}{\phi \bar{\mu}_g \bar{c}_t r_w^2}\right) + 3.23\right] \quad (8.270)$$

Estimating the AOF potential of the well requires a stabilized value of a. If the drainage area size and shape are known, the gas permeability calculated from Eq. 8.269 and the skin factor from Eq. 8.270 can be used to calculate a:

$$a = \frac{1.422 \times 10^6 T}{k_g h}\left[1.151 \log\left(\frac{10.06A}{C_A r_w^2}\right) - \frac{3}{4} + s\right] \quad (8.271)$$

Table 8.A-1 gives shape factors for various reservoir shapes and well locations. The stabilized value of a then is used in Eq. 8.262 to calculate the AOF of the well:

$$q_{\mathrm{AOF}} = \frac{-a + \sqrt{a^2 + 4b\left[p_p(p_s) - p_p(p_b)\right]}}{2b} \quad (8.262)$$

Stabilized C Method—Rawlins-Schellhardt Analysis. Although the Houpeurt equation has a theoretical basis and is rigorously correct, the more familiar but empirically based Rawlins and Schellhardt equation continues to be used and is indeed favored by many in the natural gas industry. The Houpeurt and Rawlins-Schellhardt analysis techniques are combined here to develop a version of the Rawlins-Schellhardt method for analyzing modified isochronal tests.

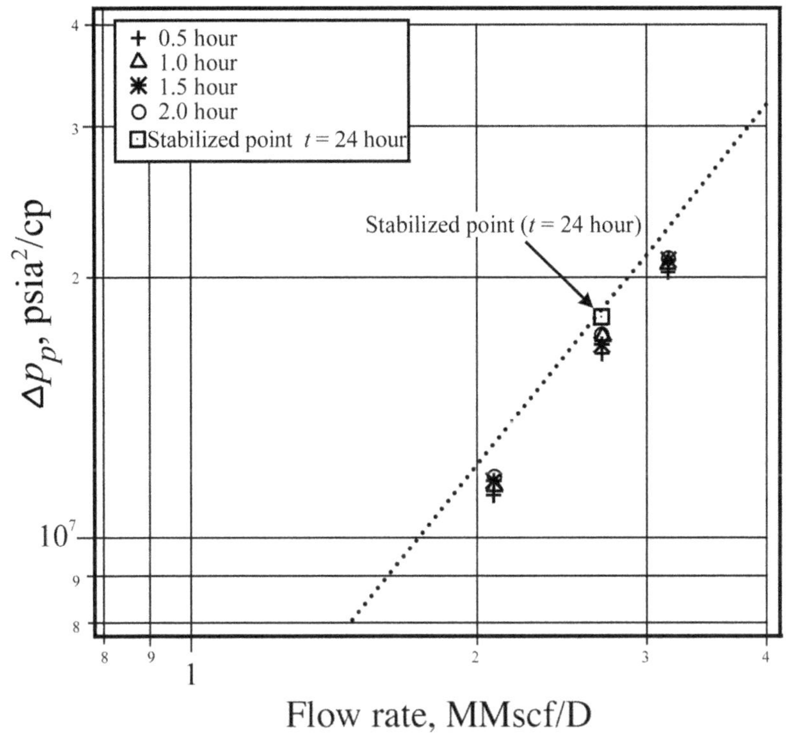

Fig. 8.125—Rawlins-Schellhardt analysis of modified isochronal test, Example 8.12.

This analysis technique, called the "Stabilized C" method,[54] is derived by equating the stabilized Rawlins and Schellhardt empirical backpressure equation with the stabilized theoretical Houpeurt equation to obtain equations for the deliverability exponent, n, and the stabilized flow coefficient, C, in terms of the Houpeurt flow coefficients, a and b.

To obtain an equation for the exponent n, take the logarithm of both sides of the stabilized Rawlins and Schellhardt empirical backpressure equation (Eq. 8.245).

$$\ln q = n \ln\left(\Delta p_p\right) + \ln C \ . \qquad (8.272)$$

n is the slope of a plot of $\ln(q)$ vs. $\ln(\Delta p_p)$. Alternatively, note that n can be expressed as the derivative of $\ln(q)$ with respect to $\ln(\Delta p_p)$:

$$\frac{d \ln q}{d \ln\left(\Delta p_p\right)} = \frac{1}{q}\frac{dq}{d\left[\ln\left(\Delta p_p\right)\right]} = n \ . \qquad (8.273)$$

Similarly, take the logarithms of both sides of the Houpeurt Eq. 8.235

$$\ln \Delta p_p = \ln\left[aq + bq^2\right], \qquad (8.274)$$

and, thus,

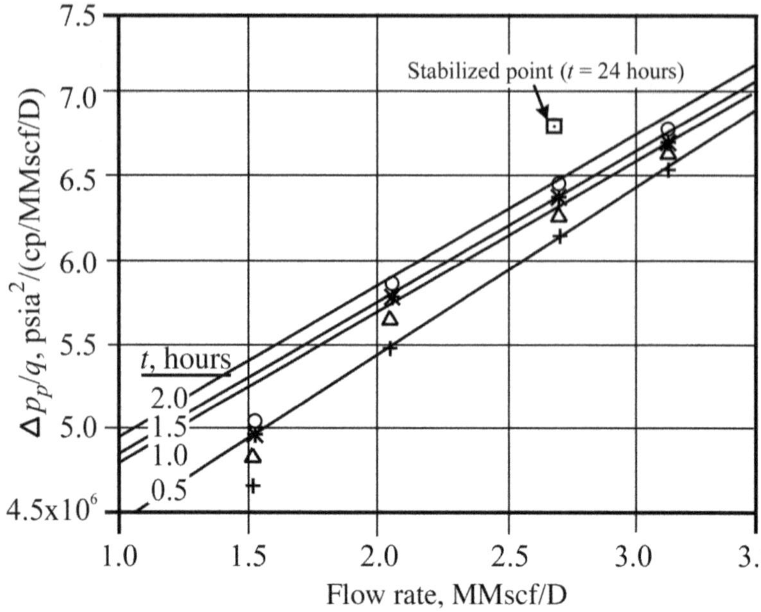

Fig. 8.126—Houpeurt plot of modified isochronal test data, Example 8.12.

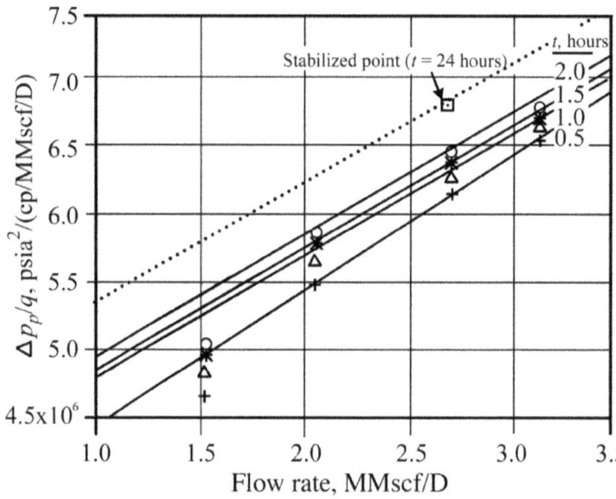

Fig. 8.127—Houpeurt analysis of modified isochronal test data, Example 8.12.

$$\frac{d\ln\left(\Delta p_p\right)}{dq} = \frac{(a+2bq)}{aq+bq^2}, \quad (8.275)$$

$$\text{or } \frac{dq}{d\left(\ln \Delta p_p\right)} = \frac{aq+bq^2}{a+2bq} = qn \quad (8.276)$$

$$\text{and } n = \frac{a+bq}{a+2bq}. \quad (8.277)$$

In Eq. 8.277, let q be the unique value q_e at which the $d \ln(\Delta p_p)/dq$ values from the Rawlins-Schellhardt and Houpeurt equations are identical. Solving Eq. 8.277 for this value of $q = q_e$,

$$n = \frac{a + bq_e}{a + 2bq_e} \quad (8.278)$$

$$\text{and } q_e = \frac{a(1-n)}{b(2n-1)} \quad (8.279)$$

Substituting in the Rawlins-Schellhardt equation and noting that, from the Houpeurt equation $(\Delta p_p)_e = aq_e + bq_e^2$,

$$q_e = C(\Delta p_p)_e^n = C(aq_e + bq_e^2)^n \quad (8.280)$$

Rearranging,

$$C = \frac{q_e}{(aq_e + bq_e^2)^n} \quad (8.281)$$

To apply the stabilized C method, it is necessary to assume that the slope, n, of the Rawlins-Schellhardt deliverability plot is constant. This assumption implies that if values of a and b can be calculated for given reservoir properties, a flow rate can be calculated from Eq. 8.279, at which the change in pseudopressures calculated by the Rawlins-Schellhardt equation is equal to the change in pseudopressure calculated by the Houpeurt equation. The substitution this flow rate into Eq. 8.281 allows calculation of a stabilized value of C and this value of C can be used to calculate a value of AOF:

$$q_{\text{AOF}} = C\left[p_p(p) - p_p(p_b)\right]^n \quad (8.282)$$

The stabilized C method is limited by the need for values of reservoir properties determined separately from the deliverability test analysis. These properties can be estimated either from drawdown or buildup test analysis or from the Brar and Aziz method.

Example 8.13: Analysis of Modified Isochronal Test Without a Stabilized Data Point. The purpose of this example is to compare results obtained from the analysis of a modified isochronal test (see **Table 8.26**) with and without an extended, stabilized data point. Calculate the AOF for the following modified isochronal test data without the extended flow point. Use both the Brar and Aziz and the stabilized C methods. Compare these results with the results obtained by using the extended flow point. This example is Well 8.[53] Only the last four flow points from the test are used in the analysis. Reservoir data are summarized here: $h = 454$ ft, $r_w = 0.2615$ ft, $\phi = 0.0675$, $T = 718$°R (258°F), $p_s \cong 4{,}372.6$ psia, $\mu = 0.023$ cp, $z = 0.87$, $c_g = 1.69 \times 10^{-4}$ psia^{-1}, $\gamma_g = 0.65$, $S_w = 0.3$, $A = 640$ acres. $C_A = 30.8828$ (assume that the well is centered in a square drainage area). In addition, the results from a drawdown test in this well indicate $k_g = 4.23$ md and $s = -5.2$.

The Rawlins and Schellhardt analysis with extended flow point gave $C = 2.426 \times 10^{-3}$, $n = 0.54$ and $q_{\text{AOF}} = 180.1$ MMscf/D. The Houpeurt analysis with extended flow point gave $a =$

1.455×10^6 psia²/cp/MMscf/D, $b = 1.774 \times 10^4$ psia²/cp/(MMscf/D)², and $q_{\text{AOF}} = 205.6$ MMscf/D.

Solution.

Brar and Aziz Method.

Step 1—Plot

$$\Delta p_p / q = \left[p_p(p_{ws}) - p_p(p_{wf}) \right] / q \text{ vs. } q$$

on Cartesian coordinates (**Fig. 8.128**). **Table 8.27** gives the plotting functions. Construct best-fit lines through the modified isochronal data points for each time. Although the data are scattered, all flow rates were used for each isochron.

Step 2—Determine the slopes of the lines, b, for each time by least-squares regression analysis. For example, at $t = 3.0$ hours, $b_1 = 1.823 \times 10^4$ psia²/cp/(MMscf/D)². **Table 8.28** summarizes the slopes for all isochrons. The arithmetic average value of the b values in Table 8.28 is

$$\bar{b} = \frac{b_1 + b_2 + b_3 + b_4}{4}$$

$$= \frac{(1.823 + 1.870 + 1.881 + 1.939) \times 10^4}{4} = 1.878 \times 10^4 \text{ psia}^2 / \text{cp} / (\text{MMscf} / \text{D}) .$$

Step 3—Using least-squares regression analysis, calculate the transient deliverability line intercepts for each isochronal line. For example, at $t = 3.0$ hours,

$$a_t = 5.677 \times 10^5 \text{psia}^2 / \text{cp} / \text{MMscf-D} .$$

Table 8.29 gives the intercepts for each isochron.

Step 4—Prepare a graph of a_t vs. log t (**Fig. 8.129**) and draw the best-fit line through data. Using all four data points, calculate m' and c' of the best-fit line of the plot of a_t vs. log t using least-squares regression analysis. The result is $m' = 3.871 \times 10^5$ psia²/(cp-MMscf/D)/cycle and $c' = 3.909 \times 10^5$ psia²/(cp-MMscf/D).

Step 5—Calculate the formation permeability to gas using the slope of the semilog straight line.

$$k_g = \frac{1.632 \times 10^6 T}{m' h} = \frac{1.632 \times 10^6 (718)}{(3.871 \times 10^5)(454)} = 6.6 \text{ md,}$$

which compares with $k_g = 4.23$ md estimated from the drawdown test analysis.

Step 6—Calculate the skin factor with Eq. 8.270.

$$s = 1.151 \left[\left(\frac{c'}{m'} \right) - \log \left(\frac{k_g}{\phi \bar{\mu}_g \bar{c}_t r_w^2} \right) + 3.23 \right]$$

$$= 1.151 \left\{ \left(\frac{3.909 \times 10^5}{3.871 \times 10^5} \right) - \log \left[\frac{6.6}{(0.0675)(0.023)(0.000169)(0.2615)^2} \right] + 3.23 \right\} = -5.0.$$

This value agrees with $s = -5.2$ estimated from the drawdown test analysis.

Step 7—Calculate the stabilized flow coefficient, a. Assume that the well is centered in a square drainage area with C_A = 30.8828.

$$a = \frac{1.422 \times 10^6 T}{k_g h}\left[1.151 \log\left(\frac{10.06A}{C_A r_w^2}\right) - \frac{3}{4} + s\right]$$

$$= \frac{(1.422 \times 10^6)(718)}{(6.6)(454)}\left\{1.151 \log\left[\frac{(10.06)(640)(43{,}560)}{(30.8828)(0.2615)^2}\right] - \frac{3}{4} + (-5.0)\right\}$$

$$= 1.227 \times 10^6 \text{ psia}^2/\text{cp}/(\text{MMscf/D}).$$

Now, calculate the AOF potential using $\bar{b}$ from Step 2 and the stabilized a value calculated in Step 7.

$$q_{\text{AOF}} = \frac{-a + \sqrt{a^2 + 4b\left[p_p(\bar{p}) - p_p(14.65)\right]}}{2b}$$

$$= \frac{-(1.277 \times 10^6) + \sqrt{(1.277 \times 10^6)^2 + 4(1.878 \times 10^4)\left[(1.049 \times 10^9) - 2{,}003.8\right]}}{2(1.878 \times 10^4)} = 205.9 \text{ MMscf/D}.$$

Stabilized C Method.

Step 1—Plot

$$\Delta p_p = p_p(p_{ws}) - p_p(p_{wf})$$

vs. q on log-log coordinates (**Fig. 8.130**). **Table 8.30** gives the plotting functions. Construct best-fit lines through the data.

Step 2—Calculate the deliverability exponent, n, for each line. For this example, use the least-squares regression analysis of all points for each isochron. For example, for t = 3.0 hours, n = 0.63. **Table 8.31** summarizes values of the deliverability exponent for each isochron. The arithmetic average slope of the values in Table 8.31 is

$$\bar{n} = \frac{0.63 + 0.64 + 0.65 + 0.65}{4} = 0.64.$$

Step 3—Calculate the theoretical value of the Houpeurt coefficient, a, using the permeability and skin factor values calculated previously with the Brar and Aziz analysis (i.e., k_g = 6.6 md, s = –5.0).

$$a = \frac{1.422 \times 10^6 T}{k_g h}\left[1.151 \log\left(\frac{10.06A}{C_A r_w^2}\right) - \frac{3}{4} + s\right]$$

$$= \frac{(1.422 \times 10^6)(718)}{(6.6)(454)}\left\{1.151 \log\left[\frac{(10.06)(640)(43{,}560)}{(30.8828)(0.2615)^2}\right] - \frac{3}{4} - 5.0\right\}$$

$$= 1.227 \times 10^6 \text{ psia}^2/\text{cp}/(\text{MMscf/D}).$$

Use the average value for the coefficient, $b = 1.878 \times 10^4$ psia²/(cp-MMscf/D), obtained from the Brar and Aziz analysis.

Step 4—Calculate the rate at which the change in pseudopressure determined with Rawlins-Schellhardt equation equals the change in pseudopressure determined with the Houpeurt equation. Use the average value for the coefficient, $b = 1.878 \times 10^4$ psia²/(cp-MMscf/D), obtained from the Brar and Aziz analysis, and the a coefficient from Step 3.

$$q_e = \frac{a(1-n)}{b(2n-1)} = \frac{1.227 \times 10^6(1-0.64)}{1.878 \times 10^4[2(0.64)-1]} = 84.2 \text{ MMscf/D} .$$

Step 5—Calculate the stabilized C value.

$$C = \frac{q_e}{(aq_e + bq_e^2)^n}$$

$$= \frac{84.2}{\left[(1.227 \times 10^6)(84.2) + (1.878 \times 10^4)(84.2)^2\right]^{0.64}} = 3.69 \times 10^{-4} .$$

Step 6—Calculate the AOF potential of the well using $\bar{n}$ from Step 2.

$$q_{\text{AOF}} = C\left[p_p(\bar{p}) - p_p(p_b)\right]^n$$

$$= 3.69 \times 10^{-4}(1.049 \times 10^9 - 2{,}003.8)^{0.64} = 218.7 \text{ MMscf/D}.$$

Table 8.32 compares the results of the analyses with and without the extended, stabilized flow points. In general, the results are comparable and illustrate the validity of the Brar and Aziz and the stabilized C methods for modified isochronal tests with no extended, stabilized flow point.

8.13 Coning

Coning is the production of an (usually) unwanted second phase simultaneously with a desired hydrocarbon phase in reservoirs with fluid contacts near the wellbore throughout much of the drainage area of a well. The term coning is used because, in a vertical well, the shape of the interface when a well is producing the second fluid resembles an upright or inverted cone (**Fig. 8.131**). Important examples of coning include production of water in an oil well with bottomwater drive, production of gas in an oil well overlain by a gas cap, and production of bottom water in a gas well.

In a horizontal well, the cone becomes more of a crest (**Fig. 8.132**), but the phenomenon is still customarily called coning. In a given reservoir, the amount of undesired second fluid a horizontal well produces is usually less than for a vertical well under comparable conditions. This is a major motivation for drilling horizontal wells, for example, in thin oil columns underlain by water.

Coning is a problem because the second phase must be handled at the surface in addition to the desired hydrocarbon phase, and the production rate of the hydrocarbon flow is usually dramatically reduced after the cone breaks through into the producing well. Produced water must also be disposed of. Gas produced from coning in an oil well may have a market, but

TABLE 8.26—MODIFIED ISOCHRONAL TEST DATA, EXAMPLE 8.13

Time (hours)	p_{wf} (psia) q =31.612 (MMscf/D)	q =44.313 (MMscf/D)	q =56.287 (MMscf/D)	q =70.26 (MMscf/D)
0 (p_{ws})	4,372.6	4,3756.0	4,344.3	4,343.6
3.0	4,274.1	4,206.5	4,111.8	3,994.6
4.0	4,266.8	4,199.3	4,103.3	3,973.3
5.0	4,262.5	4,192.0	4,097.0	3,959.6
6.0	4,258.3	4,190.4	4,093.5	3,951.9
	p_p (p_{wf}) (psia²/cp)			
0 [p_p (p_{ws})]	1.049×10^9	1.042×10^9	1.038×10^9	1.037×10^9
3.0	1.011×10^9	9.848×10^8	9.488×10^8	9.047×10^8
4.0	1.008×10^9	9.821×10^8	9.456×10^8	8.968×10^8
5.0	1.006×10^9	9.793×10^8	9.433×10^8	8.916×10^8
6.0	1.005×10^9	9.787×10^8	9.419×10^8	8.888×10^8

Extended Flow Point

p_{wf} = 3,794 psia	t = 72 hours
p_s = 4,372.6 psia	q = 77.346 MMscf/D
p_p (p_{wf}) = 8.303×10^8	p_p(14.65) = 2,003.8 psia²/cp

TABLE 8.27—PLOTTING FUNCTIONS FOR BRAR AND AZIZ ANALYSIS, EXAMPLE 8.13

Time (hours)	$\Delta p_p/q$ [psia²/(cp/MMscf/D)] q =31.612 (MMscf/D)	q =44.313 (MMscf/D)	q =56.287 (MMscf/D)	q =70.265 (MMscf/D)
3.0	1.202×10^6	1.201×10^6	1.585×10^6	1.883×10^6
4.0	1.297×10^6	1.352×10^6	1.642×10^6	1.995×10^6
5.0	1.360×10^6	1.415×10^6	1.682×10^6	2.069×10^6
6.0	1.392×10^6	1.428×10^6	1.707×10^6	2.121×10^6

also may not. In any event, production of gas in an oil well after the cone breaks through can rapidly deplete reservoir pressure and, for that reason, may force shut in of the oil well.

Several strategies may apply to wells with a potential to cone. One is to try to predict the rate at which a well will cone and produce at a lower rate as long as possible. Or, optimal economics may result by producing at a much higher rate, causing the well to cone, but increasing the cumulative hydrocarbon volume produced (and present value) at any future date. A horizontal well may be preferred to a vertical well.

Most prediction methods for coning predict a "critical rate" at which a stable cone can exist from the fluid contact to the nearest perforations. The theory is that, at rates below the critical rate, the cone will not reach the perforations and the well will produce the desired single phase. At rates equal to or greater than the critical rate, the second fluid will eventually be produced and will increase in amount with time. However, these theories based on critical rates do not predict when breakthrough will occur nor do they predict water/oil ratio or gas/oil ratio (GOR) after breakthrough. Other theories predict these time behaviors, but their accuracy is limited because of simplifying assumptions.

TABLE 8.28—TRANSIENT SLOPES, *b*, OF ISOCHRONS, EXAMPLE 8.13

Time (hours)	b (psia^2/cp/MMscf/D)
3	1.823×10^4
4	1.870×10^4
5	1.881×10^4
6	1.939×10^4

TABLE 8.29—TRANSIENT DELIVERABILITY LINE INTERCEPTS, a_t, OF THE ISOCHRONS, EXAMPLE 8.13

Time (hours)	a_t (psia^2/MMscf/D)
3.0	5.677×10^5
4.0	6.247×10^5
5.0	6.794×10^5
6.0	6.807×10^5

TABLE 8.30—PLOTTING FUNCTIONS FOR STABILIZED C METHOD, EXAMPLE 8.13

	Δp_p (psia^2/cp)			
Time (hours)	q = 31.612 (MMscf/D)	q = 44.313 (MMscf/D)	q = 56.287 (MMscf/D)	q = 70.265 (MMscf/D)
3.0	3.800×10^7	5.720×10^7	8.920×10^7	1.323×10^7
4.0	4.100×10^7	5.990×10^7	9.240×10^7	1.402×10^7
5.0	4.300×10^7	6.270×10^7	9.470×10^7	1.454×10^7
6.0	4.400×10^7	6.330×10^7	9.610×10^7	1.490×10^7

TABLE 8.31—DELIVERABILITY EXPONENTS, *n*, OF THE ISOCHRONS, EXAMPLE 8.13

Time (hours)	n
3.0	0.63
4.0	0.64
5.0	0.65
6.0	0.65

The calculated critical rate is valid only for a certain fixed distance between the fluid contact and the perforations. With time, that distance usually decreases (for example, bottom water will usually tend to rise toward the perforations). Thus, the critical rate will tend to decrease with time, and the economics of a well with a tendency to cone will continue to deteriorate with time.

Whether a cone will move toward perforations depends on the relative significance of viscous and gravitational forces near a well. The pressure drawdown at the perforations tends to

TABLE 8.32—COMPARISON OF RESULTS FROM EXAMPLE 8.13

	Without Stabilized Data		With Stabilized Data	
Parameter	Stabilized C Method	Brar and Aziz Method	Rawlins-Schellhardt	Houpeurt
n	0.64	—	0.54	—
C, (MMscf/D)/(psia^2/cp)^n	3.60×10^{-4}	—	2.426×10^{-3}	—
a, psia^2/(cp-MMscf/D)	1.227×10^6	1.227×10^6	—	1.455×10^6
b, psia^2/(cp-MMscf/D^2)	1.878×10^4	1.878×10^4	—	1.744×10^4
AOF, MMscf/D	218.7	205.9	180.1	205.6

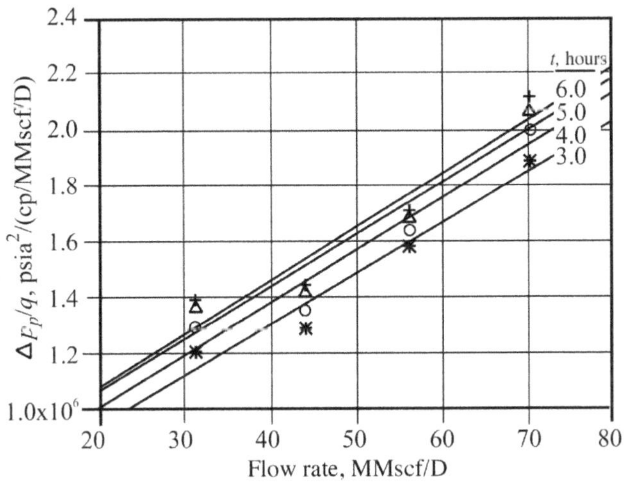

Fig. 8.128—Brar-Aziz analysis of modified isochronal test data, Example 8.13.

cause the undesired fluid to move toward the perforations. Gravitational forces tend to cause the undesired fluid to stay away from the perforations. Coning occurs when viscous forces dominate.

The variables that could affect coning are density differences between water and oil, gas and oil, or gas and water (gravitational forces); fluid viscosities and relative permeabilities; vertical and horizontal permeabilities; and distances from contacts to perforations. Coning tendency turns out to be quite dependent on some of these variables and insensitive to others.

A number of prediction methods have been published. There is no guarantee of great accuracy when using any of these methods because they all contain significant simplifying assumptions. In particular, areal and vertical variations in vertical permeability (because of flow barriers of varying extent) can cause the prediction methods to differ significantly from what actually happens in the field. Accordingly, the prediction methods are best used for quick approximations, screening, and comparison of alternatives. Reservoir simulations, based on accurate reservoir characterization, will ultimately be required.

The coning prediction method proposed by Chaperon[55] is of particular interest because of the variables it includes and because variations of the method are applicable to gas and water coning in both vertical and horizontal wells. For vertical wells, the Chaperon method calculates the critical rate for coning from the expression

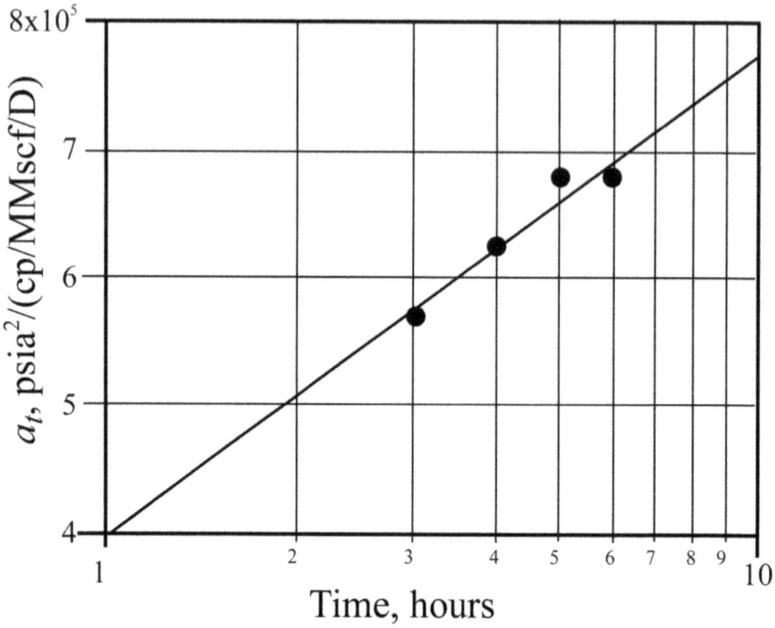

Fig. 8.129—Brar-Aziz plot of a_t vs. t.

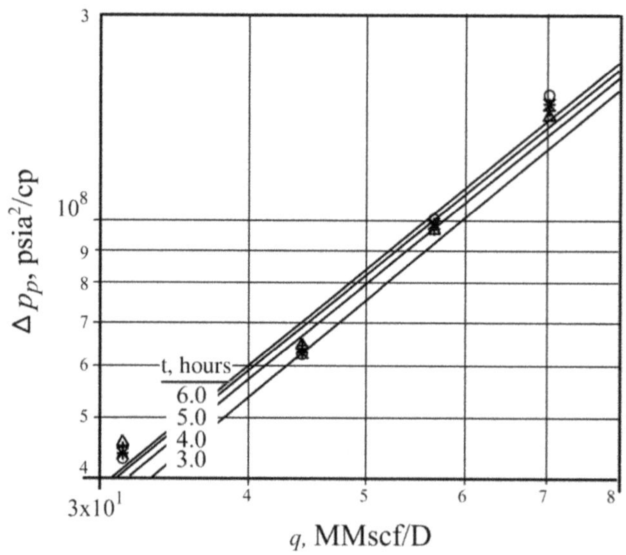

Fig. 8.130—Stabilized C analysis of modified isochronal test data, Example 8.13.

$$q_c = \frac{4.888 \times 10^{-4} k_h h_c^2 \Delta \rho q_{cD}}{B_o \mu_o}, \quad (8.283)$$

where

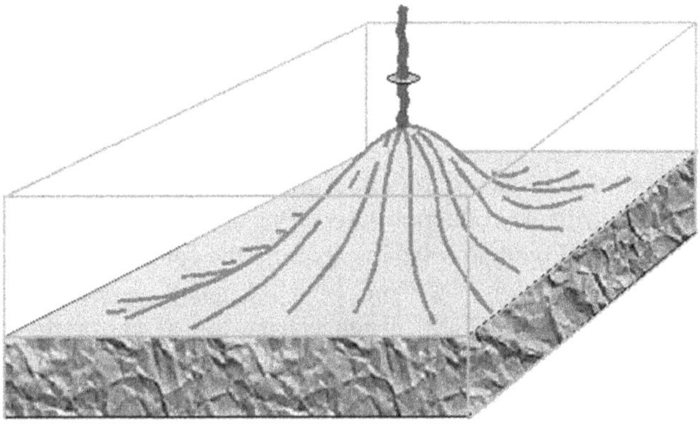

Fig. 8.131—Coning in a vertical well.

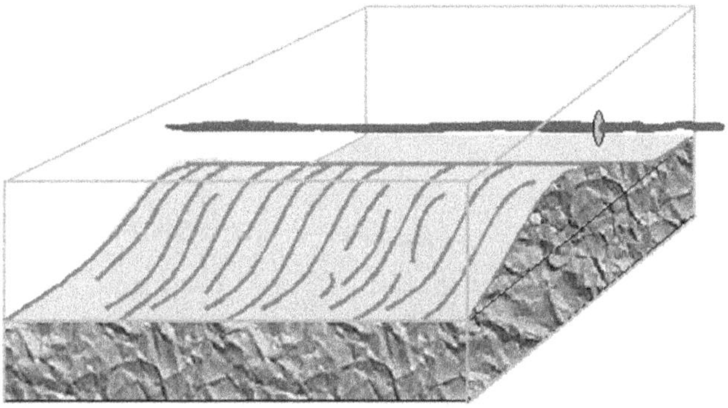

Fig. 8.132—Coning in a horizontal well.

$$q_{cD} = 0.7311 + \frac{1.843 / r_{rD}}{\sqrt{k_v / k_h}}, \qquad (8.284)$$

$$r_{rD} = (r_e / h_c)\sqrt{k_v / k_h}, \qquad (8.285)$$

$$\Delta\rho = \rho_w - \rho_o \text{ or } \rho_o - \rho_g, \text{ density difference, g/cm}^3, \qquad (8.286)$$

and h_c = distance from perforations to fluid contact, ft. For horizontal wells, the critical rate is given by

$$q_c = \frac{4.888 \times 10^{-4} L_w}{(a_H / 2)} \frac{\Delta\rho(k_h h^2)}{\mu_o B_o} F, \qquad (8.287)$$

where

$$F = 3.964 + 0.0616 a_{HD} - 0.000540 a_{HD}^2, \qquad (8.288)$$

and

$$a_{HD} = \left(\frac{a_H}{2h_c}\right)\sqrt{\frac{k_v}{k_h}} \quad \text{.......} \quad (8.289)$$

Example 8.14. Consider a square, 160-acre drilling unit in an oil reservoir overlain by a gas cap. To determine the critical production rate (at or above which coning is likely to occur) for both horizontal and vertical well alternatives, assume the following well and formation properties: k_h = 80 md, a_H = 2,640 ft, μ_o = 0.5 md, ρ_o = 0.8 g/cm³, ρ_g = 0.3 g/cm³, L_w = 1,500 ft, h = 90 ft, B_o = 1.2 RB/STB, r_w = 0.33 ft, distance from top of perforations in vertical well to GOC = 80 ft, and distance from horizontal well to GOC = 80 ft. Consider two cases: Isotropic Reservoir A, where k_v = k_h = 80 md, and Anisotropic Reservoir B, where k_v = 8 md and k_h = 80 md.

Solutions.

Isotropic Reservoir A (k_v = k_h = 80 md). For a horizontal well,

$$q_c = \frac{4.888 \times 10^{-4} L_w}{(a_H/2)} \frac{\Delta\rho(k_h h^2)}{\mu_o B_o} F .$$

Here,

$$a_{HD} = \left(\frac{a_H}{2h_c}\right)\sqrt{\frac{k_v}{k_h}} = \frac{2{,}640}{(2)(80)}\left(\sqrt{\frac{1}{1}}\right) = 16.5$$

$$\text{and } F = 3.964 + 0.0616 a_{HD} - 0.000540 a_{HD}^2$$

$$= 3.964 + (0.0616)(16.5) - (0.000540)(16.5)^2 = 4.83.$$

Then,

$$q_c = \frac{(4.888 \times 10^{-4})(1{,}500)}{(2{,}640/2)} \frac{(0.8-0.3)(80)(80)^2}{(0.5)(1.2)}(4.83) = 1{,}145 \text{ STB/D}.$$

For a vertical well, note that $r_e = \sqrt{\frac{(160)(43{,}560)}{\pi}} = 1{,}489$ ft.

$$q_c = \frac{4.888 \times 10^{-4} k_h h_c^2 \Delta\rho q_{cD}}{B_o \mu_o},$$

$$\text{where } r_{rD} = \left(\frac{r_e}{h_e}\right)\sqrt{\frac{k_v}{k_h}} = \frac{(1{,}489)}{(80)}\sqrt{\frac{1}{1}} = 18.61$$

and

$$q_{cD} = 0.7311 + \frac{1.843/r_{rD}}{\sqrt{k_v/k_h}} = 0.7311 + \frac{1.843}{18.61} = 0.830.$$

Then,

$$q_c = \frac{(4.888 \times 10^{-4})(80)(80)^2(0.8 - 0.3)(0.830)}{(1.2)(0.5)} = 173 \text{ STB/D}.$$

Anisotropic Reservoir B (k_v = 8 md and k_h = 80 md). For a horizontal well,

$$a_{HD} = \frac{2{,}640}{(2)(80)}\sqrt{\frac{8}{80}} = 5.22,$$

$$F = 3.964 + (0.616)(5.22) - (0.000540)(5.22)^2 = 4.27,$$

$$\text{and } q_c = (1{,}145)\left(\frac{4.27}{4.83}\right) = 1{,}012 \text{ STB/D}.$$

For a vertical well,

$$r_{rD} = \frac{1{,}489}{80}\sqrt{\frac{8}{80}} = 5.89,$$

$$q_{cD} = 0.7311 + \frac{1.843}{5.89} = 1.044,$$

$$\text{and } q_c = (173)\left(\frac{1.044}{0.830}\right) = 217 \text{ STB/CD}.$$

The important conclusions and lessons from this example are:

- Vertical permeability has only a modest influence on critical coning rate, at least for this situation.
- The advantage of a horizontal well over a vertical well is substantial for both isotropic and anisotropic reservoirs in this situation.
- The same sorts of calculations could be made for an oil well coning water or a gas well coning water.
- These calculations give us no information on time at which the cone will break through to the producing well nor on GOR and oil production rate following breakthrough.

While these types of simple calculations can provide some insight on the potential for coning, a finely grided simulator model could be used to more effectively predict coning behavior including timing and the benefits of a horizontal well over a vertical one.

Nomenclature

a = $\frac{1.422 \times 10^6}{k_g h}\left[1.151 \log\left(\frac{10.06A}{C_A r_w^2}\right) - \frac{3}{4} + s\right]$, stabilized deliverability coefficient, psia²-cp/MMscf-D

a = total length of reservoir perpendicular to wellbore, ft

a_h = length of reservoir perpendicular to horizontal well, ft

a_f = $\sqrt{L_f^2 + b_f^2}$, depth of investigation along major axis in fractured well, ft

a_t = $\frac{1.422 \times 10^6 T}{k_g h}\left[1.151 \log\left(\frac{k_g t}{1{,}688\phi\bar{\mu}_g\bar{c}_t r_w^2}\right) + s\right]$, transient deliverability coefficient, psia²-cp/MMscf-D

a_H = total width of reservoir perpendicular to the wellbore, ft

a_H' = modified total width of reservoir perpendicular to the wellbore, ft
A = drainage area, sq ft
A = $\pi a_f b_f$, area of investigation in fractured well, ft^2
A_f = cross-sectional area perpendicular to flow, sq ft
A_{wb} = wellbore area, sq ft
b = $\dfrac{1.422 \times 10^6 TD}{kgh}$ (gas flow equation)
b_f = $0.02878\left[\dfrac{kt}{\phi \mu c_t}\right]^{1/2}$, depth of investigation of along minor axis in fractured well, ft
b_B = intercept of Cartesian plot of bilinear flow data, psi
b_H = length in direction parallel to wellbore, ft
b_H' = modified length in direction parallel to wellbore, ft
b_L = intercept of Cartesian plot of linear flow data, psi
b_V = intercept of Cartesian plot of data during volumetric behavior, psi
B = formation volume factor, res vol/surface vol
B_g = gas formation volume factor, RB/STB
B_{gi} = gas formation volume factor evaluated at p_i, RB/Mscf
B_o = oil formation volume factor, RB/Mscf
B_w = water formation volume factor, RB/STB
$\overline{B}_g$ = gas formation volume factor evaluated at average drainage area pressure, RB/Mscf
B_{ND} = 1,422 $\mu z TD/kh$, non-Darcy flow coefficient
c = compressibility, psi^{-1}
c_f = formation compressibility, psi^{-1}
c_g = gas compressibility, psi^{-1}
c_o = oil compressibility, psi^{-1}
c_t = $S_o c_o + S_w c_w + S_g c_g + c_f$ = total compressibility, psi^{-1}
c_w = water compressibility, psi^{-1}
$\overline{c}_t$ = total compressibility evaluated at average drainage area pressure, psi^{-1}
c_{tf} = total compressibility of pore space and fluids in fracture porosity, psi^{-1}
c_{tm} = total compressibility of pore space and fluids in matrix porosity, psi^{-1}
c_{wb} = compressibility of fluid in wellbore, psi^{-1}
C = performance coefficient in gas-well deliverability equation, or wellbore storage coefficient, bbl/psi
C_A = shape factor or constant
C_D = $0.8936C/\phi c_t h r_w^2$, dimensionless wellbore storage coefficient
$(C_D e^{2s})_f$ = type-curve parameter value for the formation
$(C_D e^{2s})_{f+m}$ = type-curve parameter value for the formation plus the matrix
C_{LfD} = $0.8936\, C/\phi c_t h L_f^2$, dimensionless wellbore storage coefficient in fractured well
C_r = $w_f k_f/\pi k L_f$, fracture conductivity, dimensionless
d_x = shortest distance between horizontal well and x boundary, ft
d_y = shortest distance between tip of horizontal well and y boundary, ft
d_z = shortest distance between horizontal well and z boundary, ft
D_x = longest distance between horizontal well and x boundary, ft
D_y = longest distance between tip of horizontal well and y boundary, ft

D_z = longest distance between horizontal well and z boundary, ft
D = non-Darcy flow constant, D/Mscf
e^{-bt} = exponential decline with a constant b and elapsed time, t
E_f = flow efficiency, dimensionless
$Ei(-x)$ = $\int_x^\infty (e^{-u}/u)du$, the exponential integral
$F(u)$ = function used in horizontal well analysis
F_{CD} = $w_f k_f/kL_f$, fracture conductivity, dimensionless
g = acceleration due to gravity, ft/sec^2
g_c = gravitational units conversion factor, 32.17 (lbm/ft)(lbf-s^2)
h = net formation thickness, ft
h_D = $(h/r_w)(k_h/k_v)^{1/2}$, dimensionless
h_f = fracture height, ft
h_m = thickness of matrix, ft
h_p = perforated interval thickness, ft
h_{pD} = h_p/h_t
h_t = total formation thickness, ft
h_1 = distance from top of formation to top of perforations, ft
h_{1D} = h_1/h_t
HTR_{avg} = HTR at average drainage area pressure
J = productivity index, STB/D, psi
J_{actual} = actual well productivity index, STB/D-psi
J_{ideal} = ideal productivity index ($s = 0$), STB/D-psi
k = matrix permeability, md
$\bar{k}$ = average permeability, md
k_f = permeability of the proppant in the fracture, md
k_{fs} = permeability near the wellbore, md
k_g = permeability to gas, md
k_{gp} = permeability of the gravel in the gravel pack, md
k_h = horizontal permeability, md
k_m = matrix permeability, md
k_o = permeability to oil, md
k_r = permeability in horizontal radial direction, md
k_s = permeability of altered zone, md
k_w = permeability to water, md
k_x = permeability in x-direction, md
k_y = permeability in y-direction, md
k_z = permeability in z-direction, md
L = distance from well to no-flow boundary, ft
L_d = drilled length of horizontal well, ft
L_f = fracture half length, ft
L_g = length of flow path through gravel pack, ft
L_m = length of matrix, ft
L_p = length of perforation tunnel, ft
L_s = length of damaged zone in fracture, ft
L_w = completed length of horizontal well, ft

L_x = distance from boundary, ft
m = $162.2\ qB\mu/kh$ = slope of middle-time line, psi/cycle
m_B = slope of bilinear flow graph, psi/hr$^{1/4}$
m_L = slope of linear flow graph, psi/hr$^{1/2}$
m_s = $\dfrac{2456\sqrt{\phi \mu c_t}\, qB\mu}{k_{sp}^{3/2}}$, slope of spherical flow plot, psi-hr$^{1/2}$
m_V = slope of volumetric flow graph, psi/hr
m_{hrf} = slope of semilog plot for hemiradial flow, psi/log cycle
m_{elf} = slope of square-root-of-time plot for early linear flow, psi$/\sqrt{\text{hr}}$
m_{erf} = slope of semilog plot of early radial flow, psi/log cycle
m_{llf} = slope of square-root-of-time plot for late linear flow, psi$/\sqrt{\text{hr}}$
m_{prf} = slope of semilog plot for pseudoradial flow, psi/log cycle
M = Molecular weight of gas
MTR = middle-time region
n = inverse slope of the line on a log-log plot of the change in pressure squared or pseudopressure vs. gas flow rate
p = pressure, psi
p_{avg} = average pressure, psi
p_b = base (atmospheric) pressure, psia
p_0 = arbitrary reference or base pressure, psi
$\overline{p}$ = volumetric average or static drainage-area pressure, psi
p_a = adjusted or normalized pseudopressure, $(\mu z/p)p_p$, psia
p_{awf} = adjusted flowing bottomhole pressure, psia
p_{aws} = adjusted shut-in bottomhole pressure, psia
p_f = formation pressure, psi
p_i = original reservoir pressure, psi
p_m = matrix pressure, psi
p_p = pseudopressure, psia^2/cp
p_s = stabilized shut-in BHP measured just before start of a deliverability test, psia
p_{sc} = standard-condition pressure, psia
p_t = surface pressure in tubing, psi
p_w = BHP in wellbore, psi
p_{wf} = flowing BHP, psi
p_{ws} = shut-in BHP, psi
p_{xy} = parameter in horizontal well analysis equations
p_{xyz} = parameter in horizontal well analysis equations
p_y = parameter in horizontal well analysis equations
$p_{1\text{hr}}$ = pressure at 1-hour shut-in (flow) time on MTR line or its extrapolation, psi
p' = pressure derivative
p^* = MTR pressure trend extrapolated to infinite shut-in time, psi
p_D = $0.00708\ kh(p_i - p)/qB\mu$, dimensionless pressure as defined for constant-rate production
p_{MBHD} = Matthews-Brons-Hazebroek pressure, dimensionless
$(p_D)_{\text{MP}}$ = dimensionless pressure at match point
q = flow rate at surface, STB/D
q_{AOF} = absolute-open-flow potential, MMscf/D

q_g = gas flow rate, Mscf/D
q_o = water flow rate, STB/D
q_{Rt} = total flow rate at reservoir conditions, RB/D
q_{sf} = flow rate at formation (sand) face, STB/D
q_w = water flow rate, STB/D
r = distance from the center of wellbore, ft
r_a = radius of altered zone (skin effect), ft
r_d = effective drainage radius, ft
r_{dp} = radius of damage zone around perforation tunnel, ft
r_e = external drainage radius, ft
r_i = radius of investigation, ft
r_p = radius of perforation tunnel, ft
r_s = outer radius of the altered zone, ft
r_{sp} = radius of source or inner boundary of spherical flow pattern, ft
r_w = wellbore radius, ft
r_{wa} = apparent or effective wellbore radius, ft
r_D = r/r_w, dimensionless radius
R_s = dissolved GOR, scf/STB
s = skin factor, dimensionless
s_a = skin caused by alteration of permeability around wellbore, dimensionless
s_c = convergence skin, dimensionless
s_d = skin caused by formation damage, dimensionless
s_e = skin caused by eccentric effects, dimensionless
s_{dp} = perforation damage skin, dimensionless
s_f = skin of hydraulically fractured well, dimensionless
s_{gp} = skin factor from to Darcy flow through gravel pack, dimensionless
s_{min} = minimum skin factor, dimensionless
s_p = skin resulting from an incompletely perforated interval, dimensionless
s_t = total skin, dimensionless
s_θ = skin factor resulting from well inclination, dimensionless
s' = $s + Dq$ = apparent skin factor, dimensionless
S_g = gas saturation, fraction of pore volume
S_o = oil saturation, fraction of pore volume
S_w = water saturation, fraction of pore volume
t = elapsed time, hours
t_a = $\mu c_t t_{ap}$, adjusted or normalized pseudotime, hours
t_{ap} = pseudotime, hours
t_{bD} = dimensionless time in linear flow, hours
t_D = $0.0002637kt/\phi\mu c_t r_w^2$, dimensionless time
t_{DA} = $0.0002637\ kt/\phi\mu c_t A$ = dimensionless time based on drainage area, A
t_{eqB} = equivalent time for bilinear flow, hours
t_e = equivalent time, hours
t_{LfD} = $0.0002637\ kt/\phi\mu e_t L_f^2$, dimensionless time for fractured wells
t_p = pseudoproducing time, hours
t_{pD} = pseudoproducing time, dimensionless

t_{prf} = time required to reach the pseudoradial flow regime, hours
t_{Eelf} = end of early linear flow, t, hours
t_{Eerf} = end of early radial flow, t, hours
t_{Elf} = end of linear flow, hours
t_{Ellf} = time to end of late linear flow regime, hours
t_{Ehrf} = end of hemiradial flow, hours
t_{Erf} = end of early radial flow, hours
t_{Eprf} = end of pseudoradial flow, hours
t_p = constant-rate production period, t, hours
t_{pAD} = dimensionless producing time, hours
t_{pss} = time required to reach pseudosteady state, hours
t_{Self} = start of early linear flow, hours
t_{Sllf} = start of late linear flow, hours
t_{Shrf} = start of hemiradial flow, t, hours
t_{Sprf} = start of pseudoradial flow, t, hours
t_s = time required for stabilization, hours
T = reservoir temperature, °R
T_{sc} = standard condition temperature, °R
u = dummy variable
V = volume, bbl
V_f = fraction of bulk volume occupied by fractures
V_m = fraction of bulk volume occupied by matrix
V_w = V_{wb} = wellbore volume, bbl
w = width of channel reservoir, ft
w_f = fracture width, ft
wk_f = fracture conductivity, md-ft
w_s = width of damaged zone around fracture face, ft
WBS = wellbore storage
z = gas-law deviation factor, dimensionless
$\bar{z}$ = gas-law deviation factor at average reservoir pressure, dimensionless
Δp = pressure change since start of transient test, psi
$(\Delta p)_{\mathrm{MP}}$ = pressure change at match point
Δp_D = dimensionless pressure change
Δp_p = pseudopressure change since start of test, psia²/cp
Δp_s = additional pressure drop due to skin, psi
$\Delta p_{t=0}$ = pressure drop at time zero, psi
$\Delta p_{1\mathrm{hr}}$ = pressure change from start of test to one hour elapsed time, psi
Δt = time elapsed since start of test, hours
Δt_a = $\bar{\mu}\,\bar{c}_t\,\Delta t_{ap}$, normalized or adjusted pseudotime, hours
Δt_{ap} = $\int_0^{\Delta t} \frac{dt}{\mu(p)c_t(p)}$, pseudotime, hr-psia/cp
Δt_{Be} = bilinear equivalent time, hours
Δt_e = radial equivalent time, hours
Δt_{Le} = linear equivalent time, hours
$\Delta t_{\max}$ = maximum shut-in time in pressure buildup test, hours
ΔV = change in volume, bbl

η = $0.0002637\ k/\phi\mu c_t$, hydraulic diffusivity, ft^2/hr
η_{fD} = hydraulic diffusivity, dimensionless
λ = interporosity flow coefficient
λ_t = $\frac{k_0}{\mu_0}+\frac{k_w}{\mu_w}+\frac{k_g}{\mu_g}$, total mobility, md/cp
α = exponent in deliverability equation
α = parameter characteristic of system geometry in dual-porosity system
β = turbulence factor
β' = transition parameter
γ = Euler's constant, = 1.781, dimensionless
γ_g = gas gravity (air = 1.0)
γ_m = matrix density
ω = storativity ratio in dual porosity reservoir
μ = viscosity, cp
μ_i = viscosity evaluated at p_i, cp
μ_g = gas viscosity, cp
μ_o = oil viscosity, cp
μ_w = water viscosity, cp
$\bar{\mu}_g$ = gas viscosity evaluated at average pressure, cp
μ_{gwf} = gas viscosity evaluated at p_{wf}, cp
$\bar{\mu}$ = viscosity evaluated at $\bar{p}$, cp
ρ = density, lbm/ft^3 or g/cm^3
ρ_{wb} = density of liquid in wellbore, lbm/ft^3
ϕ_f = fraction of fracture volume occupied by pore space, $\cong$ 1
ϕ_m = fraction of matrix volume occupied by pore space
$(\phi V)_f$ = fraction of bulk volume occupied by pore space in fractures
$(\phi Vc_t)_f$ = fracture "storativity" for dual porosity reservoir
$(\phi Vc_t)_{f+m}$ = total "storativity" for dual porosity reservoir
$(\phi V)_m$ = fraction of bulk volume occupied by pore space in matrix
ϕ = porosity, dimensionless
Σs = sum of damage skin, turbulence, and other pseudoskin factors

References

1. Matthews, C.S. and Russell, D.G.: *Pressure Buildup and Flow Tests in Wells,* Monograph Series, SPE, Richardson, Texas (1967) **1.**
2. Hawkins, M.F. Jr.: "A Note on the Skin Effect," *Trans.,* AIME (1956) **207,** 356–357.
3. Wattenbarger, R.A. and Ramey, H.J. Jr.: "Gas Well Testing With Turbulence, Damage, and Wellbore Storage," *JPT* (August 1968) 877; *Trans.,* AIME, **243.**
4. Lee, W.J.: *Well Testing,* Textbook Series, SPE, Dallas (1982).
5. Joseph, J.A. and Koederitz, L.F.: "Unsteady-State Spherical Flow With Storage and Skin," *SPEJ* (December 1985) 804.
6. Horner, D.R.: "Pressure Buildup in Wells," *Proc.,* Third World Pet. Cong., The Hague (1951) Sec. II, 503–523; also *Pressure Analysis Methods,* Reprint Series, SPE, Richardson, Texas (1967) **9,** 25–43.
7. Gringarten, A.C. *et al.*: "A Comparison Between Different Skin and Wellbore Storage Type-Curves for Early-Time Transient Analysis," paper SPE 8205 presented at the 1979 SPE Annual Technical Conference and Exhibition, Las Vegas, Nevada, 23–26 September.

8. Bourdet, D. *et al.*: "A New Set of Type Curves Simplifies Well Test Analysis," *World Oil* (May 1983) 95.
9. Agarwal, R.G.: "A New Method To Account for Producing Time Effects When Drawdown Type Curves Are Used To Analyze Pressure Buildup and Other Test Data," paper SPE 9289 presented at the 1980 SPE Annual Technical Conference and Exhibition, Dallas, 21–24 September.
10. Prats, M., Hazebroek, P., and Strickler, W.R.: "Effect of Vertical Fractures on Reservoir Behavior —Compressible Fluid Case," *SPEJ* (June 1962) 87; *Trans.,* AIME, **225.**
11. Papatzacos, P.: "Approximate Partial-Penetration Pseudoskin for Infinite-Conductivity Wells," *SPERE* (May 1987) 227.
12. Cinco, H., Miller, F.G., and Ramey, H.J. Jr.: "Unsteady-State Pressure Distribution Created by a Directionally Drilled Well," *JPT* (November 1975) 1392; *Trans.*, AIME, **259.**
13. Brown, K.E.: *The Technology of Artificial Lift Methods,* PennWell, Tulsa (1984) **4,** 134.
14. McLeod, H.O. Jr.: "The Effect of Perforating Conditions on Well Performance," *JPT* (January 1983).
15. Al-Hussainy, R., Ramey, H.J. Jr., and Crawford, P.B.: "The Flow or Real Gases Through Porous Media," *JPT* (May 1966) 624; *Trans.,* AIME, **237.**
16. Agarwal, R.G.: "Real Gas Pseudo-Time—A New Function for Pressure Buildup Analysis of MFH Gas Wells," paper SPE 8279 presented at the 1979 SPE Annual Technical Conference and Exhibition, Las Vegas, Nevada, 23–26 September.
17. Lee, W.J. and Holditch, S.A.: "Applications of Pseudotime to Buildup Test Analysis of Low Permeability Gas Wells with Long-Duration Wellbore Storage Distortion," *JPT* (December 1982) 2877.
18. Spivey, J.P. and Lee, W.J.: "The Use of Pseudotime: Wellbore Storage and the Middle Time Region," paper SPE 15229 presented at the 1986 SPE Unconventional Gas Technology Symposium, Louisville, Kentucky, 18–21 May.
19. Meunier, D.F., Kabir, C.S., and Wittmann, M.J.: "Gas Well Test Analysis: Use of Normalized Pressure and Time Functions," *SPEFE* (December 1987) 629.
20. Matthews, C.S., Brons, F., and Hazebroek, P.: "A Method for Determination of Average Pressure in a Bounded Reservoir," *Trans.*, AIME (1954) **201,** 182–191.
21. Jones, S.C.: "Using the Inertial Coefficient, β, To Characterize Heterogeneity in Reservoir Rock," paper SPE 16949 presented at the 1987 SPE Annual Technical Conference and Exhibition, Dallas, 27–30 September.
22. Perrine, R.L.: "Analysis of Pressure Buildup Curves," *Drill. and Prod. Prac.,* API, Dallas (1956) 482.
23. Martin, J.C.: "Simplified Equations of Flow in Gas Drive Reservoirs and the Theoretical Foundation of Multiphase Pressure Buildup Analyses," *JPT* (October 1959) 309; *Trans.,* AIME, **216.**
24. Ramey, H.J. Jr. and Cobb, W.M.: "A General Pressure Buildup Theory for a Well in a Closed Circular Drainage Area," *JPT* (December 1971) 1493; *Trans.*, AIME, **251.**
25. Larsen, V.C.: "Understanding the Muskat Method of Analysing Pressure Build-Up Curves," *J. Cdn. Pet. Tech.* (Fall, 1963) **2,** No. 136.
26. Arps, J.J. and Smith, A.E.: "Practical Use of Bottom Hole Pressure Build-Up Curves," Reprint Paper No. 851-23-I, Tulsa Meeting, API (March 1949).
27. Cinco-Ley, H. and Samaniego-V., F.: "Transient Pressure Analysis for Fractured Wells," *JPT* (September 1981) 1749.
28. Hale, B.W. and Evers, J.F.: "Elliptical Flow Equations for Vertically Fractured Gas Wells," *JPT* (December 1981) 2489.
29. Cinco-Ley, H. and Samaniego-V., F.: "Transient Pressure Analysis: Finite Conductivity Fracture Case Versus Damaged Fracture Case," paper SPE 10179 presented at the 1981 SPE Annual Technical Conference and Exhibition, San Antonio, Texas, 5–7 October.
30. Lee, W.J.: "Postfracture Formation Evaluation," *Recent Advances in Hydraulic Fracturing,* J.L. Gidley, S.A. Holditch, D.E. Nierode, and R.W. Veatch Jr. (eds.), Monograph Series, SPE, Richardson, Texas (1989) **12.**
31. Ramey, H.J. Jr. and Gringarten, A.C.: "Effect of High Volume Vertical Fractures on Geothermal Steam Well Behavior," *Proc.,* Second United Nations Symposium on the Use and Development of Geothermal Energy, San Francisco (1975).

32. Warren, J.E. and Root, P.J.: "The Behavior of Naturally Fractured Reservoirs," *SPEJ* (September 1963) 245.
33. Gringarten, A.C.: "Interpretation of Tests in Fissured and Multilayered Reservoirs With Double-Porosity Behavior: Theory and Practice," *JPT* (April 1984) 549.
34. Serra, K., Reynolds, A.C., and Raghavan, R.: "New Pressure Transient Analysis Methods for Naturally Fractured Reservoirs," *JPT* (December 1983) 2271.
35. Barenblatt, G.E., Zheltov, I.P., and Kochina, I.N.: "Basic Concepts in the Theory of Homogeneous Liquids in Fissured Rocks," *J. Appl. Math. Mech.* (1960) **24,** 1286.
36. de Swann, A.: "Analytical Solutions for Determining Naturally Fractured Reservoir Properties by Well Testing," *SPEJ* (June 1976) 117-122.
37. Bourdet, D. *et al.*: "New Type Curves Aid Analysis of Fissured Zone Well Tests," *World Oil* (April 1984).
38. Goode, P.A. and Thambynayagam, R.K.M.: "Pressure Drawdown and Buildup Analysis of Horizontal Wells in Anisotropic Media," *SPEFE* (December 1987) 683; *Trans.*, AIME, **283.**
39. Odeh, A.S and Babu, D.K.: "Transient Flow Behaviour of Horizontal Wells: Pressure Drawdown and Buildup Analysis," *SPEFE* (March 1990) 7; *Trans.*, AIME, **289.**
40. Kuchuk, F.J.: "Well Testing and Interpretation for Horizontal Wells," paper SPE 25232 available from SPE, Richardson, Texas (1995).
41. Lichtenberger, G.J.: "Date Acquisition and Interpretation of Horizontal Well Pressure-Transient Tests," *JPT* (February 1994) 157.
42. Babu, D.K. and Odeh, A.S.: "Productivity of a Horizontal Well," *SPERE* (November 1989) 417.
43. Economides, M.J., Brand, C.W., and Frick, T.P.: "Well Configurations in Anisotropic Reservoirs," *SPEFE* (December 1996) 257.
44. Ozkan, E.: "Analysis of Horizontal-Well Responses: Contemporary vs. Conventional," *SPEREE* (August 2001) 260.
45. Houpeurt, A.: "On the Flow of Gases in Porous Media," *Revue de L'Institut Francais du Petrole* (1959) **14,** No. 11, 1468.
46. Rawlins, E.L. and Schellhardt, M.A.: *Backpressure Data on Natural Gas Wells and Their Application to Production Practices,* Monograph Series, USBM (1935) **7.**
47. Al-Hussainy, R., Ramey, H.J. Jr., and Crawford, P.B.: "The Flow of Real Gases Through Porous Media," *JPT* (May 1966) 624; *Trans.*, AIME, **237.**
48. Lee, W.J.: *Well Testing*, Textbook Series, SPE, Richardson, Texas (1977) **1.**
49. Earlougher, R.C. Jr.: *Advances in Well Test Analysis,* Monograph Series, SPE, Richardson, Texas (1977) 5.
50. Jennings, J. W. *et al.*: "Deliverability Testing of Natural Gas Wells," prepared for the Texas Railroad Commission, Texas A&M U., College Station, Texas (August 1989).
51. Cullender, M.H.: "The Isochronal Performance Method of Determining the Flow Characteristics of Wells," *Trans.*, AIME (1955) **204,** 137.
52. Katz, D.L. *et al.: Handbook of Natural Gas Engineering*, McGraw-Hill Publishing Co., New York City (1959).
53. Brar, G.S. and Aziz, K.: "Analysis of Modified Isochronal Tests To Predict the Stabilized Deliverability Potential of Gas Wells Without Using Stabilized Flow Data," *JPT* (February 1978) 297; *Trans.*, AIME, **265.**
54. Johnston, J.L., Lee, W.J., and Blasingame, T.A.: "Estimating the Stabilized Deliverability of a Gas Well Using the Rawlins and Schellhardt Method: An Analytical Approach," paper SPE 23440 presented at the 1991 SPE Eastern Regional Meeting, Lexington, Kentucky, 22–25 October.
55. Chaperon, I.: "Theoretical Study of Coning Toward Horizontal and Vertical Wells in Anisotropic Formations: Subcritical and Critical Rates," paper 15377 presented at the 1986 SPE Annual Technical Conference and Exhibition, New Orleans, 5–8 October.

SI Metric Conversion Factors

acre	×	4.046 873	E + 03	= m^2
bbl	×	1.589 873	E − 01	= m^3
cp	×	1.0*	E − 03	= Pa·s
ft	×	3.048*	E − 01	= m
ft^2	×	9.290 304*	E − 02	= m^2
ft^3	×	2.831 685	E − 02	= m^3
°F		(°F − 32)/1.8		= °C
$in.^3$	×	1.638 706	E + 01	= cm^3
lbf	×	4.448 222	E + 00	= N
lbm	×	4.535 924	E − 01	= kg
psi	×	6.894 757	E + 00	= kPa

*Conversion factor is exact.

Appendix

TABLE A-1—SHAPE FACTORS AND DURATIONS OF FLOW REGIME[49]

		Pseudosteady-State System		Infinite System
Reservoir Shape	C_A	Exact for t_{DA} >	Less Than 1% Error for t_{DA}>	Less than 1% Error for t_{DA}<
	31.62	0.1	0.06	0.10
	31.6	0.1	0.06	0.10
	27.6	0.2	0.07	0.09
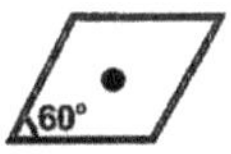	27.1	0.2	0.07	0.09
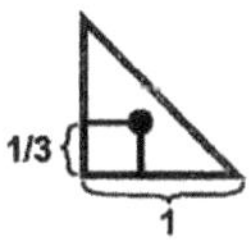	21.9	0.4	0.12	0.08
	0.098	0.9	0.60	0.015
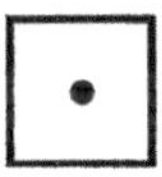	30.8828	0.1	0.05	0.09
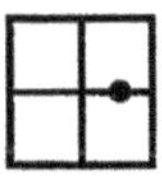	12.9851	0.7	0.25	0.03
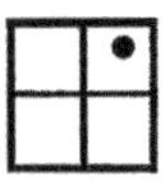	4.5132	0.6	0.30	0.025
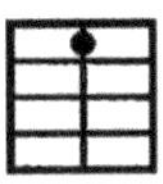	3.3351	0.7	0.25	0.01

TABLE A-1—SHAPE FACTORS AND DURATIONS OF FLOW REGIME[49] (continued)

		Pseudosteady-State System		Infinite System
Reservoir Shape	C_A	Exact for t_{DA} >	Less Than 1% Error for t_{DA}>	Less than 1% Error for t_{DA}<
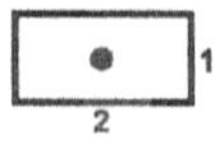	21.8369	0.3	0.15	0.025
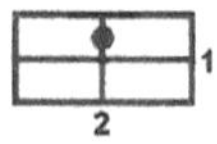	10.8374	0.4	0.15	0.025
	4.5141	1.5	0.50	0.06
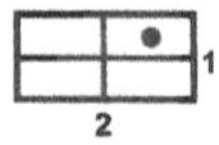	2.0769	1.7	0.50	0.02
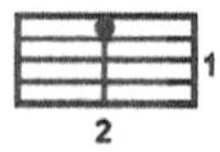	3.1573	0.4	0.15	0.005
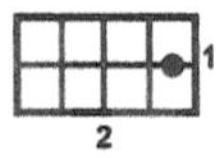	0.5813	2.0	0.60	0.02
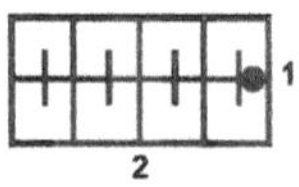	0.1109	3.0	0.60	0.005
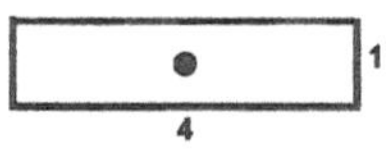	5.3790	0.8	0.30	0.01
	2.6896	0.8	0.30	0.01
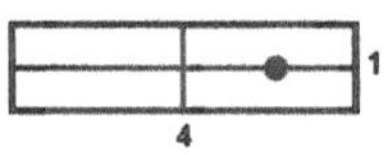	0.2318	4.0	2.00	0.03
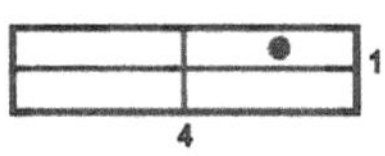	0.1155	4.0	2.00	0.01
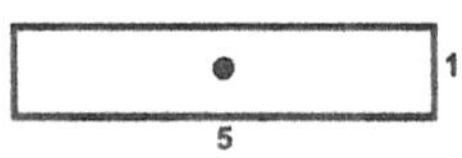	2.3606	1.0	0.40	0.025

TABLE A-1—SHAPE FACTORS AND DURATIONS OF FLOW REGIME[49] (continued)

Reservoir Shape	C_A	Pseudosteady-State System: Exact for t_{DA} >	Pseudosteady-State System: Less Than 1% Error for t_{DA}>	Infinite System: Less than 1% Error for t_{DA}<
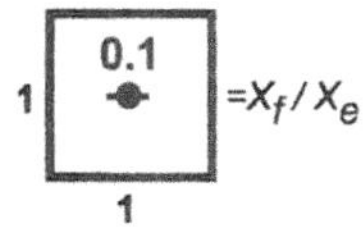	2.6541	0.175	0.08	cannot use
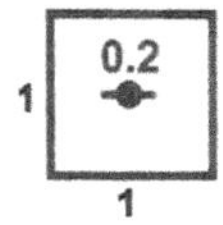	2.0348	0.175	0.09	cannot use
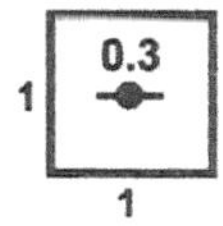	1.9886	0.175	0.09	cannot use
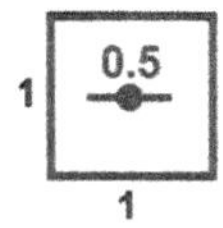	1.6620	0.175	0.09	cannot use
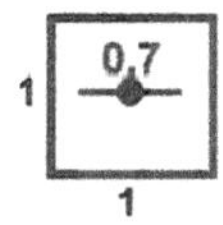	1.3127	0.175	0.09	cannot use
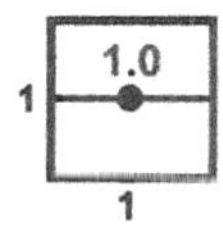	0.7887	0.175	0.09	cannot use
In waterdrive reservoirs				
	19.1	--	--	--
In reservoirs of unknown production character				
	25.0	--	--	--

TABLE A-2—EQUATIONS FOR ANALYZING FLOW IN HORIZONTAL WELLS WITH UNIFORM FLUX

Wellbore Storage

$$t_{\text{Ewbs}} = \frac{(4{,}000 + 240 s_d)C}{\sqrt{k_x k_z} L_w / \mu} \quad \text{(A-1)}$$

$$C = V_{wb} c_{wb} \quad \text{(A-2)}$$

$$C = \frac{25.65 A_{wb}}{\rho_{wb} \cos\theta} \quad \text{(A-3)}$$

$$C = qBt / 24\Delta p \quad \text{(A-4)}$$

Early Radial Flow

$$\sqrt{k_x k_z} = \frac{162.6 qB\mu}{|m_{\text{erf}}| L_w} \quad \text{(A-5)}$$

$$s_d = 1.151 \left[\frac{\Delta p_{1\text{hr}}}{|m_{\text{erf}}|} - \log\left(\frac{\sqrt{k_x k_z}}{\phi \mu c_t r_w^2} \right) + 3.23 \right] \quad \text{(A-6)}$$

$$t_{\text{Eerf}} = 1{,}800 d_z^2 \phi \mu c_t / k_z \quad \text{(A-7)}$$

$$t_{\text{Eerf}} = 125 L_w^2 \phi \mu c_t / k_y \quad \text{(A-8)}$$

Hemiradial Flow

$$\sqrt{k_x k_z} = 2\left(\frac{162.6 qB\mu}{|m_{\text{hrf}}| L_w} \right) \quad \text{(A-9)}$$

$$s_d = 2.303 \left[\frac{\Delta p_{1\text{hr}}}{|m_{\text{hrf}}|} - \log\left(\frac{\sqrt{k_x k_z}}{\phi \mu c_t r_w^2} \right) + 3.23 \right] \quad \text{(A-10)}$$

$$t_{\text{Ehrf}} \approx 1{,}800 D_z^2 \phi \mu c_t / k_z \quad \text{(A-11a)}$$

$$t_{\text{Ehrf}} \approx 125 L_w^2 \phi \mu c_t / k_y \quad \text{(A-11b)}$$

Early Linear Flow

$$\sqrt{k_x} = \frac{8.13 qB}{|m_{\text{elf}}| L_w h} \sqrt{\frac{\mu}{\phi c_t}} \quad \text{(A-12)}$$

$$s_d = \frac{L_w \sqrt{k_x k_z}}{141.2 qB\mu} \Delta p_{t=0} - s_c \quad \text{(A-13a)}$$

$$\text{where } s_c = \ln\left(\frac{h}{r_w} \right) + 0.25 \ln\left(\frac{k_x}{k_z} \right) - \ln\left[\sin\left(\frac{\pi d_z}{h} \right) \right] - 1.838 \quad \text{(A-13b)}$$

$$t_{\text{Self}} = \frac{1{,}800 D_z^2 \phi \mu c_t}{k_z} \quad \text{(A-14a)}$$

$$t_{\text{Eelf}} \approx 160 L_w^2 \phi \mu c_t / k_y \quad \text{(A-14b)}$$

TABLE A-2—EQUATIONS FOR ANALYZING FLOW IN HORIZONTAL WELLS WITH UNIFORM FLUX (continued)

Late Pseudoradial Flow (requires $L_w/b_H < 0.45$)

$$\sqrt{k_x k_y} = \frac{162.6 qB\mu}{|m_{\text{prf}}|h} \quad \text{(A-15)}$$

$$s_d = 1.151\sqrt{\frac{k_z}{k_y}}\frac{L_w}{h}\left[\frac{\Delta p_{1\text{hr}}}{|m_{\text{prf}}|} - \log\left(\frac{k_y}{\phi\mu c_t L_w^2}\right) + 1.76\right] - s_c \quad \text{(A-16)}$$

Note that s_c is defined by Eq. 8.13b.

$$t_{\text{Sprf}} = \frac{1{,}480 L_w^2 \phi\mu c_t}{k_y} \quad \text{(A-17)}$$

$$t_{\text{Eprf}} = \frac{2{,}000\phi\mu c_t (L_w/4 + d_y)^2}{k_y} \quad \text{(A-18)}$$

$$\text{or } t_{\text{Eprf}} = \frac{1{,}650\phi\mu c_t d_x^2}{k_x} \quad \text{(A-19)}$$

Late-Linear Flow

$$\sqrt{k_x} = \frac{8.128 qB}{|m_{\text{llf}}| b_H h}\sqrt{\frac{\mu}{\phi c_t}} \quad \text{(A-20)}$$

$$s_t = \frac{L_w}{b_H}\left(\frac{b_H\sqrt{k_x k_z}\,\Delta p_{t=0}}{141.2 qB\mu} - s_p - s_c\right) \quad \text{(A-21)}$$

where s_p, the partial penetration skin given in Eqs. 8.25 to 8.35, is is due to the convergence skin, s_c.

$$t_{\text{Sllf}} = \frac{4{,}800\phi\mu c_t (D_y + L_w/4)^2}{k_y} \quad \text{(A-22)}$$

$$\text{or } t_{\text{Sllf}} = \frac{1{,}800\phi\mu c_t D_z^2}{k_z} \quad \text{(A-23)}$$

$$t_{\text{Ellf}} = \frac{1{,}650\phi\mu c_t d_x^2}{k_x} \quad \text{(A-24)}$$

Calculating Partial Penetration Skin in Late-Linear Flow and Productivity Equations

$$p_{xyz} = \left(\frac{b_H}{L_w} - 1\right)\left\{\ln\left(\frac{h}{r_w}\right) + 0.25\ln\left(\frac{k_x}{k_z}\right) - \ln\left[\sin\left(\frac{\pi d_z}{h}\right)\right] - 1.838\right\} \quad \text{(A-25)}$$

$$\text{Case 1, } \frac{a_H}{\sqrt{k_x}} \geq \frac{0.75 b_H}{\sqrt{k_y}} >> \frac{0.75 h}{\sqrt{k_z}} \quad \text{(A-26)}$$

$$s_p = p_{xyz} + p'_{xy} \quad \text{(A-27)}$$

TABLE A-2—EQUATIONS FOR ANALYZING FLOW IN HORIZONTAL WELLS WITH UNIFORM FLUX (continued)

$$p'_{xy} = \frac{2b_H{}^2}{L_w h}\sqrt{k_z / k_y}\left\{F\left(\frac{L_w}{2b_H}\right) + 0.5\left[F\left(\frac{4y_m + L_w}{2b_H}\right) - F\left(\frac{4y_m - L_w}{2b_H}\right)\right]\right\} \quad \text{(A-28)}$$

where $y_m = d_y + \frac{L_w}{2}$ (A-29)

$$F(u) = -u\left[0.145 + \ln(u) - 0.137(u)^2\right],\ u < 1 \quad \text{(A-30)}$$

$$F(u) = (2-u)\left[0.145 + \ln(2-u) - 0.137(2-u)^2\right],\ u > 1 \quad \text{(A-31)}$$

Case 2, $\frac{b_H}{\sqrt{k_y}} > \frac{1.33a_H}{\sqrt{k_x}} >> \frac{0.75h}{\sqrt{k_z}}$ (A-32)

$$s_p = p_{xyz} + p_y + p_{xy} \quad \text{(A-33)}$$

$$p_y = \frac{6.28b_H{}^2}{a_H h}\frac{\sqrt{k_x k_z}}{k_y}\left[\left(\frac{1}{3} - \frac{y_m}{b_H} + \frac{y_m^2}{b_H^2}\right) + \frac{L_w}{24b_H}\left(\frac{L_w}{b_H} - 3\right)\right] \quad \text{(A-34)}$$

$$p_{xy} = \left(\frac{b_H}{L_w} - 1\right)\left(\frac{6.28a_H}{h}\sqrt{k_z / k_x}\right)\left(\frac{1}{3} - \frac{d_x}{a_H} + \frac{d_x^2}{a_H^2}\right),\ d_x \geq 0.25a_H \quad \text{(A-35)}$$

Productivity and Productivity Index (Uniform Flux Solutions)

$$q = \frac{0.00708b_H\sqrt{k_x k_z}\left(\bar{p} - p_{wf}\right)}{B\mu\left[\ln\left(\frac{A^{1/2}}{r_w}\right) + \ln C_H - 0.75 + s_p + \left(\frac{b_H}{L_w}\right)s_d\right]} \quad \text{(A-36)}$$

$$J = \frac{q}{\left(\bar{p} - p_{wf}\right)} = \frac{0.00708b_H\sqrt{k_x k_z}}{B\mu\left[\ln\left(\frac{C_H A^{1/2}}{r_w}\right) - 0.75 + s_p + \left(\frac{b_H}{L_w}\right)s_d\right]} \quad \text{(A-37)}$$

$$\ln C_H = 6.28\frac{a_H}{h}\sqrt{\frac{k_z}{k_x}}\left[\frac{1}{3} - \frac{d_x}{a_H} + \left(\frac{d_x}{a_H}\right)^2\right] - \ln\left(\sin\frac{\pi d_z}{h}\right) - 0.5\ln\left[\left(a_H / h\right)\sqrt{k_z / k_x}\right] - 1.088 \quad \text{(A-38)}$$

www.ingramcontent.com/pod-product-compliance
Lightning Source LLC
Chambersburg PA
CBHW061148240926
51846CB00006B/953
* 9 7 8 1 9 5 9 0 2 5 5 5 9 *